S0-BRB-764

WILDLIFE ECOLOGY AND MANAGEMENT

FOURTH EDITION

ERIC G. BOLEN
UNIVERSITY OF NORTH CAROLINA AT WILMINGTON

WILLIAM L. ROBINSON
NORTHERN MICHIGAN UNIVERSITY

Prentice Hall
Upper Saddle River, New Jersey 07458

Library of Congress Cataloging-in-Publication Data

Bolen, Eric G.
  Wildlife ecology and management / Eric G. Bolen, William L.
Robinson—4th ed.
     p.   cm.
  Includes bibliographical reference and index.
  ISBN 0-13-840422-4
  1. Wildlife management.   2. Ecology.   I. Robinson, William
Laughlin.   II. Title
SK355.B65 1998
639.9—dc21   98–7009
                                                   CIP

Senior Editor: Teresa Ryu
Art Director: Jayne Conte
Cover Design: Kiwi Design
Manufacturing Manager: Trudy Pisciotti
Production Supervision/Composition: WestWords, Inc.
Cover Photo: Caribou (*Rangifer tarandus*). Photo by Kennan Ward, © 1994.

© 1999, 1995, 1989, 1984 by Prentice-Hall, Inc.
Simon & Schuster/A Viacom Company
Upper Saddle River, New Jersey 07458

All rights reserved. No part of this book may be
reproduced, in any form or by any means,
without permission in writing from the publisher.

Printed in the United States of America

10   9   8   7   6   5   4   3   2   1

ISBN 0-13-840422-4

Prentice-Hall International (UK) Limited, *London*
Prentice-Hall of Australia Pty. Limited, *Sydney*
Prentice-Hall Canada Inc., *Toronto*
Prentice-Hall Hispanoamericana, S.A., *Mexico*
Prentice-Hall of India Private Limited, *New Delhi*
Prentice-Hall of Japan, Inc., *Tokyo*
Simon & Schuster Asia Pte. Ltd., *Singapore*
Editora Prentice-Hall do Brasil, Ltda., *Rio de Janeiro*

CONTENTS

CHAPTER 7

FOOD AND COVER

CHAPTER 8

WILDLIFE DISEASES

CHAPTER 15

FOREST MANAGEMENT AND WILDLIFE   317

CHAPTER 16

WILDLIFE IN PARKS AND REFUGES   338

CHAPTER 17

URBAN WILDLIFE   358

CHAPTER 18

EXOTIC WILDLIFE   383

CHAPTER 19

NONGAME AND ENDANGERED WILDLIFE   412

CHAPTER 20

ECONOMICS OF WILDLIFE   437

CHAPTER 21

CONSERVATION BIOLOGY AND WILDLIFE MANAGEMENT   450

CHAPTER 22

WILDLIFE AS A PUBLIC TRUST   474

CHAPTER 23

CONCLUSION   514

GLOSSARY   518

LITERATURE CITED   531

INDEX   585

FOREWORD

This new edition—the fourth appearance of *Wildlife Ecology and Management*—keeps pace with the avalanche of information accruing from research on wildlife populations and their management. Even a quick review of Bolen and Robinson's text brings to realization the tremendous fund of knowledge that has accumulated since Aldo Leopold's benchmark *Game Management* appeared in 1933. For many years Leopold's volume was the only text available to students of wildlife management, but it was—and remains—of historic significance because it inaugurated the modern concept of game management. Prior to Leopold's classic, game management was perceived to consist of little else than law enforcement, artificial propagation, and predator control.

The concepts broached by Leopold emphasized the importance of habitat quality, proper land use, and the ecological functions of predators—ideas that collectively represented a new environmental philosophy and ushered in the modern-day profession of wildlife management. Thereafter, the interest and growth in the profession is indexed by the membership of The Wildlife Society: From its formation in 1937 to 1997, membership has risen from a few hundred to more than 9200.

The recognition of the role played by the uses of land and water in natural-resource management gained momentum in the late 1930s and early 1940s only to suffer stagnation during World War II. With the cessation of hostilities, however, the wise use of land and water became an increasingly important doctrine in the well-being of society. No small part of this belated recognition resulted from the growing corps of professional wildlife biologists who

pointed out that our natural resources required better stewardship.

Armed with knowledge from thousands of reports and publications, which are the result of research aimed at almost every aspect of animals in relation to each other and to their environment, biologists have been at the forefront in providing society with a better understanding of wildlife in relation to land, water, and people.

That this knowledge is being applied to land and water management is evidenced in many federal and state programs involving both public and private lands. The U.S.D.A. Forest Service, the National Park Service, the Bureau of Land Management, the Environmental Protection Agency, the U.S. Army Corps of Engineers, and the Department of Agriculture have promulgated or modified many of their traditional treatments of natural-resource responsibilities to include wildlife enhancement. Most notable in its magnitude has been the Conservation Reserve Program, which is administered by the Department of Agriculture. Moreover, the growth in responsibility and stature of the U.S. Fish and Wildlife Service from its limited activity as the Bureau of Biological Survey epitomizes the increasing emphasis society has bestowed on wildlife and its management at the federal level.

But state conservation programs probably have benefited the most from the input of wildlife professionals. Prior to the 1930s, programs designed to enhance wildlife populations largely involved the artificial propagation of pheasants and bobwhite; only a few states had any interest in lands for wildlife. A review of state conservation programs today reveals the dramatic change that has occurred in the past 6 decades.

Most citizens would concur in the recognition that wildlife professionals have improved outdoor recreation, have reduced overexploitation of natural resources, and have given a more forceful meaning to the concept of a *land ethic.* The status of most wildlife populations is generally good: big game have increased to the point that certain species (white-tailed deer, for example) today pose local problems; geese, too, have become overly abundant in places; wild turkey and wood duck populations, once at precarious levels, are thriving; and trumpeter swans, bald eagles, and peregrine falcons are recovering from their perilous brush with extinction.

Also encouraging is the renewed interest by agricultural entities in recognizing the relationship between land and wildlife. New land practices are being devised and implemented to reduce water and wind erosion, which concurrently benefit wildlife. Many farmers have developed a side income from leasing or selling daily or seasonal hunting rights. As urban populations increase, along with trespass restrictions,

farm-operated hunting enterprises will be in still greater demand.

Because of the ever-growing involvement between society and the land, the knowledge disseminated in the fourth edition of this premier text on wildlife ecology is most important to the wellness of people. I know of no other text that covers so well the interrelationships and intricacies of biodiversity among vertebrates. Everyone interested in understanding wildlife values and problems will benefit from a perusal of *Wildlife Ecology and Management.* The authors, Eric G. Bolen and William L. Robinson, have produced a masterful compendium of information that embraces a wide diversity of subjects involved in the welfare of wildlife and people. To accomplish this feat, they synthesize information from more than 2500 references, a formidable review of literature that spans the breath of topics pertinent to understanding wildlife populations, their ecology, and their management.

Frank C. Bellrose
Illinois Natural History Survey (ret.)

PREFACE

Because of the continuing acceptance of *Wildlife Ecology and Management* as a basic text in its field, we are pleased to present the fourth edition. In doing so, we introduce a new feature—information boxes highlighting a selection of key figures broadly involved with the development of conservation. Other boxes profile people who work each day "in the trenches" in an arena replete with funding, political, and bureaucratic limitations and the difficult choices these present. Still other boxes describe some of the many institutions dedicated to wildlife research and management.

The new material also includes greater focus on the interactions between wolves and moose on Isle Royale, which replaces the previous text about African antelopes in Chapter 5, Population Ecology. The new information provides a prime example of the benefits reaped from a long-term research program, an approach too often stalled by shortfalls in funds or waning administrative support. We also added a section that describes a large-scale experiment in restoration ecology—a week-long release of floodwater into the Grand Canyon in 1996. Another new section outlines the role of the National Wildlife Refuge System in the United States; we also included short descriptions of federal wildlife agencies in Canada and Mexico. Additionally, the fourth edition reports the rapid spread of a "new" disease affecting house finches and possibly other songbirds, the escape of rabbit calicivirus disease in Australia, more on the bison-brucellosis controversy, and a new case of pesticide poisoning (but apparently with a happy ending). Other new material concerns fire ants, metapopulations, pending legislation popularly known as Teaming With Wildlife, and a view of wildlife management by Native Americans.

Finally, the fourth edition begins with a foreword written by a true "dean" of wildlife management, Frank C. Bellrose. His productive and far-reaching career—a work still very much in progress—spans nearly 60 years, during which he received the profession's highest recognition, the Aldo Leopold Award. We are honored to have his comments introduce this edition of *Wildlife Ecology and Management.*

OUR GOAL

The underlying purpose of *Wildlife Ecology and Management* remains unchanged from the previous editions: We wish to introduce undergraduate students to the art and science of wildlife management. These students, we hope, will include those with a variety of academic interest—"nonmajors" in university jargon—as well as those completing degrees in any of the disciplines directly associated with the conservation of natural resources, including, of course, wildlife management. For nonmajors, a class using *Wildlife Ecology and Management* may be the first, and only, time they receive formal training in a discipline related to conservation. Our presentation accordingly does not assume that students have more than modest exposure to a basic science, preferably a course or two in biology. The Glossary also remains a major element of the new edition, as does the fundamental treatment of ecology in Chapter 4.

AUTHORS' PERSPECTIVE

Our personal friendship and professional association began when we were students at the University of Maine in the late 1950s. Since then, we have taught at universities in rather dissimilar ecological settings and have studied wildlife of different types. Much of Bolen's work has been with waterfowl and wetlands in Texas and, more recently, in North Carolina, whereas Robinson has spent his career in Vermont and Michigan where he has studied deer, grouse, woodcock, wolves, and other forest-related species. These differences, we believe, strengthen the usefulness of *Wildlife Ecology and Management* for an audience of diverse interest and backgrounds.

ACKNOWLEDGMENTS

Our gratitude to others is immense. Most notably, we thank our wives, Elizabeth and Glenda, for again offering their loving support of our task. Copyright releases from many sources, including The Wildlife Society, the Society for Range Management, the Illinois Natural History Survey, and the Wildlife Management Institute, allowed us to republish several tables, illustrations, and figures, as did the cooperation of the scientists who originally conducted the research. Alda S. Ingram, Patricia M. Wagner, Suzie Piziali, and Shirley R. Burgeson provided skilled secretarial help with one or more of the four editions. Christine S. Dutton capably prepared the index, and Beth A. Pardini cheerfully attended to many organizational chores. The staff of the Randall Library at the University of North Carolina expertly assisted with many searches for material.

Our efforts in preparing the four editions of this book required the assistance and collaboration of many colleagues, and we gratefully acknowledge all who so kindly shared with us their expertise:

E. D. Ables, B. L. Allen, the late D. L. Allen, J. E. Applegate, P. Arcese, G. A. Baldassarre, B. D. J. Batt, F. C. Bellrose, W. F. Bennett, J. F. Besser, C. W. Betsill, E. P. Bristol, T. Bridges, C. M. Britton, R. P. Brooks, F. C. Bryant, S. M. Burke, B. W. Cain, B. R. Chapman, J. E. Childs, T. L. Clark, B. T. Crawford, A. T. Cringan, F. B. Cross, N. Cumberlidge, the late B. E. Dahl, M. R. Davis, D. F. Day, F. C. Dean, R. M. DeGraaf, S. Demarais, D. L. Drawe, H. E. Dregne, P. Dunne, L. D. Flake, T. M. Franklin, W. J. Foreyt, R. R. George, P. Goodman, O. T. Gorman, E. G. Gouge, K. L. Grannemann, J. W. Hardin, L. D. Harris, C. Haspel, N. L. Heitman, M. Heitmeyer, M. Herman, R. G. Hooper, J. R. Hunter, M. L. Hunter, Jr., M. L. Huster, Jr., M. Hutchins, L. R. Jahn, R. L. Jarvis, B. G. Jones, C. J. Jones, the late J. K. Jones, Jr., J. R. Karr, M. Kearsley, S. R. Kellert, W. B. King, R. E. Kirby, R. L. Kilpatrick, D. A. Klenbenow, the late W. D. Klimstra, E. Klinghammer, W. R. Kramer, J. A. Kushlan, R. F. Labisky, T. A. Langford, D. L. Leedy, J. C. Lewis, S. A. Liggett, R. L. Linder, R. M. Linn, P. Logan, W. R. London, M. P. Losito, T. E. Lovejoy III, M. P. Luttrell, C. J. Martinka, L. A. Mehrhoff, G. Merediz-Alonso, H. W. Miller, K. Miller, W. E. Moritz, E. S. Moses, P. J. Mundy, B. R. Murphy, H. K. Nelson, B. W. O'Gara, F. Opolka, R. S. Ostfeld, H. G. Packard, R. A. Parajko, R. A. Parnell, D. B. Pence, the late W. L. Pengelly, K. Penny, T. J. Peterle, G. W. Peterson, R. O. Peterson, R. D. Pettit, J. H. Rappole, B. Ralston, the late D. G. Raveling, F. A. Reid, R. J. Robel, R. R. Roth, C. E. Rupprecht, D. H. Rusch, H. Salwasser, D. E. Samuel, G. C. Sanderson, S. D. Schemnitz, J. T. Schoen, H. L. Schramm, J. L. Schuster, R. W. Seabloom, U. S. Seal, the late W. F. Sigler, C. D. Simpson, F. F. Skillern, L. M. Smith, R. E. Sosebee, R. Sparks, R. D. Sparrowe, J. Spinks, T. Stehn, J. D. Stein, F. A. Stormer, J. G. Teer, R. C. Telfair II, J. W. Thomas, D. Q. Thompson, E. T. Thorne, D. O. Trainer, Jr., J. L. Traves, G. L. Valentine, W. VandeBerg, L. W. VanDruff, B. J. Verts, D. D. Waid, R. E. Warner, R. J. Warren, L. D. Waybrant, R. P. Weaver, Jr., D. Wegner, F. Wheeler, D. H. Whiter, R. J. Whyte, A. M. Wilson, M. H. Wilson, G. W. Winegard, C. K. Winkler, B. Woodbridge, the late H. A. Wright, S. H. Young, and W. T. Zyla. To any we have overlooked, we offer our apologies as well as our thanks. We are indebted to all, but we alone bear responsibility for any remaining errors.

Eric G. Bolen
William L. Robinson

CHAPTER 1

WHAT IS WILDLIFE MANAGEMENT?

The purpose of science is not to conquer the land, but to understand the mechanisms of ecosystems and to fit man into the resources he has available on the planet on which he has evolved.

J.J. Hickey (1974)

Wildlife is a term that does not enjoy a precise or a universally accepted definition. The term implies all things that are living outside direct human control and therefore includes those plants and animals that are not cultivated or domesticated. In its fullest meaning, wildlife encompasses insects and fungi, frogs and wild flowers, as well as doves, deer, and trees.

Nonetheless, organizations concerned with wildlife generally favor the so-called "higher" forms of animal life. *The Journal of Wildlife Management,* a professional publication of The Wildlife Society, deals almost entirely with birds and mammals and usually focuses on those species prized for sport. In 1996, 65 percent of the papers appearing in the *Journal* dealt with birds and mammals, including 6 percent on threatened or endangered species. Articles on other nongame species—those not endangered—comprised 10 percent of the published titles. Others addressed research techniques (e.g., methods for census, capture, or marking animals), general conservation, and environmental law and policy. In the same year in the *Wildlife Society Bulletin,* another publication of The Wildlife Society, 59 percent of the papers concerned birds or mammals (of which 19 percent were threatened or endangered species), 28 percent dealt with techniques, and 4 percent with general topics involving more than one group of animals. Twenty-eight percent of the articles addressed ecosystem dynamics and community biodiversity; this reflected a recent broadening of the scope of wildlife management. In *National Wildlife,* published by the National Wildlife Federation—a citizen-based organization—only 53 percent of the articles in 1996 concerned birds and mammals, of which 29 percent were threatened or endangered species. Environmental law and policy were the subjects of 27 percent of the articles; 6 percent concerned fishes, amphibians, and reptiles; 3 percent featured invertebrates; 11 percent dealt with people; and 6 percent with plants (*Note:* the total exceeds 100% because of overlapping topics).

Still, birds and mammals symbolize wildlife to most people and, of these animals, game species receive a disproportionately large measure of attention. However, only a small fraction of birds and mammals are regarded as game. Wing (1951) calculated that less than 9 percent of the birds and 12 percent of the mammals in the United States are designated as game species.

In the United States, fishes were separated in name from other wildlife by a political decision made in

1940, when President Franklin D. Roosevelt combined the Bureau of Biological Survey, an agency concerned primarily with birds and mammals, with the Bureau of Fisheries. The new agency was to be called the U.S. Wildlife Service, but when fishery scientists felt they might be neglected under that name, Roosevelt added the words *Fish and* to the new agency's title and signed the proclamation (Hickey 1974). The U.S. Fish and Wildlife Service thus was born. The new name implied that fishes are somehow of a different category than "wildlife." In Canada, federal responsibilities for marine and freshwater fisheries are handled separately by the Department of Fisheries and Oceans, whereas other wildlife, particularly migratory birds, fall under the federal jurisdiction of the Canadian Wildlife Service. In the United States, commercial marine fisheries are administered by the National Marine Fisheries Service, a component of the National Oceanic and Atmospheric Administration in the Department of Commerce. Other vertebrates—amphibians and reptiles—are not managed by separate agencies and thus remain under the general category of wildlife. With the exception of threatened or endangered species, however, amphibians and reptiles generally receive little direct attention from state and federal wildlife agencies.

The meaning of *wildlife* is implied in a definition that appeared in the first issue of *The Journal of Wildlife Management* (1937): "the practical ecology of all vertebrates and their plant and animal associates." The *Journal* further stated that wildlife management "along sound biological lines is part of the greater movement for conservation of our entire native fauna and flora."

Management clearly implies the influence and application of human manipulation. In other words, management demands the use of judicious means to achieve specified objectives. In the case of wildlife, however, Hickey (1974) expressed fear that *management* holds as a premise the concept of conquering nature and suggested that attempts at dominating nature contradict the ethic of resource conservation.[1]

With that implication, wildlife management might better be termed wildlife *conservation* because in some circumstances, the best procedure may be to leave natural landscapes alone. Thus, there may be instances for which conservation decisions may not fit a strict definition of *management*.

Indeed, some people prefer leaving wildlife alone, believing that only harm comes when humans tamper with nature. However, as long as humans hunt, fish, and trap, some means are needed for responsibly regulating the harvest of wildlife populations. Hunting and fishing aside, other human influences require the presence and response of wildlife management. Urbanization and industrialization are two of the more obvious influences that commonly reduce the quality and quantity of wildlife habitat. The phenomenon popularly known as acid rain is one of many specific illustrations. A growing list of endangered species represents a major concern of wildlife management. The field of wildlife management also maintains strong links with other disciplines, among them forestry, range management, and park administration.

Although in some instances the best management decision may be no management at all, that or any other decision nonetheless remains couched in human terms and perspectives. As humans, we have, like all other animals, an inherent bias toward our own interests. However, a book about wildlife management written from the perspective of a gray wolf (*Canis lupus*) might focus on wilderness areas and place major emphasis on populations of deer (*Odocoileus* spp.), moose (*Alces alces*), and other prey. Little attention would be given to farmlands, upland birds, marshes, or waterfowl. Moreover, such a book might conclude with a section advocating control of an invading species: humans. Alternately, the perspective of a bald eagle (*Haliaeetus leucocephalus*) might emphasize fish populations, reductions of pesticides, and the preservation of nest trees. From the point of view of a sea turtle (*Chelonia* spp.), a wildlife book might highlight the management of marine algae, the preservation of beaches, and the control of gulls (*Larus* spp.) or other preda-

---

[1] *Conservation,* as a term applied to natural resources, is credited to Gifford Pinchot, a forester who served in the administration of President Theodore Roosevelt (Trefethen 1975; see also Pinchot 1947). Pinchot disliked the words *protection* and *preservation,* terms he believed implied that natural resources should be locked up and remain unused. English lacked a descriptive word for the sustained-yield use that could be achieved from forests, wildlife, and other natural resources. Pinchot thus coined *conser-* *vation* in 1907, apparently deriving his term from *conservator,* a title of an office in the British government of colonial India. The Latin roots are *servare,* to guard, and *con,* which means together. Roosevelt immediately adopted Pinchot's new word, and conservation thereafter became a priority of the Roosevelt Administration (as it would be 30 years later in the administration of another president named Roosevelt, who started, for example, the Soil Conservation Service).

tors of hatchling turtles. Try as we might, we cannot remove our own human biases and perceptions when dealing with those plants and animals we conveniently call *wildlife.*

We therefore suggest that *wildlife management is the application of ecological knowledge to populations of vertebrate animals and their plant and animal associates in a manner that strikes a balance between the needs of those populations and the needs of people.* Ecological knowledge is applied in three basic management approaches: (1) *preservation,* when nature is allowed to take its course without human intervention; (2) *direct manipulation,* when animal populations are trapped, shot, poisoned, and stocked; and (3) *indirect manipulation,* when vegetation, water, or other key components of wildlife habitat are altered.

The need for intelligent wildlife management transcends national boundaries. For example, to stem the decline of whale populations and thereafter restore their numbers throughout the world's oceans requires not only biological knowledge but also international diplomacy and cooperation. In tropical regions, where rapidly expanding human populations place ever greater demands on land for food and income production, native birds and mammals are losing habitat in which they have lived for centuries. Moreover, huge amounts of surface soils, which support humans and wildlife alike, are eroding rapidly in many regions of the world, leaving behind severely degraded landscapes. Hundreds of species of invertebrate animals face extinction each year from a combination of factors: production for food, fiber, and other goods for the immediate needs of humans; failure to consider the long-term ecological consequences of modifying the environment; and exploitative opportunities for quick profits. Thus, wildlife management also involves both knowledge about how to manage the number and consumptive habits of humans and also the careful tending of the biosphere—the environment that supports us and the other animals with which we share our planet.

A mild caveat is appropriate here: wildlife management is not basic science in the same sense as physics or chemistry, nor is it pure technology in the sense of engineering. Wildlife management draws from those fields, as well as from zoology, botany, mathematics, and several other disciplines. Despite its strong and obvious associations with the natural sciences, wildlife management also entails elements of art; that is, information from a variety of disciplines must be integrated and used skillfully with logic and imagination. A "cookbook" approach rarely suffices. The dimension of art comes into play most prominently in the design, manipulation, and evaluation of habitat (see Forman and Gordon 1986). Wildlife management thus remains both an art and a science whose practice is not dissimilar from that of psychology, anthropology, or even medicine or law. Each is a profession requiring vigorous—and rigorous—applications of skill, knowledge, and imagination.

A BRIEF HISTORY

Until the 1960s, wildlife management was primarily *game management,* the husbandry and regulation of populations of birds and mammals hunted for sport. Game management continues to be a major part of the profession of wildlife management, but wildlife managers now deal with much more than game species. Songbirds, turtles, and a full spectrum of species besides grouse and deer fall within the realm of wildlife management. Despite the anticipation of some students, wildlife management was never just "hunting and fishing—and getting paid for it!" The field now includes goals that are more complex than merely providing for the maximum harvest of game species for those who hunt and fish.

Wildlife management is changing, but its past remains relevant to the present and future. The practice of wildlife management is rooted in the intermingling of human ethics, culture, perceptions, and legal concepts. For example, a fundamental historical difference exists between the European and the North American concepts of ownership of wildlife. In most European countries, animals are the legal property of landowners, and the regulation of wildlife populations remains the responsibility of the owners. Gamekeepers or foresters who have training in wildlife management often are employed by owners of large estates. In Germany, for example, privately employed *jaegermeisters* (hunting masters) regulate hunting and closely tend the wildlife roaming the grounds. Each estate, in effect, operates within its own jurisdiction for the production and harvest of resident wildlife. By contrast, in the United States and Canada, wildlife belongs to the people of the state or province or, in the case of migratory species, to the people of the nation. These differences clearly affect the ways animals are harvested, and they define the public sector for whom wildlife is managed (see Chapter 22). Nonetheless, the basic ecological concepts governing wildlife populations remain the same, whether wildlife is publicly or privately owned.

In the 1800s, wildlife management in North America consisted mainly of adding increasingly restrictive clauses to existing hunting regulations. Some current game laws extend back to colonial times. Season lengths were shortened, bag limits were decreased, and many previously accepted methods of hunting were curtailed or eliminated. There was a general recognition that wildlife was a steadily dwindling resource that must be rationed. Regulations thus were designed to extend the period before the fateful day when the last deer, duck, and grouse might be shot. Only in the recent past have we realized that wildlife is a natural resource that is renewable—a resource that, with wise management, can be perpetuated indefinitely for the enjoyment of present and future generations.

Wildlife management emerged as a budding science in the 1930s, largely from the work, example, and writings of Aldo Leopold, Professor of Forestry at the University of Wisconsin (see McCabe 1987; Tanner 1987, Meine 1988). Indeed, the beginning of modern-day wildlife management in the United States is usually associated with the publication of Leopold's landmark text, *Game Management* (Leopold 1933a). When state and federal conservation agencies began to search for scientific information on which to base hunting policies and to improve habitat conditions, Leopold and his associates spearheaded the response. Trained biologists were needed to census deer populations, to estimate the reproductive potential and harvest rates of grouse, to find ways of producing more upland game birds on farmland, to improve wetlands habitats for waterfowl, and to restore populations where wildlife had been extirpated.

In 1937, the U.S. Congress passed the Pittman-Robertson Act, which placed a 10 percent tax on the sales of sporting arms and ammunition (see Kallman 1987). Revenues from the federal tax thereafter were distributed to the states for wildlife management and research (the rate is now 11 percent; see Chapter 22). The new revenues provided a dependable financial base that substantially augmented the scarce dollars from legislative appropriations and license fees, thereby enabling the states to expand their struggling efforts in management and research. Because "P-R" funds, like license fees, originated from the expenditures of hunters, the money understandably was directed toward the management of

game species. Great strides were made in understanding the biology and ecology of both big and small game; such information often produced new and better ways of managing selected species. The three decades following 1940 were rich in discovery, and hunters enjoyed the promise of a reasonably successful season each year. Later, however, these successes—no matter how rewarding for hunters—were viewed in a different light by other sectors of the U.S. public. New perspectives and expectations for wildlife management emerged in the 1960s and 1970s. Chief among these attitudes were an enhanced respect for nature and a growing opposition to the killing of wildlife (Scheffer 1976). These issues—reflecting a less utilitarian attitude toward wildlife—encompassed concerns for species previously neglected as "nongame," for those faced with extinction, and for wildlife exploited for commercial or recreational purposes.

Scheffer (1979) thoughtfully summarized some changes that took place in wildlife management in the United States during the latter part of this century. Chief among these is the growing number of Americans who are actively concerned about wildlife and environmental quality for reasons other than hunting and trapping. In fact, the influence of hunters and the manufacturers of sporting goods on many areas of wildlife management has diminished as the combined voices of nonhunters and anti-hunters have gained recognition in the halls of state and federal governments. Still, because wildlife programs for decades depended upon the financial support of hunters and trappers, many hunters and agencies resisted or ignored these new influences and attitudes. As Scheffer (1976) stated, "I think that hunters and trappers must accept the probability that many, if not most, Americans have a spiritual or emotional interest in wildlife which is as strong and as legitimate as their own." These are social issues, clearly indicating that the management of game species must be reconciled with a broader consideration of ecological principles and consequences. The central issue concerns the goal of *maximum* yield of a few select species (the desire of hunters and other consumers of wildlife) as opposed to *optimum* yield (which responds collectively to the demands of a broad cross section of human society and sustains the natural diversity of species within wildlife communities). In ecological terms, we should realize that management practices that maximize the yield of a single species (or crop) will inevitably disadvantage other

Photo courtesy of University of Wisconsin-Madison

Aldo Leopold
Forester, Philosopher, and Founding Father
(1887–1948)

Aldo Leopold is regarded as the "father" of wildlife management in the United States. Born in Burlington, Iowa, Leopold graduated from Yale in 1909 and began a career with the U.S. Forest Service in Arizona and New Mexico. However, his concerns went beyond forest management and included the preservation of wilderness. His efforts led to the Gila Wilderness Area in New Mexico, the first of its kind in the national forest system. In 1925, he became associate director of the Forest Products Laboratory in Madison, Wisconsin, but he left 2 years later to begin private consulting work. One of his major contracts was a survey of game populations in eight north-central states; his report won an *Outdoor Life* medal. In 1933, Leopold

completed *Game Management,* the first textbook on the topic, and began to teach at the University of Wisconsin, where he remained until his death. His appointment as professor of game management was the first such position to be recognized by a university.

In 1935, Leopold helped organize The Wilderness Society and, in 1937, cofounded The Wildlife Society. He served as president for both The Wildlife Society (1939) and the Ecological Society of America (1947). A philosopher as well as a scientist, Leopold is perhaps best known for *A Sand County Almanac,* a collection of essays largely written at his "shack," a hideaway on an abandoned farm where he developed and practiced the concept of a land ethic (that is, land is a community of life whose health is vital and not solely a commodity for economic exploitation). One particularly famous essay, *Thinking Like a Mountain,* reflects his changing attitude about the role of predators in natural systems. He also addressed the importance of biodiversity when he wrote, comparing natural communities to a fine watch, "To keep every cog and wheel is the first precaution of intelligent tinkering." Leopold may well have been the most influential figure to speak for conservation in the 20th century.

The Wildlife Society honors his memory with its highest tribute, The Aldo Leopold Award, presented each year since 1950 to a person who demonstrates distinguished service to conservation. See Flader (1974), Callicott (1987), McCabe (1987), Tanner (1987), Meine (1988), and Flader and Callicott (1991) for full accounts of Leopold's life and philosophies. The "tinkering" quote comes from journals edited by his son (Leopold 1953).

species that inhabit the same area. Such an axiom applies to forests and deer as well as to cornfields and cattle. Wildlife managers today should know not only *how* and *what* to manipulate, but also should answer *why* these actions are justified.

WHAT DOES A WILDLIFE MANAGER DO?

People involved with managing wildlife combine the talents and training of researchers, managers, and public relations experts. Problems of local importance may require one or more approaches, but in general, field biologists seek answers by following a three-step sequence: they (1) search journals and other scientific literature for parallel situations that may suggest solutions, (2) determine reasons for the difficulty using field and/or laboratory techniques, and (3) implement and evaluate remedies, the latter frequently requiring public involvement.

The designation *wildlife biologist* or *wildlife manager* encompasses a wide variety of duties. Some of these focus directly on animals and some indirectly on habitat, but most also include considerable contact with people. Throughout this edition, we feature several persons whose duties cover a broad area of wildlife ecology and management. We have asked them to highlight their duties and to mention, if they wished, their on-the-job frustrations and satisfactions.

Regardless of their assignments, wildlife biologists should know how and when to apply basic biological and ecological principles. The job is not easy, nor is the pay always attractive. Animals rarely cooperate with human aspirations for their welfare, and some sectors of the public may frustrate the best intentions of professional management. But there are other rewards, some of which reach into the depths of the human spirit: the song of spring's first wood thrush, the distant howl of a wolf wafting across a moonlit lake on a summer night, or the clean kill of a goose in the chill of a November dawn. Such experiences complement the knowledge that a wildlife manager's concern, training, and talent have helped maintain a long-standing heritage for another generation. Done well, wildlife management can strengthen the fragile web of nature and, in doing so, can enhance the quality of human existence.

SUMMARY

Wildlife is a hard term to define, but birds and mammals receive the most attention; of these, only a small percentage of species are regarded as game animals. Management implies some form of manipulation, but human perceptions usually bias our views and decisions about wildlife. Wildlife management applies ecological knowledge in ways that seek a balance between the needs of wildlife and the needs of people. In doing so, wildlife managers must apply both skills (art) and knowledge (science).

History records a long period when laws were the prime means for perpetuating animal populations. Later, wildlife was recognized as a renewable resource, and wildlife management emerged as a science when Aldo Leopold published *Game Management* in 1933. Wildlife management today considers more than game species and addresses the concept of optimum yield. Professional "wildlifers" face numerous issues that require research and management skills. These issues are addressed by (1) reviewing scientific literature, (2) finding answers with field and/or laboratory work, and (3) implementing and evaluating remedies. Political, social, and economic factors influence the ways—and the degree of success—that wildlife managers can deal with the stewardship of wildlife populations and their habitats.

CHAPTER 2

NEGLECT AND EXPLOITATION

In common with the forests, the bison herds seemed endless. But a tide of humanity drove westward, multiplying as it came. It was a characteristic biological force similar to those we habitually misjudge because of their small beginnings. If unchecked, they can take over the world while our backs are turned.

Durward Allen (1974a)

Cultures advance only when their participants learn from history. The dynamic record of human affairs thus presents each new generation with the opportunity to analyze previous mistakes. An examination of history also offers fresh opportunities for cultures to repeat previous successes. Indeed, knowledge gained from the physical and natural sciences has added steadily to advances in technology. We have been slower, however, to learn from social experiences. Human activities often have produced extensive changes—favorable and unfavorable—in populations of other animals that share our planet. With the hope that we might learn from earlier mistakes, this chapter examines the record of failings in the human treatment of wildlife.

IN THE BEGINNING

According to the King James version of Genesis 1:28, God's instructions to Adam and Eve were to "be fruitful, and multiply, and replenish the Earth, and subdue it, and have dominion over the fish of the sea, and over the fowl of the air, and over every liv-

ing thing that moveth upon the Earth." Dominion over the creatures and replenishment of the Earth implies some sort of management—or conquest—of wildlife and their environments.

Whether the story of Adam and Eve is merely symbolic of life's beginnings is unimportant here. Instead, we are concerned with the underlying attitude expressed in Genesis that portrays the Earth as the human dominion. That concept has influenced the behavior of countless generations of humans living in western Asia, Europe, and most recently in the Western Hemisphere (White 1967). For centuries, humans have enjoyed overwhelming success at being fruitful and multiplying. People have gained dominion over other living things, except for certain noxious microbes, insects, and rodents, but the human record for replenishing the Earth has been rather dismal.

Humans in the Western world have taken to heart the instruction to subdue the Earth. Wilderness and wild animals for centuries were regarded as no more than enemies to be conquered or, at best, as subordinates to be controlled (Nash 1982). Only in recent years have we started to regard ourselves as one part of an intricate web of life. In the modern view, we recognize our links not just with the physical environment but also with vegetation and other animals, including those organisms that ultimately decompose all forms of life.

As civilization spread across Europe in antiquity (ca. 1000 B.C.–A.D. 500), wild animals were gradually replaced by domestic flocks. Lions (*Panthera leo*) soon were extirpated in Europe and gray wolves (*Canis lupus*) were pushed into remote patches of mountainous and forested lands. Wild aurochs (*Bos primigenius*) were captured and bred with other cattle until the aurochs no longer existed in recognizable form. Some wild animals—such as deer, grouse, and foxes—remained in forests that were owned by nobles; these forms of wildlife were reserved for the sport of the privileged classes. Classes of lesser standing, however, regarded all property of the nobles—wildlife included—with bitter disdain. Moreover, peasants considered wildlife as competitors for forage or as predators on their herds and flocks (Leopold 1933a). Thus, in the Middle Ages the large population of peasants regarded poaching as a legitimate activity—whether for meat, for the welfare of domestic animals, or for an expression of social unrest. Even today, violation of game laws still is regarded in some parts of the world as a means of flouting authority.

Market hunting represented one of the darkest eras of exploitation of wildlife in the United States. Among the birds hunted commercially were those that today enjoy a far different status in the eyes of the public. In 1807, for example, the pioneer ornithologist Alexander Wilson planted an anonymous newspaper story that certain berries in the diet of robins (*Turdus migratorius*) rendered the birds unwholesome as human food. This fabrication for a time slowed the slaughter of robins for the Philadelphia market (Elman 1982). Robins nonetheless remained popular tablefare for Americans. In the winter of 1902–1903, a single market hunter sold 120,000 robins for the cuisine of patrons who frequented the hotels and restaurants in Texas and adjacent states (Casto 1984). Several kinds of plovers, curlews, sandpipers, and other shorebirds (Scolopacidae) also were considered delicacies and thus were shot in large numbers.

Market hunting, however, is more often associated with the exploitation of waterfowl, especially on the eastern seaboard, where immense flocks of ducks gathered on such fabled hunting grounds as Chesapeake Bay and Long Island Sound and where there was a large demand for fresh meat. Market hunters often used punt guns, which were little more than homemade cannons mounted on the bows of flat-bottomed boats. Loads equivalent to 10 or more modern shotgun shells were fired as a single charge, and one shot from such a weapon into a raft of ducks easily killed or wounded scores of birds. With two or more punt guns per boat, market hunters fired first on a raft of swimming birds, then let loose a second salvo into the remaining flock rising from the water. Because there were no legal restrictions, the only factors limiting market hunting were the skill of the hunter, the supply of powder and shot, and the seasonal availability of waterfowl (Fig. 2-1).

Day (1949) summarized some of the excesses of the era, including one dealer's shipment from Virginia of as many as 1000 ducks at a time throughout a 6-month season each year—and his marketing career spanned 30 years! Waterfowl shot on Currituck Sound, North Carolina, provided at least $100,000 each year to the local economy between 1903 and 1909. The slaughter, however, by no means was limited to the eastern states. A market hunter in Louisiana boasted of a day's kill of 430 ducks, and in 1906 two hunters in California killed 218 geese in an hour, taking a total of 450 geese for the day. In Iowa, a group of market hunters shot an average of 1000 ducks per week, or 14,000 ducks and shorebirds dur-

Figure 2-1. Waterfowl were hunted with little restraint in the days before enactment of the Migratory Bird Treaty in 1916 and the Migratory Bird Treaty Act in 1918. Among other provisions, these measures ended market hunting and became the legal foundation for modern waterfowl management. (Photo courtesy of the McCulloch Collection, Highland Heights, Ohio.)

ing a season (Musgrove 1949). The ducks sold for $6 to $15 per dozen, depending on the species, but at times the price was only 10 cents a bird. Another operation in Iowa profitably coupled market hunting with the hardware business. In some seasons, 75,000 ducks were marketed and ammunition sales reached 250,000 shells.

Some kinds of waterfowl were exploited for their plumage, among them the trumpeter swan (*Cygnus buccinator*). The skins of some 17,500 swans were sold between 1853 and 1877. Kills of such magnitude drastically reduced the population, and only 57 skins were sold between 1888 and 1897 as the availability of swans diminished over much of their former range

(Banko 1960). More than 100,000 swan skins had been marketed by the Hudson's Bay Company by the time overhunting forced trade to a close (Banko and Mackay 1964). Unfortunately, most of the market hunting for swans took place during the breeding season when the birds could be located easily. Confirmation of the Migratory Bird Treaty in 1916 and enactment of the Migratory Bird Treaty Act in 1918 brought an end to the destructive institution of market hunting for waterfowl.

Plume hunting destroyed colonies of wading birds in Florida and other states. Because of their exceptionally attractive breeding plumage, herons and egrets (Ardeidae) were ruthlessly slaughtered

Photo courtesy of the Bancroft Library, Univ. of California at Berkeley

Theodore Roosevelt
President, Hunter, and Conservationist
(1858–1919)

The man who served as the 26th president (1901–1909) of the United States established his place in history in numerous ways: as commander of the Rough Riders in the Spanish-American War; as a "trust buster;" as the first American to win a Nobel peace prize; as the builder of the Panama Canal; and, at 42, as the youngest man ever to become president. "Teddy" Roosevelt also was the U.S.'s most conservation-minded president.

Roosevelt led what he called the strenuous life, in which he enjoyed boxing, hiking, riding horses, swimming, and hunting. In the 1880s, he owned two ranches in the Dakota Territory, where he hunted bison and spent long hours in the saddle tending cattle. His western experiences produced three books about hunting; among them, *The Wilderness Hunter* (1893) included life histories of big game rather than simple narratives about trophy hunting, as was the practice at the time. Roosevelt's love of the outdoors also prompted his active political role in conservation—an involvement

strengthened by his association with Gifford Pinchot (1865–1946), the first Chief of the U.S. Forest Service.

President "Teddy" Roosevelt initiated in 1903 what was to become the National Wildlife Refuge System when he declared Pelican Island, Florida, a federal sanctuary for herons and egrets. He also withdrew 51 million ha of land from the public domain to create national forests and millions more as reserves for coal and minerals. He advised Congress that "forest and water problems are perhaps the most vital internal problems of the United States." In 1908, he convened the first White House Governor's Conference, which focused national attention to the conservation of natural resources. The conference emphasized the importance of science as the basis for management, the renewability of forest resources, and the concept of multiple use.

Roosevelt was a founding member (in 1888) of the Boone and Crockett Club, an organization of hunters dedicated to the conservation of big game and other wildlife. He also was an associate member of the Society of American Foresters. During his presidency, Teddy Roosevelt visited Yosemite National Park where he camped with John Muir (1838–1914; on right in photo), the famed advocate of redwoods and wilderness. He also maintained a close friendship with writer-naturalist John Burroughs (1837–1921), with whom he shared a lifelong interest in ornithology.

After his presidency, Roosevelt hunted in Africa and provided specimens that are still on display in the American Museum of Natural History. In 1914, he explored the River of Doubt in the Brazilian rain forest, where he contracted jungle fever. He returned to the United States weakened and prematurely aged. His legacy as an outdoorsman and proactive conservationist remains a model for future presidents. Burroughs (1906) and Cutright (1956, 1985) detail Roosevelt's life as a naturalist and conservationist; see also Pinchot (1947) and Pinkett (1970).

for the sake of women's fashions. The popularity of feathered hats boomed in about 1875, and for the next 30 years the millinery trade demanded huge quantities of egret plumes—feathers that could be obtained only from birds killed during the nesting season. Egret plumes at one time sold for $1130 per kg, and at such prices, Florida's 1877 antiplumage law remained ineffective (Wallace 1963). However, as a direct result of the campaign against plume hunting, President Theodore Roosevelt issued an executive order in 1903 declaring a small nesting area—Pelican Island, Florida—as the first federal bird sanctuary. Pelican Island itself was only a 6-ha bit of brush-covered land, but it was crowded with the nests of egrets and other species ravaged by plume hunters. More important, Roosevelt's order unleashed a flood of nominations of other sites for similar protection. In 1904, at the end of his first term, Roosevelt's decrees had created 51 wildlife refuges (Trefethen 1975). Teddy Roosevelt's reaction to plume hunting thus initiated what later became the National Wildlife Refuge System; today the refuge system includes more than 500 units in the United States and its territories.

BISON: A STORY OF NEAR EXTINCTION

Vast herds of bison (*Bison bison*) dominated the Great Plains of North America for millennia. Before 1850, people coexisted with bison; humans used the hides for shelter and clothing, the meat for food, the sinews for bowstrings, and the dung for fuel (Owen 1980; see also Barsness 1985). Whenever local segments of the herd were victims of drought or were overhunted, hard times followed for the human cultures dependent on a bison economy. Overall, a food chain of grass–bison–human persisted without interruption for thousands of years.

Matters changed rapidly in the latter half of the 19th century. Wholesale slaughter of bison began in earnest when people of European cultures, which had a long-standing ethic of subduing nature, began to settle the Great Plains. The U.S. Army waged war on the bison as a means of reducing the food supplies of the Plains Indians. Newly constructed railroads provided ready access to the herds by men equipped with efficient repeating rifles. "Buffalo Bill" Cody sometimes shot 200 bison in a day and reportedly killed 4280 bison in 18 months (Owen 1980). More than 1.5 mil-

lion shaggy hides ("buffalo robes") were sold on eastern markets in the winter of 1872–73. The thick leather also became the staple material for machine belts. Otherwise, only the tongue—considered a delicacy—was stripped from most carcasses, and huge quantities of meat were left to rot on the plains. In Kansas, the land that once teemed with life was described as a "putrid desert" (Dodge 1883). Poor shooting and inadequate preservation of the hides added to the kill; three to five animals were shot for every hide that reached the market. In 1871 and 1872, about 8.5 million bison were shot (Fig. 2-2). By 1874, the immensity of the slaughter had incensed the U.S. public, and the matter reached the floor of Congress. Congress enacted protective legislation, but the bill was vetoed by President Grant. The slaughter continued relentlessly for more than another decade. Thus, from a population sometimes estimated at 60 million in about 1860, only 150 bison remained in the wild in 1889. In 1894, a rancher in Park County, Colorado, shot the last free-roaming bison in the United States. Fortunately, a few small herds of bison still persisted on public lands (e.g., Yellowstone National Park).

Most people associate bison, or "buffalo," with the American West, but a counterpart in Europe (*Bison bonasus*) is only a little less massive, with somewhat shorter horns. Like its American relative, the European bison also barely avoided extinction. In this case, gradual reduction of the herds coincided with the transformation of their forested habitats into cultivated fields (Klos and Wunschmann 1972). Only a few hundred European bison remained in the early 19th century; these were secluded in the Bialowieza Forest in Poland. Even there, poaching continued to deplete the population, and the hunger and turmoil of World War I brought utter destruction. The last free-roaming European bison in the wilds of Bialowieza Forest was shot on February 9, 1921.

THE PASSENGER PIGEON: AN AMERICAN TRAGEDY

Some bison survived the slaughter (see Chapter 3), but not all species were so fortunate. The passenger pigeon (*Ectopistes migratorius*) probably was the most numerous bird on Earth as recently as the middle of the 19th century (Fig. 2-3). Passenger pigeons certainly were the most abundant bird in pristine North America. Alexander Wilson recorded the pas-

Figure 2-2. The slaughter of millions of bison is difficult to comprehend, but this scene helps bring the enormity of the carnage into focus. By 1889, after two decades of overexploitation, the herd of 60 million bison had been reduced to just 150 animals on the open ranges of the U. S. plains. (Photo courtesy of Panhandle-Plains Historical Museum, Canyon, Texas.)

sage of flocks that darkened the sky during migration—one flight alone was 400 km long—and was awestruck when he saw tree limbs break under the weight of the perching birds. Wilson watched for hours as a flock that he estimated at no less than 2 billion birds passed overhead at a viewing point in Kentucky. Schorger (1955) noted that Wilson once counted as many as 90 nests per tree throughout a stretch of forest 5 km wide and 67 km long. In 1871, an estimated 136 million passenger pigeons nested in a 2200-km^2 area of central Wisconsin (Schorger 1937). An immense tonnage of droppings fertilized the forests where passenger pigeons roosted (McKinley 1960). The teeming abundance of this species made it clear that fecundity alone does not account for population size; a passenger pigeon laid just a single egg. Similarly, immense abundance does not immunize a species from extinction. About 100 years after Alexander Wilson marveled at their vast numbers, passenger pigeons were gone from the Earth.

Why did this extinction occur? Millions of passenger pigeons were killed for food. In the span of 3 months in 1878, more than 1.5 million pigeons were shipped to market from a nesting area in Michigan. All told, as many as 10 million may have perished from hunting at that one colony (Blockstein and Tordoff 1985). The birds were served in fashionable restaurants in Chicago, New York, and Boston for 2 cents each. Squabs were a particular delicacy;

hence, trees in nesting colonies sometimes were felled to obtain large numbers of the young birds. Thus, the outright killing of passenger pigeons often was accompanied by the destruction of irreplaceable nesting habitats. The lack of refrigeration meant that even more birds were killed to offset losses from spoilage during transportation to market. Blockstein and Tordoff (1985) and Cobb (1987) noted that two technological developments of the 19th century were factors in the equation of extinction: railroads and the telegraph. The extensive rail network that was in place by the time of the Civil War gave professional pigeon hunters—some 1000 strong at the peak of market hunting—ready access to the major nesting colonies of passenger pigeons east of the Mississippi River. One dealer in New York City received 18,000 birds a day as a result of the new rail connections. Because passenger pigeons were nomadic, the telegraph kept hunters informed about the locations of nesting colonies. Because the railroads benefited from their association with the market hunters, the train companies likely helped transmit up-to-date information concerning places where pigeons might be harvested. Such relentless disturbance of the colonies resulted in large-scale nesting failures year after year, and passenger pigeons steadily diminished.

A subtle but significant behavioral feature added to the plight of passenger pigeons. As part of their

Figure 2-3. A rare photo of a living passenger pigeon—probably a male—taken by Wharton Huber (1877–1942). A widely traveled naturalist, Huber perhaps took the picture at Woods Hole, Massachusetts, where a few of the remaining passenger pigeons were held in an aviary. Passenger pigeons once were the most abundant bird in North America, and a single flock estimated at 2.2 billion was observed about 1810 by the ornithologist Alexander Wilson. "Martha," the last of the species, died in 1914, but the date for the photograph shown here remains unknown. (Photo courtesy of Library, The Academy of Natural Sciences of Philadelphia.)

gregarious social behavior, passenger pigeons apparently maintained their nesting colonies with a "critical mass" of breeding birds (Halliday 1980); that is, some minimum number was required as a catalyst for breeding. Hence, as the population of passenger pigeons diminished, the social efficiency of their colonial breeding also declined. By 1885, the last colony was located in Wisconsin, and only isolated reports of breeding were noted thereafter (Schorger 1955). The last passenger pigeon died in the Cincinnati Zoo in 1914 (Greenway 1967). The decline and eventual extinction of such an abundant species now seems beyond imagination. But, lacking fanfare or remedy, passenger pigeons indeed vanished forever—an event marking one of the darkest hours in environmental history.

OTHERS, TOO, ARE GONE

The Labrador duck (*Camptorhyncus labradorius*) was an attractive, small diving fowl once found along the northern Atlantic Coast of North America (Greenway 1967). The only tangible evidence of the breeding range of Labrador ducks is a set of eggs supposedly collected in Labrador and now in the care of a German museum. Johnsgard (1968) underscored how little is known about the Labrador duck: "It disappeared so swiftly and so quietly that it is difficult . . . to compose a complete obituary. We do not know for certain where it nested or exactly what it consumed, nor do we even have a record of the appearance of its downy young."

An examination of the few skins in museum collections reveals that Labrador ducks had unusual bills, characterized by soft edges bordered with a large number of comblike structures known as lamellae. Such features suggest specialized feeding habits. The diet of Labrador ducks may have included mollusks, as the birds sometimes were caught on fish lines baited with mussels, but the shape and structure of their bills also suggest surface-feeding behavior (Johnsgard 1968). The population apparently was never large, and the birds sometimes were subject to the pressures of market hunting. However, Phillips (*in* Delacour 1959) suggested that Labrador ducks were not eliminated by overshooting but by changes made by humans in the marine environment along the Atlantic Coast. These changes were associated with human settlement that consequently destroyed the food resources needed by these waterfowl. Regrettably, we shall never know what happened. Between 1850 and 1870, Labrador ducks gradually disappeared, and the last-known specimen, a male, was shot on Long Island in the fall of 1875.

The heath hen (*Tympanuchus cupido cupido*) was an eastern prairie grouse, one of three races of a once widely distributed species. The two other races include the greater prairie chicken (*T.c. pinnatus*) of the plains states and the Attwater prairie chicken (*T.c. attwateri*) of coastal Texas. The last record of a heath hen was of a male banded on Martha's Vineyard on April 1, 1931; the same bird was observed on Martha's Vineyard nearly a year later on March 11, 1932. That bird was the last-known survivor of a population that once lived in woodland clearings from New England south to Virginia and possibly into the Carolinas (Greenway 1967). As early as 1791, the New York legislature enacted protective legislation for heath hens, but the law was generally ignored because of the sporting and market values of the birds. By 1844, the population on Long Island had been eliminated. Heath hens already had been extirpated from mainland Massachusetts and Connecticut by 1830. Of 2000 birds still present on Martha's Vineyard in 1916, habitat destruction (fire) and disease claimed all but about 25 by 1925. Those losses, coupled with predation, poaching, and a sex ratio favoring males, doomed the heath hen. Unfortunately, a similar fate may await the Attwater prairie chicken if the intensive management efforts currently underway are unable to check the decline of the small population yet remaining in Texas.

A small, green parrot with prominent orange and yellow head markings once lived in the heavily timbered bottomlands primarily associated with rivers and swamps in the southeastern United States. This was the Carolina parakeet (*Conuropsis carolinensis*), the only member of the parrot family that lived in the United States. Two races apparently met along the Appalachian Mountain axis and ranged from Virginia to Florida west to the riparian forests of Oklahoma, Kansas, Nebraska, and eastern Texas, although wanderers occurred with some regularity outside this area (Greenway 1967). Coincident with expanding human settlement in the 19th century, the eastern range of the Carolina parakeet steadily diminished and was forced westward to the Mississippi River. Except for those remaining deep in the swamps of Florida, the colorful birds had disappeared by the late 1870s.

Carolina parakeets devoured quantities of cockleburs and the seeds from cypress (*Taxodium distichum*), beech (*Fagus grandifolia*), maple (*Acer* spp.), elm (*Ulmus americana*), and even loblolly pine (*Pinus taeda*) trees (Cottam and Knappen 1939). The birds were particularly destructive to orchards and efficiently stripped apples, oranges, and peaches from trees while eating only some of the fruit. Nests were located in tree cavities well above the ground, and similar sites were used for roosting throughout the year.

Because Carolina parakeets depended on mature forests, the encroaching lumber industry potentially jeopardized the continued existence of the birds. However, overshooting and capture were the more direct causes of decline. Flocks were quite susceptible to hunting because of the birds' habit of hovering over the carcasses of those already killed. Farmers also destroyed parakeets to protect orchards, and large numbers were trapped while roosting inside hollow trees. Unverified reports of Carolina parakeets persisted until 1938 (Sprunt and Chamberlain 1949), but the last authentic record is of a bird that perished in the Cincinnati Zoo in 1914—coincidentally dying in the same year and place as the last passenger pigeon.

Other American birds disappeared under the press of the spreading human population. The great auk (*Pinguinus impennis*), a flightless penguinlike bird living on the coasts and islands of the North Atlantic, was relentlessly hunted and trapped first for meat and then for eggs and feathers. Finally, collectors struck the final blow when great auks became rare and thus of greater value (Bent 1963). A pair of great auks killed at Edley Rock on the coast of Iceland in 1844 marked the end of the species (Greenway 1967). The Hawaiian rail (*Pennula sandwichensis*), a small marsh bird, was exterminated when rats (*Rattus norvegicus*) were introduced accidentally to the islands by humans. Extinction of these and a large number of other birds throughout the world occurred for many reasons, including overkilling, destruction of habitat, and introduced predators and competitors, among other causes (Greenway 1967).

Extinction is a natural process that has taken place for millennia. Fossil evidence shows that organisms arose, thrived, and disappeared long before humans appeared on Earth. New forms of life evolved as others disappeared. The ebb and flow of life thus continued in an ever-changing mosaic of species. Why then should anyone express concerns about those extinctions we are witnessing today? Are modern extinctions not simply normal occurrences in the evolution of life forms? In part, the answer is yes—a few of the extinctions in recent history probably were part of the normal march of life. However, the other part of the answer is far more alarming, for extinctions now occur at a rate much greater than in any known period in geologic time. According to Curry-Lindahl (1972), 173 species and subspecies of birds have become extinct in historic times. Of these, 157 have disappeared since A.D. 1700. Today, another 388 birds are threatened with extinction. Mammals have not fared much better: 66 species and subspecies have vanished since 1900 (Dasmann 1975). Roush (1982) estimated that the present rate of extinction occurs between 40 and 400 times the pace at which newly evolved species replace those that are lost to extinction. Most recent extinctions can be traced directly or indirectly to human activities. These losses have occurred on every continent except Antarctica.

Aside from what surely is a moral responsibility not to destroy forever those plants and animals with which we share a common environment, we should note that losses from natural communities may undermine the present and future welfare of human populations. Of particular importance are those species involved with the complex transfer and circulation of materials that are essential for life. Extinct species are lost forever, along with the intrinsic beauty, practical merits, and opportunities they offer for scientific information. In short, once extinct, the unique pool of genes we know as a species cannot be recovered. No amount of genetic engineering, animal husbandry, or wildlife management can bring back the heath hen or the Labrador duck. The clear lesson, of course, is that action should be taken long before extinction becomes an alternative.

WOOD DUCKS, WILD TURKEYS, AND OTHER NEAR EXTINCTIONS

The wood duck (*Aix sponsa*), one of the most colorful ducks in the world, almost perished in the early decades of the 20th century (Bellrose 1976). Wood ducks nest in tree cavities in wet wooded areas east of the Great Plains. A smaller population also lives along the Pacific coast. As with bison and many other species, the journals of early settlers described wood ducks in terms of seemingly inexhaustible numbers. In the 1890s, thousands of wood ducks nested in the river bottoms and lowlands of the southeastern United States. A horseman described a 33-km trip in Arkansas during which wood ducks were never out of his sight. During the same period, wood ducks ranked as the most or the second most abundant duck in Michigan, Wisconsin, and Ohio. Nonetheless, because of their attractive plumage and distribution in regions with dense human settlement, wood ducks were vulnerable to shooting during much of the year, including the summers, when fledglings and flightless molting adults were shot.

In addition, nesting habitat for wood ducks dwindled rapidly early in the present century. Swamps were drained for farmland, and lumbering claimed much of the timberlands where wood ducks nested. By 1913, George Grinnell, Edward Forbush, and other prominent naturalists had predicted the end of wood ducks. Fortunately, the birds were spared the finality of extinction, and as will be outlined in Chapter 3, the rescue of wood ducks instead became one of the better-known success stories of wildlife management.

Wild turkeys (*Meleagris gallapavo*) offer another example of a species that had fallen to alarmingly low numbers early in the 20th century. Fortunately, management efforts were successful, and turkey populations recovered in many parts of the species' former range. Early reports described the former abundance of wild turkeys: sightings between 1790 and 1850 frequently noted flocks of several hundred birds, and some observers reported seeing 1000 turkeys in a day. Turkeys were particularly abundant in the southeastern United States and in Texas and Oklahoma; estimates indicated densities of 3 to 36 birds per km². Densities of turkeys in Maine, Vermont, South Dakota, Nebraska, Colorado, and Arizona were lower, with estimates of about 0.4 birds per km². In presenting these data, Schorger (1966) concluded that about 10 million wild turkeys had lived in North America in pre-Columbian times.

Turkeys were shot and trapped for food with heedless abandon, with the result that populations in many areas soon were diminished. As early as 1708, the Colony of New York prohibited the killing of turkeys in three counties between April 1 and August 1. Thereafter, other colonies, and later states, protected wild turkeys during the nesting and brood-rearing seasons. Still, the population continued to decline and, during the following 2 centuries, turkeys were extirpated in the northern part of their range. No turkeys remained in the New England states nor in Michigan, Wisconsin, and Minnesota. Merely closing the hunting season for part of the year did not provide sufficient protection nor an adequate form of turkey management. Other restrictions followed. In 1897, Pennsylvania set a daily limit of two birds and in 1905 reduced the bag to one per day with a season limit of four birds. Soon other states also set daily and seasonal bag limits. Florida established a 10-bird season limit in one county and then reduced the season limit to five a few years later. In 1920, Kentucky set a limit of two birds per season; Texas allowed three and Virginia allowed six birds.

Such protection finally stemmed the decline of wild turkeys, and several states thereafter initiated restoration programs. Turkeys were released in parts of their former range, but many of the introduced birds simply disappeared. The reason, discovered after many failures, was that game-farm strains of turkeys were released in most of the restocking efforts. Such birds were ill-adapted for survival in the wild, and they soon perished. In a few instances, various races of wild turkey were mismatched and did not correspond to the ancestral distribution range for each race. For example, birds from southern Texas were released in Iowa, and stock from Florida was sent to Colorado. These releases also failed, but with new knowledge and renewed efforts, wildlife managers eventually restored viable populations of wild turkeys to many areas of the United States.

California condor (*Gymnogyps californianus*), the peregrine falcon (*Falco peregrinus*), Kirtland's warbler (*Dendroica kirtlandii*), and the Indiana bat (*Myotis sodalis*) are only a few of the many species that have approached extinction. As noted in Chapter 8, disease recently decimated the remaining population of black-footed ferrets (*Mustela nigripes*), perhaps the most endangered species of mammal in the world. Throughout the world, some 290 species of mammals and 388 species of birds are endangered. Each example amply demonstrates our failures in the treatment of wildlife. We have failed to recognize the plight of these animals or, worse yet, have not responded once precarious situations have been recognized. The former excuse perhaps resulted from mere ignorance, but, in light of the unchecked slaughters, the latter seems unconscionable. Today, the perilous status of many animals has been recognized, and the governments of many nations indeed have committed funds and other resources to the restoration of these species. As described in Chapter 19, biologically sound management plans are now in effect for the recovery of some threatened and endangered species. With the implementation of these plans comes new hope for the lasting coexistence of humans and wildlife.

PROBLEMS OF EXCESS: REINDEER, DEER, AND BLACKBIRDS

Not all concerns of wildlife management deal with the scarcity of animals. Problems sometimes arise from overabundance of particular species. A dynamic example resulted from the introduction of 29 reindeer (*Rangifer tarandus*) on St. Matthew Island in the Bering Sea (Klein 1968). The original herd of 24 cows and 5 bulls was released on the 332-km^2 island in 1944 and then expanded to 6000 animals by the summer of 1963 (Fig. 2-4, left). The following winter, the herd crashed in a die-off that left only 42 animals, all females except for one apparently sterile male. The animals died of starvation when the supply of lichens—the dominant plants in the diet of reindeer— was depleted by the excessive grazing pressure of the oversized herd. Overgrazing also had inflicted long-term damage on the original vegetation. Based on a comparison with a nearby island where there were no reindeer or other large grazers, no more than 10 percent (by weight) of the lichens had recovered on St. Matthew Island 22 years after the reindeer herd crashed (Klein 1987). Moreover, the new lichens were species of relatively low preference as

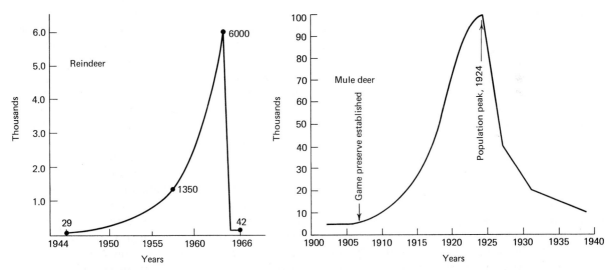

Figure 2-4. Irruption of reindeer (left) and mule deer (right) populations in situations where predators were absent or removed. After irrupting, the populations crashed when food supplies were exhausted. Data for reindeer are based on counts, whereas those for mule deer are speculative. (From Klein 1968 and Rasmussen 1941.)

reindeer forage rather than the more palatable species that once had characterized the vegetation on St. Matthew Island.

The irruption of mule deer (*Odocoileus hemionus*) on the Kaibab Plateau in Arizona is frequently cited as an example of overabundance. According to Rasmussen (1941) and Leopold (1943), mule deer in the Kaibab Forest north of the Grand Canyon apparently profited from predator control and protection from shooting during the first 2 decades of the present century. Freed from these controls, the deer population increased from about 4000 in 1905 to an estimated 100,000 in 1924. The vegetation could not keep pace with the feeding pressure of so many additional deer. Lacking enough forage, about 60 percent of the Kaibab herd starved in two winters; the population fell to 10,000 by 1940 (Fig. 2-4, right). The example of the Kaibab herd thus offers a case history of what can happen to deer populations when natural predation and hunting are eliminated. Caughley (1970) questioned the nature and magnitude of the irruption, but strong evidence suggests that the Kaibab deer herd expanded rapidly and then suffered extensive starvation.

A more recent and documented example on a smaller scale concerns an overprotected deer population in Upper Michigan (Robinson et al. 1980). In 1959, a corporation began to operate a hunting resort on 5 km^2 of coniferous lowlands in Beaver Basin on the shore of Lake Superior. A few years earlier, the last pack of wolves, the only effective natural predators of white-tailed deer (*Odocoileus virginianus*) in Michigan, had disappeared from the area. Managers at the resort nourished the herd with commercial food each winter. Deer numbers soon increased in response to the supplementary food, but hunting and other losses were inadequate to keep the well-fed herd in check—reproduction and survival simply outstripped total mortality.

By the early 1970s, the growing herd was receiving more than 80 tons of commercial deer food each year. Food and labor costs also were growing. Then, in 1974, the corporation donated the land to the National Park Service for an addition to Pictured Rocks National Lakeshore, and with the land came a winter population estimated at more than 600 deer. The Park Service reduced the food supplement to about 20 tons for each of the next two winters. In 1976, officials proposed a special hunt that would cut back the herd to a maximum of 300 deer, thereby reducing the population to a level that could be maintained by

natural browse and ending dependence of the herd on artificially supplied foods.

A local hunting club, however, claimed that the deer population had been overestimated and that a harvest of 300 would annihilate the herd at Beaver Basin. The club obtained a court injunction against the hunt and spent about $27,000 feeding the deer (against Park Service regulations) in the winter of 1976–77. In the following winter, 1977–78, the herd was neither fed nor subjected to a special hunt. As a result, starvation and predation by coyotes (*Canis latrans*), which had increased as the deer herd expanded, reduced the late-winter population to about 100 deer. By 1983, the winter herd at Beaver Basin had declined to zero. In 1996, biologists returned to Beaver Basin to determine whether deer had recolonized the area (Corrinth et al. 1997). The winter deer population was still zero, and white cedar (*Thuja occidentalis*) and maples still showed the effects of overbrowsing. However, after 14 years without browsing pressure, cedars in swamp areas and maples in uplands were regenerating with seedlings and sprouts. The heavy browsing pressure from the oversized herd had initiated the replacement of a valuable deer food, white cedar, with a marginally palatable species of browse, balsam fir (*Abies balsamea*) (Jensen 1982). The result was decimation of a deer herd, loss of recreation and meat, and long-term damage to winter deer range. Moreover, deer that now spend the warmer seasons in Beaver Basin leave in late autumn for more southern areas. Apparently, starvation had exterminated the entire deer population that once migrated westward into Beaver Basin for the winter, and despite the availability of choice winter browse, new generations of deer have not renewed the former tradition of migration.

Wildlife management did not fail at Beaver Basin because of poor biological knowledge. Instead, the failure occurred because local hunters and a judge were not convinced that a special hunt was a desirable means of balancing deer numbers with the supply of natural foods (see also Gilbert 1977).

In another example, overabundance occurs when hundreds of thousands—and sometimes millions—of blackbirds concentrate in winter roosting areas (Graham 1976). The flocks consist of four species: redwinged blackbirds (*Agelaius phoeniceus*), starlings (*Sturnus vulgaris*), grackles (*Quiscalus quiscula*), and cowbirds (*Molothrus ater*). Huge flocks of these birds winter in Arkansas, Louisiana, Tennessee, and Kentucky, where field crops and cattle feedlots provide ample food supplies (see White et al. 1985). The immense concentrations of blackbirds apparently result from the proximity of suitable roosting sites to feeding areas. In such large numbers, the birds are noisy and dirty and may damage crops and offer potential health hazards.

Several methods have been tried to reduce the flocks, including spraying the birds with detergents on cold nights (thereby causing death from exposure when the soaked plumage can no longer provide insulation), frightening the birds with noisemakers and playing recorded bird distress calls through loudspeakers, and thinning the roost trees. However, these techniques have been only partially successful and have raised the ire of many bird lovers. Also, the process of thinning the winter flocks of blackbirds has been questioned on ecological grounds because, on their summer range, these birds may reduce insect populations (Bendell et al. 1981). The difficulties of controlling some 550 million blackbirds without upsetting natural systems or human emotions thus remain a clear challenge for wildlife managers.

PREDATOR CONTROL: BOUNTIES, BAITS, AND BLUNDERS

Wildlife biologists still have much to learn about predator-prey relationships and under what circumstances predator control should be initiated. However, we do know that mistakes have been made many times in the past. The public once assumed that all carnivores threatened the continued existence of game and domestic animals. In particular, William T. Hornaday championed the view that predators were the root of considerable evil. He rallied public concern for the nation's dwindling wildlife resources with his landmark book, *Our Vanishing Wild Life* (Hornaday 1913). In his book, Hornaday condemned all killers of animals in sweeping terms, blaming cats, dogs, Italian immigrants, poor Southerners, hunters, and wild predators. A particularly forceful sentence reads: "Beyond question, it is both desirable and necessary that any excess of wild animals that prey upon our grouse, quail, pheasants, woodcock, snipe, mallard duck, shore birds, and other species that nest on the ground should be killed." We can only wonder how Hornaday, given his fateful plans for an excess of predators, might deal today with an "excess" of hunters, Southerners, or immigrants.

Predator control became a common practice in the 1930s under authorization from state and federal agencies. State and county governments offered bounties on several predators, including wolves, foxes (*Vulpes vulpes*), weasels (*Mustela* spp.), and common crows (*Corvus brachyrhynchos*). At first glance, placing a price on the head of an offending animal would seem an effective means of control. After 20 years of employing state trappers for predator control, Michigan adopted the bounty system in 1935 and offered $15 per male and $20 per female for wolves and coyotes. About 3000 coyotes were bountied each year during the 1960s and 1970s, but no reduction was noticeable in the coyote population nor was there any sign of corresponding increases in grouse, rabbits, or other wildlife populations. Trappers and hunters received millions of dollars from the public treasury for harvesting what was only part of the annual surplus of coyotes. Even drivers who accidently ran over coyotes were paid when the carcasses were turned in. The Michigan legislature finally removed the coyote bounty in 1980; since then, there has been no perceptible increase in coyote numbers nor declines in rabbits, grouse, or deer. Similarly, the number of bounties paid for weasels in Pennsylvania increased from 36,816 in the 5-year span 1915–20 to 68,423 weasels in 1930–35. After 20 years and $1,209,500 in bounty payments, however, the weasel population in Pennsylvania remained without noticeable decrease (Allen 1974a).

In addition to being generally ineffective, bounties also are subject to fraud. In New Brunswick, a 50-cent bounty on porcupines (*Erethizon dorsatum*) persisted into the 1950s. To collect a bounty, a citizen simply presented the nose of a porcupine to the town clerk, who was often a person with neither the training nor the inclination to perform close examinations of porcupine noses. As more than one entrepreneur discovered, two holes punched in each foot pad immediately produced five "noses"—and $2.50—from one porcupine. Elsewhere, bounties were paid for the ears of dogs (as ersatz ears of coyotes or wolves), and whole animals or parts thereof were carried across state and county lines for redemption. Trappers also released pregnant coyotes from traps, thereby assuring another generation of bounty payments. These practices, together with the biological shortcomings, clearly made the bounty system an ineffective and wasteful attempt at managing wildlife.

Coyotes have become the "most wanted" predators in North America. In western states, where they sometimes prey on sheep, coyotes have been killed by shooting, trapping, den digging, strychnine-treated baits, sodium monofluoroacetate (compound 1080), and cyanide guns. These methods, although favored by ranchers, have not always been acceptable to other sectors of the public. Thus, in 1970, President Richard Nixon appointed a select committee of wildlife biologists to study coyote control. The committee recommended that poisons no longer be used on federal lands (Cain et al. 1972). Two reasons were cited for the committee's recommendation: clear evidence was lacking that poisoning effectively controlled coyote numbers, and nontarget animals such as badgers (*Taxidea taxus*) too often were unintentional victims. In the mid-1960s, 1080-treated baits placed within the range of the endangered California condor produced claims that a few of the birds may have been poisoned unintentionally (McNulty 1978).

An Executive Order issued in 1972 banned any further use of poisons on federal lands, but the dispute was reopened in 1982 under heavy pressure from sheep and goat ranchers. The issues are complex and not without controversy: they involve the number of sheep killed by coyotes, the number of sheep that would die in the absence of coyotes, the effectiveness of poison in reducing sheep losses, the number of nontarget victims of poison, and the economic costs and benefits of poisoning. More about these and other issues regarding predation appears in Chapter 9.

EXOTIC WILDLIFE

The list of foreign species now residing in North America is long indeed. Among these are popular game species such as ring-necked pheasants (*Phasianus colchicus*) from Asia and brown trout (*Salmo trutta*) from Europe. On privately owned ranches in Texas, hunters pay large sums for hunting axis deer (*Axis axis*), blackbuck antelope (*Antilope cervicapra*), eland (*Tragelaphus oryx*), and other species of big game from Asia and Africa. Conversely, starlings, house sparrows (*Passer domesticus*), and brown rats (*Rattus norvegicus*)—each introduced from Europe—are well-known pests in North America. Other introductions have failed. Two large European grouse, the black grouse (*Lyrurus tetrix*) and the capercaillie (*Tetrao urogallus*), were released several times but never established viable populations in North America. North American

species also have been exported to other lands. Europeans now trap muskrats (*Ondatra zibethicus*), hunt white-tailed deer, and disparage gray squirrels (*Sciurus carolinensis*). Elsewhere, European rabbits (*Oryctolagus cuniculus*) that were released in Australia soon disrupted both the native biota and the agricultural economy in much of the island continent. The result was one of several ecological disasters associated with exotic animals that are described in Chapter 18.

Exotic plants likewise have frequently become nuisances in their new environments. On the northern prairies of the United States and Canada, two unpalatable European species—leafy spurge (*Euphorbia escula*) and spotted knapweed (*Centaurea maculata*)—have aggressively replaced native plants browsed by cattle and wild herbivores (Harris and Cranston 1979; Belcher and Wilson 1989). Similarly, the highly competitive nature of purple loosestrife (*Lythrum salicaria*) has eliminated many species of wetland plants from marshes in eastern North America (Thompson et al. 1987). An Eurasian species, purple loosestrife provides waterfowl with little cover and virtually no food compared with the native plants it displaces. Based on recent tests, three species of plant-eating, host-specific insects—a weevil and two beetles—associated with purple loosestrife in Europe may offer a means of controlling this nuisance without resorting to herbicides. If effective as biological controls, these insects may be reduce purple loosestrife to 10 percent of its current abundance over 90 percent of its range in North America (Malecki et al. 1993).

SUMMARY

Extinctions rank high among the cases of human exploitation and neglect of the biological world. The loss of those species—numbering several hundreds in recent centuries—is irreversible; with each extinction goes part of a mutually interdependent community of life and the potential discovery of new resources for human welfare. Spectacular flights of passenger pigeons will never be seen again, and the Labrador duck disappeared before we learned anything about its natural history. Wild turkeys, wood ducks, and bison each have been at such low numbers that extinction was barely averted. Conversely, overprotection of some species, particularly of deer, has led to uncontrolled population growth, starvation, and damaged habitat. The bounty system failed as a means of controlling predators. Poisoning has not produced significant reductions of predator populations, but indiscriminate use of some poisons has killed nontarget victims of other species.

Some animals have disappeared forever in the wake of human disregard for the landscape; others have become plagues; and some have stubbornly resisted control. The lesson may be that "dominion" over the creatures of the Earth may require more than mere strength and advanced technology. Instead, an ecological understanding and the application of that understanding may produce mutual benefits for humans and our fellow creatures.

CHAPTER 3

SOME SUCCESSES IN MANAGING WILDLIFE

*Every lover of nature, every man who appreciates the
majesty and beauty of the wilderness and of wild life,
should strike hands with the far-sighted men who
wish to preserve our material resources, in the efforts
to keep our forests and our game beasts, and game
birds, and game fish—indeed, all the living creatures
of prairie and woodland, and seashore—from wanton
destruction.*

Theodore Roosevelt (1905)

Theodore Roosevelt published *Outdoor Pastimes
of an American Hunter* during his tenure (1901–09)
as 26th president of the United States. Roosevelt not
only reflected on the pleasures of his outdoor experiences, but also issued a call for stemming the destructive tide that was sweeping wildlife and wild
lands from the American landscape. His plea marshaled the concerns of others. The result has been a
movement that has continued to the present day—a
movement whose goal is the conservation and
restoration of wildlife populations and wildlife
habitats—and from the context of that movement
has evolved the practice of wildlife management.

Wildlife management, as we have defined it, involves the application of ecological knowledge to
achieve a balance between the needs of humans and
those of wildlife. Ecological knowledge was still in its
infancy a century ago, and what we regard as state-of-
the-art knowledge today surely will seem primitive
and crude a century hence. Wildlife management developed as a profession in the United States during

the 1930s, but attempts at management have a much
older history.

According to Leopold (1933a), the first game law in
North America dates to 1639 when Rhode Island
closed the hunting season for white-tailed deer
(*Odocoileus virginianus*) from May to November.
Massachusetts followed with a similar law in 1694. In
1708, the colony of New York protected ruffed grouse
(*Bonasa umbellus*), heath hens (*Tympanuchus c. cupido*), and wild turkeys (*Meleagris gallopavo*) during
part of the year. Virginia enacted the first "buck law"
in 1738, which allowed the legal kill only of antlered
bucks. The concept of curbing the daily kill—known
as the bag limit—did not emerge until 1878 when Iowa
established a limit of 25 greater prairie chickens (*T. c.
pinnatus*) per day. In the 1890s, several states totally
protected passenger pigeons (*Ectopistes migratorius*).

Those laws considered the seasonal vulnerability
of each species and recognized that game animals
might be overhunted. The regulations, however,
were made without any assessment of population

sizes; nor did the laws consider the reproductive potential of each species in relation to shooting pressure. Moreover, habitat was neglected by the lawmakers of the day, and no attempt was made to preserve or restore the food, cover, and water needed by wildlife. In short, ecological knowledge and its application did not exist in the realm of wildlife management. The laws protecting heath hens and passenger pigeons were obvious failures.

RESTORATION OF BISON

The mistreatment of American and European bison (*Bison bison* and *B. bonasus*, respectively) was remedied, in part, early in the 20th century. In 1905, only a few hundred of the once vast population of American bison remained; they were found in zoos and in Yellowstone National Park. However, their pitiful numbers prompted formation of the American Bison Association at the New York Zoo. Acting under the leadership of the zoo's director, Dr. William T. Hornaday, the association prodded the conscience of the American public, with the result that bison preserves soon were established. The Wichita Game Park (now a national wildlife refuge) in Oklahoma was stocked with 15 bison from the New York Zoo, and other refuges for bison were established in Montana, Nebraska, and South Dakota. The population of bison at Yellowstone Park steadily increased without stocking.

During this period, Canada also initiated a protection program for bison. Bison Park in Alberta was established specifically for the species, and by 1920 the sanctuary protected a herd of 5000 animals. Particular concern was focused on a badly diminished subspecies of bison, the wood buffalo (*B. b. athabascae*), which occupied wooded areas in northern Alberta and the Northwest Territories. Wood buffalo were crossbreeding with the subspecies from the plains, producing hybrids that were highly susceptible to tuberculosis. Fortunately, about 200 pure-blooded wood buffalo were discovered in a remote corner of Wood Buffalo National Park in 1960, and the integrity of the subspecies now seems assured. In all, estimates indicate that the bison population in North America now exceeds 30,000 animals.

A restoration program began in Europe after poaching and the ravages of World War I had eliminated the last of the free-ranging bison herd. Luckily, 56 bison remained in zoos and private game preserves that were scattered across Europe. Dr. Kurt Priemel,

a former director of the Frankfurt Zoo, formed the International Association for the Preservation of the European Bison. The association developed a studbook that listed the names and genealogies of all pure-blooded animals remaining in Europe. With the studbook, the captive herd could be propagated without indiscriminately crossbreeding the various strains of bison. The process assured the genetic integrity and the continued evolution of the respective subpopulations from distinctive habitats in Europe.

In 1956, a small herd of bison was restocked in the Bialowieza Forest of Poland, the site where the last wild bison had been shot in 1921. By 1963, the population had grown to 57 animals, of which 34 were born in the wild. A program of winter feeding helps maintain the precious few bison in the Polish herd. Restocking also took place in the Russian part of the Bialowieza Forest. These animals in turn produced a wild herd from which biologists learned much about the natural food habits and behavioral patterns of the European bison—information never before known to science. Thus, from the dedicated efforts of concerned zoologists and wildlife managers came the successful restoration of a small, but wild, population of bison in eastern Europe that remains today. Nonetheless, Klos and Wunschmann (1972) warn against keeping all animals of a rare species in one place. The dangers of contagious disease, warfare, or fires dictate that separate populations of wild bison should be established in various locations before the species can be considered safe from extinction.

Conservationists take rightful pride in the modest recovery of bison, a success that represents an important milestone in wildlife management. Still, two difficulties persist in the management of bison herds. First, the natural predators of bison are gone from most areas. The birth rate thus exceeds the mortality rate and produces a surplus of animals. Unlike in the past, an expanding bison population can no longer wander at will across a vast landscape. Farms and ranches claim most of the countryside, and these and other uses of the land are incompatible with free-roaming herds of bison. Second, and related to the first issue, adequate food supplies are a source of concern in bison management. Expanding herds that are confined by fences soon overgraze the best of rangelands. Artificial feeding is expensive and strains the budgets of conservation agencies. An obvious solution for these difficulties—hunting—angers some segments of the public, especially when the surplus animals are shot on refuges that were established for the protection of the species. Thus, although the bison population has grown from precariously low numbers, the success of the restoration program has itself raised other issues relevant to bison management.

LEAD POISONING: ALMOST GONE

A disease known as lead poisoning once killed as many as 2 million ducks and geese each year in North American (Bellrose 1959). The malady resulted from the lead shot—estimated at 3000 tons per year—deposited in wetlands by duck hunters. Feeding waterfowl pick up and ingest the expended shot, which the birds mistake for seeds or grit. Even a few pellets may prove deadly as the lead is absorbed slowly into the blood. Local factors influence the severity of lead poisoning, including shooting pressure, firmness of the wetland bottom, water depth, and the feeding habits of the birds (see also Bellrose 1975).

The internal signs of lead poisoning include atrophy of striated muscle tissue, a distended gall bladder, anemia, fluid accumulation in the pericardial sac, atrophy of the liver and kidneys, and, especially, erosion of the gizzard's grinding surfaces. Externally, bile stains the cloacal opening, and palsy causes drooped wings and impaired movement. Feeding also is curtailed so that emaciation usually is obvious.

Other waterbirds, including sora rails (*Porzana carolina*), also died from lead poisoning (Artmann and Martin 1975). Additional evidence indicated secondary poisoning occurred in bald eagles (*Haliaeetus leucocephalus*) that fed on duck carcasses contaminated with ingested lead shot (Pattee and Hennes 1983). The issue of lead poisoning therefore stimulated a good deal of research (see Feierabend 1983)—and a lot of controversy within the public interested in hunting and conservation.

In 1976, the Secretary of the Interior ruled that duck hunters must use steel shot in certain "hot spot" areas where lead poisoning was particularly severe. Steel-shot zones were established wherever more than 5 percent of the gizzards in the local harvest of mallards (*Anas platyrhynchos*) or black ducks (*A. rubripes*) contained lead shot (Longcore et al. 1982). The zoning program reduced the occurrence of lead in the gizzards of mallards by as much as 34 percent in some areas. Overall, however, the zoning regulations did not appreciably change the potential for lead poisoning in the Mississippi Flyway (Anderson et al. 1987).

Ammunition manufacturers and many waterfowl hunters opposed the steel-shot policy (Kozicky 1975; Smith and Townsend 1981). The opposition of these groups focused on three assertions:

1. *Crippling loss.* Steel shot, because of its lesser density, lacks the killing power of lead. Thus, crippling losses—birds knocked down but not retrieved—would be greater with steel shot.
2. *Gun damage.* Steel shot, because of its hardness, will increase barrel wear and ruin shotguns.
3. *Cost.* Shotgun shells loaded with steel shot cost more than those with lead shot.

Each of these concerns was addressed in a major review of lead poisoning (Sanderson and Bellrose 1986). First, as determined by field studies, crippling losses actually diminished in areas where steel shot was required (Table 3-1). A long-term reduction in crippling loss was already underway, and the trend was *not* changed when steel shot was used (Fig. 3-1). Many hunters use larger-size shot as a means of compensating for the reduced knockdown power associated with steel pellets (i.e., selecting No. 4 steel shot instead of No. 6 lead shot).

Second, fears about barrel damage were disproven in tests conducted by arms manufacturers and gun experts using modern shotguns. No damage was detected even after firing 18,000 rounds of steel shot.

Finally, although the cost of steel shot may be 25 percent greater than the cost of lead shot, the difference may diminish somewhat with the increased production of steel loads. In any case, an average duck hunter expends only 36 shots per year; hence the increased cost per hunting season for shells loaded

TABLE 3-1. Crippling Losses Before, During, and After Implementation of Steel Shot for Waterfowl Hunting. Data Are Expressed as the Mean Number of Knocked Down but Unretrieved Waterfowl per 100 Birds Retrieved by Hunters.

Species	Implementation of Steel Shot		
	Before (1971–75)	During (1976–78)	After (1979–84)
Ducks	21.6	20.0	19.5
Geese	14.6	14.9	14.0
Coots	29.0	28.2	27.1
All Species	21.4	19.9	19.1

Source: Sanderson and Bellrose (1986).

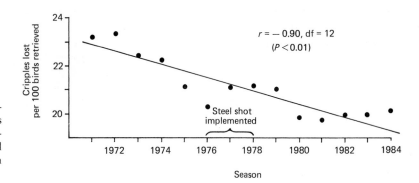

Figure 3-1. Relationship between crippling losses of waterfowl in the United States and hunting seasons, 1971–84. The relationship offers clear evidence that the use of steel shot has not increased crippling losses. (From Sanderson and Bellrose 1986.)

with steel shot is less than $5.00. Moreover, the cost of shotgun shells reflects only a small part of the total cost of waterfowl hunting, and the increased cost associated with steel shot should be assessed in terms of wildlife conservation. As Sanderson and Bellrose (1986) concluded, "the slightly higher cost of steel loads should not be a deterrent to their use, particularly in view of the dwindling populations of ducks and the keen interest of waterfowlers in perpetuating their sport."

Pressure continued for a total ban on lead shot for waterfowl hunting. More areas were added to the list of "hot spots," and by the 1985 season, steel shot was required for waterfowl hunting in all areas of Nebraska. In 1986, the Secretary of the Interior declared that, beginning with the 1991 season, lead shot could no longer be used for waterfowl hunting. When that declaration became the law of the land, the destructive course of a major wildlife disease ended at last in the United States. Even so, lead shot may have remained available to ducks for as long as 4 years after the ban started, as determined with experimental mallards feeding in some heavily contaminated wetlands in California (Rocke et al. 1977). Moreover, lead poisoning still occurs in Mexico (Estabrooks 1987, Thompson et al. 1989, Schmitz et al. 1990), although a steel-shot zone was declared for the Yucatán Peninsula in 1992. Selected areas in Canada have received similar protection (Wendt and Kennedy 1991). Scheuhammer and Norris (1995) nonetheless estimated that 200,000 to 300,000 ducks migrating from Canada may die later from ingested lead shot, and several million more may suffer from the sublethal effects of lead poisoning. Beginning with the 1999 season, Canada accordingly declared lead shot illegal for all waterfowl hunting.

Unfortunately, waterfowl also face lead poisoning from another source, fishing sinkers. On the Thames River in England, a sample of 94 dead mute swans (*Cygnus olor*) included 57 that had died from ingesting lead fishing sinkers; the digestive tracts of the lead-poisoned birds contained an average of 7 sinkers (Birkhead 1982). Elsewhere in England, 16 of 18 dead or dying mute swans on the Trent River contained an average of 11 sinkers per bird (Simpson et al. 1979). Extraordinary densities of lead sinkers occur in some bodies of water in Great Britain; one study revealed a range of 5 to 300 sinkers per square meter along the shores of some ponds and lakes (Environmental Protection Agency 1994). Lead sinkers have poisoned substantial numbers of endangered trumpeter swans (*C. buccinator*) in the western United States (Blus et al. 1989).

In addition to poisoning waterfowl, lead sinkers have poisoned significant numbers of common loons (*Gavia immer*) in several states where the species is listed as threatened or endangered, including Michigan, New Hampshire, and Vermont (Environmental Protection Agency 1994). Pokras and Chafel (1992) examined the carcasses of 31 loons from New England; 16 of the loons had died from lead poisoning, most of which was caused by lead sinkers or other kinds of lead-weighted fishing tackle.

Because of this threat to wildlife, the Environmental Protection Agency (1994) has proposed a rule under authority of the Toxic Substances Control Act that would prohibit making, processing, or importing lead or zinc-coated fishing sinkers. Substitute materials, some of which already are available, include bismuth and its alloys with tin, tungsten, or iron. Ringleman et al. (1993), while searching for a more ballistically desirable alternate to steel shot, determined that mallards (*Anas platyrhynchos*) were not poisoned by ingesting an alloy of tungsten, bismuth, and tin. Such an alloy thus may prove useful as an effective and nontoxic substitute for both lead shot and lead fishing weights.

RETURN OF WOOD DUCKS

When Congress passed the Migratory Bird Treaty Act in 1918, wood ducks (*Aix sponsa*) were given complete legal protection. The effect was heartening: Wood duck populations gradually increased. In 1938, more than 10,000 wood ducks were seen at one time feeding in wheat stubble near Havana, Illinois (Bellrose 1976). As the population grew, however, it seemed that the number of wood ducks outstripped the availability of suitable nesting sites. Instead of tree cavities, their normal nest sites, some wood ducks nested in chimneys and other atypical settings. Thus, in 1939, biologists in Illinois erected several hundred nest boxes designed specifically for wood ducks (Fig. 3-2). Nearly 52 percent of the boxes contained nests the first year; by 1942, 65 percent of the boxes were used by wood ducks (Bellrose 1955). Other states in the breeding range of wood ducks followed suit, and nest-box programs soon became the focus of wood duck management in much of the eastern United States (e.g., for Massachusetts, see McLaughlin and Grice 1952; Grice and Rogers 1965).

In 1941, wood ducks once more became legal game during the hunting season in several states. Despite a temporary setback in the 1950s, when strict hunting regulations again were imposed, wood ducks have recovered much of their former abundance. The annual harvest of waterfowl in the Mississippi Flyway included about 500,000 wood ducks during the late 1960s and 1970s; another 225,000 were shot each year in the Atlantic Flyway during the same period. There was, however, no indication that hunting was causing any decline in the population (Bellrose 1976). Today, wood ducks rank as the second- or third-most abundant species of waterfowl in the bag of eastern duck hunters.

The successful management program for wood ducks can be attributed to several factors: timely recognition of the species' perilous status; adequate legal protection; habitat improvement using artificial nesting structures; and thereafter, careful year-by-year monitoring of the population so that rigorous protection could be forthcoming, as needed, on short notice. Moreover, wood ducks responded quickly to the man-

Figure 3-2. A hen wood duck peers from an early-model nest box in Illinois. Nest boxes became a major tool in the management of wood ducks after World War II. Modifications in design and placement of next boxes later reduced losses of eggs to raccoons and other predators. (Photo courtesy of the Illinois Natural History Survey.)

agement program because they lay large clutches and readily renest if their first clutch fails. Unlike other ducks in North America, wood ducks occasionally rear two broods during a single nesting season (Fredrickson and Hansen 1983; Kennamer and Hepp 1987). See Bellrose (1976, 1980; and especially Bellrose and Holm 1994) for complete reviews of the ecology and management of wood ducks.

WILD TURKEYS: TURNING FAILURE INTO SUCCESS

By the 1930s, wild turkeys had been extirpated from much of their range. As mentioned in Chapter 2, most attempts at restocking the empty range failed because unfit breeding stock was released (Schorger 1966). After several failures, however, success was achieved when the stock consisted of wild-trapped birds of the appropriate subspecies (i.e., the stocked birds were of a genetic lineage adapted to conditions in the release area). Populations of wild turkeys thus were reestablished in much of their former range (Fig. 3-3).

Newly established populations of wild turkey also required the protection and effective enforcement of laws. Such protection did not abolish legal hunting, although it was once predicted that the successful release of turkeys in Michigan would not produce a hunting season (Schorger 1966). Nonetheless, 11 years after the first birds were stocked in Michigan, hunting began under a limited permit system that allowed the harvest of any turkey, including immature or adults of either sex. Poor nesting conditions and perhaps indiscriminate harvests in the autumn hunts caused some stagnation in turkey numbers in the late 1960s. Then, in 1970, Michigan initiated an experimental springtime hunt that was restricted to gobblers (males). A "beard"—the long tuft of bristlelike feathers on the breasts of males—enabled hunters to distinguish between gobblers and hens. Since 1970, turkey populations in Michigan have blossomed under the regime of the spring gobbler hunt. The annual harvest steadily increased from 91 birds in 1970 to 15,686 in 1996, with about one of every four hunters successfully bagging a turkey. These data, shown in Table 3-2, reflect an obvious increase in the distribution and density of the turkey population in Michigan. Today, no fewer than 40 other states also have reestablished wild turkey populations.

TABLE 3-2. Spring Harvest of Turkeys in Michigan, Gobblers Only, 1970–96

Year	Number of Turkeys Shot
1970	91
1971	96
1972	152
1973	198
1974	238
1975	349
1976	397
1977	476
1978	618
1979	627
1980	844
1981	1,033
1982	1,760
1983	1,746
1984	1,458
1985	2,016
1986	2,361
1987	3,260
1988	4,567
1989	6,195
1990	8,456
1991	9,636
1992	11,847
1993	12,931
1994	11,429
1995	13,119
1996	15,686

Source: Michigan Department of Natural Resources.

Figure 3-3. Wild turkeys have been reestablished in many parts of their former range by releasing wild birds from appropriate genetic stock. (Photo courtesy of New Jersey Division of Fish, Game, and Wildlife.)

RESTORATION OF MAMMALS IN NORTH AMERICA

A pamphlet entitled *Endangered Species, The Success of Wildlife Management in North America* (National Shooting Sports Foundation, no date) summarizes some of the lesser-known cases in which wildlife populations have been restored successfully. In 1900, estimates suggest that no more than 500,000 white-tailed deer lived in the United States. Less than a century later, in 1980, the herd numbered about 12 million deer. The distribution of the American subspecies of elk (*Cervus elaphus canadensis*) once extended from the Atlantic to the Pacific and from Canada to Mexico. By the end of the 19th century, however, the area still occupied by elk was only a fraction of its former size. Perhaps only 40,000 elk remained, most living in Yellowstone National Park. Protection of winter habitat, improved range

On the Job with...
Michael Koss
Wildlife Habitat Biologist

Mike is a wildlife habitat biologist with the Michigan Department of Natural Resources, working out of a small office in the Upper Peninsula. Most of his professional time is spent conducting operations inventories on state-owned lands—on-site inspections of large tracts of forest, which are followed by conferences with state foresters to decide on the appropriate management. Such decisions usually address the methods of harvesting timber and the regeneration strategies that should be employed after cutting. Typically, Mike advocates and defends methods that benefit wildlife, such as leaving oaks for acorn production, maintaining forest clearings where deer can graze, and opposing the establishment and perpetuation of pine monocultures favored by industrial foresters. In short, Mike integrates wildlife management with forest management, traveling 50–150 km to visit state forests in his juris-

diction. He also surveys wildlife populations, tallying drumming grouse, taking roadside censuses of singing woodcock, and counting pellet groups of deer on plots randomly located throughout his district.

Public relations is another large part of his job. On a routine day, for example, Mike might meet with conservation officers and other biologists to discuss issuing shooting permits to farmers whose crops are being damaged by deer.

Because his job is concerned with habitat, Mike often experiences a world in which the sword cuts both ways. On one hand, he finds having an influence on habitat represents a highly rewarding part of his job; yet he also finds it frustrating when he is less than a full partner in making decisions about the management of state forests. He is often outnumbered by forest managers whose goals are to maximize fiber production, but his persistence often produces reasonable compromises. His other challenges include the failure of the public at large, as well as of decision makers, to understand or implement ecological concepts in land management. Despite these frustrations, Mike enjoys his job and works hard to transfer key principles and practical information to others.

Mike advises students to continue their education beyond the bachelor's degree. In today's world, the MS degree is rapidly becoming the basic educational requirement for most entry-level positions in wildlife management. He also urges undergraduate students to "get out there and hustle"—find summer jobs relevant to the field, work as a volunteer when necessary, and undertake those activities that demonstrate a strong record of experience and leadership. He adds, "Get your foot in the door, even if it means giving away your time, and be patient. The process of getting a job usually takes a while."

conditions, and regulated hunting have produced a current population estimated at 1 million animals. Most elk are found in 10 western states, but small herds also live in Michigan, Virginia, and Pennsylvania. The population of pronghorn (*Antilocapra americana*) numbered less than 13,000 in the 1920s but increased to more than 400,000 by the early 1980s. Final mention goes to beaver (*Castor canadensis*), a species that has enriched both the history and wealth of North America. Indeed, heedless trapping nearly extirpated beaver from the United States in the 1800s. Today, however, colonies again occur across the nation, even in places where beaver lodges disappeared more than a century ago.

SOME SUCCESSES WITH BIRDS

Leafing through the pages of *Our Vanishing Wild Life* (Hornaday 1913), one finds woefully accurate predictions of extinction for several species of birds—*unless effective management activities quickly intervened*. The "candidates for oblivion" included whooping cranes, heath hens, wood ducks, and California condors (*Gymnogyps californianus*). As we have seen, wood ducks recovered dramatically with intensive management, and the future for whooping cranes is far brighter today than at any other time in recent history. Heath hens indeed are gone forever, but California condors may have gained a new lease on their precarious existence. Thanks to a successful captive breeding program at the San Diego Zoo, about 2 dozen condors have been released at two locations (e.g., Los Padres National Forest in California), thereby re-establishing wild, free-ranging populations within their former range.

Other candidates on Hornaday's list were trumpeter swans (*Cygnus cygnus buccinator*), roseate spoonbills (*Ajaia ajaja*), upland sandpipers (*Bartramia longicauda*), sage grouse (*Centrocercus urophasianus*), sharp-tailed grouse (*Pediocoetes phasianellus*), and snowy egrets (*Egretta thula*). Hornaday predicted that the sage grouse would be the first upland game bird to fall to extinction. Today, sage grouse are abundant on many western rangelands, and of the others on Hornaday's list, only trumpeter swans and roseate spoonbills remain on the current list of threatened species. That so many birds once in dire straits are today relatively safe remains a tribute to Hornaday's early call of alarm. The continued existence of these species highlights the successes of wildlife management.

ELUSIVE MEASURES OF SUCCESSFUL MANAGEMENT

An important aspect of wildlife management concerns the maintenance of wildlife populations at levels that approach neither extinction nor excess. Populations of most North American songbirds (e.g., song sparrows, *Melospiza melodia*) are secure and normally fluctuate within acceptable limits with little or no management. The same is true of most small mammals (e.g., chipmunks, *Tamias* spp.), although some species exhibit periodic irruptions or cycles (e.g., lemmings, *Lemmus* spp.) independent of human influences. Endangered species are notable exceptions that require management (see Chapter 19). For game species, however, the ability of managers to maintain populations at levels permitting a reasonable harvest each year is itself a measure of successful management. In that respect, wildlife management has been a long-standing success for many species.

Game animals such as cottontails (*Sylvilagus* spp.) or bobwhites (*Colinus virginianus*) thrive throughout most of their ranges. Indeed, the harvest of some game species numbers well into the millions each year without having a lasting effect on the size of the populations (e.g., mourning doves, *Zenaida macroura*). Much of the success in the management of these species is attributable to the biological nature of the animals themselves— they exhibit resilient population features and adaptability to human presence. Most of the more abundant species of wildlife thrive in habitats associated with human activities (i.e., habitats maintained in low or midsuccessional stages; see Chapter 4). As we have seen earlier, however, even abundant species can suddenly become scarce when they are overexploited (e.g., the passenger pigeon and the bison).

Some undertakings of wildlife management are not easily characterized in terms of success or failure. The nuances of changing attitudes often color our interpretations, as do the revelations of science. Wolves (*Canis lupus*) were extirpated 200 years ago from the British Isles, an event that was probably hailed as a triumph of progress by citizens of the day. The same, no doubt, was true when mountain lions (*Felis concolor*) no longer roamed the Appalachian, Adirondack, or Ozark mountains. Hawks, too, were enemies; thousands were killed each year. In the modern world, however, the elimination of any species from large parts of its natural range is no longer regarded as ecologically desirable, and extinction at the hand of humans clearly represents failure.

Abundance is a particularly elusive characteristic against which wildlife management might be judged. In the 1800s, the duck population in North America probably reached 400 million birds, a size about 10 times the number present in 1956. Waterfowl managers decided that the 1956 level of abundance— about 40 million birds—would be the goal of management in the decades ahead (Cooch 1969a). By 1980, the stated goal was being achieved. A higher level of abundance may be unrealistic; humans probably have altered too many wetlands in North America to restore a significantly larger duck population. The question is, should we regard maintaining 40 million ducks from an original population of perhaps 400 million as success or failure?

Successful wildlife management involves social as well as technical dimensions. From the technical standpoint, an understanding of the current status of a wildlife population is the first requirement of management. Such knowledge includes the following factors: the size of the population; its growth rate or rate of decline; the reproductive capability of the animals; and the seasonal food, cover, and water requirements of the species. Only with a grasp of these facts can managers reasonably determine the nature and extent of the biological issues, if any, confronting wildlife populations.

From the social standpoint, successful management calls for strong programs of public education, especially in cases where hunting is proposed as a means of regulating wildlife populations (e.g., doe hunts). Some sectors of the public now express vigorous opposition to all forms of hunting or trapping. Conversely, some management programs require strict protection, and the public must understand why such measures are needed. Protection may be important not just for individual species, but also for larger systems in which two or more species interact. Black-footed ferrets (*Mustela nigripes*) live in association with prairie dogs (*Cynomys* spp.), and a management program aimed at only one of these species without involving the other would be ecological folly. In still other cases, habitat preservation or modification may be an issue capturing public attention (e.g., designation of wilderness areas or prescribed burning). Finally, with the social considerations in place, the remedial phases of management can begin, whether they include modified regulations, habitat improvements, or legislative mandates.

SUMMARY

The restoration of wood ducks and wild turkeys in eastern North America, as well as of elk and pronghorn populations in western North America, illustrate clear examples of successful wildlife management. The management program for wood ducks focused on legal protection and artificial nesting structures, whereas turkeys were reintroduced in much of their former range. The rescue of bison from almost certain extinction represents a qualified success because so little of their natural habitat remains in either Europe or North America. Since 1991, only nontoxic shot may be used legally for hunting waterfowl; hence the incidence of lead poisoning that once killed as many as 2 million ducks and geese each year is no longer prevalent in the United States. Some lesser-known accomplishments of wildlife management include the restoration of upland sandpipers, sage grouse, snowy egrets, sharp-tailed grouse, and beaver populations. Successful wildlife management includes three key elements: biological and ecological knowledge of the species and its habitat; determination of the limitations, if any, facing each population; and application of a proper course of action initiated with an appropriate level of public support.

CHAPTER 4

ECOSYSTEMS AND NATURAL COMMUNITIES

Civilization is a state of mutual and interdependent cooperation between human animals, other animals, plants, and soils, which may be disrupted at any moment by the failure of any of them.

Aldo Leopold (1933a)

Each of the species we call wildlife participates in a vast network of life—a system in which nonliving elements are brought into the tissues of living organisms. These elements then undergo exchanges between plants and animals and finally again enter the physical environment. Such a network—involving the interactions of living and nonliving elements in a manner that sustains life—is called an *ecosystem.* An ecosystem therefore includes not only plants and animals but also air, soil, and water. Living organisms borrow oxygen, carbon dioxide, and nutrients from the ecosystem and then return these materials through the processes of respiration, excretion, and decomposition.

The living part of an ecosystem—at any given time and place—is known as the *biotic community,* or more simply the *community.* An ecosystem usually consists of several communities, each having distinctive groups of plants and animals. For example, a forest ecosystem may include some stands of mature trees and others of younger ages. The herbaceous cover invading a recently burned section of the forest constitutes another community in the forest ecosystem. So does the vegetation bordering the banks of a stream in the forest, and so, indeed, is the plant life in the stream itself. Various kinds of animals are associated with each of these settings, and

their presence completes each community. Ecologists know that each species in a community plays a role that may be either obvious or obscure. Nonetheless, whether dominated by bacteria, trees, amoebas, or whales, communities are identifiable associations of plants and animals living in a finite physical environment.

Ecosystems can be modified by internal or external factors; these typically operate concurrently—sometimes obviously, sometimes subtly. Some factors are natural, others are humanmade. Aging is an internal, natural factor occurring, for example, when seedlings gradually become a mature forest. Other factors include the long-term genetic responses of organisms to evolutionary adjustments in the community. Soil formation also takes vast periods of time; the process—*weathering*—results from both internal factors such as the types of parent material and vegetation and external factors such as climate. Lightning is a natural external force that may cause important modifications in several kinds of ecosystems, and so are drought and other effects of weather. In 1980, the forest ecosystem on the slopes of Mount St. Helens in Washington underwent dramatic, but natural, modifications from powerful external forces.

For the most part, however, wildlife managers are concerned with humanmade, usually external, factors bearing on ecosystems. These are too numerous to list in detail, but virtually all ecosystems have been modified by one or more of the following: mineral and energy extraction, urban development, livestock production, impoundment and diversion of water, waste disposal, forestry, and agriculture. Each has affected wildlife populations in various ways.

Wildlife management itself frequently involves the intentional manipulation of some parts of ecosystems and natural communities. For example, when water levels are adjusted to improve nesting cover for waterfowl, the changes also affect—for better or worse—the habitat for fishes, amphibians, other birds, and aquatic mammals. The interaction between waterfowl management and the management of furbearers (e.g., muskrats, *Ondatra zibethicus*) illustrates just one of these relationships (Bishop et al. 1979). In natural communities, these and other relationships often are complex, but wildlife managers should deal responsibly and knowledgeably with the interdependent nature of animals, plants, and their physical surroundings.

Much was learned in the 1960s and 1970s about the links connecting various parts of the global ecosystems. In the context of ecological relationships, our planet is small indeed. The process of tampering with one component can produce unexpected and far-reaching effects on other components. More than insects were harmed when DDT and related pesticides were applied as a means of increasing agricultural and forest production: the chemicals also produced drastic declines in bald eagles (*Haliaeetus leucocephalus*), peregrine falcons (*Falco peregrinus*), and other birds of prey (Carson 1962; Hickey 1969; see Chapter 13). Tests of atomic weapons in the 1940s and 1950s increased the risk of cancer among Inuit, even though the bombs exploded thousands of kilometers away in the South Pacific and Siberia. Radioactive particles reached the Alaskan Tundra where lichens accumulated the fallout of strontium 90 and cesium 137. These contaminants then entered the Arctic ecosystem, eventually exposing Inuit to comparatively high levels of radiation (Hanson 1967).

In 1986, a thermal explosion at a nuclear power plant at Chernobyl in the Soviet Union released high levels of cesium 137. Near the site, 31 humans were killed, and thousands more in the immediate area likely will die from the long-term effects of radiation. Other regions of the Northern Hemisphere were contaminated as well, particularly the tundra ecosystem in the Lapland region of northern Scandinavia.

As much as 10 percent of the total quantity of cesium 137 escaping from Chernobyl may have fallen over Sweden (Persson et al. 1987). After traveling 1700 km, fallout from the radioactive cloud contaminated the short Arctic food chain in Lapland: lichens to reindeer (*Rangifer t. tarandus*) to humans. Because they lack underground root systems, lichens absorb their nutrients from the air and thus become "radiation sponges" in the presence of radioactive fallout (Stephens 1987). Hence, many Lapps destroyed their reindeer, but others gamble with cancer as they continue consuming contaminated meat and milk (Anon. 1986a). In other cases, the meat was sold to fur farms or was simply buried in uninhabited regions in what essentially became nuclear waste dumps (Stephens 1987). Less certain, however, is how the Chernobyl fallout might affect migratory birds (Anon. 1985). Chernobyl lies in a migratory route where birds, wintering from eastern Africa to western India, funnel through the Ukraine to breeding grounds in northern Eurasia. Eurasian coots (*Fulica atra*) are of particular concern because these birds are common in the diet of some Asian peoples. The related American coot (*F. americana*) can accumulate high levels of cesium 137; this factor suggests the hazard that may occur in nations where Eurasian coots are eaten. In Holland, fallout from Chernobyl has been detected in the food chain of brant (*Branta bernicla*). Unfortunately, the half-life of cesium 137 is about 30 years; hence, the threat of contaminated food chains from the Chernobyl "incident" will last well into the 21st century.

Ecosystems can range in size from an area as small as a few square meters to much or all of the Earth (e.g., a desert spring to tidal zones). An aquarium is an example of a small, closed ecosystem in which humans can control the physical environment (e. g., light and temperature) as well as plant and animal life. Ponds, pastures, and woodlots also are small ecosystems. The Amazon rain forest, Lake Superior, and the Mojave Desert are specific examples of still larger ecosystems. The *biosphere*—the part of the Earth that extends from a few hundred meters beneath the surface to several kilometers into the atmosphere—is a vast ecosystem in which life flourishes. Each of these ecosystems has been influenced by humans, although some clearly are more resilient than others are to human activities. Conversely, some ecosystems require great time and effort for repair once they are degraded. Ecosystems are generally self-sustaining, but they require an external source of energy that almost always comes from the Sun. The boundaries of ecosystems, with the exception of the biosphere, are frequently hard to define. Matter as well as energy is exchanged to various degrees among adjacent ecosystems. Water, sediments, atmospheric gases, and pollutants readily travel across boundaries between forests and fields, marshes and lakes, and other ecosystems.

Cities do not qualify as ecosystems because they are not self-sustaining and daily import large amounts of matter and energy. Cities are consumers of food and fiber, but they produce little of either. Whereas raw materials may be transformed into other products in cities, almost nothing is replenished in the process; much often is expended as waste. Cities can exist because of their dependence on functional ecosystems elsewhere. Zoos also are not ecosystems, nor are animal parks (cageless zoos), where animals range over large areas with little human interference. Without imported food, herbivores in animal parks soon would overgraze their pastures, and predators quickly would overexploit their prey. Shooting preserves also lack an inherent cycle in which animals subsist without external sources of matter. By contrast, the modern basis of wildlife management operates largely in a context where elements of the total community are maintained in functional and self-sustaining ecosystems. This chapter accordingly provides a brief overview of a few basic concepts of ecosystem and community ecology. The extensive subject of ecology is treated in a wealth of other textbooks (e.g., Odum 1983; Begon et al. 1986; Colinvaux 1986; Ehrlich and Roughgarden 1987).

MATTER AND ENERGY

Plants and animals are formed of materials—matter—that occur in the physical environment of the Earth and its atmosphere. Carbon atoms form the basis of life and exist in carbon dioxide, which constitutes a small fraction of the atmosphere. Hydrogen atoms also occur in the atmosphere but are more abundant in water. Phosphorus, calcium, potassium, sulfur, iron, sodium, and all of the trace elements (e.g., zinc) forming the body of an animal are found in the crust of the Earth; these occur in one form or another in soil, dissolved in water, or both. These and other essential elements enter the body of the animal in the form of food; only oxygen is assimilated by breathing.

Because of photosynthesis, green plants have the exclusive ability of removing carbon atoms from carbon dioxide. Green plants also produce organic compounds in the form of sugars and can store the energy of sunlight in chemical bonds. Complex chemistries form fats, proteins, and a variety of carbohydrates in the tissues of green plants. Other kinds of plants—including various soil bacteria and some algae—convert nitrogen from the atmosphere into nitrites and nitrates that then can be absorbed by more-complex green plants. Thus, green plants form the crucial link between the nonliving world and essentially all forms of animal life.

However, a curious exception has been discovered recently. Chemosynthetic bacteria apparently drive the unique ecosystems that surround submarine thermal vents at isolated sites on the ocean floor. The best known of these lies in the Pacific Ocean some 2500 m underwater in the Galapagos Rift (Corliss et al. 1979; Karl et al. 1980). Because they are independent of photosynthetic plants, the feeding relationships in these highly specialized ecosystems differ from all others yet discovered. The bacteria metabolize hydrogen sulfide and other sulfur compounds that are dissolved in the hot (300°C) effluents. The vent communities include clams, crabs, anemones, fishes, and a previously unknown taxon represented solely by the vent tube worm (*Riftia pachyptila*). See Felbeck (1981) and Rau (1981), among others, for details about the nutrition of organisms associated with these sulfide-rich vent communities.

In ecosystems generally, herbivores eat plants, and carnivores eat herbivores. *Food chains* thus are established as pathways by which nutrients flow through ecosystems. Eventually, all materials return to the physical environment; respiration, excretion, and decomposition are responsible. Bacteria and fungi are the prime agents of decomposition, but a large variety of insects and other invertebrates aid in the breakdown of fallen leaves, carcasses, and other organic matter. Without decomposers, the nutrients necessary for life would remain bound up in dead plants and animals. Lacking recycling, ecosystems would soon cease to function.

Ecosystems require energy as well as matter. Living organisms build cells, tissues, and organs with energy, but energy also is required for functions other than growth. Energy is required for all metabolic processes, among them thermoregulation, digestion, and muscle activity. As we have seen, matter is continuously reused in the biosphere; infusions of new matter are rarely needed. By contrast, energy is not recycled in ecosystems. Large amounts of energy must be imported in the form of sunlight. The sun is the original source of virtually all energy at work in an ecosystem. Solar energy drives the process of photosynthesis, thereby producing chemical energy in the molecules of sugars and other compounds. These, in turn, are stored in leaves, stems, roots, seeds, and other parts of plants. Whittaker and Woodwell (1969) determined the allocation of net *primary production*—the energy incorporated into green plants by photosynthesis—in a young oak-pine forest in New York: 40 percent to roots, 25 percent to stem wood, 33 percent to twigs and leaves, and 2 percent to flowers and seeds. Herbivores such as deer (*Odocoileus* spp.), squirrels (*Sciurus* spp.), beavers (*Castor canadensis*), and porcupines (*Erethizon dorsatum*) harvest only some of the edible parts of trees above ground level. Thus, much of the vegetation remains unused or unavailable as a source of energy for most animals. Even if herbivores could consume all of the twigs, leaves, flowers, and seeds in the New York forest, fully 65 percent of the materials and energy in the trees would remain tied up in stem wood and roots. Only fire or decomposition would release the large amount of residual energy in the forest. However, little energy recycles in this forest or in any other ecosystem.

The *second law of thermodynamics* states that the transformation of energy is not 100 percent efficient. Thus, when vegetation is ingested and metabolized, not all of the energy in the plant food passes on to the herbivore. Some energy is lost as heat. The loss is equally true when animal tissues are consumed by carnivores. The process begins inefficiently because plants trap only about 2 percent of the sunlight that falls on their leaves. A great deal of solar energy, in fact, never enters an ecosystem; much sunlight is reflected back into space as light and heat. Moreover, 50,000 calories of sunlight produce only 1000 calories

of energy in a plant. Thus, as energy passes from plant to carnivore, only a small part of the original amount actually reaches the last link in a food chain. Besides heat, other calories are lost when undigested materials are excreted (e.g., the dung of bison—"buffalo chips"—fueled the fires of pioneers on the treeless plains of North America, thereby releasing the last bit of solar energy originally captured by prairie grasses). Odum (1983) described the marked reduction of available calories as energy passed through a simple food chain in a 4-ha alfalfa ecosystem (Fig. 4-1). In this example, the conversion of sunlight to alfalfa is about 0.02 percent efficient ($1.49 \times 10^7/6.3 \times 10^{10}$); the conversion of alfalfa to beef is about 8 percent efficient; and the concluding conversion of beef to human tissue is about 0.7 percent efficient.

Ecologists use 10 percent as a rule of thumb for estimating the percentage of energy converted between links in most food chains; that is, 10 percent of the energy in plants is transferred to the community of herbivores, and carnivores receive about 10 percent of the energy from herbivores. Similarly, secondary predators consuming primary predators (e.g., big fish eating little fish) also convert about 10 percent of the available energy into their flesh.

What happens to energy that is not lost as heat? Owen (1970) has described the following circumstances for a female blue-winged teal (*Anas discors*) that consumes 120 kcal daily from a diet of aquatic plants. Of that amount, 30 kcal are indigestible and pass through the duck as feces. From the residual of 90 kcal, about 80 kcal are used for maintenance activities (e.g., swimming, flying, feeding, and heart action). Some of the 80 kcal are, in fact, lost as heat in accord with the second law of thermodynamics. The remaining 10 kcal form eggs. On the particular day described, the input of calories exceeded the output, and the teal gained 2 g in weight from the surplus calories. Once an egg is laid, however, the hen loses weight. In growing animals, of course, the intake of

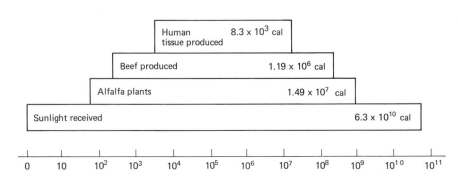

Figure 4-1. An energy pyramid showing progressively lesser amounts of available energy at each trophic level. Note that the scale is logarithmic, without which the lower levels of the pyramid would extend across several pages. (From Odum 1983.)

calories must exceed the expenditure of energy. Similarly, a growing population of animals requires a positive energy balance.

As noted earlier, food chains are the routes taken by energy and nutrients through an ecosystem. The links begin with producers (green plants), then advance to primary consumers (herbivores), and go on to secondary consumers (predators). Each of the various links in the chain is a *trophic level; trophic* means "feeding," a clear reference to the method by which energy and matter normally move through ecosystems. Grasses are the producer trophic level in a prairie ecosystem, and bison (*Bison bison*) represent the primary consumers. Wolves (*Canis lupus*) that prey on bison become secondary consumers in this food chain. Energy and matter, however, usually do not follow just one pathway through an ecosystem. Grasses are eaten not only by bison, but also by grasshoppers and other insects, and by mice, jackrabbits (*Lepus* spp.), and prairie chickens (*Tympanuchus*

cupido). Grasses also die and feed decomposers without entering any of the higher trophic levels. Moreover, the diet of prairie chickens includes grasshoppers as well as grasses. Hawks, red foxes (*Vulpes vulpes*), and coyotes (*Canis latrans*) prey on prairie chickens, jackrabbits, and mice. The flesh of all bison does not pass on to wolves or other secondary consumers. Indeed, many organisms die without their energy and matter moving to the next trophic level. In such cases, the action of decomposers shortens and redirects the food chain. Thus, instead of following a simple food chain, nutrients and energy in most ecosystems instead travel through a more complicated network, a *food web* (Fig. 4-2). In Chapter 9, we describe situations in which native species—predators—are removed from food webs; and in Chapter 18, we shall see how food webs can be disrupted when new species are introduced.

The inefficient transfer of energy from one trophic level to another produces two consequences. First,

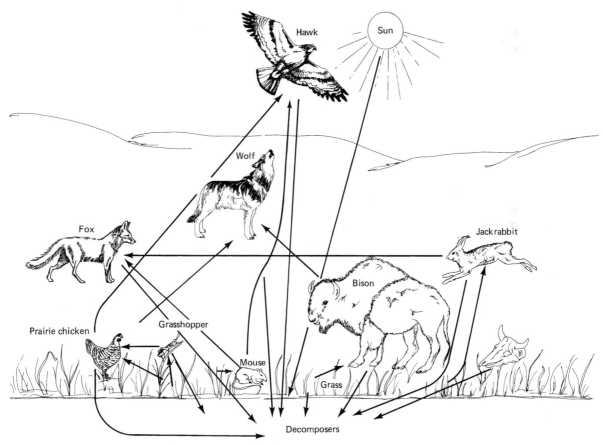

Figure 4-2. Elements of a food web on a native grassland.

less energy is available at each successive trophic level in an ecosystem. Thus, when measured per unit of area, there is more energy in green plants than in herbivores, and more in herbivores than in primary carnivores, and so on. The amount of energy, in turn, regulates the abundance of the organisms that occupy each trophic level. For example, 40 to 60 million bison once roamed North America; yet no one reasonably suggests that wolves were ever as numerous. Moreover, an average bison represents about 10 times the *biomass*—the weight of living tissues—of the largest wolf. Wildlife managers thus cannot expect larger numbers of mink (*Mustela vison*), which are predators, than of muskrats, which are herbivores—irrespective of the demands of trappers and the marketplace. The same relationship, of course, operates for wolves and deer or for any other set of predators and prey. If indeed an imbalance occurs, the predator population declines rapidly simply because the less-abundant prey population cannot supply the energy requirements for the larger number of predators. Such a relationship shows that prey populations in the long term limit the abundance of predators, not vice versa as is sometimes believed; that is, in the context of a simple food chain, deer can exist without wolves, but wolves cannot exist without deer.

Even greater amounts of energy are necessary to drive artificial ecosystems (see Chapter 23). For example, consider a breakfast of eggs. Sunlight drives production in the grainfield (along with energy for plowing, fertilizers, cultivation, insecticides, herbicides, irrigation, and harvesting); grain feeds the hen (but additional energy is required to maintain the hen house); the hen produces an egg a day (which she does nicely by herself, but large amounts of energy are needed to store, transport, and market the eggs); and, at last, each egg supplies a human with fewer than 100 calories (but only after more energy is used for cooking the meal). Thus, in terms of energy, large numbers of humans could not subsist on a diet of eggs in the same way that some cultures subsist on rice. Energy consumed en masse from the top of the food chain is produced at an unavoidable ecological price of some kind.

The second consequence is closely related to the first: *the length of a food chain is limited.* Very little energy remains after three or four steps; hence, an additional trophic level usually cannot be supported in natural ecosystems (but see Briand and Cohen 1987). A simple food pyramid reflects these relation-

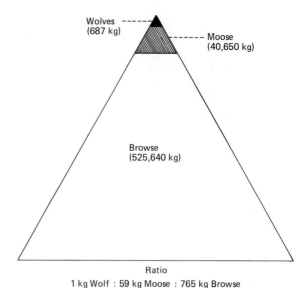

Figure 4-3. A pyramid of biomass on Isle Royale based on browse consumed by moose and moose consumed by wolves. (From Mech 1966.)

ships on Isle Royale, an ecosystem in Lake Superior where wolves subsist primarily on moose (*Alces alces*). The moose, in turn, feed on browse. Based on measurements of the biomass consumed at each trophic level, Mech (1966) showed the huge reductions occurring with each transfer in the food chain (Fig. 4-3). For each 1 kg of wolf, there are 58 kg of moose and 765 kg of browse. At Isle Royale, the conversion of energy thus involves a smaller percentage than is suggested by the 10 percent rule that is often applied to natural food chains.

RANGE OF TOLERANCE

Organisms live within a *range of tolerance,* or ecological amplitude, for each of the physical and biological components of their environment (Fig. 4-4). When either the upper or lower optimal limits of the range is exceeded, the efficiency of metabolic or reproductive processes falters and organisms begin to experience difficult circumstances. The plant or animal dies at a point where critical physiological functions are reduced or stopped. Different ranges of tolerance restrict species within certain environments, and groups of species with similar tolerances form communities within ecosystems. Environmental conditions vary greatly on the Earth; an immense variety of plants and animals accordingly thrives

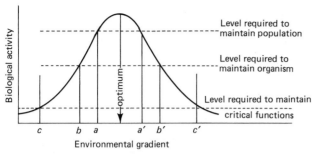

Figure 4-4. Relationship of biological activity to a gradient of environmental conditions. The range of tolerance in which an animal can survive for a short duration is from c to c'. The range in which an individual can survive for a long period of time is from b to b'; and the range in which a population can survive is from a to a'. (From Ricklefs 1979.)

from pole to pole. The thermal gradient from arctic to tropical conditions is among the more obvious environmental regimes to which organisms have adapted. The range from aridity to humidity is another important gradient. Other gradients may be somewhat less obvious but still include equally critical ranges of tolerance. For example, most plants have clear limits in their tolerance for salt, and those thriving in tidal marshes or other saline environments are known as *halophytes.* Some kinds of fishes live alternately in fresh and salt water at different stages of life, thereby showing a remarkable adaptation to salinity (e.g., Atlantic salmon, *Salmo salar*), but most fishes tolerate only slight variations in the salt content of water. Amphibians have almost no tolerance for salinity because their eggs and larvae quickly desiccate in salt water.

Ecologists use two prefixes to identify the nature of individual factors influencing the range of tolerance: *steno-,* meaning "narrow," and *eury-,* meaning "wide." Thus white-tailed deer (*Odocoileus virginianus*) are *eurythermal* species because they tolerate a wide range of temperatures; the distribution of whitetails extends from Canada to Venezuela. Conversely, polar bears (*Ursus maritimus*) live within a comparatively narrow range of temperatures in the Arctic and thus are *stenothermal* species.

The importance of these limitations sometimes was overlooked or ignored in the practice of wildlife management. In the 1940s, for example, willow ptarmigan (*Lagopus lagopus*) were released in Michigan, well south of the native range of these northern grouse. The release was doomed before it began—the range of tolerance for ptarmigan did not include

the forest environments and physical conditions present in Michigan. If conditions had been suitable, ptarmigan undoubtedly would have spread southward long ago without human transport, for physical barriers such as high mountains or large bodies of water were not blocking the way. In other words, with the exception of limitations posed by physical barriers, the range of tolerance establishes the limits of geographical distribution for willow ptarmigan and every other species.

Climographs, which depict monthly temperatures and precipitation levels, are a useful tool for determining where wildlife might thrive if managers wish to introduce new populations. Gray partridges (*Perdix perdix*) are a case in point (Twomey 1936; see also Odum 1983). In Missouri, introductions of gray partridge failed, whereas releases in Montana were successful. Figure 4-5 shows the representative climate in each state, along with the optimal conditions for the native range of gray partridges in Europe. The climograph indicates that both temperature and moisture from May to September in Missouri exceed the range of tolerance for gray partridge, thereby suggesting that those factors in some way contributed to the failure of the stocking effort. By contrast, the

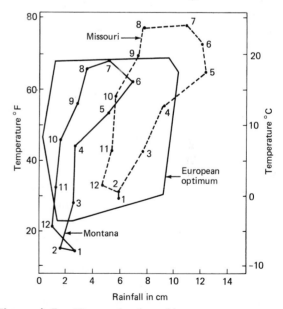

Figure 4-5. Climographs of monthly temperature versus precipitation in Havre, Montana, where gray partridge were successfully introduced, and in Columbia, Missouri, where introductions failed. Note how these compare with conditions on optimum partridge ranges in Europe. Numbers indicate months. (From Twomey 1936.)

summer conditions in Montana fell well within the range of tolerance. Temperatures during the winter months in Montana, although somewhat colder than the optimum climate in Europe, were not different enough to affect the success of the management program. Gray partridges today are well established in Montana and other areas in North America where climatic conditions are not limiting. Other factors, of course, also influence the security of wildlife populations. Nonetheless, food and cover are among the many factors for which temperature and precipitation exert strong direct or indirect influences. Climographs thus remain an important means of appraising a broad range of environmental conditions.

NICHE

Plants and animals survive only so long as they can compete successfully for resources. Each species is the product of a long evolutionary history that is governed largely by competition. Genetic differences ultimately decide competitive success. The genes of individuals that cope successfully with a given set of physical and biological conditions—their environment—are passed on to the next generation. In the process, inferior individuals are eliminated, and the less-favorable combinations of genes gradually diminish in plant and animal populations. As a result, species steadily improve the way they "fit" into their environment. The fit in some species is quite tight, and there is little or no room for dealing with changes or for colonizing different habitats. Such species can be regarded as specialists. Other species are generalists that can deal with a broader set of environmental conditions. In both cases, however, each species has a unique ability for living within its own range of tolerance. For a beaver, this means an adaptation for digesting woody food; for a wood duck (*Aix sponsa*), it means nesting in a tree cavity. Each species thus has a unique role in an ecosystem. The role is known as its *niche*. A niche emphasizes function within an ecosystem. Hence, the setting can be regarded in terms of *address* and *profession* (see Odum 1975, 1983). For bison, the setting translates into grassland and grazing. Moreover, as we shall see, niches reduce the turmoil of competition between species.

Jack pines (*Pinus banksiana*) are rather unimpressive trees in the forests of North America. Jack pines, however, produce cones that hang unopened for decades. These cones do not burn readily, even when forest fires periodically destroy the trees.

When fires occur, the cones open slowly in the heat, releasing winged seeds that the wind spreads over short distances. The seeds then germinate on the charred but cooling soil; they grow free of competition from the seedlings of other kinds of trees. Under these circumstances, a few jackpines in a forest dominated by other species can produce thousands of offspring. The result is the growth of a new forest that consists almost exclusively of jack pine. This type of forest endures until the next fire or until other kinds of trees invade and gradually overshadow the shade-intolerant pines. Jack pine thus is known as a fire-adapted species and fills an obvious niche in fire-disturbed forests—a role not shared with other trees in its community.

Where there are jack pines, there also are frequently spruce grouse (*Dendragapus canadensis*). In fact, the winter diet of spruce grouse is almost entirely composed of needles from jack pines (Crichton 1963; Robinson 1980). A large part of the winter feeding niche of spruce grouse thus depends on jack pine. Although in some locations spruce needles are substituted for jack pine, the winter diet for spruce grouse clearly originates with coniferous trees. Another species of forest grouse, the ruffed grouse (*Bonasa umbellus*), shares the same geographical range as the spruce grouse. However, the winter diet of ruffed grouse consists primarily of buds from deciduous trees (Bump et al. 1947). These associations are reflected in the kinds of trees—coniferous or deciduous—growing at sites where each species of grouse was flushed (Table 4-1). Spruce grouse were associated with sites clearly dominated by pines and spruce, whereas conifers and deciduous trees were almost equally abundant where ruffed grouse were flushed. Hence, even though spruce grouse and ruffed grouse occupy the same forest environments, the feeding niche of each is separate. Competition for winter food resources thereby is minimized (Robinson 1969).

In northern Europe, Scot's pine (*Pinus sylvestris*) is a fire species that has characteristics similar to those of jack pine. Another member of the grouse family, the capercaillie (*Tetrao urogallus*), feeds almost exclusively upon the needles of Scot's pine in winter. Thus, Scot's pine in Europe occupies the same niche as jack pine in North America, and capercaillie occupy the same niche as spruce grouse. Such species, living in different geographical areas but fulfilling similar roles in their respective communities, are known as *ecological equivalents*. Some ecological equivalents are closely related, as is the case with both the pines and grouse mentioned here, but others

TABLE 4-1. Composition of the Forest Habitat Where Spruce and Ruffed Grouse Were Flushed During the Summer Months in Michigan

Species	Percent of trees at sites with:	
	Spruce Grouse	Ruffed Grouse
Jack pine *(Pinus banksiana)*	51	8
Spruce *(Picea mariana* and *P. glauca)*	32	19
White pine *(Pinus strobus)*	8	13
Balsam fir *(Abies balsamea)*	5	15
White birch *(Betula papyrifera)*	3	12
Red Maple *(Acer rubrum)*	1	13
Quaking aspen *(Populus tremuloides)*	—	12
Sugar maple *(Acer saccharum)*	—	4
Red pine *(Pinus resinosa)*	—	2
White cedar *(Thuja occidentalis)*	—	1
All conifers	96	57
All deciduous trees	4	43

Source: Robinson (1969).

are not. Some species of kangaroos fill a grazing niche in the grasslands of Australia, and these are the ecological equivalents of bison in North America. In some communities, insects play crucial roles in the pollination of flowers, whereas birds or bats perform the same function in other communities.

Animals feeding on the same foods do not necessarily occupy the same niche. The time, location, and manner of feeding may be specialized in ways that greatly reduce or avoid competition. In eastern North America, earthworms form most of the spring diets of American robins (*Turdus migratorius*) and American woodcock (*Scolopax minor*). Nonetheless, the two birds do not compete for food because other dimensions of their feeding niches are different. Robins find worms in cleared areas, particularly in urban lawns and gardens, and they hunt by sight during the daytime. Conversely, woodcock feed in moist thickets, often at twilight, and probe for worms. The long, slender bills of woodcock are remarkably adapted for probing; a network of nerves at the bill tip senses the underground movement of earthworms (Fig. 4-6). Robins and woodcock thus maintain separate niches, even though the diets of these birds are quite similar. In Africa, Talbot and Talbot (1963) determined that wildebeests (*Connochaetes taurinus*) eat some of the same plants as topis (*Damaliscus lunatus*), but each of these antelopes selects different plant parts: wildebeest largely select green leaves, whereas topis eat mostly stalks and dried plant parts. Because of their preference for short green grasses, wildebeest normally move on to other grazing areas when sprouting grasses reach a height of about 10 cm, thereby leaving

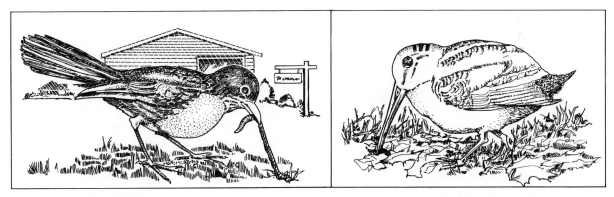

Figure 4-6. American robins (left) and American woodcocks (right) both feed largely on earthworms but find their prey in different ways and in different locations, thereby avoiding direct competition for the same food resource. Robins feed on the soil surface in open areas with short grass (lawns and golf courses) and locate their prey by sight. Woodcocks probe moist soil in woodland habitat using long bills equipped with sensitive nerves for detecting the movements of worms. Woodcock bills also are flexible and open at the tip for grasping worms below the soil surface. (Design by R. Miller.)

the maturing plants for topis and other grazing animals. Niches are by no means limited to foods and feeding. Some kinds of salmon utilize the same spawning beds but maintain separate niches because of different breeding schedules. Herons and egrets often nest concurrently in the same vegetation but do so at different heights, in effect stacking their breeding niches one above the other in a single rookery.

The ability to match habitats to the niche requirements of a species is of fundamental importance to wildlife management. Damaged or altered habitats often lack some of the resources needed for maintenance of a niche, especially for those species with small ranges of tolerance. Niches are of particular concern in regard to exotic species because two species cannot occupy the same niche. Thus, if an exotic species is introduced into a niche already occupied by a native species, one or the other will be displaced or eliminated by direct competition. After their introduction into North America, starlings (*Sturnus vulgaris*) and house sparrows (*Passer domesticus*) often preempted the nesting sites used by bluebirds (*Sialia sialis*); consequently, there now are fewer bluebirds (Zeleny 1976). In Australia, some of the native marsu-

pials apparently could not compete with European rabbits (*Oryctolagus cuniculus*); potoroos (*Potorous* spp.), no longer common, were among the native species whose grazing niche was invaded by the foreign rabbits (Tyndale-Biscoe 1973). We shall discuss more about niches and exotic wildlife in Chapter 18.

NATURAL COMMUNITIES— CHANGES IN TIME AND SPACE

Natural communities show patterns in time and space. Many features of communities clearly reflect the regional climate. For example, shrubs that are separated by bare soil and that are arranged in a widely spaced, regular pattern are typical of desert conditions. Spines, spongy cells, thickened cuticles, and other specialized features for water conservation characterize desert plants throughout the world. The species differ among desert communities, but their structural and physiological adaptations are alike (i.e., they are ecological equivalents). Ecologists thus group similar communities into larger units called *biomes* (Fig. 4-7). These units include all communi-

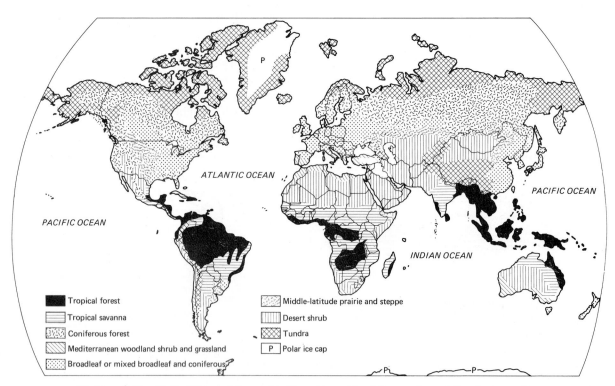

Figure 4-7.   Major biomes of the world. (From *Essentials of Geography and Development: Concepts and Processes* by Don R. Hoy, Copyright © 1980 Macmillan Publishing Co., Copyright © 1984 Merrill Publishing Company, Columbus, Ohio. Reprinted by permission of Merrill Publishing Co.)

TABLE 4-2. Biomes of North America and Some Representative Plants, Birds, and Mammals

Biome	Plants	Birds	Mammals
Tundra	Sedges (Cyperaceae) Lichens Cranberries (*Vaccinium* spp.)	Rock ptarmigan (*Lagopus mutus*) Snowy owl (*Nyctea scandiaca*) Golden plover (*Pluvialis dominica*)	Barren-ground caribou (*Rangifer tarandus*) Musk ox (*Ovibus moschatus*) Brown lemming (*Lemmus trimucronatus*)
Boreal Forest	Spruce (*Picea* spp.) Firs (*Abies* spp.)	Spruce grouse (*Dendragapus canadensis*) Boreal chickadee (*Parus hudsonicus*) Gray jay (*Perisoreus canadensis*)	Moose (*Alces alces*) Lynx (*Felis lynx*) Snowshoe hare (*Lepus americanus*) Red-backed vole (*Clethrionomys gapperi*)
Temperate Deciduous Forest	Oaks (*Quercus* spp.) Maples (*Acer* spp.)	Ruffed grouse (*Bonasa umbellus*) Black-capped chickadee (*Parus atricapillus*) Red-headed woodpecker (*Melanerpes erythrocephalus*) Blue jay (*Cyanocitta cristata*)	White-tailed deer (*Odocoileus virginianus*) Fox squirrel (*Sciurus niger*) White-footed mouse (*Peromyscus leucopus*)
Temperate Grassland	Big bluestem (*Andropogon girardi*) Little bluestem (*Schizachyrium scoparium*) Grama grasses (*Bouteloua* spp.)	Prairie chicken (*Typanuchus cupido*) Western meadowlark (*Sturnella neglecta*) Upland sandpiper (*Bartramia longicauda*)	Bison (*Bison bison*) Prairie dog (*Cynomys* spp.) Coyote (*Canis latrans*)
Desert	Sagebrush (*Artemisia* spp.) Creosote bush (*Larrea glutinosa*) Cacti (Cactaceae)	Sage grouse (*Centrocercus urophasianus*) Burrowing owl (*Speotyto cunicularia*) Road-runner (*Geococcyx californianus*)	Desert bighorn sheep (*Ovis canadensis mexicana*) Desert jackrabbit (*Lepus californicus* *mexicanus*) Desert kangaroo rat (*Dipodomys deserti*) Collared peccary (*Tayassu tajacu*)
Mediterranean Scrub Forest	Chamise (*Adenostoma fasciculatum*) Coffee-berry (*Rhamnus* spp.) Ceanothus (*Ceanothus* spp.)	California quail (*Lophortyx californicus*) Anna's hummingbird (*Calypte anna*) Brown towhee (*Pipilo fuscus*)	Mule deer (*Odocoileus hemionus*) Bobcat (*Felis rufus*) Brush rabbit (*Sylvilagus bachmani*)

Sources: Several, especially Shelford (1963).

ties of a similar type from across the globe: Mojave, Kalahari, Gobi, and Saharan deserts (among others) form a single biome. The names used for each biome vary according to individual preferences, but the terms generally reflect the dominant type of vegetation: Grassland, Deciduous Forest, Desert, Tundra, Rain Forest, Coniferous Forest, and Chaparral. Most systems further refine these designations into units such as Boreal Forest and Alpine Tundra; these are based on latitude or altitude. Distinct animal communities as well as plant communities characterize each biome. Thus, the boreal forest biome sometimes has been called the *spruce-moose biome* (Shelford 1963); however, such plant-animal designations have not been widely adopted despite their ecological merit. The spatial arrangement of biomes reflects major patterns in natural communities throughout the world (Table 4-2).

In terms of time, the composition of natural communities is realized in a sequential process known as *ecological succession,* or simply *succession.* A *pioneer community* is the first step in succession, and a *climax community* is the last in the full sequence. Some kind of disturbance, whether natural (e.g., a lightning fire, glacier, or volcanic eruption) or humanmade (e.g., plowing, lumbering, or grazing), initiates the development of a pioneer community on the exposed soil. *Primary succession* occurs where no community previously existed, as on a newly formed volcanic island. *Secondary succession* takes place where there are remnants of a previous community, as is the case in a burned-over field or a clearcut forest. In either primary or secondary succession, however, a series of different communities gradually, and sequentially, occupies a site until a climax community is established. At climax, the community is self-perpetuating in a state of *dynamic equilibrium*; that is, while there is a constant turnover of individuals within the community, the species are the same and the overall composition of the climax community remains essentially unchanged. Climax communities thus persist in both time and space, until they are disturbed.

We can see the nature of these patterns, using an example of abandoned farmland in southeastern Canada. When abandoned, tilled fields are invaded by a pioneer community of "weeds." Such plants usually are *annuals* and accordingly produce large numbers of seeds. Some of these are grasses, but most are *forbs,* herbaceous plants with broad leaves. Dandelions (*Taraxacum* spp.), goldenrod (*Solidago* spp.), and Queen Anne's lace (*Daucus carota*) are examples of forbs. Animals associated with this pioneer community might include song sparrows (*Melospiza melodia*), mourning doves (*Zenaida macroura*), meadow vole (*Microtus pennsylvanicus*), cottontails (*Sylvilagus floridanus*), weasels (*Mustela* spp.), and, at times, white-tailed deer. As succession advances, many of the annual plants are in time replaced with perennial species. Grasses become more abundant, and the vegetation is somewhat more stable than before. Still later, small trees such as gray birches (*Betula populifolia*) and speckled alder (*Alnus rugosa*) appear. The young trees shade out the grasses and other herbaceous plants. Woodcock, ruffed grouse, deer mice (*Peromyscus maniculatus*), and white-throated sparrows (*Zonotrichia albicollis*) are associated with the newer vegetation. Deer are more common than before and take advantage of the tender browse available in the low-growing shrubs. Conditions now are suitable for spruces and firs, and

their seeds germinate in what was once cropland. The young evergreens grow slowly, but they eventually overshadow the low canopy of alders and gray birches. Thus deprived of sunlight, the community of shrubs and saplings gives way to a mature spruce-fir forest. Succession now has reached its climax, in this case resulting in a Boreal Forest. Moose, spruce grouse, and red squirrels (*Tamiasciurus hudsonicus*) find their niches in this environment. Because spruces and firs can reproduce in their own shade, dynamic equilibrium is reached and the community remains essentially unchanged until an external force initiates a new cycle of succession.

Aquatic systems also undergo ecological succession, as is shown when a lake is formed by a receding glacier. The new lake is typically cold, clear, and deep, and its waters are poor in nutrients and high in dissolved oxygen. The bottom and shores are rocky and lack much aquatic vegetation. These conditions are *oligotrophic*—a term meaning "few nutrients"—and characterize lakes that are geologically young. Food webs are not complex in oligotrophic lakes because of the low levels of fertility. Trout (Salmonidae) are typical of the fish community at this stage of succession. In time, however, phosphorus, nitrogen, and other nutrients wash from the watershed into the lake, and the water slowly is enriched. Aquatic vegetation appears and gradually adds organic matter to the system as generations of plants mature and die. Decomposition of the organic matter consumes large amounts of oxygen. Moreover, the lake becomes shallower as sediments and organic debris continue accumulating as the centuries pass. Because it is shallower, the lake warms more easily during the summer and the salmonid fishes are replaced by a community of fishes whose range of tolerance includes higher temperatures (e.g., walleye, *Stizostedion vitreum,* and pike, *Esox* spp.). The lake is now *eutrophic*—"good or truly nutrient"—and the enrichment of an aquatic system follows a successional process known as *eutrophication.*

Eutrophic systems typically produce more biomass of fishes and other organisms per unit area than oligotrophic waters. The higher levels of nutrients in eutrophic systems also promote communities of greater complexity (i.e., more species) than normally occur in oligotrophic streams and lakes. Nonetheless, public regard for oligotrophic systems remains high because those systems contain relatively pure water. Also, trout and salmon fisheries command much attention from the sporting community.

Human activities can accelerate the rate of eutrophication. Abnormally high nutrient loads occur when domestic sewage is discharged into aquatic systems. Fertilizers also wash from lawns and fields into streams and lakes. Under these circumstances, the nutrient loads received in just a few years may advance eutrophication at a rate equaling many centuries of natural enrichment. Such overloads may cause biochemical havoc even in large bodies of water, and the environmental quality of the system may deteriorate into an unsavory soup. Several years ago, ecologists discovered alarming rates of eutrophication in lakes receiving waste water from residential areas; the source was phosphate-laden laundry detergents. Shortly thereafter, the phosphate levels in most detergents were reduced or eliminated. Unfortunately, however, indiscriminate releases of sewage and other wastes still cause excessive rates of eutrophication in many places.

By contrast, other human activities have reduced productive aquatic systems to nearly sterile environments. Runoff from mine wastes acidifies rivers and lakes, often reaching a point at which most forms of aquatic life soon perish. Sulfur dioxide and other emissions in the smoke from fossil fuels return as "acid rain," a phenomenon described more fully in Chapter 11. However, aquatic systems affected by high levels of acidity do not revert to the early stages of oligotrophic conditions in which a community of trout and associated organisms might flourish. Instead, the excessive acidity generally thwarts all but some primitive forms of life (e.g., microorganisms and a few species of invertebrates). The results are vastly shortened food chains and what are essentially biological wastelands.

SUCCESSION AND WILDLIFE MANAGEMENT

Succession is one of the more important concepts governing the practice of wildlife management. Within the climatic limits imposed by temperature and precipitation, each species of wildlife thrives only in those successional stages that produce rather specific arrangements for food, cover, and water. Much of wildlife management thus deals with ways of manipulating habitat for the benefit of selected species. In selecting a particular stage of succession, however, managers are thwarting the natural sequence of events. Without intervention, for example, the habitat in which bobwhite (*Colinus virginianus*)

thrive will advance to a successional stage at which the food and cover needs of bobwhite are no longer optimal. Succession thus must be arrested at a stage that coincides with the goals of management. Bobwhite, in fact, are among the many game species faring better in early successional stages than in climax communities (Table 4-3). Leopold (1933b) named four tools for setting back succession: "ax, cow, match, and plow." Although disrupting to the poetry of Leopold's list, we can add chemical herbicides to the modern arsenal of management tools. When required, succession can be accelerated with plantings and fertilizers (see Chapter 7).

The caribou is a classic example of a species associated with climax vegetation in North America (Leopold 1969). Lichens form the bulk of the winter diet of caribou; these plants grow in the understory or drape from the branches of mature trees in the Boreal Forest. Fire or other disturbances to the climax vegetation can deplete the supply of lichens and,

TABLE 4-3. Some North American Game Birds and Mammals and Their Associations with Climax and Nonclimax Communities

Climax	Nonclimax
Birds	Birds
Passenger pigeon (*Ectopistes migratorius*)	Mourning dove (*Zenaida macroura*)
Wild turkey (*Meleagris gallapavo*)	Bobwhite (*Colinus virginianus*)
Spruce grouse (*Dendragapus canadensis*)	Sage grouse (*Centrocercus urophasianus*)
	Ruffed grouse (*Bonasa umbellus*)
Mammals	Mammals
Caribou (*Rangifer tarandus*)	Elk (*Cervus elaphus*)
Musk ox (*Ovibos moschatus*)	White-tailed deer (*Odocoileus virginianus*)
Bighorn sheep (*Ovis canadensis*)	Mule deer (*O. hemionus*)
Mountain goat (*Oreamnos americanus*)	Pronghorn (*Antilocapra americana*)
Bison (*Bison bison*)	
Collared peccary (*Tayassu tajacu*)	
White-lipped peccary (*Tayassu pecari*)	
Tapir (*Tapir bairdii*)	
Brocket (*Mazama americana*)	

Source: Leopold (1969).

in turn, affect the caribou herd. Because most climax communities have been altered by humans, several of the larger species of animals associated with climax communities are no longer abundant in Europe and North America (e.g., bighorn sheep, *Ovis canadensis*). Conversely, many of the species associated with low- or midsuccessional communities are relatively abundant (see Table 4-3). Midsuccessional species seemingly have greater adaptability or tolerance than climax species have for many kinds of disturbances. Indeed, human activities have destroyed large areas of climax communities—prairies and forests, among others—so that the species associated with those communities have not fared well. However, species that have adapted to fire, insect plagues, or other disturbances persist or even increase. Passenger pigeons (*Ecopistes migratorius*), a species of climax oak forests, could not adapt to habitats modified by humans even though remnants of oaks forests remained available. The closely related mourning dove, however, has thrived in a variety of highly modified environments (e.g., urban settings; see Chapter 17). Humans are a species best adapted to successional rather than climax communities. Many human activities associated with the utilization of land, such as agriculture and forestry, continually set back succession. Agriculture wages a constant battle against the natural advances of succession. Without herbicides and cultivation, weeds and shrubs soon would eliminate the poorly adapted crops humans plant for food and fiber.

We again emphasize that the early to midsuccessional species of wildlife are most abundant today because of the many ways that succession is kept from reaching climax. The relationship clearly results from the dominant and pervasive nature of human activities. Thus, bobwhite, deer, cottontails, robins, mourning doves, and pheasants (*Phasianus colchicus*) are particularly abundant species in North America.

DIVERSITY AND STABILITY

The number of species in a community reflects richness or *diversity*. Abundance, of course, is a numerical measure of a population's size (i.e., number of individuals per species). Biomass also is an important way of measuring abundance in ecosystems. *Stability* may be defined as relative constancy in the abundance of populations.

Ecologists argue about the importance of diversity in maintaining stable communities. Simple communities generally exhibit rather unstable "boom-and-bust" characteristics. For example, the abundance of animals in tundra communities, where diversity is low, may vary greatly during the span of a decade. Keith (1963) reported average peaks in the number of snowshoe hares (*Lepus americanus*) in northern Canada as 6.7 times greater than the number at low levels. In the area near Hudson Bay, peak densities in the hare population were at an average of 242 times greater than the densities at low points in the cycles. By comparison, communities in tropical rain forests usually exhibit great diversity—there are many species, but low numbers of individuals per species. Wildlife populations in rain forests usually remain quite stable under normal conditions, although logging or other disturbances may permanently disrupt the structure of these communities.

Ecologists argue about whether diversity produces stability in natural communities, or if stability produces diversity. Proponents of the latter argument claim that physical environments with relatively constant features (e.g., precipitation, temperature, and day length) purportedly create conditions that ultimately enrich natural communities. Few experiments have systematically tested either hypothesis, but the logic behind the theory that diversity produces stability seems compelling. Many communities simplified by human influences soon produce environments of undeniable instability (e.g., desertified rangelands, some beachfront developments, and erodable slopes).

Some associations between or among members of communities are *symbiotic;* that is, close relationships are formed between two or more unrelated organisms. *Mutualism* is a common type of symbiotic association in which two organisms benefit from each other. Lichens—part algae, part fungi—are mutualistic plants that are widely distributed in terrestrial biomes from pole to pole. Some types of lichens are an important food for caribou (*Rangifer tarandus*), and because lichens readily absorb substances dissolved in atmospheric moisture, these plants are also sensitive indicators of smog and other forms of air pollutants. As noted earlier, radioactive fallout absorbed by lichens has entered Arctic food chains on at least two occasions. Some mutualistic associations are required for the survival of either member of the pair (known as *obligative mutualism*), whereas the bond may be broken in other cases without fatal results (called *facultative mutualism*).

An example of plant and animal mutualism apparently existed on the island of Mauritius in the Indian Ocean. In 1973, virtually all of the calvaria (*Calvaria major*) forests on Mauritius were gone. Only 13 trees remained, each at least 300 years old. The surviving trees still produced fruits, but their seeds did not germinate. Temple (1977) noted that the ages of the trees coincided with the extinction (by 1681) of dodos (*Raphus cucullatus*) on Mauritius. He suggested that calvaria seeds required the grinding action in the gizzards of dodos before the seeds could germinate. The dodos fed on the pulpy outer layer of plum-size calvaria fruits—a food produced abundantly in forests perpetuated by the birds' seed-laden droppings. The idea gained credence when calvaria fruits were force-fed to domestic turkeys, and, after passing through the birds' digestive tracts, some seeds germinated—perhaps the first to do so in more than 3 centuries. Thus, dodos and calvaria trees apparently had enjoyed a mutualistic relationship that ended with the extinction of one partner. The experiment with turkeys also provided a means for artificially treating calvaria seeds, thereby saving a companion species from certain extinction.

Communities may be compared to a web that has various food chains represented by separate strands. A simple web with relatively few interconnecting strands is more vulnerable to collapse if a single strand is perturbed. By analogy, in a community with few plant and animal species, the loss of just one source of plant foods would reduce the herbivore population, and the decline of herbivores in turn would affect the carnivores. By contrast, in the complex web of a diverse community, herbivores would more easily tolerate the loss of a single food source by switching to a diet of other plants, and the predators in the community would remain unaffected by the changeover. In such cases, the stability of the community is buffered by secondary, or alternate, sources of energy in the food web. Without such choices, a community is little more than an isolated—and relatively vulnerable—food chain. *Buffer species* thus are secondary foods that become of primary importance for the maintenance of stability.

The importance of diversity becomes obvious in a modern era full of frequent and strong disturbances (e.g., oil spills and tropical deforestation). In a global sense, we see that diversity is an important environmental safeguard for the stability of entire communities and ecosystems. More is involved than just a one-dimensional plea for the preservation of wildlife habitat for hunting and fishing opportunities. Reductions in the diversity of wildlife may signal threats to the life-support system on which humans depend. The global loss of diversity thus has prompted governmental studies of ways to arrest such trends (Office of Tech. Assessment 1987).

Odum (1975) noted that "Diversity in systems is undeniably a good thing. But as with most 'good things' in the real world there can be too much of it as well as too little." Low diversity may have the advantage of efficiency in exploiting energy sources; that is, if the base of a food chain is simple, a simple community of well-adapted herbivores would exploit that source more efficiently than a complex of herbivores.

Human activities, whether intentional or accidental, almost always simplify natural communities. Pollution reduces the diversity of aquatic systems, and the number of species and their abundance provide useful measures of the degree to which rivers and lakes may be polluted (see Chapter 11). Modern agricultural practices also reduce the complexity of natural communities. Dozens of native grasses in the prairie communities of North America were turned under by plows and eventually were replaced with huge blocks of single species: corn in Illinois, Iowa, and Nebraska; and wheat in the western plains. From these and similar human activities, what are known as *monocultures* are substituted for the natural diversity of the landscape. The immediate benefits of monocultures are clear: huge rates of production, efficiency of harvest, and, consequently, profitable business. Such benefits, however, present ecological risks and require large amounts of energy (e.g., gasoline for farm operations). Monocultures demand huge supplies of nutrients and other resources from external sources (i.e., labor, fertilizers, pesticides, and irrigation). Monocultures also are unusually subject to catastrophic events. Immense losses from drought, floods, hail, and outbreaks of insects and diseases are commonplace in monocultures, as are socioeconomic problems (e.g., depressed commodity prices in times of overproduction). The adage admonishing against placing "all of our eggs in one basket" has clear application to ecological systems and wildlife management.

Some highly adapted species of wildlife are not directly dependent on diversity. For example, koala bears (*Phascolarctos cinereus*) consume about 1 kg per day of vegetation from a few species of gum trees (*Eucalyptus* spp.), the sole source of food for these well-known symbols of Australian wildlife (see Degabriele 1980). Gum trees often occur in large blocks of forest in which other species are uncommon. Koalas thus exploit a simple food base with a good deal of efficiency.

Despite the advantages of diversity in many situations, wildlife management usually does not maintain or regain the diversity that once may have been present in most natural communities. A full range of species, or nearly so, may be maintained in a few, rigidly protected nature preserves, but many of these areas are small and usually are not open to hunting. Instead, some degree of compromise regarding diversity may be required or even encouraged when hunting is the prime management goal.

Some communities and ecosystems are more resilient than others. In temperate regions, where forest communities evolved in a regime of extreme temperatures and precipitation, such natural perturbations as fire or insect attacks usually are experienced without great disruption. Human disturbances also are tolerated within reasonable limits, and succession soon begins the healing process and the return of climax communities. The unusually rich soils in prairie systems for a time seemed capable of retaining their naturally high rates of production, even in the wake of abusive treatment. Resilience has limits, however, and decades of agricultural monocultures eventually exhausted the nutrients in prairie soils. Acceptable levels of crop production now require the addition of fertilizers. Desert communities do not recover quickly from disturbances nor do those in Tundra. In Arctic regions, where few nutrients and a short growing season are characteristic, land uses that might have few impacts in other ecosystems often leave long-lasting scars. Clapham (1973) noted that the trail made in 1920 by the passage of a single wagon still remained visible and unvegetated in the tundra of the Seward Peninsula, Alaska. Tropical rain forests, too, are highly vulnerable systems despite their robust appearances. Huge and highly specialized machinery today can clear large areas of rain forest in a matter of days. Clear-cutting virtually eliminates future production from the nutrient-poor soils in tropical rain forests (see Chapter 12).

Ecological circumstances such as these have direct bearing on wildlife management. Communities in ecosystems where perturbations are unusually threatening deserve the protection of special management efforts. Synthetic chemical compounds, nuclear radiation, accelerated urbanization and industrialization, and introductions of exotic organisms are among the perturbations in the modern world. The importance of these influences varies with the ecological conditions dominating each location. An Arctic environment contaminated by an oil spill may take years to recover, whereas the higher rate of biological activity in warm environments greatly accelerates the recovery period for oil spills. Wildlife managers thus adjust their goals and actions in keeping with local and regional differences in biotic communities.

SUMMARY

Humans and wildlife share a global *ecosystem* that is driven by the energy of sunlight in which matter is exchanged between living and nonliving sectors. Green plants—*producers*—capture solar energy and form the basic *trophic level* on which herbivores and carnivores—*consumers*—ultimately depend. Bacteria and other *decomposers* reduce wastes and the bodies of plants and animals into simpler materials that can be recycled. Complex *food webs* occur in most *biotic communities,* the living part of an ecosystem. Communities are organized sets of interacting plant and animal species whose *diversity* (number and complexity) varies greatly. *Niche* is the function of a species in a community. Each organism exists within a *range of tolerance* for environmental conditions. *Biomes* are large geographical units having similar vegetation.

Human manipulation of communities for any purpose, including forestry, agriculture, urbanization, and wildlife management, alters the patterns in which energy and matter flow through ecosystems. Communities vary in their tolerance for disturbance. Effective wildlife management involves recognition of the degree to which communities are disrupted when plant and animal populations are manipulated. These activities should not unduly disrupt the overall functioning of communities in which wildlife management is practiced.

CHAPTER 5

POPULATION ECOLOGY

Living as we do in a world which has been largely denuded of all the large and interesting wild mammals, we are usually denied the chance of seeing very big animals in very big numbers. If we think of zebras at all, we think of them as "the zebra" (in a zoo) and not as twenty thousand zebras moving along in a vast herd over the savannahs of Africa.

Charles Elton (1927)

A herd of zebras (*Equus burchelli*) on the African plains is indeed an awe-inspiring spectacle. Similarly, skeins of Canada geese (*Branta canadensis*) winging southward in the autumn sky or a chorus of spring peepers (*Hyla crucifer*) in the moist April woods strikes a chord deep into the human spirit. These are *populations*—groups of animals, identifiable by species, place, and time. Typically, a population rather than an *individual* is the primary focus of wildlife management. Whereas biologists regularly monitor the habitat conditions in which populations exist, they also carefully determine the size, vitality, and diversity of populations to gauge an ecosystem's health.

Population ecology has produced an extensive body of literature, much of it theoretical, and the subject continues to evolve. Ecologists attempt to explain such phenomena as population cycles and the underlying causes for the regulation of animal numbers. However, most of the theoretical concepts lie beyond the scope of this book. Instead, this chapter presents basic concepts of population ecology and describes approaches that are useful to biologists who deal with the practical aspects of managing wildlife populations.

For purposes of illustration, we shall describe in some detail research conducted on Isle Royale, a large (544 km²) island in northern Lake Superior where moose (*Alces alces*) and wolf (*Canis lupus*) populations have been studied for years (Fig. 5-1).

Figure 5-1. Research crew attaches a radio-collar to a cow moose on Isle Royale in northern Lake Superior. Signals from the transmitter help wildlife biologists determine the fates of moose, including those preyed on by wolves. This and other techniques facilitate long-term studies of predator-prey interactions at Isle Royale. (Photo courtesy of Rolf O. Peterson).

Before proceeding, however, we present some basic terminology and principles of population ecology; these provide a vocabulary and the concepts regularly used by wildlife biologists.

SOME DEFINITIONS

A *population* is defined as a group of organisms, usually of the same species, occupying a defined area during a specific time. Populations have characteristics not possessed by individual animals. For example, a population has *density,* meaning a certain number of individuals per unit area: 20 blue grouse (*Dendragapus obscurus*) per 100 ha, or 144 sugar maples (*Acer saccharum*) per ha. A population has a *birth rate,* or *natality,* defined as the number of births per thousand, per hundred, or per individual per year, and a *death rate,* or *mortality,* defined as the number of deaths per number of individuals per year. A population also has an *age structure*—that is, a distribution of numbers of individuals of various ages. Naturally, the proportion of individuals of breeding age in a population affects the birth rate and strongly influences growth. Likewise, the proportion of old animals affects the death rate. Populations also have *sex ratios* that influence the reproductive potential.

Fecundity refers to the number of eggs produced per female or to the number of sperm produced per male. Because sperm numbers rarely influence the birth rate, fecundity nearly always pertains to the number of eggs produced. *Fertility* is the percentage of eggs that are fertile. *Production* is the actual number of offspring produced, whether born or hatched, by a population during a specific period of time. Some authorities, however, are more conservative and measure production only as the number of new individuals reaching breeding age (the process is also called *recruitment* when it includes immigration); animals dying between birth and sexual maturity are not counted as recruits. Changes in such parameters as sex ratio and age distribution greatly influence production; that is, the number of offspring can vary a great deal depending on the proportion of females and animals of breeding age in a population at a given time.

THE LOGISTIC EQUATION

It has been recognized for some time that animals tend to give birth to many more individuals than will survive to breeding age (Malthus 1798; Darwin 1859).

If deaths did not offset births, the result would be an infinitely growing population. Under ideal conditions, a population for a time can show a rate of growth that is exponential—that is, it grows at an ever-increasing rate. Such conditions may occur when a small population is introduced into a new and favorable environment. No shortage of food, cover, or space exists, and no diseases, parasites, or predators affect any individuals. The birth rate is maximum, limited only by the reproductive physiology of the species; and the death rate is minimum, with deaths occurring only from old age. Such conditions have been created in the laboratory for yeast cultures and mouse populations that were provided with room to grow and plenty of food. The equation for such growth is conventionally expressed as:

$$\frac{\Delta N}{\Delta t} = rN$$

where

ΔN = change in number

Δt = change in time

r = the "per head" maximum potential growth rate

N = number of individuals in a population

As an example, suppose we have a population of 50 individuals (*N*) and each individual has the average capability of contributing one-fourth (0.25) of an individual to the population in a given unit of time (*r*). The change in number per unit time ($\Delta N/\Delta t$) would be expressed as:

$$\frac{\Delta N}{\Delta r} = rN$$

$$\frac{\Delta N}{\Delta t} = 0.25(50) = 12.5 \tag{1}$$

The answer, 12.5, is the number of individuals that is added to the population in the time interval (*t*). This number then must be added to the original population (N_t) to obtain the number in the new population (N_{t+1}), so that our new population is

$$N_{t+1} = N_t + \frac{\Delta N}{\Delta t} \tag{2}$$

$$N_{t+1} = 50 + 12.5 = 62.5$$

For the next time step, we simply repeat the process using the same r value (0.25), but a new N (62.5) to calculate the number added to the population:

$$\frac{\Delta N_{t+1}}{\Delta t} = 0.25(62.5) = 15.5$$

$$N_{t+2} = N_{t+1} + \frac{\Delta N_{t+1}}{\Delta t} \qquad (3)$$

$$N_{t+2} = 78$$

As one can see by continuing this process, an ever-increasing number of individuals is added at each time step, even though the population continues growing at the same rate (Fig. 5-2). The integrated form of the equation is: $N_t = N_0^{e^{rt}}$, where N_t represents the population at t time intervals, N_0 is the original population, r is the per head potential growth rate, e is the base of natural logarithms, and t is the time interval by which r is expressed. Such an equation is most useful for populations of organisms such as bacteria, yeast, some insects, and possibly some small mammals in which breeding and growth are continuous. For most wildlife populations, which have a distinct breeding season, growth takes place in steps, and equations (1) and (2) are more appropriately used.

Under what conditions might we expect a wild population to express exponential growth? The happy circumstances of practically unlimited food and no biological enemies of whatever size occur very rarely in nature. But there have been a few instances that have come close to such conditions (Dasmann 1964). These circumstances invariably occur when a popula-

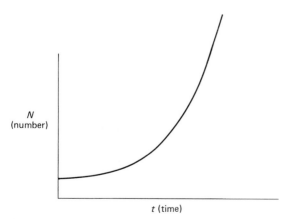

Figure 5-2. Growth of a population with unlimited food and space, $\Delta N / \Delta t = rN$.

tion is introduced into a new and favorable environment where the species was previously absent. Of the several cases reviewed by Dasmann, only one population, that of white-tailed deer (*Odocoileus virginianus*) introduced to the George Reserve in Michigan, expressed truly exponential growth (O'Roke and Hamerstrom 1948). In 1928, 2 bucks and 4 does were introduced into the 480-ha deer-proof fenced enclosure. By 1934 there were 164 deer. During the 6-year period, growth of the deer population was unrestrained by a change in birth or death rates acting under the influence of population size. In this case, the constant, r, was the only factor determining the rate of change in the deer population. Thus, for those few years, a growing population added an ever-increasing number of deer each year. Other rapid growth rates have occurred among ring-necked pheasants (*Phasianus colchicus*) introduced to Protection Island, Washington, and European reindeer (*Rangifer tarandus*) introduced to St. Paul Island off the Alaskan coast, but these populations showed less than the maximum growth that might be expected under totally favorable circumstances.

Almost any population may approach an unrestricted rate of growth if reduced to a low level, but obviously no population can increase exponentially for very long. The supply of food may not meet the demands of the ever-increasing population; space or cover availability may be limiting; predators may respond to the large numbers of prey; or disease may spread. Either birth rates decline, death rates increase, or both, so that eventually the population must stop growing. The greater the size of the population, the greater its dampening effect on the growth of the population. This effect has been mathematically defined and applied to the growth equation as follows:

$$\frac{\Delta N}{\Delta t} = rN \frac{(K - N)}{K} \qquad (4)$$

where K is defined as the maximum number of individuals the environment can sustain. As the population (N) approaches K, $K - N$ approaches zero so that when a population grows very large relative to the number the environment can sustain, its growth rate becomes nearly zero; that is, its potential growth rate (rN) is multiplied by the factor $(K - N)/K$.

For example, suppose the same population we considered earlier with an r of 0.25 now has 990 individuals and the maximum number supportable by the environment is 1000:

$$\frac{\Delta N}{\Delta t} = 0.25(990)\left(\frac{1000 - 990}{1000}\right)$$

$$= 247.5(0.01)$$

$$= 2.5$$

$$N_{t+1} = N_t + \frac{\Delta N}{\Delta t} = 990 + 2.5 = 99.25$$

Instead of the population growing by 247 individuals as it would without any limitations, it grows only by 2.5 individuals, owing to limitations placed upon it by the finite environment. Equation (4) is known as the *logistic equation.* The curve it produces is *sigmoid* (S-shaped) and is illustrated in Figure 5-3.

Populations may sometimes exceed the maximum number that can be sustained by their habitat. In such an occurrence the term $(K - N)$ is negative. Therefore, $\Delta N/\Delta t$ is negative, resulting in a decrease in numbers.

The term K often is referred to as the *carrying capacity.* One must keep in mind that carrying capacity for animals can change from time to time as food production, cover availability, water availability, and other environmental factors vary with the seasons and successive years. Factors such as territorial behavior and response to crowding may interact with these external factors so that the growth of a population may slow down before shortages of food, water, or cover are measurable in the habitat.

A factor that causes higher mortality or reduced birth rates as a population becomes more dense is referred to as a *density-dependent* factor; that is, if the

probability of an individual being born or surviving is lower as the numbers of animals in the population become higher, a density-dependent factor is acting to restrict population growth. Such factors include food supply, predation, disease, and territorial behavior. There are few density-independent factors, and they are mainly related to weather, such as cold, rain, and floods. Usually, populations in the central portions of the geographic range of their species are limited by density-dependent agencies. Near the periphery of the range, however, where habitat may be marginal and where random weather fluctuations may exceed the tolerance of nearly all animals in the population, density-independent factors may control population numbers (Krebs 1978). Such is the case with bobwhite in the northern parts of their range, where severe winters almost invariably result in a drastic reduction in quail numbers (Leopold 1931). Lack (1951) documented a similar effect on herons wintering in England.

The logistic equation is based purely on the operation of density-dependent factors. If a fatal flood or snowstorm strikes at some point, the irregularity in population growth will not be explained by the logistic equation.

The logistic equation, for the wildlife biologist, is useful in illustrating general principles of population growth as well as the theoretical effect of carrying capacity on reducing or stopping population expansion. The early portion of the sigmoid curve can serve as a theoretical model with which the manager can compare the growth rate of the population being managed. A value for r can sometimes be obtained from knowledge of natality and mortality of the species under ideal captive conditions or from potential birth rates and longevity information in the field. If population growth is close to the growth rate predicted by the early phases of the logistic equation, there is little the manager can do to increase growth of the populations.

There is a further practical application of the logistic equation. If N is set at various levels in the equation, the largest value of $\Delta N/\Delta t$ is obtained when N is half of the carrying capacity. At half the carrying capacity, an inflection point occurs in the growth curve (Fig. 5-3), the point at which population growth changes from an increasing to a decreasing rate. In managing for maximum yield of a population, it is therefore desirable to attempt to keep the population at about half the level of the carrying capacity. For example, using the previous

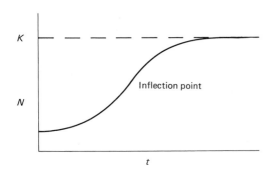

Figure 5-3. Growth of a population with a maximum number of individuals that can be sustained by the environment, $\Delta N/\Delta t = rN(K - N)/K$.

hypothetical population with K of 1000 and r of 0.25, and $N = \frac{1}{2}K$:

$$\frac{\Delta N}{\Delta t} = 0.25(500)\left(\frac{1000 - 500}{1000}\right) = 62.5$$

The maximum number of individuals that can be produced in a unit of time is about 62. At any other N, the number produced is fewer.

The factor $(K - N)/K$ in the logistic equation has no biological meaning or influence by itself. It must, in a real population of animals, represent some modification of birth rates or death rates. Therefore, the field biologist usually seeks an explanation for increasing or decreasing numbers of animals by examining the ratio of birth rates to death rates and the reasons for irregularities in either or both.

Most animals dealt with by wildlife managers reproduce seasonally, producing an annual spurt of offspring. Only if we stand back and view such a population over 50 or 100 years would these spurts become less visible. In such a long period of time, comparisons with the logistic equation could be made by wildlife biologists, but management results frequently are required in a much shorter time, and the wildlife manager must use other methods to assess the health of a population and evaluate its growth.

FIELD STUDIES

Wildlife biologists frequently encounter questions that ask, Why are the current numbers of some species so few when there were many more a few years ago? Sometimes the question is the opposite, Why are some populations larger today than in years past? These are perhaps the most basic questions of population ecology, and wildlife managers strive to provide reliable answers. Unfortunately, clear-cut responses are only rarely available. Nevertheless, there are ways to seek answers, which become more plausible when they are based on quantitative knowledge of wildlife populations.

Explanations for the fluctuations in animal populations are seldom revealed without several years of intensive fieldwork. A basic objective is to learn why animal populations fluctuate, sometimes widely, yet rarely increase to plague proportions or decrease to the verge of extirpation—a matter certainly pertinent to the management of wildlife

populations. Answers, however, usually require collecting and analyzing data for such obvious factors as the availability and quality of food, predation, competition, diseases, and climatic influences. A few examples of detailed field studies include those on snowshoe hares (*Lepus americanus*) by Keith and his coworkers (Keith and Windberg 1978, Keith et al. 1977, 1984), on African antelopes (Wilson and Hirst 1977), and on red grouse (*Lagopus lagopus scotticus*) in Scotland (Watson and Moss 1970).

Isle Royale, for several reasons, represents an excellent natural laboratory for studies of wildlife populations. First, it is more than 24 km from the nearest mainland, which essentially eliminates immigration and emigration by most land mammals. Second, the community of large mammals on Isle Royale is relatively simple: a single species of large prey (moose) and a single species of large predator (the wolf). Beavers (*Castor canandensis*), red foxes (*Vulpes vulpes*), snowshoe hares, mice, and other small mammals also occupy Isle Royale, but these play minor roles in the interactions among vegetation, moose, and wolves. By comparison, the mainland communities in this region of North America include large mammals such as black bear (*Ursus americanus*), coyote (*Canis latrans*), white-tailed deer, and a few caribou. Third, Isle Royale experiences long winters; hence, wolf and moose movements can be tracked in the snow (often from aircraft), and carcasses decompose slowly during much of the year, thereby preserving valuable evidence. In summer, volunteers systematically comb the island for moose skulls, from which age distributions can be determined. Finally, Isle Royale is federally owned and managed by the National Park Service. No hunting is allowed, and other than a few research biologists, no humans are on the island from November to March. The vegetation also has remained unmanaged for nearly a century. All told, the predator-prey regime on Isle Royale interacts in a setting where much has been learned about population ecology.

Research at Isle Royale began in 1959 when a succession of graduate students at Purdue University directed by Durward Allen initiated studies of wolf predation and behavior (L. David Mech), beaver (Philip Shelton), moose populations (Peter Jordan and Michael Wolfe), and fox predation (Wendell Johnson). In 1970, Rolf Peterson became the last in the series of Dr. Allen's students who worked at Isle Royale; now a professor himself, Peterson continues his studies of wolves to the present time. Much of the

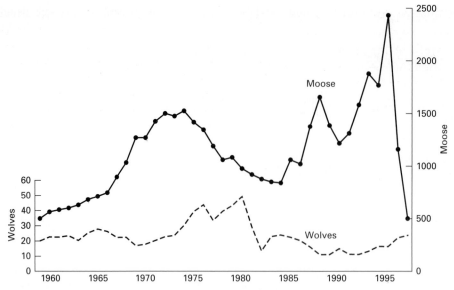

Figure 5-4. Fluctuations in wolf and moose populations at Isle Royale National Park. The wolf:moose ratio increased dramatically since 1995, and wolf predation now may limit increases in the moose population. Estimates for moose numbers between 1959–1982 were determined by population reconstruction based on recoveries of dead moose, whereas estimates for 1983–1997 are based on aerial surveys. (Based on Peterson 1997).

material in the following section is based on the early work of Mech (1966), Allen (1979), and more recently, Peterson (1995; 1997).

The results of these thorough, long-term studies provide both the public and biologists with an interesting lesson in wildlife ecology (Fig. 5-2). For the first 5 or 6 years after the research began, moose populations increased slowly; the wolves were removing approximately the number of "surplus" moose produced each year. This relationship suggested a "balance" between wolves and moose in a system free of human intervention.

Nonetheless, moose numbers continued to increase in the late 1960s and early 1970s, but the wolves showed no corresponding increase. By 1973, the moose population was estimated at 1430, while wolves remained between 15 and 20 individuals. At that level, the moose suffered from malnutrition for several winters in the late 1970s and then "crashed" by 50 percent by 1981. However, wolves apparently took advantage of the larger number of weakened moose during this period, and their numbers increased to a peak of 50 individuals. Thereafter, the moose population again increased, but the wolf population plummeted to 14 during the next 2 years, apparently because of canine parvovirus, a disease

probably brought to the island from the mainland (illegally) by a visitor's pet dog.

By the early 1990s, however, the virus disappeared from the wolf population. Moose numbers again began to increase at this time, reaching a peak of 2422 in 1995. The wolf population responded only modestly despite the recurring abundance of food and increased in the winter of 1996–97 to just to 24 individuals—not nearly enough to contain the moose population from outstripping its food supply. In the long winter of 1995–96, the moose population again crashed by 50 percent, leaving about 1200 animals; the aftershock of crash continued into May 1996, when the delayed arrival of late spring extended the period of starvation for moose. Then, with browse scarce and a second consecutive hard winter at hand (in 1996–97), the numbers of moose dropped to about 500. In this case, wolves played practically no role in controlling the number of moose, although starving moose continued to provide food for the wolf population. Wolf numbers in late winter increased from 12 in 1996 to 24 in 1997.

The lessons from the research at Isle Royale are many. First, predator-prey systems are not a neatly controlled phenomenon in which predators prudently choose their prey and adjust their reproduction to their food supply. Second, predator and prey populations fluctuate not in response to one or two simple variables (e.g., the relative abundance of each population), but they instead react to myriad variables, including diseases in either population, genetic variation, weather, and random events. Some ecologists liken these circumstances to a "discordant harmony" in a symphony (Botkin 1990).

Quoting Peterson (1995), "For ecologists, the spectre of a 'disequilibrium system,' in which the subjects are jarred unpredictably back and forth by a welter of influences, is disconcerting. Large, long-lived species such as wolves and moose can withstand many of life's vicissitudes, warding off starvation, extreme weather, shortages of prey, or surplus predators, but sooner or later a rare combination of events will turn the tide. The longer we observe a system like Isle Royale, the more shocks we see; as variability increases, our interpretations become more circumspect, more cautious, more tentative.

"This is not a sign of growing incompetence or scientific senility. It actually illustrates a classical paradox in science, wherein the longer one studies a particular living system, the less one can say with certainty about its behavior. We may show improvement in our ability to explain what has just happened, but we must ever be humble in predicting what lies immediately ahead. This is the essence of a world of 'disequilibrium.' Event in true chaos, if one can stand far enough away, patterns may be discerned, but they are unlikely to resemble the simple notions of an earlier day. Linus Pauling, a scientist whom many rank in the company of Newton, Darwin, and Einstein, said in an interview just before his death at the age of 93, 'I don't care to comment on the future of anything.'"

BIRTHS AND DEATHS

To achieve their management objectives, biologists often investigate the adequacy of existing habitat and its effects on birth and death rates. A population grows according to the simple equation:

$$r = b - d$$

where

r = actual growth rate of the population
b = birth rate
d = death rate

In some populations, animals moving in or dispersing from a population may also play a role in its growth rate. The equation then becomes:

$$r = (b - d) + (i - e)$$

where

i = immigration rate
e = emigration rate

A rate represents a change per unit time. Growth rate is the number of individuals added per individual in the population per week, per month, or per year. For example, suppose 3000 young are born each year in a population of 1000 cottontails. This represents a per-head birth rate of 3.0. During the same period, to have a stable population, there must be a per-head death rate of 3.0, which means that for every individual present in the spring population 3 must die over the course of the year. This would offset the per-head birth rate of 3.0. The population growth would then be zero ($r = 3.0 - 3.0 = 0$).

Birth rates and death rates differ with age structure and sex ratios of populations. If there are relatively many females of prime reproductive age, a population naturally will reproduce faster than one that has few females at such an age. Therefore, the wildlife biologist, to understand population growth, should consider the following seven characteristics pertaining to birth rates:

1. Age of sexual maturity of both males and females.
2. Length of the gestation period.
3. Sex ratios.
4. Whether the species is monogamous or polygamous.
5. Number of females that breed at each age.
6. Number of young per female of various ages.
7. Influence of nutritional condition on reproduction.

SEX RATIOS AND MATING SYSTEMS

Sex ratios express the relative abundance of each sex in wildlife populations. The ratios usually are expressed in one of two ways. The first tells the number of males per 100 females. The second expresses the percentage of males and females per 100 individuals in the population, with the percentage of males appearing first (e.g., in the ratio 60:40, the population consists of 60 percent males). We shall adopt the latter expression.

Sex ratios may change within populations either because of imbalanced sex ratios at birth or, more frequently, because of sex-specific mortality associated with age. Causes may be external forces such as hunting or the vulnerability of incubating females to predators. Johnson and Sargeant (1977) presented strong evidence that red foxes (*Vulpes vulpes*) killed enough hens during the nesting season to distort the sex ratio in a large population of mallards (*Anas platyrhynchos*). Some inherent mechanisms apparently cause differential mortality during embryonic development in some species, under conditions that are not always clear. Bellrose et al. (1961) recognized these age-related differences and adopted the following categories:

PRIMARY SEX RATIO.   The sex ratio at fertilization; normally 50:50 based on simple statistical probability.

SECONDARY SEX RATIO.   The sex ratio at birth or hatching; usually approximates 50:50 but may show the first indication of sex-specific mortality (e.g., 49:51). At birth, the fawns of white-tailed deer suffering from nutritional stress may exhibit a ratio favoring males by as much as 72:28 (Verme 1969).

TERTIARY SEX RATIO.   The sex ratio of juveniles; important because it indicates the proportion of each sex later entering the breeding population; for game species, hunting becomes an external influence for the first time; some inherent species-specific differences also may be present.

QUATERNARY SEX RATIO.   The sex ratio of the adult population; often clearly skewed in favor of one sex; some species of diving ducks (e.g., redheads, *Aythya americana*) are heavily imbalanced in favor of males; in most populations of large ungulates, such as deer or bighorn sheep (*Ovis canadensis*), females predominate because hunting pressure normally selects males with antlers or horns.

Waterfowl and other wildlife follow one of three types of mating systems. These systems interact with population sex ratios in ways that profoundly affect annual production. Mating systems include the following:

Monogamy

a. Seasonal. Pair-bonds established only for the current breeding season (e.g., pintails, *Anas acuta*, and other dabbling ducks)
b. Life time. Pair-bonds established for as long as both mates remain alive (e.g., coyotes, *Canis la-*

trans, and several other canids; Canada geese, *Branta canadensis,* and allied species)

Polygamy

a. Polyandry. Several males per female; extremely rare in most groups of vertebrates but occurs in a few birds (e.g., Wilson's phalarope, *Steganopus tricolor*)
b. Polygyny. Several females per male (e.g., ring-necked pheasants; elk, *Cervus elaphus canadensis*)

Promiscuity

Indiscriminate mating (e.g., cottontails, *Sylvilagus floridanus*; bobcats, *Felis rufus,* and many other felids)

The effect of quaternary sex ratios on production in monogamous species is shown in the following example; the maximum number of nests serves as an arbitrary measure of production:

Sex Ratio	Maximum Number of Nests per 100 Birds
50:50	50 (100% production)
60:40	40 (80% production)
40:60	40 (80% production)

The example illustrates the fact that monogamous species require a balanced sex ratio for the maximum production of offspring. Any deviation favoring either sex reduces production. Thus, in normal situations, hunting regulations for such species probably should not set sex-specific bag limits. However, when a sex ratio is highly imbalanced, hunting regulations can focus the harvest on the more abundant sex—for example, the harvest of ducks at times has been governed by a point system; that is, the bag limit is based on the accumulation of 100 points instead of on a set number of birds. Because the proportion of females is low in some species or populations, the point system places high values on hens and low values on drakes as a means of shifting the shooting pressure from one sex to the other.

In polygynous species, the situation is quite different. Females represent a premium for increased production. Males, within reason, become expendable, as follows:

Sex Ratio	Maximum Number of Nests per 100 Birds
40:60	60 (100% production)
50:50	50 (83% production)
60:40	40 (66% production)

Thus, in polygynous species, 100 percent production occurs at most ratios in which females make up more than 50 percent of the population (the 40:60 ratio was selected arbitrarily for comparison with production at the same sex ratio for a monogamous species, shown earlier). Indeed, the disparity favoring females might be increased to 30:70 or more for even greater production of offspring per 100 adults. Eventually, however, the point is reached where males no longer can successfully court and mate with such a large number of females. At such a point, the imbalance becomes so great that the "extra" females no longer contribute offspring and thus become an expendable surplus. Among penned ring-necked pheasants, McAtee (*in* Leopold 1933a) found that some hens remained unmated when the sex ratio reached 12:88. In most circumstances, however, male pheasants can be heavily harvested each autumn without endangering production the following spring.

AGE-SPECIFIC BIRTH RATES

The number of births per individual (or, more commonly, per 1000 individuals) in a population is called the *crude birth rate*. The crude birth rate reveals no details about what age groups actually contribute offspring to the population. Birth rates are by no means fixed within species or within populations, and variations in natality may account for large shifts in the sizes and densities of wildlife populations. The number of offspring produced during a particular period depends upon the number of females in each age class, the number of these that actually mate, and the fecundity of each age class (the same considerations may apply to males, of course, but females generally govern age-specific production). The number of offspring per female expressed by age classes is called the *age-specific birth rate*.

As an example, a population of brook trout (*Salvelinus fontinalis*) breeding in Lawrence Creek, Wisconsin, showed large age-specific differences in egg production (Table 5-1). Reproduction relied heavily on females in age classes I and II (1.5- and 2.5-year-old trout), which together produced more than 98 percent of the eggs (McFadden 1961). Three facts emerge from the details of the study. First, egg production per mature female increased with age. Yearling trout produced an average of about 425 eggs per mature female, whereas 2- and 3-year-old females each produced an average of 617 and 906 eggs, respectively. Second, because not all of the females in age class I were sexually mature, the aver-

age egg production for all females (mature and immature combined) in age class I was reduced even further (to 354 eggs). Third, even with lower egg production per female, the individuals in age class I were so numerous that the group still accounted for more than 75 percent of all egg production. These facts have management implications. In this case, Lawrence Creek is heavily fished, and few trout survive beyond age class III even though the population remains relatively stable. Thus, armed with knowledge of the age-specific reproductive rate, managers can adjust the fishing regulations in ways that will help protect the breeding population (e.g., changes in creel limits, season lengths, size limits, or a combination of these). With reduced fishing pressure, more females would reach the older age classes and thereby would contribute more to egg production. Such a shift would result in greater production from the trout population in Lawrence Creek because older females are all sexually mature and individually can produce larger numbers of eggs than younger females.

In another example, Woolf and Harder (1979) compared age-specific birth rates among white-tailed deer in three states (Table 5-2). These data revealed poor reproductive success in the Pennsylvania herd, which was densely populated and maintained with artificial food, in comparison with herds in Iowa and Ohio. Several hypotheses were explored to explain the poor reproductive performance in the Pennsylvania deer. The factors that were investigated included adrenal stress from heightened social interactions in the populous herd and reduced ovulation rates induced by chemicals in the diet of acorns. Ultimately, however, the low reproductive rates in the Pennsylvania herd were attributed to poor summer nutrition. This factor extended the lactation period into the

TABLE 5-1. Number of Eggs Produced by Brook Trout in Lawrence Creek, Wisconsin, 1955–56

Age Class (Year)	Percent Sexually Mature	Total Number of Eggs	Percent of Eggs Contributed
I	83.3	1,706,685	76.6
II	100.0	485,756	21.8
III	100.0	32,014	1.4
IV	100.0	2,318	0.1
V and VI	100.0	2,555	0.1
Total		2,229,328	100.0

Source: McFadden (1961).

autumn breeding season. Thus, fawns were weaned unusually late in the year and were unprepared to breed in the autumn. Moreover, because the older does were still lactating, their nutritional reserves were so low that ovulation and early fetal development were impaired. These conditions lowered the ovulation rates and induced early mortality in the fetuses of older does.

Woolf and Harder (1979) estimated ovulation rates using the common method of examining *corpora lutea*. These structures are the remains of ovarian follicles from which eggs are produced by sexually mature females during the current breeding season. To determine the number of corpora lutea, wildlife managers examine thin cross sections of the ovaries taken from females collected on the research area. The corpora lutea appear as small, yellowish spheres embedded in the ovarian tissues. Each corpus luteum represents the ovulation of a single egg, and the ratio of young fetuses to corpora lutea thus indicates the fertilization rate. Using these and other techniques, wildlife managers can obtain accurate measures of the reproductive contributions of each age class in a population. A table of age-specific birth rates thus becomes useful for depicting the reproductive performance of a population. A few examples are shown in Table 5-3.

As an illustration of how reproduction may vary with age structure, consider the population of Canada geese listed in Table 5-3, and two different distributions of breeding birds within the age classes. In the first case, the age structure is represented by 34 percent 1 year olds, 33 percent 2–3 year olds, and 33 percent 4+ year-old birds. Egg production from 100 females would be $(34 \times 0) + (33 \times .20 \times 4.6) + 33 \times .72 \times 6.4) = 182$ eggs. If the age ratio were to change to 40:40:20, respectively, for each of the age classes, then the number of eggs laid by 100 females would be $(40 \times 0) + (40 \times .20 \times 4.6) + (20 \times .72 \times 6.4) = 100$ eggs. Thus, with a change in age structure as described—which might easily occur in just a few years—egg production can be diminished significantly [here, by $(182 - 100)/182 = 45$ percent].

TABLE 5-2.   Comparison of Reproductive Performance in Three Populations of White-Tailed Deer

Area	Percent Fawns Pregnant	Percent Adult Does Pregnant	Corpora Lutea per Pregnant Doe	Fetuses per Pregnant Doe
Rachelwood Wildlife Preserve, Pennsylvania	0.0	93.5	1.60	1.40
Plum Brook Station, Ohio	0.0	94.8	1.95	1.77
DeSoto National Wildlife Refuge, Iowa	83.6	100.0	2.23	2.10

Source: Woolf and Harder (1979).

TABLE 5-3.   Some Examples of Age-Specific Laying and Birth Rates

Species and Source						
Brook trout	Age (years)	0.5	1.5	2.5	3.5	4.5
(*Salvelinus fontinalis*)	Percent breeding	0	83.3	100	100	100
McFadden (1961)	Number of eggs per breeding female	0	425	617	906	1196
Canada goose	Age (years)	1	2–3	4+		
(*Branta canadensis*)	Percent breeding	0	20	72		
Cooper (1978)	Number of eggs per breeding female	0	4.6	6.4		
Blue whale	Age (years)	0–3	4–5	6–7	8–11	12+
(*Balaenoptera musculus*)	Number of calves per breeding female	0	0.19	0.44	0.50	0.45
Usher (1972)						
White-tailed deer	Age (years)	0.5	1.5	2.5	3.5	4.5
(*Odocoileus virginianus*)	Percent breeding	16	68	77	81	84
Teer et al. (1965)	Number of fawns per breeding female	0.88	1.32	1.52	1.52	1.52

ADDITIVE AND COMPENSATORY MORTALITY

Animal mortality—the losses from a population—may be considered as either *additive* or *compensatory*. Numerous environmental factors, including disease, malnutrition, predation, and severe weather, act on members of a population. For example, given a population of 100 animals, food shortages and disease acting together might have the potential of removing 40 individuals during the course of several months. At the same time, however, predators also might have the potential of removing 40 animals. If these factors—starvation, disease, and predation—were *additive,* then a total of 80 animals would die from the combined action of these forces. But it is unlikely that the mortality would reach such a level. Competition for food is reduced whenever predators remove some animals from the population. As starvation lessens, so too does the incidence of disease-related mortality, and fewer animals actually die from malnutrition and sickness because of the interaction with predation. If predators remove 30 animals, only 10 might die of disease. Thus, the mortality factors act in a *compensatory* way.

Errington (1946) concluded that mink (*Mustela vison*) preyed on muskrats that were "surplus" members of a crowded population. The surplus muskrats were "social outcasts" because they could not obtain and hold breeding territories. As such, the outcasts were vulnerable to diseases and predation. If mink did not kill them, the surplus muskrats soon succumbed to disease. As Errington (1946) stated, "victims of one agency simply miss becoming victims of another" and many types of mortality are "at least partly intercompensatory in net population effect." Other histories of animal populations involving the compensatory nature of predation are described in *Of Predation and Life* (Errington 1967).

Density-dependent factors operate in a compensatory manner. By contrast, severe weather may act in an *additive* way; that is, even though a population may be reduced to low levels, a sleet or ice storm may kill a fixed proportion of the original number of animals, regardless of how many had been taken earlier by predation or disease. If, however, the storm preceded a period of food shortage, the effects of the bad weather might be compensatory, for the remaining animals likely would experience increased survival.

Hunting is a common form of mortality in populations of game species. Hunting mortality is frequently compensatory because it usually increases the life expectancy of individuals surviving the hunt, promotes higher reproductive rates, or does both. Swenson (1985) compared the reproduction of mountain goats (*Oreamnos americanus*) subjected to different levels of hunting mortality. One result of the study indicated that the summer ratio of kids to older goats (yearlings and adults) was higher after hunting had reduced the number of older goats the previous autumn. Decreased competition for winter forage explained the increased production of kids. Some caution was expressed, however, because the goats in this study were members of a population that had been introduced 10–20 years before, and the population was still expanding in logistic growth. Nevertheless, logic suggests that reduced competition in any population will increase the survival of the remaining individuals and will enhance birth rates.

Wagner et al. (1965) summarized the compensatory nature of hunting in mathematical terms in this way: The addition of a given percentage of mortality (from such factors as hunting or fishing) does not add one-for-one with the existing annual mortality from other causes (e.g., disease). Instead, the total annual mortality increases by a much smaller percentage than is measured by the actual percentage of individuals removed by hunting alone. The formula for calculating the crude annual mortality rate is

$$a = m + n - mn$$

where:

a = crude annual mortality rate

m = mortality rate from hunting or fishing

n = natural mortality rate

In a population with a natural annual mortality rate of 70 percent, the addition of 20 percent mortality from hunting would not increase the total mortality to 90 percent, but only from 70 to 76 percent:

$$a = 70\% + 20\% - (70\% \times 20\%) = 76\%$$

In other words, hunting removes some animals that otherwise would die from natural causes. Errington's ideas about surpluses do not apply here. Instead, the relationship shown in the example emerges simply because an animal can die from only one of the types of causes to which it is exposed: natural mortality or hunting mortality. The natural mortality rate remains much the same with or without hunting, although the actual number of animals dying from natural causes is less where hunting first removes a fraction of the population.

Moreover, a given percentage of harvest increases a small annual mortality rate more than a large one. For example, a 20 percent harvest raises a 40 percent annual mortality rate to 52 percent—a 30 percent increase—whereas, as we have shown, the same 20 percent harvest raises a 70 percent annual mortality rate to 76 percent, for only a 9 percent increase. Such a relationship, in part, accounts for the more visible effects of hunting species with low mortality such as big game and geese. The relationship supports Hickey's (1955) conclusion that the ability to withstand harvest is a function of each species' annual mortality rate.

LIFE TABLES AND SURVIVORSHIP CURVES

Comparisons of mortality between populations can be made by use of *life tables* and *survivorship curves*. A life table is a systematic means of describing mortality as it affects various age groups in a population. Deevey (1947) published a classic review of life tables for natural populations of animals. Murie (1944) described mortality of Dall sheep (*Ovis dalli*) in Mount McKinley National Park, Alaska. During a period of several years, he collected skulls of sheep found in the park. The horns of sheep grow in annual bursts, leaving a ring between each annual increment so that the age of a Dall sheep can be estimated by counting annual growth segments. From 608 sheep, Murie constructed the life table shown in Table 5-4.

The columns of a life table are as follows:

x = An appropriate time interval.

l_x = The proportion of animals living at the beginning of interval x. It is traditional to convert whatever sample size one has to 1000 at the beginning of the l_x column, representing 1000 animals born or hatched.

d_x = The proportion of animals dying during interval x

$1000\, q_x$ = The proportion of animals that die per interval x. It is computed as follows: $1000\, q_x = (d_x/l_x) \times 1000$.

c_x = The life expectancy expressed as the number of additional intervals an individual animal can expect to live at the beginning of interval x.

The derivation of e_x involves a few extra calculations, as follows:

$$e_x = \frac{T_x}{l_x}$$

and

$$L_x = \frac{l_x + l_{x-1}}{2}$$

or the average number of animals alive at the midpoint of an interval x.

T_x is the sum of the L_xs starting from the bottom of the column up through the desired x interval.

Biologists often are asked questions such as, "How long does a robin live?" or "How long does a mule deer live?" The answer to such questions depends on what age the animal has achieved. The American robin (*Turdus migratorius*) can live to be 7 years old, but the probability of a newly hatched robin doing so is much less than 1 percent (Farner 1945). Early mortality, in fact, is so high among songbirds that most life tables for them do not begin until late in the summer or early fall of a bird's first year. Once a robin has lived to November 1, on the average it will live another 1.37 years. More than half of them will die in the next year (Farner 1945).

Murie's life table for Dall sheep (Table 5-4) shows that life expectancy is 7.1 years for a newborn lamb. There was a flaw in Murie's data, however, in that the skulls of very young lambs that died possibly were consumed totally by scavengers or predators. Thus, young animals would be underrepresented in his life table, and life expectancy at birth would be overestimated. Once a sheep reached the age of 6 years, one may see that it can expect to live, on the average, another 3.4 years, to reach the age of 9.4 years.

A survivorship curve is constructed by plotting the l_x column of a life table against time. A classic use of survivorship curves in wildlife management was described by Taber and Dasmann (1957) for black-tailed deer (*Odocoileus hemionus columbianus*) in two habitats in California. By use of survivorship curves (Figs. 5-5 and 5-6), the researchers were able to compare survival rates of both sexes of five

TABLE 5-4. Life Table for the Dall Mountain Sheep *(Ovis dalli)* Based on the Known Age at Death of 608 Sheep Dying Before 1937 (Both Sexes Combined)[a]

x	kl_x	kd_x	kq_x	e_x
Age (years)	Number Surviving at Beginning of Age Interval Out of 1000 Born	Number Dying in Age Interval Out of 1000 Born	Mortality per Thousand Alive at Beginning of Age Interval	Expectation of Life: or Mean Lifetime Remaining to Those Attaining Age Interval (years)
0–0.5	1000	54	54.0	7.1
0.5–1	946	145	153.0	—
1–2	801	12	15.0	7.7
2–3	789	13	16.5	6.8
3–4	776	12	15.5	5.9
4–5	764	30	39.3	5.0
5–6	734	46	62.6	4.2
6–7	688	48	69.9	3.4
7–8	640	69	108.0	2.6
8–9	571	132	231.0	1.9
9–10	439	187	426.0	1.3
10–11	252	156	619.0	0.9
11–12	96	90	937.0	0.6
12–13	6	3	500.0	1.2
13–14	3	3	1000.0	0.5

Source: Deevey (1947); original data from Murie (1944).

[a]A small number of skulls without horns, but judged by their osteology to belong to sheep 9 years old or older, were apportioned pro rata among the older age classes.

populations of ungulates. All survivorship curves declined steeply in the first year, indicating high death rates of young animals. After that, considerable variation among populations occurred. The logarithmic scale is used on the abscissa to expand the lower parts of the curve. On a logarithmic scale, the removal of a constant proportion of animals woulld result in a straight declining line, such as occurs in female red deer (*Cervus elaphus*) from age 3 through 14 in Fig. 5-5.

Survival of male black-tailed deer was lower than for females because of selective hunting of males. Taber and Dasmann noted from their survivorship curves that survival of black-tailed deer was much less for both sexes in shrubland than in chaparral. By examining other population features, they found that fawn production in the shrubland deer was higher than in the chaparral deer (0.76 fawns per adult doe in shrubland to 0.53 per doe in chaparral in December). This higher production in the shrubland resulted in more intense competition for food, poorer nutrition, and higher mortality. Chaparral is a mixture of woody shrubs, while shrubland consists of scattered shrubs and herbs. In the shrubland, deer

food was more abundant, and this feature permitted deer to maintain higher densities (25 per km²) than in chaparral (11 per km²). Survival rates in the shrubland, however, were lower than in chaparral, presumably because the shrubland was fully stocked and the higher birth rates resulted in consequent higher losses through starvation. Once the carrying capacity of the shrublands had been attained, a proportion of deer that were not removed by hunting starved, so the growth rate in both shrublands and chaparral was essentially zero.

Taber and Dasmann also pointed out that dynamics of different populations of the same species may vary widely from place to place. It is therefore difficult to say that birth and death rates of a particular population are "typical" of a species; differences among populations of the same species reflect different environmental conditions.

There also is evidence that the genetic makeup of a population may change with the passage of time or in response to some environmental factor. (Such change is the basis for the theory of evolution.) Some changes can take place quite rapidly. For example, European rabbits (*Oryctolagus cuniculus*) in Australia were

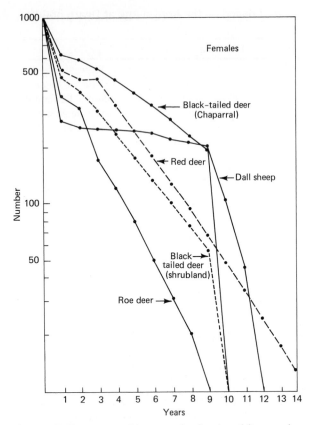

Figure 5-5. Survivorship curves for females of five ungulate populations. (From Taber and Dasmann 1957.)

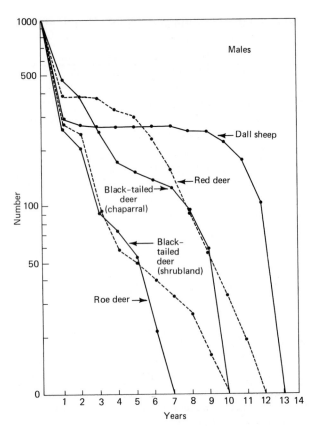

Figure 5-6. Survivorship curves for males of five ungulate populations. (From Taber and Dasmann 1957.)

in the 1950s intentionally infected with myxomatosis, a viral disease, to control their numbers. The rabbits declined rapidly. There are still rabbits in Australia—not as many, but they are genetically more resistant to the virus than the rabbits in Australia in the 1940s (Pimentel 1961). Other less-noticeable and unmeasured genetic changes possibly occur in many populations, changes that influence their birth rates and death rates (Krebs 1978) and that may be responsible for population irruptions and crashes for which no external cause may be apparent.

SOURCES OF POPULATION DATA

Obtaining accurate information about animal numbers and densities remains one of the more difficult and challenging tasks for wildlife managers. Data are gathered in the form of *censuses, estimates,* and *indices.* A census is a complete count of individuals in a

population. Some species, particularly those living in open areas, may be counted from aircraft (e.g., pronghorns, *Antilocapra americana,* or winter flocks of waterfowl). Complete counts also are possible in a few special situations. Whooping cranes (*Grus americana*) are large, white birds and thus are highly visible against a backdrop of marsh vegetation; the small population can be counted accurately in either summer or winter.

A complete count is rarely possible in areas where vegetation or topography conceals animals or where the population is quite large. In such cases, an estimate may be made on the basis of a statistical sample. A sample may be obtained by counting inanimate objects (e.g., droppings, nests, and dens) or by counting animals. In either case, the sample is taken on a plot or transect of a known size. Other information often is required before the sample can be interpreted. In the case of dens or burrows, for example, the average family size must be determined and then multiplied by the number of active burrows in the plots. Population estimates for prairie dogs (*Cynomys* spp.) are made in that manner, as are estimates based on the lodges of beaver (*Castor canadensis*). For droppings, the average number of droppings per day per animal must be known, as well as the maximum number of days since the droppings were deposited.

A common way of sampling living animals is based on a capture-recapture ratio. The estimate requires capturing, marking, and releasing a known number of animals and then resampling the population later. The population size is estimated using the proportion of the marked animals either recaptured (or resighted) in the second sample:

$$\frac{\text{No. in}}{\text{population}} = \frac{\text{no. marked \& released} \times \text{no. resampled}}{\text{no. marked in resample}}$$

The capture-recapture method, known as the Lincoln or Petersen Index (although the ratio actually is an estimate), often underestimates the population because of a higher likelihood of recapturing (or sighting) marked animals than unmarked animals (Davis and Winstead 1980).

An index is a quantitative measure of a population. However, it seldom provides exact numbers or even an estimate of the numbers or densities of animals in a population. Indices instead compare relative abundance between areas or changes in abundance from one time to another in the same area. Counts of displaying male woodcocks (*Scolopax minor*) and ruffed grouse (*Bonasa umbellus*) along established routes are commonly used indices (Mendall and Aldous 1943; Petraborg et al. 1953; Tautin et al. 1983). Others include counts of pheasants (*Phasianus colchicus*) and red foxes (*Vulpes vulpes*) by rural mail carriers (Greeley et al. 1962; Allen and Sargeant 1975).

Biologists also obtain sex and age data from various sources. The age structure of fish populations often is determined from a collection of scales; these show annual growth rings known as *annuli.* In autumn, hunters voluntarily submit wings or tails from various kinds of game birds for analysis; quail, grouse, and other birds are sampled for year-to-year changes in the ratio of juveniles to adults. Waterfowl are among the migratory birds sampled by the U.S. Fish and Wildlife Service, as thousands of duck wings are analyzed each year for age and sex ratios, by species, in what are known as *wing bees* (Fig. 5-7). At roadside check stations, state biologists obtain age and sex information from big game killed by hunters.

Tooth replacement and wear are among the features commonly inspected for age determination in deer and other large mammals. However, the methods for determining age and sex vary greatly by species. Methods range from studying the shapes of feathers among birds to determining the weight of eye lenses for some kinds of small mammals; these methods are reviewed in detail by Larson and Taber (1980).

The *Migratory Bird Harvest Information Program,* popularly known as HIP, is a new method to survey the harvest of migratory birds. Implemented in 1997, HIP requires that hunters display a HIP "certification" on their current licenses. For certification, which costs nothing, hunters merely complete a short questionnaire—just four questions—and state if they plan to hunt migratory birds, including waterfowl and webless species such as doves, rails, gallinules, woodcock, and snipe (*Capella gallinago*). Hunters also identify the types of birds they hunted during the previous year's season, with those who hunted doves forming one pool, rail hunters another pool, and so on. Each state then forwards this information to the U.S. Fish and Wildlife Service, which later selects a random sample from each pool of migratory bird hunters. Hunters in the sample are then contacted and asked to complete a more detailed survey—prepared anonymously—about the migratory birds they harvested during the season. HIP will provide wildlife managers with data to defend existing hunting practices, including season lengths and bag limits, from emotion-based legal challenges and also to protect species with declining populations. Whereas estimates of waterfowl harvests sample those hunters who buy federal duck stamps each year, HIP targets the 2 million hunters who hunt *only* doves, woodcock, and other nonwaterfowl migratory birds (i.e., hunters who do not buy federal duck stamps). By 1998, all states will participate in HIP, except Hawaii, which does not have significant populations of migratory birds.

Sampling theory must be considered when population data are collected. The subject is complex, and an extensive body of literature may be consulted (Seber 1973; Burnham et al. 1980; Sokal and Rohlf 1981). In its simplest form, however, sampling theory requires that the collected data represent the population at large. For example, the age distribution shown by scales collected from 100 fish in a lake produces a statistical estimate of the entire fish population in the lake. What is sampled (e.g., 100 fish) in theory represents what was not sampled (e.g., all of

Figure 5-7. A *wing bee* is an important source of population data for waterfowl management. Every year state and federal biologists in each flyway analyze thousands of wings mailed in by hunters. Color and other features in the plumage identify the sex and age of each species. A mallard wing is shown in the inset. (Photo courtesy of M.A. Johnson, North Dakota Fish and Game Department.)

the other fish in the population). However, one or more sources of *sampling bias* may affect the reliability of the samples, and hence the results of the analysis may not be accurate. If the fish population was sampled with a gill net, then the smaller—and thus the younger—fish likely escaped capture. Hence, scales would be collected only from the larger—and older—fish, and the analysis would reflect a population lacking young age classes. In this case, the bias results from the equipment and methods used in the field. Similarly, both age and sex data collected from deer at roadside check stations are usually biased (i.e., few does and fawns are shot),

so those data usually reflect only the adult male segment of the adult population. In this case, the sampling bias is associated with the selectivity of hunters for certain age and sex groups. However, most population data can be corrected or applied in ways that compensate for sampling bias; alternate field methods or equipment also may reduce the bias to acceptable levels (e.g., electrofishing equipment, or "fish shockers," may yield a better sample of size and age classes than gill nets provide). The important matter here, however, is that bias must be recognized and addressed before population data can produce sound management.

ORGANIZATION OF A POPULATION-MANAGEMENT PROBLEM

A wildlife manager charged with solving a problem of population ecology may be faced with a somewhat bewildering array of possible causes of an "unsatisfactory" performance by the animals he or she wishes to manage. The following outline is intended to suggest a means of organizing efforts. The objective, of course, is to identify those factors that are most responsible for preventing the further growth of the population. These factors impede births, increase deaths, or both.

A. Extrinsic Factors
  1. Density-independent (primarily weather conditions)
      a. Cause of direct mortality?
      b. Center or periphery of species range?
      c. Does weather have a substantial influence on food quality or quantity—which are density-dependent factors?
  2. Density dependent
      a. Food
          (1) Quality. Are necessary nutrients present?
          (2) Quantity. Is enough food available?
      b. Cover
          (1) Shelter from elements. Are quality and quantity sufficient?
          (2) Escape or hiding cover—for predators or from predators. Are quality and quantity sufficient?
      c. Refugia available. Are there patches of habitat in the range of the population in which animals have a high likelihood of escaping various mortality factors, such as predators, hunters, parasites, and disease?
      d. Competitors. Is there competition for resources by other species?
      e. Diseases and parasites. Are these factors influencing birth and death rates?
      f. Predators. Are predators controlling the population?
      g. Buffer species. Are other prey species present that absorb some of the impact of predation, particularly when the species being considered is at low densities?
      h. Hunting harvest. Is harvest toll replaced by the next season's production of huntable animals?
      i. Interactions among various factors. What interactions occur? Food supply–disease? Food supply–predation? Food supply–competition? Cover–predation? Buffer species–predation?

B. Intrinsic Factors
  1. Genetically stable factors
      a. Litter or brood sizes. What is the inherent potential of the species to reproduce?
      b. Longevity. How long can individuals live?
      c. Habitat selection for breeding, feeding, resting. Is it available according to inherent needs of the species?
      d. Self-limiting factors. Does the species possess self-limiting behavior such as territorial spacing or restricted breeding among selected members of a group?
      e. Dispersal. Is there an opportunity for immigration and emigration?
      f. Interactions. How do inherent features interact, such as territorial behavior–food supply, dispersal–food supply, birth rates–food supply?
  2. Genetically variable factors
      a. Birth rates. Within the physiological limits of the species, does the population show varying birth rates?
      b. Survival rates. Does a population differ genetically from time to time in the ability of individuals to withstand stress, or has there been a response to a strong selective factor such as disease or biocides?

As can be seen, answering questions about a population may be a complex task. The most fruitful and simplest approach is usually to examine the more obvious factors first, such as weather, food, cover, and the behavioral nature of the animals. Should such an approach fail to provide satisfactory answers, then more subtle interactions must be examined. Usually, the talents of a team of researchers, with various forms of expertise, must be called upon. Once the most important factors (key factors) limiting a population are identified, the manager may attempt to modify those factors. Sometimes nothing can be done, such as when the weather or climate of a particular area is simply unsuitable for the welfare of the species; but most often the application of suitable controls on key factors will result in a desired change in the population being managed.

Patuxent Wildlife Research Center

This facility, the U.S.'s first wildlife experiment station, was established in 1936 by President Franklin D. Roosevelt in the Patuxent River Valley near Laurel, Maryland. The 5260-ha site lies midway between Washington, D.C., and Baltimore, Maryland, and perhaps represents the best nature preserve for wildlife research near a major metropolitan area. Because of its size, the center provides isolated, protected areas for the study and propagation of endangered species, as well as for the experimental manipulation of wetlands, woodlots, old fields, and other types of habitat. Patuxent maintains a core of 70 scientists, supported by a staff of more than 100. Patuxent maintains 10 field stations throughout the United States. The U.S. Fish and Wildlife Service originally administered the center, but the center became part of the U.S. Geological Survey in 1996.

The Bird Banding Laboratory, a unit within the center, issues permits and bands and serves as the clearinghouse for banding data for both Canada and the United States. The lab maintains records for 56 million banded birds and 3 million recoveries of banded birds that represent 900 species and subspecies. These files are expanded each year with another 1.2 million bandings and 65,000 recoveries. Banding data are employed to monitor bird populations, to promulgate hunting regulations, and to study bird behavior and ecology. The center also coordinates the Breeding Bird Survey, begun in 1965, which annually censuses about 2900 routes at the peak of the nesting season and is an active partner in other national programs that monitor bird populations.

The Patuxent Wildlife Research Center provides curators at the National Museum of Natural History in Washington, D.C., who are responsible for the bird, mammal, reptile, and amphibian collections housed in the Smithsonian Institution. Among other tasks, the curators identify endangered animal populations, conduct systematic reviews of important taxa, and assist other agencies with taxonomic questions.

Much of the original research on eggshell thinning and other effects of DDT was conducted at the center, where a strong focus continues today on new-generation pesticides, heavy metals, and other environmental contaminants. Coordinated field and laboratory research conducted by chemists, toxicologists, and biologists examines the effects of agricultural, industrial, and other types of chemicals on wildlife and their habitats. For example, center personnel headed the investigation of waterbirds that were poisoned by selenium at Kesterson National Wildlife Refuge in California.

Studies of migratory birds and recovery programs for endangered species represent other major elements of the center's activities. Whopping cranes have been at the core of this program for many years, but bald eagles have been among the several other species housed and studied at Patuxent. The center also directs research on the Chesapeake Bay ecosystem; these studies address habitat for black ducks, forest fragmentation, wetland ecology, contaminants, and many other topics relevant to wildlife associated with North America's largest estuary.

METAPOPULATIONS

Ecologists in recent decades have studied *metapopulations*—groups of local populations of a species. Each group occupies separate patches of habitat, which are often connected by corridors through which dispersal and exchange may occur. Because each group is relatively isolated, any one of the groups within a metapopulation is prone to extirpation (often called "local extinction"), but when this happens, the vacant patch also has the capacity to be recolonized by individuals from the other patches. Changes in the distribution of New England cottontails (*Sylvilagus transitionalis*) in New Hampshire illustrate the underlying ideas about metapopulations—a case where habitat was highly fragmented by human development (Litvaitis and Villafuerte 1996).

In 1950, the distribution of cottontails spanned a more-or-less continuous area covering 60 percent of New Hampshire. By the mid-1990s, however, extensive urban development and other human-related modifications had reduced their range to only 20 percent of the state, with the remaining habitat in patches 0.2 to 15.0 ha in area. Usually only one or two cottontails in poor physical condition—and often males—populated the smallest occupied patches. Winter survival rates in small patches were low (37 percent) and about half the rate (70 percent) in larger patches where food was more abundant. However, no single population was large enough to ensure its long-term survival.

Using computer modeling, Litvaitis and Villafuerte (1996) simulated the distribution, reproduction, and survival of cottontails in a hypothetical metapopulation, a configuration of 30 patches shown in Figure 5-8. They then tested four scenarios, each populated with 200 cottontails. The first scenario consisted of the entire 30-patch landscape. Scenarios 2 to 4 consisted, respectively, of (a) 10 patches of equal size with 20 cottontails per patch, (b) 5 patches of equal size with 20 cottontails per patch, and (c) 2 large patches with 100 cottontails per patch. Variables included fluctuating temperatures, snow cover, and, based on recent trends in development, 4 percent annual loss of habitat.

Because of the large proportion of patches with small populations, the 30-patch landscape was the least likely to sustain its cottontail population (i.e., the probability of extirpation at the smaller sites was too high). In contrast, each of the three other scenarios maintained relatively secure populations. These results suggest that cottontails, which favor early successional forests, might best be managed by rotational cutting of adjacent 5- to 10-ha patches, with one-third of these cut every 10 years, thereby perpetually producing three age-classes of forests (Litvaitis and Villafuerte 1996). Such a manipulation allows cottontails to find appropriate habitat without having to travel across large areas of unfriendly terrain. Elsewhere, habitat for ruffed grouse (*Bonasa umbellus*) is currently managed using a similar practice, as described in Chapter 15.

POPULATION MODELS

When wildlife managers predict how many mallard ducks will be present in the fall population based on sample counts of breeding ducks and the number of prairie ponds in the spring, they are using a model. Similarly, an estimate of the number of deer present obtained by sampling and counting pellet groups (droppings) left by the deer is another example of the

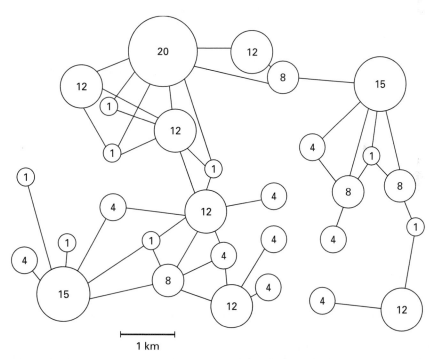

1 km

Figure 5-8. Schematic configuration—30 patches of habitat—for a hypothetical metapopulation of New England cottontails. Numbers represent the average carrying capacity for each patch; the area of each patch is proportional to carrying capacity, which assumes 1 cottontail per 0.5 ha. Lines indicate potential dispersal routes between patches. The scale applies only to the distances between patches. (From Litvaitis and Villafuerte 1996.)

use of a model. A *model*, as defined by Walters (1971), is any physical or abstract representation of the structure and function of a real system.

In the past decade, attempts have been made to construct models of everything biological, from cellular physiology to entire biomes. These representations are simulation models that trace through a period of time the changes that take place in a system, given some beginning conditions and some circumstances that effect changes in those conditions. Populations of animals lend themselves reasonably well to the methods of simulation modeling. Owing to the speed with which they perform calculations and the convenience they offer for altering variables at various stages in the operation of a model, computers offer wildlife biologists new opportunities to simulate populations of animals under study and to predict the effects of various management procedures. The basic tools for doing so are an understanding of FORTRAN or Pascal computer language and several years of data for a population, including age structure, sex ratios, birth rates, death rates, immigration, emigration, and environmental conditions that influence these factors.

Computer models have been developed to simulate wolf (*Canis lupus*)–moose (*Alces alces*) populations on Isle Royale (Jordan et al. 1971) and the projected recovery of whooping cranes (*Grus americana*) (Miller and Botkin 1974). An excellent example of a model was constructed to simulate a population of mule deer in Colorado (Medin and Anderson 1979). These authors noted that a model has three basic values: (1) it forces the researcher to think about population dynamics in new ways (conceptual value); (2) the researcher must become aware of the usefulness of various types of information necessary to construct an accurate model and, therefore, of the information necessary to understand population functions (developmental value); and (3) the model may be useful in predicting future courses of the modeled population or the effects of manipulation of the population by adjusting rates of exploitation or by altering the environment (output value).

The model of the mule deer population studied by Medin and Anderson utilized the following variables gathered over several years of study: vegetation available as food, weather, food consumption, nutrients in food, other animals as predators or competitors, age, sex, and number of deer present, natural mortality rates, hunter harvest, age structure in the hunter harvest of deer, birth rates as determined by

ovarian and fetal analyses, and the condition of the deer. Only one environmental variable, the amount of precipitation in the April–July period, seemed to have a significant effect on birth rates; it did so by affecting the amount of nitrogen available in winter forage. The density of deer in winter had an inverse influence on their birth rates. By calculating functions for these variables and feeding them into a computer, Medin and Anderson were able to simulate closely the dynamics of the deer population during the 5 years for which reliable data were available. The assumption is then that the model might be projected into the future, given a reasonable assortment of April–July precipitation values. The authors also tested the effects of different harvest strategies on the model (Fig. 5-9). Then the authors asked some "what if" questions of their model and obtained results depicted in Figure 5-10.

Students can readily see the value of a population model, provided that the model is reasonably realistic. Biologists who develop models also can recognize the frequent inadequacies of field data and assumptions of cause and effect of variables used in the model. A population model should be only one of many forms of information the wildlife manager may use. At this point in the development of population models, an attitude of informed skepticism seems appropriate.

THE HUMAN POPULATION

It is not prudent to consider management of wildlife populations without also appraising management of the human population. In 1993, there were about 5.5 billion people on Earth—a population that has been increasing at the rate of 1.6 percent per year. At such a rate, which has been declining slightly in recent decades, human numbers will double in 42 years (Population Reference Bureau 1993). More than 250,000 people are added *daily* to those already on Earth, and in fact, only 7 of the 170 nations of the world show zero or negative growth rates. The population of the United States, recently growing at a rate of 0.8 percent, reached 258 million in 1993. A glance at a newspaper suggests that many people behave as if the Earth can support an unlimited human population—one somehow unconstrained by carrying capacity (Hutchinson 1978).

For centuries human inventiveness has periodically increased the carrying capacity of the Earth, first by changing the way of living from a hunting

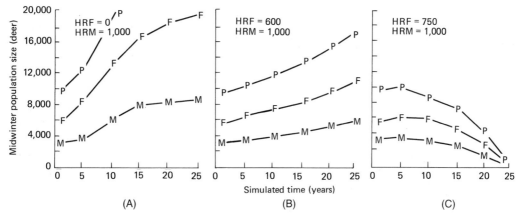

Figure 5-9. Graphs plotted from a model of simulated harvest strategies where *P* = total population size; *F* = female population size; and *M* = male population size. The annual harvest levels of females (HRF) were varied from 0 (graph A) to 750 (graph C). The annual harvest for males (HRM) was kept at a constant level of 1000 in each case. Data-based harvests were assigned for the first 3 years of simulated time; the harvest levels indicated were used thereafter. (From Medin and Anderson 1979.)

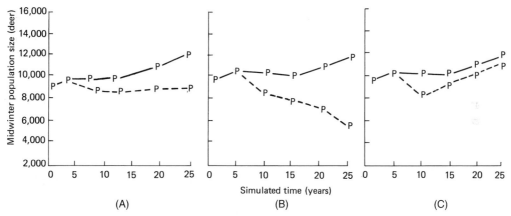

Figure 5-10. Graphs comparing a basic simulation run (where HRF = 650; HRF = 1000) with others in which specific "what if" questions were asked of the model. Solid lines show the basic simulation run; dashed lines shown a run with altered run. *P* represents the total population size. In A, the harvest rate was increased by 50 percent (HRF = 975; HRM = 1500) in year 9 only and was allowed to operate at the basic rate in all other years. In B, the harvest rate was increased by 50 percent for 2 consecutive years (years 9 and 10) and allowed to operate at the basic rate in all other years. A "catastrophic" year (year 9) is shown in C, in which the mortality rate of fawns was increased by 50 percent and the mortality rate of all older age groups was increased by 25 percent. The "catastrophic" year was followed by a year of antlered-only deer hunting (in year 10), and the basic harvest rate was allowed to operate in all other years. (From Medin and Anderson 1979.)

and food-gathering mode of existence to one of agriculture. Later, industrialization, mechanization, and rapid transportation permitted people to trade useful items for food, and fewer farmers were needed to produce human food. The "green revolution," which resulted in the development of high-yield grains, further permitted the feeding of even more people. But agricultural production during the past decade has scarcely been keeping pace with population growth, and an estimated 20–40 percent of the people in the world are underfed (Miller 1982).

Most ecologists believe that the carrying capacity of the planet for humans is rapidly being approached and that resources cannot be produced and processed fast enough to meet the demands of an ever-growing population. In some places, such as India and Bangladesh, the carrying capacity has already been surpassed.

The need to feed, clothe, and house the growing billions of people requires both more-extensive and more-intensive use of the land and waters. In doing so, humans compete with other animals. Faced with the necessity of raising crops and trees and mining energy sources of all kinds to supply starving people, many people find that arguments for providing habitat for wildlife lose their strength, and the wildlife management goal of balancing the needs of other animals and those of humans is weighted by the sheer numbers and "humane" priority of our own species.

The influence of the human population is truly global. Lead from the exhausts of millions of automobiles is found in the Antarctic ice pack, and chlorinated hydrocarbon pesticides are infused from the atmosphere into food chains of the Tundra, where pesticides have never been applied. Sulfur and nitrogen oxides emitted from power plants in one country cause acidification of lakes and soils in another. The production and use of resources to support humanity affect the resources for other animals, usually to the animals' detriment.

There is no doubt that the human population will stop growing at some point. The question is only whether human numbers will be controlled by such natural checks as starvation and territorial defense or by the intelligent application of methods to reduce the birth rate to match death rates, the latter of which have been lowered through medical treatment. The future of wildlife as well as that of humans rests upon such a choice.

SUMMARY

Wildlife management involves manipulation of populations; therefore, an understanding of population ecology is essential to wildlife managers regardless of the species being managed. Basic population attributes include density, sex ratios, age structure, natality, mortality, immigration, and emigration. The logistic equation $\Delta N/\Delta t = rN(K - N/K)$ is useful in illustrating theoretical population growth and the influence of carrying capacity. Factors acting to limit populations in the center of a species' geographic range are usually dependent upon the density of the population. Species at the periphery of their range are frequently controlled by density-independent factors. In practice, a wildlife manager should seek reasons for imbalances in birth rates, death rates, immigration, and emigration. The reasons for such imbalances may be sought in the adequacies or inadequacies of food and cover in the environment and how these limitations interact with birth, death, immigration, and emigration. Natality information is frequently easier to obtain than mortality information. Life tables are useful for comparing mortality among various populations to locate age classes affected most by mortality.

Organizing the approach to solving a problem of population management involves seeking factors of a variety of sorts that limit growth of populations. Once these are identified, managers may apply suitable modifications either directly to the population or to its habitat to bring about a desired change in the population. Models, or abstract representations of populations, may be useful to wildlife managers in illustrating which data are necessary to understand population dynamics and to predict the effects of various management measures on a population. Computers make modeling of populations a feasible management technique. Nevertheless, in actual field situations such as those on Isle Royale, a multitude of variables act on populations of predators and their prey. Such variables include weather, disease, and predator-to-prey ratio among others, and any combination of these complicates long-term predictions for the size of future populations. The principles of population ecology apply to humans as well as to other species, and the future of wildlife depends as much on management of the human population as it does on management of other species.

CHAPTER 6

ANIMAL BEHAVIOR AND WILDLIFE MANAGEMENT

I could show that none of (the) characters of instinct are universal. A little dose...of judgment or reason often comes into play, even in animals very low in the scale of nature.

Charles Darwin (1859)

Darwin recognized more than a century ago that animal behavior combines instinct and experience. Since Darwin's time, however, extensive research has greatly expanded our understanding of why animals act as they do. *Ethology,* the study of animal behavior, has emerged as a distinctive combination of biology and psychology and emphasizes inherent behavior in natural situations. *Psychology,* the more traditional study of behavior, concentrates on learned characteristics and seeks an understanding of human behavior as its ultimate goal.

The behavioral patterns of animals are often complex. The patterns result from both inherent (*innate* and *instinctive* are synonyms) traits and from learned responses to particular settings and stimuli. Distinctions between learned and inherent behavior, however, may be difficult to determine. Nonetheless, ethology has advanced rapidly in recent years, and detailed studies now appear in journals that are singularly devoted to behavioral research. Three European ethologists, Konrad Lorenz, Niko Tinbergen, and Karl von Frisch, were honored with Nobel prizes in 1973. Lorenz, in fact, is widely regarded as the "Father of Ethology."

Animal behavior is an essential component in the ecology and management of wildlife populations. Dispersal of young animals, habitat selection, court-

ship, territorial defense, flocking, daily and seasonal activities, responses to predators, renesting, migration, species recognition, and habituation are among the forms of behavior that hold strong implications for wildlife management.

Mammals generally are more adaptable than birds and other vertebrates when faced with new situations; that is, the complexity of the mammalian brain permits intelligent choices and a relatively wider range of options amid new circumstances. Most birds, although endowed with keen vision and the power of flight, have not evolved imaginative intelligence (Welty 1982; but see Manning 1967). Birds, more than mammals, generally act in ways that are determined more by genetic programming than by innovation.

This chapter describes some concepts of animal behavior and discusses their relationship to wildlife management.

HABITAT SELECTION

The selection of appropriate habitat—that is, the choice of settings that favor successful reproduction and survival—apparently is an example of innate behavior. Harris (1952) demonstrated the concept in a classic study concerning two races of deer mice. Woodland deer mice (*Peromyscus maniculatus gracilis*) inhabit forests, whereas prairie deer mice (*Peromyscus maniculatus bairdii*) normally inhabit grasslands. The experiment created two artificial habitats in the laboratory, one consisting of artificial tree trunks, fallen logs, and twigs, and the other consisting of grasslike materials. The adjacent habitats, each of equal size, were not separated by a barrier. Prairie deer mice, captured from the wild and introduced into the experimental habitats, showed a strong preference for the artificial prairie. As might be expected, woodland deer mice spent most of their time in the artificial forest. In a refined version of the same experiment, Wecker (1963) tested the preferences of prairie deer mice that had no previous experience in natural habitat. When given the choice between an artificial forest or grassland, the prairie deer mice persistently selected the artificial grassland. Those prairie deer mice that were given an early experience in a woodland, however, thereafter showed diminished selection for prairie habitat, whereas early experiences in a prairie environment generally reinforced their preference for grasslands.

Klopfer (1963) observed similar results in experiments with chipping sparrows (*Spizella passerina*). Chipping sparrows often forage on the ground near pines, and birds reared without previous exposure to such habitat were given equal opportunities of walking on pine needles or oak leaves. The result: Chipping sparrows spent 67 percent of their time on pine needles and 33 percent on oak leaves. Another group of chipping sparrows, also reared in an incubator, then were exposed only to oak leaves. Later, when given a choice between pine needles and oak leaves, the experimental birds showed a slight preference for oak leaves. Wild chipping sparrows that were captured and given the same choice spent 71 percent of their time on pine needles. The results suggest that habitat selection is instinctive, but such choices also can be modified by early learning experiences.

Although we cannot be certain, the results from the experiments with deer mice and chipping sparrows probably apply to other mammals and birds. If so, this phenomenon could explain why dispersing sharp-tailed grouse (*Tympanuchus phasianellus*), whose usual habitat is open areas with scattered trees, will colonize new habitat on a recently burned area after traveling long distances through deciduous forests. It also explains why female ducks, at the beginning of their first breeding season, choose nest sites that afford good protection from predators and access to water.

What are the management implications of habitat selection? Wildlife managers cannot judge the adequacy of habitat solely by human standards because an environment that is apparently suitable in the eyes of humans in fact may be deficient in the perception of animals. In the northern parts of their range, for example, white-tailed deer (*Odocoileus virginianus*) congregate in local areas offering shelter from deep snow, high winds, and cold temperatures. At these same sites, however, white cedar (*Thuja occidentalis*), hardwood sprouts, and other foods are often in short supply, and deer often starve to death in these "deer yards" during severe winters (Verme 1965b). Managers consequently have tried to attract deer into other areas where food is plentiful, but the animals stubbornly remain in the yards, apparently responding to an innate preference for shelter despite the shortage of food. Management plans thus must match the instinctive behavioral patterns of wintering deer; namely, winter habitat should be managed with various kinds of food-producing forestry practices on sites that border the shelter of traditional deer yards.

Most important, it should be recognized that the changes brought about by succession will produce deteriorating conditions for some species of animals, even though those changes may improve the habitat for others. Thus, if left unmanaged, many kinds of habitat will no longer suit the innate—and largely unchangeable—preferences of wildlife for specific environmental settings. Managers can better maintain a stand of pines at the expense of oak than alter the behavior of chipping sparrows.

COURTSHIP BEHAVIOR

A marsh or woodlot on a sunny spring morning is pleasantly noisy. Many kinds of animals, but particularly amphibians and birds, fill the air with a chorus of distinctive peeps, quacks, trills, drummings, croaks, and songs. Each of the calls says in effect, "Here I am; see how beautiful I sound and look," or "This is my property." The sounds deliver a highly functional two-part message to others of the same species.

The "Here I am" message—usually issued by males—functions as a mechanism for attracting mates. Courtship displays of birds and some mammals are frequently elaborate and highly ritualized (i.e., unvarying, stereotypic behavior). Courtship displays heighten intraspecific recognition, thereby reducing the chances of hybridization. The displays also stimulate and synchronize copulation, which otherwise could be thwarted if one of the partners was in a different state of sexual readiness. During the breeding season, courtship activities often belie the normally secretive nature of many animals. Courting males become unusually conspicuous. For birds, courtship activities include vocalizations, drummings, whistling of feathers, and dazzling displays of postures, plumage patterns, and colors.

An example of the implications of mating behavior and wildlife management involves black ducks (*Anas rubripes*) and mallards (*A. platyrhynchos*) in northeastern North America. The situation is complex, involving not only courtship behavior but also the evolutionary history of closely related species, habitat selection, and changes made by people in habitat. Black ducks are the highly prized quarry of waterfowl gunners in the northeastern United States and the Maritime Provinces of Canada. Although these birds are hardy and extremely wary, the black duck population nonetheless has declined steadily for the last 30 years—a period in which mallards have increased. Heusmann (1974) reviewed reasons why mallards may be replacing black ducks along the Eastern Seaboard. Geological evidence suggests that glaciers lingered along the Appalachian Mountains about 10,000 years ago, dividing the continental mallard population into two segments. Mallards in the segment lying west of the Appalachians adapted to a prairie environment that they shared with several other species of dabbling ducks. Hence, the bright plumage of drake mallards in the prairie population remained a means by which female mallards recognized the males of their species. This population of mallards winters along the Gulf of Mexico and in the river bottoms of the lower Mississippi Valley.

East of the Appalachians, however, the ducks nested in forest-related wetlands where there was no significant competition from other dabbling ducks. Hybridization was unlikely under such conditions, and the conspicuous plumage of the drakes gradually disappeared as natural selection favored birds whose plain, dark plumage offered protective coloration for males as well as for females. Black ducks thus evolved in eastern North America as "drab" mallards—nesting in forests in summer and, in winter, dwelling in the coastal marshes north and east of the winter range of the prairie mallard population. Despite these circumstances, however, the geographical separation of black ducks and mallards did not last long enough for the evolution of reproductive isolation (i.e., incompatible breeding between species).

In the past century, mallards invaded the realm of black ducks after large areas of the eastern forests were cleared (i.e., removal of the environmental barrier that formerly separated the two species). People also have provided the highly adaptable mallards living in eastern parks with handouts of bread, thereby maintaining an artificial habitat for mallards in the heartland of the black duck winter range. Hybridization thus became increasingly common when the two species came into contact, especially because pairs form on the wintering grounds; that is, female black ducks, given a choice of males, often favor drake mallards because of their more colorful plumage (Brodsky and Weatherhead 1984; see also Ankney et al. 1987). Consequently, mixed pairs form during the winter, and the drake mallards follow their mates during the spring migration northeast into the breeding grounds. The nests of such pairs, of course, produce broods of hybrid ducklings. Because mallards genetically swamp the gene pool of black ducks, the genetic integrity of the eastern species gradually has diminished. In 1969, for the first time, more mallards than black ducks were shot in the Atlantic Flyway, and the proportion has been increasing ever since.

Wildlife managers, of course, do not advocate that cleared areas in the east should revert to forest solely for the sake of preserving black ducks (in some parks, however, signs request that visitors not feed ducks). Tightened bag limits have reduced the gunning pressure on black ducks in recent years, but the population has not responded to the added protection. Indeed, maintenance of a secure niche—mallard-free environments—remains about the only hope for black ducks. Management efforts thus should center on securing wooded ponds and bogs for nesting habitat and northern salt marshes for wintering grounds (i.e., expanding specialized habitats for black ducks).

Courtship behavior often provides a useful means for estimating the abundance of animal populations (Davis and Winstead 1980). On spring evenings throughout eastern North America, for example, hundreds of biologists and amateur cooperators participate in an annual count of male woodcocks (*Scolopax minor*). About one-half hour after sunset, males begin a courtship ritual of "peenting" every few seconds while strutting in small forest clearings (a peent is a distinctive buzzing vocal sound that represents the most consistent part of the courtship display). After a few minutes of peenting, the males take off on spiral flights, ascending vertically on whistling wings and twittering loudly near the apex of a 1-minute flight. The birds thereafter silently descend to the ground and begin to peent once more.

About 10 or 11 such flights are made each evening
and again just before dawn. Cooperators drive from
point to point, pausing for 2 minutes every 0.64 km,
listening for the peent of male woodcocks. The location of each bird is recorded during a census period
that typically lasts 35 minutes (Mendall and Aldous
1943; Sheldon 1967; Tautin et al. 1983). The data
from two or three evenings along each route are for-
warded to the Inventory and Monitoring Section of
the National Biological Survey in Laurel, Maryland.
Year-to-year trends in woodcock numbers thus can
be detected over the entire range of the species so
that hunting regulations may be adjusted. The population data also may suggest where woodcock habitat
might be managed (e.g., where new clearings might
be cut in dense forests).

The abundance of other species also may be estimated using the data from courtship behavior; the
species include whistling bobwhite (*Colinus virginianus*), cooing mourning doves (*Zenaida macroura*),
crowing ring-necked pheasants (*Phasianus colchicus*), drumming ruffed grouse (*Bonasa umbellus*),
gobbling turkeys (*Meleagris gallopavo*), and barking fox squirrels (*Sciurus niger*). The results of such
surveys, although generally useful, must be applied
with caution. For one thing, the surveys focus almost entirely on males, and variations in sex ratios
therefore can introduce distortions in the estimated sizes of the populations. Other variables
also may affect the surveys, including wind, rain,
time of day, and stage of the breeding cycle (Davis
and Winstead 1980). Much of the variation, however, can be reduced if the procedures under which
the surveys are conducted are standardized (e.g.,
specific starting and ending times, uniform routes,
and specified weather conditions). Courtship surveys of breeding males thus have become useful
monitors for bird populations, particularly when
those data are analyzed in conjunction with sex ratios and brood surveys.

REPRODUCTIVE PHYSIOLOGY
AND BEHAVIOR

Animal reproduction normally follows a rather rigid
sequence that involves the interactions of photoperiod, weather, endocrine glands, and sex-specific behavior. *Photoperiod*—the relative length of days and
nights—stimulates the release of hormones from the
pituitary gland, which in turn stimulate the enlarge-
ment and respective functions of ovaries and testes.

The sex organs, in addition to producing eggs or
sperm, also secrete sex hormones. In female birds, estrogens secreted by the ovaries ready the oviduct and
shell gland for egg production and physiologically
prepare the female for a long period of incubation. In
female mammals, estrogens prepare the uterus for
pregnancy. In males, testosterone secreted by the
testes initiates the intense courtship and territorial
behavior typical of the breeding season.

Most wild birds and mammals require a suitable
environment for completing the reproductive process. Many species do not breed in captivity because
one or more crucial environmental elements are
missing, and the physiological and behavioral mechanisms associated with breeding thereafter fail. Likewise, in nature, animals that do not obtain territories
or adequate food or shelter may experience a physiological shutdown of breeding activities. Even though
last year's nest still may be available, the courtship
behavior of golden eagles (*Aquila chrysaetos*)
nonetheless involves ritualized nest-building activities, and reproduction will stop unless sticks are carried and passed between the adults (Hamerstrom
1970). Ring doves (*Streptopelia risoria*), even under
laboratory conditions, will not incubate eggs without
first building a nest. Doves provided with ready
made nests and eggs build another nest on top of the
old one before incubation begins (Lehrman 1964).
The lesson from these examples is that the reproductive needs of some animals may include subtle environmental features that are not readily apparent to
human observers.

Renesting—that is, laying a replacement clutch
after an earlier nest is destroyed—requires that
nesting females reverse their physiological and behavioral activities (i.e., incubation reverts to the
process of laying eggs). Accordingly, the interval
between the destruction of the first clutch and the
beginning of the second nest—the renesting interval—
lengthens in proportion to the stage of incubation,
and the number of eggs in renests is usually smaller
than in the original nest (Sowls 1955; Coulter and
Miller 1968). Thus, when large numbers of early
nests are destroyed by floods, storms, or other
causes, most of the broods may originate from the
smaller clutches of renests, and production may be
reduced accordingly (Cringan 1970). Also, if nests
are destroyed late in the season, there may not be
enough time for renesting, which again reduces
total production. Renesting nonetheless is an important adaptation for birds that normally nest only
once each year. Without renesting, the adversities

of predation, flooding, and other nesting disasters would be far greater.

TERRITORIAL BEHAVIOR

Individuals—usually males—of many species claim and defend specific sites against other members of the same species. The defended area is known as a *territory* (Noble 1939). A territory is usually part of a larger unit called the *home range,* an area in which an individual animal conducts its normal daily activities. Resting, foraging, and watering areas, including the travel routes to each, together form the home range. The home range may be shared with other individuals (e.g., water holes), but the territory is exclusive. Territories in some species are of no value as a source of food or even as a nesting or denning site. Instead, territorial behavior represents a type of biological symbolism in which males express their social dominance over other males. In effect, territorial displays proclaim, "This is my property," a declaration that does not necessarily imply a physical structure such as a nest. The territories of bobwhites, for example, often are no more than the space around a single fencepost, from which males individually perch and imperiously issue their calls of ownership. Indeed, territorial behavior many times is initiated before females are present, although the same displays usually play a role in courtship later in the breeding season; that is, the same displays function not only as a warning sign for other males, but also as an advertisement for a mate with whom the territory will be shared. Territoriality is best understood in birds because they generally are more conspicuous and more readily observed than mammals and other vertebrates. However, territorial behavior occurs widely among many kinds of animals.

Territorial defenses at times involve violent behavior. The headlong battering of bighorn sheep (*Ovis canadensis*) is a familiar example. The massive horns of bighorn sheep are remarkably adapted for just such behavior and illustrate one of many examples in which physical structures and behavior have evolved together. Combat among bighorns, in fact, is highly ritualized and has little difference in function than the postures of species that do not involve physical contact at all. Combat between two male robins (*Turdus migratorius*) is a common sight each spring in yards across North America. These conflicts last only a few moments and, like bighorn sheep, robins rarely harm each other. One individual usually retreats before much physical damage is done, and after territorial boundaries have been established among neighbors, further strife is uncommon. Overall, territorial defense involves more bluffing than fighting.

Territories are proclaimed with some combination of visual displays, threats, sounds, and scent. Wolves (*Canis lupus*), for example, mark the boundaries of their territories with urine. By howling, wolves also notify other packs that the territory is occupied (Peters and Mech 1975; Harrington and Mech 1979). Smell and hearing thus are important senses in the lives of wolves; these features illustrate the principle that both sending and receiving mechanisms must be in place before a behavioral message can be transmitted appropriately. The songs of birds, nearly always performed by the males, announce to other males that territories have been established and that potential intruders should move elsewhere. Some birds proclaim their territories with nonvocal sounds. For example, ruffed grouse perch on logs and rapidly beat their cupped wings, creating a drumming sound that can be heard nearly 1.5 km away (Bump et al. 1947). In a forest, where vision is limited to a few meters, many birds proclaim their territories with sound, whereas sight becomes a more viable means of communication in the relative openness of a grassland. Thus, prairie chickens (*Tympanuchus* spp.)—another type of grouse—inflate colorful air sacs and perform an elaborate dance as prominent parts of their display. Similarly, fishes normally living in muddy waters lack bright coloration, whereas those dwelling in clear water often bear colorful markings. These differences illustrate another principle, namely that behavior of the various species has evolved in ways suited to their surroundings.

Most species of grouse and many other nonmigratory birds establish territories in autumn. This explains why hunters frequently hear ruffed grouse drumming in October, an activity that is usually associated with the spring mating season. Such birds maintain close year-round ties with their territories. Migratory birds, however, necessarily establish (or renew) territories each spring after reaching their breeding grounds, and some species also set up territories on their wintering grounds (see Rappole and Warner 1978, 1980).

Territory sizes vary widely among species. Herons (Ardeidae) and some other colonial-nesting birds defend a circle only 1–2 m in diameter immediately around the nest (Wynne-Edwards 1962). Some grouse, such as sharp-tailed grouse, prairie chickens (*Tympanuchus cupido*), sage grouse (*Centrocercus urophasianus*), and black grouse (*Lyrurus tetrix*), perform their courtship and territorial defense on a common dancing ground known as a *lek*. Here the birds gather at dawn and display a behavioral repertoire of dancing, calling, and rattling of tail feathers. Each male claims a parcel of several square meters from which other males are excluded. Females stroll into the lek and usually mate with those dominant males holding centrally located territories (Hamerstrom 1939). After an hour or two of intense territorial defense and mating activity, the whole flock may fly off and feed together quite amicably. Among spruce grouse (*Dendragapus canadensis*), males establish, occupy, and defend territories against males, and females defend a second set of territories against other females (Herzog and Boag 1977).

Pochards (*Aythya* spp.) exhibit moving territories; males in such species as ring-necked ducks (*A. collaris*) defend a circle about 2 m in diameter around their mates (Mendall 1958). Dabbling ducks (*Anas* spp.) typically defend part of a stream or pond, including a loafing spot and all the air space above the territory. A drake of the same species flying through the air space will be pursued by the resident male until the chase reaches the territorial boundary. The nests of most ducks are located outside the defended area, and mated pairs often feed outside of their territories with others of the same species.

In some animals, however, the entire home range may be defended, at least during the breeding season. Songbirds spend nearly all of their time within territories that are established and maintained by males. Territorial boundaries do not always touch one another, and the gaps remaining between actively defended territories may be occupied by nonbreeding, nonterritorial individuals. Hensley and Cope (1951) conducted a field experiment in Maine in which territorial birds of various species were eliminated from a 16-ha area. About 79 percent of the 154 original holders of territories were shot, as were their replacements. As soon as the owner of a territory was killed, a new bird appeared and claimed the same territory. Nearly 530 adult birds were removed in about 5 weeks from the same study area. The new birds apparently came from a "floating" population of nonterritorial individuals that occupied gaps and other areas lacking territories. Thus, for such species, census techniques based on territorial displays may underestimate the total population. Mech (1977b) speculated that gaps between wolf territories served as a refuge for white-tailed deer, in which deer populations may recover after experiencing periods of heavy wolf predation.

The energy spent defending and maintaining territories must be repaid with increased survival or productivity of the territorial animals. If this did not occur, territoriality would not persist as a behavioral feature. For some species, territorial behavior assures adequate physical spacing; males are thereby available to females and copulation can proceed without undue interference from competing males (Lack 1966). Holders of territories are more likely familiar with the immediate sources of food and shelter and thus may function more efficiently than transient animals. Once established, territories usually are maintained with a minimum of effort—in birds, often by song alone—thereby avoiding waste of energy that otherwise might be expended in continual disputes over mates, food, or other resources.

Territories set a limit on the sizes of breeding populations. According to Wynne-Edwards (1962), a territory represents a wealth of resources in a place where an animal can reproduce and rear a family in relative prosperity. Individuals that do not hold territories frequently do not reproduce and suffer higher mortality rates than owners of territories (Jenkins et al. 1963). Territorial behavior thus can restrict breeding, thereby regulating the rate at which populations expand. The numbers of pintail ducks (*Anas acuta*) breeding on a Canadian marsh or the population of red grouse (*Lagopus lagopus scoticus*) on a Scottish moor are thus limited not directly by space, food, or nest sites, but by the territorial behavior of the birds. The lesson is that breeding animals do not tolerate crowding.

The size of territories is not constant from year to year. Territory size typically becomes smaller when resources are plentiful, which suggests a means by which wildlife managers might influence the sizes of populations in territorial species. With dabbling ducks, males usually defend loafing sites and the surrounding water and air space. Hence, when several small ponds are blasted in a marsh, or when crooked rather than straight stream channels are dredged, the number of potential duck territories can be increased. In general terms, territory sizes may be reduced and higher population densities of desired species may be achieved by intermingling resources in ways that enrich the habitat (i.e., methods that improve interspersion). This is what Leopold (1933a) called *edge effect;* that is, more resources are available per unit of area when there are more edges between cover and food types (see Chapter 7). Such areas, thus enriched, can support more individuals holding territories.

SEXUAL SEGREGATION

Many species of animals are sexually dimorphic (i.e., males and females are readily distinguishable from each other). In some sexually dimorphic species, sex also tends to occupy separate habitats during much of the year, although the reasons for this form of segregation are not always clear. Bleich et al. (1997), using radio telemetry, determined that predators may be involved in the habitats selected by mountain sheep (*Ovis canadensis nelsoni*). Females and lambs, both more vulnerable than males, tend to occupy steep slopes where, although food was of lower quality, there are fewer predators. Conversely, the larger and stronger males select areas with higher-quality food during most of the

nonbreeding season, despite the greater abundance of predators in those areas (i.e., large size and horns apparently help intimidate potential predators). For sexual segregation in moose (*Alces alces*), see Miquelle et al. (1992).

CIRCADIAN RHYTHMS

The term *circadian* means "approximately one day." Circadian rhythms refer to activities of animals that show a regular pattern that occurs about every 24 hours (Fig. 6-1). Most animals of a particular species typically feed at certain times of the day or night and remain inactive at other times. Such behavior may offer management opportunities. Brood counts of ducks, for example, are best conducted near dawn and dusk when the birds feed actively (Mendall 1958). The home-range size of animals, as determined by radio telemetry, must be computed during the period of greatest activity. In Wisconsin, for example, Smith et al. (1981) showed that home-range size of coyotes (*Canis latrans*) can be described only if the radio locations are plotted when the coyotes are active. Because the coyotes in the Wisconsin study were active mainly between 6 P.M. and 6 A.M.,

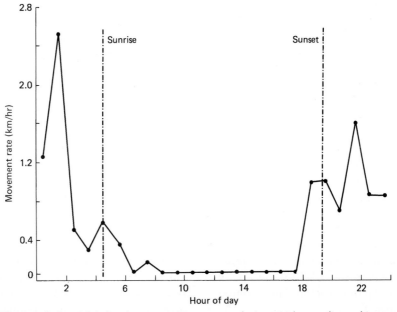

Figure 6-1. Adult female coyote movement rate during a 24-hour radio-tracking period. This daily activity pattern was typical for 19 full-day activity graphs plotted for adult and juvenile coyotes in northern Wisconsin. (From Smith et al. 1981.)

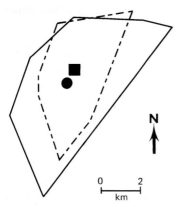

Figure 6-2. Adult female coyote home ranges determined from 48 locations during four nights (solid line) and 30 daylight locations obtained on different days (interrupted line). Daylight center of activity is represented by the square, night center of activity by the circle. (From Smith et al. 1981.)

radio locations taken only during the day (when most biologists work) seriously underestimated the actual areas used by the coyotes (Fig. 6-2). For a reliable estimate of home-range sizes, biologists had to locate coyotes during the night.

DISPERSAL

Most young animals leave the home range in which they were reared and wander to new locations—a phenomenon called *dispersal.* The distances that dispersing animals travel vary among species. Many small mammals move less than 1 km (Allred and Beck 1963), but some carnivores may disperse great distances. A red fox (*Vulpes vulpes*) dispersed 160 km (Phillips et al. 1972) and a wolf traveled 886 km from northern Minnesota to central Saskatchewan (Fritts 1983). Normal dispersal distances, however, are considerably less.

Because they are young, inexperienced, and traveling in unfamiliar terrain, dispersing animals often experience higher mortality rates than resident animals. Ruffed grouse are well known for their autumn "crazy flights" in which some young birds fly into fences and windows, wounding and killing themselves (Bump et al. 1947). The mortality rate for the dispersing grouse, however, is probably lower than it would be if the young birds had not dispersed but instead remained in their parents' home range competing for food and mates.

The effects of dispersal include the following: (1) maintenance of genetic variability within the species; (2) repopulation of depleted areas; and

(3) colonization of new areas when suitable habitat becomes available. Thus, most suitable habitat eventually is occupied unless some barrier prevents dispersal. Such barriers occur in the form of intervening mountain ranges, deserts, oceans, rivers, or other large tracts of uninhabitable range.

Howard (1960) proposed two types of dispersal. The first is "innate dispersal," a tendency of young animals to leave their natal areas and thereby prevent inbreeding. The second is "environmental dispersal," which is a behavioral response to stresses such as shortages of food or space, as described by Poole (1997) for food-stressed lynx in Canada.

Islands, to which dispersal of land animals is limited, generally have fewer species than areas of equivalent size on the adjacent mainland. The probability of extinction of a species is higher on an island than on a continent. Humans sometimes create "islands" of wildlife habitat by surrounding parks with cities and small woodlots with expanses of intensively farmed fields. Unless animals can disperse from larger populations into smaller areas, some species may be extirpated from the smaller units of habitat (see Chapter 21).

RESPONSES OF WILDLIFE TO HUMANS

Behavioral responses to the presence of humans are important factors in the survival of animals. Dodos (*Raphus* sp.), great auks (*Pinguinus impennis*), and Steller sea cows (*Hydrodamalis stelleri*) did not respond successfully to humans and suffered extinction. Other species, however, such as Norway rats (*Rattus norvegicus*), ring-necked pheasants, and starlings (*Sturnus vulgaris*), have benefited from living in association with humans. In this section we will consider only the immediate behavioral responses of wild animals to the presence of humans. We will not deal with the responses of animals to the broad array of habitat modifications brought about by humans.

Normal behavioral responses, such as an animal's avoidance of danger, have been used by biologists for estimating numbers. King (*in* Leopold 1933a) calculated the average flushing distance of birds on either side of a line of known length as a means of estimating ruffed grouse populations. The average flushing distance is doubled and then multiplied by the length of the line transect to determine the area sampled. The count of birds within that area provides a simple way to determine density (Fig. 6-3). Hahn (1949) later modified the technique for white-tailed deer. The modification employs a standard 3.3-km transect and the average flushing distance is determined by two people; one person walks at right angles from the baseline at 90-m intervals with a white handkerchief dangling to simulate a deer's tail; the other observer notes when the "tail" is no longer visible from the transect. After the average distance is determined at 90-m intervals along both sides of the transect, the plot size is permanently established for future censuses. Thereafter, only one person need walk the transect and record the numbers of deer encountered to arrive at density estimates. If the transects in either the King or Hahn technique are located in representative habitat, then the sampling procedure compensates for the normal variation in cover types (small clearings, dense brush,

etc.). Burnham et al. (1985) determined that line transects are more accurate for estimating population density than are methods that use fixed, long, narrow plots.

Most wild animals respond to humans by hiding or fleeing, but others may develop a tentative trust, particularly if rewards of food or shelter are a part of the bargain. Bald eagles (*Haliaeetus leucocephalus*) are among the animals that flee from humans. Some evidence indicates that human activities near nests lower the reproductive success of eagles (Juenemann 1973; Grubb 1976; but see Newman et al. 1977). Thus, because bald eagles are classified as endangered in most of the United States, human activity near nests is prohibited on federally managed lands (e.g., national parks). On their wintering grounds in Washington and Michigan, eagles avoid humans, particularly when

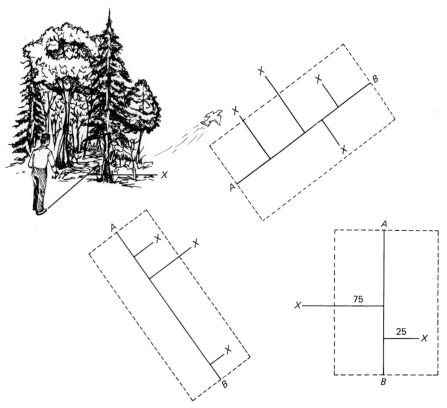

Figure 6-3. Schematic diagram of three King census plots for ruffed grouse. Line transects are shown between *A* and *B;* the lines need not be of the same length. Plot sizes, in part, will vary accordingly. Locations of grouse flushed by an observer walking the line are indicated by Xs. If the baseline, lower right, is about 2175 m in length, and two grouse are flushed at 25 and 75 m, respectively, then the average flushing distance in this sample is 50 m, making the plot's width 100 m. The plot's area is thus 217,500 m² (or 21.75 ha, as shown by the dashed outline) and the density is about 1 grouse per 11 ha. In a forest of 1 km², the grouse population based on this plot would be estimated at 9.2 birds, but several plots normally should be sampled.

humans approach from unaccustomed locations (Stalmaster and Newman 1978; Jensen 1980). Avoidance behavior is more pronounced in adult eagles than in juveniles. Stalmaster and Newman (1978) suggested that 90 percent of the eagles wintering in Washington would be protected if human activities were restricted within 250 m of streams, and further seclusion would result if buffer zones of vegetation were established. Knight and Knight (1984), however, after observing eagles on two rivers in Washington, recommended that human activities be restricted within 450 m of waterways, thereby protecting 99 percent of the bald eagles wintering in such sites.

Birds wintering in subtropical and tropical areas also are subject to human disturbances. Tests in Florida revealed the flushing distances for 16 species of water and shorebirds and, based on these data, humans in boats, cars, and all-terrain vehicles, as well as walking, should maintain a "buffer zone" of at least 100 m to keep from disturbing these birds (Rodgers and Smith 1997).

In Nevada, Leslie and Douglas (1980) described how desert bighorn sheep (*Ovis canadensis nelsoni*) developed a peculiar dependence on human activities. Bighorns did not summer in the River Mountains until a sewage-leach field was established in the 1940s. The Southern Nevada Water Project, a piped-water system constructed in 1971 between Lake Mead and Las Vegas, also provided several pools from coolant water that was discharged from pumps. The sheep, responding to the new but artificial sources of water, started to spend the summer in the River Mountains. In 1978, new work began among the old pipelines, and the bighorns often avoided water holes near the construction sites. As a result, forage near other water holes probably received greater feeding pressure. Despite the artificial origin of the water sources and some habituation of sheep to humans, the study recommended that human activities at the water sources should be curtailed in favor of protecting the bighorns. Such sites represent the most critical component in the habitat of desert bighorn sheep.

In some instances, problems are created for both animals and humans when wildlife habituates to people. The interactions between bears and humans offer the most dramatic example. Black bears (*Ursus americanus*) are usually wary animals, and in the wild they enjoy an omnivorous diet of fruits, insects, honey, and some meat. Black bears in wild country rarely approach humans, but those feeding

regularly at garbage dumps and along roadsides soon lose any fear of humans. As a result, black bears can cause extensive property damage by breaking into buildings, cars, ice chests, and tents and, at times, have injured or killed people in the process (Baptiste et al. 1979; Tate 1980).

Grizzly bears (*Ursus arctos horribilis*) are dangerous, whether or not they have had any previous experience with humans (Craighead and Craighead 1971). Discarded and stored food at campsites encourages the approach of grizzlies and may further reduce their fear of humans. Between 1970 and 1986, six back-country hikers in Glacier National Park were killed by grizzlies, and several others have been injured (Martinka 1982 and personal communication). The attacks have produced opposing views: extermination of the bears at one extreme (Moment 1970) and the total exclusion of humans from grizzly country at the other.

Management goals currently encourage returning bear populations in parks to a more natural state. The methods include closing garbage dumps, educating visitors about the behavior patterns of bears and about how to respond when bears are encountered, and restricting human use of well-known haunts (see Chapter 16).

The habituation of songbirds to winter feeders is the result of benign human intentions. Mallards and Canada geese (*Branta canadensis*) often remain much farther north than would be the case if food were not supplied artificially. Handouts of food in parks maintain wintering populations of mallards in the northern cities of Calgary (Sugden et al. 1974), Minneapolis (Cooper and Johnson 1977), Sault Ste. Marie (Jensen 1980), and Montreal (Reed and Bourget 1977). Canada geese may experience better survival by remaining in the north than by migrating southward and experiencing additional hunting pressure (Raveling 1978).

IMPRINTING AND PARENTAL CARE

Imprinting is a type of permanent learning that takes place during a relatively brief period of responsiveness early in the life of some animals. Imprinting behavior was first described by Spalding (1873) and was later brought to the attention of a wider audience in a delightful book, *King Solomon's Ring*. In that book, Lorenz (1952) described how young precocial birds such as ducklings and goslings become attached to the first object that moves nearby.

The imprinting process takes place within a short time, usually when the young birds are ready to leave the nest. In normal circumstances, the moving object is the mother bird, but the strength of the attachment was proven experimentally when broods were imprinted on other objects, among them a cardboard duck pulled on a string, Lorenz's dog, and Lorenz himself.

The imprinting period lasts only a few hours, but the imprint endures for the life of the bird. Thus, when males later reach sexual maturity, they choose a copy of the mother figure as a sexual partner. Schutz (1965) reared ducklings of various species with their own parents, with parents of other species, and with brood mates of other species. Of 28 mallard ducklings reared with mallard parents, all subsequently paired with their own species, but of 34 male mallards reared with foster siblings or foster parents, 22 attempted to pair with a member of their foster species. Imprinting behavior thus establishes both a bond between mother and offspring and a means by which males recognize the females of their species.

Young mammals of some species also imprint and, in some cases, parents imprint on their offspring. Mother goats (*Capra aegragus*) apparently imprint on their kids (Klopfer et al. 1964). If nannies cannot lick and smell newborns within an hour of birth, the kids are rejected. If, however, nannies are permitted contact with their newborn for a few minutes during the first hour, the parent-offspring bond is assured. Similar bonds may be established in other species, especially those such as wildebeest (*Connochaetes taurinus*) whose calves might become separated from their mothers in a vast herd.

The implications of imprinting in wildlife management are varied. In dense breeding populations of ducks and geese, particularly in city parks, the likelihood of brood mixing among different species is higher than under wild conditions. Consequently, the probability of confused matings is higher; indeed, ducks in parks frequently hybridize. Birds and mammals reared for stocking must be handled carefully so that the young animals do not imprint on humans. Those that do will likely experience low rates of survival after release.

Female whooping cranes (*Grus americana*) raised in captivity frequently imprint on humans, and if so, they may not respond later to the courtship rituals of males. To prevent imprinting in these situations, young cranes are exposed only to humans dressed in cranelike costumes (Fig. 6-4). Handlers also wear these costumes to induce adult

Figure 6-4. Staff at the International Crane Foundation, Baraboo, Wisconsin, wear cranelike costumes when teaching juvenile whooping cranes how to find food. The costume is used to prevent the young birds from imprinting on humans, and prior to their release into the wild, the hand-reared whooping cranes never saw humans out of costume. (Photo courtesy of D. H. Thompson, International Crane Foundation.)

females into ovulation by simulating the courtship dance of male whooping cranes—a frenetic ritual of flapping arms, calling, and throwing grass over the shoulder. The females then are artificially inseminated and fertile eggs are produced. Conversely, imprinting may be used intentionally, as is done when young masked bobwhite (*Colinus virginianus ridgwayi*) are imprinted on a foster parent before the brood is released (see Chapter 19). Imprinting also makes it possible to move hen wood ducks (*Aix sponsa*) and their broods into unoccupied habitat (Capen et al. 1974).

The secretive nature of wild animals makes it difficult to study maternal behavior. Recent studies of white-tailed deer, however, have revealed somewhat complex behavioral relationships among does and their fawns, including fawning territories, the age of the does, and ability of the deer to move their fawns out of danger (Ozoga et al. 1982; Ozoga and Verme 1986). These studies used radio-marked does and fawns that were maintained in a 252-ha enclosure for a 5-year period. The results indicated that does established and defended a fawning territory against other does, particularly during the first month of the fawns' life. Experienced does returned each year to the same territory, while young does established territories near their mothers. Does more than 4 years old had smaller territories than younger does, but the older does moved their fawns farther within those territories. Older does also spaced twin fawns farther apart than younger does. When disturbed, the fawns of older does were moved greater distances, sometimes beyond the territorial boundaries.

These strategies, apparently developing in association with experience, effectively reduced predation. During the 3 years of the study when black bears were present in the enclosure, does 4 years old or older lost 32 percent of their fawns, while the fawn loss was 58 percent for younger does. In contrast, the losses were 13 and 4 percent, respectively, during the 2 years when bears were absent. For management, the findings indicate that a breeding population of younger deer will prove less productive than an older population. In addition to giving birth to fewer fawns, younger does also lose a higher proportion of fawns to predators.

Older mothers also may provide more milk for their young. Bison (*Bison bison*) calves born to cows more than 4 years old have a nutritional advantage over calves from younger mothers. Compared with calves nursing younger cows, calves with older mothers had longer feeding bouts and more often nursed until satiated (Green 1986). Because of these differences, we can reasonably assume that calves with older mothers may be better nourished and, if so, they also may enjoy better rates of survival.

MIGRATION

Baker (1978) defined *migration* as the simple act of moving from one spatial unit to another. Orr (1970), however, believed that *migration* is a periodic movement involving a round trip. An even more limited definition would add that *migrations* are regular movements to and from breeding areas. Whatever definition is used, it is clear that many species migrate across land and water as a means of survival. Migrations may not encompass the great distances we often envision, and indeed, distance is not a component of the definition. A population moving 50–100 km is just as migratory as one moving 5000–10,000 km. In each case, migration is an equally important biological feature.

The spectacular migrations of birds have attracted the most attention of biologists, but other kinds of wildlife also migrate. Some insects are migratory; of these, the monarch butterfly (*Danaus plexippus*) has gained the most attention (Brower 1985). Atlantic salmon (*Salmo salar*) are among the many species of migratory fishes. Some mammals migrate, including several species of bats and whales.

Elk (*Cervus elaphus canadensis*) and some species of grouse (Tetraonidae) migrate relatively short distances up and down mountainsides in what is called *altitudinal migration*. More typically, however, we envision migration as north-south movements known as *latitudinal migration*. Arctic terns (*Sterna paradisaea*) are the long-distance champions; their migratory route covers nearly 16,000 km—one way—between the polar regions of the Northern and Southern hemispheres. Gradients exist in the migratory behavior of some birds. Barn owls (*Tyto alba*) are migratory in the northern part of their range in the United States but remain relatively sedentary in the south (Stewart 1952).

Lincoln (1935) classified the migratory routes across North America into four flyways, based on a large-scale analysis of banded birds and their movements. Each of the flyways—Atlantic, Mississippi, Central, and Pacific—corresponds to a major route followed each year by millions of waterfowl and other birds (Fig. 6-5). The boundaries between these migratory routes are not always well defined, and several species regularly cross from one flyway to another, for example, canvasbacks (*Aythya valisneria*) and tundra swans (*Cygnus columbianus*). Bellrose (1968) thus refined the flyways into subunits known as *corridors,* which more accurately describe waterfowl migrations in North America. Nonetheless, the four flyways serve as administrative units for waterfowl managers (see Hawkins et al. 1984). Each unit is managed separately with hunting regulations that are tailored each year to match brood production, shooting pressure, and other variables.

BIRDS

Waterfowl offer many useful examples of the relationships between migratory behavior and wildlife management. Two of these illustrations are briefly noted here, one dealing with a distinctive migrational schedule, the other with nesting behavior.

Blue-winged teal (*Anas discors*) are the earliest of the autumn migrants. They begin their southward journey from North America in late summer. By October, most blue-winged teal are in the southern United States, Mexico, or Central America. Blue-winged teal accordingly escape much of the hunting pressure experienced by species that migrate later in the year. Consequently, relatively few teal are har-

ATLANTIC FLYWAY

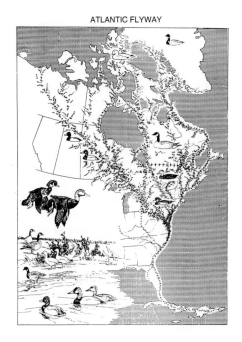

MISSISSIPPI FLYWAY

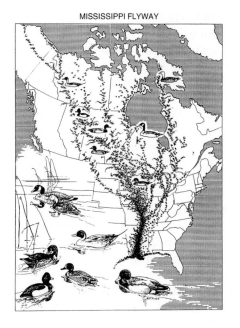

CENTRAL FLYWAY

PACIFIC FLYWAY

Figure 6-5. Four flyways—Atlantic, Mississippi, Central, and Pacific—cross North America. Whereas many birds faithfully stay within these routes, some species or subpopulations regularly cross from one flyway to another. The flyways serve as administrative units for managing migratory waterfowl. (Maps courtesy of U.S. Fish and Wildlife Service.)

vested during the regular hunting season, which is scheduled to coincide with the peak migration of other waterfowl in late autumn. Managers thus proposed a special 10-day season in September exclusively for teal, and after considerable deliberation, an

experimental season was authorized in 1965 (Martinson et al. 1966). The experimental status later was removed, and the early teal season then became an established policy in many states of the Central and Mississippi flyways. The special teal season carries some risk that other species may be shot, especially in the northern parts of the flyways, but a vigorous education program for hunters reduces the kill of "mistake birds." Current regulations thus require that hunters identify ducks in flight, and those who cannot do so risk stiff penalties.

Behavior known as *migrational homing* describes the return of adult hens to the same nesting area year after year. First-year hens also exhibit migrational homing, for they return to the place where they were raised the year before. Sowls (1955) demonstrated that homing behavior is strongest in pintails, gadwalls (*Anas strepera*), and shovelers (*A. clypeata*); homing behavior is somewhat less strong for mallards and blue-winged teal. Each year, the nests of these species are located near the nesting sites that had been selected the previous year. Moreover, resident birds return first each spring, thereby claiming prime nesting cover before other ducks arrive. Migrational homing thus reflects strong attachments to specific nesting areas. The phenomenon underscores that local production depends largely on birds with previous experience at each site.

But what if the local waterfowl population experiences heavy shooting pressure early in the hunting season? Hochbaum (1955) described breeding marshes as "burned out"—a reference to gunpowder—when overhunting has decimated the population to levels where few ducks survive and return the following year. In these situations, the number of breeding ducks falls below the carrying capacity of the nesting habitat. Conversely, when hunting seasons are set later in the autumn, an influx of other migrants dilutes the local population and lessens the percentage of local birds in the bag. Thus, more of the local breeding population survives and migrational homing replenishes the marsh with nesting birds the following year.

Strongly developed homing tendencies may limit another behavior known as *pioneering*—defined as the search for new nesting sites. Because wood ducks rarely pioneer, potential breeding habitat may remain unused. Thus, an otherwise suitable nesting area in Wisconsin lacked a breeding population of wood ducks for more than 50 years until hand-reared birds were stocked at the site and a new tradition of homing behavior thereafter assured future nesting activities (McCabe 1947). On the other hand, well-developed pioneering abilities have permitted large-scale range expansions for such species as the ring-necked duck (*Aythya collaris*) (Mendall 1958).

One benefit of migration is that birds can exploit resources on a seasonal basis, thereby allowing for the recovery of food supplies in the off season. The costs of migration, however, include a large expenditure of energy and, for some species, crucial dependence on a few resting and feeding sites along the route. Such sites may support fully 80 percent of the breeding population for some shorebirds (Myers et al. 1987). Those areas are usually patches of coastal wetlands and beaches, without which shorebird populations could not maintain their current numbers. Protection of these sites has begun with the cooperative efforts of 23 state and provincial wildlife agencies, the U.S. Fish and Wildlife Service, the Canadian Wildlife Service, and the Peruvian government. The result is the Western Hemisphere Shorebird Reserve, a network that identifies and protects critical migratory, wintering, and breeding sites from destructive human activities. Similarly, many state and federal refuges were selected because of their strategic locations in the migratory routes of waterfowl and other birds. DeSoto National Wildlife Refuge on the shores of the Missouri River between Nebraska and Iowa, for example, is a key staging area in the migratory path of snow geese (*Anser caerulescens*).

REPTILES

Some reptiles make seasonal voyages of hundreds or even thousands of kilometers. Sea turtles of five genera migrate long distances between their breeding beaches and their feeding grounds (Carr 1965). For example, one population of green sea turtles (*Chelonia mydas*) feeds in the "turtle-grass pastures" off the Brazilian coast. Then, in April, these turtles begin to arrive at their breeding grounds on tiny Ascension Island, 2300 km offshore in the Atlantic. After mating, the males remain at sea while the females haul up on the beaches, deposit their eggs in the sand, and leave. The young that hatch face a hazardous trip across a few meters of sand where gulls prey on large numbers of the fledgling turtles. Once in the water, the survivors apparently drift westward with the currents toward South America. In 2–3 years, the young turtles mature and return to Ascension Island, using some as yet unexplained navigational ability. Similar migrations of sea turtles to specific coastal beaches or remote islands occur throughout the tropical and subtropical oceans.

On a lesser scale, timber rattlesnakes (*Crotalus horridus*) are among the many serpents that migrate seasonally between winter dens known as *hibernacula* and their breeding and foraging areas. Based on studies of marked individuals, most rattlesnakes remain within 3 km of their hibernacula, but a few move almost twice that distance (Brown 1987). Because of what may be a form of imprinting, timber rattlesnakes migrate each year to and from the same hibernacula, some of which may have been in continuous use for hundreds and perhaps thousands of years. Housing developments and other human activities, however, have destroyed the natural hibernacula of these and other snakes. Accordingly, artificial hibernacula have been designed and installed for the management of snakes (Fig. 6-6). Unfortunately, roadways often cross the migratory pathways to natural hibernacula, thereby adding another hazard for what usually are already highly persecuted animals. Orr (1970) and Baker (1978) provide further discussions of other reptilian migrations.

MAMMALS

Among mammals, only four groups—bats, cetaceans (whales and porpoises), pinnipeds (seals and sea lions), and large hooved herbivores—undergo regular migrations. Altitudinal migrations, however, occur in deer and some other mammals living in mountainous areas.

Scandinavian lemmings (*Lemmus lemmus*)—hamster-sized rodents—sometimes are reported as making mass "suicidal" migrations in which they leap from cliffs into the sea and drown. Baker (1978) noted that lemmings indeed migrate, but for short

distances down mountainsides from their periodically crowded breeding grounds. In Norway, Clough (1965) observed that lemmings migrate individually and not by orienting with each other en route (lemmings are generally intolerant of one another). Lemming migrations apparently proceed from areas of high population density to those that have lower densities. Because most of the travelers are young animals, *dispersal* perhaps is a more specific term for describing the movements of lemmings. In any event, small streams, large rivers, or other water barriers are encountered in these movements. Baker (1978) suggests that lemmings do not readily take to the water unless they can see the opposite bank or unless conditions become so crowded that the risk of death by remaining becomes higher than that of swimming for new terrain. Lemmings entering the water swim with the intention of reaching the opposite shore, but some drown when the expanse of water proves too extensive. Despite the dramatic reports of such events, however, mass drownings are not a common occurrence in lemming populations. Moreover, because suicide has no survival value, neither lemmings nor any other species with such tendencies could long persist, and any genes allegedly promoting suicide scarcely could be passed on from one generation to the next.

Several kinds of bats (*Myotis* and *Eptesicus* spp.) summer throughout New England, but banding studies have shown that most hibernate in winter in caves in Massachusetts, Connecticut, and Vermont (Griffin 1940, 1945). In the central United States, Indiana bats (*Myotis sodalis*) may migrate up to 500 km,

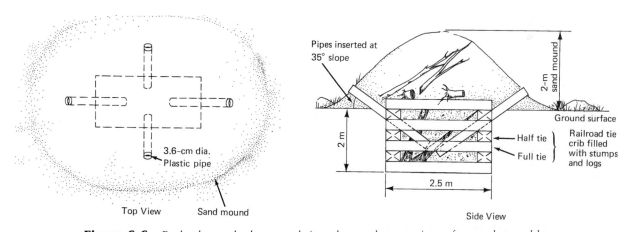

Figure 6-6. Rocky dens and other natural sites where snakes overwinter often are destroyed by housing developments and other human activities. These sites, known as *hibernacula*, may be replaced using artificial dens constructed of railroad ties and plastic pipe. The design shown here is used in New Jersey. (Courtesy of R. T. Zappalorti, Herpetological Associates.)

apparently following watercourses to their roosting caves (Hall 1962). In Europe, bats of the genus *Myotis* move an average of only 32 to 80 km between summer and winter, but other bats (*Nyctalus* and *Pipistrellus*) banded in Europe migrate 750 to 1160 km, respectively. Management strategies for bats, whether for the protection of an endangered species (e.g., the Indiana bat) or for controlling rabies in such groups as the free-tailed bats (Molossidae), must take into account their migratory behavior.

Of some 85 species of whales, about a dozen are listed as threatened or endangered, and virtually all of these exhibit some form of migratory behavior. Orr (1970) has discussed the migrations of whales in detail, of which some highlights are presented here. The baleen whales, which feed by straining water through comblike plates (baleen, or "whalebone") made of the same material as human fingernails, find immense food resources in the polar seas. Small, shrimplike crustaceans called *krill* are the primary food for baleen whales. Newborn whales, however, cannot tolerate the cold temperatures of Arctic and Antarctic waters, and baleen whales thus calve in zones nearer the equator. Young whales, nourished by fat-rich milk, quickly develop an insulating layer of blubber and thereafter accompany their mothers back to the polar feeding grounds. Baleen whales thus undergo north-south migrations each year in both the Northern and Southern hemispheres. Indeed, the rather predictable nature of these migrations has enhanced the success of whalers from the days of Nantucket to the present time.

Studies of humpback whales (*Megaptera novaeangliae*) that were marked in five Antarctic zones revealed that the populations in each area maintain separate summer ranges and migratory routes. The migratory route of gray whales (*Eschrichtius robustus*) extends from the Arctic to their primary breeding grounds in Scammon's Lagoon in Mexico. Part of the route parallels the shoreline of southern California where thousands of people marvel at the leisurely but determined passage of these great marine mammals. Because their migratory route lies largely within the territorial waters of the United States, gray whales could be protected from the undue exploitation that once threatened the future of the species. Unfortunately, other baleen whales migrate in international waters where the protective measures adopted by one nation do not preclude whaling by other nations.

Toothed whales feed on various aquatic organisms ranging from mollusks to seals but have somewhat different migratory habits from baleen whales. Sperm whales (*Physeter macrocephalus*) move about in response to availability of their primary food, cuttlefish, which abound where cold and warm currents mix. Females and young bulls rarely leave the tropical waters in such places as the west coasts of Africa and South America, but in summer the older bulls may venture as far north as the Bering Sea and then return again in the autumn to join the females in the warmer oceans. Other toothed whales, including porpoises, tend to follow the fish on which they feed.

A few seals, such as the harbor seal (*Phoca vitulina*) and the Weddell seal (*Leptonychotes weddelli*), are nonmigratory. Others, however, such as northern fur seals (*Callorhinus ursinus*), may migrate as far as 4800 km between their summer breeding areas in the Pribilof Islands and the southern reaches of their winter feeding range in southern California. A similar pattern occurs among those fur seals migrating along the Asian coast of the Pacific. Harp seals (*Phoca groenlandica*) have been the subject of heated controversy because their newborn pups are killed for their fur. These seals migrate from their foraging range throughout the Arctic and Subarctic seas to three principal breeding or "pupping" grounds: the White Sea in northwestern Russia, the Norwegian Sea east of Greenland, and the North Atlantic near Newfoundland.

Large terrestrial herbivores also migrate. Various species of antelope migrate across the dusty plains of Africa, usually moving from place to place in search of food and water. Because rainfall and the responses of the vegetation vary in location and quantity from year to year, so do the movements of the animals. Thus, the migrations of African mammals are irregular and may not follow the more predictable patterns of many other mammals.

As shown by Talbot and Talbot (1963), wildebeest (*Connochaetes taurinus*) illustrate the irregular migrations of African herbivores. Wildebeest usually drink every day—but *must* drink every 2-3 days—and generally stay within 8 km of water. For forage, they prefer grass less than 25 cm tall. During the wet seasons, wildebeest continually move about, grouping and regrouping on open plains. They remain in these areas as long as water and green grass are available. When conditions become dry, however, wildebeest leave the plains to ebb and flow into the surrounding bushlands. The movements are irregular, depending on available water and forage, but a roughly circular migration pattern results. Wildebeest in the southern portion of the range begin to move northwesterly, and those in the more northern plains leave in a southwesterly direction. The animals at times may retrace part of their route, but most important, these movements may be diverted by the occurrence of rain during the dry season. Under those circumstances, wildebeest move toward rainstorms, sometimes responding to the sound of thunder. Hence, wildebeest migrations vary each year.

In northern North America, some large herbivores show migratory tendencies. In Michigan, Verme (1973) showed that white-tailed deer move up to 64 km (averaging about 14 km) between their widespread summer ranges and the centers of winter concentration known as deer "yards." Moreover, deer wintering in specific yards usually occupy the same summer ranges, with the result that there is little intermingling among the herds. In nearly all cases, the summer ranges of the deer were located within a 180° arc, rather than in a sunburst pattern radiating from the winter yard (Fig. 6-7).

Altitudinal migrations occur among big game in western North America. In the mountains of the Pa-

cific Coast, black-tailed mule deer (*Odocoileus hemionus columbianus*) move down the western slopes in late summer, whereas the typical race of mule deer (*O. h. hemionus*) migrates down the eastern slopes. The differential nature of these migrations thus segregates the two populations during the autumn breeding season, thereby maintaining the distinctiveness of each race (Orr 1970). By contrast, where the coastal areas and inland valleys are not separated by well-defined topographical features, the two races of mule deer mix and interbreed (Cowan 1936).

Moose (*Alces alces*) also migrate in some areas. In British Columbia, moose summer in subalpine forests at elevations of 1500–2100 m but slowly descend to winter ranges at 700–800 m (Edwards and Ritcey 1956). The availability of food apparently is the prime force motivating these movements. Interestingly, this altitudinal migration of less than 2 km between the summer and winter ranges produces an ecological change equivalent to a distance of 560 km— the latitudinal equivalent of moving between the southern Arctic Tundra and the Boreal Forest. Construction

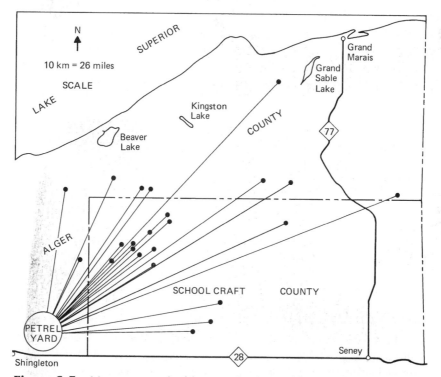

Figure 6-7. Movement records of deer tagged in the Petrel Yard near Shingleton, Michigan, and recovered by hunters in the latter half of November. The longest distance traveled from this yard was 48 km and the average distance was 21 km for the 24 animals. Note the distinctive summer range dispersal pattern for this herd. A similar tendency for directional movement, or zone of occupancy, was evidenced by the recovery of tagged deer from other yards. (From Verme 1973.)

of a 1440-km pipeline across Alaska raised concerns for the migrations of big game. In response, the pipeline was alternately buried and elevated to a height of 3 m so that migrating animals might continue their movements. Thus, when migrating moose began to encounter the completed pipeline in the 1970s, only 14 of 1082 (1.3 percent) attempts at crossing ended with the animals turning back. A somewhat larger number moved along the pipeline before crossing, but most of the moose eventually migrated successfully (Klein 1979).

Caribou (*Rangifer tarandus arcticus*) and reindeer (*R. t. tarandus*) migrate almost continuously in the northern realms of North America and Eurasia. Their movements extend for as much as 500 km between summer ranges in the Arctic Tundra and winter quarters among the scattered trees at the edge of the Boreal Forest and then back again. Klein (1979) reported that the 6000 caribou in the central herd have not negotiated the Alaska pipeline as well as moose. Although bull caribou apparently cross the pipeline without difficulty, cows and calves definitely avoid the structure. Caribou migrations across the tundra also were interrupted where a road and associated traffic ran parallel to the pipeline (Curatalo and Murphy 1986). Where there was no roadway, however, the pipeline did not pose a serious deterrent for caribou migrations. Farther south, in forested habitat, moose and caribou have not been significantly deflected by the Alaska pipeline. Of 7900 caribou in the Nelchina herd, only 4 did not cross the pipeline (Carruthers and Jakimchuk 1987). The caribou in the Nelchina herd thus crossed regardless of whether the pipe was buried or was elevated at its normal height of 2.4 m or higher. Both caribou and moose have increased in forested areas in the years since the pipeline was constructed (Eide et al. 1986). Traffic and other human activities associated with the Alaska pipeline—rather than the physical presence of the pipeline itself—apparently represent the primary obstacle in the migrations of large northern mammals.

MANAGING MIGRATORY ANIMALS

Migratory wildlife often present difficult management challenges, not the least of which occur when such species cross state, provincial, or international boundaries. In those situations, management requires coordination between the political units so that a species is not overexploited when it falls under the temporary jurisdiction of one state or nation. On the

open seas, for example, the endangered status of several species of whales and sea turtles undoubtedly is related to the difficulties of establishing mutually agreeable regulations among the various nations.

The Migratory Bird Treaty signed in 1916 by the United States and Great Britain (on behalf of Canada) effectively protected all migratory birds crossing the borders of both nations. The treaty and its enabling legislation, the Migratory Bird Treaty Act (1918), established appropriate regulatory authority, especially for the determination of waterfowl harvests. Market hunting was outlawed, and the federal government gained regulatory authority for the management of migratory birds. Previously, state governments harvested waterfowl in an uncoordinated system of regulations that often included spring hunting seasons. A similar treaty was signed with Mexico in 1936, thereby bringing migratory birds under a coordinated umbrella of management encompassing all of North America.

In 1987, the United States joined what is popularly known as the "Ramsar Convention." The thrust of the treaty is indicated by its full name, the *Convention on Wetlands of International Importance, Especially as Waterfowl Habitat,* which originally had been adopted in 1971 by 23 nations at Ramsar, Iran (Anon. 1971). Today, there are about 93 signatory nations. For the purposes of the treaty, the term *waterfowl* includes ducks, geese, shorebirds, and all other birds ecologically dependent on wetlands. Signatory nations promote the conservation of wetlands in general, and each country officially lists at least one wetland of importance within its territory. Four national wildlife refuges are among the six wetlands listed thus far by the United States. All told, more than 300 areas exceeding 19 million ha now are enrolled on the official list.

Also in 1987, an international agreement was signed for the protection and management of 180,000 caribou migrating in the region of the Porcupine River between the Yukon and Northwest territories in Canada and the Arctic National Wildlife Refuge in Alaska (Anon. 1987). The Porcupine herd lives in Canada for most of the year but migrates to calving grounds on the coastal plain of Alaska. The treaty establishes an International Porcupine Caribou Board that will manage the herd. Two issues are of immediate concern to the board: subsistence hunting by native peoples and the proposal by the United States for opening the refuge to gas and oil development.

At the other end of the spectrum, the International Whaling Commission (IWC) for many years had little success in limiting harvests of the rapidly dwindling stocks of whales. Critics noted that some IWC delegates represented commercial whaling interests and consequently sought harvests that were inconsistent with the welfare of future whale populations. Even then, Japanese and Russian whalers consistently violated the IWC quotas for sperm and bowhead whales (*Balaena mysticetus*) (Stevens 1980; Weber 1982). In 1982, however, after membership had expanded from 14 to 38 nations, the IWC passed a resolution calling for the elimination of commercial whaling by 1986. The vote was 25 to 7, with 5 abstentions and 1 absence. The moratorium has been less than successful, however, because Norwegian whalers continue taking some whales for "research purposes" (Bush 1993). Likewise, the moratorium has been ignored consistently by the Japanese, whose arguments have been presented eloquently by Kalland and Moeran (1992). Among these are accusations that the IWC membership was padded by adding such nonwhaling nations as Antiqua, Belize, Costa Rica, Monaco, and Santa Lucia. The IWC dues and expenses for the delegates of these nations were paid by antiwhaling activists because they would vote for the moratorium and thereby curry favor from friendly larger countries. Other arguments declared that Western nations harbor prejudice against the Japanese; also, that the United States and United Kingdom use the IWC meetings as a venue for appeasing animal rights groups back home by lashing out against a cruel and inhumane activity in which they no longer take part. Thus, the conservation of whales—a resource held in common by all nations—has been lost in a milieu of international bickering.

The International Union for the Conservation of Nature and Natural Resources (IUCN), based in Switzerland, consists of 438 member organizations in 103 countries. The objectives of IUCN include promoting the conservation of wild places and wild animals and plants in their natural environments. IUCN continually assesses world environmental problems and promotes research relating to their solutions. Among these problems are the suitable protection of migratory animals. Maintenance of a close working relationship between the IUCN, the United Nations, and other intergovernmental bodies such as the Council of Europe, the Organization of African States, and the Organization of American States holds promise for the management of migratory birds and mammals. The success of international cooperation for the con-

servation of migratory wildlife may well serve as a test of the civility of people in the world.

TOO MANY GEESE

In 1997, a committee of wildlife biologists assessed an embarrassment of riches—the exploding populations of three species collectively known as "white geese" (Batt 1997; see also Ankney 1996). Chief among these are lesser snow geese (*Anser caerulescens caerulescens*), which now number more than 3 million birds—a 300 percent increase since 1969. The geese have benefited from an abundance of agricultural foods on their wintering grounds in Texas, Louisiana, and other locations in the Central Flyway. By spring, the birds are in top physical condition and realize exceptional rates of reproduction after returning to their nesting areas in the Canadian Arctic. Refuges, which often supply additional food resources, also increase the winter survival of lesser snow geese. The birds accordingly derive an energy and a nutrient subsidy during both the winter and the spring migration period. All told, these conditions have effectively removed the limits that winter carrying capacity once placed on lesser snow geese. Largely for the same reasons, greater snow geese (*A. c. atlantica*) in the Atlantic Flyway also are increasing, as are Ross geese (*A. rossii*) in western North America.

The burgeoning numbers of white geese have produced massive habitat damage in the Arctic Tundra where the birds nest: Hundreds of thousands of geese today lay bare large areas of ground. When foraging, lesser snow geese "grub" at roots and tubers belowground; hence, they quickly denude their feeding areas in what are often called "eat outs." Thus, with the grazing pressure from such large numbers of geese, the exposed soils erode and develop into mudflats that will take decades to recover, if ever, their former vegetation (Kerbes et al. 1990; Kotanen and Jefferies 1997). The damage is greatest along the western shore of Hudson Bay, where the concentration of breeding birds and steadily deteriorating habitat has been likened to a "snow goose ghetto."

Biologists accordingly fear that an ecological catastrophe is imminent (i.e., a population crash accompanied by irreversible habitat degradation). Efforts to curtail the geese with longer hunting seasons and larger bag limits have failed because the birds, usually in large flocks, are wary and do not decoy easily. Moreover, some hunters do not regard snow geese as good table fare and therefore seek other game. In short, the harvest of snow geese currently lags far behind the growth of the population.

Recommendations for dealing with the explosion of lesser snow geese challenge the orthodox view of migratory bird management. These include proposals to (a) legalize the use of electronic calling devices, (b) legalize baiting at special times and places, (c) encourage Native Americans to harvest more birds and, also, to eliminate the prohibition on taking eggs, and (d) sanction additional hunting in and around refuges where snow geese concentrate. While drastic, such measures represent the tough choices now facing managers responsible for the long-term health of snow geese in North America.

SUMMARY

This chapter outlines several aspects of ethology that are relevant to wildlife management. Birds and mammals select suitable habitat innately, but the process can be modified by learning experiences early in life. Courtship behavior often increases the conspicuousness of animals during the breeding season, thereby providing managers with a means for estimating the size of animal populations. Reproductive physiology is tied closely to behavior. Animals must have suitable physical surroundings and mates that perform the appropriate behavioral rituals before reproduction can succeed. Under atypical conditions, courtship behavior may be directed at inappropriate mates; the result is hybridization or reproductive failure. Ducks and some other birds exhibit renesting behavior if their first nests are destroyed (i.e., the hens lay another clutch); this process lessens the effects of predation and other difficulties.

Territorial behavior is a means of spacing animals in their habitat. Thus, if habitat is limited, breeding also may be limited. In rich and diverse habitats, territories are smaller than they are in poor habitats. Knowledge of circadian rhythms, particularly of the time of day or night in which animals are most active, offers wildlife managers the means of determining movements and other activities of animals. Dispersal—the tendency of young animals to leave their parents and natal range—is important in the colonization of newly available habitat and in nourishing depleted populations.

Because of imprinting, the early experiences of young animals may determine the level of attachment to a mother figure, the future choice of mates, and to some extent the choice of habitat. Good management practices assure that young animals are afforded appropriate opportunities for imprinting on the correct parents and surroundings. The ways animals respond to humans have important management implications; sensitive species must be isolated from human activities, and habituated animals may become pesky or dangerous. Various animals migrate, some for thousands of kilometers, as a means of survival. Migratory movements may involve seasonal changes in altitude or latitude. Migrations frequently take animals across international boundaries, thereby complicating management activities. The welfare of migratory animals often requires cooperation among nations and is best shown by international treaties for the protection of migratory birds.

CHAPTER 7

FOOD AND COVER

Cover is a magic word in wildlife management. It is, indeed, often a magic wand with which wild animals and birds are made to populate places formerly uninhabitable. ...It seems desirable that we should seek to analyze the complex nature of cover more carefully.

Charles Elton (1939)

No animal survives without food, and few can exist without cover. In places where water is relatively abundant, availability of food and cover strongly influences the kinds and numbers of animals occupying an area. A driving force in evolution is the acquisition and conservation of nutrients and energy. Those animals that are best adapted for extracting energy from their surroundings and for using that energy to best advantage are most likely to survive. Less efficient individuals are displaced into suboptimal environments where few survive.

Food is the source of nutrients and energy, and good cover prevents the loss of energy by providing shelter from extremes of wind and temperature. Cover also offers protection from predators. Wildlife managers normally are concerned with the year-round habitat for resident species, but a desirable mixture of food and cover may be a seasonal consideration for migratory species. Habitat requirements for migratory waterfowl, for example, include nesting and brood cover in the spring and summer months, but these features are not important during winter.

FOOD

The energy in food serves as fuel for the metabolic processes of animals. The nutrients in food support growth and maintenance of body structures. A wealth of literature describes the food habits and dietary requirements of scores of species. However, our purpose here is not to list food requirements species by species but instead to describe and illustrate some generalities that pertain to the ecology and management of wildlife foods, nutrition, and feeding (see Robbins 1983 for a detailed treatment of these subjects).

DIGESTIVE SYSTEMS OF BIRDS AND MAMMALS

Birds and mammals have evolved a wide range of adaptations for utilizing potential foods that occur in their surroundings. Birds and mammals have developed ways of eating and digesting lichens, fungi, leaves, stems, tubers, roots, nectar, seeds, fruits, and nuts from the plant kingdom. From the animal kingdom, all but a few of the largest and most powerful species may serve as food for predators. Some animals are highly specialized and feed within a limited range of foods. For example, the snail kite (*Rostrhamus sociabilis*), a rare species of hawk in the United States, feeds almost entirely on apple snails (*Pomacea paludosa*); these birds have evolved precisely curved bills for extracting the snails from their shells (Stieglitz and Thompson 1967). Similarly, crossbills (*Loxia* spp.) are uniquely adapted for extracting seeds from cones. Mammals specialized for a narrow range of foods include anteaters (Myrmecophagidae) and moles (Talpidae). Other species, such as crows and jays (Corvidae), bears (Ursidae), and raccoons (*Procyon lotor*), have adapted an *omnivorous* strategy; their diets include a wide variety of plant and animal foods.

The process of converting plant and animal materials into usable food occurs in the digestive system (Fig. 7-1). In birds, the digestive tract includes a relatively long *esophagus,* a part of which in most species is an expandable chamber known as a *crop* [Fig. 7-1(a)]. Crops are storage areas that are usually larger in herbivores than in carnivores. Plant eaters consume more food but at slower rates than meat eaters, so storage capacity is an important feature for herbivores. While foods are stored in the crop, they undergo practically no digestion.

Beyond the esophagus is the stomach, consisting of two parts—a soft glandular stomach known technically as the *proventriculus* and a muscular gizzard called the *ventriculus.* Enzymes secreted in the glandular stomach begin the process of chemical digestion. The inner walls of the gizzard are lined with a tough, corrugated surface adapted for grinding. Most species of birds ingest grit, which aids the grinding action of the gizzard. Food thus is ground into small fragments, thereby speeding up the action of the digestive enzymes. The action of the gizzard in many carnivorous species is modified so that bones, fur,

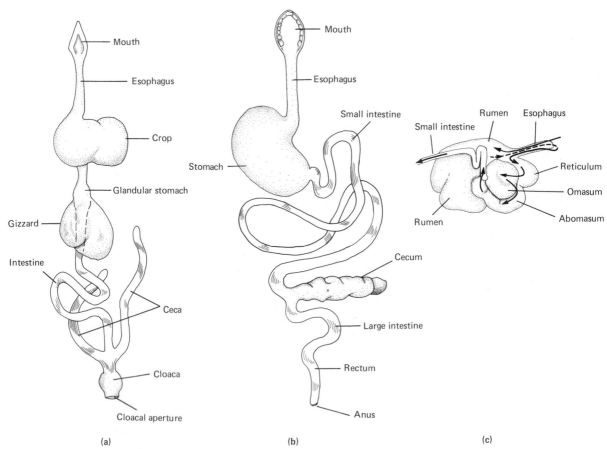

(a)   (b)   (c)

Figure 7-1.   Digestive tracts: of a bird (a, generalized); a mammal (b, generalized); and a ruminant stomach (c).

feathers, and other relatively indigestible materials are rolled into pellets and then are regurgitated from the digestive system. Such pellets accordingly contain identifiable remains of prey; they provide biologists with a ready means of analyzing the food habits of owls and other carnivorous birds (see Korschgen 1980 for a full discussion of food-habits techniques). Gizzards function as powerful crushers in some birds. Wild turkeys (*Meleagris gallopavo*) swallow nuts whole but can crack a pecan in an hour and can fragment a hickory nut in about 8 hours; to do so requires a pressure of 52–150 kg (Schorger 1960).

Ground food, mixed with digestive enzymes, passes from the gizzard into the intestine, where nutrients are absorbed into the bloodstream. Blood vessels lead from the intestine into the liver—a large organ in birds—which stores excess sugars and fats, synthesizes some proteins, and secretes bile into the intestine. Additional digestive enzymes are secreted into the intestine from the pancreas. The intestine is short in meat-eating birds and is comparatively long in those birds that eat seeds and other plant materials. These differences reflect an adaptation to the high fiber in diets of vegetation.

A pair of elongated, blind tubes called *ceca* arise near the posterior end of the intestine in most birds. Ceca provide an important environment for bacterial decomposition of fibrous foods and also for absorbing water and digested proteins (Welty 1982). Because birds, mammals, and other vertebrates do not produce enzymes capable of breaking down the molecular bonds of fibrous compounds, these animals depend on gastrointestinal bacteria for splitting and reducing these molecules into absorbable compounds. Grouse, with their winter diets of fibrous foods—buds and conifer needles—are among the birds with well-developed ceca. In one study, the cecal lengths of spruce grouse (*Dendragapus canadensis*) increased from about 27 to 40 cm between September and February, coincident with the high content of fiber in the winter diet (Pendergast and Boag 1973). Similar changes occurred simultaneously in the lengths of the intestines. Spruce grouse experimentally maintained on low-fiber diets did not respond to the same extent, thereby suggesting that diet rather than some other factor indeed stimulated the elongation of these organs. Miller (1975) and Paulus (1982) showed similar changes in the gut morphology of waterfowl resulting from high fiber diets.

Food materials move into and out of the ceca about once every 24 hours (Welty 1982). McBee and West (1969) determined that bacterial fermentation in the ceca provides from 15 to 30 percent of the daily energy requirements of willow ptarmigan (*Lagopus lagopus*); these data probably are similar in other species of grouse. Cecal droppings are usually soft and dark, in contrast with the drier, whitish droppings coming from the intestine (see the discussion on vitamins later in this section).

In birds, the *cloaca* is the final passageway for food. The cloaca is a short chamber that accepts several materials: undigested foods from the intestines and ceca, uric acid from the kidneys, and eggs and semen in season from the ovaries and testes, respectively. These materials leave the body through the *cloacal aperture.*

There is considerable variation among birds regarding their feeding schedules. Many species feed early in the day and again in the evening, but those feeding on carrion sometimes eat only once every few days. Food passes through the digestive tract of birds at rates that vary with the kind of food. Berries appeared in the droppings of blackcap warblers (*Sylvia atricapilla*) within 12 minutes, although the entire meal of berries took 2 hours to clear the tract. Dyed oats fed to a goose (*Anser* sp.) appeared in 4 hours; the entire meal cleared in 17 hours (Ziswiler and Farner 1972). Because different foods are digested at various speeds, studies of food habits may be subject to biases and misinterpretation of the evidence remaining in crops and gizzards. Swanson and Bartonek (1970) fed known amounts of food to blue-winged teal (*Anas discors*) and examined the digestive tract of the teal at selected intervals thereafter. Foods such as the soft bodies of snails were seldom detectable after 10 minutes, whereas several kinds of seeds remained identifiable for several hours or days. The results of these experiments suggest that only foods in the esophagus (crop) should be analyzed, thereby avoiding the risk of bias associated with differential rates of digestion. Also, to avoid postmortem digestion, alcohol should be forced into the upper gastrointestinal tract immediately after the birds are collected in the field.

Beginning with the mouth, the digestive system of mammals consists of the pharynx, esophagus, stomach, small and large intestines, and anus [Fig. 7-1(b)]. Unlike birds, most mammals have teeth, and all lack either crops or gizzards. Teeth, of course, initiate the digestive process by physically altering food into sizes that can pass into the pharynx and beyond. Chewed foods are usually difficult to identify, but new methods use the species-specific cellular structure of plant and animal materials as the basis for determining the food habits of mammals (Korschgen 1980).

As with birds, the length of the mammalian digestive tract varies with diet. However, unlike birds, the length of the tract does not vary with seasonal changes in food habits. Shorter tracts are associated with those species of mammals eating meat or other highly digestible foods. The vampire bat (*Desmodus* spp.) has a stomach that is a simple, straight, and highly expandable tube for storing and digesting blood (Glass 1970). In contrast, herbivores have some combination of complex stomachs, long ceca, and long intestines. These features again reflect the high fiber content in the diet of plant-eating mammals. The stomachs of cud-chewing herbivores—animals known as *ruminants*—are subdivided into four main chambers [Fig. 7-1(c)]. Members of the deer family (Cervidae) are examples of ruminants. Beavers (*Castor canadensis*), collared peccaries (*Tayassu tajacu*), and rabbits and hares (Leporidae) are herbivores lacking the ruminant-type of stomach; these species instead each have a simple stomach and a single, well-developed cecum. Bacteria in the stomachs of ruminants or in the ceca of other mammalian herbivores produce enzymes that help digest fibrous foods. Microbial fermentation provides 60 to 70 percent of the energy requirements of ruminants (Annison and Lewis 1959). Whereas most mammals can eat a variety of foods, herbivorous mammals often experience a period of adjustment when changing from one kind of food to another. During this period, the composition of bacterial populations in the digestive system adjusts to the new diet.

ENERGY

Energy is released when large food molecules are broken apart. Once released, some of the energy is converted into heat, which is used for the maintenance of body temperature in birds and mammals. Some energy also is transformed into mechanical energy for muscle contraction, for the transport of materials through membranes, and for assembling new and different kinds of cells (e.g., feathers, teeth, skin, liver, and other tissues).

How much energy does an animal use? The level varies with the size of the animal. In warm-blooded animals, small species require proportionately more energy than larger species because the greater ratio of surface area to volume permits more heat to escape. The formula of Brody (1945) for active mammals is generally used for energy analyses in wildlife: daily energy requirement (kcal) = $140 \times$ (body wt in $kg^{3/4}$). Thus, a masked shrew (*Sorex cinereus*) weighing 5 g

(0.005 kg) needs 2.63 kcal per day; an active 68-kg human needs 3300 kcal, and a 544-kg brown bear (*Ursus arctos*) needs 15,800 kcal. The shrew needs about 526 kcal per kg of body weight, whereas the bear requires only 28 kcal per kg. These figures vary with the degree of insulation of the coat, with the temperature of the environment, and with the cover or shelter surrounding the animal, as will be discussed later.

It would seem that birds, with higher body temperatures and higher metabolic rates, also would have greater energy demands than mammals. King and Farner (1961), however, noted that the equation describing energy demands in kcal per kg of body weight is indistinguishable statistically between birds and mammals—with one notable exception: Energy requirements are substantially higher for birds weighing less than 10 g than for mammals of equivalent size. For example, two species of hummingbirds weighing about 4.0 and 3.5 g require 1400 and 1600 kcal per g per day, respectively, compared with the far smaller requirement of 526 kcal for a 5-g shrew.

CARBOHYDRATES

Carbohydrates include cellulose, starches, and sugars, which are various compounds of carbon, hydrogen, and oxygen. Simple sugars are metabolized quickly in the digestive system and therefore are a source of quick energy: simple sugars yield 4.2 kcal per gram. Larger molecules, such as cellulose and lignin—the woody material in trees—contain a great deal of energy (of which we take advantage when burning wood for fuel). However, the abundant calories in wood are not available to most animals. Only a few species, such as beavers, porcupines (*Erethizon dorsatum*), and lagomorphs (Leporidae), have a digestive system and intestinal bacteria capable of digesting and extracting energy from wood. Carbohydrates abound in all parts of plants eaten by animals.

FATS

Fats (and oils) consist of carbon and hydrogen atoms. In comparison with carbohydrates, fats are composed of fewer oxygen atoms. Fats contain more than twice the energy per unit of weight as carbohydrates (9.5 versus 4.2 kcal per gm), but the digestive system extracts the energy at a slower rate. Fat deposits in the body serve as storage depots for energy. The gastrointestinal tract of many grazing and browsing species is not adapted for digestion of fats. Fats and oils are broken apart physically by bile produced in the liver, stored in the gall bladder, and secreted into the small intestine. Deer, lacking a gall bladder, continually secrete bile into the intestine in small but sufficient quantities for the emulsification of the few fats normally occurring in their diet. Larger amounts of fats, however, cannot be fragmented into digestible units in deer because too little bile is available. Fats are found in small quantities in the vegetative parts of plants (e.g., leaves and stems) but are often abundant in seeds such as corn, beans, and peanuts.

PROTEINS

Proteins contain nitrogen in the form of amino acids (NH_2 groups) in addition to carbon, hydrogen, and oxygen. The energy available in proteins is about equivalent to the energy present in carbohydrates, but the amino acids in protein build nucleic acids that are essential for the reproduction of cells. Moreover, nucleic acids help construct enzymes, substances that are necessary for practically every chemical reaction in the body. Proteins, like fats, are not abundant in vegetative parts of plants, although proteins are concentrated in tips of growing stems. Seeds such as beans, grains, and nuts contain the highest concentration of proteins in plants. Legumes, such as clover, alfalfa, beans, and peas, fix atmospheric nitrogen (with the help of bacteria in their root nodules) and thus generally are good sources of nitrogen in the diets of animals. Because protein often is scarce in plants, the abundance of protein in the diets of herbivores provides a measure of food quality.

VITAMINS

Vitamins are complex molecules that function as enzymes in the body; small but essential amounts are required. In some locations, vitamins may be naturally deficient and thus may prevent the growth and vigor of wildlife populations. In most cases, however, the diets of wild animals contain the required amounts of vitamins. A few groups of animals, such as lagomorphs (rabbits and hares), conserve the vitamins produced by bacteria in the digestive tracts by consuming their own feces. That process, *coprophagy,* effectively recycles vitamins. Lagomorphs and other animals practicing coprophagy produce two types of feces: dry pellets from the intestine that

are not ingested, and moist droppings from the *cecum* that are ingested immediately. Animals experimentally denied access to their feces experience reductions in vitamin K and biotin, thereby decreasing the digestion of other foods and causing reduced growth rates (Schmidt-Nielsen 1979).

MACRONUTRIENTS

Macronutrients include several minerals and dissolved ions. Among these are sodium, potassium, phosphorus, calcium, magnesium, and sulfur. Phosphorus and calcium are essential for the formation of bones, teeth, and eggshells. Other macronutrients are necessary for such functions as the transmission of nerve impulses, muscle contraction, blood coagulation, and maintenance of proper osmotic conditions. Minerals are contained in ash in the gross analysis of foods (Table 7-1).

MICRONUTRIENTS

Micronutrients are elements that are present in animal tissues in minute quantities. Their total makes up less than 0.01 percent of the body. The functions of some micronutrients are known, but our knowledge of the role played by many others remains incomplete. Iron forms the core of the hemoglobin molecule that transports oxygen in the blood (among other functions), and copper aids in the manufacture of hemoglobin and contributes to the functions of enzymes. Other micronutrients include cobalt, nickel, zinc, vanadium, chromium, molybdenum, manganese, silicon, tin, arsenic, selenium, fluorine, and iodine (Schmidt-Nielsen 1979).

ECOLOGY AND EVOLUTION OF FEEDING BEHAVIOR AND DEFENSE

Feeding relationships represent an evolutionary arena in which there is an ongoing contest for survival between the eaters and the eaten. Survival of a species depends on a sufficient number of individuals obtaining enough nutrients and thereafter producing enough offspring to replace at least those that are consumed by other organisms. Natural selection favors efficient feeders as well as efficient escapers; that is, as the eaters evolve toward increased feeding efficiency, the eaten concurrently develop means of escape and defense that enhance their own survival. Plants have developed characteristics for resisting

TABLE 7-1.   Composition of Some Foods Eaten by Wildlife

Food	Water (% wt)	kcal/ 100 g	Protein	Fat	Carbohydrates and Lignin	Minerals (Ash)
			Nutrients in g/100 g dry wt			
Plant						
Grass (Graminae)	80	220	13.0	6.0	73.0	8.0
Jack pine *(Pinus banksiana)* needles	55	524	8.9	11.8	77.0	2.5
White cedar *(Thuja occidentalis)* twigs and leaves	54	237	3.3	4.4	91.3	2.0
Animal						
Invertebrates						
Snail (Gastropoda)	79	219	52.0	1.0	0.0	47.0
Crab (Decapoda)	26	170	33.0	2.0	9.0	56.0
Insect *(Notonecta)*	25	374	56.0	4.0	24.0	16.0
Vertebrates						
Fish (perch, *Perca flavescens*)	80	426	76.1	3.5	0.0	9.0
Bird egg (chicken, *Gallus gallus*)	73	430	38.0	31.0	0.0	31.0
Whole bird (chicken)	76	446	57.0	24.0	0.0	19.0
Whole mammal (pig, *Sus scrofa*)	75	448	89.0	10.0	0.0	1.0

Sources: Ullrey et al. (1968); Pendergast and Boag (1971); Bardach et al. (1972); Brambell (1972); Gurchinoff and Robinson (1972); Watt and Merrill (1973).

herbivores, and herbivores have developed means of coping with the defense mechanisms of plants. Predators have developed keen senses, speed, and stalking abilities, while prey species have developed watchfulness, agility, camouflage, and hiding abilities. Selective pressures are applied mutually between the eater and the eaten. These processes are known as *coevolution*—the joint evolution of two or more groups of organisms with close ecological relationships but that lack any exchange of genes. Reciprocal selective pressures operate in the process of coevolution, making evolution of one group partially dependent on evolution of the other (Pianka 1983).

Smith (1970) described an example of coevolution between pine squirrels (*Tamiasciurus* spp.) and conifer trees in northwestern North America. Conifer seeds are the main food of the squirrels, and squirrels can remove virtually all cones from a tree. The trees, however, reduce the effectiveness of squirrels and thereby enhance their own survival in several ways: (1) by producing cones that are difficult for squirrels to open; (2) by producing cones with fewer seeds; (3) by increasing the thickness of the seed coat; (4) by producing less-nutritious seeds; (5) by dropping seeds early in the season before young squirrels

begin to feed on seeds; and (6) by periodically producing almost no cones in some years. Some of these adaptations apparently reduce the potential number of young trees that might be produced. With fewer seeds per cone and periodic skips in seed production, however, the evolutionary strategies would seem to be self-defeating; similarly, so does the production of seeds that have a low nutrient content, for these could produce weakened seedlings. In fact, however, such features reduce the feeding efficiency of the squirrels and thus prevent consumption of the entire seed crop. When they occur, the cone "failures" decimate the squirrel populations, a process that increases the survival rate of the seed crop in the following year. Such fluctuations in the cone crop occur regardless of favorable weather. Also, squirrels, to survive, have developed behavioral means of dealing with the defense mechanisms of the conifers; these include (1) careful selection of cones so that squirrels avoid those with few seeds and thus conserve their foraging energy, and (2) stockpiling cones as a food source for lean years.

Other plants have developed physical and chemical defenses against herbivores. For example, African acacias (*Acacia* spp.) produce long spines on twigs after the tip is browsed, whereas unbrowsed twigs have shorter spines (Janzen 1978). Chemical defenses are more common, however, and only in recent years have ecologists realized and appreciated the variety and ubiquity of these mechanisms (Janzen 1978; Bryant and Kuropat 1980). Some of the chemicals in plants that repel animals include tannins, oils, resin, alkaloids, terpenes, and terpenoids (Pianka 1983).

Bryant and Kuropat (1980) noted that food choices may be made primarily on the basis of nutritional quality or, as increasing evidence suggests, because browsing animals may avoid plants that contain certain chemicals. In the latter instances, individuals that continue eating plants with chemical defenses may experience reduced fitness. Vegetation developing in the absence of large herbivores (e.g., on oceanic islands) may lack adequate defenses against browsing animals. Thus, red deer (*Cervus elaphus elaphus*) introduced into New Zealand caused widespread destruction of native shrubs (Howard 1964; see also Chapter 18). In Europe, where red deer are native, many plants are resistant to the effects of browsing.

Chemical defenses of plants frequently poison the microflora in the gastrointestinal tract of herbivores, thereby inhibiting crucial digestive processes (Bryant and Kuropat 1980). Some chemicals such as sesquiterpenes are fatal to domestic sheep and, presumably, to wild herbivores as well. Wild herbivores, however, probably evolved the means of avoiding such plants (Pianka 1983). Thus, herbivores are confronted with two questions when feeding: First, is the plant toxic, and second, does the plant contain sufficient digestible nutrients? Research only recently began to investigate the nature of chemical defenses in plants. We predict that our current knowledge about the quality of wildlife food will be much revised in a few years.

QUALITY OF FOOD

Plants have evolved myriad organs, structures, and forms for coping with the vagaries of their environments. Tubers, seeds, bulbs, roots, leaves, stems, flowers, and fruits are just a few of the better known parts of plants. For herbivores, the nature and availability of plant materials vary from place to place, species to species, season to season, and, indeed, on the structure of a single plant. Herbivores thus select—in time and space—those parts of a plant from which they can extract a living. Diets of barely digestible food may carry herbivores through a lean season, but, ultimately, foods of good quality must be available for the growth and reproduction of animals.

Some animals supplement their diets with bones. Porcupines and many kinds of mice gnaw on shed antlers. Desert mule deer (*Odocoileus hemionus crooki*) and desert bighorn sheep (*Ovis canadensis mexicana*) eat dried mammal bones (Krausman and Bissonette 1977; Warrick and Krausman 1986). Bones contain both calcium and phosphorus, but because calcium normally is abundant in the diet of desert animals, bone-chewing probably provides a means of supplementing phosophorus and other minerals.

The diet of carnivores varies little in quality. Nutrients available in the flesh of one species of animal are quite similar to those in the flesh of another species. A few animals have developed some repulsive chemical features (e.g., musk glands in shrews and poison glands in toads) and thus are seldom eaten by predators. A few others, such as porcupines, turtles, and armadillos (*Dasypus novemcinctus*), have specialized dermal features that hinder predators. But these animals are exceptions. By and large, any food deficiencies of carnivores are those of quantity rather than quality.

The nutritional quality of vegetation as wildlife food usually depends on soil fertility—that is, on the abundance of nutrients in the soil. Numerous studies showing a positive relationship between the size of animals, population densities, and soil fertility are cited in Chapter 12. For example, in Missouri, the size and bone strength of adult cottontails (*Sylvalagus floridanus*) varied in different parts of the state according to soil fertility (Crawford 1950). In Michigan, cottontails that live on vegetation growing on sandy, sterile soils doubled their numbers from spring to fall, whereas those rabbits living 60 km away on fertile soils increased their numbers fivefold during the same period (see Allen 1974a). Cottontails on the sandy soils also required a larger winter range than the other rabbits. The relationships between good soil and high-quality plant food for herbivores is now well established. In the words of Wilde (1958), "Nothing will come of nothing, and soils deficient in nutrients cannot become productive, regardless of whether they are planted to rye, pine, or grouse." Unfortunately for wildlife, however, agriculture has preempted nearly all of the richest soils on which many species of animals once prospered. Biologists thus often face the task of managing wildlife populations under less than optimal conditions on lands in agricultural areas (see Chapter 13).

Nutritional studies have verified the importance of an adequate diet for growth and reproduction in white-tailed deer (*Odocoileus virginianus*). French et al. (1955) showed that antler development, as well as body size, was greatly influenced by nutrition. Yearling bucks on good diets developed antlers with 6-8 points, whereas their siblings on poor diets produced small, unbranched antlers ("spike bucks") or those with a single fork.

That poor nutrition adversely affects the physical condition of animals is not surprising, but poor nutrition also affects reproduction. Verme (1965a) compared fawn production of penned white-tailed deer kept on experimental diets. Twenty-seven does that were given a highly nutritious diet during the autumn breeding season produced an average of 1.74 fawns per doe, whereas those that were maintained on poor diets during the same period produced only 0.95 fawns per doe. Further, 68 percent of the fawns from the undernourished does were males versus 36 percent in the case of the well-nourished does (Table 7-2). Fawn production from the well-nourished does in this experiment was equivalent to production on good range in Michigan, but production from the poorly nourished does fell below that occurring on poor deer ranges. As shown in roan and sable antelopes (*Hippotragus equinus* and *H. niger*), good nutrition also enhances resistance to diseases that otherwise can be fatal in malnourished animals (Wilson and Hirst 1977).

Anderson and Stewart (1969) examined soil micronutrients and macronutrients as a means of explaining why pheasants (*Phasianus colchicus*) were consistently abundant in some areas of Ohio and scarce in others. The concentrations of macronutrients—calcium, magnesium, phosphorus, potassium, and sodium—were higher both in the soil and in the internal organs of pheasants in places where pheasants were abundant. Some micronutrients—barium, manganese, and titanium—also were more abundant where pheasant numbers were high. In poor pheasant areas, however, other micronutrients—chromium, lead, nickel, vanadium, and zirconium—were twice as concentrated in the organs of the birds. The study suggested that the soil nutrients may limit pheasant numbers in two ways: by scarcity (of the macronutrients, calcium, magnesium, and potassium) and by abundance (toxic concentrations of such micronutrients as chromium).

TABLE 7-2. Relation of Autumn Nutritional Levels to Deer Reproductive Patterns

	High Nutrition	Low Nutrition
Does examined	27	22
Does not bred	0	2
Does bred but unproductive	3	1
Fawns produced	47	21
Litters (1:2:3 fawns)	2:21:1	17:2:0
Fawns per doe	1.74	0.95
Fawns per pregnant doe	1.96	1.11
Male fawns	16[a]	13[a]
Female fawns	29	6
Percent males	35.6	68.4

Source: Verme (1965a).
[a]One set of twins could not be sexed.

Selective feeding behavior may partially compensate for the scarcity of some nutrients. Animals at times demonstrate "hungers" for specific nutrients that may be deficient in their bodies (Hainsworth 1981). Laboratory rats presented with a "cafeteria" of unnatural foods, including dextrose (for carbohydrates), casein (for protein), cod liver oil and olive oil (for fats), and yeast (for vitamins), selected a nutritionally balanced diet from the available foods and showed normal growth and development. When yeast was removed from the cages, vitamin B became deficient and the growth rate of the rats diminished. However, the rats partially offset the deficiency by consuming their own feces (coprophagy), which intestinal bacteria had enriched with vitamin B.

Gurchinoff and Robinson (1972) found that wild spruce grouse selected the needles of jack pines (*Pinus banksiana*) that possessed significantly higher protein and mineral content than those of neighboring trees. Thus, grouse could detect differences in nutritional content in their food, but the sensory mechanism remains unknown. Hohf et al. (1987) offered penned spruce grouse boughs from lodgepole pines (*P. contorta*) in a feeding experiment. Two kinds of branches were tested: those from trees that were browsed in the wild by spruce grouse and branches from randomly selected trees. Given equal access to each kind of branch, the penned birds preferred the needles from the same branches that the wild spruce grouse selected as food. Analysis showed that needles from those trees contained more protein than the others, but no chemicals were identified that might have repelled the birds from the uneaten needles.

Gullion (1966) noted that ruffed grouse (*Bonasa umbellus*) often select the flower buds of aspen (*Populus* spp.) from trees that appear less vigorous than others. Perhaps such trees can no longer produce resins or other chemical defenses (Bryant and Kuropat 1980). Dasmann (1981) cited studies suggesting that deer avoid browsing on healthy sagebrush (*Artemisia tridentata*) and instead feed on plants of lesser vigor. The digestibility of healthy sagebrush may be inhibited by an abundance of plant oils.

When penned white-tailed deer were offered a choice of hardwood browse cut before leaf fall (in September), immediately after leaf fall (in October), and at monthly intervals throughout the winter, they chose the October-cut and fresh-cut twigs over those cut in September (Alkon 1961). The main factor in the choice apparently was the higher moisture content of the preferred twigs; there were few differences in the availability of nutrients. Water content of the September-cut twigs was below 40 percent, whereas the moisture content in twigs cut later was consistently above 40 percent. When given a choice of equally nutritious foods, domestic animals prefer those with higher moisture content, probably because moisture aids digestion. Thus, if cuttings are planned for an area where deer winter, the animals will obtain greater benefits from slash cut after the leaves fall (i.e., cuttings made in late autumn or during the winter).

Young animals invariably require more protein than adults. Milk supplies the protein requirements for juvenile mammals. For young carnivores, however, the demand for protein often is supplemented with partially digested food regurgitated by their parents. For young herbivores, milk is needed for a relatively long period of time. Milk supplies, however, are not always assured. In white-tailed deer, the nutrient level in the milk remains the same regardless of the nutritional condition of the doe, but the *quantity* of milk production declines with malnutrition (Youatt et al. 1965). Fawns born to does that have struggled through a difficult winter may have little or no milk available and thus often may die of starvation. Hence, the debilitating effects of winter malnutrition commonly carry over into the next generation of deer.

Young birds are generally voracious consumers of protein (Kear 1972a; Fig. 7–2). As shown in Fig. 7-3, mallard ducklings (*Anas platyrhynchos*), for example, feed almost exclusively on insects and other arthropods, but the young birds become vegetarians at 2 months of age (Chura 1961). Finches (Fringillidae), which as adults are seed-eaters, feed insects to their young. Adult yellow-bellied sapsuckers (*Sphyrapicus varius*) catch flying insects, dip them in tree sap, and then feed their young the sticky mixture. Pigeons and doves (Columbidae) feed their young a secretion of protein-rich "crop-milk." Only a few birds, such as geese and oilbirds (*Steatornis caripensis*), mature without a diet of animal protein. Goslings instead obtain protein from young grasses, and young oilbirds eat protein-rich nuts (Kear 1972a; Welty 1982).

Sex-specific differences in diet occur in some animals. For example, Beier (1987) found that female white-tailed deer wintering on good range in southern

Figure 7-2. Protein is essential for the growth and development of young birds and usually is obtained from a diet of insects and other invertebrates. Hence, these mallard ducklings require more protein than their mother. (Photo courtesy of Northern Prairie Wildlife Research Center, U.S. Fish and Wildlife Service.)

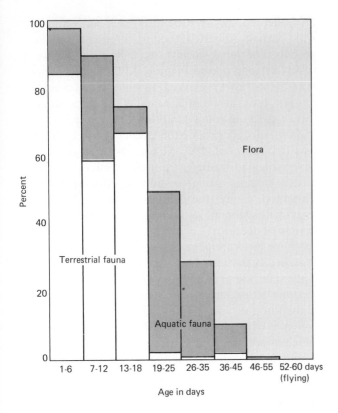

Figure 7-3. Age-related changes in the diet of mallard duck-lings. Note how insects and other invertebrates (white area) initially dominate the diet, which later begins to include more plant materials (shaded, at right) as the young birds mature. Also note the shift to aquatic fauna (dark area) as the ducklings lose their down and gradually acquire feathered plumage, thereby facilitating subsurface feeding activities. (From Chura 1961.)

Michigan consumed more grass and less browse than males. The feces of does also contained larger quantities of nitrogen, indicating a regimen of better quality in comparison with the diet of bucks. In northern Wisconsin, however, other studies have shown that adult bucks consistently dominate does when competing for scarce winter food (Kabat et al. 1953). As a result, the bucks probably are better nourished. Differences in food availability may explain these circumstances. When food is adequate, bucks may not contest feeding opportunities with does, but competition between the sexes is initiated when food is scarce. Female pintails (*Anas acuta*) switch from seeds and other plant materials to a diet dominated by animal matter during the nesting season (Krapu 1974a,b; Krapu and Swanson 1975). The shift provides hens with the protein and calcium necessary for egg formation—demands that are not placed, of course, on drakes.

When available, natural mineral springs attract wildlife, as do the surrounding soils (for more about "salt licks," see Chapter 12). Moose, white-tailed deer, bighorn sheep, and muskox (*Ovibos moschatus*) are among the big-game species frequenting such areas. Sodium, calcium, magnesium, and potassium represent the principal minerals available in natural springs, but the groundwater also may include sodium bicarbonate, which ruminants may seek to alleviate excess stomach acid (Bechtold 1996).

QUANTITY OF FOOD

Beginning with 1900, Leopold et al. (1947) documented about 100 cases where herds of white-tailed deer or mule deer (*Odocoileus hemionus*) expanded to such numbers that their food supply was threatened or they caused excessive crop damage. In about half of those cases, disaster was averted when humans altered the habitat or initiated hunting. In the other cases, however, overbrowsing was followed by mass starvation.

What does *overbrowsing* mean? How can an animal consume more food than is available in the habitat? The answer is that woody plants grow only seasonally, sending out new twigs in the spring and early summer. The new growth contains most of the digestible nutrients. If animals eat too much of the new growth, the damaged plants cannot regenerate additional twigs (Fig. 7-4). In some valuable food plants, such as white cedar (*Thuja occidentalis*), regeneration is impaired when as little as 20 percent of the new growth is removed (Aldous 1952). Other plants can tolerate up to 100 percent browsing, although pressure of that magnitude year after year eventually will decrease production of new tissues. The ultimate result is starvation and low birthrates—unless the herbivores are reduced by other means. Food shortages were implicated as the factor initiating declines of snowshoe hares (*Lepus americanus*) in northern Alberta, and reductions in the prey population of hares thereafter diminished the numbers of lynx (*Lynx canadensis*) (Keith 1974; see also Chapter 9).

RESPONSES TO FOOD SHORTAGES

Some birds can live without food for extended periods of time. Several species of hummingbirds (Trochilidae) in Peru and European swifts (*Apus apus*) survive short-term food shortages by entering a state of hypothermic torpor for a few hours to sev-

Figure 7-4.　Deer faced with food shortages must reach higher and higher for browse (left). The result of such feeding pressure year after year is a browse line, below which almost no new annual growth is produced (right). Fawns are the first to starve under such conditions. (Photos courtesy of U.S.D.A. Forest Service.)

eral days. Poorwills (*Phalaenoptilus nuttallii*) in the western United States may hibernate for periods as long as 88 days (Jaeger 1949). However, even without highly specialized physiological adaptations such as these, other birds can withstand temporary food shortages while maintaining normal body temperatures. Under experimental conditions, Jordan (1953b) kept wild mallards alive for about 3 weeks without food, but the surviving birds lost 43 percent of their body weight. When returned to a normal diet, however, the mallards regained the weight loss in 28 days. Gerstell (1942) determined that wild turkeys and ring-necked pheasants (*Phasianus colchicus*) can survive 1 and 2 weeks, respectively, of severe winter weather without food.

Hibernation is a fairly common phenomenon in mammals. Watts et al. (1981) defined *hibernation* as "a specialized adaptive seasonal reduction in metabolism concurrent with the environmental pressures of food unavailability and low environmental temperatures." The length of the hibernation period and the amount of reduction in body temperature vary among mammals. During hibernation, the body temperatures in some rodents, such as ground squirrels (*Spermophilus* spp.), may drop to a few degrees above freezing, whereas black bears (*Ursus americanus*) maintain body temperatures at 31°C or higher, only about 7°C less than normal (Rogers 1981). Hibernation saves a bear about 600 kcal per day, about half of the energy that is expended at

normal rates of body temperature and activity (Watts et al. 1981). White-tailed deer in northern areas exhibit lowered metabolic rates in midwinter without denning or prolonged sleeping. Under these conditions, deer travel less and eat less even when adequate food is available (Silver and Colovos 1957). Nevertheless, deer frequently die as a result of food shortages during severe winter weather.

We earlier outlined the large-scale starvation of mule deer on the Kaibab Plateau (Chapter 2). Another widely cited case involves the catastrophic loss of brant (*Branta bernicla*) in the wake of a sudden food shortage. Brant winter on the Atlantic Coast from New Jersey to North Carolina, where they feed almost exclusively on a marine plant known as eelgrass (*Zostera marina*). Because of what probably was a parasitic disease, eelgrass nearly disappeared from the Atlantic coastline in 1931 and 1932. Large dislocations resulted in those animal communities dependent on the once widespread beds of eelgrass (e.g., see Stauffer 1937). For brant, the loss of their primary food was calamitous: the population along the Atlantic coast fell to 10 percent of its former size (Cottam et al. 1944). At some concentration points, the winter census in 1933-34 revealed only 2 percent of former numbers, and the hunting season for brant was closed in light of the abrupt decline. Before 1932, eelgrass formed 85 percent of the winter diet of brant, whereas afterward the same food constituted only 9 percent of the diet. The brant instead switched to a diet dominated by algae, which is undoubtedly of lower quality and palatability than eelgrass. Brant numbers have since recovered, but the shock of the eelgrass failure amply illustrates the severity of food shortages on wildlife populations.

At present, perhaps the most widely publicized case of food shortage concerns giant pandas (*Ailuropoda melanoleuca*) and the depletion of bamboo in China. Pandas feed almost exclusively on bamboo (of several genera, but primarily *Fargesia* spp.) and daily consume up to 18 kg of the leaves and stems. The large volume of food is necessary because pandas, which evolved as carnivores, digest only 17 percent of the bamboo, and the animals thus spend two-thirds of the day feeding just to maintain a small surplus of calories (Schaller 1986). Curiously, the individual plants of many kinds of bamboo flower, set seed, and then die all at once. The flowering cycles have a frequency of 40 to 120 years, depending on the species, and several years of growth are required before the new generation of seedlings again becomes part of the panda diet. In former times, pandas could feed on other kinds of bamboo whenever one species died en masse. Under present conditions, however, farming has greatly reduced the once vast bamboo forests, and the remaining habitats for pandas often include only one or two species. Thus, pandas currently face critical food shortages whenever a flowering cycle occurs, and at least 138 of the endangered pandas—long the international symbol of conservation—died of starvation in the mid-1970s (Schaller 1986). In 1983, another flowering cycle killed bamboo over a wide area, and pandas again starved. Today, about 1000 pandas survive in relatively small areas—including 12 refuges—representing only a fraction of their former range. Hence, management considerations include replanting extirpated species of bamboo on existing refuges where only one species of bamboo is available and establishing new refuges where several species of bamboo are present.

PHYSICAL CONDITION AND NUTRITION

Among humans, excessive reserves of fat usually indicate poor physical condition. Indeed, athletes desire and train for a low percentage of body fat. Such a measure of fitness may be unique among humans in wealthy countries where food is plentiful and fat can be replaced easily. Few studies have measured the fitness of wild animals in an environment where food shortages and competition for food often are commonplace. Wildlife biologists generally assume that

large reserves of body fat are useful indicators of good condition and survival. Hepp et al. (1986) compared the body weights of mallards with hunting mortality. To discount the differences between large and small mallards, the weight of each mallard was divided by the wing length of the same bird—the ratio is one of the more commonly used measurements known as a *condition index*. From a total of 5610 mallards banded and measured in autumn along the Mississippi River, 234 birds were recovered in the same season or succeeding winter. The recovered mallards—those shot by hunters—often were birds with low condition indexes at the time of banding, thus indicating that mallards in poorer condition were more vulnerable to hunting than those with larger fat reserves. The data were similar for all sex and age groups. The results suggest that mallards in poorer condition may be driven from suitable feeding areas by stronger birds, thereby forcing the less-hardy birds into places where they would more likely encounter hunters.

FOOD MANAGEMENT

At first glance, it might seem that managing food resources for wildlife is a two-step process: determine the *food habits* (i.e., diet) for each species; then supply those foods. However, food management is not so simple an activity. Food habits usually are determined by examining the digestive tracts (stomach analysis) or droppings (scat analysis) of animals. There are many degrees of resolution by which food habits can be described (Table 7-3). The lowest degree of resolution consists simply of a list of foods eaten by the target population. Such lists have some value, of course, but they do not distinguish between foods eaten frequently and those taken only rarely, between items of high nutritional content and those of poor quality, and between those of high availability and those of scarcity. Various ways have been devised for describing the preferences animals may have for certain foods, each involving some ratio of components in the diet to the availability of each food in the habitat. For example, if acorns constitute only 5 percent of the seeds, browse, and other foods available in an area but make up 50 percent of the diet, then acorns represent a preferred item.

The amount and detail of food-habits information should be tied to the objectives of management. Managers evaluating habitat for the possible reintroduction of an extirpated species may need no more than a comparison of the locally available foods with the

TABLE 7-3. A Comparison of Methods of Analysis of Wildlife Foods, Progressing from Simple to Complex

Method	Advantage(s)	Disadvantage(s)
List of foods in diet	Quick	No information on relative value of different food items
Frequency of occurrence: $\frac{\text{(No. of animals eating item)}}{\text{(No. of animals examined)}}$	Easy and quick to give rough idea of foods eaten by what proportion of individuals	Little information on quantities of various items eaten No information on digestibility
Percent volume: $\frac{\text{(Volume of item)}}{\text{(Total volume of all items)}} \times 100$	Fairly quick; requires only identification, graduated cylinder, and water displacement	No information on digestibility May overrate the value of items with much water or air space
Percent weight: $\frac{\text{(Weight of item)}}{\text{(Total wt. of all items)}} \times 100$	Fairly quick, requires only identification and a balance	May overrate value of items with high water content
Percent dry weight: $\frac{\text{(Dry wt. of item)}}{\text{(Total dry wt. of all items)}} \times 100$	Can be done with identification, balance, and drying oven Eliminates water as a factor	Provides no information on digestibility of foods or on chemical makeup of food
Nutrient analysis: Percent dry weight of protein, fat, carbohydrates, and minerals	Gives information on fiber, digestible carbohydrates, protein, fats, and minerals eaten	Requires special equipment Does not reveal digestibility of foods for animal eating them
Digestibility: $\frac{\text{(Dry wt. of material eaten)} - \text{(Dry wt. of material voided)}}{\text{(Dry wt. of material eaten)}} \times 100$	Provides information on amounts of food digested and therefore relative values of food items	Requires animals to be penned Usually only one food item at a time is analyzed, losing possible effects of food mixtures
Food preference: $\frac{\text{(Proportion of item in diet)}}{\text{(Proportion of item available)}}$	Gives information on food selection	Should be combined with nutrient analysis and condition of animals to obtain most information

known dietary requirements of the animals in question. If few of the local foods match with the dietary needs of the species, then there is little likelihood that the reintroduction program will succeed. In situations where a wildlife population is not thriving, however, a detailed analysis may be required, including studies of nutritional conditions and the possibility of dietary competition with other species.

In some circumstances, food plants may be propagated for wildlife (Fig. 7-5). Plantings may range from oak (*Quercus* spp.) for the long-term production of acorns, to a planting of ryegrass (*Secale cereale*) for the short-term production of green fodder. Yoakum et al. (1980) offered several recommendations. Plantings should be limited to those species that are adapted to local climate, soils, and moisture conditions. For economic reasons, plantings should be coordinated with other land-use activities such as forest cuttings, controlled burns, road construction, or surface mining. Sharecropping arrangements often can be made in which local farmers plant and tend food patches, and then harvest part of the crop, leaving a contracted portion for wildlife. However, plots of less

than 2 ha of unharvested grain are not effective for increasing game. The usefulness of wildlife food plantings has not received thorough scientific evaluation; little data are available to demonstrate whether food plantings actually increase wildlife populations.

Seasonal changes in food requirements are sometimes overlooked in discussions of feeding ecology. As mentioned earlier, the diet of female pintails changes dramatically during the breeding season (Krapu 1974a,b; Krapu and Swanson 1975, 1977). Adult pintails are largely vegetarians for most of the year, but their nutritional needs change when developing egg follicles deplete the fat reserves of breeding females. The percentage of animal food (invertebrates) in the spring diet rapidly increases to 56 percent. Thereafter, as eggs are laid, the demand for protein increases again and invertebrates make up 77 percent of the diet. The sudden surge of protein in the diet probably replaces that which is lost during the laying period. Proteins are not stored readily in the body; thus, sudden losses must be replaced by feeding rather than by drawing from stored reserves. After the clutch is completed, invertebrates make up

Figure 7-5. Food patches may increase the availability of nutritious food for herbivores, provided that the size, distribution, and choice of crop are appropriate for local conditions and for the food habits of wildlife present in the area. Shown here is a food patch of cereal grains and legumes next to a forest edge in Ohio. (Photo courtesy of U.S.D.A. Forest Service.)

only 29 percent of the food, and the hens begin to replace their fat reserves by returning to a diet of seeds.

The protein needed by nesting pintails is by no means present year-round. The invertebrates—fairy shrimp, larvae of midges and blackflies, snails and earthworms—are available as duck food mainly after spring floods. Krapu (1974b) suggested that the timing of nesting in pintails may vary with the occurrence of flooding and the consequent availability of the high-protein foods required for egg production. Thus, nesting may be delayed in those years when flooding occurs late in the spring. In Nevada, female redhead ducks (*Aythya americana*) selectively consume the eggs of largemouth bass (*Micropterus salmoides*), thereby capitalizing on a seasonally available source of protein coincident with the birds' breeding season (Noyes 1986). The bass eggs contained nearly 64 percent protein and were available in 8–10 locations per ha. It is interesting that male redheads apparently increase the foraging abilities of their hens by distracting the bass guarding the nest.

Gullion (1966) observed that many studies of food habits are based on the contents of crops or stomachs obtained in autumn when samples are readily avail-

able from hunters. However, conclusions stemming from such studies may not provide the best information for management. Thus, because a large variety of food usually is available in the autumn—a pattern reflected in the stomach analyses—the inference might be that food is not a key factor in the regulation of wildlife populations. Food-habits studies accordingly might better focus on the time of year when food is scarce. As an example, food studies conducted during spring, summer, and autumn reveal that ruffed grouse eat almost everything available: leaves of clover (*Trifolium* spp.) and strawberry (*Fragaria virginiana*), seeds of grasses (Graminae), fruits of hawthorn (*Crataegus* spp.), chokecherry (*Prunus virginianus*), and apple (*Malus* spp.), as well as a variety of insects. With such an adaptable appetite, ruffed grouse thus appear free of food shortages. The picture of abundant food resources changes in winter, however, when the berries have disappeared, insects are no longer available, and snow covers the low-growing herbaceous vegetation. Ruffed grouse turn in winter to a diet of tree buds, especially those of aspen. But grouse do not eat just any aspen buds—the buds of the male flowers are the most nutritious. In 15 minutes of intensive feeding each evening, a grouse can obtain enough energy and nutrients from the male buds to last 24 hours. Such efficient feeding permits the minimum expenditure of energy for gathering food and greatly reduces the time grouse are exposed to predators and chilling winter winds. Hence, Gullion (1966) asserted that the availability of male aspen buds, whose abundance may vary from year to year, remains a primary factor in the survival of ruffed grouse.

In light of the numerous cases of starvation in overpopulated deer ranges cited by Leopold et al. (1947), a seemingly humane and obvious management option might focus simply on bringing food to starving animals. Trucks or planes rushing hay or grain to a snowbound herd create a dramatic scene, as does the heartwarming vision of starving fawns digging into a bale of alfalfa. Despite the attractiveness of such a remedy, however, artificial feeding presents some serious drawbacks, as was shown at Beaver Basin, Michigan.

The case at Beaver Basin, Michigan, is not unique, and the example well illustrates the biological, economic, and political difficulties generated when deer or other animals are supplied with large quantities of unnatural food (Robinson et al. 1980). In the early 1950s, a private club owning about 500 ha of winter deer range along the south shore of Lake Superior began to feed white-tailed deer. Swamps of mature cedar at the site provided good shelter but rather meager food for the few hundred deer wintering in the area. Typically, the deer fed through the winter, then left in the spring for their summer range, bore fawns, and returned to Beaver Basin in the fall. Wolves (*Canis lupus*), the natural predators of deer, were extirpated in the late 1950s, and hunting was not regulating the deer population. By the early 1970s, the club was feeding about 40 tons of commercial deer rations each winter and another 40 tons during the rest of the year. Signs of starvation were evident even while the program poured food into the bottomless pit of the ever-increasing deer herd.

In 1974, the owners of the club gave the property to the National Park Service as an addition to Pictured Rocks National Lakeshore. Along with the land came about 650 deer, which by then were heavily dependent on commercial food. A field survey showed that natural browse could support only about 150–200 deer. The program of supplemental feeding, normally discouraged by Park Service policy, was continued for two more winters; 20–23 tons of commercial rations were supplied while managers

sought a better solution for dealing with the oversized deer herd. Deep snows, windfalls, and inadequate cover prevented the success of a plan for luring the starving animals onto baited snowmobile trails leading to a nearby browsing area.

In 1976, the Park Service and the Michigan Department of Natural Resources approved a special winter hunt for removing up to 300 of the estimated 650 deer at the site. At the urging of a local hunter's group, however, the hunt was stopped by a court order issued on the grounds that the deer population had been overestimated and that the hunt, therefore, would cause irreparable damage. The deer thus were fed for another winter.

In the following two winters, 189 and 87 deer, respectively, died primarily from malnutrition. By 1980, the winter population in Beaver Basin was down to about 165 deer, and none remained in 1983. Furthermore, overbrowsing and natural succession had changed the composition of the vegetation in the deer yard. Between 1975 and 1979, the availability of white cedar diminished from 151 kg per ha to 6 kg per ha. Balsam fir (*Abies balsamea*), a species of marginal palatability for deer, increased from 3 kg per ha to 10 kg per ha in the same period (Jensen 1982).

The lesson from Beaver Basin is that artificial feeding, if not accompanied by suitable cropping of the herd, can become financially and ecologically expensive. Unfortunately, the long-term implications are seldom considered by those who initiate feeding programs designed to "save the animals." Doman and Rasmussen (1944) warned long ago that winter losses increase as deer become more dependent on supplemental foods and less on native forage. But, as at Beaver Basin, the hard facts of such a message are not always accepted. Moreover, a diet suddenly enriched by unnatural foods may disrupt the digestive physiology of starving animals. The four-chambered stomach of deer and other ruminants contains a microbial flora that aids digestion. The flora changes slowly when the diet undergoes normal alterations between seasons, but a sudden shift from rich to poor food—or even from poor to rich food—may cause illness and sometimes death (Moen 1978). Wobeser and Runge (1975) thus described severe rumenitis in white-tailed deer feeding on wheat and barley during winter. The digestible carbohydrates in the enriched diet directly or indirectly killed 9 of 108 deer that died during the winter. Deer accustomed to a diet of browse accordingly may be injured or killed when their winter diet includes grains.

Artificial feeding programs also raise philosophical questions about the degree to which the health and perpetuation of natural communities should depend on humans. Animals subsisting on a regime of easy handouts eventually may develop undesirable characteristics that may not benefit the long-term interests of the species. Among the more obvious risks is diminished fear of humans, even to the point of adopting an aggressive attitude. Artificial feeding may involve nontarget species as well. As an example, mallards adapting to handouts in parks are apparently displacing the wary population of black ducks (*Anas rubripes*) in northeastern North America (Heusmann 1974; see also Chapter 6). With game species, artificial feeding also sharpens the issues raised about the ethics of sport hunting. Hunters breed, shelter, and feed wildlife with increasing frequency. Although such practices are often legal, intensive feeding programs may enter the gray area between wildlife management and game farming.

Insuring an adequate supply of food for deer and other herbivorous wildlife on a sustained basis is not an easy task. Nutritional requirements must be known; the rate of production must be determined as well for the plants forming the diet. The number of animals involved also is of obvious importance, but useful population data are not obtained easily. Instead, managers often judge the condition of the vegetation as a barometer of wildlife populations in relation to the food supplies. Forage availability and consumption normally are measured for such purposes (Aldous 1941; Beals et al. 1960). A browse line offers a clear illustration of these relationships: the higher the line, the heavier the pressure for food. Managers also assess the fat reserves of the animals as an indirect but useful indication of forage conditions. Fat deposits around the kidneys and in the bone marrow are examined commonly for this purpose. Thus, with the combined information from habitat and animals, managers can reasonably determine if the population is too large for the available food supply. If the population is too numerous, then the following courses of action are available:

1. *Do nothing.* Disadvantages to this course of action frequently include starvation and long-term damage to the vegetation and perhaps to other parts of the habitat (e.g., soil erosion). The public image of wildlife management very likely will be damaged as well.

2. *Reduce the animal population by (a) increased hunting or (b) introduction of predators.* Liberalized hunting regulations—usually involving an increase in the bag limit—above and beyond the normal restrictions often bring forth outcries from the public. In some cases, public criticism emerges when females are added to the bag as a means of reducing the size of wildlife populations. Introduction of predators also may cause negative reactions from stock owners and some hunters. Increased hunting offers the advantages of adding recreational opportunities for humans while reducing the level of suffering from starvation for the animals. The advantages of adding predators include the restoration of a somewhat more natural system and the enhancement of aesthetics within the wildlife community.

3. *Live-trapping, removal, and relocation.* This method usually gains public approval and improves the visibility of wildlife management. The disadvantages, however, include high operating costs and locating an area where animals can be released without generating another crowding problem. Moreover, this remedy usually offers only a temporary solution.

4. *Artificial feeding.* Whereas this response has broad public appeal, feeding programs also encourage further expansion of the population, thereby creating ever-increasing demands for both artificial and natural foods. Artificial feeding may be justified in cases where a food shortage is temporary or where hunting later will curtail continued growth of the population.

5. *Habitat modification.* This option increases production of natural foods, usually by setting back succession or by planting self-perpetuating vegetation. Slow response time and the lack of public enthusiasm are among the disadvantages, but cost effectiveness and permanence—relative to artificial feeding—are overriding advantages.

COVER

Earlier, we proposed that cover prevents the wastage of energy by protecting animals from adverse weather (shelter) or from predators and other enemies (concealment). (From the standpoint of survival of an individual animal, the transfer of its own energy into the body of a predator may be regarded as waste.) In some cases, cover also may help predators obtain food, as when a bass hides in aquatic vegetation waiting for a passing minnow.

Giles (1978) and Dasmann (1981) correctly noted that the term *cover* has been used loosely in wildlife literature. *Good woodcock cover* usually implies the entire habitat, including feeding grounds (some of which may have very little vegetation for concealment or shelter), resting areas, and breeding habitat that may consist of very densely vegetated terrain. In plant ecology, the term *cover* refers to the percentage of ground concealed by a vertical projection of the overlying vegetation. Here, however, the term will refer to any physical or biological features or arrangements of features that provide shelter from weather or concealment from or for predators. Note also that cover requirements may vary not only with function but also with time (i.e., daily or seasonally). Burrows in soft snow can provide nocturnal cover for grouse; a mature conifer swamp is winter cover for deer; and a rocky ledge overhanging a game trail provides hunting cover for a mountain lion (*Felis concolor*).

COVER AS SHELTER

The body temperatures of birds and mammals are regulated internally (i.e., they are *homeothermic* animals), and their activities are largely independent of ambient temperatures. Nonetheless, homeotherms often require cover as means of preventing excessive buildup or loss of heat. Conversely, fishes, reptiles, and amphibians—*poikilothermic* animals—lack internal heat-producing mechanisms, and their rates of activity are more dependent on ambient temperatures

than are those of birds and mammals. Reptiles, for example, may maintain a suitable body temperature by moving in and out of the shade.

Unfortunately, the function of shelter as a component of cover has not been widely studied. The wind-chill index developed by the U.S. Armed Forces and now widely used by weather reporters (partly to exaggerate how cold it is) illustrates the benefits of shelter from wind (Table 7-4). A wind of 32 km per hr at a temperature of 0°C produces an effect on exposed flesh of −14°C in a calm. Thus, with such a wind, an animal finding shelter in which the wind is 0 km per hr effectively raises its environmental temperature 14°C. A coat of fur or feathers (or blubber, in the case of seals and whales) reduces the effects of cold temperatures and wind, and the insulating capacity of these coverings may define the limits of thermal stress and, in turn, may limit the geographical distribution of many species.

Robinson (1960) tested the effects of removing shelter on the survival of white-tailed deer fawns. The physical condition of one group of fawns, kept in a 0.4-ha enclosure that had been clear-cut, was compared with the condition of fawns kept in two similar-size pens, one with moderate cover and the other with a dense canopy of conifers. Available natural foods were removed from each pen, and all fawns were fed submaintenance levels of commercial deer food. The moderate and dense canopies measurably reduced wind and retarded heat loss at night, producing warmer temperatures than in the clear-cut pen. The deer, however, demonstrated no significant differences in physical condition, primarily because

TABLE 7-4. Temperature Equivalents Caused by Wind Effects

Wind Speed (km/h)	Actual Thermometer Reading (°C)									
	10	*5*	*0*	*−5*	*−10*	*−15*	*−20*	*−25*	*−30*	*−40*
0	10	5	0	−5	−10	−15	−20	−25	−30	−40
8	9	4	−1	−7	−12	−17	−23	−28	−34	−44
16	4	−2	−8	−14	−20	−27	−33	−39	−45	−57
24	2	−5	−11	−18	−25	−32	−38	−45	−52	−65
32	0	−7	−14	−21	−28	−36	−43	−50	−57	−71
40	−1	−9	−16	−24	−31	−39	−46	−54	−61	−76
48	−2	−10	−17	−25	−32	−40	−48	−55	−63	−78
56	−3	−11	−19	−26	−34	−42	−50	−58	−65	−81
64	−3	−11	−19	−27	−35	−43	−51	−59	−66	−82
64+	Little additional effect									
Danger for humans:	Little danger for properly clothed person			Moderate danger				Great danger		

Source: Department of the Air Force (1974).

those in the clear-cut pen found shelter among the remaining small trees where temperature and wind conditions were comparable to those in the more densely vegetated enclosures. In nature, however, deer must move about as they forage for food, and those in a clear-cut area thus would have been severely affected by the shortage of shelter.

In the northern parts of their range, white-tailed deer traditionally seek conifer swamps for winter cover, sometimes migrating up to 50 km from their summer range to reach such habitat. Given a choice between available cover and available food, the deer choose cover (Verme 1965b). Such behavior suggests that food does not compensate for the loss of energy occurring in poor shelter because adequate winter browse may be gained only when deer wade through deep snow while being exposed to high winds and predators. During midwinter, the metabolism of white-tailed deer slows and they enter a sort of semi-hibernation period (Moen 1978). Deer thus spend less time moving about in winter and reduce their intake of food, even when abundant food is available. Hence, for deer, the quality of shelter may be a key factor in the conservation of energy. The best winter cover for white-tailed deer consists of a closed canopy of balsam fir and mature spruces (*Picea* spp.). The physical characteristics of such stands reduce wind chill and intercept falling snow (Fig. 7-6).

Birds and mammals may find shelter among others in their flock or herd. The value of huddling is nowhere more pronounced than among emperor penguins (*Aptenodytes forsteri*), which breed in one of the most severe climates on Earth. At the approach of winter, emperor penguins walk 50–100 km across sea ice to the permanent ice shelf where a breeding rookery is established. Each female penguin lays a single egg, which is cradled on the feet of the male. She then returns to the sea to feed, leaving incubation duties to the male for a 60-day period in a regime of high winds and −30 to −40°C temperatures. Counting travel time to and from the rookery, male penguins may not feed for up to 100 days; their weight accordingly diminishes from about 35 kg to about 20 kg (LeMaho et al. 1976; Pinshow et al. 1976). Based on laboratory studies, the energy used walking to and from the rookery costs the male pen-

1. Dense stands of even-age swamp conifers.

2. All-age mixed swamp conifers.

3. Even-age swamp conifer-hardwood stands.

4. Two-storied cut-over timber.

Figure 7-6. Comparison of the effects of dense conifer cover (1) with decreasing amounts of crown cover (2–4) in reducing cold wind and snow depth. (From Verme 1965b.)

guin only about 1.5 kg, but exposure to cold for a lone penguin during the long fasting period expends about 25 kg—a loss (70 percent) that would be fatal for a 35 kg penguin. However, under similar conditions, huddling penguins each use only about half the energy expended by a lone penguin, and the weight loss per bird thus is reduced to an acceptable level (15 kg, or about 40 percent). The difference, then, between life and death for incubating male penguins lies in their huddling behavior. Moreover, a reduction in population size would force a larger proportion of penguins to the outside of the huddle, thereby taking a greater toll on the fat reserves of birds on the periphery. Thus, for management purposes, populations of emperor penguins probably cannot be maintained in small numbers.

COVER AS CONCEALMENT

The value of cover as a hiding place for wildlife is more often studied than is its value as shelter. Within their range, snowshoe hares are the favorite prey of almost every medium-to-large carnivore and undoubtedly can use all the concealment available. In Utah, Wolfe et al. (1982) found a strong correlation between the horizontal density of cover up to 2.5 m high and the abundance of snowshoe hare droppings, which provided an index of the hare population. Horizontal density of cover is measured with a *cover board,* a board marked into appropriate sections and held vertically at a sample point (Fig. 7-7). An observer stands at a stipulated distance from the board and records the proportion of the board obscured by vegetation. In the Utah study, hares chose a cover density of at least 40 percent in the first 1.5 m above the snow level. Such cover occurred in subalpine fir (*Abies lasiocarpa*), Engelmann spruce (*Picea engelmanni*), and aspen-conifer edges. By contrast, the cover requirements of hares did not occur in mature Douglas fir (*Pseudotsuga menziesii*) and pure stands of aspen (*Populus tremuloides*). At the peak of the population cycle in Minnesota, snowshoe hares occupied some cover types affording less protection than those used at low densities (Fuller and Heisey 1986). When hare numbers are high, competition for good cover apparently forces some individuals into areas where the risk of predation is increased.

The cover requirements for ruffed grouse are now well known, largely from the studies of Gullion (1966, 1972) in Minnesota. The dense regrowth of small stems found in regenerating stands of young aspen of-

Figure 7-7. A cover board is used to determine the density of cover, shown here at Moosehorn National Wildlife Refuge in Maine. The proportion of the board obscured at various heights is recorded as a quantitative measure of concealment. (Photo courtesy of U.S. Fish and Wildlife Service.)

fers both young and adult grouse ideal protection from predators. In such stands, the low undergrowth is usually sparse, and thus the cryptically plumaged grouse can see approaching predators. Moreover, the thicket of aspen stems severely restricts the maneuverability of either avian or mammalian predators. Other plants such as alder (*Alnus* spp.) can provide similar physical protection for grouse, but the leaves and buds of mature aspen offer the added benefit of high-quality food. As aspen trees mature, however, the stands become less dense and offer progressively poorer cover for the birds. Gullion (1972) suggested that good ruffed grouse habitat should include three age classes of aspen within a distance of a few hundred meters, the area usually occupied by a bird. The youngest age class may be too immature to support grouse but serves as a replacement stand; a middle-age class (8–20 years old) provides good cover; and an older class (20–40 years) produces winter food. Tall conifers should be discouraged because of their potential for hiding hawks and owls. Overall, aspen cover does not hide grouse from predators. Instead, the cover helps grouse see predators and impedes the access of predators.

Cultivated plants are an essential component in the cover requirements for ring-necked pheasants. In Wisconsin, virtually no pheasants occurred in counties with less than 20 percent cultivation; pheasant densities increased in counties where up to 55–70 percent of the land was plowed each year, but diminished at higher levels of cultivation (Wagner et al. 1965). Such data clearly define the range of cultivation providing optimum habitat for pheasants in Wisconsin. In those counties with more than 55 percent of the land in cultivation, pheasants were more abundant where wetlands were also more abundant (see Fig. 11-5, Chapter 11). The wetlands provide a permanent grassy and brushy type of cover where pheasants find protection from both severe weather and predators.

In South Dakota, Hanson and Progulske (1973) recorded the cover types used by radio-tracked pheasant hens and broods during two summers and autumns. Available cover included corn, small grain, residual cover, pasture, summer fallow, alfalfa, shelterbelts, ditches, and spoil pits. Although alfalfa fields constituted only about 5 percent of the area, more than 13 percent of the pheasant locations occurred in alfalfa. Ditches, small grain fields, and shelterbelts also were used more frequently than their percentage of occurrence in the area, while the least-frequented cover types were pastures, summer fallow, and spoil pits. Cornfields were used only slightly more commonly than expected by random movements.

A partial explanation for declines in pheasant populations in the recent decades is unfolding in Illinois (Warner 1979, 1981, 1984). Crops were rotated in the World War II era, and approximately 80 percent of the pheasant chicks survived from hatching to early autumn. By the 1970s, however, the survival rate in summer had declined to about 50–55 percent (Warner et al. 1984). The poor survival rate involves an interaction of food, cover, agricultural practices, and herbicides.

Pheasant chicks require a high-protein diet during the first several weeks of life. Whitmore et al. (1986) determined that 2-week-old pheasant chicks require about 967 to 2676 mg of food per day—a need the young birds meet almost exclusively with insects 3 mm or larger in size. Fields of oats and other small grains, along with forage legumes such as sweet-clover (*Melilotus officinalis*), supported high densities of the larger-size insects. Moreover, such fields were well dispersed in rural landscapes. These cover types, along with weeds in and bordering the fields, hosted an abundant and diverse insect fauna for foraging broods (Hill 1976; Warner 1979; Whitmore et al. 1986). After World War II, however, production of row crops—mostly corn and soybeans—and the associated usage of herbicides gradually expanded over pheasant ranges in the Midwest. Prime habitat for pheasant broods thereby diminished as the newer farming regime replaced the old. In other words, because they now were largely weed-free and thus contained fewer insects, row-crop monocultures only partially provided the dietary needs of young pheasants.

Although more research is needed before the exact mechanisms can be identified, some interactions between increased brood mortality and intensive row-cropping methods are now clear. On farms with small grains and/or hay, Warner (1984) determined that the field-to-field movements of hen pheasants and their broods were relatively restricted during the time the young birds foraged for insects. Conversely, broods moved much more widely in habitats dominated by row crops, apparently because attractive foraging cover was scarce. Hence, the more extensive movements of pheasant broods in corn and soybean fields heightens energy demands and increases the probability that broods might encounter predators or other causes of mortality.

Nonetheless, some management opportunities for enhancing the survival of young pheasants may be possible even though there may be no substitute for the interspersion of small grains and forage legumes in the Midwest. Since 1983, however, some improvement in brood survival has occurred on those small tracts of cover established in conjunction with farm programs that require reduced production of row crops (R. E. Warner, personal communication). Curtailed or reduced use of pesticides may help restore an insect fauna that meets the dietary needs of pheasant broods. For example, farming practices that integrate mulching may provide adequate foraging habitat if pesticides are not used. Buffer strips of small grains or a mixture of grasses and legumes planted along riparian zones also may provide brood habitat and help reduce soil erosion.

Similar circumstances befall gray partridge (*Perdix perdix*), once the most abundant and widespread game bird in Britain. In fields where pesticides kill much of the insect population, partridge broods moved an average of 102 m between roost sites (an index of day-to-day travels for food) and accordingly experienced 40 percent mortality. In fields where a 6-m-wide strip around the edge was left unsprayed, however, the broods moved only 43 m between roost sites and experienced just 2 percent mortality. Rands and Sotherton (1985) thus concluded that chicks with access to the small, but critical, unsprayed strips could find adequate insect foods with less travel and consequently showed much higher survival rates. Moreover, the unsprayed strip did not affect overall crop yields.

Some states offer economic incentives for farmers to leave strips of cover for wildlife along roadsides. Pheasant hunters in some states also are assessed extra license fees for purchasing vacant lands, which then are managed for pheasant cover and public hunting. Additionally, the U.S. Government administers the Conservation Reserve Program as part of the Food Security Act of 1985. The measure encourages farmers to leave blocks of uncultivated land, which should improve food and cover conditions for pheasants and other species (see Chapter 13).

Unlike an animal that might flee from poor cover, a nest of eggs lacks such an option when a predator approaches. Thus, for most birds the quality of nesting cover is of paramount concern. In some birds, such as shorebirds (Scolopacidae) and nightjars (Caprimulgidae), the nest is placed on the ground virtually without cover. A suitable background is chosen by the female, however, so that her speckled eggs are visible only to the keenest eye. In addition, the incubating female, also cryptically colored, usually orients herself toward the sun so that her shadow is minimized and visual blending with the background is almost complete.

Other birds, including most waterfowl, lay eggs that are not cryptically colored. Nests are partially concealed by the camouflaged plumage of the incubating female and, when she is away, by a covering of down. Nonetheless, the vegetation surrounding the nest is the primary source of concealment for most waterfowl nests, and the influence of nest cover on nest success has been studied at length. Results based strictly on the density of vegetative cover have not been consistent. Of 88 duck nests studied in South Dakota, predators claimed 29 percent in heavy cover, whereas 50 and 55 percent, respectively, were destroyed in medium and light cover (Schranck

1972). Similar results occurred when simulated nests of chicken eggs were placed in each type of cover. The results suggest that the density of cover was of greater influence on restricting the movements of mammalian predators than on obscuring their vision. In another study, Dwernychuk and Boag (1972) observed the fate of a large number of natural duck nests and simulated nests associated with several avian predators: gulls (*Larus delawarensis* and *L. californicus*), crows (*Corvus brachyrhynchos*), and magpies (*Pica pica*). No mammals were present on the islands where the study was conducted. Nest cover was classified as (1) completely visible; (2) more than half visible; (3) less than half visible; and (4) completely invisible. These are the results, measured by percentage of nest success in each cover type:

Nest type	Nest success (percent) by cover type			
	1	2	3	4
Natural	80	61	78	75
Simulated	0	41	44	66

There was no relationship between concealing cover and nest success in the natural nests, but there was in the simulated nests. In this case, human disturbance may have attracted predators to the natural nests, thereby negating any influences of the protective cover. Lacking human disturbance, however, avian predators probably claim few nests except when the eggs are partially exposed and lack the screening protection of overhead vegetation. Under natural conditions, incubating hens place a protective layer of down over their clutches before leaving their nests. Otherwise, duck eggs are exposed only when an incubating hen is suddenly flushed from her nest or before incubation begins (i.e., during the laying period). Some workers (e.g., Mendall 1958) carefully replaced disturbed vegetation near duck nests under study and found no indications of higher rates of predation in comparison with nests that were not subject to human disturbances.

Besides the effectiveness of density, cover also may be improved if it consists of a mixture of several elements rather than one or a few kinds of vegetation. Bowman and Harris (1980) released hungry raccoons (*Procyon lotor*) that were trained to eat bobwhite eggs in a pen containing a randomly placed quail nest. The density of the cover inside the pen was varied using several kinds of plants found in southeastern forests, for example, slash pine (*Pinus elliotti*) and bracken ferns (*Pteridium aquilinum*). Based on tests with three different densities and mixtures of cover, nests survived longer in heavier mixed cover than in lighter cover. The results suggested that a mixture of cover plants in nesting habitat may be an effective deterrent to nest predation.

Snow is another cover type of importance to many kinds of wildlife (see also Chapter 11). For some species, snow is a salvation, for others a detriment. Many northern animals take advantage of snow cover for both concealment and shelter. Indeed, some species have evolved remarkable adaptations for living in snowy environments (e.g., white winter pelage in ermines, *Mustela erminea*). Formozov (1946) provided an encyclopedic review of the ecology of snow, which includes numerous examples of the benefits snow offers wildlife.

Because air is trapped in the crystalline structure of snowflakes, snowfall forms an insulating blanket over the ground. The insulative capacity, however, varies inversely with the density of the snowpack, and so the thermal protection of snow varies accordingly. Typical temperature profiles for January snow in Michigan are shown in Figure 7-8. Under more than 20 cm of snow, temperatures near the ground are seldom lower than a degree or two below freezing when, at the surface of the snowpack, the ambient temperatures may be much colder.

Red-backed voles (*Clethrionomys gapperi*), deer mice (*Peromyscus maniculatus*), shrews (*Sorex* spp.), and other small mammals spend most of the winter in snow tunnels along the surface of the ground and seldom venture to the top of the snow. In this subnivean world, these mammals find safety from predators as well as nearly constant and moderate temperatures. Trips to the surface of the snowpack are necessary only on occasion, usually to find seeds or to climb over hard-packed snow such as occurs on a snowmobile trail. In winters with deep snow, small mammals continue breeding throughout the coldest months, whereas breeding does not occur in snow-free winters (Formozov 1946; Kott and Robinson 1963). In Eastern Europe, Formozov (1946) reported that voles thrived, whereas owls perished in a regime of deep snow during the winter of 1928–29.

Although few birds seek shelter underground, several species of grouse roost at night in snow; these include sharptail (*Tympanuchus phasianellus*), blue (*Dendragapus obscurus*), black (*Lyrurus tetrix*), ruffed, and spruce grouse. For these species, concealment from predators may rank equally with the insulating value of snow. Blohkin (*in* Formozov 1946) attrib-

uted a die-off of black grouse in Russia to a cold, snowless winter. Also, when snow crust prevents ruffed grouse from burrowing, the resulting thermal stress reduces the fat reserves of the grouse (Gullion 1970). In the following spring, the eggs of females surviving such conditions may produce chicks with reduced vitality.

Great gray owls (*Strix nebulosa*) are among the few predators that can prey on snow-burrowing rodents. These northern raptors apparently can hear mice moving beneath the snow and can pluck unsuspecting rodents from their tunnels without relying on visual cues (Master 1979). Conversely, snow interferes with the hunting patterns of other owls, and deep snow can hinder the movements of some larger mammals. Coyotes (*Canis latrans*) can be tracked and captured by humans when the snow is deep and soft (Ozoga and Harger 1966). Deep snows, along with cold temperatures, confine white-tailed deer into sheltered areas (deer yards). Nelson and Mech (1986) determined that wolves capture significantly more deer during winters with deep snow than in winters with shallow snow. Managers use the depth and compaction of snow, along with wind and temperature measurements, for evaluating the severity of winter on white-tailed deer (Verme 1968).

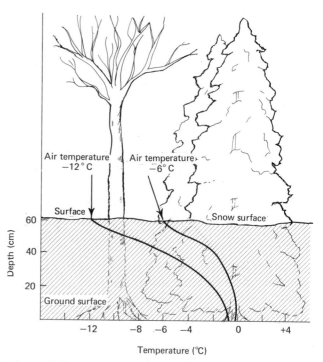

Figure 7-8. Typical temperature profiles in 60 cm of snow cover in January in Michigan.

EDGES AND EDGE EFFECT

Hunters and bird-watchers often search the places where forest meets field and where water meets shore. Many animals frequent such places more commonly than they do the interiors of forest or field and the open waters of oceans or lakes. The apparent increase in abundance of wildlife at places where two habitat types meet is known as *edge effect.* In his description of edge effect, Leopold (1933a) noted that the exact reasons for the increase of animal life near edges were not known, but he speculated that two factors were involved. These were (1) the attraction of simultaneous access to more than one environmental need, such as food and shelter, and (2) a greater variety of vegetation, including species common to each of the adjacent habitats plus some plants characteristic of only the edge. The origin of an edge may be *inherent* (natural) or *induced* (caused by humans or created by short-term phenomena). Inherent edges may be caused by abrupt changes in soil or changes in topography. In eastern North America, an inherent edge between deciduous forest and coniferous forest frequently occurs where uplands meet lowlands. Similarly, inherent edges are found in the Great Plains between grasslands and the woodlands that border rivers. Induced edges may occur where floods or fires have reversed succession or where humans have created a patchwork of tilled land interspersed with fencerows and woodlots (Thomas et al. 1979a). Aerial photographs are among the tools available for quantifying edges in assessments of wildlife habitat (Brooks and Scott 1983).

Lay (1938) was the first to test the hypothesis that the numbers of animals and species of animals are greater at an edge between a forest and a clearing than in either the clearing or the forest alone. As is true of many subsequent investigations of edges and wildlife, the study focused on birds. The result was that edges of clearings in eastern Texas averaged 16.5 birds of 6.5 species, whereas the birdlife in the forest interior averaged 8.5 individuals of 4.6 species. Other studies obtained similar results, but with some refinements. Strelke and Dickinson (1984) found a significantly higher number and diversity of breeding birds only in the first 25 m of woodlands adjacent to clearcuts, but no corresponding increase of birds on the clear-cut side of the induced edge. Gates and Gysel (1978) found that nest densities were distinctly higher on both sides of a forest-field edge compared to the densities farther into each habitat. However, nest parasitism by brown-headed cowbirds (*Molothrus ater*)

and predation also were significantly higher in nests along the edge—factors that lowered the reproductive success of birds nesting at the edge. Kroodsma (1984), studying five species of birds breeding along powerline corridors in Tennessee woodlands, determined that edge effect was more pronounced in those areas containing patches of blackberry (*Rubus* spp.). The results indicated that the attraction of birds to a single species—blackberry, in this case—was characteristic of disturbed environments. Narrow corridors showed less edge effect than wide corridors. In addition, edge effect diminished as the frequency of saplings more than 2 m tall increased within the corridors. The latter served as perches for male songbirds. As the saplings became more prevalent in the corridor, more birds perched in the interior of the corridors. Hence, the influx of birds into the corridor diminished the ratio of birds on the edge to birds in the interior of the corridor. Some reports of edge effect thus may result from the presence of singing posts near the forest edge and not necessarily from an actual increase in the densities of breeding birds; that is, birds are counted on their singing posts at edges, but their territories may extend well into the interiors of either wooded or cleared areas.

Studies of edge effect on deer have been conducted by several researchers (Willms 1971; Scott 1982; Hanley 1983). Williamson and Hirth (1985) studied browsing by white-tailed deer in clear-cuts adjacent to hardwood forests in Vermont. Clearcutting frequently stimulates abundant regeneration of woody sprouts of value as deer browse. The study determined that the distribution of various browse plants was homogeneous throughout the clear-cut areas, but that deer preferred feeding near the edge of the forest cover. The exception to that situation arose where two species of highly preferred browse—pincherry (*Prunus pennsylvanica*) and red maple (*Acer rubrum*)—were sought farther out into the openings than other species. Deer, in fact, moved as far as 100 m into the open areas for pincherry and red maple browse, perhaps because such species may provide a level of nourishment sufficient to offset the additional risk of predation (as well as offsetting the added energy demands of reaching the choice browse). Large clear-cuts in which the centers are more than 100 m from shelter thus would be less useful for deer than smaller clearcuts (Williamson and Hirth 1985). In this instance, edge effect clearly presents a case where two resources—food and cover—occur in proximity instead of reflecting responses to a larger number of species near the edge.

Among waterfowl, however, the amount of interspersion of water and land strongly influences the number of breeding ducks per wetland. Most ducks are not hatched and reared on large marshes but instead are raised on small wetlands where the area of surface water required to produce one brood is much smaller. The prairie pothole region of the northern Great Plains thus is one of the greatest duck-producing areas in the world. Bennett (1938) noted that 95 percent of all duck nests are on land within 200 m of water. Large marshes have proportionately less edge than small prairie ponds. For example, a round 1-ha pond would have 19.5 ha of nesting range within 200 m of its shore and a ratio of shoreland area to water of almost 20:1. A 100-ha circular pond would have 832 ha of shoreland area within 200 m of shore for a ratio of slightly more than 8:1. In other words, 100 1-ha ponds offer more nesting habitat per area of water than a single 100-ha pond of the same shape. Irregularly shaped water areas further increase the water-to-shoreland ratio and thereby potentially increase duck production (Fig. 7-9). Furniss (1935) surveyed duck broods on potholes in Saskatchewan and found that smaller potholes produced more broods per hectare (fewer hectares per brood) than larger ones; Bennett (1938) found a similar relationship in Iowa (Table 7-5). The most efficient wetlands for duck

TABLE 7-5.   Brood Production by Size of Water Area

Size of Water Area in Hectares	Hectares per Brood	
	Saskatchewan	*Iowa*
0.2	1.1	0.7
0.4	1.6	1.5
0.8	1.6	1.7
1.2	1.5	2.1
1.6	2.1	5.3
2.4	3.2	—
3.2	3.2	3.7
4.0	4.0	4.0
4.9	4.9	—
5.1	—	5.0

Sources: Saskatchewan data, Furniss (1935); Iowa data, Bennett (1938).

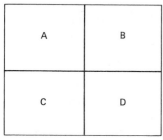

Habitat diversity with little interspersion and poor edge effect. Four plant communities (letters) meet just once in this case.

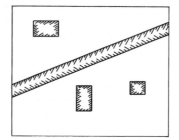

Straight ditch and regular sided patches of vegetation (shaded areas) add little edge effect.

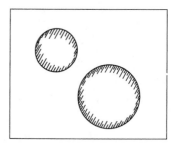

Large ponds (shaded areas) have relatively little shoreline edge per surface area of water.

Habitat diversity with good interspersion and edge effect. Four plant communities now have many more contact points, without reducing the total area occupied by each.

Meandering ditch and irregularly shaped patches greatly increase the amount of edge for wildlife use.

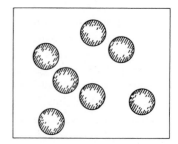

Numerous small ponds create a large proportion of edge with no loss of total surface area.

Figure 7-9. Schematic diagrams comparing poor interspersion and edge effect (top row) with greatly improved habitat conditions (bottom row).

production thus are small ponds. Unfortunately, such ponds also are the most vulnerable to drainage for agriculture and residential development.

The American woodcock (*Scolopax minor*) is a bird that uses the ground beneath small dense trees for daytime cover and feeding, small clearings for display flights, and larger clearings for night roosting. The closer such areas are to one another the more they are favored by woodcock. An interspersion of such habitats, as recommended by Sepik et al. (1981), creates considerable edge, as clear-cuts through young forest are made in narrow strips or blocks (Fig. 7-10).

It should not be assumed that edge inevitably brings about an increase in species richness or in the numbers of animals. For example, a plowed field maintained with bare topsoil amid weeds would not be productive habitat for wildlife compared with natural vegetation, even though edge would be increased along the border of the plowed area. Most studies of edge ecology have been conducted at the edges of forests or marshes using birds or, rarely, mammals as indicators of edge effect. A broader range of studies using other types of edges and other kinds of animals and plants would help bring the

Figure 7-10. This opening in the forest was created as a displaying area for male woodcocks. (Photo courtesy of Greg Sepik, U.S. Fish and Wildlife Service.)

concept of edge effect into better focus. Some birds, such as ovenbirds (*Seiurus aurocapillus*), require extensive tracts of forest and disappear when woodlots become too small (Robbins 1979). Spotted owls (*Strix occidentalis*) also depend on large unbroken stands of old growth forest. With clear-cuts, however, the owls lose parts of their foraging areas, a factor that may restrict their distribution (Fosburgh 1986). Ruffed grouse crossing the edges of good cover experience greater predation and hunting pressure than those well within good cover (Gordon Gullion, personal communication). Hunters thus search edges because the chance of seeing and shooting birds in such places is better than in the interior of good cover, where birds may in fact be more abundant but less vulnerable. A paradox of habitat management for ruffed grouse is that the desired mixture of aspen (stands of three age classes within a few hundred meters) creates edges. The amount of edge, however, can be minimized in such situations if the clear-cuts are made square or round rather than long and narrow.

Compared with food, less is known about cover as a requirement for wildlife. The properties of cover for shelter and concealment require considerably more study before management recommendations for most species can be made with confidence.

Food and cover are two of the foremost components of wildlife *habitat,* for which Hall et al. (1997) proposed a standard definition:

habitat is "the resources and conditions present in an area that produce occupancy—including survival and reproduction—by a given organism. Habitat is organism-specific; it relates the presence of a species, population, or individual (plant or animal) to an area's physical and biological characteristics. Habitat implies more than vegetation or vegetation structure: it is the sum of the specific resources that are needed by organisms....Thus, migration and dispersal corridors and the land that animals occupy during breeding and non-breeding seasons are habitat."

SUMMARY

Two of the basic essentials of wildlife habitat are food and cover. Food supplies nutrients for the substance of the body and energy for vital processes. Herbivores select plants that contain adequate digestible nutrients and a minimum of chemical and mechanical defenses. Soil provides the nutrients in plants, and the quality of wildlife food thus is directly related to soil fertility. For herbivores, the amount of protein in plants is frequently used as an indication of food quality, although some mineral deficiencies also may limit the growth and condition of grazing and browsing animals. Proteins are especially important for the growth and development of young birds and occur in a diet of insects and other invertebrates. Milk normally provides the protein needs of young mammals.

Habitat manipulations, rather than artificial feeding programs, are the best means of managing food shortages. Appropriate use of fire, timber harvest, and other techniques alter succession, thereby providing a renewable source of food for herbivores, but management also should include adequate cropping of the animal population.

Cover protects an animal from extremes of temperature and wind and provides concealment from predators or for predators. Needs for shelter vary from species to species depending on the insulation provided by feathers, fur, or blubber. Some animals overwinter in dense stands of trees; others burrow into snow; and some huddle.

Habitat management—of forests, marshes, and fields alike—for the benefit of wildlife usually aims for the interspersion of food and cover, thereby creating abundant edges between these two essential elements. Creation of edges appears beneficial for many kinds of wildlife such as waterfowl, woodcock, deer, and some songbirds, but other species may be more vulnerable to predation along edges. Species such as ovenbirds and spotted owls require large tracts of unbroken woodland habitat.

CHAPTER 8

WILDLIFE DISEASES

Disease in a wildlife population is rarely a simple, one-cause, one-effect situation. Usually it is the product of profound changes in the environment.
Lars Karstad (1971)

We shall consider a wildlife disease as a disturbance to the normal function or structure of an animal. Wildlife diseases may result from a broad array of causative agents and may be assigned to the following categories: infectious, parasitic, toxic, physiological, nutritional, congenital, and degenerative. The advent of environmental pollution highlighted the importance of toxicity. Compression of populations into restricted areas such as zoos, parks, and preserves may encourage physiological and degenerative diseases. Small, isolated populations may suffer congenital anomalies such as inbreeding infertility or other forms of inbreeding depression. Infectious, disease-spreading agents are known as *pathogens,* which include bacteria, viruses, rikettsias, parasites, and fungi (Fig. 8-1).

Instead of attempting an exacting clinical review of the many diseases affecting wildlife, we will select only a few examples that illustrate the ecological relationships—real or potential—that exist between pathogenic agents and wildlife populations. We will assess how disease-producing agents interact with habitat conditions, how some wildlife populations may be influenced, and how management sometimes may be deployed to offset disease-related adversities. In large measure, we will be concerned with *epizootiology,* the "how" and "why" of diseases in either their *enzootic* (chronic) or *epizootic* (eruptive) states in wildlife populations.

WHY STUDY WILDLIFE DISEASES?

At least four reasons compel wildlife managers to address the issue of diseases in animal populations. First, either domestic or wild animals may serve as *reservoirs* or as *vectors* for pathogens that ultimately affect each other or, indeed, humans. In 1924–25, mule deer (*Odocoileus hemionus*) in Stanislaus National Forest were slaughtered when foot-and-mouth disease ravaged livestock in California. The herd was decimated when more than 22,000 deer were shot in the months following the discovery of the disease. Of these, about 10 percent showed lesions associated with foot-and-mouth disease (Keane 1926). Similarly, even larger numbers of African wildlife were slaughtered as a means of eradicating the reservoir of parasites transmitted by tsetse flies (Jahnke 1976; see Chapter 14). Such measures are, of course, a drastic treatment for protecting livestock, but wildlife is not always the villain in such relationships. DeArment (1972) found no evidence that two serious livestock diseases, leptospirosis and brucellosis, were carried by wildlife, despite the claims of ranchers. During the same 10-year period that cattle were stricken, neither disease was detected in more than 1600 blood samples taken from the pronghorn (*Antilocapra americana*) population in the same region. The findings of that study undoubtedly saved many

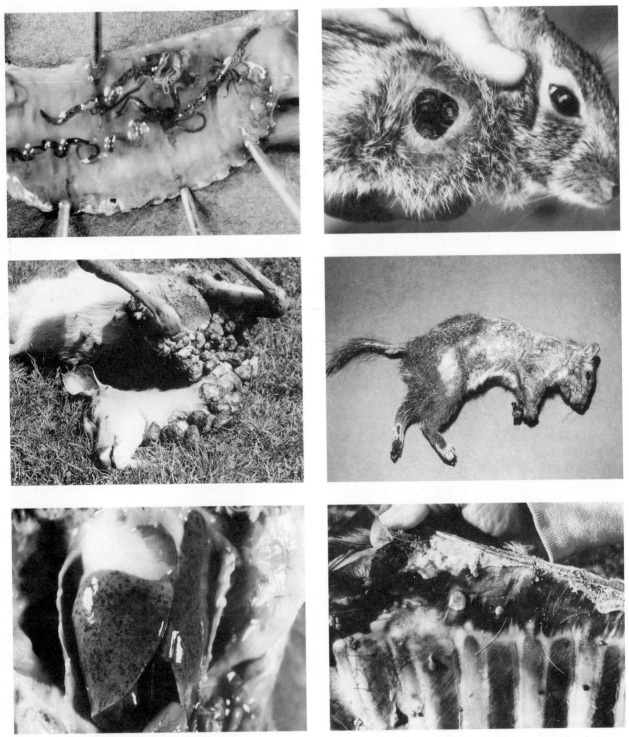

Figure 8-1.   Pathogens produce external and internal infections. Upper photos show gapeworms in the opened trachea of a pheasant (left) and botfly larvae in the skin of a cottontail (right). In the middle row, fibromas cover the body of a white-tailed deer (left), and sarcoptic mange infects the skin of a squirrel (right). Lesions on the liver of a mallard indicate duck virus hepatitis (bottom left), and those in the pleural cavity of a white-tailed deer result from bovine tuberculosis (bottom right). (Photos courtesy of M. Friend, U.S. Fish and Wildlife Service.)

hundreds of pronghorns from needless destruction. More recently, Kingscote and Bohac (1986) found no evidence of either leptospirosis or brucellosis in more than 200 pronghorns examined in Alberta and suggested that a barrier prevents the transmission of leptospirosis between pronghorns and cattle.

Another case involving brucellosis is more complex. Some ranchers in Montana suspect that bison (*Bison bison*) in Yellowstone Park act as reservoirs for brucellosis. The infectious disease presents an economic concern for the livestock industry. Infected cattle suffer abortions, and the *Brucella* bacteria can be transmitted to humans as undulant fever. Although often testing positive for antibodies, bison seldom exhibit clinical signs of brucellosis, a phenomenon that may be an evolutionary adaptation of long standing (Meagher 1973). Brucellosis entered North America with imported cattle, so the bacteria originally traveled from livestock to bison and elk. In cattle, brucellosis causes the spontaneous abortion of calves, reduced milk production, infertility, and other reproductive disorders. No effective treatment or cure exists for animals infected with brucellosis, but the disease can be prevented to some degree with a vaccine. On average, the vaccine is 65 to 75 percent effective in cattle but less so in bison.

The issue centers on those bison that move from the park to adjacent grazing lands. Ranchers feel the bison may infect their cattle with brucellosis, despite the lack of evidence to support such a claim. Transmission of the disease from bison to cattle, while possible, is less easily accomplished than from cattle to cattle (Davis 1986). After thoroughly reviewing the status of bison as reservoirs of brucellosis, McCorquodale and DiGiacomo (1985) concluded that these and other North American ungulates play little role in the transmission of the disease to cattle. Similarly, Kingscote et al. (1987) found little evidence for the exchange of several microbial diseases between elk (*Cervus elaphus canadensis*) and cattle on shared range in Alberta. However, these researchers recommended periodic monitoring because of the potential for epizootics.

The threat, whether real or imagined, of Yellowstone bison spreading brucellosis to livestock in Montana nonetheless erupted in 1996 when the summer herd in Yellowstone National Park numbered about 3500 animals. Deep snows the following winter reduced the availability of food and, with sanction from state and federal agricultural agencies, about 1000 bison were shot after they followed plowed roads and snowmobile trails out of the park and into nearby public lands and ranches in Montana. Together with other deaths from the hard winter, the killings reduced the Yellowstone bison herd to about 1300. In effect, livestock officials have taken unwelcome jurisdiction over wildlife, in this case free-ranging bison—a prominent icon for the conservation of U.S. wildlife (Keiter 1997; see also Peacock 1997 for a popular account).

Other means for controlling diseases—or the vectors of diseases—may impair the ecological state of natural systems. To control mosquitoes, for example, tidal marshes along the eastern seaboard were drained with ditches, thereby destroying much of the wetland vegetation and decreasing wildlife habitat. Among other effects was the invasion of noxious woody species into the saltmarsh communities (Bourn and Cottam 1950; Miller and Egler 1950) and decreased use of habitat by several groups of birds (Clarke et al. 1984).

A second reason for addressing wildlife diseases concerns the density of animal populations. As habitat dwindles in both quality and quantity, wildlife populations become more concentrated. Quite probably, many animals are stressed so that they are predisposed to diseases beyond former levels. Unfortunately, there are indications that efforts to manage wildlife populations sometimes heighten disease-related mortality. The severity of infectious diseases, as we shall see, may be density dependent, and when management successfully increases densities of animal populations, a greater proportion of the population becomes infected when disease strikes. For example, the consequences of building high-density waterfowl populations on intensively managed refuges in one case probably induced additional mortality. Geese and other waterfowl on a refuge in Missouri experienced a winter epizootic of avian cholera (*Pasteurella multocida*) when the birds concentrated on small, ice-free areas kept open by pumping water (Vaught et al. 1967).

Third, diseases may cause serious losses in already small populations of endangered species. In 1984, what was believed to be an insect-borne virus killed 7 of 39 whooping cranes (*Grus americana*) held in captivity by the U.S. Fish and Wildlife Service at the Patuxent Wildlife Research Center. Besides causing the immediate loss of crucial breeding stock—the dead birds represented 18 percent of the captive flock—the disease threatened the future of the captive-breeding program. Moreover, at least one immature whooping crane raised in Idaho was weakened severely by avian tuberculosis and then died from salmonellosis (Stroud et al. 1986). Because the pathogen causing avian tuberculosis, *Mycobacterium avium,* can persist in soil for months or even years, the disease may threaten the restoration program for whooping cranes described in Chapter 19. In the late 1970s, canine parvovirus (CPV) was discovered in wolves (*Canis lupus*) in Minnesota (Mech et al. 1986), and the disease persisted there and on Isle Royale in Lake Superior (Peterson and Krumenaker 1989). During the course of a 12-year study in Minnesota, wolf pups generally experienced lower survival in years when CPV was more prevalent in adults, but the population nonetheless increased during the period (Mech and Goyal 1993). These results suggested that established wolf populations apparently can withstand CPV, but the disease may limit isolated groups or those colonizing new areas, such as those in Wisconsin, Michigan, Montana, Idaho, and Washington. In 1985, an epizootic of canine distemper decimated most of the only known colony of black-footed ferrets (*Mustela nigripes*). The colony, located near Meeteetse, Wyoming, was estimated at 59 ferrets before the epizootic, but only 6 ferrets were located later in the year (Williams et al. 1988). Because black-footed ferrets may be the rarest mammal in the United States, and possibly in the world, these losses were a serious setback for the conservation of an endangered species. Price (1985), after determining the efficacy and safe use of a vaccine for avian cholera, suggested that endangered birds such as the Aleutian Canada goose (*Branta canadensis leucopareia*) might be immunized when the geese are trapped for banding. High levels of lead present in the tissues of urban rock doves (*Columba livia*) are a potential threat for peregrine falcons (*Falco peregrinis*) living in cities on the Atlantic seaboard. In cities, the endangered falcons prey heavily on rock doves ("pigeons"), which contain an average of 4.6 ppm of lead. In contrast, rock doves living elsewhere contained only 0.33 ppm of lead (DeMent et al. 1986). Rock doves themselves are relatively resistant to the toxic effects of lead and offer a means of monitoring accumulations of lead in urban settings (Dement et al. 1987). Fortunately, however, no evidence yet suggests that urban-dwelling peregrine falcons have experienced secondary poisoning from the high dietary levels of lead in rock doves.

Fourth, diseases are a part of the whole spectrum of issues facing wildlife managers; that is, diseases are just as much a part of the management puzzle as are food habits, population dynamics, and habitat requirements. In fact, diseases usually are related directly to each of these subjects. Parasites in the abomasal chamber of the rumen of white-tailed deer (*Odocoileus virginianus*) may offer managers a way of estimating the health of deer herds. The intensity of these infections theoretically varies with the density of the herd in relation to carrying capacity. Eve and Kellogg (1977) thus suggested that counts of abomasal parasites were an "early warning device" for detecting overstocked deer ranges. The usefulness of this evaluation, however, may vary between summer and winter and the method should be refined for local situations (Demarais et al. 1983). Indirectly, weather conditions, soil, water, and other environmental settings influence diseases and their epizootiology. Karstad (1971) investigated a disease that was killing large numbers of caribou (*Rangifer tarandus*) calves in Newfoundland. Up to 50 percent of the calves were dying from sizeable abscesses on their necks. No pathogen was evident as the source of the infections, despite careful examinations of the dead animals. Subsequent fieldwork determined that the abscesses were caused from wounds inflicted by lynx (*Felis lynx*). Although the calves often were defended successfully by their mothers, the attacking lynx usually were able to wound the calves before being driven off, thereby initiating an infection that later killed the young animals. The severity of the losses coincided with the population cycle of lynx, and hence means were taken to reduce the lynx population on the calving grounds during peak years. Migratory birds face a continual series of diseases year round along all parts of their north–south axis of movement, thereby underscoring the fact that disease biology and management must be addressed across state, provincial, and national lines (Friend 1984). The incidence and severity of wildlife diseases thus are not isolated matters often associated with a single time and place.

Brucellosis, as noted earlier, is a disease of economic importance in livestock, but it also may curtail reproduction in some species of big game. In cattle, *Brucella abortus* causes abortions in the latter half of pregnancy, sterility in cows, and pathological changes in the genital tracts of bulls. Because about 50 percent of the female elk in some herds in Wyoming are infected, the loss of elk calves represents a serious impediment for management (Thorne et al. 1979, 1981). However, because the infected herds in Wyoming are fed artificially in winter, the accessibility of the elk permits administration of an immunizing vaccine. In 1985, about 490 cow and calf elk—70 percent of the herd feeding at one management area—were immunized with vaccine-loaded "biobullets" shot from an airgun (Thorne and Anderson 1985).

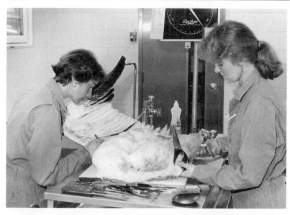

National Wildlife Health Center

A catastrophic loss of more than 40,000 waterfowl from duck plague at a national wildlife refuge in 1973 triggered creation of this facility. As a result of recommendations from a blue-ribbon committee appointed to evaluate events such as this epizootic, the center was established in Madison, Wisconsin, in January 1975. Originally operated as a facility of the U.S. Fish and Wildlife Service, the center became part of The National Biological Survey in 1993 and was transferred to the U.S. Geological Survey in 1996. The facility presently consists of two laboratory-office buildings situated on nearly 10 ha of land, much of which was restored to native prairie. The center also operates a field station in Hawaii and another in Bozeman, Montana. All told, the National Wildlife Health Center represents the largest and most comprehensive program—nationally and internationally—devoted exclusively to the diseases of free-ranging wildlife.

The center's mission is to provide the lead role in addressing wildlife health issues for species falling under the stewardship of the U.S. Department of the Interior. Specifically, the center monitors the status of health and diseases in wildlife populations, detects changes in disease patterns, determines the impact of disease on wildlife resources, and investigates disease ecology. When epizootics occur, the center also provides on-site assessments, training, and other types of assistance designed to prevent further loss of wildlife. In sum, the center provides an integrated program involving disease diagnosis, research, health assessments, field response to epizootics, and training programs to address wildlife health issues in a broad spectrum of species and ecosystems (e.g., sea turtles, bison, and migratory birds). The work which is conducted under both field and laboratory conditions entails essentially all levels of detail—from molecular to gross examinations. These efforts are collaborative and involve clients from federal, state, and private sectors.

Since its inception, the National Wildlife Health Center has conducted disease assessments for more than 70,000 specimens, which involved mortalities resulting from new types of pesticides, biological toxins, oil and mining operations, electrocution, and lead poisoning, as well as detecting the occurrence of previously unreported infectious diseases. Health issues related to the recovery of endangered species represent a major component of the center's program. The staff issues a quarterly report of wildlife mortality, which is distributed internationally. The center also published *Field Guide to Wildlife Diseases, Volume 1. General Field Procedures and Diseases of Migratory Birds,* which is being revised and will be followed by two additional volumes. Center personnel also teach a course, "Diseases of Wildlife," at the University of Wisconsin and offer instruction at other institutions. The center also provides on-site training for veterinary students and for visiting scientists from other nations, including India, Russia, Spain, Finland, and Egypt.

Some diseases are a part of natural mortality, but this factor should not preclude their identification, nor should it deter efforts to minimize the impact of diseases on wildlife populations. Other diseases, such as lead poisoning, are unnatural in the sense that they are caused by humans. In any case, pathogens striking wildlife are not only a matter of concern for animal health. For some species, pathogens also represent lost recreation, diminished aesthetics, and waste of food and fiber. A well-illustrated handbook, *Field Guide to Wildlife Diseases,* outlines general field procedures and describes the diseases of migratory birds (Friend 1989).

Finally, the public may become passionately involved with ways biologists deal with wildlife diseases.

A case in point concerns a flock of 350–400 semidomestic mallards (*Anas platyrhynchos*) and muscovy ducks (*Cairina moschata*) living on a canal in Venice, California. Residents regarded the birds as pets; hence when some of the birds developed an infection of duck virus enteritus (DVE), the citizens protested the state's decision to exterminate the flock. DVE, also known as duck plague, is unusually virulent, and state biologists feared the dreaded disease might spread to wild waterfowl in the Pacific Flyway. A single epizootic of DVE in 1973, for example, killed more than 40,000 mallards in South Dakota. Lacking any other form of treatment, the state biologists hoped to contain the disease by destroying the Venice flock. But when biologists arrived on the scene, the residents formed a human chain and blocked access to the birds. A superior court judge also issued a temporary restraining order halting the roundup. The order later was removed after representatives of public and private conservation agencies described the threat posed by leaving the birds in the Venice canal. The flock thereafter was humanely destroyed, but not before residents moved some birds from the contaminated site at Venice to wetlands elsewhere in the Central Valley, including areas near Kern National Wildlife Refuge, in effect releasing a potential time bomb. Luckily, no further epizootics of DVE have yet occurred in the Pacific Flyway, but in hindsight, a better job of public relations might have avoided some of the unfortunate aspects of this event.

PERSPECTIVES

As we have stated, some diseases are natural phenomena, and their occurrences should not necessarily be viewed with alarm. Others indeed severely affect wildlife populations (Table 8-1). Managers, to be effective as well as informed, must remain watchful for situations where action may lessen the impact of disease on wildlife populations or on those who use wildlife resources.

For example, botfly infections of gray squirrels (*Sciurus carolinensis*) can adversely affect the behavior of hunters. Larvae of the squirrel botfly (*Cuterebra emasculator*) are subcutaneous parasites that form grotesque, but nonlethal, infections in gray squirrels (Fig. 8-2). These may affect significant percentages of their host populations (see Atkeson and Givens 1951 and Parker 1968 for examples in Alabama and Virginia, respectively). Because of the influence these parasites might have on the attitudes of hunters, Jacobson et al. (1979) conducted a survey of hunters in Mississippi where botflies sometimes infect more than 50 percent of the gray squirrel population. The estimates from this survey, when extrapolated on a statewide basis, indicated that no less than 60,000 squirrels were discarded by hunters because of the unsightly infections. The survey also examined the sociological impact of parasites on hunters; among these results were that some hunters quit squirrel hunting (8 percent), others hunted elsewhere (9 percent), and some reduced the time spent hunting squirrels (26 percent). In all, more than half of the hunters reacted in some fashion to the parasitized squirrels. In light of the parasite's life history, a delayed open-

Figure 8-2. Unsightly botfly larvae caused hunters to discard large numbers of squirrels. A simple remedy was developed by wildlife managers who delayed the hunting season by one month, thereby allowing time for most of the parasites to leave their hosts before the season opened. (Photo courtesy of H. A. Jacobson, Department of Wildlife and Fisheries Sciences, Mississippi State University.)

ing of the hunting season—from September until October—diminished this problem in squirrel management. Gray squirrels generally are rid of the objectionable infections later in the autumn when the botfly larvae emerge from their hosts and pupate in the soil. In North Carolina alone, the delayed season probably saved 880,000 gray squirrels from being discarded by hunters (Jacobson et al. 1979).

Similarly, human exposure to tularemia is reduced when the opening date of the rabbit season is late enough in the year so that 10 or more nights of freezing temperatures have passed before hunting begins (Yeatter and Thompson 1952). The primary vectors of tularemia, ticks, drop from rabbits with the onset of cold weather, and the disease dissipates accordingly. A week later, most of the infected rabbits have died, leaving a relatively disease-free population available for safe hunting. In this case, the opening date becomes a management tool for lessening the hazard of tularemia to human health (Table 8–2).

Diseases also should be viewed in an ecological context. They are not phenomena isolated in nature. Several pathogens offer useful insights about the workings within and between wildlife populations and about natural selection. Predation may increase on disease-stricken populations where large numbers of weakened prey are predisposed to attack. Nearly one-third of the adult moose (*Alces alces*) killed on Isle Royale

TABLE 8-1. Selected Diseases of Wildlife[a]

Common Name	Type	Pathogen	Principal Targets	Characteristics
Avian cholera	Bacterial	*Pasteurella multocida*	Most birds, especially waterfowl	Poor coordination; hemorrhages on pericardium, necrotic foci on liver; unusually rapid mortality
Tularemia ("rabbit fever")	Bacterial	*Francisella tularensis*	Mammals, especially rodents and lagomorphs	Lethargic or spasmodic behavior; swollen lymph nodes; necrotic foci on liver and spleen; high mortality
Brucellosis ("abortion disease")	Bacterial	*Brucella* spp.	Mammals, especially caribou, bison, and other ungulates	Lameness from infected joints; scrotal enlargement in males; uterine thickening and edema in pregnant females; aborted fetuses; stunted or weakened calves; little adult mortality
Sylvatic plague	Bacterial	*Yersinia pestis*	Mammals, especially rodents	Locally chronic; enlarged spleen; hemorrhagic nodules ("buboes") in lymph system in acute cases; necrotic foci on liver, spleen, or lungs; death within 3–5 days when virulent
Duck virus enteritis ("duck plague")	Viral	*Herpesvirus*	Waterfowl	Bloody feces, cloaca, and nares; slow movements and reduced wariness; internal hemorrhaging and lesions, especially on intestines; rapid mortality often preceded by convulsions
Infectious hemotopoetic necrosis (IHNV)	Viral	*Rhabdovirus*	Fishes, especially trout	Prominent in hatcheries; anemia, lethargy; distended abdomen; hemorrhages in muscles and internal organs; high mortality in water less than 10°C; clinical signs and mortality stop in water 15°C or higher
Aspergillosis	Fungal	*Aspergillus fumigatus*	Many birds, including quail, grouse, pheasants, and waterfowl	Gasping; emaciation; plaques resembling bread mold form in bronchi, lungs, and air sacs; variable mortality
Sarcoptic mange	Parasitic arthropod	*Sarcoptes scabiei*	Mammals, especially foxes and other canines	Erratic behavior in advanced stages (wandering in open, little fear of humans, etc.); vigorous scratching; mites inflame skin, causing loss of hair; thick, crusty skin often infected with *Staphylococcus* bacteria; epizootic levels perhaps in 20-year cycles; high fatality

[a]Comprehensive treatments of these and other diseases appear in Davis et al. (1970), Davis et al. (1981), Page (1976), Wobeser (1981), and Davidson et al. (1981).

TABLE 8-2.   Relationship Between Freezing Temperatures, Opening Date of the Rabbit Hunting Season in Illinois, and the Rate of Tularemia in Humans

Days[a]	Rate of Tularemia per 100,000 Humans
After opening of rabbit season:	
30–39	14.1
20–29	14.2
10–19	11.1
0–9	4.3
Before opening of rabbit season:	
0–9	1.0
10–19	0.4

Source: Yeatter and Thompson (1952).

[a]The numbers in the first column mark periods in which, on average, the 10th night with a freezing temperature occurred for the autumn. Hence, if it took 30 to 39 days after the season opened before freezing temperatures had occurred for 10 nights (i.e., a warm autumn), tularemia was recorded in about 14 of every 100,000 humans (top line). If 10 nights of freezing temperatures had occurred 10 to 19 days before the season opened (i.e., a cooler autumn), the rate of infection dropped to 0.4 (bottom line).

suffered from a necrotic inflammation of the mandible; the disease is known as periodontitis or "lumpy jaw" (Fig. 8-3). This and other debilitating conditions, including heavy tick infections, predisposed moose to wolf predation (Mech 1966). Other studies of big game also linked the frequency of predation with disease-related settings (Murie 1944; Crisler 1956). Cheatum (1951) discovered a correlation between lungworm infections and "winter kill" in deer dying during severe winters. Malnutrition predisposed the deer to a weakened condition whereby lungworm-induced pneumonia eventually proved fatal.

Moose are particularly susceptible to tapeworm infections; as intermediate hosts, moose are infected when eggs of proglottids dropped in the feces of carnivores are consumed along with forage or drinking water. These develop into golfball-size cysts in the lungs of moose, and the infection can reach 68 percent of the herd (Ritcey and Edwards 1958). Completion of the parasite's life cycle depends on ingestion of the cysts by a carnivore; thereafter, the larvae mature in the carnivore's intestinal tract and the cycle is renewed. Because the infections in moose are initiated by feeding and drinking, older moose have heavier infections than younger animals. Thus, older moose are steadily debilitated by ever-increasing lung infections and become more susceptible to predation. As suggested by Mech (1966), this situation raises an interesting philosophical relationship between parasitic infections and predation. For the parasite to continue as a biological entity in nature, it *must* mature and reproduce in carnivores. In terms of natural selection, it thus becomes beneficial for older, more heavily parasitized moose to fall victim to wolves. Indeed, the

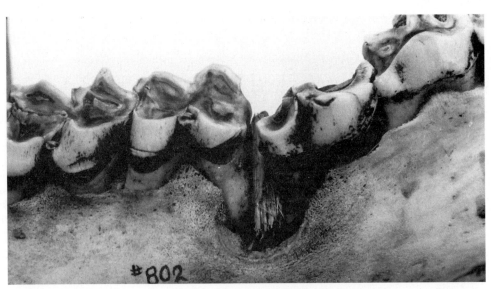

Figure 8-3.   Close-up of a mandibular infection commonly known as lumpy jaw. Moose on Isle Royale, debilitated by this disease, often were predisposed to predation. (Photo courtesy of R. O. Peterson.)

older age classes of moose no longer exert much influence on the herd's reproductive potential and seemingly represent an "expendable" segment of an otherwise healthy population. Furthermore, the very existence of this pathogen and its obligate life cycle are evidence of predation's natural role in animal populations.

As outlined by Holmes (1982), the ecology of predation is particularly noteworthy in regard to diseases. Wolves "test" their prey for signs of vulnerability. Hence, wolves normally cull individuals having debilitating infections. Predation in this instance can reduce the impact of the pathogen (i.e., diseased prey are quickly removed, leaving a healthy population). In the absence of predation, however, sick animals are not culled and pathogens may assume a more important role in population control. Conversely, mountain lions (*Felis concolor*) ambush their prey and seem less likely to select heavily infected prey. The culling effect of mountain lions and other ambush predators thus may be less in comparison with that of wolves.

Foraging experiments have indicated that starlings (*Sturnus vulgaris*) are more likely to consume prey infected with parasites than nonparasitized prey of the same species (Moore 1983). The parasites altered the behavior of the prey in ways that predisposed them to starling predation (e.g., infected prey occurred more frequently in exposed areas and thus were more readily encountered by predators). In this case, the infected prey serve as the intermediate host in the life cycle of a parasitic worm (i.e., the parasite must pass from the intermediate host to the final host before completing the transition from egg to adult and does so when infected prey are consumed by terminal hosts). Such results have several implications. First, prey populations may consist of two types of individuals, those that are vulnerable to predation (parasitized) and those that are less so (not parasitized). Thus, predation rates for the same species may differ between sites because of variable levels of parasitism in the prey populations. Second, some species of prey may be relatively uncommon in predator diets but, because parasitized individuals are more vulnerable, a large percentage of the predator population still may become infected with parasites. Third, parasitism that increases the vulnerability of prey likely influences the diet of predators; that is, parasitized prey may dominate the diet of predators in one community, whereas elsewhere the same species of prey may be consumed less frequently if they are free of parasites. Finally, we see the evolutionary benefit to the parasite of increasing the vulnerability of its intermediate host: Parasitized prey are consumed more often by the terminal host, thus facilitating completion of the parasite's life cycle.

Cowan (1947) cautioned that examination of carcasses often is limited to skeletal remains so that the presence of many debilitating diseases in prey likely is underestimated. The viscera of prey and any pathogens contained therein seldom remain available for study after predators (and scavengers) have finished feeding. As we have noted, parasitism may increase the vulnerability of prey to predators. Conversely, parasitism increased among white-tailed deer that were freed of coyote (*Canis latrans*) predation (Kie et al. 1979). Knowledgeable biologists should correctly identify such relationships so that mortality will not be assessed in error, thus thwarting the possibility of initiating effective management responses.

Many wildlife populations experience only enzootic levels of disease when other factors hold their densities in check. With excessive densities, however, diseases may reach epizootic levels. Thus, hunting may be crucial for maintaining healthy wildlife populations. Numerous attempts to protect wildlife fully from reasonable hunting pressure only assured that other mortality factors, among them disease, came into play as regulators of population density. When left to their own workings, infectious diseases eventually are self-regulating in concert with their hosts. As epizootics peak, density-dependent mechanisms in the host population return the pathogen to enzootic levels. As mortality or resistance progresses, infectious contact diminishes within the host population and the disease dissipates accordingly.

Some management situations are inherently conducive to epizootics. None seems more susceptible than those occurring in hatcheries. With the advent of fish culture late in the 19th century, the epizootiology of fish maladies assumed new dimensions. Fishes in hatcheries are held in closed systems, thus placing virtually the entire stock in jeopardy should a pathogen appear. Furthermore, hatcheries receive stock from many sources, which offers a high risk of introducing pathogenic agents that are otherwise foreign to local environments. Unfortunately, hatchery workers once discounted disease-related mortality simply by obtaining more eggs than otherwise needed rather than by developing appropriate management techniques. Exchanges of fishes and their eggs became so commonplace that few hatcheries remained free of disease (Bowen 1970). Eastern brook trout (*Salvelinus fontinalis*) were exposed to furunculosis (*Bacterium salmonicida*) when rainbow trout (*Oncorhyncus mykiss*) were brought to eastern hatcheries for propagation and

TABLE 8-3.   Evolution of Management Responses to Wildlife Diseases

Level	Response Characteristics	Stage	Perspective
1	Recognition that disease exists; acceptance of disease as a natural event	Awareness	Fatalistic
2	(a) Desire to respond to die-offs; no action taken because none obvious (b) Generate concern to others	Concern	Frustration
3	(a) Responses to die-offs in an unplanned manner (b) Plans developed and utilized in responding to die-offs (c) Improved method developed for combatting die-offs (d) Control efforts evaluated for effectiveness	Control	Fire fighting
4	(a) Reaction procedures well organized, effective within limitations of capabilities; prevention of future outbreaks given attention (b) Short-term research carried out for disease control; long-term research initiated for disease prevention (c) Disease concepts integrated as part of routine wildlife management; integrated program of research and disease control underway	Prevention	Problem solving

Source: Friend (1981).

stocking. Major symptoms of furunculosis include blisters and ulcers penetrating deeply into body tissues, as well as hemorrhaging of the swim bladder, large intestine, and peritoneum; the bacteria collect in the spleen, liver, and kidneys and may destroy these organs (Lagler 1956a). The disease spread rapidly in hatcheries, and only with the development of sulfa drugs in the 1940s did furunculosis come under partial control. Prolonged use of sulfonamides, however, may lead to the development of drug-resistant strains of the bacteria. Even in more modern times, infectious pancreatic necrosis showed reverse movement, traveling from northeastern hatcheries to western North America. This is a viral disease that is difficult to detect in its early stages. Once the disease is established, however, fishes with infectious pancreatic necrosis typically swim in spiral corkscrew movements and thereafter experience high rates of mortality. Today, significant advances in fish culture have curbed the incidence of many diseases, but the very nature of hatchery operations still provides environments predisposed to epizootics of major proportions.

DISEASES AND HABITAT

In earlier times, the pathogens attacking wildlife populations were regarded somewhat passively by wildlife managers. Wholesale applications of remedies scarcely seemed possible even if vaccines or other treatments were available. Little could be accomplished in a wild, free-roaming population by clinical ministrations to the few individual animals that might come to hand. Diseases thus were dismissed as regrettable but unmanageable misfortunes befalling wildlife populations (i.e., a fatalistic view that diseases were "an act of God"). That attitude reflected the first of four phases in the management history of wildlife diseases (Friend 1981). Later, as human regard for diseases matured, attitudes changed and an era of concern began (Table 8-3). Managers expressed alarm when diseases killed wildlife but still lacked the knowledge to take much action. In the third phase, control measures were initiated as a means of stemming epizootics (e.g., pick up and disposal of dead and dying animals). The last phase, prevention, still lies ahead.

Experts in epizootiology now believe that habitat conditions influence the course of many wildlife diseases. Whereas pathogens are not likely to be eradicated completely, the severity and frequency of their actions on wildlife populations often may be limited by human intervention. If so, greater control over disease becomes an attainable goal of wildlife management. Despite the attractiveness of this concept, however, the epizootiology for many pathogens remains unclear, thereby dampening the full application of disease management for some species of wildlife. Nonetheless, the ecological settings for some pathogens and their victims are well enough known so that habitat manipulations sometimes can function as "wildlife medicine." In time, and with more knowledge, additional wildlife diseases undoubtedly will succumb to intense management practices.

Avian botulism has been linked with habitat conditions since the earliest report of massive waterfowl deaths in 1876 at Owens Lake, California. Kalmbach (1968) compiled a full history of the disease in wild birds. Botulism is known throughout much of the world, including Europe, South Africa, New Zealand, and Australia (Woodall 1982; Galvin et al. 1985). In the United States, botulism in Utah and California still accounts for 100,000 or more deaths of waterfowl in some years (Fig. 8-4). Among the first hypotheses about causes suggested that a mineral or organic constituent in the water was responsible for the epizootics (Clarke 1913). Later, the disease became known as *alkali poisoning,* after Wetmore (1915, 1918) concluded that water-soluble salts provided the toxic agent. Because the disease primarily occurred in western states, it also has been called *western duck disease.* Finally, the toxin produced by *Clostridium botulinum* type C was isolated from both sick ducks and mud samples, indicating that the disease was a form of food poisoning known as botulism (Giltner and Couch 1930). Kalmbach and Gunder-

son (1934) subsequently believed that dead organic matter, shallow water, high temperatures, and an alkaline environment were related to production of the toxin. These conditions, acting in combination, formed the basis of the "sludge-bed hypothesis" wherein decaying organic matter provides the medium in which the bacterium produces its toxin in the absence of dissolved oxygen.

Bell et al. (1955) offered an alternate hypothesis to the sludge-bed concept. Instead of believing that toxin production occurs directly on newly exposed or flooded mudflats, they suggested that wetland invertebrates were the prime transmitters of *botulinum* toxin to waterbirds. Known as the *microenvironment concept,* this hypothesis contends that the bacteria essentially produce their own environments in the carcasses of aquatic and semiaquatic invertebrates. Thus, when water levels recede, aquatic species of invertebrates die and the bacteria flourish on this medium rather than in soil or water. Conversely, if previously dry areas are flooded by shallow water (e.g., by wave action), then terrestrial invertebrates are killed, leading to the aftermath of toxin production. In addition, maggots thriving on the carcasses of

Figure 8-4. Epizootics of botulism claim many thousands of waterfowl annually. Shown here are carcasses of ducks that died on the mudflats of wetlands surrounding Great Salt Lake. (Photo courtesy of Utah Division of Wildlife Resources.)

decaying waterbirds also accumulate botulism toxin (but are themselves immune). In turn, the maggots are ingested by previously unaffected birds, thereby causing more deaths and continued production of maggots (Figs. 8-5 and 8-6). The maggot cycle continues until the sources of toxin are no longer present (i.e., when the invertebrate larvae mature); until the toxin breaks down, losing its virulence; or until stabilized water levels or other habitat factors limit the toxin's availability (Hunter et al. 1970).

Some epizootics of botulism may result from human activities. Malcolm (1982) noted an association between botulism and instances of ducks and other birds colliding with a power line strung over a large wetland in Montana. Collisions with the wires killed at least 4100 birds of 55 species in 17 months; of these, ducks of 14 species accounted for 44 percent of the victims. Maggots feeding on the rotting carcasses contained botulism toxin and helped perpetuate an epizootic that killed 5200 more birds, of which 71 percent were ducks. It is unlikely that the power line will be moved away from the wetland (clearly the best solution) but, at the minimum, the carcasses of the power-line casualties should be picked up frequently (see also Anderson 1978; Rusz et al. 1986; Faanes 1987). Interestingly, turkey vultures (*Cathartes aura*) are highly resistant to type C toxin; this factor presumably represents an adaptation for feeding on the carcasses of birds killed by botulism (Kalmbach 1939). Other species of scavenging birds may have similar resistance (Woodall 1982).

Methods for controlling botulism once relied on frightening birds with aircraft or fireworks, thus scaring them from areas of high risk or ongoing epizootics (Rosen and Bischoff 1953). Direct treatment of sick birds also is possible, either with injections of antitoxin or by placing these birds in holding pens with clean water. However, neither method is very practical. Better results of far wider scope are achieved when water levels are manipulated to eliminate conditions favorable for production of *botulinum* toxin.

Wherever possible, water levels should remain constant during the botulism season (usually during the autumn migration). The shallow edges of the land-water interface in wetlands are prime sites for toxin production. Unfortunately, these same areas are attractive to loafing or feeding waterfowl (Hunter et al. 1970). Where water can be controlled in separate units, water from a larger number of units can be concentrated in one or two units. Instead of all the management units suffering from unfavorable water levels—and botulism—a smaller number are

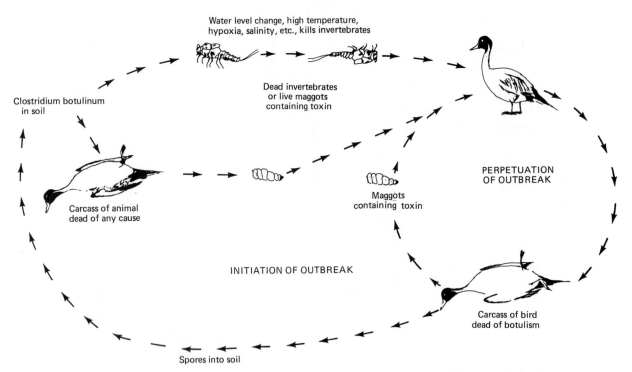

Figure 8-5.   Pathways in the epizootiology of botulism. Note the key role of fly larvae and other invertebrates in the botulism cycle. (From Wobeser 1981, redrawn by B. L. O'Kelley.)

Figure 8-6. Duck carcass floating among hundreds of maggots (upper). A cycle of botulism infections develops when the toxin-bearing maggots are ingested by healthy birds, as shown by the presence of maggots (circled) in an opened duck gizzard (lower). (Photos courtesy of W. E. Clark, California Department of Fish and Game.)

stabilized while the others are drained completely. Much can be gained if waterfowl habitats with a history of epizootic botulism are developed with water-management systems (dikes, pumps, channels, etc.) to reduce losses during the critical season. Regrettably, only a relatively small number of waterfowl habitats are adapted to this form of management; most of these are state or federal refuges, whereas unmanaged wetlands still experience uncontrolled outbreaks of botulism.

Soils are fundamental components of wildlife habitat (see Chapter 12). Relationships between soil and wildlife often are indirect, as is illustrated by certain helminth parasites and their infections of white-tailed deer and livestock sharing the same ranges. *Fascioloides magna* is a large fluke that normally infects the livers of white-tailed deer. In deer, the flukes cause little damage despite their large size and robust encapsulations (Fig. 8-7). In cattle, however, the flukes often produce significant damage and hence become a pathogen of considerable economic importance (Foreyt and Todd 1976). Deer thus act as reservoirs for flukes infecting livestock, as do elk (Kingscote et al. 1987).

Like other trematode parasites, deer-liver flukes require intermediate hosts. In Texas, *Lymnaea bulimoides* is the sole intermediate host (Olsen 1949). This snail, appropriately known as the deer-liver fluke snail, requires shallow surface water to complete its own life cycle; rainwater temporarily held at the soil's surface suffices for this purpose. A survey of snail populations revealed that *Lymnaea bulimoides*

was absent on sandy soils where surface water drained rapidly. By contrast, nearly all transects on heavy clay soils were populated with snails (Foreyt and Samuel 1979).

These circumstances translate rather directly into the rates of helminth infection in both white-tailed deer and cattle. Animals grazing on sandy soils remain largely free of the parasites, whereas those on clay soils become infected. Thus, cattle grazing in the same clay pastures as deer are constantly exposed to fluke infections. However, because the snails generally are absent on sandy soils, there is little or no passage of the fluke's larval stages to either deer or cattle in sandy environments.

In a similar vein, Prestwood and Smith (1969) linked the spotty distribution of parasitic nematodes in deer to the availability of appropriate gastropods. Deer on sandy soils with pine forests lacked infections, whereas those on soils with subclimax or climax deciduous forests were parasitized. The presence of wallowing species may exacerbate fluke infections. For example, the rooting behavior of feral hogs (*Sus scrofa*) creates depressions that hold water for considerable periods of time even in dry environments. Thus, their wallows create and enlarge areas of ideal snail habitat, indirectly increasing the potential of fluke infections among susceptible hosts (Foreyt et al. 1975).

Little management, other than treating cattle with efficacious anthelminthics, is possible on clay pastures. On sandy ranges, however, it seems unwise to develop artificial sources of water where the snails then might survive. Where windmills are necessary, water should be confined to troughs that are periodically treated with molluscicides. Fencing, too, might better follow soil types so that fluke infections remain confined to a few units of the cattle herd where intensive treatment can be administered economically.

Gross changes in land use also have increased contact between species that have differential responses to parasitic diseases. Prior to the 20th century, there was little contact between moose and deer in northern Minnesota. Thereafter, logging, homesteading, and fires created favorable habitats for deer and brought large numbers of deer into the range of moose (Karns 1967). Moose populations soon declined, probably because of the meningeal parasites carried by deer (see following section for more details).

Habitats sometimes can be "improved" to the point where diseases may be enhanced. Changes include inadvertent as well as directed alterations of wildlife environments, as shown by actions affecting nomadic

Figure 8-7. Large liver flukes infect white-tailed deer in some regions of the United States. Infected deer usually are not harmed, but the same parasite infecting cattle often is damaging and, hence, causes significant economic losses. (Photo courtesy of M. Friend, U.S. Fish and Wildlife Service.)

grazing animals in Africa. Deep pits left after roads were graveled in Etosha National Park, South West Africa, initiated epizootics of anthrax. When the gravel pits filled with water, they attracted a variety of wildlife. Additional water ("boreholes") was established to hold grazing animals in areas opened for tourists. As predicted, the permanent water concentrated the animals, but their grazing pressure degraded the grassland into weeds and bare soil. Whereas anthrax apparently always had been present in the park, the watering sites made by humans proved highly effective incubators for *Bacillus anthracis,* producing deadly epizootics among the crowded populations (Ebedes 1976). The combination of stress from overgrazing and the permanent source of infection effectively overcame the level of anthrax immunity developed naturally by zebra (*Equus burchelli*) and other species. Following this hard-learned lesson, elimination of the artificial water reduced the sources of virulent infection and initiated recovery of the nearby vegetation.

DISEASES AND POPULATIONS

The matter of disease effecting regulatory controls on wildlife populations is a question that stirs debate among biologists (e.g., Holmes 1982; May 1983). Data usually are insufficient to assess definitively the impact of pathogenic agents on populations because of understandable difficulties in sampling and because of the interactions of disease with predation, weather, competition, and the availability and quality of food. Epizootics are, of course, rather easily documented (e.g., rabies and sylvatic plague), but whether such occurrences actually regulate populations in the long run is another question. One can understand the theoretical notion that infectious diseases, as well as predation or other limiting factors, may act in a density-dependent fashion, but proving such a relationship is difficult when other variables may be acting simultaneously (see Anderson and May 1982 for a review of issues related to infectious diseases and populations).

The case for density-dependent responses of diseases frequently has been championed to explain cyclic behavior in certain wildlife populations. Periodic and often rapid declines in the numbers of red grouse (*Lagopus lagopus scoticus*) in Scotland were attributed to the parasitic nematode, *Trichostrongylus tenuis,* early in the 20th century (Lovat 1911). The larvae of the parasite, upon hatching, climb to the tips of

heather, where they are ingested by red grouse; the parasite then infects the bird's digestive tract. On reaching maturity, the parasite's eggs pass in the bird's feces back onto the heathlands. Thus, the denser the grouse population, the greater is the proportion of infected heather-tips. This density increases the number of parasites entering each bird. The intensity of infection rises as the number of grouse increases until many birds are eventually killed by their parasite loads. Thereafter, there is a reduction in both the grouse population and in the incidence of parasites in the heather. However, Lack (1954, 1966) presented evidence in rebuttal, suggesting instead that it is the number of grouse in relation to their food supply that is critical; in this view, strongylosis is lethal only to those birds already weakened by starvation.

The naturalist Ernest Thompson Seton (1929) proposed that the declining phase in the 10-year cycle of snowshoe hares (*Lepus americanus*) was precipitated by devastating "plagues," seemingly triggered by a density-dependent mechanism. Conversely, Keith et al. (1984) have found that the combination of food shortages, cold, and predation initiates declines in snowshoe hare populations. In any event, mortality rates for adult snowshoe hares change significantly during the course of the cycle. During 4 years of decrease, the rates varied between 64 and 72 percent, reaching 97 percent in the fifth year, whereas the rate dropped to 59 percent in the first year of the upward phase; results of similar magnitudes were found in the mortality rates of juveniles (Green and Evans 1940). Green and his associates (1938a; 1938b; 1939) described a nontransmissible condition, "shock disease," as the cause of wholesale deaths of hares in the 10-year cycle. These cases involved hypoglycemia with pathological manifestations focused on the spleen and liver, conditions that Keith (1963) frequently discovered in his survey of the literature dealing with snowshoe hare mortality. Chitty (1959) nonetheless doubted that shock disease occurs naturally in wild populations, and the matter rests largely unresolved even as the 10-year cycle continues in hares across North America.

As mentioned earlier, one type of pathogen sometimes predisposes its host to other diseases. Nematodes of the genus *Prostostrongylus* are transmitted prenatally as larvae in bighorn sheep (*Ovis canadensis*), maturing in the lungs of lambs within 30–45 days of their birth (Hibler et al. 1974). Infected ewes thus transfer the parasites to their unborn young when the larvae migrate through the placenta. The parasite also is transmitted when an infected animal coughs up the larvae, swallows, and introduces the parasites to the digestive tract where they are passed in the feces. Larvae survive for at least 15 months in fecal pellets (Honess 1942). Small, inconspicuous land snails of several species harbor the larvae for additional development; they reach maturity in the bighorn sheep when the snail is ingested passively with the host's forage. Lungworms thereafter foster bacterial and viral lung infections in bighorns, often leading to excessive mortality (Marsh 1938). Bighorn lambs seem particularly susceptible to these infections, especially when they are stressed by other events. Woodard et al. (1974) associated exceptionally high bighorn mortality from disease with cold, wet weather; ewe:lamb ratios dropped from 100:83 and 100:72 in June to 100:17 and 100:22 in September, respectively, in two consecutive years of study. Such losses in the juvenile age class restrict the herd's growth and create an overaged population that functions as a reservoir for future lungworm infections. Buechner (1960) more fully described the impact of these relationships on bighorn populations in what is known as the *lungworm-pneumonia complex* (see also Post 1971; Uhazy et al. 1973).

Fortunately, measures have been developed to control lungworm infections in bighorn lambs by treating their mothers. The treatment interrupts transmission of lungworm larvae from pregnant ewes to their unborn lambs. When 52 ewes were treated experimentally with one of four drugs and then were released on Pikes Peak, Colorado, 80 percent later produced offspring that survived, whereas only 5 percent of the herd at large successfully reared lambs (Schmidt et al. 1979). Based on these tests, the most effective drugs then were mixed with apple pulp and were set out at feeding stations. The results were no less dramatic than those which had occurred with the experimental animals. Lamb survival exceeded 64 percent in the bighorn sheep herds treated with anthelminthic drugs. Managers realized, however, that they were treating only the symptoms of a larger problem: overcrowded ranges where the animals were exposed repeatedly to lungworm infections. Shortly after the Colorado herds were treated with drugs, an either-sex hunting season for bighorns was initiated to help reduce the population to levels more appropriate to each area's carrying capacity (Wishart 1978).

The magnitude of disease losses among wildlife populations is difficult to assess, but sometimes it may be of considerable proportions. The extent of nonhunting mortality still is largely unknown for many species, but diseases may be responsible for the majority of such losses. For example, Friend (1976) estimated that nonhunting losses of waterfowl may exceed by twofold the legal harvest annually, with disease alone accounting for most of the nonhunting mortality. Life tables for mourning doves (*Zenaida macroura*) suggest that overall mortality rates are about equal where the species is hunted and where it is not (Austin 1951; Hickey 1952; Winston 1954). If so, this discovery suggests the importance of diseases and other types of natural mortality in the dynamics of wildlife populations whereby hunting and other forms of cropping largely replace mortality from other causes (see Harris and Kochel 1981 for population models of compensatory mortality).

Among mourning doves, trichomoniasis is widespread in geographical occurrence, and its virulent strains may cause death within 4 days of infection. This disease is a parasitic infection caused by the protozoan, *Trichomonas gallinae,* attacking the upper digestive tract of doves and pigeons (Stabler and Herman 1951). Lesions (cankers) may develop externally on the head and neck and in the mouth, throat, and crop. The disease spreads rapidly to nestlings that are fed regurgitated "crop milk" from their infected parents. Among adults, the organism is spread when infected birds, unable to swallow food because of the lesions in their throats, drop the now-contaminated seeds only to have other birds feed on the infectious food materials (Kocan and Herman 1971). Transmission among adults drinking from common water sources also may be a factor in the disease's epizootiology (Kocan 1969).

In Alabama, a severe outbreak of trichomoniasis apparently curtailed reproduction of mourning doves as diseased adults generally lacked the gonadal development necessary for breeding; the age ratio in the affected population was altered, with juveniles constituting only 10 percent of the sample (Haugen 1952; Haugen and Keeler 1952). Interestingly, the courtship behavior of billing and mutual feeding among doves and pigeons further enhances transmission among adults. On nesting, infected birds are almost certain to pass on the parasites to their young. Trichomoniasis also has been suggested as contributing to the extinction of the passenger pigeon (*Ectopistes migratorius*) (Stabler 1954).

Unfortunately, there are no successful treatments for trichomoniasis epizootics in wild dove populations. Relatively effective treatment for penned birds is possible with some chemotherapeutic agents, including copper sulfate administered in drinking water

(Jaquette 1948). However, liver damage develops when the optimal dosage is administered to infected birds. Other treatments include the registered drugs, Enheptin and Emtryl (Stabler and Mellentin 1953; McLoughlin 1966), but these and other treatments render drinking water unpalatable unless the birds are restricted to a single source (i.e., confined in pens). A water-borne therapeutic agent nonetheless seems the best means of treatment, but until a palatable and widely applied compound is developed, epizootics of trichomoniasis may remain regulators of local or even regional dove populations.

Saumier et al. (1986) demonstrated the reduced reproductive success of American kestrels (*Falco sparverius*) that were experimentally infected with parasitic roundworms, *Trichinella pseudospiralis.* Compared with uninfected controls, infected kestrels exhibited delayed egg laying, produced fewer eggs (4.9 vs. 7.1), and experienced greater egg breakage (29 vs. 1.6 percent because infected birds did not lay their eggs in nests) and embryo mortality (40 vs. 4.7 percent). The infected birds produced an average of only 0.6 hatchlings, whereas the controls produced 2.1 offspring. The experiment clearly indicated how parasitic infections may impair the reproductive fitness of kestrels and thus may illustrate an important pathogenic factor that affects the recruitment of fledglings into raptor populations.

A parasitic nematode, *Parelaphostrongylus tenuis,* offers an interesting contrast in its effect on two species of big game. (In a confusing array of taxonomic distinctions, this parasite at times has been included in the genera *Protostrongylus* and *Pneumostrongylus.*) As with other helminth parasites, snails and other gastropods serve as intermediate hosts for the parasite's larval stages. Deer, likewise moose, are infected when they incidentally ingest larva-bearing snails along with their forage (Lankester and Anderson 1968). *P. tenuis* mature in the meninges of the brain, but in deer little damage develops, either because the worms have impaired development or because of natural immunity in these tissues (Anderson 1963; Anderson and Strelive 1967). In moose, however, these parasites severely damage the central nervous system, causing paralysis and death. Thus, the same parasite is essentially harmless in deer but in moose produces a fatal disease known as *blind staggers* or *moose sickness* (Anderson 1965; Bindernagel and Anderson 1972).

The differential pathogenicity of *P. tenuis* in deer and moose apparently plays a regulatory role where their respective populations overlap. Saunders (1973)

recorded small moose populations where infections of deer were high. Moose densities dropped to about 2 per 5 km² where *P. tenuis* infections in deer reached 60 percent. Conversely, moose densities doubled to about 10 per 5 km² where the infection rate in deer was 14 percent (Fig. 8-8). Karns (1967) associated infections in moose with high densities of deer and recommended that deer populations be kept at minimal levels—less than 5 per km²—in areas managed for moose. Gilbert (1974) also correlated deer densities with the occurrence of moose infections in Maine and suggested that hunting tends to remove diseased animals from the population, thus lowering the percentage of infected moose left in subsequent years. Reintroductions of moose into ranges where they have been extirpated should be weighed in light of current deer densities, or the restocking may fail (Severinghaus and Jackson 1970). In 1985 and 1987, 59 adult moose from Ontario were released in upper Michigan. To reduce the probability of infection by *P. tenuis,* a release site was selected where the density of white-tailed deer was relatively low (2–3 per km²). *P. tenuis* infections nonetheless killed about 25 percent of the introduced moose within 2 years. However, because natality exceeded mortality, the surviving moose population had grown

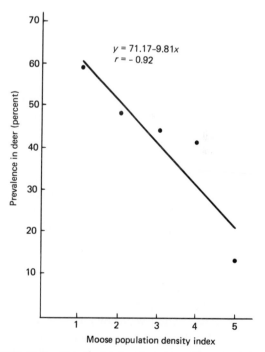

Figure 8-8. Relationship between rates of *P. tenuis* infection in white-tailed deer and moose densities. (From Saunders 1973.)

to about 200 animals by 1994 (Michigan Department of Natural Resources, file data). Moose sickness also appears to be moving westward with the extension of white-tailed deer range (Coady 1982).

Finally, the moose-deer disease relationship offers another example of ways that ecological conditions influence the epizootiology of a disease. In Nova Scotia, snow depth isolates deer from moose in winter, reducing interspecific transmission of the fatal nematodes from deer to moose (Telfer 1967). However, when conditions forced moose into winter deer habitat at lower elevations in New Brunswick, the moose died before spring, apparently from contact with deer and the parasites they transmit (Kelsall and Prescott 1971). In effect, the parasite has become a weapon of competition between deer, moose, and perhaps other large herbivores. Holmes (1982) reported that several attempts to reintroduce elk and caribou have failed in areas now occupied by white-tailed deer, apparently because of mortality associated with meningeal worms.

The malarial parasite, *Leucocytozoon simondi,* is transmitted by blackflies (Simuliidae). One species, *Simulium rugglesi,* is commonly associated with the spread of the pathogen in waterfowl. Shewell (1955) and Bennett (1960) suggested that the blackfly vectors may feed exclusively on waterfowl. Thus, infections coincide largely with the flies' geographical distribution, an area that, unfortunately, covers much of the breeding range of ducks and geese in northeastern North America. Leucocytozoonosis occurs in waterfowl of several species and age classes, but its pathogenic effects are most severe in juveniles. Indeed, losses of young birds may reach catastrophic proportions. Herman (1968) summarized studies in which mortality among ducklings reached 71 percent, and losses of goslings in epizootics occurring at 4–5-year intervals is a serious limiting factor among Canada geese (*Branta canadensis*) nesting in some settings (Herman et al. 1975). O'Roke (1934), in an early assessment of leucocytozoonosis, found that 100 percent of the black ducks (*Anas rubripes*) he examined in Michigan were infected, and Trainer et al. (1962) located the parasite in 64 percent of 6 species of waterfowl examined in Wisconsin.

Infections of *Leucocytozoon* follow a schedule that supplies blackflies with ample numbers of parasites for infecting new individuals, particularly young birds (Herman 1968). Adult waterfowl are carriers of the parasite and serve as a natural reservoir for new infections. On breeding areas where blackflies occur, the pathogen spreads throughout the waterfowl population when flies inject infected salivary fluid as they obtain a blood meal. At this point, the infection is limited to adult birds that may not show advanced parasitemia in what is known as the *prepatent* period (Fig. 8-9). When the number of parasites peaks, the *patent phase* of the schedule begins. This phase is followed by a *latent period* when it may be impossible to locate *Leucocytozoon* in blood smears from infected birds; there may be only one parasite per tens of thousands of normal blood cells, making detection of the disease unlikely at such times. However, during the latent period, a *relapse* often occurs when the level of parasitemia again increases, providing the blackflies with a second supply of parasites. The relapse phase occurs at a time when many infected birds are nesting, so the flies now are able quickly to infect ducklings. The coincident timing of the relapse with nesting was confirmed by Chernin (1952), when he exposed latent-phase ducks to a photoperiod that advanced egg laying by one month; both male and female ducks exhibited the relapse phase one month early, coinciding with their reproductive efforts under the artificial-light regime.

Blackflies largely restrict their feeding activities to sites near the shoreline and at distances less than 1.8 m above the surface (Bennett 1960; Fallis and Bennett 1966). Blackflies thus match the ecological setting of most ducks and geese for their nesting and brood cover. Clinical signs of *Leucocytozoon* infections in ducklings include listlessness, weakness, and loss of appetite; death may occur within 24 hours of infection (O'Roke 1934). Goslings infected at 2–3 weeks of age often survive, but those infected within a week of

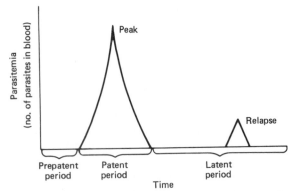

Figure 8-9. Schematic representation of parasitemia in *Leucocytozoon* infections. Relapse phase in adults often coincides with nesting, thus supplying blackfly vectors with ample parasites to infect newly hatched waterfowl. (From Herman 1968.)

hatching usually die (Herman 1968). This phenomenon underscores another example of age-related susceptibility in the ecology of diseases and animal populations. As summarized by Cook (1971), the parasites invade the lungs, heart, gizzard, and intestines, with the severest involvement occurring in the host's spleen and liver; anemia develops for the duration of the disease's patent phase. Control of *Leucocytozoon* rests on control of the blackfly vectors, but, unfortunately, there is presently no means of large-scale regulation of their population.

Introduced diseases also contributed significantly to the extinction or range restriction of endemic species of forest birds in Hawaii, particularly for the honeycreepers of the family Drepaniidae (Warner 1968). A potential reservoir for avian malaria, *Plasmodium,* undoubtedly existed in Hawaii for some time as migratory birds regularly visited the islands and poultry had been introduced as well. What was lacking for the transmission of avian malaria, however, was an appropriate insect vector. The necessary vector unfortunately was introduced in 1826 when the tropical-subtropical form of the night-flying mosquito (*Culex pipiens*) escaped from the water kegs of a ship anchored at Maui. Thereafter, the mosquitoes spread rapidly throughout the lowland

---

On The Job With...
Ronnie R. George
State Wildlife Agency Administrator

Like most administrators in state wildlife agencies, Ron rose through the ranks. However, before he began his professional career, Ron worked as a ranchhand while he completed (in 1965) a bachelor's degree in science; he then served as a jet pilot in the U.S. Air Force. Later, he enrolled in graduate school and earned (in 1972) a M.S. degree in range and wildlife science for which he completed a thesis on waterfowl parasites.

With his formal education completed, Ron worked for a short time for what is today the Natural Resources Conservation Service and then spent 9 years with the Iowa Conservation Commission, first as a wildlife management biologist, then as a wildlife research biologist. In 1981, he became program leader for migratory shore and upland game birds with the Texas Parks and Wildlife Department. In 1992, he advanced to deputy director of the Wildlife Division, the position he currently holds.

Along the way, Ron qualified as a Certified Wildlife Biologist and held several offices, including president of the Texas Chapter of The Wildlife Society. In 1990, he received the National Wildlife Health Research Center's Outstanding Publication Award for his part in a study of mycotoxin poisoning of sandhill cranes. In 1996, he was a guest lecturer in the College of Wildlife Resources, Northeast Forestry University in Harbin, China. He has authored or coauthored more than 50 publications that appeared in professional journals or popular media.

His current responsibilities involve supervision of personnel in the Wildlife Division of the Texas Parks and Wildlife Department and coordination of the wildlife research program. These duties are broad in nature and extend to the staff's work with wildlife habitat on both private and public lands. For example, he helped plan the management of a historic bison herd scheduled for release on public land the Texas High Plains. Ron also played a role in developing and implementing wildlife management plans on 10 million acres of private land in Texas.

Ron finds particular satisfaction in guiding research projects undertaken annually by the field staff, but he is often overwhelmed by the volume of information that flows into his office from all quarters. He advises students to become acquainted on-the-job with professionals and to volunteer to help with administrative and field activities. To prepare for a job, Ron also recommends that students become proficient at public speaking, writing, photography, and computers and develop a working knowledge about farming, ranching, and forestry operations. He noted that his agency often receives 200+ applications for every entry-level position, so more than just a full transcript of biology courses is usually necessary to survive the intense competition for jobs.

areas of Hawaii, breeding in brackish water along the coast and in more typical surface-water habitats. However, mosquito populations are largely unsuccessful above elevations of 600 m. Furthermore, birdpox virus ("bumblefoot") arrived with the domestic fowl of early European settlers, along with the introduction of another insect vector. The wake of these introductions brought havoc to the endemic birds of Hawaii. Today, the surviving species of honeycreepers are restricted to elevations where the insect vectors are absent.

An epizootic of rinderpest in Africa late in the 19th century represents the most infamous case of an introduced disease that produced calamitous results on native wildlife (Holmes 1982). The viral disease probably was introduced with oxen during the Italian military occupation of Ethiopia. Rinderpest killed vast numbers of wild ungulates as well as cattle and also has produced secondary effects. First, the deaths from rinderpest eliminated the reservoir for a deadly parasite carried by tsetse flies (*Glossina* spp.). The setting served as an example that led to large-scale slaughter of wildlife elsewhere in Africa (Ford 1971). Staggering numbers of wildlife were shot in futile attempts to mimic the devastation of rinderpest (i.e., remove the reservoir herds; see also Chapter 14). Second, changes wrought by rinderpest produced anomalies in the distributional patterns of wildlife that persisted for years (Spinage 1962).

Even casual relocations of animals may lead to epizootics. In the 1930s, a sturgeon (*Acipenser stellatus*) from the Caspian Sea was released in the Aral Sea. Gill parasites carried by this single fish depleted stocks of native sturgeon (*A. nudiventris*) for more than two decades afterward (Bauer and Hoffman 1976).

What appeared to be a new viral disease among white-tailed deer was confirmed in 1955 (Shope et al. 1955; Fay et al. 1956). The severity and pathological signs of the malady prompted the name *epizootic hemorrhagic disease.* Except for states in the extreme northeast and southwest, occurrence of the disease largely parallels the distribution of white-tailed deer in the United States and Canada. Other species, including pronghorns and mule deer, sometimes are infected, but epizootic hemorrhagic disease is most evident in white-tailed deer. Based on carcass counts after epizootics, deaths among white-tailed deer outnumber those of mule deer at a ratio of about 23:1, with losses of whitetails reaching 60 percent of a single herd (Hoff and Trainer 1981). Mortality rates of infected animals approach 90 percent, and thousands of deer may succumb during outbreaks.

Few organs are immune from the hemorrhages, but the heart, liver, kidneys, lungs, gastrointestinal tract, and spleen commonly exhibit lesions of varying sizes (Fig. 8-10). Karstad et al. (1961) determined that the

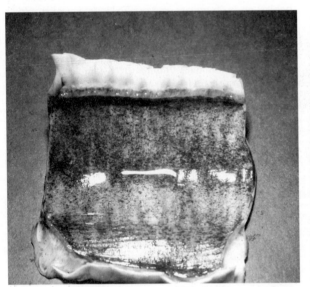 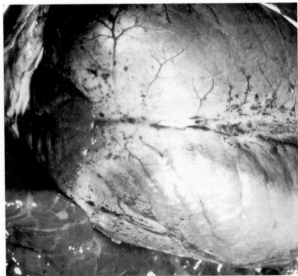

Figure 8-10.   Hemorrhages in the opened trachea (left) and on the heart of white-tailed deer (right) characterize epizootic hemorrhagic disease. Many other internal organs also may show hemorrhages of various sizes. (Photos courtesy of M. Friend, U.S. Fish and Wildlife Service.)

hemorrhaging results from malfunction of the blood's clotting mechanism and from degenerative changes in the walls of blood vessels.

External symptoms include loss of appetite, weakness, salivation, fever, hemorrhages in the oral cavity, and, sometimes, the occurrence of blood in the feces and urine. Coma and death follow within 3–36 hours after these symptoms develop. The disease is sudden in its onset, often breaking out among large numbers of deer in single nonrecurring epizootics. Known outbreaks of epizootic hemorrhagic disease have occurred in August, September, and October; the epizootics terminated abruptly with the first frost (Trainer and Karstad 1970). In North Dakota and South Dakota, long periods of excessively hot, dry weather precede epizootics, but initial outbreaks of the disease occur when there are abrupt changes in temperature and barometric pressure accompanied by high humidity (Hoff et al. 1973).

Arthropod vectors had long been suspected in the transmission of epizootic hemorrhagic disease, but it was not until the 1970s that midges of the genus *Culicoides* finally were associated with the disease's epizootiology (Jones et al. 1977; Foster et al. 1977). Effective control or treatments are unknown at present. Trainer and Karstad (1970) speculated that oral vaccines someday might be administered to deer in foods supplied during winter or droughts.

DISEASES AND BIOLOGICAL CONTROLS

The native Australian fauna has been supplemented by a host of exotic species, but none has proved more troublesome than the European rabbit (*Oryctolagus cuniculus*). After some earlier and unsuccessful attempts to establish rabbits, a successful transplant was achieved on the southeastern coast near Geelong, Victoria, in 1859. They spread rapidly, reaching the west coast of Australia 16 years later and ultimately occupying more than half of the Australian continent (Frith 1973). The explosion of rabbits quickly led to severe overgrazing and to widespread wind and water erosion of the soil (Fig. 8-11). Conflicts with the sheep industry were inevitable as the rabbits consumed forage otherwise allocated to livestock. The networks of rabbit warrens also were pervasive disturbances in the rangeland system. Despite widespread efforts to control the rabbits with fences and poisons, little population regulation was effected. The exotic rabbits continued registering their impacts on the landscape well into the 20th century.

Then, in 1950, after long experimentation, the mammalian disease myxomatosis was introduced into the rabbit population. Whereas myxomatosis is a poxvirus whose pathogenic effects are self-limiting in its natural hosts (*Sylvilagus* spp.), this disease is especially destructive in European rabbits (Yuill 1970).

Figure 8-11. Burgeoning numbers of European rabbits devastated Australian rangeland until they were controlled, at least for several years, with a viral disease known as myxomatosis. (Photo courtesy of E. Slater, Division of Wildlife Research, CSIRO, Australia.)

Marsupials and a variety of other native and domestic species virtually are unaffected, so myxomatosis remained a virulent, host-specific control agent for an unwanted exotic species within the modern Australian fauna (Bull and Dickinson 1937). Myxomatosis is spread by a variety of arthropods, particularly by mosquitoes; transmission occurs when the infected skin tissues (not blood) are carried from rabbit to rabbit on the mouthparts of the vectors (Fenner et al. 1952; Fenner and Woodroofe 1953). Nodular tumors develop on the body, followed by internal hemorrhagic necrosis of the intestine, liver, and other organs. Fenner and Ratcliffe (1965) presented a full account of the disease's epidemiology, but suffice it to note here that no more than 10 to 20 percent of the rabbit population survived the initial impact of myxomatosis (see also Frith 1973). Nonetheless, some rabbits developed resistance to myxomatosis—and the virus itself undergoes changes in its virulence—and the rabbit population rebounded to an estimated 300 million by 1996.

Currently, a disease called rabbit calicivirus disease (RCD) again decimated rabbit populations in Australia. Rabbits infected with RCD experience extensive internal blood clotting, lapse into a coma, and die within 36 hours. RCD first appeared mysteriously in China in 1984, where it killed millions of rabbits as it spread in Asia and into Europe. In 1995, RCD was undergoing tests on remote island off the Australian mainland to confirm its host-specificity, but despite elaborate precautions, the disease escaped to the mainland. Whereas no field studies have yet been published, RCD apparently is killing only its intended targets, but the wholesale elimination of rabbits nonetheless may produce some unwanted results. For example, food chains may collapse, thereby causing declines in predator populations of wedge-tailed eagles (*Aquila audax*) and other native species, that have grown dependent on rabbits—but perhaps not before the predators first switch to livestock and pets. Also, with fewer rabbits competing for grass, Australian ranchers may increase their flocks and herds, thereby continuing soil erosion and other environmental damage at the same rate as occurred with a burgeoning rabbit population. The effect of millions of rotting rabbit carcasses remains uncertain, but it seems likely to include plagues of blowflies (Calliphoridae). Conversely, some species of native plants, long suppressed by the rabbits, may again flourish on Australia's rangelands and begin to restore the much-depleted seed bank. In short, the Australian Outback perhaps may respond to a massive reduction of rabbits just as it would to the end of a long drought.

A novel experiment concerns attempts to immunize desert bighorn sheep against the viral disease known as bluetongue, or "soremuzzle," using natural vectors. Bluetongue infects both wild and domestic ruminants, including white-tailed deer, pronghorns, elk, and domestic sheep and goats (Trainer and Jochim 1969). The disease inflames the mucous membranes of the gastrointestinal tract, especially affecting those of the mouth and nose. The animal's tongue swells, becomes discolored and blue, and typically hangs loosely from the side of the mouth 24 hours before death occurs (Fig. 8-12). A variety of clinical and pathological symptoms occur that may forecast death 7–8 days after exposure (Trainer 1970).

Because much of the original range of desert bighorns subsequently was grazed by domestic sheep, the suggestion arises that bluetongue carried by these flocks more than likely contributed to reductions in bighorn populations (Robinson et al. 1967). Thus, reintroductions of desert bighorns may require large blocks of land where domestic livestock are excluded, a requirement that is obviously expensive—financially and politically—to implement. Alternatively, ways to rid bighorn ranges of the disease-carrying vectors would be needed—a virtual impossibility. However, Robinson et al. (1974) presented experimental evidence that a com-

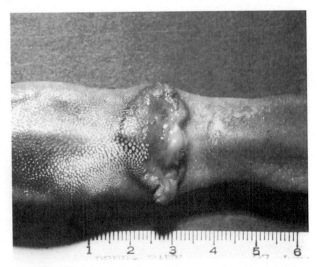

Figure 8-12. Large lesion on the tongue of a white-tailed deer resulting from the viral infection bluetongue. Bluetongue infects several species of wild and domestic ruminants but has been particularly disastrous for desert bighorn sheep. (Photo courtesy of M. Friend, U.S. Fish and Wildlife Service.)

mon arthropod vector might be used to immunize desert bighorns. Gnats (*Culicoides varipennis*) are the natural vector of the bluetongue virus and, when given a blood meal that includes bluetongue vaccine, they become "mobile syringes" seeking out and vaccinating animals that cannot otherwise be approached by humans. Results of experiments on captive bighorns were similar, whether the animals received injections by needle or received the vaccine from the treated gnats. Robinson et al. (1974) concluded that "where bluetongue is a proven problem in bighorn sheep, vaccination is indeed possible, even mandatory in the management of this species." Such imaginative methods as this undoubtedly will be required to restore desert bighorns in much of their former range.

Figure 8-13. Conjunctivitis in a house finch infected with *Mycoplasma galliseptium,* a pathogen currently producing an epizootic in this and perhaps other species of wild songbirds in eastern North America. The clinical signs include swollen eyelids and nasal and ocular exudate. (Photo courtesy of Southeastern Cooperative Wildlife Disease Study, College of Veterinary Medicine, University of Georgia.)

LYME DISEASE AND OTHER "NEW" DISEASES

Diseases not previously known or recognized periodically arise in the course of history, as illustrated by the scourge of AIDS that currently claims many thousands of human lives each year. Another is hantaviral pulmonary syndrome (HPS), which emerged in the American southwest; the disease has since appeared in California, North Dakota, and a number of other states (Hughes et al. 1993). Hantaviruses are carried by small rodents, chief among them deer mice (*Peromyscus maniculatus*), but also by other species such as pinon mice (*P. truei*), cotton rats (*Sigmodon hispidus*), and cliff chipmunks (*Tamias dorsalis*) (Stone 1993; Childs et al. 1994). Humans apparently contract the hantavirus by inhaling or ingesting dust contaminated with urine or feces from infected rodents. The pathogen causing HPS is virulent: Of nearly 70 cases reported by early 1994, 39 (56 percent) resulted in death, and no cure has yet been developed.

Conjunctivitis in songbirds caused by the microorganism *Mycoplasma gallisepticum* also represents a disease not previously detected in wildlife. The infection causes respiratory disease in domestic poultry and other pen-reared birds, but in house finches (*Carpodacus mexicanus*) it produces conjunctivitis characterized by swollen, crusty eyes (Fig. 8-13). The disease was first observed in house finches in the mid-Atlantic states in 1994 (Luttrell et al. 1996), but it has now spread elsewhere across the eastern half of North America. The rapid spread of conjunctivitis may be enhanced by the propensity of house finches to assemble at bird feeders, where sick birds may spread the in-

fection to other birds or contaminate the feeders (Fischer et al. 1997). House finches in eastern North America also migrate (Belthoff and Gauthreaux 1991) and thereby could disseminate the pathogen across a large geographical area. Unfortunately, conjunctivitis was detected recently in another species, the American goldfinch (*Carduelis tristis*), but its distribution in this species remains unknown.

Lyme disease is another disease seemingly new to North America. While it is rarely fatal, Lyme disease is now the most common arthropod-borne disease of humans in the United States; according to the Centers for Disease Control and Prevention (1996), about 16,460 cases were reported in 1996. Up to 10 percent of the human population in some locations has been infected with Lyme disease (Daniels and Falco 1989, Barbour and Fish 1993, on which much of this discussion is based; see also Ginsberg 1993).

The pathogen is a spirochete bacterium, *Borrelia burgdorferi,* which is transmitted in the northeastern United States by the deer tick (*Ixodes scapularis,* which at times has been known as *I. dammini*), although elsewhere other species of *Ixodes* also serve as vectors. The disease gained national attention in 1975 when an unusually high incidence of "juvenile rheumatoid arthritis" was mistakenly diagnosed in Lyme, Connecticut.

Hunters, hikers, and others venturing into woodlands often are among the victims of Lyme disease, but infections may occur anywhere there are infected ticks (e.g., lawns). The disease thus is a threat in both rural and urban areas. Focal outbreaks have occurred near Boston, Philadelphia, and New York (Steere 1994). In its advanced stages, the symptoms of Lyme disease include attacks of arthritis, abnormal heart rhythm, and neurological difficulties. A circular red rash—technically known as *erythema migrans*—may develop around the site where the tick was attached, but such evidence is not always produced nor is a rash necessarily indicative of Lyme disease. Symptoms may not appear for several weeks or months after the disease is contracted. The true incidence of the disease is difficult to determine, as (a) other diagnoses may be tendered for actual cases of Lyme disease, and (b) Lyme disease may be wrongly diagnosed for other ills. Because Lyme disease is so difficult to diagnose, it often may go untreated. With correct diagnosis, however, acutely infected victims respond to treatment with orally administered antibiotics; for chronic infections, antibiotics are administered intravenously. Unfortunately, the illness does not ordinarily produce immunity; hence Lyme disease may be contracted a second time.

Ecological changes may be responsible for the recent emergence of Lyme disease, especially in the northeastern United States. Conditions favorable to deer developed when abandoned farmlands reverted to the early stages of forest succession. With more deer—the keystone host for adult deer ticks—the prevalence of Lyme disease also increased. This scenario suggests that Lyme disease may have infected humans in earlier centuries (i.e., before large-scale deforestation), but its presence may have been masked by the occurrence of more serious maladies. Today, Lyme disease may be as much of a threat to the residents of wooded, suburban communities as to persons visiting forests for recreational purposes. The fear of Lyme disease also has affected public attitudes about deer and nature preserves.

Whereas *Ixodes scapularis* is responsible for more than 80 percent of the cases of Lyme disease in North America, other ticks also may serve as vectors (e.g., *I. pacificus* is the principal vector in the western United States; *I. ricinus* and *I. persulcatus* transmit the disease in Europe and Eurasia, respectively). In fact, all terrestrial mammals and as many as half of the birds occurring in Eastern Deciduous Forests are suitable hosts for various stages of the deer tick, which therefore is poorly named. Note, however, that deer are not reservoirs for the spirochetes. Deer serve only as favored hosts for adult ticks, and ticks dropping from deer present no risk of further infec-

tion (i.e., such ticks are at the end of their life cycle and will not feed again; moreover, most larvae produced by these ticks hatch without spirochete infections). In other words, deer are dead-end hosts for the spirochetes and cannot transmit the pathogens to ticks. Instead, several other species of wildlife—birds as well as mammals—are among the reservoirs for the spirochetes; one of the more common and efficient of these is the white-footed mouse (*Peromyscus leucopus*) (Lane et al. 1991). Raccoons (*Procyon lotor*) and striped skunks (*Mephitis mephitis*) are among other species that often serve as reservoirs (Fish and Daniels 1990).

The life cycle of ticks includes two immature stages: larvae, which hatch from eggs and have only six legs, and nymphs, which develop from the larvae and later become adults (nymphs and adults each have eight legs). Although larvae, nymphs, and adults may become infected with the spirochetes, only infected nymphs and adults transmit the disease to humans or other vertebrates (i.e., larval ticks are not important vectors of Lyme disease). There are several pathways by which the spirochetes may be transmitted:

- After hatching, the larvae may obtain a blood meal from a mouse or other infected host, thereby ingesting spirochetes. The larvae molt and spend the winter as nymphs, which seek hosts again when they emerge the following summer. Spirochetes in the larvae persist and survive in the nymphs, and when the nymphs feed after emerging, they transmit the spirochetes to another host.

- Alternately, if the larvae are not initially infected, they may acquire spirochetes after becoming nymphs if they happen to feed on previously infected hosts. In short, either the larvae or the nymphs may acquire the pathogens, but of these, only the nymph can transmit the spirochetes to a vertebrate host (i.e., the larvae do not feed a second time and therefore normally can only acquire—not transmit—the pathogens). Humans are usually infected by bites from nymphs in the spring or summer (del Rio et al. 1996).

- The spirochetes survive and persist in an adult tick produced from either a larvae or a nymph that previously obtained a blood meal from an infected vertebrate. When such a tick—a female seeking a blood meal necessary for egg production—feeds on an uninfected vertebrate, the spirochetes are transmitted. Adult ticks are twice as likely to be infected with the spirochetes because they have had two chances to acquire the pathogens, once as larvae, and again as nymphs. However, adult ticks are larger, and in the case of human epidemiology, they are more likely to be removed from a person's body within a few hours. Therefore, adult ticks are not as likely as nymphs to infect humans.

Field biologists recently discovered that ear tags may double the rates at which white-footed mice may be infested with *Ixodes* ticks (Ostfeld et al. 1993). The tags either reduced the ability of the mice to groom themselves, or the perforation itself attracted ticks to the tagged ear (Fig. 8-14). The results suggest that factors—ear tagging, in this case—that increase the likelihood of a larval tick obtaining its first blood meal from a reservoir species also may increase the incidence of Lyme disease.

Prevention of Lyme disease remains difficult. One approach is to control the vectors with an appropriate pesticide. However, the two-stage life cycle of *Ixodes* ticks spans at least two years, and each stage potentially involves their distribution over a large area (i.e., the larvae are carried about on a host animal during the first year, and additional movement occurs when the nymphs are transported in the second year). Thus, while ticks are susceptible to a number of pesticides, several applications covering a large area may be necessary before the tick population is effectively suppressed, and such treatments may raise environmental concerns. Nonetheless, residents in areas with an established history of Lyme disease may choose to spray their yards each year.

An imaginative technique was designed to reduce ticks harbored by white-footed deer mice. Cotton, treated with an insecticide, is inserted in paper tubes and supplied to the mice as nesting material, thereby eliminating ticks from a major reservoir species during the period when Lyme disease is transmitted. This method reduced ticks in one study, but the technique was not effective in three other tests conducted elsewhere, perhaps because other reservoir species were present at those locations. Greatly reducing the deer herd seems like a tempting solution, but virtually all deer must be removed to control the local tick population, and the presence of other hosts might counteract the effectiveness of such a drastic measure.

Various means of personal protection currently remain the primary means of preventing Lyme disease. These measures include (a) inspecting for ticks after outdoor activities and promptly removing any ticks with tweezers, grasping the tick as close to the skin surface as possible and pulling straight back with a slow steady force (avoid crushing the tick), (b) wearing light-colored clothing (on which ticks are more visible), (c) using insect repellents regularly (e.g., DEET or permethrin), and (d) inserting pant cuffs inside the tops of socks. Showering after visiting wooded areas also may lessen the chance of infection. Deer ticks do not always begin to feed right away; hence, they may be washed off before imbedding themselves.

WILDLIFE DISEASES AND HUMANS

Wildlife populations serve as reservoirs or carriers for some diseases affecting humans. Should we tenuously include commensal rodents as wildlife, then foremost of these diseases was the devastating Black Death that gripped Europe in the 14th century. Scholars of medieval history estimate that fully one-third of the population in Western Europe—about 20 million people—succumbed during an epizootic of bubonic plague (Tuchman 1978). In Paris, the death rate was 800 per day, and by the time the disease had run its course, the toll had reached 50,000—half the city's population. Fortunately, bubonic plague never again inflicted such widespread damage on human populations, although statistics collected by the World Health Organization in 1967 indicate that more than 5000 cases occurred under wartime conditions in Vietnam.

Conversely, sylvatic plague directly concerns species clearly falling within the realm of wildlife (e.g.,

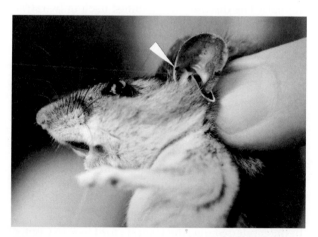

Figure 8-14. Arrows points to one of several deer ticks, the vectors of Lyme disease, attached to a white-footed mouse. Recent evidence indicates that infestations of deer ticks may increase on mice marked with ear tags, as shown here. (Photo courtesy of P. Klose.)

bobcats, *Felis rufus*, badgers, *Taxidea taxus;* see Hopkins and Gresbrink 1982; Tabor and Thomas 1986). This form of plague is caused by the same pathogen as bubonic plague (i.e., *Yersinia pestis;* see Table 8-1), but transmission stems from wild rodents, not from commensal species; fleas are the principal vectors in either instance, however. More than 230 species or subspecies of wild rodents are associated with natural infections of sylvatic plague (Olsen 1981).

The epizootiology of sylvatic plague is complex and is associated with the pathogen's persistence (e.g., in burrows) and with the structure and social systems of the reservoir population. Sylvatic plague also is localized and is discontinuously distributed, factors that strongly suggest that specific relationships between the reservoir species and their environment have much to do with the disease's occurrence. Four basic components seem involved, although not all may be of equal importance or in effect at one time. These are (1) fleas, (2) soil, (3) hibernating rodents, and (4) varying degrees of susceptibility among rodent populations to *Y. pestis* (Olsen 1981).

Cases of sylvatic plague may develop when humans visit prairie dog (*Cynomys* spp.) towns or other communities of wild rodents, but these are isolated exposures, not human epizootics. Nonetheless, mention of plague evokes strong currents of fear when cases are reported in the public media. Such publicity sometimes leads to overreactions, as when prairie dog towns are sprayed aerially with massive amounts of pesticides. The intent of such treatments is to destroy fleas and other potential vectors, but too often the environment also is contaminated with persistent chemicals. Barnes and Kartman (1960) described a more localized approach: Bait boxes, open at both ends and treated inside with insecticide, are placed in areas of high human exposure to rodents (e.g., campgrounds and parks); rodents, attracted to the bait, are dusted as they enter the box. Not only are the animals cleansed of fleas, but flea populations that remain underground are treated when the dusted rodents return to their burrows.

Tularemia also infects humans, but this disease, although highly debilitating, is not contagious among humans. Contact with ticks or other arthropods that previously had fed on infected rabbits is the most common means of transmission. Sometimes persons skinning infected rabbits or muskrats (*Ondatra zibethicus*) may contract tularemia, but the disease is not encountered often by humans.

Rabies, like plague, is another disease that strikes fear among humans—and with good reason. If not treated immediately, rabies is virtually 100 percent fatal in humans. The pathogen, a *Rhabdovirus*, attacks the brain and the spinal cord. Furthermore, the symptoms are horrifying. Human victims become uncontrollably thirsty, but when given water they begin to convulse and to elicit terrified responses to the sight of water (thus the human disease's other name, *hydrophobia*). A series of vaccinations can prevent rabies. However, the efficacy of the vaccinations is dependent on treatment soon after exposure; by the time clinical signs develop, it is too late to save patients from death. Accordingly, animals that bite humans must be sacrificed immediately for critical examinations of their brain tissues in order to confirm the presence of rabies. Otherwise, bite victims must undergo the expensive series of vaccinations without benefit of any diagnosis (other than the bite itself). Because it often is impossible to examine the biting animal, persons may undergo treatment when they actually are not infected.

Rabies virus is transmitted from the salivary glands of infected animals; from there it spreads when an infected animal bites another animal or a human. A few cases of rabies also have developed when humans breathed the air in bat caves (Warrell 1977). In some parts of the world, particularly India and the Philippines, as many as 25,000 people die of rabies each year, but only 1 or 2 deaths per year occur in the United States (Centers for Disease Control 1993). Rabies has been eliminated in Great Britain, although an epizootic on the European mainland has renewed concern about pets reintroducing rabies (Anderson et al. 1981).

Rabies likewise is nearly always fatal in nonhuman animals. The disease may manifest itself in two strikingly different forms of behavior: *furious* and *paralytic* (or "dumb") rabies. In the furious form, the typical "mad-dog" signs are expressed, with the victim frantically running about, biting and snapping at other animals or even at sticks and other inanimate objects. Hooved animals bite and lash out with their legs. In the less-common paralytic form, animals become semiparalyzed; they drop their jaws and cannot bite or produce sounds.

Many species of wildlife are implicated in the spread of rabies. In India, jackals (*Canis aureus*) are important hosts; red foxes (*Vulpes vulpes*) and badgers (*Meles meles*) are frequently infected in Europe. In the United States, the Centers for Disease Control (1993) reported the following cases in wildlife for 1992: bats, 647; foxes, 397; raccoons, 4311; skunks, 2334; and another 223 cases in other species. Woodchucks (*Marmota monax*) also have been reported with rabies in the mid-Atlantic states, perhaps resulting from competition with raccoons for dens in areas of high rabies activity (Fishbein et al. 1986). An epizootic recently spreading along the mid-Atlantic coast perhaps originated from raccoons translocated from Florida in 1977 (Beck 1984; Rupprecht and Smith 1994). Whatever the origin may be, rabies in raccoons has increased dramatically in the mid-Atlantic region, including metropolitan Washington, D.C. (Table 8–4). The epizootic in raccoons eventually spilled into the dog and cat populations, posing a clear threat to humans. In Ontario, Canada, a program of rabies prevention targets the strain transmitted by raccoons (Rosatte et al. 1997). In addition to trapping and destroying rabid animals, an eradication zone 4 km in diameter is established around any location where a rabid raccoon is encountered. A second area, a live-trap and vaccinate zone, extends another 4 km beyond the first. The program is publicized widely and includes campaigns to vaccinate cats and dogs and to discourage humans from feeding and domesticating raccoons. See Bacon (1985) for a full review of the dynamics of rabies in wildlife populations.

Rabies is particularly dangerous in urban settings. Raccoons head the list of carriers in cities because these animals adapt well to urban areas (see Chapter 17). Other species of urban wildlife, such as striped skunks (*Mephitis mephitis*), also are potential transmitters of rabies but are less likely to be tolerated by humans. Moreover, raccoons pose a greater threat of transmitting rabies to dogs and therefore to people. Inoculation of pet cats and dogs clearly remains the best means of preventing rabies among humans because pets often are the link between humans and wildlife. The number of human cases of rabies corresponds almost directly with the number of cases in dogs and, in turn, with the extent with which the dog population has been immunized (Beck 1984). Dogs used for hunting should always be inoculated. Fully 75 percent of 427 raccoons Hubbard (1985) tested for rabies in a rural area of Virginia were infected. A shortage of rainfall seemed linked with the epizootic in Virginia. The drought reduced the habitat, with the result that raccoons crowded into a still-suitable watershed where intraspecific competition probably induced physiological stress and lowered resistance to rabies infections. Rabies also occurs in wildlife associated with rangelands (Young 1984).

Nearly all mammals are susceptible to rabies, but the disease predominates in certain reservoirs, as exemplified by species of carnivores and bats. For example, 30 of the 42 species of insectivorous bats have been reported with rabies in the United States (Constantine 1993). In recent years, several humans

and several millions of livestock have died in Latin America from rabies transmitted by vampire bats (*Desmodus rotundus*) (Sikes 1981; Acha and Alba 1988). Transmission of rabies by bloodsucking arthropods has not been demonstrated (Bell et al. 1957).

Because of an understandable desire to control rabies, extreme measures occasionally are advocated to eradicate wildlife reservoirs. An epizootic of rabies in central Europe featured a high incidence in red foxes; 70 percent of the more than 16,800 cases reported in Europe in 1979 occurred in foxes (Anderson et al. 1981). In Europe, foxes were slaughtered with guns, poison, and gas, but only in Denmark have such means been successful (Lloyd 1977). A fox-free zone is maintained along the German-Danish border, but only with considerable expense and unrelenting effort; vacant territories are quickly colonized by immigrating foxes (Kaplan 1977). After examining a population model for foxes, Anderson et al. (1981) concluded that culling methods will achieve little results except where foxes live in poor habitat.

Nonetheless, culling programs have been tried locally in the United States where, interestingly, the reservoir species may differ by region (Sikes 1981; Smith 1989). Foxes are responsible for rabies epizootics in the Appalachian Mountains. Skunks are the primary reservoir in the midwest, even though foxes are present in dense populations. In the southeast, rabies transmitted by raccoons is troublesome in Florida, Georgia, Alabama, and South Carolina, but the disease seems almost absent in the fox and skunk populations in these states. Foxes thus were the target of rabies control in Kentucky (Lewis 1966), whereas skunks were controlled in Ohio (Schnurrenberger et al. 1964). Strychnine-treated baits were used in these efforts but, in Kentucky, 135 dogs, 25 cats, 202 birds and rodents, and 129 other animals, in addition to 65 foxes, accepted the poisoned baits. Overall, the cost of control measures reached $208 per fox. At this rate, further reduction of costs would be required before continuation of the program could be justified economically (Lewis 1966).

Lloyd (1977), upon considering widespread fox control, concluded as follows:

The expense of depressing fox numbers nationally would be enormous and the results could not be guaranteed. In addition to the unwarranted economic, ecological, and social effects of such a measure, there is no justification for the slaughter of a wild animal on such a scale as a precautionary measure, against an event which is only remotely and locally probable. There is a better case for reducing the numbers of

TABLE 8-4. Documented Cases of Rabies Occurring in Raccoons in the Mid-Atlantic Region of the United States, 1977–92. The Epizootic Eventually Included Cats and Dogs, Thereby Increasing the Threat of Rabies Infections in Humans

	1977	1978	1979	1980	1981	1982	1983	1984	1985	1986	1987	1988	1989	1990	1991	1992	1993	1994	1995
West Virginia	1		8	14	22	43	89	27	15	30	38	103	21	17	17	23	35	38	64
Virginia		3	4	7	102	645	545	158	102	139	253	366	148	129	167	203	213	251	271
Maryland					7	118	735	964	672	588	384	338	295	382	467	413	501	412	326
Pennsylvania						26	81	281	285	409	240	543	493	405	219	208	229	222	219
Washington, D.C.						5	158	12	4	29	42	13	12	8	23	16	16	4	13
Delaware											8	36	21	37	143	162	100	38	53
New Jersey													18	408	787	579	332	194	219
New York														84	666	1355	2369	1284	846
Totals	1	3	12	21	131	837	1608	1442	1078	1195	965	1399	1008	1470	2489	2959	3795	2443	2011

Source: Centers for Disease Control.

ownerless stray cats and dogs, and for imposing some restraint to the wanderings of dogs let out for the day while their owners are at work.

Vaccination thus seems a better method than culling, although as the density of foxes increases, a higher proportion of the fox population must be vaccinated (Anderson et al. 1981), but field work in Western Europe suggests that rabies can be controlled in red foxes when oral vaccines are properly distributed in attractive baits (Winkler and Bogel 1992). An effective delivery system remains a major obstacle for vaccinating wildlife against rabies. Recently, biologists experimented with a technique that may also help stem the rabies epizootic sweeping across the mid-Atlantic states (Rupprecht et al. 1992). The tests were designed to determine if raccoons would accept vaccine-treated baits scattered over a wide area of Pennsylvania (see also Winkler et al. 1975 regarding foxes). However, an effective live vaccine sometimes can *cause* rabies; thus, with genetic engineering, biologists inserted a gene from the rabies virus into the genetic material of another virus. The product is a vaccine powerful enough to immunize raccoons but different enough from the rabies virus that it will not produce the disease. The vaccine still is not approved for use, however, but the delivery system was perfected in the interim. As currently designed, the baits are wax ampules filled with the vaccine, which are placed inside a cylinder of fishmeal and fish oil bound by a polymer. To test the delivery system, however, tetracycline was used in place of the vaccine. Tetracycline stains teeth and glows under the illumination of ultraviolet light. Hence, with the cooperation of trappers and hunters, a large number of raccoons were examined. Field studies revealed that nearly 70 percent of the animals had chewed open the test baits. Thus, pending approval of the recombinant vaccine, biologists may have a means of immunizing large numbers of raccoons against rabies.

SUMMARY

Wildlife diseases originate from infectious and noninfectious pathogens such as viruses, bacteria, parasites, and toxic materials that impair the normal body structure or function. Diseases exert direct and indirect influences on wildlife and often are associated with complex ecological settings before and after they strike animal populations. Animals weakened by disease often are more vulnerable to predators, and wildlife may interact with domestic livestock or humans by serving as reservoirs of debilitating pathogenic agents. Whereas knowledge about wildlife diseases is increasing rapidly, the ability of managers to limit diseases in wildlife populations generally lags well behind. Nonetheless, diseases remain a part of the puzzle of wildlife management, especially where unfavorable habitat conditions may be involved.

Epizootiology concerns the ecology of diseases and asks the questions *how* and *why*. Diseases are enzootic in their chronic form and epizootic when they erupt.

Ecological conditions involving habitat quality and quantity commonly are associated with wildlife diseases, as is illustrated by epizootics of botulism in waterfowl. Diseases usually involve several components of an ecosystem, such as invertebrates or soil conditions, as well as the infected animals. Knowledge of these factors and the way they interact with disease transmission is necessary for a full assessment of management options.

Many diseases seem density dependent, affecting larger proportions of animal populations at high densities than at low densities. Regulation of wildlife populations by disease agents nonetheless is difficult to demonstrate conclusively. Disease sometimes has been cited as controlling natality, as a causative agent in population cycles, or as affecting age structure in wildlife populations. Exotic diseases may have devastating results on the viability of an isolated, endemic fauna (as is illustrated by avian malaria on some species of Hawaiian birds). At the same time, host-specific diseases offer options for controlling exotic pests.

Lyme disease represents a "new" malady for humans in the northeastern United States and other areas where its primary vector, the deer tick, occurs along with white-footed mice and other species that serve as reservoirs for the spirochete pathogen, *Borrelia burgdorferi*. The disease is not readily diagnosed, but once identified, antibiotics offer an effective treatment. Control of the vector populations is difficult, although careful applications of pesticides on lawns in urban areas may reduce infection rates. Various types of personal protection (e.g., tucked-in pant legs) are recommended for persons frequenting outdoor areas where deer ticks occur.

Tularemia and sylvatic plague are two wildlife diseases of concern to humans, but rabies elicits the greatest fear. Rabies occurs in skunks, foxes, and a variety of other wildlife that may come into contact with people. Reducing stray cats and dogs may be more effective, however, than attempts to control rabies by reducing wildlife populations.

CHAPTER 9

PREDATORS AND PREDATION

All living things are destined to die and be recycled as a part of the flow of energy through the life community. Which is to say, a creature must feed, and sooner or later it will be fed upon.

Durward L. Allen (1979)

Predators are animals that survive by killing and eating other animals. Every ecosystem—whether aquatic or terrestrial—contains predacious species, although certain environments may lack some kinds of predators. Mammalian predators, for example, seldom occur in the native fauna of oceanic islands. Nonetheless, mountain lions (*Felis concolor*), bobcats (*F. rufus*), gray wolves (*Canis lupus*), hawks and eagles (Falconiformes), and alligators (*Alligator mississipiensis*) are familiar predators, many of which are persecuted because of their role in nature. Predators generally bear an unsavory reputation in Western culture. A big, *bad* wolf, for example, is the villain in "Little Red Riding Hood" and "The Three Little Pigs." The stigma is thus passed on from adults to children in nursery stories, generation after generation. We often forget—or overlook—that such "friendly" animals as bullfrogs (*Rana catesbeiana*), swallows (Hirundinidae), dragonflies (Odonata), and robins (*Turdus migratorius*) are also predators. In fact, some seemingly innocuous species are especially voracious predators (e.g., shrews, Soricidae). Among the many species of predacious birds, the American woodcock (*Scolopax minor*) is one of the few regarded as game. Mink (*Mustela vison*) and ot-

ters (*Lutra canadensis*) are valuable furbearers as well as predators.

Humans generally have little regard for the function of predation in wildlife communities, and predatory species—especially birds and mammals—have been much maligned in our efforts to bring ecosystems under human control. Humans often reserve for their own judgment decisions about who may kill whom. Curiously, predation in aquatic systems generally goes on without human condemnation, perhaps because the bloodshed occurs out of sight in a medium foreign to our own environment. In fact, most aquatic predators are valued as game species. Salmon and trout (Salmonidae), for example, are voracious predators that are held in high esteem, but carp (*Cyprinus carpio*) and other herbivorous fishes are largely scorned as "trash fish." Such perceptions, however, are reversed in the human regard for terrestrial wildlife (i.e., terrestrial predators are "varmints," whereas herbivorous birds and mammals include highly prized game species). Predators and predation, perhaps alone in the spectrum of natural phenomena, clearly remain among the least understood and most unappreciated components of natural communities (see Kellert 1985).

PREDATOR BEHAVIOR AND PREY SURVIVAL

In Chapter 7, we described the coevolution of plants and herbivores, but the evolutionary view of feeding relationships bears repetition here because the principles also concern predation. For plants, the survival of a species depends upon the evolution of an effective defense against overutilization and destruction by herbivores. Herbivores, in turn, must gain their nutritional welfare from the same plants that have evolved defensive mechanisms (e.g., thorns). Thus, the Earth's biota today is replete with many species of plants and herbivores that have met and survived such evolutionary challenges. Herbivores, however, also must develop means of resisting carnivores or face premature death. Insects resist or avoid insectivores with protective shapes and coloration; some resemble sticks or leaves, whereas others mimic stinging or distasteful species. Vertebrate prey species also have developed protective coloration as a means of avoiding detection. Others defend themselves against attack with their hooves or other structures, but alertness, swiftness, and an affinity for remaining in protective cover are the foremost means by which most herbivores avoid predation. The survivors, of course, form the breeding population, thereby passing on to the next generation those genetic traits especially adapted for escaping predators. Similarly, the survival of predators depends on specialized behavioral and physical features for capturing enough prey. The special hunting traits of predators include speed, agility, claws, sharp teeth, strength, keen senses of smell and vision, and a sense of geometry in determining angles of pursuit.

Predators hunt their prey in various ways, but these methods are by no means as efficient as we might believe. Owls (Strigiformes), accipiter hawks (Accipitrinae), flycatchers (Tyrannidae), frogs (Ranidae), cats (Felidae), rattlesnakes (*Crotalus* spp.), and largemouth bass (*Micropterus salmoides*) are among the predators that hunt by ambush. Others, such as falcons (Falconinae), cheetahs (*Acinonyx jubatus*), and striped bass (*Roccus saxatilis*), swiftly pursue and overtake their prey. Wolves, coyotes (*Canis latrans*), buteo hawks (Buteoninae), and chain pickerel (*Esox niger*) employ methods intermediate between ambush and swift pursuit. Regardless of the method, predators usually hunt over large areas and generally select physically weakened prey or those in a vulnerable position relative to escape cover.

Most predators normally expend a large investment of time and effort per unit of food. Prey animals may find cover or may outrun, outfly, outmaneuver, or fend off a predator. In fact, predators of large animals make successful kills rather infrequently. Thus, of 124 moose (*Alces alces*) approached and "tested" by wolves on Isle Royale in Lake Superior, only 9 were killed, a success rate of about 7 percent. In such circumstances, strength and condition—functions often associated with age—were the prime factors that enabled moose to repel a pack of attacking wolves (Allen 1979). Likewise, Rudebeck (1950, 1951) recorded a success rate of only 7.6 percent in 688 attacks made by predacious birds in Europe, including two species of falcon (*Falco* spp.), a hawk (*Accipiter* sp.) and an eagle (*Haliaeetus* sp.). The success rate varied by species from 4.5 to 10.8 percent. Other studies have shown similarly low capture rates.

Predators often select individual prey animals that show some oddity in color, behavior, or location. Thus, a minnow impaled on a fishhook and thrown back into a lake among thousands of minnows of similar size and shape will be attacked more often by largemouth bass or pickerel than other minnows in the lake. Fishing with minnows would be quite fruitless if the appearance and behavior of hapless baitfish were not different. Krischik (*in* Morse 1980) studied such relationships in the laboratory, using juvenile bluegills (*Lepomis macrochirus*) as prey and largemouth bass as predators. Some of the bluegills in the prey population were normal, whereas others were infected with parasites that produced abnormal coloration and physical changes that affected swimming behavior. Bass could prey freely on bluegills in either condition, but the experiment showed that the odd color and behavior placed the infected bluegills at greater risk of predation. Pielowski (1959) conducted a similar predator-prey experiment with goshawks (*Accipiter gentilis*) and rock doves (*Columbia livia*). When goshawks hunted in a prey population dominated by white birds, most of the kills were dark-colored rock doves. Conversely, when the situation was changed so that dark-colored rock doves were more common, the goshawks usually attacked the smaller percentage of white birds in the experimental flock. These results indicated that odd birds—in this case, those that were unusually colored—in the population were singled out as prey. It is not hard to imagine that the erratic behavior of diseased animals quickly attracts predators or that predation often limits the successful introduction of hand-reared wildlife. Balcomb et al. (1984) reported that a type of pesticide produced spasms, coiling, and disorientation in earthworms, thereby increasing the chances that robins also would be poisoned after preying on worms showing such unusual behavior. The attention that predators pay to unusual behavior presumably led to the development of distraction displays, such as the feigning of a broken wing by female birds when their nests or broods are threatened.

Conspicuousness also contributes to the vulnerability of prey. Dice (1947) and later Kaufman (1974) studied the success of owls searching for mice of different colors placed on contrasting backgrounds. In each study, the conspicuous mice (e.g., dark mice on a light background) were more vulnerable than those where contrast was lacking (e.g., dark mice on a dark background). Such evidence further supports the idea that predation is a major force that promotes the selection of fit individuals (i.e., predators remove unfit individuals). As an environmental force, predation also plays a clear role in habitat utilization. In the temporary absence of predators, for example, ducks began to nest in obvious locations lacking protective cover (Balser et al. 1968).

Predators frequently are more successful when part of the prey population is forced outside of its normal habitat. The outcast individuals thus are predisposed to predation (and other forms of mortality) because of their unfavorable location. As such, they represent a less-fit segment of the population (Errington 1967). These individuals are removed quickly. Conversely, the stronger individuals remain in the more suitable habitat, thereby favoring survival, successful breeding, and passage of their dominant traits to the next generation. In the bluegill-bass experiment described previously, normal individuals outranked infected bluegills in the social hierarchy of the prey population and forced the odd individuals into marginal habitat (Krischik, *in* Morse 1980). Likewise, wolves more often killed white-tailed deer (*Odocoileus virginianus*) at the fringes, rather than in the interior, of winter deer yards (Nelson and Mech 1981).

Because the success rate of capturing prey normally is so low, predators seldom enjoy an easy livelihood. Like other kinds of organisms, however, predators have evolved ways of coping with environmental hardships. Many predators, such as canids and felids, are strongly territorial. This characteristic assures adequate hunting grounds for established individuals and their families. In some pack animals such as wolves, behavioral mechanisms restrict breeding to one or two dominant females in the pack, thus increasing the chances that pups will have enough food (Mech 1970). Even with territorial defense and limited breeding, however, many young wolves die of starvation in times of low prey availability (Allen 1979), and older individuals dispersing into new areas are often unsuccessful at establishing territories (Fritts and Mech 1981). Similar fates undoubtedly befall other predators.

Whereas the immediate effect of predation is the death of individual animals, we also have seen the significance of predation in the light of evolutionary development: a steady strengthening of the genetic heritage of the survivors within a prey population. Nonetheless, various factors influence the limitations that predation may place on the size of prey populations. Because predation is an age-old phenomenon, the continued presence of both predator and prey populations confirms that a natural coexistence has endured for millennia without direct human intervention (e.g., no predator control). In a highly modified world, however, the argument may arise that predators, as well as herbivores and scavengers, deserve the attention of management (Berryman 1972). Such management may include protecting predators from human persecution (e.g., owls and hawks and endangered species), establishing controlled harvests (e.g., furbearers such as mink), manipulating habitat (e.g., improving cover for ground-nesting birds or placing predator guards on nest boxes), or directly reducing predator populations by lethal and nonlethal means (e.g., trapping and sterilants). The management of predators thus has important implications across the breadth of wildlife management, some of which are addressed in other chapters. However, because predation on wildlife presents a somewhat different management problem then predation on domestic animals, these aspects are presented separately in this chapter.

PREDATION IN NATURAL COMMUNITIES

A search of the wildlife literature will uncover contradictory conclusions about the limitations, if any, that predators exert on the abundance and density of prey populations. One study may indicate that predators do not control prey populations, whereas another suggests that the removal of predators caused the explosion of ungulate herds in Yellowstone National Park. We also are told not to shoot hawks and owls because these birds are important regulators of rodent populations. One of the foremost figures in the study of predator-prey relationships thus stated, "When reexamining questions of social intolerance and population effects of predation in the animal kingdom, I am not surprised that few pat answers come to mind" (Errington 1967). The subject of predation clearly remains a focus for heated debate.

Unfortunately, many uncontrollable variables interact and thereby confound field studies, whereas controlled experiments in the laboratory are conducted under overly simplified conditions. The results in one case are seldom clear-cut, and those in the other may be of little management value. Nonetheless, in this chapter we shall review studies that have shaped our knowledge of predation and improved the management of predator and prey populations.

In Europe, predators have been regarded as enemies, in keeping with the prevalent Western attitude that nature is a force to be conquered (see Nash 1982). Wolves were extirpated in the British Isles during the 1700s and in the past century have been isolated in a few small areas on the European continent (Pimlott 1972). Full-scale predator control remains a central part of the intensive management of game in Europe (see Chapter 22). The strong bias against predators was one of the European traditions imported into North America, and during an era of exploitation lasting for three centuries, many predator populations were persecuted relentlessly. The large predators—wolves and mountain lions (*Felis concolor*)—were primary targets of control. Vital habitat also was destroyed both for predators and

their prey. The distribution of wolves in North America was greatly reduced (Fig. 9-1), and aside from Alaska, their range in the United States now includes only 5 percent of the area occupied in 1700 (Mech 1970). Mountain lions remain on no more than 30 percent of their former range (Grzimek 1975).

The general plight of wildlife in North America was recognized in *Our Vanishing Wild Life,* a book by William Hornaday (1913), director of the New York Zoological Park. Hornaday presented an urgent and effective plea for the protection of wild animals, but his concern scarcely included carnivores. Underlying much of his management philosophy was the conviction that the "relentless pursuit and destruction of the savage-tempered, strong-jawed, fur-bearing animals is in part the salvation of the ground birds of to-day and yesterday." Therefore, Hornaday believed that "scores upon scores of species long ere this would have been exterminated" without predator control, including the elimination of predatory birds. "A Species To Be Destroyed" was the terse caption for pictures of a Cooper's hawk (*Accipiter cooperii*) and a sharp-shinned hawk (*A. striatus*) in Hornaday's book. Indeed, even a rodent was included in Hornaday's indictment of predators. Beneath a picture of a red squirrel (*Tamiasciurus hudsonicus*) appears the phrase, "A Great Destroyer of Birds." Hornaday also praised the bounty on weasels (*Mustela* spp.) in Pennsylvania and lauded the work of the United States Bureau of Forestry, whose agents in 1910 rid from 15 western states 6487 coyotes, 241 wolves,

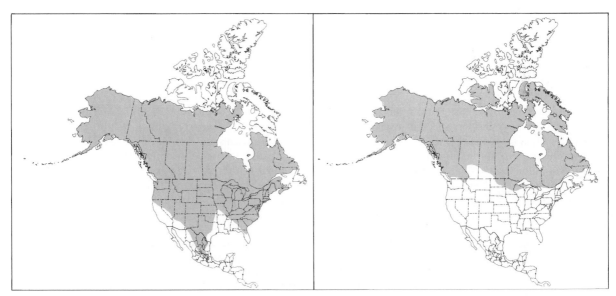

Figure 9-1.   Former (left) and present (right) distributions of gray wolves in North America. (From Mech 1970.)

88 mountain lions, 870 bobcats (*Felis rufus*), 213 bears (*Ursus arctos* and *U. americanus*), and 72 lynxes (*Felis lynx*). "Buffalo Jones," a government hunter who protected young elk (*Cervus elaphus*) in Yellowstone National Park, received similar commendation from Hornaday. Jones killed about 40 mountain lions and claimed that he found the skulls of 9 elk calves near the den of an old male mountain lion.

The professional qualifications and convincing literary style of Hornaday likely influenced and reinforced American attitudes and official policy toward predators. Bounties and predator-control agents were common features of wildlife management in the first 50 years of the 20th century. Hornaday's viewpoint was human oriented; he championed an outlook that projected inherent qualities of *good* or *bad* to animals of all kinds. *Bad*, of course, was a designation Hornaday clearly reserved for predators. Animal populations were not regarded, as they are now, as members of a broad community in which each species, including humans, interrelates with other species—a community in which humans may not always be the central focus.

The wisdom of looking at predators in the "old way" was examined by Leopold (1949) in *Thinking Like a Mountain* (see also Flader 1974 for a book-length analysis of Leopold and the evolution of his thoughts about predators). In his dramatic essay, Leopold recalled shooting at a pack of wolves early

National Museum of American History, Smithsonian Institution

William T. Hornaday
Zoo Director
(1854–1937)

Hornaday was the first American to translate an interest in zoos and museums into wildlife conservation. While attending college, Hornaday taught himself taxidermy and mounted specimens for museum displays at what is now Iowa State University. He later joined a commercial firm that specialized in the preparation of scientific specimens. His new position included collecting trips to such far-flung areas as India, South America, and the East Indies, and he soon established himself as an expert field naturalist as well as a skilled taxidermist. In 1882, he became chief taxidermist for the U.S. National Museum in Washington, D.C., where he prepared a famous diorama of six bison positioned around a waterhole—a format known as a great habitat exhibit and later copied by other museums. Hornaday later helped establish the National Zoological Park in Washington, D.C., and served 2 years as its director before resigning in 1890 over a disagreement. Hornaday entered private business for 6 years but then was invited to become the first director of the fledgling New York Zoological Park (Bronx Zoo). In this position, he emerged as a world-famous zookeeper and a leading patron of wildlife conservation. He remained director until his retirement in 1926.

Although Hornaday freely shot animals for his taxidermy work, he nonetheless opposed unrestrained sport hunting and battled for laws requiring daily bag limits. A brief quote illustrates his uncompromising language: "The idea that in order to enjoy a fine day in the open a man must kill a wheelbarrow load of birds is a mistaken idea; and if obstinately adhered to, it becomes vicious!…one can fill his day and his soul with six good birds just as well as with sixty." He also worked vigorously to outlaw the commercial exploitation of plumage from wild birds and lobbied successfully for enactment of the Bayne Law, which prohibited selling meat from game animals. Hornaday pursued his objectives with little concern for the opinions or feelings of others; an obituary described him as being among the "most acrimonious of conservationists." He blamed predators—weasels and hawks, as well as larger carnivores—for the decline of other species, but he also regarded European immigrants and poor southerners as destructive agents of wildlife.

Although his many crusades included the protection for fur seals and pronghorns, Hornaday is best remembered for his unrelenting efforts to save bison from extinction. In particular, he supplied 15 animals from the Bronx Zoo in 1907 to stock a new refuge in Oklahoma, which today maintains a thriving herd of nearly 500 bison. Another shipment later established a herd in the Black Hills of South Dakota. Meanwhile, as president of the American Bison Society, he worked hard and successfully to establish a national bison refuge in Montana. Hornaday wrote 15 books about wildlife, of which *Our Vanishing Wild Life* (1913) is perhaps the best known.

in his career, an opportunity not to be missed in those times of ardor for a predator-free environment. After the fusillade, a wounded pup hobbled off and an old female lay dying. Leopold wrote:

We reached the old wolf in time to watch a fierce green fire dying in her eyes. I realized then, and have known ever since, that there was something new to me in those eyes—something known only to her and the mountain. . . . I thought that because fewer wolves meant more deer, that no wolves would mean hunters' paradise. But after seeing the green fire die, I sensed that neither the wolf nor the mountain agreed with such a view.

Since then I have lived to see state after state extirpate its wolves. . . . I have seen every edible bush and seedling browsed . . . [and] . . . the starved bones of the hoped-for deer herd.

I now suspect that just as a deer lives in mortal fear of its wolves, so does a mountain live in mortal fear of its deer.

THEORETICAL PREDATOR-PREY SYSTEMS

Theoretical ecologists have developed models of predator-prey relationships, and some of these illustrate principles that are useful in wildlife management. Lotka (1925) and Volterra (1926) developed equations that assumed a certain potential rate of increase for the prey population in the absence of predators and a decline in predator populations in the absence of prey. For example, the rate of increase might be calculated for a population of cottontails (*Sylvilagus floridanus*) in the absence of foxes (*Vulpes vulpes*), or the decline in the fox population could be assessed with a shortage of cottontails (i.e., the rise and fall of predator and prey populations as the numbers of each change). Figure 9-2 illustrates the mathematical and ecological logic of the Lotka–Volterra equations.

The theory of the Lotka–Volterra equations seems sound. Predators have more food when prey populations are high, and thus the numbers of predators increase as their death rate diminishes. The increase in predators then places more pressure on the prey population, which declines accordingly. As the prey population declines, however, predators find less food and their death rates increase until there are so few predators that the prey numbers again begin to increase, and so on.

Are there any natural populations that show such oscillations? A classic example of predator-prey fluctuations is shown in the fur records of the Hudson's Bay Company, in which the pelts of snowshoe hares (*Lepus americanus*) peaked in 10-year cycles (MacLulich 1937). Lynx pelts showed similar peaks and troughs, which generally lagged behind those of the hares (Fig. 9-3). The assumption that predator-prey interactions are responsible for the cycles, however, can be questioned. Perhaps the most telling argument

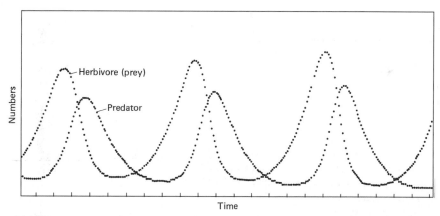

Figure 9-2. A graphical representation of the Lotka–Volterra equations depicting predator-prey oscillations. The differential equations are $dH/dt = (r_h - aP)H$ and $dp/dt = (bH - r_p) P$, where H = numbers of herbivores (prey), r_h = natural rate of increase of herbivores in the absence of predators, a = decrease in herbivore population growth rate per predator present, P = number of predators, r_p = growth rate of predators in the absence of prey (negative because under such conditions deaths would exceed births), and b = per head increase in predators per prey animal available. Students who have access to a computer may wish to program these equations and experiment with different starting levels and different rates of increase. Equilibrium is attained when $H = r_p/b$ and $P = r_h/a$. (Courtesy of P. I. Pavlik.)

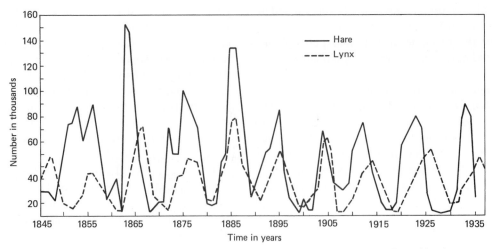

Figure 9-3. Changes in the abundance of the lynx and the snowshoe hare, as indicated by the number of pelts received by the Hudson's Bay Company. This is a classic case of cyclic oscillation in population density. (Redrawn from MacLulich 1937.)

against the theory of interaction is that hare populations on islands, where there were no lynxes, fluctuated in general synchrony with hare population on the mainland. Moreover, a ratio of 1 lynx per 4500 hares prevailed in some years of peak hare populations. Even the hungriest of lynxes could not reduce a hare population of such magnitude (Keith 1963).

The Lotka–Volterra equations assume a constant food supply for the prey population. The equations also assume that all other factors, such as disease, climate, and other predators, exert a constant effect on the prey population. We must conclude, therefore, that a system in which the number of prey depends directly on the number of primary predators, and vice versa, grossly oversimplifies what occurs in nature.

LABORATORY STUDIES OF PREDATOR-PREY SYSTEMS

A number of laboratory studies have been conducted that use a single prey and a single predator population. Huffaker (1958) conducted a particularly interesting and informative study of mites living on the surfaces of oranges (wildlife students may find experiments using microscopic arthropods unexciting, but counting mites on oranges as compared with deer and wolves in forests presents the experimenter with many logistic advantages while revealing similar principles). Oranges in each experiment were placed in various spacings and arrangements, thereby altering the movements of the mites within their habitat.

In the absence of predators, the populations of herbivorous (or orangivorous) mites (*Eotetranychus sexmaculatus*) grew in a sigmoid fashion and then fluctuated widely about a mean of 3000 mites per orange. When Huffaker stocked the oranges with both predatory mites (*Typhlodromus occidentalis*) and herbivorous mites, and both kinds of mites could move from orange to orange, the prey and then the predators began to increase. Thereafter, however, the flourishing population of predators overate and eliminated the prey population. The lack of food then caused the extinction of the predators.

The next experiment was conducted in a complex environment of single oranges and clumps of oranges separated by varying distances. Tiny wooden posts inserted in the oranges offered the prey mites anchors for their silken threads. With a breeze for assistance, the prey population could disperse and colonize new oranges a step ahead of the predacious mites. The complexity of such an environment helped sustain the predator and prey populations through three fluctuations (Fig. 9-4).

Although Huffaker (1958) conducted his experiments with microscopic invertebrates, the results are pertinent to wildlife management. The outcome of placing 10 deer and 1 mountain lion together in a pen is quite predictable: extermination of the deer, just as predatory mites eliminated their prey from a single orange. In a larger pen, however, deer have a greater probability of surviving. If the area within the pen is not only large but also complex—including areas where an ambush might be avoided—the ad-

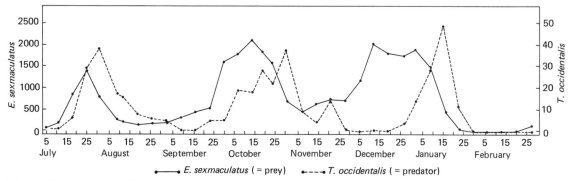

Figure 9-4. Three oscillations in density of a predator-prey relation in which the predatory mite, *Typhlodromus occidentalis,* preyed upon the orange-feeding six-spotted mite, *Eotetranychus sexmaculatus.* These oscillations were observed after mites were given a complex environment permitting emigration and immigration of mites from the surface of one orange to another. (From Huffaker 1958.)

vantages for deer are increased even more. The conclusion seems clear: mobility, adequate food, and suitable cover assure the survival of prey species. Conversely, predation may be excessive in unsuitable environments (e.g., in large areas with poor food or cover or in highly fragmented habitats of otherwise good quality).

FIELD OBSERVATIONS OF PREDATOR-PREY SYSTEMS

Errington (1956, 1967) strongly influenced the thoughts of many wildlife ecologists about the role of predation in animal communities. Much of his work concerned muskrats (*Ondatra zibethicus*), social interactions within muskrat populations, and the role of mink. These studies indicated that mink predation was greater when muskrat numbers were high. However, all muskrats were not equally vulnerable to predation because of social differences that arose within the dense population. Those muskrats that became social outcasts—most of which would die from starvation or disease—were almost exclusively the ones killed by mink. Indeed, Errington regarded most such muskrats as "walking corpses," a reflection on their predisposition to a more immediate cause of death (i.e., predation before starvation). Errington thus felt that social interactions in crowded populations were more significant than predation as a means of limiting numbers of prey.

From Errington's work we learn that the total effect of predation cannot be assessed simply by counting the number of animals killed by predators. Mortality factors such as disease, starvation, and predation frequently tend to be *compensatory* rather than *additive* (see Chapter 5). For example, if a certain percentage of animals dies from predation, then the percentage of animals dying from starvation will be diminished in comparison to that of a predator-free population. Conversely, if a certain percentage starves to death, then a smaller percentage of animals will die from predation. Such an effect is said to be compensatory—that is, one cause of mortality largely replaces another—whereas, additive mortality involves summing the percentages dying from each cause of death.

Some insights about the ways predators may control prey populations can be seen in the responses of small mammals that prey on the larvae of pine sawflies (*Neodiprion sertifer*) in southern Canada (Holling 1959). The eggs of pine sawflies hatch in the spring after overwintering on the ground; the larvae then creep up pine trees and begin to feed on the needles. In early June, the larvae drop from the trees and then spin cocoons that remain in the ground litter until late September. Three species of small mammals prey on the cocoon-wrapped larvae at this time: masked shrews (*Sorex cinereus*), short-tailed shrews (*Blarina brevicauda*), and deer mice (*Peromyscus maniculatus*). Each species, however, extracts the sawfly larvae in a different way, and the opened cocoon cases thus present clear evidence about which species of predator consumed the larvae. This evidence, plus the unopened cocoons of living sawfly larvae, permitted a unique opportunity for detailed sampling of a predator-prey community in a natural setting.

Holling (1959) found that the density of prey influenced both the feeding behavior of the predators and the number of predators. In Figure 9-5a, we see that

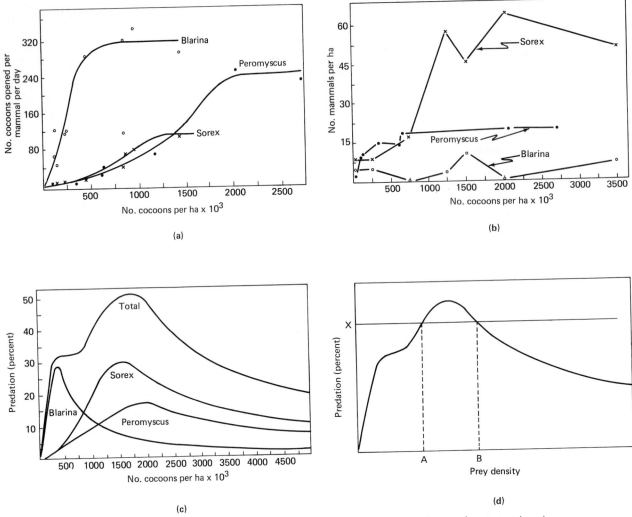

Figure 9-5. Responses of three species of small mammal predators to density of a prey species, pine sawfly cocoons: (a) functional responses; (b) numerical responses; (c) functional and numerical responses combined; and (d) theoretical model showing relationship between prey density and regulation of prey population by predation, where *x* = an arbitrary level of predation. (From Holling 1959.)

an average predator consumes more prey per day when more prey are available—up to a certain limit, probably to the point where the predator simply can eat no more. The tendency of predators to shift their diets toward an abundant prey is known as a *functional response*. The functional response of *Blarina* was relatively steep and that of *Sorex* more gradual. The *Sorex* population was satiated at a lower level than *Blarina*, no doubt because *Sorex* are much smaller animals than *Blarina*.

A *numerical response* occurs when the numbers of predators increase with an increase in the density of the prey population. The numerical response of the

three small mammal predators to densities of sawfly cocoons developed in two different ways (Fig. 9–5b). *Sorex* and *Peromyscus* populations responded to increased densities of sawfly larvae with increases in their own numbers, again up to a certain point. The *Blarina* population, however, did not show a numerical response and apparently was held in check by something other than food limitations. *Peromyscus* and *Sorex* populations responded numerically until some factor other than food limited additional growth.

The combination of functional and numerical responses determines the total effect of predation on a prey population at a particular density. For example, at a density of 250,000 cocoons per ha, each *Sorex* consumes an average of about 10 cocoons. At that density of cocoons, there are approximately 12 *Sorex* per ha, and these predators removed about 120 cocoons per ha. To calculate the total number of cocoons eaten per day by the predators, one multiplies the number of individuals of each species of predator by the number of prey consumed daily by each predator. For the pine sawfly–small mammal community, the effect of each predator and the cumulative effect of all species is shown in Fig. 9-5c for the 100-day period when cocoons were available to predators. Note that each predator population contributes in various ways (i.e., at different prey densities) to the losses experienced by the prey population.

If the reproductive output of a prey population exceeds the losses to predators, then predation is not itself a limiting factor. For example, reproduction may increase a prey population by 35 percent each year, but predators may remove only 20 percent, thereby presenting no direct limitation on the continued growth of the prey population. If, however, reproduction remains unchanged in the prey population but predators remove 40 percent, then predation indeed will cause a decline in the prey population. Thus, predation can control and reduce prey populations at certain densities (between A and B in Fig. 9-5d). At low and high densities of sawfly larvae, however, predators did not remove enough larvae to prevent an increase in the prey population (Holling 1959).

Even though the small mammal–sawfly community represents a somewhat specialized predator-prey system, the interplay of functional and numerical responses of predators to various densities of prey nonetheless illustrates sound principles. These responses have important applications when managers consider whether predators are responsible for the decline of prey populations. Questions about the effects of predation on prey populations cannot be answered without addressing such issues. Yet the simplicity of the small mammal–sawfly study should not cause us to overlook other factors related to predation. In other situations, prey animals are usually more active than insect cocoons, escape cover is usually available for prey, and most wildlife communities include alternate sources of food for predators, each interacting with their own food supplies. Leopold (1933a) therefore proposed five factors that

must be considered when one pieces together the puzzle of predator-prey interactions:

1. Density of the prey population.
2. Density of the predator population.
3. Characteristics of the prey, including reactions to predators, and nutritional condition.
4. Density and quality of alternate foods available to the predator.
5. Characteristics of the predator, such as its means of attack and food preferences.

Hamerstrom et al. (1985) conducted a long-term study in Wisconsin of interactions between northern harriers (*Circus cyaneus*) and their primary prey, meadow voles (*Microtus pennsylvanicus*). An extremely close relationship existed between the number of harrier nests and the abundance of voles, except during a period (1964–70) of heavy DDT applications (Fig. 9-6). The results in this case strongly suggest that the size of the predator population depends on the abundance of prey. Harriers, however, are migratory and do not exert year-round pressure on the vole population in Wisconsin. An unusual finding from this study concerned the influence of prey density on the mating systems of the predators. Harriers are primarily monogamous, but polygyny frequently occurred in years when voles were abundant. Polygyny apparently prevails only when voles are abundant because male harriers bring food to incubating females. Thus, one male probably could not supply several females with food when voles were scarce. Additionally, subadult harriers bred more often during periods of vole abundance. Thus, the abundance of the prey—voles—governed not only the abundance of the predator, but also influenced the adoption of specific mating systems.

Hornocker (1970) studied mountain lions and their prey in the Idaho Primitive Area for 4 years. He found that the physical condition of elk, except of large antlered bulls, bore no relationship with their susceptibility to lion predation. Mountain lions hunt by ambush rather than by pursuit. Thus, the physical condition—good or bad—of a deer or an elk was of little value in avoiding mountain lions. Elk and mule deer (*Odocoileus hemionus*) populations each increased during the study, indicating that lions were not limiting the size of prey populations. The research indicated, however, that lion predation was a significant factor that dampened violent oscillations in the numbers of deer and elk. The removal by lions

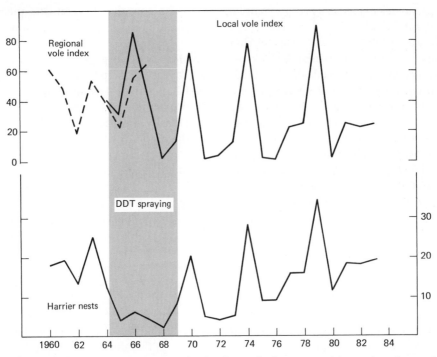

Figure 9-6. Relationship between the abundance of voles (prey) and the number of nests of northern harriers (predators). A close tie is indicated except when DDT was applied heavily (shaded) and disrupted nesting efforts in the harrier population. (From Hamerstrom et al. 1985.)

of 200–300 or more prey animals per year reduced range damage in the study area. Predation in this case averted overstocking before other factors began to limit the deer and elk herds. In the absence of predators, elk and mule deer have overpopulated their ranges and have suffered large-scale starvation (Cahalane 1948; Rasmussen 1941).

Perhaps the most famous study of predator-prey relationships has been conducted on Isle Royale, a national park in Lake Superior, where wolves, moose, and beaver (*Castor canadensis*) exist in a relatively undisturbed environment (Mech 1966; Peterson 1977; Allen 1979). Moose established themselves on Isle Royale in the early 1900s and, by 1925, had replaced woodland caribou (*Rangifer tarandus*) as the dominant herbivore. In a setting lacking predators, the moose population grew to an estimated 1000–3000 animals by 1920, overbrowsed their range, then crashed to several hundred in the mid-1930s. Two fires in 1936 renewed the browse supply, and the numbers of moose increased once, then crashed again in the mid-1940s. Wolves crossed the ice to Isle Royale about 1949. In 1959, Durward Allen and a research team from Purdue University began what was

to become a long-term study of predation in the natural laboratory of Isle Royale.

Moose and wolf numbers remained relatively stable between 1959 and 1968. The wolves, most of which traveled in a single large pack of 15–22 animals, regularly encountered and "tested" moose. When tested, a healthy moose lowered its head and faced the pack, threatening and ready to attack with its forefeet any wolf that might charge. Faced with such a dim prospect, wolves bluffed for a few minutes and then retreated. However, a moose that fled and tried to outrun the pack in the snow frequently would be attacked from behind, brought down, and killed. Physical weakness apparently influenced the success of these tests because disproportionately high numbers of old moose were killed between 1959 and 1968 (Table 9-1). Healthy moose, including cows protecting their calves, successfully defended against most attacks. Thus, wolves were taking old and presumably diseased moose and consuming practically all of the available meat on each carcass. Browse for moose was sufficiently plentiful on Isle Royale during this period, and Mech (1966) concluded that wolf predation was maintaining the

Photo courtesy of Elva Hamerstrom Paulson

The Hamerstroms,
Frances (1907–) and Frederick (1910–1990)
Field Biologists

The wildlife profession can honor few husband-wife teams, but of these few, the Hamerstroms reign supreme as field biologists and as partners in a 59-year marriage. After earning a B.A. degree at Harvard, Frederick—"Hammy"—began his wildlife career learning artificial propagation of game birds in New Jersey. He later received a M.S. degree in 1935 working with the pioneering wildlife biologist, Paul L. Errington (1902–1962), at what is now Iowa State University. In 1937, the two men subsequently published the first paper in the first volume of *The Journal of Wildlife Management.* Frances—Fran, with a soft *a*—was very much a part of these early studies of pheasants, quail, and raptors, and she, Hammy, and Errington wrote a paper about owl predation that won The Wildlife

Society's first publication award (in 1940). In 1941, Hammy earned a Ph.D. under Aldo Leopold while Fran earned a master's degree, thereby becoming the only woman to earn a graduate degree under Leopold's tutelage.

During the Depression years, the Hamerstroms began their long-term work with prairie chickens in Wisconsin, but they also studied other kinds of grouse as well as deer, sandhill cranes, and furbearers and continued their lifelong interest in raptors. Hammy, along with Fran and another author, again won a publication award (in 1957) from The Wildlife Society for *A Guide to Prairie Chicken Management.* Hammy was a gifted technical writer, editor, and critic and applied those skills to his own work as well as those of Fran and many others. Northern harriers were a special interest of Fran, and her two-decades-long study of the species, its nesting ecology, and the abundance of prey led to a book about these relationships (Hamerstrom 1986). After retiring, the Hamerstroms continued working with raptors, especially Harris hawks and ospreys, during winter trips to Texas and Mexico.

Of special note is Fran's engaging book, *Strictly for the Chickens* (1980), about her life and experiences with Hammy while they studied prairie chickens in Wisconsin. She also wrote several books for children. Although born into Boston society, Fran and Hammy lived much of their lives in a well-worn farmhouse, where they established a research headquarters, raised a family, and warmly welcomed and informally trained a generation of students.

See Anderson (1991) and McCabe (1991) for memorials to Hammy; see also Bolen (1992). For Errington, see Scott (1963).

---

moose population at levels below those at which food became limiting.

Social dominance restricted breeding in the large wolf pack. Only the dominant male and his mate produced offspring, and the pups survived roughly in proportion to vacancies created by deaths of other members of the pack. Overall, wolves, moose, and browse seemed in "balance" on Isle Royale.

Beginning with the winter of 1969, however, weak links developed in the browse–moose–wolf food chain. Unusually deep snow accumulated during the winters of 1969, 1971, and 1972. Moose congregated in the conifers along the edges of Isle Royale, thereby depleting the balsam fir (*Abies balsamea*) browse in these lowland areas. Some moose starved, and calves born in the spring following a severe winter began life in poor nutritional condition. Wolves responded in

TABLE 9-1. Age Composition by Percent of Moose Killed by Wolves on Isle Royale, 1959–68 (177 Moose) and 1969–76 (351 Moose)

	Percent Killed	
Age	*1959–68*	*1969–76*
Calves	24.8	42.2
1–5 Years	9.6	18.2
6–10 Years	32.8	23.4
11–15 Years	32.8	16.2

Source: Allen (1979).

various ways to the weakened condition of their principal prey. Thus, between 1969 and 1976, the proportion of calves taken by wolves increased from about 25 percent to 42 percent, and moose in the prime of life became more vulnerable than before (Table 9-1).

At times, wolves resumed hunting before the pack had finished consuming an earlier kill. New packs formed as the numbers of wolves increased. By the winter of 1979–80, approximately 50 wolves on Isle Royale were apportioned among five different packs, a duo, and several loners (R. O. Peterson, personal communication). Social strife developed, and some wolves attacked and killed other wolves in 1980 and 1981. The numbers of moose declined in the 1970s as the wolf population increased. Then, in the winter of 1980–81, wolf numbers dropped suddenly from 50 to 30. Researchers found three dead wolves in malnourished condition and, by the winter of 1981–82, the wolf pack on Isle Royale was down to 14 animals. Small gains and losses occurred thereafter, and 15 wolves remained by the winter of 1986–87. Meanwhile, moose numbers increased rapidly from about 600 in 1981–82 to 1400 in 1986–87, a response probably associated with a series of moderate winters and better calf survival because of improved nutrition and lower wolf predation. R. O. Peterson (personal communication) noted the improved physical vigor of the moose population during the mid-1980s and the high proportion of middle-aged moose that lacked arthritis, a disease that increases the vulnerability of older moose to wolves. Biologists monitoring the Isle Royale community predict, however, that successful wolf attacks will increase as the moose population ages. Consequently, the wolf population will enlarge, possibly bringing about a predator-prey cycle with a periodicity of about 38 years (Peterson et al. 1984).

Simple generalizations still cannot be made about predator-prey relationships, even after more than 25 years of intensive research on Isle Royale. An undeniable value of the research is that the work has been conducted in an environment—a national park—where humans have not complicated the setting, as has happened in studies conducted in Alaska and British Columbia (see following). One lesson, however, from Isle Royale is that 10 years of study may not be enough to determine the full range of ecological settings that may develop, even in relatively simple natural communities; that is, had the work stopped after a decade of study—as it might well have, considering the length of most research programs—the dramatic shift in the age structure of the prey would have remained undetected (Table 9-1). The unpredictable nature of weather and its influence on the wintering behavior of moose and the effects of moose behavior on their own food supply were critical factors in the changing predator-prey equation on Isle Royale. If a fire rejuvenated browse

production, for example, the new food supplies undoubtedly would affect the numbers and physical condition of the moose population (see Fig. 15-3, Chapter 15). Such events, in turn, would influence the hunting success of wolves and their numbers. Finally, the individual personalities of the leaders of wolf packs may be significant. The leader's degree of dominance over the members of the pack, its abilities regarding travel routes and attack strategies, and its regulation of mating among other wolves in the pack may play an important role in the coherence and hunting success of the pack, much as leadership in human societies may determine the fate of a large group of people.

Long-term studies in northern Minnesota also have yielded valuable information on wolf behavior and the effects of wolf predation on prey populations (Mech et al. 1971a, b; Mech and Frenzel 1971; Mech 1973; Mech and Karns 1978; Nelson and Mech 1981). In Minnesota, the diet of wolves includes beaver and moose, but white-tailed deer are the major prey. Wolf predation intensified the losses of deer during severe winters when the restricted availability of browse already was reducing the deer population. Deer that survive times of heavy predation do so by occupying the seams—vacant areas—between wolf territories (Mech 1977b; Nelson and Mech 1981). Wolves trespass more often into the territories of other packs when prey populations are at low densities. For such transgressions, the invaders at times have been attacked and killed by the owners of the territories (Allen 1979; Smith et al. 1987).

Studies of snowshoe hare populations in central Alberta have shed some light on the role of predation in cyclic populations (Meslow and Keith 1968; Rusch and Keith 1971; Keith and Windberg 1978; Keith et al. 1977, 1984). Besides the snowshoe hares, the components included vegetation, ruffed grouse (*Bonasa umbellus*), great horned owls (*Bubo virginianus*), lynxes, and their interactions during a 17-year period in which the hare population crashed twice. Highlights of these studies, summarized by Keith and Windberg (1978), emphasized that a decline in a hare population begins when high numbers of hares produce a shortage of aspen (*Populus* spp.) twigs and other winter food. The food shortage results in high mortality of young hares, lowered reproduction the following summer, and consequent reduction of the size of the hare population. The predator-to-prey ratio thus increases so that owls and lynxes exert a greater effect on the hare population by killing a larger proportion of those that remain available. Predation intensifies and continues depressing the hare population. As hares become scarcer, however, the predators turn increasingly to ruffed grouse, thereby initiating a decline in the grouse population. Eventually, the overall shortage of both species of prey diminishes the size of the predator populations, mainly from emigration and reduced survival of immature owls and lynxes. The scarcity of hares thereafter permits regeneration of the vegetation and increased reproductive success in the remaining hare population (a female hare may produce 15–18 young per year in a good year and half that in poor years). More hares also survive the winter because aspen twigs again are available. When the number of hares then reaches the point where the demand for food outstrips the supply of vegetation, the hare population begins to decline and another cycle begins.

In the situation just described, predation plays its greatest role by intensifying and prolonging the reduction in hare numbers and not by initiating the decline nor by preventing the recovery of the depressed prey populations. A secondary prey species—known as a *buffer* species—also was involved in the cycle in Alberta. In this case, the buffer species was ruffed grouse, which absorbed some predatory pressure when the hares were at low numbers. Conversely, hares absorbed the pressure when grouse were at low numbers.

Based on 60 radio-collared and 31 ear-tagged lynx (*Lynx lynx*) in the Northwest Territories, Canada, Poole (1997) determined that young lynx dispersed whether or not their primary prey (snowshoe hares, *Lepus americanus*) were adequate. However, adults as well as young lynx dispersed when snowshoe hares declined to very low densities. Dispersing lynx traveled up to 930 km; none returned to their original home range. Nine lynx did not disperse and instead remained near their established home ranges, and each of these died of starvation, thereby illustrating the consequences of staying home when food is limited. Studies of bobcats also verify that some adults may disperse in response to prey shortages (Knick 1990). In general, both bobcats and lynx remain attached to their established home ranges until food shortages become critical (usually in winter), then leave to search for vacant areas with adequate food resources—and often fail.

In Sweden, Erlinge et al. (1984) studied the interactions among nine species of predators—five mammals and four birds—and three species of prey, predominantly field voles (*Microtus agrestis*) and rabbits (*Oryctolagus cuniculus*). In this 6-year study, predator and prey populations tended to dampen the cycles or regulate each other, thereby contrasting with other studies in which predators deepen and prolong population cycles of prey species (e.g., Keith et al. 1984 and others) or show essentially no impact on prey populations (Errington 1946). In the Swedish study, predators took a number of voles about equal to the annual production. Two types of predators were identified: generalists and specialists. Generalists, such as red foxes, ferrets (*Mustela putorius*), and domestic cats (*Felis catus*), found buffer species when vole numbers reached their low points each year. Generalist populations accordingly remained fairly constant throughout the study. Specialist predators, such as eared owls (*Asio otus*) and ermines (*Mustela erminea*), depended heavily on voles. Populations of these predators fluctuated more widely, as they were apparently sensitive to small changes in the number of voles each spring. The rabbit population crashed after a severe winter, and increasing predation on voles by generalist predators caused declines in the populations of specialist predators. The conclusion was that rabbits acted as a buffer for the voles, permitting the existence of a relatively constant number of generalist predators during periods when voles were scarce. When rabbit numbers declined, heightened competition between generalists and specialists for the limited numbers of voles caused a reduction in the numbers of specialists, and the vole population stabilized. Voles generally follow a cyclical pattern in northern Sweden, however, where there are few buffer species.

FIELD EXPERIMENTS WITH PREDATOR-PREY SYSTEMS

Predators sometimes have been removed from local areas in experiments designed to assess their impact on prey populations. Beasom (1974) removed 188 coyotes and 120 bobcats from a 2186-ha area in south Texas and compared fawn production of white-tailed deer on the experimental area with a nearby control area with similar vegetational and topographic features. The results gathered during a 2-year period showed a striking increase in the fawn:doe ratio in the area from which predators had been removed (Table 9-2).

Such results suggest that deer production could be increased significantly under a regime of effective predator control. Beasom (1974) stated that "It seems probable that white-tailed deer densities could be increased in this area with intensive predator control

TABLE 9-2.   Fawn:Doe Ratios on an Experimental Area From Which Predators Were Removed and From a Control Area Without Predator Removal in South Texas

Year	Experimental Area	Control Area
1971 (drought year)	0.47	0.12
1972 (normal year)	0.82	0.32

Source: Beasom (1974).

TABLE 9-3.   Observed Fawn:Doe Ratios in White-Tailed Deer Populations Inside a Predator-Free Exclosure and Outside the Exclosure

Cohort and Date	Outside Exclosure	Inside Exclosure
1972 cohort		
September 1972	0.49	0.47
1973 cohort		
September 1973	0.29	0.66
March 1974	0.31	0.47
1974 cohort		
September 1974	0.47	0.63
December 1974	0.46	0.51
March 1975	0.43	0.45
1975 cohort		
September 1975	0.32	0.45
December 1975	0.34	0.40
March–April 1976	0.31	0.31
1976 cohort		
September–October 1976	0.40	0.61
1977 cohort		
September 1977	0.60	0.73
December 1977	0.66	0.55

Source: Kie et al. (1979).

efforts if other compensating sources of mortality did not become immediately operative." The part of that sentence concerning compensating mortality turned out to be important.

Kie et al. (1979) conducted a longer study on the Welder Wildlife Refuge in south Texas, in which 14 coyotes and 2 bobcats were removed from a 361-ha exclosure. (Deer could jump the fence surrounding the exclosure—moving in or out—but only a few actually did so. However, by design, the fence retarded the entrance of predators.) Fawn:doe ratios were compared between white-tailed deer populations inside the exclosure and outside on an adjacent control area where predators had not been removed. In 1972, before predators had been removed, fawn:doe ratios between the experimental and control areas were nearly identical (Table 9-3). In the absence of predators, however, fawn:doe ratios increased inside the exclosure for a few years, and the density of deer increased from about 35 to 84 animals per km². Thereafter, deer in the exclosure began to show signs of malnutrition; fawn:doe ratios in winter and early spring again were similar to those in the control area. Such conditions in the deer herd in the exclosure indicated that winter mortality of fawns was compensating for the earlier losses to predators (Table 9-3). After 3 years, deer densities began to decline inside the exclosure, and after 5 years, many adult deer— apparently weakened by malnutrition—succumbed to parasites. By 1978, the density of the deer population in the exclosure returned to the same level as in 1972—before predators had been removed (Fig. 9-7). Removing predators simply increased the numbers of deer for a short time. Then other causes of mortality, namely malnutrition and parasites, returned the herd to its original density.

Predator control, however, has increased the nesting success of ducks in experimental areas. In North Dakota, Duebbert and Lokemoen (1980) controlled red foxes, striped skunks (*Mephitis mephitis*), raccoons (*Procyon lotor*), and badgers (*Taxidea taxus*) in a 259-km² area

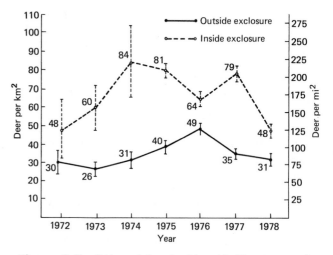

Figure 9-7.   Estimated deer densities with 95 percent confidence intervals. Estimates for 1972–74 derived from ground transect counts in summer; estimates for 1975–78 derived from helicopter counts in January. Numbers equal mean values in deer per km². (From Kie et al. 1979.)

for a 2-year period with strychnine baits, trapping, and shooting. A 51-ha plot of natural prairie near the center of the area was searched intensively for duck nests, and the fate of each nest was determined. Ninety-four percent of 756 duck nests hatched in the near absence

TABLE 9-4. Comparison of Four Measures of Duck Reproduction on Treated Sites (Predators Removed) With Untreated Sites (No Predators Removed) on the Agassiz National Wildlife Refuge, Minnesota

	Success of Simulated Nests	Success of Natural Nests	Breeding Pair Counts (Mallards, Pintails, Gadwalls)	Number of Ducklings Produced
Treated	531/654 (81%)	146/247 (59%)	3176	7571
Untreated	239/699 (34%)	32/112 (29%)	3140	4858

Source: Balser et al. (1968).

of mammalian predators. By contrast, only 56 percent of 320 duck nests hatched on nearby areas where predators were not controlled. Very high nest densities in this study—up to 3 nests per ha—were attained in the predator-controlled area. The nests of some mallards (*Anas platyrhynchos*) were as close as 2 m to each other, which suggests that even higher nest densities might be achieved without territorial strife.

In a similar study, mammalian and avian predators on the west half of Agassiz National Wildlife Refuge in Minnesota were poisoned and trapped during a 3-year period (Balser et al. 1968). No predators were controlled on the east half of the 15,400-ha refuge. Then, in a succeeding 3-year period, the practices were reversed (i.e., predators were removed on the east half *versus* no control on the west half). Waterfowl production was assessed by (1) numbers of breeding pairs, (2) numbers and sizes of broods, (3) fates of natural nests, and (4) fates of simulated duck nests (i.e., several chicken eggs were placed in appropriate cover; these "hatched" for purposes of the experiment if the artificial clutch survived a normal laying and incubation period). The results clearly indicated that predator control increased duck production (Table 9-4). Sargeant et al. (1984) examined remains of ducks that were found at the dens of red foxes and estimated that foxes may kill as many as 900,000 adult ducks per year in midcontinental North America. The majority of the remains were those of females, which for mallards offered a hypothesis explaining the imbalanced sex ratios present in mallard populations (see Johnson and Sargeant 1977). Sargeant et al. (1984) concluded that management schemes that curtail fox predation have the potential of increasing duck production.

These studies indicate that localized predator control can increase waterfowl production, at least under some circumstances. Obviously, predator control would be out of place on sites where maintenance of natural communities is the primary management goal. If, conversely, maximum production of ducks for sport shooting is the goal of

management, then predator control might be justified. The costs of such programs, however, must be carefully considered. If predator control is practiced for increased duck production, the cost of such control should be balanced against the increased number of ducks actually shot by hunters. For example, if predator control costs $1 for each additional duckling that hatches, but only 1 of every 2 ducklings reaches flying age, then the cost rises to $2 per duck available during the hunting season. Moreover, if only 1 of every 5 additional flying-age birds is shot, then the cost of predator control reaches $10 per bird.

Additional study also is needed to determine the limits to which breeding ducks can be crowded into available habitat and whether the high local densities of duck nests achieved by predator control can be sustained for many years. The beneficial aspects, if any, of predator control may be of short duration. Predators normally reinvade nesting areas; in one study, the effects of predator removal lasted only 9 months after control measures stopped (Duebbert and Kantrud 1974). The long-term and perhaps unanticipated effects of predator control also should be examined. In the absence of predators, for example, Balser et al. (1968) noted that ducks began to nest in the open. Hence, if predator control stopped for economic or social reasons, then virtually all production might be lost during the next nesting season as predators reinvade the area.

These and other studies have underscored the interaction of predation and cover. Duck nests concentrated in the remaining areas of suitable cover may suffer inordinately high rates of predation. The same number of nests distributed randomly over a larger area probably would suffer fewer losses. Hence, where intensive uses of the land leave only narrow strips of nesting cover at the edges of fields and wetlands, predators can concentrate their searches and destroy more nests. In effect, the remnants of linear cover become "predator lanes." Conversely, blocks of dense cover act as barriers to predators and reduce the linear edge of cover available as travel lanes for mammalian predators. Thus, lacking either the finances or the desire to reduce predator populations, management that is geared toward improved nesting cover will itself significantly enhance breeding conditions for waterfowl (Duebbert 1969; Duebbert and Kantrud 1974; see also Schrank 1972).

An unexpected source of predation on ducklings comes from northern pike (*Esox lucius*). Solman (1945) estimated that about 10 percent of the ducklings produced in northern Saskatchewan are eaten by pike. About 3.8 percent of 759 pike stomachs that contained food included remains of ducklings. The calculated loss was one duckling per 0.57 ha of water in June and July, a rate that perhaps doubled for the entire brood season. Pike thus may kill more than 1 million ducklings on the deltas of the Athabasca and Saskatchewan rivers (Solman 1945). Lagler (1956b) studied predation at Seney National Wildlife Refuge in Michigan, where both pike and duck broods were commonplace but with far different results. More than 1200 pike stomachs were examined, but only 3 (0.2 percent) contained ducklings. Duck broods also were observed for 5535 "duckling minutes" without the occurrence of pike attacks. Furthermore, the vulnerability of ducklings was tested when a tethered mallard duckling was towed across a pool from which pike were caught on fishing lures immediately before and after the experiment. However, the duckling was not attacked. Lagler (1956b) concluded that pike predation was an extremely minor cause of duckling mortality at Seney. He suggested that the greater availability of buffer species such as bullheads (*Ictalurus nebulosus*), bluegills (*Lepomis macrochirus*), and several species of minnows at Seney might account for the lessened vulnerability of ducklings to pike.

From the studies reviewed here, the question as to whether predators control prey populations has not been answered clearly. The reason is that, in some situations, predators *do* control prey populations and in others they *do not*. The U.S. Fish and Wildlife Service (1978) reviewed some 58 North American studies of large predators, including bobcats, lynxes, mountain lions, coyotes, wolves, bears, and golden eagles (*Aquila chrysaetos*). Prey species included deer, pronghorns (*Antilocapra americana*), and wild sheep (*Ovis canadensis*). In 31 studies, predators were regarded as having a limiting or regulating effect on their prey populations, and in the other 27 studies, predators did not control or limit prey populations. Clearly, we cannot yet generalize about

those circumstances in which predators control natural populations of prey. Predation remains just one of several factors that bring deaths into balance with births. Predation thereby contributes to the elastic stability exhibited by natural communities.

Public perceptions of predators and their prey vary widely. Prejudices still exist in European and American cultures in which wolves and bears are portrayed as ferocious killers of humans, as well as causes for diminished populations of deer, elk, and other big-game species. Hunters commonly blame hawks, owls, foxes, and coyotes for depressed quail, grouse, and rabbit populations. On the other hand, predators also are regarded as taking only sick, disabled, and old prey, thereby benevolently culling unfit individuals and improving conditions for the remainder of the prey population. In this view, predators prevent overpopulation and starvation in prey species. Moreover, one well-known book, *Never Cry Wolf* (Mowat 1963), and a film by the same name suggested that wolf diets include large proportions of rodents and other small mammals—a notion that is completely unsupported by scientific evidence.

WOLF CONTROL IN ALASKA

As arguments about coyote control continue in the western United States (see following), an equally intense controversy has developed in Alaska over the killing of wolves. Wolf control in certain areas has been carried out under the auspices of the Alaska Fish and Game Department as a means of enhancing the recovery of moose and caribou populations. Justification and rationale for removing wolves in recent years have been found in an analysis and scientific review of interactions among wolves, their prey, and humans in Alaska (Gasaway et al. 1983). In that study, wolf, moose, and caribou populations were evaluated in a 17,000-km² experimental area in which wolves were shot and trapped in 1976–79. Those populations then were compared with populations in several nearby control areas where wolves were not removed. Estimates of wolf populations in the experimental area had increased from about 100 in the early 1960s to about 240 in the early 1970s. Moose populations declined from a high of about 23,000 in 1965 to about 3000 in 1975, when the first wolves were removed. Wolves in all of the study areas fed primarily on moose; an examination of wolf stomachs revealed 95 percent moose remains, 3 percent Dall sheep (*Ovis dalli*), and 2 percent small birds and mammals. The Alaskan study provides at least provisional answers to several ecological questions about relationships between large-bodied predators and prey.

Some of these questions and their answers include the following:

Did reproduction in the surviving wolves increase when the wolf population was thinned, thereby negating the control program? (See the density-dependent response shown for coyotes in Table 9-6.) No. At removal rates of 38 to 61 percent of the autumn populations, neither the percentage of reproductively active females nor the number of pups per female changed significantly. Based on estimates, the wolf population did decline from about 239 to 80 animals during the four-year removal period.

Did prey populations increase after wolves were controlled? Yes. Cow and calf moose survival more than doubled, and the mortality rates for adult moose outfitted with radio collars declined from 20 percent in the 3 years before wolves were removed to 6 percent in the 3 years afterward.

If prey populations increased in the area where wolves were removed, did prey populations also increase in the areas where wolves were not removed? Yes. Prey living in areas adjacent to the experimental area apparently benefited from the removal of wolves whose ranges overlapped into both areas. Prey populations remained depressed in a study area that was not immediately adjacent to the area where wolves were removed.

Was wolf predation the primary cause of moose mortality? Yes. Wolves killed all 20 moose wearing radio collars that died from causes unrelated to human activities. Six radio-collared moose were killed by hunters during the study.

Was hunting by humans a contributing factor in the decline of moose? Yes. In 1970–74, hunting mortality was 10 percent and apparently equaled the annual recruitment rate of yearlings. At such a rate, the moose population could not increase, even in the absence of wolf predation. A reduction in hunting kill was recommended.

Was severe weather a factor in the moose decline? Yes. The severe winter of 1970–71 reduced the prey:predator ratio, thereby increasing the impact of wolves on moose.

Did low numbers of moose cause a decline in wolf numbers? Yes, but the decline in wolves occurred more slowly than the drop in moose numbers. The effect intensified, however, as the moose:wolf ratio continued falling.

Would moose populations recover without humans interfering with the wolf population? Yes, but both predator and prey numbers would probably remain very low for a decade or more.

With such convincing arguments in favor of wolf control, why does controversy surround current proposals of the Alaska Fish and Game Department for initiating similar measures in other areas? Many protectionists fear that wolves may be extirpated by an overly enthusiastic control program, as has happened in most of the United States. Some oppose killing any animals, and others simply may want as little human meddling as possible in one of the few natural communities remaining in North America. Gasaway et al. (1983), however, noted that hunting already interferes with the prey population in Alaska; thus, humans have an ecological obligation to adjust the number of predators. The biologists contend that a simultaneous reduction of hunting and predators will accelerate the recovery of the depressed prey populations and, eventually, of the predator population as well. The process will thereby enrich the entire community. In their words,

The "balance-of-nature" concept is firmly entrenched in the public's mind, and it has underlain the teachings of some university courses in wildlife management and ecology for at least two decades. The balanced system envisioned is one that generally remains near an equilibrium through sensitive regulatory feedback mechanisms. A "prudent" predator is required in this system, that is, one that will consume its prey such as to maximize its own food supply while at the same time minimizing the possibility that the prey population will be unable to maintain itself and serve as food for the future.

However, wolves and probably most other predators are not quite so prudent as the theory demands, and the wide oscillations in predator and prey populations are more the rule than are systems balanced near equilibrium (Gasaway et al. 1983).

Ballard et al. (1997) recently investigated the complexities of wolf-prey relationships in northwestern Alaska. In this 6-year study, 86 wolves in 19 packs were tracked with radio telemetry, from which the authors determined that the prey of wolves consisted of moose (42 percent) and migrating caribou (51 percent). However, none of the radio-collared wolves in this study actually followed caribou on migration. Instead, the wolf packs usually maintained year-round territories with an average size of 1868 km², although apparently when prey populations reached extremely low densities within their territories, a few packs occasionally did follow migrating caribou.

Although the data and arguments for predator control outlined here appear convincing, the full story may not yet be at hand. We have learned from the studies on Isle Royale, for example, that various ecological settings (e.g., snow depth) greatly influence the nature of predation. Moreover, it seems possible that long periods of scarcity of both herbivores and predators have occurred in systems free from human interference and that the recovery of vegetation may depend on prolonged periods of animal scarcity.

Since 1985, policies governing wolf control in Alaska have required consideration of the following: determination of a desired level of wolf and prey populations in each management area; permission for hunting wolves from airplanes only if the Board of Game determines that the prey population is severely depressed and that big-game harvests are below long-term averages; that scientific evidence indicates that the cause of low prey populations is wolf predation; and that alternate management methods are not likely to bring about the recovery of depressed prey populations. The board also must act in accordance with the public interest, considering consumptive and nonconsumptive uses of wildlife, and must hold meetings where the public may comment.

Under those policies, wolf control continues in some places (Reardon 1986), whereas control measures planned for other areas have been stopped. One of the latter areas is Unit 20E in eastern Alaska, where field studies conducted by Gasaway and reported by Reardon (1986) revealed low densities of moose and caribou. About half of the 600 moose calves born each spring are killed in their first summer by grizzly bears (*Ursus arctos horribilis*). About 90 additional calves are killed by wolves, and another 90 die of other natural causes. The ratio of moose to the combined predator population of bears and wolves is about 2:1. An earlier recommendation for controlling wolves in Unit 20E thus was rejected because the majority of the moose calves were killed by bears. Similarly, a proposal for removing both bears and wolves lacked favor with the local populace, suggesting the complicating influence of public attitudes toward various types of predators (i.e., bears *versus* wolves in this case).

Wolf control has been a common practice in Canada for a century. With a recent public interest in environmental matters, however, provincial governments are experiencing more pressure for justifying the killing of wolves. In British Columbia, the Ministry of Environment and Parks (1985) detailed the history of wolf management in that province. The history includes bounties, poisoning programs, protection of wolves in some areas, bag limits for wolves in other areas, and, recently, the resumption of control programs in local areas where prey populations were depressed. In British Columbia, as in Alaska, two factions of the public—the protectionists and the consumptive users of wildlife—battle for acceptance of their respective viewpoints. Such arguments are part of the democratic way of life in which decisions are based upon both public will and biological knowledge. Gasaway et al. (1983) commented as follows: "Acceptance of the dramatic natural changes that occasionally occur in predator and prey numbers would simplify the future management of some wildlife resources such as wolves and their prey."

PREDATION ON DOMESTIC ANIMALS

Natural selection has endowed most wild herbivores with the genetic features of alertness, agility, speed, and stealth. Most individuals thus escape predation. Domestic animals, on the other hand, are products of artificial selection, having been bred over the course of centuries for meat, milk, wool, hides, eggs, and docility. On the whole, domestic animals are no match for wild predators, and it becomes almost inevitable that conflicts arise between the interests of livestock owners and those of wolves, coyotes, hawks, and other predators.

The issue is not new. Predator-prey conflicts led years ago to the extirpation of wolves from most of Europe and the United States. Similarly, tigers (*Panthera tigris*) vanished from much of Asia. In modern times, however, the western United States has emerged as the classic arena of confrontation between predators and livestock interests. At one time, it was simply assumed that predators must be eliminated from western rangelands, and all methods of control were regarded as both necessary and acceptable. In the 1960s, however, concerns about wildlife and ecology captured the attention of the American public. There was interest in the role predators play as selective agents on prey populations and in the effects of poisons on nontarget species. Furthermore, the inhumaneness of some methods of predator control attracted much public notice. Demands arose for scientific evaluations of predator-control methods and programs. The legitimate interests of citizens in these matters was fueled by a related issue: privately owned livestock graze on public lands administered by the Bureau of Land Management and the U.S.D.A. Forest Service. Indeed, federal lands in some parts of the west provide up to 50 percent of the grazing lands for sheep (U.S. Fish and Wildlife Service 1978).

Predators of many kinds—golden eagles, foxes, bobcats, mountain lions, and wolves—at times may be of local importance, and cattle, goats, and domestic fowl are among the livestock killed by predators. Nonetheless, in the American West, attention remains sharply focused on sheep and coyotes (Fig. 9-8).

EXTENT OF THE PROBLEM

Data compiled by the U.S. Fish and Wildlife Service (1978) provide an overview of coyote damage in 15 western states; some findings are highlighted here. Losses from predation are not distributed evenly among all sheep ranchers. Fully 45 percent of the ranchers reported no loss of lambs to coyotes, and 67 percent reported no loss of adult animals. But coyotes claimed more than 20 percent of the lambs

on 10 percent of the ranches (Table 9-5). By age class, 8.1 percent of the lamb population and 2.5 percent of the adult sheep population were killed by coyotes. Other causes of mortality, known and unknown, claimed 15.1 percent of the lambs and 7.9 percent of the adult sheep. Assuming an average rate of 4 percent loss to predation, 276,000 fewer lambs were marketed each year because of coyote predation. At an average value of $52.40 per 48-kg lamb, coyotes thus caused a $14 million loss of lambs. An average loss of 1.5 percent of the ewe population

TABLE 9-5. Percentage of Sheep Losses to Coyotes Reported by Western Sheep Producers

Percent Lost to Coyotes	Percent of Producers Losing	
	Lambs	*Adult Sheep*
0	45	67
0.1–10	32	30
10.1–20	13	2
>20	10	1

Source: U.S. Fish and Wildlife Service (1978).

Figure 9-8. Coyotes are the primary target of predator control in the western United States. State, federal, and private wildlife agencies are conducting studies of coyote ecology, searching for ways of reducing coyote predation on sheep and other livestock. Shown here is a group of young coyotes inspecting the remains of a sheep. (Photo courtesy of G. E. Connolly, U.S. Fish and Wildlife Service.)

adds $5 million to this sum, thereby bringing the total loss to $19 million per year attributed to coyote predation.

CONTROL METHODS

Coyotes are controlled with four basic methods (Wagner 1972):

1. *Trapping,* nearly always with steel leg-hold traps using either bait or scent as an attractant. Coyotes rarely enter any type of box trap.
2. *Den hunting* ("denning"), in which pups are killed at den sites, often by asphyxiation using carbon monoxide cartridges thrown into the den. Dens sometimes are excavated and the pups are then destroyed by various means.
3. *Shooting,* in chance encounters or by luring with distress calls; aircraft hunting is particularly effective.
4. *Poisoning,* with any of several substances, including strychnine (now largely abandoned because of its nonselectivity) or compound 1080 (sodium monofluoroacetate) placed in baits. A device known as a coyote-getter also administers poison by firing a cyanide-loaded pistol cartridge directly into the mouth of a coyote pulling on the lure-tipped muzzle of the weapon. A spring-loaded M-44 has replaced the cartridge-fired coyote-getter in recent years.

Since 1964, two special scientific overviews of predation and predator control have been commissioned by the U.S. government (Leopold et al. 1964; Cain et al. 1972).

THE LEOPOLD REPORT.   This first report was prepared at the request of the Secretary of the Interior Stewart Udall by his five-member Advisory Board on Wildlife Management, chaired by A. Starker Leopold of the University of California. Two basic tenets of the report were that (1) all native animals are resources of inherent interest and value to the people of the United States; therefore, husbandry of all forms of wildlife is a proper responsibility of the government; and that (2) local control of wildlife populations is an essential part of management policy where crops, livestock, or human health and safety are damaged or jeopardized.

Before the Leopold Report, predator control was conducted by the Branch of Predator and Rodent Control (PARC) of the U.S. Fish and Wildlife Ser-

vice. However, the Leopold Report stated that these duties frequently were conducted without adequate and critical appraisal of the biological or financial need for control.

The Leopold Report made six recommendations:

1. That the Secretary of Interior appoint an Advisory Board on Predator and Rodent Control, with representatives from the livestock industry, conservation groups, and scientists.
2. That PARC reassess its own goals.
3. That PARC operations be based upon specific criteria that substantiate the need for control, that costs of control be shared by local and state funds, and that PARC efforts in the eastern United States be greatly curtailed.
4. That research efforts be greatly expanded to develop means of minimizing animal damage, that control be as specific to offending animals as possible, and that nonlethal control methods be explored.
5. That PARC find a new name that would reflect its function as an organization that does more than eliminate "undesirable" species.
6. That strict legal controls be placed upon poisons, particularly compound 1080, to prevent ecological abuses.

As a result of the Leopold Report, PARC was renamed the Division of Wildlife Services and a wildlife biologist was appointed chief of the division (directors of the former unit had often lacked biological training). Also, compound 1080 was used more selectively than before. Nonetheless, the public remained dissatisfied with predator-control policies.

THE CAIN REPORT.   In 1971, the Nixon administration established another committee: the Advisory Committee on Predator Control, chaired by Stanley A. Cain, a former Assistant Secretary of the Interior. The committee was charged to review the entire question of predator control and to evaluate both direct and indirect effects, including environmental impact and alternatives to methods that were then practiced.

The continued use of compound 1080 was specifically addressed in the Cain Report. Subsequently, President Nixon issued an executive order prohibiting the use of all poisons for predator control on federal lands and by federal agencies. The United States Environmental Protection Agency thereafter restricted the availability of chemical toxicants for state and private use in animal control. Since 1973, poisons have been used sparingly.

The Cain Report and the policies it generated were attacked by ranchers and criticized by some biologists. Howard (1974) regarded coyote control as necessary for the economic welfare of livestock growers; his views made several points. Among these were that coyotes could not be controlled effectively with repellants, chemosterilants, or other nonlethal methods in their current state of development. Moreover, he believed that 1080 baits and M-44s set by trained personnel could be individually selective for troublesome coyotes. He also argued that Federal policies adopted in 1972 transferred predator control from trained agents to untrained landowners, which he believed was an ecologically undesirable change in responsibility. Howard (1974) thus contended that "poisons are not the problem. It is how they are used that creates the hazard." O'Gara (1982) noted that some scientists and environmentalists have not accepted coyote predation as an economic burden for some livestock producers—a position that polarized the issue between ranchers and some segments of the public. Such divergence can be detrimental when cooperative efforts are needed for solving other environmental issues involving the management of western lands. Moreover, because their position has been dismissed by

Photo courtesy of Rob and Bessie Welder Wildlife Foundation

Rob and Bessie Welder Wildlife Foundation

The Welder Wildlife Foundation was established by the generosity of Texas rancher Robert "Rob" H. Welder (1891–1953). In his will, Mr. Welder designated a 3157-ha parcel of his large ranch as the site for a nonprofit foundation dedicated expressly to wildlife research and education. Income from earlier on-site oil and cattle operations support the foundation's activities. A headquarters building contains staff offices, a 20,000-volume library, a 72-seat lecture hall, student study, reception area, laboratory, and a small museum. Other facilities include a student dormitory, bunkhouse, necropsy laboratory, animal holding pens, and a rotunda designed for social events. Public tours are conducted once a week, and groups may schedule visits at other times.

Although active in several conservation-related activities, the Welder Foundation's foremost endeavor is to sponsor wildlife and range research conducted by graduate students from universities across the United States. Much of the research is undertaken locally, but some projects are completed elsewhere. Students working at the foundation are provided free on-site housing. To date, more than 250 students have received M.S. or Ph.D. degrees with foundation support. Research conducted by the staff and students has produced about 1200 scientific papers. The conservation education program includes teacher workshops, wildlife camps, field days, and outdoor activities for public school students.

Habitats at the Welder Foundation include several communities, of which a mesquite-grassland community is the most extensive. Others include chaparral-grassland, riparian woodlands, and bunchgrass prairies. The area also includes three relatively permanent lakes, edged by marshes. The foundation is located near Corpus Christi, Texas, in a transition zone where a mix of eastern and western species, as well as those with northern and southern affinities, creates an unusually rich biota (e.g., about 1400 species of native plants and 380 species of birds).

White-tailed deer, wild turkeys, bobwhites, coyotes, bobcats, and collared peccaries are among the prominent species of wildlife available for research. Several species of ducks and other waterbirds seasonally visit the wetland areas, and the riparian woodlands offer significant habitat for migrant songbirds. The Welder deer herd, which averages about 900 animals, is perhaps the most thoroughly studied deer population in North America. In addition to detailed investigations of food habits, diseases and parasites, and behavior, the herd's response to coyote predation has been emphasized. Management activities demonstrate the compatibility between wildlife and ranching operations in the midst of an active oil field.

See Kie et al. (1979) for the influence of coyote predation on the Welder deer herd. Box (1961) and Drawe et al. (1978) describe the local vegetation.

much of the public, ranchers may resort to illegal means when combatting coyotes.

EFFECTIVENESS OF CONTROL METHODS ON COYOTE NUMBERS

A program of Animal Damage Control (ADC) was organized by mandate of federal legislation enacted in 1931. In 1986, Congress transferred all ADC activities from the U.S. Fish and Wildlife Service in the Department of the Interior to the Department of Agriculture. The transfer removed the responsibility for predator control from an agency primarily concerned with wildlife conservation and placed those responsibilities instead in the hands of an agency whose concerns were more closely aligned with the interests of ranchers and farmers. All previous legal constraints, however, such as the restricted use of poisons, remained in force under the terms of the transfer. The ADC program kills 70,000–85,000 coyotes annually in 13 cooperating western states. These efforts, along with other control measures, remove perhaps 18–29 percent of the coyote population each year (U.S. Fish and Wildlife Service 1978). Connelly and Longhurst (1975) examined the effects of control with a simulation model. Their conclusion was that although predator control somewhat depresses coyote densities, coyote populations are resilient; increased reproduction and pup survival compensate for the losses. Coyote numbers may be reduced substantially only at the highest levels of control. Indeed, coyotes could be exterminated in about 50 years—but only if 75 percent of the population were removed each year.

Field data from Texas suggest that coyote populations compensate for lower densities by producing larger litters (Knowlton 1972). Coyote litters averaged 7.0 pups per litter in areas where control was intensive and coyote densities were relatively sparse. In contrast, litter size averaged 3.6 pups in areas with light control and higher densities (Table 9-6). Thus, control measures are both thwarted and made more costly by the compensatory nature of reproduction in coyote populations.

Studies of coyote densities before and after control measures are criticized because few data exist for comparable situations in which predator control was not carried out. Wagner (1972) analyzed coyote populations, as determined by coyotes killed per work-hour of effort, in eight western states before

TABLE 9-6.   Comparison of Average Litter Sizes of Coyotes from Seven South Texas Counties in Relation to Control Efforts

Intensity of Control Effort	County	Number of Pregnant Females Examined	Average Number of Uterine Swellings per Female
Intensive	Uvalde	10	6.2
	Zavala	8	8.9
	Dimmit	12	6.4
Moderate	Jim Wells	21	5.3
	Hidalgo	11	3.7
Light	Jim Hogg	17	4.2
	Duval	11	2.8

Source: Knowlton (1972).

and after the use of compound 1080 (Fig. 9-9). From these data it seems that coyote densities were lower after the application of compound 1080 in the more northern and colder states, where the relative scarcity of winter food presumably increased the rate at which coyotes ingested poisoned baits.

Thus, compound 1080 apparently can reduce coyote numbers, at least under certain environmental conditions. However, such a conclusion does not indicate the associated effect, if any, on sheep populations. The critical point remains whether fewer sheep die when and where coyotes are controlled. Accordingly, Wagner (1972) compared *total* sheep losses (from predation and other causes) before the application of compound 1080 (1940–49) with those that occurred during a period when compound 1080 was used widely (1950–70). The comparison showed no detectable differences in total losses of sheep as a result of using compound 1080 for coyote control (Table 9-7). However, the suggestion of compensation again arises, this time with regard to the mortality rate experienced by sheep populations; that is, about 9.5 percent of all sheep die each year—from many causes—with or without effective coyote control. Other forms of mortality apparently compensate for coyote predation, and the broad-scale application of compound 1080 as a means of protecting sheep from coyotes must be seriously questioned, as it was in the Cain Report.

Nonetheless, individual case histories often show that ranchers who suffer high losses of sheep can be

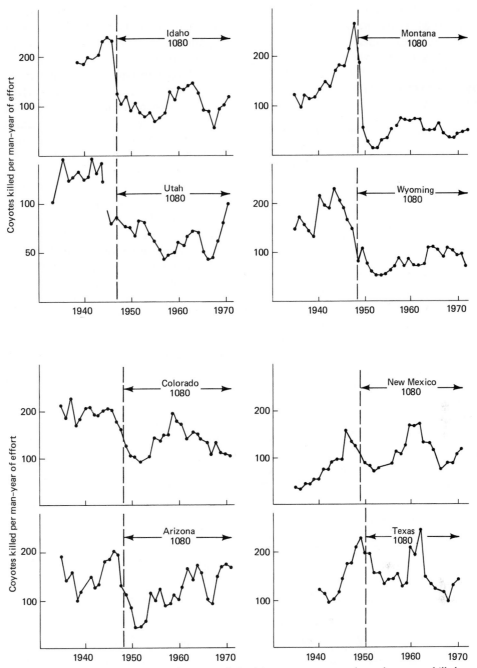

Figure 9-9. Annual coyote population trends in eight western states as shown by coyotes killed per man-year, based on records from the Division of Wildlife Services, U.S. Fish and Wildlife Service. (From Wagner 1972.)

helped when predators are removed (U.S. Fish and Wildlife Service 1978). Therefore, if predator control seems necessary, responses made on an individual basis seem more effective than widespread measures.

EFFECTS OF COYOTE CONTROL ON OTHER ANIMALS

Other animals are killed in the course of controlling coyotes (Table 9-8). The U.S. Fish and Wildlife Service (1978) contends that such losses are insignificant in comparison with fur harvests. For example, ADC activities killed 2298 bobcats compared with a fur harvest of 24,900 bobcats. The ADC kill of coyotes, however, represents only about 25 percent of the total kill of coyotes. Quite possibly, then, the 2298 bobcats may represent only 25 percent of those killed in the total control program. If so, about 9000 bobcats becomes a realistic estimate of those killed incidentally by various means of coyote control.

O'Gara (1982) reviewed the characteristics of compound 1080. It is tasteless and highly toxic to canids and rodents but is less so to other animals. Raptors,

vultures, mustelids, bears, rabbits, and humans are less than one-tenth as sensitive as coyotes are to compound 1080 poison. Because it is water soluble, compound 1080 does not accumulate in food chains as fat-soluble substances (this is not the case with DDT and other chlorinated hydrocarbons). The compound breaks down in soil and in water in a matter of weeks. Under experimental conditions, Burns et al. (1986) determined that secondary poisoning—that is, animals poisoned after scavenging on the carcass of a previously poisoned animal—with compound 1080 was not a serious threat to dogs, striped skunks, or black-billed magpies (*Pica pica*). However, although compound 1080 may be less lethal to other wildlife than to coyotes, the effects of sublethal dosages remain largely unknown (Robel 1982). For example, sublethal doses of compound 1080 produced permanent reproductive damage in laboratory animals. Therefore, the fact that the poison remains in the body only a short time is irrelevant.

PUBLIC POLICY AND PREDATOR CONTROL

Public acceptance remains an important element in determining the methods wildlife managers use for achieving various goals. The U.S. Fish and Wildlife Service (1978) accordingly prepared a chart of the public's response to each of 11 methods of coyote control (Fig. 9-10). Nonlethal methods clearly are preferred over lethal methods, and such traditional methods as trapping, denning, and using slow-acting poisons are among the least acceptable.

Nass et al. (1984) identified ecological and herd management practices associated with the loss of lambs and kids to predators in five western states.

TABLE 9-7. Mean Annual Loss of Sheep from All Causes in Western States

| Location | Percent Sheep Loss | |
	1940–49 (Before 1080)	1950–70 (During 1080 Program)
Texas	10.3	11.3
New Mexico	10.5	10.8
Other Western States	7.9–11.7	7.0–10.3
Mean	9.6	9.4

Source: Wagner (1972).

TABLE 9-8. ADC Take of Carnivores, in FY 1976, by Method

Species	Leg Traps	M-44	Called and Shot/ Shot	Dens	Snares	Aircraft	Dogs	Live Traps	Totals
Badger	1,247	0	20	0	15	0	0	0	1,282
Bear, black	1	0	4	0	77	0	17	1	100
Bobcat	1,924	0	63	2	146	144	19	0	2,298
Coyote	31,581	5,328	5,347	5,226	3,187	33,626	204	0	84,499
Dog, domestic	418	3	161	0	0	0	0	0	582
Fox, red/gray	2,635	216	195	222	232	175	0	9	3,684
Mt. lion	5	0	1	0	2	0	41	0	49
Raccoon	1,376	5	9	11	41	0	9	296	1,747
Skunk, striped/spotted	3,711	60	98	0	10	0	0	1,509	5,388

Source: U.S. Fish and Wildlife Service (1978).

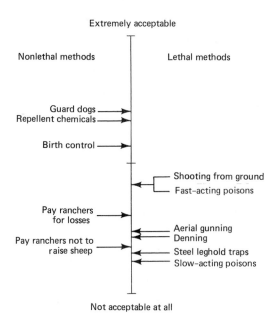

Extremely acceptable

Nonlethal methods   Lethal methods

Guard dogs
Repellent chemicals

Birth control

Shooting from ground
Fast-acting poisons

Pay ranchers
for losses

Aerial gunning
Denning
Pay ranchers not to
raise sheep   Steel leghold traps
Slow-acting poisons

Not acceptable at all

Figure 9-10. Acceptability of coyote damage-control methods expressed as national averages. (From U.S. Fish and Wildlife Service 1978.)

Low losses of livestock—less than 5 percent—were associated with low populations of natural prey and low predator populations and ranching operations that used feeder lambs and range sheep. Higher losses were associated with combined sheep-goat operations, small ranch sizes, high predator populations, a local history of predation, and the nearness of brush cover for predators.

Aversive conditioning is a behavioral concept that has potential promise for limiting predator damage. The method calls for injecting sheep carcasses with distasteful chemicals—lithium chloride, for example—that sicken coyotes eating the treated meat. Thus conditioned, coyotes thereafter avoid sheep in the same manner as one avoids any situation involving an unpleasant sensory experience (e.g., encounters with striped skunks). Experiments with aversive conditioning thus far have produced mixed results, and field application of such a system awaits more research and development. Guard dogs (Green and Woodruff, 1990), electric fencing (Linhart et al. 1982), flashing strobe lights and sirens (Linhart et al. 1984), and even llamas (*Lama peruana*) serving as "watchdogs" (Anon. 1981; Lye 1981) are among other deterrents that may prevent coyotes from attacking sheep.

If predator control is undertaken for economic reasons—and there seems to be no other justification—

then the results also should be evaluated in economic terms. Evaluations might include the worth of lambs that otherwise would not have been marketed and game birds or deer in the bag that would not have been there without predator control. Such savings, of course, should then be weighed against the costs of predator control.

The coyote-sheep controversy illustrates that wildlife management often involves more than dealing with wild animals for recreational purposes. Domestic animals, economics, public relations—and indeed, politics—are parts of the complex setting in which wildlife managers must operate. Watt (1982) presented a flow chart showing the interactions of 11 different elements in the coyote-sheep complex; the chart included economic factors such as tariffs and competition from synthetic fibers (Fig. 9-11). Competition reduces the price of wool, decreases profit margins, and ultimately forces ranchers into additional methods of predator control. The flow chart also notes the lack of adequate knowledge about such matters as the densities of other herbivores and predators and their interactions with sheep and coyote populations. Regrettably, however, managers seldom have complete information before decisions must be made.

Federal policies in the United States concerning predator control have been challenged in recent years. The policies may be summarized as follows (Cain et al. 1972):

Animal damage control should be designed to ensure the maintenance of native wildlife and habitats. Control methods should reduce animal depredations as selectively as possible, and control should be directed at the depredating individual or local depredating population.

Goals of control should include protection of human health and safety from wildlife-borne diseases; from birds or other animals threatening human safety, such as birds in the vicinity of airports; and from economic losses in residential and industrial situations resulting from mice, bats, and nuisance birds.

Forests, range, agricultural crops, and livestock should be protected from direct damage by wildlife and from wildlife-borne diseases when economic and social benefits are judged to offset all costs of control.

An animal-control program will be conducted when and where there is a demonstrated need, as determined by the U.S. Fish and Wildlife Service after a careful review of all available evidence.

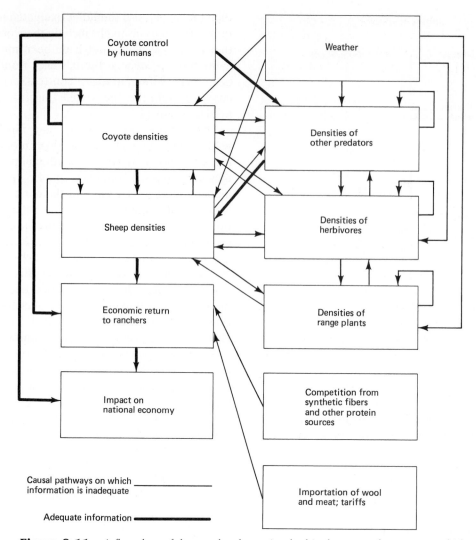

Figure 9-11.   A flow chart of the causal pathways involved in the coyote-sheep system, which generates the sheep rancher-conservationist controversy. The thin lines represent causal pathways on which the information developed by research to this point is inadequate. The thick lines represent causal pathways on which the available information is adequate or approximately adequate. (From Watt 1982.)

An annual work plan related to land use, planning, and zoning will be developed for animal control in each cooperating state, and will be coordinated with plans of the Forest Service and the Bureau of Land Management so that the plans will be consistent with their multiple-use concept.

The Wildlife Society (Position Statements Committee 1975) has taken the following position:

The policy of The Wildlife Society, as regards control of animal damage, is to:

1. Support only those animal control programs that are justified biologically, socially, and economically.
2. Encourage continuing research designed to improve methods of (a) accurately assessing the damage caused by wildlife, (b) controlling and preventing animal damage, especially by nontoxic means, and (c) measuring the effectiveness of damage-control programs.
3. Recommend that efforts of control be the minimum required to bring damage within tolerable limits.

4. Support the use of only the most efficient, safe, economical, and humane methods to control depredating animals, and advocate effective lethal control only when other methods are unsatisfactory.

5. Urge that all control programs directed at wildlife species be closely regulated by state or federal laws.

SUMMARY

Generalities about predator-prey systems are elusive, but ecological knowledge indicates that extermination of any element in a community is unwise. The coexistence of prey and predators for centuries indicates that, given proper food and cover, natural prey populations can withstand predation and that predator populations can survive low prey densities. The effect of predators on prey populations in nature may depend on the complexity of the community in which they live. In the absence of predators and hunting, many prey populations overpopulate their ranges and ultimately decline rapidly from starvation and disease. In complex communities that have abundant food and cover and several species of predators and prey, predators moderate the occurrence of drastic peaks and troughs in prey populations. Recent ecological evidence from relatively simple natural communities, however, indicates that predators may depress declining prey populations to very low levels and maintain those levels for extended periods of time. In such situations, the effects of severe weather and hunting on prey numbers are additive. Control of predators for the recovery of depressed prey populations, especially those of game animals, is proposed by some wildlife managers and some segments of the public. Other segments of the public oppose such actions.

Protection of livestock from predators may be warranted in some local situations, but broad-scale programs of predator reduction have not proven effective. Overall losses of sheep were not reduced during a period when coyotes were poisoned with compound 1080. The results suggest that other factors maintain a relatively stable mortality rate irrespective of coyote predation. Two reviews made by special commissions produced important changes in federal programs concerned with predator control (e.g., a ban on compound 1080). Because of a complex setting of biological, sociological, psychological, and economic factors, predators and predator control present continuing challenges for wildlife managers.

CHAPTER 10

HUNTING AND TRAPPING

They lunged their spears....Realization came slowly to the exhausted men. In the sudden silence, the hunters looked at each other. Their hearts beat faster with a new kind of excitement....They did it! They killed the mighty mammoth!...[They] had killed the gigantic creature no other predator could.

Jean M. Auel (1980)

The evolutionary and cultural development of humans has been shaped by their dependence on other animals (Eltringham 1984). Fossils dating back some 500,000 years show that *Homo erectus* killed large mammals with spears and then cooked and ate the flesh of their quarry. Ancient cultures indeed relied on animals for many essentials of human existence: meat for nourishment, skins for clothing, sinews for rope and thread, fat for fuel, antlers for tools, horns for drinking vessels, and bone for tools, weapons, and needles.

Each man was a hunter in early human cultures. Paintings on the walls of caves and bones in middens reveal the importance of wild animals as prey and food of Stone Age humans (Brodrick 1972). A man's duty was to obtain those essentials, and a successful hunter was respected by his family and the community. Between 9000 and 6000 B.C., however, some wild animals were brought under human control, and with the domestication of cattle and goats, fewer people thereafter were required for hunting and farming (Cole 1972). As human efficiency at farming and livestock production increased through the centuries, a complex economic system developed, and most people found themselves ever further removed from the land. In the late 20th century A.D.,

most people, without struggle or bloodshed, procure meat nestled on a styrofoam plate and wrapped in clear plastic. The role of the hunter and the trapper is being challenged (Clarke 1958; Amory 1974; Scheffer 1976; Grandy 1986), and wild animals are regarded by many as having legal rights similar to those of humans. Yet in 1996, as many as 22 million people in the United States of all ages went hunting (Ference 1997). Moreover, the traditionally male-dominated sport of hunting is rapidly gaining popularity with women (see following).

This chapter addresses the evolution of sport hunting, attitudes of hunters and antihunters, the ecological effects of hunting on animal populations, the role of hunting and hunters in wildlife management, and the role of trapping and its effects on populations of furbearers.

TO HUNT OR NOT TO HUNT

Hunting has been challenged from several quarters. Those who speak against hunting are frequently articulate and respected citizens such as the late physician-philosopher Albert Schweitzer; the late Joseph Wood Krutch, a New York drama critic and writer-turned-naturalist; and Cleveland Amory, New York writer and president of The Fund for Animals, an animal protectionist group. Along with these leaders are several citizens groups, including The Friends of Animals, New York City; The Society for Animal Rights, headquartered in Pennsylvania; Greenpeace, of British Columbia; and the Humane Society of the United States, Washington, D.C. (the latter is not to be confused with the American Humane Association, which maintains a neutral stance on hunting). The arguments of these individuals and organizations are primarily philosophical and moralistic in nature.

Schweitzer (1965) based his philosophy on "reverence for life." In his works, "Ethics is nothing else than reverence for life. Reverence for life affords me my fundamental principle of morality, namely, that good consists in maintaining, assisting, and enhancing life, and that to destroy, to harm, or to hinder life is evil." Regarding hunting, Schweitzer felt that humans have no right to inflict suffering and death on another living creature unless it is unavoidably necessary. In his view, humans should express horror for thoughtlessly causing death and suffering. Nonetheless, Schweitzer kept a gun for killing snakes and predatory birds (Clarke 1958). In "A Damnable Pleasure," Krutch (1957) launched a more direct attack on hunting, condemning the sport as "the perfect type of that pure evil for which metaphysicians have sometimes sought." One of his frequently quoted statements is, "When a man wantonly destroys one of the works of man we call him Vandal. When he wantonly destroys one of the works of God we call him Sportsman." Grandy (1986) argued that modern-day hunting in North America is an anachronism that is "not romantic, not meat hunting, and not necessary." He asked if the killing of birds and mammals for fun will continue as a socially acceptable form of human pleasure.

In *Man Kind?*, Amory (1974) cites numerous examples of what most hunters would call atrocities. One example takes place on a game ranch in Texas where a hunter rides afield in a car, makes a pretense of offering food to a big game animal from Africa, then kills his exotic trophy without getting out of the car—unless, as Amory says, the hunter first wants to pet the animal. He borrowed a description of hunters (from an article in the *Saturday Evening Post*) as "paunchy and redfaced, slow of foot, and carrying hip flasks." Hunters, according to Amory, rarely pursue wounded game, and the woods thus are filled with cripples after the hunting season. The quotation also notes that hunters are "the dirtiest of outdoor users, leaving behind a landscape littered with bottles and aspirin tins."

Cohen (1977) asked student hunters taking a class in wildlife management, "Why do you hunt?" A few samples illustrate the antihunting tone of the author's analysis. The answer that hunting may be instinctive was dismissed quickly as "a twisted pseudointellectual response." Students who claimed they hunt for aesthetic reasons could not satisfactorily answer—at least in Cohen's opinion—a follow-up question about why they did not simply take pictures or hike in the woods, and a student who said that he gets a thrill from the kill was called "sick."

Organizations in the United States sympathetic toward hunting include the National Rifle Association, the National Wildlife Federation, The Wildlife Management Institute, and numerous local and state hunter's groups. Responses of hunters to antihunters have ranged from immoderate insults, similar in terms to the antihunting articles by Krutch (1957) and Cohen (1977), to philosophical musings on the role of the human as a predator in natural communities.

In "Hunting—An American Tradition," Adams (1982) laments that hunters were widely respected only two or three decades ago, but today they "endure a constant barrage of negative remarks." He stated that the idea that killing is morally wrong is ludicrous and wrote, "You show me a person who doesn't directly or indirectly kill on a regular basis and I'll show you a bleached, well-weathered pile of human remains. Every living creature takes life to stay alive, and if it doesn't, it quickly starves and dies a slow, agonizing death." Adams's advice to hunters defending their sport against an adversary is to "immediately launch a direct frontal attack. Be polite, if possible, because the poor snook has probably never examined his gut-level reaction in a sensible, rational way. He simply has been programmed by the widespread antihunting 'morality' pervading modern urban life. With luck, you might be able to shake him up enough to bring him to his senses." Little is accomplished for either wildlife or humanity when some antihunters and some hunters resort to name calling and insults.

Anthony (1957) offered a hunter's response to Krutch (1957) in an essay, "But It's Instinctive." He argued that humans are predators, both biologically and historically, and thus exhibit mental processes conditioned by the fact of such behavior. Whereas complete agreement is lacking, many biologists believe that human behavior has a genetic basis, just as does our anatomy (i.e., human dentition and digestive system are adapted for meat as well as for vegetable foods).

Hutchins and Wemmer (1986) compared the philosophies of "animal rights" and wildlife conservation; a summary of their main points follows: The animal-rights ethic maintains that humans should cause no pain, suffering, or death to creatures that are *sentient*—animals that are capable of experiencing pain. Nonhuman animals should be accorded the same ethical concerns as fellow humans. In this view, the integrity of ecosystems, communities, and animal populations is secondary to the primary rights of individual animals. That view contrasts with the land ethic of Leopold (1949), which contends that "a thing is right if it tends to preserve the integrity, stability, and beauty of the biotic community. It is wrong when it tends to do otherwise." Hunting is permissible in the land ethic of Leopold, but it is not in the animal-rights ethic. Conflicts between animal-rights and land ethics thus focus on the ways humans control animal populations, especially in the absence of natural predators. The land ethic indeed encourages hunting and trapping, thereby preventing overuse of vegetation and die-offs of starving animals. The two philosophies also differ on the issue of killing exotic species that threaten the existence of native species. Philosophical articles in defense of hunting have been published by Clarke (1958), Zern (1972), Hutchins and Wemmer (1986), and Robinson (1986).

While he was bow hunting for deer, Olson (1980) mused on the role of humans as hunters. He recalled a Hopi Indian ceremony in which a hunter fasts 3 days before hunting. If the hunter is worthy, a deer comes forward and the gift of life is exchanged: the life of the deer for the life of the hunter, who also will ultimately go to the Earth Mother. Olson also remembered a friend who ate only wild meat, spurning meat from the supermarket because, she claimed, such an anonymous environment insulated the buyer from the reality of an animals's death. As Olson pondered these thoughts, a small buck approached his stand. With a hunter's excitement, he killed the deer. Yet he wondered about his feeling of euphoria. Was it from the "fun" of killing or the satisfaction of participating in the natural system? He would use the hide and meat. "Is it," he asked, "more humane to munch steak or soybeans and feel nothing at all?" Olson further observed that among hunters, there are two kinds of kill, one that dulls and another that sensitizes. "The former," he says, "is tragedy, and the latter is sacred celebration, as old as time. Every living being kills life. Some firsthand."

Geist (personal communication) expressed a slightly different viewpoint: "I hunt no more to kill animals than I garden to kill cabbages." Eaton (1986) similarly examined an analogy between managing wildlife for hunting and tending and harvesting a garden. He wrote, "The people who most love roses, who spend hours pruning them, kill them; the people who most love vegetables are those who grow, kill and eat them; the men who most love waterfowl…who invest in their habitat and protection, also kill and eat them. This is a paradox only to men and women who suffer from centuries of separation from nature, and thus from their true nature."

Elton (1939) offered the interesting analogy that most athletic games involve the elements of predator, prey, and cover. American football, for example, is a ritualized version of prey (the team with the ball) attempting to reach cover (the end zone) before being captured (tackled) by the predator (the defensive team). In baseball, the ball is prey to the batter (who becomes prey upon hitting the ball), who attempts to reach the first base (cover) before the fielding team (the predator) gets the ball to the base. The excitement that people feel for such events probably is rooted in an innate human interest in predator-prey relationships, an interest that is very likely tied to the survival value of eating while avoiding being eaten. Hunting and fishing thus are merely more primitive expressions of our predatory nature. Of course, the predatory behavior of humans can be regulated, just as it is with rules for athletic contests—with hunting regulations for seasons, bag limits, and equipment limitations.

The strength of human predatory nature probably varies among individuals, depending upon both genetics and experience. If so, we may expect that nonhunters may never understand hunters, just as football fans may never fully appreciate grand opera. Many people do not wish to hunt, which may be fortunate in view of the present world population of humans. Kellert (1978) analyzed attitudes and characteristics of hunters and antihunters. Whereas both groups are interested in the future of wildlife, misunderstandings often mar the efforts of each for productive conservation and management. In any case, 29 percent of the American public harbored strong objections to hunting. Women were particularly against hunting, as were residents of large urban areas. Few antihunters had raised animals or had enjoyed associations with farmers. Of all groups identified by the survey, antihunters were among the least knowledgeable about animals. Based on the attitude scale shown in Table 10-1 (see Kellert 1976a,b), antihunters were divisible into those holding primarily humanistic values about animals and those with strong moralistic beliefs.

Antihunters with humanistic attitudes strongly identified themselves with the feelings of individual animals. Members of The Friends of Animals and The Humane Society of the United States clearly reflect the humanistic attitude. In large measure, this attitude developed from associations with pets; proponents transferred such emotions as fear and pain from pets to wild animals. Objections to hunting, then, arise from a conflict between loving animals and killing them for food and recreation. In this view, the welfare of a species or population is subordinate to feelings for individuals, and recreational hunting remains a deliberate infliction of pain and death without reason. As might be expected, these feelings are heightened when attractive animals—waterfowl and deer, for example—are considered.

The moralistic antihunter objects to hunting for more subtle reasons (see Schweitzer 1965 and Krutch 1957 for examples). Hunting in this case is considered contrary to human social values and represents an unnecessary—and even evil—exploitation of nature. Shooting animals is regarded as degenerative behavior that violates the sacred values humans attach to life. Hence, because wild animals possess the same will to live as humans, hunting reflects as much an antihuman attitude as an antianimal posture. In sum, moralistic antihunters identify hunting with ethics and social behavior, whereas humanistic antihunters transfer their own emotions to individual animals.

Hunters, including persons who once hunted but may no longer do so, constituted 37 percent of the population sampled by Kellert (1978). About half (17 percent) actually hunted in the past 5 years, with only 5.5 percent hunting regularly during that period. The primary reasons for hunting, in order, were (1) for meat, (2) for sport, and (3) for contact with nature. These responses can be associated with three of the attitudes appearing in Table 10-1, but some degree of overlap among these categories must be appreciated. "Meat hunters" represent the utilitarian attitude, associating the practical value of animals and indifferently connecting hunting with outdoor experiences. This view predominated among 44 percent of the persons who hunted in the past 5 years.

"Sport hunters" were associated with the dominionistic attitude. About 38 percent of all active hunters associated hunting with competition or the mastery of

their contact with animals. This context may be similar to an athletic contest in which skill, endurance, and effort are values associated with victory. Elton's (1939) analogy, cited earlier, apparently applies to the "sport" hunter. Hunters in this group were not especially knowledgeable about animals, nor did they exhibit much interest or emotional attachment for wildlife. Guns were important to hunters with dominionistic attitudes, as was the ritualistic nature of hunting where costuming and other forms of pageantry are emphasized. For these hunters, game animals represented trophies, opponents in a contest, and tangible measures of human skills.

"Nature hunters" accounted for 18 percent of those hunting within the past 5 years. However, this group made up the largest part (35 percent) of those hunting frequently. As a group, they were under 30 years of age and of higher socioeconomic status than other hunters. Nature hunters also exhibited more knowledge about animals than other hunters and reported more involvement with camping and other outdoor activities. These hunters held high naturalistic values, but their affection for animals—a humanistic attitude—was general and not especially high. In part, naturalistic hunters were escaping the rigors of modern living, replacing these with a fundamental contact with nature at a level reminiscent of humanity's ancestral origins as hunters. In so doing, however, nature hunters were confronted with the rationalization of killing versus their other attitudes about wildlife. This conflict of attitudes was assuaged, in part, by including death as a natural part of life in nature (namely, an unending drama; see Olson 1980).

Kellert (1978), in reporting these results, concluded with recommendations that may minimize conflicts between hunters and antihunters when policies for wildlife management are formulated. These are as follows:

1. Place greater educational emphasis on the ecologistic attitude as a way of establishing a dialogue between hunters and antihunters. This attitude possesses viewpoints compatible with the perceptions of both groups. Moreover, this attitude discourages antihunting sentiment based on anthropomorphic notions derived from intuitive emotion and opposes hunting practiced without regard for the needs of game animals in their natural habitats.
2. Encourage greater governmental recognition of different attitudes toward animals as reflecting multiple uses and satisfactions derived from wildlife.

TABLE 10-1. Categories Used to Classify Attitudes of the American Public Toward Wildlife

Attitude	Key Identifying Terms	Highly Correlated With	Most Antagonistic Toward
Naturalistic	Wildlife exposure, contact with nature	Ecologistic, humanistic	Negativistic
Ecologistic	Ecosystem, species interdependence	Naturalistic, scientistic	Negativistic
Humanistic	Pets, love for animals	Moralistic	Negativistic
Moralistic	Ethical concern for animal welfare	Humanistic	Utilitarian, dominionistic, scientistic, aesthetic, negativistic
Scientistic	Curiosity, study, knowledge	Ecologistic	None
Aesthetic	Artistic character and display	Naturalistic	Negativistic
Utilitarian	Practicality, usefulness	Dominionistic	Moralistic
Dominionistic	Mastery, superiority	Utilitarian, negativistic	Moralistic
Negativistic	Avoidance, dislike, indifference, fear	Dominionistic, utilitarian	Moralistic, humanistic, naturalistic

Source: Kellert (1976a, b).

Management should be carried out for different types of hunters. There should also be an explicit recognition of the potentially legitimate moralistic and humanistic attitudes concerning cruelty and animal welfare.

3. Diversify wildlife funding sources to include contributions from antihunters and other potential nonconsumptive users of wildlife resources. If nonconsumptive users desire increased attention from fish and wildlife agencies, they should share some of the financial burden.

4. Increase attention and financial allocation for nongame research and recreational programs.

WOMEN HUNTERS

Hunting is commonly regarded as a male-dominated outdoor sport; yet according to surveys conducted in 1996, women in the United States are hunting in larger numbers than ever before (Ference 1997). Indeed, about 13 percent of the 15.2 million licensed hunters over 16 years of age are women. Women hunters increased by almost 49 percent between 1989 and 1996. At such stunning rates, women are expected to represent a force of nearly 3 million hunters by the turn of the century. As a result of this surge, a nonprofit organization—the Women's Shooting Sports Foundation—now facilitates the participation of women in hunting and other shooting-related activities.

Three reasons apparently account for much of this dramatic change in the demography of U.S. hunters. First, more women are joining their male companions in outdoor recreation (e.g., family outings,

which sometimes include adolescent children). Second, more women are learning to shoot competently as a means of self-protection, and hunting is a natural extension of such proficiency with weapons. Finally, the manufacturers of sporting arms started to produce rifles, shotguns, and archery equipment designed expressly for women.

THE EFFECTS OF HUNTING ON POPULATIONS OF ANIMALS

Hunting by primitive humans may have caused extinctions of many large mammals and birds. Fisher et al. (1969) described the impact of Stone Age humans on birds and mammals. During the ages in which hunting was the main livelihood of humans, a progression of inventions continually improved the efficiency of hunters. Such devices as bolas, bows and arrows, boomerangs, blowpipes, atlatls and spears, traps, snares, decoys, and nets made humans into master predators. Most creatures on land and in shallow waters fell prey to human appetites. Large herbivores were especially desirable sources of food, and populations of many species were decimated. Several kinds of large carnivores—the rivals of human predators—also were affected.

A number of hypotheses have been advanced to explain widespread extinctions of large animals in North America about 10,000–11,000 years ago. Martin (1971, 1973) proposed that the growth and the spread of the human population were responsible for the extinctions—a theory sometimes known as "Pleistocene Overkill." According to this view, humans arriving in North America encountered abundant populations of large mammals—the North American megafauna—that had not evolved adequate defenses against human predation. The human population expanded across the hemisphere and, during the next 1000 years, such species as mammoths (*Mammuthus* spp.), ground sloths (*Nothrotheriops shastense*), dire wolves (*Canis dirus*), and cave bears (*Ursus spelaeus*) disappeared forever. Thus, beginning with the Clovis culture, a blitzkrieg of overhunting soon eliminated 32 genera of large animals from North America (Martin and Klein 1984; Lewis 1987; Martin 1987).

Some disagreement exists, however, about the total impact of primitive cultures on wildlife populations. Brodrick (1972) believed that humans were relatively rare during Paleolithic times and therefore may not have exerted as great an effect on wildlife as others believe. Rapid climatic changes also may have been responsible for extinctions when animals who had adapted to warm climates did not adapt to the advance of glaciers and the accompanying changes in vegetation. Nevertheless, many large birds and mammals were nearing extinction, and it seems likely that humans played a role in the demise of at least some species of the North American megafauna.

Rates of extinction vary with geographical regions. Island faunas (because of their small populations, limited potential for immigration, and often specialized behavior and structure) are more vulnerable than continental faunas. In North America, Fisher et al. (1969) estimated that about 50 mammals and 40 birds became extinct in the 3000 years preceding 1600, a rate of about 3 per century. Since 1600, however, the extinction rate has increased to 19 species per century, and we have more information about the causes of extinction. Hunting and human-destroyed habitat were responsible for about 57 percent of the extinctions of birds and 62 percent of the extinctions of mammals since 1600.

Some qualification is needed, however, about hunting in modern societies. Just as humans have developed more-efficient ways of killing animals, so also have humans developed more sophistication about the disastrous results of overhunting. Fisher et al. (1969) noted that both tribal and urban societies gradually restricted the efficiency of hunting with self-imposed regulations. As a result, hunting gradually became more ritualized as human survival depended less on wildlife for food and fiber. Sport hunting, as currently practiced and managed, does not pose a threat to the existence of game species. In fact, many leaders in the movement for the conservation of wildlife were avid hunters, perhaps none more prominent than President Theodore Roosevelt and the "father" of wildlife management, Aldo Leopold. Likewise, most of today's leaders are also hunters. Wildlife managers are professionally responsible for the welfare of wildlife populations and are charged with related missions: protection of the resource against overharvest and maintenance of high-quality sport hunting. Despite the current anti-hunting campaigns of some organizations, we probably can assume that sport hunting will continue for some time as a factor in the ecology and management of many wildlife populations.

Investment overkills are types of hunting that still threaten wildlife (Fisher et al. 1969). Such hunting is done for profit and often requires a financial return on an investment in expensive equipment. The perilous status of many whales has resulted from investment overkills. Other examples include some African animals, from which profits are gained from the sale of ivory, hides, rare trophies, and live animals for zoos. Some relief from the overkill of whales occurred in 1982 when growing pressures from concerned nations and conservation groups yielded concessions from The International Whaling Commission (McCloskey 1982). International agreements also have declared the trade in endangered animals and their products illegal in dozens of countries.

HARVEST AND HUNTING

Hunting represents a highly visible form of mortality for wildlife. Each autumn, guns and ammunition are advertised in full-page "specials," and newspapers and magazines later feature pictures of hunters displaying game birds or admiring big-game trophies. During the hunting season, thousands of deer are transported from woods to cities on the roofs and fenders of vehicles. The attention of the public is thereby called to wildlife, and concern develops for the potential impact of hunting on animal populations. The argument frequently is advanced that hunting merely removes a "surplus" of animals that likely would die anyhow. Is there validity in such an argument?

Are hunting and trapping essential to wildlife management? Can wildlife be managed without hunting or trapping? Hunting is not a part of managing nongame species. Habitats in healthy communities can be manipulated in ways that can maintain populations at certain levels of numbers, and where healthy communities include game species, hunting or no hunting may make little difference. But where humans have simplified communities by altering vegetation, removing buffer species, and, in particular, eliminating predators, the job of managing wildlife without hunting or trapping may be quite difficult and even impossible. Under such circumstances—which are often the case—the goals of preventing extermination at one end and preventing a plague of animals at the other can be met most effectively with some sort of regulated harvest.

The various causes of natural mortality among animals tend to be compensatory. For example, if a population is reduced by predation, then fewer animals normally die from disease or starvation. As a result, the number of remaining animals tends to be relatively constant by the time the breeding season arrives. Populations of robins (*Turdus migratorius*) do not grow infinitely even though the birds are not hunted. In most years, the losses of robins from predation, disease, and other causes of death more or less equal production from the preceding nesting season. Hence, the question arises: does hunting mortality act in a compensatory manner in game species, or should hunting mortality be treated separately from other causes of deaths? Such are the concepts and questions wildlife managers address when deciding whether a species can be harvested by humans. Managers also are concerned with what level of hunting pressure a game population can sustain.

The general theory of harvesting animals is based on the following premise: without harvest, growth and recruitment of the population are balanced by natural mortality; therefore, the average growth rate of a population at carrying capacity is zero (Dixon and Swift 1981). Hunting reduces the population, but the loss also increases the growth rate. The increase in growth is the consequence of higher birthrates and lower death rates, which result from decreased competition for food and other resources (i.e., with fewer animals, there are more resources per individual). Consequently, the accelerated growth rate provides a surplus of animals beyond the number required for replacing the losses—a surplus that can be harvested.

In Chapter 5, we noted from the logistic equation that a maximum number of individuals is added when the population is at one-half of its carrying capacity. Wildlife populations maintained at such levels thus produce the maximum number of animals that could be harvested each year. *Maximum sustained yield* is defined as the largest average harvest that can be taken continuously from a population

under existing conditions. It therefore appears that maximum sustained yield is attainable when (1) a population is kept at a level of about half of its carrying capacity and (2) the harvest takes the annual production of the population (Holt and Talbot 1978). The formula for maximum sustained harvest (*H*) is one-fourth the maximum population (*K*) times the intrinsic rate of growth (*r*):

$$H = \frac{Kr}{4}$$

At first glance, maximum sustained yield seems a desirable management goal for harvesting wildlife, and for a time it was, but a number of reasons later modified that goal. Because the carrying capacity is not constant in most environments, maximum population sizes are not easily determined. Also establishing harvest rates with the logistic equation requires assumptions that are not often met in actual populations (Dixon and Swift 1981). Specifically, the equation assumes (1) that the rate of increase responds immediately to changes in population density, (2) that the age-and-sex structure in the population is stable, and (3) that the harvest is spread equally throughout the population.

In addition to such biological flaws, other considerations temper the use of the logistic equation for determining harvest levels. Wildlife in North America is public property, for example, but some sectors of the public are unwilling to have their animals managed purely for maximum harvest (Dixon and Swift 1981). Hence, optimum yield has replaced maximum sustained yield as a management concept for most game birds, game, furbearing mammals, and fishes. In the United States, *optimum yield* is defined by the Fishery Conservation and Management Act of 1976 as the quantity of fish (1) that will provide the greatest overall benefit to the nation, with reference to food production and recreational opportunities, and (2) that is prescribed as such on the basis of maximum sustained yield as *modified by any relevant economic, social, or ecological factors*. By consensus, the same concepts apply to the harvest of terrestrial wildlife.

Wildlife managers, however, have yet to quantify the ecological, social, and economic influences affecting the harvest of wildlife. Dixon and Swift (1981) described how ecological features might be incorporated into a model for the harvest of furbearers. The model consisted of three interacting species of furbearers: two predators and one prey

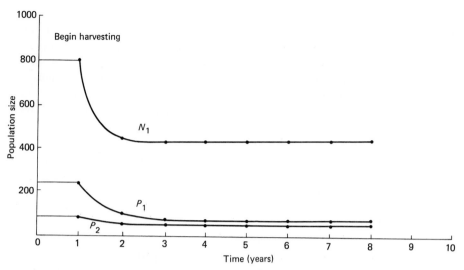

Figure 10-1. Simulated interaction among an *r*-furbearer prey *(N₁)*, an intermediate furbearer predator *(P₁)*, and a *K*-furbearer predator *(P₂)*, where N_1 and P_1 are harvested. $N_1(o) = 808$, $P_1(o) = 242$, $P_2 = 81$. (From Dixon and Swift 1981.)

(Figure 10-1). In the model, the prey furbearer is an *r*-selected species (having a high reproductive rate, short life span), one predatory furbearer is a *K*-selected species (having a low reproductive rate, long life span), and the other predatory furbearer has intermediate characteristics. Without harvest, the prey maintains a population of 808 animals, the "intermediate" predator, 242 animals, and the *K*-selected predator, 81 animals. With maximum harvest of only the prey species, equilibrium is reached at 404 animals, the intermediate predator numbers 121 animals, and the *K*-selected predator population contains 40 animals. Table 10-2 shows the outcome under conditions of maximum harvest for various combinations of animals in the model. Such models still are quite simplistic and do not consider food availability for the prey population or weather conditions that can cause year-to-year fluctuations in the populations of all species.

Social factors such as the pressures from antitrapping groups, demands for more animals for public enjoyment, and competition among trappers further complicate the establishment of optimum yield. Economic factors such as fur prices and operating costs (e.g., trap and gasoline prices) constitute another influence on the harvest rates of furbearers. Dixon and Swift (1981) noted that most trappers and other users of natural resources understand and use short-term economic returns as a basis for increasing or decreasing exploitation. When prices are high, pres-

TABLE 10-2. Harvesting Regimes and Resulting Predicted Equilibrium Populations for One-Prey and Two-Predator Species

Harvested Species	Equilibrium Populations		
	N_1	P_1	P_2
None	808.3	242.5	80.8
N_1	404.1	121.2	40.4
P_1	880.6	132.1	88.1
P_2	819.5	245.8	41.0
N_1, P_1	440.3	66.0	44.0
N_1, P_2	409.7	122.9	20.5
P_1, P_2	894.0	134.1	44.7
N_1, P_1, P_2	447.0	67.0	22.4

Source: Dixon and Swift (1981).

sures for heavy short-term harvests are strengthened by the uncertainty of the long-term economic future.

Setting harvest rates by means of imposing hunting and trapping regulations remains an inexact science. Hunting regulations probably have received more attention than any other game management tool, but Weaver and Mosby (1979) noted that little factual information exists about the impact of hunting regulations on game populations. In general, managers set the regulations and then observe the population for possible effects. Regulations typically are conservative; that is, season lengths and bag limits are strict, so that any error favors underharvest rather than overharvest. Further restrictions may be recommended if the population shows a consistent decline, whereas stable or increasing numbers suggest that a larger harvest might be possible without reducing the breeding stock. We now will consider some examples of studies that have assessed the effects of hunting on various game species.

MALLARD

Mallards (*Anas platyrhynchos*) are adaptable, migratory ducks with a high reproductive potential. Females breed at 1 year of age and lay 9–12 eggs in any of several types of nesting cover (e.g., fields and wetlands). About 12–18 million mallards leave their breeding grounds in northern North America each autumn and become targets for several million duck hunters. Mallard populations have been studied in detail (e.g., Cooch 1969a; Anderson and Burnham 1978; Cowardin and Johnson 1979; Nichols et al. 1984), but little is certain about the ways hunting affects the annual survival rates for this important game duck. There is no doubt that liberal hunting regulations (bag limits of 3–7 per day, long seasons, and early opening dates) increase the bag of mallards compared with the numbers shot in restrictive sea-

sons (Table 10-3). In some years, up to 25 percent of the mallards are shot, and hunting accounts for 50 percent of all mortality. Nonetheless, analyses of band recoveries show that the survival rates for mallards, from September 1 to August 31 of the following year, are no different whether hunting regulations are liberal or strict (Table 10-4). Such results indicate that other forms of mortality compensate for hunting mortality.

Nichols et al. (1984) reviewed and analyzed numerous population studies of mallards and other waterfowl to test whether mortality is compensatory or additive. Such information is crucial for determining bag limits and hunting season lengths. For example, if the hunting kill is additive, then restrictive regulations will assist in the recovery of depressed populations. If it is compensatory, however, liberal regulations, to a point, will have little influence on the survival of birds until the following breeding season. In general, the review found evidence in nine studies that hunting mortality was compensatory among various age groups and sex groups of mallards. In only one study was mortality additive, and that was only among young females. However, Nichols et al. (1984) did not wholeheartedly accept

TABLE 10-3. Summary of Waterfowl Hunting Regulations and Mallard Harvest Estimates for Selected States in 1962 and 1970[a]

Flyway/ State	1962 (Restrictive)				1970 (Liberal)			
	Opening Date	Season Length	Mallard Bag Limit	Mallard Harvest[b]	Opening Date	Season Length	Mallard Bag Limit	Mallard Harvest[b]
Pacific								
Washington	Oct. 13	75	4	177	Oct. 10	93	6	311
California	Oct. 13	68	5	168	[c]	[c]	7	372
Central								
Colorado	Nov. 9	25	1	12	Oct. 17	90	[d]	115
North Dakota	Oct. 12	25	1	59	Oct. 3	70	5	203
Mississippi								
Minnesota	Oct. 13	25	1	114	Oct. 3	45	4	299
Arkansas	Dec. 6	25	1	41	Nov. 27	45	4	542
Louisiana	Nov. 30	25	1	54	Nov. 7	55	[e]	500
Atlantic								
New York	Oct. 20	45	2	24	Nov. 16	50	4	83
Pennsylvania	Oct. 20	50	2	20	Oct. 10	60	3	66

Source: Anderson and Burnham (1978).
[a]The estimated size of the continental mallard population was 7.6 million in 1962 and 11.6 million in 1970 (Pospahala et al. 1974: 70–71).
[b]In thousands.
[c]Variation by zone within the state.
[d]Point-system season.
[e]Total bag limit was 6 per day.

TABLE 10-4.   Estimates of Average Annual Survival Rates for Mallards in Years of Restrictive and Liberal Waterfowl Hunting Regulations.

Age-Sex	Number of Reference Areas	Average Survival in Restrictive Years	Average Survival in Liberal Years	Difference (Liberal–Restrictive
Adult males	12	0.6573	0.6366	−0.0207
Adult females	11	0.5500	0.5587	0.0087
Young males	10	0.5175	0.5204	0.0029
Young females	9	0.5263	0.5371	0.0108
All combined	–	0.5628	0.5632	0.0004

Source: Anderson and Burnhan (1978)

the compensatory mortality hypothesis and the implications for setting waterfowl regulations accordingly (i.e., adoption of liberal seasons). They noted that data for species other than the mallard are extremely limited and that other species may not respond to hunting losses as the mallard does. Furthermore, they argued that data for survival studies have been collected without a specific experimental design; therefore, the liberal seasons in some cases frequently coincided with large duck populations. They suggested that conclusive evidence about the relationship between hunting mortality and annual survival will require a long-term experiment with large numbers of banded ducks and hunting regulations, varied by design over a large geographic area. The factors that limit the number of mallards surviving through the winter, spring, and summer following the hunting season remain unidentified.

Figure 10-2 illustrates how hunting mortality would influence survival rates under additive and compen-

satory conditions, respectively. Under additive conditions, survival declines linearly, beginning with the shooting of the first bird (Fig. 10-2, left); that is, the survival rate diminishes whenever a bird is shot. With compensatory conditions, however, survival is unaffected by hunting mortality until a certain point is reached (*C*), beyond which survival declines linearly (Fig. 10-2, right).

A distinction should be made between compensatory mortality and compensatory population growth. *Compensatory mortality* exists when one mortality factor replaces another mortality factor in the population. With regard to hunting, therefore, the concept of compensatory mortality states that if hunting does not kill a certain proportion of animals, about the same proportion will die of starvation, predation, or disease. Natality is not a part of the equation. *Compensatory population growth,* on the other hand, may involve an increase in natality and survival of a new cohort that offsets losses from mortality. Hunting losses, if not

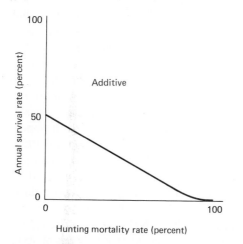
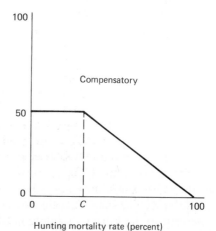

Figure 10-2.   A theoretical comparison of the effect of hunting mortality on total annual survival. With *additive mortality,* an annual survival rate of 50 percent is reduced immediately and proportionately, starting with the death of the first bird shot during the hunting season (left). With *compensatory mortality,* the annual mortality rate of 50 percent is unaffected by hunting mortality up to a point (*C*), beyond which hunting mortality becomes additive (right). (Modified from Nichols et al. 1984.)

compensated by reductions in other forms of mortality, may in some cases be compensated by increased births and survival of animals produced during the next breeding season. Thus, populations may be somewhat self-regulating because of both compensatory mortality and compensatory natality or because of either one or the other. Further research is needed both to identify the causes of nonhunting mortality and to determine when these factors operate.

AMERICAN WOODCOCK

Woodcock (*Scolopax minor*) are migratory shorebirds occurring in the thickets of eastern North America. These popular game birds sit tight for pointing dogs, and their erratic flights present sporting targets. The clutch consists of only four eggs, which arouses concern about how much mortality woodcock populations can sustain without declining. In the decade beginning in 1965, the total harvest of woodcock doubled to an annual kill of about 1.6 million birds; yet measurements of the breeding population and of age-and-sex ratios in the harvest showed that there was little change in woodcock numbers during the period (Owen et al. 1977). Hunting regulations for woodcock—a migratory species—are set by the U.S. Fish and Wildlife Service in consultation with representatives of state wildlife agencies. Woodcock are managed in two geographical units known as the Eastern Region and the Central Region, which are separated along state lines more or less parallel to the axis of the Appalachian Mountains.

Woodcock populations are assessed each spring by counts of males singing on their breeding grounds and by surveys of hunters during the autumn migration (see Chapter 5). Tautin et al. (1983) reported good correlation between the two types of surveys. Some field evidence, however, suggests that there may be as many as three nonsinging males for every singing male and that changes in numbers of the silent males may go undetected. Until the 1980s, daily bag limits for woodcock had been set at 5 birds for decades, and lengths of the hunting season were traditionally liberal (Owen et al. 1977). From the late 1970s through the early 1980s, however, woodcock numbers began to drop. The harvest declined by an estimated 26 percent in the Eastern Region and by 14 percent in the Central Region. Such changes were reflected in the counts of breeding males in the Eastern Region, but not in the Central Region where breeding males increased by 20 percent (Wood et al. 1985). Beginning in 1985, eastern states responded by reducing the daily bag limit from 5 to 2 or 3 birds and by shortening the hunting season for woodcock (e.g., 21 days in Pennsylvania). The exact causes of the population declines remain unknown, but the growing popularity of woodcock hunting in the southeastern United States may be a factor. Wood et al. (1985) proposed that increased hunting pressure on the wintering grounds in the southern Atlantic and Gulf Coast states may be responsible for reduced woodcock numbers.

Based on counts of courting males, woodcock populations continued to decline at about the same rate even after the bag limit was reduced in 1985 to 2 or 3 birds per day in the Eastern Region. However, woodcock populations declined more rapidly than before in the Central Region, where the daily bag limit continued at 5 birds per day (Bruggink 1997). Some form of additive mortality may be occurring in the Central Region, possibly the result of lost habitat, but the exact cause of the continued decline is unknown. In 1997, the daily bag limit in the Central Region was reduced from 5 to 3 birds and the season length was shortened from 60 to 45 days. Further monitoring of woodcock populations, their habitat, and their mortality perhaps will identify the causes underlying their continued decline.

WILD TURKEY

Hunting regulations were set cautiously after wild turkeys (*Meleagris gallopavo*) were reestablished successfully in much of the United States. Weaver and Mosby (1979) analyzed the effects of varying season lengths and bag limits by comparing population data for turkey flocks in two areas of Virginia. In a study area in the Central Mountains, the turkey population was estimated at 19,600 in 1963 and 29,400 in 1976, representing a gain of 50 percent. In an area of the Eastern Piedmont, turkeys numbered 20,700 in 1963, but the population declined by more than 13 percent to 18,100 birds in 1976. The harvest generally increased in the Central Mountain population; the average kill of 1126 birds per year for 1959–62 increased to 1794 for 1964–68 and reached an average of 2271 turkeys per year for 1969–76 (Fig. 10-3). Conversely, harvests in the Eastern Piedmont declined from an average of 1075 birds for 1951–62 to 379 turkeys per year for 1971–76 (Fig. 10-4). The results show that a reduction of the harvest to levels of about 2–3 percent of the autumn population did not halt a decline of turkeys in the Eastern Piedmont. Conversely, harvests of 8–10 percent did not prevent steady growth in the Central Mountain population. Therefore, hunting

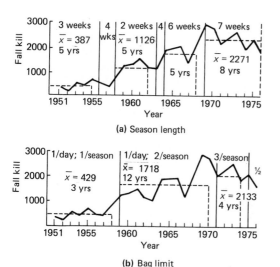

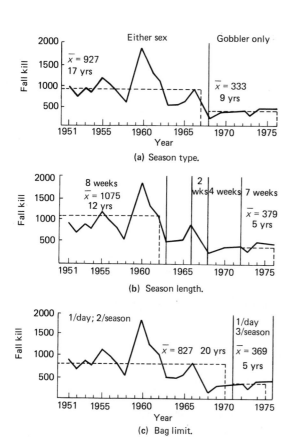

Figure 10-3. Reported fall wild turkey kill and changes in (a) season lengths and (b) bag limits in the Central Mountains Game Range, Virginia, for the period 1951–76. (From Weaver and Mosby 1979.)

Figure 10-4. Reported fall wild turkey kill and changes in (a) season types, (b) season lengths, and (c) bag limits in the East Piedmont Game Range, Virginia, for the period 1951–76. (From Weaver and Mosby 1979.)

was not a factor causing a decline or preventing an increase, respectively, in these turkey populations.

Turkey populations in good habitat may prove resilient to large reductions. Rush (1973) described the effects of live trapping and removing turkeys for stocking in other areas. No detectable decline in the resident population could be detected, even though 38–43 percent of the autumn population of 350–400 birds was removed each year during a 10-year period. As shown in Chapter 3, turkey numbers in Michigan have continued growing concurrently with increased harvests of gobblers.

One might ask, if turkeys can withstand heavy hunting pressure, why were they so scarce 50 years ago? Factors apart from hunting apparently play a major role in determining turkey numbers, especially habitat conditions, disease, and weather. Therefore, beyond adjusting harvest rates, a full range of ecological conditions must be addressed in the management of wild turkeys.

DEER

Late in the 19th century, overhunting and habitat destruction had reduced the white-tailed deer (*Odocoileus virginianus*) population in North America to less than 500,000 animals (Downing 1987). By the 1980s, however, the population of whitetails numbered more than 12 million, and a western relative, the mule deer (*O. hemionus*), rebounded from scarcity in the early 1900s to about 5.5 million animals (Mackie 1987). Deer populations were restored, in part, by imposing strict hunting regulations, including short seasons. In many cases, a limit of 1 antlered deer per hunter per year—a provision widely known as "the buck law"—also was enacted. Because 1 buck often breeds with several does, an imbalanced sex ratio favoring females has little impact on reproduction. Under good range conditions about half of the female white-tailed deer breed in their first autumn, giving birth to a single fawn the following spring. Virtually all females 1.5 years of age and older reproduce each year, averaging 2.0 fawns each.

Under the buck law, deer numbers had rebounded by the 1940s. The recovery occurred in the absence of natural predators—for the most part, wolves (*Canis lupus*) and mountain lions (*Felis concolor*) that had been extirpated in the early 1900s—and during a period when abandoned farms and cutover forests were reverting to young woodlands. An entire generation of hunters therefore had grown up with the buck law and witnessed its success.

New problems arose, however, as the deer herds increased in density (Leopold 1943). In their northern range, white-tailed deer traditionally spend the winter congregated in coniferous lowlands called deer yards, where they find protection from the elements (Chapters 6 and 7). Under such conditions, large numbers of deer place excessive pressure on the food supply, resulting in malnutrition, starvation, and reduced reproduction. Deer numbers and harvests stabilized or declined despite continuation of the buck law. In response, biologists recommended liberal hunting regulations as a means of reducing the number of deer entering the yards, thereby relieving overbrowsing and increasing reproductive success. These recommendations frequently were met with strong objections from hunters who were unfamiliar with the concepts of carrying capacity and limiting factors. The hunters feared an overharvest and consequent decline of their favorite species of big game.

Eventually, however, biologists gradually were granted opportunities to refine and liberalize hunting regulations for deer, but only after two decades of controversy (Day et al. 1971; Cronin 1979; Gilbert 1977). The current approach usually employs a variety of regulations within each state: more liberal harvests in those areas with high deer numbers and a potential shortage of winter browse, and stricter hunting regulations in areas with smaller herds. The methods used for a desirable geographic distribution of the deer harvest include the allocation of permits for 2 or more antlerless or antlered deer per hunter, lengthened seasons in areas with dense deer populations, and shortened seasons for bucks only in less-populated areas. Hunters take more than 150,000 deer each year in New York (Underwood and Porter 1983). In Texas, the harvest regularly reaches 350,000 deer and at times exceeds 400,000 (Texas Parks and Wildlife Department files).

In addition, hunters' dollars have been used for the development and maintenance of deer habitat. Managed forests thus have played an important role in the revival and sustenance of large numbers of deer in many states. In Michigan, for example, $2 from each deer license now goes into the improvement of deer habitat on state lands. In 1987, Michigan hunters harvested 240,000 deer with guns and 60,000 with bows and arrows (Michigan Department of Natural Resources files). Thus, 300,000 deer were harvested in 1987 from an autumn population estimated at somewhat more than 1 million deer. By comparison, the average harvest in the early 1960s was about 100,000 deer (Bennett et al. 1966). Biologists in Michigan believe that the 1987 bag of 25–30 percent of the autumn population can be sustained, provided that habitat conditions remain favorable.

In 1983, a symposium described several examples where the sex and age ratios of deer populations were manipulated by hunting (Caesar Kleberg Wildlife Research Institute 1983). Without restrictions, hunters on a large private ranch in Texas generally overharvested adult bucks, resulting in a posthunt sex ratio of 1 buck per 11.6 does (Adams 1983). Implementation of a permit system limited the harvest to only 33 percent of the antlered deer and required that hunters kill at least 2 does per buck. Although the permit system initially was unpopular with hunters, the technique changed the sex ratio to 1 buck per 4.9 does and increased the size of the bucks in the harvest—results that later improved hunters' acceptance of the regulations. In contrast, the body size and antler growth of white-tailed deer were not improved in another experiment in Texas when only 25 percent of the buck population was harvested (Gore et al. 1983). In Ontario, Euler and Smith (1983) determined that there is no danger of overharvesting when the kill of antlerless deer is held below 30 percent. To achieve a suitable harvest of black-tailed deer (*O. h. columbianus*) and thereby reduce forest and crop damage on an 8000-ha refuge, De Calesta (1983) found that it was necessary to loosen hunting restrictions and schedule hunts conveniently for applicants. The result was a heavier harvest and a deer population with larger body weights, balanced age ratios, and healthy fawns and yearlings.

Medin and Anderson (1979) used a computer model for estimating the effects of varying harvest strategies on a population of mule deer in Colorado. The model considered the interactions of weather conditions and food quality, which in turn affected the condition and reproductive rates of the Cache la Poudre River deer herd. Between 650 and 1526 deer normally were harvested from the herd of 10,000 animals. Various levels of doe harvests were tested, keeping the harvest of males at 1000. The model indicated that up to 650 does could be harvested without causing a decline in the herd. The herd diminished, however, when the harvest reached 700 or more does. One way of potentially providing a high yield of deer was to double the harvest rate of does every second year (i.e., 1300, 0, 1300, 0, and so on). According to the model, such a harvest could produce a greater increase in the population than would result from a constant harvest of 650 does each year. Computer models contain imperfections, however, and should be tested continually against field data (Medin and Anderson 1979; Underwood and Porter 1983).

MANAGING FOR THE HUNTER

Despite large deer harvests in many states, a significant number of hunters do not bag a deer every year. In Michigan, more than 750,000 hunters bag 300,000 deer; thus only about 1 hunter in 3 or 4 is successful (hunters legally may take more than 1 deer). Indeed, if each hunter bagged a deer—thereby removing a total of about 750,000 deer from a population of 1 million—the annual harvest would be biologically impossible to sustain. Perhaps wildlife managers should consider that a maximum level of hunter satisfaction may be a more desirable goal than maximum harvests. A few studies have addressed questions about the relationships between hunter satisfaction and game harvests.

Kennedy (1974) categorized factors contributing to the satisfaction of deer hunters as "hunt rewards" and "extrahunt rewards." Hunt rewards include game-dependent satisfactions such as seeing animals and their signs and bagging a deer. Those goals are managed traditionally with habitat improvement, hunting regulations, and law enforcement. Extrahunt rewards include companionship, cooking and eating outdoor meals, and camping. Such goals are managed by controlling the number of hunters, providing maps and hunting information, and developing desirable camping facilities.

In Maryland, fellowship played a more important role among hunters than did bagging a deer (Kennedy 1974). In Michigan, however, the satisfaction of hunters was directly related to the bag of antlered deer (Langenau et al. 1981). About 67 percent of the hunters bagging deer reported good or very good hunts, whereas only 19 percent of those who did not bag a deer shared the same opinions. Nonetheless, two-thirds of the hunters who reported poor or very poor hunts returned a second year, which suggests some degree of satisfaction with extrahunt rewards.

Results from the Michigan survey support the belief that hunter satisfaction is dependent on the density of legal game, but wildlife managers nonetheless face a dilemma. How can more hunters be satisfied if deer already are produced and harvested at maximum rates consistent with local soil and vegetation conditions? There are three choices: (1) a small proportion of a large number of hunters can be satisfied; (2) the total number of hunters can be reduced so

that the percentage of satisfied hunters increases; or (3) all hunters may be provided with extrahunt benefits (e.g., the development of better camping facilities and the distribution of maps and other information). Where deer numbers are not at their maximum, Langenau et al. (1981) concluded that financial resources might best be put into increasing deer densities, thereby increasing the number of satisfied hunters (i.e., more deer are seen and bagged).

In addition, cooperation between hunters and wildlife managers undoubtedly improves hunter satisfaction. Many national and local hunters' organizations now contribute money and labor directly toward land acquisition and habitat management, often working in the field with biologists. The resulting sense of accomplishment in seeing increased wildlife production very likely offsets failure to attain a bag limit on every hunt.

Stocking game for shooting is a somewhat controversial practice. In some cases, pen-reared birds are set out for hunters to shoot about an hour later, but this procedure does not quality as wildlife management. Conversely, stocking birds or mammals to create permanent breeding and huntable populations is a legitimate form of wildlife management, and these programs frequently involve exotic species, as discussed more fully in Chapter 18. Despite the apparently logical nature of stocking, augmenting local breeding populations with stocked game animals is quite expensive and seldom achieves desired results. With radio telemetry, for example, Leif (1994) determined the fates of 159 pen-reared female pheasants (*Phasianus colchicus*) and 44 live-trapped wild female pheasants released in the same area of South Dakota. Only 8 percent of the pen-reared birds survived, whereas 55 percent of the wild pheasants survived. The pen-reared pheasants also raised only three broods per 100 females, but the wild pheasants produced broods at the rate of 34 per 100 females. Moreover, the concentration of hand-reared hens attracted predators, thereby causing higher than normal losses of nests for both wild and pen-reared pheasants in the release area. Based on these results, pen-reared hens released as breeding stock in fact may negate the survival of *all* pheasants in the local population.

HUNTING BY NATIVE AMERICANS

For the most part, hunting in North America is based on the use of efficient weapons developed during the past two centuries. The result for most American hunters has been an outdoor activity that combines European influences, often ceremonial as well sporting in nature, and North American motives that focus on recreation and meat. In recent years, an emphasis on shooting large animals appears to be increasing, producing detailed record keeping and a scoring system for trophies that encourages competition among hunters.

Conversely, Native Americans, at least in some areas, are developing greater interest in restoring their cultural hunting traditions. For example, in the Yakima tribe in Washington State, hunters represent a small and select fraction of the community in which some tribal members hunt as a form of public service to obtain meat for others in the tribe and not as a personal pursuit (McCorquodale 1997). Meals in which wild game is shared are central features of funerals, memorials, name-giving rituals, and weddings. Such ceremonies may take place throughout the year, such as when a young man kills his first animal—an event regarded as a rite-of-passage into adulthood. Celebration of this event requires not only that the meat be given to others within the tribe, but also that his weapon be given away during the ritual.

Interest in reviving age-old hunting traditions without regard for season, however, brings with it opposition from some other quarters of the hunting public—those who question the fairness of one group hunting under a different set of rules from the other. Nonetheless, McCorquodale (1997) argues that "There is nothing inherently moral about 2- or 3-week hunting seasons or 1-animal bag limits." Further, he regards regulations as amoral and simply tools to achieve desired management objectives. He also noted that subsistence hunting by Native Americans is "need-driven," whereas recreational hunting is "opportunity-driven," and that opportunity-driven hunting regulations inherently promote greater participation (i.e., an attitude exemplified by "we'd better get out and shoot a deer because we have only 2 weeks to do so"). Moreover, the number of tribal hunters tends to be small, with less than 1000 who hunt elk in Washington State in contrast to 31,000 recreational elk hunters in an equivalent area.

McCorquodale (1997) asserts that "the recreational hunting culture has been unable to affirm subsistence and ceremonial hunting by Native Americans as a legitimate resource use [that] betrays an attitude of intolerance toward other cultural views of hunting. Native subsistence and ceremonial hunting should be viewed as a legitimate allocation of the big game resource, which is shared with recreational hunters and which, with few exceptions, poses no threat to the recreational allocation of big game." However, few recreational hunters have accepted the merits of these agruments.

Allocations of fish stocks have produced similar controversies in recent decades. In these cases, Native Americans claim their rights to take fish by any means, at any time, and in any numbers, whereas sport anglers protest the methods and fairness of such unregulated harvests (which indeed may be stipulated in treaties dating to the 19th century). Resolutions to conflicts such as these will require thoughtful tolerance and reason as well as close monitoring of the shared fish and wildlife populations so that neither approach threatens the continued vitality of these resources.

TRAPPING AND FURBEARERS

Trapping for some is a profession, but for most trappers, it remains a hobby. Boddicker (1981) presented a profile of the "typical trapper" in America as male, 35 years old, a blue-collar worker, at least a high school graduate, earning $11,000 per year, of which $1596 (14.5 percent) came from trapping. Such persons spent about $500 on trapping and used 56 leghold traps or snares for an annual catch of 112 furbearers, mainly muskrats (*Ondatra zibethicus*), raccoons (*Procyon lotor*), and red foxes (*Vulpes vulpes*).

Even more than hunting, trapping has been the subject of heated controversy for the past two or three decades (Mitchell 1982). Opponents of trapping argue (1) that traps cause animals undue suffering; (2) that taking an animal's life for the sake of fashion is immoral; and (3) that trapping kills or maims many nontarget species. Since 1973, Florida, Massachusetts, Rhode Island, and New Jersey have banned or severely restricted the use of leghold traps. Referenda calling for a ban on leghold traps failed in Ohio in 1977 and in Oregon in 1980. Since then, antitrapping groups have concentrated their political efforts on townships, counties, and other local governments rather than on initiating statewide campaigns.

Arguments in defense of trapping include the following: (1) that harvest data from trappers help wildlife managers keep track of furbearer populations; (2) that trapping provides an annual crop of furs that otherwise would be lost to other forms of mortality; (3) that fur sales help the economic welfare of states and individuals; (4) that rabies, distemper, and other diseases are suppressed by trapping; and (5) that the inevitable death of an animal probably is less painful for a trapped animal than it would be from starvation, predation, or other "natural" causes (Woodstream Corp. 1977). Smith and Brisbin (1984), in part, refuted the argument that trapping data provide a means of monitoring furbearer populations. Instead, the study showed that the probability of capturing striped skunks (*Mephitis mephitis*), gray foxes (*Urocyon cinereoargenteus*), and raccoons varied considerably from year to year. Moreover, total trapline captures did not correlate well with the estimated populations of those species. Therefore, further study seems necessary before wildlife managers can determine whether changes in furbearer populations may be reflected in fur harvests. Some of the controversies and arguments about trapping are summarized in Table 10-5.

TABLE 10-5. A Summary of Statements and Arguments About Trapping

Statement	Protrapping		Antitrapping	
	Argument	*Solution*	*Argument*	*Solution*
Trapping is cruel.	All animals must die. Death in a trap is no more cruel than starvation, disease, or predation.	Frequent checks of traps. Continue research and development of humane traps.	Animals caught in traps suffer pain and loss of limbs.	Stop trapping; use box traps.
Traps are nonselective.	Traps properly set and without bait catch only intended animals most of the time.	Use bait only where necessary.	Many nontarget animals are killed and wasted.	Outlaw baited traps.
Only traps that kill quickly should be used.	No chance for release of nontarget animals.		Prevents lingering deaths	
Trapping controls disease.	Controlling numbers of animals prevents overpopulation and spread of disease. Special trapping may reduce rabies outbreaks.		There is little supporting evidence that fur trapping controls disease.	
Trapping prevents overpopulation.	Unharvested populations overeat food supply and damage habitat.		Many species that are not trapped do not experience destructive overpopulation.	Establish natural predators to control numbers.
Fur harvests help to monitor size of populations.	Sound management is based upon number of pelts taken.		Strict management is necessary if trapping is outlawed. Fur harvests may reflect price of fur rather than number of animals present.	
Furs are an economic resource.	Natural furs represent a useful and renewable resource.	Continue regulated trapping.	Furs are not essential but a luxury.	Stop trapping.

TRAPPING IN ONTARIO

Nowhere is fur trapping as carefully managed as in Ontario, Canada. The fur industry has been an important part of the life and economy of Ontario since 1534 when Jacques Cartier first traded with Native Americans. Monk (1981) described the experience of managing fur resources in Ontario, which we summarize here.

For four centuries, the Canadian fur market has sold the pelts of muskrat, fox, wolf, mink (*Mustela vison*), lynx (*Felis lynx*), fisher (*Martes pennanti*), marten (*M. americana*), otter (*Lutra canadensis*), snowshoe hare (*Lepus americanus*), and red squirrel (*Tamiasciurus hudsonicus*). But beavers (*Castor canadensis*) have been the backbone of the fur trade, and ups and downs in beaver populations have served as a barometer of the trapping industry in Ontario.

The early days of fur trading were characterized by violent battles between rival fur companies and fights over trapping territories. Lives were lost, and trading posts were besieged. Furbearer populations—especially those of beaver—steadily declined as a result of the competition for furs. In the first half of the 19th century, the major fur trader, the Hudson's Bay Company, imposed a limit of 650 beavers from northern areas as a means of protecting the remaining population. Nonetheless, beaver populations continued declining. In the latter half of the century, federal and provincial governments enacted laws that closed the trapping seasons for beaver, otter, fisher, marten, and muskrat between May 1 and October 31 each year. But the numbers of furbearers still declined.

In 1892, the trapping season for beaver, otter, and fisher was closed entirely for 5 years. In 1895, deputy game wardens were appointed, but the numbers of furbearers still fell. The deputies were not salaried, and the laws largely were ignored. Little was known about habitat, distribution, productivity, and harvest rates. A new effort was made in 1916 to control the excessive harvest of furbearers. Beavers and otters could be trapped only under license ($5) and coupons for 10 beavers and 5 otters were issued with each license; each pelt required an attached coupon.

Beaver trapping in southern Ontario was closed in 1922, and the number of game preserves where hunting and trapping were prohibited was increased from 7 to 11. Beaver and otter seasons were shortened the following year in northern Ontario, and trapping was illegal within 1.5 m of a beaver lodge. Additional areas were closed to beaver trapping as the 1920s progressed, but beaver and otter populations declined further because of inadequate law enforcement. Fur resources were still being overexploited.

In 1938, the provincial government closed beaver trapping throughout Ontario. Stiff penalties were exacted for violations. One fur buyer was fined $15,395 for illegally trading 1000 beaver pelts, which were confiscated and sold by the government for $14,000, a sizeable sum at the time.

Strict enforcement of laws produced hard feelings between the government and people who were used to living off the land. Fur buyers, long accustomed to buying and selling as they pleased, also resented government interference, but, by the late 1940s, furbearer populations were recovering. The success of strict laws, tainted with distrust of government, brought about new approaches to managing trapping—approaches that placed considerable responsibility on the trappers themselves.

Registered trap lines were begun in 1947. Thereafter, individuals trapping on provincial lands, by license, had exclusive rights to the furbearer harvest on such lands. An annual quota for fisher, marten, lynx, and beaver was established by negotiations with each trapper. Trappers also formed the Ontario Trappers Association, which sponsored fur auctions where buyers bid competitively for pelts.

Areas depleted of animals were restocked from the Chapleau Game Preserve, a large, centrally located area that had been closed to trapping. Annual beaver harvests rose from about 23,000 in early 1940s to about 120,000 in the 1950s, and to 150,000 in the 1960s. In 1979–80, 209,000 beavers were harvested.

For the 1979–80 season, Ontario sold 17,772 trapping licenses. Many trappers operate on private lands, but there were 2691 registered traplines, some of them shared. A minimum of 75 percent of the negotiated quota was expected, but the maximum number could not be exceeded. Most beaver quotas are set at the rate of 1.5 beaver per colony, and aerial surveys determine the number of colonies. The beaver population in Ontario is estimated at 2 million and seems to be stable. Trapping quotas for other species are mandated according to population trends.

An organized program of trapper education has been instituted along with the carefully regulated harvest system. About 300 two-day workshops are conducted throughout the province under government and private sponsorship. Sessions are conducted by experienced trappers, enforcement officers, and wildlife biologists. Courses about the conservation of furbearers, consisting of 40 hours of classroom instruction and 10 hours of field demonstrations, are given at several community colleges. The courses cover not only the biology and trapping of furbearers, but also address the ethics and conduct of modern trappers.

Monk (1981) concluded that the key to sound furbearer management in Ontario is good communication, thereby establishing strong liaisons among trappers, resource managers, and enforcement officers. Although unfortunate and all too common, the near-collapse of a resource and the forceful intervention of government were necessary before appropriate conservation measures were adopted. The Ontario experience thus offers a lesson for others about ways in which future crises might be avoided in the harvest of furbearers.

TRAPS

Three types of traps are commonly used for capturing furbearers: steel or "leghold" traps, snares, and killing mechanical traps. Leghold traps are by far the most popular among trappers and the most unpopular among antitrappers. Leghold traps are portable, relatively inexpensive, and can be set in a variety of places. Unless a leghold trap is set in or near water in such a way that a trapped animal drowns, the victim is held alive until the trapper arrives and dispatches the animal with a bullet or club.

Snares are simply wire loops that catch the necks or feet of animals. Foot snares seemingly cause less damage than leghold traps (Mitchell 1982), but neck snares usually strangle their victims. Despite the lingering trauma endured by animals caught alive in leghold traps and snares, many animals unintentionally caught in such devices can be released.

Killing mechanical traps include the Vital and the Conibear. These kill when large clamping bars strike the head or neck region. Although killing mechanical traps are humane, they must be set in special conditions (a narrow runway or a sheltered "cubbyhole") and their deadliness eliminates the option of releasing alive an unintended victim.

The major objection to trapping—inhumane treatment of animals—has produced constructive attention from several quarters. The Canadian Association for Humane Trapping promotes the research and development of traps that minimize the pain suffered by captured animals. Another agency, the Federal-Provincial Committee for Humane Trapping, gives

governmental impetus for the development of humane trapping methods. Gilbert (1976, 1981) tested the killing thresholds and improvements of the Vital and Conibear traps and concluded that humane (quick-killing) traps theoretically can be developed for practically any species. The remaining drawback, however, is that trap design alone does not prevent the capturing and killing of nontarget species. To resolve that problem, the behavior of target and nontarget species must be understood before a trapping system can be developed. There must be knowledge about species-specific set locations, placement of baits, and a selective means of setting the trigger mechanism.

TRAPPING AND HUNTING EDUCATION

The educational program is one of the important features of the successful furbearer management in Ontario. Boggess and Henderson (1981) noted that hunters, trappers, and their antagonists all agree that more education is needed. In the United States, about half the states currently require hunter-safety courses, but only New Jersey and Washington require trapper education courses. The extension service at Kansas State University conducts an educational program for trappers and hunters. The program encourages selective, ethical, safe, and humane trapping and hunting, and improved understanding of wildlife and the environment. Boggess and Henderson (1981) emphasized that successful programs involve knowledgeable citizens from the local community. Attempts also have been made to educate the general public about wildlife ecology, management, hunting, trapping, and the moral and philosophical aspects of hunting and trapping.

Because decisions about natural resources are ultimately political decisions, an informed public is essential if the mission of wildlife management is to go forward. The legitimacy and sincerity of the arguments of antihunters and antitrappers, as well as those of hunters and trappers, must be recognized and calmly addressed. Shouting matches will not enhance the views of either side in these issues; nor will the public long embrace positions founded on falsehoods and inaccuracies. Mitchell (1982) therefore quoted Marietta Lash of the Canadian Association for Humane Trapping, a group committed to a moderate viewpoint: "How can you possibly get someone around to your way of thinking if you're forever telling him he's the scum of the earth?"

SUMMARY

Hunting and trapping, once regarded as essential human activities, persist as recreational pursuits in modern societies. Opposition to hunting and trapping has grown in recent years, and opponents hold humanistic and moralistic viewpoints. Hunters view their sport from a utilitarian, dominionistic, and naturalistic standpoint. Regulated sport hunting has not endangered any species with extinction. Because wildlife harvests probably will continue, management should adopt a strategy of *optimum yield,* which considers ecological, social, and economic factors as well as consumption of the resource. Ecological theory suggests that maximum harvests may be attained by trimming a population to half of carrying capacity. In practice, however, harvests are set conservatively, and the game populations are monitored to observe the impact of hunting regulations.

Champions of animal rights particularly question the morality of trapping because of the presumed suffering of trapped animals. Trapping has been outlawed in a few states, whereas a strong program of trapping management has evolved in Ontario. Humane trap designs, trapper education, and strict enforcement of trapping regulations may achieve biological goals and improve ethical standards. Management of hunting and trapping today involves the management of controversy. Some degree of mutual understanding will occur when antagonists understand that both consumptive and nonconsumptive users of wildlife share a common goal: the welfare of animal populations.

CHAPTER 11

WILDLIFE AND WATER

The creeks overflow: a thousand rivulets run 'Twixt the roots of sod;
the blades of marsh-grass stir; Passeth a hurrying sound of wings that
westward whirr; Passeth, and all is still; and the currents cease to run;
And the sea and the marsh are one.

Sidney Lanier, "The Marshes of Glynn" (1878)

Water is indispensable to all organisms; yet it is unevenly distributed—in time and space—across the Earth's surface. Water also is a simple chemical compound, and unlike most others, it readily appears in liquid, gaseous, and solid forms that act in an immense variety of ecological settings. It occurs in grades essentially free of minerals as well as in those laden with salts and other materials. Of the immense amount of water on our planet, fully 97 percent occurs in oceans, about 2 percent in glaciers, and less than 1 percent in the combined volume of aquifers, rivers, and lakes.

Many ecological relationships between water and wildlife are direct and obvious; others are more subtle. All are essential. Water management touches the distribution, quality, and quantity of water not only in arid regions, but also where water may be plentiful.

SOME PROPERTIES OF WATER

A high *heat capacity* is among the more important properties of water in relation to living matter; that is, water is able to absorb a great deal of heat without becoming much warmer. For example, if equal weights of iron and water were frozen to absolute zero ($-273°C$) and then subjected to equal amounts of heat, the iron will have melted at 1298°C, whereas the ice would have just reached its melting point (0°C). Only liquid ammonia has a greater heat capacity than water, making ammonia-based compounds useful in refrigerators, air conditioners, and other cooling devices. Because water forms a high percentage of an animal's body weight, its heat capacity helps stabilize body temperatures even under extreme environmental conditions. Water content reaches 67 percent by weight in the bodies of mule deer (*Odocoileus hemionus*) and 82 percent in pronghorn (*Antilocapra americana*) (Knox et al. 1969; Wesley et al. 1970). Homeotherms—those so-called warm-blooded animals with self-regulating body temperatures—take special advantage of this property; birds and mammals live under the most extreme temperature regimes on Earth. Even poikilothermic, or cold-blooded, species such as fishes survive under a wide range of climatic regimes because the temperature of their habitat (i.e., water) changes relatively little between tropical and polar climates. Adaptions of many kinds, of course, delimit most animals to more specific ranges on land or in water, but water's high heat capacity is fundamental to each species' overall tolerance for ambient temperatures.

Water also is an universal solvent that alters the chemical state of an unusually large number of other substances. This property makes essential nutrients available for life processes while concurrently serving as the medium of their transport. Therefore, dissolved materials from fundamental sources such as bedrock may be transferred to the living components of ecosystems everywhere. Indeed, Leonardo da Vinci noted that "water is nature's carter," clearly referring to its importance as a life-giving transportation system (MacCurdy 1939).

Water's action as a solvent may change the carrying capacity of wildlife habitat. Inundation of soils otherwise above the water table leads to a rapid, if temporary, increase of available nutrients that may greatly expand both the numbers and growth rates of plants and animals at the flooded site. Conversely, water may dissolve or carry harmful materials. In modern times, the solvent properties of water have fostered disasters that were caused by humans. Pollution of many kinds has ruined entire aquatic systems and destroyed the biotic communities they supported.

Closed watersheds continually receive an ever-larger concentration of salts that steadily accumulate to the point where many forms of life are harmed or excluded. This phenomenon is accentuated where evaporation exceeds precipitation. The Great Salt Lake in Utah and the Dead Sea bordering Israel are classic examples of closed watersheds naturally affected by salt accumulations, but immense numbers of lesser sites are similarly affected throughout the world. In Kenya, after the traditional breeding area of lesser flamingos (*Phoeniconaias minor*) suddenly was flooded, the huge flock—perhaps 2 million birds—quickly moved en masse to Lake Magadi, a shallow lake with no outlet that is fed by saltwater springs. The intense heat caused rapid evaporation, further concentrating the salts in the lake. After the birds hatched, the legs of the young flamingos built up anklets of encrusted salt deposits as they waded in the mineral-laden water (Fig. 11-1). Few could survive the burden of the encrustations. Fortunately, a group of conservationists happened on the scene and began the arduous task of removing the deposits from the doomed birds. In all, some 27,000 young flamingos were freed from their shackles (Williams 1963). The flamingo colony returned to their original breeding areas the following year, avoiding the lethal waters of Lake Magadi.

Similarly, heavy encrustations of sodium salts killed or immobilized about 300 Canada geese (*Branta canadensis*) and other waterfowl on a hypersaline lake in western Saskatchewan (Wobeser and Howard 1987). Some of the birds were encrusted with at least 3 kg of salt crystals. The dead geese showed

Figure 11-1. High concentrations of dissolved salts in the water of Lake Magadi in Kenya formed heavy encrustations on the legs of young flamingos (left). Conservationists chipped the mineral shackles from the debilitated birds, saving thousands from sure death (right). Fortunately, the large flamingo colony returned to a less-threatening breeding habitat the following year. (Photos courtesy of Alan Root, with permission of the National Geographic Society, © 1963.)

evidence of acute muscle degeneration and had aspirated lake water. The encrustations apparently occurred when the lake water cooled rapidly, resulting in the supersaturation and crystallization of the dissolved salts. Most of the immobilized geese (155 birds) survived after being captured and relocated on nearby freshwater wetlands.

Human disturbances also can increase the salinity of water, of which the case at Mono Lake, California, has attracted wide attention (Sun 1987). The city of Los Angeles diverts the streams feeding Mono Lake, which not only lowers the lake's water level, but also further concentrates the already saline water. Mono Lake, famous for its mineral towers, lacks fishes but produces immense numbers of brine shrimp and brine flies that feed hundreds of thousands of migrant birds traveling in the Pacific Flyway. If the current lake level falls by 3 m, the brine flies will lose 40 percent of their habitat. Further drops will impair reproduction of the brine shrimp, again impairing the food supplies for birds. Lower lake levels also mean that predators could gain access to birds nesting on the lake's islands. Mono Lake already has fallen about 11 m, and conservationists worry that continued losses will trigger the collapse of the lake's food chain.

SOME ECOLOGICAL INFLUENCES OF WATER

Light penetration into water interacts with depth and clarity to support the basic productivity necessary for aquatic food chains. Birge and Juday (1929) recorded light penetrations of 67 percent full intensity and 10.5 percent full intensity at 1-m and 10-m depths, respectively, in the usually clear waters of Crystal Lake, Wisconsin. However, wave action, erosion, or other factors may inhibit light penetration, causing diminished productivity in disturbed systems. Robel (1961b) reported a negative relationship between water turbidity and the biomass of submersed vegetation important as food for waterfowl (Table 11-1). Carp (*Cyprinus carpio*) and other so-called rough fishes contribute to turbidity in soft-bottomed marshes, often to the point where vegetation may be affected and the carrying capacity for waterfowl may be reduced (Chamberlain 1948). Water clarity increased from a transparency of 15 cm to more than 90 cm after rough fishes were removed from a North Carolina lake (Cahoon 1953).

Water exerts a powerful physical force on the landscape. The natural phenomenon of geological erosion

TABLE 11-1. Relationship of Water Turbidity to Submersed Plant Production

Relative Turbidity Rating	Water Turbidity (Colorimeter Units)	Plants (kg/ha)	
		Range	*Mean*
Low	0–50	1680–3585	2297
Medium	51–100	18–1064	448
High	101–150	6–560	140

Source: Robel (1961b).

has created the canyons, valleys, escarpments, and other features of the landscape that have much to do with wildlife habitat. Topography, of course, thereafter may influence the amount and quality of water available to fishes and other wildlife. The same water supporting trout and other cold-water organisms at high elevations later produces very different aquatic communities as it moves to lower elevations. Oxygenation and clarity as well as temperatures are transformed as water velocities respond to topographical gradients.

Rivers meandering across flat terrain cut new channels so that water eventually flows in newer, self-made courses (Fig. 11-2, upper). Old river channels thus separated from the new riverbed are known as oxbow lakes in North America and as billabongs in Australia. Horseshoe Lake Game Refuge in southern Illinois was developed around a 485-ha oxbow of the Mississippi River. Some 30,000 Canada geese, about half the region's winter population, once used the refuge each year, establishing the oxbow lake as the most important winter goose habitat in the Mississippi River Valley (Hanson and Smith 1950). More recently the refuge has supported an even larger winter population of about 100,000 geese (Fig. 11-2, lower). The numerous billabongs developing from the old river systems in Australia are crucial breeding, feeding, and refuge habitat for ducks and other waterbirds. Like oxbow lakes, billabongs become progressively shallower, develop aquatic vegetation, and lose many of their original riverbed characteristics. Mature billabongs are among the most important waterfowl habitats in Australia (Frith 1959, 1967, 1973). Whereas these and other processes are natural phenomena of a maturing landscape, water also erodes immense amounts of soil when humans are careless. Sheet, gully, and riparian erosion each has forceful implications on soil, vegetation, wildlife, and human resources (see Chapter 12).

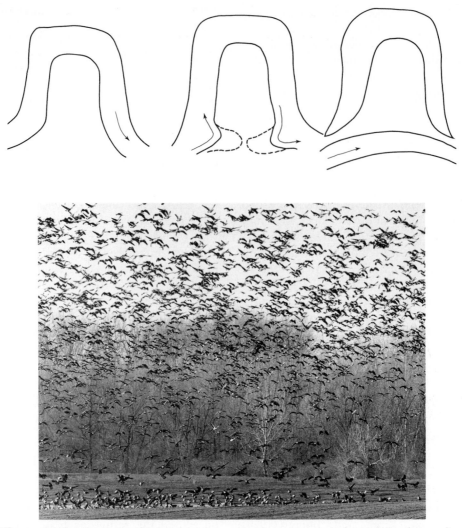

Figure 11-2.   Formation of oxbow lakes is a continual process typical of meandering, low-gradient streams and rivers. Water currents (arrows) gradually carve new channels that bypass and isolate the oxbows (top). Horseshoe Lake Game Refuge in Illinois is managed for Canada geese and other wildlife on an oxbow formed on the upper Mississippi River (bottom). Note the river-bottom forest remaining in the background. (Photo courtesy of Illinois Department of Conservation.)

Large-scale patterns in rainfall distribution are influenced by air currents moving across mountain ranges. Warm, moisture-laden air moving upward over mountainous terrain cools at the higher elevations and can no longer retain its moisture. Rain or snow falls on the windward side of the slope. However, little moisture remains by the time the air currents have passed over the mountains, and a dry "rain shadow" develops on the lee side of the range. The Great Basin of North America, encompassing much of Nevada, Utah, and southern Idaho, lies in the rain shadow of the North American cordillera, in part, creating an arid environment suitable for chukar partridges (*Alectoris chukar*).

In the form of ice, water during the glacial epochs shaped lakes, rivers, and terrain as the ice sheets advanced—and then retreated—across much of North America. The numerous prairie potholes that dot the northern plains claim glacial origins. Some were formed by gouging, and others resulted when ice chunks calving from the glacier's leading edge were forced downward under the immense weight of the advancing glacier. When the glaciers eventually retreated, myriad water-filled potholes remained that became the heartland of breeding habitat for waterfowl in North America. At least 50 percent of the North American duck population is produced in this region—aptly known as America's "duck factory"—

even though these wetlands make up only 10 percent of the total breeding habitat (Smith et al. 1964).

On May 29, 1986, an ice dam formed by a surge of Hubbard Glacier closed off Russell Fiord, Alaska, and impounded about 5.3 km³ of fresh water. The new lake stratified; a layer of dense salt water lay beneath an accumulating layer of fresh water. The lake level rose nearly 30 cm per day and, had the water continued rising, drainage patterns in local rivers would have changed radically, no doubt impairing the migrations of Pacific salmon (*Oncorhynchus* spp.) and threatening the economy of neighboring villages. When the ice dam broke open on October 8, 1986, the discharge was about 14 times the amount of water cascading over Niagara Falls each second. In places, erosion removed the alluvial soils down to bedrock and produced a 200–300-m retreat in the shoreline (Mayo 1986). Until then, however, marine life was trapped behind the ice dam, largely depending on food and oxygen in a layer of salt water buried beneath the deepening layer of fresh water. About 170 seals and sea lions (Pinnipedia) were among the marine mammals confined in "Lake Russell." Rescue efforts, organized by the California Marine Mammal Center, received widespread coverage in the news media, but only a few of the entrapped animals were captured and released before the ice dam ruptured and Russell Fiord returned to a marine environment (Gage 1987; Roletto 1987). Geologists, however, believe that Hubbard Glacier again will close Russell Fiord, forming an even larger freshwater lake and once more producing a rapidly changing environment for marine wildlife.

Bruemmer (1987) has described the ecological importance of permafrost, the stratum of icebound ground that underlies about one-quarter of the Earth's land surface. Thick layers of permafrost develop wherever annual freezing exceeds annual thaw. In Barrow, Alaska, permafrost is about 400 m thick, and in Siberia, maximum depths of nearly 1500 m are known. In summer, the Arctic becomes waterlogged and green with a rich carpet of vegetation. About 900 species of vascular plants occur in areas with permafrost; yet the average precipitation (20 cm) is less than occurs in the Mojave Desert. Permafrost explains the paradox of a watery land with an arid climate. The scant rainfall and snowmelt cannot seep into the frozen soil, and the water therefore remains in the root zones of the vegetation. Without permafrost, most of the Tundra would become a lifeless desert. With permafrost, the food web includes herbivores such as caribou (*Rangifer taran-*

dus), muskox (*Ovibos moschatus*), and Arctic hares (*Lepus arcticus*).

Some appreciation of the shaping force of water can be gained when estimates of soil carried annually by some of the world's great rivers are considered: the Yellow River, 2080 million tons; the Ganges River, 1600 million tons; the Amazon River, 400 million tons; the Mississippi River, 344 million tons; the Colorado River, 149 million tons; and the Nile River, 122 million tons (Holeman 1968). The Yellow River in China and the Colorado ("red," in Spanish) in the United States, among others, derive their names from the colored burden of soil they carry. Some of these loads are, of course, human-caused, resulting from abusive practices upriver, but much of the burden is the normal result of water's geological erosion. In the form of glacial ice, water discharges even larger amounts of sediments. Friedman and Sanders (1978) estimated that glaciers in Antarctica discharge 35–50 billion tons of sediments annually at a rate of 2500 to 3570 tons per km² of ice (versus about 13 billion tons or 260 tons per km² of watershed for the world's rivers combined).

Sedimentation in such vast amounts often has produced large deltas such as those at the mouth of the Mississippi and Nile rivers. The delta at the mouth of the Mississippi River is expanding at a rate of about 90 m per year (Shirley and Ragsdale 1966). These deposits are rich, carrying with them a nutrient-base of continental proportions. The marshes of the lower Mississippi Delta in Louisiana, for example, are renowned for the populations of muskrats (*Ondatra zibethicus*) and other wildlife they support (O'Neil 1949; Lowery 1974a,b). Conversely, when the Aswan High Dam impeded the Nile's nutrient discharge, the Mediterranean Sea's sardine fishery quickly collapsed. Before completion of the Aswan High Dam, Egyptian fishers harvested 16.3 million kg of sardines from the shallow waters just beyond the Nile Delta, but shortly after the dam began to restrain the nutrient–enriched floodwaters of the Nile, the sardine harvest dropped by 97 percent (Owen 1980).

1996 EXPERIMENTAL FLOOD IN THE GRAND CANYON

In 1963, completion of the Glen Canyon Dam altered the downstream flow of the Colorado River through the Grand Canyon. Previously, flows peaked each summer after the snowpack melted in the Rocky Mountains and sent a torrent rushing downstream, but the huge dam "smoothed" out the seasonal changes in the river's flow through the Grand Canyon. Water that accumulated behind Glen Canyon Dam formed an immense reservoir, Lake Powell. The river water that flowed through Grand Canyon—once warm and laden with silt and sand—now was cold and relatively clear (i.e., new features of the water released downstream from the depths of Lake Powell). Pre-dam water temperatures of about 27°C in summer dropped to a nearly constant post-dam average of about 8°C. Likewise, the dam reduced sediment loads in the river water by 87 percent (Bureau of Reclamation 1995).

These changes affected the communities in and bordering the Colorado River in the Grand Canyon (Carothers and Brown 1991). The post-dam thermal regime no longer matched the requirements of several native fishes, included the endangered humpback chub (*Gila cypha*), whose reproduction and recruitment are impaired by the colder water (e.g., limited survival of eggs and larvae; see Hamman 1982; Kaeding and Zimmerman 1983; Marsh 1985). With cold water now flowing in the main river, suitable habitat increased for the trout that were released (in 1965) in the tailwater below the dam and in some of the Colorado's tributaries within the Grand Canyon. The trout population accordingly prospered—and increased predation on juvenile humpback chub. Moreover, because of the reduced sediment loads, sandbars no longer developed along the riverbanks, thereby eliminating much of the backwater habitat—protected, low-velocity, shoreline environments—that previously provided nursery areas for young chub and other native fishes. Nonetheless, humpback chub survive in the Grand Canyon because they reproduce successfully in the warmer tributaries flowing into in the canyon (e.g., the Little Colorado River). Today, the chub population in the Grand Canyon represents the largest aggregation still remaining in the Colorado River system.

Control of the water flowing into the Grand Canyon from Lake Powell also eliminated repeated scouring of the riparian communities bordering the river. Previously, yearly floods washed away much of the vegetation and reduced plant invasions. Each flood initiated another round of secondary succession, and the scour zone therefore remained largely free of permanent vegetation. However, completion of Glen Canyon Dam stabilized the flow and ended scouring, which set the stage for the development and invasion of other vegetation. Among these was the rapid expansion of saltcedar (*Tamarix* spp.), an exotic shrub that is well known for its ability to cope with drought and salinity. Saltcedar generally outcompetes native

vegetation and, in doing so, typically produces dense monotypic stands. As it proved in the Grand Canyon (and elsewhere), saltcedar is an excellent example of an exotic species that thrives in altered ecosystems. Despite its undesirable qualities, saltcedar—because it is the only tree-size vegetation available—provides nesting habitat in the Grand Canyon for the endangered southwestern willow flycatcher (*Empidonax traillii extimus*) and other birds that depend on riparian habitats (Unitt 1987; Brown and Trosset 1989).

An experiment in restoration ecology took place in spring 1996 when a large volume of water was released for 1 week from Lake Powell into Grand Canyon. One of the objectives of this experiment was to rebuild sandbars along the margins of the Colorado River in the Grand Canyon and thereby create and rejuvenate backwater habitats for humpback chub and other native fishes (i.e., flooding amplifies sediment transport). The week-long experimental flood increased the average volume and surface area of sandbars by 176 percent and 48 percent, respectively (Parnell et al. 1997a). The flood also produced a twofold increase in the surface area of backwater environments and created eight new backwater sites (Parnell et al. 1997b). The flood also buried large amounts of organic matter, which increased nutrient concentrations in the aquatic systems for more than a year (Parnell et al. 1997a). Eventually, however, the sandbars again began to erode and, in the absence of additional flooding, many of the newly created backwater habitats proved temporary (Valdez 1997). Nonetheless, the experimental flood proved that important habitat can be restored in the Grand Canyon.

WATER, DISTRIBUTION, AND ISOLATION

The configurations of water, past and present, play a role in the geographical distribution of organisms. Perhaps the best known example is Wallace's line, a zoogeographical boundary that coincides with a deep trench in the seabed between the small islands of Bali and Lombok in the East Indies (Fig. 11-3). The narrow trench remained an effective barrier to faunal dispersal even when sea levels were lowered during the glacial ages of the Pleistocene. Conversely, the islands between Bali and the Asian mainland were surrounded by much shallower seas; the islands joined as a contiguous landmass with Asia when the ocean receded. Faunal dispersal occurred between the mainland and the islands up to and including Bali, but not beyond. The gap between Bali and Lombok, although less than 32 km in width, still marks the terminus for the Oriental fauna, even for such mobile groups as birds.

Fisher (*Martes pennanti*) and some other mammals living in the montane forests of northern Utah are barred, in part, by the Green River from moving eastward into similar habitats in Colorado; conversely, the range of Abert's squirrels (*Sciurus aberti*) does not extend westward from Colorado into northern Utah for the same reason (Findley and Anderson 1956). However, for reasons that remain unclear, the Columbia River has not been a similar barrier to mammalian distributions; most of the species on the northern side of its course also are present on the southern side (Gordon 1966). Nonetheless, some mammals have extended their ranges westward or eastward across the Cascade Mountains because of the gorge cut through the mountains by the Columbia River. Two species of western squirrels moved eastward through the gorge, whereas white-tailed deer (*Odocoileus virginianus*) and muskrats moved westward into Washington and Oregon along this "highway" through the mountains.

The swimming abilities of terrestrial animals clearly influence their predisposition to cross water and hence expand their ranges; the width and velocity of rivers are other factors in this matter. Some animals confronted with water barriers successfully expand their distributions by "rafting;" that is, where tangles of vegetation and soil break away from one side of a river, small animals thereon may be transported to the opposite side, but rafting as a means of transporting larger animals probably is rare. Ice bridges are another means for animals to cross water barriers (Banfield 1954). Wolves (*Canis lupus*) reached Isle Royale by crossing 29 km of ice on Lake Superior, thereafter establishing a predator-prey interaction with the island's moose (*Alces alces*) population (Mech 1966; Allen 1979). Fuller and Robinson (1982) recorded the movements of deer and predators across ice on the St. Mary's River between Ontario and Michigan; deer seldom crossed more than 1000 m of ice, whereas the predators did so more readily. Lack (1954) noted that eiders (*Somateria mollissima*) that breed on small Arctic islands delay nesting until after the surrounding ice melts, after which Arctic foxes (*Alopex lagopus*) cannot reach the islands and destroy their nests; a similar defensive mechanism is employed by gulls nesting on islands in Finland. More than 250 goose nests on Arctic islands were destroyed when foxes crossed ice bridges during a late spring (Ryder 1967). However, because geese normally abandon breeding areas where successful nesting is thwarted repeatedly, it seems probable that such large losses are rare occurrences (Owen 1980).

Water barriers also serve as isolating mechanisms leading to the development of new forms by adaptive radiation. The isolating function of water may be either real, in terms of distance or physiology (e.g., intolerance to salt water), or psychological. Hawaiian

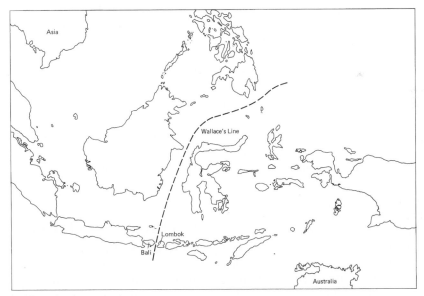

Figure 11-3. Location of Wallace's Line in the East Indian archipelago. The line coincides with a deep ocean trench and marks the southern extension of the Oriental fauna from the Asian mainland into the archipelago.

geese (*Branta sandvicensis*) apparently evolved from Canada geese that were isolated, perhaps by the misadventure of a Pacific storm, from their North American homeland. The species is the only goose endemic to an island and, in isolation, it has adapted uniquely to a terrestrial lifestyle in the lava fields of Hawaii (Weller 1980; Kear and Berger 1980). Trappers apparently introduced blue foxes—a color phase of the Arctic fox—onto some Aleutian Islands to assure inbreeding and, hence, perpetuation of the more valuable blue-phase pelage of these animals for commercial purposes. Ground-nesting birds, among them Canada geese, have been virtually eliminated from certain islands where the foxes were released (Murie 1959).

Perhaps the most heralded instance of adaptive radiation concerns the water-related isolation of certain birds in the Galapagos Islands. Charles Darwin (1845) immortalized these birds in his journal as "a most singular group of finches." Known today as Darwin's finches (Geospizinae), their common ancestral stock arriving from South America subsequently evolved into 14 species that have the lifestyles of warblers, woodpeckers, and, of course, ground-feeding finches (Lack 1947). The water between the islands was of sufficient distance to allow time for the development of reproductively isolated populations before additional colonizers invaded from other islands in the Galapagos archipelago. Conversely, Mayr (1942) described species and subspecies of land birds in the Solomon Islands that easily could cross the small water gaps separating the islands, but do not, apparently because of the strong psychological barrier presented by the water itself.

Fishes have obvious problems crossing land barriers, although walking catfish (*Clarias batrachus*) are remarkably adapted for overland movements. The distributions of many fishes are limited by their inability to tolerate either salt water or fresh water, whereas others, such as Atlantic salmon (*Salmo salar*), seasonally visit each type of regime. Vast geological changes also have produced unusual distributions—and problems—for certain species. Among these is the former occurrence of immensely large lakes that subsequently disappeared, leaving only remnant waters with isolated fish populations. In late Pliocene-Pleistocene times, a lake of nearly 52,000 km^2 with a maximum depth of about 300 m covered much of what is now Utah. Known as Lake Bonneville, it drained northward via the Snake River into present-day Idaho, ultimately connecting to a vast network of river drainages. Subsequent geological and climatological

events diminished the lake so that, today, only a few isolated lakes dot the ancient lakebed. No less than seven endemic species of fishes evolved in the isolated remnants of Lake Bonneville and their immediate drainages (Sigler and Miller 1963).

Springs in desert regions are as isolated as oceanic islands, but, unlike islands, their faunas are derived from past ages rather than by chance invasion and subsequent adaptive radiation (Minckley and Deacon 1968). The Devils Hole pupfish (*Cyprinodon diabolis*) perhaps is the best-known case of highly isolated, relict species threatened by civilization. This species—one of several with restricted distributions in the family of killifish—evolved in a single spring in Nye County, Nevada (Miller 1948). Unfortunately, demands for irrigation water lowered the spring's level to the point that spawning habitat along with pool's periphery was no longer inundated, and the population at times has numbered no more than 125 breeding adults. More recently, the spring has been protected and an artificial spawning platform was installed just below the water's present level. Also, a population of Devils Hole pupfish now is maintained artificially in a refugium located near the foot of Hoover Dam.

Water-influenced environments also may effectively have an impact on animal distributions. Historically, pronghorns failed to cross the Missouri River eastward from Nebraska into Iowa, not so much because of the river itself but because the riparian forests bordering the river were unsuitable habitats. Farther north in the Missouri's drainage, the forests diminished and pronghorns ventured successfully across the river into the eastern parts of the Dakotas and beyond (Jones 1964). Conversely, riparian forests extended the range of white-tailed deer westward into the prairie states; forested river corridors reach like fingers into grasslands where the deer otherwise would not have found suitable habitat in pre-settlement times (Fig. 11-4). Similarly, woodchucks (*Marmota monax*) moved westward across the Flint Hills using the riparian corridors along tributaries of the Kansas River (Choate and Reed 1986).

WATER AND WILDLIFE POPULATIONS

Water's impact on wildlife populations takes many different forms. For many animals, successful breeding is directly or indirectly related to water, usually in the form of precipitation. For example, the har-

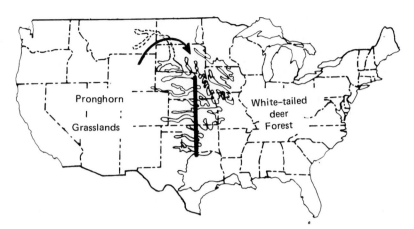

Figure 11-4. Generalized map of riparian forest intrusions into North America's grassland interior. Eastward dispersal of pronghorns was blocked by the forest barrier in Nebraska and elsewhere (solid bar), but movements across the upper Missouri River were successful (arrow) as riparian forests thinned to the north (Jones 1964). Conversely, eastern populations of white-tailed deer extend their westward range into grassland regions along the forest corridors.

vest of cottontails (*Sylvilagus floridanus*) in New Jersey on the first day of the hunting season was related to the combined amounts of rain falling during March and September. Higher amounts of rainfall in these months diminished the number of cottontails killed per hunter (Applegate and Trout 1976). In this relationship, the March rainfall probably influenced the survival rates among the first litters of the year, whereas rainfall in September likely governed the late-summer breeding production from juvenile cottontails.

Several species of common game ducks, particularly in the genus *Anas,* depend largely on semipermanent wetlands for nesting habitat. Much of the North American population of mallards (*A. platyrhynchos*), pintails (*A. acuta*), and other ducks breeds on the prairie potholes of the northern Great Plains. Potholes are capable of producing more than half of

the game ducks available for hunting. However, as is typical of a prairie ecosystem, the northern plains commonly experience droughts. The result is a regime of wet–dry periods that directly affects the nesting habitat available for waterfowl. With drought and the lack of spring runoff, the potholes are diminished temporarily in their carrying capacity, and breeding waterfowl experience reduced production (Rogers 1964) or are forced to move elsewhere (Smith 1970; Derksen and Eldridge 1980). With adequate water, however, the waterfowl population increases rapidly. Therefore, a census known as the *July Pond Index* was developed to assess the availability of prime breeding habitat for ducks. Crissey (1969) subsequently plotted numbers of breeding ducks against the numbers in July ponds available the previous year (Fig. 11-5). The strong statistical relationship that resulted underscores the

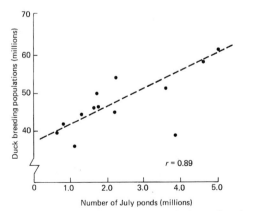
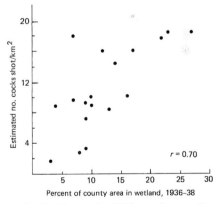

Figure 11-5. Two examples of relationships between wetland habitats and wildlife populations. On the left, the number or prairie potholes (*July Pond Index*) in Canada and the size of the continental breeding population of ducks the following spring are shown for a 10-year period (1955–65). A positive relationship also existed between the percentage of wetland areas (on the right) in selected Wisconsin counties and pheasant densities (as measured by the estimated harvest of cocks). (From Crissey 1969 and Wagner et al. 1965.)

value of habitat assessment as a useful indicator of wildlife populations and, of course, the immense value of potholes as waterfowl breeding habitat. Also shown in Figure 11-5 is a similar correlation between the area of wetlands in Wisconsin and pheasant (*Phasianus colchicus*) densities.

Bobwhites (*Colinus virginianus*) are widely distributed in the United States and are among the most popular of game birds. Millions are harvested each year by hunters, making bobwhite management of paramount importance for many state conservation agencies. This concern has resulted in comprehensive studies of bobwhites (Stoddard 1931 is a classic, but, more recently, Klimstra and Roseberry 1975) and hundreds of other research reports. The volume of data available clearly illustrates that bobwhite populations are highly responsive to climatic patterns, particularly as measured by rainfall. In the heartland of South Carolina's best bobwhite habitat, for example, drought reduced the number of juvenile quail per adult female to 4.9 versus an 11-year average of 9.1 and a maximum of 13.2 (Rosene 1969). Similarly, Kiel (1976) found a linear relationship between rainfall and the age ratios of bobwhite on the King Ranch in Texas (Fig. 11-6). Bobwhite populations thus have remarkable shifts in abundance, responding with "boom or bust" years. Of four types of weather patterns monitored along with bobwhite populations for 25 years, Stanford (1972) found that quail are most severely reduced by droughts and high temperatures. The following condensed list reflects the variety of effects produced by limited rainfall.

1. Pairing, covey breakup, and nesting are delayed, limited, or fail to occur at all.
2. Smaller clutches are laid and incubated; high nest abandonment occurs among normal-size clutches.
3. Females are emaciated and die incubating; eggs spoil from high ground temperatures before incubation begins; desiccation traps hatching chicks in their eggshells; uneven hatching prompts females to leave their nests prematurely with fewer chicks.
4. Large percentage of adults lack broods in the summer; above-average number of males alone care for small broods.
5. The normal second hatching peak (in August) does not occur.

For California quail (*Lophortyx californicus*), McMillan (1964) found only 35 percent young in the harvest when rainfall averaged 16 cm, whereas 70 percent young occurred in years when rainfall averaged 30 cm. The difference between good and poor reproduction for California quail is based largely on the annual increment produced from renesting efforts; in years of low rainfall, renesting is limited and the overall production of young is reduced greatly.

The western limits of bobwhite distribution are irregular and fluctuate locally with precipitation and perhaps other environmental factors (Fig. 11-7). Rainfall patterns in the Rolling Plains of Texas, for example, vary annually from about 18 cm to more than 127 cm and cause major fluctuations in bobwhite populations. In describing these conditions, Jackson (1969) emphasized that this problem for wildlife managers is not unlike a similar situation in agricultural management. Dry years inflict crop losses, too, but these failures ultimately have stimulated better methods of land management. This lesson should not be missed in quail management, as repetition of outdated techniques such as predator control, stocking, and closed or reduced hunting are ineffective responses when rainfall shortages temporarily depress quail densities. The flux of quail densities at the periphery of their distributions also was suggested in Oklahoma where Schemnitz (1964) studied a sympatric population of bobwhites and scaled quail (*Callipepla squamata*). During 3 consecutive years of drought, bobwhite rapidly declined, with juveniles making up less than 48 percent of the autumn population. In contrast, young scaled quail made up nearly 75 percent of the fall population dur-

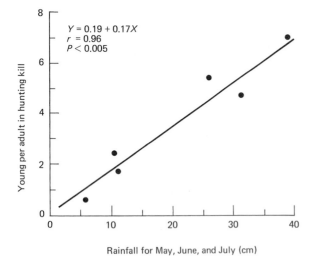

$$Y = 0.19 + 0.17X$$
$$r = 0.96$$
$$P < 0.005$$

Figure 11-6. Relationship between rainfall in late spring–early summer and the ratio of young to adult bobwhites bagged in south Texas. (From Kiel 1976.)

ing the same period, and the population increased. On a long-term basis, the distribution of these species, as reflected by their densities, swings back and forth with extremes in precipitation. Bobwhite breed with a minimum of 40 cm of rainfall (Robinson 1956), whereas scaled quail experience successful reproduction with as little as 15 cm (Sehemnitz 1964).

Production of wild turkeys (*Meleagris gallopavo*) likewise depends on seasonal rainfall. Remarkable differences in juvenile-adult age ratios—from 0 to 576 poults per 100 adult hens—were recorded for a 10-year period on the King Ranch in Texas. Beasom and Pattee (1980) explained nearly all of this variation with models primarily based on soil moisture storage and rainfall from the preceding late summer and autumn months. Egg production, rather than nesting activities or poult survival, apparently is the

critical phase in the reproductive cycle determining productivity in wild turkeys. Whereas the causative agent directly affecting egg development is unknown, the influence of late summer and autumn rainfall prior to the breeding season triggers a mechanism involving vegetational quality. Similarly, Francis (1967) formulated an equation based, in large measure, on precipitation and available soil moisture that accurately predicted the productivity of California quail.

Teer et al. (1965) found that population densities of white-tailed deer in Texas were related to the precipitation occurring in the previous year. In drought years, the relationship was especially close and lessened only when rainfall again occurred in average or above-average amounts (Fig. 11-8). Sowls (1961) reported strong indications of reduced reproduction among javelina (*Tayassu tajacu*) in Arizona following uncommonly dry years. Such relationships suggest that precipitation controls the population of some game species. The significance of control, however, is manifested largely under extreme conditions (e.g., drought) rather than when precipitation falls at or near its long-term mean. Barring such extremes, most species seem geared to normal precipitation patterns. For example, jackrabbits (*Lepus californicus*) have two peaks for reproduction during the year, apparently coinciding with the two rainy periods normally occurring in Arizona (Vorhies and Taylor 1933).

PHYSIOLOGICAL AND BEHAVIORAL RESPONSES

That some wildlife populations may respond to water in positive or negative ways has been shown in the preceding discussions. How this happens is not always clear, although the *why* seems more certain, namely, it makes little sense for eggs to hatch successfully if the habitat's carrying capacity is so diminished by water shortages that there follows little or no chance for the young to survive. Even waterfowl nesting on dry land—that otherwise might successfully produce and hatch eggs in years of drought—require aquatic habitat for rearing their broods. Reproductive efforts for water-dependent species, if proceeding routinely in dry years, thus would remain only fruitless exercises, costly in energy, and contrary to the overriding dictum of natural selection.

Returning to quail for the moment, we find that these game birds apparently have evolved a trigger

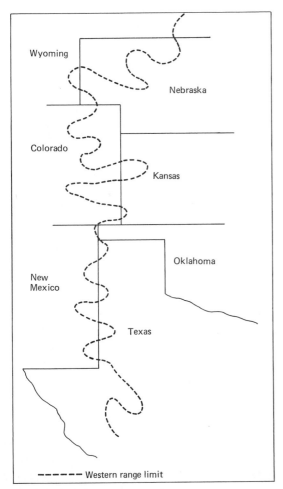

Figure 11-7. Approximate western limits of bobwhite. The irregular nature of the boundary reflects year-to-year changes in precipitation. (From Jackson 1969.)

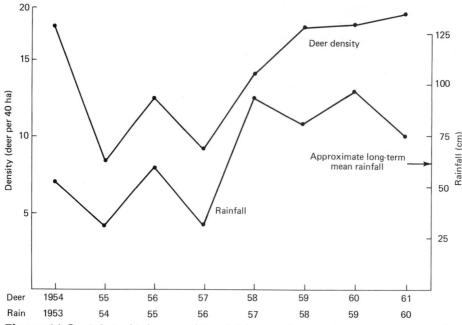

Figure 11-8. Relationship between white-tailed deer populations and precipitation the preceding year in the Llano Basin, Texas. (From Teer et al. 1965.)

mechanism for breeding that is dependent on rainfall. Lehmann (1946, 1953) found a high correlation between rainfall and bobwhite reproduction and suggested that reduced amounts of vitamin A in dry years limit breeding. Carotene is the precursor of vitamin A and normally is contained in green foliage. As the succulence of green vegetation usually is a function of precipitation, the amount of carotene varies with rainfall. So, in droughts, carotene levels are reduced (Beck 1942) and vitamin A ultimately is not available to quail (Nestler et al. 1949). With a dry winter and spring, bobwhite remained in coveys and did not breed until rainfall occurred. Winter mortality also was above normal, and the vitamin A reserves in their livers were greatly diminished under the stress of drought. Similar results were recorded between Gambel's quail (*Lophortyx gambelii*) populations and rainfall (Swank and Gallizioli 1954); these also seemed related to the availability of vitamin A. Hungerford (1964) recorded only 19 to 28 percent juveniles in the autumn population when the vitamin A stored in adult livers that was collected the preceding spring was less than 400 micrograms. However, the proportion of juveniles jumped to 74 percent in a year of abundant rainfall when more than 1600 micrograms of vitamin A per liver were stored. The reproductive organs of Gambel's quail with less than 550 micrograms of stored vitamin A

failed to develop during the normal breeding season, but adult birds were not harmed by the deficiency. In experiments with penned bobwhites, Nestler (1946) showed that fewer juveniles survived in broods of hens that were fed diets deficient in vitamin A. Interestingly, broods from these hens experienced higher mortality rates even when they were fed large dietary supplements of vitamin A after hatching. These results correspond with field studies indicating that juvenile Gambel's quail are more vigorous and experience greater survival rates in wet years than in years of drought (Sowls 1960).

A linkage between rainfall, plant growth, and reproduction was indicated by an analysis of plant hormones known an phytoestrogens in the green foods of California quail (Leopold et al. 1976). The physiological action of phytoestrogens inhibits reproduction in some domestic mammals (Labov 1977) and may offer another explanation for the "boom or bust" pattern of quail production and rainfall. Whereas California quail will pair and attempt to nest even in dry years, these efforts result in as few as 25 young per 100 adults. In one study, the crop contents of adults collected before and during the breeding season in a dry year revealed reduced volumes of foods, but the food contained detectable amounts of phytoestrogens. The plants were no less available, but their growth was stunted and poor. In the following year, when rainfall was generous, large amounts of green foods were consumed from the vigorous growth of forbs and grasses. Phytoestrogens in these samples were virtually absent, and 325 young per 100 adults subsequently were produced. In an accompanying experiment with penned quail, birds receiving a diet containing phytoestrogens initiated laying 2 months later and produced 81 percent fewer eggs than did those fed a phytoestrogen-free diet. Based on these studies, Leopold et al. (1976) concluded that the presence or absence of phytoestrogens in the foliage of annual vegetation may well coordinate quail reproduction with the prospective food resources later available for broods.

In Australia, the seasonal flooding in the Murray-Darling drainage normally stimulates breeding in local fish populations. These fishes respond to a temperature threshold, below which breeding is inhibited. Above the threshold, however, flooding is required to induce breeding activity (Lake, *in* Frith 1973). Unfortunately, water impounded behind Murray-Darling dams is thermally stratified, so that cold water released from the bottoms of the dams depresses the temperature and inhibits breeding downstream for considerable distances.

Another striking relationship occurs among some species of Australian waterfowl. Frith (1959, 1967) reported that breeding in grey teal (*Anas gibber-*

ifrons) develops directly in response to rising water levels. In inland Australia, where most of these birds normally nest, there is no regularly defined breeding season. Grey teal initiate courtship and nest whenever the billabongs and other depressions are rapidly filled; breeding might occur in midwinter one year and midsummer the next and may be skipped altogether in years when water supplies are inadequate. As shown in Figure 11-9, gonadal development closely parallels the pattern of the steadily rising water, although the nutritional status of the birds affected this response (Braithwaite and Frith 1969). Sexual activities begin almost immediately, with virtually every male at once seeking a mate in a highly compressed period of courtship. Eggs may be produced in as few as 10 days after the water begins to rise, and nests occur in every conceivable site throughout the affected area. If the rise in water levels is not great, only the local population breeds, but if a major flood develops, immense flocks of nomadic teal appear and breed also. Flooding that is initiated from rainfall or melting snow occurring at distant sites suffices to initiate breeding: when the water travels downstream, the flooding stimulates breeding of grey teal even in the absence of local rainfall. Local rainstorms that are insufficient to change the water levels fail to trigger breeding. Recently, Crome (1986) suggested that the massive

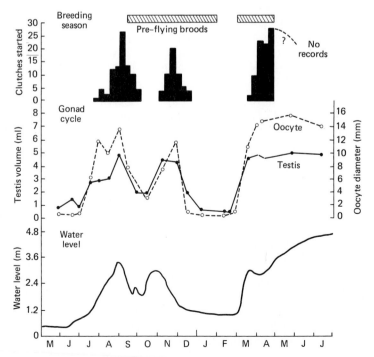

Figure 11-9. Relationships among water levels, gonadal development, nesting, and brood production for grey teal in Australia. (From Frith 1959.)

production of invertebrates when the dry wetlands refill is responsible for the onset of breeding in grey teal, rather than the rainfall and rising water levels per se.

Wildebeest (*Connochaetes taurinus*) and perhaps some other grazing species in Africa exhibit a remarkable response to rainfall. Because a premium exists for fresh grass, wildebeest rapidly abandon their grazing areas for those where rain stimulates new growth. Talbot and Talbot (1963) found that every major movement of wildebeest herds was associated with rainfall. Herds of 100,000 animals may turn in their tracks to move toward a rainstorm, covering distances of 50 km or more a day. The sound of thunder or the smell of rain may trigger these movements, but sight seems the primary sense attracting wildebeest to a distant storm. Rainfall on the African plains is localized during the dry season, and dark columns of clouds 80 km distant are seen easily. The stimulus of rain is strongest when the plains are dry; at such times, wildebeest even leave areas of adequate forage when rains fall elsewhere on the plains. Once there, the wildebeest mill about for a day or two until fresh grass appears. Occasionally, mistakes occur, and wildebeest move long distances only to find that the rain fell on wooded hills instead of on the plains. On balance, however, this response remains an adaptation that gives wildebeest the best chance of finding fresh grass in a semiarid environment.

Some animals can manufacture water within their digestive systems. Reliance on *metabolic water* has freed some animals from the necessity of drinking (e.g., kangaroo rats, *Dipodomys* spp.). The manufacturing process, simply stated, is the recombination of the carbohydrates available in the diet. Seeds are a common source, with each molecule of a simple carbohydrate ($C_6H_{12}O_6$) potentially producing 6 molecules of H_2O for the animal's needs. Metabolic water contributes 18 percent of the total daily water requirements for kit foxes (*Vulpes macrotis*), and because kit foxes live in areas devoid of free water, the balance is obtained as preformed water in the bodies of their prey (Golightly and Ohmart 1984). Kit foxes thus require at least 175 g of prey daily to meet their needs for preformed water, and more prey is needed for this purpose than for energy demands. Animals producing metabolic water often have concomitant adaptions for water conservation. These include highly effective kidneys or other mechanisms that wring virtually all usable water from the urine before it is eliminated and behavioral patterns that reduce water loss (Schmidt-Nielsen 1964; Dantzler 1982).

Sand grouse (*Pterocles* spp.) inhabit some of the most arid regions on Earth, including the Sahara and the Kalahari Desert in Africa. Because the nests of sand grouse are often located several kilometers from the nearest water, ornithologists have long questioned how the chicks drink. In answer, Cade and Maclean (1967) described a remarkable adaptation whereby the male sand grouse carries water to the young in its feathers. The male's abdominal plumage is highly specialized for this purpose and, in fact, holds more water than a paper towel or a cellulose sponge of equal weight. Water in the feathers will, of course, evaporate in flight, but even so, tests indicate that a sand grouse beginning a 32-km flight with 25 g of water can deliver 10–18 g to the brood per trip. On arrival, the male stands erectly, allowing the young birds to strip water from the specialized feathers. In the past, the small number of natural water sources undoubtedly restricted sand grouse populations, but more recently, artificial water developments have enhanced the birds' distribution and numbers in parts of their overall range. Sand grouse are obligate drinkers surviving in water-poor environments and thus demonstrate model examples of the ways evolution involves adaptive compromises (Maclean 1985).

It was once thought that nasal glands rinsed away salt water entering the sensitive nasal membranes of seabirds. Later, research determined that the glands secrete highly concentrated saline fluids, functioning in effect as an extrarenal mechanism for eliminating salts. Technau (1936) showed that the development of the nasal (or salt) glands in seabirds was highly correlated with each species' association with seawater. Not only do seabirds have large nasal glands when compared with terrestrial species, but the glands' development also varies within subspecies of seabirds, depending on their use of salt water. Terrestrial birds have rudimentary nasal glands but are unable to secrete salts. Among aquatic species, the development and function of the glands depend on each individual's exposure to salt water. In ducks, for example, nasal glands of individuals emanating from freshwater marshes show little development, whereas those of the same species raised on saline lakes have well-developed, functional glands (Cooch 1964). There may be racial differences in other cases. Among mallards, a resident race living on the coast of Greenland exhibits a well-developed nasal gland, whereas those of the European race are much reduced (Stresemann 1934). Therefore, the development and function of nasal glands in birds depends on two factors, a primary genetic factor (i.e., marine versus terrestrial species) and a secondary environmental effect of salt stress (Schmidt-Nielsen and Kim 1964). Individual aquatic birds possessing the genetically adapted nasal glands may show temporary function when confronted with saltwater environments, only to have the gland regress when moving on to freshwater habitats (and vice versa).

Cooch (1964) examined the possibility that impairment of nasal glands in ducks was related to avian botulism (see Chapter 8). In a series of experiments, it was determined that sublethal dosages of *Clostridium botulinum* (type C) toxin can cause mortalities in ducks drinking salt water (Table 11-2). Ducks drinking fresh water, but given the same dosages of toxin, showed improved survival. With little or no toxin, salts in the water were excreted by the birds' nasal glands. Massive dosages of the toxin, of course, remain lethal, but more moderate exposure to the toxin seems to impair the functional ability of the gland and leads to death when the osmolarity of the blood plasma can no longer be regulated. In effect, these experiments suggest that Wetmore's early (1915) implication of "alkali poisoning" may describe botulism-induced mortality more accurately than supposed.

The physiological compatibility—or incompatibility—of some organisms with regional precipitation regimes has implications for some management strategies. Wild turkeys belong to a single species but exhibit six reasonably well-defined races (Aldrich 1967). A constellation of environmental factors undoubtedly contributed to the formation of these races, but rainfall patterns influence at least the Rio Grande race and its distribution. An isohyet of 81 cm of rainfall coincides well with the eastern edge of this race's distribution, beyond which, in wetter areas, it

TABLE 11-2. Percent Survival of Pintails Subjected to Five Doses of *Clostridium botulinum* (Type C) Toxin in Association with Three Levels of Water Salinity

Water Type	Toxin Doses[a]				
	2000	*1500*	*1000*	*500*	*250*
Fresh	67	67	100	100	100
5 percent NaCl	17	50	50	83	100
10 percent NaCl	0	33	50	67	83

Source: Cooch (1964).

[a]Measured in mouse lethal doses.

does not survive and another race occurs. Glazener (1967) suggested that the efforts of past decades to transplant Rio Grande turkeys from Texas into the southeastern states were doomed from the onset. These restoration efforts simply used the "wrong" race, one ill-adapted to the rainfall regime of the wetter release sites. On the other hand, successful transplants in Kansas, Oklahoma, and New Mexico were made within the precipitation limits tolerated by Rio Grande turkeys. In fact, some of these and other transplants, especially those in Nebraska and North Dakota, were established well outside the race's original distribution.

The Russian ecologist Formozov (*in* Pruitt 1959) formulated a classification for mammals and their relationships with snow: *chionophobes*, those avoiding snow; *chioneuphores*, those adapted to snow; and *chionophiles*, those highly specialized or restricted to snow. Pruitt (1959) considers New World caribou, unlike Old World reindeer, as chionophiles because they spend two-thirds of their annual cycle in snow and because of their behavioral and morphological adaptations to snow conditions. In fact, "snow caribou" seems a more appropriate name for the species than barren-ground caribou. Caribou in Canada are confined in winter by snow that has certain characteristics. Soft, thin snow is preferred, and, as harder snow conditions develop, the animals progressively travel to sites where winter forage is more accessible. Snowfall and topography interact so that the winter migration routes of caribou likely are governed by these features, indicating a behavioral reaction to favorable pathways through snowfields, perhaps not unlike the flyways of birds. Within their wintering habitat, caribou progressively forage more on upland areas as snow depths increase on lower-lying sites (Fuller and Keith 1981). The energy expended by caribou digging ("cratering") in snow for lichens increased proportionately with the density of the snow cover. About twice as much energy was required for cratering in dense snow with a thin crust, compared with a light, uncrusted snow, and it increased fourfold in snow compacted by a snowmobile (Fancy and White 1985). Therefore, snow depth and density interact on the energy caribou expend with feeding; the energic cost of obtaining food must remain low or be offset by the energy derived from the forage.

The altitudinal migrations of moose in British Columbia also are correlated strongly with snow depths. Spring migration upward to higher elevations is marked by rapidly decreasing depths of 46 cm or less, whereas downward movements in winter coincide with increasing accumulations of snow at higher elevations. Under these conditions, the winter range of moose is dynamic (Edwards and Ritcey 1956). Snow depths, density, and hardness are important features interacting with the winter activities of moose, but unless snow depths exceed 70 cm, moose experience little or no serious restrictions of movement (Coady 1974). Telfer and Kelsall (1984) have discussed the adaptations, including track loads, of some large North American mammals for survival in snow.

Ice also is important in the breeding behavior and management of several polar species. For example, harp seals (*Phoca groenlandica*) whelp in the rough ice near the interior of ice floes where there is some degree of shelter. Female harp seals mate again soon after whelping, but the embryos do not begin development for 11 weeks afterward. Both the breeding schedule and the phenomenon of delayed implantation are responses to ice patterns; mating takes place when the herd is congregated on the floes, and the pups are born a year later when ice conditions again are optimal. The pups' white coat, lasting only for about 2.5 weeks after birth, is a valuable fur, promoting passionate controversies about harvesting the young animals while they are still nursing their mothers. The pelts' whiteness makes it difficult to photograph the pups aerially against a background of snow and ice for inventories governing the annual harvest. However, the white pelt absorbs ultraviolet radiation so that special film produces a black image of the pups against a white background and a useful census of production now is possible (Fisheries & Environment 1977).

WATER, DISASTERS, AND HARD TIMES

Water-related disasters befalling animal populations, even if occasional, may require management action, as occurs in the aftermaths of flooding, blizzards, storm tides, or droughts. Virtually nothing can be done, of course, about the disaster per se, but harvest regulations may be tightened when large segments of wildlife populations are killed or fail to reproduce because of a short-term calamity (Spencer et al. 1951).

Severe winter drought contributed to massive losses in Texas when 21,000–30,000 deer died between January and August 1962 within part of a single county (Marburger and Thomas 1965). Densities before the die-off were 26 deer per 40 ha and about 15 per 40 ha afterward, indicating a 44 percent reduc-

tion in the population. With the return of normal rainfall, the deer population recovered in 1 year's time to about the same level that existed prior to the die-off. Spevak (1983) recorded decreasing species diversity within a community of desert rodents during a drought, followed by recovery to predrought levels when rainfall patterns returned to normal; densities also decreased for nearly all of the rodents during the drought. In 1982–83, an 11-month drought reduced kangaroos (*Macropus* spp.) by 40 percent across more than 1 million km² of eastern Australia (Caughley et al. 1985). Such droughts and their impact on kangaroo numbers may be intrinsic to the ecology of kangaroos, thereby counterbalancing much of the rapid population growth that occurs between droughts.

Heavy snowfall can limit populations of deer and other ungulates (Edwards 1956). Deep snows may entrap animals for extended periods, and starvation follows when local food supplies are exhausted. In the Adirondack region of New York, Severinghaus (1947) recorded mortalities of 21 deer per 2.6 km² when snows were deep, long-lasting, and late in the year. Deep snows lasting 5 weeks significantly affected deer mortality, and those exceeding 10 weeks caused severe losses. Snow depths in excess of 50 cm retard movements, especially among fawns. In Utah, snows up to 90 cm in depth caused mortalities of 9 to 42 percent, depending on the available forage, within segments of that state's mule deer herd. Such conditions often bring forth public pressure for artificial feeding programs, but this shortsighted remedy usually further elevates deer populations above the carrying capacities of winter ranges. In fact, herds kept in proper balance with winter forage supplies suffered only slightly higher than normal mortality in deep snow. Conversely, on depleted ranges, losses were magnified several times over (Robinette et al. 1952). The condition of the snow, particularly its density and water content, interacts with the amount of snow on deer mortality. Light, fluffy snows are troublesome because the legs of deer penetrate deeply, making movement extremely difficult and energetically expensive. Conversely, wetter snow packs may develop a crust that enables deer to move rather easily on the snow's surface.

Snows accumulating in burned-over or clear-cut forest limit elk (*Cervus elaphus canadensis*) distributions in winter; with 60 cm or more of snow, elk seek out coniferous cover or sites at lower elevations even at the expense of leaving better forage behind (Gaffney 1941; Martinka 1976; Leege and Hickey

1977). In the central Rocky Mountains, mule deer follow a snow-imposed rotation in their grazing pattern, moving on to other areas as the snow depth increases above 46 cm; in severe winters, more than 90 percent of the winter range may be excluded because of deep snows. The concentrations of deer eventually crowding into sites with less snow (usually southern-facing slopes) severely deplete forage supplies, and as many as 381 carcasses were located in about 5.2 km² of overbrowsed winter habitat (Gilbert et al. 1970).

Snow depths affect the movements and interactions of both predators and prey. Einarsen (1948) found a herd of nearly 1000 pronghorns concentrated on 26 km² of land during a severe Oregon winter and described a scenario where pronghorn vulnerability to coyote (*Canis latrans*) predation was increased by deep snows. Moose on Isle Royale seek dense cover along the lake edges when snows are deep, and wolf (*Canis lupus*) packs run on the ice bordering the shoreline at these times. So Peterson (1977) located significantly more moose kills near shorelines when snows were more than 50 cm deep, but at lesser snow depths, more kills occurred elsewhere on the island. Snow depths also influenced the age structure of moose killed by wolves. Losses of calves increased from 30 percent to 47 percent when snow depths exceeded 76 cm (Peterson 1977). Within the adult segment of the Isle Royale moose population, the vulnerability of "prime-aged" animals changed with snow depth (Peterson and Allen 1974). Normally, adults between 1 and 6 years of age avoid significant wolf predation, but kills of adult moose in these younger age classes increased dramatically with deeper snowfall (Table 11-3). Snow depths also have important implications for wolf ecology as well as for prey populations (Nelson and Mech 1986). In Minnesota, wolf kills of white-tailed deer increased in winters with greater snow depth. Deep snows influenced the vulnerability of deer by acting as a physical impedance for escape and by reducing fat reserves (i.e., restricted feeding and increased energy costs from travel). Therefore, to a large extent, the food supply of wolves is determined by an external factor, snow depth. Wolves breed in midwinter and whelp in spring; hence, litter size and pup survival may be affected by the vagaries of winter snowfalls.

The results of severe winters, including deep snows, on the age-specific mortality of deer may be pronounced. Of 323 black-tailed deer (*Odocoileus hemionus sitkensis*) starving in southeastern Alaska, 56 percent were fawns, 8 percent were adults less than 5 years old, and 36 percent were older adults (Klein and Olson 1960). The heavy losses of fawns were further reflected in changes in the age ratio before and after a critical 15-day period when snow depth increased to a maximum of 210 cm; the ratio dropped from more than 45 fawns:100 adults to about 15:100 when warmer weather and rains finally ameliorated the severe conditions (Table 11-4).

Snow depths in the winter preceding the birth of Dall sheep (*Ovis dalli*) were correlated strongly with lamb:ewe ratios about one month after the lambing season in Mount McKinley National Park (Fig. 11-10). The inversity of this relationship probably results from a combination of reduced rates of pregnancy, natality, and/or neonatal survival in years of heavy snowfall (Murphy and Whitten 1976).

Snow cover during the nesting season in Spitsbergen was associated with the percentage of young barnacle geese (*Branta leucopsis*) present in the autumn population; a good correlation existed even though

TABLE 11-3. Percent Age Distribution of Adult Moose Killed by Wolves in Relation to Snow Depth

Adult Age Classes	Snow Depth		
	Less than 51 cm	*51–76 cm*	*More than 76 cm*
1–3	5.9	18.2	29.7
4–6	8.8	10.2	13.5
7–9	38.2	25.0	27.0
10–12	35.3	30.7	13.6
13–16	11.8	15.9	16.2

Source: Peterson and Allen (1974).

TABLE 11-4. Changes in Age Ratios (Fawn: Adult) of Black-Tailed Deer with Snow Depth

Date	Fawns per 100 Adults	Snow Depth (cm)
December 15	48	25
February 26	57	46
March 1	46	76
March 8	29	178[a]
March 10	36	137
March 13	33	117
March 16	50	71
March 19	13	58
March 20	8	53
March 21	17	48
March 23	14	41
April 10	14	30

Source: Klein and Olson (1960).
[a]Maximum snow depth of 210 cm occurred March 7.

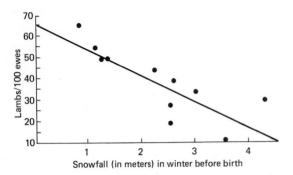

Figure 11-10. Relationship between total snowfall and lamb-to-ewe ratios for Dall sheep the following spring in Mount McKinley National Park, Alaska. (From Murphy and Whitten 1976.)

the population data were obtained 4 months after the nesting season ended (Owen and Norderhaug 1977). In this case, snow cover influenced the proportion of adults nesting successfully; brood size was not affected. Hence, the percentage of juvenile geese varied from less than 3 percent in years with prolonged snow cover to 25 percent in years when snow did not persist as long (Owen 1980). Similarly, the date when snow cover was 50 percent cleared correlated with several parameters associated with the breeding activities of Canada geese nesting on the Tundra of the Northwest Territories (MacInnes et al. 1974). Among these was a strong association between the persistence of snow cover and the percentage of eggs failing to hatch.

In Alaska, Hansen (1961) underscored the effects that storm tides might have on nesting black brant (*Branta nigricans*); these and other waterfowl often adapt to rising water by elevating their nests, but the hatching success of eggs was limited under flood conditions. Whereas most other waterfowl attempt second nestings when their first nests are destroyed, brant are physiologically unable to renest (Barry 1962), and large-scale flooding may cause major population depressions. A torrential rain of 61 cm in 5 hours followed 2 years later by the heavy rains of Hurricane Carla flooded 3000 ha of prime habitat of Attwater prairie chickens (*Tympanuchus cupido attwateri*) on the Texas Gulf Coast; prior to these floods, the site had the highest-known density of this endangered species, whereas none was seen there in the years afterward (Lehmann and Mauermann 1963). Cultivation of the better-drained lands at this site forced the prairie chickens into lower areas, thereby magnifying the impact of these floods on their population. Fire, as well as cultivation, also may

influence the effects of flooding on ground-nesting birds. Low, wet sites escape fire, so that when bobwhites were attracted to these remaining patches of nesting cover, their nests were flooded by the next heavy rain (Stoddard 1931).

Changes in the thermal structures of the Pacific Ocean at equatorial latitudes initiate what can become one of the most disastrous events on Earth. South Americans call the event *El Niño* ("The Child") because of the coincidental timing with the Christmas season; episodes are irregular but recur at an average of about 4–5 years. The phenomenon is an anomalous incursion of warm water in the eastern Pacific that may unleash droughts, floods, and other powerful forces across the globe. The results sometimes produce widespread suffering and loss of human and animal life but always affect the abundance and distribution of marine organisms on the Pacific coast of South America.

The precise causes of El Niño are unclear, but the mechanics of an episode include surges of warm water—known as Kelvin waves—flowing toward the western coast of South America, as well as a deep surface layer of warm water developing offshore. Normally, the currents flow in the opposite direction, leaving only a shallow layer of warm water at the surface along the western coastline of the Pacific basin (Fig. 11-11). The usual pattern also includes an upwelling of nutrient-enriched cold water off the coasts of Peru and Ecuador. Upon reaching the surface, the enriched cold water supports an immense base of plankton and, in turn, a huge pyramid of marine life. With El Niño, however, the upwelling is contained within the thick layer of warm water, thereby interrupting the flow of nutrients to the surface and collapsing the food web.

The consequences of El Niño may disrupt socioeconomic conditions as well as biological communities on both land and sea. In one episode, the catch of anchovies (*Engraulis ringens*) in Peru and Chile fell from 13 to 2 million tons, thereby sharply reducing foreign trade and initiating a wave of high unemployment (Avaria 1985). Conversely, tropical species of marine organisms for a time may invade the coastal zones as a consequence of the switch from cold to warm water during an El Niño. These include several species of sharks, marlin, and other fishes, as well as sea turtles; these events may produce temporary changes in the fishing and dietary habits of Peruvians (Caviedes 1984). Anglers as far north as California have been confused by catches of unfamiliar species, and millions of small tropical crabs, (*Pleuroncodes*

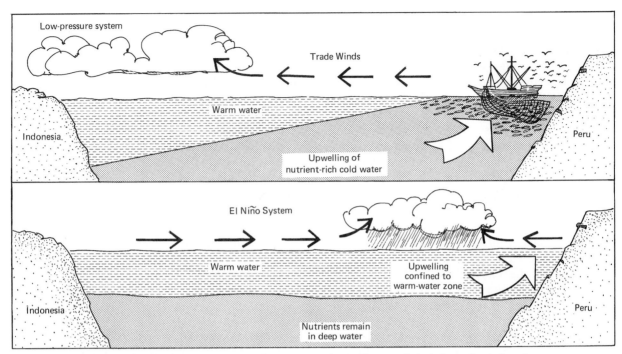

Figure 11-11. The equatorial weather pattern in the Pacific normally includes trade winds and surface currents headed toward a low pressure system lying over Indonesia. Warm water is driven westward so that the upwelling along the coast of South America consists of nutrient-rich cold water. The upwelling supports a huge pyramid of marine life (top). With El Niño, the pattern is altered so that a thick layer of warm water accumulates along the coastline and thereafter contains the upwelling. Denied the flow of nutrients to the surface, the food web collapses when plankton no longer flourish (bottom.) The El Niño of 1982–83 was unusually severe, including a strong reversal in the direction of the trade winds. (From Avaria 1985.)

planipes) clogged the pipes at the San Onofre Nuclear Power Plant during a recent El Niño (Thayer and Barber 1984). Because of food shortages, some seabirds moved as much as 1600 km from their normal range (Aid et al. 1985). The reproductive success of seabirds as far north as Oregon was reduced because of the abnormally high water temperatures associated with a recent El Niño (Hodder and Graybill 1985). The stress of El Niño also alters the depositional pattern of new shell material on some species of mollusks. Because the shells expand in increments similar to tree rings, it seems possible to detect El Niño events in the archaeological record from shells unearthed in ancient middens (Rollins et al. 1986).

The El Niño of 1982–83 was particularly severe; it caused major perturbations in much of the world, including epidemics of typhoid fever, fires, crop losses, mudslides, and invasions of snakes and insects (see Gill and Rasmusson 1983; Caviedes 1984). Barber and Chavez (1983) reported as much as a twentyfold decrease in the biomass and productivity of phytoplankton resulting from the diminished upwelling of

nutrients. At the time, anchovy stocks still were depleted from an earlier El Niño and the overfishing that followed (Idyll 1973; Bardach and Santerre 1981). Thus, the consequences of the 1982–83 episode were catastrophic for marine organisms dependent on these and other food fishes.

Colonies of Humboldt penguins (*Spheniscus humboldti*) in Peru suffered reduced breeding success and increased mortality as a result of the 1982–83 El Niño; the population declined 65 percent—from 6000–8000 birds prior to 1982 to 2000–3000 birds in 1984 (Hays 1984). Seabirds of several other species also experienced wholesale reproductive failure in the 1982–83 episode, as witnessed at Christmas Island (Schreiber and Schreiber 1984). Instead of reaching an expected breeding population of about 10,500 pairs, frigate birds (*Fregata* spp.) gave up nesting and departed for areas where small fish, squid, and other foods still flourished. Abandoned nests and dead or starving nestlings were all that remained on Christmas Island from the impact of El Niño. Torrential rains associated with El Niño also thwarted nesting efforts. A variety of terns (*Sterna* spp.), shearwaters (*Puffinus* spp.), and boobies (*Sula* spp.) experienced similar fates. In this instance, the adult birds responded to the adversities of El Niño by dispersing, thereby channeling the limited food resources into their own survival rather than continuing the energy-demanding activities of nesting and rearing young. Because most seabirds are long-lived species, the strategy of skipping a breeding season for a short term is an appropriate evolutionary adaption to food shortages. In extreme situations, however, adult birds sometimes starve and their carcasses begin to litter the beaches (Barber and Chavez 1983). Overall, only 150,000 seabirds, or less than 2 percent, of an estimated population of 14 million returned to their nesting sites on Christmas Island at the end of the 1982–83 El Niño (Caviedes 1984). Major declines in seabird populations also occurred in the Galapagos Islands from the same episode (Valle and Coulter 1987). On Isla Daphne Major, the exceptional rainfall produced thick vegetation, which greatly diminished the suitability of some nesting sites formerly used by boobies—a habitat limitation posing long-term effects (Gibbs et al. 1987).

Similar evidence for the stress of El Niño on the food web can be found among marine mammals. According to Barber and Chavez (1983), all of the pups of the Galapagos fur seal (*Arctocephalus galapagoensis*) born during 1982 died before the end of the 1982–83 episode; juvenile mortality also occurred during the same period in three other species of seals and sea lions. The seal pups probably died from the lack of adequate milk resulting from the poor foraging success of their mothers. As an indirect measure of the poor food supplies, the female seals foraged at sea for an average of 5 days during the El Niño, whereas 1.5 days is the typical foraging period.

Reptiles in the Galapagos Islands also were vulnerable to the extreme conditions of the 1982–83 El Niño (Merlen 1984). The surge of flood waters displaced giant tortoises (*Testudo elephantopus*) from highland habitat on the Tortoise Reserve on Santa Cruz Island. The circumstances perhaps explain how these plodding terrestrial animals colonized other islands in the Galapagos archipelago: The violence of the El Niño floods swept the tortoises into the ocean. Also, because of changes in water temperature, large areas of algae died on the Galapagian coast, thus eliminating the staple food for marine iguanas (*Amblyrhynchus cristatus*). Starvation soon killed thousands of iguanas.

RESERVOIR EFFECT AND MANAGEMENT

Organic matter accumulating in the upper horizons of exposed soil normally undergoes oxidation; this chemical process is enhanced further by the actions of microbial and other biological agents. The net result—a steady release and recycling of nutrients—is essential for virtually all food chains in terrestrial systems. In newly flooded aquatic systems, however, the impounded waters may be enriched rapidly by the sudden incorporation of plant and soil nutrients. The reaction is not unlike the explosive organic activity within a hay infusion. Production at virtually all trophic levels usually is increased substantially, although local factors such as water temperature affect the rates at which productivity accelerates. The initial flush of nutrients may enrich the waters of larger impoundments for several years. Thereafter, further oxidation of soil organic matter is curtailed and the system eventually stabilizes at a lower nutrient level. Such dynamic enrichment of water following impoundment is known popularly as the *reservoir effect*.

Reservoir effects occur naturally on many ephemeral wetlands, among them the prairie potholes and playa lakes of the North American Great Plains. These wetlands are maintained by precipitation that is characteristically unpredictable; they accordingly undergo frequent but irregular wet-dry fluctuations promoting high biological activity after drought periods end. A microcosm involving the reservoir effect important to waterfowl develops when melting snow forms temporary wetlands on the northern prairies. Blue-winged teal (*Anas discors*) and other ducks feeding on the invertebrate populations irrupting in the temporary wetlands derive the protein sources critical for successful breeding (Swanson et al. 1974).

The seasonal flooding of river-bottom forests is another natural reservoir effect. These habitats remain rich, productive ecological zones because of the wet-dry cycles they experience. Among the benefits for wildlife is the availability of mast—primarily acorns—for wood ducks (*Aix sponsa*) foraging in shallowly flooded timber (Brakhage 1966a). Pin oak (*Quercus palustris*) is among the more valuable producers of mast in the Mississippi Delta country, and flooding of this species in its dormant season does not affect acorn production. Pin oak stands are subject to water and other management practices leading to improved hunting success (Merz and Brakhage 1964). In addition to waterfowl, such species as swamp rabbits

(*Sylvilagus aquaticus*) are associated with lowland hardwood forests, and their distribution in Missouri suggests that forests of more than 100 ha may be necessary to sustain sizable populations (Korte and Fredrickson 1977a). A rich community of wildlife makes seasonal or permanent use of lowland forests and adjacent uplands. However, impoundments stabilizing water levels in bottomland forests interrupt the normal flooding regime and destroy the community; trees begin to fall after 3 years of flooding, reaching a maximum at 8 years (Yeager 1949). Whereas waterfowl utilization of permanently flooded bottomlands may increase for a time, the effect is temporary and declines as the artificial system matures.

Regrettably, no more than 25 percent of North America's original lowland hardwood forest remains today, a result of wholesale drainage and flood-control projects. Although the area of forest lost directly to these activities may not be great, drainage promotes conversion of forested bottomlands into farmland (Fredrickson 1978). Runoff from these croplands further limits the remaining forest system when siltation reduces root aeration, a primary mortality factor in lowland forests (Featherly 1940).

That basic productivity can be increased temporarily when land is periodically flooded suggests that water levels might be manipulated to accrue certain advantages. Indeed, the concept is not new, as periodic drying of ponds in Europe was considered an essential element of fish culture in 1883 (Neess 1946). Production in the ponds responded to the flush of nutrients available first to microorganisms and, ultimately, to the fishes after each dry period in the management rotation. This strategy is well known to rice farmers as a fundamental method for increasing yields. During the dry, fallow period, the soil is enriched naturally before reflooding and the planting of the next crop. Thus, reservoir effects influence both the plant and animal components within aquatic systems.

In the 1930s, reservoirs constructed by the Tennessee Valley Authority initially furnished excellent fishing (Cahn 1937). Increased fish yields, both in numbers and in poundage, from new impoundments supported active sport and commercial fisheries (Ellis 1937, 1942). Turner (1971) recorded a doubling of fishing success (measured as the catch per work-hour) between the pre- and postimpoundment fishery in Kentucky. Fishing success was greatest in the year immediately after impoundment; it then declined in each of the following 4 years. The weights of many species steadily increased in the same time period (Table 11-5). Jenkins and Morais (1971) reported a negative relationship between the ages of 103 reservoirs and the biomass of bass and sunfish harvested; as the impoundments aged, catches decreased. In some impoundments, the nutrient flush and subsequent increase in benthic fish foods are supplemented by the cover afforded to young fish from submerged trees and shrubs (Hoffman and Jonez 1973).

The deliberate, seasonal drying of wetlands is known as *drawdown*; it, too, relies on the reservoir effect. Wildlife managers have applied drawdowns for the production of waterfowl food and cover plants, control of succession, and improvements of the interspersion of vegetation in wetlands (Kadlec 1962). These management goals depend on the control of water supplies into and from the wetland, a feature of intensive management at many state and federal wildlife refuges. Elaborate water-control structures and diversion schemes have been developed for such manipulations; these usually are individually tailored to meet a variety of local conditions and needs (Addy and MacNamara 1948; Seamans et al. 1959).

Fredrickson and Taylor (1982) described the benefits of drawdown management and the production of moist-soil vegetation for many kinds of songbirds, shorebirds, and mammals, as well as waterfowl. The scheme they proposed recognizes four categories of water depth: deep (more than 15 cm); medium (15 cm); shallow water and mudflat; and dry. Ideally, as many units as possible should be developed, each held at a water depth attracting different groups of

TABLE 11-5. Average Weights in Grams of Selected Fishes Harvested Before and After Impoundment, Rough River, Kentucky

Species	Preimpoundment	Postimpoundment (Years)			
		1	2	3	4
Black bass	520	540	560	610	630
Sunfish	20	40	50	50	50
Catfish	150	180	350	560	710

Source: Turner (1971).

wildlife. The moist-soil management plan offers an alternative to the common practice of producing row crops as wildlife food; grains are suitable food only for waterfowl and a limited number of other species, whereas the diversity of foods encouraged by drawdown management supports an equally diverse community of wildlife. Row crops also offer little cover and often are nutritionally incomplete, whereas natural vegetation generally overcomes these deficiencies. Furthermore, drawdown management maximizes production of most invertebrates, thereby providing a ready source or proteinaceous food required by many kinds of birds. Midge (Chironomidae) larvae respond favorably to water-level manipulations and, as a major food of renesting pintail hens, increase the renesting efforts of these birds (Krapu 1974b). A population of midge larvae in a Utah salt marsh managed for waterfowl was 18 times larger after a drawdown cycle than that of an adjacent site inundated continuously for 3 years (McKnight and Low 1969).

Green-tree reservoirs are managed units of bottomland forest that are temporarily flooded during the fall and winter months; the object is to make mast crops on the forest floor available to wintering waterfowl (McQuilkin and Musbach 1977). This system duplicates the natural flooding experienced in hardwood bottomlands in much of the south and midwest, but because flooding is controlled, the practice assures that water is applied in years of diminished rainfall. The idea originated in Arkansas's pin oak bottomlands among duck hunters wishing to maintain their legendary hunting in dry years (Rudolph and Hunter 1964). Green-tree reservoirs offer two management values; first, unlike permanently flooded reservoirs, the trees are not killed, so that timber resources are not lost; and second, the acorn crops important as foods for wintering waterfowl remain available each year regardless of rainfall. Dramatic results were achieved, and the concept spread to other states; a duck population on one southern refuge immediately doubled and ultimately expanded from 21,000 to as many as 100,000 birds (Rudolph and Hunter 1964). Whereas several species of waterfowl respond to green-tree reservoirs, mallards and wood ducks are the principal targets for this water-management practice in the Atlantic and Mississippi flyways. In the northeast or other regions where timber is not adapted to seasonal flooding, preliminary management schemes suggest that waterfowl use may be improved when impoundments are created, even though flooded trees may be killed (Thompson et al. 1968; see also Cowardin 1969).

The dynamic nature of green-tree reservoir management also affects the invertebrate foods necessary for breeding ducks. Drobney and Fredrickson (1979) recorded a shift from 33 percent invertebrate foods in the fall diet of wood duck hens to 54 percent for prelaying hens, 79 percent for laying hens, and 43 percent for postlaying hens in a large block of water-managed hardwood forest in southwestern Missouri. Depressions within green-tree reservoirs in Illinois remaining permanently flooded harbored 10 times more biomass of invertebrates in the spring than in the autumn (Hubert and Krull 1973).

BEAVER, WATER, AND WILDLIFE

Beaver (*Castor canadensis*) are renowned for their manipulation of water. Their dams alter the movement of water and, in doing so, produce significant changes in the local ecosystems (Naiman et al. 1986). Trees flooded by beaver impoundments generally die, although in southern bottomland forests the distribution of bald cypress (*Taxodium distichum*) often seems associated with the ponds beaver create (Fredrickson 1978). For a time, the dead trees offer habitat for cavity-nesting birds, fungi, and a host of wood-boring insects. In the absence of foliage, the amount of light penetrating through the canopy is increased greatly. The impounded water is warmed, both from the increased amount of solar radiation and from the reduced rate of stream flow. Siltation also is more rapid behind the dam—what once might have been a gravelly streambed becomes a mud-bottomed pond. Subtle but profound differences also occur in such basic phenomena as nitrogen cycling in streams where beaver have impounded water. Naiman and Melillo (1984) showed that nitrogen accumulations were about 1000 times greater in a subarctic stream after beaver dams modified the environment.

Immense numbers of beaver once were widely distributed in North America (see Naiman and Melillo 1984: Howard and Larson 1985). Estimates suggest that 60 million beaver ranged over 15.5 million km² at a density of about 4 animals per km², with a carrying capacity varying between 0.9 and 1.25 colonies per km of stream. As much as 30–50 percent of the smaller streams within the current range of beaver may fall under the direct influence of their activities.

Marked differences in the biota develop between beaver impoundments and unaffected downstream communities. A comparison in California showed that beaver ponds supported a less-diverse assemblage of bottom organisms than did the more heterogeneous habitats in the stream (Gard 1961). However, both the numbers and the biomass of organisms in the impoundment were significantly greater. The reservoir effect, described earlier, likely explains much of the increased production for both invertebrates and fishes in beaver ponds. Knudsen (1962) reported increased production of brook trout (*Salvelinus fontinalis*) after new beaver ponds were established, but trout habitat in the ponds deteriorated a few years later. Active beaver ponds on a small stream in Colorado provided at least 200 people with fishing recreation each summer; each person stayed an average of 2 days and caught 5 trout per day (Neff 1957).

Some studies have uncovered few differences in the numbers of trout between beaver ponds and comparable lengths of unaltered stream, but as much as a fourfold increase in trout populations has been recorded for beaver ponds in New Mexico (Huey and Wolfrum 1956). Despite these differences in fish numbers, comparative research consistently has shown that significantly larger trout (individual length and/or weight) occur in beaver ponds (Rutherford 1955; Gard 1961). Such differences may be dramatic when expressed in total weight. On the basis of a 4-year average, Gard (1961) found that trout biomass in beaver ponds reached 217 kg per ha, whereas only 36 kg per ha occurred elsewhere in the same stream. The composition of the trout community also may be altered under the influence of beaver impoundments (Table 11-6). In at least some instances, beaver dams present complete or partial obstacles to trout movements within streams. Upstream movements are usually more restricted than those downstream, and this limitation may have consequences when trout are denied access to their spawning habitat (Rupp 1955).

TABLE 11-6. Composition of a Trout Community Before and After Removal of Beaver Dams

Species	Composition (Percent)	
	Before	*After*
Brown trout	74	31
Brook trout	23	16
Rainbow trout	3	53
Totals	100[a]	100[b]

Source: Gard (1961).
[a] 103 fish totaling 3.8 kg
[b] 19 fish totaling 1.2 kg

Beaver impoundments often provide nesting habitat for waterfowl. Production estimates vary between 1.6 (Knudsen 1962) and 3.0 (Brown and Parsons 1979) ducklings per ha on beaver impoundments in Wisconsin and New York, respectively. Beard (1953) identified six major components of marshlands created by beaver that favor waterfowl. These are (1) interspersion of cover with water, (2) composition of cover types, (3) water depth, (4) amount and types of food resources, (5) freedom from human disturbances, and (6) the proximity of nesting cover to brood habitat. With these components at their best, the average number of young produced each year reached 11.4 ducklings per ha. Collins (1974) attributed the increase in duck populations during a 20-year period in Ontario to the increased number of active beaver ponds. In Maine, beaver ponds offer black ducks (*Anas rubripes*) nesting and feeding habitat and also serve as effective waterfowl refuges during the autumn migration (Hodgdon and Hunt 1955). Also, black duck broods showed definite patterns correlating with active beaver ponds, largely because of suitable cover produced by stabilized water levels (Table 11-7).

Beavers are, of course, a manageable resource in their own right. But prices for beaver pelts fluctuate somewhat cyclically in the marketplace so that trapping efforts also vary. In off-years, beaver populations expand, often causing timber loss, flooded roadways, and diversion of drainage systems. At other times, high fur prices place considerable pressure on beaver populations.

The beaver's own practice of water management may affect other wildlife such as otter (*Lutra canadensis*) or ruffed grouse (*Bonasa umbellus*), but the implications of their activities on two groups—trout and waterfowl—clearly suggest that orderly management is possible. For example, the age of beaver ponds is an important influence on waterfowl production, with newer impoundments normally having greater use than older ones (Benson and Foley 1956; Brown and Parsons 1979). Regularly conducted aerial surveys thus permit determinations of each pond's status so that older ponds might be identified (see Bogucki et al. 1986). Arner (1963) found that the temporary drainage of beaver impoundments, followed by the seeding of millet (*Echinochloa crusgalli*), produced large amounts of food for ducks wintering in Alabama. Impoundments managed in this way created substantial income for landowners when they were leased for duck hunting.

Because of the unique influences beavers exert on the landscape (e.g., raised water tables, upstream

TABLE 11-7. Relationships Between Active and Inactive Beaver Ponds with Black Duck and Other Waterfowl Brood Utilization

Condition of Beaver Pond	Location of Broods of Six Species (Percent)	Broods/ha		Ponds Used by Broods (Percent)		Ponds With >33 Percent Cover (Percent)
		Black Duck	*All*	*Used*	*Not Used*	
Active	81	0.98	1.85	53	47	53
Inactive	19	0.22	0.27	27	73	8

Source: Renouf (1972).

sedimentation, nitrogen accumulation, and transfer of biomass from terrestrial to aquatic systems), their widespread removal undoubtedly altered watercourses throughout North America (Naiman and Melillo 1984; Naiman et al. 1986). Brayton (1984) thus described how beaver restored severely eroded riparian habitats on streams in Wyoming. Biologists stocked the eroded areas with beaver, and because the sites were denuded of suitable vegetation, logs and branches were cut and delivered to the beavers for contructing their dams. Three years later, the silt load declined by 90 percent, erosion stopped as vegetation again stabilized, and the banks and, not incidentally, various kinds of wildlife returned to the restored streamside communities. At one site, spring floods washed away the dams, so stronger building materials were added to the delivery of logs: old truck tires. The beavers immediately incorporated the tires into the log dams and, as a result, the strengthened dams withstood the spring floods. These examples of "beaver engineering" understandably have attracted the interest of land managers concerned with restoration of damaged riparian habitats.

ALLIGATORS AND MARSH ECOLOGY

Just as bison (*Bison bison*) once dominated the ecology of the American plains, so have alligators (*Alligator mississipiensis*) assumed an influential role in southern marshes (Craighead 1968). Nowhere are these relationships more prominent than in the Everglades, where alligators shape the structure and survival of plant and animal communities. As always, many factors are involved, but even a cursory look at the ecology of 'gator holes underscores the importance of a single species in maintaining the integrity of a larger, complex biota.

'Gator holes overlay natural depressions in the limestone floor of the Everglades. Alligators clear marsh vegetation from the depression and then move the debris to the rim, forming a pool surrounded by a levee of plants and mud (Fig. 11-12). Willows (*Salix caroliniana*) are the first woody plants that invade the elevated rim, but bald cypress characterizes the trees that eventually grow on the levees (Craighead 1968). Understory vegetation on the levees consists of grasses, sedges, ferns, and other plants tolerant of shade. Of the herbaceous plants, flag (*Thalia geniculata*) is conspicuous, reaching heights of 3 m in the fresh water inside the 'gator hole (these plants are intolerant of the saltwater tides sometimes flooding the marshes outside the rim protecting 'gator holes). Indeed, the large bananalike leaves of flag turn brown at the beginning of the winter dry season and, because they stand out against the green background of other marsh vegetation, unfortunately may guide poachers to alligator dens. The vegetation developing on the levees offers diversity otherwise lacking in much of the Everglades' "river of grass," and hence several kinds of birds, mammals, and reptiles find food, cover, and breeding habitat on the levees.

Perhaps of greater importance, however, is that 'gator holes become miniature refuges of fresh water during the winter dry season, maintaining not only alligators but also other creatures dependent on water (Fig. 11-13). As many as 23 species of fishes concentrate in 'gator holes when the water in the marshes recedes each winter, as do large numbers of crustaceans; recorded densities reach 1600 fishes and crustaceans per m^2 (see review by Williams and Dodd 1978; also Kushlan 1979; Kushlan and Hunt 1979). Such a readily available prey base establishes food chains for predators, whereas the organic waste left by the predators, in turn, maintains a nutrient base for organisms serving as food for the fish populations.[1] Thus, a largely self-sustaining system develops at a critical season, and when the winter drought ends, organisms harbored in 'gator holes rapidly repopulate the surrounding marshes.

A particularly interesting relationship exists between 'gator holes and wood storks (*Mycteria americana*). These large wading birds have developed a breeding cycle coinciding with the normal dry season each winter and hence with the prey concentrated in 'gator holes. Large quantities of food are required by nestling storks, and the adult birds capitalize on the dense fish populations in 'gator holes to meet the demands of their young (Kahl 1964). For example, wood storks and other wading birds reduced the biomass of fishes concentrated in dry-season ponds by 76 percent and their numbers by 77 percent, perhaps assuring survival through reduced competition for the remaining fish stocks during the rigors of the dry season (Kushlan 1976). Conversely, in years of exceptional rainfall, wood storks postpone or cease breeding altogether because fish populations then are dispersed throughout the marshes and cannot be harvested efficiently as food. Wood storks and other marsh birds attracted to 'gator holes are prey for alligators, but, as one biologist remarked, for every bird eaten by an alligator, another 10 birds survive on food and water in 'gator holes during the dry season (Craighead 1968).

These interactions have clear importance for the management of wetlands inhabited by alligators. If poaching significantly reduces alligator populations, then an entire chain of events is interrupted to the detriment of a large number of organisms in a carefully arranged ecosystem. In short, protection of one resource, alligators, becomes a means of assuring continuation of an entire wetland community adapted to seasonal patterns of rainfall.

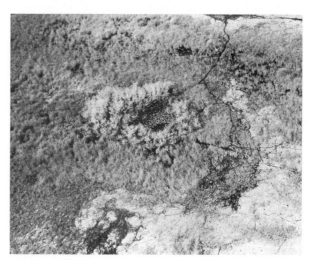

Figure 11-12. Alligators maintain water-filled depressions known as '*gator holes* in the Everglades and other southern wetlands (top). During periods of drought, 'gator holes serve as miniature refuges for freshwater fishes, wading birds, and other kinds of water-dependent wildlife (bottom). Note the influence of the 'gator hole on the vegetation immediately surrounding the open water. (Photos courtesy of T. C. Hines, Florida Game and Fresh Water Fish Commission; and J. A. Kushlan, Everglades National Park.)

[1] A South American crocodillian known as the caiman (*Caiman crocodilus*) apparently plays a similar role in nutrient cycling (Fittkau 1970, 1973). In this case, caimans feed primarily on fishes in the nutrient-deficient waters in parts of the Amazon drainage. In turn, organic materials from the caimans (e.g., food scraps, feces, and carcasses) form a nutrient base on which the fish populations depend. The subtlety of this relationship between predator and prey became evident when local anglers and hide hunters killed caimans in the belief that fishing would improve, when, in fact, the fish populations diminished with the disappearance of the caimans. A thriving caiman population simply produced better fishing for all concerned.

OIL, WATER, AND BIRDS DON'T MIX

Among the pollutants contaminating water resources and the wildlife associated with aquatic habitats are agricultural chemicals, heavy metals, and a complex of industrial wastes (e.g., polychlorinated biphenyls). All of these can, and do, affect wildlife, particularly birds (see Longcore and Samson 1973; Heinz 1976a,b; Chupp and Dalke, 1964; Hays and Risebrough 1972; Gasaway and Buss 1972;

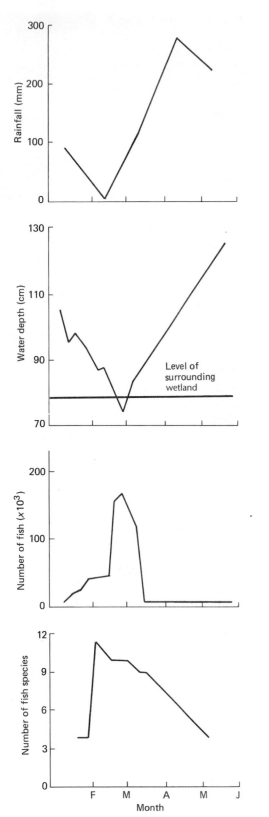

Figure 11-13. Dynamics of water depth, fish numbers, and composition in a 'gator hole during the normal dry season (February, March, April, May, June) in a Florida marsh. (From Kushlan 1974.)

Ohlendorf et al. 1978). In recent years, however, public attention has been captured by the sensational impacts, real and potential, of large-scale oil spills. Supertankers, far larger than any battleship, carry immense tonnages of petroleum, and modern capabilities for off-shore drilling have developed sizeable oilfields around the globe. In marine environments alone, some 6 million tons of oil are introduced each year (Wilson and Hunt 1975). Misfortunes besetting tankers and other oil facilities have produced some major wildlife disasters, among them the Santa Barbara (California, 1969) and Ixtoc I (Mexico, 1979) blowouts and the wreckage of the *Torrey Canyon* (Cornish coast of England, 1967) and the *Argo Merchant* (New England coast, 1976).

On March 24, 1989, the largest tanker spill in U.S. history occurred when the *Exxon Valdez* rammed Bligh Reef and spilled 240,000 barrels of crude oil into Prince William Sound on the coast of Alaska. The spill contaminated more than 1900 km of shoreline, including parts of one national forest, four national wildlife refuges, and three national parks. Oil from the spill eventually reached shorelines as far as 965 km from Bligh Reef. Unfortunately, Prince William Sound—one of the largest tidal estuaries in the United States—is the site of valuable salmon fisheries and an abundance of other wildlife. More seabirds and marine mammals were killed as a result of the *Exxon Valdez* grounding than by any other spill ever recorded. According to data published by the Committee on Merchant Marine and Fisheries (1993), about 1000 dead sea otters (*Enhydra lutris*) were recovered, but as many as 5500 may have been killed from exposure to the oil. Among seabirds, common murres (*Uria aalge*) suffered the greatest losses, with the toll of dead estimated at 300,000 birds. The carcasses of about 150 bald eagles (*Haliaeetus leucocephalus*) were tallied, and the total loss may have been four times greater. Moreover, the spill occurred just prior to the most biologically active season of the year; hence, the loss of breeding stock reduced recruitment and further diminished the populations.

Based on recent evidence, the bald eagle population in Prince William Sound recovered and returned to its pre-spill size by 1995 (Bowman et al. 1997). Similarly, various seabirds, including the species suffering the greatest harm—pigeon guillemots (*Cepphus columba*)—had recovered by 1991 (Murphy et al. 1997; see also Day et al. 1997). Other aspects—and opinions—of the *Exxon Valdez* oil spill appear in Well et al. (1995), although Senner (1997) cautions readers that many of the reports therein are based on work sponsored by Exxon Corporation.

Although large numbers of waterbirds are lost in oil spills, a positive correlation does not necessarily exist between the size of the spill and the number of birds affected (Perry et al. 1978). Instead, location of the spill and the time of year are more critical factors associated with high mortality—circumstances all too well illustrated when the *Exxon Valdez* ripped open in Prince William Sound.

Reactions of birds confronted with oily water seems species-specific, with surface-active taxa such as diving ducks (Aythyinae) or penguins (Spheniscidae) more susceptible than others (Custer and Albers 1980). In 1985, for example, oil spills in the heavily traveled shipping lanes around the Cape of Good Hope threatened colonies of jackass penguins (*Spheniscus demersus*). In the worst spill, about 1180 penguins—probably an underestimate—were oiled when the bulk carrier *MV Kapadistrias* ran aground near the birds' feeding grounds (Kerley 1986). The number of penguins oiled in this spill exceeded the total population of smaller breeding colonies elsewhere on the coast of South Africa. Conversely, fewer than 100 Cape gannets (*Sula capensis*) in a nearby colony of 140,000 birds were oiled, thereby illustrating the vulnerability of penguins to oil spills.

Unfortunately, no effective means have been developed to treat oil-damaged birds on a large scale. In fact, most oiled birds cannot be captured easily for even cursory treatment until they are already debilitated by starvation, toxicity, or exposure (Fig. 11-14). Methods currently employed rely in part on removing oil with solvents or detergents on a one-bird-at-a-time basis that usually requires holding the birds for long periods afterward (Smith 1975; Williams 1977, 1978). Even so, only a small percentage (5–10 percent) of the treated birds survives this costly process, and of these, perhaps only a fraction thrives after they are released (Hay 1975).

Oil in aquatic environments directly affects birds in one of two ways (or, more likely, with both acting simultaneously). The first result is a loss of insulation when plumage is fouled with oil. More than most terrestrial species, aquatic and semiaquatic birds rely on their plumage as a protective medium against heat loss. Even small amounts of oil, especially on a bird's underside, effectively render the plumage useless for this purpose; less than 1.0 g of oil has caused the death of ducks (Hunt 1961). Hartung (1967) found that ducks undergo rapid increases in their metabolic rates to overcome heat losses when their plumage is oiled. Because oiled birds usually stop feeding, they must draw on their body reserves, and when these are exhausted, they experience greatly accelerated starvation.

Second, oil is ingested as the birds preen their plumage. If, as reported by Hartung and Hunt (1966), ducks acquire about 7.0 g of oil on their plumage when exposed to spills, they ingest about 1.5 g the first day and about half of the total on their feathers within 8 days after exposure. Subsequent effects vary somewhat with the type of oil (e.g., diesel, lubricating, etc.), but each of the oils tested induced at least lipid pneumonia, gastrointestinal irritations, fatty livers, and adrenal cortical hyperplasia and indicated that toxicity is a definite factor in the mortality of oiled birds (Hartung and Hunt 1966).

The interaction of heat loss and starvation with toxicity in settings where birds may already be experiencing environmental stresses from food shortages, cold, disease, or even migration is not difficult to imagine. Hunt (1961) reported the deaths of 12,000 wintering and transient ducks on the lower Detroit River where more than 60,500 liters of waste oil entered the system daily (see also Hunt and Cowan 1963).

An indirect effect of oiling concerns reproduction. Ducks ingesting 2.0 g of lubricating oil ceased laying immediately and did not resume egg production until 2 weeks later (Hartung 1965). Furthermore, fertile mallard eggs exposed to small amounts of mineral oil experienced 68 percent less hatching success than untreated eggs; this experiment was designed to simulate eggs contaminated by oil washed up on shore

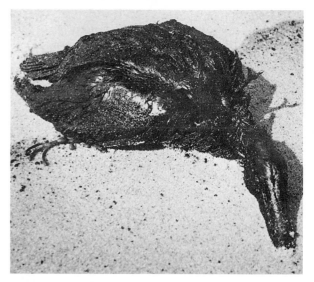

Figure 11-14.   Oil-covered birds, such as this almost unrecognizable duck, often are the pitiful aftermath of spills from disabled tankers or blowouts of offshore wells. Only a small percentage of oiled birds are recovered for treatment, and, of these, only a few survive despite the care they receive. (Photo courtesy of U.S. Fish and Wildlife Service.)

(as might occur among gulls, terns, and other shore-birds nesting near the water's edge). In still another experiment, this one simulating the application of oil to eggs from the breast of an incubating bird, Hartung (1965) found that none of the eggs hatched, in turn suggesting that prolonged incubation of the dead eggs lessened the chances for a second nesting attempt. Whereas a coat of oil might interrupt the normal gaseous exchange through an eggshell and thus induce death of the embryo, Albers (1977), Hoffman (1978), and Szaro et al. (1978) determined that it was the toxic components of oils that drasti-cally reduced the hatching success of eggs oiled on as little as 20 percent of their surface. The toxic action works rapidly, as 82–94 percent of the embryos were killed within 96 hours after exposure. Lewis and Malecki (1984) showed that weathered No. 2 fuel oil (common home-heating fuel) remained as toxic as fresh oil to the embryos of gulls (*Larus* spp.). Only after a month of weathering was the composition of the oil altered enough to pose little or no threat to the hatching success of gull eggs.

Birds also are trapped in oil pits and sumps con-structed near oil fields, refineries, and petrochemical factories. Most of the victims are ducks and other waterbirds, which apparently mistake the oil for water, although songbirds are killed in summer (Flickinger 1981). Oil pits in arid zones are particu-larly dangerous, especially during droughts and in winter when freshwater ponds are frozen. Oil sumps in the San Joaquin Valley of California claim an esti-mated 150,000 birds each year (Banks 1979), and in New Mexico, Glover (*in* Flickinger and Bunck 1987) estimated an annual loss of 225,000 birds in crude oil pits. After placing carcasses of various-size birds in oil pits, Flickinger and Bunck (1987) concluded that the rates of sinking and disappearance of the birds were related positively to body size and that the car-casses disappeared more rapidly in summer when oil temperatures were hot and more slowly in winter when the oil was cooler. Therefore, frequent counts of birds killed at oil pits in winter may overestimate mortality, but the losses of songbirds in summer may be greater than reported. Depending on body size and season, bird carcasses in oil pits can be counted with reasonable accuracy every 1 to 3 weeks.

WATER AND RAW SEWAGE

Few streams are exactly alike in their unaltered state, but it is possible to illustrate some generalized results experienced when a stream is polluted by raw sewage.

Bartsch and Ingram (1959) described a model stream with a flow of 2.8 m³/second and a water tempera-ture of 25°C receiving sewage from a community of 40,000 persons. Upstream from the point of contami-nation, the water is high in dissolved oxygen (DO) and low in its biological oxygen demand (BOD). On entry, however, the sewage causes radical changes in these parameters of water quality (Fig. 11-15, upper). BOD increases markedly, reflecting the depletion of dissolved oxygen by the demands of an immense bacterial population in a *zone of degradation*. This is followed by a *zone of active decomposition* where cil-iates and other somewhat higher forms adapted to sludge deposits and oxygen famine exist. Down-stream, nearly 80 km distant, a *recovery zone* begins, marked by a replenishment of DO and a greatly re-duced BOD resulting from the stream's ability to reaerate itself. Here the bacteria-eating ciliates give way to rotifers and microscopic crustaceans, and a full range of aquatic life again begins (Fig. 11-15, lower). However, the stream's recovery takes 8 days of flow and covers a distance of 154 km from the point of contamination.

The theoretical model of the foregoing example largely was realized when the fish fauna was exam-ined upstream and downstream from a sewage dis-posal plant in Illinois (Larimore and Smith 1963). Pollution reduced both the density and the composi-tion of the fish community downstream from the sewage plant, although, in this instance, local condi-tions permitted recovery within a lesser distance than is shown in the model (Table 11-8).

The zones of biological degradation, decomposi-tion, and recovery are more or less collapsed into a single ongoing process in sewage-treatment ponds and lagoons. The degree to which sewage materials recover depends on the type and extent of treatment (e.g., aeration, filtration, BOD in subunits), but vir-tually all of the systems are water-based operations. Although not planned for such purposes, construc-tion of sewage-treatment plants often creates wildlife habitat for a number of aquatic species. For water-fowl, the lagoons are unusual habitat because the water is often deep and the edges lack emergent veg-etation and, in fact, may be covered with rock or other hard-surfaced materials (Swanson 1977). Instead, the attractive component seems to be the abundant invertebrate food supplies available in nutrient-enriched ponds. Uhler (1956, 1964) perhaps was the first to describe waterfowl usage and production on sewage lagoons, underscoring the abundance there of midge larvae and other invertebrates—moreso than in natural wetlands. Midges are particularly well-known

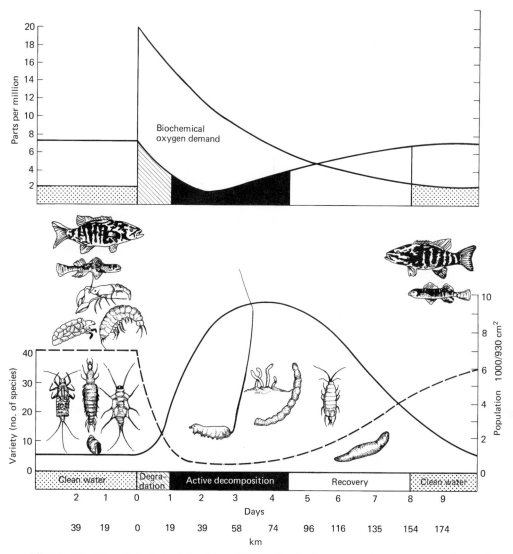

Figure 11-15. Biological relationships within an idealized stream receiving raw sewage. Dramatic changes in dissolved oxygen, biological oxygen demand, and the kinds and numbers of organisms take place downstream from the point of pollution indicated by 0 km. (From Bartsch and Ingram 1959.)

sources of protein required by nesting hens and ducklings (Krapu and Swanson 1975; McKnight and Low 1969), and in sewage-treatment lagoons in Missouri, their numbers—sometimes exceeding 16 per cm^2—made up more than 94 percent of the total insect population (Kimerle and Enns 1968).

Maxson (1981) recorded waterfowl usage of a sewage lagoon in North Dakota and noted that this habitat served migrating and premolting birds as well as those raising broods. More than 60 waterfowl broods were recorded on the 263-ha lagoon each year, or one brood per 183 m of shoreline. Conversely, sewage environments possibly may promote

avian diseases (Moulton et al. 1976), feather-wetting from detergent accumulations (Choules et al. 1978), or poisoning from blue-green algal toxins (Olson 1964). However, Maxson (1981) did not observe significant mortality of either adults or ducklings from these or other factors. Swanson's (1977) summary of brood usage of lagoons showed a range of 2.9 to 19.3 ducklings per ha of surface. Dornbush and Anderson (1964) found that lagoons had more than twice as many broods per ha as natural wetlands in South Dakota. Therefore, within the confines of their primary purpose, sewage lagoons may offer a variety of management and research opportunities,

TABLE 11-8. Numbers of Species and Density for a Fish Fauna Upstream and Downstream from a Sewage Disposal Plant in Illinois

Sampling Distance (km) in Relation to Sewage Disposal Plant	Fish Fauna	
	No. of Species	Density per 83.5 m^2
Upstream from plant:		
11.3	10	150
6.4	10	167
1.6	11	67
0.8	11	332
Downstream from plant:		
2.0	2	2
3.6	3	14
6.4	2	7
10.0	4	30
19.3	15	397

Source: Larimore and Smith (1963).

including those associated with urban wildlife management (see Chapter 17).

"ACID RAIN:" A CHANGING ENVIRONMENT

Combustion of fossil fuels, particularly coal, leads to chemical reactions in the atmosphere. Acids form when sulfur dioxide (SO_2) and nitrogen oxides (NO_x) released from smelters, power plants, or other industrial installations combine with atmospheric moisture (Fig. 11-16). Environmental damage occurs when the moisture, now infused with sulfuric and nitric acids, falls as precipitation.

In principle, any acid-sensitive environment subject to acid precipitation may be harmed. Cowling and Linthurst (1981) warned that acid precipitation may damage terrestrial ecosystems more than is supposed. For example, rain falling in forests is intercepted by several tiers of foliage before reaching the ground, so contaminated rainwater can alter the structure and function of leaves well before the rainwater can be buffered, possibly by a neutral soil. Addison and Jensen (1987) analyzed the complex setting regarding air pollution for forest ecosystems. Nonetheless, the effects of acid precipitation currently are most apparent in lakes. The acidity steadily increases as runoff from watersheds accumulates in lacustrine systems. Prevailing air currents dictate the deposition of acids formed in the atmosphere so that lakes far removed from the sources of combustion are affected. Hence, oxide-bearing smoke from industries in Ohio or New Jersey eventually may damage lakes in Canada.

Because both rainfall and snowfall act as vehicles, the phenomenon is described correctly as acid precipitation. However, the term *acid rain* has been used widely in the public media and now seems established beyond recall. In fact, *acid rain* was the term initially used in 1872 when British chemist Robert Angus Smith noticed that rainfall in industrial England was eroding the surfaces of buildings (Begley 1987). Unfortunately, acid rain is a subtle form of pollution that initially did not attract the same notoriety as did smog. Today, the biological implications of acid rain strain international politics as well as aquatic environments.

The formation of acid rain has been accelerated by recent economic development and, ironically, by some environmental efforts promoting clean air. The vast U.S. coal reserves promise to alleviate, in part, the nation's dependence on foreign oil, and numerous industries accordingly have altered their fuels. New smokestacks, heightened to reduce pollution at ground level, actually disperse the oxide-bearing gases high and wide into the atmosphere. Treatments for one set of ills, it seems, have become the cause of another.

That rain and snow were gaining widespread acidity was first noted in Europe and, later, in eastern North America (Gorham 1955; Oden 1976; Likens and Bormann 1974). Changes in hydrogen ion concentration are measured exponentially by a pH scale of 0 (acid) to 14 (alkaline) around the neutral point of pH 7. A decrease from pH 5 to pH 4, for example, represents a tenfold increase in acidity. Precipitation relatively free of contamination normally would register about pH 5.7 and values below this point define acid rain (Cowling and Linthurst 1981). Therefore, the apparently small decline in pH to 4.0–4.5 recorded for 1972–73 in eastern North America actually represents an alarming increase in acidity of the region's precipitation (Cogbill 1976).

Acid rain also contains higher amounts of lead, cadmium, mercury, and other heavy metals than does unaffected precipitation (Elgmork et al. 1973; Schlesinger et al. 1974). The transfer of the methyl form of mercury from aquatic systems into fish-eating wildlife is particularly efficient because methylmercury is readily absorbed and slowly eliminated in birds and mammals (Wiener 1987). Moreover, methylmercury biomagnifies in aquatic food chains, so the tissues of fish-eating species such as otter (*Lutra canadensis*) are contaminated at levels greater than are found in the fishes on which they

Figure 11-16. Sulfur dioxide and nitrogen oxides released into the atmosphere from the combustion of fossil fuels form sulfuric and nitric acids in the atmosphere. The result is acid rain, which has affected aquatic ecosystems far removed from the original sources of pollution. (Photo courtesy of Elizabeth D. Bolen.)

feed (Wren et al. 1986). Although conclusive data are lacking, acid rain potentially threatens human health because heavy metals and other contaminants increase in drinking water. With great acidity, for example, the concentration of dissolved aluminum in water increases massively, and aluminum has been implicated in the pathogensis of several human disorders, including kidney disease, Alzheimer's disease, and amyotrophic lateral sclerosis ("Lou Gehrig's disease") (Research and Monitoring Coordinating Commission 1986).

Regional and local properties of watersheds affect the impact of acid rain. If limestone is the dominant parent material, then the acids often are neutralized as they pass through the ecosystem. On the other hand, where parent materials lack this buffering effect, watersheds are susceptible to increased acidity from the contaminated precipitation. Unfortunately, the latter is true in much of eastern North America where granitic bedrock predominates (Fig. 11-17) More than 50 percent of the high-elevation lakes sur-

veyed in 1975 in New York's Adirondack Mountains had pH values of less than 5.0, whereas earlier (1929–37) only 5 percent of these lakes were that acidic (Schofield 1976). Lakes in Nova Scotia originally surveyed in 1955 all exhibited increased acidity after 21 years of exposure to acid rainfall (Watt et al. 1979), and in Ontario, the acidity of lakewater gained at rates of up to 0.16 pH units per year (Beamish and Harvey 1972). Loucks (1980) reported estimates suggesting that more than 48,000 lakes in Ontario alone will experience significant acidification by the year 2000.

Whereas acid rain can damage all levels of aquatic food chains (beginning with single-celled organisms), fish populations are often the first vertebrates stressed by increased acidity. McNicol et al. (1987), for example, reported fewer species and reduced biomass among minnows (Cyprinidae) in lakes with a pH of 5.5 or less. Conversely, yellow perch (*Perca flavescens*) are among the few fishes relatively tolerant to acidification. Brown trout (*Salmo*

Figure 11-17. Regions in North America where parent material does not buffer acid rain (diagonally lined area) and the breeding range of the black duck (dashed outline). A large overlap of the two distributions occurs in the Northeast (dotted area). (From West 1980 and Johnsgard 1975.)

trutta) populations now are gone in about one-third of the lakes in parts of southern Norway (Wright and Snekvik 1978), and several species have disappeared from some lakes in Ontario (Beamish et al. 1975; Beamish 1976). Fish losses may be dramatic, as when snowmelt or heavy rains suddenly flush aquatic systems with acidic water (Wright et al. 1976; Schofield and Trojnar 1980), but more often the increased acidity inhibits reproduction. Thereafter, fish populations gradually diminish as older cohorts expire. Egg fertility, hatching, and juvenile survival are each limited by reductions in pH (Runn et al. 1972; Beamish et al. 1975; Daye 1980).

In addition to loss of eggs and fry, fishes of some species fail to spawn when acid rain lowers the pH of lakewater. The acidity also seems linked with skeletal deformities and with major reductions in the biomass of fish populations (Beamish et al. 1975). Spawning may be limited or terminated when the calcium in the serum of female fishes reaches such levels that ovarian maturation no longer occurs. Reproductive failure of this type apparently occurs when pH ranges between 5.2 and 4.7. As shown in Fig 11-18, the same stress probably demineralizes the skeletons of fishes, leading to deformities when pH drops below 5.0 (Beamish, 1972; Beamish et al.

Figure 11-18. Deformed white sucker *(Catostomus commersonnii)* from a lake near Sudbury, Ontario, where contamination from acid rain was first discovered in North America. Low pH in the lakewater interfered with normal calcium metabolism, resulting in skeletal deformities such as the one shown here. (Photo courtesy of R. J. Beamish.)

1975). So when acid rain lowered the pH of lakewater below critical levels, fishes of several species experienced poor reproduction, reduced numbers and size, and physical deformities (Fig. 11-19). Increasing acidity also may disrupt the form and function of gill filaments (Fig. 11-20).

Acid rain also may inhibit the sense of smell in salmonids and therefore may modify the normal migratory behavior of Atlantic salmon. Acidification of pH 4.5 or less kills adult salmon, their eggs (pH 3.5), and immature forms (pH 4.0), but reductions in salmon populations also occur in rivers where the pH is above lethal concentrations (Daye and Garside 1977; Watt et al. 1983). Hasler (1960) determined that the migration abilities of salmon are controlled largely by olfaction; hence, the upstream migrations of spawning salmon may be thwarted in rivers with increased acidity; that is, if low pH alters the manner in which salmon return to their home streams, adults may no longer locate suitable spawning sites and the population will diminish accordingly. Royce-Malmgren and Watson (1987) demonstrated that the olfactory responses of juvenile salmon were impaired when the pH was lowered from 7.6 to 5.1. The effects were reversed when the pH returned to 7.6. These findings have important management implications, especially for programs designed to reintroduce

salmon; in acidified rivers, stocked fish may not find their way back to sites where they were released.

Other wildlife directly dependent on aquatic systems also may be affected by acid rain. Freda (1986), for example, noted the high mortality of amphibian embryos reared in acidic water. As we have seen, acidification clearly damages fish faunas, and so fish-eating birds also are affected. In Sweden, Eriksson (1986) reported declines in ospreys (*Pandion haliaetus*) in areas with acidified waters; this probably occurred because nestling survival was affected by the reduced foraging success of adults. However, other associations are somewhat more subtle. Longcore et al. (1987) evaluated the risks to which birds may be exposed as a result of acidified wetlands. For example, mergansers (*Mergus* spp.) and loons (*Gavia* spp.) normally return to natal areas for nesting, and if these waters become barren because of acidification, future breeding efforts may be fruitless; that is, the adults may continue nesting at sites where there is little chance of successfully raising a brood. Acidity also eliminates clams and snails, each an important source of dietary calcium required by birds during egg laying. Calcium shortages therefore may affect such species as black ducks (*Anas rubripes*). Fish-eating birds can obtain calcium from the bones of fish—but only as long as fish are available.

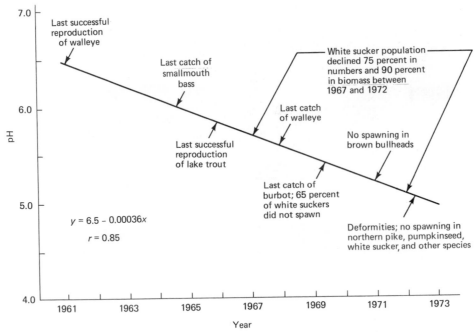

Figure 11-19.   Summary of influences of decreasing pH from acid rain on fish populations in George Lake, Ontario. Species include walleye *(Stizostedion v. vitreum)*, smallmouth bass *(Micropterus dolomieu)*, white sucker *(Catostomus commersonnii)*, brown bullhead *(Ictalurus nebulosus)*, burbot *(Lota lota)*, northern pike *(Esox lucius)*, pumpkinseed *(Lepomis gibbosus)*, and lake trout *(Salvelinus namaycush)*. (From Beamish et al. 1975.)

Acidity also affects the abundance and diversity of the invertebrate fauna required by young birds as a source of protein. Compared with less acidic wetlands in Maine, brood survival for ring-necked ducks *(Aythya collaris)* was lower on wetlands where pH was less than 5.5 (Blancher and McAuley 1987). The differences appeared in broods of older ducklings in which the growth of feathers and body mass required high levels of dietary protein. However, invertebrate numbers may increase and become more available to waterfowl when acidity reduces fish populations (i.e., fish and waterfowl may compete for invertebrate foods). Hence, contrary to the situation with ringneck broods, Pehrsson (1979) found that more foods were available for mallard ducklings on acid lakes than on those unaffected by acid rain. Ericksson (1976) found that alternate prey foods for juvenile goldeneyes *(Bucephala clangula)* may be more difficult to catch and also may be less palatable. McNicol et al. (1987) noted that increased acidity may influence predator-prey relationships between fish and waterfowl and their foods.

Some invertebrates are more sensitive than others to acid environments. Mayflies (Ephemeroptera), which are four times more common than any other

food in the diets of young black ducks, are among the sensitive groups and so may be eliminated when wetlands become acidified (Bell 1971; Reinecke 1979). Such data seem significant because nearly all of the breeding range of black ducks lies within the northeastern region affected by acid rain (Fig. 11-17). Therefore, how do black duck broods fare in acidified wetlands, and in what ways does the presence of competitors—fish—interact with the availability of invertebrate foods?

DesGranges and Hunter (1987) addressed such questions with data pooled from three sets of experiments (e.g., comparisons of various combinations of lakes and experimental wetlands with and without fish and acidity). Data from these experiments suggested that the quality of lakes as habitat for either fish or black ducks declines steadily with increasing acidity. However, in acidified lakes, fish are affected by both direct toxicity and diminished food supplies and thus are more sensitive to acidity than ducklings. Virtually all fish disappear below pH 5.0, and the sharp decline in fish density means more food for ducklings, at least for a time. Eventually, perhaps at pH 4.5, so few invertebrates survive that ducklings are deprived of food and succumb to the

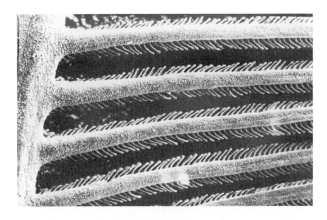

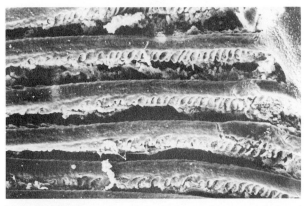

Figure 11-20. Enlarged views of normal (top) and damaged (bottom) gill filaments. The damaged filaments were taken from a fish living in an aquatic system polluted by acid rain. (Photo courtesy of C. L. Schofield.)

effects of acidity. The study concluded that acidification has a negative effect on black duck broods and could be disastrous in extreme cases. Therefore, concern for waterfowl is yet another reason for controlling acid rain (DesGranges and Hunter 1987; see also Hansen 1987).

WATER DEVELOPMENTS AND WILDLIFE

The uneven distribution of fresh water necessarily gained the attention of early civilizations. Campsites were associated with available water of suitable quality, but as civilization advanced, methods to move water were developed. At first, crude wells were constructed, but subsequently dams and aqueducts altered the distribution of water for human advantage. The technology of modern civilization developed the large-scale wonders of Hoover Dam and the Cal-

ifornia water scheme, among countless others. The Murray-Darling river system is Australia's largest, draining more than 1,036,000 km²; yet the average depth of runoff in the system is only 1.3 cm annually. However, 84 major dams built on the system divert about half the natural flow of the rivers for irrigation and urban uses. The region's mammalian fauna, already largely displaced by livestock, was little affected by the widespread irrigation scheme, but wholesale displacement of birds adapted to semiarid conditions did occur. These birds were largely replaced by introduced species such as starlings (*Sturnus vulgaris*) and house sparrows (*Passer domesticus*) that are adapted to an agricultural regime. However, the Murray-Darling irrigation scheme provides some benefits for waterbirds during dry periods, and depressions filled by tailwater runoff have become the principal habitat for the otherwise declining populations of freckled ducks (*Stictonetta naevosa*). In the main, however, control of the flow in this river system will lead to diminished waterfowl populations, in part because wildlife requirements were not part of the planning for the irrigation scheme (Frith 1973).

Water also has been diverted for other reasons, most notably to "reclaim" wetlands for agricultural production and other developments. Wetland reclamation has occurred on both large and small scales, involving everything from the establishment of major cities such as Leningrad and Mexico City to the farming of prairie potholes. In the case of prairie potholes, the figures are staggering. Estimates of the original number of potholes in Prairie Canada (Manitoba, Saskatchewan, and Alberta) vary from 6.7 million (Lynch et al. 1963) to 10 million (Cooch 1969b). Some 1.2 million more potholes occur in the United States. A recent summary underscored the drastic impact of agricultural and other activities on these wetlands in the United States (Sanderson 1976). By 1950, only 50 percent of the potholes remained, and during the period 1943–61, more than 405,000 ha of pothole habitat were lost. Some 64,000 potholes were drained in the Dakotas and Minnesota in 2 years, and losses continue at a rate of more than 2 percent per year. As a direct result, waterfowl production on the prairies, once estimated at 15 million ducks annually, now totals only about one-third that number.

In some areas, however, blasting offers a useful management tool for creating or reclaiming small wetlands (Fig. 11-21). Blasting may be used effectively in bogs and marshes where succession has advanced to the point that vegetation encroaches on open water, thereby reducing the carrying capacity of the site for

Figure 11-21. Water developments for wildlife include blasting potholes in bog areas for the creation and improvement of waterfowl habitats (top). The potholes shown here are about 1.5 m deep and 9 m long (bottom). (Photos courtesy of U.S.D.A. Forest Service.)

Figure 11-22. Section of streambed in Vermont undergoing straightening in a process known as channelization (top). Although channelization is designed to reduce flooding, it frequently disrupts wildlife habitat by degrading streamside communities (bottom). (Photos courtesy of Massachusetts Cooperative Wildlife Research Unit.)

waterfowl and other wildlife (Scott and Dever 1940; Provost 1948; Strohmeyer and Fredrickson 1967).

In many areas of the United States, streambeds have been straightened as a means of reducing floods. This process, *channelization*, often destroys riparian zones bordering the stream and so alters the size and composition of wildlife communities associated with streamside vegetation (Fig. 11-22). Possardt and Dodge (1978) recorded 636 birds of 62 species associated with a stream in Vermont, but in channelized sections of the same stream only 387 birds of 53 species were present. In particular, warblers and vireos were reduced because of their dependence on insect foods gleaned from under- and overstory vegetation. In Iowa, Stauffer and Best (1980) predicted that removal of woody vegetation in a riparian community would eliminate 78 percent of 41 species of breeding birds even though herbaceous cover remained; another 12 percent would de-

crease in numbers, and 10 percent would increase under these conditions. Conversely, a much greater percentage of birds could continue breeding if channelization were conducted so that a part of the riparian canopy remained intact. Geier and Best (1980) developed similar predictions for nine species of small mammals confronted with channelization and other disturbances of riparian habitat.

Entire watersheds and the wildlife population therein also may be degraded by channelization (see

Simpson et al. 1982 for an extensive review). For one example, Shapiro et al. (1982) summarized the aftermath of channelization of the Kissimmee River basin in Florida. In all, nearly 16,500 ha, or 78 percent, of the marshland in the basin were drained, producing devastating effects on productivity and faunal associations in these wetlands. Sport fisheries were reduced and probably other segments of the fish fauna were limited as well. Waterbirds declined by 93 percent from prechannelization levels. Furthermore, wholesale disruption of the aquatic ecosystem reduced the prey base available for bald eagles, producing a 74 percent reduction in the number of active eagle territories in the basin even though suitable nesting sites remained in place after channelization was completed. In another case, Erickson et al. (1979) found that a 40-km channelization project in North and South Dakota stimulated a nearly fourfold increase in the rate of wetland drainage. In other words, the presence of a channel prompted landowners to drain wetlands that otherwise would have remained intact as valuable wildlife habitat. Channelization also contributed to the dislocation of blackbirds (Icteridae) in Tennessee (White et al. 1985). When streams were channelized and the bottomland forests destroyed, as many as 10 million birds started to feed in agricultural areas. If future land-use patterns are similar, then human-bird conflicts of this type can only become more intense.

A classic case of biological disruption resulting from manipulation of water occurred when the Welland Canal was enlarged in 1932. The canal bypassed the formidable barrier of Niagara Falls between lakes Ontario and Erie. Its reconstruction permitted sea lampreys (*Petromyzon marinus*) access to the upper Great Lakes and their tributaries (only 3 adult lampreys were recorded in Lake Erie prior to 1932, suggesting that the original canal, completed in 1829, itself was something of a barrier for marine lampreys). The adults breed in rivers, depositing about 60,000 eggs per female in their single breeding effort before dying. Filter-feeding larvae hatch from the eggs and remain in their natal streams for several years before transforming into parasitic adults. Equipped with suckerlike mouths, adult lampreys attach themselves to fishes and rasp through their victim's skin with rings of concentric teeth. They feed on body fluids and blood of a variety of hosts, but lampreys have been particularly devastating to the lake trout (*Salvelinus namaycush*) fishery in the Great Lakes (Applegate 1950). Production of lake trout in Lake Michigan under the impact of lam-

preys fell from nearly 1.95 million kg in 1935 to 155,000 kg in 1949 (Hile et al. 1951). As lake trout populations collapsed, complex changes occurred in other fish populations, resulting in alterations of basic predator-prey relationships and an unstable fishery for all species of commercial value (Smith 1968, 1972a,b). Millions of dollars have since been spent controlling lampreys using mechanical traps, electric wires, and, more recently, selective larvicides (Applegate and King 1962). The latter have successfully controlled sea lampreys in Lake Superior.

Despite these and other disruptions, other kinds of water development have proven beneficial for wildlife even though their construction was for other purposes. Among these are flood-prevention lakes (Grizzell 1960; Nord 1963; Day 1964). These are impoundments of various sizes designed to reduce floods by confining excess water in the upper reaches of watersheds. Construction was authorized by federal legislation (most recently, by the Flood and Agricultural Act of 1962) and is carried out by the Natural Resource Conservation Service. The impoundments, however, remain the property of the owner. Depending on the region, warmwater or coldwater fisheries often may be initiated in flood-prevention lakes with stocking programs. For waterfowl, the attractiveness of these impoundments seem self-evident under the presumption that any new wetland areas have some habitat value. However, because the impoundments remain in private ownership, no recommendations were formulated by public conservation agencies for managing these habitats. Hobaugh and Teer (1981) accordingly studied 55 flood-prevention lakes in Texas to determine features that might influence their value as waterfowl habitat. The lakes they surveyed supported more than 42,000 wintering and migrating ducks, especially where the impoundments had large areas of surface and abundant aquatic vegetation. Because lakes with clear water attracted more ducks than those with turbid water, practices for reducing erosion were recommended, including livestock fencing and the seeding of annual grasses on disturbed sites around each lake. Furthermore, the importance of the watershed projects as waterfowl habitat demonstrated that wildlife values should be incorporated in the planning and construction phases of each project, rather than have them emerge as random or incidental benefits afterward.

In part, the losses of prairie potholes as crucial waterfowl habitat are offset by the existence of ordinary stock ponds. Stock ponds, of course, are not designed primarily for waterfowl or other wildlife requiring aquatic habitat, but their presence nonetheless has produced benefits. Bue et al. (1952) studied waterfowl production on stock ponds in South Dakota, finding that more than 20 ducks were raised per pond. The implications of these data are clear when they are expanded to include all of the 1850 stock ponds then available in a single county: Production varied from 36,000–43,000 ducks in each year of the study. Later, Ruwaldt et al. (1979) estimated that 88,600 stock ponds made up 14 percent of the total wetland area in South Dakota, again indicating the importance of these manufactured structures as auxiliary habitat.

Perhaps of greater importance is what might be done if stock ponds were managed for wildlife in keeping with the concept of multiple use. Hudson (1983) suggested that only slight modifications would be needed to increase waterfowl production on stock ponds. For example, much might be done to increase the amount of shoreline length per area of surface water in the design and construction of stock ponds without compromising their primary purpose. Mack and Flake (1980) found that shoreline length had a strong relationship to the presence of waterfowl broods on stock ponds in South Dakota. Furthermore, stock ponds should cover more than 0.5 ha with about 40 percent of the area less than 61 cm deep, indicating that minor shifts in the initial selection of the site may provide these features (Hudson 1983). Better grazing practices and therefore more vegetation suitable for nesting probably would produce even more broods on stock ponds. Stock ponds also represent an underutilized resource for riparian and other species. Menasco (1986) recorded 115 species of wildlife, including 3 endangered species, on ponds in the Tonto National Forest in Arizona. With fencing, stock ponds can be managed as riparian communities while still providing water for livestock. Ekblad and Crockford (1983) suggested the construction of "no-freeze" water troughs as another improvement in regions where winter temperatures cause the ponds to ice over. Water is piped through a pond's embankment to a trough partially buried for insulation in the backslope. Regulated by a valve, the water runs continuously and hence does not freeze. Because the trough is a separate source of water, the pond can be fenced to reduce the risk of cattle breaking through the ice and drowning, while also protecting wildlife cover on the shoreline (see Chapter 14).

Specific water developments for wildlife encompass a variety of structures, many of which are one of a kind, designed to match a local setting. For exam-

ple, seeps may be impounded to collect water for desert bighorn sheep (*Ovis canadensis*) as Halloran and Deming (1958) described for refuges in Arizona and Nevada. Other kinds of water developments increase the carrying capacity of arid and semiarid habitat for various kinds of big game (Fig. 11-23). Horizontal wells are particularly useful where declining water tables no longer make vertical wells or other surface-water developments practical or possible for livestock and wildlife (Welchert and Freeman 1973; Bleich et al. 1982; Gartner 1986). In California, for example, Weaver (1973) noted the extirpation of bighorn sheep from arid regions where springs and seeps no longer reached the surface. Additionally, horizontal wells flow by gravity, are unlikely to become contaminated, and require little maintenance and no operational costs. Development costs are minimal except when drilling equipment must be hauled by helicopter to remote areas. Sites for horizontal wells are selected on the basis of (1) former occurrence of springs or seeps, (2) suitable geological formations, especially dike or contact aquifers, and (3) presence of indicator plants known as phreatophytes. Some common phreatophytes include willow (*Salix* spp.), reed (*Phargmites australis*), and salt cedar (*Tamarix* spp.). McKenzie (1985) has provided a review of solar-powered and other systems for pumping water on rangelands.

The original self-filling water basins were designed for desert quail and led to their popular identification as "gallinaceous guzzlers" (Glading 1943, 1947). Guzzlers consist of (1) a collection basin having at least one side gently sloped or stepped for access, (2) a cover to reduce evaporation, and (3) an apron serving as a watershed for the collection basin (Fig. 11-24). Most guzzler units are fenced so that the installation cannot be damaged by livestock.

The surface area of the apron is dependent on (1) the capacity of the collection basin and (2) the minimum amount of rainfall expected at the site each year. A surprisingly small artificial watershed is needed because virtually all of the precipitation falling on the apron is collected (Yoakum et al. 1980). For example, in a region receiving only 2.5 cm of rainfall annually, a 2650-liter collection basin can be filled to capacity by 105 m^2 of runoff surface; a circular apron only 11.5 m in diameter fulfills this requirement.

Roberts (1977) noted that distribution patterns and consumption rates should be determined if wildlife is to benefit from guzzlers. His summary showed that the home ranges of mule deer are larger in arid environments so that managers space guzzlers

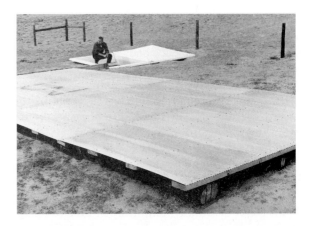

Figure 11-23. Many kinds of wildlife drink at water structures, including javelina (top) and mule deer (bottom). (Photos courtesy of U.S. Fish and Wildlife Service.)

Figure 11-24. Two examples of wildlife "guzzlers." The top photo shows a structure built for pronghorn in Wyoming, but the collection basin was designed so that rabbits, songbirds, sage hens, and other wildlife also could gain access to the water. Note that half of the plastic basin is covered to reduce evaporation and that the site is fenced to exclude livestock. The guzzler (bottom photo) was designed for desert bighorn sheep in Nevada. In this case, the apron covers the collection basin lying beneath it and, hence, serves double duty as a watershed and as a retardant for evaporation. Water collected in this unit is discharged by a float valve into a small drinking trough adjacent to the apron. (Photos courtesy of J. D. Yoakum, U.S.D.I. Bureau of Land Management.)

at intervals of about 1.5–5.0 km, depending on local conditions. Water consumption also varies according to the setting. Mule deer in Oregon consume an average of 3 liters per visit, whereas their average consumption in Arizona is nearly twice as much. Hence, water developments must be designed in a manner appropriate to each environment. For instance, where drifting snow can be expected, proper placement of the apron will catch the melt water and greatly enhance the unit's function, whereas, lacking such a natural situation, snow fences may be used to enhance drifting in the watersheds of guzzlers.

Guzzlers, while effectively improving habitats where water is a limiting factor, sometimes may produce new management concerns. Vegetation surrounding guzzlers may be overutilized and damaged by grazing species attracted to these or other water

developments. In fragile desert communities, recovery of overgrazed vegetation may be extremely slow, if it occurs at all. Mule deer moving to guzzlers in Arizona severely hedged palatable shrubs within 180 m of the structures to the extent that many plants were killed (Elder 1956). Predation rates also may be higher than normal at artificial watering sites. If this seems likely, guzzlers should be constructed near appropriate escape cover. Also, fenceposts should be pointed on top to discourage avian predators from perching near guzzlers where small animals might be expected to visit. The location of guzzlers or other water developments should not encourage poaching (Halloran and Deming 1958). Water developments situated near roads, for example, may prove unwise. Finally, once guzzlers are functional, other factors besides water availability will become limiting, thereby requiring new assessments of the desired size and density of wildlife populations, including measurements of hunting pressure.

SUMMARY

Water is not distributed uniformly in time or space; yet it is essential for life. Ecological relationships concerning water are direct and indirect, including its physical force on the landscape and hence on wildlife habitat. Distributions of many animals are limited by water either by physiological or psychological means or by water acting as a physical barrier.

Wildlife populations may increase or decrease dynamically in keeping with precipitation. The number of water-filled potholes on the northern prairies greatly influences the size of North American waterfowl populations even though these wetlands are a small percentage of all waterfowl habitat. For other animals, such as bobwhites, precipitation influences breeding by regulating the quality of their food. Therefore, precipitation has both external (habitat) and internal (nutrition) effects on reproduction and hence on the year-to-year abundance of wildlife.

Wildlife living in arid and semiarid environments have behavioral, physiological, and physical adaptations permitting their existence under harsh conditions. Some adaptations concern the initiation of breeding. Animals with specialized tolerances to water regimes may not thrive if they are introduced into areas that have wetter or drier climates.

With extreme conditions, such as drought or flood, wildlife populations may experience high mortality. Deep accumulations of snow reduce winter survival for deer and other big game animals, often by increasing predation as well as by restricting the availability and procurement of adequate food. Winter conditions sometimes influence the breeding success of wildlife the following spring and the movements and distribution of animals.

A phenomenon known as reservoir effect greatly increases biological productivity after land is flooded, but the effect lasts only a few years if the site remains continuously under water. Most populations of aquatic organisms respond dramatically when a river is impounded but thereafter stabilize without further increases in biomass. Where water levels can be manipulated, wildlife managers produce a wet-dry-wet regime with drawdown management. Drawdowns create several water depths, including exposed soil, thereby establishing seasonally important food and cover for wildlife. Beaver impoundments produce similar results, and periodic removal of old beaver dams can benefit trout and waterfowl habitat.

Alligators conserve water in depressions that function as miniature refuges during dry periods. Vegetation and mud piled on the edges of 'gator holes create diversity and hence produce new ecological conditions and habitat for various kinds of wildlife. Interactions within 'gator holes illustrate the importance of a single species—alligators in this case—in the function of larger biological communities.

Water often is polluted by ill-advised activities or disasters. Sewage released into streams quickly degrades aquatic life; considerable time and space are required before recovery occurs. Birds and other wildlife exposed to oil spills seldom survive despite treatment.

"Acid rain" forms when atmospheric moisture combines with sulfur and nitrogen oxides released from combustion of fossil fuels. Lakes in northeastern North America and northern Europe are especially vulnerable to acidification because of air currents and regional soil chemistry. The full impact of acid rain on terrestrial ecosystems remains unclear, but ample evidence demonstrates that acidification has severely harmed fishes and other aquatic organisms. Other information also suggests that acid rain may be disrupting duck and waterbird populations.

Water developments involving drainage, channelization, and large-scale impoundments usually produce undesirable effects on wildlife populations and their habitat. Stock ponds and flood-control lakes, however, are water developments of proven benefit to waterfowl, even though few of these have been designed as wildlife habitat. With planning, many kinds of water developments can benefit wildlife without interfering with their primary purpose. Structures designed to collect and store water for wildlife are known popularly as gallinaceous guzzlers. These consist of an apron serving as an artificial watershed that diverts precipitation into a collection basin protected from evaporational losses. Such devices may benefit wildlife locally but also may encourage overbrowsing and poaching in the vicinity.

CHAPTER 12

WILDLIFE AND SOILS

The pattern of life, natural or managed, must fit patterns of soil fertility.
William A. Albrecht (1944)

Soils affect the basic nature of aquatic and terrestrial environments. The distribution and abundance, as well as the quality, of organisms are influenced by the soils associated with ecosystems. Some systems are poor and nearly sterile, whereas others are immensely rich. In the thoughts of Allen (1974a), farmers gauge their land in yields of crops, but it makes little difference if we measure the productivity of land in bushels of wheat or in coveys of quail. Each is in some way a product of the soil and its capacity. These relationships may be complex, but they are no less true for songbirds than for quail, for minnows than for bass, for Africa than for North America.

In this chapter, we shall examine both direct and indirect interactions between soil and wildlife. The focus highlights the fundamental importance of soil as a component of wildlife habitat. Each soil type has an inherent capability for producing biomass in some form. But unless these capabilities—or limitations—are understood, the best management of farm or forest, plain or mountain, remains unfulfilled. In short, soil management is an investment for the future, and its renewability as a natural resource must be maintained by stewards of the land. Abuse of the soil resource clearly diminishes the capacity of an ecosystem to function as fully as it should.

SOME FEATURES OF SOIL

Soils are classified by texture, based on their content of sand, clay, and silt (Fig. 12-1). These materials vary from coarse grains to particles resembling powder. Because of the different nature of each textural group, many kinds of wildlife have developed associations with certain soil conditions. Many of these relationships are indirect and usually concern vegetation and so form a link with wildlife seeking food and cover (Fig. 12-2). For others, the associations are more direct.

Sidewinder rattlesnakes (*Crotalus cerastes*) offer a clear illustration of an adaption to the loose, granular texture of desert sands; movement would be difficult for sidewinders without their peculiar but effective twisting and sliding means of locomotion. Some evidence suggests that at least two species of desert rodents locate more buried seeds as the soil moisture in sand increases (Johnson and Jorgensen 1981). In their native range in Europe, gray partridges (*Perdix perdix*) show an association with sandy or other well-drained soils, but their distribution in the United States is not so uniformly related to soil texture (Dale 1942, 1943). "Buffalo wallows" formed by the dust-bathing of bison (*Bison bison*) support a flora

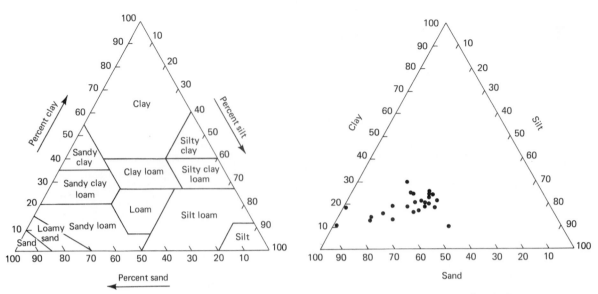

Figure 12-1. Classification guide for soil texture employed by the U.S. Department of Agriculture (left). As an example, a soil with 30 percent clay, 40 percent silt, and 30 percent sand is classified as a clay loam. If more than 20 percent of the soil materials are larger than 2.0 mm, the classification is modified accordingly (i.e., as gravelly clay loam). The distribution of pocket gophers in eastern Kansas (right) coincides with loams low in clay and silt and high in sand. (From Downhower and Hall 1966.)

Figure 12-2. Soil type often has direct bearing on vegetation. The sharp ecotone between a prairie community (foreground) and the dense brush coincides with an abrupt transition from a sandy soil to clay. Soil-plant relationships of this type directly influence wildlife habitat. (Photo courtesy of R. L. Urubeck.)

unlike the surrounding grasslands, thereby increasing habitat diversity on the plains. Soil texture and soil moisture are among the differences between wallows and nearby sites (Polley and Wallace 1986). The texture of soils on streambanks may influence the channel profile, which in turn influences water temperature, stream velocity, and cover availability for fishery resources. Streambanks with high silt-clay contents generally form deep, narrow channels, whereas those with more sand develop wide and shallow channels (Bohn 1986). In Australia, Wood (1980) studied a rabbit (*Oryctolagus cuniculus*) population located on an area of small sand dunes where the length of the breeding season—from 4 to 14 months—was related to soil moisture. Soil texture also limits the natural distributions of some animals and sometimes influences the ways in which wildlife and their habitats are managed.

Pocket gophers (Geomyidae) and moles (Talpidae) have elongated skulls, small eyes, reduced tails and ears, and short fur capable of lying forward or backward in keeping with their fossorial habits. These adaptations clearly have evolved for life underground, and it is not surprising that soil texture strongly influences where they might occur (Fig. 12-3). Soil texture was second only to the availability of food in determining the distribution of the plains pocket gopher (*Geomys bursarius*) in Kansas (Downhower and Hall 1966; see also Miller 1964). In Minnesota, most burrows of Richardson's ground squirrels (*Spermophilus richardsonii*) were located in loams instead of in other soils (Laundre

Figure 12-3. Soil texture influences the distribution of many species of fossorial animals such as pocket gophers. (Photo courtesy of M. H. McGaugh and R. L. Hendricksen.)

and Appel 1986). Clay loams were too wet for the squirrels, whereas the higher sand content of other soil types in the study area may have limited the stability of the burrows. Similarly, pine snakes (*Pituophis melanoleucus*) burrowed in a narrow range of sandy soils in the Pine Barrens of New Jersey; the limits were defined by a soil soft enough for digging and, at the other extreme, not soft enough to collapse the burrow or nest (Burger and Zappalorti 1986). The safety of the eggs and, later, of the emerging hatchlings is of obvious importance in relation to soil type for pine snakes and other kinds of burrowing reptiles. Gophers were found only in soils with low clay content and more than 40 percent sand (Fig. 12-1). Soil texture also limited the distribution of pine voles (*Microtus pinetorum*) in Pennsylvania (Fisher and Anthony 1980). These rodents are semifossorial in their habits and lack specialized morphological adaptations for digging or living permanently in burrows. Nonetheless, soil texture peculiar to their distribution suggested methods limiting the injury pine voles cause to fruit trees (root and bark damage). New orchards might better be located on soils unsuitable for pine voles. In established orchards, poison and other direct controls can be concentrated only on those soil types preferred by the animals instead of more widespread and costly treatments on areas where soil texture already acts as a deterrent.

On the other hand, prairie dogs (*Cynomys ludovicianus*) show less selectively and dig their burrows in several kinds of soil (Osborn 1942). King (1955) reached a similar conclusion but noted that only exploratory tunneling occurred in rocky ground or loose sands; no successful burrows were established in soils of these textures.

Crawford and Bolen (1976) found correlations between the amount of deep sandy soil and populations of lesser prairie chickens (*Tympanuchus pallidicintus*) in west Texas. The relationship provided indirect evidence of the reliance of these birds on native vegetation adapted to these soils. Denney (1944) and Crawford (1946) called attention to several other soil-wildlife associations and their relationship to management; Wilde (1946), Cronemiller (1955), and Allan et al. (1963) proposed ways that soil data might be applied to managing wildlife habitat. Managers thereby often can locate areas suitable for habitat improvements directly from soil maps or, conversely, they can rule out sites where management might be inappropriate (Fig. 12-4).

In aquatic environments, the ecological conditions influencing the occurrence of lead poisoning include

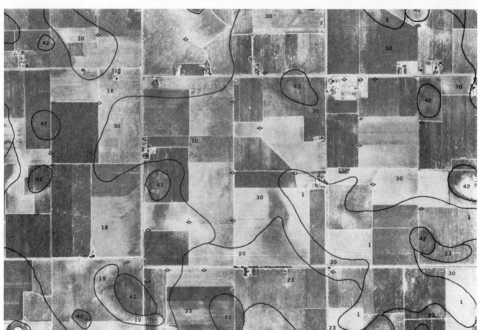

Figure 12-4. Many kinds of wildlife habitat can be located on soils maps for management purposes. Playa lakes are crucial wetlands for more than 1 million ducks overwintering on the Southern High Plains of Texas (top). The lake basins—wet or dry—are located easily on soils maps because of their unique clay soils, indicated here by the circular patterns on a Soil Conservation Service survey map (bottom). (Top photo courtesy of High Plains Underground Water Conservation District No. 1, Lubbock, Texas.)

soil texture. Waterfowl feeding in silty-bottomed wetlands largely avoid contact with expended shot-gun pellets, whereas the pellets remain more available on clays or other hard-bottomed substrates where shot does not settle rapidly (Bellrose 1959; but see Longcore et al. 1982 for an interesting exception). In terrestrial environments, however, the availability of lead shot seems independent of soil texture, and upland birds may mistakenly ingest pellets lying on the surface of moist soils. Pheasants (*Phasianus colchicus*) ingest lead shot (Hunter and Rosen 1965) and quail also may experience lead poisoning (Campbell 1950; Westemeier 1966). Locke and Bagley (1967) suggested that lead poisoning in mourning doves (*Zenaida macroura*) may occur more frequently than supposed. Doves collected from game management areas in the mid-Atlantic states contained lead shot and/or elevated lead concentrations in their livers; up to 11 percent of the samples showed these traits, indicating that doves have a high degree of exposure to lead (Kendall and Scanlon 1979). More than 50 percent mortality occurred in mourning doves experimentally fed as few as two pellets, and females ingesting only one pellet hatched fewer eggs (Buerger et al. 1986). Hence, management may be required on all soil types in terrestrial habitats where accumulated shot might pose a hazard to feeding birds. Two days after the hunting season opened for mourning doves on a public hunting area in Tennessee, Lewis and Legler (1968) recorded a 300 percent increase in shot on the soil surface, or a total of about 107,500 pellets per ha. If hunting areas such as this also are managed with food patches—and lead shot still is used—then disking or plowing soon after the hunting season is over may be required to reduce the risk of lead poisoning in birds seeking seeds or grit.

Unusual soil conditions at a wildlife refuge in Texas once caused the deaths of about 500 ducks (O'Neill 1947). The birds died of exhaustion after being shackled by sticky balls of mud weighing between 400 and 740 g. Gray partridge chicks also have died when their feet were encumbered by mudballs formed from wet clayey soils (Yeatter 1934).

With accelerated losses of wetlands, biologists have tried to improve the remaining areas as a means of enhancing waterfowl and other wildlife habitat. One technique, dynamiting, is used to blast heavily vegetated areas so that open-water "potholes" are created for feeding and other benefits. An evaluation of these structures 20 years after they were blasted revealed that soil type greatly influenced their longevity as waterfowl habitat (Strohmeyer and Fredrickson 1967). Potholes blasted in coarser-textured soils retained greater depths over time than those with finer, more erodable soils, even though their original depths were greater in finer soils. Accordingly, soil texture should be considered when planning wetland blasting programs.

Soil chemistry also interacts with wildlife. Some soils are alkaline, some are acidic, and still others are neutral in their pH. The percentage of soil samples with a pH above 6.0 correlated positively with the litter size of cottontails (*Sylvilagus floridanus*) in five soil regions of Alabama; litters increased from 3.6 to 4.7 with increasing pH (Hill 1972). Soils in some

places are laden heavily with salts of many kinds. Areas with high evaporation, low precipitation, and high water tables have soils encrusted with soluble salts that support sparse vegetation of unusually high salt tolerance known as *halophytes.* We have noted in Chapter 11 that the granite-based soils of the northeastern United States lack the buffering capacity to offset the devastation of acid-bearing precipitation.

Soils vary in their nutrient content. Phosphorus, nitrogen, potassium, and other major nutrients are present in various amounts and may be deficient for the welfare of plant and animal communities. Farmers, of course, regularly add fertilizers to enhance crop production, at times providing better nutritional values for wildlife as a beneficial side effect. However, wildlife managers seldom are able to fertilize soils solely for the crops of animals they produce. Fertilizers are expensive and often remain outside the economic realm of wildlife management. In general, most wildlife habitat is associated with low-quality soils that are unsuitable for cultivation. Albrecht (1944) was among the first to herald associations between soil fertility and its effects on wildlife. He stressed the importance of nutrients passing from the soil into the diets of animals instead of continuing the old belief that wildlife can thrive merely with ample vegetative bulk irrespective of its nutritional qualities. Indeed, most trophy-sized big game come from sites rich in nutrients, especially where calcium and phosphorus are abundant (R. J. Robel, personal communication).

Phosphate deficiencies in some soils of Hawaii affect livestock and, presumably, the exotic pronghorn (*Antilocapra americana*) population there as well. After introduction, the pronghorn herd on Lanai split into smaller units and frequently ventured into habitats other than grasslands. Among other symptoms, insufficient phosporus promotes easily broken bones, disturbances in sexual functions, and appetite irregularities. In the first hunting season, more than one-fifth of the pronghorns shot as game in Hawaii indeed had bones so brittle that they broke in the hands of hunters. The animals also did not reproduce as expected, and this, along with their wandering into atypical habitats (for food?), suggests that pronghorns cannot prosper on Hawaii's phosphate-deficient soils (Kramer 1971).

Some minor elements also are necessary for the normal growth and development of plants and animals. These are known as *micronutrients,* and include boron, cobalt, copper, zinc, and iron, among others. Most serve as catalysts for complex physiological processes, and thus only small amounts need be present in soils to satisfy these functions. When deficiencies occur, however, organisms commonly exhibit altered growth patterns or other abnormalities (see Chapter 7).

When there is less than 0.05 ppm of selenium, for example, in the diets of grazing animals, they regularly acquire nutritional muscular dystrophy or "white muscle disease" (Allaway and Hodgson 1964). Selenium may be deficient in several types of soil, but this most often occurs in those of volcanic origin (Church 1971). Mountain goats (*Oreamnos americanus*) in British Columbia showed symptoms of white muscle disease after they were captured, and some died shortly after they were confined in pens (Hebert and Cowan 1971a). An analysis of 13 species of forage plants from their range showed that 11 species contained less than 0.05 ppm of the element and that two salt licks also had little selenium. Strangely, no evidence of white muscle disease occurred in the mountain goat herd before these individuals were captured. This suggested that the stress of capture and confinement may have triggered the disease. If so, Hebert and Cowan (1971a) speculated that stresses from heavy hunting or predator pressure could lead to an outbreak of white muscle disease in the herd still living on soils with marginal amounts of selenium.

Selenium also can be too abundant for the welfare of wildlife, thereby illustrating the principle that "the dose alone determines the poison." Irrigation runoff supplies the water for the Kesterson Wildlife Refuge in California, but the soils in the surrounding San Joaquin Valley are mineral-rich deposits from an ancient seabed. The refuge in effect is a settling basin for irrigation water laden with selenium and other toxic elements (e.g., cadmium, arsenic, lead, mercury). The ever-increasing concentrations of selenium eventually produced abnormalities in the embryos of waterbirds: missing legs, feet, wings, or eyes, as well as other kinds of disfigurements. Fully 41 percent of the nests of aquatic birds at Kesterson contained at least one dead embryo, and 20 percent included at least one embryo or chick with an obvious abnormality (Ohlendorf et al. 1986a,b; see Ohlendorf et al. 1986c for a related study of wintering waterfowl). Reproduction in waterfowl and other birds breeding at Kesterson thus remains poor because of the high rate of embryonic mortality. In response, a three-part management plan was initiated to reduce the hazard: Waterfowl are hazed and frightened away; selected habitat on the refuge was made unattractive; and safer habitat elsewhere in the area was improved to attract the displaced birds (Zahm 1986; see also Mosher 1985 for the legal

and political concerns of selenium and other contamination of national wildlife refuges). Unfortunately, selenium continues to be a hazard for ducks and other aquatic birds in central California, even after contaminated drainage water was diverted elsewhere for 9 years (Paveglio et al. 1997). Selenium contamination in the watershed of the San Joaquin River in California also exceeds those concentrations that can impair the development and maturation of chinook salmon (*Oncorhynchus tshawytscha*) (Hamilton et al. 1986).

Another illustration was found in a Minnesota deer population associated with iodine-deficient soils (Seal 1979). This herd experienced a high incidence of goiter. Thyroid glands from these deer were about three times heavier than normal, and their thyroxine levels were one-half to one-third of normal. When members of the herd were given iodized oil, a significant increase in thyroxine levels followed. Testis size also increased in response to the hormonal stimulation of the additional thyroxine. This study suggested the potential of management geared to specific soil deficiencies as opposed to more general nutritional shortcomings related to poor soil fertility.

SOME INFLUENCES OF SOIL ON WILDLIFE

Fossorial animals, as we have already seen, depend on soil features for the integrity of their tunnels and dens. Some soils may crumble easily, whereas others may resist digging by even the strongest of the burrowing species. Of 50 red fox (*Vulpes vulpes*) dens examined in central New York, all but four were dug in fine sandy soil (Sheldon 1950). Soil structure also influences the distribution and internal features of fox dens in Eurasia. Dens in clay soils were shallow with many branches, whereas those in sandy soils were deep and lacked extensive branching (Kolosov 1935). Moss (1940) found that the woodchuck (*Marmota monax*) burrows closely coincided with the occurrence of sandy loam on a study area where loam and fine sandy loam also occurred; 100 of 115 burrows were restricted to sandy loam, and given the accuracy of the soil map used in the study, the remaining burrows also may have been located in soils of that texture. Bank-dwelling muskrats (*Ondatra zibethicus*) exhibit some selection for soil types, usually choosing clay-based soils for their burrows whenever possible. On more porous soils, however, muskrat burrows can damage the integrity of earthen dams on farm ponds (Beshears and Haugen 1953).

Soils consisting of more than 75 percent sand and less than 7 percent clay characterize the streamside banks where kingfishers (Ceryle alcyon) excavate their nesting burrows (Brooks and Davis 1987). The high percentage of sand not only aids excavation, but also may improve drainage for both rainwater and the wastes of the 6–7 nestlings. Gopher tortoises (*Gopherus polyphemus*) are most often associated with dry soils with high sand content. In autumn, however, gopher tortoises move to soils on wetter sites where the clay content is higher, perhaps because burrows in such soils remain moister during the dry winters and the risk of desiccation is lessened (Means 1982).

Some relationships extend beyond a single species. For example, soil texture influences the locations of gopher colonies, as noted earlier, but the conspicuous nature of gopher mounds in turn may help hawks locate profitable hunting areas (Gilmer and Stewart 1984). Burrowing owls (*Athene cunicularia*) nest in the burrows of prairie dogs (*Cynomys* spp.) and, because the owls enlarge the openings, those burrows located in sandy soils often are selected. Presumably, the sandy soils facilitate enlarging the passageways (MacCracken et al. 1985). Rattlesnakes (*Crotalus* spp.) often find refuge in the tunnels of burrowing mammals, and the association between prairie dogs and black-footed ferrets (*Mustela nigripes*) has important management implications (Hillman 1968; Hillman et al. 1979). Similarly, Florida mice (*Podomys floridanus*) and a race of gopher frog (*Rana areolata aesopus*) live in the burrows of gopher tortoises, which show strong associations with sandy soils during the summer in Florida (Blair and Kilby 1936). Interspecific relationships such as these no doubt extend to marine ecosystems, although our knowledge of these associations is still meager. Gray whales (*Eschrichtius robustus*) gain most of their nourishment from amphipods living in very fine sand on the floor of the Bering Sea. When feeding, the whales gouge pits up to 4 m long, 2 m wide, and 0.4 m deep. Such activities profoundly disturb a large area of the seafloor and inject immense volumes of sediment into the water column each year—more than twice the annual sediment load of the Yukon River. However, the upheaval of the seafloor winnows clay and silt from the sand substrate, and the disturbed areas become favored habitat for new colonies of amphipods. Moreover, the disturbance helps recycle nutrients otherwise trapped in the sediment, thereby suggesting that the feeding whales may be a significant factor in the enrichment of the Bering Sea for a large part of the marine community (Johnson and Nelson 1984).

Wild pigs (*Sus scrofa*) rooting in the deciduous forests of Great Smoky Mountains National Park mixed the upper soil horizons and reduced leaf litter and other ground cover. The disturbances nearly eliminated southern red-backed voles (*Clethrionomys gapperi*) and northern short-tailed shrews (*Blarina brevicauda*) from sites where the uprootings were particularly severe (Singer et al. 1984). These and other disruptions of native ecosystems by wild pigs are a source of controversy because environmentalists want the animals eliminated, whereas hunters and state wildlife agencies encourage maintenance of the population (see Wood and Lynn 1977; Wood and Barrett 1979).

Soils, when frozen, may inhibit or alter certain activities. Woodcock (*Scolopax minor*) normally probe for earthworms in soft, moist soils, but when these suddenly freeze, the birds seek other foods in decaying stumps or logs (Mendall and Aldous 1943). In other instances, plants and animals are adapted to frozen soil. Ground remains permanently frozen below its surface at high latitudes or altitudes even in summer. Permafrost indeed characterizes much of the Arctic Tundra where cold-adapted species dwell for some or all of their lives. Arctic foxes (*Alopex lagopus*), faced with the difficulties of digging in permafrost, instead rear their pups in dens excavated in sandy slopes (Macpherson 1969). Better drainage at these sites apparently keeps the soil unfrozen and accessible for denning activities. Drainage also seems the major factor determining the location of burrows used by hibernating Arctic ground squirrels (*Spermophilus parryii*) (Mayer 1955). A characteristic pattern of hexagons develops over permafrost when temporarily thawed soil moves into shallow fissures in a process known as *solifluction.* The intensity of this movement and the instability of the soil further limits the occurrence of fox dens (Macpherson 1969).

More than 60 years ago, Leopold (1931) suggested that successful pheasant populations in the north-central states largely coincided with the area covered by the Wisconsinan glacier. This was the last ice sheet in the series of glaciers once extending over much of North America, and the drift it deposited indeed differs from the soils left from earlier glacier events (Anderson and Stewart 1969, 1973). In southern Ohio, for example, the distribution of sizable pheasant populations was limited to "ribbons" paralleling glacial outwashes along streams; few or no pheasants otherwise occurred on the unglaciated soils in this region (Leopold 1931). The glaciated-soil hypothesis suggested a second idea, namely that soil minerals, because of their surpluses or deficits, may affect pheasant distributions and abundance. In particular, available calcium seemed the most important mineral in relation to self-maintaining pheasant populations (Dale 1954, 1955). Juvenile and hen pheasants are capable of differentiating the calcium content available in grit, and hens indeed select calcareous grit during the nesting season (Harper 1963, 1964; Sadler 1961). Thus, even where calcium is limited (as in soils not associated with the Wisconsinan glacier), pheasants still may prosper by selecting grit from the small amounts of available calcium (Harper and Labisky 1964). However, as Labisky (1975) emphasized, the causal effects of soil elements on the hardiness of pheasant populations are anything but resolved and no longer seem exclusively peculiar to glacial or nonglacial soils.

Hanson and Jones (1976) analyzed soil elements as a means of determining the nesting sites of geese in the high Arctic. They determined the concentrations of 12 elements in the feathers of geese shot during the hunting season for comparisons with the same elements present in the soil-plant complex at known nesting sites. Calcium proved the most useful feather element for distinguishing between the various nesting colonies of lesser snow geese (*Anser caerulescens*). Magnesium, along with calcium, interacted strongly with the other elements studied and showed a frequent number of statistical relationships between the feather and soil minerals. Similar trials conducted on the feathers of other species of waterfowl also showed promise, but results from wing bones were not consistent (Devine and Peterle 1968; Kelsall and Calaprice 1972). Analytical techniques, however, for determining feather profiles are not standardized (see Edwards and Smith 1984 for recommendations). Nonetheless, among other uses, feather profiles may offer management approaches for assessing differential production, harvest, and distribution among goose populations nesting in remote settings. Only a small proportion of geese are banded each year, and most of these are caught on wintering grounds after birds from several nesting colonies already have mixed. By comparison, feather samples from the many geese shot each year potentially offer more information than bands about the origins of these birds.

Likewise, hair from moose (*Alces alces*) in Alaska showed variations in its mineral content. In this instance, the differences were seasonal and varied in proportion to the animals' body condition, suggesting a means of monitoring the welfare of big game and, indirectly, their habitat (Franzmann et al. 1975). Concentrations of magnesium, copper, and most of the other eight elements studied peaked in the fall and reached their lowest levels in winter, thereby paralleling the timing of fat deposition and food availability for moose (see also Kubota et al. 1970). There also were marked differences between the hair of winter-killed calves and those of live moose from the same area. Like feathers, hair samples are physiologically stable and are easily collected and stored.

Ruminants, among them elk (*Cervus elaphus canadensis*), mule deer (*Odocoileus hemionus*), and bighorn sheep (*Ovis canadensis*), are attracted to so-called salt licks. Heimer (1972) reported that about 1500 Dall sheep (*Ovis dalli*) travel up to 19 km to a large lick in Alaska. Licks are natural formations of mineral-bearing soils, sometimes associated with saline springs (see Fraser et al. 1980). Large amounts of geological materials often are removed after long use. Knight and Mudge (1967) described a lick where more than 28,000 m^3 of sediment were removed by the combined activities of consumption and trampling-induced erosion. Some droppings of animals visiting licks may be composed almost entirely of salt-bearing clay or other soils (Geist et al. 1983).

In general, sodium salts are preferred, but calcium and magnesium salts also attract big game (Chapman 1939; Stockstad et al. 1953; Holl and Bleich 1987). Sodium salts were present in all of the natural licks sampled by Cowan and Brink (1949). Because sodium is essential for many body functions in animals, yet is not required by most plants, herbivores actively seek alternative sources of minerals in what is known as a *salt drive*. Salt drive among white-tailed deer (*Odocoileus virginianus*) includes all age and sex classes except nursing fawns and occurs during the spring and summer months but seldom in winter (Weeks and Kirkpatrick 1976). With elk, a conspicuous salt drive also developed late in May and peaked in June, 2–3 weeks after they began to forage on fresh vegetation (Dalke et al. 1965). Likewise, mountain goats sought sodium from natural licks during the spring months (Hebert and Cowan 1971b). Weeks and Kirkpatrick (1976) suggested that the high intake of water and potassium associated with a spring diet of succulent forage creates a temporary sodium imbalance and therefore initiates the salt drive early in the growing season. The moisture content of forage available to bighorn sheep was correlated with the use of mineral licks in California (Holl and Bleich 1987). The spring salt drive also may help replace skeletal minerals metabolized during winter (Geist

1971). Jordan et al. (1973) and Aumann and Emlen (1965) suggested that the availability of sodium may control the productivity and population size in such diverse herbivores as moose and rodents. Besides apparently providing big game with mineral supplements, licks also may serve as centers of social interactions. Animals perhaps develop an acquired habit of visiting such sites. Jones and Hanson (1985) have produced a comprehensive report on the role mineral licks play in the physiological ecology of North American big game and livestock.

Salt also attracts many kinds of rodents. Mice, voles, and other small rodents crave and consume mineral salts, as shown by rapid disappearance of the huge number of antlers shed each year. Adult porcupines (*Erethizon dorsatum*) have strong salt drives in summer, and 71 percent leave the forest for human settlements in search of sodium (Roze 1985). Salt drives for woodchucks and fox squirrels (*Sciurus niger*) peak in the spring and again in autumn for the squirrels (Weeks and Kirkpatrick 1978).

These associations offer several management opportunities. In earlier times, the dietary benefits of salt artificially supplied to big game remained the primary objective, but Dalke et al. (1965) cited additional opportunities. Some of these included using sulfur additives to control mange and ticks and, by distributing salt blocks in various arrangements, reducing crop depredations, controlling diseases and parasites by reducing the number of animals at natural licks, altering unfavorable grazing patterns, and manipulating hunting pressure. For these and other reasons, many tons of block salt and bagged rock salt have been distributed by pack animals and air drops in Idaho and other western states, although the effectiveness of this management practice has not been demonstrated. However, salt blocks were no longer provided for wildlife in Canada's national parks after Samuel et al. (1975) discovered contagious ecthyma or "soremouth" in bighorn sheep and mountain goats. Similarly, natural licks can be coated with creosote when it is desirable to force abandonment of such sites. Cowan and Brink (1949) noted that natural salt licks inordinately subject big game to predators. Indeed, they found more mountain goat carcasses near salt licks than elsewhere in Jasper National Park. Nematodes and other parasites may have better chances of spreading among big game concentrated at salt licks. Because of their craving for salt, mountain goats in Olympic National Park paw at the ground where hikers have urinated, thereby increasing soil erosion (Hutchins and Stevens 1981; see also

Chapter 18). Salt blocks placed near hiking trails might alleviate much of that damage. Losses of bighorn sheep lambs in New Mexico were associated with aberrant movements to mineral licks (Watts and Schemnitz 1985). Ewes and their lambs left the relative safety of the Big Hatchet Mountains and traveled across 4 km of desert enroute to the licks. Lamb mortality during these trips was a prime factor limiting the expansion of the bighorn population. When artificial licks were placed in the mountains, movements of ewes and lambs across the desert dropped significantly. Conversely, Wiles and Weeks (1986) found that white-tailed deer in Indiana seldom moved more than 1.5 km for salt and therefore doubted that the distribution of deer would be altered greatly by a program of salt management.

Sodium chloride (NaCl) applied to de-ice roads in Ontario probably contributes to the frequency of traffic accidents involving moose (Fraser and Thomas 1982). The accidents peak during the spring and early summer when the salt drains off into roadside pools, attracting moose to the roadways. The accident rate does not correlate with increased vehicular traffic but instead coincides with the time of year moose experience their greatest salt drive. Although other materials might be substituted for salt as a means of de-icing roads, they are too costly, suggesting that alternate sources of salt placed at a distance from highways may divert moose and thus may effectively reduce the collision rate. This procedure actually was employed many years ago when salt blocks were used to attract deer away from salted highways in Michigan. The collision rate dropped by 87 percent as a result (Leopold 1933a). Deer in New Hampshire probably obtain much of their dietary sodium from salts applied to roadways, either in runoff or in roadside vegetation influenced by the runoff. In fact, deer in New Hampshire could obtain almost pure NaCl simply by licking road surfaces, and the continued use of salt for de-icing highways in New Hampshire precludes the possibility of NaCl shortages for deer populations in the area (Pletscher 1987). Infant porcupines, while not subject to salt drive, sometimes are orphaned when their mothers are killed looking for sodium salts along roadways (Roze 1985). Unfortunately, salts spread on snow-covered roadways also may attract some kinds of songbirds, and at times several hundred birds have been killed by traffic (Meade 1942).

The attraction of band-tailed pigeons (*Columba fasciata*) to calcium-rich mineral deposits has managerial as well as biological implications (March and Sadleir 1970, 1972). Adult pigeons apparently require calcium supplements during and after their lengthy breeding season—in females for egg production and in both sexes for production of crop milk. Accordingly, band-tailed pigeons regularly visit mineral deposits where, in autumn, they often are subject to concentrated hunting pressure. However, age ratios derived from pigeons shot at mineral deposits may show a low proportion of juveniles and yield misleading production data, presumably because young-of-the-year do not have the same calcium requirements as adults and hence visit mineral-rich sites less frequently. Conversely, age ratios derived from pigeons shot on their feeding grounds include more juveniles and therefore may better represent annual production estimates. The association of adult band-tailed pigeons with mineral deposits also suggests that these sites may be useful locations for banding activities and taking censuses of breeding populations. Finches, particularly crossbills (*Loxia* spp.), pine siskins (*Carduelis pinus*), and evening grosbeaks (*Coccothraustes vespertinus*), also are attracted to both natural and artificial sources of salt. However, no explanation is at hand for the extradietary need for salt in these birds. Whatever the relationship, salt may prove useful for luring large numbers of finches for public viewing in parks, backyards, and along nature trails (Bennetts and Hutto 1985). In addition to three species of finches, Fraser (1985) also recorded repeated visits by two species of butterflies to salt-laden sites along roadways.

Soil fertility long ago was linked with the size of furbearers and several species of small game (Crawford 1950). Weights of raccoons (*Procyon lotor*) in Missouri correlated closely with the fertility ratings for the counties where the animals were harvested. The relationship held, irrespective of sex and age classes. Perhaps the most telling influence of soil fertility concerned cottontails. Based on a sample of more than 175,000 live-trapped cottontails, animals from areas with high soil fertility were 33 percent heavier than those from poorer soils (but see Williams 1964). Other results showed that the femurs of cottontails from the more fertile soil types were larger, had higher specific gravities (i.e., were less porous), and were broken less easily (Table 12-1). Larger amounts of calcium and phosphorus in the stronger and larger bones clearly reflected the availability of these elements in the soil where the cottontails were collected. The amount of body fat in mourning doves in Illinois also varied with soil fertility. Fatter doves occurred where the soils were rich, whereas doves were leaner on sandy soils that had little fertility (Hanson and Kossack 1963).

A general pattern emerges that better quality soils support larger bodied animals. For deer in North America, this relationship was shown decades ago by Einarsen (1946) and Cheatum and Severinghaus (1950). The linkage, of course, is through the quality of forage consumed by the animals, and a pattern of "better soil–bigger deer" thus emerges (but see Klein and Strandgaard 1972 for a curious situation in Denmark). Likewise, species more closely associated with farmland—and thus with richer soils—also show this relationship. The harvest of bobwhites (*Colinus virginianus*) in Missouri averaged about 1 per 2.4 ha on a fertile soil type, but only 1 per 877 ha on stony soils (Crawford 1946).

Turkey (*Meleagris gallopavo*) numbers and soil types were related in Missouri (Dalke et al. 1946), but this relationship did not hold in West Virginia, presumably because farming on the better soils largely had displaced turkeys to the stony soils on mountainsides (Uhlig and Bailey 1952). Similarly, the high fertility of certain soils in parts of Missouri encouraged intensive farming, thereby reducing populations of prairie chickens (*Tympanuchus cupido*) in those areas (Christisen 1985). The weights of bobwhites collected from glaciated soils in Ohio significantly exceeded those from nonglaciated soils (Schultz 1948). However, bobwhite densities were less on the glaciated soils where farming also was more inten-

TABLE 12-1. Summary of Physical Properties of 450 Cottontail Femur Bones in Relation to Soil Fertility

Soil Fertility[a]	Avg. Weight (g)	Avg. Length (cm)	Avg. Thickness of Bone Walls (mm)	Avg. Maximum Diameter (cm)	Avg. Breaking Strength (kg)	Avg. Volume (cc)	Avg. Specific Gravity
High	4.172	8.03	0.82	0.76	20.2	3.98	0.854
Medium	3.841	7.87	0.74	0.72	16.6	3.48	0.814
Low	3.403	7.32	0.68	0.69	12.4	3.21	0.786

Source: Crawford (1950).

[a]Soil fertility determined by calcium, nitrogen, phosphorus, and potassium contents, and crop yields.

sive. Better farming conditions in Illinois were associated with high levels of fat in mourning doves (Hanson and Kossack 1963). The pelts of muskrats from streams influenced by rich alluvial soils differed from those with less-turbid waters and gravel substrates. Inferior pelts, in terms of their size and fur quality, originated from clear, almost sterile streams, whereas the largest and best pelts were associated with watersheds having rich soils. In gathering those data, Crawford (1950) noted an interesting exception that showed the influence of fertility in another way. Muskrat pelts from two forks of the same low-turbidity stream ranked quite differently in their size and quality. However, the fork where the better grade of muskrat fur originated was enriched by the organic wastes from a commercial fish hatchery. Like muskrats, the fur of opossums (*Didelphis virginiana*) also showed direct correlations with the soil fertility of watersheds where the animals lived (Crawford 1950).

Wishart and Bider (1976) found no differences in woodcock habitat, including the abundance of earthworms, based on soil types in southwestern Quebec. However, the area of exposed soil surface did influence woodcock use of the sites. On the heavily used habitats, an average of 87 percent of the soil was free of matted vegetation and debris, perhaps indicating that this factor increased the contact between predator and prey. If so, it points out the ecological difference between abundance and availability—earthworms protected from the probings of woodcock by debris on the soil's surface remained unavailable as food, irrespective of their abundance.

Kantrud and Kologiski (1982) found differences in the composition of dominant birds breeding among major soil types in the northern Great Plains, but the densities of the bird populations were less related to soil classifications. Increased mean annual soil temperatures seemingly depressed the richness of the avifauna more than decreased soil moisture or organic matter. Soils heated by both artificial and natural causes are associated with a fungus causing avian encephalitis and may be natural reservoirs for other pathogenic microorganisms (Tansey and Brock 1973). Among the timber rattlesnakes (*Crotalus horridus*) living in the Pine Barrens of New Jersey, gravid females prefer basking on exposed sandy soil, whereas males and nongravid females largely remain in forested areas (rattlesnakes are on the New Jersey list of endangered species). However, because the sandy sites most often occur on the edges of unpaved roads

in the area, substantial numbers of gravid females concentrate in areas of high human activity. Thus, a specific and crucial segment of the rattlesnake population is subject to higher rates of accidental deaths, wanton killing, and illegal collecting (Zappalorti and Reinert 1986). Perhaps as a management practice, basking areas could be bulldozed on sandy soils where human activities are less intensive.

A condition known as "velvet horn" occurs among white-tailed deer living on a certain soil type in central Texas (Thomas et al. 1964a, b, c, 1970). About 2 percent to 9 percent of the bucks shot in this area retain velvet-covered antlers during the autumn and winter months instead of developing the hardened antlers of normal bucks (Fig. 12-5). These animals also have small, dysfunctional testicles that are no more than 25 percent of the size of normal bucks, and their social behavior is subordinate to other deer of both sexes. Unfortunately, the precise cause of velvet horn remains unknown, but nearly all of the velvet-horned bucks were shot on granite-gravel soils (Table 12-2). Whatever the cause may be, this association does not impair the deer herd's reproductive performance. Doe-fawn ratios for the herd living on granite-gravel soils were the same as for other herds nearby, and

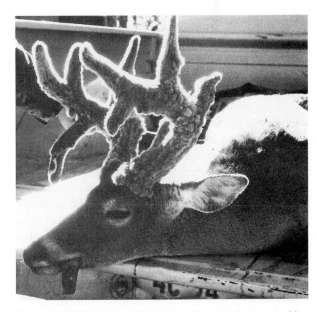

Figure 12-5. White-tailed deer with velvet horn, an oddity associated in some unknown way with granite-gravel soils in central Texas. Antler tissues of bucks with this condition remain soft and slowly erode until new growth builds on the stump left from the previous year. Examinations have not detected any pathogenic agents for velvet horn nor is there any evidence that bucks ever recover from the condition. (Photo courtesy of Texas Parks and Wildlife Department.)

TABLE 12-2. Association of Velvet-Horned Bucks with Granite-Gravel Soils Based on Deer Shot by Hunters in Central Texas.

Year	Total Velvet-Horned Bucks Killed	Velvet-Horned Bucks Killed on Granite-Gravel Soils		Velvet-Horned Bucks Killed on Other Soil Types	
1959	143	130	(90.9%)	13	(9.1%)
1960	406	372	(91.6%)	34	(8.4%)
1961	441	392	(88.9%)	49	(11.1%)
1962	328	311	(94.3%)	17	(5.2%)
Totals	1318	1205	(91.4%)	113	(8.6%)

Source: Thomas et al. (1964a).

just as many does were pregnant. So the disorder remains an oddity associated with a peculiar soil type but is of no consequence to deer management.

Striking parallelism between the pelage colors of small mammals and the background color in their environments suggests another influence of soil on wildlife. Pale- and dark-colored rodents live, respectively, on the white sand and dark lava beds in the Tularosa Basin of New Mexico as well as in other deserts of the southwestern United States. However, the color of these soils does not directly influence the pelage color of the rodents. Instead, the matching pelage color seems the result of selective predation and long periods of reproductive isolation (Benson 1933; Dice and Blossom 1937). In this case, reduced predation favors retention of advantageous characteristics among the survivors so that the surviving rodents transmit genetic materials that steadily enhance each generation's protection (i.e., a larger proportion of the appropriate color survive). In time, and given the normal genetic variation in pelage colors, dark pelage emerges in those mammals living on the lava and pale pelage dominates the populations dwelling on the white sands. Experimental evidence later indicated the strength with which soil color may protect desert rodents from predation. Dice (1947) covered half a room with dark soils and the other half with light soils and then added populations of pale- and dark-colored mice—and an owl. When the owl was forced to hunt by sight, more protectively colored mice (i.e., dark mice on dark soil) escaped predation than did conspicuously colored mice (i.e., dark mice on light soil). In fact, mice matching their background soils in the experiment enjoyed more than a 20 percent advantage escaping predation. The magnitude of this selective advantage, if applied to natural populations, would soon result in the genetic dominance of one color over others with respect to specific soil types (see also Kaufman 1974).

A TROPICAL PARADOX

Perhaps no soils have been more misjudged than those underlying tropical rain forests. The lush vegetation that they support led early explorers and colonists to believe that these soils, once cleared, would yield equally productive crops. Regrettably, this myth continues, and about 250,000 km^2 of rain forest are lost each year (BioScience 1981), often with governmental encouragement (Fearnside 1983). After a few short years of crop production, the deforested soils lie wasted and capable of sustaining little more than poor-quality pastures where cattle can forage for less than a decade before the grazing system collapses (Nations and Komer 1984). Moreover, in part because of the peculiar circumstances outlined later, there is little chance that successional processes again will revegetate these soils with rain forest (see also Gomez-Pompa et al. 1972). The consequences surely will prove as destructive for human communities as for wildlife (see Raven 1981; Caufield 1984). Deforestation of rain forests adjacent to the Panama Canal, for example, has increased soil erosion to the point where siltation may reduce the capacity of a reservoir providing water for operating the canal (where the passage of each of 12,000 ships per year flushes 43.3 million imperial gallons from the canal to the sea). Moreover, deforestation reduces evapotranspiration, and lesser amounts of water vapor return to the atmosphere. Reports thus indicate that rainfall may have been reduced by as much as 10 percent in central Panama since 1900, perhaps posing another threat to the operation of the canal (Borrell 1987).

It is easy to understand how the fertility of rainforest soils might be misjudged. Huge trees tower above one or more dense understories of still other trees in seemingly endless tracts of impenetrable rain forest. The biomass of this vegetation is awesome,

Figure 12-6.   Lush vegetation in tropical rain forests belies the fragility of crucial plant-soil relationships. Rain forests typically have multilayered canopies (left), leafy understories, and rich ground litter (center). Note the buttressed base of the large tree, an indication of the shallow root system. Most rain-forest plants find nutrients in the forest floor instead of penetrating deeply into the infertile soils. Once cleared, tropical rain forests produce a legacy of ruined soils, temporary croplands, and poor forage for livestock (right). Destruction of these ecosystems limits the winter habitat of migratory birds and causes the extirpation or extinction of resident species. (All photos from southern Veracruz, Mexico, courtesy of J. H. Rappole and E. G. Bolen.)

with a dry weight averaging 450 tons per ha compared to 300 tons for temperate deciduous forests (Whittaker and Likens 1975). Likewise, the *leaf-area index* (i.e., surface area of leaves above 1 m^2 of the forest floor) is 8 and 5, respectively, for the two types of forests. Other measures of luxuriance might be cited for tropical rain forests, but the foregoing adequately illustrates the dominance of these environments among the world's forests. Even lacking such data, as did the colonists, just a glance at a virgin rain forest transmits an impression of high soil productivity (Fig. 12-6; see also White 1983). There was little evidence that clearing would produce a legacy of degraded soils and a regime of impoverished agriculture. Instead of high fertility, rain-forest soils are leached almost completely of nutrients. High rainfall washes from the soil any reserve of nutrients that otherwise might accumulate as sources of fertility. Where then does the rain forest gain its nutrients? Jordan (1982) summarized the remarkable functions

of litter accumulations on the floor of Amazon rain forests as a dynamic reservoir of nutrients. Thus, litter—not the mineral soil—largely maintains the rich vegetation.

Litter mats in tropical rain forests vary between 15 and 40 cm in thickness and contain more than half of the smaller root systems in the forest (Stark and Spratt 1977). This layer of humus intercepts and absorbs the nutrients entering the rain-forest ecosystem; virtually none enters the mineral soil underneath. As the litter decomposes, the root systems in the mat gain access to the nutrients. It is likely that algae in the litter also trap nutrients in rainfall and store these until the algae decompose, again providing direct access to the root systems. Further, acidity and other conditions in the litter inhibit bacteria that utilize precious nitrogen. Many leaves are coated with algae and lichens that absorb the nutrients in rainfall, and when the leaves fall, their decomposition releases even more nutrients to roots in the litter zone.

Sanford (1987) determined that apogeotropic roots—those that grow *up* the trunks of neighboring trees—of some species in the Amazon rain forest cycle nutrients directly from plant to plant; that is, such roots absorb and transport nutrients from precipitation flowing down the stems of nearby trees. These structures and their unique function seemingly represent another adaptation to the nutrient-poor soils of tropical rain forests. A measure of this system's efficiency can be determined by the ratio of recycled nutrients to the total amount available so that a value of 1.0 represents maximum results. Amazon rain forests achieve a rating of 0.7 despite their poor soils, surprisingly similar to the efficiency of forests elsewhere growing on rich soils (Jordan 1982).

The leaves of many tropical plants seem adapted in ways that resist attack from insects and other herbivores, again conserving nutrients within the vegetation instead of passing them to consumers (Jordan and Uhl 1978; Janzen 1974). Finally, much of the available nutrient supplies in tropical rain forests are bound in the trees themselves. Thus, when tropical forests are cut and the litter cycle is destroyed, only a nutrient-deficient soil remains that can support neither farm nor forest. Indeed, surveys of Amazonian soils, while still far from complete, suggest that no more than 1 percent of the area may be suitable for conventional agriculture (Goodland and Irwin 1975).

The soils of many tropical regions contain laterites, materials rich in secondary oxides of iron, aluminum, or both (see Alexander and Cady 1962). Lateritic soils form in association with heavy rainfall and luxuriant vegetation. But what happens when the forest cover is removed from lateritic soils? Simply stated, the soil hardens into ironstone. The processes involved are complex but involve exposure to sunlight and the ensuing chemical changes this exposure brings to the hydrated iron oxides. The point here, however, is that deforested laterites harden into untillable pavement, too often creating a wasteland within a few years. Thus, considering present technology, clearing lateritic soils of rain forests in the hope of gaining additional agricultural production is fruitless, and in fact it destroys the integrity of these soils as a renewable resource. Indeed, some economies produce only building materials from hardened laterites, reflecting the term's Latin origin from *later,* or brick (Russell 1966). The results of continued deforestation and exposure of lateritic soils in tropical environments were expressed dramatically by Goodland and Irwin (1975) in *Amazon Jungle: Green Hell to Red Desert?*

What does deforestation and the subsequent degradation of tropical soils mean? Raven (1981) estimated that some 3 million species of plants and animals occur in the tropics, whereas temperate regions of the Earth harbor about half that number. Myers (1980) suggested that nearly half of all species on Earth apparently occur in the moist tropical forests that occupy only 6 percent of the Earth's land surface. The biological richness of the tropics is staggering. Ecuador, about the size of Colorado, harbors more than 1300 species of birds, roughly twice the combined total for the United States and Canada. Almost as many species of frogs and toads can be found in a few square kilometers of rain forest on the eastern slopes of the Andes as occur in all of temperate North America (Raven 1986). In Peru, a survey of insects in the rain-forest canopy uncovered 12,000 kinds of beetles (Coleoptera) in a total of 41,000 species—all on a plot no larger than 1 ha (Lewin 1986). Unfortunately, only one-sixth of the tropical species have been studied or even catalogued. By comparison, fully two-thirds of the temperate biota are known to science.

The richness of tropical systems is by no means confined to terrestrial life. About as many kinds of fishes—nearly 5000 species—swim in the Amazon drainage as exist in the Atlantic Ocean (Raven 1981). Thus, many forms of tropical life likely will vanish even before they are discovered. Construction of the Transamazon Highway alone likely caused the extinction of a number of species, probably without any awareness on the part of the builders (Myers 1976). Many of these undoubtedly were so-called lesser organisms (e.g., invertebrates); yet these surely functioned importantly in the food webs of higher species in tropical forests, not to mention their inherent worth as biological entities of the Earth's biota. Nor are such losses without implications for human welfare: The struggle of tropical plants against insect predators led to the evolution of many natural chemicals, of which quinine and several others serve as important medicines and drugs (Jackson 1983). Tropical forests also were sources of invaluable dyes, oils, gums, and foodstuffs (e.g., coffee and bananas), to name only a few commodities of global importance. Nonetheless, only a few of the immense number of tropical species have been tested for their usefulness, and countless products will go undiscovered as rain forests vanish forever.

Many of the rivers draining Amazonian rain forests reflect the poverty of tropical soils. Primary productivity is so low in the clearwater and blackwater rivers draining these soils that the large biomass of fishes they support instead depends on the adjacent flood-plain as an energy source. During periods of flooding, fishes move into the inundated forests and feed on seeds, fruits, invertebrates, and detritus (Goulding 1980). In fact, fully 75 percent of the commercial fish catch may originate in flooded forests. These rela-tionships, again underscoring the fundamental nature of impoverished tropical soils, clearly suggest that fish populations will decline as the floodplain rain forests are destroyed. Carr (1982) also stressed the depen-dence of rivers and estuaries on tropical forests and urged that these ecosystems, while distinct, should be managed ecologically using the principles of water-shed management.

A large number of North American songbirds, es-pecially species of warblers, overwinter in tropical forests even though they breed in the temperate biomes of North America. In fact, about 51 percent of the 650 birds in the North American avifauna win-ter in the tropics, and these spend one-half or more of their life cycle in tropical communities (Rappole et al. 1983). Many biologists once believed that mi-grant birds might not require specialized habitats in winter, implying that the loss of tropical forest did not seem important to the needs of these species. However, Rappole and Warner (1978, 1980) pointed out that migrant birds indeed are constrained by food resources, competition, and other exacting eco-logical features in the tropics. In fact, migrants are no less dependent on habitat quality and quantity in winter than they are during the breeding season (Rappole and Morton 1985, 1986). Accordingly,

many songbirds familiar to North Americans will di-minish in numbers as forests—and soils—far to the south are ruined (Table 12-3).

Regrettably, similar fates seemingly await a large number of resident species. The dependence of tropi-cal mammals on the forest canopy is emphasized by data from a Panamanian rain forest. Some 70 percent of the mammalian biomass is contained in arboreal species, with monkeys and sloths alone representing more than half of the total biomass at this site (Eisen-berg and Thorington 1973). Jaguars (*Panthera onca*) seem better able to cope with clearings, although they avoid open pastures. However, their travels in Brazil encompass large areas, including forest, where their natural prey occurs. Female jaguars range over at least 25–28 km^2 and males move in an area more than twice that size (Schaller and Crawshaw 1980). Be-cause large cats often have social systems correlated with their environments, jaguars faced with shrinking habitats and declining numbers may encounter limi-tations lying beyond their social tolerances.

Farnworth and Golley (1974) suggested that many tropical vertebrates are remarkably sedentary. Gaps of suitable habitat as small as a few hundred meters apparently bar the movements of some birds and mammals. Unfortunately, preliminary studies fur-ther suggest that blocks of uncut rain forest left in the midst of a clear-cut are poor refuges for wildlife. D'Aulaire and d'Aulaire (1986) reported that birds displaced from the clear-cut area at first crowded into the "island," but after much turmoil with the res-ident birds, the population crashed, leaving only half

TABLE 12-3. Estimated Decline in the Populations (× 1000) of Migratory Birds Wintering in the Lowland Rain Forests Associated with the Tuxtla Mountains in Southern Veracruz, Mexico. Note the Relationship of These Data to the Loss of Habitat.

Species	Territories per km^2	1500	1960	1975	1985
Wood Thrush (*Hylocichla mustelina*)	150	225	112	74	34
Black-and-White Warbler (*Mniotilta varia*)	40	60	30	20	9
Ovenbird (*Seiurus aurocapillus*)	40	60	30	20	9
Kentucky Warbler (*Oporornis formosus*)	70	105	53	35	16
Hooded Warbler (*Wilsonia citrina*)	60	90	45	30	14
Forest—Percent Remaining		100	50	33	15

Source: Rappole and Morton (1986).

of the original number of birds. A year later, only 18 of the 39 species were left in the 1-ha block. Moreover, trees died four times faster in the small blocks than in the undisturbed rain forest, which indicated that habitat conditions for wildlife could not be sustained in the would-be refuges. Other interactions in the small area contributed to the decline in species. White-lipped peccaries (*Tayassu pecari*) lacked foraging space in the 1-ha blocks and quickly disappeared, but so did three types of frogs. Without peccary wallows, the frogs lacked puddles in which to breed. Such changes also occur with somewhat less severity on blocks of 10 or 100 ha, but the implication is obvious: continued cutting and fragmentation of tropical forests will be accompanied by wholesale reductions and extinctions of species.

Deforestation accordingly poses interference with movements, breeding opportunities, and other crucial activities of many animals. For other species, deforestation simply means the inexorable loss of essential habitat. A single example will suffice. Populations of a small monkey, the golden lion marmoset (*Leontideus rosalia*), once ranged over 6500 km² of

Brazilian rain forest, but now no more than 600 of these animals remain in only 550 km² of that habitat (Myers 1976). One can ponder the dismal future of white-lipped peccaries, brocket deer (*Mazama americana*), and an immense fauna of other species as tropical communities are reduced to wasted soils.

Despite an uncertain future, methods are available for stemming the destructive tide of tropical deforestation (Nations and Komer 1983; Rubinoff 1983; Office of Technology Assessment 1984). These include strategies for preserving rain forests as well as for changes in farming (i.e., elimination of slash-and-burn agriculture). One of these is the *chinampa* system, which intensively uses lands already cleared, thereby halting further expansion into additional forest. Small canals around each farm plot not only provide a simple irrigation/drainage system, but aquatic vegetation grown in the waterways also provides "green manure," and fish in the canals offer a source of protein. No machinery, insecticides, or fertilizers are needed, and evidence at hand indicates that a *chinampas* of only 2000 m² can produce food and cash crops for a family of five. Agroforestry is

Tall Timbers Research Station

Tall Timbers is a nonprofit research station founded in 1958 near Tallahassee, Florida. The station enjoys an international reputation for its management-oriented research, which places heavy emphasis on fire as an ecological force that influences forest management and wildlife populations. Much of the station's research work and demonstration areas are located in habitat dominated by longleaf pine. The original research station consisted of 1133 ha, but gifts and donations have subsequently expanded the facility to nearly 1619 ha.

The focus placed on prescribed burning results from the premise that lightning fires have played an evolu-

tionary role in shaping the biological landscape on the Coastal Plain bordering the Atlantic Ocean and the Gulf of Mexico. Prescribed burning has become a commonplace tool in this region for the management of bobwhite, a concept pioneered by Herbert Stoddard (1889–1970), who served as vice president and, later, president of Tall Timbers. Many of the research projects are long-term studies and thus are of particular value for understanding ecological processes. Among these are detailed records for 84 research plots burned at different intervals since 1959.

In addition to research, Tall Timbers is active in conservation education, which is addressed with seminars, field days, and training programs for landowners and managers, as well as with technical and semitechnical publications. The staff demonstrates techniques for managing land in ways that promote sound conservation while allowing economic uses of natural resources. In other words, Tall Timbers works to accomplish a sustainable balance between people and natural systems.

The year-round staff at Tall Timbers includes wildlife biologists, foresters, and plant ecologists, as well as seasonal interns and visiting scholars and naturalists. In addition to its other holdings, the station library includes a database consisting of more than 10,000 sources of information concerning fire ecology. Every other year, the station hosts a scientific conference, from which the papers are published as the *Proceedings of the Tall Timber Fire Ecology Conference*.

another technique, which features farms of both trees and crops. As many as 75 species of crops have been combined on 1-ha plots, and a farm family may exist for a generation on no more than 10 ha of cleared land. While encouraging, the technology of these methods must be complemented with political and financial constraints if tropical forests are to persist. The threat is particularly acute where massive logging and ranching operations devour large areas of rain forest each year.

In 1987, an imaginative financial arrangement emerged that will save large blocks of tropical forest in selected countries. Conservation groups will pay the debts of foreign nations in exchange for the protection of rain forests. The exchanges are known popularly as "debt-for-nature" swaps. Bolivia was the first nation to negotiate such a swap and will protect about 1.5 million ha in the Amazon Basin in exchange for $650,000 in debt relief, although the sponsoring group—Conservation International—will pay only $100,000 to the owners holding the loans (Walsh 1987). The swap will work so long as indebted nations are willing to abandon their development plans and lien-holders are willing to write off poor loans for a fraction of their face value (about $0.15 on the dollar in the agreement with Bolivia). Costa Rica and Equador are working on similar swaps. The idea has sparked interest in Congress, where bills are pending that will urge the World Bank and similar institutions to pursue debt-for-nature swaps as a matter of policy. Other proposals under study include changing the U.S. tax code so that tax credits are offered to commercial banks that negotiate loans with debtor nations willing to adopt conservation measures.

DESERTIFICATION

Mabbutt (1981) refers to desertification as the spread or intensification of desert conditions in and around arid lands, involving lessened biological productivity, accelerated soil deterioration, and impoverishment of human livelihood systems. About 80 percent of the world's agricultural land in arid and semiarid regions is desertified to some degree, affecting 700 million people (Dregne 1983).

Desertification only rarely involves sand dunes creeping over better kinds of soil, although that situation sometimes happens. More often, desertification develops in times of drought where already vulnerable arid and semiarid lands have been abused. These lands then permanently degrade into deserts. Symptoms of desertification include (1) declining water ta-

bles, (2) salinization of soil and water, (3) reduction of surface water, (4) high rates of soil erosion, and (5) degradation of native vegetation (Sheridan 1981). With these systems in mind, it is easy to see how badly wildlife might fare as desertification advances.

Perhaps no area serves as a better example of desertification than does the region bordering much of the Mediterranean Sea. After centuries of deforestation and overgrazing, the soils of this once productive region are now largely eroded and infertile. This degradation sometimes has been known as *Mediterraneanization* because the process is so common in that region. Desertification in the Mediterranean region dates at least to 300 B.C. and the observations of Plato in *Critias*: "The annual supply of rainfall was not lost, as it is at present [after the forests were cut], through being allowed to flow over a denuded surface to the sea." Previously, when forests abounded, rainfall instead was stored in the soil and was discharged gradually in rivers and springs "with an abundant volume and wide territorial distribution" (Anon. 1956). The glory of Greece diminished with the abuse of its forests and soils. History similarly records the cedars of Lebanon—once forests—where there are none today, and the fertile plains of Iraq and Iran, where now are found only sandy wastelands dotted with the ruins of abandoned townsites. The geographical crescent from North Africa through Asia Minor to Greece and Spain remains a catastrophic example of exploitive management imposed on vulnerable landscapes (Burger 1978).

Talbot (1957) reported desertification of the Gir Forest in India, the home of the last wild population of the Asiatic lion (*Panthera leo persica*). Up to 80,000 head of livestock—a more accurate estimate could not be made—ravaged the forest, leaving little growth beneath the canopy. The forest retrogressed first into thorn scrub and then into desert. Since the 1880s, the Gir Forest has shrunk from 3260 km^2 to 1300 km^2, and wildlife has been virtually eliminated even though hunting has been almost nonexistent there for decades. The lions, as might be expected, began to kill livestock as their natural prey vanished with advancing desertification. So maintaining the remnant lion population has become a political concern as well as a biological challenge as the Gir Forest continues shrinking. In Africa, desertification has destroyed wildlife habitats, leading to reductions in the distribution and abundance of several species of antelopes and other native herbivores (deVos 1979). Nor are freshwater fish populations immune to desertification. Increased turbidity, siltation, and salinity adversely affect these fisheries, as do changes in the volume of stream flow.

Native faunas contribute little to desertification unless human influences already have initiated the process (Cloudsley-Thompson 1977). Desert vegetation normally withstands the damage of locusts, recovering quickly from these or other periodic insect attacks. But when arid soils are irrigated, locust and grasshopper populations may explode suddenly, causing enormous damage. Under these conditions, desert soils no longer have any kind of protective cover, and desertification advances. Locust swarms indeed seem one of the threats resulting from—not causing—desertification. Favorable breeding and swarming conditions are created as more land becomes impoverished. Deforestation and overgrazing also are linked with irruptive locust populations (Rivnay 1964). Under natural conditions, larger herbivores exert little influence on desertification. Gazelles and antelopes, for example, are nomadic, and their grazing pressure is distributed over large areas without lasting environmental damage. However, when these animals are confined in fenced pastures, they soon overgraze the vegetation, and desertification indeed may occur (Cloudsley-Thompson 1977).

Thus, the relatively simple ecological systems involving native animals in desert biomes do not influence desertification unless human modifications assist the process. The logical point for management again is based on soil relationships. Ormerod (1978) stressed that African nations should assess the capacity of their soils to withstand exploitation and thereafter set limits for economic development below which ecological degradation will not occur. As shown later, that concept has equal application elsewhere.

Some 650,000 km^2 of productive land on the Sahara's southern fringe has become desert in this century (Tolba 1979). But desertification plagues the arid regions of the United States, Canada, and Mexico, as well (Table 12-4). Many public and privately owned

lands in Texas, New Mexico, Arizona, Colorado, and other western states have been degraded, including those on Native American reservations and under the jurisdiction of the Bureau of Land Management (Sheridan 1981).

Along with other kinds of habitats, desertification has ruined many riparian communities in the western United States (Ohmart and Anderson 1982). Desert streams, tapped for irrigation, suffer return flows of higher salinities, and their overall flow often is less dependable. Wildlife populations are impoverished as the vegetation bordering desert streams degrades. Johnson et al. (1977) predicted the extirpation of nearly half of the 166 species of birds breeding in a riparian zone as desertification advanced. Similar results are anticipated for the mammalian fauna.

SOME INFLUENCES OF WILDLIFE ON SOILS

Soils are subject to many biological influences. Among these are a range of activities associated with wildlife. Some are obvious, whereas others are more subtle and require the insight of ecological linkages between and among animals, water, vegetation, and soils. For example, the burrows of land-based iguanas (*Conolophus* sp.) in the volcanic ash on the Galapagos Islands influence soil development and eventually form the microhabitat where plants reinvade the barren landscapes created by volcanic eruptions (Hendrix 1981). Similarly, sea lions (Otariidae) transport pebbles long distances—in their stomachs!—thereby introducing new geological materials to beaches far removed from the source of the stones (Fleming 1951).

Burrowing animals are well-known movers of soil. Prairie dogs, gophers, and other rodents move and mix as they tunnel, as do foxes, coyotes (*Canis latrans*), badgers (*Taxidea taxus*), and other denning species. Such excavations may be extensive, particularly for those species that dwell in colonies. Most of

TABLE 12-4. Desertification in North America Measured in Thousands of Hectares

Nations	Irrigated Land			Rangeland			Dry Cropland		
	Total	Desertified	Percent	Total	Desertified	Percent	Total	Desertified	Percent
United States	15,500	1,650	11	235,000	188,000	80	30,000	15,000	50
Canada	300	60	20	10,000	7,000	70	5,000	3,000	60
Mexico	3,750	1,125	30	100,000	96,000	96	7,500	6,700	89

Source: Dregne (1983).

the prairie-dog burrows Smith (1955) examined in Kansas were 1.5 to 2.4 m deep, with about 4.3 m of lateral tunneling, but others reached depths of 4.6 m with 11 m of lateral tunneling. At their entrances, prairie dog burrows are marked by mounds 0.3 to 0.9 m high and from 0.9 to 3.0 m in diameter. Given these statistics, one can envision the amount of soil moved in the Kansas study area where 6344 burrows were counted on 46.5 ha. Besides the protection they offer, burrows have important social features for prairie dogs. The spatial arrangement between burrows essentially remains unchanged from generation to generation so that their location has a stabilizing effect on the colony (King 1955).

More exact estimates are available for the amount of soil moved by pocket gophers (*Geomys bursarius*). Downhower and Hall (1966) calculated that a single gopher transports 2025 kg of soil to the surface each year. In Texas, Buechner (1942) estimated that gophers brought 807 kg to the surface per ha in a tall grass area, but the amount jumped to 15,864 kg/ha on an overgrazed grassland. Nutrients may be deficient in new gopher mounds because subsoil deposited at the surface has been leached as well as tapped by root systems (Spencer et al. 1985). However, northern pocket gophers (*Thomomys talpoides*) accelerated the recovery of the blast-damaged ecosystem at Mount St. Helens by transporting to the surface the nutrient-rich soils buried under a layer of relatively sterile volcanic ash (Anderson 1982; see also Franklin et al. 1985). Furthermore, the excavations of rodents and badgers completely converted the surface soils from silt loam to loam on some sites in Colorado (Thorp 1949). Ground squirrels and gophers may bring 7 to 9 kg of subsoil to each square meter of surface area, for a total of about 67,200 to 89,600 kg/ha (Taylor 1935; Thorp 1949).

Invertebrate animals also move and mix soil. Earthworm casts, which formed in 2- to 3-cm layers on the surface of an undisturbed prairie in Texas, equaled about 24,000 kg/ha (Dyksterhuis and Schmutz 1947). This influence was impaired, however, when the grasslands were disturbed. In Africa, the soil-building properties of earthworms vary with climate; the weight of earthworm casts per ha increased about 50 times during the rainy season compared with a hot dry regime (Evans and Guild 1947). Lunt and Jacobson (1944) found high amounts of nitrogen, phosphorus, and other beneficial materials in earthworm casts; surface soils with casts thus are richer than those lacking casts. Lee (1985) presents a

full review of the role earthworms play in soil formation. Crayfish, mound-building ants and termites, and some beetles also move large amounts of subsoil to the surface (Thorp 1949). McColloch (1926) estimated that 95 percent of all species of insects dwell in soil at some time in their life cycles. Bryson (1931) suggested that all insects burrowing 15 cm or more below the surface effect an interchange of soil. For example, ants in New Mexico transport 2-cm layer of soil to the surface per 100 years, which may influence long-term soil processes (e.g., water-holding capacity) and the composition of plants in semiarid communities (Whitford et al. 1986).

Animal carcasses as well as molted feathers and hair add large amounts of organic matter to soils. Decay of these materials continues largely unnoticed, but their contribution is better realized when the biomass of just two groups of mammals is considered. Taylor (1935) estimated that more than 2.1 million rabbits and rodents occupied 20,200 ha of desert grasslands in Arizona, a biomass of 198,870 kg. This equals the addition of 9.8 kg of organic matter per ha to the soil with each turnover in the rabbit and rodent population. Similarly, Koford (1958) calculated that prairie dog carcasses contribute 12 kg/ha of organic matter to the soil each year, much of it enriching the deeper strata when the animals die in their burrows. Insects, earthworms, and countless microorganisms return even larger amounts of biomass to the soil.

The excreta of both small and large animals represent still additional nutrient and organic matter. Jackrabbit (*Lepus californicus*) droppings averaged more than 89 kg/ha on rangelands, exceeding by about 30 times the weight of the jackrabbit population living on the same area (Vorhies and Taylor 1933). One can only wonder about the tonnages of "chips" produced over the centuries by successive generations of some 60 million bison.

Large amounts of excrement also are deposited on the vegetation and substrate beneath heron colonies. In swamps, the excrement falling into the water is diffused but nonetheless enriches the growth of algae and other forms of aquatic life (Dusi 1979). Similar amounts of excrement falling on upland soils may be toxic. The pH of soils in a plantation in Alabama supporting a large heron colony increased 1.2 times; soil phosphates increased more than 60 times under the influence of the birds' droppings (Dusi 1978). Potash, magnesium, and calcium also increased severalfold. These changes in soil chemistry killed the trees after a single nesting season. When the dead trees fell, the birds moved to nearby trees in the plantation, eventually killing them as well (see also Young 1936). Similar destruction of nesting vegetation from the excreta of a large heron colony in Delaware forced half of the birds to nest at another site the following year (Wiese 1978). At the new site, 60 percent of the shrubs were defoliated and 8 percent were killed after a single nesting season. Toxic levels of soil nutrients and pH in a Minnesota heron colony also affected the underlying vegetation; as nest densities increased and hence more excreta was deposited, fewer plants survived (Weseloh and Brown 1971). Hicks (1979) recorded significant changes in phosphate and nitrate levels in the soils under a roost of more than 250,000 crows (*Corvus brachyrhynchos*), but the vegetation in the roost was not affected by the birds' excreta.

Many rodents store plant materials below the soil's surface. Squirrels (*Sciurus* spp.) bury nuts, and many of these caches remain unused. Denning species deposit immense amounts of vegetation below ground for either food or nest materials, and much of this is incorporated into the mineral soil. A population of 100,000 kangaroo rats (*Dipodomys* spp.) introduces about 180,000 kg of plant materials to the subsoil each year (Taylor 1935). Dwellings above ground also accumulate large amounts of organic matter. Wood rats (*Neotoma* spp.) construct large "houses" of sticks and other plant materials, and when these decompose, the mixture of compost and droppings produces 10 to 20 sacks of rich fertilizer per dwelling (Streator 1930; see also Greene and Murphy 1932; Greene and Reynard 1932).

Water impounded by beaver (*Castor canadensis*) dams gradually deposits a silt overburden on the inundated area, significantly influencing aquatic life in the new pond (see Chapter 11). When beaver dams are removed, a deep accumulation of muck and peat usually is exposed. However, revegetation of these soils may not proceed according to the expected course of succession. Soils submerged for several years are deoxidized and saturated with hydrogen sulfide and other marsh gases. The hydrogen sulfide acts on ferric compounds so that a surplus of soluble ferrous iron develops in these soils. This, in turn, ties up phosphorus in a form unavailable to many higher plants. Furthermore, the toxic nature of hydrogen sulfide and the action of ferrous iron may injure root systems and the soil fungi necessary for tree growth. Wilde et al. (1950) demonstrated that the root systems of seedlings developed poorly on previously flooded soils, with little or no evidence of symbiotic soil fungi. Accordingly, a new forest may be delayed

until these soils are reinoculated gradually with fungi carried by birds, runoff, and other ecological phenomena. Therefore, removal of beaver dams for the purpose of reclaiming forest environments may not achieve the expected results quickly. In areas underlain with limestone or other carbonate bedrock, beavers at times also have helped produce sinkholes by providing more water for the underground drainage system (Cowell 1984).

Hooved animals compact soils. Depending on the number of animals and their size, otherwise permeable soils may become so compacted as to prevent the percolation of water. Most vegetation suffers in such a regime, as does the soil itself. Barren, eroding soils are commonplace around waterholes or salt licks frequented by grazing animals. These circumstances are localized, however, and generally do not represent normal soil conditions for unconfined wildlife populations. Most herds of grazing species are more or less nomadic, so that soil compaction and trampling are minimal as the animals move about.

Only in a few instances have native wildlife populations caused serious soil erosion in the United States. Up to 4000 elk wintering in the Gallatin River Valley of Montana may concentrate in 1800 ha for as long as 5 months during hard winters. They trample and paw the south-facing slopes and wind-bared ridges, damaging the vegetation and compacting the soil. These conditions accelerate soil erosion from high-intensity rainstorms occurring in the following summer. Sediments washed from the winter elk range contribute to the lessening of water quality in the Gallatin River and its tributaries, affecting both irrigation practices and the insect fauna available for trout. Two habitat variables contribute significantly to the severity of erosion: ground cover provided by both growing plants and litter, and soil bulk density, an indicator of soil compaction and percolation. (Slope is also important, but this factor was a constant in the elk studies.) After testing various combinations of treatments, including seeding, a strong relationship emerged between the interaction of cover and bulk density. Soil porosity increased in proportion to the amount of ground cover; consequently, increases in both reduced significantly the amount of erosion. Even under a regime of high-intensity rainfall, management for ground cover of at least 70 percent and soil bulk densities no greater than 1.04 g/cc restored and maintained soil stability on the Gallatin winter elk range (Packer 1963).

Soil erosion is more common in situations where animals have been introduced outside their native ranges. Mountain goats released in Olympic National

Figure 12-7.   Red deer and other exotic animals have destroyed the understory part of the canopy in some forests in New Zealand (top). General abuse, including the presence of exotic animals, produced barren habitats and immense losses of soil on much of New Zealand's mountainous terrain (bottom). (Photos courtesy of W. E. Howard.)

Park paw the soils of delicate plant communities; thereafter, erosion steadily expands the pawings into gaping scars (see Fig. 18-8, Chapter 18). Erosion also is rampant in New Zealand, where burgeoning populations of red deer (*Cervus e. elaphus*) and other ex- otics have destroyed native vegetation (Fig. 12-7). Numerous mammalian herbivores now are established on mountainous terrain characterized by immature, highly erodable soils. With the loss of protective plant cover, New Zealand's high-intensity rainfall has severely damaged large areas and has created substantial instability in the original island communities (Howard 1964).

In Africa, cattle are herded in tight units because of predators. Intensive herding thus concentrates grazing and contributes greatly to soil erosion on African grasslands (Wright 1960a). So enmassed, cattle soon graze plants to the root level and cut the sod with their hooves. Hence, a combination of human culture, grazing, and predation acts to destroy the environment on which all depend. In contrast, the native ungulates spread over the range, distributing their grazing pressure far more evenly and effectively. Predation—or rather the removal of this influence—also affected soil conditions in Itasca Park, Minnesota. When wolves (*Canis lupus*) were eradicated from the park, beaver and deer populations increased beyond the site's natural carrying capacity, destroying certain kinds of vegetation and, in turn, interrupting the soil regime in the park (*fide* Taylor 1935). Both the African and Minnesota episodes illustrate the balance among abiotic and biotic components of ecosystems and their evolution as a single interacting complex.

In a sense, highly specialized "soils" actually may be created by some animals. The most profound instance concerns the accumulations of guano deposited by generations of bats. Caves generally are sterile environments, lacking either sunlight or the organic materials necessary for energy transformation and a life-support system. Bats, however, provide an influence that overcomes these deficiencies. Horst (1972) described an ecological system developing in an energy base of bat guano in Sonora, Mexico. The cavern happened to be an abandoned mine, but the circumstances probably are little different from those in most bat caves (see Poulson and White 1969).

Three species of bats jointly roost in the cavern; of these, two feed on nectar and/or pollen, and the other feeds on insects. All of the bats return to the cavern after feeding, with the resorptive phase of their digestion ending in defecation on the floor of the mine's innermost recesses. With this organic "soil" in place, the foundation of one or more food chains was established. Cave-dwelling cockroaches are the primary consumers in the chain, subsisting entirely on the guano or on moribund bats. The roaches live throughout the cavern but reach densities of more than 100/m² near the guano deposits. Because of seepage from the cavern's interior, a pool of permanent water exists at the mouth of the mine, and here a large number of frogs subsists on the cock-roaches. Whereas sunlight entering the mine's entrance supplies photic energy for aquatic vegetation and also food for the tadpoles, the nutrient base for this vegetation remains directly associated with the guano deposits. The fauna of this miniature ecosystem also includes freshwater cave crabs that feed on dead bats and frogs, but they forage in the guano as well. With this supply of food at hand, animal visitors from outside the cave seek foods from the guano-based resources. Snakes, raccoons, and turtles entered the cave and surely consumed one or more of the prey species available therein. In larger caves, the deposition of bat guano undoubtedly reaches several tons per year. Extractions of guano for fertilizer give an indication of the accumulations possible. About 91 million kg of bat guano had been removed from Carlsbad Caverns by 1928 (*fide* Taylor 1935). Many bat caves also harbor large populations of beetles scavenging on moribund bats, and the heaps of guano beneath roosting sites swarm with these beetles (R. J. Baker, personal communication). In all, guano "soil" serves as a soillike energy source sustaining a system of consumers not otherwise possible in sunless cave environments.

FERTILIZATION

Widespread fertilization of wildlife habitat currently lies beyond the capabilities of most management agencies. Wildlife occasionally may gain some secondary benefits in situations where other considerations are foremost, as when rangelands are fertilized (see Carpenter and Williams 1972). However, costs generally prohibit treatments of large areas when wildlife is the only consideration. But in the long run, management will become more intense on the shrinking reserves of wildlife habitat, undoubtedly bringing greater focus on the feasibility and benefits of soil fertilization. Such thoughts actually were heralded when Mendall and Aldous (1943) proposed that fertilizers may improve woodcock habitat; nutrients applied to depleted soils might increase earthworm numbers, thereby restoring the major food of woodcock on sites selected for intensive management.

Some research already has suggested the usefulness of fertilizers as management tools. Longhurst et al. (1968) noted dramatic results when fertilizers were added to nutrient-deficient soils in California. On plots observed for 4 years, deer consumed an average of 1280 kg/ha of fertilized herbaceous vegetation, whereas only 315 kg/ha of forage was removed on unfertilized sites. In the Black Hills, nitrogen fertilizer nearly doubled the consumption by deer of range grasses, and the fertilized forage also was used earlier in the season and grazed for a longer period of time than unfertilized vegetation (Thomas et al. 1964). McGinnies (1968) also recorded increased utilization of fertilized grass by mule deer. In central Pennsylvania, the soils of oak forests are deficient in several nutrients; deer in this region are smaller than average, and trophy-sized antlers rarely develop. The principal species of browse increased in their nutrient content after nitrogen fertilizers were applied, and at least one food plant received heavier browsing as a result (Wood and Lindzey 1980). Japanese honeysuckle (*Lonicera japonica*), an important food source for white-tailed deer in the southeastern United States, responded to nitrogen fertilizers with higher plant production and increased crude protein content in leaves, but fruit production diminished (Segelquist and Rogers 1975). These results appear significant because crude protein often is deficient in southern forages, perhaps accounting for low productivity and poor physical development in some deer herds. The increased cover produced by fertilized honeysuckle likely improves the habitat of many small birds and mammals as well, although the reduction in fruit may affect some species. Similarly, the crude protein content of wavyleaf oak (*Quercus undulata*) was increased with nitrogen fertilizers, and mule deer used this forage more than unfertilized oak for 2 years after treatment (Anderson et al. 1974). Abell and Gilbert (1974) also reported minimal crude protein and phosphorus in some species of browse utilized by white-tailed deer in Maine. Fertilizers increased the nitrogen content of the plants, but the results were inconclusive for phosphorus.

Likewise, elk selectively sought fertilized forage by pawing through snow that covered treated plots on the Olympic Game Range in Washington (Brown and Mandery 1962). In Oregon, wintering Roosevelt elk (*Cervus elaphus roosevelti*) preferred pastures of perennial ryegrass (*Lolium perenne*) managed by a combination of haying and fertilization. The treatments produced higher quality forage that met the minimal nutritional requirements of the elk herd; this forage should also overcome the previous deficiencies that curtailed the herd's reproductive rates and calf survival (Mereszczak et al. 1981). After a single treatment with fertilizers, pronghorns spent more time and consumed more range forage in winter for the next 3 years, even though the increased nutritional qualities of the plants diminished after a

few months (Barrett 1979). Forage production also was maintained at higher levels for 3 years after the single fertilizer treatment. Whereas the reduction of the protein content somewhat detracts from the usefulness of this management procedure, the continued production of additional forage on the fertilized sites and the preference of pronghorns for the treated vegetation suggests that fertilizers may have merit as a management procedure.

Smaller animals also respond to fertilized soils. White-tailed jackrabbits (*Lepus townsendi*) favored forage on sites fertilized with nitrogen and phosphorus (Johnston et al. 1967), and the fecal matter of red grouse (*Lagopus lagopus scoticus*) accumulated nearly three times more rapidly during winter on fertilized heather than elsewhere, indicating a clear pattern of selective foraging (Miller 1968). On fertilized heather, red grouse also reared larger broods and experienced a fivefold increase in spring densities if the area also was protected from grazing (Watson and O'Hare 1979). As part of a study of food quality and its effects on breeding densities, Ash and Bendell (1979) found that urea and ammonium nitrate each increased the nitrogen content in four important food plants on the summer range of blue grouse (*Dendragapus obscurus*). Rodent damage to Pacific silver fir (*Abies amabilis*) increased after the trees were fertilized with nitrogen, again indicating that the added nutrients increased the palatability of the vegetation (Gessel and Orians 1967). White-fronted geese (*Anser albifrons*) wintering at the Wildfowl Trust showed 42 percent more usage of fertilized plots compared with untreated sites and, when used in combination with mowing, fertilizers increased usage by 87 percent (Owen 1975).

Increased palatability of forage seems the key response to fertilization, although the actual mechanism for this response is not understood fully. Fertilizers undoubtedly enhance the nutritional qualities of forage, but this factor alone does not seem to explain why palatability is affected. Several plants high in nutrients remain unpalatable as forage, indicating that the nutrient content of plants is not always correlated with palatability (Longhurst et al. 1968). Thomas et al. (1964) suggested that fertilizers increase the succulence of plants and hence their palatability. They found that grasses fertilized with nitrogen contained about 90 percent more moisture than unfertilized plants. Additionally, nitrogen fertilizers often induce larger-sized plant cells without proportionately increasing the amount of cell-wall material, thereby enhancing overall succulence with a favorable ratio of water content per cell. Fertilized plants, because of their more rapid growth and earlier maturity, remove water from the soil at higher rates than unfertilized plants, thus effectively using the available soil moisture (McKell et al. 1959; Burzlaff et al. 1968). Jones and Handreck (1967) suggested that nitrogen supplements may reduce the silica content of plants, thereby indirectly increasing the palatability of fertilized plants. Olfaction seems the primary sense used by deer for selecting preferred browse, but it remains doubtful that smell alone enables animals to detect nutritious vegetation (Longhurst et al. 1968). Whatever the mechanisms may be, ample evidence indicates that fertilizers often improve the palatability of forage (Carpenter and Williams 1972).

The responses of wildlife to fertilized vegetation, described earlier, indicate several management possibilities. Fertilization changed the foraging patterns of white-tailed deer in Pennsylvania, indicating that the browsing pressure of the herd might be shifted from site to site following a planned schedule within each management unit (Wood and Lindzey 1980). The migration routes of elk were altered using fertilizers so that crop damage was reduced (Brown and Mandery 1962). Similar possibilities exist for mule deer (Holechek 1982). When used on selected sites, fertilized browse attracts numbers of deer away from overused winter ranges. Nitrogen fertilizers are particularly useful for improving winter browse by simultaneously improving the palatability and nutritional quality of vegetation in underutilized areas. Conversely, fertilizers may be used to attract livestock, thereby relieving key sites for deer from additional foraging pressure. As shown, animals often respond to fertilization with improved reproduction and survival. In several of the studies mentioned, fertilizers increased carrying capacities for wildlife in a variety of habitats, figuratively transforming enriched soil conditions into healthy animal populations.

Toth et al. (1972) reported significant increases in seed production when nitrogen-bearing fertilizers were applied to smartweed (*Polygonum* spp.) communities in New Jersey. One or more species of smartweed occur on most freshwater wetlands in the United States, and their seeds are major foods for many kinds of waterfowl. More seeds were produced, but the nitrogen and crude protein contents of the seeds remained essentially unchanged with fertilization. The costs, based on the additional amounts of seed produced, indicated that fertilizers were a practical means of increasing the carrying capacity of waterfowl habitats vegetated with smartweeds.

In special instances, local conditions supply natural fertilizers not otherwise available. One concerns "fairy rings," the growth of fungal mycelia outward from a central point in an ever-expanding circle (Rogers and McAllister 1969). Grasses growing at the circumference of the circle are enriched when the older fungae die and release nitrogen into the soil. Indeed, soils collected on the perimeter of a 30-m fairy ring in South Africa were far richer in nitrogen than those either inside or outside of the perimeter. A circular band of conspicuously greener vegetation thereby developed on the circumference compared to the grasses either inside or outside the ring, even though the species of grass might have been the same throughout. Such luxuriance attracted grazing animals, presumably because of the increased palatability of the enriched vegetation.

SUMMARY

Soils are a fundamental component of ecosystems and directly or indirectly influence the distribution, abundance, and quality of biological communities. Wildlife, like other products of the land, fares in proportion to the capabilities and limitations of soil fertility. Vegetation usually serves as the link between soil and animal associations, but there also are direct soil-animal interactions. Some species of wildlife are adapted to certain kinds of soils. Soil texture can limit the distribution of tunneling animals, and soil nutrients can affect the welfare of animal populations as shown by natural situations and by experiments with fertilizers. Wildlife, in turn, can influence soil conditions by adding organic matter, by mixing soil, and sometimes by altering vegetation or enhancing erosion. Both wildlife and human populations have been impoverished by soil depletion. Ill-advised land management of tropical forests and arid lands provides striking examples of soil depletion and hence diminished capacities for continuing biological productivity.

CHAPTER 13

WILDLIFE AND FARMLANDS

> *The earth is given as a common stock for man to labor and live on....The small landowners are the most precious part of a state.*
>
> **Thomas Jefferson (1785)**

The Jeffersonian ideal of an agrarian America lasted little more than a century after his death in 1826. Expanding industry and urbanization steadily changed the economic landscape following the Civil War. For small landowners, the tempo of change peaked at the end of World War II. The family farm and a way of life began to vanish. Today, agribusiness has replaced the historical image typically associated with tillage of the land (Thomas et al. 1982). Farmers are now producers who represent only one segment of a larger complex that includes manufacturers and suppliers of agricultural equipment and services, processors and distributors of farm products, and technical and financial assistance. Nonetheless, farms and farming—the bond between humans and soil—have made a lasting imprint across America and the world. We shall look at that imprint in this chapter from the viewpoint of wildlife management. As we shall see, a shelterbelt can reduce erosion on a windswept farm while also serving as cover for wildlife.

AGRICULTURE: A BRIEF HISTORY

Agriculture has experienced four revolutions (Laetsch 1979). The first of these started about 10,000 years ago when wild plants were domesticated. This happened worlds apart—in northern China, in the central valley of Mexico, in Iraq, and in the Andes of South America. Millet, corn, wheat, and potatoes have been the modern results, but many others have been added. With the domestication of food plants came the birth of human civilization (but see Diamond 1987). People were freed from the tenuous existence of gathering nuts and berries or hunting for animals. They instead became wedded to the land, initiating not only farming but also permanent settlements and an array of social developments. In the course of these events, several kinds of wildlife also evolved a measure of dependence on farming. Some of those associations created management challenges that still lack full solutions.

The second revolution was the product of exploration and discovery. This era waited until the 15th century, when Columbus and other adventurers set forth and found new lands. Later, Magellan, Cabot, Pizzaro, and others began to uncover the character of the Earth, its wealth, and its people. Sea routes and trade soon developed and with these came a worldwide exchange of agricultural commodities. Corn and potatoes, long cultivated in the New World, were introduced into Europe. Coffee, spices, and sugar, among others, commanded trade among far-flung nations. Today, international commerce in agriculture has reached major proportions, sometimes with indirect bearing on wildlife management.

The Industrial Revolution spawned the third agricultural revolution. Steam power, fossil fuels, and mechanization initiated huge increases in crop production. Cotton gins, reapers, and other equipment replaced arduous human and animal labor. Railroads carried farm products to far-off markets. Farm mechanization continued, with the tractor and combine becoming commonplace after World War II. We shall look at several of the ways that mechanization has affected, directly or indirectly, the status of farm-related wildlife.

Technology founded the fourth revolution. Scientific progress, particularly in chemistry and genetics, brought forth herbicides, fertilizers, and insecticides, along with new varieties of crops. Plant and animal genetics improved on nature. Hybrid crops and livestock were matched to the Earth's soil and climatic regimes, replacing less-productive types and again increasing yields of food and fiber to new highs. Human populations spiraled during this same period. The world population numbered about 3 billion in 1960 but jumped to 5.7 billion by 1994—and 9 out of 10 births occur in developing countries. As might be expected, the race between agricultural technology and a rapidly growing human population profoundly influences wildlife management.

WHAT'S HAPPENED TO FARMS AND FARMLAND?

Populations of many game and nongame animals rise and fall with the ebb and flow of farming. Their fates go hand in hand with the dynamics of tillage, crops, economic patterns, and, indeed, the very existence of agriculture. Historically, the initial conversion of forest and grassland to farmland brought with it many changes in numbers and kinds of animals as a direct result of the gross alterations of those environments. However, the original conversion was not always a negative influence on wildlife habitat. Small farms cleared from eastern forests offered more "edge" and often a supply of emergency foods. Cottontails (*Sylvilagus floridanus*), bobwhite (*Colinus virginianus*), and later, pheasants (*Phasianus colchicus*) generally thrived on farms during an era of axe, hand labor, and horse-drawn plow. Later, when small farms carved from woodlands were abandoned during periods of economic troubles, the tilled land gradually reverted to shrubs and other vegetation favoring white-tailed deer (*Odocoileus virginianus*).

Large numbers of farms were abandoned during the hard years of the 1930s, but even later, abandonment continued at an accelerated pace (Burger 1978). Deer populations in eastern states grew in response to the newly available habitat. In New York State alone, at least 100,000 ha of farmland were abandoned between 1950 and 1960. However, the carrying capacity for deer presumably will lessen as succession progresses toward a mature forest. One sees evidence of such changes where stone fences, once separating farms and fields, extend through land now revegetated with developing forests (Fig. 13-1).

Mechanization brought more rapid alterations and a new era of farm production. In 20th-century America, vast tracts of land have been cleared of native vegetation and broken to the plow. Powerful tractors and other machinery have contributed to farm efficiency to the point that less than 4 percent of the workforce now directly produces the food and fiber for the American population, and the size of farms has increased 200 percent from an average of 61 ha in 1875 to 183 ha by 1980 (Thomas et al. 1982). Under this regime, an agricultural monoculture generally has replaced diversity. The result is often "clean farming." Vance (1976) aptly demonstrated the decline in farm-related wildlife populations during the period 1939–74, as clean farming swept across Illinois. Hayfields were replaced by soybeans and other cash crops, and the average size of each field more than doubled. These changes not only furthered a monoculture but also largely eliminated the beneficial edges provided by fencerows between the once-smaller fields (Fig. 13-2). An interspersion index, measuring the availability of edge, dropped from 351 to 115 as a result of large fields replacing smaller units without the intervening fencerows. In all, grasslands and bushy fencerows were reduced by 84 percent between 1939 and 1974. In the process, greater prairie chickens (*Tympanuchus cupido*) were eliminated, and bobwhite and cottontail populations were reduced by 78 percent and 96 percent, respectively. Taylor et al. (1978) likewise recorded reduced densities of pheasants as land-use patterns in Nebraska changed toward clean farming. In Iowa, where 94 percent of the land is farmed, fencerows with a continuous row of trees and shrubs provided habitat for up to 36 species of birds per 10 km, whereas fencerows lacking woody plants contained no more than 9 species of birds in the same distance (Best 1983). Unfortunately, the species-rich fencerows of woody vegetation are being removed at a faster rate than those having only herbaceous vegetation (Vance

Figure 13-1. A stone fence, once separating fields on a Vermont farm, today stands without purpose in a regrowth of hardwood forest. The setting reflects regional changes in agricultural patterns during the 19th century. Wildlife communities undergo corresponding changes in abundance and composition as abandoned farmland reverts to mature forest. (Photo courtesy of R. W. Fuller.)

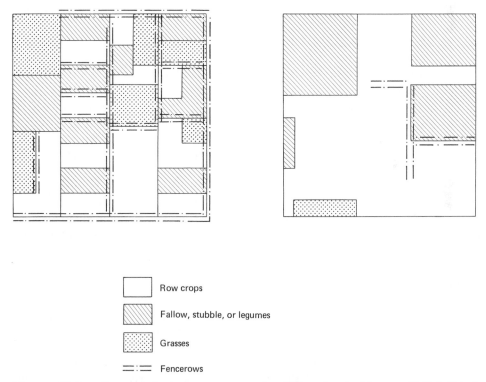

Figure 13-2. Changes in land use on a square mile of Illinois farmland between 1939 (left) and 1974 (right). Compare field sizes, diversity, and the amount of fencerows. Major differences in this example included more than a sevenfold increase in soybeans and virtual elimination of hayfields, illustrating the overall trend toward monocultures and clean farming. (From Vance 1976.)

1976). Fencerow width provided the most important variable for predicting the abundance and diversity of birds nesting in fencerows in Michigan. Accordingly, Shalaway (1985) recommended that fencerows be at least 3 m wide and include a few snags for cavity-nesting birds. In most areas, an unplowed strip along a fence will produce good cover in 1 year and small shrubs in 2–6 years.

Southwood (1972) described similar trends in Britain. Increased field sizes, monocultures, and the removal of hedgerows interacted with animal populations of several kinds. In particular, major reductions of Britain's renowned hedgerows seem associated with faunal disruptions, although these relationships are complex. Nonetheless, hedgerows produced 2.4 to 4.7 nests of gray partridges (*Perdix perdix*) per km, whereas an average of only 0.9 nests was found in the same distance where wire fences separated fields. Forman and Baudry (1984) concluded that hedgerows also function as corridors by which birds and other species move across agricultural landscapes, again noting that the removal of hedgerows continues in Europe and the United States.

A national inventory of the 39 states within the range of bobwhites found that clean farming was the major cause of declining quail populations during the decade spanning World War II (Goodrum 1949). Similarly, Leedy and Dustman (1947) correlated a serious loss of nesting habitat for pheasants in Ohio with a 24 percent decline in hayfields and an increase in areas devoted to row crops. This switch in land use contributed significantly to declines of one-third of the pheasant population between pre- and postwar years. In Iowa and Minnesota, clean farming and changing land-use patterns were more related to pheasant numbers than to the length of the hunting season (George et al. 1980). In Illinois, changes in crops after World War II apparently subjected pheasant chicks to greater mortality, as shown by declining brood sizes between 1946 and 1981 (Warner 1984; Warner et al. 1984). Brood size was correlated positively with plantings of hay and small grains but correlated negatively with corn and soybeans. Intensive farming has eliminated nesting cover in important breeding areas for waterfowl, suggesting low production of ducks in the years ahead unless farmers are offered incentives for maintaining wildlife cover (Higgins 1977; Cowardin and Johnson 1979; Sugden and Beyersbergen 1984). Barn owls (*Tyto alba*) increased in Ohio as agriculture developed, but the subsequent conversion of grass-associated agriculture to intensive row-crop farming diminished the availability of foraging habitat, and the owl population thereafter decreased steadily (Colvin 1985).

The loss of fencerows and other habitat to intensive agriculture also contributed to marked reductions in songbird populations (Graber and Graber 1963). Hedgerows planted in Nebraska in the early 20th century probably contributed to the increase in range and abundance of woodchucks (*Marmota monax*), but their abundance decreased in the 1950s and 1960s when many hedgerows were removed (Jones 1964; see also Choate and Reed 1986). Allen et al. (1982) found that cottontails preferred brush habitat irrespective of the intensity of farming, suggesting that brush cover must be restored where clean farming is practiced if cottontail habitat is desired. An increase in irrigation marked the shift from small farms to corporate monocultures in central Wisconsin, but white-tailed deer did not adjust well to these changes. Such findings suggest that the distribution and amount of winter range for deer may be restricted with the continued spread of irrigated agriculture in the Midwest (Murphy et al. 1985).

The switch from horse- to tractor-drawn machinery produced other effects on wildlife besides clean farming. Fields once allocated to horses as pastures or for their food production were no longer needed, and with this change came even less diversity in the farm landscape. Tractors presented still other difficulties: with their increased speed, power-driven machines began killing wildlife. Leedy and Dustman (1947) found that mowing equipment increased pheasant mortality by 60 percent. Nighttime mowing, made possible by headlights, heightened mortality fivefold over the daytime rate. Hens and juvenile pheasants especially were vulnerable to power mowers. Cottontails also were killed by mowing equipment. On the other hand, pheasant mortality in wheat fields decreased as combines replaced binders. Wheat is harvested about 10 days later when combines are used, and more nesting pheasants escape death with the additional time to complete incubation. Also, cutting bars on combines are higher than those on binders and hence pass over nesting pheasants more readily. Dabbling ducks nesting in hayfields experience relatively low vulnerability to mowing equipment because they generally rise swiftly and nearly vertically from their nests when flushed by mowing machines and hence avoid the cutting bar (Labisky 1957).

Coincident with these new uses of the land came other technologies supporting agricultural development. Water was managed with irrigation schemes of many kinds, and potent fertilizers and pesticides enhanced crop production. Today, U. S. agriculture dominates world production, but its status is dynamic and remains subject to a range of economic restraints and political manipulations (e.g., world trade, price supports, and embargoes).

In the course of managing agriculture production in the United States, several programs were designated to reduce surpluses of wheat and other crops by removing land from production (Harmon and Nelson 1973). The Agricultural Act of 1956, popularly known as the Soil Bank, was among the more influential of these in terms of farm management and wildlife. The Soil Bank allowed farmers to enter into 5- or 10-year agreements for retiring land contributing to crop surpluses. A major provision of these agreements was that the idled land must be protected by adequate plant cover. Farmers could enroll until 1960, and, except in a few instances, the agreements expired by 1969. The response of pheasant populations in South Dakota illustrates what happened throughout much of the bird's range. At the Soil Bank's peak, nearly 725,000 ha of land lay uncropped in South Dakota—an excellent cover for nesting pheasants. The pheasant population nearly doubled. Dahlgren (1967) reported that the state's pheasant population increased from 4–6 million to 8–11 million birds during the Soil Bank era. Similar increases in pheasant numbers occurred in other states as a direct result of the program (Schrader 1960; Bartmann 1969; Erickson and Wiebe 1973).

After its expiration, the Soil Bank was replaced with other incentives for controlling surplus farm production. Unfortunately, the new programs did not always require protective cover for the idled lands. Pheasants thus benefited only when farmers provided adequate cover and suffered when they did not. In Wisconsin, Gates and Ostrom (1966) found that just over 4 percent of their study area was retired from grain production and, of this, only half was covered with habitat suitable for pheasant nesting. Nonetheless, at least a 10-percent increase in pheasant production was attributed to this area alone. Joselyn and Warnock (1964) calculated that up to 39 percent of the pheasant nests on Illinois farms were located in cover that would have been in crops ill-suited for nesting had an idle-soil program not been in force.

The Food and Agriculture Act of 1965 established the Cropland Adjustment Program (CAP) as an-

other means of offsetting surplus farm production. Farmers agreed to plant grasses and legumes instead of crops for periods of 5 or 10 years. Some 16 million ha, or an area equivalent to all of Ohio and half of Pennsylvania, were eligible under terms of the program (Jaenke 1966). The dense cover produced by CAP yielded significant results. Duebbert (1969) reported 61 duck nests of 7 species on about 50 ha of CAP land in South Dakota, far in excess of the nest densities in nearby fields still in production. Moreover, the hatching success was an astounding 79 percent, compared to only 30 percent on nearby farmed lands. In all, CAP lands produced about 8 ducklings per ha (see Duebbert and Lokemoen 1976).

The tangle of ground litter formed by the vegetation on CAP lands served two major functions. First, it was desirable as nesting cover. Mallards (*Anas platyrhynchos*), in particular, seemed attracted to this cover as they nested there in greater proportions than elsewhere. Second, the dense cover offered ideal concealment so that predation was reduced and more nests hatched (Duebbert 1969). Subsequent research on CAP lands also determined that the benefits of prime nesting cover exceeded those of predator control. CAP lands with no predator control produced six times the number of ducklings per ha than did habitat of lesser quality where predators were reduced (Duebbert and Kantrud 1974). Because estimates suggest that only 2 percent of all waterfowl production originates on federal and state lands, privately owned lands clearly remain the mainstay of waterfowl production (Hochbaum and Bossenmaier 1965). The results achieved on CAP lands suggest the potential that good cover may afford waterfowl nesting on private land (see Livezey 1981 for the management of retired cropland on public refuges).

In 1985, Congress passed the Food Security Act (the "Farm Bill"), which offers strong remedies for the ills of crop surpluses and soil erosion. The measure also provided important benefits for wildlife. In one segment of the program, known as the Conservation Reserve Program (CRP), farmers voluntarily remove highly erodible land from production for 10 years under an agreement with the U.S. Department of Agriculture. The agreement carries obligations for both parties. Farmers must establish and maintain permanent cover on the CRP land covered by the agreement; the cover may include trees, shrubs, forbs, legumes, and native or introduced grasses, but the vegetation cannot be harvested commercially; that is, mowing or grazing is prohibited during the

10-year period, but hunting is permitted according to the wishes of the farmer. In some cases, water-control improvements that help maintain the cover crops or manage small wetlands for wildlife may supplement the program. Shelterbelts are another choice available in the list of approved practices. For its part, the government guarantees the landowner a rental payment annually for 10 years and shares up to 50 percent of the cost of establishing the cover. The rental payments are established when the farmers offer a bid that is accepted by the government; bids averaging nearly $100 per ha were approved for payment in the first year of the program. CRP clearly offers new hope for the agricultural community, for soil and water conservation, and for wildlife management, particularly if the cover crops are targeted for the needs of selected species (e.g., nesting cover for pheasants and waterfowl; see George et al. 1979; Duebbert et al. 1981; Klett et al. 1984). More than 12.6 million ha of highly erodible croplands were taken out of production and placed in CRP by 1992, mostly in the Great Plains states (Kurzejeski et al. 1992).

Changes are likely to occur; contracts for CRP lands began to expire in 1995. Kurzejeski et al. (1992) examined the attitudes of Missouri farmers about both the current program and their future involvement with CRP. About 12 percent of the farmers claimed wildlife habitat as the primary reason for enrolling land in the program, although 37 percent listed wildlife as having some importance in their decision. Other reasons included concerns about soil erosion, realization of more profit than otherwise possible, and personal retirement. However, 57 percent of the landowners were not aware of the option for wildlife management—a choice that includes planting vegetation especially for wildlife food and cover. As for the future, only 6 percent of the farmers said they would retain the present vegetative conditions if the CRP was discontinued in 1995, but 55 percent would renew their contracts if the program were to continue for another 5 years. The study also noted a lesson from history: When lands held idle in the Soil Bank returned to crop production in the 1970s, upland game-bird populations thereafter declined. So to avoid repeating these circumstances, landowners interested in wildlife conservation should be identified and encouraged to continue participating in CRP.

Other segments of the Food Security Act also are of major importance to conservation. Two of these are known popularly as the "Swampbuster" and the "Sodbuster" provisions. Farmers lose whatever benefits for federal subsidies (e.g., price supports, crop insurance, and low-interest loans) they may have or wish to have, if they drain wetlands or plow previously untilled land that is highly erodible. Both restrictions prevent marginally suited land from coming into production and furthering larger crop surpluses; in doing so, both measures also preserve wildlife habitat. The Swampbuster provision, by stemming the tide of wetland destruction for additional crop production, is a much-needed improvement in national farm policy. Yet another segment of the act enables the Secretary of Agriculture to acquire and retain easements in wetlands, uplands, or highly erodible lands for conservation, recreation, and wildlife purposes for periods of at least 50 years. In return for granting the long-term easements, farmers with land held in security against loans from the Farmers Home Administration (FmHA) have part of their debts canceled. (In 1987, the link between FmHA and the U.S. Fish and Wildlife Service was strengthened further: Future loan applications will be assessed with regard for wetland protection. Moreover, when lands held in the FmHA inventory are sold to private interests, the deeds will be restricted in ways that will preserve or restore wetlands on the properties.) Thus, with enactment of the Food Security Act of 1985, the federal government served clear notice that it no longer will be a partner to the conversion of wetlands or highly erodible lands into cropland. It also took a bold step in furthering the conservation of soil, water, and wildlife resources.

FARM CROPS AS WILDLIFE FOOD

Farms provide sources of food for many kinds of wildlife. In some cases, little or no harm accrues to either crops or wildlife. In others, crop depredations are severe and pose difficult situations for wildlife managers. As we shall see, crops can expose wildlife to harmful pesticides and, on occasion, to at least one infectious disease.

Bobwhite adapt well to most forms of agriculture. They consume many kinds of foods, among them large numbers of seeds from both wild and cultivated vegetation. One biologist estimated that more than 5 million seeds are consumed each year by a single bobwhite (*fide* Rosene 1969). The variety of these foods is no less amazing. Among the many

On The Job With...
Melissa "Missy" Sparrow
Wildlife Biologist

Missy Sparrow has capped her undergraduate degree with a M.S. degree, completing a thesis on woodcock populations. She worked as a seasonal biologist at the Northern Prairie Wildlife Research Center in Jamestown, North Dakota, before being hired as a Wildlife Biologist with the Wisconsin Department of Natural Resources. Missy specializes in assisting private landowners and spends most of her time helping citizens apply for state funds to share the cost of restoring wetland and grassland habitats on their property. She also helps citizens establish breeding populations of ring-necked pheasants on privately owned land.

On a typical day, Missy will visit with landowners to determine whether they qualify for one of Wisconsin's cost-sharing programs for habitat restoration. That done, she might continue working on budgets and work plans for the next 2 years and begin to prepare a grant proposal for funding grassland projects for the following year. Sales of pheasant stamps provide the pool of grant money, but like scientists everywhere, Missy has to apply for the funds with a detailed proposal. Her workday often ends with a meeting with her field employees to review the progress of wetland- or grassland-restoration projects in her district. When she has time, Missy might join others conducting a prescribed burn at a grassland site.

Wetland restoration is especially rewarding for Missy. Sites are often revitalized simply by destroying drainage pipes and by plugging ditches, but others require greater efforts. Either way, Missy enjoys the more-or-less instant results produced by her program. She once saw several ducks and a pair of geese at a wetland where the restoration work was completed only the day before—and she admits that the experience sent her heart into overtime. For Missy, these special moments overshadow the frustrations of dealing with overwhelming paperwork and the pressure from politically powerful people to undo or reverse her efforts to increase wildlife habitat.

Missy's advice to students who aspire to become wildlife biologists: Be willing to work for years at seasonal jobs, which may help you get a permanent job. "Be willing to work anywhere," she advises, and accept duties that may be "something you don't really want to do." Another of Missy's observations concerns an important difference between research and management positions. "They are," she says, "two different worlds." Wildlife managers typically deal far more with the public; therefore, these positions are, in her words, "becoming more oriented toward management of people than of wild animals."

She points out that a M.S. degree is "a good idea if you want to advance in your career," although she adds that even an advanced degree won't lead to financial success, especially if you are stationed in an area with a high cost of living. A "second income in a family might be advised in order to lead a comfortable life."

studies of quail foods is a list of 302 species of plants (Korschgen 1948), and even lengthier lists have resulted from more extensive investigations (Stoddard 1931).

Virtually all studies of bobwhite food habits show the influence of crops, although there are regional differences. Preference trials indicate that bobwhites generally prefer the seeds of grasses over those of legumes, and especially those of annual grasses (Michael and Beckwith 1955). In the northern range of bobwhite, corn is common, whereas sorghum, wheat, rye, and several legumes predominate elsewhere. Crops are especially important in the winter diet, as they offer bobwhites nourishment during a season of when natural foods may be scarce.

Therefore, farming provides critical food resources for a major species of game bird. Bobwhite prospered under most agricultural regimes until clean farming started to claim crucial habitat (Goodrum 1949). Our point here, however, is that populations of an important game bird often were sustained by crops without inflicting any damage. The seeds they consumed were the waste of crop production and therefore do not represent losses to farm income. Edminster (1954) aptly concluded that, of all the game birds, bobwhites probably enjoy the highest esteem of farmers.

Waste corn and small grains also become a mainstay in the fall and winter diets of pheasants. However, pheasants sometimes eat newly sprouted corn or peck at ripe tomatoes, causing localized damage to these crops. Dambach and Leedy (1948) witnessed 3 pheasants digging up 19 corn sprouts in 20 minutes during a survey of 139 Ohio cornfields. They detected some damage of this type in 12 percent of their sample, but extensive losses were limited to 4 cornfields bordering long-established pheasant refuges. In all, the results suggested that damage to freshly planted corn is negligible even in the best pheasant-producing areas in Ohio. However, damage to sprouting corn may be more troublesome in states with larger pheasant populations. If so, repellents may be necessary. A carbamate insecticide tested in South Dakota proved effective for this purpose (West et al. 1969). Unprotected fields lost as much as 33 times more corn sprouts than fields treated with the repellent.

In general, however, pheasant depredations seem exaggerated and the bird's food habits remain compatible with most farming operations (Edminster 1954). The same can be said of prairie chickens and mourning doves (*Zenaida macroura*). However, farmers often experience other problems indirectly related to their crops. Attractive populations of game birds thriving on farmlands may encourage unauthorized trespass and other discourtesies by thoughtless hunters.

The situation regarding bobwhites and pheasants contrasts with the depredations by deer, raccoons (*Procyon lotor*), and several kinds of rodents and birds. Javelina (*Tayassu tajacu*) and coyotes (*Canis latrans*) are known to damage watermelons and other succulent crops. However, losses of unharvested wheat and other cereal grains to waterfowl in Canada remain the most severe example of crop depredation (see Gillespie 1985a about this problem in New Zealand). These losses occur on the northern prairies when waterfowl begin to concentrate prior to their autumn migration—just at the time of harvest. Unlike damage from insects or hail, crop depredations from waterfowl assume a different perspective among farmers. Hail is an "Act of God," and insects are a hazard triggering remedies clearly aimed at pests. But ducks are not pests. They are instead a natural resource rigorously protected by international treaty. Farmers accordingly seek responses to their plight from wildlife managers (Stephen 1975).

The first reports of serious depredations began in the 1940s when mechanization brought major changes in farming practices (Bossenmaier and Marshall 1958). Initiation of new methods of harvesting grain immeasurably exacerbated the possibilities for waterfowl damage (MacLennan 1978). Wheat and other small grains are cut in strips and are left to dry in windrows. With dry weather, 8–10 days of ripening are required before the grains are harvested, and during this period crops are susceptible to waterfowl depredations. Losses include both consumption of the grain and trampling damage (Stephen 1975). Jordan (1953) found that mallards, on the average, consume about 73 g of small grains daily during the autumn, but this amount increased to 82 g under the demands of cooler temperatures. Other ducks and geese consume proportionately smaller or larger amounts, respectively, according to their body sizes. Other factors affect waterfowl depredations. Among these are the size of the waterfowl populations, field size, proximity of the crops to water, and weather. In years when crops require longer periods to dry, their exposure to field-feeding increases, and so do depredations. In a short harvest period, losses might reach 7 million kg, but in a longer harvest season, this figure could jump to about 380 million kg (MacLennan 1978).

Colls (1951) summarized the dilemma. On the one hand, farming destroys waterfowl habitat when wetlands are converted into croplands. On the other hand, waterfowl destroy crops. Furthermore, as waterfowl habitat becomes more restricted, there are greater chances of depredation in fields near the remaining wetlands. This, in turn, prompts more endeavors to restrict waterfowl habitat. Waterfowl themselves are "poor ambassadors" for wetland preservation in grain-producing regions (Leitch 1951). Farmers, when asked to help preserve valuable wetland habitat, sometimes do not cooperate because of the damage ducks cause to crops. Unfortunately, these confrontations still continue (Miller 1976). MacLennan (1978) estimated the value of a single year's destruction of grain at $30–$40 million in Canada as an extreme—but real—loss and suggested that $12–$15 million represented a long-term average. If the average loss were evenly distributed, each farmer would lose about $100. But that is not the case. A small percentage of the farmers actually bear the majority of the depredations, and these individuals each may lose $2000–$5000 each year. In Saskatchewan, where about one-third of Canada's duck population nests, more that 35 percent of the depredations occur on less that 1 percent of the area farmed (Atkinson 1975). As suggested earlier, these croplands usually are located near the best waterfowl habitat.

Management solutions vary in their nature and effectiveness. Tactics designed to scare field-feeding waterfowl often are only temporarily effective and, at best, merely transfer the problem to another field. Specially designated fields are left in "lure crops" under the sponsorship of government; here the birds are neither hunted nor scared so that the damage is deflected from other fields during the harvest (Fig. 13-3). In comparison with wheat, barley may be the best lure crop because it matures faster, costs less, and can be eaten more efficiently (Clark et al. 1986). Bait stations also are deployed to attract ducks away from crops. Leitch (1951) noted that development of some wetland seed plants helped reduce crop depredations in Alberta. Overall, however, the seasonal abundance and availability of wheat and other grains simply prove too attractive for large flocks of foraging waterfowl to resist, and attractive marsh vegetation often remains unutilized when crops mature (Horn 1949; Bossenmaier and Marshall 1958). Although the long-term solution to crop depredations presumably involves some type of habitat modification—either in the fields or in the wetlands—managers are far from reaching this goal. Combinations of scaring tactics, feeding stations, and

lure crops seem the best management techniques currently available. These, plus government-sponsored insurance programs, offer a degree of relief for farmers suffering severe losses. Ennis (1981) suggested that prevention of depredation is cost effective; every $1 invested in prevention saves $4 of compensation to farmers experiencing crop damage. Funds obviously are needed to develop new ways of diverting waterfowl from unharvested fields. Also needed are new farming practices that would make crops less susceptible to depredation.

Waterfowl depredations also assume dimensions lying somewhat beyond the scope of basic wildlife management. Wheat and other grains produced in North America have become world commodities. Sales represent not only income for farmers, but also a means for governments to offset their national trade deficits. National priorities accordingly are geared for more production. For better or worse, these sales also may be used as "chips" in international politics, particularly in deals with hungry nations whose socioeconomic systems may differ from those in North America. Some persons also have raised the question about who should have the fundamental responsibility for the food security of the world (Arkinson 1975). North Americans currently produce a surplus of grain and other foods at a time

Figure 13-3. Depredation management with a lure crop on the prairies of North Dakota. The smaller print on the sign (insert) reads, "The purpose of this lure crop is to allow waterfowl to feed, thereby preventing damage to other crops in the area." (Photo courtesy of S. D. Fairaizl.)

when other peoples face starvation. In short, are we obligated to feed present and future human populations or to nurture ducks? This issue may force a moral dimension on our land-management decisions, perhaps excluding wildlife values as an unconscionable luxury. Finally, who pays for crop depredations caused by waterfowl? Canadian farmers suffer the financial losses; yet U.S. hunters enjoy a harvest of more than 70 percent of the ducks and geese produced in Canada. Under these circumstances, should waterfowl hunters in the United States contribute to Canada's burden? Canadian farmers might just as easily produce fewer ducks and still meet the needs of Canadian hunters without experiencing severe crop damage. In other words, why should Canadian farmers produce waterfowl for both Canadian *and* U.S. duck hunters? There are no easy answers to these issues, and none has become established policy, but responses surely will be needed before long.

Blackbird is a collective term often employed for one or more species of dark-colored birds that flock to croplands. Grackles (*Quiscalus quiscula*), cowbirds (*Molothrus ater*), and red-winged blackbirds (*Agelaius phoeniceus*) are lumped conveniently in this group to which we tenuously can add crows (*Corvus brachyrhynchos*) and starlings (*Sturnus vulgaris*) for purposes of our discussion. These birds damage regionally important crops, among them mature and sprouting corn and rice. A survey of 24 states producing 97 percent of the corn crop in the United States indicated that more than 16 million bushels, or 0.16 percent of the total harvest, are lost because of birds (Stone et al. 1972). In terms of 1970 dollars, this damage amounted to $9.3 million, with Pennsylvania and Indiana experiencing losses of $1 million or more each, closely followed by Ohio, Illinois, Wisconsin, and Minnesota (see also Wakeley and Mitchell 1981). Such damage varies greatly, however. Farmers in some areas experience heavy losses, whereas others have little or no damage. Overall, White et al. (1985) found that blackbirds, even in numbers of a million or more, caused only minor economic impact and suggested that the strategy of reducing roosting populations is not an effective means of controlling agricultural damage. Indeed, many of the earlier methods for controlling bird damage proved relatively ineffective (Fig. 13-4). Regrettably, farmers sometimes have turned to toxic materials instead of nonlethal repellents (see Stone et al. 1984).

Figure 13-4. Various kinds of noisemakers, such as the propane-operated exploder shown here, have been used to scare blackbirds and other wildlife from fields. Such means often lose their effectiveness when the animals become habituated to the disturbance and continue feeding on crops. (Photo courtesy of U.S. Fish and Wildlife Service.)

Some nonlethal chemicals repel animals because the materials produce an objectionable taste or odor. Other materials produce a behavior known as aversion conditioning when chemically treated food makes animals sick. Animals thereafter avoid the same food—even when it is no longer treated—because of the unpleasant association (Rogers 1974, 1978). Because they are nonlethal, these materials are acceptable to conservationists and yet help to repel troublesome birds from corn and other crops (Schafer and Brunton 1971). Corn seed, rice, berries, and small cereal grains have been protected using methiocarb (Guarino 1972; Besser 1973; Stone et al. 1974; Conover 1982), although it seems less effective and more costly when applied to ripening field corn (Joyner et al. 1980). Methiocarb limited blackbird damage to sprouting corn to 0.3 percent, whereas untreated fields nearby experienced 44 percent damage (Stickley and Guarino 1972). The nature of aversion conditioning is such that only part of a field may require treatment with a chemical repellent—methiocarb, in this case—yet produces the same effectiveness against birds as treatment over the entire area (Avery 1985). Aversion conditioning using methiocarb effectively repelled Canada geese (*Branta canadensis*) from golf courses: It also might be useful for protecting winter wheat or other crops from geese (Conover 1985; see also Kahl and Sampson 1984).

Another compound, 4-AP (4-aminopyridine, or Avitrol), combines lethal and nonlethal approaches to blackbird depredations. This chemical does kill some birds, but before they die, they exhibit squawking, erratic flight, and other irregular behaviors that frighten unpoisoned birds from the croplands. Fields treated with 4-AP distributed at bait stations experienced 56 percent less damage than untreated fields (Stickley et al. 1972). De Grazio et al. (1972) modified the bait-station method by spreading cracked corn treated with the same chemical throughout fields damaged by blackbirds. This procedure decreases the chances for pheasants to consume hazardous amounts of the baits as they might at a feeding station. Overall, 4-AP lessened damage by 85 percent, and flocks of blackbirds were repelled after no more than 1 percent of the population consumed the baits. However, note that Dolbeer (1981) suggested that the costs of blackbird control often may exceed the value of lost corn crops. He accordingly proposed an economic model for determining cost-efficient management decisions when blackbirds threaten cornfields (see also Somers et al. 1981). Recently, the performance of 4-AP has been improved by the addition of a stabilizing ingredient, and favorable cost:benefit ratios now may be obtained under most conditions (Besser and Hanson 1985).

Reflecting tapes present a new means of repelling blackbirds from crops (Dolbeer et al. 1986a). The tapes are 11 mm wide, red- and silver-colored on opposite sides, and flash in sunlight. Under certain wind conditions, the tapes also produce a roaring noise. The tapes are suspended across fields at intervals of 3 to 7 m. In Ohio, only 3.2 percent of the ears in cornfields taped at 3-m intervals were damaged, compared with 6.3 percent damage where tapes were spaced at 7-m intervals. Damage in fields with no tape amounted to 17.2 percent of the ears. The results were similar for millet and other crops and included field trials conducted in Bangladesh, India, and other developing nations (Bruggers et al. 1986). Reflecting tapes seem especially useful on small fields of high-value crops but also may deter birds from visiting polluted or toxic sites or from roosting in urban or residential area (see also Blokpoel and Tessier 1984). Depending on the interval between tapes, the costs of the materials and labor are about

$23 to $84 per ha in the United States, but savings of about 5 to 8 times these costs may be realized in protected fields (Dolbeer et al. 1986a).

Other control measures include the possibility of sterilizing blackbirds with compounds such as Ornitrol (Lacombe et al. 1987). The effectiveness of Ornitrol on spermatogenesis in red-winged blackbirds seems dependent on the time when the compound is sprayed on corn. Under field conditions, Ornitrol should be applied early in the spring when the testes of males are not yet fully developed.

Genetically resistant strains of corn and other crops offer the ultimate protection against depredations by wildlife (and, of course, from attacks by insect pests and plant diseases). Rapid advances in such "high-tech" fields as gene splicing may produce marvels within the next decade. Meanwhile, horticultural experiments with 25 cultivars of sweet corn indicated that the range of damage by birds varies nearly fivefold between the most- and the least-damaged cultivars (Dolbeer et al. 1986b). The research identified those physical features (e.g., husk weight) that should be bred into new lines of sweet corn to reduce damage by blackbirds (but see Bollinger and Caslick 1985). Damage to sunflower crops may be reduced by breeding plants with seeds having an objectionable taste to birds, but more research is needed before these features are fully effective (Dolbeer et al. 1986c). In the long run, crops protected by such means hold promise for an environment uncontaminated with pesticides and other life-threatening chemicals for humans and wildlife.

Deer frequently cause excessive losses of soybeans and other crops (for deer-related losses of alfalfa, nursery plants, and Christmas trees, see Palmer et al. 1982; Scott and Townsend 1985a). Various methods are used to reduce this damage, but timing often is crucial in the application of these measures. Soybeans suffer heavy damage—as much as 80 percent of the harvest—if grazed during the first week of their emergence (deCalesta and Schwendeman 1978). After the first week, deer remove a smaller percentage of the plants with far less crop damage. Unfortunately, by the time deer have damaged soybeans, it is usually too late to initiate control measures assuring good yields. Control methods must be in effect prior to the emergence of soybeans in order to achieve satisfactory results. Further, because damage is greater near the edges of soybean fields bordered by woodlands, effective control procedures can be concentrated along the borders rather than throughout the fields. As with unfenced orchards, owners of soybean fields sometimes are authorized to shoot deer as a means of reducing crop depredations. However, by the time state biologists can confirm the damage and issue permits, the period of maximum damage usually has passed. Shooting deer after the fact no longer has any economic justification. Repellents and scaring techniques properly timed to coincide with the first week of sprouting are better ways of dealing with deer depredations in soybean fields. Deer repellents show considerable variation in their effectiveness, however, and many of the commonly used types offer little protection (Harris et al. 1983; Palmer et al. 1983; Conover 1984).

Surveys of farmers' attitudes about crop losses sometimes reveal influences that should govern wildlife policies (see Scott and Townsend 1985b). Farmers in New York generally seem quite tolerant of crop damage caused by deer and, in fact, 79 percent recognized the aesthetic values of deer visiting their fields (Brown et al. 1978). As with any estimates of crop damage, it is possible that losses are exaggerated or that other kinds of animals beside deer contributed to the damage. Even so, no more than 2 percent of the farmers in New York consider their losses substantial or severe. Only when estimates exceeded $3000 did a clear majority of farmers consider the damage unacceptable. Nearly half considered losses between $1000 and $3000 as tolerable. These findings suggest that management policies may be based on some erroneous assumptions; namely, many wildlife managers suppose that farmers experiencing crop losses want fewer deer. They also may judge the losses of a few farmers as representative of all farmers in the region. Contrarily, the results outlined earlier indicate that immediate financial losses often are tempered by the sociological benefits that deer and other wildlife convey to farmers. At the same time, the few farmers complaining of serious crop damage very likely do need assistance, and these warrant quick responses from management agencies.

Changing cultural practices have influenced crop depredations, as we have seen in the case of waterfowl damage on Canadian grainfields. Another example merits note, not only because it increased susceptibility of orchards to depredation, but also because the "problem" itself seemingly abetted a cure. Deer may be troublesome in apple orchards. Mature apple trees largely escape damage when their limbs exceed the reach of browsing deer. Repeated browsing, however, can reduce young trees to stunted, mis-shapen stems that are useless for production. Financial losses in these situations may be severe, and shooting under some type of permit system offers the only feasible relief. Fencing usually is too expensive to protect large orchards (but see Palmer et al. 1985). In recent years, however, more orchards have been planted with dwarfed varieties of fruit trees. These are even more susceptible because deer can reach the twigs of mature trees, but the dwarfed varieties permit high-density plantings and simplify picking the fruit. Those features make deer-proof fencing far more feasible than it is with standard-sized trees (Table 13-1). The dwarfed varieties thus represent a cultural system producing more fruit per hectare while offering better opportunities for controlling browsing damage (Caslick and Decker 1979). Additionally, Porter (1983) found that single-strand electric fences to which flags of aluminum foil coated with peanut butter are attached at 10-m intervals also reduced browsing damage in orchards. The flags provide both visual and odorous stimuli, thereby encouraging nose-to-fence contact. The electric shock behaviorally conditions the deer to avoid the fence and thus reduces browsing damage. The fence is cost effective on orchards of 5 ha.

Under certain conditions, some crops may adversely affect wildlife. Soybeans are grown in many areas frequented by Canada geese, and at times, these impact in the esophagi of feeding birds. Estimates of mortality vary with the size of the flock, but 3100 geese died in an extreme outbreak in Illinois (Jarvis 1976). Some geese die suddenly, whereas others starve slowly when the impacted soybeans prevent passage

TABLE 13-1.   Feasibility, as Measured by Savings, of Fencing a 20-ha Orchard with a 1829-m Perimeter Against Deer Depredation with Different Densities of Apple Trees

Density (Trees/ha)	Annual Damage Estimate @ $0.50/Tree (Dollars)	Annual Cost of Deer-Proof Fence[a] (Dollars)	Potential Savings/Year (Dollars)
299	3,025	2,024	1,001
539	5,450	2,024	3,425
1,112	11,350	2,024	9,326
1,957	19,800	2,024	17,776

Source: Caslick and Decker (1979).

[a]The annual cost figure ($2024) includes depreciation, interest, maintenance and repairs, real property taxes, and insurance but does not include the original costs of fencing ($7500) and labor ($2300).

of food into the stomach and erode the lining of the esophagus (Fig. 13-5). Durant (1956) described the hemorrhaging and necrosis of impacted esophagi and determined that dry soybeans expanded 2.5 times in volume when moistened, easily causing fatalities. Lead poisoning may contribute to the incidence of impacted esophagi in geese and other waterfowl (Hanson and Smith 1950). Other factors affecting the incidence and the severity of this malady include the migration patterns of geese, the availability of corn or other foods, and most important, the amount of rainfall before and during the autumn feeding period (Jarvis 1976). Only dry soybeans impact, and with adequate precipitation, the dampened beans swell before they are ingested. Geese prefer corn and will feed on soybeans primarily when the corn harvest is delayed. Early migrations of geese into areas producing soybeans increase the length of the feeding period and thus the incidence of impaction. Two of these factors—rainfall and the harvest of corn—enable managers to anticipate impaction mortality each year; responses include disking harvested soybean fields, providing artificial supplies of corn, and hazing geese from soybean fields.

Aspergillosis is a fungal infection of the respiratory tracts of birds exposed to molding crops (Fig. 13-6; see also Table 8-1, Chapter 8). Bellrose et al. (1945) described an epizootic among wood ducks (*Aix*

sponsa) feeding on recently flooded corn; about 10 percent of the feeding birds seemed affected, and these likely spread the infection to others in the vicinity. About 2000 Canada geese died in an epizootic of aspergillosis on a refuge in Missouri, apparently from ingesting infected grains that were grown, ironically, as part of a waterfowl feeding program (McDougle and Vaught 1968). When a blizzard pre-

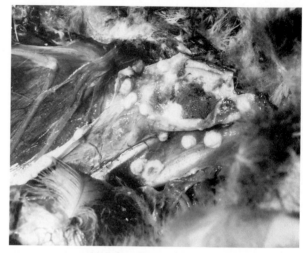

Figure 13-6. Molding crops are a common source of aspergillosis infections in the respiratory tracts of birds. Shown here are fungal growths on the lungs of a Canada goose. (Photo courtesy of M. Friend, U.S. Fish and Wildlife Service.)

Figure 13-5. Esophagus of a Canada goose fatally impacted with soybeans. (Photo courtesy of R. L. Jarvis.)

vented mallards from feeding in fields, they instead fed on silage spread over the snow for cattle, and more that 1000 birds thereafter succumbed to aspergillosis presumably obtained from this source (Neff 1955). These cases represent single, local sources of contamination, and prevention by hazing or by supplying alternate sources of attractive food sometimes may be feasible (Wobeser 1981). Even with a single source of unfit food, however, timely detection of the infections remains difficult, permitting an epizootic well before the source is located.

Restoration of diminished Canada goose populations remains an outstanding achievement of wildlife management. Winter inventories indicated that their numbers nearly doubled in the 2 decades following 1955, reaching about 3 million by the start of the 1974 hunting season (Bellrose 1976). This was accomplished with several kinds of management but also included a role played by corn. A dynamic situation resulted, not only in the buildup of goose numbers, but also in a changing pattern of their winter distributions. As mentioned previously, new types of farm machinery were developed rapidly after World War II. Mechanical corn pickers were among these, and, whereas they can harvest corn more rapidly than hand labor, the machines also left more corn as field waste. Depending on several factors, among them moisture content, as much as 15 percent of the corn remains on the ground after harvest (Jugenheimer 1976). With these additional food resources at hand on public and private lands—and with regulated hunting—Canada geese again thrived in the Mississippi and Central flyways. Fall and winter food reserves were assured in Illinois, Minnesota, and other northern states, far from the bird's traditional wintering grounds. In response, some subpopulations stayed through winter. Fewer geese, so it seemed, traveled south, bringing cries of "short stopping" from hunters in southern states as their goose hunting became less rewarding. Indeed, the continental Canada goose population was building and most of the increase was occurring in the northern states. The number of birds wintering in Rochester, Minnesota, for example, increased from 250 in 1951 to 8650 by 1966–67 (Gulden and Johnson 1968) and to 20,000 by 1970 (Raveling 1978).

Closer examination of the "short-stopping" phenomenon revealed that other factors were affecting the winter distribution and abundance of Canada geese. Raveling (1978) determined that hunting mortality was greater for those geese continuing southward in the Mississippi or Central flyways.

These segments of the goose population were subject to about five times more shooting pressure than those staying in Minnesota. Quite simply, the birds were increasing their vulnerability by flying south. Annual mortality in the Minnesota flock was 19 percent, whereas it reached 50 percent for those continuing their migration southward. With this differential mortality, the northern subpopulations increased, while those wintering in the south steadily diminished. What appeared to be short stopping actually was a dynamic change in mortality rates among subpopulations influenced, in part, by the availability of corn.

EROSION, SEDIMENTATION, AND WILDLIFE

Wind and water erode unprotected soil, taking a heavy toll on a basic natural resource in all regions of the United States (Fig. 13-7). The losses are staggering. Inventories estimate that more than 6 billion tons of soil erode each year from nonfederal lands in the United States (U.S. Department of Agriculture 1981). The greatest losses occur on croplands, although pastures, forests, and rangelands also experience soil erosion (Table 13-2). Whereas it may take 1000 years for natural forces to form 2.5 cm of enriched topsoil, erosion may claim nearly the same amount every 30 years, resulting in what might be called a soil deficit. Water erosion alone claims a national average of about 11.6 tons per ha on croplands each year. The implications are obvious: A life-supporting resource steadily erodes with each particle of soil swept from productive areas. Indeed, accelerated soil erosion may be the underlying reason for the fall of civilizations. In the view of Carter and Dale (1974), civilizations have not lasted more than 30 generations except in the valleys of the Nile, Tigris-Euphrates, and Indus rivers where recurring flooding naturally restores soil fertility (only time will tell whether completion of the Aswan High Dam in 1968 will thwart the flood-based fertility on the Nile). Elsewhere in the world, human abuses so accelerate erosion that soil fertility cannot be renewed naturally.

Accelerated sedimentation—the result of erosion—is a creeping form of environmental degradation in aquatic environments. To be sure, natural geological forces produce turbidity in some aquatic systems, but there is little evidence that sediments of that origin are lethal to fishes (Wallen 1951; Phinney 1959).

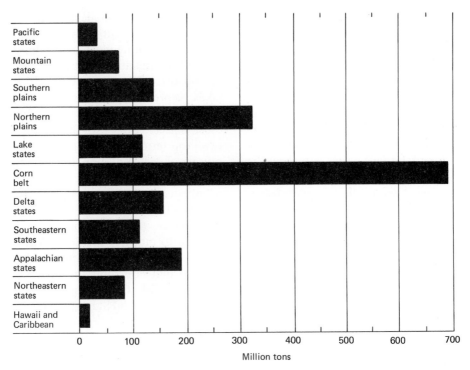

Figure 13-7.   Regional losses of soil per year for croplands in the United States. Estimates for sheet and rill erosion only. (From U.S. Department of Agriculture 1981.)

TABLE 13-2.   Gross Annual Soil Losses from Wind and Water Erosion on Nonfederal Land in the United States

| | Hectares (Millions) | | | | |
Land Use	>5 Tons/Ha	5–12 Tons/Ha	12–34 Tons/Ha	34+ Tons/Ha	Total Erosion (Billions of Tons)
Cropland	64.2	46.0	37.7	19.4	2.82
Pastureland	42.5	5.7	3.8	2.0	0.35
Forestland	132.4	10.5	4.7	2.0	0.44
Rangeland	114.8	22.5	16.2	11.6	1.71
Other	—	—	—	—	1.10
Totals	353.9	84.7	62.4	35.0	6.42

Source: U.S. Department of Agriculture (1981).

With chronic soil erosion, however, the carrying capacity of lakes and rivers is destroyed almost without notice or the fanfare of obvious fish kills. Some fishes, of course, are adapted to turbid waters. For these, reduced eye size, compensated by enhanced development of other sense organs and modifications in contour, body form, fin development, and color, evolved under the selective pressures of turbidity (Hubbs 1940). Species adapted to turbid waters have silt-tolerant eggs and often scatter these instead of building nests. However, for species adapted to clear water, erosion leading to high levels of turbidity poses a number of difficulties. The intensity of farming—along with deforestation—has so increased erosion that excessive turbidity sometimes chokes or scours the sensitive gill elements of many fishes. With increasing turbidity, fishes increase their ventilation rates as a means of overcoming respiratory deficiencies so that their oxygen demands might be met (Horkel and Pearson 1976). There is, of course, a limit to which many fishes can adapt to these circumstances, and they die when that limit is exceeded. For others, bottom sediments smother their eggs. Silt-laden water also reduces the activities of juvenile bass (*Micropterus* spp.) and interferes with the social structure of sunfish (*Lepomis* spp.) (Heimstra et al. 1969). The complex reproductive behavior of many species relies on visual cues and therefore can be affected by unusual turbidity. Because little or no aquatic vegetation can thrive in highly turbid waters, food chains and habitat conditions also have changed drastically for many fishes. Predacious species rely on their sight for seeking food, and reduced visibility in turbid water diminishes their effectiveness and decreases the probability of a game fish seeing a lure.

It is not surprising, then, that fish faunas have changed in keeping with degrading environments. The shift increases the populations of small and large "rough" fishes such as spotfin shiner (*Notropis spilopterus*) or carp (*Cyprinus carpio*) while decreasing the numbers of walleye (*Stizostedion vitreum*) and other game species that usually are of better food value (Trautman 1957). As an illustration, Peters (1967) studied the effects of sedimentation introduced by irrigation from farmlands bordering spring-fed trout streams in Montana. Trout were abundant and egg survival was high only at the lower ranges of sedimentation, whereas increasing sediment loads were accompanied by fewer trout, more rough fish, and high egg mortality (Table 13-3).

Sedimentation also influenced key waterfowl habitat in the Illinois River Valley—first positively, then negatively. The situation helps to illustrate the ecology of natural sedimentation versus the aftermath of accelerated soil erosion. As described by Bellrose et al. (1979), the geological events forming the valley gradually built natural levees parallel to the river, thereby separating most of the channel from the adjacent bottomlands. Until the 1930s, water trapped behind these levees created a series of bottomland lakes, representing some of the finest waterfowl habitat in North America. Since then, however, soil erosion accelerated with the adoption and spread of row-crop farming. The results produced an unnatural rate of sedimentation, the formation of an unstable layer of silt in the river bed, and the steady losses of aquatic vegetation serving as valuable food for waterfowl.

Turbidity, as a function of sedimentation, contributed to much of this loss. Wave action and fish movements stir the silty bottom sediments into a suspension that interrupts the penetration of sunlight, disrupting fish faunas as well as waterfowl food plants (Jackson and Starrett 1959). Nearly 70 percent of the fish species in the Illinois River have declined or disappeared as vegetation and other features of the aquatic system were degraded (Karr et al. 1985). Sedimentation also filled the bottomland lakes and marshes at rates in excess of normal geological schedules. The capacity of Lake Chautauqua, one of the bottomland lakes bordering the Illinois River (and a national wildlife refuge) was reduced by more than 18 percent in less than 24 years; about 76,400 tons of sediment are deposited each year (Bellrose et al. 1979). As a result, sediments originating from the accelerated erosion of farmlands on the upper reaches of the Illinois River are bringing wetland habitats downstream into an irreversible and premature extinction. Widespread adoption of conservation tillage will reduce erosion, thereby stemming much of the harmful sedimentation, but reclamation

TABLE 13-3. Effects on a Trout Stream of Sediment from Agricultural Practices

	Station	Average Monthly Sediment Concentration (ppm)	Maximum Daily Sediment Loads (Tons/Day)	Average Number of Fish		Mortality of Trout Eggs (Percent)
				Trout	*Rough[a]*	
Upstream	1	18	1.2	216	7	3
	2	79	32	197	17	22
	3	167	35	55	225	54
	4	186	61	8	1201	70
Downstream	5	319	2700	6	378	47

Source: Peters (1967).

[a]Includes three species of sucker and one chub.

of the damaged river system is virtually impossible. In summing up their review of this setting, Havera and Bellrose (1985) state, "The decline in the ecological integrity and productivity of the Illinois River floodplain ecosystem should serve as an example of man's misuse of our natural resources.... The Illinois Valley is not the only loser in the misuse of agricultural lands.... Our nation's productivity is also being washed away."

AGRICULTURAL CHEMICALS AND WILDLIFE

Despite long-term losses of topsoil, technology has produced phenomenal increases in farm production. Whereas 1.0 ha was required to sustain one person in food and clothing in 1930, the ratio is now 0.4 ha per person (*fide* USDA Economic Research Service). Production of food and fiber on U.S. farms has no equal, in part because of the intensity of technologically oriented management. Chemicals have been instrumental in this achievement. These include fertilizers and groups of compounds designed to kill weeds and insect pests. The latter, together with rodenticides and a few other specialized chemicals, are known collectively as pesticides.

DDT, perhaps the best-known insecticide, was discovered in 1874 but remained shelved until its effectiveness against insects was realized decades later (Cottam 1969). DDT sprang into prominence during World War II, prompting widespread applications of this and other pesticides on croplands after the war. In the wake of their use, however, came a belated awareness that some pesticides produced harmful side effects. Many kinds of wildlife frequently bore the brunt of a well-intentioned technology that failed to envision its total consequences.

A voluminous literature has developed about pesticides and their effects on wildlife, including the introduction of scientific journals devoted to environmental toxicology (e.g., *Pesticide Monitoring Journal* and *Bulletin of Environmental Contamination and Toxicology*). Many of the studies are experimental, using captive animals fed diets contaminated with known amounts of agricultural chemicals. Other studies take place in the field and report the incidence of pesticides in birds and mammals exposed directly or indirectly to treated croplands. Some of these studies were triggered by serious die-offs of wildlife, whereas others marked the potential

for damage. Korschgen (1970) traced Aldrin applied to cornfields through the soil and into the food web of earthworms, insects, seeds, and ultimately into a variety of vertebrates. Similarly, Meeks (1968) followed DDT through the food web of a freshwater marsh in Ohio, and Herman and Bulger (1979) intensively studied the impact of DDT on nontarget organisms in forests of Oregon. In addition to the environmental damage resulting from the regular use of pesticides, accidents involving pesticides also cause ecological disasters. In 1986, 30 tons of insecticides, herbicides, and fungicides spilled from a warehouse into the Rhine River near Basel, Switzerland, causing the ecological death of a major river for years to come (*Time* 1986). Regrettably, some pesticides at times have unintentionally poisoned wildlife even when they are applied properly.

Unusual habits occasionally predispose some species of wildlife to pesticide poisoning. For reasons that remain unclear, black-billed magpies (*Pica pica*) routinely ingest cow hair. Magpies thereby were poisoned for 3 months after cattle were treated externally with a pesticide used to control dermal parasites (Henny et al. 1985). Although other factors may be involved, a 10-year decline in magpie populations in several western states coincided with the widespread application of the pesticide. Fortunately, this form of poisoning can be eliminated if the pesticide is applied in other ways (e.g., injection, food additives, or capsules), thereby eliminating the harmful residues on cow hair. Other kinds of special circumstances frequently concern pesticide poisoning and wildlife. Aerially sprayed chemicals, for example, may drift into fish hatcheries and kill the entire stock. Although poisoned elsewhere, animals dying from pesticides may enter key areas, thus bringing lethal materials into the habitat of important species of wildlife (e.g., into refuges). In fact, the safety of two endangered species—ocelot (*Felis pardalis*) and jagaurundi (*F. yagouaroundi*)—was compromised when birds poisoned by a highly toxic pesticide died in a national wildlife refuge in Texas (White et al. 1982).

Large die-offs of lesser snow geese and other birds have been associated with pesticides by Flickinger (1979) and others. Scott et al. (1959) summarized the virtual elimination of wildlife after Dieldrin was applied to farmland in Illinois. Japanese beetles were the targets, but pheasants, a variety of songbirds, cottontails, muskrats, and fox squirrels also were killed. From 1961 to 1975, agricultural pesticides were responsible for nearly one-quarter of all fish kills for which the cause could be determined, including the loss of 109,000 fish when pesticide-contaminated runoff entered an Alabama fish hatchery (U.S. Environmental Protection Agency 1979). However, the harm some pesticides inflict on wildlife often is insidious, taking effect in ways that are not obvious. Eggs with shells so thinned by DDT that they break during incubation are more subtle aftermaths of contamination than a field of poisoned geese, but the result is no less devastating. Other sublethal effects of DDT and its chemical relatives on birds have been shown from experimental research. Among these are delayed migratory conditioning (Mahoney 1975), delayed ovulation (Jefferies 1967), increased thyroid weight and activity (Jefferies and French 1972), sterility (Ratcliffe 1969), and irregular behavior (James and Davis 1965; Heinz 1976c), possibly including increased vulnerability to predators (McEwen and Brown 1966). The irony of these findings is reflected in a 1965 report of the Presidential Panel on Environmental Pollution: as little as 1 percent of the pesticides applied to agricultural lands in the United States actually may hit their intended targets.

Some chemicals produced results far beyond the expected. Corn treated with the herbicide 2, 4-D accumulated more nitrogen that, in turn, promoted a population explosion of aphids feeding on the crop. Additionally, the treated corn was 26 percent more susceptible to corn borers, and the females of this insect were a third larger and produced a third more eggs (Oka and Pimentel 1974). Honeybees often were killed by insecticides, affecting not only the honey industry but also the pollination of fruit and vegetable crops, as well as uncultivated vegetation (Pimentel et al. 1980). Food chains, many of which involved wildlife, also were interrupted by pesticides. Reductions in insect populations on grasslands treated with insecticides were followed by decreases in small rodent populations dependent on insect foods (Barrett and Darnell 1967). The sources of food for Atlantic salmon (*Salmo salar*) were altered by DDT (Keenleyside 1967). In England, Potts (1977) suggested that insecticides so reduced insect populations that this loss of food produced declines in the gray partridge populations. Hamerstrom (1979) linked the nesting of northern harriers (*Circus cyaneus*) with the cyclic nature of rodent populations, but when DDT was applied to the study area, the normal rodent-hawk cycle was interrupted and far fewer harriers nested successfully (see also Hamerstrom et al. 1985). All told, pesticides have extraordinarily influenced ecosystems by reducing bio-

logical diversity, interrupting food chains, modifying energy transfer, reducing the quality of soil, water, and air, and lessening the stability and resilience of both natural and managed environments (Pimentel and Edwards 1982).

Some pesticides were banned in the United States after their harmful effects were discovered. DDT was banned for use in the United States in 1972, but this restriction did not prohibit its continued manufacture for export to other nations. Indeed, black-crowned night herons (*Nycticorax nycticorax*) nesting in Idaho experienced impaired breeding success because of their exposure to DDT on wintering areas in Mexico (Findholt and Trost 1985). Similarly, the organochlorine residues were several times higher in migratory insectivorous birds in the diet of peregrine falcons (*Falco peregrinus*) than in resident prey species, suggesting that the poisons are acquired from prey wintering in countries where harmful organochlorines are still used. The eggshells of the falcons nesting in the Rocky Mountains thus remain abnormally thin, and improved reproductive performance in the population cannot be expected until organochlorine contamination is reduced in the prey base (Enderson et al. 1982).

Scores of troublesome insects, including mosquitoes, houseflies, and lice, as well as agricultural pests, are no longer susceptible to some chemical controls, having gained resistance after generations of exposure. Of about 2000 insect and other arthropod pests, nearly 400 species have evolved resistance (Georghiou and Taylor 1977). Resistance in some of these species increased 25,000 times. Chancellor (1978) supplied evidence that some plants also have developed resistance to herbicides. Adaptive resistance seems limited to those species with high rates of reproduction and short life cycles. These organisms produce so many generations of offspring that the probability of encountering genetic resistance is realized—and these generally are the pests for which pesticides were intended. Only a few survivors are enough to establish new and resistant populations. Conversely, few vertebrates develop any degree of resistance to toxic chemicals. Some rodents have developed 12 times the normal tolerance to Endrin (Webb and Horsfall 1967; Webb et al. 1973). Resistance to chlorinated hydrocarbons in mosquito fish (*Gambusia affinis*), yellow bullhead (*Ictalurus natalis*), and some species of frogs living near heavily treated cottonfields was documented by Ferguson (1963), Ferguson and Bingham (1966), and Culley and Ferguson (1969). Resistance of mosquito fish to Endrin and Toxaphene increased more than 520 and 375 times, respectively. Ferguson (1967) noted the dangers resistant organisms pose to food chains and, indeed, to humans when they are consumed by nonresistant species occupying higher trophic levels.

Most insecticides are not species-specific; that is, they kill all insects—harmful and beneficial species alike. Many insects are predators of harmful species, and the benefits of natural pest control are diminished when predators are removed. Further, because of basic ecological relationships between predators and prey populations, the predacious species take far longer to recover from insecticides than do the damaging, herbivorous species (Croft and Brown 1975).

Recent approaches to crop protection recognize the role of beneficial insects and take means to maximize their influence. Overall, the result has been a combination of chemical, cultural, and biological methods known as integrated pest management (Huffaker 1980). This approach reduces the amount and frequency of insecticide applications. It also uses altered cultural practices and the development of insect-resistant crops as means of reducing the harmful insecticide contaminations so prevalent in the past. Biological control of pests takes advantage of a pest's natural vulnerability, to which it cannot adapt genetically. For example, some pest populations succumb rapidly when management enhances their exposure to species-specific diseases or parasites. Some harmful insects are lured into traps by species-specific sex attractants known as pheromones. Others are sprayed with nonlethal materials that cause interruptions in their life cycles so that they remain as juveniles and never mature into egg-laying adults. Other materials change the way insect colonies function. For example, fenoxycarb is an insect growth regulator that greatly reduces infestations of imported fire ants (*Solenopsis invicta*). Workers carry fenoxycarb-treated baits into the colonies, where they are dispensed as food to other ants, including the queen. Although not toxic, fenoxycarb nonetheless prevents the queen from laying any more worker eggs—only those that produce sterile nonworkers. Thereafter (a) the existing workers, forced to labor harder to secure food, rapidly die without any replacements, (b) the queen dies without the care of the worker caste, and (c) and no females are available to start new colonies elsewhere. Colonies treated with fenoxycarb lack the requisite infrastruc-

ture and thus collapse. In field tests, fenoxycarb eliminated 60 percent of the colonies and reduced the population by 67–99 percent within 13 weeks (Banks et al. 1988; see also Collins et al. 1992 for seasonal trends in the application of fenoxycarb).

Rands (1985) described a cultural approach that lessened the impact of pesticides on the production of gray partridges. Pesticides reduce the abundance of insect foods on which the partridge broods depend, and the survival of chicks thus is impaired in fully sprayed fields. Gray partridges in Britain in fact have been declining since 1945, largely because of chick mortality (Potts 1980). However, when 6-m strips around the perimeters of grain fields were left unsprayed, the chicks found enough insects, and the mean brood size often doubled in comparison with those in fully sprayed fields. Moreover, the unsprayed strips had little or no effect on the total yield of grain (Rands 1985; see also Rands and Sotherton 1985).

A news release from the U.S. Department of Agriculture (1987) announced the development of a genetic technique that may control tobacco budworms (*Heliothis virescens*), insects that damage cotton, vegetables, and tobacco. Crosses with a related, but harmless, species created hybrid females that not only produce sterile male offspring but whose female offspring also continually pass on the trait for male sterility. This form of biological control is especially promising because tobacco budworms are rapidly developing resistance to chemical pesticides. The initial test with the hybrid females successfully introduced male sterility into the budworm population and reduced the number of insects by 75 percent.

SOME KINDS OF INSECTICIDES

The "families" of insecticides increase each year as new compounds are formulated. However, for our purposes, we shall consider only three of the better-known types of the several described by Rudd (1964).

Chlorinated hydrocarbons, also known as organochlorines, embrace a number of insecticides used widely to control agriculture and forest pests. The best known is DDT, but Chlordane, Heptachlor, Endrin, Aldrin, Dieldrin, and Toxaphene also belong to this group. Some chlorinated hydrocarbons such as Toxaphene and DDT now are banned or restricted in the United States, but others still are used. News of waterfowl contaminated with Endrin made national headlines in 1981 (*Time* 1981).

Chlorinated hydrocarbons attack the central nervous system. Tremors, tonic contractions, convulsions—

and usually death—occur in cases of acute toxicity. Repeated low-level ingestion leads to accumulations in fatty tissues, including those of the liver and heart. Storage of chlorinated hydrocarbons in fatty tissues is significant because they may later be released rapidly when the stored fat is mobilized for energy (Van Velzen et al. 1972). For example, Babcock and Flickinger (1977) described the death of geese mobilizing Aldrin-contaminated fat under the stress of migration. Chlorinated hydrocarbons are especially deadly to aquatic organisms. Toxaphene, in fact, once was used by fishery biologists to reclaim ponds and lakes overpopulated with undesirable kinds of fishes. Even when used for this purpose, however, Toxaphene often killed other animals. Lennon et al. (1970) cited a 4-year fish reclamation project in Nebraska where each aerial application of Toxaphene was accompanied by 15–100 percent losses of waterfowl.

Chlorinated hydrocarbons may persist for years. Almost 40 percent of the DDT applied to a field in Maryland was present 17 years later (Nash and Woolson 1967). Soils treated with a single application of Aldrin were 95 percent free of residues after 1 year, but no further reduction was detected after an additional 5 years (Korschgen 1971). However, most croplands are treated year after year so that residues of Aldrin or other chlorinated hydrocarbons may steadily accumulate in soils. Whereas a single application of DDT had little initial effect on birds breeding in a forest habitat, repeated applications over a 4-year period led to a 26 percent reduction in the population by the spring of the fifth year; the numbers of some species decreased by 44 percent (Robbins et al. 1951). On the other hand, a single application of Aldrin killed 25–50 percent of a pheasant population within 1 month of its application and severely reduced reproduction of the survivors (Labisky and Lutz 1967).

A critical feature of persistent insecticides is their increasing concentration in each succeeding level of the food chain. This process, known as *biomagnification*, is particularly common in ecosystems treated with chlorinated hydrocarbons. Even at authorized rates of application, passage of these materials through the ecosystem typically ends with excessive accumulations among organisms at the higher trophic levels. Predators such as brown pelicans (*Pelecanus occidentalis*) and falcons (*Falco* spp.) are unusually susceptible to biomagnification. Direct mortality may result from these accumulations of toxic chemicals, or, as described earlier, indirect dysfunctions may

TABLE 13-4.   Biomagnification of a Chlorinated Hydrocarbon in an Aquatic System. These Data Are Maximum Estimates of Accumulations in Organisms Occupying Producer and Consumer Trophic Levels, as Indicated. Numbers Shown Are Multiples of the Original Contamination in Water of 0.02 ppm

Organisms	Trophic Level	Multiple of Original Contamination
Plankton	Producer	265
Frogs	Secondary consumer	250
Small fishes	Secondary consumer	500
Predacious fishes	Tertiary consumer	85,000
Fish-eating birds	Tertiary consumer	80,000

Source: Rudd (1964) based on fieldwork of Hunt and Bischoff (1960) and others.

inhibit successful reproduction. An example of biomagnification, based on an aquatic ecosystem in California that was treated with a chlorinated hydrocarbon, shows the concentrations eventually reaching several kinds of wildlife (Table 13-4). Whereas the original application rate was only 0.02 ppm, levels in fishes and western grebes (*Aechmophorus occidentalis*) were magnified more than several thousand times. The population of western grebes was affected significantly. About 1000 pairs of these birds nested in the area prior to its treatment, but no young were produced for the next 12 years (Hunt and Bischoff 1960; Rudd 1964).

Soil type is among the factors influencing the persistence of chlorinated hydrocarbons. A comparison between organic soils and silty loams for Heptachlor residues showed that 27 percent and 4.5 percent, respectively, remained after 6 months. Unfortunately, organic soils, because of their earthworm populations, are favorite feeding sites for woodcock (*Scolopax minor*). Even when Heptachlor was applied at recommended levels, earthworms accumulated enough of the chemical to kill 10 of 12 woodcock within 53 days (Stickel et al. 1965). An even more telling case concerned the contamination of woodcock in New Brunswick (Wright 1960b, 1965). Fully 86 percent of the woodcock in the fall migration were contaminated from DDT applied on their breeding grounds. These birds received further exposure—this time to Heptachlor—on their wintering grounds so that the breeding populations the following spring produced

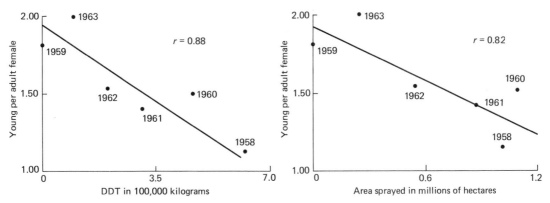

Figure 13-8. Relationship between breeding success of woodcock in New Brunswick and DDT applications. The number of young per adult female is correlated strongly with the amount (left) and area (right) of DDT applications. (From Wright 1965.)

contaminated hatchlings. The result was indicated in the decreased numbers of young woodcock produced per adult female (Fig. 13-8).

When penned pheasants were fed Dieldrin, no mortality occurred in the first generation, although their egg production was decreased (Atkins and Linder 1967). However, the offspring of these birds experienced mortality, further loss of egg production, decreased fertility, and impaired behavior (Baxter et al. 1969). Up to 37 percent of the Dieldrin ingested by the first generation of pheasant hens was transferred to the yolks of their eggs (Lamb et al. 1967). This transfer proved an important means by which hens eliminated Dieldrin from their own tissues, but a second generation was contaminated, contributing to the result mentioned earlier. However, Dieldrin does not seem to cause eggshell thinning in pheasants (Dahlgren and Linder 1970).

For avian predators such as prairie falcon (*Falco mexicanus*), contamination with DDT and other chlorinated hydrocarbons had disastrous effects on the thickness of their eggshells (Enderson and Berger 1970). Chlorinated hydrocarbons interrupted the normal transport of calcium to the oviducts of contaminated birds by inhibiting the action of the enzyme carbonic anhydrase. With less than normal amounts of calcium, the eggshells of these birds were thin and broke during incubation. Hatching success dropped rapidly in proportion to the thickness of eggshells (Table 13-5).

The ban on applying most kinds of chlorinated hydrocarbons in the United Stated initiated the recovery of some species harmed by eggshell thinning, including bald eagles (*Haliaeetus leucocephalus*). Production of brown pelicans in California began to improve when DDT decreased in the food chain

(Anderson et al. 1975). In a 6-year period, fledgling rates improved from just 4 young pelicans from 1125 nests (0.004 young per nest) to 1185 young from 1286 nests (0.922 young per nest). A fivefold decline of DDT metabolites in the eggs of ospreys (*Pandion haliaetus*) nesting in Connecticut and Long Island coincided with the return to almost normal production; fledgling success increased from a low of 0.4 young per nest to 1.2 young per nest (Spitzer et al. 1978).

Organophosphates are less persistent than chlorinated hydrocarbons in soils and other components of the environment. However, they generally are highly toxic and have caused significant mortality in birds, including a single event when more than 1450 Canada geese died (White et al. 1982; see also Grue et al. 1983 for a useful review). Organophosphates have been linked with secondary poisoning. About 400 birds of prey died after they consumed rodents and birds initially poisoned with an organophosphorous compound (Mendelssohn and Paz 1977). Secondary poisoning also killed Franklin's gulls (*Larus pipixcan*) that fed on cicadas (Cacadidae), which had been poisoned unintentionally with an organophosphate (White and Kolbe 1985). Barn owls (*Tyto alba*) experimentally fed quail poisoned with an organophosphate were themselves affected, although not lethally (Hill and Mendenhall 1980). Secondary poisoning is a hazard dependent upon the number of poisoned prey consumed, the body parts consumed, and the degree to which the prey was initially contaminated.

A serious case of secondary poisoning developed in Argentina when Woodbridge et al. (1995) discovered 700 dead Swainson's hawks (*Buteo swainsonii*) in the winter of 1994–95. The hawks died after consuming

TABLE 13-5.   Comparison Between Production of Young and Eggshell Thickness in Prairie Falcons

Eggshell Thickness Index[a]	Prairie Falcon Nests		Number of Young Fledged	
	Total	*Percent Fledging One or More Young*	*Total*	*Per Pair*
More than 1.75	21	75	54	2.6
1.45–1.75	31	50	51	1.6
Less than 1.45	10	10	3	0.3

Source: Enderson and Berger (1970).
[a]Index based on Ratcliffe (1967).

grasshoppers contaminated with monocrotophos, an acutely toxic organophosphate insecticide widely used in Argentina. Swainson's hawks feed almost exclusively on grasshoppers during the winter but prey heavily on small mammals after they migrate to their breeding grounds in western North America. In winter, these hawks also concentrate in flocks of several thousand, whereas they disperse widely in pairs during the breeding season. In 1995–96, a more-extensive field survey revealed a count of 5000 victims, which provided the basis for an estimated loss of 20,000 Swainson hawks in Argentina—and once again, monocrotophos was implicated as the causative agent (Fig. 13-9). If such losses continued, drastic reductions in the breeding population would be obvious in North America within 5 to 10 years. Indeed, the pesticide-related mortality in Argentina may have represented about 5 percent of the total population of 450,000 Swainson's hawks (Goldstein et al. 1996). Fortunately, because of rapid international collaboration, new government restrictions were implemented and a major manufacturer of monocrotophos recalled stocks of the insecticide, which curtailed further damage (Stafford 1997).

Malathion and Parathion are the best known of the organophosphate group. These and other organophosphates inhibit the enzyme acetylcholinesterase in the nervous system, causing disruption in the transmission of impulses. Death usually occurs from asphyxiation when the respiratory center in the brain fails (see Hill and Fleming 1982). Human fatalities have occurred from external contact with small amounts of Parathion. Unlike chlorinated hydrocarbons, organophosphates seldom accumulate in fatty tissues.

Unfortunately, the acute toxicity of some organophosphates at times has been misused for the deliberate poisoning of wildlife. Carson (1962) estimated that farmers intentionally poisoned 65,000 birds with pesticides. More recently, Stone et al.

Figure 13-9. Pesticides killed thousands of Swainson's hawks in Argentina, as shown here by a sample of dead birds collected during the winter of 1995–96. Fortunately, rapid reactions by the chemical industry, farmers, and private and governmental agencies stopped further losses. (Photo courtesy of Brian Woodbridge, U.S.D.A. Forest Service.)

(1984) described two cases in New York where, in total, farmers intentionally killed more that 8000 redwinged blackbirds and other birds. Hawks of three species died from secondary poisoning in these events. In Texas, about 11,000 birds of 12 species were killed by bait poisoned with highly toxic organophosphates (Flickinger et al. 1984). Agricultural chemicals also have been used for the deliberate poisoning of wildlife in Europe (Hamilton et al. 1981). Intentional poisoning is difficult to detect because it can be masked by what otherwise seems to be a legitimate use of the pesticide. Thus, Diazinon, applied for the control of lawn-damaging pests, poisoned about 700 brant (*Branta bernicla*) on a golf course in New York where the grazing birds probably were regarded as a nuisance (many of the dead birds were salvaged for scientific uses; see Vangilder et al. 1986). Diazinon also has killed Canada geese, wigeon (*Anas americana*), and many other birds on golf courses and lawns elsewhere (Zinkl et al. 1978; Stone and Gradoni 1985; Littrell 1986). Stone et al. (1984) believe that education, highly publicized prosecutions, and substantial fines may reduce the incidence of deliberate poisoning of wildlife.

Carbamates are a relatively new family of pesticides. Some impetus for carbamate production stemmed from the larger number of insects developing resistance to other insecticides. Most carbamates in use have shorter life spans than chlorinated hydrocarbons and so reduce the chances of biomagnification. Like organophosphates, carbamates inhibit acetylcholinesterase activity in the nervous system (Hill and Fleming 1982). Fenoxycarb, in part, inhibits the metamorphosis of immature insects into adults, and in water, the chemical is degraded by sunlight with a half-life of 5 hours (Environmental Pesticide Fact Sheet, 1986). Until then, however, fenoxycarb is highly toxic to aquatic invertebrates and is registered only for the control of fire ants on turf and nonagricultural lands. The toxicity of carbamates varies widely, with some being rather harmless, whereas others are deadly in small amounts. Temik is highly toxic, but Sevin has a safe environmental record because of its low toxicity to vertebrates (Stickel 1974). A single granule (about 0.6 mg per bird) of Furadan killed 4 of 5 experimental birds within 24 hours (Balcomb et al. 1984). Rice seed legally treated with Furadan killed several species of songbirds and sandpipers; the carbamate apparently was applied expressly to kill birds feeding in newly planted rice fields (Flickinger et al. 1986).

Furadan, also known as Carbofuran, is a systemic insecticide; this means that it is applied to the soil but then is absorbed through the roots into most plant tissues, giving crops internal protection against insect attacks. Flickinger et al. (1980) recorded the effects of Furadan used in rice fields after Aldrin was suspended by the Environmental Protection Agency in 1974. The effects included rapid mortality of fishes, frogs, and crayfish, but five times fewer birds were killed than with Aldrin, in part because Furadan was applied after the peak of spring migration. In stronger formulations, however, Furadan killed several thousand waterfowl in British Columbia, California, and Oklahoma, indicating that only the lowest strength possible of this carbamate should be used, and then only after migration is over (Flickinger et al. 1980).

Secondary poisoning also can occur with some carbamates. Red-shouldered hawks (*Buteo lineatus*) died after feeding on birds poisoned in a cornfield treated with Furadan. Expansion of the mortality data to include the total area treated each year with Furadan indicated that several thousand hawks of various species might be harmed (Balcomb 1983). Earthworms poisoned in fields treated with Furadan became spasmatic and so attracted the attentions of feeding robins (*Turdus migratorius*); such data suggest that huge numbers of robins may die each year in this fashion (Balcomb et al. 1984).

HERBICIDES

In general, herbicides seemingly represent a far less-direct threat to wildlife than insecticides. Laboratory tests with many of the better-known herbicides indicate that these cause little or no mortality in birds under conditions simulating rates of field application (Hill et al. 1975). These materials are less toxic than insecticides to terrestrial animals and do not seem to biomagnify. However, there may be delayed effects that are not expressed for some time after initial exposure. Controversy surrounds the cancer-forming potential of some herbicides, as witnessed by the incidence of this malady in people exposed to Agent Orange in Vietnam. However, even if some herbicides are carcinogenic to humans, it may not necessarily mean that wildlife is also threatened. The life span of most wild animals is short enough that they die from other causes before lethal cancerous tissues have time to develop. The matter remains speculative at this time, however.

Still, recent tests suggest that some widely used herbicides produce significant mortality in some animals. Embryos in mallard eggs exposed to Trifluralin experienced nearly 50 percent mortality when treated with normal application rates of this herbicide (U.S. Fish and Wildlife Service 1982a). Either direct spraying on eggs or transferral of the herbicide on the plumage of the incubating adults to their eggs may pose hazards. When applied at recommended rates, 2, 4,5-T did not significantly impair mallard embryos, but another herbicide, Paraquat, produced high rates of mortality and impaired embryonic growth when applied at one-half the level recommended for field use (Hoffman and Eastin 1982).

As with insecticides, a large number of herbicides entered the marketplace after World War II. Only 14 herbicides were sold in the United States before 1940. By 1963, the number had increased eight times (Ennis 1964). Unlike insecticides, however, herbicides must be more specifically tuned to their targets or they will kill crops along with the undesirable weeds. Most herbicides are designed to kill either grasses or broadleaved plants, but not both. Even so, species tolerances to herbicides vary greatly, and further specificity may be possible once tolerances are determined for target and nontarget vegetation. Factors such as soil temperature, stage of plant development, and season also influence the effectiveness and selectivity of herbicides.

Herbicides may kill or impair vegetation that is of benefit to wildlife. Klebenow (1970) found that herbicide treatments reducing sagebrush (*Artemisia tridentata*) also killed understory vegetation important to sage grouse (*Centrocercus urophasianus*). Therefore, wildlife habitat may suffer in quality or quantity when some herbicides are used indiscriminately. Attempts to rid sandy soils of shinnery oak (*Quercus havardii*) in west Texas with high application rates of herbicides also impaired production of the range grasses that were desired as replacement vegetation. At lower rates of application, however, prairie grasses thrived and most of the oaks were killed, but the patches that remain may be crucial habitat for prairie chickens and other wildlife (Jones and Pettit 1984). Adjustments in the rates at which herbicides are applied thus may improve range conditions for both cattle and wildlife. In Oregon, clear-cut areas are sprayed with herbicides to suppress invasions of brush, thereby favoring the regrowth of commercially valuable conifers. The treatments accordingly reduced the complexity of the vegetation, but the density and diversity of the avian community remained largely unchanged (Morrison and Meslow 1984). However, MacGillivray's warblers (*Oporonis tolmiei*) were unusually sensitive to short-term defoliation of deciduous shrubs, and Wilson's warblers (*Wilsonia pusilla*) declined two or more times in density even after altering their foraging behavior in re-

sponse to the modified habitat. For management purposes, the study nonetheless revealed that retention of even small amounts of deciduous tree cover on sprayed clear-cuts would result in near-normal bird communities. So, if left unsprayed, the borders of logging roads, creek edges, and steep slopes would maintain crucial habitat for most songbirds without affecting overall timber production.

FERTILIZERS

Ferilizers are applied to farmlands in various ways and at different levels of concentration depending on the crops and soils involved. Most fertilizers pose no harm to wildlife. However, fertilizers in granular form may resemble seeds or grit and may offer potential hazards for wildlife ingesting large numbers of granules. Fredrickson et al. (1978) summarized instances of moribund birds associated with fertilizer poisoning, including the death of nearly 4500 juvenile pheasants held in pens. Because huge amounts of granular fertilizers are applied each year in the United States—204,000 tons in South Dakota alone—experiments were conducted with fertilizers to determine their effects on pheasants. Breeding hens were force-fed fertilizers in capsules, and chicks ate mixtures of fertilizers in their foods. The results showed no influences on reproduction, behavior, or survival, leading to the conclusion that granular fertilizers normally do not affect unconfined pheasant populations (Fredrickson et al. 1978).

FARMING FOR WILDLIFE

As described earlier, humans often have modified the condition of many soils throughout the world. Desertification of lands bordering the Mediterranean is a good example, but many others exist. The Dust Bowl of the 1930s, triggered when prolonged drought struck the North American interior, was the aftermath of years of negligence and mismanagement (Worster 1979; Fig. 13-10).

For all its tragedy, the Dust Bowl sired new ideas about soil and water relationships. The Soil Conservation Service, established in 1935, was the immediate federal response to abused land (now the Natural Resources Conservation Service). Contour plowing, terracing, alternate and strip cropping, and other means of stabilizing soil resources were advanced under the sponsorship of this agency (Fig. 13-11). Biologists for

Figure 13-10. The Dust Bowl of the 1930s resulted when abused farmlands in the Great Plains fell under prolonged drought. Gritty storms of eroded soil were commonplace in parts of Texas, Oklahoma, Kansas, and Colorado, but clouds of western soil at times reached the eastern United States. (Photo courtesy of The Southwest Collection, Texas Tech University.)

many years have suggested the benefits of mixed crops, hay, and woody cover for cottontails. Shelterbelts, thickets, or other kinds of cover available about every 200 m offer ideal habitat for cottontails on farmlands (Hendrickson 1947). These measures promoted soil stability. But soil by the millions of tons still erodes each year, undermining a multitude of relationships involving water, vegetation, and wildlife as well as agricultural production (see Table 13-2).

SHELTERBELTS

Shelterbelts, or *windbreaks*, were adopted widely in the 1930s as a means of protecting soil against wind erosion. The U.S.D.A. Forest Service alone planted more than 200 million trees in nearly 30,000 km of shelterbelts in 6 states between 1935 and 1942 (Droze 1977). In recent years, however, many shelterbelts have been removed despite evidence of improved yields when crops are protected from wind (Ogbuehi and Brandle 1981; Brandle et al. 1984; but see McMartin et al. 1974). Others are aged and dying, often without being replanted. The Dust Bowl

is not within memory of today's generation of young farmers, and shelterbelts may seem only a relic of another era. With fewer shelterbelts, conservation of soil and wildlife habitat share mutual degradation. Shelterbelts function as islands or corridors of trees in a matrix of cultivated land (Fig. 13-12). Griffith (1976) estimated that less than 3 percent of the area in the Great Plains is covered by woodland, thereby emphasizing the potential of shelterbelts as crucial wooded habitat for many species of wildlife (see also Emmerich and Vohs 1982). Today, shelterbelts are included in the cover types offered by a new conservation program—the Food Security Act of 1985, discussed earlier—in which erodible lands are protected with stable vegetation. So more of these important habitats may be forthcoming on U.S. farmlands in the decade ahead.

Shelterbelts are ecological units offering opportunities for managing game and nongame in otherwise treeless environments (May 1978; Martin and Vohs 1978; Podol 1979; see especially Johnson and Beck 1988 for a thorough review). All told, 17 species of birds nested in shelterbelts sampled by Yahner

Figure 13-11. Contour farming promotes soil and water conservation (top). Likewise, alternate strips of crops help stabilize and protect soil from erosion (bottom). (Photos courtesy of Natural Resources Conservation Service.)

(1982a), and of these, mourning doves nested at a density of 20 nests per ha. Emmerich and Vohs (1982) recorded 15 species of birds using at least 25 percent of the shelterbelts they studied in South Dakota. Shelterbelts on the northern prairies are important nesting areas for merlins (*Falco columbarius*) and Swainson's hawks (*Buteo swainsoni*) (Fox 1971; Gilmer and Stewart 1984). A variety of small mammals occupies shelterbelts, but few of these are agricultural pests (Yahner 1982b, 1983a). Other species benefiting from shelterbelts include bobwhite, pheasants, and cottontails (Stormer and Valentine 1981). Shelterbelts extended the range of fox squirrels (*Sciurus niger*) westward into the southern Great Plains (Trippensee 1948) and connected the ranges

of two color phases of the common flicker (*Colaptes auratus*) (Oberholser 1974). The geographical center of breeding range for Mississippi kites (*Ictinia mississippiensis*) shifted westward when shelterbelts provided new nesting and foraging habitat (Love and Knopf 1978; Glinski and Ohmart 1983). Because they create mosaic patchworks and increase "edges" in open landscapes, shelterbelts may afford Mississippi kites with more nesting sites, enhanced habitat for prey, and increased feeding opportunities (Glinski and Ohmart 1983; Love et al. 1985; see also Forman and Baudry 1984).

Other benefits may accrue from shelterbelts. Ferber (1974) reported estimates that birds consume about 118 kg of insects per 0.8 km of shelterbelt

Figure 13-12. Shelterbelts provide corridors and islands of woody cover in a matrix of intensive farming (top). Note the exposed and wind-blown soils in the fields lacking shelterbelts in the lower center of the photo. Nest of a lark sparrow (*Chondestes grammacus*) illustrates the value of shelterbelts for wildlife associated with woody habitat (bottom). (Photos courtesy of Natural Resources Conservation Service.)

each year. Larger shelterbelts also offer opportunities for hunting. More pheasants were killed in Colorado with less effort near woody plantings than in habitat without shelterbelts (Lyon 1961); strips of woody cover also enhanced quail hunting in Texas (Jackson 1969).

Not all species of wildlife benefit from shelterbelts, however, and distinctions should be made between what is or is not essential habitat (Podoll 1979). For example, shelterbelts are prime habitat for cottontails but they seldom offer the same degree of security for deer unless the sites provide year-round habitat requirements. Prairie chickens actually avoid shelter-belts during certain times of the year. Great horned owls (*Bubo virginianus*) and other avian predators at times may increase the mortality of pheasants near windbreaks, but herbaceous vegetation in these habitats can be managed in ways that reduce predation (Petersen 1979; Snyder 1985). With these distinctions in mind, shelterbelts offer considerable opportunities for managing a large number of species.

Because shelterbelts are artificial habitat, many of the features desirable for wildlife habitat may be planned from the outset. Podoll (1979) summarized the major factors influencing the value of shelter-belts as wildlife habitats as follows: (1) species of food and cover plants selected, (2) the density and arrangement of plants within the shelterbelt, (3) the width of the shelterbelt, and (4) the right-angle orientation of the shelterbelt in relation to prevailing winds. The arrangement and thickness of shelterbelt cover seem crucial, particularly where severe winter weather may be expected. Wandell (1949) cited losses of several hundred pheasants in North Dakota in some shelterbelts, whereas in the same area a better-situated shelterbelt successfully protected 300–400 pheasants from the rigors of winter. Shelterbelts established on the edges of small lakes in heavily cultivated regions may protect wintering waterfowl, thereby conserving their energy reserves during cold, gusty weather (Bennett and Bolen 1978). Stephen (1975) proposed that shelterbelts located next to grain fields may reduce crop depredations because field-feeding ducks generally prefer large, open spaces unencumbered by tree growth. Simple considerations such as the spacing of trees can favor certain species. Spaces of 5–6 m between spruce trees, for example, enhance robin and mourning dove nesting habitat while retaining the primary benefits of the shelter-belt as a wind barrier (Yahner 1982a). Shelterbelts of 8 rows occupying about 0.6 ha next to croplands were recommended, and, as individual trees age and die, some should be left as nesting and foraging sites for birds requiring snags (Yahner 1983b).

Capel (1988) reviewed the layout and design criteria for shelterbelts, of which a few are highlighted here. In northern areas such as the Dakotas and Manitoba, the greater demands for thermal protection and the adversities of drifting snow require at least 8 and as many as 20 rows of trees and shrubs in a belt 33 m or greater in width. Shelterbelts of 2 or 4 rows, however, are sufficient in Texas. The basic design for wildlife includes 5 rows, of which the north or west side consists of conifers (for winter wind and snow protection), an inside row of one or more species of tall, deciduous trees (for vertical structure, nesting cover, and additional wind protection), and then 2 rows of shrubs or short trees on the leeward side (for food production). However, because the width of shelterbelts remains the dominant feature associated with avian diversity and nesting success, additional rows of shrubs further improve the benefits for wildlife. Besides their usefulness for shelter and food for wildlife, the species of trees and shrubs must be selected for their tolerance to conditions such as drought and temperature. Field offices of the Agricultural Extension Service or the Natural Resource Conservation Service can recommend species suitable for local and regional conditions.

Diversity of both the over- and understory vegetation should be encouraged. About 60 percent of the birds using shelterbelts in Minnesota were most often seen on or near the ground, indicating the importance of the understory in shelterbelt management (Yahner 1982c). Grazing is particularly damaging to the understory in shelterbelts as it drastically reduces much of the habitat available for wildlife (Capel 1988). Shelterbelts on grazed lands therefore should be fenced as protection from livestock. As shelterbelts mature, other management can be employed for maximum results, including selective thinning, some weed control, and placement of food patches nearby (Stormer and Valentine 1981). Strips of alternate vegetation bordering fields with shelterbelts offer additional diversity in farm habitat and contribute to the conservation of soil and water resources (Fig. 13-13). Swihart and Yahner (1982) found more cottontails in shelterbelts where debris added structural complexity to these habitats. Whether managers are renovating existing shelterbelts or are planting new ones, development of as much vertical stratification as possible in the structure of the vegetation will improve their usefulness as wildlife habitat (Yahner 1983a,b).

ODD AREAS AND ROADSIDES

Farms often have small areas unsuitable for cultivation, sometimes known as *odd areas*. An aggregate of some 4 million ha of odd areas once was available in the United States (Anderson and Compton 1971), but new techniques of land reclamation have reduced this area. Although they are individually of small size, odd areas may be important habitat for wildlife. Corners of fields where drainage is poor, for example, produce little income; yet they offer several species of

Figure 13-13. Protective strip of rye grass borders shelterbelt and crops. In addition to providing further diversity in farm habitat, the rye grass matures before the critical period of water stress, thereby reducing competition for water between crops and the trees in the shelterbelt. Rye grass also produces more stems than other cereal grains. (Photo courtesy of Natural Resources Conservation Service.)

farm-related wildlife sources of food and cover. The importance of such patches lies in their juxtaposition to large cultivated areas where songbirds, cottontails, quail, and other species often cannot gain a foothold. For example, where center-pivot irrigation is employed, the corners of fields not covered by the circular pattern of water distribution remain available as potential pheasant habitat (Guthery et al. 1980). Center-pivot sprinklers water a circle of about 53 ha, leaving an aggregate of nearly 3 ha in the four corners of each irrigated field available for wildlife management (Fig. 13-14), but newer systems water these previously fallow corners. In wetlands, however, center-pivot sprinklers require construction of travelways (i.e., earthen ramps on which the wheels of the sprinkler move), which act as predator lanes and thereby decrease the nesting success of marsh birds (Peterson and Cooper 1987).

Odd areas need not be managed in many cases—protection from intensive cultivation often is enough to produce benefits, although management sometimes can enhance the carrying capacity of these sites. Farmers may be concerned that odd areas are sources of weed infestation and, if so, limited treatment with herbicides often curbs this threat. In some

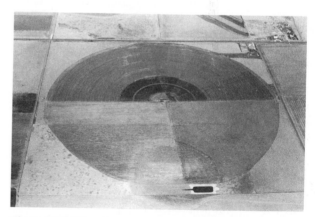

Figure 13-14. Pivot-irrigation systems offer opportunities for conserving wildlife habitat in the corners lying outside of the watered area. Instead of trying to produce crops in the unwatered corners, these areas might be left in native vegetation and, therefore, may support some wildlife on otherwise intensively cultivated farmlands. (Photo courtesy of D. D. Drbal.)

cases, food patches are created when corn or other grains are planted and left unharvested for wildlife. Sod-forming grasses or other undesirable vegetation may limit the success of these without judicious treatments with herbicides (Hamilton and Buchholz

1953; Buchholz and Bayer 1960). A combination of shelterbelts and food patches is a particularly effective management on farmlands. But with or without management, patches of uncultivated vegetation provide wildlife with essential habitat within farming regimes otherwise devoted to a monoculture.

Roadsides offer significant potential for habitat management, not only on secondary roads in farmland but also along major highways and railroad rights-of-way (Oetting and Cassel 1971; Page and Cassel 1971). On farmlands, the trend toward larger fields, fewer fencerows, and a monoculture of row crops has caused major reductions in habitat for many species of desirable wildlife. Few game birds seem more affected than pheasants. Studies of pheasant nesting in managed roadside habitats were initiated in an intensively cultivated area in Illinois (Joselyn et al. 1968; Joselyn and Tate 1972). Selected roadside areas were seeded with a grass-legume mixture and left unmowed until the peak of hatching had passed. More pheasant nests per area were established on these sites than on other types of roadside cover or, indeed, on any of the other habitat available throughout the study area. Hayfields, in particular, are important nesting habitat for pheasants, but as the area of hayfields is reduced, the contributions of roadside environments become more crucial. About 9 percent of the Illinois study area was hayfield at the start of the study, but hayfields amounted to no more that 3 percent of the area 8 years later. Coincident with the reduction in hayfields, the percentage of pheasant nests established on roadsides increased from 7 percent to 45 percent during the same period, further indicating the importance of roadsides as supplemental nesting cover. On a long-term basis (1967 to 1984), roadside management amplified the increase in regional pheasant populations and moderated declines associated with changes in land use and severe winters (Warner and Joselyn 1986; see also Warner et al. 1987).

Cooperation of farmers remains a key element in roadside management. To enhance such associations in Illinois, farmers and state conservation officials entered into cooperative agreements. The state undertook the seeding program and farmers agreed not to mow until August 1 of each year. Up to 89 percent of the farmers complied with their agreements during the first 4 years, but thereafter the compliance dropped to 63 percent (Warner and Joselyn 1978). Most of the farmers supported the roadside management program, although some were concerned about the appearance of unmowed roadsides and weed con-

trol. Some of these shortcomings were addressed by reminding farmers with a newsletter describing the need for their continued cooperation. Also, the farmer's prideful concern for the public's reaction to an unkempt roadside was reduced with signs advising motorists about the program and its merits for pheasant management. Roadside management also provides favorable habitat for numerous songbirds and small mammals, not just for pheasants. This illustrates a relationship often neglected when wildlife management is perceived only as game management. In reality, habitats that have been improved for a few game species produce favorable conditions for dozens of nongame species. So diversity is improved in the overall wildlife community. Further, for each pheasant nest in a managed roadside, several nests of other species could be expected. One might hope that future measurements of improved habitat consider *all* species benefiting from management activities.

Recent evidence suggests some caution, however, about intensive roadside management for wildlife. O'Neill et al. (1983) found that lead from automobile emissions concentrated in the soils, vegetation, small mammals, and some insects along roadsides. The concentration of lead diminished as the distance from the road increased, and the amount of lead in small mammals increased in proportion to the volume of traffic. Because lead may impair reproduction, increase mortality, and cause renal abnormalities in wildlife, roadside management should be limited to thoroughfares where traffic volume is less than 7000 vehicles per day. Bats (Chiroptera) feeding near a major parkway on average contained more lead than small terrestrial mammals (Clark 1979). The stomach contents and carcasses of barn swallows (*Hirundo rustica*) feeding over the same parkway also contained greater amounts of lead than swallows foraging elsewhere. Nonetheless, Grue et al. (1984) concluded that lead in automotive emissions does not pose a serious threat to birds feeding on flying insects over highways. Harrison and Dyer (1984) found accumulations of lead along roadways in a national park and calculated that some mule deer (*Odocoileus hemionus*) could ingest harmful amounts of lead if only 1.4 percent of their daily forage was consumed from roadsides. Continued use of unleaded fuels eventually will reduce lead concentrations in roadside environments.

TILLAGE

Traditional methods of preparing seedbeds for row crops often begin with fall plowing, followed by one or more disking treatments. After planting, the fields again are cultivated; consequently, the land is tilled repeatedly several times each year. Under this regime, tilled lands are exposed to wind and water erosion until crops again help anchor the soil. Years of tillage often may form a hardened plow pan, a compacted layer of relatively impenetrable soil just beneath the bite of the plowshare.

Conservation tillage—known in some of its various forms as *no-till, minimum tillage,* and *stubble mulching*—is a recent concept that overcomes some of the disadvantages outlined above. "Stubble" and other residue from the previous crop are left on the soil surface, thereby forming a cover that reduces wind and water erosion. The cover may be as much as 79 percent on untilled fields compared with 6 percent on tilled fields (Sloneker and Moldenhauer 1977). New crops are planted directly through the residue with seed drills, without plowing or other disturbance of the soil and ground cover. The net results of conservation tillage include less soil erosion, more water infiltration, and reduced runoff, as well as reduced operating costs. Increased chemical treatments, however, may be required for weed and insect pests. Conservation tillage also may increase the density of rodent populations, thereby increasing damage to newly planted crops (Johnson et al. 1985; Castrale 1985). Nonetheless, estimates suggest that conservation tillage can reduce soil loss and fuel costs by as much as 90 percent and 80 percent, respectively (Soil Conservation Service 1980; see also Gebhardt et al. 1985; D'Itri 1985; Sprague and Triplett 1986). Some form of conservation tillage was practiced on 33 percent of all cropland in the United States in 1986 (Conservation Tillage Information Center 1987).

Conservation tillage has definite advantages for soil protection (Table 13-6). The practice also benefits wildlife. Prairie chicken populations in west Texas are influenced by the amount and intersper-

sion of native and cultivated habitats (Crawford and Bolen 1986). Within limits, cultivation increased the carrying capacity for prairie chickens, but additional conversion to row crops thereafter reduced the habitat's suitability. However, positive correlations between the acreages protected by crop residues and the seasonal populations of prairie chickens suggested that conservation tillage partly overcame the intrusion of additional cultivation.

In Nebraska, Nason (1982) also described the advantages of conservation tillage for pheasants and other birds nesting in spring-fallowed croplands. He likened the method to the close association between pheasant abundance and the undisturbed lands during the era of the Soil Bank and predicted that ground-nesting birds will prosper as conservation tillage becomes more widely adopted. However, the relative abundance of insects and other arthropods seems unaffected by conservation tillage, and the practice apparently offers no advantage over tilled fields during the spring when pheasant chicks require proteinaceous foods (Basore et al. 1987).

About half of the northern prairies of North America now are under intensive cultivation. This same region provides the breeding habitat for more than 50 percent of the continent's waterfowl population. Unfortunately, sharp conflicts, including those about wetland drainage and crop depredations, exist between agricultural interests and those concerning the maintenance of large waterfowl populations. Some of these issues must be addressed by one or more levels of governmental policy makers, but other issues involve tillage and other day-to-day farming operations.

On the northern prairies, the stubble of harvested grain crops may be left over winter. Early migrants, particularly pintails (*Anas acuta*), use stubble fields for nesting as little other cover is available until later in the season. However, many pintail nests are destroyed when these fields are cultivated. Milonski (1958) reported that 72 percent of the first nests of pintails were in stubble fields, but about half of these were destroyed by farming operations before they hatched. Pintails, after losing their first nests in stubble, often renested in hayfields, but many of these also were lost when the hay was mowed. Densities of waterfowl nests on untilled land may be 12 times those on croplands and may yield 16 times as many ducklings (Higgins 1977). However, most fields are tilled, and the larger rates of duck production are not realized with current management practices. Conservation tillage thus offers new opportunities for vastly

TABLE 13-6. Water Runoff and Soil Erosion for Fields in Wisconsin Treated with One Kind of Conservation Tillage Compared With Conventional Plowing and Fallowing

Parameter	Conventional Plowing	Fallow	Conservation Tillage
Runoff (mm)	29.5	50.3	24.4
Runoff (%)	7.0	12.0	5.8
Erosion (tons/ha)	6.0	17.5	3.7

Source: Onstad (1972).

improved duck production while offering economic benefits to farmers (e.g., reduced fuel costs) and environmental advantages (e.g., soil and water conservation) to a broader sector of society. Nearly four times as many ducks were produced on farms where conservation tillage was practiced than on conventionally tilled farms (Cowan 1982). Croplands managed with conservation tillage not only expand the habitat base for the entire nesting season but also provide optimal nesting cover with relatively low rates of predation. Winter wheat and fall rye are particularly suitable for conservation tillage and improved duck production because both crops are seeded in the fall, thereby avoiding the risk of destroying nests with seed drill during the spring months.

Basore et al. (1986) recorded 12 species of birds with an average density of 36 nests per 100 ha in fields managed with conservation tillage, compared with only 3 species and a density of 4 nests per 100 ha on tilled fields. Killdeer (*Charadrius vociferus*) and vesper sparrows (*Pooecetes gramineus*) were two of the major species, as were pheasants and mourning doves. Predation rates were high in this study, but Basore et al. (1986) calculated that the continued switch to conservation tillage in Iowa would offer habitat for about 25,000 pheasant nests per year by 1988. In addition to cover, some types of conservation tillage leave large amounts of waste corn and soybeans available as food for wildlife, whereas the amount and availability of waste grains diminish rapidly as the intensity of plowing increases (Warner et al. 1985).

Best (1986), however, speculated that conservation tillage may produce "ecological traps" because nesting success may be lower than is needed for the replacement of breeding stock. If so, birds nesting in such places form so-called sink populations. In this view, the attractive nature of the nesting cover lures birds from other areas, but farming practices (e.g., chemical applications) thereafter may severely reduce breeding success. In other words, the birds may be better off nesting elsewhere instead of breeding in fields managed by conservation tillage (see also Wooley et al. 1985). The potential for more and more sink populations increases as productive habitats are converted into croplands managed with conservation tillage. Nonetheless, such relationships remain unproven, and the "Sodbuster" and conservation reserve provisions of the Food Security Act of 1985 in any case may offset much or all of this threat.

Labisky (1957) studied fields where duck nests reached densities of about 1 nest per ha. The area was associated with a waterfowl refuge but was farmed privately. Virtually all active duck nests in hayfields were destroyed by mowing unless protective measures were employed. These measures included dragging ropes across the field so that incubating hens were flushed and thereafter marking the location as a site to avoid when mowing. This left an "island" of unmowed hay surrounding the nest, but it also attracted predators. Some evidence tentatively suggested that larger islands lessened the incidence of predation, but the size of the predator population may have an overriding influence. Delayed mowing seems the best management procedure, but the delay may reduce the quality of the hay and therefore may diminish the financial reward for farmers.

Because of this, the chronology of some farming activities often adversely coincides with the nesting seasons of several birds. Weigand (1980), for example, concluded that haying and other farm operations neutralize gray partridge production in the United States. Higgins (1977) also projected a dismal future for prairie-nesting waterfowl. Virtually all nesting studies conducted on the intensively farmed prairies indicate that too few nests hatched to maintain waterfowl populations at desirable levels. Poor nesting cover resulting from intensive cultivation, coupled with nest destruction caused by farm machinery and predators, pose severe limitations for future populations of several species.

Rodgers (1983), however, determined that a subsurface cutting blade, used in lieu of surface tillage for weed control, can save up to 53 percent of the bird nests located in wheat stubble. Adults continued incubation on 89 percent of the nests in the undercutting treatments, and no deaths or injuries were observed. The undercutting method provides both wildlife and agronomic benefits because the surface litter continues to protect the soil while controlling weeds and successfully maintaining many bird nests. Because at least 18 species of birds nest in wheat stubble and wheat is fallowed on some 33 million ha of North America, the undercutting method potentially conserves an immense number of nests from destruction each year.

As we have seen, corn has become a staple in the diets of several kinds of wildlife. Many state and federal refuges produce corn solely as fall and winter food for waterfowl, but these efforts achieve maximum benefits only if as much of this food as possible is consumed. It is one matter to leave large amounts of corn in a field and quite another to have it used efficiently. So how might refuge managers enhance the availability of the corn they produce? Let us look at private farming operations before returning to that question. Contrary to the goals of refuge management, the private farming sector removes as much corn as possible. Any of the crop left unharvested represents lost income. Despite this, mechanical corn pickers often leave sizable amounts of waste corn. Waste corn amounted to nearly 4 percent of the harvest, or 364 kg per ha, on farms in the Texas Panhandle (Baldassarre et al. 1983). So regardless of corn's abundance on either refuges or private lands, the same question again arises, namely, how to make this resource available most effectively for field-feeding waterfowl.

Postharvest tillage greatly affects the abundance and availability of waste corn. Plowing turns under 97 percent of the leftover ears and kernels, whereas disking claims 77 percent of this waste (Baldassarre et al. 1983). Both methods disturb the soil and require additional energy and labor costs. Burning, however, circumvents these drawbacks and provides the maximum availability of waste corn. Cornstalks and other litter present a physical impediment for feeding waterfowl so that burning the litter significantly increases the birds' access to waste corn. Whereas the surface litter is removed, the rootstalks remain unburned and continue binding the soil against erosion. Burning thereby increases availability of waste corn, irrespective of its original abundance. Waterfowl respond quickly, selecting newly burned fields in preference to others. Prescribed burning accordingly presents managers of either private or public lands with a cheap, effective tool for manipulating field-feeding waterfowl populations. Feeding pressure and distribution may be managed according to specific objectives, the size of the waterfowl population, and the acreage of corn produced. The corn resource may be apportioned over a longer or shorter period by an appropriate burning schedule, and spatial relationships also may be devised. For example, it may be desirable to disperse birds in the case of an epizootic or to create a patchwork of burned fields in order to distribute hunting pressure more evenly. Also, unburned fields can be held in reserve to meet the sudden demand for extra food when unusually cold weather strikes.

"WILDLIFE PARTNERS" CALENDARS

Calendars, designed especially for the purpose, may motivate farmers and landowners to adopt conserva-

tion practices (Messmer et al. 1996). Farmers are more likely to implement practices beneficial to wildlife if the communication is personalized, so calendars offer a simple and inexpensive means of providing information. The management practices deal with nesting structures, food plots, buffer strips, wetland restoration, and delayed haying until pheasant nests have hatched. Other practices concern conservation education and funding. The calendars, provided free, feature attractive wildlife artwork, which are a means of recognizing as well as motivating landowners' conservation efforts. Moreover, the calendars provide an efficient and accessible means of transmitting time-specific information and therefore do not require several mailings throughout the year (e.g., conservation practices related to plowing or harvest appear in the calendar at the appropriate seasons). When the effectiveness of the calendars was evaluated, 98 percent of the respondents reported adopting at least one of 46 specific conservation practices, some of which represented ideas new to farmers (Messmer et al. 1996). "Wildlife Partners" calendars thus represent a means for conservation agencies to distribute useful information with expectations for achieving desirable results.

SUMMARY

Increased farm efficiency enables production of immense quantities of food and fiber in the United States, but such efficiency often diminishes wildlife. Clean farming reduces habitats required by many species of farm-related wildlife by replacing diversity with monocultures. The Soil Bank and other government programs helped curtail surplus farm commodities, leading to improved habitat conditions for pheasants and some species of ducks when fields were removed from crop production. Similarly, the Food Security Act of 1985, with its Swampbuster and Sodbuster provisions, also holds great promise as a source of wildlife habitat. Wildlife often consumes waste crops without harming farm production, but crop losses in Canada caused by waterfowl remain a difficult conflict still lacking satisfactory management.

Water and wind erosion waste immense quantities of soil each year on croplands, lessening capabilities for production of plant and animal biomass. Sedimentation resulting from soil erosion has diminished the quality and quantity of aquatic habitats for wildlife. Pesticides improved agricultural production but often produced harmful side effects on wildlife.

Chlorinated hydrocarbons are persistent and commonly accumulate in ever-increasing concentrations as they move upward in food chains, and many of these chemicals have been restricted from further use. Instead of relying only on poisons, biological controls and other forms of integrated management can be used to reduce agricultural pests.

Shelterbelts, roadsides, and odd areas on farms provide wildlife habitat, and when needed, these can be improved with management. Conservation tillage saves fuel and labor costs while protecting soil and offers secondary benefits for farm-related wildlife. Whenever possible, farming activities should be scheduled for minimal disruption of wildlife, but full integration of farming operations and wildlife management still poses many challenges. Calendars represent a convenient and effective means of advising farmers about conservation practices.

CHAPTER 14

WILDLIFE AND RANGELANDS

And I will send grass in thy fields for thy cattle that thou mayest eat and be full.

Deuteronomy 11:15

Rangelands occupy about 47 percent of the world's land area (Williams et al. 1968). They characteristically are unsuited for cultivation but instead produce forage for livestock and wildlife. Rangelands also are associated with water, timber, and other natural resources. In the United States, rangelands occupy about one-third of the country—much of it public land—mostly in the 17 states west of the Mississippi River, but also in the southeastern states (see Campbell and Keller 1973).

The potential for effective utilization of all range resources initiated the practice of range management, a discipline integrating biological, physical, and social sciences (Stoddart et al. 1975). Range management is a biological science because it addresses interactions between vegetation and animals; physical because of the roles topography, climate, soil, and water play on rangeland utilization; and social because of the demands people place on the goods and services produced by rangelands. The practice of wildlife management is confronted by these same considerations, each requiring knowledgeable use of scientific principles tempered by sound judgment. Of paramount importance to either discipline is that management of livestock, wildlife, and other natural resources conserves the integrity of the rangeland system without permanent damage.

One generally envisions a range as a vast grassland somewhere "out West," grazed by roaming herds of cattle. Whereas such a scenario is part of the picture, it is a restricted view of a much larger canvas of rangeland types. Arctic Tundra, grazed by caribou (*Rangifer tarandus*), is no less a rangeland, nor are the hot desert-scrub grasslands and their bands of sheep. Grasslands in the Great Plains of North America are usually categorized by their height; tall, mid-, or short, although other categories are recognized in other regions. Grazing also occurs in both deciduous and coniferous forests that are open enough to support herbaceous vegetation. These are found from the higher elevations of the Rocky Mountains to the coastal plains of the Gulf states. Savannas and coastal marshes also are valuable grazing resources. In all, rangelands encompass a broad sweep of ecological settings, but each has in common with the others the utilization of forage by grazing animals.

GRASSES

Grasses are by no means the only forages of interest to range managers. Shrubs, forbs, and even trees fall within the realm of grazing and livestock management. Nonetheless, the production of useful grasses remains a central theme, and some knowledge of grasses and their characteristics is essential for wildlife managers concerned with rangeland systems.

Agrostology is the study of grasses. Most agrostologists recognize Gramineae as the family of grasses, although some favor the designation Poaceae for the group. In all, more than 6000 species of grasses have been organized into 6 subfamilies and 26 tribes (Stebbins and Crampton 1961). However, several other taxonomic systems have been proposed, but most systems for identifying grasses rely on the complex morphology of each species' inflorescence.

The growth of grasses involves complex relationships, but a generalized overview will underscore how grasses respond to grazing. Further, we can compare how grazing may damage other kinds of plants.

Plants produce new growth from specialized tissues known as meristems. In shrubs and most other vegetation, new shoots grow from the tips of older stems, but in grasses, the location of the meristem producing new shoots remains at the base of the plant. Growth in grasses thus is initiated upward from the base. When upper leaves are grazed, grasses generate new herbage from the meristem remaining near ground level, replacing the consumed tissues. In fact, modest grazing stimulates regrowth and, in doing so, produces repeated yields of forage each growing season. Without grazing, many grasses mature into rank, unpalatable vegetation no longer suitable as prime forage. The same capacity for regrowth results after fire or mowing removes the upper parts of most grasses.

Conversely, the meristem tissues producing new shoots in most other kinds of plants remains at the exposed tip of their stems. For these, cropping of the terminal herbage usually impairs further regrowth of a grazed shoot. New growth must await activation of a dormant bud on another stem, which may not occur until the following spring. This is why some woody plants appear "hedged" when their twigs are overbrowsed repeatedly (Fig. 14-1). Many grasses also produce rhizomes, stolons, and tillers as a means of establishing new plants. In sum, major differences between grasses and other plants are the location and vulnerability of their growing points.

Additionally, the high ratio of vegetative to reproductive tissues enables grasses to withstand grazing. In general, a large area of grasses is devoted to photosynthetic activity and therefore to carbohydrate production. There is variation among species in this relationship, however. Grasses that decrease with heavy grazing generally have more reproductive shoots in proportion to the number of vegetative shoots (Branson 1953). Reproductive

Figure 14-1. Palatable shrubs, such as Mexican cliffrose (*Cowania mexicana*), become "hedged" when their exposed growing points are subject to repeated browsing. As shown here, unbrowsed cliffrose develops into fully formed shrubs (top), but a hedgelike form results when deer overbrowse these plants (bottom). (Photos courtesy of North Kaibab Ranger Station, U.S.D.A. Forest Service.)

shoots are adapted for seed production rather than for their tolerance to repeated herbage cropping (Hyder 1972).

Because most grasses grow rapidly, their leaves are able to transport carbohydrates produced by photosynthesis to other parts of the plant, either for continued growth or for storage. Young leaves thus import sugars from older, developed herbage, as does the root system. Grasses vary in their rates of carbohydrate production and storage, with those species that do so quickly showing greater resistance to grazing pressure. Species that accumulate carbohydrates earlier in the growing season are better adapted for spring grazing, whereas those that are slower to produce mature leaves are less able to restore carbohydrates to their growing points (Hyder and Sneva 1959, 1963). So a critical aspect of grassland management is the timing of grazing pressure. Grazing is scheduled ideally to coincide with the end of the active growth period so that new herbage production is stimulated by the removal of the older growth. However, old herbage may be deficient in its nutritional qualities, forcing a trade-off in the design of grazing schemes between maximum regrowth and forage quality (Dahl and Hyder 1977).

Adaptations of grasses to growing seasons influence grazing management as well. Those species that produce most of their growth during fall, winter, or early spring are *cool-season grasses*, whereas those that grow rapidly in the summer months are known as *warm-season grasses*. Ideally, grasslands subject to continuous grazing pressure should contain both warm- and cool-season species.

THE ANIMAL UNIT

The balance between the number of animals consuming range vegetation and the ability of the vegetation to withstand foraging is crucial to management. Ranchers can control the size of their herds, but this alone does not account for the foraging pressure of wildlife sharing the same rangeland. How then do we try to standardize measures for both livestock and wildlife?

The Animal Unit (AU) recognizes the various kinds of livestock—sheep, goats, and cattle—as well as the several kinds of wildlife dependent on range vegetation. One AU equals the live weight of a cow and a calf, or 454 kg, under the assumption that animals of this weight consume a constant amount of forage. The average weights of all other grazing animals are converted to this standard. Thus, 9.6 pronghorns (*Antilocapra americana*), 5.8 mule deer (*Odocoileus hemionus*), 1.9 elk (*Cervus elaphus canadensis*), or 7.7 white-tailed deer (*Odocoileus virginianus*) equal 1 AU (Stoddart and Smith 1955). Conversion to AUs not only permits assessments of foraging pressure but also allows a common base for economic decisions. For example, grazing fees on public lands are based on an AU allotment, giving ranchers opportunities to herd sheep (5/AU), goats

(6/AU), or cattle, or some mixture of these. The AU allotment is based on the overall grazing capacity for each section of range, but it may be filled with different species or classes (e.g., steers, cows, and calves) of livestock allocated according to the AU equivalent for each.

Unfortunately, the AU does not consider different types of foraging pressure; that is, some species might require browse, whereas others subsist primarily on grasses. For example, about 10 pronghorns equal 1 AU, but because cattle generally select grasses and pronghorns greatly favor forbs and browse, the AU equation fails to represent the potential carrying capacity of ranges where both animals forage. Hoover et al. (1959) calculated that 105 pronghorns would actually consume the same amount of grass as one cow, concluding that "all the antelope in Colorado [then about 9000] would not eat enough grass to feed 100 head of mature cattle." Thus, the AU provides only an estimate of foraging pressure from all animals, irrespective of the specific requirements of each.

MANAGEMENT OF RANGE VEGETATION

The practice of range management is as diverse as the types of rangelands and the animals and plants growing on them. In some instances, water development for livestock or wildlife may be the primary management consideration. In general, range management addresses the manipulation of vegetation with treatments such as spraying or by regulating grazing pressure. No matter how intensively managed, however, range vegetation cannot maintain its integrity if it is abused by grazing animals.

Range vegetation is managed to improve the quality and/or quantity of forage available for the production of livestock and, sometimes, for improving wildlife habitat. Practices are used to promote certain classes of forage—usually grasses but sometimes forbs or shrubs as well. Further, certain plants are desired because of species-specific differences in their nutritional values, palatability, or adaptability to local conditions. These often are seasonal in nature so that some plants are favored for spring or summer grazing, whereas others are desirable as winter forage. To accomplish these goals, desirable species sometimes may be seeded but, more often, it is the removal of undesirable species that dominates the management regime (both practices may be undertaken in tandem, although sometimes the desirable plants flourish by themselves once the noxious vegetation is treated). Where overgrazing has deteriorated a range, removal of the excess animals usually will initiate some recovery of the vegetation, but such recovery often takes long periods of time and may be economically unacceptable. Scifres and Polk (1974) recorded desirable changes after spraying a shrub-infested range that took place in about one-fifth of the time required for the same amount of natural recovery. Thus, undesirable vegetation has become the target of mechanical, chemical, and biological methods of management.

Invasions of woody species, particularly shrubby plants, have reduced the carrying capacities on many of the world's rangelands. The immensity of the problem is well illustrated in Texas, where woody plants infest some 36 million ha, or 82 percent, of the state's rangelands (Smith and Rechenthin 1964). Woody plants on more than half of this area are so dense that little or no restoration of grass is possible without intensive management (Fig. 14-2). Left untreated, mesquite (*Prosopis* spp.) and other shrubs continue to limit livestock production as their distribution and densities increase.

Diverse theories abound as to why shrub communities have expanded dramatically in the last century, but most of these implicate human influences rather than precipitous changes in climate or other naturally occurring phenomena. Overgrazing and suppression of wildfires figure prominently in most explanations of increasing shrub densities and invasions. In any case, shrubs (more commonly known as "brush") became the primary target of management practices on most rangelands. Today, however, brush management rather than brush eradication is the goal. Brush is a natural component of most rangeland and offers shade and browse for livestock. Complete removal of brush is as unsound economically as it is ecologically. Land managers now are adopting a broader view of rangeland as an integrated system of plant communities. For wildlife, recognition of this concept coincidentally follows the long-standing principle of habitat interspersion and its creations of "edge."

A primary objective of range management on brush-infested ranges is to reduce the amount of woody biomass so that grasses and other herbaceous vegetation are favored. Water—already a precious resource on most rangelands—also is conserved when the foliage of moisture-demanding brush is reduced and more sunlight reaches the herbaceous understory. Furthermore, dense brush physically interferes with many day-to-day ranching opera-

tions. Working cattle from horseback is almost impossible when brush impedes visibility, accessibility, and movement.

Conversely, elimination of brush is unfavorable for many species of wildlife that now supplement the economic returns on many ranches. A leasing system that permits daily or seasonal fee-hunting currently adds considerable income to private ranching operations, particularly in light of rising operational costs, periodic droughts, and changing livestock markets. In some regions of the West, hunting revenues may equal or exceed the income per hectare from livestock production, suggesting that the ranching industry will adopt new ways of managing land for both livestock and wildlife (see Ramsey 1965; Teer and Forrest 1968; Teer 1975). Steuter and Wright (1980) defined the trade-off between brush cover and its relationship with deer densities against livestock production requiring other types of range vegetation. This relationship indicated that deer numbers will diminish when and if managers remove brush on a large scale. However, if herbicides are applied in alternating strips instead of spraying large blocks, then as much as 80 percent of brush-infested rangelands may be treated (Beasom and Scifres 1977). Further, selective treatments of specific range sites within large pastures may overcome the negative impact of herbicides on deer habitat while concurrently improving forage production for livestock (Beasom et al. 1982). Box and Powell (1965) suggested that brush be managed as forage for both livestock and wildlife and not as woody weeds. Mowing, for example, removes the physical barrier of dense brush, promotes water conservation, and generates regrowth of succulent, highly available browse (Table 14-1). One year after mowing, brush regeneration also was suitable for bobwhite (*Colinus virginianus*) and turkey (*Meleagris gallopavo*) nesting and as cover for deer fawns. Within 3–5 years, adult deer and javelina (*Tayassu tajacu*) found the regrowth high enough for their cover requirements, and all the while, palatable forage was available to both cattle and wildlife.

Wildlife potentially affected by range improvement techniques includes, among many others, pronghorns, quail, and deer. Of the several species of exotic grasses commonly used to revegetate ranges in Texas, only kleingrass (*Panicum coloratum*) appears to provide adequate forage for livestock *and* seed acceptable to bobwhites (Pitman and Holt 1983). A host of songbirds and other nongame species also depends on rangeland habitats, and techniques that alter the composition and structure of range vegetation influence

Figure 14-2. Various kinds of management are used where dense woody vegetation limits livestock production on rangelands (top). In this case, a heavy chain drawn by two bulldozers mangled mesquite and other brush (bottom). (Photos courtesy of N. E. Adams.)

TABLE 14-1.   Preference and Forage Values for Selected Species of Brush Following Treatment with Mechanical Methods

	Soil Type	Treatment Method					
		None	*Mowed*	*Roller Chop*	*K-G Blade*	*Root Plow*	*Root Plow and Rake*
Preference Value[a]							
	Heavy clay	1,330	12,755	12,670	9,660	4,000	8,820
	Fine sandy loam	4,721	—	19,205	25,575	915	1,000
Forage Value[b]							
	Heavy clay	25	196	160	108	6	13
	Fine sandy loam	165	—	255	260	9	6

Source: Box and Powell (1965).
[a]Percentage utilization × frequency of utilization.
[b]Preference value × density of the plant.

these animals as well. Recall that populations of "lesser species" often are keys in food webs and other ecological patterns, such as seed dispersal, of rangeland systems (see McAdoo et al. 1983). For example, when forage-depleted ranges are seeded artificially with grasses and a monoculture subsequently develops, the numbers of rodents and lagomorphs in the new community often are reduced, presumably affecting the prey base available to raptor populations (Howard and Wolfe 1976; but see Guthery et al. 1979 for other results).

MECHANICAL METHODS

Range management has at its command several ways to treat brush. Each of these, however, must be assessed prior to use for the type of response desired. Individual differences in the tolerance and adaptation of woody plants and the soils on which they occur have much to do with the selection of treatments. This seems particularly true where mechanical means are used. Shredding, roller chopping, root plowing, and chaining are among the more common mechanical treatments on brush-infested rangelands (Fig. 14-3). Box (1964) found that 98 percent of the mesquite growing on fine sandy loam in south Texas was killed by root plowing, whereas on heavy clay soil, about 19 percent died when subjected to the same treatment. Conversely, other plants that were rather obscure in the pretreatment composition increased dramatically: Two species of cacti increased 600 percent after root plowing the sandy-loam rangeland.

With such dramatic changes in vegetation, it is clear that equally profound changes likely will occur in the composition and density of animal populations on treated rangelands. The results include both the obvious cover and food requirements for native and domestic animals as well as some less obvious results (e.g., nutritional and other changes, even in the same plants, between pre- and posttreatment vegetation). For example, five of six species of brush showed increased crude protein content after mowing, a factor that contributes to their increased preference values and forage ratings for deer and cattle following this treatment (Powell and Box 1966). However, mechanical treatments of this type usually enjoyed short-lived results as the succulent sprouts springing from the cut stems mature and steadily develop more woody growth. At another extreme, woody plants such as big sagebrush (*Artemisia tridentata*) contain large amounts of essential oils that, although seemingly of nutritional benefit as a prime source of energy, may be indigestible and perhaps harmful in the rumens of mule deer (Nagy et al. 1964; but see Welch and Pederson 1981 for other results). Such findings emphasize the subtle nature of nutrition in the management of woody vegetation as forage.

Mowing, instead of killing brush outright, effectively alters the growth form, or physiognomy, of the plant community. Woody species such as mesquite or huisache (*Acacia farnesiana*) can develop into trees with sizable trunk diameters, whereas others such as agarito (*Berberis trifoliolata*) and catclaw (*Acacia greggii*) remain shrubby and seldom exceed 2–3 m in height at maturity. Mowing reduces the canopies of shrubs and trees alike but, for sprouting species, the root system remains viable and generates new growth; sprouting occurs at the root crown lying below the cutting level of the mowing equipment. For mesquite, Wright and Stinson (1970) reported that about 25 percent of the removed top growth was replaced with new sprouts by the end of

Figure 14-3. Roller chopping (top left) and bulldozing (top right) are two means of mechanically treating brush-infested ranges. With root plows, a heavy blade is towed through the soil, severing brush below ground (middle row). Heavy chains dragged behind bulldozers are another common means of reducing brush cover (bottom row). (Photos courtesy of Texas Agricultural Extension Service.)

the first growing season. Mowing also replaces armored stems or foliage, a common feature of many brushy plants, with regrowth that initially is less protected by spines and thorns. In sum, mowing reduces brush canopies and generally produces regrowth of the same species but, in doing so, succulent forage is regenerated at a height where it is accessible to browsing animals. However, complete removal of the screening cover provided by brush may negate the benefit to deer of newly available forage resources. White-tailed deer otherwise commonly adapt to changes in food availability wrought by mechanical brush removal if adequate cover remains on the treated area (Quinton et al. 1979).

Chaining also removes gross amounts of cover, often disrupting the behavioral patterns and distribution of deer. This was indicated when chaining reduced the canopy cover of bottomland deer habitat from 20 to 4 percent but otherwise did not alter the cover of forbs or grasses (Darr and Klebenow 1975). Previously, white-tailed deer numbers were four to five times greater in the bottomlands than in the surrounding uplands. After the bottomlands were chained, however, deer densities dropped to about half of their former numbers. Movements to and from the chained areas were limited to morning and evening and the animals attempted to use any clumped cover remaining in the chained area for concealment. They otherwise retreated to unchained sites during the daytime and when venturing into the chained areas tended to remain near the edge of the cleared area.

In Colorado, O'Meara et al. (1981) determined the aftermath of chaining on nongame populations. Whereas 10 species of breeding birds occurred on an unchained area, only 4 species were observed on similar sites chained 15 years earlier. Further, the density of the bird population was reduced by half on the chained sites. Conversely, small mammals were more abundant on the chained areas but included fewer species compared to unchained habitats. Management procedures minimizing the adversity of chaining on nongame animals include (1) using lighter weight chains and thus improving the survival of some shrubs and smaller trees, (2) selectively retaining trees with cavities, and (3) limiting the widths of cleared areas to 200 m.

Root-plowing, a method that also drastically reduces the cover of woody vegetation, seems equally inhibitory to deer movements and foraging behavior (Davis and Winkler 1968). Deer in south Texas moved onto root-plowed areas at night in groups of 5–30 animals but made little use of the forage available there and then returned to the cover of untreated brush during the day. The stress of drought-induced shortages of forage was not enough to force deer into the open root-plowed sites where escape cover was absent. Further, not all deer left the brush cover even at night, and in the brush cover, they remained active during daytime. Drought was more devastating to the availability of forage on root-plowed areas than in stands of native brush. On root-plowed sites, drought and overstocking of cattle produced severe shortages of forage so that only the untreated brushlands could support either deer or livestock. After 25 or more years, root-plowed brush in south Texas showed long-term reductions in both the density and diversity of browse plants preferred by white-tailed deer and increases in those of lesser preference (Fulbright and Beason 1987). These results again strongly suggest that brush should be managed in small units or strips rather than in large blocks.

HERBICIDES

Aerial applications of herbicides are widely used to reduce undesirable plants on rangelands (for example, see Hylton et al. 1972 for sagebrush control with chemicals). Some are selective or only partially effective on certain species, whereas other herbicides have a broader spectrum of effects. The most widely used herbicides are those that attack broad-leaved plants without affecting grasses. Unfortunately, many herbicides injure forbs as well as woody species (Blaisdell and Mueggler 1956). Such results are of mixed value for wildlife, depending on the importance of forbs as food. Two months after spraying solid blocks of rangeland in south Texas, Beasom and Scifres (1977) determined that forb production was reduced from about 200 kg/ha to nearly 30 kg/ha; recovery took more than 2 years. Species diversity also was affected; 15 months after spraying, only 13 of 30 forbs were present on the treated rangeland, and others were reduced in density by as much as 75 percent. Meanwhile, numbers of white-tailed deer were reduced by 60 percent on the sprayed site, in clear response to the loss of forbs (Fig. 14-4). Other research has shown that white-tailed diets contained as much as 68 percent forbs by volume (Chamrad and Box 1968; Drawe 1968), confirming the importance of forbs as a necessary component of rangelands managed for deer. Similarly, javelina populations were significantly reduced on sprayed areas because of the nearly complete elimi-

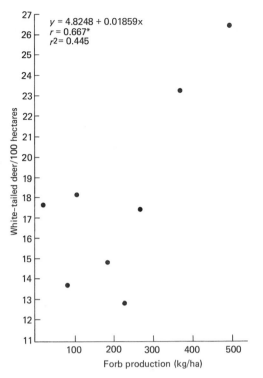

$y = 4.8248 + 0.01859x$
$r = 0.667*$
$r2 = 0.445$

Figure 14-4. Relationship between dry weight of forbs and deer densities after herbicides were sprayed on brush communities on south Texas rangelands. (From Beasom and Scifres 1977.)

nation of prickly pear (*Opuntia* spp.), a food dominating 78 percent of their diet by volume with 100 percent frequency of occurrence (Jennings and Harris 1953). Turkeys were not affected in this instance, reflecting their more general diets and opportunistic feeding habits.

Based on the evidence at hand, efforts to reduce sagebrush with herbicides have severely restricted sage grouse (*Centrocercus urophasianus*) populations. In Montana, Wallestad (1975) reported that a 31 percent loss of sagebrush habitat near a strutting ground coincided with a 63 percent decrease in courting males, and Peterson (1970a) described a strutting ground with a long-term average of 54 courting males that diminished to only 3 birds within 2 years of spraying. Nesting and brood habitats also were depleted by spraying. In Idaho, Klebenow (1970) found only 1 nest per 89 ha on sprayed sites, whereas nesting densities reached 1 nest per 26 ha on untreated rangeland. Spraying removed enough shrub cover to inhibit nesting, as shown by a comparison of cover at treated and untreated areas and nest sites (Table 14-2). Broods were more common on unsprayed sites where forbs generally remained more abundant. However, some individual species such as dandelions (*Taraxacum* spp.) occurred more frequently on the sprayed habitats and, as these plants are important food for juvenile sage grouse, some compensation for the loss of other foods was achieved (Klebenow and Gray 1968; Peterson 1970b). Nonetheless, a recovery time of about 5 years after spraying was judged necessary before nesting resumed and 10 years before the original carrying capacity returned fully (Fig. 14-5). Martin (1970) found parallel results on sagebrush rangelands in Montana; herbicides killed 97 percent of the sagebrush and reduced forb coverage by half, and only 4 percent of all sage grouse observations occurred on the sprayed area. Human changes in vegetational composition clearly were responsible for the reduction of sage grouse in both Idaho and Montana. In Wyoming, Johnson (1969) estimated that sagebrush requires 14 to 17 years to recover fully from herbicidal treatments, suggesting that the spraying likely causes a similar impact on that state's grouse populations.

With planning, however, sagebrush may be treated with herbicides without wholesale damage to sage grouse habitat. Sage grouse favor nesting sites between 1650 and 1770 m elevation and begin to concentrate

TABLE 14-2. Comparisons of Cover in Sagebrush Communities (Sprayed and Unsprayed) with Sage Grouse Nesting Cover

Treatment or Location	Shrub Cover (percent)				Basal Area of Forbs (Percent)
	Sagebrush (Artemisia spp.)	Bitterbrush (Purshia tridentata)	Horsebrush (Tetradymia canescens)	Total	
Unsprayed	13.5	0.1	0.4	15	3.2
Sprayed	5.8	1.9	0.6	7	2.4
Nest sites	16.4	0.9	0.7	18[a]	3.0

Source: Klebenow (1970).

[a]Minimum shrub cover at any nest was 21 percent ($N = 32$).

Figure 14-5. Sage grouse, shown here in courtship, are affected by indiscriminate applications of herbicides on large, unbroken tracts of sagebrush. Only when key areas or strips within sagebrush communities remain unsprayed do these habitats retain their carrying capacity for sage grouse. (Photo courtesy of U.S. Fish and Wildlife Service.)

their broods in June near swales, drainages, or other mesic locations as adjacent habitats dry (Klebenow 1970). Accordingly, Braun et al. (1977) proposed management recommendations that would minimize damage to sage grouse populations whenever ranges with sagebrush are treated. These include (1) no treatments within 3 km of strutting grounds, nesting habitat, or brood areas, (2) applying treatments in irregular patterns no more than 30 m wide and leaving untreated areas of at least the same width, (3) leaving 100-m strips of untreated sagebrush along the edges of meadows and drainages, and (4) overall, avoiding consideration of any treatment wherever sagebrush cover is less than 20 percent of the total range community.

Herbicides at times may produce selective results benefiting wildlife. In northern transition zones between prairie, deciduous, and coniferous forest types, spraying reduced low-quality deer foods and increased supplies of better browse species (Krefting and Hansen 1969). Deer were attracted to the sprayed plots for winter forage and for summer bedding and fed four to five times more there than on adjacent unsprayed sites. Counts of pellet groups showed increased deer usage on the sprayed areas, especially in

aspen (*Populus tremuloides*) areas, 8 years after treatment. Krefting et al. (1956) also stimulated productive regrowth of highly preferred deer browse when mature plants were treated with herbicides. Spraying remains an efficient tool for managing many other types of habitat where dense, often monotypic communities preclude desirable amounts of diversity and interspersion of food and cover vegetation. For example, herbicides may improve habitat for lesser prairie chickens (*Tympanuchus pallidicinctus*) on ranges where shinnery oak (*Quercus havardii*) forms dense canopy cover. Prairie chicken numbers decreased as the cover of shinnery oak increased (Cannon and Knopf 1981). Hence, some reduction of the shrubby cover may result in larger numbers of prairie chickens, as suggested by experiments with a common herbicide, Tebuthiuron (Doerr and Guthery, 1983). Tebuthiuron applied at 0.4 kg/ha effectively controlled shinnery oak and increased grass production, yet did not reduce the availability and diversity of forbs required by lesser prairie chickens.

FIRE

What maintains the integrity of grassland systems? Why have shrubs and trees not long ago replaced grasses in the course of plant succession? In part, the answer lies in soil characteristics and precipitation regimes, but as woody plants are scattered in grassland communities, other forces also must exert strong influences on grasslands.

Because ecologists have learned that fire favors grasses over woody plants, they attribute this force as much as any to the continued maintenance of many grasslands. This happens in at least three ways (Cooper 1961):

1. The growing points of most grasses are protected; they lie near or below ground level so that fires do not often kill the meristematic tissues essential for regrowth. Conversely, the growing points of woody plants are exposed well above ground level where they are damaged easily by fire.
2. Grasses regrow rapidly after burning and most produce seed within the same growing season. Woody plants, by comparison, take several years to produce seed.
3. For grasses, fire removes only 1- or 2-year growth above ground, but for woody plants, burning causes a loss of fiber, representing many years' accumulation. So the rate of development and maturity after burning greatly favors grasses.

Therefore, one easily sees the implication for management of grassland communities and the grazing animals they support. If fires are repeatedly suppressed, woody vegetation indeed may develop and largely replace grasses. Coupled with overgrazing, long-term fire suppression enhances deterioration of productive grasslands into shrub-dominated ranges of diminished value.

Fire also increases the palatability of many grasses; that is, stands of old grasses may be rank and strawlike and therefore spurned by grazing animals. After burning, however, regrowth of the same grasses is tender, usually enriched with protein, and highly palatable.

For effective management, fires must be prescribed to fit local conditions (Wright and Bailey 1982). In California chaparral communities, for example, sprouting and nonsprouting species are abundant. If the interval between burns is too short—less than 15 years—fire will favor the sprouting species because nonsprouting plants have inadequate time to produce seed between fires. Conversely, long fire-free intervals favor nonsprouting species. If only one noxious nonsprouting species is present in a community, such as one-seed juniper (*Juniperus monosperma*), two fires within 10 years will keep this species suppressed for 30 or 40 years. Frequent fires in communities dominated by sprouting species such as honey mesquite (*Prosopis glandulosa*) will reduce the cover of this and other shrubs but maintain a continuing occurrence of shrubs in the community. In all, prescribed burning requires considerable ecological knowledge.

GRAZING AND WILDLIFE

Grazing animals understandably influence the integrity of range vegetation. Bison (*Bison bison*) and pronghorn are the more obvious grazers coming to mind in considering once pristine rangelands, but jackrabbits (*Lepus* spp.), deer, prairie dogs (*Cynomys ludovicianus*), numerous other rodents, insects, and many birds also foraged on native rangelands. Although these animals certainly exerted some influence, rangeland vegetation coevolved with foraging wildlife into a dynamic equilibrium between consumers and producers (Stoddart et al. 1975). With the human introduction of domestic livestock, however, a new force was exerted on rangeland vegetation, and the equilibrium of the past changed quickly.

Livestock altered the composition of rangeland vegetation. Plants with high forage values became less abundant, creating a grazing disclimax whose development remains proportional to the stocking rate and its duration. Livestock also alter the physiognomic aspects of rangeland vegetation. Tall grasses may be held at lesser heights and densities so that ground cover and other physical features of the community are affected. For example, tall grasses used as bedding sites by pronghorn fawns also are desirable range forage for livestock. If grazing reduces this cover to the point that fawns no longer have adequate concealment, fawn survival may be jeopardized by higher rates of predation (Tucker and Garner 1983). Influences of this kind also were illustrated by studies of Mearns quail (*Cyrtonyx montezumae mearnsi*) on ranges in Arizona (Brown 1982). Larger amounts of seeds and other quail foods were produced when more plants were grazed by livestock. Indeed, production of important quail foods more than doubled when grazing intensity increased from 50 to 91 percent of the range vegetation. But quail were uncommon or even absent in those pastures where plants were heavily grazed. It became clear that the heavier grazing pressure, while producing more quail food, concurrently reduced escape cover below a critical threshold and so effectively eliminated quail from these ranges. The study demonstrated that no more than 50 percent, by weight, of the available forage could be utilized by grazing animals in order to maintain Mearns quail at optimum numbers.

Managers appraise the composition of rangeland vegetation to determine *range condition*. This appraisal is based on how much the current vegetation deviates from its potential at the range site. In other words, what percentage, if any, of the present community still represents climax vegetation? With few exceptions, climax grasses are preferred livestock forage and therefore are known as *decreasers*. Somewhat lower quality plants are known as *increasers* because their percentage expands with grazing. A balance is sought between decreasers and increasers so that, overall, range vegetation remains productive and useful forage. Range condition declines from excellent into classes of good, fair, or poor, as the percentage of climax species decreases, accompanied by a reverse pattern in the increasers (Table 14-3). With long periods of excessive grazing pressure, *invaders* become more prevalent in the community, reaching a point where the range's former carrying capacity for livestock and wildlife becomes severely reduced.

The welfare of livestock, of course, coincides with range condition, but so do many features of wildlife

TABLE 14-3.   Sample Calculation of Range Condition Based on the Composition (Percent Cover) of the Climax Community Weighed Against Current Composition of Vegetation

Species or Group	Climax Vegetation (Percent Cover)	Current Vegetation (Percent Cover)	Current Proportion of Climax (Percent Cover)
Sideoats grama (*Bouteloua curtipendula*)	100	10	10
Perennial threeawn (*Aristida* spp.)	5	10	5
Texas grama (*Bouteloua rigidiseta*)	5	5	5
Forb increasers	10	5	5
Woody increasers	5	20	5
Hairy tridens (*Erioneuron pilosum*)	0	15	0
Annuals	0	35	0
		100	30[a]

Source: Dyksterhuis (1949).

[a]Four classes of range condition can be recognized based on the current percentage of climax vegetation: excellent, 75–100 percent; good, 50–75 percent; fair, 25–50 percent; and poor, 0–25 percent. In the example above, the range condition is at the low end of the fair class.

TABLE 14-4.   Percentage of Yearling Bucks Harvested with Spike Antlers in Relation to Range Conditions in Two Ecological Regions of Texas

Range Condition	Percentage of Yearlings with Spikes	
	South Texas Plains	*Edwards Plateau*
Good	34.5	26.7
Poor	58.4	59.7

Source: Gore and Harwell (1983).

associated with rangelands. For example, range condition influences antler development of white-tailed deer. Larger percentages of yearling bucks with spike antlers—instead of multibranched antlers—were harvested from ranges in poor condition, whereas the percentage decreased when range conditions improved (Table 14-4). In part, this may be explained by enhanced nutrition of deer on ranges in better condition. The diets of white-tailed deer on ranges in excellent condition were 29 percent higher in crude protein and 27 percent higher in phosphorus than those feeding on poor ranges (Bryant et al. 1981). Further, yearling deer maintained on experimental diets of low nutritional quality produced an average of 0.62 fawns, whereas those fed high-quality diets produced 1.63 fawns per doe (Verme 1969).

Livestock and wildlife also may compete directly for food. Items in the diets of each may reflect strong competition for specific plants, but data are more often reported by forage classes: grasses, forbs, and browse. Distinctions between these groups occasionally are unclear, primarily for some species of broad-leaved herbaceous plants having woody stems (known as *suffrutescents* or "half shrubs"); examples include Gardner's saltbrush (*Atriplex gardneri*) and winterfat (*Ceratoides lanata*). A wise manager also is aware that food habits may be seasonal. Regional differences also occur in the diets of animals of the same species. Furthermore, close study of interspecific foraging behavior may show that parts of the same plants are consumed differentially. If so, competition is less than it might seem initially. Finally, a number of food-habits techniques are available, each having some amount of inherent bias for determining diets accurately. Stomach contents are commonly reported by volume and by frequency of occurrence, but to be truly useful, these data should be accompanied by some measure of forage availability, yield, and nutritional value.

Studies of competition among sheep, goats, cattle, and white-tailed deer indicated strong overlaps in diets, depending on season (McMahan 1964; Bryant et al. 1979). In winter, each of these ruminants competed for browse, with deer and goats showing the greatest degree of competition for browse and mast in all seasons. Sheep and deer each made heavy use of forbs, particularly in summer, when 65 percent and 68 percent of their respective diets were dominated by forbs. However, Bryant et al. (1979) suggested that competition for forbs also occurred in winter and early spring. Cattle and goats did not compete strongly with deer for forbs (McMahan 1964). Grasses were used primarily by deer only in the spring when the plants were succulent, but if forbs and browse were absent at other seasons, grasses then were eaten in larger than usual amounts.

Photo courtesy of Northern Prairie Wildlife Research Center

Northern Prairie Wildlife Research Center

Established in Jamestown, North Dakota, in 1965, this facility recently gained a new affiliation. Previously, the Northern Prairie Wildlife Research Center was operated by the U.S. Fish and Wildlife Service, but the center now is operated by the Biological Resources Division of the U.S. Geological Survey. The center's research mission also has changed from one primarily focused on waterfowl west of the Mississippi River to a mission expanded to include research, monitoring, and inventory of the entire Great Plains biota, but with an emphasis on the ecology of waterfowl and other grassland birds, wetland biology, predator ecology, and the management of grassland habitats. Results obtained by the Northern Prairie Wildlife Research Center guide the management programs of the National Park Service, U.S. Fish and Wildlife Service, and other agencies in the Department of Interior.

In addition to office and support buildings, the 260-ha facility includes an aquatic laboratory, holding pens for predators, experimental ponds, a herbarium, dormitories for visiting scientists, and an advanced geographical information system. A herbarium of more than 6000 specimens is housed at the center, and a database that reports the fates of more than 85,000 waterfowl nests from 1966 to the present is available for study. The permanent staff numbers about 45, and up to 65 employees have temporary and seasonal positions. Graduate students are frequently based at the center.

The Northern Prairie Wildlife Research Center also maintains a 1073-ha study area at Woodworth, North Dakota, which is dedicated to prescribed burning and other types of land-management research. Another site, the Cottonwood Lake Study Area, is a fully instrumented wetland complex where hydrology, water chemistry, and other features are monitored to determine the physical and biological functions of prairie wetlands. The center specializes in field, rather than laboratory, research.

The center's research staff is organized into a Section of Northern Plains Ecology that includes a Grassland Ecosystem Initiative Team with an associated field station at the University of Minnesota, a Wetlands Restoration Initiative, and a Wildlife/Landscape Ecology Team. Also at Jamestown is a Section of Support Services that consists of statisticians, remote-sensing staff, a library and a librarian, and a computer staff that operates the center's home page (http://www.npwrc.org). A Section of Central Plains/Ozark Plateau Ecology is based in Columbia, Missouri; a field station at the University of Arkansas supports this program. Overall, the Northern Prairie Wildlife Research Center is dedicated to the development and dissemination of new information about the resources of the grasslands and adjacent areas in the central United States.

Competition for grasses otherwise was greatest among cattle, sheep, and goats.

The inferences of these data become clear when the performances of deer herds are weighed against the accompanying stocking rates of domestic animals (Table 14-5). In much of the Southwest, and especially in the thickly populated deer range in central Texas, summer is the critical season. Extensive losses of deer may occur when summer rains fail and forage is diminished. If livestock already have depleted most of the forage, the few remaining food resources are subject to intense utilization. Deer first compete with goats for browse in winter and then are forced into a second competitive season in summer with

TABLE 14-5. Average Results from White-Tailed Deer Maintained Experimentally for 5 Years on Rangelands With and Without Livestock (Sheep, Goats, and Cattle)

Livestock Stocking Rate	Ha/Animal Unit	Deer Herd	
		Adult Mortality (Percent)	*Change in Size (Percent)*
Heavy	3.2	40	−43
Moderate	6.5	14	9
Light	9.7	14	6
Deer only[a]	—	5	32

Source: McMahan (1964).

[a] 2.6 ha per deer.

sheep for forbs. Competition for grasses exists largely between cattle and sheep, without major involvement of deer. Continuous livestock grazing, when it includes sheep and goats, therefore presents a highly competitive arena for deer.

Warren and Krysl (1983) demonstrated that grazing and hunting pressure interacted with the food habits and nutritional status of white-tailed deer. On an area with 20–25 percent harvest of deer, few exotic big game, and regulated grazing (cattle only), deer experienced better nutritional condition at a much earlier age than on another area where exotics were common, livestock of several kinds grazed continuously, and the harvest of deer was only 5–10 percent of the population. Comparisons of the diets of deer also reflected the better range conditions on the managed area.

Bobwhite apparently spread westward after livestock were established on western rangelands, likely because of the physiognomic changes grazing wrought to the vegetation (Edminster 1954). The same alterations greatly expanded mule deer populations (Holechek 1982). Tall, dense grasses were replaced in part by annual forbs and woody species. California quail (*Lophortyx californicus*) benefit from moderate grazing in humid regions that otherwise would support vegetation less suited for the birds (Leopold 1977). Moderate grazing also provided Attwater's prairie chickens (*Tympanuchus cupido attwateri*) with favorable nesting, escape, and winter cover, whereas ungrazed pastures were not used by the birds (Kessler and Dodd 1978). Ungrazed vegetation develops into thick mats of cover avoided by prairie chickens. In short, grazing *can* be a positive factor in the distribution and abundance of wildlife when the composition and physiognomy of range vegetation are appropriate. Undergrazed grasslands, as suggested above, once may have limited some wildlife populations, but in modern times overgrazing now seems far more critical.

When grazing replaced farming in northwestern Florida, the pressures of booming cattle prices and drought led to overgrazing that subsequently diminished the food and cover available to quail. Quail populations dropped when these changes in land use occurred despite attempts to manage quail with food plantings (Murray 1958). Likewise, Klimstra and Scott (1957) noted that overgrazing markedly reduced bobwhite nesting on pastures in Illinois. On southern pastures, Lay (1954) suggested that declines in bobwhite coincided with the increase in livestock and stated that "maximum numbers of quail and maximum numbers of livestock are incompatible."

Nonetheless, grazing can be manipulated for the betterment of both bobwhite and cattle. Bobwhite in Illinois usually nested in grazed pastures, likely reflecting the open ground cover that grazing produced (Klimstra and Scott 1957). Furthermore, heavy grazing of selected "spots" of small size can create lower successional stages favorable for bobwhites; these should be grazed before the growing season begins so that the vegetation can respond during the same year (Jackson 1969). A simple way to accomplish spot grazing is to move the winter feeding stations for cattle periodically, thus forming a limited patchwork of heavily grazed areas within a more uniform rangeland type.

OVERGRAZING

Overgrazing represents not only eventual depletion of forage resources, but also may work against current income (Table 14-6). Based on a 19-year study of grazing intensity in Colorado, moderate grazing yielded more profits than either heavy or light stocking rates (Bement 1969). As vegetation retrogresses with continued overgrazing, crucial soil and water relationships also fail, ultimately reducing the rangeland's carrying capacity for grazing animals, even after stocking rates are lessened. Rates of water infiltration on watersheds in South Dakota showed proportional relationships with grazing intensity (Rauzi and Hanson 1966); that is, infiltration was lowest with heavy grazing and highest with light grazing. Heavily grazed watersheds also had the highest rates of runoff.

Range deterioration is not without other measures of ill health. Taylor et al. (1935) recorded significant increases in insect numbers on overgrazed range-

TABLE 14-6.   Returns and Production per 260 ha for Three Levels of Grazing Pressure on Rangelands in Colorado

	Grazing Pressure		
	Heavy	*Moderate*	*Light*
Number of livestock	53	47	27
Gain per head in 5 months (kg)	78	96	105
Gross return ($)	661	1027	724
Operating costs ($)	188	163	97
Net profit ($)	473	864	627
Profit/gross return ratio (%)	72	84	87

Source: Johnson (1953).

TABLE 14-7. Insect Populations (1000s per 0.4 ha) on Overgrazed and Properly Grazed Rangelands

Range Condition	Beetles (Coleoptera)	Flies (Diptera)	Bugs (Hemiptera)	Leafhoppers (Homoptera)	Bees, Ants (Hymenoptera)	Grasshoppers (Orthoptera)	Total
Overgrazed	118	30	100	214	140	180	782
Properly grazed	50	28	8	52	28	20	186

Source: Taylor et al. (1935).

lands (Table 14-7). Typically, the number of species declines, and this loss in diversity is accompanied by a larger number of total insects (Smith 1940). Ironically, forage-consuming insects such as grasshoppers show dramatic increases in numbers on overgrazed ranges (Nerney 1958). Grazing also may influence—one way or another—the abundance and diversity of lizard communities (Jones 1981), birds (Wiens 1973), and small mammals (Grant et al. 1982).

Black-tailed prairie dogs and jackrabbits often are held responsible for depletion of range vegetation. Hansen and Gold (1977) found that the overall diets of prairie dogs and cattle overlapped by 64 percent, but many of the plants were grazed at different times of the year, thereby reducing the degree of direct competition. Nonetheless, prairie dogs and jackrabbits are common targets of control programs on western rangelands. Jackrabbit "drives" often gain national attention in the news, as has the poisoning of prairie dogs, causing emotional responses in some sectors of the public (see McNulty 1971). Human responses aside, the ecological issues may be complex and may directly or indirectly involve other wildlife. In particular, endangered black-footed ferrets (*Mustela nigripes*) are intimately associated with prairie dogs so that the fates of the two species remain linked (Hillman 1968; Hillman et al. 1979).

Large populations of jackrabbits and prairie dogs actually may be the result of overgrazing, not the cause of range depletion. These animals are associated with the early stages of succession in grassland ecosystems and therefore are "weed species" (Fig. 14-6). Bison herds apparently created favorable habitat for prairie dogs and jackrabbits in earlier times. Today, excessive numbers of cattle can produce the same or even more disturbances to grassland communities. Control of either prairie dog or jackrabbit populations may require no more than better management of livestock grazing. Some ecologists claim this can be achieved simply with barbed wire, meaning that fences used to regulate grazing pressure will promote better range conditions and hence fewer prairie dogs or jackrabbits. In Arizona, Taylor et al. (1935) studied

heavily grazed pastures separated only by barbed wire. The jackrabbits could forage freely on either side of the fence, according to their preferences, but cattle continued grazing only on one side. The result was that jackrabbits were three to four times more abundant on the heavily grazed pastures.

When cattle were removed from a wildlife refuge in Oklahoma where prairie dogs were protected, the grass cover increased and the prairie dogs abandoned the site (Osborn and Allan 1949). In Kansas, prairie dogs never established colonies on a range revegetated with tall grasses even though five colonies were begun on overgrazed ranges nearby (Smith 1955). Koford (1958) observed that prairie dogs rarely start new colonies where ranges are in good to excellent condition. More recently, Uresk et al. (1982) compared the numbers of prairie-dog burrows before and after fencing cattle from an active town site in South Dakota. The density of active burrows was twice as great where livestock continued grazing. These authors mentioned that the costs of controlling prairie dogs with poisons did not produce financial rewards and instead recommended reducing livestock numbers as a means of natural control. In Kansas, elimination of grazing during the growing season helped reduce a 44.5-ha prairie dog colony to less than 5 ha after 4 years (Snell and Hlavachick 1980). Some of the interactions between livestock and prairie dogs undoubtedly are related to habitat. In shortgrass environments, prairie dogs may be the primary influence on range vegetation, whereas the intensity of cattle grazing seems the dominant regulator in tallgrass prairies (Hansen and Gold 1977).

Overgrazing, particularly during dry periods, was identified by Jackson and DeArment (1963) as an important factor influencing population levels of lesser prairie chickens. In a heavy grazing regime, highly palatable tall grasses are replaced by shorter species of lesser quality as habitat for prairie chickens. Conversely, better grazing management—for example, grazing systems that periodically rest rangelands and use more moderate stocking rates—

Figure 14-6. Prairie dogs are associated with the early stages of plant succession and so thrive on heavily grazed ranges where livestock suppresses taller grasses (top left). When grazing pressure is reduced or eliminated, prairie dogs commonly abandon their colonies as the vegetation recovers. Range vegetation surrounding a burrow abandoned for 1 year (top right) and 3 years (bottom) shows recovery resulting from deferred grazing during the growing season. (Photos courtesy of Natural Resources Conservation Service.)

contributed to the increasing numbers of lesser prairie chickens in Colorado (Hoffman 1963). On southwestern ranges, densities of grassland gamebirds probably depend on survival during the winter and therefore on the amount of residual vegetation available for food and cover (Brown 1978). If, after summer and fall grazing, too little residual vegetation is left, grassland birds will fare poorly during the winter and experience erratic changes in their abundance the following year.

GRAZING AND TROUT

Unfortunately, relationships between grazing and fisheries have not always been integrated in resource management, with the result that streamside habitats and fish populations have been damaged (Meehan and Platts 1978). Riparian zones make up a small percentage of rangeland areas; yet because cattle and other livestock prefer the succulence and diversity of riparian vegetation, these zones receive dispropor-

tionate grazing pressure (Ames 1977; Thomas et al. 1979). Besides damaging aquatic and streamside communities, this often means that forage available on nearby upland ranges remains unused.

In Montana, Gunderson (1968) found that a section of ungrazed stream had 76 percent more cover for trout (Salmonidae) than did a grazed section of the same stream. Undercut banks, overhanging vegetation, and other conditions desirable for trout were important in this evaluation. Furthermore, trout 15 cm or more in length were 27 percent more numerous and 44 percent heavier in the ungrazed section. Platts (1982) summarized 20 studies of grazing and its influence on fish populations and habitat; all but one concluded that grazing degraded streamside environments and hence the local fishery. The exception involved a grazing plan protecting streamsides on a well-managed sheep allotment. Sheep otherwise may severely degrade streamside habitats. A section of stream bordered by a heavily grazed meadow was about five times as wide and only one-

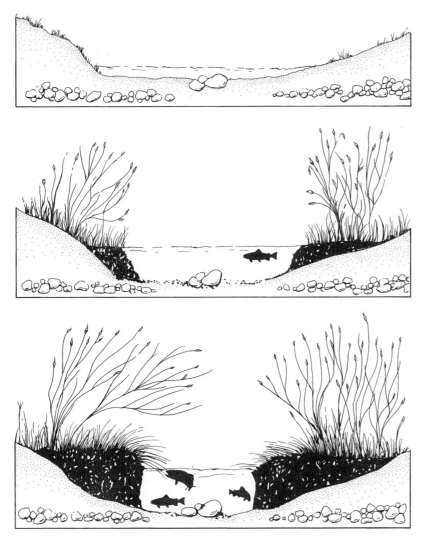

Figure 14-7. Schematic representation of stream conditions subject to heavy grazing (top). Banks are grazed and trampled, leading to increased stream width, shallow water, and poor habitat for trout. Shallow water may be warmed enough by direct sunlight to limit a trout fishery. After 2–3 years without grazing, vegetation begins to recover as stream banks again develop their structure (middle). The habitat is improving for trout as food, cover, and spawning conditions become more favorable. Conditions after 5 to 10 years without grazing offer excellent trout habitat (bottom). Overhanging banks and deeper water have been restored as the vegetation recovered. Sedimentation also is reduced significantly. (From Bowers et al. 1979.)

fifth as deep as a section where the meadow received light or no grazing (Platts 1981). Recovery from heavy grazing may take up to 10 years and then only after most or all of the livestock are removed (Fig. 14-7).

Bowers et al. (1979) reported that trout production increased by an average of 184 percent where grazing was light or eliminated and suggested that production could be increased by about 200 percent if management decisions were made to optimize trout habitat. They presented some management recommendations beneficial for trout production, including (1) implementing grazing systems that will create and/or maintain good trout habitat, particularly those that protect streamside areas until the fall months, (2) fencing easily damaged areas on the most important trout streams, and (3) using salt or water developments to attract livestock to other parts of the range. However, Bryant (1982) found that neither salt nor alternate water sources appreciably reduced livestock utilization of riparian zones in Oregon and that cattle continued grazing streamside vegetation during the summer months unless these sites were fenced.

In sum, grazing systems and other management practices designed to meet the requirements of a livestock operation and maintenance of range vegetation also should consider their effects on riparian zones and the fisheries dependent on these environments (Meehan and Platts 1978). Too often, however, streamside studies of grazing and its influence on trout fisheries have not been rigorously conducted, and the results of these may not always be as instructive as supposed (Platts 1982).

GRAZING SYSTEMS

Several systems have been developed to supplement the traditional method of *continuous grazing* (Kothmann and Mathis 1974; Kreitlow and Hart 1974; Merrill and Taylor 1975; Kothmann 1980; Merrill 1980; Bryant et al. 1982). These include *deferred-rotation systems* of several types but these always include one or more pastures in some stage of resting while others are grazed. The length of time each pasture in a deferred system is rested or grazed depends on the season, the kind of vegetation, and, of course, on the class and density of livestock. Other grazing schemes are known as *short-duration systems*. These include high intensity–low frequency systems that exert heavy grazing pressure for short periods of time. Another, even more complex short-duration scheme is the Savory grazing method (Savory 1978). In this case, the range is divided into pastures resembling spaces between the spokes of a wagonwheel, with a common water source at the hub. The grazing period in each pasture is short, no more than a few days, before the livestock are moved to the next cell in the series. This and other grazing systems may permit large increase in stocking rates compared with continuous grazing.

The major differences between deferred and short-duration grazing systems are the ratios of (1) area grazed to ungrazed at any one time and (2) the length of grazing and resting periods (Steger 1982). With deferred rotation, at least one-half of the area is grazed and the grazing period at least equals the resting period. With short-duration systems, less than half the area is grazed at any one time and the grazing period is shorter than the resting period. Kothmann (1974) defined an array of terminology for grazing management, to which students should refer.

Anderson and Scherzinger (1975) described a grazing plan for cattle that improved the quality of winter forage for elk. Cattle grazed the vegetation in late spring and early summer, but they were removed before the end of the growing season. This strategy allowed time for the plants to regrow and cure as forage of high nutritional quality. Without cattle, the ungrazed vegetation remained rank and of low quality. In sum, the cattle stimulated the regrowth of forage later used by elk. The winter elk population jumped from about 320 animals to 1190 in the 10 years after the grazing plan was implemented. Furthermore, blue grouse (*Dendragapus obscurus*) numbers also increased under this grazing regime.

Dual use of rangelands in Utah by mule deer and sheep produced similar results (Smith et al. 1979). In this case, sheep grazed in the late spring, reducing the amount of herbaceous plants, but subsequent regrowth of this same vegetation increased the proportion of green plants available to wintering deer. Concurrently, the temporary removal of the herbaceous vegetation freed moisture and nutrients for bitterbrush (*Purshia tridentata*)—a major browse species—and increased its availability to the deer herd. Bitterbrush itself can be managed as forage for both sheep and big game if livestock grazing is terminated before the twigs begin to grow rapidly. After bitterbrush twigs have produced about one-third of their annual growth, even moderate use by sheep will curtail further growth and hence the forage production of this important shrub (Jensen et al. 1972).

The density of white-tailed deer in Texas was highest in a seven-pasture, high intensity–low frequency grazing system when compared to either continuous or four-pasture, deferred-rotation systems (Reardon et al. 1978). Each of the pastures was grazed by a mixed herd of cattle, sheep, and goats. The seven-pasture system was grazed for 3 weeks and then rested for 18 weeks. Deer responded with a clear preference for the short-duration system, showing stronger preferences in proportion to the frequency of the rest periods. Further, the seven-pasture system also supported the highest stocking rate of livestock and produced high rates of hunter success and hunting revenues.

After careful study of the effects of grazing systems on habitat requirements of deer, as well as quail and turkey, Bryant et al. (1982) recommended a Merrill deferred-rotation system in Texas for the optimum production of both livestock and wildlife. In contrast, they also suggested that continuous yearlong grazing was rarely of value to most wildlife. Pastures in southeastern pinelands often degrade under continuous grazing. When these are managed with short periods of intensive grazing followed by long rests, vegetation important to deer and game birds increased while saw-palmetto (*Serenoa repens*) and other troublesome plants decreased (Moore and Terry 1979).

These studies indicate that both livestock and big game can share rangelands profitably when an appropriate grazing regime is employed. Moreover, the effective stocking rate, measured by AUs, of livestock plus wildlife can be greater than with wildlife or livestock alone.

Some game birds also may respond favorably to grazing systems. Bobwhite densities were highest on a high intensity–low frequency system, followed by continuous grazing, and then a four-pasture deferred-rotation system (Hammerquist-Wilson and Crawford 1981). The response seemed related to the increased amounts of forbs and bare soil, together with fewer grasses, produced by the "high-low" system. These conditions favored the movements and feeding activities of the quail. Kiel (1976) also suggested that thick stands of grasses produce few seeds suitable as quail food, impede the birds' movements, and reduce the production of winter forbs and insects. The needs of other game birds may be tied similarly to grazing systems. Merrill (1975) found that a four-pasture deferred-rotation system produced better nesting habitat for turkeys than did continuous grazing. A rotation system also provided the flexibility needed to assure nesting and brood cover for waterfowl on western rangelands (Gjersing 1975). When pastures in the system were rested or grazed only during the first part of the year, brood production increased the following spring. Conversely, late-season grazing reduced the amount of residual cover, and brood production the following spring decreased. On the average, the rotation system produced about four times as many broods as did season-long grazing. Baker (1978) compared rates of nest predation between two grazing systems and continuous grazing using simulated turkey nests. Both grazing systems had higher rates of nest survival than continuous grazing. Nest survival was greater in pastures deferred for 41 days than in those deferred for 10 days.

At first glance, grazing systems might seem to reduce the success of ground-nesting birds, presumably because the temporarily concentrated livestock might trample more nests. However, trampling losses were similar between continuous grazing and a short-duration system tested with simulated ground nests (Koerth et al. 1983). Cattle seem to move more in the larger pastures required by continuous grazing than in the smaller cells of grazing systems, thereby offering at least as much opportunity for trampling nests as in a short-duration grazing system. Further, each pasture under continuous grazing is stocked all the time, whereas those in the short-duration system were grazed only 6 percent of the time. Hence, during a 40-day period—approximately the combined laying and incubation period for turkey, quail, and similar birds—only 87 percent of the total pasturage was grazed in the short-duration system, whereas all of the area remained utilized under continuous grazing. The rates of nest losses per week of grazing also

were alike statistically in this experiment. All told, Koerth et al. (1983) suggested that short-duration grazing systems should not increase trampling losses of ground nests.

As might be expected, each grazing system varies somewhat in its influence on wildlife. Each may increase or decrease the abundance and quality of food and cover for wildlife or simply affect the social interactions between wildlife and livestock. Elk preferred pastures managed with a deferred-rotational system in Oregon largely because this plan, rather than season-long grazing, reduced the disturbance by cattle even though there was no direct competition for forage (Skovlin et al. 1976). Holechek et al. (1982) summarized their review of grazing and wildlife relationships with the caution that the balance between defoliation and recovery must be assessed for any grazing system. Excessive defoliation, even for relatively short periods of time, may not benefit either wildlife or livestock. Some vegetation may not recover from a period of heavy grazing with an equal period of rest. In fact, assessments of grazing pressure should emphasize the amount of residual vegetation, not the amount of forage removed by grazing animals. Because plant production on western ranges varies with precipitation, the amount of forage utilized each year is not as important as the amount of vegetation left at the end of the grazing period. In droughts, even moderate grazing of the current year's production may leave too little residual vegetation to maintain the health of the range. Fortunately, it is far easier to measure the amount of residual vegetation than to measure the amount removed. By doing so, managers have a common point of reference for determining grazing pressure and thus the welfare of the range ecosystem (Holechek et al. 1982).

RANGE FIRES AND WILDLIFE

Wildfires exerted dynamic perturbations on vegetation and the organisms associated with plant communities long before humans tamed fire for domestic use. Fires were surprisingly frequent in prehistory. Kilgore and Taylor (1979) reported the incidence of fires in the Sierra Nevada based on scarred tree rings of giant sequoias (*Sequoia sempervirens*) and ponderosa pine (*Pinus ponderosa*). Giant sequoias were subject to fires every 10 to 20 years before 1875, with individual trees experiencing fires at intervals of 3 to 35 years. Fires coursed through ponderosa pine communities with even more frequency, averaging a fire every 6 to 9 years. Such frequencies are conservative estimates as they include only those fires severe enough to damage the cambium layer; lesser fires passed unrecorded over the centuries. Fires in grasslands and other herbaceous communities cannot be estimated by the same technique, but one can safely assume that fire was a regular phenomenon in these types. Wright and Bailey (1982), after considering a variety of historical data, believed that prairie fires occurred at frequencies of 5 to 10 years. Where the topography of grasslands is more dissected, fires probably were limited to 10- to 30-year intervals. As we have seen, grasslands indeed may owe their very existence to recurring fires. Other communities contain biological components dependent on fire (i.e., fire-adapted species) in ways that fulfill critical requirements in their life histories.

Ecological history experienced dramatic changes when humans regularly began to use fire: Uncontrolled, human-caused fires were more common. In Alaska, for example, the intrusion of miners, trappers, and others who deeply penetrated a virgin land initiated more fires than would have occurred naturally, and about 80 percent of the evergreen forests

there have burned since the turn of the last century (Leopold and Darling 1953b). Conversely, efforts were at least partially successful in suppressing wildfires. Advancing settlement brought with it organized efforts to prevent fires and to put out those that did occur. Roadways acted as fire breaks, as did lanes cleared expressly for that purpose. Fire protection is known to everyone familiar with "Smokey the Bear," smoke jumpers, and forest rangers. Finally, Chapman (1932, 1936, 1944) and Stoddard (1931, 1962) are credited, respectively, with initially determining the ecological role of fire in the management of southern pine forests and bobwhites. Their work was a pioneering assessment of fire ecology (sometimes accompanied by cries of heresy), a forerunner of today's acceptance of practical fire management in many vegetational types. Now, prescribed burning has become an ecological tool for forest, range, and wildlife management (Fig. 14-8).

We should emphasize that fires ordinarily do not kill significant numbers of wildlife. Most animals escape the ravages of fires and only occasionally are carcasses found afterwards; some of these may be of animals already dead or dying before the fire. Howard

Figure 14-8. Prescribed burning offers management opportunities for improving range conditions for livestock and wildlife. These include controlling undesirable vegetation and improving the palatibility of coarse grasses by stimulating regrowth of succulent foliage. Most fires are started on the ground (left) but helicopters ("helitorches") may be used in some circumstances (right). (Photos courtesy of C. M. Britton.)

et al. (1959) concluded that there was little chance of wild vertebrates becoming entrapped in fires; virtually all mammals and birds they observed simply avoided a fire set in an annual grass-brush community in California. Their experiments determined that rodents in simulated burrows 5–18 cm deep died only when temperatures there exceeded 59°C (surface temperatures reached 149°C), with these losses occurring primarily beneath fallen logs. Interestingly, some forms of wildlife may themselves deter fires. For example, California quail, rodents, and cottontails (*Sylvilagus floridanus*) maintain trails or remove enough herbaceous vegetation from the bases of brush clumps so that the cleared areas become effective fire breaks. However, fires occurring during the nesting season may destroy nests, eggs, or fledglings, and prescribed burning clearly should be scheduled to avoid that possibility.

The major result of fire remains its impact on habitat conditions. Left alone, wildfires may accrue either positive or negative outcomes. Fires create dramatic changes in the environment, benefiting some species of wildlife at the expense of others. In northern forests, this relationship is clear between the responses of moose (*Alces alces*) and caribou, respectively (see Chapter 15). In grasslands, fire also influences wildlife populations. Among game birds, bobwhite show the most remarkable responses to fire management. The oft-cited standard that quail could not achieve greater densities than 2.5 birds per ha remained widely accepted until prescribed burning was practiced on lands otherwise managed for timber and grazing. Stoddard (1931 and later) established the role of fire in the management of bobwhites, and densities of 5 birds per ha eventually were attained with effective use of this tool. Winter burns on rangelands for

brush control in southern Texas also initiated short-term increases in several measures of bobwhite abundance (Wilson and Crawford 1979). These were pronounced in the first year after the burning, but diminished as the vegetation recovered and the amount of bare ground decreased (Table 14-8). Feeding, roosting, and travel were enhanced for quail on the newly burned ranges. Reid (1953) also recorded that 1- and 2-year-old burns produced greater amounts of quail food than older burns.

Prairie chickens also respond to fire management. Booming grounds (leks) may be abandoned when residual vegetation from the previous year is so dense or tall that the birds' courtship activities are inhibited; grasses and forbs on booming grounds should be maintained at heights of less than 15 cm (Anderson 1969a). Prairie chickens in Illinois preferred recently burned areas for their booming grounds, and new sites were selected by the birds in response to late winter burnings; inactive booming grounds also were reoccupied after they were burned (Westemeier 1972). Cannon and Knopf (1979) suggested that burning may encourage prairie chickens to move into new habitats or to recolonize their historic ranges. Stimuli associated with newly burned sites may cause the birds to move into burned areas, offering an alternative to transplanting as a means of reestablishing prairie chickens within their former range. The additional habitat also may permit more males to take part in breeding activities.

Prescribed burning has proven equally valuable for management of prairie chicken nesting habitat. In Illinois, Westemeier (1972) found nest densities of 3.8 ha per nest versus 2.4 ha per nest between unburned and burned grasslands; these data were

TABLE 14-8. Bobwhite Abundance 2 Years After a Rangeland Burn (Winter) in Southern Texas

Bobwhite Census Method	Area	Two-Year Index of Bobwhite Abundance Following Burning			
		Spring	*Fall*	*Spring*	*Fall*
Along transects	Burned	9.0	15.0	10.8	7.8
	Control	3.0	8.0	4.5	9.0
Trapping	Burned	0.5	3.3	1.3	0.3
	Control	0	0	0	0
Flushes with dog	Burned	7.0	5.0	0.8	0.1
	Control	4.0	4.0	0.2	1.8
Vocalizations	Burned	849	95	378	56
	Control	215	105	260	136

Source: Wilson and Crawford (1979).

recorded 2–4 years after burning. Because much of the native Illinois prairie has been replaced with cultivated, cool-season grasses, prescribed burns of these in August give better nesting results, whereas on warm-season prairie grasses, burning in March is best. Thus, prescribed burning must be tailored by season to account for differences between altered or natural habitats. Similarly, Chamrad and Dodd (1972) found that the beneficial effects of prescribed burning on habitat managed for Attwater's prairie chickens varied with grazing. Fire enhanced the habitat only on ungrazed grasslands, presumably because either fire or grazing separately can maintain those physiognomic features of the vegetation attractive to the birds.

Accidental fires created new strutting grounds for sage grouse (Dalke et al. 1963), leading to the suggestion that prescribed burns of 0.4–4.0 ha might be useful in otherwise homogeneous sagebrush communities (Klebenow 1972). However, because some brush cover also is critical (15–20 percent cover seems optimal), a burning program should be on a long rotation (up to 20 years). On the birds' spring and summer ranges, staggered burning dates will create a mosaic of habitat with a full array of food and cover requirements. Forbs are particularly important as food for sage grouse broods, and these plants respond well to burning. Conversely, prescribed burning is not recommended on winter ranges as the birds then rely on mature sagebrush for their food and cover needs (Klebenow 1972).

The image of a mourning dove (*Zenaida macroura*) nest brings to mind a flimsy platform of twigs rather insecurely placed in a tree. Thus, if fire were used to remove the woody vegetation, the apparent conclusion would seem that doves might fare poorly whenever rangelands were burned. Evidence to the contrary suggested the adaptability of mourning doves nesting on burned-over rangelands (Soutiere and Bolen 1972, 1976); the birds instead nested on the ground even when some trees still were available. Nesting densities were greatest in the current year's burn and decreased each year thereafter, suggesting that the amount of ground cover was an important habitat characteristic. A comparison of nest success between ground nests and those in trees showed nest success at 21 percent and 15 percent, respectively. Wind damage caused the diminished success of tree nests. Predation rates were similar, and production from the ground nests largely offset any losses resulting from removal of the trees.

FENCING

Pronghorns, although often called "antelope," are actually a species indigenous to North America not closely related to the true antelopes (Bovidae) of the Old World. In his explorations of the American southwest, Francisco Vasquez de Coronado was likely the first Western man to record encounters with pronghorns (Leister 1932). Some three centuries later, the developing livestock industry entered the pronghorn's realm, and among the inevitable changes that ensued was the fencing of rangelands.

Fencing controls livestock movement without need of herders, prevents intermixing of separately owned stock, and may promote the proper usage of forage when herds are manipulated between pastures. Usually, barbed-wire fences posed little threat to pronghorns; the strands were widely spaced and often loose enough for safe passage of pronghorns through or under the fence, and yet they still confined cattle. With sheep, however, woven-net fences were required, and pronghorns no longer could move through these barriers for forage or winter cover. Although physically able to do so, pronghorns by nature do not jump over confining obstacles, so net fences became serious barriers for animals adapted to a free-roaming existence (Fig. 14-9). Large numbers have died, particularly in severe weather when entire herds were cornered by sheep-proof fences (Knipe 1944; Buechner 1950; Hoover et al. 1959). In 1882, during the first winter after a 97-km drift fence was completed in the Texas Panhandle, a blizzard forced large numbers of pronghorns into a pocket along the fence where, unable to proceed further, some 1500 were killed by settlers (Haley 1936). In another instance, drought reduced a population by 40 percent, largely because net fences confined pronghorn herds to forage-depleted ranges shared with livestock where they were forced to consume low-quality or toxic vegetation (Hailey et al. 1966). Other effects of fencing on pronghorns, including injury and starvation, are mentioned by Gross et al. (1983).

Spillett et al. (1967) studied 22 types of fencing arrangements that might confine sheep and yet not hinder pronghorn movements. They recommended a net fence of 81-cm maximum height or a 66-cm net fence topped by a 10-cm gap and a single strand of barbed wire (see also ZoBell 1968). As these are shorter than the fences normally used by sheep ranchers, the considerable savings for posts and wire costs favor adop-

Figure 14-9. Pronghorns once roamed freely on the grassland ranges of North America, but some kinds of livestock fences later interfered with their movements. Wildlife managers thus were challenged to develop ways of confining livestock without hindering the travel of pronghorn herds. (Photo courtesy of U.S. Fish and Wildlife Service.)

tion of fencing that simultaneously meets livestock and wildlife needs. Mapston et al. (1970) designed a horizontal pass that relied on the broad-jumping abilities of pronghorns. Livestock were confined by grill-work, similar to a cattle guard, overlying a shallow pit, whereas pronghorns easily jumped across the same barrier. A grill about 2 m² is sufficient. The grill-work should be elevated about 25 cm and installed with earthen ramps on each side. Topography is important in locating pronghorn passes. Sites should be selected where livestock exposure is minimal and pronghorn exposure is highest, although no site or structure will be completely effective for confining all types or classes of livestock (Gross et al. 1983). Fence corners and hilltops are good locations for pronghorn passes, whereas sites near salt licks, water sources, livestock trails, and roadways should be avoided.

Elsewhere, other fencing situations sometimes cause other conflicts between livestock and wildlife interests. When fencing on cattle ranches in Zambia was destroyed by roaming wildebeest (*Connochaetes taurinus*) and zebra (*Equus burchelli*), game officials were forced to shoot numbers of the offending animals (C.D. Simpson, personal communication). Between 1927 and 1958, nearly 32,000 elephants

(*Loxodonta africana*) were shot in Uganda to protect agricultural operations, whereas only 8170 were killed by licensed hunters during the same period (Brooks and Buss 1962). In Botswana, large numbers of wild animals are killed by fences, and the damage they cause is reported to exceed $100,000 annually; yet even more fencing was planned to isolate cattle from the potential reservoir of foot-and-mouth disease (Riney 1967). Wholesale removal of elephants, giraffes (*Giraffa camelopardalis*), and other large browsing species to protect fencing subsequently increased the amounts of shrubby vegetation on African rangelands so that the brush now must be controlled to encourage herbaceous forage for livestock (Stoddart et al. 1975). In Kenya, barbed-wire fences on rangeland surrounding Nairobi Royal National Park block the traditional lines of movement for game crossing the sanctuary's boundaries, threatening the park's future value (Darling 1960).

Fences around small wetlands eliminate the trampling and foraging of cattle on shoreline vegetation important as food sources or nesting cover for waterfowl. Kirsch (1969) found reductions in breeding pairs, nest densities, and nest success for waterfowl nesting on grazed pastures and concluded that shoreline grazing was harmful to waterfowl production. Grazing damage, including trampling, is somewhat selective by species and community; aquatic communities may not be affected, whereas some shoreline food plants such as smartweed (*Polygonum* spp.) may fail to recover from trampling even after cattle are removed (Whyte et al. 1981). In all, the effects of grazing on wetlands are a function of site, animal pressure, plant zonation, and composition, with these factors varying locally and regionally. However, the benefits of fencing may be short-lived if cattail (*Typha* spp.) later dominates littoral zones in dense, monotypic communities; food plants are crowded out and duck nests in these areas seem especially vulnerable to skunks (*Mephitis mephitis*) and other predators attracted to the dense cover (Keith 1961).

Management potentials nonetheless exist for these situations. Mundinger (1976) described a rotational grazing system that increased brood production by 50 percent on grazed wetlands. Fences permanently protecting parts of wetlands should leave the deeper sites accessible to livestock so that water is available even during dry periods, whereas fencing the remaining shoreline maintains undisturbed zones that are potentially available for waterfowl and other wildlife (Whyte and Cain 1981).

High-speed thoroughfares necessarily are bordered by fences that keep livestock off roadways. Wildlife, particularly deer, pose the same problem for traffic safety but require fences that are higher and remain flush with the ground's surface (Falk et al. 1978). Deer often seek the forage available on roadsides or medians; they sometimes successfully penetrate so-called "deer-proof" fences and then are entrapped on the roadway by the same fence designed to prohibit their access. One assessment indicated that deer-vehicle accidents represent an average cost of $648 per incident, including property damage, injury, and loss of life, thus giving highway engineers and wildlife managers an economic basis for judging the cost-effectiveness of safety devices and/or reducing deer herds (Hansen 1983). Reed et al. (1974) found that one-way gates along a high (2.4 m) fence bordering an interstate highway in Colorado effectively allowed mule deer to leave the roadway. Fenced roadways that cross the migratory paths of large mammals pose still another problem. Mule deer migrations between summer and winter ranges are interrupted by deer-proof fences unless other means allow their safe passage across busy highways. A special underpass located on a well-established migration route permitted about 61 percent of the local deer population to migrate safely beneath an interstate highway, but a somewhat larger and more open underpass likely would have even better effectiveness (Reed et al. 1975).

BURROS AND RANGELANDS

Spanish explorers of the 16th century introduced domestic burros (*Equus asinus*) into North America. Later, burros became the mainstay of prospectors and others who required dependable, sturdy pack animals for the rugged terrain of the American West. Burros are the domestic descendants of the African wild ass, a species of three races endemic to the arid zones between Algeria and Somalia. Given this heritage, burros adjusted readily to range environments in the western United States when they escaped or were turned loose at the end of the mining era (Fig. 14-10). Burros eventually established large feral populations in California, Nevada, and Arizona (see Chapter 21).

The versatility of burros to cope successfully in various ecological settings is reflected in their diets. In one area, burros consumed 4 percent grasses, 30 percent forbs, and 61 percent browse (Woodward and Ohmart 1976), whereas in another, the composition—almost reversed—was 61 percent grasses, 11 percent forbs, and 28 percent browse (Hansen and Martin 1973). Moreover, burros remain in good health, despite seasonal deficiencies in the quality of their forage, without suffering the nutritional consequences affecting native ruminants (see Hanley and Brady 1977a; Urness and McCulloch 1973).

Figure 14-10. Feral burros adapted readily to ranges in the western United States. As shown here, burros subsist on desert ranges where vegetation is sparse, often damaging fragile plants and disrupting associations within native communities. (Photo courtesy of R. F. Seegmiller.)

Burros, with their cecal digestive systems, enjoy an ecological advantage over ruminants in sparsely vegetated habitats (Seegmiller and Ohmart 1981). Bighorn sheep (*Ovis canadensis*), cattle, and other ruminants require foods whose particles are small enough to pass through the recticulo-omasal orifice. As cellulose and other fibers increase in the diet's content, the time required to reduce these to the appropriate size in the rumen is lengthened, depriving the ruminant animal of adequate nutritional benefits. Conversely, burros are not similarly limited and can utilize a diet high in fiber. They and other equines can ingest more forage with a more rapid rate of passage (compared to ruminants) at the expense of reduced digestion of cellulose. Thus, when supplies of low-fiber foods—usually preferred forage—are exhausted, burros continue feeding on a diet of high-fiber foods, whereas ruminants cannot. In short, a larger percentage of the total vegetational biomass (both high and low fiber species) is available to burros than to ruminants.

Burro populations largely are free of natural regulation, and the only effective control fell to ranchers armed with rifles. In the ranchers' view, burros competed with livestock for forage, and shooting presented a direct measure for reducing their numbers. Without control, burro populations can increase 20–25 percent within 18 months (Woodward and Ohmart 1976; see also Ruffner and Carothers 1982). In Death Valley, aerial censuses found that burro populations jumped from 1426 animals in 1978 to 2500 in 1981, indicating that even more rapid rates of expansion are possible (Allen et al. 1981).

Collectively, these observations give clear indication that burros experience few of the ecological restraints typical of most native or domestic animals in arid environments.

The damaging impacts of burros on public land in the United States have been assessed in detail (Carothers et al. 1976). Overgrazing and trampling by burros altered both composition and densities of vegetation and small rodent communities, compared with a burro-free site in an otherwise similar setting. Without burros, 28 species of vascular plants occupied 80 percent of the area; with burros, 19 species of plants covered only 20 percent of the plot. Species diversity of small rodents also was higher in the absence of burros, and the density of these mammals was about four times greater than where burros occurred. As discussed further in Chapter 18, burros also have displaced bighorn sheep and, without removal of the burros, few bighorns will ever again occupy Death

Valley National Monument and other public lands in the western United States (Buechner 1960).

Palo verde (*Cercidium* spp.) is a staple in the diets of burros; these trees are browsed throughout the year but especially in late summer and fall when grasses and other vegetation are less available. Palo verde provides a rich source of phosphorus and carotene at that time (Hanley and Brady 1977a). The feeding behavior of burros is particularly destructive to palo verde and other woody plants. Burros break off large branches but consume just a small part, leaving the remainder untouched (Fig. 14-11). Only the large end of the branch is chewed, along with some of the bark. In contrast, bighorn sheep forage at the tips of palo verde branches, consuming only the new growth produced annually. Observations of burros feeding on palo verde revealed that the animals browse for about 30 minutes at a time, removing an average of 23 limbs per feeding period (Seegmiller and Ohmart 1981).

Because of the movements and activities of burros, their influences on range vegetation vary largely in proportion to habitat preferences dictated by terrain and the proximity of water. Burros typically concentrate in dry washes where their heavy utilization of forage is pronounced; a comparison of washes with high and low utilization showed vegetational differences of 2.8 percent and 8.6 percent in canopy cover and density ratings of 252 and 721, respectively (Hanley and Brady 1977b). These effects varied with distance from water, with the heaviest browse utilization occurring near permanent water and the lightest utilization at distances of more than 2.5 km from water. Riparian habitats, in particular, receive heavy browsing pressure during the summer months.

GAME RANCHING

Unfortunately, the term *game ranching* may have two very different meanings. First, game ranching sometimes refers to those pay-for-hunting enterprises where big game (usually exotic species) is harvested for sport (Attebury et al. 1977). Large sums are involved, with trophy animals commanding fees exceeding $1000 per head. Further mention of exotics and the economics of this form of game ranching is made in Chapters 18 and 20.

The second meaning of game ranching concerns the husbandry of native animals *in situ* for the production of meat and other products. In this case, game ranching is a variation of range management

Figure 14-11.   Large stems of palo verde are broken by feral burros, even though only a small part of the branches is consumed, leaving the plants severely damaged. (Photo courtesy of R. D. Ohmart.)

that otherwise concerns the husbandry of domestic livestock on rangelands.

The basic principle of game ranching is that native species are far better adapted to local conditions than are domestic livestock. In short, attempts to raise cattle, sheep, or goats on many rangelands would be better forgone in favor of direct management of the same ranges for native species. The model best illustrating game ranching is the herding of reindeer (*Rangifer tarandus*) in northern Scandinavia by Lapps. Similar reliance on cattle in such a climatic and ecological realm would be uniformly disastrous for animals and people alike. Regrettably, however, the time-proven lesson of the Lapps has not been fully learned by other peoples; ill-suited livestock are still "forced" into environments where they may be no more than marginally successful.

The concept of game ranching is developing in Africa but has not yet been widely adopted in the United States (Lambrecht 1983). Broader adoption of game ranching probably rests upon three criteria:

(1) a pool of native species possessing large body sizes, behavior, and other features associated with livestock production, (2) national demand and need for protein in human diets, and (3) the availability of rangelands for the conversion of forage into meat and other animal products. Had North American grasslands remained open range, one could speculate that bison, and not beef, might have become the staple meat in the United States. Nonetheless, enterprising ranchers in the United States might develop domestic strains from their herds of exotic big game (Dasmann 1968). Stocks of these, when returned to their native rangelands overseas, may well initiate profitable ranching operations and ready sources of red meat. Nilgai antelope (*Boselaphus tragocamelus*) are well established on rangelands in southern Texas (Sheffield et al. 1971), and if selectively bred for their latent domestic values, this stock might become the nucleus of a ranching industry in parts of Asia. Eland (*Tragelaphus oryx*), the largest species in the antelope family, already is at least semidomesticated and

seems likely to offer people of several African nations a new source of protein.

Cattle and the large pool of antelopes in Africa are related taxonomically in the family Bovidae, but comparisons between cattle and antelopes often show striking differences in their respective efficiencies and tolerances under African range conditions. The history of cattle in Africa dates to well before the Christian era, but even this long period of adjustment has not adapted them fully to the stresses of heat, disease, and drought experienced in Africa.

Zebu cattle (*Bos indicus*) have been the traditional breed of the Masai, but even with their tolerances for African conditions, calf crops seldom exceed 50 percent and annual mortality is about 15 percent. By comparison, the most advanced management in Africa yields calf crops of 80 percent and losses of about 5 percent (Skovlin 1971). Furthermore, zebu characteristically have slow growth rates and delayed maturity (i.e., mature cows weigh 272 kg and calve at intervals of 18–24 months).

Water shortages during the African dry season interact strongly with the availability of forage for cattle. During the dry season, the ranges nearest permanent water are severely overgrazed, and the cattle must be driven farther away for forage. At first, a full day's trip each way (covering some 16 km) is required, and the cattle have water only every other day on a rotation between feeding and watering; as the forage becomes less available, an even longer cycle without water occurs. Cattle experience considerable stress during the normal dry season, but in prolonged droughts, only minimal forage is obtained with ever-increasing walking distances, and cattle die in massive numbers. Three-quarters of 450,000 head likely died in a single district of Kenya in 1961 (Payne and Hutchinson 1963). Even when new waterholes are constructed, the buildup of cattle may quickly destroy the range. In Senegal, Riney (1967) reported the deterioration of previously ungrazed grasslands into desert 4 years after water was developed for cattle.

In contrast, antelopes are particularly water efficient, having adapted various ways of conserving the water they consume. They excrete smaller amounts of body wastes per unit of body weight, and even these have less water content than those of cattle. Sweating also is reduced compared with cattle, and species such as the eland, oryx (*Oryx gazella*), and Grant's gazelle (*Gazella granti*) raise their body temperatures during the day so that sweating is not initiated until an upper limit is reached. For example, the temperature of a 500-kg eland can rise from 39°C to 42°C during the daytime, thus conserving 5 liters of water (Kyle 1972). The food habits of the native antelopes also favor efficient water intake; by feeding at night, when the relative humidity is at its highest, the animals' forage contains larger amounts of moisture than during the daytime. Some browse plants contain more than 50 percent water at night, even during droughts. An eland eating the leaves of *Acacia* can derive 5.3 liters of water per 100 kg of metabolic body weight, an amount corresponding to its needs for maintaining a constant body weight (Kyle 1972). At the extreme, the large-bodied addax (*Addax nasomaculatus*) virtually never drink, but instead obtain all their moisture from the forage they consume; this species is adapted admirably for arid rangelands where cattle cannot exist.

Cattle were excluded from nearly 40 percent of Africa by tsetse flies (*Glossina* spp.) and the lethal trypanosome parasites they carry. This subject is mentioned later, but suffice it to note here that native wildlife are not susceptible to nagana (or n'gana), the animal form of the disease known in humans as sleeping sickness. Conversely, only with expensive methods can cattle remain free of nagana on many African rangelands. Other diseases affecting cattle in Africa include those borne by ticks (e.g., East Coast fever, anaplasmosis, and redwater) that require weekly dipping or spraying to keep livestock disease-free; a large variety of still other diseases (e.g., anthrax, rinderpest, and contagious bovine pleuropneumonia) also require constant vigilance (Skovlin 1971).

Perhaps the strongest argument for game ranching lies in the natural efficiency of native species to utilize diverse range forage. In short, they convert range vegetation into meat more efficiently than cattle. The foods and feeding habits of native animals are either diverse or adapted to ways that maximize utilization of the available forage. Many species also produce higher weights of usable meat compared with cattle. Eland, for example, produce more than 75 kg of lean meat per 100 kg of carcass, whereas cattle yield only 55 kg, or 73 percent as much (Lambrecht 1983).

Maximum production of meat per area is achieved when large animals of several species share a common rangeland without competition or degradation of the vegetation. Several species of African big game show remarkable adaptations in this regard. Some are grazers, whereas others forage on browse, so that nearly all forms of vegetation are consumed. Conversely, cattle thrive on a narrower selection of grasses and generally ingest these only when they are succulent or properly cured. Furthermore, even when the same plant is utilized by two or more different species of native animals, either spatial (e.g., having stems high or low above the ground) or temporal (e.g., having different growth stages) separations occur. Giraffes obviously are able to forage well above the level of other animals so that the browse on a single tree is used concurrently without interspecific competition. Ecological partitioning also is shown by four grazing species eating the same kind of grass. Zebra select the upper stems of the grass, taking lignified material that antelopes are unable to digest readily. Topi (*Damaliscus lunatus*) then eat the lower stems, and wildebeest graze the leaves. With these materials removed, the grass produces new shoots from its base that are selected by Thomson's gazelles (*Gazella thomsoni*). Thus, a naturally balanced system has evolved for the native species—both plants and animals—that cannot be matched by the much narrower feeding niche of cattle and other domestic livestock.

In sum, these features strongly indicate that animal husbandry in Africa is better practiced by game ranching than by similar attempts with cattle, sheep, or goats. Capital investment is minimal and many native species offer rapid financial returns because a larger percentage of their populations can be removed annually. Relatively rapid growth rates and the high stocking densities possible with native

species clearly suggest the benefits of game ranching in Africa (Table 14-9). The underlying principle, however, may well be applied elsewhere, especially where transportation and marketing facilities already may be established. From the standpoint of wildlife management, game ranching need not exclude fee-hunting. Trophy animals might still be harvested under field conditions and in natural settings. Further, the financial rewards from a well-managed game ranch should assure not only perpetuation of the stock but also maintenance of wildlife habitat.

AN AFRICAN SAGA

Tsetse flies throughout history have shared much of Africa's native rangelands with vast herds of big game. Several species of flies are involved so that, collectively, they infest a broad continental belt south of the Sahara. Their stinging bites are annoying, but because they also are vectors of trypanosome parasites, tsetse flies are in fact deadly insects. Some of the parasites kill people and nearly all are lethal to livestock. However, a natural immunity largely protects native wildlife from what is known as sleeping sickness in humans and nagana in animals. The net effect was that the flies and parasites they carry effectively prevented occupation of the African rangelands by domestic livestock—and without their herds, few pastoral peoples settled the lands (Petrides and Swank 1958). Tsetse flies thus repre-

TABLE 14-9. Selected Comparisons of Growth Rates and Standing Crop Weights Between African Antelopes and Livestock

Species/Herd	Weight in kg		
	Gain/Day	*Adult Males*	*Approx. Standing Crop-kg/km²*
Eland (*Tragelaphus oryx*)	0.33	726	
Wildebeest (*Connochaetes taurinus*)	0.24	208	
Kongoni (*Alcelaphus buselaphus*)	0.23	150	
Topi (*Damaliscus lunatus*)	0.20	132	
Domestic cattle	0.14	454	
Domestic sheep	0.05	45	
Mixed herds of wild ungulates			12–17 thousand
Cattle only			3–5 thousand
Cattle, sheep, and goats			2–3 thousand
Sheep and goats			0.4–2 thousand

Source: Kyle (1972).

sented a biological barrier to human encroachment, leaving much of the African landscape essentially as a pristine wildlife preserve.

Parts of this setting yielded to the handiwork of modern people. Pressures to establish cattle on the rich grazing lands led to mass killings of wildlife so that the sources of trypanosome infections might be eliminated. Entire herds of zebra, bushbuck (*Tragelaphus scriptus*), warthog (*Phacochoerus aethiopicus*), and other large animals—reservoirs for the parasites—were slaughtered. Even a partial recounting of the kill is staggering. In Zululand during 1929–30, some 26,000 head were shot solely to maintain a disease-free buffer zone around a game reserve; in Tanzania, 8000 rhinos (*Diceros bicornis*), gazelles, giraffes, and lions (*Panthera leo*) were killed in one experimental unit (McKelvey 1973); and in Uganda, more than 161,000 animals fell to rifles in the name of tsetse-fly control (Jahnke 1976).

Next, the insecticide era began. Vast areas of the African range were sprayed to control tsetse flies. More recently, work on vaccines suggests that cattle, and possibly humans, might be immunized temporarily against the parasites, and efforts toward an effective biological control of the flies may be promising (see McKelvey 1973). Thus, a disease of continental proportions may soon be eliminated and with it the sufferings of a large human population. Few clear-thinking persons would argue against the control of a dreaded disease, but the ecological aftermath may not be as bright for African wildlife. The heavily shot game populations have, or will, recover as other methods are employed to fight nagana, but the underlying conflict remains no less prominent than before: settlement and land use versus wildlife habitat. Dynamic changes seem sure to follow as extensive grazing lands are opened to cattle and their herders. The problems are ecologically complex, and include the cultural values of the indigenous human population. Briefly outlined, these are as follows:

1. More cattle will be stocked, eventually leading to overgrazing. Sophisticated grazing systems are not yet widely adopted in Africa and the new ranges will surely deteriorate under continuous grazing pressure as others have in the past (see Heady 1960). The heaviest grazing pressure centers on the same waterholes where wildlife drink. Unfortunately, cattle are not efficient users of either grass or water in comparison to native wildlife, and overexploitation of these resources is almost certain, especially in periods of drought.

2. Cattle are a measure of status and wealth for many native peoples, and grazing schemes that might otherwise sustain their forage may be countermanded by local cultures (Maloiy and Heady 1965). Cattle, although providing products for human consumption, also are living bank accounts that increase with every calf. Additionally, cattle have become part of the social life of many native people. In Zambia, for example, cattle are (a) paid to a bride's father by the future son-in-law, (b) slaughtered to commemorate a girl's eligibility for marriage or to mourn deaths, and (c) given as recompense for hostile acts so that revenge is avoided (Larson 1966). In such cultures, a premium is placed on the largest possible herds.

3. Native wildlife may be viewed as competitors with cattle for forage and water, and human tolerance for sharing the range with wildlife may diminish. For example, the Masai in Kenya attach fundamental and spiritual significance to their cattle, and they perhaps will abandon their traditional laissez-faire attitude toward native animals (grazers and predators) as disease-free rangelands are opened (Darling 1960). Furthermore, direct measures to control wildlife often are carried out unwisely. When 1000 warthogs were shot in a single operation in Senegal, the meat was left to rot because the natives were Muslim, and no scientific use was made of the animals for research (Riney 1967).

SUMMARY

Sizable areas in the United States are managed for grazing animals. Whereas shrubs and forbs may be valuable forage, a large part of range management is concerned with grasses. Grasses withstand grazing because their growing points lie near or below ground level, and hence partial removal of their foliage does not impair regrowth. Indeed, moderate grazing in the appropriate season stimulates new production.

A cow and a calf, or an assumed body weight of 454 kg, equal 1 Animal Unit (AU) for purposes of assessing forage consumption. Body weights of other animals are converted to this standard to assess total foraging pressure on rangelands. For example, 5.8 mule deer equal 1 AU. However, Animal Units do not fully equate grazing animals with browsing animals.

Range management involves many practices designed to improve or maintain the quantity and quality of vegetation for livestock and wildlife. These include mechanical and chemical means of managing dense or undesirable vegetation, but each must be used wisely or wildlife food and cover may be damaged. Prescribed burning also can be a useful management tool on many rangelands to improve conditions for livestock and wildlife.

Range vegetation coevolved with native grazing animals, but livestock may disrupt the equilibrium between forage production and consumption. Livestock also may alter the composition and physiognomy of range communities, often at the expense of wildlife. Some plants decrease with grazing, whereas others increase or invade. Livestock also may compete with wildlife for food, but competition varies among regions, seasons, and kinds and classes of animals. Deer, for example, face stronger seasonal competition from sheep and goats than from cattle on ranges in the southwestern United States. Without some grazing, however, range vegetation may become too dense for birds such as bobwhite or prairie chickens.

Overgrazed ranges deteriorate, producing undesirable vegetation, soil and water losses, and poor habitat for many kinds of wildlife. However, prairie dogs and jackrabbits generally thrive on overgrazed ranges because they favor early stages of plant succession. Heavy grazing may ruin streamside habitat crucial for trout by trampling banks, removing overhanging vegetation, and reducing stream depth.

Grazing systems provide periods of rest for range vegetation. Depending on the length of the rest periods and the area of ungrazed units, the systems are known as either deferred or short-duration grazing systems. Either method offers advantages over continuous grazing on many ranges, but the balance between foliage production and recovery remains crucial in determining the best grazing system for individual ranges.

Grasslands, in part, are maintained by recurring fires that inhibit woody species. Few animals are killed by fires, but wildfires often may destroy their habitats and hence drastically reduce wildlife populations. However, prescribed burning may improve range conditions for wildlife, but fire ecology is complex and requires careful study before fires are set for management purposes. Bobwhite and prairie chickens are among the species for which prescribed burning has proven an effective management tool.

Pronghorn herds often declined when livestock fences limited their movements on western ranges. Specially constructed passes, when properly located along fencelines, reduce this hazard, as do certain types of fences. In Africa, fences often are damaged by wildlife, leading to harsh measures for controlling herds of large animals. Livestock may damage vegetation on the shorelines of small wetlands, thereby reducing food and cover for waterfowl unless fences protect these habitats.

Burros adapted well to ranges in the western United States after they were released by prospectors, but they now damage range communities on some public lands.

Game ranching concerns the management of native species of grazing animals for meat production. The concept is better developed in Africa than in the United States. Native animals have greater efficiency compared with livestock for converting range vegetation into meat; they also are more water efficient. Nevertheless, cattle ranching is likely to continue in Africa, encouraged by tradition and control of sleeping sickness, and will likely cause reductions in native mammal populations.

CHAPTER 15

FOREST MANAGEMENT AND WILDLIFE

*The importance of nontimber values [of forests] is dramatized by
the provocative assertion that modern civilization could get
along without wood, but not without forests.*

Samuel T. Dana (1968)

Many people think of the forest as a place abounding
in wildlife. That is often true, but the presence of trees
does not automatically ensure the presence of animals.
Wildlife may, in fact, be quite scarce in some forests.
The abundance of wildlife in woodlands depends upon
the quantity and quality of food and cover available
for the needs of various kinds of animals, and these
conditions in turn depend largely on the way forests are
managed. This chapter reviews the methods of forest
management and relates these to the needs of wildlife.

Prior to about 1900, forests in North America were
regarded as virtually inexhaustible, and there was no
real forest management. Forests were exploited without regard for the future. Loggers of the day commonly purchased land, cut the timber, and moved on,
leaving behind a barren and often burned landscape.
Taxes on the abandoned land often went unpaid,
and the land reverted to state and federal ownership.
But two events gradually closed the era of western-
moving "cut-out and get-out" forestry in the late
1800s. First, the Pacific Ocean was looming closer as a
visible western boundary of what was once regarded
as an unlimited forest. Second, concerned citizens took
action. In 1875, the American Forestry Association
was organized in Chicago for the promotion of forestry
and timber culture. In 1891, Congress authorized the

president to set aside and proclaim forested lands as
public reservations before the lumbermen arrived
(Allen and Sharpe 1960). President Benjamin Harrison promptly designated 10,000 km² in Wyoming and
Colorado as forest reserves; he eventually reserved
52,600 km² as federal forests.

In the early 1900s, the management of federal
forests was transferred, despite severe opposition,
from the Department of Interior to the Department
of Agriculture, and the name *National Forest* was applied to what were formerly called *federal forest reserves*. At the same time, universities such as Cornell,
Yale, and Michigan began to graduate academically
trained foresters such as Gifford Pinchot and Aldo
Leopold. From 1905 to 1907, President Theodore
Roosevelt enlarged the national forest lands to
404,860 km². Roosevelt acted with the advice of Pinchot, then chief of the U.S. Forest Service, who is
credited with coining the term *conservation* (Trefethen 1975).

These sudden and rather exciting developments
for forest conservation in the United States heralded
an era in which emphasis was placed on the *sustained
yield* of wood products. The goal of sustained yield
was met with new harvest methods and management
approaches, including planting and fire protection.

Photo courtesy of Jack Ward Thomas

Jack Ward Thomas
Former Chief, U.S.D.A. Forest Service
(1934–   )

Dr. Thomas began his career as a manager and research biologist for the Texas Parks and Wildlife Department (1957–66). His assignments centered on the dense deer herd in the Llano Basin of the Edwards Plateau and resulted in publications concerned with population dynamics, diseases, and the soil-related condition known as velvet horn. He then joined the U.S.D.A. Forest Service in West Virginia and began to research the effects of even-aged forest management on deer, turkeys, and other wildlife. In 1969, he transferred to Massachusetts to establish the first research unit in the Pinchot Institute of Environmental Forestry, where his studies focused on wildlife habitat in urban and suburban environments—pioneering work that cemented urban wildlife and especially "backyard management" to the framework of the profession.

In 1973, Thomas was appointed chief research wildlife biologist and director of the Range and Wildlife Habitat Laboratory in La Grande, Oregon; here, he led a team of specialists who developed a system that considered all species in the practice of forest management. The results were published in the award-winning book *Wildlife Habitats in Managed Forests—the Blue Mountains of Oregon and Washington* (1979). Becoming involved in the heated controversy between the conservation of northern spotted owls and logging in old-growth forests, he chaired a series of committees charged with developing a scientifically credible conservation strategy to deal with the preservation of biodiversity in the old-growth forest ecosystems in the Pacific Northwest. Thomas has written and edited more than 350 publications, including another award-winning book *The Elk of North America: Ecology and Management* (1982).

In December 1993, Thomas was the first wildlife biologist appointed chief of the Forest Service. During his tenure, logging on national forests was reduced by more than 50 percent, and he integrated ecosystem management into the agency's planning processes. When some members of Congress moved to privatize the national forests, Thomas helped thwart these ill-advised efforts. He retired as chief in November 1996 after serving during the most tumultuous periods in Forest Service history since the tenure of the first chief, Gifford Pinchot. Thomas then occupied the Boone and Crockett Club chair at the University of Montana; in addition to teaching, he oversees a research facility—the Theodore Roosevelt Ranch—operated by the club.

Thomas was president of The Wildlife Society (1977–78) and received the Aldo Leopold Memorial Medal (1991) for distinguished service to wildlife conservation. During his career, Thomas demonstrated an array of talents in the stewardship of natural resources, including research and management of wildlife in both urban and exurban environments, leadership in public policy and agency administration, and, currently, university teaching.

Foresters began to guide the nation's forests into a condition of continuous production.

In the 1950s, the rapidly growing human population began to turn from the cities to the forests, seeking recreation and a contact with nature that had been lost somewhere in the rush to suburbia. The Forest Service initiated *multiple-use management*, meaning that in addition to timber and pulpwood production, attention would be paid to camping, hunting, skiing, fishing, birdwatching, and other recreational uses of the National Forests. The Multiple Use Act of 1960 officially sanctioned the goal of managing the National Forests for a variety of purposes. Campgrounds were built and maintained, and wildlife biologists joined the Forest Service to recommend ways that forest animals might be managed. But multiple-use management made only token gestures toward managing the total forest ecosystem. Major emphasis still centered on wood production.

The Wilderness Act of 1964 mandated absolute preservation in a natural state of large tracts of national forests, national parks, and other federal lands in remote areas. The act clearly recognized the values of forested lands for uses other than timber production. About 36,000 km² were set aside in which roads, timber harvests, commercial developments, and buildings have been prohibited. Later, an additional 28,000 km² were added under the Endangered American Wilderness Act of 1978, and another 244,000 km² were designated as wilderness areas under the Alaska Lands Act of 1980. About 2 percent of government-owned lands in the United States outside of Alaska now is designated as wilderness. In these areas, human activities are limited to "quiet" pursuits such as hiking, camping, fishing, and, in some places, hunting. Most timber and mining interests continue to oppose wilderness designations.

A new wave of public opinion was rising in the 1960s and 1970s. People became interested in ecology and the total environment, including herbaceous plants, fungi, fishes, amphibians, birds, and mammals. The public saw the forests not only as places where trees are grown for harvest, but also as places where wildlife can live and where solitude helps replenish the human spirit (Allen and Sharpe 1960). The laws setting aside wilderness areas were augmented with new laws that protected endangered species, the first of which was enacted in 1966. In 1969, Congress passed the National Environmental Policy Act, which required preparation of an Environmental Impact Statement on any action involving federal permits. In 1976, the National Forest Management Act was enacted, requiring the Department of Agriculture to ensure consideration of the economic and environmental aspects of managing renewable resources. To meet the objectives of multiple-use management, the act addressed a list of concerns: silviculture; forest protection; outdoor recreation; range, timber, watershed, wildlife, fish, and wilderness preservation—and the diversity of plant and animal communities.

Such laws, backed up by public sentiment, caused no small amount of turmoil among professional foresters whose training had focused on the principle of sustained production of wood fiber. Thomas (1979), in his Preface to a superb treatment about the new role of wildlife management in national forests, summed up the setting as follows:

> Until just a few years ago, … many resource management professionals were secure in the assurance given them in their basic forestry and wildlife management classes that good timber management is good wildlife management.

Times have changed. Laws have changed. Public demands and politics have changed. The public forests have been thrust into the forefront as a prime supplier of wood products and recreation. It seems likely that these forests will become ever more intensively managed to meet the nation's burgeoning demands for wood, water, recreation, grazing, and wildlife. As for good timber management being good wildlife management, wildlife biologists and foresters alike have found that it is not necessarily so!

Forest managers are now under increasing pressure to account for wildlife in their management activities, particularly land-use planning. That means all wildlife—not just species that are hunted or are aesthetically pleasing or classified as threatened or endangered.

Most of the laws enacted since 1970 pertain primarily to federal lands, but many states also have adopted laws and policies for state forests that parallel those for national forests. Furthermore, although owners of private commercial forests do not operate under the same legal constraints as national and state forest managers, private owners are concerned about wildlife for reasons of public image. As a result, private forest managers often institute practices that involve good wildlife management, provided these can be accomplished without severe economic strain.

Small tracts of forests owned by private individuals are probably the most difficult to manage. These include farm woodlots and tracts of 16–65 ha used frequently by their owners as recreational lands. On federal, state, and corporate lands, central organizations screen and approve management information, usually with the advice of professional foresters and wildlife biologists. But owners of small private tracts do not have an organizational structure or the trained professionals to make management decisions. In addition, the land parcels in private ownership often are so small that management plans, unless they are coordinated with the plans for adjacent lands, may have unfavorable effects on both forest and wildlife production. Table 15-1 indicates the ownership of commercial forestlands in the United States. In the northeast, 44 percent of forested land is in the ownership of farms or other privately owned nonforest industries, and in the southeast, the total is 73 percent. By contrast, in the west, only 21 percent of forestland is in the same categories.

SOME BASICS OF FOREST MANAGEMENT

Forest management involves a multitude of practices, including economic and social considerations as well as the application of silvicultural methods to the land. Silvicultural activities involve regeneration, tending, and harvesting the forest. Included are planting, insect

TABLE 15-1.   Ownership of Commercial Forest Lands in the United States by Major Regions in Millions of Hectares

| | East | | West | | |
	North	South	Rocky Mountains	Pacific Coast	Total
Private					
Forest industries	7.20	14.48	0.85	5.00	27.53
Farm	18.37	23.16	3.31	2.38	47.27
Other	30.89	31.47	1.70	3.30	67.36
Total	56.46	69.11	5.91	10.68	142.16
Public					
Federal	4.96	5.86	16.55	15.44	42.81
State	5.32	1.02	0.89	2.34	9.57
Local	2.41	0.30	0.03	0.19	2.93
Total	12.68	7.18	17.48	17.97	55.31
All ownership (total)	69.14	76.29	23.39	28.65	197.47

Source: Yonce (1983).

control, pruning, thinning, fire control, and cutting. The method by which trees are harvested is by far the most influential practice that determines the character of the forest. A sustained yield of marketable wood can be attained with two basic approaches: *even-aged management* and *uneven-aged management.* Each approach is determined by the harvest method. Even-aged management means that all trees in a stand are approximately the same age and therefore usually of the same size. A comparison of the features of even-aged and uneven-aged management is shown in Table 15-2.

Some definitions are appropriate here:

Clear-cut. The removal of all trees from an area.

Shelterwood cut. The removal of all trees from an area except for several large trees that provide shade for developing seedlings. The large trees are removed a few years later, after the seedlings have been established.

Seed-tree cut. Similar to a shelterwood cut, but the large trees are left as a source of seeds rather than for shade.

Single-tree selection cut. Individual trees having certain characteristics are marked and removed.

Group-selection cut. Small groups of trees are marked and removed. As the group gets larger, the cut begins to resemble a clear-cut.

Rotation time. The number of years between the time a tree or stand is cut and the time it is replaced by another harvestable tree or stand.

Cutting cycle. The number of years between cuts on a particular forest area. (In a selection cut, the cutting cycle may be 10 years, but the rotation time may

TABLE 15-2.   A Comparison of Characteristics of Forests Managed by Even-Aged and Uneven-Aged Methods

Characteristic	Even-Aged	Uneven-Aged
Harvest method	Clear-cut Shelterwood Seed tree	Single-tree selection Group selection
Type of trees	Usually shade intolerant (shelterwood cut may produce even-aged stands of shade-tolerant species)	Shade tolerant
Stand appearance	Uniform tree height in each stand, except for a few years after shelterwood or seed-tree cut. Often aesthetically unattractive, especially the first few years after cutting	Great variation in heights of trees, although if group selection is used with groups larger than 0.4 ha, appearance is like small even-aged stands
Forest appearance	A patchwork of stands of various ages	Aesthetically acceptable. A large expanse of uniformly mixed sizes of trees
Type of wildlife use	Mobile species adapted to early successional and mixed successional stages	Species adapted to mature forest conditions

be 120 years; that is, some trees may be removed every 10 years, but it may take 120 years for a new tree to replace a harvested tree.)

Shade tolerant. Refers to trees that can reproduce and grow in shade.

The question of how clear-cutting can produce a sustained yield of wood may be puzzling. Suppose a corporation owns 100 large units of timberlands. In 1990, one unit is clear-cut. The clear-cut is followed by the establishment of seedlings. In 1991, another unit is clear-cut, in 1992 another, and so on, with one unit cut per year for 100 years. By 2090, the trees on the original unit, which were cut in 1990, are 100 years old and ready for cutting. Some clear-cut stands of Douglas-fir (*Pseudotsuga menziesii*), jack pine (*Pinus banksiana*), and aspen (*Populus* spp.) have rotation times as short as 40 years. Thus, for these species, the same area of forest might be clear-cut, then regenerate, reach harvestable age, and be clear-cut again within the working career of a forester.

A SYSTEM FOR CONSIDERING NEEDS OF FOREST WILDLIFE IN FEDERAL AND STATE FORESTS

With the new public interest in managing forests for wildlife as well as for trees—and for all forms of animals, not just game animals—wildlife managers have found themselves wanting for information. The question as to how a 16-ha clear-cut will affect indigo snakes (*Drymarchon corais*), red-backed salamanders (*Plethodon cinereus*), or downy woodpeckers (*Picoides pubescens*) might be answered with no more than a shrug of the shoulders.

On a large forest in a temperate region, there may be 15–20 species of amphibians, 10–20 species of reptiles, 75–100 species of breeding birds plus migrants passing through, and 30–40 species of mammals, making a total of 130–200 species. Determining the habitat needs for these species in order to give each one fair treatment in forest management would be an extremely tedious process. It also would be difficult to synthesize all the factors into a forest management plan.

Thomas et al. (1979a) developed a systematic approach, adapted from Haapanen (1965), using categories of breeding and feeding habitats of the various animals. These categories are called "life-form associations," of which 16 were described for the 327 species of vertebrates living in the Blue Mountains of Washington and Oregon (Table 15-3).

The value of using the system of life-form associations lies in the ready assessment of the benefits or detriments associated with each forest management practice. In the Blue Mountains, for example, a practice in which old and dead trees are left standing would benefit the 50 species that are members of life forms 13 and 14 (Table 15-3). Among those species are the common flicker (*Colaptes auratus*), the red-breasted nuthatch (*Sitta canadensis*), and the northern flying squirrel (*Glaucomys sabrinus*).

Application of the life-form association classification system involves considerable effort. Forest animals must be inventoried, and then the breeding and feeding habitats of each species must be determined from the literature. (Concise regional guides are available from the U.S.D.A. Forest Service so that field biologists can find such information.) Then a list of life forms is prepared, along with the species occupying each category. Finally, the influence of cutting is estimated for each of the various life forms. These efforts may be necessary to assure that proper recognition is given to wildlife as part of the forest community.

CLEAR-CUTTING AND WILDLIFE

Will a clear-cut help or harm wildlife? Someone reading the November–December 1972 issue of the *Journal of Soil and Water Conservation* might become confused. One article is entitled "Clear-cutting: Beneficial Aspects for Wildlife Resources" (Resler 1972), whereas the following article is "Clear-cutting: Detrimental Aspects for Wildlife Resources" (Pengelly 1972). For herbivores, the primary benefits of clear-cutting include the new growth of nutritious forage that normally begins the year after cutting and lasts up to 20 years (Regelin and Wallmo 1978). Although clear-cutting may provide food for large mammals such as deer (*Odocoileus* spp.), elk (*Cervus elaphus*), and moose (*Alces alces*), the practice concurrently destroys cover for these animals, both for concealment and for protection from cold temperatures and deep snow. Among other detrimental effects, clear-cutting also produces long unbroken rows of slash that impede the travel of large animals.

Whether a clear-cut will help or hurt wildlife depends upon the kind of wildlife, how large the clear-cut area will be, and how long a period of time it will take for reforestation. Those wildlife species that need openings may be helped, whereas those that need dense trees will be harmed. Those that require shrubby growth for browse may be

TABLE 15-3. Life-Form Descriptions

Life Form	Reproduces	Feeds	Number of Species[a]	Examples[b]
1	In water	In water	1	Bullfrog
2	In water	On the ground, in bushes, and/or in trees	9	Long-toed salamander, western toad, Pacific treefrog
3	On the ground around water	On the ground, and in bushes, trees, and water	45	Common garter snake, killdeer, western jumping mouse
4	In cliffs, caves, rimrock, and/or talus	On the ground or in the air	32	Side-blotched lizard, common raven, pika
5	On the ground without specific water, cliff, rimrock, or talus association	On the ground	48	Western fence lizard, dark-eyed junco, elk
6	On the ground	In bushes, trees, or the air	7	Common nighthawk, Lincoln's sparrow, porcupine
7	In bushes	On the ground, in water, or the air	30	American robin, Swainson's thrush, chipping sparrow
8	In bushes	In trees, bushes, or the air	6	Dusky flycatcher, yellow-breasted chat, American goldfinch
9	Primarily in deciduous trees	In trees, bushes, or the air	4	Cedar waxwing, northern oriole, house finch
10	Primarily in conifers	In trees, bushes, or the air	14	Golden-crowned kinglet, yellow-rumped warbler, red squirrel
11	In conifers or deciduous trees	In trees, in bushes, on the ground, or in the air	24	Goshawk, evening grosbeak, hoary bat
12	On very thick branches	On the ground or in water	7	Great blue heron, red-tailed hawk, great horned owl
13	In own hole excavated in tree	In trees, in bushes, on the ground, or in the air	13	Common flicker, pileated woodpecker, red-breasted nuthatch
14	In a hole made by another species or in a natural hole	On the ground, in water, or the air	37	Wood duck, American kestrel, northern flying squirrel
15	In a burrow underground	On the ground or under it	40	Rubber boa, burrowing owl, Columbian ground squirrel
16	In a burrow underground	In the air or in the water	10	Bank swallow, muskrat, river otter
		Total:	327	

Source: Thomas et al. (1979a).

[a]Species assignment to life form is based on predominant habitat-use patterns in Blue Mountains of Oregon.

[b]Scientific names are deleted for space considerations.

helped for many years after the clear-cut, but only if adequate shelter also is available. Information presented in Table 15-4, collected from cutover and virgin stands of Douglas-fir in northern California, shows that some species of birds require trees, others require shrubs, and still others require openings for feeding and breeding (Hagar 1960). Of 33 species of songbirds on a 4-ha area of aspen in northern Utah, DeByle (1981) found that five species declined or disappeared after clear-cutting, and three species increased or appeared. In a before-and-after study of an area with 12 circular clear-cuts of 1.2 ha, Scott et al. (1982) found an increase in chipmunks (*Eutamias minimus*) on clear-cuts and no decline of other small mammals. The study also reported a small decline in the numbers of ruby-crowned and golden-crowned kinglets (*Regulus calendula* and *R. satrapa*, respectively), but no other measurable changes in the

TABLE 15-4. Comparison of Distribution and Abundance of Species of Birds in Cutover and Uncut Douglas-Fir Forests in Northern California

Distribution	Number of Species	Examples
Found in forest but not in cutover area	7	Spotted owl (*Strix occidentalis*), pileated woodpecker (*Dryocupus pileatus*), Hammond flycatcher (*Empidonax hammondii*), red-breasted nuthatch (*Sitta canadensis*)
Found in cutover area but not in forest	20	California quail (*Lophortyx californicus*), golden-crowned sparrow (*Zonotrichia atricapilla*), lazuli bunting (*Passerina cyanea*), Allen's hummingbird (*Selasphorus sasin*)
More abundant in forest than in cutover area	12	Hermit thrush (*Hylocichla guttata*), golden-crowned kinglet (*Regulus satrapa*), raven (*Corvus corax*)
More abundant in cutover area than in forest	13	Red-breasted sapsucker (*Sphyrapicus thyroideus*), red crossbill (*Loxia curvirostra*), varied thrush (*Ixoreus naevius*)
Similar abundance in both cutover area and forest	3	Screech owl (*Otus asio*), mountain chickadee (*Parus gambeli*)

Source: Hagar (1960).

species or numbers of birds were observed. If the clear-cuts were larger, however, the changes probably would be greater.

The size of a clear-cut has important bearing on its value to wildlife. For deer and elk, proximity to cover is a paramount consideration. Wallmo (1969) found that clear-cut strips of about 20, 40, 60, and 120 m in width in forests of lodgepole pine (*Pinus contorta*) and spruce (*Picea* spp.), when alternated with uncut strips of the same width, doubled the number of mule deer (*Odocoileus hemionus*) on a study area in Colorado. Reynolds (1966) also noted that clear-cuts in blocks of less than 8 ha in size or in strips of less than 300 m in width benefit deer and elk in Arizona. (Very few clear-cuts in western forests, however, have been of such small dimensions.) In eastern North America, Krefting (1962) similarly recommended shelterwood cuts or clear-cuts in strips up to 20 m wide for white-tailed deer (*Odocoileus virginianus*).

Robbins (1979), however, warned of the problems of fragmenting forests (see Chapter 21). Some songbirds, particularly those that occupy the interiors of forests and have large home ranges, disappear when woodlots reach some minimum size. For Kentucky warblers (*Oporornis formosus*), the critical minimum was about 33 ha, but ovenbirds (*Seiurus aurocapillus*) were nearly absent in forests of less than 11 ha. Birds with large home ranges probably occupy special niches in the forest and require large areas to obtain their resources. Such birds thus are among the rarer species in forests and woodlots. Consequently, fragmenting the forest is of greater detriment to rarer species than it is to common ones. Noon et al. (1979) also noted the disappearance of rare species of forest songbirds on clear-cut areas.

Robbins (1979) suggested that creating numerous openings is not good management for an intact forest community: "Some of these recommendations are counter to traditional game management approaches of opening up the forest to provide more shrub growth for deer or ruffed grouse. Openings do provide more food for certain species, especially for support of seed-eating birds through the winter, but such openings as are desired for management of special interest species can be put in areas where they will not jeopardize a forest interior ecosystem."

This statement suggests that wildlife managers may have to decide on one of two approaches: management for *featured species* or for *species richness*. The former entails maintenance or improvement of habitat for one or a few species; in this case, the food and cover needs of other species in the same habitat remain of minor concern. Such management is frequently applied to game animals and endangered species. In managing for species richness, the objective is to maintain or create suitable habitat for the entire community of plants and animals. Species-richness management is desirable in wilderness areas and, possibly, in the management of productive forests.

SNAGS

Forest birds have evolved under conditions in which trees were subject to natural competition for sunlight and nutrients. As a result of competition, many trees die but remain standing for several years. These dead and dying trees, called *snags,* provide nesting, feeding, and perching sites for a wide variety of birds (Fig. 15-1). In North America, 55 bird species nest in tree cavities (Evans and Conner 1979), including the wood duck (*Aix sponsa*), common goldeneye (*Bucephala clangula*), peregrine falcon (*Falco peregrinus*), kestrel (*F. sparverius*), barn owl (*Tyto alba*), chimney swift (*Chaetura pelagica*), most woodpeckers (Picidae), chickadees (*Parus* spp.), nuthatches

(*Sitta* spp.), wrens (Troglodytidae), and bluebirds (*Sialia sialis*).

Until recently, modern forest management removed snags as a means of reducing competition for sunlight and eliminating breeding sites for insects and fungi. The absence of snags, however, also deprived many insectivorous birds of an essential component of their environment. Following removal of conifer snags with a heavy timber harvest in Arizona, the population of cavity-nesting birds declined by 52 percent. On an adjacent plot where snags were left standing, bird populations increased by 23 percent, and by 31 percent on an uncut control plot (Scott 1979).

Thomas et al. (1979b) and Evans and Conner (1979) reviewed literature on the role of birds in controlling insects. They concluded that birds often suppress outbreaks of insects. Takekawa et al. (1982) suggested that avian predators may have a value of $14.50 per ha for controlling outbreaks of spruce budworms. Most passerine birds that depend on snags for nesting sites are insectivorous. Thomas et al. (1979b) found that the lack of suitable nesting sites is

Figure 15-1. Snags provide nesting cavities and sources of food (e.g., insects) for woodpeckers and other birds and dens for squirrels and raccoons (left). Snags should be left standing during and after logging operations, as indicated here by the regrowth of younger trees. The U.S.D.A. Forest Service protects selected trees on National Forests (right). The sign reads, "Wildlife Tree–This tree saved for wildlife food and shelter. Please help protect it." (Photos courtesy of Elizabeth D. Bolen and the U.S.D.A. Forest Service.)

frequently a limiting factor on cavity-nesting birds and, in European forests where snags had been removed, that forest managers erected nesting boxes for birds as a method for controlling insects. Some 400,000 birdhouses were set on one 140,000-ha area in Bavaria for that purpose. Thomas et al. (1979b) suggested that the cost of leaving some natural snags would be considerably less than the cost of constructing and setting out birdhouses.

Regardless of their role in insect control, songbirds are a part of the forest community, a fact that seems reason enough to justify preservation of suitable habitat. (In National Forests, protection of birds is mandated by regulations under the National Forest Management Act.) Evans and Conner (1979) recommended leaving 0.1-ha clumps of standing trees within each 2-ha area treated with clear-cuts or shelterwood cuts. In northeastern forests, this prescription leaves between 800 and 1200 standing trees per 40 ha and provides habitat—including snags—for most of the smaller cavity-nesting species, but not for larger species such as hairy (*Picoides villosus*) and pileated (*Dryocopus pileatus*) woodpeckers. To satisfy the larger home-range requirements of the latter species, Schroeder (1982) recommended about 1 snag per 5 ha, with snag trees averaging about 54 cm in diameter in eastern North America and 76 cm in the west. Forest stands should be at least 70 ha in area in the east and 200 ha in the west. Bull and Meslow (1977) suggested that other woodpeckers usually will benefit if the minimum requirements are met for pileated woodpeckers. Evans and Conner (1979) recommended that uncut areas should be concentrated along stream edges where the trees also provide erosion control, protect fish habitat, and offer cover for game species.

DEADWOOD AND FUEL

Foresters frequently view logs rotting on the forest floor as a wasted resource. In some British and European forests, downed woody material is promptly picked up and used for firewood, giving the forests a neatly manicured, parklike appearance. Dead and down woody material, however, may play several roles in forest ecology, such as providing a base for growth of new trees ("nurse logs"), harboring fungi that aid in nutrient cycling, as well as providing habitat for wildlife (Fig. 15-2). In the Blue Mountains of Oregon, Maser et al. (1979) found 179 vertebrates, including 5 amphibians, 9 reptiles, 116 birds, and 49 mammals, making some use of logs on the forest floor. The study noted, however, that there is insufficient information as to how many logs and other woody material should be left for wildlife. In the absence of such information, the study recommended that prudent management should leave many logs and the surrounding natural materials for the benefit of wildlife.

Wood has become increasingly popular as heating fuel since the sudden rise of the price of fossil fuel in the mid-1970s. The total impact of collecting dead and dying trees or thinning stands for firewood remains to be assessed, but as Carey and Gill (1980) put it, "From the chickadee's point of view, such action may be tragic." They presented a system by which various woody species are rated according to value for songbirds, upland game birds, furbearers, and game mammals. The system encourages collectors of firewood to take those species that are of only fair value to wildlife but of high value as fuel. In the northeastern United States, such species include the hickories (*Carya* spp.), ashes (*Fraxinus* spp.), and

Figure 15-2. Decaying logs may fulfill several roles in forest ecology, including nutrient recycling and as sources of food and shelter for wildlife. The logs shown here litter the floor of a deciduous hardwood community in the eastern United States. (Photo courtesy of Elizabeth D. Bolen.)

black locust (*Robinia pseudoacacia*). At the other end of the scale, trees having high wildlife value and lower fuel value should be saved. These include aspens, spruces, and birches (*Betula* spp.). Carey and Gill (1980) suggested a rule of thumb for maintaining a diversity of wildlife in stands where firewood is cut: Harvest those species of trees that are most abundant, and thereby retain the greatest variety of species in the forest community.

FOREST FIRES AND WILDLIFE

Until recently there was only one acceptable thing to do with a forest fire—put it out. Attitudes toward fire in North America had been influenced strongly by fires that occurred in 1871 in Peshtigo, Wisconsin, claiming 1152 human lives (Holbrook 1945); and in 1894 in Hinckley, Minnesota, killing 418 people. Walt Disney's famous movie *Bambi*, made late in the 1930s, depicted a forest fire as a terrifying enemy of wildlife. About the same time, the U.S.D.A. Forest Service conceived of *Smokey*, the friendly and wise talking bear in a ranger's hat that admonished the public to prevent forest fires. The campaign was effective, and the exclusion of fire quickly became a basic tenet of forest protection.

Ecologists now understand that fire has been a natural part of the environment for centuries, as evidenced by fire scars buried deep within the soil and among the annual growth rings of trees (Ahlgren 1974). For example, two fire scars on a tree that is 100 years old indicate an average of one fire every 50 years. Examinations of very old trees thus offer opportunities to determine the frequency of fires before Europeans settled in North America. Other evidence about the history of fires comes from the type of plant cover. Curtis (1959) noted that mature vegetation may be either grassland or forest in areas on the eastern edge of the Great Plains where annual precipitation is between 50 and 75 cm (e.g., parts of Wisconsin and Minnesota). However, forests normally develop on those areas where fires occur irregularly, whereas grasslands develop in areas that are burned frequently (see Chapter 14).

In other areas, particularly the pine forests in the southeastern United States, fire protection produced an unanticipated effect (Wagner 1978). After the initial logging of longleaf pine (*Pinus palustris*), hardwood species of lower economic value replaced the pines. Foraging hogs (*Sus scrofa*) and fires were blamed for suppressing reproduction of the pines, but protection from these influences did not reestablish healthy pine forests. Instead, scrubby broadleaved trees shaded out the pine seedlings.

The life history of longleaf pine suggests adaptation to fire. A sprouting seed produces a root and a ground-level bud, from which sprouts a grasslike tuft of needles. For 7 years, the seedling remains in this "grass stage" while storing nutrients in a stout, growing taproot. A ground fire at this stage of development burns the needles but does not destroy the root or bud. Thereafter, the young tree draws on nutrients stored in the taproot, springs to a height of 3–5 m in a few years, and develops a thick fire-resistant bark. The tree continues growing at a more typical rate, provided that other trees do not compete for sunlight and nutrients.

Complete fire protection, originally advocated by the U.S.D.A. Forest Service in longleaf pine stands, favored the competing hardwood species and thereby suppressed the more economically desirable pines. Long periods of fire suppression also permitted large accumulations of combustible ground litter, so the fires that eventually occurred were larger and hotter than those that previously had occurred more frequently (whether set by local residents or from lightning)—and the hot fires generally destroyed all species, including the pines.

The effects of fire suppression in the southeast also had an impact on wildlife. For example, bobwhite (*Colinus virginianus*) habitat became less suitable in the absence of periodic fires. Stoddard (1939) found that fire prevented the development of dense tangles of brush, which generally are avoided by bobwhites, and stimulated the growth of grasses, legumes, and other food plants. When prescribed burning eradicates all understory vegetation, however, nongame birds are affected detrimentally (Wood and Niles 1978).

A general ecological effect of fire is to set back succession. Many wildlife species depend upon periodic destruction and renewal of the forest, with various species occurring at different stages in succession. In commercial forests, succession is set back when trees are harvested, but in forests with little or no potential for wood harvest, fire is the most frequent large-scale destructive force.

In July 1976, lightning started a fire in a remote portion of the Seney National Wildlife Refuge in Michigan. Fires on the refuge normally burn out on the sandy ridges surrounded by wet bogs, but a severe summer drought dried the bogs and the fire spread. By September, 30,000 ha had burned and the fire continued spreading from the refuge to adjacent commercial forests despite the efforts of firefighters. For a 3-year period afterward, Anderson (1982) studied the effects of the fire on wildlife by comparing vertebrate populations in burned areas with those in similar unburned areas nearby. At sites where the fire was the hottest, new habitat was created for sharp-tailed grouse (*Tympanuchus phasianellus*), black-backed woodpeckers (*Picoides arcticus*), and black bears (*Ursus americanus*). Amphibians and reptiles probably were killed in the fire, as indicated by populations after the burn that consisted of younger age groups; these likely represented animals that had immigrated after the fire. In areas where the fire was relatively cool, trees were not destroyed and ground vegetation had reestablished itself by the following spring, leaving few traces of the fire.

Bendell (1974) reviewed literature on direct effects of fire on birds and mammals. In some cases, bears (*Ursus* spp.) and moose have been seen swimming rivers to escape fire, whereas in other cases, animals have reacted without apparent alarm. In Alaska, despite many observers, no birds and mammals were reported fleeing hastily from a large fire on the Kenai Peninsula. A family of swans (*Cygnus* sp.) and a moose were seen feeding in a small lake while fire burned to the shore, and a small band of caribou (*Rangifer tarandus*) rested while surrounded by fire and then moved into the burn as the fire subsided. A few small mammals such as cotton rats (*Sigmodon* spp.) have been known to perish in forest fires, as have some large African mammals such as elephants (*Loxodonta africana*), lions (*Panthera leo*), and warthogs (*Phacochoerus* spp.) in tall-grass fires.

But the bulk of evidence reviewed by Bendell (1974) indicates that few birds or mammals perish in fires. Small animals in burrows only 10–20 cm below ground surface experience temperatures that are tolerable, and studies of tagged mice of various species show that they survive fires. Deaths do occur, however, from fires during the breeding season, when there are eggs, nestling birds, and relatively helpless young mammals.

Moose thrive in early successional stages, whereas caribou live in older climax forests. Edwards (1954) described the changes occurring after a fire burned 520 km² in northern British Columbia. Young woody growth became abundant, whereas arboreal lichens nearly disappeared. Moose, which browse heavily on tender woody vegetation, had been practically nonexistent in the area before the fire but became abundant during the following 27 years. Caribou, which depend largely on lichens, declined from abundance to scarcity.

Scotter (1970a) compared the density of droppings of moose and barren-ground caribou in forests of various age classes in northern Canada and demonstrated the changes in proportions of the two species as coniferous forests mature after a fire. Figure 15-3 clearly shows that moose reached their greatest abundance in forests 11–30 years old, whereas caribou numbers were low in early successional stages and then gradually increased with the age of the forest. Klein (1982) concluded that fire destroys lichens and renders areas unsuitable for caribou for a period of about 50 years. Without fire, however, lichen production may stagnate in stands more than 100 years old. Thus, over the long term, fire rejuvenates the growth of lichens and maintains ecological diversity in the boreal forest.

In northern Minnesota, Irwin (1975) observed that moose and deer commonly shared summer and autumn range on large burns of recent origin. The increased forage on these areas permitted an increase in the moose population despite the presence in the deer population of the parasitic nematode (*Parelaphostrongylus tenuis*), which is often fatal when transmitted to moose (see Chapter 8).

Because of the growing recognition that fire benefits many species of wildlife, controlled burning now is used more frequently as a management tool (Fig. 15-4). Perhaps the most publicized species for which burning is conducted is the Kirtland's warbler (*Dendroica kirtlandii*). This endangered species requires stands of jack pines that are 1.5–6.0 m tall and that range from 7–20 years of age. Without pines of this size, Kirtland's warblers lack suitable nesting habitat. Jack pine is a fire-adapted species, whose cones remain closed on the tree for decades, but open soon after being heated by forest fires. The U.S.D.A. Forest Service and the Michigan Department of Natural Resources, under the recovery plan for this bird, conduct a regular program of burning to promote the development of jack pine stands of the proper structure (Buech 1980). Further discussion of Kirtland's warbler management is found in Chapter 19.

Aspen forests seldom persist for more than 80–100 years, unless a disturbance such as cutting, wind, or fire sets back succession. DeByle (1985) reviewed the potential of fire for maintaining aspen stands in the interior western United States. In this region are more than 2.86 million ha of aspen, most of which are older than 60 years. These stands, left undisturbed,

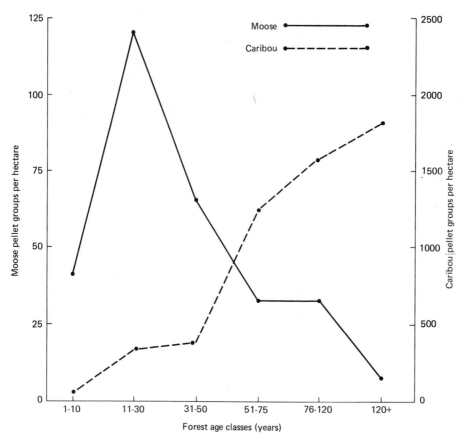

Figure 15-3.   A comparison of droppings of moose and caribou in sample plots of various ages indicating preference of moose for early successional stages and of caribou for more mature forests. (From Scotter 1970a.)

will be replaced by conifers in the climax community. Cutting or burning the old aspen, however, will proliferate stands of rapidly growing young aspens that spring from the living root systems of the older trees. An average of 25,000 stems per ha may arise from a mature aspen stand of 500 stems per ha. Wildlife and forest managers must choose whether to permit natural succession to proceed to conifers, which favor such species as southern red-backed voles (*Clethrionomys gapperi*), red squirrels (*Tamiasciurus hudsonicus*), and martens (*Martes americana*), or to cut or burn to favor deer, elk, moose, snowshoe hare (*Lepus americanus*), and ruffed grouse (*Bonasa umbellus*). The decision cannot be deferred until the conifers replace the aspens and the aspen roots die.

DeByle (1985) suggested that if the decision is made to retain aspens and no commercial market is available for aspen wood, burning is a feasible alternative to cutting. The size of the burns can be selected, and burns can be made in patches at intervals to encourage a diversity of density and age categories. Units of 2–24 ha on a rotation of 60–100 years would provide a mixture of even-aged stands to meet the food and cover needs of several wildlife species.

Fire suppression in the subtropical Florida Keys probably has reduced habitat quality for Key deer (*Odocoileus virginianus clavium*), an endangered species whose population has declined recently. Carlson et al. (1993) therefore measured the nutrient content of browse following natural and prescribed fires. Crude protein, phosphorus, and digestibility—measures of browse quality—were improved by fires. Additionally, fire was an important agent for maintaining the presence of early successional species of endemic herbaceous plants. The research produced recommendations that parts of the Keys should be managed with fire to provide high-quality browse and early successional vegetation important in the diet of Key deer. However, to maintain the natural diversity in the local biota, management plans for the Keys also should en-

Figure 15-4. A prescribed burn sets back succession on wildlife habitat in New York. The burn provides grass and other herbaceous plants as food for deer and grouse and opens up the cover for many songbirds. For safety, the burn originates from a staffed fireline. (Photo courtesy of M. E. Richmond, U.S. Fish and Wildlife Service.)

courage the development of more mature successional stages (e.g., prescribed burns scheduled for designated areas at intervals of 5, 15, 25, and 75 years).

In national parks, wilderness areas, and other places where commercial timber production is not the management goal, federal agencies have developed a viable policy for dealing with forest fires. Fires started by lightning are permitted to burn so long as they pose no danger to human life, structures, or commercial forests. This policy recognizes the historic role of fire in forest and other ecosystems. It acknowledges that plants and animals dependent on fire are not stable entities that can be preserved in an unchanging form but instead are dynamic communities continually undergoing natural processes of maturity, destruction, and rejuvenation.

We emphasize that wildlife managers are not advocating the abandonment of fire suppression on commercial forests or uncontrolled burning anywhere but in those areas set aside for their value as natural ecosystems. But listeners to Smokey the Bear in recent years will notice a slight change of tune. Instead of once familiar all-inclusive condemnation of fire, Smokey now says that fire, *except in the hands of experts*, can destroy valuable timber and wildlife.

OLD-GROWTH FORESTS— A SPECIAL CASE?

Along the Pacific coast from northern California to southern Alaska, there are forests with trees up to 1000 years old. The species include redwoods (*Sequoia sempervirens*), western hemlock (*Tsuga heterophylla*), Sitka spruce (*Picea sitchensis*), and Douglas-fir. These are uneven-aged forests; gigantic trees stand beside young trees, and sunshine reaches the ground only in places where old trees have fallen. Such stands have ecological characteristics that are not readily replaced or mimicked by management techniques.

The complexity of food webs in old-growth forests has been described in detail (Meslow et al. 1981; Baker 1986; Kelly 1986). Included are rotting logs (often regarded by commercial foresters as wasted wood), shelter snails, centipedes, and numerous other invertebrates. These animals feed upon the fungi that decompose the logs and thereby assist in returning nutrients to the soil. Some fallen trees, resting in streams, shade the eggs and fry of salmon. Flying squirrels (*Glaucomys sabrinus*), several species of bats, tree voles (*Arboremus longicaudus*), and western redbacked voles (*Clethrionomys occidentalis*) are among the small mammals that spread the spores of fungi, especially truffles and other species whose spores are produced underground and depend entirely on animals for their dissemination. The spores cling temporarily to the pelts of the rodents that eat the truffles, and then they drop off as those animals travel about. In addition to their role as decomposers, some fungi act as mycorrhizae, combining with the root hairs of living trees and increasing the ability of the trees to absorb water, nitrogen, and minerals. Predators on the small mammals include martens, goshawks (*Accipiter gentilis*), and northern spotted owls (*Strix occidentalis caurina*). The latter hunt and nest almost exclusively in the old-growth forests.

Bald eagles (*Haliaeetus leucocephalus*) also nest and roost in these old-growth stands, particularly along rivers and estuaries where fishes are available as food. The large trees intercept snow and provide shelter from wind for wintering mule deer. Wallmo and Schoen (1980) found that Sitka black-tailed deer (*Odocoileus hemionus sitkensis*) used old-growth stands considerably more in both summer and winter than stands that were 1–150 years old. Figure 15-5 compares the undergrowth in an old-growth stand with the forest floor in a regenerated even-aged forest. A "rain" of broken twigs and branches falls to the ground from the crowns of the old-growth trees,

Figure 15-5. Comparison of an old-growth stand (left) with a 100-year-old even-aged stand (right). Note the uneven-aged trees and abundant shrubs in the old-growth forest. Such sites are favored by black-tailed deer. The even-aged stand regenerated after old-growth forest was destroyed; the trunks are of similar size and there is little understory vegetation. Black-tailed deer are not plentiful in such stands. Both sites are in coastal Alaska. (Photos courtesy of J. W. Schoen.)

and shrubs growing in the patches of sunlight create a sufficient supply of browse. Many other species of birds and mammals, whose living requirements are not fully understood, also are a part of the old-growth forest community. Unfortunately for these species of wildlife, old-growth forests also provide high-quality lumber. About 5 million of the original 6 million ha of old-growth forest have been cut in the last century (Baker 1986).

Perhaps 2000 pairs of spotted owls remain, but their numbers are declining rapidly as old-growth forest disappears (Fosburgh 1986). Loss of these owls could eliminate their regulatory effect on rodent populations, bringing about "boom and bust" cycles (Kelly 1986). How this disturbance would affect fungi and forest growth, if at all, remains unknown at present. The spotted owl has been designated by the U.S.D.A. Forest Service as an *indicator species*, which means that the presence of these owls is used as a barometer of the health of old-growth forest communities. Current federal policy requires maintenance of "viable" populations of all species of native wildlife on lands managed by the U.S.D.A. Forest Service and the Bureau of Land Management (Meslow et al. 1981). This process entails maintaining enough individuals throughout the range of each species to perpetuate naturally self-sustaining populations. Depending on the geographical area, the goal for spotted owls will require maintaining 600–1800 ha of old-growth timber per pair of birds and also preserving a habitat "network" of connecting corridors between the larger stands of old growth (Fosburgh 1986).

Studies of juvenile northern spotted owls outfitted with radio transmitters emphasize the essential role of older, closed-canopy forests to the birds' dispersal (Miller et al. 1997). Young owls that are forced to disperse through open stands of saplings and clear-cuts suffered greater mortality than did those that traveled through older forests or closed stands of saplings. Furthermore, the owls dispersed for shorter distances in areas where clear-cuts were abundant in comparison with those that dispersed into forests with closed canopies. These results suggest that the natural dispersion and survival of young spotted owls are lessened when older forests are heavily cut. Hence, to promote the dispersal and safety of young owls, forests should be managed to reduce clear-cuts and to maintain a broad matrix of stands with different age classes.

About 2.3 million ha of old-growth forest remain in the Tongass National Forest of Alaska. Clear-cutting is scheduled to remove large blocks of this forest, and plans call for regeneration of even-aged stands that will be harvested on a 90–125-year rotation (Matthews and McKnight 1982). To ensure local employment in the forest industry, an agreement has been made by the U.S.D.A. Forest Service to harvest about 4.5 billion board feet per decade. In 1980, Congress with the Alaska Lands Act allocated $40 million annually to assist in meeting this harvest level on lands where the low volume of timber makes logging unprofitable without a subsidy. Less than 30 percent of the old-growth stands will be preserved intact (Hoopes 1982).

The even-aged stands that regenerate after harvest do not attain the characteristics of present old-growth stands, even after 100 years. Crown closure is complete, preventing sunlight from reaching the forest floor, and bole sizes are relatively uniform. Forage in the understory is sparse. In addition, Matthews and McKnight (1982) note that removal of 70 percent of the forest will leave remaining stands isolated, preventing the migration of some animals from stand to stand. In western Oregon, several species of wide-ranging carnivores, including grizzly bear (*Ursus arctos horribilis*), wolf (*Canis lupus*), lynx (*Felis lynx*), wolverine (*Gulo gulo*), and fisher (*Martes pennanti*), already have been extirpated or reduced to low numbers in the remaining stands of old growth that have been fragmented and isolated by clear-cuts (Harris et al. 1982). In addition, with the exception of bald eagles and mule deer, we have little detailed information about the requirements of wildlife living in old-growth forests. We do know, however, that such forests are irreplaceable and that attempts to simulate the ecological characteristics of old-growth stands will fail with a 100-year rotation. In fact, Hoopes (1982) contended that old-growth forests are not harvested but instead are mined. Wildlife researchers need time to evaluate the ecological interactions of wildlife in the northwestern old-growth stands, but it appears that economic pressures for cutting these stands will prevail before sufficient ecological data can be gathered and interpreted.

PRIVATE WOODLANDS

In harvesting private woodlands, numerous concessions can be made so that many species of wildlife are not left wanting for suitable habitat. Guljas (1975) and Hassinger et al. (1981) have pointed out how timber, pulp, and firewood cuts can be made with consideration for wildlife needs. In planning a tree harvest, the landowner should consider the value of each tree both for wood production and for wildlife use. Guljas (1975) described numerous means of managing for both timber production and wildlife in the eastern United States, where the majority of forest lands are privately owned. Many of these recommendations could be modified for other geographical areas.

Guljas (1975) recommended against grazing sheep and cattle in woodlots. The activities of large animals may be detrimental both to timber production and wildlife, as ground vegetation is reduced, preventing reproduction of trees as well as depriving small animals of food and shelter. Hooves trample the ground, compacting the soil and reducing water and air supplies to tree roots.

The edges between field and forest usually produce a growth of shrubs. In the eastern United States, edges with shrubs such as sumac (*Rhus* spp.), sassafras (*Sassafras albidum*), and dogwood (*Cornus* spp.) harbor nearly double the number of birds and 1.4 times the species of birds than woodlands that lack shrubby edges (Guljas, 1975).

Snags should be spared, along with trees providing a variety of mast, such as acorns, walnuts, and hickory nuts, and some small trees furnishing low cover. The boundaries of clear-cuts should be uneven, and trees on stream banks should be left standing. Stream banks should be protected by building bridges for machinery. After the harvest ends, loading sites for logs should be fertilized and seeded with herbaceous plants that provide cover for the exposed soil and food for wildlife (Figs. 15-6 and 15-7).

Many owners of small forests sell their timber "on the stump" to a contractor who harvests and markets the wood. Hassinger et al. (1981) prepared a sample contract by which landowners can protect wildlife values when their timber is sold in this manner (Fig. 15-8).

FOREST MANAGEMENT FOR RUFFED GROUSE— AN EXAMPLE OF FEATURED-SPECIES MANAGEMENT

A number of forest wildlife species are of special interest because of their endangered status or their value as game animals. Scientific research conducted over the past few decades has revealed much about managing habitat for animals such as wild turkey (*Meleagris gallopavo*), white-tailed deer, bobwhite, and ruffed grouse. The last species, a popular and challenging game bird with a transcontinental range, is chosen here as an example of featured-species management. As noted earlier, management for featured species differs from management for species richness, but the former also may benefit other wildlife. Such is the case with forest management for ruffed grouse.

Material for this discussion originates almost entirely from the work of Professor Gordon Gullion of the University of Minnesota, whose long-term research yielded a picture of interactions involving two major players—ruffed grouse and aspen trees—and several minor players, including snow, goshawks, and humans.

Ruffed grouse are nonmigratory birds of northern deciduous and coniferous forests. In spring (and sometimes in autumn), males advertise their presence

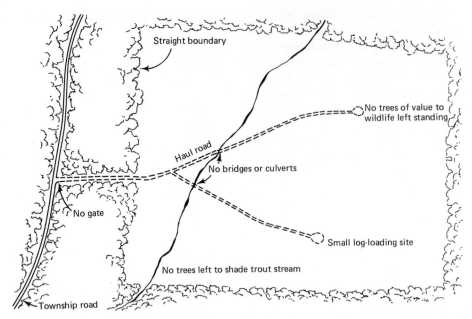

Figure 15-6. Clear-cut without considerations for wildlife. The map illustrates a clear-cut conducted for short-term economic benefits. No shade trees were left along the stream, snags and den trees were removed, and loading sites for logs were not seeded with cover and food plants. Note that the straight boundaries of this clear-cut leave a minimum of edge. (From Hassinger et al. 1981.)

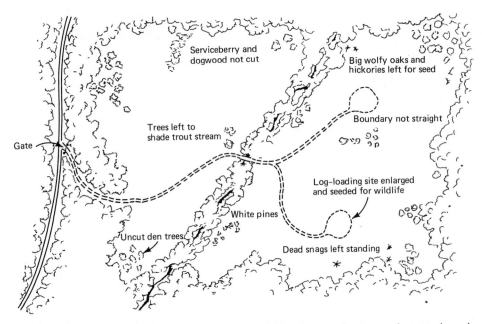

Figure 15-7. Clear-cut with considerations for wildlife. The map illustrates a clear-cut planned to leave streamside trees for shade and erosion control. Snags, den trees, and mast-producing trees also remain after the cut, and loading sites were seeded. Note that the boundaries of this clear-cut are irregular and create a good amount of edge. (From Hassinger et al. 1981.)

I, Mark Hamilton of Masten, Pennsylvania (Purchaser) agree to purchase from Woodrow Meristem of Pomfret Center, Pennsylvania (Seller) the trees described below.

I. Location of Sale: The 42-acre woodland is in Derry Township, Tioga County, Pennsylvania, at the intersection of PA Route 804 and Legislative Route 7221, as shown on the attached map.

II. Trees to be Cut: Cut all designated trees and/or trees marked with yellow paint. Reserve all hemlock, hickory, dogwood, serviceberry, and black gum. Additional trees of special wildlife value to be left are marked with blue paint. Also not to be cut are any trees within 100 feet of Brougher Run except those marked with yellow paint by Seller.

III. Conditions of Sale:

    A. The Purchaser agrees to the following:

        (1) To pay the Seller the sum of $16,350 for the above designated or marked trees, and to make payment in advance of cutting.

        (2) To waive all claim to the above described trees unless they are cut and removed on or before one calendar year from the date on this contract. In the event Purchaser is, due to circumstances beyond his control, unable to complete the sale in the time allowed, the Seller and Purchaser may agree on an extension of time for this contract.

        (3) To construct a log-loading site approximately one-half acre in size in the southeast portion of the tract at a location agreed upon by the Seller and Purchaser.

        (4) To do all in his power to prevent and suppress forest fires on, or threatening, the sale area.

        (5) To avoid unnecessary injury to all trees not designated to be cut.

        (6) To repair damages caused by logging to ditches, fences, bridges, roads, trails, or other improvements damaged beyond ordinary wear and tear.

        (7) Not to assign this Agreement in whole or in part without the written consent of the Seller.

        (8) All tops and slash from felled trees within 25 feet of the adjoining highway will be removed. No slash will be left across or on the public road, cleared field, or Brougher Run. Tops may be left on skid trails to prevent erosion.

        (9) To leave standing all marked property boundary trees.

        (10) Purchaser will take precautions to prevent soil erosion and other conditions detrimental to the property resulting from logging operation. Should such conditions occur, they will be corrected by the purchaser. He also will remove all oil cans, paper, and other trash resulting from the operation.

        (11) To furnish to Seller 20 pounds of perennial rye grass seed, 2 pounds of timothy seed, and 6 pounds of innoculated birdsfoot trefoil which Seller will apply to the log-loading site and roads upon completion of this timber sale.

        (12) To maintain public liability and workman's compensation insurance policies for the duration of this contract.

    B. The Seller agrees to the following:

        (1) To guarantee title to the forest products covered by this Agreement, and to defend it against all claims at his expense.

        (2) The property boundary lines shown to the Purchaser by the Seller are correct as located on the attached map. The Seller will save harmless the Purchaser from all trespass claims originating as a result of errors in the boundary line location made by the Seller.

        (3) To allow the Purchaser to make necessary logging-road improvements such as bridges and gates which shall be removed or left in place as agreed upon by the Seller and the Purchaser. Trees designated for cutting may be used to construct such improvements.

        (4) To grant freedom-of-entry and right-of-way to the purchaser and his employees on and across the area covered by this Agreement, and also other privileges usually extended to purchasers of timber which are not specifically covered, provided they do not conflict with specific provisions of this Agreement.

    C. In case of dispute over the terms of this Agreement, we agree to accept the decision of an Arbitration Board of three selected persons as final. Each of the contracting parties will select one person, and the two selected will select a third to form this Board.

        Signed this _____ day of _____, 19____

Witness:

_____   (Signed) _____
                                                        Purchaser

_____   _____
                                                          Seller

Figure 15-8. A sample contract for a timber sale in which written provisions are included for the protection and development of wildlife habitat. Similar contracts can be prepared for any timber sale. (From Hassinger et al. 1981.)

and territories with a ritual known as "drumming" (i.e., males stand on logs and produce powerful drumming sounds by rapidly beating their cupped wings). In May, after locating a drumming male and mating, a mature female lays a clutch of 8–15 eggs that hatches in 23 days. The precocial young leave the nest as soon as their down is dry, and the female accompanies the brood for 6–10 weeks, after which the young disperse to seek their own ranges. In winter, ruffed grouse feed heavily on tree buds, and, if the snow is soft and deep, the birds burrow underneath the surface to escape low temperatures and chilling winds.

A number of factors affect the reproduction and survival of grouse, thereby influencing the size of the population. These include quantity and quality of food available, snow conditions, predation, and hunting. The timing at which these factors are present or absent has great bearing on grouse numbers.

In much of their range, ruffed grouse find food, shelter from weather, and escape cover almost entirely in habitat dominated by aspen trees (Gullion 1970, 1972, 1982). Two species, quaking (small-tooth) aspen (*Populus tremuloides*) and big-tooth aspen (*P. grandidentata*), produce buds in winter that contain all the necessary minerals, fats, proteins, and carbohydrates to nourish ruffed grouse through the winter. Furthermore, the density of male flower buds

on mature aspen trees, a preferred and most nutritious food, is usually sufficient so that a grouse can fill its crop in just a 15-minute feeding period, thereby limiting exposure to predators to only a minimal period. Grouse also can subsist on buds of hazel (*Corylus* spp.), maple (*Acer* spp.), and birches (*Betula* spp.), but securing these foods requires that the birds move about, thereby expending more energy and increasing their exposure to predators. Mature aspens thus provide a high quality and quantity of food. Grouse can subsist on a wide variety of food in spring, summer, and autumn, and so food does not limit their numbers in those seasons. Figure 15-9

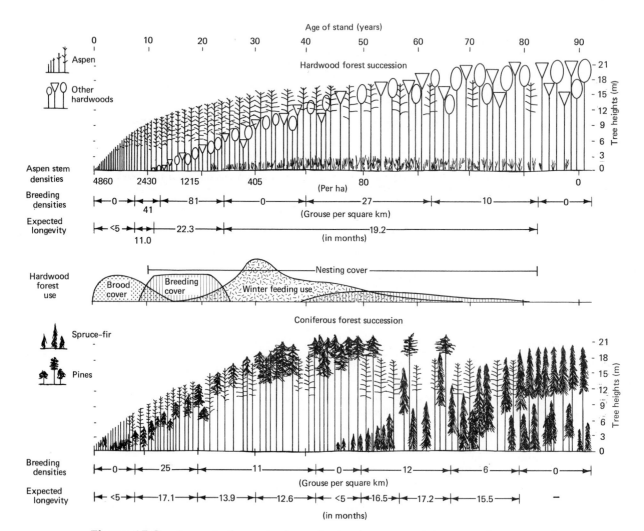

Figure 15-9. A portrait of a growing forest, showing the sequence of changes as a forest ages and how these changes influence ruffed grouse use and densities. The hardwood forest succession represents a fairly accurate portrayal of the natural sequence, but the coniferous forest succession is faithful only to the age of about 40 years—the remainder of this sequence is compressed for the sake of convenience. (From Gullion 1972.)

shows deciduous and coniferous forest succession with aspen—the pioneer species—gradually being crowded out by climax species.

Aspen saplings, which grow in profusion (up to 30,000 stems per ha) after a clear-cut or a fire, provide excellent cover for young grouse. The young birds can move about in the dense cover, but the high density of vertical stems prevents effective pursuit by hawks.

Nesting sites and drumming logs frequently are found in aspen stands of intermediate age (8–25 years old), although mature stands (40+ years) also provide drumming logs and nest sites. Although grouse sometimes seek out coniferous cover in winter, Gullion's studies in Minnesota have shown that conifers are not necessary for grouse, provided that the snow is deep (about 25 cm) and soft enough to permit burrowing. Conifers, in fact, may offer predators the advantage of concealment.

Good grouse cover offers protection against hunters as well as against other predators. Gullion (1982) contends that most grouse are shot not in good grouse cover but in marginal cover. These are frequently young birds dispersing through unfamiliar and less-than-ideal habitat where they are more vulnerable to the gun than birds that have established territories in good cover.

So ideal ruffed grouse habitat consists of abundant aspens, including saplings, pole-size trees, and mature trees, all within the annual foraging range of a bird. Forest management can create these conditions by clear-cutting in a pattern in which a patchwork of four age-classes of aspen are present. Figure 15-10 illustrates how this may be accomplished using 4-ha blocks, the size recommended for large commercial forests. In smaller forests, blocks down to 0.4 ha may be desirable if they are economically feasible. The blocks should be nearly square in shape to minimize

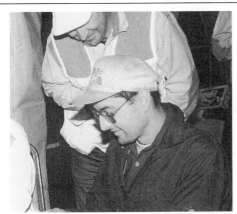

Photo Courtesy of Dan Desseker

On the Job with...
Dan Desseker
Forest Wildlife Biologist

Dan Desseker is a forest wildlife biologist with the Ruffed Grouse Society. He is stationed at Rice Lake in northern Wisconsin—a town known for experiencing a record low temperature of −53°C several years ago. The Ruffed Grouse Society is a private organization with over 25,000 members who are dedicated to improving the environment for ruffed grouse, American woodcock, and other species of forest-related wildlife. The society achieves its goals by encouraging landowners—whether private, state, or federal—to practice management that encourages a variety of forest types, including the regular presence of young stands of timber.

Dan earned both B.S. and M.S. degrees in Wildlife Management. He spends much of his time "on the job" communicating by letter and telephone with land managers and other professional biologists across North America. He encourages and advises these contacts about management practices that produce early and mid-successional forests. Dan also attends and participates in state and regional conferences that deal with forest management. In the course of these activities, he routinely delineates and advocates the position of the Ruffed Grouse Society. On a recent typical day, Dan opened, read, and attempted to respond to a 10-day accumulation of mail, and he participated in a conference call to Washington, D.C., regarding hunting regulations.

According to Dan, the most rewarding aspect of his job stems from being able to provide information to policy makers, such as state and federal legislators, as well as to other professionals in the field of forest management. He finds it particularly gratifying to see his efforts result in prime habitat for grouse and other early successional species of forest wildlife. He points out, in response to those who might be more concerned with large trees in mature forests, that "baby trees and the animals associated with them need love, too." Conversely, his greatest frustrations arise from the lack of adequate time to address all of the conservation and management issues that need attention.

Dan's advice for college students who hope to enter the wildlife profession: Earn a master's degree, develop public-speaking skills, take a course in marketing—and read Edgar Allen Poe.

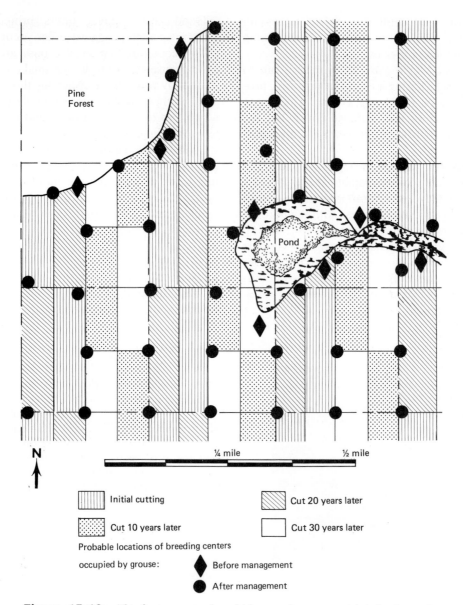

N

¼ mile   ½ mile

▯ Initial cutting   ▨ Cut 20 years later

▨ Cut 10 years later   ☐ Cut 30 years later

Probable locations of breeding centers

occupied by grouse: ◆ Before management

                        ● After management

Figure 15-10. This figure suggests how fairly extensive commercial timber harvesting could be done in aspen or hardwood forests in a manner that should substantially benefit ruffed grouse and other forest wildlife species. Each of these rectangular strips consists of 4 ha, and 64 ha could be cut from 2.5 km² in each operation, spaced at 10-year intervals. On the other hand, cuttings could be at 5-year intervals, cutting half as much each time. (From Gullion 1972.)

edge. This form of management has produced breeding densities of about 1 pair of grouse per 5 ha in the Mille Lacs area of Minnesota.

This example of featured-species management for ruffed grouse obviously produces an aspen monoculture. However, aspen growth benefits several other species of wildlife, including deer, woodcock (*Scolopax minor*), and a variety of songbirds. The presence of aspen stands in different age-classes creates diversity in the undergrowth. The long-term effects of maintaining an early stage of succession has not been studied, however, and research of this type will be needed before the picture of grouse management is complete. Furthermore, the management of forests for wildlife will require compromises between those who desire environments for particular species and those who advocate a full spectrum of species richness.

SUMMARY

Human interest in forests in North America has evolved in three stages: from exploitation, to management for wood production, to conservation of the entire forest ecosystem. Recognition that forests hold values for uses other than wood production ushered in the present era of conflicts among wood producers, wildlife managers, and wilderness preservationists. In the United States, legislation such as the National Forest Management Act, the Multiple-use Sustained Yield Act, and the Wilderness Preservation Act requires that wildlife be considered in the management of National Forests.

Some methods of harvesting trees can enhance the habitat for some kinds of wildlife, but the needs of all animals cannot be met in the management plans for a single forest. Some management may be designed to enhance species richness, whereas other management may be directed toward one or a few featured species.

Fire, now recognized as a part of the natural environment, has profound effects on the forest community. Fires normally do not kill many animals, but vastly altered habitat resulting from fires usually changes the abundance and composition of wildlife communities. The general effect of fire is to set back succession to early stages, favoring such large mammals as deer and moose and acting against caribou for the first 100 or more years after a fire.

Evidence from old-growth forests in the Pacific Northwest contradicts the commonly held belief that forests in early successional stages produce more wildlife than climax forests. Old stands of uneven-aged trees provide a mixture of shelter and food for black-tailed deer, nesting sites for eagles, and breeding and hunting areas for spotted owls that are not available in second-growth stands. Only 10–20 percent of the original old-growth forests remain in North America. Conservationists hope to preserve the complex floral and faunal communities in old-growth forests and use the spotted owl as an indicator species.

Management for a featured species—the ruffed grouse—largely consists of clear-cutting aspen stands, thereby setting back succession and forming a patchwork of different age-classes. The dangers of a monoculture in featured-species management may be partially avoided by incorporating stands of various age-classes in the forest.

At present, forest management includes conflicting recommendations from forest managers, game managers, and nongame managers. Resolution of these conflicts will require not only more biological data, but also imaginative management approaches coupled with social and political acumen.

CHAPTER 16

WILDLIFE IN PARKS AND REFUGES

> *Beyond the countryside lie the lands largely uninhabited or undisturbed by man. These remote mountains, forests, and deserts include the remnants of primeval America. Here, the ... park lands offer opportunities for memorable outdoor experiences in surroundings of superlative natural beauty. But here, too, population growth and technological change are combining to destroy these opportunities.*
>
> **President's Council on Recreation and Natural Beauty (1968)**

We shall define a *park* as an area that is designated primarily for the purpose of human recreation. This definition requires qualification in some national parks, as we shall see, and it excludes such areas as national and state forests and wildlife refuges where, although recreation also occurs, the primary management goal is something other than human recreation. Parks are administered under various political jurisdictions, including governments of villages, cities, counties, states or provinces, and nations. This chapter deals with the ecology and management of animals in large nonurban parks and refuges. These areas are generally under jurisdiction of state, provincial, or federal governments. Wildlife in urban parks is considered in Chapter 17.

Textbooks on park management and outdoor recreation devote considerable attention to such topics as management of vehicular traffic, campground design, swimming facilities, location of concessions, trash collection, and sewage disposal. However, the management of wild animals in parks receives only passing mention. Most considerations of wildlife in parks deal only with conflicts between humans and wild animals (Kellert 1980a). Whereas problems with such animals as bears (*Ursus* spp.) are an important part of wildlife

management in parks, other species and the natural communities to which they belong also deserve the attention of park managers and biologists. Large herbivores, for example, frequently overpopulate parks where hunting and natural predators have been eliminated. The natural dynamics of ecological succession also foil attempts for preserving natural communities in parks, and the movement of animals in and out of parks illustrate that ecosystems in parks interact with highly modified environments outside.

This chapter reviews some special problems of managing wildlife in parks and refuges, including interactions between wild animals and visitors, overpopulation of protected animals, natural succession, management functions, and interaction between animals in parks and surrounding lands.

ENJOYMENT OF WILDLIFE BY PARK VISITORS

Although it seems intuitively clear that wildlife enhances recreational values for park visitors, there is little quantitative information to support that assumption. Such data would be useful for describing the value of wildlife and for comparing wildlife values with other values associated with outdoor experiences (e.g., quality of camping facilities and pleasant scenery). Kellert (1980a) noted that an understanding of the satisfactions associated with people–wildlife interactions would aid intelligent park planning and management. Additionally, safe and relatively natural opportunities for human interactions with wildlife offer the public an educational means for gaining a better understanding of animals and ecological relationships. The sort of information needed, Kellert noted, concerns the types of visitors and their preferences for contact with various animals, links between types of habitats and wildlife-related opportunities, and species-specific tolerances of wildlife to humans.

Using interviews and questionnaires, Brown et al. (1980) assessed the recreational experiences enjoyed by 312 back-country hikers in Colorado. The study quantified the various factors or segments contributing to the overall satisfaction of a hike through wild country (Table 16-1). Hikers placed the highest psychological value on achieving a relationship with nature; this aspect scored +3.2 of a possible 4.0 points. Following that, among the psychological attributes of hiking were the escape from pressure (+2.9) and the benefit of exercise (+2.6). The same hikers also

TABLE 16-1. Some Values Obtained by Cluster Analysis of Perceived Contributions to Recreational Experiences of Back-Country Hikers in the Weminuche Area of Colorado

	Value[a]
Psychological Attributes:	
Relationship with nature	3.2
Escape physical pressure	2.9
Exercise	2.6
Freedom	1.9
Achievement	1.6
Reflection on personal values	1.3
Wildlife Values:	
Large wildlife (bighorn sheep, deer, mountain goats)	2.7
Small wildlife (beaver, ptarmigan, and other birds)	2.7
Good fishing	2.4
Naturally reproducing fish	2.4

Source: Brown et al. (1980).
[a]Scale: highest value = 4.0; no value = 0.0; strongest negative value = –4.0.

scored wildlife values using another set of questions. The value of seeing large animals, such as bighorn sheep (*Ovis canadensis*), mountain goats (*Oreamnos americanus*), and deer (*Odocoileus* spp.), scored +2.7, as did the value of seeing ptarmigans (*Lagopus leucurus*), beavers (*Castor canadensis*), and smaller animals. The interest-in-wildlife values thereby compared favorably with those of exercise (+2.6) and outranked such values as the personal achievement of hiking (+1.6) and the opportunity to reflect on personal values (+1.3). Evaluations of this sort may be different for families using campgrounds, for hunters, or for birdwatchers, but the appreciation of wildlife by these or other types of park-users remains unstudied. Nonetheless, the study of hikers supports, in quantitative terms, the premise that wildlife plays a significant role in the recreational value associated with parks and other wild areas.

The popularity of shows featuring trained killer whales (*Orcinus orca*) and dolphins (*Tursiops* spp.) illustrates the high level of public interest in animals. Such shows inform thousands of people about the behavior and intelligence of these animals, but the public learns little about the ecology of marine mammals and their natural environment. Unlike marine aquaria, many parks have the resources to combine recreational experiences with education, thereby stressing the overall ecological setting in which animals live in the wild. To perform these functions, viewing facilities should be designed in ways that ensure minimum disruption of the animals' normal activities. In addition, an ecological understanding of what visitors see should be enhanced by the availability of self-guided tours, educational materials, and informed naturalists. Many national parks in the United States and Canada have visitor centers with informative displays and daily lecture programs emphasizing the ecological relationships of wildlife.

Agreeing that wildlife in parks should be managed for human benefits and public education, Shaw and Cooper (1980) described how park managers can enhance the ways visitors appreciate animals. The method is based on a system of guidelines that develop appropriate opportunities for viewing wildlife in parks (Fig. 16-1). Some species have little or no tolerance for humans, however, and species such as mountain lions (*Felis concolor*) and wolves (*Canis lupus*) cannot be viewed easily. Nonetheless, an extremely popular program at Algonquin Park, Ontario, treats visitors to nightly "wolf howls" in which wolves answer the calls of park guides. Other species

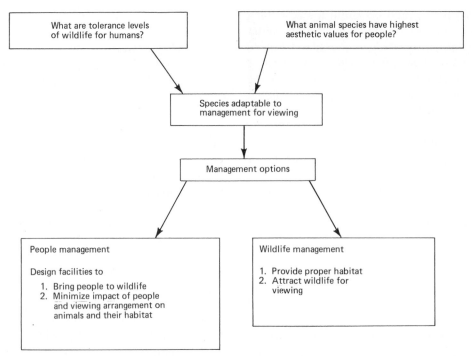

Figure 16-1. Management guidelines for developing wildlife viewing opportunities in parks. (From Shaw and Cooper 1980.)

have seasonal needs for seclusion, whereas at other times people may approach the same animals rather readily. Some endangered species at times may be tolerant of human intrusions. For example, boat tours and a well-placed tower produce good views of whooping cranes (*Grus americana*) on their wintering grounds in Texas (Fig. 16-2). However, similar arrangements on the nesting grounds of whooping cranes probably would be disruptive.

Figure 16-2. Thousands of tourists view whooping cranes each winter from this tower at Aransas National Wildlife Refuge, Texas. (Photo courtesy of U.S. Fish and Wildlife Service.)

Because of their unique features, certain animals are more aesthetically appealing than other animals (Shaw and Cooper 1980). In general, large mammals and birds are more popular than smaller species, and colorful animals attract more attention than drab forms. Interactions between parents and their young also are of great public interest, and predators have special appeal to most people. Whatever the species may be, however, they should be viewed in settings that are as natural as possible. Kellert (1980a) thus did not recommend such practices as those carried out in Yellowstone National Park in the early 1900s when bison (*Bison bison*) were rounded up daily in locations where people could conveniently watch the animals feed. The presence of natural habitat remains the essential difference between viewing animals in wild areas and in zoos; thus habitat management may be necessary in parks where viewing wildlife is a major objective.

Various designs are available for managing people who are viewing wildlife. These range from such simple things as roadside turnouts and picnic tables near water to observation towers and specially designed buses (Figs. 16-2 and 16-3). Some visitor centers are built underwater with windows offering a fish-eye view of aquatic life. The U.S. Army Corps of Engineers has such an arrangement at Bonneville Dam,

Figure 16-3. Tour vehicles reduce the impact of visitors viewing wildlife at Everglades National Park. Vehicles keep visitors in a group and allow interpretive comments from guides. (Photo courtesy of R. Frear, Everglades National Park.)

on the Columbia River in Oregon, where visitors can watch the ascent of salmon, trout, and other species in the fishway (Fig. 16-4).

Providing opportunities for viewing wildlife in natural settings presents park supervisors and wildlife managers with a major difficulty: attaining a proper balance between human–wildlife interactions without impairing the environment and disrupting the behavior of the animals (Shaw and Cooper 1980). Therefore, educating large numbers of people about wildlife while avoiding the disturbance of natural processes remains a paradox of park management.

HUNTING IN PARKS

Recreational hunting may be appropriate in parks offering adequate areas of undeveloped land or water. Hunting currently is permitted on 41 areas under jurisdiction of the U.S. National Park Service (Williamson 1985). In parks where natural predators are absent, hunting may offer a desirable means of preventing overbrowsing by abundant populations of deer and other large herbivores. A small number of expert marksmen may fulfill—manageably and effectively—the immediate objective of reducing the number of animals. But if the objective also includes providing additional outdoor recreation, a reasonable alternative may be a public hunt controlled by permits (provided that the animals are not excessively tame). Another consideration addresses the effect of hunting on the behavior of wildlife; that is, hunting may

increase the wariness of animals, thereby reducing the opportunities for park visitors to see and photograph wildlife. Much of the following discussion is based on guidelines Fogg (1975) developed for planning and managing hunting in parks.

In many ways, hunting in parks can be managed just as it is on other lands. Harvest levels and methods may conform with regulations set by state agencies, and habitat can be manipulated for the production of game species. Food patches, impoundments, and small openings often benefit nongame as well as game animals and add to the enjoyment of park visitors during the spring and summer months. Such manipulations, however, should conform to the general purpose and plan of the park.

Because of potentially serious conflicts between hunting and other recreational uses of parks, however, the management of hunting requires special attention. A minimum of 160 ha of open land or water area should be available for hunting. For safety reasons, restrictions may be necessary regarding the density of hunters and the types of weapons they are permitted to use. Rifles might be restricted to large parks, and shotguns and archery equipment may be allowed in smaller parks. Hunting seasons should coincide with times of low use by nonhunters, and boundaries of hunting areas should be defined clearly. Adequate buffer zones are needed between hunting zones and picnic areas, campgrounds, and ski trails.

Fogg (1975) concluded that hunting in parks remains a controversial issue, especially in densely populated parts of the world. It is a topic that requires thorough discussion, including public participation, before management recommendations are adopted.

NATIONAL PARKS

National parks are usually of grand scale. These areas encompass scenic and environmental features of national or international significance rather than those limited to local or regional importance. Yellowstone was the world's first national park, established by the United States government in 1872. Since then, the National Park System has grown to include nearly 300,000 km^2 (including 178,000 km^2 added in Alaska). The system consists of more than 40 parks, as well as more than 25 national monuments, historic sites, recreational areas, near-wilderness areas, seashores, and lakeshores. The concept of national parks has been so successful in the United States that more than 100 other countries now have adopted the idea.

Figure 16-4. At this visitor center inside Bonneville Powerhouse and Dam on the Columbia River, Oregon, a glass window constructed beside a fish ladder offers outstanding views of salmon ascending the river. Besides serving as a public attraction, the same facility is used for counting salmon during their migration to spawning areas upriver. (Photo courtesy of U.S. Army Corps of Engineers.)

In 1916, the U.S. Congress formalized the purpose of national parks in the United States with three mandates:

First, that the national parks be maintained in absolutely unimpaired form for the use of future generations;

Second, that they are set apart for the use, observation, health, and pleasure of the people; and

Third, that the national interest must dictate all decisions affecting public or private enterprise in the parks.

The National Park Service has experienced difficulty in reconciling the first objective of *preservation* with the second of *recreation*. As Fuller (1969) pointed out regarding a similar quandary for parks in Canada, "It seems never to have been spelled out whether pleasure was to come from the natural magnificence of the area or from artificial embellishments. The problem is still with us."

Preservation generally prevails as the motive for managing wildlife in national parks, but the traditional meaning of preservation was revised by a committee reviewing the management of U.S. National Parks (Leopold 1963). After a year of study, the committee proposed a specific goal for managing national parks, as follows: "To preserve, or where necessary re-create, the ecologic scene as viewed by the first European visitors. . . . Protection alone is not adequate to achieve this goal. Habitat manipulation is helpful and often essential to restore or maintain animal numbers.... A greatly expanded research program oriented toward management needs to be developed."

That report set the stage for a new perspective on preservation—preservation of representative ecosystems and preservation by active management. It also admitted that the goal of preserving natural communities would require more scientific research. Thus, national parks in the United States have become sites for widely acclaimed research on wildlife (e.g., the long-term studies of wolves at Isle Royale National Park; see Mech 1966; Peterson 1977; Allen 1979).

The degree of manipulation that should be carried out, however, remains a delicate subject, and it varies from nation to nation and from park to park. Large herbivores, for example, are commonly cropped or culled for human consumption on the national parks in Uganda (Ssemwezi 1974). In Kenya, Olindo (1974) reported that surplus elephants (*Loxodonta africana*) are shot as a means of curtailing tree damage and the destruction of habitat for small animals. The park system in India, on the other hand, includes eight areas where inner sanctuaries of at least 300 km^2 remain undisturbed for Bengal tigers (*Panthera tigris tigris*). Game cropping in such areas is left to the tigers (Sankhala 1974).

Lamprey (1974) noted that any philosophy of park management is flawed if it considers national parks only as animal sanctuaries and "attaches more value to animal life than to plant life." Unfortunately, natural and artificial changes in natural communities are not always easy to distinguish. Ecological succession was not understood when the original concept of preservation was established for national parks in the United States. So the question of whether natural succession is desirable actually is a matter of defining the objectives of each park.

Lamprey (1974) cited an example in which the character of temperate and tropical woodlands is influenced by the frequency and intensity of grass fires. Fire has been a natural part of those environments for centuries. In the absence of fire, the woodlands become denser and thicker and therefore unsuitable as habitat for some kinds of birds and mammals. As a management practice, fire suppression clearly influences both the flora and fauna in these woodland communities. Conversely, an intensive fire may eliminate the woodland, again changing—but in greatly different ways—the character of the community for many years afterward. Park managers thus should address the question of what the community structure will be with or without fires and thereafter decide whether natural fires should be suppressed. A policy of fire suppression in a park established for the preservation of savanna species likely will eliminate the very species the park intended to preserve.

In Africa, elephant populations in parks pose a difficult management situation. Elephant populations have been compressed into limited areas on several national parks because of habitat destruction in the surrounding areas. As happens with overabundant populations of other herbivores, elephants overeat their normal food supply and thereafter decline in numbers. In this case, however, there is a time lag of several years during which elephants uproot trees for food on the upper branches and, in so doing, severely damage the park environment, even to the point of turning woodlands into grasslands (Lamprey 1974).

The "elephant problem" has been dealt with in a variety of ways. Authorities shot elephants in several parks (Ssemwezi 1974). In another park, management was intentionally passive: no controls were applied and the elephants damaged the environment, thereafter dying by the thousands. In the process, the elephants and a coincident drought created a grassland, which resulted in an increase in grazing ungulates. In other areas, fire suppression slowly regenerated the woodlands and thereby offset the more dramatic and sudden damage caused by elephants (Lamprey 1974). Park managers do not yet know what the "right" way is—or if there is a "right" way at all—for handling large numbers of elephants. Research remains the only means for answering such questions about wildlife management in national parks.

In 1962, at the First World Conference on National Parks, an international committee chaired by F. Bouliere of France prepared a seven-point statement on management of national parks. Highlights are as follows:

1. Management is any activity directed toward achieving conditions consistent with the plan of the park. It may involve manipulation of flora and fauna or protection from modification.
2. Few national parks are isolated from influences of surrounding areas. Interactions of parks with their surroundings include immigration and emigration of animals, the spread of fire, and the flow of water and air, including pollutants.
3. Climax communities such as Rain Forests and Tundra need no modification.
4. Successional communities, such as grasslands, must be managed. Fire, for example, is an essential tool for maintaining open savanna and prairies.

5. Where some animal populations, because of loss of predators, immigration, or compression of habitat, threaten a desired environment, population control should be practiced.
6. Management should be based upon scientific research, and both research and management should be undertaken only by qualified personnel.
7. Management based upon scientific research is desirable and essential to maintain some biotic communities in accordance with the objectives of a national park.

OVERPOPULATION OF ANIMALS IN PARKS

Elimination of both natural predation and hunting has created overpopulation of some animals in many parks (Leopold et al. 1947; Bendt 1962). A consistent management policy has not been developed for handling overabundant herds of deer or elk (*Cervus elaphus canadensis*) in parks. As with elephants, the animals first denude the vegetation and then starve to death. The elk herd at Grand Teton National Park migrates seasonally in and out of the protection offered by the park. After years of controversy, Congress finally authorized a measure for controlling the Grand Teton elk herd. Public Law 787 stated that "qualified and experienced hunters licensed by the state of Wyoming and deputized as park rangers by the Secretary of the Interior" could shoot elk in the park. Between 1951 and 1961, 5866 deputized hunters removed 1610 elk (Bendt 1962). Artificial feeding of elk wintering on the National Elk Refuge in the Yellowstone-Grand Teton area of Wyoming has been a common practice (Fig. 16-5).

Park employees also periodically reduce the bison herd at Yellowstone National Park. Up to 400 animals per year were removed at times between 1936 and 1966. Such "herd reductions" kept the bison population at a size consistent with the available range conditions (Meagher 1973).

Public outcries usually ensued, no matter what controls were applied to animal populations in Yel-

Figure 16-5. Many of the elk wintering on the National Elk Refuge near Jackson, Wyoming, spend other seasons on lands administered by the National Park Service. (Photo courtesy of J. Wilbrecht, U.S. Fish and Wildlife Service.)

lowstone and other parks in North America. Shooting raised the ire of persons with moralistic and humanistic views of animals. On the other hand, those with utilitarian attitudes were incensed if nothing was done and animals starved in parks.

Deer are overcrowded on Angel Island, a 260-ha state park in San Francisco Bay (White 1981a,b). Angel Island became a state park after the site was abandoned by the military in the 1950s. Each year, hundreds of thousands of city dwellers from the Bay Area hike, bicycle, sunbathe, and picnic at the park. Black-tailed deer (*Odocoileus hemionus columbianus*) offer an added attraction for visitors. By 1966, a growing population of more than 100 deer had overeaten the vegetation on Angel Island, and only unpalatable plants remained within the reach of the hungry animals. Deer begging tidbits from picnickers soon became commonplace. In November 1966, park rangers shot 50 deer as a means of averting starvation, but public opposition prevented further reduction of the herd and a widespread die-off followed.

The few survivors of 1966 gradually increased and brought the deer population back to the boom phase of the cycle. By 1975, 227 deer were counted on Angel Island, but in the following year the carcasses of 56 dead deer provided ample evidence of renewed starvation. The San Francisco Society for Prevention of Cruelty to Animals (SPCA) began to feed the deer, with permission from the California Department of Parks and Recreation. Supplemental feeding began in the fall of 1976 and continued into 1977. The success of the feeding program was not evaluated, but the presence of 60–80 deer carcasses suggested that the effort did not totally prevent starvation, and the deer population apparently declined further.

A browse line appeared again in 1980, when the population was estimated conservatively at 150 deer. At that time, Dr. Dale McCullough of the University of California at Berkeley suggested that coyotes (*Canis latrans*) might be introduced as an experimental means of controlling the deer population on Angel Island. It was hoped that a natural predator–prey system might gain public acceptance, but sentiments expressed to the California Department of Fish and Game (DFG) indicated otherwise, and the coyote plan was scrapped. Shooting once more was proposed as a means of thinning the herd, but the idea was rejected when the SPCA in San Francisco filed a lawsuit against the DFG. Negotiations between the DFG and SPCA then produced a plan for removing deer from Angel Island: deer would be live-trapped and released on the mainland into a low-density population of the same subspecies.

In August and September 1981, with $20,000 from the SPCA, the DFG trapped and relocated 214 deer on a recently burned location on the California mainland. Radio telemetry studies showed that 85 percent of the translocated deer died in their new environment within a year, more than half within the first 3 months. Causes of death included malnutrition, poaching, and predation by coyotes and domestic dogs. The researchers, led by McCullough, believed that the poor nutritional condition of the deer and their inexperience outside the controlled environment at Angel Island were responsible for the poor survival rate (Anon. 1983).

About 40 deer remained on Angel Island after the removal program. (The estimate of a total of 150 deer made in 1980 indeed was conservative.) The California DFG and Department of Parks and Recreation then began work on a management plan to prevent renewed overpopulation of deer on Angel Island. In view of past experience, however, it seems unlikely that any plan will satisfy all segments of the public. The Angel Island experience is not unique, and if the management plan for Angel Island proves successful, it could serve as a model for controlling herbivores in other parks.

Wildlife managers attempt to reduce the growth rates of rapidly expanding animal populations, especially when their burgeoning numbers may cause long-term damage to food supplies (e.g., overbrowsing). Some animal populations also are nuisances, and when dealing with these and similar situations, wildlife managers traditionally increase the death rate, often by encouraging more hunting, with poisons, or by introducing predators. Recently, however, Fayrer-Hosken et al. (1997) tried immunocontraception to control an elephant (*Loxodonta africana*) population expanding by 5 percent each year in Kruger National Park, South Africa. They assessed the reproductive condition for elephants tranquilized with darts shot from helicopters. Pregnant females were not treated, but other females were marked and injected with an immunocontraceptive, porcine zona pellucida (pZP). Four weeks later, the marked elephants received a booster shot administered by a dart shot from the air. Based on preliminary results, the method effectively limits conception and therefore may represent a management tool that, although expensive, may appease those who oppose shooting to cull growing elephant herds (see also Kirkpatrick et al. 1996 for a review of immunocontraception in wildlife).

Several questions remain about the fate of animal populations in parks. Should we consider die-offs of elephants in Africa and elk in Wyoming as natural phenomena that have occurred throughout the ages, or have die-offs resulted from human interference with natural processes? Shooting presents numerous difficulties: determining the necessity for shooting; deciding how many animals should be shot and when and where to shoot, who should do the shooting, and what should be done with the meat and hides (and ivory, in the case of elephants); and dealing with opposition from the public. Arrangements should be made for collecting scientific data from animals killed when populations are thinned in parks (e.g., for analyses of stomach contents, parasite loads, body condition, and reproductive status). Only long-term research will help determine the ways of managing parks for what we now regard as excess numbers of animals.

BEARS

On the evening of August 13, 1967, two young women in Glacier National Park were killed by grizzly bears (*Ursus arctos horribilis*) in separate unprovoked attacks. The sensational publicity surrounding that tragedy attracted worldwide attention. Management policies for bears in national parks thus came under criticism from several sources. One critic proposed that grizzlies should be eliminated from national parks so that visitors would not be subjected to the potential threat of attacks (Moment 1970). Before such extreme measures are adopted, however, the history of grizzly bears and their interactions with humans in parks should be examined.

Grizzlies killed a total of 5 park visitors during the first 97 years that national parks existed in Canada and the United States, and only 25 people had been injured before 1960 (Herrero 1970). In the 1960s, injury rates increased from about 1 per 3 years to about 5 per year. The increase in bear attacks was correlated with increased numbers of visitors, but the probability of a person being attacked nonetheless remained quite low: 1 injury per 1.5 million visitors in parks where grizzlies occur (Herrero 1970).

In the 1950s and 1960s, sows with cubs were involved with 37 (82 percent) of the 45 attacks where the sex and age of the bears could be identified. Of all known attacks, 56 percent occurred in campgrounds and 31 percent happened to hikers in back country. The frequent attacks in campgrounds apparently stemmed from an association with humans that developed when the bears fed at garbage dumps and when visitors deliberately offered bears handouts (Craighead and Craighead 1971).

McArthur (1980) reviewed the changing patterns of bear attacks in recent decades. In Glacier National Park, concerted management efforts in the 1960s diminished the access of bears to human food, and the rate of attacks thereafter fell from 1.7 per million visitors to 0.2 per million. But after 3 years (1969–71) without grizzly attacks, 6 injuries occurred in 1972–75 (1.04 per million visitors) and 5 in 1976–78 (1.02 per million visitors). Of the 11 injuries in Glacier Park in the 1972–78 period, only 2 resulted from females with young, 6 from single adult bears, and 3 from subadults. The number of attacks increased in circumstances not related to food—namely in encounters with bears in remote areas—a pattern verified by Martinka (1982). According to McArthur (1980), the increased aggression of adult and subadult bears cannot yet be explained; nor have the events been identified that elicit bear charges toward humans. In other words, grizzlies are unpredictable.

It is debatable whether the potential for bear attacks has diminished with the removal of human-generated food sources. Bear attacks still occur, and given the nature of grizzlies, future attacks can be eliminated only if bears are exterminated from parks or if people are completely removed from grizzly habitat. The question of whether bears should be eliminated was examined in a newspaper poll conducted about a year after the fatal attacks in Glacier Park. Of 3420 responses, only 104 (3 percent) favored exterminating grizzlies in national parks (Herrero 1970). Accepting the danger of bears is probably the price visitors must pay for enjoying a truly natural environment. Indeed, in the early 1980s 2 more persons died from grizzly attacks in Yellowstone Park (Chase 1986a). The injury rate of 1 per 800,000 to 1 million visitors per year certainly compares favorably with a rate of 1 automobile injury per 100 people. Herrero (1970) concluded that humans must enter the domain of grizzly bears as cautious and alert visitors, thereby relinquishing the human role as the tamer and reducer of wilderness. In such a setting, humans may not be the dominant species, and people become more truly a part of nature. Providing such a quintessential experience is the highest purpose that national parks can serve.

Since 1968, grizzly bear management in the Greater Yellowstone Park Ecosystem has been marked by conflicts between various researchers and the National

Park Service (NPS). At the heart of the controversy are three issues: NPS decisions for eliminating garbage dumps where bears obtained much of their food; the NPS policy of removing bears from places where they may threaten humans; and declining grizzly populations (Craighead 1979; Knight and Eberhardt 1985; Chase 1986a,b). The controversy includes those who want to protect grizzly bears as well as the voices of those who oppose the preservation of bears or who rank bears with low priority in park management. Suggestions include retaining wilderness for grizzly habitat by reducing road access and timber harvests in the areas surrounding Yellowstone Park. Conversely, an employee of a forest products corporation said, "From my point of view, people are more important than grizzly bears" (Brown 1985).

In 1983, the final Environmental Impact Statement was approved for managing grizzlies in Yellowstone.

The management objectives were to preserve and maintain natural populations of bears as part of the park's native fauna and to provide for the safety of park visitors. Major recommendations included (1) eliminating access to all human garbage; (2) educating visitors with literature and warnings from rangers; (3) closing some areas of high grizzly density to visitors; (4) live-trapping and removing bears that persist in human activity areas (Fig. 16-6); (5) killing only those bears deemed dangerous by the park superintendent; (6) monitoring bear numbers and distribution; (7) providing supplemental food (carcasses of large herbivores) when natural foods are not widely available; and (8) coordinating bear management with the U.S. Forest Service and state agencies.

In 1986, after several years of work by a federal-state committee, the Interagency Grizzly Bear Guidelines were published, covering lands managed by the National Park Service, Forest Service, Fish and

Figure 16-6. Grizzly bears visiting campgrounds and other areas of human activity in parks are trapped, anesthetized, and released in remote areas. The procedure is only partially successful, however, as the bears frequently return to the original site. Radio collars give researchers valuable information about the movements of bears. (Photo courtesy of Wyoming Game and Fish Department.)

Wildlife Service, Bureau of Land Management, and various state wildlife agencies (Interagency Grizzly Bear Committee 1986). The guidelines describe procedures for dealing with grizzlies in five different management situations, ranging from places where grizzlies do not occur to bear "population centers" where grizzlies are abundant. The management activities are wide ranging and include using bear-proof garbage containers, closing logging or mining roads to public use, and conducting logging in some areas during winter when the bears are hibernating.

The Wildlife Society (1986) approved a position statement on management and conservation of brown bears, including the grizzly and the Alaska brown bear (*Ursus arctos middendorfi*). The statement recognizes increasing conflicts between bears and human activities and the diminishing range and numbers of bears. The Wildlife Society recommends several measures, including (1) encouraging coordinated efforts among state, federal, and provincial agencies to include bears in management plans; (2) reducing human use of areas frequented by brown bears; (3) eliminating domestic sheep grazing on public lands in the Greater Yellowstone Ecosystem; (4) prohibiting the sale of all bear parts (e.g., claw jewelry); (5) enforcing strict management of hunting; (6) developing an accurate means of measuring population trends, including recruitment and mortality; (7) reducing vehicle access into bear habitat; (8) reintroducing bears into suitable but currently unoccupied habitat, especially in wilderness areas and national parks; and (9) disseminating accurate information to the public on bear conservation and management.

One overriding difficulty is that grizzly bears are a K-selected species (i.e., long-lived species with low birthrates), and there are not enough bears to permit experiments with various management measures. The margin for error therefore is small, and serious and long-lasting consequences might result from a failed management effort.

Grizzlies are not the only dangerous creatures occurring in parks, but they receive the most publicity. Bison and mule deer (*Odocoileus hemionus*) have killed park visitors, and black bears can be dangerous clowns. A popular cartoon describes the adventures of a lovable bear begging from tourists in "Jellystone Park," but all black bears are not as good-natured as this comic-strip counterpart. In Great Smoky Mountains National Park in 1977–78, black bears injured 24 of the 17.7 million visitors (Tate 1980). The injury rate in this case—1 per 700,000 visitors—is slightly greater than the injury rate caused by grizzlies in western parks. Tate (1980) studied the attack behavior of black bears that were panhandling along roadsides in the Great Smokies Park. She found that a bear was more likely to attack humans as a feeding session grew longer and as the number of different feedings increased. Older bears were more likely to attack than young ones, and males were more aggressive than females. Hot weather and rain also increased the likelihood of bear attacks on tourists.

These observations provide some basis for predicting behavior of black bears in particular settings, but the management implications are unclear. As with grizzlies, the lure of food undoubtedly overcomes a bear's tendency to avoid humans, and each handout encourages more begging behavior. The concurrent roles of national parks—recreation and preservation—are highly strained when it comes to dealing with panhandling bears. Park rangers cannot easily stop tourists from tossing snacks to bears, but with vigorous enforcement of the "no-feeding" rule and intensive education of park visitors, black bears may regain their reputation as secretive and resourceful denizens of the forest. Neither the image nor the reality of black bears as pitiful, subservient, and sometimes rebellious roadside beggars has any place in park management.

PRESERVATION, HUMAN POPULATIONS, AND PARK DEVELOPMENT

In North America, the arguments for and against resource preservation in parks, particularly in large national parks, are waged between two groups of relatively well-fed people. Such is not the case, however, in central and eastern Africa where hungry people look on parks as a source of food. In Tanzania, Miller (1982) described the setting where thousands of wildebeest (*Connochaetes taurinus*) flourished in the protection of Serengeti Park, while outside the fence starving African families watched in hunger. In such a context, wildebeest may be considered a wasted and desperately needed source of human food, especially because many of the animals are killed and eaten by lions (*Panthera leo*) and other predators. Hence, the question arises as to whether concerns for human welfare should override those for preserving natural systems. Must the concept of nature preserves such as the Serengeti Park be abandoned in favor of providing food for an ever-increasing human population?

Miller (1982) says the concept need not be abandoned, but perhaps national parks in developing countries should contribute tangible resources to local economies. He claims that arguments for "pure" preservation in parks are flawed because human influences already may have disrupted natural systems. Fences prohibit the natural movement of animals; and roads, campgrounds, and concession areas in parks violate and displace natural environments. Therefore, culling excess animals in parks for human food might be similarly justified on the grounds of serving the public interest. The existence of national parks in developing countries can be justified further when important watersheds are protected in ways that assure the availability of clean water for industrial and domestic consumption. Parks also may serve as reservoirs for plant and animal species that can be propagated for use outside the park.

Other parks in Africa are not under the same pressures as those so visible in the Serengeti. In Kruger National Park in South Africa, however, human intrusions have been a part of wildlife management. Material in this section is based on Smuts (1982), who described the interactions of humans, carnivores, and herbivores in African parks.

In the central district of Kruger Park, continued declines of zebra (*Equus burchelli*) and wildebeest prompted a detailed search for causes. Grazers adapted to shortgrass, such as the zebra and wildebeest, seem more vulnerable to lion predation in wet years, when grass grows tall. The general decline of these species halted temporarily during the dry years of the late 1960s but resumed with the series of wet years between 1973 and 1977. The decline involved interactions among weather, food, and cover and the respective sizes of the predator and prey populations. In some animal populations in Africa, the ratio of predators to prey animals was 1:1000; predation had little impact on the prey population in such cases. In Kruger Park, however, the ratio is 1 lion per 110 prey animals. The lion population of 700 in 1975 required about 2500 wildebeest per year. The potential annual recruitment into the wildebeest population was only 3300 calves, leaving an excess of only 800 available for wild dogs (*Lycaon pictus*) and other predators. Predation obviously was accentuating the decline of wildebeest and zebra in Kruger National Park during the 1970s.

Park managers had two choices: Let nature take its course or intervene and curb the declining numbers of zebra and wildebeest. In keeping with the current emphasis in national park management, laissez-faire would seem the simplest choice, but the history of the central district of Kruger Park indicated that human influences may be responsible for the unfavorable predator-prey ratio.

Between 1933 and 1977, water developments in the park opened new grazing areas for native herbivores; 70 windmills and 23 dams were built during the period. As a result, wildebeest and zebra herds broke into smaller groups and reduced their migratory behavior. Such changes, along with controlled burning and cropping of other species, initially increased the size of the zebra and wildebeest herds. The same changes, however, also increased the vulnerability of the herds to predators. Proportionally more animals are on the fringes of small herds and thus more are exposed to predators, and newborn calves in smaller herds also are more noticeable to hyenas (*Crocuta crocuta*). The new water supplies also stimulated the growth of tall grasses, thereby offering lions better hiding cover and increased hunting efficiency. Furthermore, humans cropped 3500 buffalo (*Syncerus caffer*) and 7500 impala (*Aepyceros melampus*) between 1968 and 1977, which reduced the buffering effect of these species.

For a management goal, Smuts (1982) recommended that wildebeest and zebra herds once more should be formed into large mobile aggregations rather than remain dispersed and sedentary. Water manipulation, prescribed burning, and reduced culling of elephants and buffalo in years of good rainfall were the prime means of encouraging formation of the larger herds. Because the predator-prey ratio seemed artificially high, the systematic removal of lions and hyenas also was part of the management plan. However, immigration soon fills vacant territories, especially with lions, and defeats the purpose of the removal program. Therefore, to minimize the replacement of lions by immigration or reproduction, the management plan recommended that the lion population should be cropped gradually, taking all from a single pride rather than a few individuals from several neighboring prides. Predator control, however, also should include careful monitoring as protection against increases in other predators or increases in prey populations beyond the carrying capacity of the habitat. The management program in Kruger Park now is well entrenched and seems indispensable for the maintenance of confined plant and animal populations. Prescribed burning and the other practices focus, as much as possible, on providing adequate habitat conditions and minimize the direct manipulation of animal numbers.

Parks in India have experienced pressures similar to those in Africa. The last stronghold of lions in Asia is a 1265 km^2 tract in India known as the Gir Forest (Desai 1974). Besides lions, 25 species of mammals, 14 species of reptiles, and numerous rare birds occupy the Gir Forest. The "forest" actually consists of several habitat types, ranging from closed teak forest to thorny brush to open grasslands. In 1963, the lion population numbered 266 animals, the descendants of about 12 lions saved when shooting was stopped in 1889. By 1968, however, the population was down to 177 lions. In recent years, the Gir has attracted growing numbers of tourists.

The Gir Forest is not a declared nature preserve. Instead, the site has been managed for teak production and for cattle grazing, of which the latter has posed the greater threat to wildlife. About 4800 Maldharis and 17,000 of their cattle occupy the Gir, and another 47,000 cattle move into the forest during the monsoon season. The Maldharis are impoverished pastoral people who barely subsist by selling ghee processed from milk. Overgrazing and overcutting have marred the recent history of the Gir. In the 1960s, these abuses, along with export of cattle manure from the forest, reduced plant production to about 15–30 percent of former levels in areas occupied by the Maldharis. Estimates suggest that domestic livestock consumed 85 percent of available grass in the forest. Plant production on the Gir was almost nil.

Under these conditions, livestock constitute about 90 percent of the lion diet in the Gir. Large native herbivores, normally the prey of lions, were all but eliminated from the stripped forest. Desai (1974) noted, "It is a vicious and paradoxical position that lions feed upon cattle, and cattle in turn are gradually destroying the habitat of the lion."

In India, national parks are declared by the individual states, and a large part of the Gir Forest soon may become a national park in the State of Gujarat. Meanwhile, the government of Gujarat decided to "restore the ecological balance in the Gir and to ensure the survival of the lion and other wildlife in perpetuity" (Desai 1974) with the following steps:

1. To reduce overgrazing by closing the forest to cattle immigrating during the monsoon season.
2. To resettle the Maldharis and their livestock on "government wastelands" outside the forest and prevent their reentry.
3. To construct and maintain a fence and patrol road around the forest.
4. To provide an alternate food supply for lions until such time as the native ungulate population increases and replaces cattle as the mainstay of the lion diet.

Costa Rica is a country with several new national parks. The most successful park, Santa Rosa, features a historical site where a battle for national sovereignty was waged. Surrounding the park are examples of dry tropical woodland, savanna, mangrove swamp, and gallery forest. The public readily accepted designation of the battleground as a national reserve, but an educational program was needed before the public was convinced of the ecological values of the surrounding areas (Boza 1974). Another new park in Costa Rica is Poas Volcano, an extremely scenic location on an active volcano where a cloud forest is inhabited by spectacularly beautiful birds known as quetzals (*Pharomacrus mocino*). Another park protects the nesting beaches of sea turtles (Cheloniidae), another a coral reef, and still other parks protect caverns, grasslands, lakes, and archaeological sites.

National parks were not established easily in Costa Rica (Boza 1974). Opposition came from lumbering interests, hunters illegally taking rare wildlife, and apathetic government institutions. Support came from local municipalities where tourism would flourish (but conflicts arose when the municipalities wanted to maximize tourist revenues without giving priority to long-term park goals), from international groups (such as the International Union for Conservation of Nature and the Audubon Society), and from local conservation groups. When she was the nation's First Lady, Señora Karen de Figueres was a most influential ally of park development in Costa Rica.

The initial strategy in Costa Rica was to manage a few parks in ways that would gain public acceptance. The first few parks included sites of distinct national interest, and once public support was assured, the park system was expanded to include areas of representative flora and fauna. Costa Rican parks suffer from budget limitations, and international volunteers from the U.S. Peace Corps, the British Voluntary Service Organization, and the Caribbean Conservation Corporation assist with development and patrol duties.

In North America, some government lands are designated for timber production and grazing, whereas others are set aside for preservation and recreation. In most African countries, however, parks are the only designated use of government lands. Miller (1982) suggested that third-world nations promote "ecodevelopment" outside the parks. These programs include establishing firewood plantations, creating extension programs with farmers, and promoting rural education. Without proper rural education and development, poor and starving people ultimately will invade the parks for food, fuel, and shelter. As the human population continues growing worldwide at a rate of 1.7 percent per year—thus doubling each 41 years—parks are becoming islands of nature surrounded by an increasingly human-modified environment.

The President's Commission on Americans Outdoors (1987) prepared a report containing elements of a vision for the future in the United States. The vision included a nationwide network of greenbelts of trails, rivers, and abandoned railroads linking urban and rural areas; protection of 2000 segments of rivers and their banks; and a populace with an ethic of respect for nature. Foremost among the recommendations is a call for a national trust that would provide a minimum of $1 billion per year for land acquisition and park development. After reviewing the report, Pritchard (1987) noted, "These are difficult times for budgets. Yet we are foolish if we sell the backyard to pay off the mortgage. We need to continue investing in our future." The executive summary of the report concludes, "We have to create . . . opportunities by preserving and nurturing the natural world before we lose it forever. If we pay our debt to the great outdoors, we will be repaid many times over."

REFUGES

The origins of National Wildlife Refuge System, as noted in Chapter 1, date to 1903 when President Theodore Roosevelt proclaimed Pelican Island in the Indian River, Florida, a federal wildlife sanctuary. His action protected a large rookery where plume hunters were slaughtering nesting egrets. Whereas Roosevelt's intervention represents a significant milestone—and he later created many more—federal refuges were added in the following decades only when Congress approved site-specific funding for each refuge. Two of the better-known refuges acquired in this fashion include the National Bison Range in Montana (1908) and the Bear River Migratory Bird Refuge (1928) in Utah.

A more systematic approach for establishing federal refuges did not emerge until later. The Dust Bowl of the 1930s ushered in an era in which many refuges were added to the federal system, especially to protect the dwindling numbers of waterfowl and their wetland habits. A number of big-game refuges also were added during these years; in particular, these "game ranges" added large areas of desert, grassland, mountains, and coniferous forest to the landscapes not otherwise protected at the time (Fig. 16-7). Offshore islands—even entire chains of oceanic islands—eventually were added to the refuge system, primarily to protect sea birds or marine mammals. Despite its

Figure 16-7. Visitors enjoy the scenery as well as the wildlife at the National Elk Refuge near Jackson Hole, Wyoming. This refuge protects the winter range of elk but also provides the public with selected outdoor activities on a year–round basis. (Photo courtesy of U. S. Fish and Wildlife Service.)

Photo courtesy of Phil Norton

On The Job With . . .
Phil Norton
Refuge Manager

Phil Norton's career as a refuge manager began in 1968. After finishing his undergraduate degree, he started as a biological aide at Buffalo Lake National Wildlife Refuge (NWR) but soon was promoted to assistant manager at a succession of other western refuges. He later spent 3 years on the NWR staff in Washington, D.C., and then returned west to begin duties as zone supervisor for federal refuges in Utah, Colorado, and Kansas. In 1986, wanting to return to the field, Phil assumed his current position in New Mexico as complex manager for Bosque del Apache NWR and its satellites. In 1995, the National Wildlife Association and the National Audubon Society presented Phil with the Paul Kroegel Award—named for the warden of Pelican Island, the nation's first wildlife refuge—and selected him Refuge Manager of the Year.

Bosque provides winter habitat for sizable populations of waterfowl and sandhill cranes, but the refuge depends on irrigation and drainage systems. To renovate these, Phil secured a $1 million grant; he also revitalized the farming operations and implemented moist-soil management to supplement food production for wildlife. Saltcedar, which had invaded the riparian zone on the 20,240–ha refuge, was another management target, and the monoculture is now being steadily converted into stands of cottonwood and willow, the historical riparian vegetation at Bosque. In essence, Phil spends his time developing projects and obtaining the resources necessary to complete the work.

Community involvement is another of Phil's goals. For example, every November the refuge and the nearby town of Socorro, New Mexico, jointly hold a Festival of the Cranes, which attracts about 10,000 visitors. The activities include cultural and historical displays, crafts, conservation-related workshops (e.g., birding classes), and field trips. Phil also keeps Bosque user-friendly, in part, with signs that explain to visitors the "why" as well as the "how" of management. These and other activities have produced a strong union between the refuge staff and the local and regional citizenry. Indeed, Friends of the Bosque, a citizen-support group, helped obtain $4 million to improve refuge buildings, equipment, and other facilities. Citizen volunteers also serve as tour guides and help with on-site projects such as boardwalk construction, building renovations, and road repairs.

Despite the many rewarding aspects of his job, of which "developing an area for wildlife and watching people respond" is paramount, Phil also experiences frustrations. In particular, these concern the revolving door of politically appointed leaders in a bureaucracy for "whom process is more important than results." For Phil, each new corps of bureaucrats seem "to reach further into the day-to-day operation" of refuges than did the previous set.

growth, however, this vast network of wildlife habitat was not officially united into one land-management system until 1966, when President Lyndon B. Johnson signed the National Wildlife Refuge System Administration Act.

The largest single addition to the refuge system occurred in 1980 when more than 21.8 million ha of land in Alaska became part of the federal system; this acquisition resulted from the Alaska National Interest Lands Conservation Act. For the most part, however, federal refuges are now acquired with funds from either the Migratory Bird Conservation Fund or the Land and Water Conservation Fund (see Chapter 22), although Congress itself can still authorize and fund refuges directly. The U.S. Fish and Wildlife Service administers the acquisition, operation, and management of all federal refuges.

Today, the National Wildlife Refuge System includes more than 500 refuges and includes sites ranging in size from less than 1 ha (Mille Lacs National Wildlife Refuge in Minnesota) to almost 81,000 ha (Yukon Delta National Wildlife Refuge in Alaska). Federal refuges occur in all 50 states and 5 territories. More than 80 percent of the system's 38 million ha are in Alaska. In many cases, federal refuges are managed in groups known as complexes, in which one or more satellite refuges falls under a central administration housed at a larger refuge.

Contrary to popular belief, federal refuges do not eliminate the equivalent of county property taxes. Instead, federal revenue-sharing funds partially or fully reimburse the county school system for the revenue "lost" when the lands were removed from the local tax base. Current policy also precludes using the force of public law—eminent domain—to acquire refuge lands from unwilling sellers.

STATE WILDLIFE REFUGES

Virtually every state maintains a network of wildlife refuges, but most of these sites are designated as game management areas. Such a distinction reflects that these state "refuges," for the most part, were acquired with the assistance of Pittman-Robertson funds (described in Chapter 22), and therefore are areas open to public hunting, as required by the original legislation. In addition to hunting, these areas also offer other types of wildlife-dependent recreation and locations for field research, as noted later. State-owned game management areas, like their federal counterparts, provide prime habitat for a large number of nongame species on a year-round basis.

MANAGEMENT OVERVIEW

State and federal refuges are managed, often intensively, but we cannot review here the multitude of these practices. Suffice it to mention that fire is a common tool at locations with grass or forest cover, as is the manipulation of water levels on refuges where wetlands represent the dominant habitat (Fig. 16-8). Some wetlands are drained in spring to promote production of natural foods on the exposed mudflats,

which are then reflooded when migrating shore birds and waterfowl return in autumn. In some cases, native vegetation is established to provide nesting cover, but in others corn and other grains are planted to supply large amounts of high-calorie foods for migrating waterfowl and to reduce damage to crops on privately owned farms nearby. Timber is selectively harvested on some refuges and grazing is allowed on others. Whereas these practices at times may generate earnings, income remains secondary to the primary focus of improving habitat conditions for wildlife (e.g., cutting that promotes stands of mast-producing trees in bottomland hardwood forests). However, not all refuges are managed proactively. Some units in the system—designed as Wilderness and Research Natural Areas—are left in their natural state.

Refuges also are convenient sites for monitoring the status of waterfowl and other migratory birds. For example, each year refuge personnel trap, band, and release thousands of ducks. Later, when the bands are recovered, the data provide managers with vital information about the population ecology of each species (e.g., mortality rates). Also, waterfowl and other migratory birds are censused each year on refuges located in key staging and wintering areas (i.e., major concentration areas in each of the four major flyways, described in Chapter 6).

SOME FUNCTIONS OF REFUGES

Each refuge fulfills at least one primary function (i.e., the original reason for its designation as a refuge). These functions are not mutually exclusive, but for

Figure 16-8. Many state and federal refuges are managed for waterfowl and shorebirds. As shown here, water-control structures are built and operated to provide optimum habitat for waterbirds, especially in the more arid parts of the United States. (Photo courtesy of U. S. Fish and Wildlife Service.)

convenience they are listed here separately. For the most part, the list focuses on federal refuges, but many state-owned sites share similar functions.

ENDANGERED SPECIES. The National Bison Range, mentioned earlier as one the nation's oldest federal refuges, was established exclusively to protect some of the precious few bison (*Bison bison*) that survived the abuses of the 19th century. In 1937, Aransas National Wildlife Refuge began to protect the wintering grounds of whooping cranes (*Grus americana*), and in 1972, a grassland area in Texas was acquired to safeguard the dwindling population of Attwater prairie chickens (*Tympanuchus cupido attwateri*). Established in 1957, the National Key Deer Refuge in Florida protects its namesake subspecies of white-tailed deer (*Odocoileus virginianus clavium*). Today, federal refuges provide habitat for about 60 species of endangered animals.

REST STOPS FOR MIGRATORY BIRDS. Many federal refuges, along with a good number of state management areas, were acquired to provide resting stops for migrating waterfowl. Most federal refuges were withdrawn from lands held as public domain, but others were purchased using funds derived from sales of "duck stamps" (i.e., the Migratory Bird Conservation Fund), which have added about 1.6 million ha to the refuge system. More than providing hunters with access to migrating waterfowl, these areas assured that crucial wetland habitat would remain protected and available for future flights of ducks and a host of other species. Refuges are indeed reservoirs of biodiversity for a rich constellation of organisms, of which only a few species are ever targets for hunters.

The underlying concept for these refuges recognizes that waterfowl need resting areas about 320 km apart in each flyway. These refuges provide the birds with food and cover, much of it made available as the result of intensive management, as noted earlier. Although a bit fanciful, their distribution has been likened to a "string of pearls" stretching along the north-south axis of each flyway. A considerable number of refuges serving as rest stops provide visitors with unusual opportunities to witness spectacular numbers of waterfowl.

PROTECTION OF UNIQUE OR HIGHLY PRODUCTIVE NATURAL SYSTEMS. Refuge lands and waters protect a broad spectrum of the nation's biological diversity. Recognition of this function of refuges has developed in parallel with the ever-increasing pressures exerted by a growing human population. Many refuges were acquired specifically because of their unique and/or productive fish, wildlife, and plant resources. Special designations, such as Wilderness, Research Natural Areas, Wetlands of International Importance (as designated by the RAMSAR Convention, Chapter 6), and Shorebird Reserves, among others, give added importance to refuges. Many conservationists indeed believe that the refuge system eventually must include representative examples of all the major biotic communities in the United States. At present, however, this concept is not a specific charge of the National Wildlife Refuge System, but it points out that the system continues to evolve and mature.

WILDLIFE-DEPENDENT RECREATION. Of the estimated 30 million people who visit federal wildlife refuges each year, the great majority pursue nonconsumptive activities. These include hikes, picnics, photography, and boating, as well as the simple enjoyment of the outdoors and an opportunity to observe wildlife. Refuges are exciting places for field trips for school children and senior citizens alike. Several refuges preserve archaeological or historical sites as well as wildlife habitat. Many refuges accordingly provide visitor centers, interpretive trails and drives, observation towers, photographic blinds, and other facilities designed to serve and educate the public (Fig. 16-9). Refuges located in or near large urban centers are particularly valuable resources for public education. San Francisco Bay National Wildlife Refuge, for example, includes an environmental center with two classrooms, an auditorium, dip nets and other field equipment, and a library of environmental materials, including lesson plans and curricular materials for schoolteachers—all available at no cost to visitors—in addition to protecting more than 6800 ha of wetland habitat in the heart of a major metropolitan area.

Consumptive outdoor recreation, including hunting, is allowed on many federal refuges, and where permitted, up to 85 percent of the area of each refuge is open to hunting. Fishing is generally permitted on refuges with lakes, streams, and other suitable bodies of water. Some waterfowl refuges maintain blinds for duck hunters, who use these on a first come, first served basis or on the basis of a daily lottery. Anglers and hunters must have valid state licenses and adhere to current bag limits and other regulations when pursuing their sport on federal refuges (Fig. 16-10). Regardless of whether it is consumptive or nonconsumptive in nature, however, outdoor recreation enjoyed by the public on refuges

Figure 16-9. Visitors centers at refuges offer fine opportunities to learn about wildlife, including hands-on experiences (left), whereas nature trails and other sites provide outdoor classrooms (right). (Photos courtesy of U. S. Fish and Wildlife Service.)

Figure 16-10. Hunting is allowed on many federal refuges, some of which provide blinds for waterfowl hunters. Daily harvests are checked and tallied by refuge personnel, as shown here at Sacramento National Wildlife Refuge in California. (Photo courtesy of U. S. Fish and Wildlife Service.)

shares a common ground—the maintenance of healthy wildlife populations.

RESEARCH. Because they provide secure habitat, refuges are sites where important research has been, and continues to be, conducted. Many refuge-based research projects are long-term, again because the day-to-day, season-to-season risks are minimal in comparison with other, less-secure areas. Diked management units on many waterfowl refuges permit studies where water levels can be manipulated to determine the responses of desirable as well as undesirable aquatic and semiaquatic vegetation. A 6-year study of nest predators—still perhaps the most thorough of its kind—was conducted on a federal refuge in Minnesota (Balser et al. 1968). Similarly, a federal refuge bordering Great Salt Lake in Utah for many years served as the principal site for field and clinical studies of avian botulism. A full account of the hundreds of research projects conducted on refuges would include studies on all classes of vertebrates, many taxa of invertebrates, and vegetation ranging from algae to trees. The ecological relationships would run a gamut of food habits to predation, cover requirements to disease, and reproduction to parasitism. Suffice it to conclude that most of this research could not have been completed successfully in other locations.

1997 ORGANIC ACT

Unlike any other federal land-management agency in the U.S. government, the U.S. Fish and Wildlife Service operated for years without the guidance of an organic act—a legislative charter that describes the agency's mission on the land it administers. In other words, while the U.S. Fish and Wildlife Service indeed managed its refuges for wildlife, other interests sometimes claimed the commercially valuable resources often available on refuge lands (e.g., timber). Then, in 1991, an organic act was introduced that specifically designated wildlife as the highest priority on federal refuges. The proposal granted that recreation and even such profit-making activities as mining on occasion might play secondary roles, but only if these enterprises did not interfere with the first and primary function—protection and management of wildlife resources. Despite its seemingly self-evident merits, the bill faded from Congress's legislative agenda by 1994. More ominous, however, were the continued attempts of some special-interest groups to make mining, lumbering, grazing, and other activities *legally mandated purposes* for operating federal refuges. In such a scheme, wildlife no longer represented the paramount purpose of refuge management and instead became only one of several co-equal uses for the lands and waters in the National Wildlife Refuge System.

Fortunately, a sound organic act for the federal refuge system finally reached the desk of President Bill Clinton, who signed the bill on October 9, 1997. The measure, while not perfect in every respect, nonetheless significantly reduced many threats to the wildlife resources protected by the refuge system. This legislation—the National Wildlife Refuge System Improvement Act of 1997—includes several key provisions:

- Defines the mission of the system, which is "to administer a national network of lands and waters for the conservation, management, and where appropriate, restoration of the fish, wildlife, and plant resources and their habitats within the United States for the benefit of present and future generations of Americans."
- Defines compatible wildlife-dependent recreation as "legitimate and appropriate general public use" and establishes hunting, fishing, wildlife observation and photography, and environmental education and interpretation as "priority public uses" where these are compatible with the mission and purpose of individual national wildlife refuges.

- Requires the secretary of the interior to ensure and maintain the biological integrity, diversity, and environmental health of the refuge system.
- Retains the authority of the refuge manager to determine compatible public uses (e.g., hunting and grazing) and whether these will be allowed; the law established a formal process to determine "compatible use."
- Requires each refuge to develop a comprehensive conservation plan within 15 years and to update the plan at least once every 15 years thereafter.

PROBLEMS

Despite its obvious benefits, the federal system of wildlife refuges faces several difficulties, some of which are highlighted here (for an overview of these in a popular format, see Williams 1996 and an earlier report by Leopold et al. 1968).

Inadequate funding has been a chronic difficulty for the National Wildlife Refuge System. A comparison with two other land-management agencies illustrates the point. The U.S. Fish and Wildlife Service receives about $5 per ha per year to manage its lands, whereas the corresponding figures for U.S.D.A. Forest Service and National Park Service are about $17 and $32 per ha, respectively. Because of limited funding, about 43 percent of the refuges lack on-site staff and many others are understaffed. The replacement cost for dikes, road, buildings, and other structures and facilities totals more than $4 billion, but only 1 percent of that amount is spent for maintenance each year, whereas the private sector normally annually spends 2 to 4 percent to maintain its capital investment. Consequently, the backlog of repair and maintenance costs today is in excess of $500 million.

Perhaps one of the great challenges facing natural-resource managers today is the trend toward increasingly broad responsibilities in an environment filled with highly complex issues. Formerly, managers were concerned primarily with a relatively small group of so-called game species and, under those conditions, most refuges were managed rather simply by "staying home and tending the farm." Today, with a growing interest in biodiversity, refuge staff face a far-greater burden: understanding the trade-offs associated with various types of management and dealing with conflicting public opinions and constituencies as to the "proper balance" between species. To complicate matters even more, these issues arise at a time when basic resources—air, soil, and water—are under increasing stress. For example, water shortages, water quality, and urban development have direct bearing on the wildlife populations at many refuges. These concerns require that refuge managers take a broad view of resource issues and, to resolve these, they necessarily become actively involved with a host of organizations, agencies, and individuals. Several factors make resource issues difficult to resolve: the problems are often highly technical; negotiations frequently involve several entities and occur in a politically charged arena; and the "fixes" are generally expensive and therefore not readily implemented.

Finally, refuges are not always managed as natural ecosystems. In particular, waterfowl often become highly concentrated on refuges, especially where crops are raised solely to feed overwintering flocks. The outcome may be a change in distributional patterns and, at least in some cases, an increased threat of epizootics or crop damage on neighboring farms. In the recent past, some refuges seemed to be competing with each other for waterfowl instead of managing the birds across a network of refuges. Artificial feeding—typically baled hay in winter—on some big-game refuges also countermands any intent for a naturally functioning ecosystem. Indeed, ungulate populations on some refuges may build to levels exceeding carrying capacity, which in the absence of sufficient predation often means hunting becomes a much-needed management tool to deal with these herds. At Great Swamp National Wildlife Refuge in New Jersey, however, antihunting forces routinely attempt to block the managed harvest of an overpopulated deer herd.

SUMMARY

Wildlife in parks enhances the recreational experiences of most visitors. If people and animals are managed properly in parks, wildlife may advance public understanding of ecological relationships. Managers face a major difficulty in most parks where the public can enjoy wildlife: a demand for developing viewing opportunities without disrupting the environment and behavior of the animals. Wildlife management in parks may include habitat modifications and even hunting in larger parks. In national parks, the primary management objective is preservation of nature for recreation and education. Preservation in some cases may involve active management, particularly to recreate or maintain plant and animal communities characteristic of early succession, such as grasslands and savannas.

Animal populations rarely are manipulated in the national parks of some nations, including the United States, whereas other nations regularly intervene. Only long-term research in parks will determine what is "natural" and what is not.

Bears have created special management problems in some parks, although attacks on humans occur at a rate of only 1 per 700,000–1 million visitors. In some developing countries, wildlife in parks represents a potential source of food for hungry people. The future of wildlife in parks, and perhaps elsewhere, ultimately depends on decisions concerning the growth and control of the human population.

Since its beginnings in 1903, what is today known as the National Wildlife Refuge System has expanded to include more than 500 federal refuges. Some refuges protect endangered species, but others secure key habitat for big game and waterfowl. Others protect sites with unique biological features, including wilderness areas. Refuges offer useful locations to conduct wildlife research. Refuges also are important sites for wildlife-dependent recreation, including photography, birding, nature study, and, often, hunting and fishing. At the state level, refuges are commonly identified as game management areas, most of which fulfill the same functions as federal refuges.

In 1997, the National Wildlife Refuge System Improvement Act at last provided the system with a well-defined mission statement in which wildlife conservation and management are paramount. Grazing, boating, and other "compatible uses" remain secondary considerations.

Funding for refuge maintenance lags behind other federal land-management agencies. Refuge personnel today face complex environmental issues, many of which have political as well as biological implications. On some refuges, overcrowding and other highly unnatural conditions may longer represent the best interests of wildlife management.

CHAPTER 17

URBAN WILDLIFE

Wouldn't we be wise to bring nature and wildlife areas into the city where the pleasures they bring can be enjoyed on a daily basis? Man has an inherent metaphysical and spiritual need for nature and he should not retreat from it. Most urban people want open spaces and trees and a contact with nature as part of their daily lives.

Lorraine C. Smith (1975)

Urban growth is rampant worldwide. In 1992, more than 2.3 billion people—or 43 percent—of the world's population lived in cities (Population Reference Bureau 1992). By the year 2000, nearly half of humanity will reside in cities, and more than half of the largest cities will be in developing nations where economic stresses remain especially burdensome. Urbanization also has marked the growth of the United States. About 75 percent of the U.S. population resides in metropolitan environments, representing about 212 million humans dwelling on 16 percent

of the land (Table 17-1). In the process, concepts and concerns such as urban sprawl, green spaces, smog, concrete jungles, and urban renewal have crept into the lexicon of the 20th century.

The realization of an urbanized society portends challenging directions in the theory and practice of wildlife management. The challenge includes using wildlife to remind urbanites that a world of nature still exists (Allen 1974b). Indeed, wildlife managers may no longer enjoy the luxury of self-imposed interests centered only on farm or forest. Human needs

TABLE 17-1. Metropolitan Statistics for the Resident Population of the United States, 1940–90

	1940	1950	1960	1970	1980	1990
Metropolitan Areas:						
Number of SMSAs[a]	168	168	212	243	318	268
Population (× 1000)	69,535	84,854	112,885	139,419	169,405	197,467
Percent change over previous period	—	22.0	33.0	23.6	21.5	11.8
Percent of total U.S. population	52.8	56.1	63.0	68.6	74.8	79.4
Percent of U.S. land area	7.0	7.0	8.7	11.0	16.0	19.0
Nonmetropolitan Areas:						
Population (× 1000)	62,135	66,472	66,438	63,793	57,100	51,243

Source: U.S. Bureau of the Census (1993).

[a]Standard Metropolitan Statistical Areas (SMSAs) are integrated economic and social units with a large population nucleus.

and development seem a part of this evolution. As long ago as 1970, people of the United States indicated strong preferences (95 percent) for trees and grass as components of their surroundings (*Life*, January 9, 1970). These preferences outscored such features as a nearby shopping center (84 percent), a kitchen with modern conveniences (84 percent), or the proximity of good schools (81 percent). The implications seem clear: The human spirit desires—and perhaps needs—landscapes with at least some degree of naturalness.

Wildlife seems to be a part of this experience. A survey of households in a Canadian city revealed strong support for the presence of wildlife and a desire for management within the city (Gilbert 1982; see also Brown et al. 1979; Witter et al. 1981). Nearly half of the residents surveyed were willing to pay a municipal tax for management activities. As might be expected, television provided the primary source of wildlife-related information and entertainment. Unfortunately, however, few urbanites could identify the public agencies responsible for wildlife management. Urbanites also did not recognize the ecological value of diversity or the suitability of selected environments as wildlife habitat, and few pictures of animals were identified correctly by more than half of the residents in the sample. (There are, however, no comparable studies measuring these traits in rural residents; perhaps animal identification, habitat associations, and the value of diversity are not familiar to any segment of the population.)

The findings of this survey indicated that management agencies need more visibility with the urban public and that municipal planning for wildlife programs will receive firm support. Furthermore, urban residents may be more responsive than supposed to a broad interpretation of the meaning of *wildlife*. *Game*, *furbearers*, *big game*, and other designations used by wildlife managers may be meaningless terms for residents who largely associate with the animals seen in backyards. Residents of metropolitan areas in New York, for example, ranked butterflies and songbirds well above waterfowl and pheasants as animals they would rather see near their homes (Brown et al. 1979). Therefore, managers should adopt a more holistic approach to the principles and practice of wildlife management when dealing with urban communities (see George et al. 1974 for a general discussion of updating the training of wildlife managers).

Evidence of the intangible relationships between humans and nature is difficult to document in quantitative terms. Nonetheless, there is a relationship. The National Council for Therapy and Rehabilitation Through Horticulture, created in 1973, was established to further the treatment of physically and mentally impaired people. The premise is simple and direct. Contact with some form of nature—plants, in this case—is therapeutic and beneficially influences the course of human development. The renowned physician Karl Menninger noted that such contact "is one of the simple ways to make a cooperative deal with nature for a prompt reward" (NCTRH brochure, undated). In a study of patients recovering from the same kind of surgery, Ulrich (1984) found that those assigned rooms with a view of a natural setting had shorter postoperative stays, received fewer negative comments in the evaluations of nurses, and took fewer strong analgesics in comparison with patients in rooms with a monotonous view of a brick wall. Other findings suggest that animals enrich the lives of elderly people and others requiring specialized care (Anderson 1975). Social scientists also have found that pets remarkably reduce several kinds of human trauma, even among the criminally insane (Curtis 1981). If we extend these extreme situations to everyday life, the suggestion emerges that contact with plants and animals carries some ill-defined but deep-seated meaning for humans. So for many urban-bound residents who otherwise have minimal contact with nature, management programs enhancing human interactions with wildlife may be especially beneficial in ways not before appreciated.

Other reasons for urban wildlife management address more tangible considerations. Taxpayers, in one way or another, support many of the costs associated with wildlife management. As the majority of U.S. citizens dwell in cities, it follows that these taxpayers should receive some measure of return on their investment. The responses assume two directions, one inward and the other outward. First, creative ways are needed for making cities as hospitable as possible for wildlife. Many people decry the aftermath of urbanization on wildlife habitat and hence on wildlife populations, but little action follows unless hazards to human safety and health are at issue. Parks, cemeteries, schoolyards, and other open spaces are prime sites for urban wildlife management. However, few of these places actually have been managed for wildlife; their primary purposes lie in other realms. But what if ways are developed to add wildlife habitat as a secondary purpose for urban open spaces? Therein lies the inward direction, namely to facilitate appropriate human exposure to wildlife within urban communities.

Outwardly, urbanites also must be instilled with appreciation for wildlife, even though most wild creatures cannot exist in cities. The voting strength within state legislatures now resides in urban centers. In time, this voice may no longer be supportive if urbanites maintain only a "vacation interest" in wildlife management (Davey 1967). Urbanites, of course, do not inherently lack appreciation for wildlife. Most of the members of national conservation organizations live in cities, and bird guides and other nature books are sold widely in city bookstores. Zoos improve the contact between humans and animals, but for all their benefits, zoos do not manage populations of wild animals; they create stable environments for individuals. Other means are needed to gain urban support for management programs in far-off places. Greater attention to urban wildlife management thus may prove useful in educating large numbers of citizens about the basic principles and values of wildlife management. The process could stimulate greater support of the full range of national and global wildlife programs (Adams et al. 1987). Therefore, urbanites must be educated about herons and deer, marshes and forests, if wildlife management is to gain the fullest national support for its far-flung activities.

Until recently, urban wildlife management largely consisted of ridding cities of vermin and troublesome animals. Dedicated biologists have struggled for years with these problems, often in the face of adverse publicity when certain kinds of animals were controlled. Commercial firms, of course, also deal with many forms of undesirable animals, usually insects, rats, and mice, but, otherwise, urban environments usually have been "written off" as locales unsuitable for practicing wildlife management. No management strategy purports to create and maintain pristine environments within a city, although innovative methods can establish habitats suitable for more than the monotony of house sparrows (*Passer domesticus*), domestic pigeons (*Columba livia*), and wandering household pets. A schoolyard or park endowed with appropriate vegetation, feeding stations, or shelters might easily attract a variety of songbirds while concomitantly improving the quality of life for human inhabitants. As with many other aspects of wildlife management, its practice in urban settings relies heavily on creating and maintaining habitat diversity. Relatively small areas of appropriate habitat may attract songbirds. Evenden (1974) noted only house sparrows, pigeons, and starlings (*Sturnus vulgaris*) in downtown Washington, D.C., but nearby he found 19 species of birds among the oases of spring gardens surrounding the White House; 17 species were recorded there during the winter. As the White House grounds occupy about 8 ha, these censuses represent a richness of more than 2 species per hectare in an otherwise highly urbanized environment.

Realization of the opportunities that city environments afford wildlife management is relatively new to the profession. Based on a survey conducted in 1985, only 9 percent of the universities with wildlife programs offered courses dealing specifically with urban wildlife and, at most other schools, about 8 percent of the class time in other courses is devoted to the subject (Adams et al. 1987). A greater percentage of western (36 percent) than eastern (29 percent) universities reported some research activities with urban wildlife, which indicates that interest in the subject is not concentrated in the urban East. In 1983, just six state wildlife agencies reported having urban programs, and only half of these included research activities (Lyons and Leedy 1984).

The full scope of urban wildlife management leads to functional networks with other disciplines, among them sociology, architecture, psychology, landscape design, law, medicine, and urban planning (see Stenberg and Shaw 1986; Adams and Leedy 1987, 1991). In short, gaps between design and management—or between art and science—are not opposing ideologies, but instead can be blended for the betterment of metropolitan landscapes and urban living (Twiss 1967). The Wildlife Society adopted a statement of program elements—Guidelines for Implementing Urban Wildlife Programs under State Conservation Agency Administration—in 1986. Later in the same year, the statement was endorsed by the International Association of Fish and Wildlife Agencies. Like the undercurrents of city life itself, the issues are complex, but therein lies the challenge of a new dimension for wildlife management.

URBAN WILDLIFE RESOURCES

Wildlife managers, in keeping with the more traditional responsibilities of their state and federal agencies, have spent untold work-hours enumerating wildlife on farmlands, forests, prairies, and wetlands. These efforts have produced an impressive volume of literature. In contrast, few studies have assessed the wildlife resources in cities, although long ago Pearson (1915) noted the sanctuary value of urban cemeteries for songbirds. Thomas and Dixon (1974) have since made more exhaustive surveys in the cemeteries of greater Boston; these findings underscore the promise of maintaining wildlife in cities.

Cemeteries exist within the confines of most cities. To the extent that they are permanently located, cemeteries represent—along with urban parks—habitat immediately available to wildlife (Fig. 17-1). In greater Boston, cemeteries represent about 35 percent of the remaining open space. Surprisingly, 95 species of birds were discovered, including hawks, herons, and game birds. Further, many species nested in cemeteries. Common species such as starlings, robins (*Turdus migratorius*), and blue jays (*Cyanocitta cristata*) accounted for 36 percent of the 34 nesting species, but 27 percent of 1195 nests found were those of flickers (*Colaptes auratus*), catbirds (*Dumetella carolinensis*), song sparrows (*Melospiza melodia*), pheasants (*Phasianus colchicus*), and mockingbirds (*Mimus polyglottos*). Many of these species successfully produced broods.

The Boston study also uncovered 20 species of mammals residing in urban cemeteries. These included raccoons (*Procyon lotor*), striped skunks (*Mephitis mephitis*), red foxes (*Vulpes vulpes*), woodchucks (*Marmota monax*), muskrats (*Ondatra zibethicus*), cottontails (*Sylvilagus floridanus*), opossums (*Didelphis virginiana*), and gray squirrels (*Sciurus car-*

Figure 17-1. Many species of wildlife, among them squirrels, find suitable habitat in urban cemeteries. (Photo courtesy of L. W. VanDruff.)

olinensis). Gray squirrels in another urban cemetery were as numerous as in any nonurban population (Hathaway 1973). In this case, the cemetery's vegetation was maintained at the best stage of succession for squirrels. Coupled with the presence of diversified shrubbery, the setting produced a steady supply of food throughout the year. Several red foxes also living in the cemetery contributed to the natural regulation of the squirrel population.

Cemetery settings also provided a diverse pattern of human activities, beyond the expected high incidence of family visitations to gravesites (Thomas and Dixon 1974). Many visits were recreational; some included the feeding of wildlife. Legally or otherwise, bowhunters apparently hunt successfully for pheasants and cottontails in cemeteries and other urban environments in Minneapolis–St. Paul and Milwaukee (Stearns 1967). The wildlife scenarios seen by cemetery visitors often resembled those of more rural settings. For example, in a downtown Boston cemetery surrounded by tall buildings, observers watched a kestrel (*Falco sparverius*) deliver meals to its nestlings. Boston cemeteries accordingly ranked fifth among the 10 categories of zoos and other sites visited by schoolchildren on organized field trips, which again indicates the potential of these habitats for both wildlife and human values (Thomas and Dixon 1974). Managers can enhance these opportunities by recommending vegetation carefully selected for the food and cover needs of wildlife. Such recommendations naturally need to consider the aesthetic values of plants as ornamentals for gravesites and also the kinds and sizes of hedges and other vegetation bordering roadways, paths, and ponds in cemetery settings (Fig. 17-2).

Lussenhop (1977) compared bird populations in Chicago cemeteries with those of surrounding urban areas. Interestingly, this study indicated that indigo buntings (*Passerina cyanea*) and other native songbirds spilled over from the cemeteries into the adjacent urban habitat, rather than urban birds such as house sparrows spilling into cemeteries. Furthermore, in cemeteries where there are few nesting sites, the songbirds nested in the adjacent city but foraged in the cemetery, thereby suggesting that bird refuges in cities might attract more wild native species than was previously suspected.

Other studies have described urban wildlife populations of various kinds. Raccoons are one of the more common species that adapt to urbanization (Cauley and Schinner 1973; Schinner and Cauley 1974). Within the city limits of Cincinnati, raccoon

Figure 17-2. Cemetery landscaping and arrangement can enhance wildlife habitat in urban environments. Vegetation along walls and other "edges" provides excellent habitat for birds and other small animals without lessening the site's tranquility (top), whereas a highly regimented and closely manicured cemetery offers little value for wildlife (bottom). (Photos courtesy of Elizabeth D. Bolen.)

densities often exceeded those reported in more natural habitats. Densities in the urban population never were less than 1 raccoon per 3.6 ha, whereas maximum densities of 1 per 4.7 to 6.5 ha characterize populations elsewhere. Hoffman and Gottschang (1977) found 160 raccoons residing in a 234-ha residential area. Raccoons in these studies denned or sought refuge in churches, inhabited and empty houses, garages and other buildings, and hollow trees, some of which were close to central business districts. Storm sewers offered a network of travel routes safe from traffic and dogs. Some animals turned up in places befitting a raccoon's agility and ingenuity, including the ventilation ducts of a high-rise office building and an unused sofa on the third floor of a house. Urban raccoons fed on many kinds

of food, reflecting their use of gardens, birdfeeders, and garbage as well as more natural food sources. Besides cantaloupe, corn, watermelon, and sunflower and millet seeds, raccoon droppings contained cellophane wrappers, string, rubber bands, and aluminum foil. The diet of opossums collected in Portland, Oregon, included 9 percent each of garbage and pet food (Hopkins and Forbes 1980). Opossums in urban Corvallis, Oregon, moved shorter distances per night than in rural habitats, probably because the year-round availability of garbage, pet food, and other food sources reduced the need for exploring new areas (Meier 1983).

Peregrine falcons (*Falco peregrinus*) regularly nested on the window ledges of skyscrapers in eastern cities before pesticides later decimated their numbers. High above the busy streets below, window ledges on the upper floors of tall buildings are among the urban sites serving as analogs for settings—sheer rocky cliffs, in this case—otherwise regarded as natural habitat for wildlife (Bolen 1991). Twenty-two young falcons fledged in a 13-year period from an office building in Montreal (Herbert and Herbert 1965, 1969). Bent (1938) wrote that it was not unusual for peregrines to winter in large eastern cities where pigeons formed a readily available staple in their diet; the Customs House tower in Boston, the Post Office tower in Washington, and the City Hall tower in Philadelphia each has been favorite winter quarters for peregrine falcons. In Canada, efforts to reestablish peregrine falcons included 34 birds hacked from buildings in Edmonton. Other peregrines were hacked from cliffs along rivers, but the released birds showed a clear preference for urban nesting habitat, perhaps because they are attracted to the city by the availability of prey species such as pigeons (Dekker and Erickson 1986). The prey of an urban peregrine falcon in Baltimore included 91 percent pigeons, with the balance consisting of 11 species of other birds and 1 mammal (Barber and Barber 1983). Today, city environments play an important role in the restoration of peregrine falcon populations (Fig. 17-3).

An increase in merlins (*Falco columbarius*) wintering in the north coincides with their adoption of urban centers in Canada as breeding habitat (James et al. 1987). The size of the merlin population in Calgary, Edmonton, Regina, and Saskatoon was significantly correlated with the numbers of Bohemian waxwings (*Bombycilla garrula*), suggesting that the availability of prey also played an important role in the urbanization of merlins. The numbers of waxwings in cities have increased, proba-

Figure 17-3. This female peregrine falcon, affectionately known as Scarlett, selected the 32nd floor of an insurance building in Baltimore as her nesting site. Ledges on tall buildings in cities resemble cliffs, and pigeons offer a ready supply of food. In addition to fledging 4 of her own offspring, Scarlett raised 17 young as a foster parent in a program designed to restore urban populations of peregrine falcons. Scarlett died in 1984 at age 7, but during her lifetime she became a widely publicized symbol of urban wildlife management. (Photo courtesy of J. C. Barber, The St. Paul Companies, formerly United States Fidelity and Guaranty Company.)

bly as a response to the planting of fruit trees that supply winter food for waxwings. Crows (*Corvus brachyrhynchos*), normally associated with rural environments, now nest regularly in cities (Knight et al. 1987), and cuckoo-shrikes (*Coracina* spp.) are among the native birds in Australia adapting to urban life (Debus 1985). In suburbia, screech owls (*Otus asio*) typically range over 4–6 ha, whereas those in rural settings occupy about 30 ha, or about one-fifth the density of the suburban birds (Gehlbach 1986). Screech owls often catch earthworms on wet pavement; therefore, those owls living in suburbia experience more collisions with vehicles. However, because of increased predation, the mortality rate for nestling screech owls is 2.7 times greater in the rural population. Studies elsewhere also report higher rates of nest predation in rural versus urban bird populations

(Tomialojc 1980). Many other birds become established urban visitors when suitable habitat is available. The checklist of birds recorded in New York City, and especially in Central Park, includes more than 200 species (Bull 1964, 1974).

Given a spacious landscaped setting, university campuses offer suitable habitat for birds, despite heavy human traffic. Swank (1955) reported an extremely heavy density of nesting mourning doves (*Zenaida macroura*) on a campus where the walks and streets were lined with high-canopy trees of several species. He recorded 648 active dove nests on 33 ha, with 77 percent production of young in a single nesting season. The doves gathered nesting materials from street gutters, and the campus sprinkler system provided a dependable water source during periods of short rainfall. Mourning doves in Berkeley, California, are heavily dependent on bird feeders, and the urban habitat carried a population of about the same magnitude as nearby rural areas. Based on a study of doves banded on a campus, as well as in residential areas of Berkeley, there was little difference between the adult mortality rates for the urban doves and those that were hunted (Leopold and Dedon 1983). Collisions with vehicles, windows, or wires were the most common causes of mortality for doves in Berkeley. On a diversified campus in Cleveland, Dexter (1955) recorded 11 amphibian species (of 30 occurring in Ohio), 8 reptiles (of 37), 143 birds (of 329 for the Cleveland area), and 22 mammals (of 51).

Chipmunks (*Tamias striatus*) are amusing rodents that adapt well to residential settings. Ryan and Larson (1976) frequently observed them in yards with stone fences, piles of debris, and drainage systems, as well as in wooded sites on or near house lots. The chipmunks responded to artificial food sources, thereby presenting themselves for human observation. Urban populations of gray squirrels may be larger than those elsewhere, reaching densities of 12–15 animals per ha (Flyger 1959, 1970).

Autumn populations of gray squirrels in a 3.3-ha park across from the White House in Washington, D.C., reached 51 animals per ha, the highest density reported for the species (Manski et al. 1981). Large quantities of supplemental food provided throughout the year and the availability of nest boxes buoyed the squirrel population to this level. Red squirrels (*Tamiasciurus hudsonicus*) and fox squirrels (*Sciurus niger*) also are common in cities within their respective ranges. Because large trees are part of older residential sections of many cities, the exceptional canopy cover provides tree squirrels with nearly ideal habitat conditions. Flyger (1974) accordingly noted that it is incorrect to consider squirrels as having adapted to urban environments; squirrels instead have taken advantage of superior habitats provided by humans. The prevalence of urban squirrel populations tends to reflect conditions no less desirable for human habitation (e.g., tree-lined streets, parks, etc.).

Gill (1966, 1970) and Gill and Bonnett (1973) have reported on a remarkable population of about 400 coyotes (*Canis latrans*) within Los Angeles. These animals may be larger and hardier than their nonurban counterparts. Coyotes are numerous enough in Los Angeles to warrant public calls for control. Leach and Hunt (1974) cited a news article referring to "marauding bands of coyotes who are walking the streets forcing people to keep their dogs, cats, and guinea pigs in their homes so that they will not be eaten by hungry coyotes." Other complaints noted the disturbance of coyotes howling at night; the howls can rise above the urban din of cars and aircraft. Andelt and Mahan (1977) monitored the movements of a radio-collared coyote in Lincoln, Nebraska, and determined that 70 percent of the time the animal was in a well-developed 14-ha residential area even though about half of its range included nonresidential sections of the city. The relatively small range of this coyote seemed related to the diversity of habitat and to the abundance of food. Coyotes adapt well to food sources associated with humans. About 17 percent of the materials in the droppings of coyotes living in a suburb of San Diego was associated with garbage (MacCracken 1982). Rabbits and rodents dominated their diet, but the occurrence of house cats, apples, and melon seeds reflected associations between coyotes and human habitation. Red foxes sometimes locate their dens in wooded areas within some of the larger towns and cities in Kansas; natural prey and garbage both were used as food in these instances (Stanley 1963). In New Mexico, gray foxes (*Urocyon cinereoargenteus*) adapted to development by becoming active at night in residential areas, but they avoided residential neighborhoods altogether when the density of houses exceeded 128 per km² (Harrison 1997).

Some cities harbor larger wildlife. In Winnipeg, a city with a population of more than 500,000, a herd of some 200 white-tailed deer (*Odocoileus virginianus*) flourishes at a density (30 deer per km²) in excess of nonurban herds elsewhere in Manitoba (Shoesmith and Koonz 1977). Good winter cover and food supplies from trees and shrubs maintain the herd's vitality. Some animals probably move along a river drainage through the city, but fences, busy streets, and other artificial barriers make the herd essentially sedentary; however, these animals by no means are semidomesticated. In Madison, Wisconsin, an urban deer population threatened vegetation in a university arboretum; the animals therefore required control. Between 1958 and 1982, 162 deer were removed from the arboretum and another 54 were removed in the winter of 1982–83 (Ishmael and Rongstad 1984). Deer-vehicle accidents in the densely populated counties around Chicago have increased in the last decade and have caused an average loss of $1306 per accident (Witham and Jones 1987a,b). The Chicago-area herd was thinned (by both lethal and nonlethal means), in part to reduce collisions, and the venison was distributed to more than 500 missions and soup kitchens that feed the indigent of Chicago. Urban deer populations continue to stimulate new management strategies, as evidenced by several recent papers (e.g., Nielson et al. 1997). Herds of fallow deer (*Cervus dama*) and red deer (*C. e. elaphus*) live within a circle 64 km in diameter centered in mid-London, as do a number of other species (Hurcomb 1969). Within the vast city of Tokyo, Kishida (1934) found nondomestic mammals of 10 families and listed 3 additional families of cetaceans encountered in Tokyo Bay. In Miami, manatees (*Trichechus manatus*) commonly are viewed from city bridges, where these endangered animals are safer from unlawful persecution than elsewhere in Florida because of close scrutiny by interested citizens and conservation groups (Moore 1956). The density of magpie (*Pica pica*) nests was not significantly different between highly urbanized parts of Seoul and a rural area in Korea (Lee and Koo 1986).

Throughout history, sites along water were preferred for human settlements. Transportation and power for commerce were keys in the location of first camps and villages and then towns and cities. In short, urbanization prospered along the rivers and larger lakes of the United States (as it has elsewhere in the world). Indeed, data summarized by Leedy et al. (1981) indicate that 70 percent of the 415 cities in the United States with a population of 50,000 or more are located on the edge of a river, lake, bay, or ocean. Thus, many urban areas offer significant potentials for fishing and other forms of aquatic recreation, but lack of public access and pollution often have destroyed these opportunities (Duttweiler 1975). Of more than 3700 urban lakes in the United States, about 80 percent are significantly degraded; yet these offer potential recreational and aesthetic values to more than 94 million metropolitan residents (Council on Environmental Quality 1979). Managers nonetheless have created fishing where water conditions are still acceptable (Fig. 17-4). Limited habitat for trout in a highly urbanized area bordering the Huron River in Michigan is managed by special regulations. After trout are stocked, varying limits are imposed each month on the number and size of the legal catch. At times, all fishes must be released. Under this intensive management regime, each fish was caught more than two times, and the 10 to 1 benefit-to-cost ratio represented a justifiable and efficient expenditure of public funds (Carl et al. 1976). Several thousand trout are stocked in park lakes in Milwaukee as part of a stepped-up fisheries project for cities in Wisconsin (Wisconsin Department of Natural Resources 1985).

In 1970, inner city residents of St. Louis spent more than 100,000 hours angling for fishes stocked in five urban lakes. Many of the persons using this resource were under 16 or more than 60 years old, and 77 percent of those fishing consumed the fish they caught. Local residents thought that the program had social benefits, but there was no measurable influence of urban fishing activities on crime rates (Ikeda 1971).

Duttweiler (1975) concluded that sport fishing is an outdoor activity highly adaptable to urban settings. Various types of programs, mostly under local sponsorship, therefore have evolved to serve city res-

idents. These include stocking urban waters, "kids-only" and senior-citizens fishing, fee fishing, fishing day camps, clinics and interpretive services, and pier construction. Indeed, *urban fisheries* became one of the most prominent "buzz words" in the vocabulary of biologists, thereby taking fishing to people instead of using the reverse approach of the past (Sullivan 1980). Maintaining free-flowing urban streams with vegetated borders, instead of lining the banks with concrete or enclosing them in conduits, offers a first step for urban planners and developers. A full range of opportunities and concepts for developing urban fishing is offered in Leedy et al. (1981) and Allen (1984).

Urban-related developments may lead to special situations benefiting the management of aquatic organisms. Water heated by a nuclear power plant increased the catch rates for 18 species of fishes during the winter and spring months (Marcy and Galvin 1973). Artificially warmed water also increases the growth rates and production of many aquatic animals. For this reason, shrimp are reared commercially in pools fed by the heated water of a power plant in Corpus Christi, Texas. Hartman (1979) found that water heated by factory cooling systems provided favorable winter habitat for manatees during harsh periods of winter weather. However, artificial sources of warm water have encouraged manatees to expand their winter range northward on the Florida coast. Thus, because manatees cannot survive cold water temperatures, power plant shutdowns pose a serious risk for manatees relying on artificial warm water refuges during winter. Shane (1984) accordingly urged that no new sources of artificial warm water effluents accessible to manatees should be created north of the species' historical winter range limits.

URBAN MONOCULTURE

The prime enemy working against wildlife in urban environments is not so much the crowded human community itself as it is the monoculture created by humans. Two features of downtown or even many suburban areas work against the influx and establishment of wildlife. These are the lack of sufficient vegetation (quality and quantity) and environmental uniformity. Both features promote a monoculture that limits development of a native fauna within the city's confines. Indeed, few cities were designed with corridors of vegetation that permit animals to colonize

Figure 17-4. Opportunities for fishing may be realized in association with other uses of urban ponds. Water quality is a major consideration in the management of urban fisheries. (Photo courtesy of Elizabeth D. Bolen.)

any habitats remaining in their inner recesses. In fact, urbanization usually progressed only with the constraints of zoning laws and construction codes. Regard for other types of city planning associated with environmental quality is a recent development, which evolved after urbanization already had claimed many natural resources. Walcott (1974) found that the birds breeding in a residential area were reduced from 26 to 9 species after a century of advancing urbanization.

Urban vegetation, except for the diversity found in some parks and cemeteries, often consists of only a few species. Much of it is widely spaced and in poor condition. Smog or other disturbances, including the thoughtless carving of trees, greatly reduce the vitality of downtown plant life. Trees surrounded by pavement may receive inadequate water, and many are trimmed heavily. The vegetation often is of exotic origin, selected for its ornamental properties rather than for its suitability as food and cover for native animals. At its extreme, downtown vegetation may consist only of potted shrubs placed on sidewalks. Ground litter is absent, as is the humus supporting many organisms essential for food chains. Finally, spraying can further reduce urban bird populations either directly or indirectly (see Hickey and Hunt 1960).

The city also is the architectural zenith of environmental uniformity. By definition, a city is a crowded collection of structures and pavement offering little more than superficial variation in outward appearance. Storefronts and office buildings are a nearly uniform facade of glass and masonry blended into vertical walls. In between, roadways and sidewalks form a continuous horizontal pavement. The results are sterile "concrete canyons." Urbanization of residential districts also promotes monocultures, unless careful planning precedes construction of new housing developments. A study of bird populations during urbanization found that species diversity decreased as residential development progressed (Geis 1974). Overall, bird populations were of higher densities, but, unfortunately, most of the increase resulted from unwanted starlings and house sparrows. Thus, monocultures and socioeconomic settings are reflected in the diversity of bird communities. Congested downtown areas and low-income neighborhoods have few species, whereas spacious residential areas reflecting high living standards enjoy many more kinds of birds (Barnes 1966; Williamson 1973). DeGraaf and Wentworth (1981) further refined bird communities in functional units known as guilds, based on

their association with urban and suburban areas, respectively.

At the other extreme, far-sighted planners sometimes have incorporated aesthetic relief within dense enclaves of human habitation. Central Park in New York is one of these; this 349-ha oasis of countryside lies within the heart of a metropolis packed with some 8 million humans; yet it harbors, in season, about 260 species of birds (Arbib et al. 1966). Fredrick Law Olmsted and Calvert Vaux created—with brilliant vision for the future—the park from wasteland in the late 19th century. Ironically, birds and other wildlife in Central Park today seem safer than many humans venturing into this urban "wilderness." Tanner (1975) also traced the political and developmental history of Jamaica Bay Wildlife Refuge, a treasure of marshland bordered by two metropolitan airports and urban Brooklyn and Queens. Thanks to the planning of Robert Moses and the management zeal of Herbert Johnson, the refuge regularly hosts some 250 species of birds and another 50–60 species of infrequent visitors. On the southern end of Lake Michigan, near the industrial goliaths of Gary, Indiana, and Chicago, lie the Indiana Dunes, a scenic and pristine shoreline with its own scientific history. There H. C. Cowles (1899) initiated his classical studies of vegetation dynamics, thereby introducing concepts of community succession to the ecological sciences. That outstanding parks and recreational areas replete with wildlife are found within or near highly urbanized environments (see Tanner 1975) indicates that wildlife management can and should occur in cities as well as in rural settings.

URBAN ZONES

VanDruff (1979) separated the urban complex into three zones based on the intensity of development and their potentials for wildlife management. First, and least developed, is the *rural-urban interface*, where it is still possible to preserve areas of native habitat. Ponds, wooded sites, and marshes occurring at the city's edge can be scheduled for green spaces before development encroaches and eliminates or alters their values as wildlife habitat. Other benefits also may accrue from preservation of this zone. For example, a wetland preserved during urban expansion in the Minneapolis–St. Paul metropolitan area retained sediments and nutrient loads contained in surface runoff, thereby improving the quality of water entering a nearby lake (Brown 1985). Planning is crucial, for the rural-urban interface experiences the most rapid urbanization of any in the city. The rural-urban interface therefore is the first line of defense for ecological diversity because, in part, existing natural environments can be maintained as the city expands.

Suburbia is the second zone; it represents the most widespread type of urbanization. One- or two-family homes, along with schools, churches, and other facilities such as hospitals, parks, and cemeteries, characterize suburbia. Wildlife still may exist here as remnants of former natural communities, dwelling in patches of habitat interspersed among human habitations. However, some settings may be dominated by large numbers of a few species (Fig. 17-5). In suburbia, wildlife managers find good opportunities for reaching residents with conservation education programs. *Backyard management* becomes a reality in suburbia when residents are informed about ways and means of enhancing wildlife. Because the wildlife community in suburbia still may contain a good deal of diversity, management with nest boxes, feeding stations, and plantings should be directed toward all components of the community. Efforts within suburbia for managing larger open spaces may lead to neighborhood wildlife sanctuaries as well. Conversely, the potential for wildlife to become nuisances also is high, and managers must be equally responsive to these concerns. A study of public attitudes toward deer in a suburban area illustrated some of the complexities faced by wildlife managers (Decker and Gavin 1987). The matrix of concerns included positive elements such as the aesthetic values of deer and negative elements such as shrub and garden damage, deer-vehicle collisions, and the threat of disease. The study emphasized that the major task for managers in suburban areas is to formulate management activities that are consistent with public attitudes.

Finally, *metropolitan centers* mark the core of the urban complex. This is the inner city or downtown area, where tall buildings, busy commerce, and crowds largely replace all remnants of former natural systems. *Wildlife* usually is limited to exotic pests, and control of starlings, rats, pigeons, and house sparrows may be required. Rubbish accumulations and other sanitary problems often provide these pests with a stronghold of resources. Besides eliminating these situations, managers face the task of introducing as much diversity as possible. Virtually any addition of vegetation will help, including street-side plantings, corner plazas, and rooftop and window gardens. Above all, managers working in metropolitan centers must allow for maximum amounts of human interaction in planning inner-city improvements for wildlife.

Figure 17-5. Urban settings often are dominated by large numbers of a few species such as these starlings. (Photo courtesy of Elizabeth D. Bolen.)

MULTIPLE-USE MANAGEMENT IN THE CITY

Multiple use is an accepted concept for many areas where wildlife resources are meshed with timber, mining, and other uses of the landscape. However, the concept may reach its maximum application in urban environments where human pressures are at their greatest for space and other resources already in short supply. Room for landscaping is limited, or perhaps even nonexistent, in many cities, and even those areas already landscaped may suffer debilitating abuse. Wildlife otherwise capable of existing in marginal urban habitats may require some degree of insulation from the heavy domestic traffic of pedestrians, vehicles, and pets. The innocent play of children might easily cause abandonment by nesting song-birds, and a child's inquisitive examination of a nest may lead to "collection" of the eggs. At the other extreme, vandalism of feeder stations or nest boxes, along with the purposeful destruction of animals, must be considered in any urban management program.

In other cases, the habitat may be "too good" for compatibility with human activities. Tragic accidents at airports have occurred when aircraft hit birds living adjacent to runways (Solman 1968). Some species of gulls (*Larus* spp.) and even birds such as long-billed curlews (*Numenius americanus*) coexist with the noise and jet-age activities of large urban airports. Thompson (1977) described a burrowing owl (*Athene cunicularia*) community living in the abandoned dens of California ground squirrels (*Spermophilus beecheyi*) at the Oakland Municipal Airport. The owls made foraging flights to an adjacent golf course and used divots from the course as nesting material. Jackrabbits (*Lepus californicus*) also are commonly associated with the closely cropped turf of many airports.

Multiple-use problems are numerous for the urban wildlife manager, but they may not be altogether unrelated to those of his or her counterparts working outside the city. For example, many lakes in urban parks are stocked with domestic ducks, and these

Figure 17-6. Wild ducks often are attracted to lakes in city parks, where they provide enjoyment for urbanites of every age. However, the birds also may become dependent on artificial foods of low nutritional quality, may enhance outbreaks of diseases, and may hybridize with other species. (Photo courtesy of H. K. Nelson, U. S. Fish and Wildlife Service.)

habitats also may serve as refuges for wild ducks (Fig. 17-6). Opportunities for genetic mixing are manufactured; the domestic birds attract wild ducks, and pairing develops in highly artificial situations. In Massachusetts, Heusmann (1974) reported increasing contact and hybridization between mallards (*Anas platyrhynchos*) and black ducks (*A. rubripes*) in urban parks, thereby contributing to the genetic swamping of the black duck's gene pool and, in part, to its possible demise as a biological entity. Furthermore, artificial feeding of mallards wintering in parks may lead to their undue dependence on bread or crackers, perhaps a critical shortcoming because of rapidly increasing food costs (Heusmann and Burrell 1974). It also seems likely that artificial feeding and shelters may increase the survival of individuals that otherwise would be eliminated by the rigors of natural environments. Populations of wintering birds using backyard feeders in urban Fort Worth, Texas, showed higher-than-usual frequencies of bill and leg abnormalities (Sharp and Neill 1979). Gray squirrel populations supported by artificial foods in the small, isolated parks in downtown Washington, D.C., frequently include individuals with abnormal pelage (Flyger 1974).

Still other multiple-use situations are unique to management in cities. Banks (1976) drew attention to the large number of birds killed by collisions with buildings and broadcasting towers. Such collisions are even more prevalent today because of the common practice of constructing modern buildings with large areas of reflective glass (Fig. 17-7). The birds see the reflections of sky or nearby vegetation and fly headlong into the images. Most birds killed in window collisions die from brain damage and ruptured blood vessels rather than broken necks, as commonly supposed; skeletal fractures indeed occur only rarely (Klem 1990a). Windows at private residences also account for large numbers of collisions. Klem (1990b) recorded annual kills of 26 and 33 birds, respectively, at two houses. Overall, he estimated the deaths of at least 98 million birds each year from window collisions in the United States. Unfortunately, the common practice of placing silhouettes of falcons or owl decoys on or near windows does not greatly reduce collisions. Instead, collisions are reduced or eliminated by placing netting or narrow strips of cloth over the entire surface of the window. The ultimate remedy lies in the design of buildings: windows installed at an angle so as to reflect the ground instead the surrounding habitat or sky (i.e., not in the usual verticle position) or using nonreflecting glass (Klem 1991). Avian predators such as sharp-shinned hawks (*Accipiter striatus*) may exploit prey attracted to bird feeders erected near windows (Klem 1981). When fleeing, cardinals (*Cardinalis cardinalis*) and other songbirds often struck the windows and were seized by the hawks. Whereas such tactics probably reinforced the return of sharp-shinned hawks to areas with window casualties, the predators also fell victim to windows where they hunted. Thus, windows pose a hazard for certain predators that exploit easily available prey. Industrial pollution is another concern, especially because many factories and processing

Figure 17-7. Modern buildings constructed with dark-tinted plate glass exteriors present birds with an illusion of unobstructed habitat. (Photo courtesy of D. Klem, Jr., Biology Department, Muhlenberg College.)

Figure 17-8. Industrial pollution threatens the health of wildlife. Shown here are herons and egrets assembled in a highly industrialized area on Nueces Bay in the Port of Corpus Christi, Texas. (Photo courtesy of D. H. White, U.S. Fish and Wildlife Service.)

plants commonly are located near wetlands that attract numerous species of wildlife (Fig 17-8).

The challenges that expanding urbanization place on wildlife managers can be illustrated by Winnipeg's urban deer herd (Shoesmith and Koontz 1977). Some 200 deer reside in Winnipeg's metropolitan area on habitat that diminished by 90 percent between 1926 and 1976. Most of the remaining habitat is included in a 285-ha nature park; undeveloped industrial areas provide additional wooded habitat. However, urbanization is closing in around the park. Housing developments and high-rise apartment buildings have isolated the park from other wooded areas. Furthermore, a major freeway may cross through the park and another roadway bordering the areas was expanded to accommodate truck traffic to a new railroad yard. The railroad yard itself will interrupt a major travel route for deer moving between the park and other forested sites. Yet another block of existing deer habitat in Winnipeg may become the location of a federal radar facility. Thus, only with thoughtful planning for access routes between the remaining habitat will Winnipeg's free-ranging deer herd remain intact.

Wildlife managers confronted with these and other realities must reach deeply into their intuition and imagination, as well as into their training, to master the challenges of multiple use in urban environments. Urban wildlife habitat may be developed creatively in more places than supposed. In one experiment, waste water was applied to the dense vegetation screening an outdoor theater. The plantings, invigorated by the effluent, offered food and cover for rabbits, pheasants, and a number of other kinds of wildlife (Stearns 1967). Railroad rights-of-way and city golf courses are other areas where multiple-use concepts can be applied. Golf courses in northern states, for example, lie unused for much of the year and might easily be managed as winter habitat for several kinds of animals.

ASPECTS OF URBAN MANAGEMENT

As we have seen, urban and suburban environments can and often do provide a richness not otherwise thought of when local conditions match, even minimally, the needs of many forms of wildlife. Some conditions are coincidental (e.g., garbage for food), whereas others are the result of careful planning and thoughtful attention to a mixture of human settlement and wildlife habitat (e.g., parks and greenbelts). In addition, the urban wildlife resource offers both consumptive and nonconsumptive opportunities for human enjoyment, not the least of which are favorable economic benefits and social conditions. Any thoughtful person quickly sees the relationships between one's own psychological and physical welfare and the presence of wildlife. Social conditions are difficult to quantify for such purposes, but other means illustrate the stark reality of the coexistence between humans and wildlife: The livers of gray squirrels living in low socioeconomic areas of a city contained significantly higher levels of lead than did those of squirrels from middle or high socioeconomic neighborhoods (McKinnon et al. 1976). Indeed, because they are sensitive indicators of lead and other pollutants, squirrels seem useful as monitors for many types of environmental hazards in urban areas (Bigler and Hoff 1976).

Management is necessarily intensive in most cases, especially when it is attempted long after urbanization already has formed a bleak monoculture. Many areas accordingly may have to be "written off" as essentially hopeless for wildlife management, although even the worst "concrete jungles" may one day give way to urban renewal and the promise of revitalization. But opportunities remain elsewhere in the city, and they are greatest when wildlife resources can be incorporated into the planning of housing developments, shopping centers, and other construction associated with urbanization of the landscape. Collaboration between wildlife agencies and municipal planners clearly is required; at the least, city planners should gain familiarity with wildlife resources and their management (Greer 1982). For their part, wildlife managers must adopt different goals and use different procedures in accordance with the intensity of urbanization (VanDruff 1979).

In urban management, soil conditions are particularly important because of their limitations in terms of both quality and quantity. Construction activities typically mix topsoil with subsoil, remove topsoil altogether, or introduce a large amount of debris into much of the stratum. Minerals necessary for plant development often are limited or even absent, and the soils may suffer from poor drainage, high compaction, and excessive heating. Patterson (1974) therefore stressed that soil sampling was an important priority in the management of urban vegetation. Composite samples may be collected for pH; available magnesium, phosphorus, and potash; and texture and organic matter analyses. Slope, drainage, and other features of the landscape also are part of these assessments. With these data at hand, site planning can proceed (McCormack 1974). Much information, in fact, can be computerized to determine soil ratings for urban uses and capabilities (Rogoff et al. 1980). Suitable vegetation may be found to match the existing soil conditions, or, alternately, the soil may first need treatment. The latter usually is an expensive process but may be necessary if the plantings are to succeed. If drainage is poor, the topsoil may have to be removed temporarily and the proper drainage gradient established in the subsoil before the topsoil is replaced. Excessive heat or dryness may require the installation of subsurface irrigation equipment. If so, it should be installed with minimal disturbance to the topsoil. If necessary, loam topsoil acquired elsewhere may be added to make the site suitable.

Once the site has been readied properly and consideration given to both the macroclimate and the microclimate, the selection of plants can begin (Patterson 1974). Whereas regional temperature and precipitation regimes are obvious, the urban microclimate is more subtle. In effect, wind tunnels are created between corridors of buildings. This phenomenon may have much to do with the selection of vegetation and its location, as do such other factors as lighting, radiant heating, and exhaust fumes. These and other conditions peculiar to the urban setting determine the selection of plants that must be hardy and compatible with city life. Some plants, for example, have root systems that clog underground pipes; others have relatively short lifespans or inordinately brittle limbs that pose special maintenance problems.

It is beyond our scope to present exhaustive lists of plants suitable for urban wildlife. Many local and regional factors influence these selections, but it is important that the vegetation match the goals of an urban management plan. Food and cover requirements for the target species of wildlife must be understood if the effort is to succeed. Studies of bird habitats in Tempe, Arizona, for example, suggest that native vegetation planted in parks, residences, and other urban areas has much potential for offsetting the rapid loss of riparian vegetation in the Southwest. Some plants offer outstanding habitat values (see Martin et al. 1951; DeGraaf and Witman 1979). Hackberry (*Celtis* spp.) trees alone provide persistent fruits for 40 species of birds (Barnes 1966). Many kinds of ornamental trees also produce fruits attractive to a variety of birds, but exotic plants may support fewer kinds of insects than native trees, thereby increasing the ecological handicap of insectivores in urban areas (Beissinger and Osborne 1982). DeGraaf and Wentworth (1981) recorded a striking lack of insectivorous birds in urban environments where natural vegetation was absent. Obviously, an appropriate mixture of vegetation must be at hand to meet the daily and seasonal needs of animals, a tenet no less true of urban environments than of wildlands.

Beyond these considerations lies the arrangement of the urban habitat. Vegetation, once selected, should be arranged to facilitate adequate patterns of interspersion attractive to wildlife. Symmetrical arrangements pleasing to humans may not necessarily serve the best interests of animals. Hooper et al. (1975) believed that clumping of vegetation was an important habitat feature attracting birds in a suburb near Washington, D.C. About as many nesting sites occurred in relatively small amounts of clumped vegetation as in a larger area of dispersed foliage in woodlots. Urban vegetation created or preserved as "islands" favors many kinds of songbirds (Beissinger and Osborne 1982). Furthermore, residential landscapes should include the same layers of canopy cover as found in undisturbed forests to attract a full spectrum of birds into suburban areas. Patchy networks of multilayered shrubs and trees thereby complement open spaces as ideal urban habitat. Lacking either vertical or horizontal structure, residential vegetation largely remains a landscape of lawns attracting only ground-gleaning birds.

Whereas it is desirable to have as much food and cover as possible available from a carefully selected array of vegetation, urban management may require additional features to assure the stability and diversity of wildlife in the city. Among these are nest and roosting boxes for birds and mammals, and various types of feeding stations.

Birdhouses are an obvious tool, both for backyard and greenspace management. They are available commercially in a variety of sizes and materials, or they may be made easily by even novice carpenters who have only modest equipment at hand. Construction of birdhouses often lends itself to community projects tackled by scouting groups and supported by civic clubs. Birdhouses left in place year round serve as winter roosts for some species, but heat escapes if the entrance hole is near the top. Special winter-roost boxes invert the entrance position, placing the hole near the floor of the structure. Another type of specialized nesting box is a multichambered "bird hotel" designed to accommodate the gregarious purple martin (*Progne subis*).

Shelters for squirrels also are constructed easily (Fig. 17-9). These might be made from nail kegs fitted with a top and entrance hole, although other kinds of structures serve the same purpose. Burger (1969) used folded tires instead of wooden boxes, and reported increases of 65–100 percent in squirrel populations 1–2 years after the artificial dens were erected (see also Barkalow and Soots 1965; Peterle and Sadler 1965). Squirrels and other wildlife were attracted to nest boxes erected in a metropolitan area

Figure 17-9. Simple structures adequately serve as shelters for squirrels and other wildlife in urban environments. (Photo courtesy of L. W. VanDruff.)

in Knoxville, Tennessee, with only 1 of 50 boxes remaining unused during a 3-year study (Fowler and Dimmick 1983).

Artificial burrows for rabbits and other ground-denning wildlife are useful, especially where the soil is highly compacted or mixed with brick fragments and other such debris. These burrows are simple boxes, placed beneath the ground's surface and fitted with drainage tiles as entrance tunnels (2 per burrow, set at 45°). The dimensions of the boxes are not critical and no floor is necessary, but drainage patterns are a major consideration in locating the burrows, as is the proximity of suitable ground cover near the entrances. If vandals or sightseers present no problems, the top of the box can be hinged to remain flush with the ground's surface for inspection and maintenance of the burrow. Rock or log piles can be decorative in their arrangement and yet provide suitable quarters for chipmunks or other wildlife adapted to this form of shelter. Rotting logs also provide natural feeding stations for birds foraging on invertebrates. Where ponds are available and large enough to harbor waterfowl, artificial nesting platforms or nest boxes may enhance breeding activities. In some instances, only small piles of clean, loosely packed hay on the shoreline are required; in others, platforms in the water are desirable. Old washtubs, for example, make satisfactory nest platforms for geese (Brakhage 1966b).

Predation is always a potential problem with artificial structures, and every reasonable means should be employed to reduce this hazard.

Feeding stations for birds can be simple structures that protect food from undue exposure. Winter and George (1981) found that feeding stations enhanced the abundance, distribution, and diversity of birds in wooded residential areas. However, interactions between feeding stations and vegetation alternately may favor native or exotic birds. After comparing wooded and unwooded sections of Chicago, Guth (1981) recommended increasing vegetation if higher densities and more species of native birds are desired. In the absence of trees, starlings and other exotic birds are the primary beneficiaries of feeding stations both in summer and winter. With adequate vegetation, winter feeding benefits native birds. So to encourage only the operation of feeding stations without also developing a vegetated environment produces few benefits for native birds. Feeding stations pose certain risks for birds, including the attraction of predators to the concentration of prey, the danger of flying into windows and other accidents, and disease (Brittingham and Temple 1986). Platform feeders may encourage diseases because birds stand in the food and likely contaminate the seeds with fecal matter.

There are several points to consider in constructing and erecting structures for wildlife. First, the materials must be durable enough to be long-lasting and to withstand the local climate. Any extra expense required to obtain good materials will reap returns by increasing the lifespan of the structure. Second, the structure must be matched carefully with the desired species. Nesting boxes with large entrance holes may be usurped quickly by starlings, and boxes placed too high or too low will preclude the use by some kinds of songbirds. Most local chapters of the National Audubon Society can provide detailed information on the proper dimensions of nest boxes and entrance holes for a variety of birds. Third, the structures should blend with the environment whenever possible. Whereas it is not necessary to "fool" cavity-nesting birds by covering the exterior of nest boxes with bark, concealment reduces the chances of vandalism while simultaneously lessening the degree of artificiality. Finally, a maintenance plan is needed to assure continued utilization of the structures. Normal wear in time will require repairs, but more important, the boxes need an annual inspection to remove unwanted debris or to clear away any growth that may inhibit their continued use. Maintenance of nesting boxes is best accomplished late in the winter, prior to the breeding season and at a time when foliage is at a minimum.

Roaming housecats and dogs may pose problems of predation and harassment for urban wildlife, especially because wildlife usually is concentrated in small areas. More than 60 percent of the urban residents surveyed by Yeomans and Barclay (1981) believed that pets would frighten wildlife that was attracted to their yards. Predation by cats on birds was a major concern expressed in the survey, and many residents accordingly attempted some means for controlling cats roaming residential neighborhoods. Studies conducted in an English village, in part, confirm these fears (Crucher and Lawton 1987). Whereas small mammals formed about 64 percent of their overall diet, domestic cats nonetheless accounted for 50 percent of the mortality experienced by the local sparrow population in the more urbanized interior of the village. If these data are applied to the total population of domestic cats in Britain—estimated at 5 million—then domestic cats account for the loss of some 20 million birds each year (Harrison 1992). Worse still, this estimate may be conservative; the data are based only on birds the cats actually brought home and do not account for prey consumed or left elsewhere, which may equal as much as half of the total kill. Note, too, that the cats in the English study were fed regularly with pet food (i.e., domestic cats hunt for reasons other than hunger). Unfortunately, declawing or bell collars do not reduce the hunting prowess of domestic cats. Confining the animals inside homes remains the only reliable way of reducing their effect on birds. The diets of rural cats generally include a small proportion of birds (e.g., McMurry and Sperry 1941; Eberhard 1954; and Toner 1956). Nonetheless, the overall toll may be large; Harrison (1992) cited a study which estimates that farm cats in Wisconsin alone may kill 19 million songbirds and 140,000 game birds each year (see also Coleman and Temple 1993). Fitzgerald (1988) offers a useful review of the impact of domestic cats on prey populations. At a residential area in Brooklyn, New York, free-ranging cats, excluding pets, reached densities of nearly 5 per ha, and their distribution depended on the availability of shelter provided by abandoned buildings rather than on the abundance of food (Calhoon and Haspel 1989). Cat populations in Baltimore, Maryland, varied from 1.5 to 7.4 per ha (Oppenheimer 1980).

Dogs are even more numerous than cats in the United States. Estimates differ, but Feldman (1974) noted that 46 percent of all households have at least 1 dog, for an estimated total of about 40 million canine pets; some 500,000 dogs are owned in New York City alone. In Baltimore, 32,400 to 54,000 dogs, or up to one-half of the canine population in that city, roam freely (Beck 1973). The hazards posed by free-roaming dogs are manifold, including biting, property damage, and traffic accidents. That vast amounts of canine wastes are deposited each day in large cities adequately illustrates the pollution and health hazards of large dog populations (e.g., 75 tons of feces and 340,000 liters of urine daily in New York City). Whereas discussions of unleashed dogs harassing deer and other wildlife in rural environments occur in the literature (Lowry and McArthur 1978), comparable data addressing this problem in cities are wanting. Nonetheless, urban residents wishing to manage their neighborhoods should be advised about ways to minimize the influence of pets so that wildlife is not placed at a disadvantage (Yeomans and Barclay 1981).

Opportunities for educational and action programs for wildlife in the cities run a full gamut. In some places, citizens in cooperation with museums, zoological and botanical gardens, nature clubs, and other institutions already have such programs in motion. Yet more can be accomplished. Urban wildlife management seems an unparalleled opportunity for blending theory and practice with education by demonstration. Even simple attempts might influence young minds. What might transpire in a classroom when birds visit a window-ledge feeding station? Even with house sparrows, the phenomenon of social ranking might be observed and studied as one aspect of ethology. The possibilities grow as one thinks of a simple capture and marking program using these highly accessible birds as models for behavior and population analyses. Experiments with food selection (by color and shape) and preferences are as easily executed at a window feeder as are observations of plumage development. One need only reflect upon the original and classic studies of song sparrows made by Margaret Morse Nice (1937, 1943) in her yard and neighborhood in Columbus, Ohio, to comprehend that valuable scientific contributions are possible in urban settings.

Neighborhood committees bent on improving both living conditions and real estate values need information on landscaping that also attracts wildlife. Officials responsible for parks and recreation similarly should be advised of habitat improvements for public lands within their cities. Inner cities are often dynamic locales. Buildings, and indeed entire city blocks, are razed for new projects; fires gut tenements and warehouses; and freeway-construction cuts swaths across downtown neighborhoods. These and other events continually create opportunities for spacious greenbelts and parks where once there was little to please the human eye or spirit. Urban renewal projects alone can revitalize areas of decayed neighborhoods that imaginative planning might develop into new dwellings; these areas can provide tasteful surroundings for people and new space for birds and small mammals. Communities setting aside land for industrial parks may easily include attractive landscaping consistent with the needs of wildlife. Gill and Bonnett (1973) and Leedy et al. (1978) offer fuller treatment for these and other urban situations where nature might be a planned component of a city's system. We see in these examples, however, that such opportunities really require little more than a thoughtful commitment to a better environment—for humans and wildlife alike.

One novel idea concerns the development of rooftops as minirefuges for wildlife. The flat rooftops of office, factory, and apartment buildings represent the greatest area of vacant space in the inner core of most cities. With imaginative planning, these offer myriad opportunities for wildlife habitat. Nighthawks (*Chordeiles minor*) commonly breed on the roofs of urban buildings. In Detroit, the home ranges of nighthawks varied inversely with the number of flat roofs per hectare (Armstrong 1965). Other birds nest on roofs, among them killdeer (*Charadrius vociferus*) (Fisk 1978). Based on samples collected in urban Canada, Ankney and Hopkins (1985) projected that many thousands of killdeer pairs nest on roofs. Whereas the nests of killdeer and many other roof-nesting birds generally experience little predation from mammals, chicks of precocial birds must leave roofs for food. Therefore, roofs enclosed by parapet walls prevent chicks from jumping; chicks jumping from high buildings may be killed by the fall.

High above the street in the Kensington district of London, a 0.6-ha rooftop garden with pools, trees, and lawns attracts several kinds of wild birds, including nesting mallards (Fig. 17-10). Herons hunted in the garden when the pools contained fish. Whereas the Kensington garden is exceptional, it illustrates what could be accomplished when old, inner-city housing is replaced with flat-roofed houses and walled roof gardens. The gardens give added insulation and provide places where inner-city residents can grow flowers and vegetables. Roof gardens thus represent an excellent way of extending green spaces into deprived urban areas (Anon. 1986b). The idea offers builders and city officials several possibilities for tax advantages and public-relations benefits.

Sewage-treatment lagoons offer other opportunities for urban wildlife management. Maxson (1981), for example, noted that sewage lagoons offer food resources, dependable water supplies, and relative isolation from disturbances, all of which enriched local bird populations in North Dakota. Similar observations have been made throughout the world (e.g., Bell 1985). In some locations, breeding activities of waterfowl and other birds can be enhanced by installing nesting structures. Several kinds of waterbirds—along with mink (*Mustela vison*), raccoons, and other mammals—were among the species listed in another survey of sewage-treatment lagoons (Dornbush and Anderson 1964). Swanson (1977) recorded densities of about 3 to 10 ducklings per ha on sewage lagoons in North Dakota and summarized other data reporting densities exceeding 19 ducklings per ha. In comparison with natural wetlands in South Dakota, sewage lagoons were at least twice as productive; an average of about 14 ducklings per ha was estimated for a sample of 36 lagoons (Dornbush and Anderson 1964). More should be learned about the design and construction of sewage lagoons for ways that will provide better habitat and yet remain compatible with the primary purpose of waste treatment. Shoreline development for nesting and brood cover is chief among these requirements. At present, vege-

tation suitable as cover is discouraged in the operation of sewage lagoons (Barsom 1973), but ongoing research suggests that some species of wetland plants absorb certain kinds of toxic substances, thereby offering the twofold promise of cleaning waste water and improving waterfowl habitat in sewage lagoons.

Mortality also should be monitored as part of the management program for sewage lagoons. Botulism has killed waterfowl on sewage lagoons in Utah (Mouton et al. 1976), and the feather-wetting properties of detergents may prove lethal when ducks and other waterbirds are chilled (Choules et al. 1978). If these or other agents cause inordinate and chronic mortality for waterbirds, then means should be taken to discourage further use (e.g., by removing nesting cover). However, illegal shooting probably remains the most common cause of mortality for waterfowl on sewage lagoons (Maxson 1981).

Mallards are particularly adaptable to many urban situations (Greer 1982). Among these environments are residential areas developed on coastlines where artificial channels create a maze of waterways (Fig. 17-11). Figley and VanDruff (1982) found mallards using ornamental vegetation and boats, docks, and other structures associated with a residential community for nesting habitat on the New Jersey coast. Nesting success was high, but a variety of disturbances caused the loss of about 40 percent of all broods. Nonetheless, the area produced about 600 mallards in one season, providing residents with significant enjoyment. The urban population of ducks also contributed to hunting, with about 230 of the 800-bird population bagged in a single hunting season.

Buildings, pavement, and the loss of vegetation in urban areas reduce the ground's natural storage capacity for water. After heavy rains, large volumes of water flow into storm drains leading to nearby streams. The surge of runoff, however, can flood and pollute downstream areas, increase erosion and sedimentation, and damage wetland communities (Adams et al. 1986; see also Brown 1985). In recent years, however, the runoff from urban development has been controlled with excavated drainage basins. Such basins, in turn, offer opportunities for creating wetlands of benefit to wildlife, including nesting and brooding areas for mallards (Adams et al. 1985). Moreover, homeowners strongly favored having wildlife and wetlands nearby (Adams et al. 1984; see also Shaw et al. 1985). Adams et al. (1986) recommended the following guidelines for optimizing the design of stormwater basins for wildlife: (1) grade gently sloping sides (about a 10:1 ratio), which will

Figure 17-10. A one-of-a-kind rooftop garden in London attracts wild birds, including mallards and herons, and suggests opportunities for managing flat-topped buildings typical of the downtown area in most cities. (Photo courtesy of J. C. Dale.)

Figure 17-11. Mallards readily adapt to urbanized recreational developments with waterfront homes and so experience close contact with humans. Urban mallards were studied in detail by Figley and VanDruff (1982) at one such site in Ocean County, New Jersey (top). A flock of mallards raft on a dredged channel in another developed waterfront on the New Jersey coast (bottom). (Photos courtesy of W. K. Figley and Elizabeth D. Bolen.)

encourage development of marsh vegetation and be safer for children; (2) include water depths of not more than 61 cm for 25–50 percent of the surface area, with about 50–75 percent at a depth not less than 1.1–1.2 m; (3) achieve a 50:50 ratio of emergent vegetation to open water; (4) construct small islands on larger impoundments and establish vegetation suitable for nesting cover; (5) select designs that permit regulation of water levels, including complete drainage; and (6) locate permanent impoundments near existing wetlands whenever possible, which will enhance the wildlife values of the stormwater basins. Such designs seem crucial because urbanization produces significant changes in the richness of waterbird communities (Bezzel 1985).

A growing concern for every citizen is the needs of the aging—none of us escapes the process. The world of nature is as fascinating to the old as to the young. Elderly people are enrolling en masse in special

classes and workshops, and as they form an enlarging block of voters, training in the ecological sciences proves not only popular but enlightening for a responsible, tax-paying citizenry. Such classes of course need not be for the elderly alone, but the mechanisms for reaching this identifiable audience often are already in place in urban communities.

Civic clubs are always looking for speakers, and an urban wildlife specialist can become a welcome guest, perhaps stimulating sponsorship for a wildlife project (see Proulx 1986). Rapid technological advances in mass communications have produced a wealth of informative media available for presentations; many of these are colorful and dynamic portrayals of wildlife. Publishers of newspapers are aware of the public's interest in wildlife and welcome material appropriate to their readership. Many papers print a regular column featuring wildlife and natural history.

This discussion by no means fully treats the opportunities available for wildlife educational programs in urban communities. For example, on-site educational programs for visitors to urban duck ponds offer ideal means for shaping the attitudes of urbanites toward wildlife and other forms of natural resource management (Hardin and VanDruff 1978). We do emphasize, however, that the modern wildlife biologist needs training in communications as well as in the sciences (see Gilbert 1971). Speaking well in an orderly and coherent fashion, having a familiarity with visual media, and writing effectively for a defined audience are basic requirements for many professions, but nowhere do they seem more essential than for the biologist entering the realm of urban wildlife management.

URBAN WILDLIFE AS PESTS

The presence of wildlife in urban or suburban environments can pose difficulties as well as benefits. As we have seen, the variety of wildlife occurring in cities often exceeds what might be expected. Many animals are more adaptable than supposed to some form of association with dense human habitation (Fig. 17-12). Raccoons and opossums are clear examples, but deer and coyotes (and often "coydogs") are among the larger species sometimes adapting to urban life. In Florida, a population of feral swine (*Sus scrofa*) living in a rapidly expanding urban area uprooted natural vegetation in an ecological preserve and damaged the greens and fairways on a golf course (Brown 1985). Night hunting at bait stations

Figure 17-12. Raccoons often become nuisances in residential areas, where they fare on readily available sources of food and cover. As shown here, a raccoon has raided a bowl of pet food in a suburban garage (left), eventually leading to its capture in a backyard patio (right). (Photos courtesy of Texas Rodent and Predatory Animal Control Service and Patricia A. Chamberlain.)

proved an efficient means of eliminating the swine. Bats, squirrels, snakes, and many kinds of birds regularly are subjects of human ire or fear. Exotic birds such as starlings and house sparrows are well-known nuisances. Many species damage property or pose threats to human health. Weber (1979) reported that pigeons can transmit more than 30 diseases to humans (among them a bacterial malady resembling appendicitis) and another 10 to domestic animals. Raccoons pose a serious threat in the transmission of rabies on the eastern seaboard, including Washington, D.C. (Beck 1984; Lipske 1985). So too much of a "good thing" or, in the case of animal-borne diseases, even a little of a "bad thing" in urban settings quickly attracts attention and calls for action. However, the public may not always speak in unison about the ways and means in which urban wildlife is managed.

Starlings and pigeons are common targets of pest control in cities. An array of measures is used to rid or at least reduce urban populations of both species. Trapping is time-consuming and usually removes only a few birds from much larger residual populations. A wetting agent, PA-14, is the only chemical registered with the Environmental Protection Agency for the lethal control of blackbirds and starlings. When sprayed on roosting birds, PA-14 permits water to penetrate and saturate the feathers, thus destroying the insulative properties of the plumage and causing fatal hypothermia at low ambient temperatures (Stickley et al. 1986). Poisons can be effective but often trigger objections from citizens sympathetic to

animal rights or concerned about harm to nontarget organisms. However, some poisoning programs are relatively selective. One method uses poisons placed on window ledges where pigeons roost; special perches sometimes are employed to administer the poison. When the birds alight on the perch, they absorb the poison on contact. Other methods are designed to poison only a few birds, relying on the erratic behavior of the treated individuals to scare away the remainder of the flock (see Chapter 13).

Blokpoel and Tessier (1984) strung overhead wires and monofilament lines as a means of excluding ring-billed gulls (*Larus delawarensis*) from public places in Toronto. The technique effectively eliminated gulls from the sites but did not deter pigeons. *Biosonics* is another nonlethal means for controlling urban birds while minimizing public outcries or fears about misdirected poisoning (Frings and Jumber 1954). Tapes of distress or alarm calls of the target species are recorded and played at intervals near roosting sites. In theory, the birds respond to the alarm and disperse from the roost. In practice, many birds may return, especially when a few individuals remain in place and recall the dispersed flock. The technique may be more effective if the unfrightened, residual birds are removed by other means before they can influence their frightened counterparts. Block (1966) temporarily reduced an urban population of starlings from 10,000 birds to a few hundred using tape-recorded distress calls. Johnson et al. (1985) recommended that a combination of sounds or stimuli be used for dispersing unwanted birds instead of using only a single stimulus.

Physical barriers retarding the presence of unwanted animals often are overlooked in the design of urban buildings. Open beams in airport hangars commonly offer attractive roosting sites for pigeons and other birds hazardous to aircraft. Wide window ledges and decorations on the outside of many buildings likewise provide nesting or roosting sites for starlings, house sparrows, or pigeons. Attempts to modify existing perches on buildings generally produce limited results, again indicating that the original building design remains foremost as a method of displacing unwanted, urban-adapted birds (Fig. 17-13). Architectural styles indeed have played a key role in regulating the nesting habitat of starlings and house sparrows (Geis 1974). These birds found ample nesting sites under unboxed eaves or where widely spaced louvered vents were left unscreened. Almost any type of structural opening on the exterior of residential structures attracts sparrows and starlings (Fig. 17-14). Conversely, where the structures were modified, indexes of starling and house sparrow populations were lower, whereas those of other birds were higher (Table 17-2).

The ultimate means of reducing established urban pests likely revolves around some type of species-specific, biologically based methodology. Further refinement of biosonic methods remains one interesting avenue for controlling starlings and other urbanized birds. Host-specific pathogens may be another. Sex attractants and other types of "chemical messengers" known as *pheromones* also have potential applications for controlling rodents—and perhaps other vertebrates—just as they have been successful with insects (Christiansen 1976). Such methods, when coupled with proper sanitation and construction, encourage minimal disruptions to urban environments.

Whereas control of vermin infesting cities seldom lacks public support, measures directed toward other animals may be viewed somewhat more skeptically. Urbanites, in particular, have more emotional attachment for wildlife than their rural counterparts (Kellert 1976a,b). Human compassion for some animals thus poses difficult public relations issues for biologists charged with controlling urban pests.

An analysis of more than 7500 complaints made by residents in 12 Texas cities indicated a range of concerns about urban wildlife (Table 17-3). Structural

Figure 17-13. Attempts to discourage pigeons and other birds from perching on window ledges generally are unsuccessful when the modifications are not part of the original building design. As shown here, pigeons were not inhibited by a wire thicket laid on the window ledge after the building was finished and, in fact, the birds used the wire as support of their nests (left). Far better results are obtained when buildings are designed and constructed with features such as steeply sloping window ledges (right). (Photos courtesy of Patricia A. Chamberlain and J. L. Stair.)

damage was foremost in this regard, including the damage resulting from skunks burrowing under houses, raccoons and opossums denning in attics, and squirrels chewing internal wiring. Some animals accumulate combustible materials inside dwellings, presenting unwelcome fire hazards. Analyses from other geographical regions, if available, undoubtedly would show a similar range of damage to homes and businesses. Any list of troublesome species probably would be dominated by Old World rodents, but other animals on such a list will vary from region to region (Table 17-4). The range of armadillos (*Dasypus novemcinctus*), for example, generally is limited by their intolerance to cold temperatures but, even so,

Figure 17-14. House sparrows, starlings, and pigeons take advantage of buildings with unscreened louvers, vents, or other open structures suitable for nesting and roosting. Note the accumulation of nesting materials in open tiles above the house sparrow (left). A detached drainpipe offered a starling access to shelter for a nest under eaves (right). (Photos courtesy of Patricia A. Chamberlain and Elizabeth D. Bolen.)

TABLE 17-2. Population Indexes for Starlings, House Sparrows, and Other Birds in Relation to Construction of Homes, Townhouses, and Apartment Buildings

Site	Population Index		
	House Sparrows	*Starlings*	*Other Birds*
Homes with trees:			
Modified construction[a]	0.12	0.29	3.18
Unmodified construction	1.78	1.52	2.63
Townhouses near woods:			
Modified construction	1.54	3.64	4.00
Unmodified construction	5.23	5.50	0.50
Apartments:			
Modified construction	0.12	0.12	1.25
Unmodified construction	7.25	4.50	0.75

Source: Geis (1974).

[a]Modified construction inhibited nesting starlings and house sparrows, whereas open eaves and lattice work (unmodified construction) provided attractive nesting places and, in turn, bolstered populations of these birds. Note that where starlings and house sparrows are abundant other bird numbers are low.

TABLE 17-3. Distribution of Complaints Concerning Wildlife Damage Made by Residents of 12 Cities in Texas[a]

Type of Damage	Number of Complaints	Percent
Structural	4953	65
Yard, plants, trees	1966	26
Personal property	252	3
Garden (vegetables and fruits)	225	3
Pets and livestock	113	1
Electrical, plumbing, phone, or other utilities	88	1
Personal injury	84	1
Total	7681	100

Source: Chamberlain et al. (1982).

[a]Data are based on claims of actual or potential damage including fear of attack.

TABLE 17-4. Animals Named as Damaging or Threatening Property or Human Safety by Residents of 12 Texas Cities

Common Name or Group	Number of Complaints	Percent
Rats and mice[a]	3282	42.7
Raccoons	834	10.9
Squirrels	753	9.8
Skunks	514	6.7
Pocket gophers	474	6.2
Opossums	367	4.8
Armadillos	356	4.6
Birds	302	3.9
Moles	193	2.5
Bats	189	2.5
Snakes	169	2.2
Other[b]	248	3.2
Total	7681	100.0

Source: Chamberlain et al. (1982).

[a]Category includes 98 percent Old World rats and mice and 2 percent native species.

[b]Among these, coyotes, beavers (*Castor canadensis*), deer, nutria (*Myocastor coypus*), prairie dogs (*Cynomys* spp.), foxes, rabbits, and ground squirrels (*Spermophilus* spp.) were subjects of complaints made by urban residents.

their distribution has expanded in the southeastern United States (Humphrey 1974). Armadillos are linked with several diseases of concern to humans, including leprosy. In urban and suburban settings, armadillos dig in gardens, cemeteries, golf courses, athletic fields, and parks, thereby initiating management for limiting what otherwise is regarded as a delightful and innocuous species (Chamberlain 1980).

In Australia, a free-flying population of sacred ibis (*Threskiornis aethiopica*) at a wildlife sanctuary increased dramatically when the birds adapted to urban conditions (Underwood 1994). The birds began to scavenge food from picnic areas, rubbish bins, and dumps outside the sanctuary. Additionally, the breeding biology of the urbanized ibis changed in significant ways: a greatly lengthened nesting season; atypical nesting sites, including trees near picnic areas far from water; increased number of successful nests per adult per season; and greater production per nest (1.3 chicks compared with about 0.6 elsewhere). These conditions soon produced an unusually large ibis population, which in turn produced problems such as excessive fouling under roosting trees, offensive odors, harassment of visitors for handouts, dead vegetation, and a potential disease hazard. Initial attempts at relocating some birds to a site 160 km distant failed when most of the birds later returned. Other management strategies therefore were employed, including restricting the availability of food both inside and outside the sanctuary, harassing ibis nesting at undesirable sites, removing nests and eggs, disrupting roosting ibis with spotlights, and substituting artificial eggs in active nests (the latter prolongs incubation and so precludes further nesting attempts). These measures brought an immediate response. Many birds dispersed to other areas, and for those that remained, both the length of the breeding season and the rate of nesting success returned to normal. As a result, the breeding population dropped from 900 to about 100 birds. The management program thus reduced the troublesome aspects of an overabundant but otherwise highly attractive urban wildlife resource.

Ironically, the success of management programs for restoring Canada goose (*Branta canadensis*) populations in North America fostered new concerns. Expanding populations of geese now burden city parks, golf courses, and residential areas (Fig. 17-15). Canada geese occurred on 42 percent of the eastern golf courses surveyed by Conover and Chasko (1985) and, of these, the birds were considered as nuisances on the majority (62 percent). The survey also determined that the average flock in the mid-Atlantic states averaged 250 geese per golf course, of which a high proportion remained during the summer months (i.e., many nuisance problems stemmed from resident flocks). More than half of the difficulties on golf courses originated within recent years, suggesting that problems may be increasing. Nelson and Oetting (1981) also described the magnitude of this latter-day

Figure 17-15.   Canada goose populations in urban environments often attract favorable public reactions (left); yet continued expansion of urban goose populations poses a number of difficulties for wildlife managers. Golf courses are one of the sites where geese are not always welcome (right). (Photos courtesy of H. K. Nelson and Elizabeth D. Bolen.)

management problem. All told, several hundred thousand Canada geese can be considered as urban, and the number is increasing. Sizable flocks occur in Seattle, Toronto, Cleveland, Chicago, Boston, Wilmington, Denver, Indianapolis, Nashville, and Minneapolis, just to name some of the more prominent sites. In Minnesota, the urban Canada goose flock is estimated at 30,000–40,000 birds, excluding another 25,000 migratory birds. Newspapers often feature photographs of broods parading within city limits. In some cases, news about urban geese is not always welcome, as when city water supplies are fouled or when the birds become hazards at airports. In Maryland, a frustrated golfer clubbed to death a Canada goose with his putter when the bird interfered with his game. Even small populations of urban geese serve as decoys for larger numbers of wild birds, with the result that large flocks can build quickly. Furthermore, diseases may be transmitted more readily when migratory and nonmigratory geese mix in highly modified urban environments. Uncontrolled feeding with such nutritionally deficient foods as bread also may predispose the birds to contagious diseases.

Management of urban goose populations still is in its infancy, however. One difficulty, mentioned earlier, is the polarization of public attitudes. For the most part, legal hunting is not possible and the birds remain free of shooting pressure so long as they remain within city limits. One also can imagine the controversies developing when eggs and nests are destroyed as means of controlling goose numbers in cities. Aversive conditioning holds some promise as an effective, yet acceptable method for managing nuisance geese. Conover (1985) sprayed grass with methiocarb, a nonlethal chemical that sickens geese grazing on the turf. The results indicated that nuisance geese reduced their use of treated areas by 71 percent for 8 weeks after the site was sprayed. About 1 ha of turf can be treated at 3 kg/ha for $110, a cost acceptable to many golf course managers (see also Conover and Chasko 1985). Until methiocarb gains clearance from the Environmental Protection Agency, however, state and federal biologists stage goose "round ups" and relocate the birds in more suitable environments. These are costly operations that often may offer only a temporary solution, especially when part or all of the captured birds are migratory and later return. Cooper (1987) demonstrated that adult females often return to the original capture site, whereas young birds may be moved shorter distances (e.g., 32 km) from the capture site without significant returns.

Guidelines currently set forth for managing urban Canada geese in the northeastern United States focus on relocating troublesome birds (U.S. Fish and Wildlife Service 1980). These recommendations include (1) releasing birds a minimum of 800 km from the capture site, preferably to the south and outside of watersheds where the geese originated; (2) releasing geese where hunting is permitted or at least where hunting pressure offers some means of population regulation (potentials for crop damage on private land should be minimized when selecting release sites); (3) examining trapped geese for disease, malnourishment, or irregular plumage, and culling affected birds before they are relocated (with the states donating and receiving captured birds sharing this responsibility); and (4) sterilizing all birds released on areas where hunting does not afford a feasible measure of control. Eventually, however, translocation will be self-limiting when states and provinces accepting geese fulfill their stocking goals (Cooper 1987). Translocation guidelines also stress public education. The public should be informed that urbanized geese are subsidized artificially and often are, or become, distinct from the migratory populations of far higher esteem as a natural resource. Artificial feeding should be discouraged, and when possible, urban habitats should be manipulated with hedgerows or other structures that reduce their attractiveness to geese.

Cooper and Keefe (1997) recently compared various means of dealing with troublesome urban geese. Fencing and harassment with dogs were the only effective short-term methods of reducing goose damage, but these methods merely redistribute the birds and cause new problems elsewhere. In contrast, sport hunting, egg destruction, and relocation can limit population growth on a long-term basis, but each of these has serious drawbacks (e.g., limited application and/or excessive cost, as well the refusal of states to accept any more birds). Trapping and processing urban geese for human food may be an economically feasible alternative, however, as shown by a pilot pro-

gram in which goose carcasses were readily accepted as ground meat, boneless breast, or whole-bird products at food-distribution centers in Minneapolis and St. Paul, Minnesota.

SUMMARY

Until recently, little thought has been directed to urban wildlife; yet most people in the United States live in metropolitan areas where human associations with nature seem especially valuable. Future support of all wildlife management may depend, in part, on increasing human–wildlife interactions in urban environments.

Several species of wildlife are well adapted to urban habitats, whereas others fare well in open areas associated with human activities. Cemeteries, for example, often harbor well-developed wildlife communities where breeding and predation take place amid dense human settlement. However, urbanization typically promotes monocultures, and hence relatively large populations of few species may characterize urban landscapes of pavement and manicured lawns. Starlings and house sparrows are particularly abundant in urbanized environments. With planning and management, however, many urban settings can be developed into diversified habitats where a larger range of native wildlife might flourish.

The goals and activities of urban wildlife management vary with the degree of urbanization. Components of the urban complex include the rural-urban interface, suburbia, and metropolitan centers, each having different management potentials. Managers must use creative means to incorporate wildlife habitat into a multiple-use scheme compatible with other urban activities. Furthermore, well-planned educational programs should complement urban management. City residents need to understand ecological concepts governing wildlife in their parks and backyards and in rural and wilderness environments. Managers also should be prepared to control nuisance or hazardous wildlife in ways that meet with public acceptance.

CHAPTER 18

EXOTIC WILDLIFE

It may have begun with Noah, but, wherever it started, the whole idea of rearranging the earth's wild creatures still seems irresistible. Man, the supreme meddler, has never been quite satisfied with the world as he found it.

George Laycock (1966)

A popular television commercial once admonished, "It's not nice to fool Mother Nature." Nonetheless, humans have long considered the rearrangement of wildlife as their prerogative for improving the landscape and the order of nature. In particular, birds, mammals, and fishes have been introduced for aesthetic, economic, and recreational reasons, but an awesome variety of invertebrates, herpetofauna, and plants also have found their way to other continents at the hands of people. Such designs have not been uniformly wise, nor has "Mother Nature" often been fooled.

Some basic distinctions should be made concerning exotics—those organisms introduced into places they have not previously occupied. Many North American garden plants, crops, and livestock are exotic in origin and, although these indeed may have caused some ecological disturbances, the wildlife manager's concern with exotics lies primarily with those species that are intended to establish wild populations for recreational purposes.

Accidental introductions of exotic organisms—those escaping from ships, for example—are another but sometimes related matter. Some of these, such as

Old World rats (*Rattus* spp.) and imported fire ants (*Solenopsis invicta*), represent serious pests that directly or indirectly affect native wildlife. The unfortunate result of exotic rats escaping from ships onto oceanic islands often was compounded when yet another exotic, the mongoose (*Herpestes auropunctatus*), was released in anticipation of eliminating the foreign rodents. Such hopes were short-lived. Mongooses actually brought additional problems to island faunas. Along with other exotic mammals, they contributed to the near-elimination of Hawaiian geese (*Branta sandvicensis*) (Kear and Berger 1980). In fact, all mammals now in the Hawaiian chain were introduced except for one species of bat (Kramer 1971). On Caribbean islands, mongooses were introduced to control snakes as well as rats. The snakes were only partially controlled, but after depleting the more easily available rats (in sugarcane fields), mongooses preyed on native mammals, ground-dwelling birds, lizards, and other island fauna, nearly eliminating several species. Within 20 years of its introduction into the Caribbean, the mongoose was regarded as the worst of all pests; some evidence suggests they also carried rabies (deVos et al. 1956).

Bird populations on Guam were ravaged after brown tree snakes (*Boiga irregularis*) reached the island in the early 1950s, probably as stowaways on ships carrying military cargo (Fritts 1988; Engbring and Fritts 1988). The snakes—which may exceed 2 m in length—occurred islandwide by 1970 and, at peak populations, approached densities of 100 per ha (Rodda et al. 1992). In the course of this irruption, the snakes completely eliminated several species of once-common birds, including some found nowhere else (e.g., Guam rail, *Rallus owstoni;* bridled white-eye, *Zosterops conspicillatus*; and Guam flycatcher, *Myiagra freycineti*) (Savidge 1987). When other species of birds were no longer abundant, the snakes began to prey on lizards, again causing marked reductions or extirpations of several species (Rodda and Fritts 1992). Guam's native bats also fell prey to the exotic snakes (Wiles 1987). Brown tree snakes are mildly venomous, attack when threatened, and commonly invade human dwellings; snakebites now are widely reported on Guam (Fritts et al. 1990). Moreover, Guam has experienced numerous and expensive power outages as a result of brown tree snakes climbing on electrical lines (Fritts et al. 1987). If these snakes reach other areas, particularly Hawaii, the unfortunate story at Guam almost surely will be repeated. In 1991, the Brown Tree Snake Control Act became law and directs the secretary of agriculture to initiate a program of suppression, control, and eradication.

The zebra mussel (*Dreissena polymorpha*), a mollusk from southern Russia, escaped into the Great Lakes from the water ballast of a foreign ship probably in 1985 or 1986 and thereafter spread rapidly into the watersheds of the Hudson, Tennessee, Ohio, Mississippi, St. Lawrence, and Arkansas rivers among others (Ludyanskiy et al. 1993). The immense populations of these small bivalves already have caused serious economic and ecological impacts. They befoul marine equipment and close industrial and municipal pipelines; densities as high as 750,000 zebra mussels per m² were reported from the intake canal of a power plant in Detroit. Zebra mussels filter algae and other microorganisms so effectively that water clarity is remarkably improved, but this action also interrupts food chains and removes chlorophyll from aquatic systems. In effect, they reduce primary productivity, thereby disrupting the normal course of energy flow in aquatic systems. Because they colonize virtually any hard surface underwater, zebra mussels attach themselves to the shells of other species and thereby quickly overpower native mollusks.

Zebra mussels also absorb and concentrate contaminants. Based on field data and experiments conducted in Holland, cadmium and pesticides were transferred from zebra mussels to tufted ducks (*Aythya fuligula*) in such quantities that the embryos in their eggs showed teratogenic effects (De Kock and Bowmer 1993). Hence, a similar risk is present in North America for ducks and other wildlife that consume significant amounts of zebra mussels. Mitchell and Carlson (1993) recorded an average of 260 zebra mussels in the upper gastrointestinal tracks of 19 lesser scaup feeding in Lake Michigan; the maximum number was 987. Indeed, zebra mussels have become a primary food in the diets of several species of waterfowl in the lower Great Lakes and now serve as an effective means of transferring persistent organic contaminants to higher trophic levels (Mazak et al. 1997). The availability of zebra mussels in Lake Erie also has altered the migratory movements of lesser scaup (*A. affinis*) and other species of pochards (Wormington and Leach 1992). In all likelihood, zebra mussels will occur in almost all parts of the United States and southern Canada by the 21st century; yet at present there is little indication that any safe and effective method of control is forthcoming (Ludyanskiy et al. 1993).

Imported fire ants also produced harmful responses following their escape into the United States. In this case, massive amounts of Heptachlor and other pesticides were sprayed to control these stinging pests, in part, because biologists reasoned that the ants might threaten bobwhites (*Colinus virginianus*). At least two mechanisms may be involved: (a) the ants may kill hatching chicks and (b) the ants may reduce the abundance of invertebrate foods seasonally required by bobwhite (Allen et al. 1995). Whereas Heptachlor itself apparently harmed bobwhite populations before it was banned, recent experimental evidence indicates that imported fire ants indeed reduce the survival of quail chicks as well as cause weight loss in the young birds (Giuliano et al. 1996). Mortality increased when chicks were exposed to as few as 50 ants for 60 seconds. Imported fire ants accordingly may adversely impact bobwhite populations wherever these species come into contact (see also Allen et al. 1993, 1994).

Additional evidence suggests that imported fire ants also may reduce recruitment in white-tailed deer populations (Allen et al. 1997). Where fire ants were controlled with chemicals, recruitment in the deer population—measured by doe:fawn ratios—was about twice as high in comparison with sites where the ants were left untreated. The reduction in fire ants on the treated areas presumably lessened contact between newborn fawns and fire ants, which are attracted to mucous membranes such as the eyes and mouth. A newborn fawn's natural defense is to freeze in place, but this behavior in fact may increase attacks from fire ants. Moreover, when forced into repeated movements to new bedding sites, fawns may experience greater predation from coyotes (*Canis latrans*). Because does remain faithful to fawning sites, they do not avoid these areas even when they are heavily infested with fire ants; that is, deer have not had adequate time to develop avoidance behavior to fire ants. Overall, the results suggest that fire ants contribute to both direct and indirect mortality of newborn fawns.

Imported fire ants exhibit two types of mating behavior: monogynous (single queen) and polygynous (multiple queens). These behaviors in turn translate into differences in densities. Monogynous colonies are territorial and their densities are limited by territorial behavior (i.e., less than 100 mounds per ha). Conversely, polygynous colonies lack intraspecific territoriality and reach much greater densities (e.g., up to 1400 mounds per ha, each with 200,000 or more individuals). Habitats infested with polygynous colonies thus pose far greater threats to wildlife, and their continued spread in Texas and other areas is a cause of concern (Porter et al. 1991). Imported fire ants currently infest more than 101 million ha in the United States and are expected to cover almost 25 percent of the United States before reaching their range limits (see Vinson and Sorensen 1986).

Nonetheless, a favorable degree of biological control was achieved in Australia when the rapid spread of exotic cactus (*Opuntia* sp.) was curbed by importing moths (*Cactoblastis cactorum*) whose larvae severely damaged the plants. Some conservation agencies anticipate that grass carp (*Ctenopharyngodon idella*) from eastern Asia will reduce dense growth of noxious aquatic vegetation—including some exotic species such as hydrilla (*Hydrilla hydrilla*) and water hyacinth (*Eichhornia crassipes*)—choking recreational waters in the southern United States (Mitzner 1978; Kilgen 1978; Pierce 1983). The impact of grass carp introductions remains controversial, as does the desirability of hydrilla (see Johnson and Montalbano 1987). However, the recent development of a sterile hybrid may increase the use of grass carp as a management tool for controlling unwanted aquatic vegetation (Pierce 1983). Another

carp (*Cyprinus carpio*), widely introduced into North America during the 19th century, subsequently reduced native vegetation valuable as waterfowl food (Threinen and Helm 1954; Tyron 1954; Robel 1961a) and disrupted the faunal balance of many aquatic systems.

In still other instances, organisms originally intended for domestic uses (e.g., pets, beasts of burden, etc.) escaped or were released (McKnight 1964). Burros (*Equus asinus*) have become troublesome exotics as their numbers have increased in the western United States. Now well established in large feral populations, burros (and wild horses) have been the focus of both passionate controversy and practical resource management (see Chapters 14 and 22).

European mute swans (*Cygnus olor*) have escaped from captivity in Michigan, New England, and on Chesapeake Bay in Maryland (Reese 1975; see also O'Brien and Askins 1985). The population in Maryland, starting from one pair of mated swans escaping in 1962, is experiencing unrestricted growth—42 percent in the first 10 years after escaping—and the several hundred birds in the population likely will expand to nearly several thousand (Reese 1980). A large sedentary population of aggressive waterfowl thus may have a detrimental effect on both the ecology of Chesapeake Bay and the large numbers of native waterfowl overwintering in the same habitat (Reese 1975).

Only a few foreign species have successfully colonized North America without human assistance. Perhaps the most dramatic example is the cattle egret (*Bubulcus ibis*) that made its way to the New World from Africa late in the 19th century (Sprunt 1955). Cattle egrets reached the northeastern coast of South America, likely driven by an Atlantic storm, and then gradually dispersed northward, arriving in Florida by 1942. Once in the United States, they spread along the Atlantic and Gulf coasts and then extended into the interior, eventually reaching the West Coast (Fig. 18-1). Cattle egrets associate with grazing livestock and exploit a food resource of insect-rich pastures and fields, but the impact of their nesting habits on native wading-bird colonies has not yet been determined fully.

The status of today's pests scarcely was perceived when colorful plants or delightful songbirds found new homes in North America. Several species of exotic fishes jeopardize the integrity of aquatic systems in many parts of the world (see Courtenay and Stauffer 1984). Walking catfish (*Clarias batrachus*), now established in Florida, are one of the more

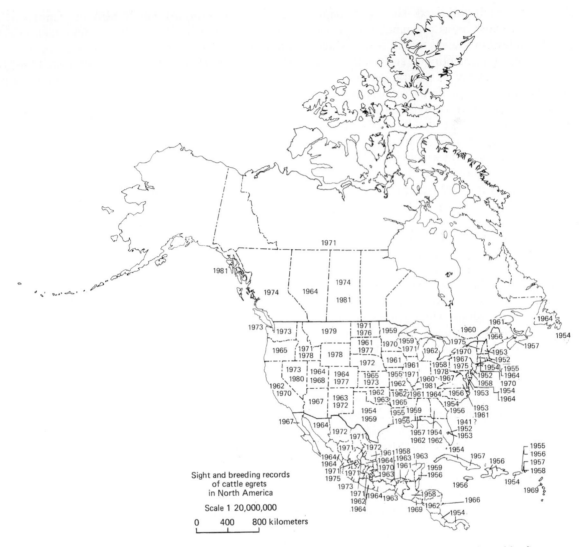

Figure 18-1. The advance of cattle egrets across North and Central America is indicated by first sightings (upper date or single date) and first breeding records (lower date or those marked with stars). Some records from Mexico represent band recoveries. (From Telfair 1983.)

publicized exotics in North America. As its name implies, this species expands its range by "walking" overland from one body of water to another, thereby thwarting local efforts to confine its distribution (Fig. 18-2). In still other cases, exotic animals were imported and released—usually after great expense—as a means of improving hunting or fishing. Introduction of such species is virtually worldwide, having been attempted whenever outdoor recreation seemingly needed improvement. There are now Canada geese (*Branta canadensis*) in New Zealand, muskrats (*Ondatra zibethicus*) in Europe, rainbow trout (*Oncorhyncus mykiss*) in Chile,

raccoons (*Procyon lotor*) in the former Soviet Union and, as we shall see, a global collection of exotic game in North America.

THE CASE FOR EXOTIC GAME

Proponents of exotic introductions sometimes argue that foreign species effectively fill so-called vacant niches. They believe it is as prudent to introduce game from other lands as it is to manage native species for hunting and fishing. Where native species are absent, perhaps an exotic can be found instead.

Figure 18-2. Walking catfish are difficult to control because of their ability to move overland between aquatic systems. This exotic species now infests the St. John's River Watershed, the largest in Florida. (Photo courtesy of P. Shafland, Florida Game and Fresh Water Fish Commission.)

In this view, any foreign opportunities for outdoor recreation should not be ignored just because a native species is not already fulfilling this function. Chukar partridges (*Alectoris chukar*) already are established on the arid, rocky slopes of Utah and Nevada, presenting challenging hunting in habitats where none existed previously. Also, there may be places where native game, although present, is not plentiful enough for the demands of hunters. Relatively few North American waterfowl reside in the southeastern United States, even though Florida and other states in this region have large wetland environments. Lacking either abundance or variety of native waterfowl—and therefore, duck hunting—the southeastern states have considered several species of South American ducks as exotic introductions (Weller 1969).

Foreign species also might replace those native animals that have not coped well with environmental changes. Ring-necked pheasants (*Phasianus colchicus*) clearly are better adapted to farming than greater prairie chickens (*Tympanuchus cupido*). Heavy grazing or logging pressure also may present regional situations where an exotic might provide better hunting than a native species ill-adapted to these or other manufactured alterations. Some exotic fishes may survive better than native species in reservoirs, thermally heated waters, or otherwise modified aquatic systems. In sum, three habitat situations arise for considering introductions of exotic game: (1) natural habitats lacking any suitable game, (2) natural habitats with limited abundance of native game, and (3) modified habitats no longer producing sufficient native game and/or those where habitat restoration is not practical. Bump (1963a) found these situations in about 20 percent of the United States.

Financial rewards are another reason exotic game is released, although this incentive usually is limited to lands held in private ownership. Schreiner (1968) and later Attebury et al. (1977) described the financial returns for trophy-class exotics stocked on game ranches. Gross profits varied from about 40 percent to 80 percent, depending on the species. However, the average return on capital investment was less than 2 percent if land costs were included in the economic analysis. Without land costs, the average return on capital reached 10 percent. The addition of exotics increased returns by 20 percent where native wildlife also was hunted for economic gains.

Special situations also seem favorable for exotics. Some species of exotic fishes may control noxious vegetation (Stanley and Lewis 1978); others seem more responsive to intensive management for commercial enterprises (e.g., aquaculture).

Exotic introductions, as we shall see, generally parallel the movements of humans throughout history. But the major impetus for releasing foreign game in the United States came after World War II, when the sales of hunting licenses soared. Soldier-hunters added to the fervor for exotic game when they returned home with tales of animals they had encountered on foreign soil (Bump 1951). As pressures mounted for more outdoor recreation, it became clear that attempts to introduce potential game species on a "hit-or-miss" basis were ill-advised and should be administered by some central authority. Thus, in 1948, a State-Federal Cooperative Foreign Game Program was established by the U.S. Fish and Wildlife Service. The program was guided by three objectives (Bump 1968):

1. To provide a fund of ecological and life history data for individuals or agencies wishing to evaluate exotic animals for release in the United States.
2. To discourage exotic introductions when these data suggested that releases might be unwise.
3. To fill vacant or understocked habitats with foreign species as an alternate course of action following appropriate testing and trial introductions.

A PATTERN OF SUCCESS?

The history of exotic wildlife in North America does not precisely record the early attempts that our ancestors made to import wildlife. Ocean traffic during the colonial period did not enter ports regulated by quarantine stations. Nonetheless, Benjamin Franklin's son-in-law, Richard Bache, was one of the first to introduce exotic wildlife in North America. Bache released European game birds in New Jersey late in the 18th century, no doubt hoping to duplicate the gentlemanly estate-shooting of Western Europe (Phillips 1928). His efforts failed, and exotic game waited another century for its place in North America.

The centerpiece of exotic game in North America clearly remains the ring-necked pheasant. Originally from Asia, pheasants had long been established in Western Europe as sporting birds; they perhaps reached Great Britain in the baggage of Roman legions (Delacour 1951). It logically followed, given the longstanding success and excitement of pheasant hunting in England, that immigrants would want to experience pheasant shooting in the American colonies (the birds are sometimes, although erroneously, known as "English" ring-necked pheasants). Repeated attempts at introduction failed, however, and not until 1881 were pheasants successfully established—not on the East Coast nor with English stock, but in Oregon with pheasants transplanted directly from China.

Credit for the nucleus of the North American pheasant population rests with Owen N. Denny. As a U.S. consul stationed in Shanghai, Judge Denny shipped pheasants for release in Oregon's Willamette Valley. His first transplant failed, but a second effort in the same year established a population that 10 years later enabled an annual harvest estimated at 50,000 birds (Laycock 1966). The rest is sporting history. Pheasants were rapidly transplanted into several adjoining states, eventually establishing strongholds in the midwestern states of South Dakota and Nebraska, where millions were shot annually. Paradoxically, the ring-necked pheasant is the "state bird" of South Dakota.

The presumed success of the ring-necked pheasant in North America has become the prime example of those justifying continued trials with exotic game animals. The species surely has provided millions of hunter-days of outdoor recreation and literally tons of table meat. Revenues generated from pheasant hunting also are immense, reaching millions of dollars annually. Some ethical considerations about exotic wildlife are mentioned later, but we should also examine the biological implications—real and potential—of the pheasant as a notable "success story."

Other galliforms, particularly prairie grouse, once occupied most of the range where pheasants later became established in large numbers. Among the grouse, prairie chickens are adapted to native grasslands for their food and cover requirements. With the advent of farming, the native grasslands steadily gave way to the plow and domestic grasses. Today, cereal grains of several types blanket yesteryear's plains and prairies, with the wheat and corn belts marking much of this conversion. It is doubtful that prairie chickens could have long maintained large populations in a regime so completely altered.

Nonetheless, some evidence of interspecific competition with pheasants suggests that prairie chickens do not cope effectively with the exotic birds. For example, pheasants physically dominate greater prairie chickens where the two species share common ranges. Persistent daily attacks eventually drive the native birds from their long-established booming grounds and cause pronounced reductions in the prairie chicken population (Sharp 1957). In 506 observations of booming grounds in Illinois, Vance and Westemeier (1979) recorded pheasants intruding on 104 occasions (21 percent), and of these, cock pheasants harassed prairie chickens 45 times (43 percent); the pheasants dominated the prairie chickens in 78 percent of the encounters. In one encounter, a single pheasant apparently dominated the entire booming ground and its population of 80 prairie chickens. The Illinois study also revealed instances of nest parasitism (i.e., pheasant eggs laid in prairie chicken nests) that led to reduced hatching success (24 percent for parasitized nests versus 51 percent for unparasitized nests) and, possibly, to increased rates of nest abandonment and predation.

Such strife seems especially critical where management efforts focus on saving the populations of prairie chickens still remaining within the established range of ring-necked pheasants. Reintroductions of prairie chickens, necessarily involving small numbers of birds, into otherwise suitable habitats may fail if pheasants subsequently share the same sites. On smaller booming grounds, pheasants easily disrupt the courtship activities of prairie chickens to the extent that breeding may be delayed or prevented; females are almost always driven from the booming grounds by intruding pheasants so that some hens may not mate (Vance and Westemeier 1979). Harger

(1956) also described details of an aggressive encounter between a cock pheasant and displaying prairie chickens that caused disruption of the chickens' booming grounds. Anderson (1969b) concluded that the similarities in breeding habitat and timing of courtship (both daily and seasonal) between pheasants and prairie chickens, coupled with the aggressive behavior of pheasants, would adversely affect prairie chickens in Wisconsin. Such contacts between these species would, of course, be exacerbated when pheasant densities were high and minimized when they were low. Continued stocking of pheasants within the remaining distribution of prairie chickens accordingly seems undesirable.

Grange (1948) offered another perspective about releasing pheasants in the range of prairie grouse; namely, he deplored the stocking of pheasants because of the incidental losses hunting imposed on grouse populations during the hunting season for pheasants.

There also may be reciprocal dynamics between populations of the exotic pheasant and native galliforms; as pheasants gain in density, the others wane. Sharp (1957) suggested that the invasion of pheasants in Nebraska brought about a drastic decline in prairie chickens that later was reversed when the pheasant population crashed. Errington (1945), in his classic study of bobwhites, stated that quail competed with pheasants on a bird-for-bird basis in winter. Even at low densities, pheasants competed significantly with larger populations of bobwhites.

Little evidence exists to support any premise that pheasants may be declining permanently in North America. In most instances, the unsuccessful nature of transplanted species eventually destined for failure becomes obvious in rather short periods of time. Large numbers of coturnix quail (*Coturnix coturnix*) from the Old World, for example, were released repeatedly in the United States without further evidence of their existence a year later (Labisky 1961). Their rapid disappearance occurred despite careful planning, study, and the birds' extremely high rate of reproduction (Stanford 1957). Three species of exotic partridges stocked for "put-and-take" management on shooting preserves also showed poor postrelease survival (Byers and Burger 1979). Such a short-lived existence indeed marked the initial introductions of pheasants, but thereafter additional releases of the transplanted birds eventually flourished. Today, large fluctuations in pheasant populations are commonplace, reflecting the changing agricultural practices

discussed in Chapter 13 (see Erickson and Wiebe 1973; Labisky 1976; Vandel and Linder 1981). In the recent past, irruptions in pheasant populations have been most prevalent only in areas with poor or marginal habitat where their numbers previously were quite low (Labisky 1975). At best, the matter remains speculative, but the pheasant as an exotic species may have passed its peak in North America.

REASONS FOR FAILURE

Bump (1951, 1963a, 1968) proposed that future attempts to introduce exotic game should be enlightened by failures of the past. The reasons vary in detail for each case but their general nature, summarized below, includes a range of ecological, physiological, and behavioral features.

1. *Failure to evaluate the characteristics of the exotic species.* This includes food habits, breeding biology, and similar species-specific features, as well as climatic and habitat requirements. In the past, introductions of exotics often proceeded without proper research or even in spite of known inadequacies.

2. *Failure to release sufficient numbers of individuals in good health to assure successful establishment of the exotic population.* Releases of too few animals, even those in good health, may be negated by mortality or reduced encounters for mating. For game birds, at least 200–300 wild-trapped individuals or about 1000 hand-reared individuals should be released in a single area for each of at least 3 years. Any number of animals released in poor physical condition obviously reduces the odds for success.

3. *Failure to condition hand-reared animals properly before they are released.* Many captive animals rapidly become domesticated and cannot cope with predators and other environmental limitations normally encountered after their release. Failure to avoid humans may be significant.

4. *Failure to manage the exotic population.* This includes the methods of release, the follow-up after release, and, when necessary, taking prompt action to correct mistakes. Species with gregarious habits should be released in units meeting their social requirements. Unless dedicated fieldwork follows the release of exotics, little evaluation of the program is possible. This commitment may be long-term, perhaps as much as 25 years, before the relative terms of success or failure can be assigned accurately.

Releases of foreign grouse in the United States between 1883 and 1950 failed for one or more reasons listed above. These included 23 trials of 4 species, but only one trial was based on any scientific background, and none of the attempts met the minimum requirements for an adequate test (Bump 1963b). Many of the reasons for failures with exotics also apply to reintroductions of native species. Lewis (1961) reported the failure of a stocking program designed to restore wild turkey (*Meleagris gallopavo*) populations in Missouri. During an 18-year period 14,000 turkeys were released, but because these were hand-reared stocks, the effort failed and was discontinued. Similarly, hand-reared turkeys introduced into Nebraska failed, either dying rapidly or becoming no more than "glorified barnyard fowls" (Suetsugu and Menzel 1963). Only when wild birds were trapped and released were turkey transplants successful in these states.

SOME CONCERNS ABOUT EXOTICS

Opposition to exotic wildlife focuses on a number of biological questions, among them the possibilities of hybridization and competition. Unfortunately, these possibilities are seldom known in advance.

Normally, native species of similar heritage and distribution (e.g., sympatric congeners) long ago developed behavioral or other means that maintain their respective genetic integrities. In unaltered settings, hybridization among native congeners is avoided largely by daily or seasonally competitive functions leading to one or more isolating mechanisms. Hybrids between native species remain more noteworthy for their rarity than for the possibility that such crosses might occur. However, if one of these species is introduced into a foreign fauna, its genetic compatibility with a local—but related—species still may function, and high rates of hybridization may result in its new environment (Fig. 18-3). Thus, mallards (*Anas platyrhynchos*) introduced into Australia now hybridize commonly with the native grey duck (*Anas superciliosa*) to the point that the native species has been swamped genetically by the exotic (Frith 1967). A similar situation has developed in New Zealand, where the proportion of pure grey ducks dropped to 4.5 percent of the total mallard–grey duck population, a level probably too low for maintenance of the species (Gillespie 1985a).

When ibex (*Capra ibex*) were extirpated in the High Tatra Mountains of Czechoslovakia late in the 19th century, new stocks were found to restore the

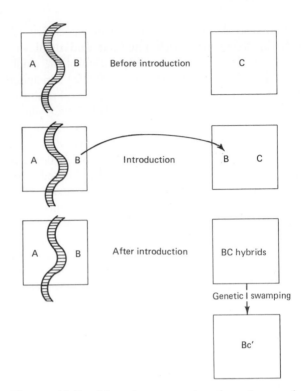

Figure 18-3. Schematic representation of genetic swamping after the introduction of an exotic species. In this example, A, B, and C are related species naturally occupying separate distributions. A and B, although sharing the same distribution, are ecologically and/or biologically maintained as species by one or more isolating mechanisms (wavy lines) and normally do not hybridize. C exists alone. However, when B is introduced into the distribution of C, the absence of long-established isolating mechanisms between B and C may lead to their frequent hybridization. In some instances, B may swamp C so that the genetic integrity of C is threatened, as indicated by the symbol Bc'.

original herd. Unfortunately, two related species (*C. hircus* and *C. nubiana*) from warmer climates in the Old World also were introduced in the following years. Turcek (1951) recorded the results interbreeding brought to the reestablished population of native ibex. As might be expected, the hybrids were of a different color, and they developed variable horn configurations. More unexpectedly, however, the crosses between the exotic and native species produced hybrids whose breeding season shifted to late summer. Normally, the native ibex breed in midwinter so that the young are born in April or May, when conditions are favorable for their survival. The two species, having been introduced from warmer climates, however, each had breeding schedules producing young in the winter. Under the influence of the *ibex* × *hircus* × *nubiana* crosses, the hybrid kids thereafter were born at a time when snowstorms and freezing weather claimed their lives generation after generation. Only four ibexlike animals (all females) survived.

Turcek (1951) also described the aftermath from the release of a Siberian race of roebuck (*Capreolus capreolus*) in Czechoslovakia. The larger Siberian animals crossed with the smaller race of native roebuck, but when the crosses involved females of the native race, the fetuses were so large that their birth killed the mothers. Similarly, after two races of elk (*Cervus elaphus*)—one from Asia and another from North America—were released to "freshen up" the bloodlines of a third race in Poland, the hybrids produced antlers with fewer points and smaller spreads than before (Lindemann 1956). Given these unexpected results, Polish game managers thereafter spent many years selectively weeding out animals with these characteristics. A genetic success apparently occurred, however, when sable (*Martes zibellina*) in western Siberia developed heavier and more valuable fur following the introduction of another race of the same species, but this remained the only improvement among many introductions of furbearers in Siberia (Lindemann 1956).

Competition under field conditions may assume a number of dimensions. Food is a resource obviously subject to competition, especially when it is limited (e.g., forage during a drought), but tree cavities or other breeding sites also may be preempted. The loss of natural cavities to nesting starlings (*Sturnus vulgaris*) and house sparrows (*Passer domesticus*)—both exotics—has markedly reduced bluebird (*Sialia sialis*) populations (Zeleny 1976). The competition may be highly aggressive; Gowaty (1984) determined that house sparrows will kill adult and nestling bluebirds and build nests over their bodies. Muscovy ducks (*Cairina moschata*) introduced into the range of hole-nesting waterfowl might displace species

such as wood ducks (*Aix sponsa*) or black-bellied whistling-ducks (*Dendrocygna autumnalis*) (Weller 1969; Bolen 1971). When equal numbers of sika deer (*Cervus nippon*) and white-tailed deer (*Odocoileus virginianus*) were confined together experimentally for 9 years, the greater dietary adaptability of the exotic sikas greatly outstripped the whitetail population. The original herd of 6 whitetails eventually fell to 3 animals, while the sika population grew to 63 individuals (Armstrong and Harmel 1981).

Sika deer also were released on the eastern shore of Maryland in 1916 (Flyger 1960). During the ensuing years, the sika herd increased and expanded its distribution at an average of 0.8 km/yr and thereafter was harvested in greater proportions than white-tailed deer in Maryland's best deer-hunting area (Table 18-1). Because of the decreasing density of white-tailed deer, the legal bag limit for sikas was increased in 1976—and the limit for white-tailed deer decreased—to curtail growth of the sika herd (Feldhamer et al. 1978). Elsewhere in Maryland, the percentage of sika in the annual deer harvest increased from 72 to nearly 83 percent in the 5-year period ending in 1982 (Keiper et al. 1984). Compared with white-tailed deer, sika deer forage on a greater variety of plants, including vegetation in fresh- and saltwater marshes, and are better food competitors in overbrowsed habitats (Keiper 1985). Additionally, Davidson and Crow (1983) suggested that differential susceptibilities to infectious pathogens may be important factors in the interactions between white-tailed and sika deer (see Davidson et al. 1985, 1987 for related studies).

In the western United States, the behavior of feral burros at water holes so greatly disturbs these sites that desert bighorn sheep (*Ovis canadensis*) may no longer frequent many of the sources previously

TABLE 18-1. Five-Year Harvest Data for Sika and White-Tailed Deer in Dorchester County, Maryland, 1973–77 (Data Refined for Seven Election Districts Where Sika Deer Are Most Numerous)

	Entire County			Seven Election Districts		
Year	Total Number of Deer Harvested	Sika (Percent)	White-tailed (Percent)	Total Number of Deer Harvested	Sika (Percent)	White-tailed (Percent)
1973	1503	16.3	83.7	973	24.8	75.2
1974	1495	21.3	78.7	883	33.6	66.4
1975	1259	27.3	72.7	828	39.6	60.4
1976[a]	1445	36.4	63.6	921	55.8	44.2
1977	1539	41.7	58.3	981	64.1	35.9

Source: Feldhamer et al. (1978).

[a]Increased legal bag limits for sika and decreased limits for white-tailed deer were initiated in 1976.

available to them (McKnight 1958). This is an example of *interference competition*, where the resource is not necessarily in short supply but becomes limited functionally because of interference with its utilization.

Marti (1993) described the plight of white-headed ducks (*Oxyura leucocephala*) in Spain, where they have faced various forms of competition from ruddy ducks (*O. jamaicensis*), a North American species that escaped from an aviary in England. The Spanish population of white-headed ducks—the only one in Western Europe—reached a low of just 22 birds in 1977 but slowly recovered as a result of stingent conservation measures. Wetland drainage and other human disturbances contributed to the threatened status of the species throughout its range, which includes other parts of Eurasia, but encroachment by ruddy ducks may represent the greatest threat to survival of the isolated population in Spain. The two species occupy similar niches; therefore, they compete for habitat in shallow wetlands with submerged vegetation and beds of reeds. Moreover, because male ruddy ducks are more aggressive than their Old World counterparts, they often acquire mates from the population of the native species. Hybrids from these matings are fertile, and the outlook is uncertain for maintaining the current stock of about 800 white-headed ducks in Spain. In England, the exotic population of ruddy ducks is increasing by about 10 percent a year and now numbers about 3500 birds. Ruddy ducks also continue expanding in other parts of Europe, where they now breed in the Netherlands, France, and Iceland.

Competition, of course, more often may favor a native species and lead to the demise of an exotic, but the "risk" is not of the same magnitude. When a native species is displaced by a successful exotic, its continued existence at former levels of abundance is far from certain, as with the eastern bluebird. Conversely, the small numbers of exotics represent only expendable segments of larger populations elsewhere; such species suffer little, if any, jeopardy if the introduction fails. Ironically, however, the exotic population of blackbuck antelope (*Antilope cervicapra*) now in North America may exceed the numbers remaining in its native range. Whereas the blackbuck antelope perhaps was once the most abundant big game animal on the Indian subcontinent, overhunting and conflicts with agriculture have reduced the large herds of thousands to scattered groups of 50–100 animals (Ramsey 1968).

Diseases and parasites represent other risks from imported wildlife, but these dangers have been reduced in recent times. In the United States, imported stocks now are held in quarantine and then are confined in zoos or other propagation facilities; only the offspring are released. These precautions apply to legally imported animals. However, they may be circumvented by unscrupulous purveyors of exotic wildlife.

Because many families of animals have wide representation throughout the world, it is likely that any disease or parasite introduced by exotics would find alternative hosts with at least as much susceptibility to the "new" pathogen (Parker 1968). Members of the deer family, for example, are found in Asia and Europe as well as North America, and mixing of these species easily might spread a pathogen to another relative. Domestic animals have even a greater degree of similarity throughout the world, a factor that immensely increases the odds for epizootics when livestock are moved without appropriate safeguards. As described in Chapter 8, some pathogens affect both wildlife and domestic animals. When cattle from Asia were introduced into Africa, they brought rinderpest ("cattle fever") that subsequently killed large numbers of Cape buffalo (*Syncerus caffer*) (deVos et al. 1956). In this and other cases, the host may be related, as with bighorn and domestic sheep or domestic and wild turkeys. In others, the taxonomic relationship is less close, as with cattle and deer.

Our knowledge of diseases often is incomplete, and introductions of exotics may further compound this situation. At one time, a bacterial disease then known as psittacosis ("parrot fever") was believed to be confined primarily to the parrot family and was of little danger when parakeets, cockatoos, and parrots were transported widely as pets. (These birds belong to the family Psittacidae; hence the former name for the disease.) Regrettably, this was untrue. More than 130 species of wild birds have been infected (Burkhart and Page 1971). Some species contract the disease, whereas others may be only carriers. Losses may be severe in poultry, and potentially all birds seem susceptible to some degree, as are mammals, including humans. Wobeser and Brand (1982) cited an outbreak where the human fatality rate reached 23 percent. Ironically, Wobeser (1981) contracted an acute case of what is now known as chlamydiosis while writing his acclaimed book on waterfowl diseases. Chlamydiosis sometimes is called ornithosis, again recognizing that the disease is more or less of universal occurrence in all birds, not just parrots.

Another twist of fate involved a disease of rainbow trout. The native range of rainbows occurred in the western United States, but when stocks were cultured for release into eastern rivers, they carried the pathogen causing furunculosis to eastern fish hatcheries. Whereas rainbows generally seemed immune to furunculosis, the disease proved devastating to brown trout (*Salmo trutta*) that were recently imported from Europe, as well as to eastern brook trout (*Salvelinus fontinalis*) (Bowen 1970). A gill parasite carried by a single sturgeon (*Acipenser stellatus*) that was released into the Aral Sea subsequently caused high mortality and depleted stocks of the native sturgeon (*A. nudiventris*) for more than two decades (Bauer and Hoffman 1976). Large-mouth bass (*Micropterus salmoides*) introduced into England from North America carried with them the trematode, *Urocleidus principalis*. Unlike the case of the sturgeon, however, this gill parasite seems host-specific and has not spread among other fishes native to the British Isles (Maitland and Price 1969). Still, introduction of bass and other fishes in the past proceeded in ignorance of the possibilities for contaminating native species with new pathogens.

One must remember that an exotic population released into the wild is no longer an "experiment." It is a fact accomplished, no longer under complete control. Whereas one might hope to hit the target, an unconfined exotic population is not unlike a bullet— once fired, it cannot be called back. A misplaced shot may have wide-reaching implications. Furthermore, imported wildlife may escape despite the best intentions to keep it confined. Texas law requires that certain species of exotic big game must be contained by deer-proof fencing, but about 34 percent of these animals later roamed free (Armstrong and Wardroup 1980). Animals, of course, can and do travel across boundaries, and whereas the boundaries may be between farms and ranches in some cases, they may also be between states and nations where the "value" of a neighbor's exotic might not be appreciated equally. Weller (1969) discussed this and other problems for exotic waterfowl where long-distance migrations might be expected.

Nor have all exotic introductions coincided with so-called vacant niches. In Nevada, about 5000 francolins (*Francolinus* spp.) and sand grouse (*Pterocles exustus*) were released precisely where the state's highest densities of Gambel's quail (*Lophortyx gambelii*) already occurred, placing the exotics in direct competition with the native species (Gullion 1965). Fortunately, the quail persisted and the exotics apparently disappeared.

The far-reaching consequences resulting from introductions of foreign species into new ecosystems seldom have been established beyond those of an obvious and direct nature, such as releasing a disease. Examination of biological communities and their structure is difficult under the best of circumstances, and to do so as a means of tracing the role of an introduced species represents an awesome task indeed. Soil and water relationships may be involved, along with behavioral or other subtle interactions within the community of organisms. The perturbations caused by burros on soil, vegetation, and small mammal faunas in the Grand Canyon are one example (see Chapter 14). However, the introduction of peacock bass (*Cichla ocellaris*) into Lake Gatun, Panama, even better demonstrates the disruptions an exotic may inflict on various trophic levels (Zaret and Paine 1973).

Peacock bass—so named because of a large spot at the base of their tails—are native to the Amazon River and its tributaries in South America. They reach 2 kg in weight and 50 cm in length, and because of their size, fighting abilities, and tastiness, they seemingly offered exceptional sport-fishing opportunities in Panama when introduced in 1967. The first introductions occurred in the upper reaches of the Chagres River, an arm of Lake Gatun (the lake is part of the Panama Canal). Within 2 years, peacock bass entered the lake, gradually spreading across its expanse. These movements were not haphazard, however, as the fish moved in a wave whose leading edge was composed of subadults. The structure therefore is quite different between the advancing and established segments of the population. Peacock bass are voracious predators, and as they spread across Lake Gatun, they effectively eliminated six of the eight common native fishes previously inhabiting the waters and drastically reduced another. Among these was a 90-percent reduction in the numbers of silversides (*Melaniris chagresi*), a species important as food for tarpon (*Megalops atlanticus*). Several species of waterbirds also relied on the formerly abundant schools of silversides, and all of these consumers either declined or stopped foraging altogether where peacock bass dominated the lake (Fig. 18-4). Further, reduction in the populations of forage fishes increased the numbers of zooplankton on which they fed (some forms of zooplankton also changed when the feeding pressure on them was lessened). The increased numbers of zooplankton, in turn, eventually may decrease the amount of algae in Lake Gatun. In the years immediately following the

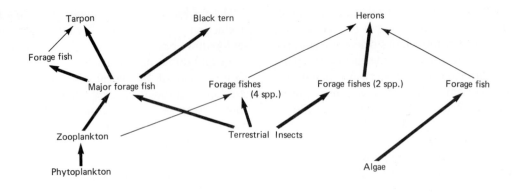

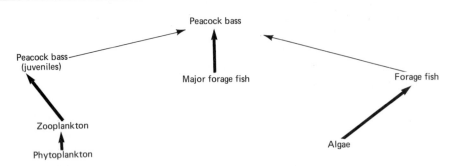

Figure 18-4.   Effect of peacock bass on food webs in Lake Gatun, Panama. Upper food web reflects alternate pathways of energy to final consumers in a diverse aquatic system lacking the exotic bass. Lower food web shows the loss of several forage fishes and the replacement of native consumers by peacock bass. Heavy lines indicate major relationships. (From Zaret and Paine 1973.)

entry of peacock bass into Lake Gatun, mosquito populations also increased severalfold, presumably because the small forage fishes that eat mosquito larvae had been eliminated or reduced in numbers. Coincidentally, the type of malaria that infects humans suddenly changed when the incidence of *Plasmodium falciparum* increased from 5 to 60 percent; *Plasmodium vivax* decreased proportionately. Since their introduction, peacock bass thus have exerted ecological perturbations in virtually all of the trophic levels in Lake Gatun (Zaret and Paine 1973).

In Hawaii, the successful colonization of Kalij pheasants (*Lophura leucomelana*) is a symptom of a degraded ecosystem because the birds largely depend on other exotics for food and cover (Lewin and Lewin 1984). Kalij pheasants, native to much of southern Asia, were liberated in North America early in the century and again between 1962 and 1976, but no sustained breeding populations resulted from these releases. In Hawaii, however, 67 birds released in 1962 founded an abundant and widespread population, and Kalij pheasants became legal game

in 1977. The success of the species may prompt transplants elsewhere in Hawaii, but further expansion may be unwise. Kalij pheasants enhance the establishment of exotic plant pests, may prey on rare species of land snails, and serve as potential disease reservoirs.

Another concern has to do with the profession of wildlife management. If we seek exotic wildlife to supplement native species, might this not be an admission that we have already failed in our task at home? Introductions of coturnix quail within the range of bobwhites hardly would seem necessary if populations of the native bird provided sufficient hunting success; yet coturnix were released by the thousands in 19 states between 1956 and 1958 (Laycock 1966). Coturnix introductions failed in every instance, but one might wish that an equal amount of funds and effort had been expended on the management of bobwhites. Conversely, we might ponder the ecological consequences on bobwhites had coturnix quail succeeded. Continued abuse of habitat for native wildlife is regrettable enough without being compounded by the hope that some magical exotic species might be found to occupy a diminished environment. In ethical terms, wildlife administrators and others in leadership positions might well discourage hunters from the notion that better recreation awaits the discovery and introduction of the "perfect" game animal. In fact, more hunting does not necessarily result as the aftermath of exotic introductions. Two species of European grouse—capercaillie (*Tetrao urogallus*) and black grouse (*Lyrurus tetrix*)—have been proposed for release in northern forests of the United States; yet their normal densities in Europe are no more than those maintained by native ruffed grouse (*Bonasa umbellus*) at their lowest levels (Gullion 1965). Furthermore, these European species also experience population fluctuations, so, at the ebbs, hunters would locate few of the exotic birds for sport. In fairness, however, it should be pointed out that wildlife administrators sometimes have been pressured into supporting exotic game programs when, in reality, their own feelings were of a contrary nature (Gullion 1965).

Figure 18-5. Axis deer are large, spotted cervids from Asia that are currently the most abundant species of exotic big game in Texas. Shown here are axis and white-tailed deer on a privately owned ranch managed for trophy hunting. (Photo courtesy of R. J. Whyte.)

NORTH AMERICA'S OUTDOOR ZOO

Exotic introductions—past and present—are not restricted geographically; nearly every region of North America has been implanted with exotic wildlife at some time. However, the U.S. Southwest has been the most active region in these efforts, particularly with big game introduced from the drier parts of Africa, Asia, and Europe. Release of exotics has been highlighted in the public game policy in New Mexico, and stocks of several species command sizeable hunting fees on large, privately managed ranches in Texas (Fig. 18-5). A survey completed in 1980 reported 72,000 exotic big game animals in Texas, but the population jumped to 120,000 animals by 1985 (Armstrong and Wardroup 1980; Traweek 1985). Of the 59 species represented, all except two have increased in abundance. The three most-abundant species are 38,000 axis deer (*Axis axis*), 18,700 blackbuck antelope, and 15,000 nilgai antelope (*Boselaphus tragocamelus*). Only 67 percent of the animals in the 1980 survey still were confined within deer-proof fences. Except for two special circumstances, exotic animals are not classified as big game in Texas. Instead, current law regards exotics as domestic livestock, which means they are exempt from control by the Texas Parks and Wildlife Department and can be hunted at any time.

Any listing of exotic game species released in North America would be lengthy if all attempts—failures and successes alike—were included. The question of "success" often is debatable, and, depending on one's own criteria and evaluation, a verdict for some species still may be undecided. For example, descendants of 93 gemsbok (*Oryx gazella*) originally released on a military post in New Mexico eventually spread into parts of a national monument. On one hand, these animals are hunted and seemingly represent an economic asset in deserts that have few large herbivores; on the other hand, gemsbok present park managers with new dilemmas in keeping with their charge to maintain a native environment (Reid and Patrick 1983). The following examples suggest the variety of exotics currently established to some degree in North America.

CHUKAR PARTRIDGE (*ALECTORIS CHUKAR*). The chukars most commonly established in North America originated in India, Pakistan, and Afghanistan. However, as several subspecies likely have been introduced at one time or another, the birds now in North America actually may be an amalgamation. After numerous stocking attempts in 42 states and 6 Canadian provinces, chukars are now hunted in 10 western states and British Columbia. Nevada and Utah are at the center of the species' distribution in North America. Other chukar populations are isolated in Hawaii, New Zealand, and Mexico.

The earliest introduction of chukars occurred in 1893 when 10 birds were released in Illinois (Cottam et al. 1940). However, the first successful transplants did not take place until 1935 in Nevada. Additional transplants later established chukars throughout Nevada, and a hunting season was opened there in 1948 (Christensen 1970).

Chukar range is found in rugged mountainous terrain in arid and semiarid climates. Sparse grass–forb communities, sometimes mixed with low shrubs and occasional conifers on rock-strewn slopes, represent ideal habitat. These conditions in the birds' native range in Asia essentially are duplicated in parts of the Great Basin region of the western United States (Fig. 18-6). Nonetheless, chukars seem to have more exacting requirements than first supposed. Introductions of more than 7600 birds failed completely in New Mexico, despite the gross similarity of that state with Nevada (Gullion 1965). Cheatgrass (*Bromus tectorum*), itself also an exotic, is highly important in the chukar diet, and the distribution of this plant on western rangelands likely is responsible for the chukar's success. Such habitat largely lies outside the immediate ecological realm of native gamebirds occurring in the western United States. In fact, chukars thrive on overgrazed rangelands no longer suitable

Figure 18-6. Chukar partridges introduced from Asia found suitable habitat on talus slopes and rocky outcrops in Utah and Nevada. (Photo courtesy of D. Erickson, Nevada Department of Wildlife.)

for sage grouse (*Centrocercus urophasianus*). Thus, chukar partridges provide sport hunting in a large area of western North America where native game birds are either ecologically excluded or have been extirpated.

BARBARY SHEEP (*AMMOTRAGUS LERVIA*). The Barbary sheep is native to the Atlas Mountains of northern Africa. The sheep first were imported for zoos and parks in the eastern United States in 1900, and as their numbers rapidly increased, the surpluses were given to private ranches for release (Ogren 1965). Wild populations are now well established in California, New Mexico, and Texas. Two separate herds also occur in Mexico (Rangel-Woodyard and Simpson 1980).

In California, the free-roaming Barbary population originated in 1953 when 85 animals escaped from the privately owned zoo of newspaper magnate William Randolph Hearst (Barrett 1980). The herd apparently stabilized at about 400 animals and is subject to privately controlled hunting. Whereas releases have been made at several places in New Mexico, the largest herd there stemmed from 57 Barbary sheep freed in 1950 on state-owned lands along the Canadian River; the stocking program was authorized when the state's efforts to manage bighorn sheep were disappointing and a substitute was desired for the native species (Ogren 1965). Morrison (1980) estimated that 1750 Barbary sheep currently reside in New Mexico, and since 1955, when the species was declared a game animal, the legal harvest produced a 39 percent hunter success rate. The Texas population of Barbary sheep includes both free-roaming and confined animals. Releases in 1957 and 1958, 44 animals in total, established a wild population in Palo Duro Canyon in the Texas Panhandle after releases of mule deer (*Odocoileus hemionus*) there failed; later estimates placed this population at 1400 to 1600 animals (Dvorak 1980). Between the first hunt in 1963 and 1978, 959 of the exotic sheep were harvested by hunters issued 3011 permits (32 percent hunter success rate). Ramsey (1970) estimated that 1300 Barbary sheep were confined on 40 privately owned ranches in Texas, primarily to offer supplemental income for landowners leasing hunting rights. Overall, about 5100 Barbary sheep occur in Texas, New Mexico, and California, but a proposal to release the species on federal lands in Nevada was rejected by the Bureau of Land Management (Yoakum 1980).

Ogren (1965) discussed the potentials for competition between Barbary sheep and native wildlife.

Mule deer, the staple big game species in much of the western United States, may be jeopardized where the two species come into contact. The exotic sheep outproduce mule deer and utilize some of the same food plants. Browse, particularly oaks (*Quercus* spp.) and mountain mahogany (*Cerocarpus montanus*), is sought by both species, but Barbary sheep overall have far greater plasticity in their diets than mule deer. In Palo Duro Canyon, Texas, Krysl et al. (1980) found a 74 percent index of dietary similarity between mule deer and Barbary sheep, with browse showing the greatest amount of overlap. They warned that continued expansion of these big game populations would decrease forage availability. Thereafter, with more interspecific competition for the remaining forage, mule deer are likely to be displaced. Accordingly, for mule deer and Barbary sheep to remain sympatric, their combined populations must be carefully regulated below forage carrying capacities.

Greater concern, however, perhaps rests where Barbary sheep interact with bighorn sheep. Desert bighorns are no longer thriving, despite numerous and intensive management practices by both state and federal agencies. In fact, their diminished populations often provided the very impetus to replace them with Barbary sheep.

Pence (1980) lamented that examinations of Barbary sheep to detect infectious diseases and parasites occurred only after the fact of their releases. Bluetongue is paramount in this regard. Robinson et al. (1967) believed that this disease, transmitted by contact with domestic sheep, may have played a dominant role in the disappearance of desert bighorns from parts of their range. Thus, when positive titers of bluetongue also were detected in Barbary sheep (Trainer and Jochim 1969; Hampy et al. 1979), another source of infectious contact was established for desert bighorns.

Many characteristics of Barbary sheep diets are essentially like those of desert bighorns. The foods of both species in North America show considerable, and perhaps critical, overlap. Of 49 foods important to desert bighorns, 37 also occur in the diets of Barbary sheep in Texas and New Mexico (Simpson et al. 1978). The relative proportions of grasses, forbs, and browse ingested by both species also are similar, especially during the winter months when stresses may be acute.

Competitive interaction between desert bighorns and Barbary sheep was examined in conceptual terms by Seegmiller and Simpson (1979). Among other considerations, they noted that Barbary populations, unlike those of desert bighorns, can increase rapidly. Barbary sheep commonly bear twins and occasionally triplets, whereas the native sheep produce a single lamb each year. In at least one instance, a female Barbary sheep produced two births within a 7-month interval. Barbary sheep also have greater dispersal abilities than desert bighorns. Simpson et al. (1978) summarized records showing dispersals of at least 145 km as Barbary populations increased.

In all, a growing body of evidence suggests wholesale incompatibility between Barbary sheep and desert bighorns—and when and where their ranges eventually overlap, only the Barbary may emerge. If so, the strong force of interspecific competition is illustrated in judging the merits of exotic game introductions. Moreover, in some regions, this may be a case where a clear choice must be made for either Barbary or bighorn sheep—one or the other—but not for both.

REINDEER (*RANGIFER TARANDUS*). For many years, the taxonomic status of reindeer and caribou was unsettled. The animals are circumpolar in distribution and were treated as several species or, alternately, as one species with several races. Banfield (1961), in a widely accepted revision of *Rangifer*, claimed that the complex is a single species of several races, but that matter is beyond our interest here except that Old World reindeer were introduced into the range of the New World caribou.

The domestication and husbandry of reindeer in northern Eurasia are centuries old, with the animals providing virtually the entire livelihood for Lapps. Until late in the 19th century, domestic reindeer remained in Eurasia. Then, between 1891 and 1902, 280 reindeer were brought to Alaska by government officials hoping to provide Inuit with a dependable source of food and fiber; in later years, five other attempts followed to develop a reindeer industry in Canada (Scotter 1970b).

The goal to establish reindeer-based economies for native peoples in Alaska and Canada failed. Inuit, hunters by culture, did not adapt to the pastoral way of life that herding demanded (Lantis 1954). Arguments about ownership and overall mismanagement also took their toll, and by 1964, the last native-owned herds in Canada failed as a ranching operation (Scotter 1970b). But doomed as these projects were, reindeer remain an established latter-day component of North America's arctic and subarctic fauna.

Leopold and Darling (1953b) offered some useful observations about the introduced intrusion of reindeer into the native realm of caribou. In Alaska, after the reindeer population reached a zenith of about 600,000 animals in the early 1930s, their numbers steadily diminished to 252,000 in 1937 and to 155,000 in 1941. About 26,700 reindeer remained by 1952. This setback, blamed on predation and poor management, largely discounted a more fundamental problem. Constant overgrazing rapidly deteriorated the range and eventually exhausted the requisite supply of lichens. In fact, reindeer herds on islands—lacking either predators or herding problems—followed the same pattern as befell the mainland population. On Nunivak Island, the reindeer herd of 22,000 collapsed to less than 5000 as the range was depleted. Such losses of habitat affected caribou as well as reindeer. Unfortunately, the disappearance of caribou in much of western Alaska was related directly to the excessive grazing of reindeer. Furthermore, the remaining patchwork of suitable habitats fragmented the large herds of caribou into smaller bands and, in doing so, apparently further reduced their productivity.

Similarly, Klein (1968) described the expansion of 29 reindeer released on St. Matthew Island in 1944 into a herd of 6000 by the summer of 1963. At this level, the herd reached a density three times greater than the recommended carrying capacity of about 30 animals per 5 km^2. During the following winter, a crash die-off diminished the herd to 50 animals, nearly all of which were females, and an environment virtually stripped of lichens.

Social, cultural, and economic factors still influence the management of reindeer herds in North America (Klein and White 1978). In recent years, an Asian market has developed for velvet-covered reindeer antlers because of the alleged aphrodisiac and medicinal properties attributed to the velvet. Because this market has greatly increased the cash income of local herders, economic incentives have arisen again for increasing the reindeer industry. Furthermore, segments of the caribou herd are declining in the western Arctic, placing limits on subsistence hunting by Inuit. At the same time, reindeer meat is increasing in importance. Thus, land managers are still facing a complicated situation as efforts mount to expand reindeer husbandry in areas within the range of caribou.

BROWN TROUT (*SALMO TRUTTA*). The original distribution of brown trout extended continuously across Eurasia from the northern coastal countries of Europe (including the British Isles and Iceland) to the Black, Caspian, and Aral seas and their tributaries in western Asia; brown trout also occur on the Mediterranean islands of Corsica and Sardinia and along the coast of Algeria in northern Africa. Today, introduced populations are found on every continent except Antarctica (MacCrimmon and Marshall 1968).

Brown trout were likely candidates for transplanting, as they seemingly are more tolerant of a wider regime of environmental conditions than many other salmonids. Waters impounded behind dams, for example, might become too warm for native trout, whereas these waters often remain suitable for brown trout. Because of their hardiness, brown trout also adapt well to hatchery conditions and to the rigors of stocking. Further, the species is wary and grows rapidly, offering skilled anglers opportunities for trophies weighing more than 14 kg; brown trout of 3–5 kg are commonly landed. Those released in the streams of Chile are renowned for their great size, and a brown trout netted in Utah exceeded 17 kg (Sigler and Miller 1963).

Brown trout first reached North America in 1883 when eggs from Germany were shipped to a hatchery in New York (Laycock 1966). Fry from these and subsequent shipments of eggs from Germany and Scotland were released in Michigan, New York, Maine, and Minnesota. Their distribution in North America today includes virtually every aquatic system with suitable environments (e.g., lakes and streams in 45 of the 50 United States), as well as in some systems where, because of unfavorable conditions, they are stocked under put-and-take management (Catt 1950; MacCrimmon and Marshall 1968).

Results of brown trout introductions in North America and elsewhere in the world have brought about various ecological and biological reactions, primarily from competition. Some of the interactions concern other fishes and, perhaps, a bird. Populations of blue ducks (*Hymenolaimus malacorhynchos*) are less abundant now in New Zealand than a century ago when trout initially were released. Brown trout consume large volumes of aquatic invertebrates, the staple summer diet of blue ducks, leading Kear and Burton (1971) to speculate that competition for these foods limited the ducks as exotic trout populations expanded in New Zealand. Indeed, based on the current distribution of these species, it seems more than coincidence that blue ducks thrive in streams where trout have been excluded (see Kear 1972b).

Brown trout are strong competitors with other fishes, and their introductions have wrought changes in the distribution and abundance of native fish populations. MacCrimmon and Marshall (1968) cited evidence that brown trout have displaced brook and rainbow trout. In Alberta, Nelson (1965) reported that brown trout spread at least 241 km from the point of release, contributing to the marked decrease of Dolly Varden (*Salvelinus malma*) in the Kananaskis River system. Brown trout have displaced some populations of land-locked salmon (*Salmo salar*) in Maine; the predominance of large brown trout overwhelmed other fish populations. Similarly, the brown trout in Convict Lake, California, were described as a "burden on the biological economy of the lake" (*fide* Fenderson 1954).

The difficulties of catching brown trout often result in a disproportionate number of larger fish. At first glance, that might seem desirable, but the presence of the larger individuals usually reduces overall fishing success. One study compared the yield of brown trout versus native brook trout based on residual populations of each species at the end of the fishing season. Only 0.4 brown trout were landed for each 1.0 fish of the same species remaining after the season closed, but, among brook trout, fully 3.0 fish were caught for each 1.0 remaining (Cooper 1951). In other words, the native species yielded 7.5 times more fish than did the exotic species. A trade-off may be involved with brown trout introductions; namely, should a few trophy-sized brown trout or smaller yet more-catchable native trout remain the goal of management? Responses, of course, will vary among both managers and anglers, but the question should be considered whenever future introductions

of brown trout are contemplated for waters already inhabited by native species of trout.

Brown trout, more than other species of trout, prey on fish. When they exceed 22 cm, their diets increasingly include smaller fishes. Partly for this reason, brown trout are not stocked in Wisconsin where native brook trout are being encouraged (Brynildson et al. 1977). Furthermore, brown trout have more than average ability to reproduce. Because they spawn in the autumn when streams usually are clear and stable, brown trout may produce larger numbers of fry than native rainbow trout; the latter, breeding in the spring, often encounter flooding that may result in less-favorable spawning conditions (Sigler and Miller 1963).

On the other hand, artificial situations creating unstable or simplified environments often favor invading species (Elton 1958). For example, the seasonally unstable nature of reservoirs impounded for hydroelectric power may increase brown trout populations over those of other species (see Nelson 1965). Still other lakes and streams provide only marginal habitat for native trout and salmon. In these waters, brown trout may represent the most reasonable alternative for maintenance of a productive salmonid fishery (Pierce 1981).

NUTRIA (*MYOCASTOR COYPUS*). Notions of quick riches led to the introduction of nutria, a large aquatic rodent endemic to South America (Fig. 18-7). Its heavy fur seemingly made the nutria an attractive candidate for commercial fur farming, and beginning in the 1930s, large numbers of breeding pairs from

Figure 18-7. Nutria are large aquatic rodents from South America introduced into North America for their fur value. They now occur commonly in coastal wetlands from Texas to Maryland but also have colonized some inland states. (Photos courtesy of J. A. Chapman.)

Argentina were brought to the United States (Presnall 1958). Unscrupulous sellers of breeding stock advertised widely that one could "retire on a half acre" devoted to nutria farming; promoters asked $800 for females and $150 for males at a time when nutria pelts usually sold for less than $4 each (Laycock 1966). The rodents escaped or were released soon afterward (usually when would-be fur ranchers failed to gain rapid profits), establishing wild populations virtually everywhere they found freedom.

Nutria spread into wetlands along the Gulf Coast after escaping from Avery Island, Louisiana, in the aftermath of a hurricane in 1940 (Harris 1956). Nutria occurring in Texas descended from those escaping in Louisiana, although others were released to establish outdoor fur farms; still others were sold and released to control aquatic vegetation (Swank and Petrides 1954). In Maryland, nutria were introduced in the late 1930s or early 1940s and now are found in about 75 percent of that state's marshlands (Willner et al. 1979). Other populations exist on the West Coast of the United States (Howard 1953) as well as in local settings well inland (e.g., Ohio; Petrides and Leedy 1948). Ashbrook (1948) described one case of accidental escape and two others where nutria were released intentionally by fur farmers in Washington.

Nutria rapidly destroy wetland vegetation, competing in this way with muskrats and waterfowl. Willner et al. (1979) reported that nutria, in part, severely damaged coastal marshes, including more than half of the primary wetlands in a national wildlife refuge in Maryland. In particular, nutria consume large quantities of 3-square bulrush (*Scirpus olneyi*), a staple food of southern muskrats. Assessments of competition between nutria and muskrats are not always clear. Atwood (1950) believed that nutria occupy mainly freshwater marshes along the Louisiana coastline, whereas muskrats prefer brackish wetlands. If so, competition seems minimized by each species' preference for one habitat or the other. Conversely, Swank and Petrides (1954) suggested that nutria compete with muskrats for food and cover, citing a report that nutria already have been detrimental to muskrat populations in Washington and Oregon. Croplands near marshes also are damaged by nutria. In California, Schitoskey et al. (1972) reported that 11 percent of the croplands near nutria habitat was damaged. Sugarcane and cornstalks were nipped, whereas entire plants of alfalfa, rice, and ryegrass were consumed. In Oregon, nutria girdled trees in orchards and forests (Kuhn and Peloquin 1974).

Nutria can be especially damaging where rice is grown commercially. Their burrows riddle levees surrounding irrigated fields. Nutria also create large openings in rice fields when they cut the plants for food and platforms (Harris 1956).

Evans (1970) accordingly described poisons and other means of controlling nutria. Baits treated with zinc phosphide were recommended, especially where nutria densities are high. However, nonselective poisons or even rodenticides should be used with caution where nutria share habitats with muskrats. Severe winter weather causes drastic reductions in nutria populations—as much as 90 percent—but cold generally represents a limiting factor only at the northern edges of the species' distribution (Willner et al. 1979). In Louisiana, where nutria are well established, alligators (*Alligator mississipiensis*) may exert some biological control on dense nutria populations. Remains of nutria were found in 56 percent of the alligator stomachs examined by Valentine et al. (1972), but this occurred only when nutria reached peak numbers. At lower densities, nutria occurred in only 5 percent of the alligator stomachs.

Trapping may control nutria populations at moderate to low densities. But even with commercial trapping, the vagaries of fur prices at times may preclude the desired amount of control. In England, where leghold traps are illegal, an experimental 5-year program of live-trapping reduced nutria numbers by 99 percent on a 30-km^2 wetland (*fide* Willner 1982). Ultimately, effective and safe control of nutria may be accomplished only if a host-specific disease or parasite is implanted where the animals are troublesome. Biological control of this sort may be difficult to realize, however, because of the potential for simultaneously damaging muskrat populations with the same pathogen.

Despite these considerations, the fur of nutria has commercial value, and so the animals are of importance to some trapping economies. Trappers in Maryland harvest large numbers of nutria at prices approaching $10 per pelt (Willner et al. 1979). Since the 1970–71 trapping season, harvests of nutria in Louisiana have exceeded 1 million animals, reaching a value of about $9 million in 1979–80 (*fide* Willner 1982). Whatever the pros and cons of nutria might be, they are now well established in much of North America. Given this fact, managers now face the challenge of balancing nutria populations as a fur resource against their real or potential overabundance as pests.

TRANSPLANTS WITHIN NORTH AMERICA

Native species of wildlife also have been moved within the confines of North America. In their new settings, native species can function as exotics, potentially offering the same range of problems outlined for foreign transplants. As noted earlier, rainbow trout transplanted from western states transmitted a parasitic disease to eastern brook trout (Bowen 1970). Conversely, wild turkeys also have been widely relocated, apparently without harming ecological relationships within their new homes. Stock from Texas has since established turkey populations in Nebraska, Kansas, and North Dakota (Glazener 1967).

Elk (*Cervus elaphus canadensis*) transplanted from the Rocky Mountains initiated populations elsewhere in the United States, either where local races of elk earlier had been extirpated or where elk seemed desirable additions to the regional fauna. A private landowner released 44 Rocky Mountain elk in the Guadalupe Mountains on the Texas–New Mexico border in 1928, replacing the extirpated Merriam's elk (Davis and Robertson 1944). This population increased to 400 animals and then declined to 100 when Guadalupe National Park was created and artificial water sources for livestock were removed (Moody and Simpson 1977). In Washington, the descendants of transplanted Rocky Mountain elk now apparently threaten the delicate alpine vegetation on Mount Rainier National Park (Allen et al. 1981). Virginia opossums (*Didelphis virginiana*) brought to western Oregon in the early 1930s by workers in the Civilian Conservation Corps escaped or were released when the corps disbanded in 1939. The opossums in Portland reproduce at a higher rate than other populations elsewhere (Hopkins and Forbes 1979).

In 1939 and again in 1941, hunters released eastern cottontails (*Sylvilagus floridanus*) from Ohio and Illinois into western Oregon, apparently because the native brush rabbit (*Sylvilagus bachmani*) offered little sport (Graf 1955). The introduced rabbits expanded their range locally during the next 40 years, but, fortunately, they have not become agricultural pests or hybridized significantly with brush rabbits. Nor have eastern cottontails become the popular game animals in Oregon that once was anticipated (Verts and Carraway 1980, 1981).

The original distribution of mountain goats (*Oreamnos americanus*) was limited to northwestern North America. However, between 1924 and 1929,

6 goats and their 4 offspring escaped from their pen in the Black Hills of South Dakota, eventually establishing a disjunct population of about 300–400 animals (Harmon 1944; Jones et al. 1983). Because of overgrazing and declines in the herd's general welfare, South Dakota initiated a hunting season for mountain goats in 1967 to forestall a population crash. Nonetheless, overcrowding and heavy lungworm infections continued plaguing the Black Hills goat herd (Turner 1974).

In Washington, the historical distribution of mountain goats included the Cascade Range but not the nearby mountains of the Olympic Peninsula (Webster 1925). Indeed, the geographical isolation of the peninsula prevented natural colonization even though habitats there otherwise seemed suitable for mountain goats. This changed in 1925 when four animals from British Columbia were released into the unique montane environments of the Olympic Peninsula then administered by the Forest Service; by 1929, a total of about 12 mountain goats had been introduced. In the years since, this nucleus increased to a population of 500–700 animals in a region now largely within Olympic National Park (Moorehead et al. 1981). The park was created in 1938, and with it came cessation of legal hunting. With a capability of doubling in numbers about every 9 years—and no hunting—the Olympic population of mountain goats steadily increased, reaching a size inimical to the fragile environments that the park was designed to protect. Thin soils and delicate vegetation in the park's alpine zones today show increasing disturbance under pressure of the unchecked herd (Fig. 18-8). The harsh conditions at higher altitudes, including short growing seasons, low soil moisture, high winds, and frequent freeze-thaw cycles, contribute to the sensitivity of the alpine regime; vegetation there is slow growing and easily disturbed. Furthermore, several species of native plants evolved in the relative isolation of the Olympic Mountains, and 11 of these grow where mountain goats now occur; another 43 plants exposed to the burgeoning herd are listed as endangered by state conservationists. Mountain goats in the park already have reduced the abundance of palatable forage, and many of the native plants show signs of grazing disturbance (Pike 1981). Mosses and lichens, crucial to the stabilization of alpine soils, have been eliminated in many areas.

As shown in Figure 18-9, mountain goats paw wallows up to 10 m in diameter, and as many as

Figure 18-8. Individual goat wallows cause craterlike scars in Olympic National Park (top). At some sites, the introduced herd of mountain goats has cleared large areas of vegetation and initiated extensive erosion (bottom). (Photos courtesy of M. Hutchins.)

Figure 18-9. Mountain goats have remained unchecked by hunting since their introduction into Olympic National Park (top). Vegetation and thin soils in the alpine zones are disturbed easily by pawing animals (bottom). (Photos courtesy of M. Hutchins.)

70 wallows, eroding to depths of 1 m, have disrupted the integrity of a single mountain meadow in Olympic National Park (Pfitsch 1980). Humans also may accentuate the physical disturbances of mountain goats in a curious way. Because there are no natural salt licks in Olympic National Park, the urine of hikers and campers creates salt-rich patches attracting the animals. Thus, in their craving for salt, mountain goats paw and trample these sites so that further erosion develops in the montane regime (Hutchins and Stevens 1981).

Happenstance led to the transformation of a saltwater game fish into freshwater trophies and, thereafter, to its relocation elsewhere in reservoirs far removed from salt water. The saltwater distribution of striped bass (*Morone saxatilis*) stretches along the eastern coastline from the St. Lawrence River to Florida, and westward to Louisiana on the Gulf of Mexico (Karas 1974). Surf anglers prize the species for its sporting value and as tablefare. As an anadromous species, striped bass spawn in coastal river systems, for the time tolerating brackish or even fresh water before returning to marine environments. In 1941, a spawning population of striped bass was trapped upriver when the Santee-Cooper River system in South Carolina was closed from the sea by two dams. By 1950, resident striped bass were schooling in the Santee-Cooper Reservoir, and anglers began to experiment with ways to catch the "new" freshwater gamefish (Stevens 1957). More important, striped bass were completing their life cycle entirely in a freshwater environment (Fig. 18-10). This prompted the interest of biologists from inland states where large freshwater reservoirs lacked a gamefish large enough to utilize the abundant shad populations choking those fisheries (Bailey 1974). Further, the addition of striped bass to reservoirs in southern states apparently posed little competition with other fishes.

Based on the circumstances in South Carolina, stocks of striped bass were released in the reservoirs of many southern states and in California. However, these introductions generally failed to establish self-sustaining populations. Successful spawning developed only where the transplanted stocks had access to large rivers with a sufficient water velocity to suspend the species' semibouyant eggs for 36–75 hours (Pflieger 1975). This means that striped bass populations maturing in reservoirs also must have long stretches of moving water available to fulfill their spawning requirements. Otherwise, the eggs settle to the bottom, where they are smothered by silt. As few rivers maintain their previous velocities after they are impounded, the continuation of striped bass fisheries in freshwater reservoirs remains largely dependent on repeated stockings of hatchery-produced fingerlings (Fig. 18-11).

Striped bass relocations deserve one further comment. Between 1879 and 1881, fish from the New Jersey coast were transported by rail to California, where they were released in San Francisco Bay. Although this transplant did not involve adaptation to fresh water, it did establish a sizable commercial fishery yielding many tons of striped bass annually (Surber 1957).

SOME NORTH AMERICAN EXPORTS

Exotic transplants are by no means limited to those brought to North America; several species have been exported from North America as well. Many of these were selected for export because of their recreational values.

Oceanic islands have been particularly attractive for the importation of foreign species. In the absence of any previous geological connections with continents, oceanic islands developed depauperate faunas that were highly susceptible to human intervention. Land mammals, particularly hooved species, were absent in pristine times, and native bird faunas usually were highly specialized. Moreover, in the eyes of the settlers, the fauna of oceanic islands lacked species suitable as game animals. Such settings were ripe for the "correction" of nature's shortcomings. As a result, even some of the most remote islands now commonly feature a biota of international proportions.

New Zealand, like Hawaii and most other oceanic islands, has suffered from a host of exotic introductions. Cervids from both Europe and North America were stocked as game species, among them a lot of elk presented in 1905 as a gift by President Theodore Roosevelt. The ecological havoc subsequently befalling New Zealand's native vegetation and soils from the burgeoning cervid populations is now well established (Howard 1966). By 1930, elk and their European counterpart, red deer (*Cervus e. elaphus*), were declared "vermin;" more than 1 million were killed by government agents between 1931 and 1968 (Roots 1976). Hunters also were provided free ammunition for the same purpose. Recently, farmers in New Zealand have domesticated red deer as sources of meat and other products. Up to 20,000 animals from the wild population are captured each year to

Figure 18-10. Striped bass trapped upriver by the Santee-Cooper Reservoir in South Carolina established a breeding population existing entirely in fresh water. Stock from this population was used to initiate other fisheries, but most of the transplants did not result in self-sustaining populations. (Photo courtesy of *South Carolina Wildlife Magazine*.)

supplement domestic herds, thereby offering some relief for the overpopulated countryside (Finkelstein 1983).

Canada geese were released in Europe as early as the 18th century, but they were not introduced successfully into New Zealand until 1905. There, according to Delacour (1954), the birds were distributed widely and produced remarkable numbers of young without first experiencing long periods of adjustment to either the Southern Hemisphere or to local ecological conditions. Several races were released, including the giant subspecies (Yocom 1970), eventually resulting in a racially mixed complex in much of New Zealand. But the habits of the expanding Canada goose population proved troublesome. The birds grazed in sheep pastures, making no friends of herders wishing to conserve forage for their livestock. Croplands, too, were gleaned, as were smaller vegetable gardens. In brief, Canada geese quickly be-

came pests in New Zealand. After additional releases in 1920, a limited hunting season was initiated; this was followed by a regular goose season. In 1931, however, all restrictions were removed and Canada geese could be hunted year-round by any means.

In 1959, a group of 56 pronghorns (*Antilocapra americana*) from Montana was transported to the Hawaiian Islands; of these 44 survived shipment.[1] Four animals were given to the Honolulu Zoo, with the balance scheduled for release on some 317,000 ha of grassland (another two died during the interval). However, when the remaining herd was released, the thirsty animals immediately struck out for the ocean, thinking it was a lake and a source of fresh water. A day later, the herd was wandering aimlessly on the beach, still thirsty and cut off from the island's interior by thick thorn forest. A team of volunteers was organized to recapture the animals. In the process, some pronghorns swam out to sea until the surf forced them back to shore. Others were frightened enough to scatter into the forest, and at least two more died on the beach, presumably from drinking salt water. Even after the survivors of this episode were returned to the grassland, more deaths occurred when some wandered off, never to be seen again. Others died when their eyes were punctured by the thorny vegetation of the forest. In all, 18 pronghorns survived, but even these suffered an intestinal disorder from the marked change in their diet. Nonetheless, this pitiful nucleus regained its health and formed a scattered herd of about 150 animals nearly a decade later. Kramer (1971), in reporting these events, concluded that the release of pronghorns into Hawaii must still be considered as "a trial introduction." The herd has failed to increase at normal rates since its introduction and little effort has been made to determine why. Certainly, any future assessment of the pronghorn herd in Hawaii must be colored by the ecological wisdom of releasing a temperate plains animal into a semitropical island environment.

Although the impacts of exotic wildlife often may be more pronounced on oceanic islands such as New

---

[1]Big game often experience high rates of mortality during transport. Pimlot and Carberry (1958) reviewed records for moose (*Alces alces*) transported within North America; on the average, 1 out of every 2 died. Based on observations of confined populations, Bailey and Franzmann (1983) determined that introduced moose suffer higher mortality rates than resident animals of similar age, presumably because the introduced animals are faced with unfamiliar surroundings. Only 7 percent of the calves produced by introduced moose survived their first winter, compared to a survival rate of 28 percent for calves born in resident herds.

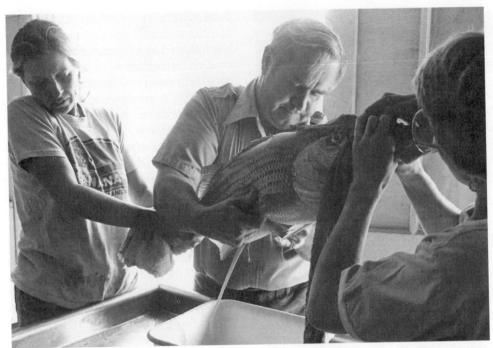

Figure 18-11. Hatchery operations are needed to sustain striped bass fisheries in most freshwater environments. At the Bonneau Fish Hatchery in South Carolina, striped bass are stripped of milt and eggs to supply more than 100 million fry each year to 30 states, New Zealand, South Africa, and the former Soviet Union. (Photo courtesy of Santee Cooper Country.)

Zealand and Hawaii, North American species also have been released on other types of landforms. Some of these have been viewed as beneficial introductions, whereas others again seem of far lesser value than originally intended. These contrasts are illustrated in the release of some common North America mammals into other regions of the world.

The British Isles are continental islands—they once were contiguous with the European landmass and thus share much of the fauna found in western Europe. Nonetheless, the deciduous forests of England lacked an ecological counterpart of the North America gray squirrel (*Sciurus carolinensis*), and the North American species initially was released there in 1828. This and subsequent introductions were not successful, but 10 animals gained a foothold on the estate of the Duke of Bedford in 1890 (Laycock 1966). As the squirrels flourished, stock from this nucleus was transplanted to still other parts of England. Further importations from North America finally were curtailed early in the 20th century, but by then gray squirrels were firmly established on English soil. The value of the species as a game animal in North America was not realized in its new home. Neither did the squirrels become the delightful scamps of urban settings that

they were in the cities of eastern North America. Unlike their habits in North America, gray squirrels in England often fed on bark—not nuts—and irreparably damaged great numbers of trees with their girdling activities. Why this trait developed in the exotic population is not understood, but the habit of barking trees in England certainly made gray squirrels a nuisance in woodlot and park alike.[2] In any case, the damage was sufficient to effect a bounty on gray squirrels. Authorities even supplied squirrel hunters with free ammunition, but these controls proved fruitless and later were discontinued (Roots 1976).

---

[2]Changes in food habits are not unknown for other exotic species. When raccoon dogs (*Nyctereutes procyonoides*) were brought from eastern Asia as an addition to the furbearing fauna of western Siberia, they no longer fed on fishes and crustaceans as before, but instead preyed on game birds, hares, and poultry. Furthermore, their value as a furbearer in this region never was realized (Lindemann 1956). An article appearing in *Audubon* magazine (March 1983) noted that fur farms in Illinois and Wisconsin already housed about 50 raccoon dogs before this species was placed on the prohibited list by the U.S. Department of the Interior. In 1981, Canada also prohibited further importations of raccoon dogs and persuaded fur farmers to eliminate their current stocks and instead return to raising foxes.

The damage inflicted by gray squirrels was not limited to English treelife. A native squirrel likewise experienced difficulties as gray squirrels spread in England. As is often the case, a strict cause-and-effect relationship is not easily demonstrated, but endemic red squirrel (*Sciurus vulgaris*) populations diminished as the exotic North American species occupied England. The red squirrel is the only native tree squirrel in England, and lacking an aggressive nature, the species apparently was displaced by the larger gray squirrel. Red squirrels generally inhabit coniferous forests rather than deciduous trees, so competition for food and other resources with gray squirrels was not readily apparent. There remains the possibility that gray squirrels released in England carried a disease (coccidiosis?) that proved lethal to red squirrels, but such a relationship has not been proven. Nonetheless, using 10-km^2 grids, a national census of the ranges occupied by the two species showed the concurrent decline of red squirrels as gray squirrel populations expanded (Shorten 1946, 1953). After 15 years of occupancy by gray squirrels, fully 66 percent of the grids formerly occupied by red squirrels no longer contained the native species.

North American furbearers have been no less unpopular as transplants in other lands. Muskrats were stocked widely in Eurasia shortly after 1900 because of the economic value of their pelts. Within a few years, however, the aquatic rodents' troublesome activities (mainly tunneling in drainage systems and damaging crops) reached the point where government controls were authorized. England eventually was able to rid itself of muskrats. Elsewhere, muskrats remained valued only in Finland and the former Soviet Union (Laycock 1966).

Economic incentives also fostered the introduction of raccoons into the former Soviet Union. In czarist times, the Russian government encouraged the nation's rich fur trade, including experiments with furbearers imported from other countries (Redford 1962). These policies were continued by the Soviet government after the czarist empire fell in 1917 and eventually led to the introduction of nine species of exotic furbearers (Safonov 1981). Among these were 22 raccoons released in Central Asia in 1936. Additional transplants of both zoo and wild stocks continued until 1958, extending raccoon populations to 18 regions in the former Soviet Union (Aliev and Sanderson 1966). By 1964, the introduced raccoon population was estimated as more than 4 million animals, with recorded densities of 1.6 per km^2 in suitable habitats. Somewhat later, however, Safonov (1981)

found raccoon populations stable only in the Caucasus and Byelorussia—and there seems little chance for further range expansion. Deep, loose snows for much of the year may explain the ecological limitations working against further expansion of raccoon populations in the former Soviet Union. Harvests have not exceeded 35,000 pelts per year. As a management procedure in regions where natural cavities are limited, Russian biologists have encouraged raccoons by deploying artificial den structures like those described by Stuewer (1948). Interestingly, raccoons in the former Soviet Union no longer harbor most of their North American parasites; instead, their parasites now are those common to Asian wildlife (Aliev and Sanderson 1966).

Final mention can be made of the introduction in 1958 of large-mouth bass into Lake Atitlan, a large tropical lake in Guatemala. For at least 3 centuries, the native fishes and crabs in Lake Atitlan provided local natives with a productive source of food and commerce. Within 15 years of their release, however, the bass population decimated the native fishes as well as the crab population, and only swimmers equipped with diving gear can harvest the few large bass now in the lake. As a result, the native lost a critical source of protein, along with the income once generated by the native fish populations (Zaret and Paine 1973). Furthermore, the exotic bass population apparently extirpated six species of amphibians from Lake Atitlan and competes for food with an endangered grebe (*Podilymbus gigas*), perhaps preying on the chicks as well (LaBastille 1974).

FOR BETTER OR WORSE?

Exotic animals present a continuum of ecological and socioeconomic values (the same can be said for exotic plants). Old World rats fall at the bottom of the scale, remaining universally regarded as vermin. House sparrows are less noxious but generally are considered undesirable additions to the North American avifauna. Like Old World rats, house sparrows also are associated with human activities in urban and rural settings. Sika deer illustrate another gradation in the continuum. They were introduced exclusively for their sporting value, and while they are seemingly well established in Maryland and Texas, it may be too early to render a firm decision on their desirability. As we have seen, experimental evidence indicates that sika deer can displace white-tailed deer in a few years. Although still speculative, the same relationship seems plausible between Barbary sheep and desert bighorn sheep. Red-whiskered bulbuls (*Pycnonotus jocosus*), colorful birds native to India, escaped from captivity in 1960 and now thrive in suburban Miami without apparent disruption to the local community. Although considered agricultural pests in other areas, the birds do not imperil citrus crops and seem unlikely to spread beyond the influence of Florida's tropical climate and vegetation (Carleton and Owre 1975). However, another species of bulbul (*P. cafer*) introduced in Hawaii is spreading rapidly (Williams and Giddings 1984). Finally, pheasants and brown trout are as entrenched in North America as starlings and carp, but unlike the latter, these species are heralded widely as welcome additions. Nonetheless, even carp have some commercial value, as do nutria, adding another dimension to the role of exotics.

These examples illustrate a full range of human esteem for exotic animals, and herein lies the crux of dealing with their place in wildlife management.

These illustrations should be tempered by intent. Clearly, no one wanted Old World rats or even red-whiskered bulbuls as free-roaming "wildlife." These and many others simply escaped when given the opportunity. On the other hand, pheasants, carp, and starlings were released intentionally. A conscious decision by a few people—even a single person in some cases—was made on behalf of all sectors in our society. Starlings were released in New York's Central Park in 1890 by Eugene Schiefflin, a man driven to introduce all of the birds mentioned in Shakespeare's plays (Laycock 1966). "Nay, I'll have a starling shall be taught to speak nothing but Mortimer"—no more than that dramatic line from *Henry IV* sparked the successful release of birds that now cover most of North America. Even so, sporting values—real or potential—remain a powerful impetus for releasing exotics, especially where nature seemingly fails to provide a suitable native species for rod or gun.

What conclusions can be reached about the role of exotic wildlife? Regrettably, there are few pat answers, even when we confine our question to intentional releases of species with clear sporting values. Like beauty, views about exotics lie in the eyes of the beholder. Ranchers reaping $1000 or more per animal certainly appreciate exotic big game and, obviously, so do the hunters paying for such trophies. Any pheasant hunter would give a similar endorsement. Yet it is clear that many exotics also have disrupted entire biological systems after their release. A large body of evidence, some of which we have examined, strongly suggests that the ecological price often exceeds the expected benefits of exotic introductions.

Social goals are part of the picture. If we assume, for the moment, that more hunting and fishing opportunities are the only objectives of wildlife management, then exotics indeed might reasonably satisfy the demand for additional outdoor recreation. With some modest constraints, sika deer might be hunted year round without diminishing the resource they represent. Their trophy value and adaptability to some environments in the United States are well established. What, then, is the trade-off if sika deer are encouraged at the expense of white-tailed deer? Perhaps none if consumptive recreation is the only objective. This is, of course, a simplified look at a complex matter, but it illustrates the attainment of a specific objective. Further, our simple "this for that" example obviates the need for testing, as our objective is not to save whitetails but instead to promote sika deer. We would not even be too concerned if our exotic failed. If it did, others might be tried, or at worst there still would be whitetails to hunt. Coturnix quail failed despite exhaustive study. Without testing, starlings released in New York City expanded across North America (Kessel 1953).

Conversely, a broader range of sociological values may be judged of higher value than hunting. Most cultures recognize both the aesthetic and tangible values of their native biota. Native species distinguish the coinage, stamps, seals, and even flags of many geopolitical units throughout the world. National parks also are a direct reflection of the heritage and value society attaches to native communities. Presumably, these and other manifestations indicate a commitment endorsed by a large segment of the public. If so, is it then reasonable to conclude that most citizens might oppose exotic introductions made only on the behalf of sporting interests? One also can suppose, a priori, a preference among most Americans, including hunters and anglers, for bluebirds and bass instead of starlings and carp. Thus, what sector of society is best served when exotic introductions are considered, and who decides?

These questions aside, managers must face the fact that an array of exotic species now occurs in North America. For these it is no longer a question of whether they should have been released—they are here, and resource managers must consider them carefully in the planning and execution of their duties. In some cases, this may mean developing ways and means to limit the undesirable features of an exotic species. Nest boxes for wood ducks have been designed to reduce competition from starlings (Grabill 1977), but ways for reducing competition for forage between exotic and native ungulates still are unclear unless food is supplied artificially. Methods designed to remove or eliminate troublesome exotics also may be necessary, but such actions often bring disfavor from groups concerned with animal rights (Hutchins et al. 1982). Conversely, pheasant populations follow trends in agriculture, and with modern farming practices, their numbers no longer reach former levels of abundance. In this instance, managers seek ways of increasing pheasant numbers by developing roadsides or other "new" habitats (see Chapter 13). In all, the management of established exotics focuses on manipulating the strengths and weaknesses of each species in its new home.

GUIDELINES AND POLICIES

Three separate actions currently guide or control the importation of exotic biota into the United States. Two of these were enacted, respectively, by the U.S. Department of the Interior and by presidential executive order. A third is the policy statement of The Wildlife Society that, while not legally binding, represents the official position of a prominent organization of wildlife biologists (Labisky et al. 1975). Each of these policies highlights modern-day concerns for additional disruptions of native biota by imported wildlife. Whereas not limited to big game alone, these policies in the main reflect an emphasis on exotic ungulates.

U.S. Department of the Interior recommendations as a basis of public-land policy for exotic wildlife were adopted in 1966 by Secretary Stewart Udall. These remain in effect and include eight conditions for considering exotic introductions. In abbreviated form, these are as follows:

1. Critically determine that a need exists with desirable ecological, recreational, and economic impacts.
2. A definite niche is available and unsuited for a native species.
3. Introductions should not be considered if they threaten the reduction or displacement of native populations, nor should existing or proposed land uses be in conflict with an exotic transplant.

4. Introductions should be preceded by ecological studies of both the animal and the habitat proposed as the release site.
5. Disease relationships require special study as well as the steps assuring appropriate quarantine leading to disease-free stock.
6. Species with close relatives in the United States should be avoided to preclude hybridization with native wildlife.
7. Small-scale experiments and a thorough evaluation of these should precede larger introductions.
8. Before an exotic is released, methods for controlling its abundance and expansion must be available.

These criteria led to eight recommendations that became active policy; again abbreviated, they state as follows:

1. No decision regarding the importation of an exotic species should be made until a full assessment, based on thorough research, is at hand.
2. No exotic game should be permitted in national parks or other public lands devoted to the preservation of native biota or within a suitable buffer zone surrounding such areas.
3. Exotics should be excluded from the vicinity of rare or uncommon native species unless potential conflicts between them are proved unlikely.
4. Exotic grazers should be excluded from federal lands primarily devoted to, and fully stocked with, domestic grazers unless interspecific competition is insignificant.
5. In the absence of full coordination, exotic big game should be excluded from areas where timber production, farming, intensive recreation, or other land uses have primary value.
6. Exotics may be introduced on federal lands under permit, including a commitment by the appropriate state agency to control the size and distribution of the population consistent with the habitat's carrying capacity.
7. Exotic big game spreading into federal lands from other types of ownership should be treated as trespassing livestock, with the responsible party held liable for damages and the expense of control.
8. Public-land policy regarding exotics should be reviewed periodically and, when necessary, revised to reflect better the real and potential impacts of exotic wildlife on U.S. environments.

Jimmy Carter, 39th president of the United States, signed Executive Order 11987 on May 24, 1977, which, in part, (1) restricted federal agencies from introducing exotic organisms onto land they administer, (2) encouraged the prevention of exotic introductions by other levels of government and by private citizens, and (3) restricted federal support of exotic transplants outside of the United States. The limitations apply unless either the Secretary of the Interior or the Secretary of Agriculture determines that the introduction will not have an adverse effect on natural ecosystems.

In 1975, the governing council of The Wildlife Society approved a three-point policy concerning exotic introductions, as quoted:

1. Support the introduction of exotic species only after competent scientists have demonstrated that (a) the exotic can potentially satisfy a specific recreational or biological need in the ecosystem to which it will be introduced, (b) the exotic is ecologically suitable for introduction into the new ecosystem, (c) the exotic will not be deleterious to desirable species (native or exotic) or cause any deterioration of the ecological complex, and (d) the exotic has satisfied all appropriate quarantine requirements upon entry.
2. Urge that no state, provincial, or national agency shall introduce, or permit to be introduced, any exotic species into any area within its jurisdiction unless such species can be contained exclusively within that jurisdiction or unless adjoining jurisdictions into which the species could spread have sanctioned the introduction officially.
3. Exclude from the provisions of this policy the importation of exotic species by officially recognized scientific and educational organizations and the interinstitutional exchange of such species, provided that the exotics are maintained in captivity at all times. (Labisky et al. 1975.)

Each of these policy statements clearly indicates that exotic transplants in the past too often transpired without full recognition of the form and function of native ecosystems. Large sums of money have been expended attempting to correct past mistakes, and some environments still remain severely disrupted under the continuing influences of exotic organisms. Hindsight is far clearer than foresight, but the lessons of history sometimes are lost in the present. Public policies often are political matters, subject more to immediate pressures for results than to long-range implications. More often than not, decisions to import alien wildlife succumb to the politics of expediency. However, the above policies represent commitments to the future, and those believing otherwise might well recall the admonition of Aldo Leopold (1938) that exotic wildlife "has depleted the game funds of 48 states for a half-century and has served as a perfect alibi for postponing the practice of game management." Regrettably, Leopold's insight largely went unheeded for much of another half-century.

SUMMARY

Humans are responsible for intentionally or accidentally introducing many organisms into new environments throughout the world. Many introductions of exotic plants and animals have failed, others have become pests, and still others have improved outdoor recreation. Pheasants remain the most successful example of an exotic game animal introduced into the United States.

Reasons for exotic introductions are based on "filling" habitats where native species are no longer successful or are limited in abundance or were absent in the first place. Economic incentives also promote introductions of big game animals, especially in the southwestern United States.

Concerns about exotics include potentials for spreading pathogens, habitat modification, and hybridization or competition with native species. Competition for food between exotic and native animals seems especially crucial, but other forms of competition exist. Once released, exotics are no longer an experiment, and ways to control troublesome exotics seldom are effective. In some cases, exotic introductions suggest we have failed to manage native species adequately and instead have relied on searching elsewhere for replacements. Chukar partridges, Barbary sheep, reindeer, nutria, and brown trout are additions to the North American fauna, each representing the "pros and cons" of exotic introductions.

Native animals also have been transplanted within North America, again with debatable results in some cases. The sporting, aesthetic, or economic values of North American animals led to their export to other continents. Among these, gray squirrels have displaced native squirrels in England, but raccoons seem a valuable addition in parts of the former Soviet Union.

In all, exotic wildlife presents a continuum of values to humans, ranging from vermin to trophy big game. In proposing new introductions, managers must carefully assess a broad spectrum of social goals and not just those of hunters and anglers if wildlife management is to assume its fullest meaning. Policies have been established for considering further introductions of exotic animals and these presumably will preclude repetition of past mistakes. Nonetheless, large numbers of exotic animals already are well established throughout the world, and managers must cope with their presence.

CHAPTER 19

NONGAME AND ENDANGERED WILDLIFE

If a species becomes extinct, its world will never come into being again. It will vanish forever like an exploding star. And for this we hold direct responsibility.
David Day (1981)

Wildlife management for many years was equated with game management. Species perceived to be of lesser status—particularly songbirds—were relegated to either the concerns of Audubon chapters or, more often, to benign neglect. Although many birds, reptiles, amphibians, and other species that generally lack status as game animals have been studied for many years, the intent was to understand their basic biology and only rarely to manage their numbers or habitats.

It remained until the ecological awakening of the 1960s for the idea of matching management principles, long practiced solely with game species, with the needs of nongame animals (Bolen et al. 1980). The environmental movement initially arising during the presidency of John F. Kennedy spurred legislative action for the next decade, in parallel with the earlier era of conservation aroused by President Theodore Roosevelt. Not only did the mandates by legislative bodies foster attention to nongame species and a broad array of environmental issues, but so, coincidentally, did the public conscience. Membership in citizens' conservation groups rapidly increased, in turn, leading to even more action and reevaluations (Scheffer 1976). The impact was considerable and far reaching. No longer would massive schemes for moving soil and water progress unchallenged without prior consideration of the ecological and biological implications

of these actions. The nation's economic development came face-to-face with a quest for both qualitative and quantitative environmental standards in confrontations that often brought forth difficult choices and ill feelings. Nongame or endangered species were often at the heart of these matters.

Today, the wildlife profession has affirmed its commitment to nongame animals and their management. State and federal managers now are routinely assigned nongame responsibilities, both in basic research and in the management of populations and habitat. Attention to nongame species does not, and should not, exist in isolation; rather, nongame rightly is addressed as part an ecosystem replete with other species managed as game. Graul et al. (1976) formalized this concept with a plan integrating species management and ecosystem preservation. First, distributions are determined for all vertebrates undergoing known or suspected declines; emphasis is placed on those "indicator" species with narrow environmental tolerances and hence on those most susceptible to ecological disturbances. This eliminates immediate concern for animals with widespread distribution or those not declining in numbers. Second, the indicator species then are associated with specific ecosystems, thereby focusing the management plan on key environments. Third, management is implemented on those habitats within the key ecosystems where several indicator species occur. The ecological foundation for this approach assumes that the maintenance of the indicator species at desired levels concurrently means that the ecosystem also has been preserved. This approach, while not perfect, offers three advantages: (1) wildlife managers not yet trained in the complexities of ecosystem management can carry out the work because they are familiar with individual species, (2) it is adaptable to any level of funding or availability of personnel, whether increasing or decreasing, and (3) it allows wildlife agencies to initiate ecosystem management using existing data concerning the distribution and habitat requirements of specific animals. Graul and Miller (1984) later discussed the strengths and weaknesses of the ecological-indicator approach in comparison with other strategies for managing ecosystems.

DEFINITION

There are no universal biological criteria that define either game or nongame animals. The appropriate definition lies in the political arena and not within the spheres of the biological sciences or sporting image. Game animals are designated as such by legislative action (often guided by no more than tradition). Whereas mourning doves (*Zenaida macroura*) are prized gamebirds in the western United States, they are legislatively designated as songbirds in many eastern states. By subtraction, then, those species not listed as game are nongame. But the definition still is not simple. A nongame classification does not necessarily imply full protection and, indeed, may be the opposite. Mountain lions (*Felis concolor*) are a case in point. In Arizona, this species is considered a game animal, whereas in Texas it is a "varmint," and it is legal to kill one at any time by any means.

Still another complicating factor concerns those species designated as threatened or endangered. Whooping cranes (*Grus americana*), California condors (*Gymnogyps californianus*), and Kirtland's warblers (*Dendroica kirtlandii*), just to name a few, are endangered birds that clearly are not listed as game animals. They are nongame as well as endangered. Remnant populations of the greater prairie chicken (*Tympanuchus cupido*) are threatened in Wisconsin and Illinois but are hunted in Kansas, underscoring the geographical considerations of management goals for this and other species. The apparent convenience of placing such animals into neat administrative boxes often is fraught with overlapping lines of

management and budgetary authority. Species such as the bobcat (*Felis rufus*) may be assigned in some states to a furbearer management program or to an animal damage control unit as well as to an endangered species team.

Another attempt to contrast game and nongame animals involves whether they are used consumptively (hunted or trapped) or nonconsumptively (by birdwatchers, nature photographers, etc.). This is a convenient breakdown for many purposes, but most game species also are enjoyed by birdwatchers and photographers. Brocke (1979) therefore urged that professional biologists and the agencies they represent stress game and nongame *values*, not species. This proposal, including the simple administrative categories of urban, rural, and wilderness management, has merit; it transcends the choking array of organization boxes, budgets, and personnel that agencies often employ to meet their responsibilities (Fig. 19-1). Management efforts within each category address the animal systems of each, whatever the status of the species involved.

ECONOMIC VALUES

Expenditures, either direct or indirect, for wildlife by the U.S. public are immense. The dollar figures for nongame birds alone run into millions annually; such data are a powerful force partially justifying nongame management and research programs.

Based on purchases by urban dwellers in 5 large U.S. cities (Table 19-1), some 20 percent of U.S. households buy about 27 kg of birdseed annually (DeGraaf and Payne 1975). At the then-current value of $0.40/kg and with 15 million households actively feeding birds, the annual retail sales of birdseed in 1974 were estimated at $170 million (see also George et al. 1982). Other kinds of economic impact also are of significance: bird field guides ($3 million), commercially produced birdhouses and feeders ($15 million), gift books about birds ($4 million), Audubon Society memberships ($3.1 million), binoculars ($115 million, about half of all binocular sales), and cameras and camera equipment ($187 million if only 5 percent of the $3.7 billion spent annually on photography is attributed to nature photography).

Therefore, about $500 million is spent annually on the nonconsumptive "use" of wildlife, primarily in appreciation of nongame birds. This is a conservative estimate but it nonetheless represents considerably more

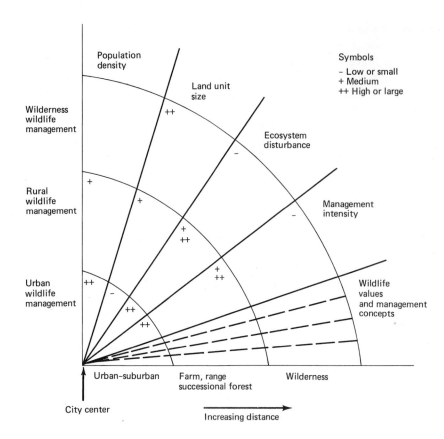

Figure 19-1. A conceptual scheme for wildlife management dealing with a continuum of conditions. Wildlife values (hunting, birding, scientific, etc.) occur in each management category but in differing intensities. (From Brocke 1979.)

TABLE 19-1.   Purchases of Birdseed by Households in Five Major U.S. Cities, 1972

City	Number of Households	Households Feeding Birds (Percent)	Average Annual Purchase per Household (kg)
Milwaukee	442,804	19.4	29.2
Cleveland	656,487	24.7	26.1
St. Louis	750,164	19.8	29.2
New York	3,949,454	15.1	22.3
Boston	861,024	23.8	31.6

Source: DeGraaf and Payne (1975).

than the $300 million spent by duck hunters in 1974 seeking a consumptive use of waterfowl resources (all data exclude food, transportation, and lodging costs).

Endangered species that can be viewed without jeopardizing their safety often render significant economic impacts to regional economies. The Corpus Christi Chamber of Commerce estimates that visitors to Aransas National Wildlife Refuge spend in excess of $1 million annually in transportation, lodging, and other activities associated with viewing whooping cranes. The Colima warbler (*Vermivora crissalis*) also is a specialty species, in part attracting bird-watchers to Big Bend National Park. Throngs of people expend considerable sums in the process of observing gray whales (*Eschrichtius robustus*) migrating along the California coast or watching the undisturbed grazing of bison (*Bison bison*) in several national parks and wildlife refuges (Fig. 19-2). A collage of species at such places as the Everglades, Yellowstone, or Hawk Mountain also attracts large numbers of people each year. Untold millions are spent by visitors to these and numerous other places where animals live and interact with each other in natural settings.

Even species otherwise considered as game animals may yield important economic returns in a nongame capacity. An Illinois refuge managed for Canada geese (*Branta canadensis*) operated with a budget of $175,000; using an average population of 200,000 birds, the operating costs were $0.88 per living bird and $6.00 per goose shot by hunters. Against these cost figures were returns to the county of $11 spent per bird by persons viewing the birds and $45 spent

Figure 19-2. Bison are one of many attractions at several national parks and wildlife refuges. At Theodore Roosevelt National Park, above, a free-roaming herd of nearly 300 bison attracts an average of 610,000 visitors each year. (Photo courtesy of U.S. Fish and Wildlife Service.)

per bird by hunters (Arthur 1968). This represents a twelvefold return for nonconsumptive users versus a sevenfold return by consumers. Further, the same birds generating the higher rate of return remained alive potentially to incur their economic value another time; the birds shot by hunters were a one-time product by comparison (although hunters, too, enjoy seeing flocks of geese, not just the ones they shoot). In short, the aesthetic values of this goose population brought a higher monetary return than did the consumptive use of the resource, but both uses were compatible with management of the goose refuge.

AN UNFINISHED AGENDA

In 1975, the Wildlife Management Institute published a comprehensive examination of the funding base needed for nongame management, and it proposed model legislation to meet those needs. The report briefly spurred legislative action, but the bills languished in committee without reaching the floor of Congress.

Renewed legislative efforts were successful when the Fish and Wildlife Conservation Act became law in September 1980. Under the act, states could plan and/or implement nongame projects funded by a 4-year appropriation of $20 million. The act also charged the

U.S. Fish and Wildlife Service to study alternate ways of funding the program after the initial funding expired. Regrettably, the funding authorized by the act was never appropriated, although the study of alternate funding was completed. Among the 18 potential sources of revenue listed in the report was a proposal to levy user fees on various kinds of recreational equipment and other items (U.S. Fish and Wildlife Service 1984). However, neither this nor any of the other proposals was adopted, and while Congress later renewed the act, funding once again was not appropriated.

A new effort is now at hand: *Teaming with Wildlife* (TWW), a proposal to protect nongame species before they become endangered. TWW, which flows from the initiative of the 1980 Fish and Wildlife Conservation Act, targets nongame management as a formal element of the larger issue of preserving biodiversity. Also known as "The Wildlife Diversity Funding Initiative," TWW mimics the highly successful Pittman-Robertson and Dingell-Johnson legislation described in Chapter 22 (i.e., they each collect user fees). Hence, by expanding on the financial footing already provided by "P-R" and "D-J," species that are neither hunted nor fished also will benefit from responsible wildlife management.

If enacted, TWW will levy a modest fee—0.25 to 5 percent—on the manufacturer's price of such products as birdseed, field guides, canoes, sleeping bags, tents, and other products associated with outdoor recreation. The cost to consumers is small because the fee is *not* based on the retail price. For example, if a manufacturer sells a $100 tent to a retailer for $50, the 5 percent fee increases the manufacturer's price by $2.50, and the consumer now pays $102.50 for the tent. The amount of the fee will appear on the manufacturer's invoice to the retailer, and green tags will inform purchasers that a portion of the retail price will be used for conservation. Funds collected by TWW will be allocated to each state, using a formula based on population $\left(\frac{2}{3}\right)$ and land area $\left(\frac{1}{3}\right)$. Each state must match its allocation with nonfederal dollars at a ratio of 25 percent state to 75 percent federal. The program has a natural constituency: In 1991, 108.7 million people in the United States spent $59 billion feeding birds, hiking, camping, hunting, fishing, and in other types of outdoor recreation (Franklin and Reis 1996).

TWW hopes to raise $350 million each year to fund three target areas:

Conservation: emphasis placed on preventing species from becoming endangered (e.g., conduct inventories to determine the status of plant and animal populations; take action to prevent population losses; conserve habitat; restore native species to their original ranges).

Education: emphasis placed on fostering an ethic of responsible stewardship for wildlife resources (e.g., interpret nature along trails, campgrounds, picnic locations, and other viewing areas; establish and maintain nature centers; offer wildlife and outdoor recreation programs; issue activity guides for school and communities groups).

Recreation: places emphasis on enhancing outdoor recreation (e.g., provide access areas for hikers, photographers, boaters, bird-watchers, and other outdoor enthusiasts with trails, blinds, observation towers, and other accommodations; establish a nationwide network of wildlife viewing areas; publish and distribute guides for wildlife habitat in backyards).

STATE FUNDING FOR NONGAME MANAGEMENT

Several states have established specific funding sources for nongame programs. Colorado was the first state to do so when it provided a checkoff on state income-tax forms. Taxpayers voluntarily could contribute a part of their tax refund for nongame management by marking the appropriate box on the tax form (the idea mimics the checkoff for presidential campaigns appearing on federal income-tax forms). In the first year, 1978, $350,000 was raised for nongame in this manner, and by 1980, the amount increased to $650,000 when nearly 12 percent of Colorado's taxpayers contributed an average of $5 of their income-tax refund to nongame programs. Checkoffs in the United States produced a total of $9.4 million in 1986. By 1994, 36 states permitted voluntary funding for nongame management from state income-tax refunds (Fig. 19-3).

Because the checkoff idea was successful, however, the tax forms in some states later included a number of competing options—cultural activities and prevention of child abuse, among others—with the result that revenues for nongame often diminished (Stranahan 1987). In Colorado, for example, checkoff income dropped to $398,000 in 1985 after reaching a peak of $692,000 in 1982. In periods of economic difficulties, persons due tax refunds may be less willing to contribute funds for any purpose and instead request full payment. Because effective management requires continuity, the uncertainty of year-to-year funding remains a major drawback of checkoff programs (Harpman and Reuler 1985; see also Mangun 1986). Nonetheless, tax forms reach 100 percent of the potential contributors, cost wildlife agencies nothing to produce, and yield far greater returns than the T-shirts, decals, stamps, and paintings that have been marketed for nongame funding (Applegate 1984).

Some states fund nongame programs in other ways. In Washington and North Carolina, sales of special license plates support nongame management and research. A $4 surcharge imposed on new vehicles registered in Florida brought in $2.2 million for nongame in 1985. Part of the state sales tax in Missouri is dedicated to conservation programs, including the management of nongame.

In New Jersey, contributors to the checkoff program included equal groups of hunters as well as nonhunters, including antihunters (Applegate 1984). Nearly 50 percent of the hunters surveyed in Virginia agreed that nongame could be managed using license fees (Moss et al. 1986). Such findings seem significant because they indicate an important overlap in the consumptive and nonconsumptive users of wildlife (see Connelly et al. 1985). So, to some degree, funds normally expended exclusively for the management of game species may be spent legitimately on a wider spectrum of wildlife (i.e., nongame).

Figure 19-3. Roadside sign encourages checkoff contributions for the nongame program in New Jersey. Taxpayers voluntarily dedicate part of their state income-tax refunds to nongame research and management. By 1994, checkoffs were available on the tax forms in 36 states. (Photo courtesy of New Jersey Division of Fish, Game, and Wildlife.)

Numerous surveys describe the characteristics of contributors to state checkoff programs (e.g., Applegate 1984; Carothers and Knight 1984). After completing a survey of donors and nondonors in Oregon, Manfredo and Haight (1986) recommended that wildlife managers bolster checkoff programs by (1) clarifying the meaning of "nongame;" (2) describing projects supported by the funds so that potential donors will understand what happens with their money; (3) soliciting opinions as well as donations, thereby increasing public involvement with nongame management; and (4) working with professional tax preparers so that their clients become more aware of the checkoff option (see also Applegate and Trout 1984).

STATUS AND CONCERNS OF NONGAME MANAGEMENT

Thompson (1987) provided an overview of nongame programs in the United States. With the exception of South Dakota and Vermont, all states have adopted formal programs for nongame and endangered species. Statutory authority has extended broadly to vertebrates in most cases but has been more limited in the case of invertebrates and plants. Budgets averaged $292,000 per program, but these varied considerably (range: $30,000 to $1.1 million). Tax checkoffs largely were responsible for generally improved

budgetary situations, but 13 states reported net reductions in funding. An average of 5 persons per state worked on nongame and endangered species. According to the survey, the greatest needs concern developing reliable funding sources, expanding statutory authority for nongame management, and integrating nongame programs into routine wildlife operations. In England, societies and trusts usually assume the initiatives for wildlife conservation, including concerns for species lying well outside the realm of game management (Fig. 19-4).

Various surveys reflect the considerable interest and involvement of the U.S. public in wildlife (e.g., Shaw and Mangun 1984). Nonetheless, Jackson (1982) noted that the public commitment of time and money for nongame wildlife seems less than expected and therefore identified four areas of concern:

1. *Nongame* is a poor name and lacks identity. Biologists define other wildlife in positive terms (e.g., *big game*, *songbirds*, and *sport fish*), but the term *nongame* lacks any descriptive power for public appreciation. Similarly, Kellert (1984) contended that broad, adequately funded programs for nongame will not be established easily until the artificial and negative distinction between game and nongame species is replaced with an emphasis on *comprehensive* wildlife management. *Nongame* also lacks any biological identity (e.g., songbirds

Photo courtesy of Pete Dunne

Cape May Bird Observatory

Cape May Bird Observatory (CMBO) is the keystone facility of the New Jersey Audubon Society, an independent organization founded in 1897. Cape May, at New Jersey's southern tip, is among the foremost sites in North America for watching the autumn migration of birds, especially hawks, shorebirds, and seabirds, but also for watching migrating butterflies. The significance of the location stems from a combination of wind and geography, which together funnel millions of birds each year to Cape May, where they concentrate before continuing southward across Delaware Bay.

Alexander Wilson (1766–1813), the "father of American ornithology," praised the bird life at Cape May, as did John James Audubon (1785–1851) and the 20th century's preeminent artist-naturalist, Roger Tory Peterson (1908–1996). Accordingly, CMBO was established in 1975 to promote the understanding and appreciation of the needs of resident and migrating birds so that they are not undermined by human activities. CMBO's program centers on research, environmental education, bird conservation, and recreational birding.

In 1997, CMBO opened its showpiece, the Center for Research and Education, which operates 7 days a week and features displays, a 170-seat lecture room, a bookstore, offices, and a viewing deck surrounded by 10.5 ha of saltmarsh and uplands. The center affords access to the wealth of biological resources along New Jersey's shoreline on Delaware Bay. A second facility, the Northwood Center, lies at the tip of the Cape May Peninsula—the epicenter for birding activities. Also open all week, the Northwood Center provides visitors with travel and birding-related information and a store where field guides and other equipment are sold. CMBO is the centerpiece of ecotourism in southern New Jersey: Birders spend about $31 million annually in Cape May County, where 700 jobs also originate from recreational birding.

CMBO supports applied research, including censuses and ecological studies of migrating raptors, winter surveys of bald eagles, and work designed to ensure protection of North America's second-greatest population of migrating shorebirds (e.g., immense numbers of red knots and ruddy turnstones). Because the shorebirds feed heavily on the eggs of horseshoe crabs breeding in Delaware Bay, the research agenda also includes these unusual invertebrates.

More than 400 species of birds, of which 200+ have been seen in a single day, provide the core of recreational birding at CMBO. At the peak of migration, 1.5 million shorebirds pass through Cape May, as well as 60,000 hawks and more than a million cormorants, sea ducks, loons and other seabirds. Workshops, nature weekends, field trips, and other activities occur throughout the year. The "World Series of Birding" is a high-profile event originated by CMBO staff; each May, teams of 3–5 birders compete for 24 hours to find the most species of birds in New Jersey. CMBO publishes a newsletter, *The Kestrel Express,* and a journal, *The Peregrine Observer,* to keep its members informed about birding events at Cape May.

are nongame to some people; reptiles are nongame to others).

2. Nongame laws are inconsistent and are selectively enforced. Large sectors of the public sometimes view those species that are "obscure, slimy, or scaly" as retarding progress (e.g., on highways or dams). Snails and spiders enjoy little public value, but public appreciation rises with complexity so that toads, turtles, birds, and mammals gain progressively more appreciation. Thus, for species with weak public support, an approach that advocates habitat protection may be a better strategy than one that enacts species-specific laws.

3. Public attitudes about the uses of wildlife are polarized. For example, the moralistic/humanistic views typical of antihunters clash with the ecologistic/ dominionistic/naturalistic attitudes associated with hunters (Kellert 1976a,b; see also Table 10-1). Overall, only 7 percent of the U.S. public holds three strongly ecologistic attitudes, which is the sector that actively supports wildlife conservation and management.

4. Interest in nongame wildlife is perceived as influencing antihunting. Some sectors of the public suspect that nongame enthusiasts are aligned with the antihunting movement. Therefore, with greater

Figure 19-4. In England, tunnels were constructed under roadways where large numbers of toads otherwise might become traffic victims. "Toad tunnels" reflect a mature concern for the welfare of all wildlife and gained widespread recognition in the world press. (Photo courtesy of Times Newspapers Ltd., London.)

funding, nongame management might gain undue influence and diminish the importance of hunting-oriented management. Although such views are erroneous, the perception nonetheless weakens support for nongame management.

OUT OF AFRICA: VULTURE RESTAURANTS

A uniquely conceived plan for vultures offers a prime example of nongame management in a part of the world known for its many species of big game. Eight species of vultures occur in southern Africa, and of these, only the white-backed vulture (*Gyps africanus*) is not considered either rare, vulnerable, or endangered. For the others, especially the Cape griffon (*G. coprotheres*), sites artificially supplied with large animal carcasses—known as *vulture restaurants*—have enhanced the conservation of these large and often unappreciated birds (Mundy 1983; Friedman and Mundy 1983).

The concept of vulture restaurants originated in the 1966 as a means of feeding bearded vultures (*Gypaetus barbatus*) in South Africa. Similar feeding areas later were established in Europe for a dwindling population of griffon vultures (*Gyps fulvus*) and in the United States for an endangered species, the California condor. By the 1970s, vulture restaurants had become a management tool for the Cape griffon, a species listed as threatened by African biologists (Mundy et al. 1992).

Vulture restaurants are operated for four somewhat overlapping reasons (Butchart 1988). First, the stations provide the birds with a source of safe and available food. Because some African cultures believe certain body parts of vultures possess medicinal or magical properties, livestock carcasses are sometimes poisoned as a means of obtaining birds. In other areas, encroaching brush has limited accessibility to naturally occurring sources of food, so the restaurants, which are located in open areas, receive exceptional utilization. In still other areas, food sources may be scarce, and the restaurants help remedy such shortages. Additionally, the availability of food offers a means of luring Cape griffons back into areas they earlier abandoned.

Second, the restaurants provide the birds with a crucial nutrient while illustrating the complex workings of some food webs. A calcium deficiency was suggested when biologists discovered skeletal deformities in nestling Cape griffons breeding in some ranching areas of South Africa where livestock carcasses were available for scavenging. Missing from the scene, however, were spotted hyenas (*Crocuta crocuta*), which have been eliminated from much of their range in South Africa. Feeding hyenas crush the bones of large carcasses, thereby leaving small bone fragments that adult griffons transported to their nestlings. In the absence of spotted hyenas, however, small bones were no longer available. The diet of the young birds therefore lacked sufficient calcium, resulting in osteodystrophy, the abnormal skeletal development of the nestlings. The incidence of these deformities dropped from an average of 17 percent to 2.5 percent when crushed bones were supplied in the food provided at vulture restaurants in areas where spotted hyenas had been eliminated (Richardson et al. 1986).

Third, vulture restaurants have sparked opportunities for tourism and conservation education. Blinds situated near the restaurants provide fine setups for photography and observation of vultures and other scavengers. Display boards advise visitors to these sites about the ecology and conservation of the birds. Public opinion about vultures improves after visitors observe the birds at close quarters.

Fourth, the sites are useful for research. For example, young vultures color marked before fledging may be resighted at the restaurants as they mature, thereby allowing scientists to monitor the movements and behavior of individual birds.

Vulture restaurants require some degree of management (Butchart 1988). The birds are not very discriminating in their choice of foods, and carcasses from virtually any large animal are suitable forage. However, domestic pigs should not be provided regularly because of the high fat content of their flesh. Management is enhanced when the restaurants are located near sources of carrion (e.g., feedlots). Animals euthanized with barbiturates should not be used because these drugs remain viable in body tissues for some time, and secondary poisoning kills the birds. As might be expected, most humans do not enjoy handling carcasses, especially those that are partially decomposed; hence the work often is unpleasant, and volunteers may be hard to find. Where needed, adequate amounts of crushed bones are easily supplied by a worker with a hammer; a nestling's calcium requirements are fulfilled by about three dozen small (8 g) bone fragments consumed during a 60-day period (Richardson et al. 1986). Vulture restaurants should be located in open areas where visibility is good and where these large birds can land and take off without obstruction. Such locations also should be managed to minimize human disturbances, and for obvious reasons, powerlines should be avoided when selecting sites for restaurants (Mundy et al. 1992). Finally, the availability of carrion may attract feral dogs and other species of undesirable scavengers. Hence, when funds permit, a fenced area of at least 50 × 50 m is recommended.

ENDANGERED SPECIES

Species judged as endangered with extinction are listed by state, federal, and international agencies as well as by some private organizations. The most-

cited of these lists is the *Red Data Book*. This volume, continually updated, is issued by the International Union for Conservation of Nature and Natural Resources (I.U.C.N) located in Morges, Switzerland. "Red," of course, is symbolic of the danger that these species, both plants and animals, presently experience throughout the globe. The *Red Data Book* was first issued in 1966 by the I.U.C.N.'s Survival Service Commission as a guide for the formulation of policy, preservation, and management of the species listed. Information for endangered mammals and birds is more extensive in the *Red Data Book* than for other groups of animals and plants because of the prominence in the world's biota, but coverage also is given to less-prominent organisms facing extinction.

The I.U.C.N. found that endangered species usually reach their dire status because of one or more of five main factors (Fisher et al. 1969).

NATURAL CAUSES.   Extinction is recognized as a natural biological process consistent with the concepts of evolution. Based on the fossil record, birds have a mean species life span of about 2 million years and mammals of about 600,000 years. Extinction by natural causes may mean the actual death of a species or the evolution of the stock into one or more newer forms. Overspecialization, competition, sudden climatic change, or catastrophic events such as earthquakes or volcanic eruptions are among the natural causes of a species' death.

HUNTING.   Unwise and unregulated hunting has contributed to the extinction of some species (e.g., passenger pigeons, *Ectopistes migratorius*). This category also includes the diminishing populations of nontarget species that mistakenly bear the brunt of hunting, trapping, poisoning, or other lethal actions. Special cases concern those animals considered as exceptional trophies or of commercial value (e.g., whales, spotted cats, and certain ungulates with exceptional horn or antler development) and those possessing allegedly mystic or medicinal properties (e.g., Javan rhinoceros, *Rhinoceros sondaicus*).

INTRODUCED PREDATORS.   Introduction of exotic species to ecosystems often plays havoc with natural systems, especially when a predator is introduced. Native species seldom are adapted to cope with such a new and often devastating component of their environment. Sometimes the exotic predator in fact was introduced to control another exotic species, thereby compounding the biological effects on the native ecosystem (e.g., mongooses, *Herpestes auropunctatus*, introduced to prey on rats, *Rattus* spp., in Hawaii). The exotic predators regularly turn to the native fauna in their search for prey.

NONPREDATORY EXOTICS. Nonpredatory exotics are primarily the agents of competition and disease, although other effects may be attributed to their introduction. Goats ruined much of the habitat in Hawaii, and Asian cervids (e.g., sika deer, *Cervus nippon*) effectively outrival populations of white-tailed deer (*Odocoileus virginianus*) in central Texas. Throughout much of North America the now-widespread starling (*Sturnus vulgaris*) successfully preempted many of the cavity nesting sites once used by eastern bluebirds (*Sialia sialis*).

HABITAT MODIFICATION. Habitat destruction and disturbance are the ultimate destroyers of wildlife. Humans, of course, are the prime mover—and remover—of soil, water, and vegetation. Drainage schemes, reservoir construction, industrial and agricultural development, and deforestation, among a host of other human-induced changes, too often are the basis for the threatened status of plants and animals.

Assessment of these agents against the birds and mammals listed in the *Red Data Book* strongly indicates that humans have been the prime cause of extinction since A.D. 1600 and, unfortunately, continue to exert a negative impact. Only a quarter of the species meeting with extinction in modern times apparently have died out naturally, and humans seem responsible for the balance, including those currently threatened (Table 19-2). Species limited to island ecosystems are especially vulnerable to the presence of civilization, but the data in Table 19-2 include ample numbers of continental species as well.

What features may predispose certain animals to becoming endangered? Two or more features may be involved, and of course no single species possesses all of the following characteristics, although polar bears (*Ursus maritimus*) come uncomfortably close to a composite example. The list includes:

- Species with narrow habitat requirements or restricted distribution, especially if limited to islands or isolated bodies of water. For example: Puerto Rican parrot (*Amazona vittata*), pahrump killifish (*Empetrichthys latos*), Indiana bat (*Myotis sodalis*), and golden-cheeked warbler (*Dendroica chrysoparia*).
- Species of economic importance, especially those crossing national boundaries or living in international territory. For example, blue whale (*Balaenoptera musculus*), Atlantic green sea turtle

TABLE 19-2. Percent Assessment of Causative Agents Leading to Extinctions Since A.D. 1600 and the Current Causes of Enthreatenment for Birds and Mammals

Cause	Birds			Mammals
	Large Nonpasserines	Small Passerines	Total	
Leading to extinction:				
Natural	26	20	24	25
Hunting	54	13	42	33
Introduced predators	13	21	15	17
Nonpredatory exotics	—	14	4 } 76 human	6 } 75 human
Habitat modification	7	32	15	19
Total causes of extinction	100	100	100	100
Leading to enthreatenment:				
Natural	31	32	32	14
Hunting	32	10	24	43
Introduced predators	9	15	11	8
Nonpredatory exotics	2	5	3 } 68 human	6 } 86 human
Habitat modification	26	38	30	29
Total causes of enthreatenment	100	100	100	100

Source: Fisher et al. (1969)

(*Chelonia mydas*), ocelot (*Felis pardalis*), and Atlantic salmon (*Salmo salar*).

- Species of large size, especially predators, or those intolerant of humans (or vice versa). For example, grizzly bear (*Ursus arctos horribilis*), bald eagle (*Haliaeetus leucocephalus*), Asiatic lion (*Panthera leo persica*), and gray wolf (*Canis lupus*).
- Species that have limited numbers of offspring per breeding or long gestation/incubation periods or that require extensive parental care. For example, Mountain gorilla (*Gorilla gorilla beringei*), Mississippi sandhill crane (*Grus canadensis pulla*), Abbott's booby (*Sula abbotti*), and California condor.
- Species with highly specialized physical, behavioral, or physiological adaptations or high genetic vulnerability. For example, manatee (*Trichechus manatus*), Auckland flightless teal (*Anas aucklandica*), red wolf (*Canis rufus*), and giant panda (*Ailuropoda melanoleuca*).

A BRIEF HISTORY OF NONGAME AND ENDANGERED SPECIES

Although concern for endangered species usually is identified initially with the formal enactment of the Endangered Species Preservation Act in 1966 and its successors (discussed later), other, and often less heralded, actions tracing back into the early years of the 20th century also affected nongame and endangered species. Among these is establishment of Pelican Island National Wildlife Refuge on the Indian River near Sebastian, Florida, the first unit of a now extensive federal refuge system. President Theodore Roosevelt responded to the pleas of conservationists, among them the noted ornithologist Frank Chapman, with an executive order protecting Pelican Island from the ravages of poachers. The island's shrubby undergrowth of mangroves attracted nesting populations of wading birds whose plumage was the target of the lucrative millinery feather trade. Herons and egrets (Ardeidae) were slaughtered for their plumes, steadily diminishing their numbers as well as interrupting the breeding activities of the survivors. Anglers, believing their catches of commercial fishes were reduced by brown pelicans (*Pelecanus occidentalis*), also visited the island to break the birds' eggs or to destroy their young. Therefore, with a stroke of his pen on March 14, 1903, Roosevelt created not only the first national wildlife refuge, but also formally endorsed the conservation of beleaguered species.

Passage of the Migratory Bird Treaty in 1916 (following several years of frustrated attempts for federal regulations) gave international recognition to both migratory game and nongame birds (Article I). Some birds, previously subject to hunting, such as the wood duck (*Aix sponsa*), were given protection by closed seasons, but the treaty also afforded protection for shorebirds and, notably, for whooping cranes. Article VIII of the treaty led to the enactment of the Migratory Bird Treaty Act in 1918 in the United States and to its Canadian counterpart, the Migratory Birds Convention Act in 1917. These acts brought the tenets of the treaty into national law: Each country issued regulations and provided the policing powers for their enforcement.

The decades that followed witnessed actions that in one way or another affected wildlife. Most of these circumstances were directed to the needs of game species, but benefits also were reaped by nongame animals. Expansion of the National Wildlife Refuge System (funded in large measure by hunters purchasing "duck stamps") set aside habitat for waterfowl, but numerous other species of aquatic and semiaquatic creatures were housed concurrently in these same wetlands. The dual traumas of economic depression and widespread drought in the 1930s fostered heavy federal investment of money and personnel; the Tennessee Valley Authority, Soil Conservation Service, and Civilian Conservation Corps were among the programs involved with the nation's recovery. Each was geared to land and water management that often had significant impact on the creation or maintenance of wildlife habitat.

As an upwelling of World War II, the pesticide popularly known as DDT came into widespread use. This chemical was formulated in 1874 but had languished unused until its impact as a cheap and effective insecticide was recognized more than a half-century later (Cottam 1969). Allied troops were quickly freed of body lice, and civilians of war-torn nations were saved from plagues of typhus and other arthropod-borne diseases with liberal applications of DDT. Following the war, DDT was released for public use on farms and forests to attack a host of insect pests. Thousands of tons were produced and applied annually; landscapes of every type were dusted or sprayed with the newest weapon in civilization's age-old fight against pestilence.

Songbirds were among the first nontarget species to suffer from DDT and its sister chemicals in the family of chlorinated hydrocarbons (Hotchkiss and Pough 1946; Robbins and Stewart 1949), but early reports mention other animals as well (see Linduska 1948). Documentation was especially forceful when the mortality of robins (*Turdus migratorius*) and other songbirds was related to DDT used to control Dutch elm disease (Fig. 19-5); robins suffered about 90 percent mortality within 17 days of the application (Hickey and Hunt 1960; Hunt 1960).

These accounts dealt only with direct losses of songbirds; more was to come. Other types of nongame birds were experiencing the latent effects of DDT. These were the birds of prey, including the peregrine falcon (*Falco peregrinus*), bald eagle, and brown pelican. Few deaths occurred, but populations of these carnivores were rapidly declining nonetheless. In this instance, pesticides accumulated through the food chain and inhibited the transfer of calcium to the female's oviduct so that the bird's eggshells were unnaturally thinned and broke during incubation (Hickey and Anderson 1968). With so few young surviving, even the normal rate of adult mortality quickly reduced populations to alarmingly low numbers.

These and related events largely concerned nongame species. As such, they were of considerable interest to professional biologists and some amateur bird-watchers, but overall public awareness was minimal, perhaps even lacking. It remained for Rachel Carson to arouse widespread concern with the publication of her best-selling book, *Silent Spring*. In 1962, this courageous author sent shockwaves across the nation and into the halls of Congress and industry. *Silent Spring*, its title grimly reflecting the loss of

breeding birds, signaled to the U.S. public the critical stage to which its own environment had degraded, and it did so with more impact than anything published before or since (Trefethen 1975). A modern era of environmental concern was born, and nongame species had been its midwife.

Three legislative responses highlighted the new era. First, the National Environmental Policy Act (NEPA) without doubt was the capstone of the movement inspired by *Silent Spring*. NEPA became—and remains—a powerful tool for maintaining environmental quality when federal actions are proposed. It requires consideration of endangered species as well as other wildlife in making assessments before initiating actions with environmental implications (see Chapter 22). The other two responses were the Endangered Species Act, including its successors, and the Fish and Wildlife Conservation Act of 1980.

ENDANGERED SPECIES LEGISLATION

Whereas legislation designed to protect certain animals has a long history in the United States, the first act specifically addressing species threatened with extinction was the Endangered Species Preservation Act of 1966 (Johnson 1979). As we have noted, the first federal refuge was created to protect wading birds. Other measures included protection of bison when the National Bison Range was established in 1908 and the Bald Eagle Protection Act of 1940 as a means of guarding the national symbol of the United States. But

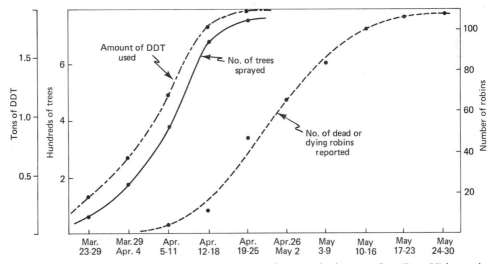

Figure 19-5. Cumulative lag between application of DDT and robin mortality. (From Hickey and Hunt 1960.)

Photo courtesy of Magnum Photos, Inc.

**Rachel Carson
Writer and Conservationist
(1907–1964)**

A shy but courageous woman, Rachel Carson played a key role in establishing the modern era of environmental awareness. She wrote several books on marine biology including *The Sea Around Us* (1951), but *Silent Spring* (1962)—a thoughtful and exquisitely written exposé of the harmful effects caused by DDT and other pesiticides—brought Rachel Carson into the national limelight. *Silent Spring* produced an immediate and angry uproar from chemical manufacturers and the agricultural industry, whereas conservationists praised the book. The interest and controversy of her message generated a national audience and a runaway bestseller. She did not advocate an end to pest control but instead urged safer applications of pesticides and, es-

pecially, development of other methods for controlling pests (e.g., biological controls). She hoped both individual citizens and governments would address what she felt was the most important and neglected aspect of all—the effects of pesticides on future generations. In large measure, the U.S. Environmental Protection Agency was established because of the public concern that resulted from *Silent Spring*—truly one of those few books that changed the course of history.

Rachel Carson was born in Springdale, Pennsylvania, and earned degrees from Pennsylvania College for Women (B.A.) and Johns Hopkins University (M.A.). Several universities later awarded her honorary doctorates. In 1936, she became an aquatic biologist with the Bureau of Fisheries (later included in the U.S. Fish and Wildlife Service) and in 1949 began duties as editor-in-chief of the agency's publication. However, in 1952, she resigned her government job and devoted herself to writing full-time about conservation, ecology, and nature. Her concern for accuracy gave her writing credibility; yet her rich prose merged science and literature in ways rarely achieved.

Her books and articles won many awards, including medals from the John Burroughs Memorial Association, the National Audubon Society, the Garden Club of America, and the New York Zoological Society. In 1963, the National Wildlife Federation recognized Rachel Carson as Conservationist of the Year. The Rachel Carson National Wildlife Refuge (in Maine) was named in her honor, as was the Rachel Carson Estuarine Reserve in North Carolina. See Graham (1970), Brooks (1972), and Lear (1997) for more about the life, writings, and influence of Rachel Carson.

the 1966 legislation remains an explicit statement concerning endangered species and their protection.

The 1966 legislation authorized the secretary of the interior to determine the wildlife facing extinction in the United States. Further, the act sponsored research on those species, authorized acquisition of terrestrial habitat using funds from the Land and Water Conservation Act, and formalized the National Wildlife Refuge System as a means of protecting endangered species. In 1969, the Endangered Species Conservation Act expanded the original legislation to include all species of vertebrates and some invertebrate groups as well. Importation of endangered species or their products was prohibited, and addition of foreign species to the United States list was authorized by this legislation (Fig. 19-6). Another section led to formation of the Convention on International Trade in Endangered Species of Wild Fauna and Flora, popularly known as

CITES. Members of CITES, now about 125 nations, recognize the status of endangered species by listing them in one of three appendixes (see Johnson 1980):

Appendix I. species endangered by trade.
Appendix II. species that are, or may become, threatened by trade.
Appendix III. species that, in the view of a member nation, should be subject to regulation within that nation as a means of conservation.

Each member nation of CITES is required to appoint a management authority for issuing import or export permits for species listed in the appendixes and to appoint a scientific advisory authority. In the United States, the secretary of the interior is the management authority operating through the U.S. Fish and Wildlife Service. The Endangered Species Scientific

Figure 19-6. A roomful of furs, leather goods, and other items confiscated under the importation and trade restrictions of the Endangered Species Conservation Act of 1969 represents only a small sample of the illegal products seized each year by federal agents. Jewelry and souvenirs made of ivory are among the items often confiscated from tourists returning from Africa and Asia (inset). (Photos courtesy of Steve Hillebrand, U.S. Fish and Wildlife Service.)

Authority (ESSA) originally advised the management authority (Brown 1978), but this role later was placed elsewhere when further reorganization was authorized.

In 1973, the Endangered Species Act again was modified, extending coverage to plants as well as animals. Only insect pests are excluded from protection. In letter and spirit, the 1973 act recognized endangered species as components of ecosystems and stressed that the integrity of ecosystems must be maintained because of scientific, aesthetic, educational, historical, and ecological values. It is important that distinctions were made between *threatened*

and *endangered.* In the new definitions, *endangered species* are those faced with extinction in all or much of their distribution, whereas *threatened species* are those that seem likely to become endangered. The term *rare* was dismissed because it could include thin populations of otherwise secure species at the edge of their distributions. The 1973 act also recognized separate populations and subspecies as *species,* thereby offering protection for vertebrates that otherwise might be excluded. For example, the Florida panther (*Felis concolor coryi*) is endangered, with perhaps no more than 50 animals of this subspecies of mountain lion remaining in Florida (Hendry et al.

1980), whereas other subspecies of mountain lions elsewhere in the United States are not necessarily diminished in numbers. Therefore, if subspecies had not been recognized in the 1973 act, the Florida population of mountain lion would not have qualified for protection because mountain lions in the western United States were relatively abundant.

The listing process may be initiated by any individual, agency, or group submitting a petition to the secretary of the interior; requests for delisting a species are initiated the same way. The weight of the supporting evidence supplied by the petitioner is judged by the U.S. Fish and Wildlife Service or, if a marine species is involved, by the National Marine Fisheries Service. The process of listing or delisting a species requires public notification in the *Federal Register* and contact with state governments and must be completed within 1 year. Criteria specified in Section 4 for determining a threatened or endangered species include (1) habitat destruction, (2) overexploitation, (3) threatened eradication from disease or predation, and (4) inadequate regulations for protection. The list of endangered species must be reviewed every 5 years by the Department of the Interior.

Emergency powers also are granted to the secretary of the interior for declaring a species as endangered. This procedure was used when a small herd of about 20 woodland caribou (*Rangifer tarandus*) living in northern Idaho was placed on the endangered list by the secretary of the interior. Woodland caribou once occurred in New England and the Great Lakes region, but the small herd in Idaho represents virtually all of this species still remaining in the United States outside of Alaska. Poaching, wildfires, and collisions with vehicles threaten the herd, and inbreeding seems the reason for reduced calf survival.

Section 7 of the 1973 legislation stirred several controversies, of which the endangered snail darter (*Percina tanasi*) received the most public attention. Language in Section 7 required all federal agencies to consult with the Department of the Interior about proposed actions that might jeopardize threatened or endangered species. However, the department can only recommend changes in these actions and not render final decisions about continuing, modifying, or stopping an action associated with other federal departments (Bean 1983). This language forced the issue between completion of the Tellico Dam on the Little Tennessee River and the welfare of an endangered species. Impoundment of the river seemed likely to ruin the existing habitat then known for the

snail darter. Arguments in the halls of Congress and in the public media pitted what was seemingly an insignificant little fish against the alleged benefits of completing the multimillion-dollar Tellico Dam.

In the course of the debate, little public mention was made of the limited economic importance of the dam; snail darters remained the highly visible culprit even though the project was not economically viable. Then-Secretary of the Interior Cecil D. Andrus pointed out that the long-range benefits of the dam were negligible and scarcely justified the immense cost of its construction. In a letter dated July 16, 1979, to the chairman of the Senate Committee on Appropriations, Andrus wrote, "The annual benefits of the project are $6.52 million, compared with annual costs of $7.24 million." Moreover, Andrus pointed out that the Tellico Dam failed to meet government standards for safety and that completion of the dam would not add appreciably to overall production of electricity by the Tennessee Valley Authority. Nonetheless, Congress later specifically exempted Tellico Dam from the Endangered Species Act, and the dam was completed. Meanwhile, segments of the snail darter population were transplanted to other waters, and some previously undiscovered populations of snail darters later were found in other rivers, giving additional hope that the species will remain part of the North American fish fauna.

The snail darter controversy, along with some other conflicts between endangered species and economic developments, led to passage of the Endangered Species Amendments Act of 1978 (Johnson 1979). Section 7 was modified as a compromise for handling future conflicts, in part by requiring all federal agencies to assess the status of any species listed in the action area and thereafter to consult with the Department of the Interior about alternate actions to protect endangered species. If no agreement is reached on an alternate plan, the agency involved must apply for an exemption from a review board which then reports to a cabinet-level committee, popularly known as the "God Squad." Exemptions may be granted by the committee if (1) there are no acceptable alternatives to the proposed actions jeopardizing endangered species, (2) the action is in the public interest and does not violate any international agreement, (3) the benefits of the actions exceed the benefits of the alternatives, and (4) the actions are of regional or national significance and reasonable mitigation and enhancement procedures for the species are established (Fig. 19-7). Ironically, the committee decided to block completion of the Tellico Dam fol-

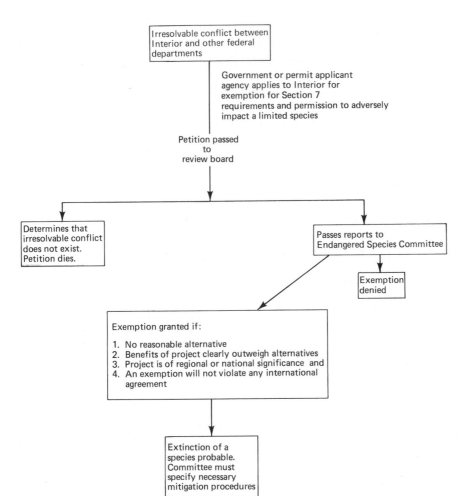

Irresolvable conflict between Interior and other federal departments

Government or permit applicant agency applies to Interior for exemption for Section 7 requirements and permission to adversely impact a limited species

Petition passed to review board

Determines that irresolvable conflict does not exist. Petition dies.

Passes reports to Endangered Species Committee

Exemption denied

Exemption granted if:

1. No reasonable alternative
2. Benefits of project clearly outweigh alternatives
3. Project is of regional or national significance and
4. An exemption will not violate any international agreement

Extinction of a species probable. Committee must specify necessary mitigation procedures

Figure 19-7. Flow chart outlining the exemption process to Section 7 of the Endangered Species Act of 1973. The process, prompted partially by the controversy over snail darters and the completion of Tellico Dam, was authorized by the Endangered Species Amendments Act of 1978. (From Johnson 1979.)

lowing these criteria and thus protected the snail darter population on the Little Tennessee River, but, as indicated, this decision was overridden by an exemption granted by the Congress.

TRIAGE: THE CRUSH OF DECISION

It seems tragically clear that not all species facing extinction can be saved or even assisted. Human assault on natural systems has been, and remains, awesome. Not enough funding exists to muster campaigns for the biota rapidly disappearing from our planet. Some estimates suggest that one species currently is lost each day, and by the end of the 1980s, the rate may increase to one per hour (Myers 1979). Thus, only a small percentage of endangered plants and animals can be scheduled for recovery activities, and by selecting these few, we necessarily deny efforts for assisting many others.

How may such decisions by reached? Previously, choices about endangered species were made largely on some arbitrary measure of public appeal and on a more-or-less glamorous image projected by the species in question. Whooping cranes and California condors thus have captured considerable public attention, but little publicity was given to the dusky seaside sparrow (*Ammodramus maritimus nigrescens*) or Bachman's warbler (*Vermivora bachmanii*). Among mammals, it remains far easier to generate concern for ocelots than for saltmarsh harvest mice (*Reithrodontomys raviventris*), and with few exceptions, fishes, amphibians, and reptiles enjoy little more support than the many invertebrates and plants threatened with extinction. Lovejoy (1976) accordingly proposed a system known as *triage*, an environmental adoption of the method for making the difficult choices that faced medics treating large numbers of soldiers wounded in World War I. With triage, wounded soldiers were separated into three groups: (1) those who likely would die despite medical attention, (2) those with light wounds who likely would recover without attention, and (3) the remainder—those who likely would survive if given intensive care.

Several factors influence triage decisions (Lovejoy 1976; Myers 1979). Among these is the relative difficulty in aiding the organisms involved. Some organisms require huge tracts of habitat, and acquiring these today as refuges or parks may prove impossible. It seems safe to assume that few grizzly bears would still exist in the 48 contiguous states had not Yellowstone Park been established in 1872; yet acquisition of such a huge park today would be a remote possibility as a means of conserving that species. Conversely, relatively small areas may suffice for protecting red-cockaded woodpeckers (*Picoides borealis*) in the pine forests of the southeastern United States; breeding colonies and foraging habitat may consist of no more than 4 ha and 40 ha, respectively (Hooper et al. 1980).

Genetic issues also may be involved. When populations reach drastically low numbers, the desired amount of genetic variability usually is reduced, and so the chances increase for poor reproduction or for the emergence of undesirable traits because of repeated inbreeding. Hawaiian geese (*Branta sandvicensis*) suffered poor fertility, at times reducing the effectiveness of a captive breeding program designed for their recovery (Kear and Berger 1980). Red wolves were completely extirpated in the wild, in part because they crossbred with coyotes (*Canis latrans*) and thereafter became genetically swamped by the far larger coyote population (Pimlott and Joslin 1968; Riley and McBride 1972). Thus, when an endangered species is so low in numbers as to influence desirable genetic relationships, it may be necessary to consider carefully whether such a species can be saved no matter what management efforts might be applied.

Other factors include economic relationships with humans. It is easy, of course, to identify those species or groups of species of known benefit to humans. Primates clearly are of great value for medical research, whereas few socioeconomic values seem assignable readily to Devils Hole pupfish (*Cyprinodon diabolis*) or Houston toads (*Bufo houstonensis*). But it is far more difficult to predict what other species might become useful in the future and hence to save those that are threatened with oblivion.

Finally, triage might consider the biological uniqueness of species. Some animals, such as aardvarks (*Orycteropus afer*), are the sole representative of a taxonomic group (Tubulidentata), whereas others, such as Darwin's finches (Geospizinae) together represent a "singular group" of well-known scientific importance. Application of these and perhaps other criteria at best is a regrettable process. Triage thrusts crushing decisions upon mortal beings about the continued existence of other creatures. Yet, just as it was for medics examining wounded soldiers, the reality of the situation seemingly makes triage a necessity. Without it, our skills and funds may be dissipated, thwarting any chances of effectively rescuing at least some endangered species.

MANAGEMENT OF ENDANGERED SPECIES

The perils of extinction were highlighted by the enactments of the legislation outlined earlier. Recognition by Congress set into motion commitments to conserve the remaining stocks of threatened and endangered species. Now equipped with a funding base, federal and some state agencies marshaled biologists into recovery teams charged to assess the status of certain species and to design ways and means to halt their decline. Some efforts, of course, were initiated long before the passage of specific legislation, but now the efforts were concentrated by legal mandate.

At least one difference in the biological features of many endangered species may require special consideration. Most game species are *r*-selected, meaning that they have high intrinsic rates of population growth. However, endangered species often are *K*-selected: They have adapted an evolutionary strategy that favors lower rates of population growth with a higher efficiency in their use of resources. *K*-selected species also seem more specialized than *r*-selected animals, likely a cofeature of their higher efficiency of resource utilization. Such differences may dictate that the broad-brush approach often taken by game managers requires thoughtful refocus to provide effective management for endangered species (Temple 1977a).

Recovery plans for endangered species have no single pattern of action. They include, on a species-by-species basis, habitat preservation and management, captive breeding programs, census requirements, control of predators and competitors, supplemental feeding, and other techniques tailored to meet the needs of each species (see Temple 1977b). Only a few of these can be outlined here, but the selection underscores the scope of management addressing these challenges.

On the Job with...
Janet Schultz
Forest Plant Ecologist

Janet Schultz works in the Hiawatha National Forest in the Upper Peninsula of Michigan. She began her career with the U.S.D.A. Forest Service in 1987, previously earned B.S. and M.S. degrees, and currently is working toward a Ph.D. In her job, Jan is responsible for surveying 4000 km² of forest for rare plants and significant community types. Aside from field work and coordinating management and research, she spends much of her time responding to mail and phone calls from citizens and city, state, and federal agencies with questions about plants and plant ecology. Jan conducts workshops designed to train people who are interested in topics such as plant identification, habitat restoration, and how to use native plants for landscaping. She also helps identify areas of future research within the U.S.D.A. Forest Service and spends considerable time attending meetings and writing reports.

Her current research projects include developing a means for expressing the viability of rare plant populations discovered on the Hiawatha National Forest. In part, she uses the age structure of the rare species to determine the status of these populations, and she assesses the ecological significance of the sites where these plants occur, ascertains threats, and provides mitigation measures. Jan also analyzes the occurrence and distribution of rare plants at seven other national forests in the Great Lakes Region, and she develops strategies for controlling noxious plants such as purple loosestrife, especially where this and other invasive and persistent species threaten ecologically sensitive areas. Additionally, she coordinates the Research Natural Area program, which identifies and establishes sites within Hiawatha National Forest for special protection and research.

Jan Schultz feels that her job has a beneficial effect on wildlife—both plants and animals—by serving as a catalyst for change and by developing "ecological sensitivity" for the regionally and globally significant forest communities in Michigan's Upper Peninsula. Her frustrations include the struggle with individuals and organizations that cannot accept those changes in resource management that are designed to address today's ecological concerns. On occasion, she also lacks sufficient time or personnel to finish projects or to do them as well as she wishes. Jan regards the most rewarding part of her job as learning about the diversity of plant and animal life in the region's ecosystems and her activities that ultimately effect positive management of natural environments in the Great Lakes Region. She advises students not to lose sight of why they chose a profession in natural resource management. To avoid becoming overwhelmed by the workload, Jan suggests "finding renewal by periodically getting out in the field."

WHOOPING CRANE (*GRUS AMERICANA*). No other species in North America serves as a greater symbol of conservation than the whooping crane (Fig. 19-8). The dire status of whooping cranes received formal recognition in 1918 when the meager population still remaining received complete legal protection under the Migratory Bird Treaty Act. Legal protection of this sort was largely passive, however, and for many years thereafter whooping cranes seldom benefited from any type of active management designed to increase their numbers.

Whoopers apparently were never abundant, although they once nested across much of the northern prairies and into the forests beyond. A small sedentary population occurred in Louisiana. Most of the interior population wintered on the coast of the Gulf of Mexico, but others wintered in the mid-Atlantic region between New Jersey and Florida. Unfortunately, agricultural development adversely affected whooping cranes, and their breeding grounds were unprotected from the steady turn of the plow across the North American heartland. Many whooping cranes also were shot, especially in such places as the Platte River in Nebraska, where these and other birds concentrated during migration. Allen (1952) suggested that no more than 1400 whoopers survived by the mid-1800s as a result of these activities.

Despite the protection from further shooting afforded by the Migratory Bird Treaty Act, the meager population of whooping cranes steadily declined in the wake of unabated habitat destruction and limited enforcement of the new law. Then, in 1937, the

Figure 19-8. Whooping cranes symbolize wildlife conservation and management in North America. Shown here are three adults wintering in coastal marsh habitat at Aransas National Wildlife Refuge in Texas. Birds in this flock breed at Wood Buffalo National Park, a remote site in northern Canada more than 4000 km from the Texas coast. (Photo courtesy of U.S. Fish and Wildlife Service.)

Aransas National Wildlife Refuge was established in Texas. This crucial step protected about 22,000 ha of coastal habitat that forms the core of the wintering grounds for whooping cranes. Nonetheless, only 23 birds remained by 1941 (Allen 1952). Of these, 15 migrated between their winter quarters in Texas and some then-unknown breeding area in the far north, 6 constituted a nonmigratory population in Louisiana, and 2 were held in captivity. The location of the nesting grounds for the migratory population of whooping cranes remained a mystery until 1954, when nests were discovered in the remote reaches of Wood Buffalo National Park in the Northwest Territories of Canada (Allen 1956). The habitat in the park did not require further protection, and at last, studies could begin on the rare bird's breeding habits. Moreover, management options no longer were limited solely to the wintering area in Texas, but even with this advantage, the population remained unsteady.

In one year, six or eight young birds might be produced, only to have two or three (or none) reach Texas the following season. Such production barely kept pace with the loss of older birds, and the migratory flock wintering in Texas generally hovered between 25 and 40 birds for many years—the resident population in Louisiana was gone by 1950.

In 1967, a captive breeding program was initiated using eggs collected from the wild population and the few whooping cranes held in zoos, but few of the young birds survived to an age where they could be released. Then an imaginative technique, known as the *foster-parent program* was proposed. Research indicated that whooping cranes' eggs and young would fare well under the parental care of sandhill cranes (*Grus canadensis*), a related and far more abundant species (Drewien and Bizeau 1978). Thus, in 1975, 14 whooping crane eggs were placed in the nests of sandhill cranes nesting a Gray's Lake National Wildlife Refuge in Idaho. Of the 14 transplanted eggs, 9 hatched and 6 young whooping cranes reached flying age in the first year of the experiment. Of these, four survived the fall migration to Bosque del Apache National Wildlife Refuge in New Mexico, the traditional wintering area for the sandhill cranes acting as foster parents.

The program continued each year thereafter, but production from the transplanted eggs remained low, largely because of the poor survival of the hatchlings (i.e., most eggs hatched under the incubation of the foster parents, but the chicks seldom reached fledgling age). The newly created flock of whooping cranes eventually peaked in 1985 at 33 birds, of which 12 were young, but production dropped to four and one young, respectively, in the next 2 years. No young have been produced since, and only eight adults remained by the winter of 1992–93. More important, none of the whooping cranes produced during the 15-year history of the foster-parent program ever nested, thereby negating hopes for establishing a new breeding population. Aspirations also were dashed for starting other flocks by the same means in northern Michigan or southern Ontario (Brownlee 1987). Meanwhile, however, the flock wintering in Texas and nesting in Wood Buffalo National Park reached 182 birds in 1997, of which 30 were young-of-the-year.

The foster-parent experiment was abandoned in 1990 in favor of establishing a nonmigratory population of whooping cranes on the Kissammee Prairie in central Florida. In this case, juvenile whooping cranes raised from the captive flocks housed at Patuxent Wildlife Research Center at Laurel, Maryland, or the International Crane Foundation at Baraboo, Wisconsin, were introduced using a "soft-release" technique. This involves a gradual transition from captivity to free-ranging living conditions. At first, the birds are confined in fenced, predator-proof enclosures in which food and water are provided, but after a period of acclimation lasting about 1 month, they are free to fly in and out of the pen whenever they wish. A test group of 25 whooping cranes was released in this fashion in 1993 to determine the birds' response to the habitat in Florida and to assess the soft-release technique. Bobcats (*Lynx rufus*) preyed heavily on the initially released cranes, but predation was lessened when those released later were conditioned to roost in water, and the area near the pens was mowed to reduce the cover available to bobcats. As of 1998, the Florida population consisted of 65 whooping cranes, and although some of these have constructed nests, none has yet produced offspring.

The long-term goals call for reclassifying whooping cranes from endangered to threatened when two additional self-sustaining, wild populations, each consisting of at least 25 breeding pairs, are established. The two new populations proposed in the recovery plan, one of which may result from the efforts in Florida, may be either migratory or nonmigratory (U.S. Fish and Wildlife Service 1994). Captive flocks are being established in Canada to supplement to those already quartered in the United States. The recovery plan sets a target of 1000 birds for the population nesting at Wood Buffalo National Park. Such goals should be in place for 10 consecutive years before the status of whooping cranes is "downlisted" from an endangered to a threatened species, which is unlikely before the year 2020. The ultimate goal, of course, is to completely "delist" whooping cranes—official recognition that the species is no longer at risk.

EASTERN TIMBER WOLF (*CANIS LUPUS LYCAON*).

Controversy rages whenever predators are mentioned, particularly for the larger species that may conflict with civilization's many enterprises. None is more bitter than the proposals addressing the eastern timber wolf. Wolves have been branded as villains in folklore for centuries, instilling in people from childhood a prejudice that pervades much of modern society. For conservationists, wolf management remains something of a no-man's-land between ardent protectionists and those bent on extermination of the last unconfined animal.

The eastern race of gray wolf once ranged widely in the United States east of the Mississippi, but its current distribution is restricted to about 5 percent of its former U.S. range. Minnesota, where there are about 1200 wolves, is the only state where substantial numbers of eastern timber wolves still remain; others persist in eastern Canada. In 1997, about 24 animals made up the highly celerated wolf populaion at Isle Royale, along with another 130 in northern Michigan and 150 in northern Wisconsin. [Note: Biologists released 31 wolves of another subspecies (*C. l. irremotus*) in Yellowstone National Park in 1995 and 1996, but this newly established population was supplemented when more wolves immigrated from Canada, thereby bringing the park population to 83 in early 1998. Similarly, 11 Mexican wolves (*C. l. baileyi*) were introduced in Arizona in 1998.]

Incompatibility between humans and wolves led to the predator's widespread extirpation. Centuries of European folklore had fostered antiwolf attitudes among settlers so that the normal thing to do to a wolf was to kill it. Some conflicts also concerned farm animals as an available source of prey, subsequently giving rise to bounties and poisons as sanctioned means of controlling wolves. A secondary but associated impact was the loss of the wolf's natural prey. As human settlement cleared the eastern forests, habitats for moose (*Alces alces*), white-tailed deer, and other mammals were reduced beyond their ability to support prey populations in sizes large enough for the needs of wolves.

The Recovery Plan includes three kinds of management zones in Minnesota, each with a different status for the current wolf population. The first is some

26,000 km² of wilderness areas where wolves continue to enjoy full legal protection from any form of human exploitation consistent with their pending status as a threatened species. The wilderness habitat is extensive enough to support a population of about 1 wolf per 26 km² and includes an established national park, a scenic canoe area, and portions of a national forest. A second management zone of 54,000 km² will permit regulated control of wolves to maintain their numbers at about 1 animal per 130 km²; this zone lies in a more-settled area of northern Minnesota where wolves might inflict livestock damage. Finally, the remainder of Minnesota represents a zone currently without wolves.

The zoning measures are designed to achieve several goals, among them (1) preserving wolves in Minnesota, (2) decreasing livestock losses, (3) minimizing the public's negative attitude toward wolves and the Endangered Species Act, and (4) increasing local support for the wolf's role in our environment. These goals are complemented by the formulation of a vigorous educational program, implementation of habitat improvements fostering larger prey populations of deer—perhaps including reintroduction of woodland caribou—and continuing support for the protection of wilderness areas (Mech 1977c).

Reestablishing resident wolf populations in other states lying within the species' former range is another recommendation in the Recovery Plan, although many difficulties confront such intentions. Not the least of these is the lack of significant blocks of wilderness in such states as Wisconsin and Michigan. Human attitudes also remained a critical concern. The small population still struggling in the Upper Peninsula of Michigan (Hendrickson et al. 1975) was supplemented by the release of four radio-collared wolves in 1974 (Weise et al. 1975). However, all were killed within 8 months, two by hunters and one each by trapping and an automobile. Further, examination of three other wolves killed in Michigan revealed that, while otherwise in good physical condition, they were carrying previously fired .22 caliber bullets, birdshot, and/or evidence of trapping damage to their legs and feet. This suggested that without an educational program designed to reform public attitudes toward wolves, the direct killing of wolves, rather than shortages of either habitat or prey, may limit the chances for successful restocking (Robinson and Smith 1977).

In the early 1990s, the Michigan Department of Natural Resources (DNR) organized educational meetings throughout the state. Citizen volunteers and DNR biologists informed the public about the ecology and behavior of wolves. These meetings apparently were effective—reinforced, perhaps, by $1100 fine levied in 1995 on a person who shot a wolf. Prior to the deer season, DNR personnel and citizen volunteers also visit hunting camps to inform hunters about the presence of wolves and to seek their cooperation. Further, coyotes cannot be shot legally during the gun season for deer, thereby reducing the probability of a hunter mistaking and shooting a wolf for a coyote. In 1997, about 130 wolves—up from virtually none in the 1970s—occupied Michigan's Upper Peninsula, and humans intentionally killed few of these.

KIRTLAND'S WARBLER (*DENDROICA KIRT-LANDII*). The breeding population, about 750 pairs in 1997, of Kirtland's warblers now nests not only in central Michigan, but also in the Upper Peninsula, where 19 breeding pairs were discovered (Fig. 19-9). After breeding, the birds migrate from Michigan to winter quarters in the Bahama Islands (Radabaugh 1974).

The welcome increase from a low of 167 pairs in 1987 to 750 pairs in 1997, together with the expanded breeding range, is cause for optimism. Nonetheless, a population of just 1500 birds is by no means secure, as might be envisioned by comparing this to the town of Eaton, Indiana, as representing the only human population in existence worldwide.

Some aspects of habitat management for Kirtland's warblers are discussed in Chapter 15, but the successful intrusion of brown-headed cowbirds (*Molothrus ater*) into the warbler's breeding habitat has heightened management problems. The cowbirds are obligate brood parasites, no longer building nests or incubating their own eggs. Instead, they lay their eggs among the clutches of other nesting birds (Table 19-3). Further, the female cowbird returns to remove some, but not all, of the host's eggs; she ably discriminates her own eggs from those of the host (Mayfield 1960). Whereas nest parasites such as the European cuckoo (*Cuculus canorus*) have evolved remarkable adaptations for this habit, brown-headed cowbirds are not egg mimics, nor do they evict their nestmates after hatching; they succeed simply by outstripping their host's own young for food and space (Friedmann 1963). In short, the hosts faithfully nourish nestling cowbirds, unknowingly raising the young of another species, usually at the expense of their

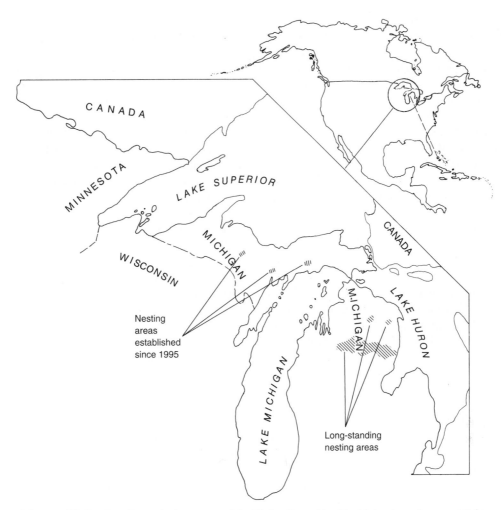

Figure 19-9. Breeding and winter range of the Kirtland's warbler. The birds migrate between Michigan and the Bahama Islands (dashed line in small map). Nesting areas formerly were limited to a few sites largely within a three-county region in the Lower Peninsula of north-central Michigan, but the breeding range expanded to the Upper Peninsula in the mid-1990s.

TABLE 19-3. Contents of 75 Kirtland's Warbler Nests Parasitized by Brown-Headed Cowbirds

| Warbler Eggs/Nest | Number of Nests Based on Cowbird Eggs/Nest | | | | Nests | | Egg Numbers | | Cowbird Eggs/Nest |
	1	*2*	*3*	*4*	*Total*	*Percent*	*Warbler*	*Cowbird*	
0	0	1	2	0	3	4	0	8	2.6
1	1	4	1	0	6	8	6	12	2.0
2	11	7	2	1	21	28	42	35	1.7
3	14	10	3	0	27	36	81	43	1.6
4	8	6	1	0	15	20	60	23	1.5
5	1	1	0	0	2	3	10	3	1.5
6	1	0	0	0	1	1	6	1	1.0
Totals	36	29	9	1	75	100	205	125	1.67[a]

Source: Mayfield (1960).

[a]Mean clutch size for 75 Kirtland's warbler nests, above, was 2.73 eggs. Therefore, cowbird eggs make up 38 percent of the average combined clutch: $1.67/(1.67 + 2.73) = 0.38$.

own. Moreover, Brittingham and Temple (1983) have shown that fragmentation of eastern forests has contributed to the incidence of cowbird parasitism; as forests become more patchy, cowbirds gain further access to the nests of other species from intervening clearings (see Chapter 21).

Brown-headed cowbirds apparently expanded their range eastward from the continental grasslands when the forested areas of the north-central states were cleared for agriculture and may have reached the breeding grounds of the warblers by the 1870s (Mayfield 1960). Since then, their population and parasitic effects on the warblers have increased; 48 percent of Kirtland's warbler nests were parasitized before 1950, whereas 67 percent were similarly affected since that time (Mayfield 1972). The consequences of cowbird parasitism are particularly severe for a host species already reduced in numbers. Pressures exerted on the warblers by the parasitic birds seem density independent; that is, the cowbird population can independently maintain itself by parasitizing nests of other songbirds while the warbler population continues to diminish. Therefore, cowbirds annually exert a constant, if not increasing, toll on the nesting efforts and production of Kirtland's warblers.

Mayfield's studies (1960, 1961) of cowbird-warbler interactions clearly indicate the impact of nest parasitism: The presence of nestling cowbirds reduces by more than one-half the chances that any warblers will fledge from a warbler's nest. With 55 percent of their nests parasitized, Kirtland's warblers could produce 60 percent more young annually if the interference of cowbirds were removed. Sampling, however, continued to show that 60–70 percent of the warbler nests are victims of cowbird parasitism (Mayfield 1973).

Direct control of brown-headed cowbirds seemed the only plausible measure to offset their detrimental influence on Kirtland's warblers. A trapping program was initiated in 1972 (Mayfield 1977). Large wire traps, baited with sunflower seeds, were deployed in the core nesting areas; their effectiveness was improved when a few entrapped cowbirds were left inside as decoys. More than 17,000 cowbirds were removed in the first 6 years of the program, with dramatic results. The rate of production of Kirtland's warblers increased, exceeding that of other species of warblers in North America. Parasitism dropped to 5 percent or less, and the number of fledgling warblers increased to more than 4 per pair of adults (from 2 nesting efforts per season) versus the rate of less than 1 fledgling per pair annually before the trapping

program. By 1981, more than 33,500 cowbirds had been removed, and the 10-year average fledging rate for Kirtland's warblers was 2.76 young per nesting attempt (Kelly and DeCapita 1982). The trapping program continues as an effort to increase the warbler population to at least 1000 pairs.

MASKED BOBWHITE (*COLINUS VIRGINIANUS RIDGWAYI*).   The masked bobwhite is a subspecies of the familiar bobwhite, one of the more abundant and prized game species occurring in a wide range of habitats in North America (Aldrich 1946). Masked bobwhites, however, occurred in a more limited area within the confines of the Sonoran plains; their original range included south-central Arizona and northwestern Mexico. Ligon (1952) expressed fears that masked bobwhites already were extinct in the early 1950s, but a wild population was discovered 14 years later in Mexico. They were extirpated completely from Arizona by 1900. Their known distribution lies solely in two areas in Sonora, Mexico, with the larger population of these estimated at about 1000 birds (Tomlinson 1972).

Males are distinctively marked by a red breast, not unlike a robin, and black head plumage with only traces of white coloration, but females are quite similar in general appearance to typical female bobwhites. Because their habitat requirements are linked with grassland communities, masked bobwhites are susceptible to the effects of overgrazing. Unfortunately, overgrazing has been the normal procedure for ranching in the range of masked bobwhite, and large areas suffered heavy reductions of desirable grasses while simultaneously experiencing the encroachment of woody shrubs. Populations of masked bobwhites correspondingly diminished as these changes occurred (Tomlinson 1972).

Attempts to reintroduce masked bobwhites with hand-reared and wild birds began as early as 1937 and continued periodically until 1974. Unfortunately, none of these efforts established self-sustaining populations, as few of the quail survived for even the first year after their release. Ellis et al. (1977) summarized the reasons for these failures: (1) some releases were made outside the bird's historical range, (2) many of the hand-reared quail were not physically suitable for transplanting, and (3) habitat conditions at the release sites, even when within the previous range of the species, were less than suitable.

Two methods have been adopted to enhance a newer propagation program for masked bobwhites (Ellis et al. 1977). The first of these uses a "covey box" designed to safely regroup released birds while they are getting accustomed to their new environment. "Call birds" are confined to the covey box, but the balance of the covey is released outside; the covey reenters the box via a one-way entrance in response to the calling of the confined birds. The covey is gradually exposed to humans, bird dogs, and hawks as part of a 24-day training schedule (Fig. 19-10, left). As training progresses, the quail learn to avoid predators and when to hold or flush. The birds also are evaluated during the training period for body and plumage condition and for their stamina and use of cover, especially when avoiding hawks.

The second method employs male bobwhites from Texas as foster parents for masked bobwhite chicks. Male bobwhites readily adopt the chicks and can be vasectomized to prevent hybridization with masked bobwhites (Ellis and Carpenter 1981). Experiments proved male bobwhites better foster parents than other species of quail of either sex; the results also were better than with masked bobwhites raised in the covey box. Masked bobwhite chicks are placed in warmed brooding chambers; thereafter, a male bobwhite is added to the group and the heat is turned off. If he begins brooding the chicks, adoption has occurred; if not, another male bobwhite is

selected. After 2 days, adopted broods and their foster parents are placed in larger pens designed to instill following behavior through dense vegetation. The entire group is released in the wild 5–10 days later (Fig. 19-10).

SUMMARY

Public awareness and the professional biologist's responses to nongame management are relatively recent and fostered by legislative action. "Nongame" is a political designation and not a biological reality. Species as varied as mountain lions and mourning doves are given equally variable recognition by legislative bodies. The economic impact of nongame animals is immense, reaching hundreds of millions of dollars annually.

Endangered species are those facing extinction, whereas threatened species are those potentially en route to endangered status. Endangered species are listed in the *Red Data Book* issued by the International Union for Conservation of Nature and Natural Resources. Reasons for their alarming status are (1) natural causes, (2) unregulated hunting, (3) introduced

Figure 19-10. Juvenile masked bobwhites are released from a covey box during a training program acclimating the birds to new surroundings (left). Another release method relies on vasectomized bobwhite males acting as foster parents for broods of masked bobwhites (right). (Photos courtesy of D. H. Ellis, U.S. Fish and Wildlife Service.)

predators, (4) nonpredatory exotics, and (5) habitat modification. Humans are responsible for most of the extinctions since A.D. 1600 and for most of the current problems.

Historical actions to protect and conserve nongame wildlife included the Migratory Bird Treaty Act (1918), the National Environmental Policy Act (1969), the Endangered Species Act (1966 and later), and the Fish and Wildlife Conservation Act (1980).

Management of endangered species includes intensive efforts with captive breeding and foster parent programs, habitat preservation, control of predators and competitors, and other tailor-made approaches based on Endangered Species Recovery Plans.

CHAPTER 20

ECONOMICS OF WILDLIFE

How can you buy or sell the sky, the warmth of the land? The idea is strange to us ... And what is there to life if a man cannot hear the lonely cry of the whippoorwill or the arguments of the frogs around a pond at night? I have seen a thousand rotting buffaloes on the prairie, left by the white man who shot them from a passing train. I am a savage and I do not understand how the smoking iron horse can be more important than the buffalo that we kill only to stay alive. This we know, the earth does not belong to man; man belongs to the earth.

Attributed to Chief Seattle (1854)

Chief Seattle spoke those words when the lands of the Suquamish Tribe in the Washington Territory were transferred peaceably to the U.S. government. Native Americans had difficulty conceiving of land and wildlife as commodities to be bought and sold.

A hundred years later a white man in Manitoba wrote, "There come unforgettable evenings in late August when a soft calm settles over the marsh. The breeze is spent and the reeds duplicate themselves in perfect reflections. The sky is clear; yet with sundown comes a refreshing chill. The tinkle of water from the canoe paddle and the rustle of the wren in the tules are the only sounds. Here is the world as God made it" (Hochbaum 1955).

Harmon (1976) contrasted those words with a passage from a book written by another white man about the benefits of draining marshes for agriculture: "Drainage permits better timing of seasonal cultivation, lowers the costs of cultivation, and improves seed germination." A marsh becomes a grain-field. Here, lamented Harmon, is "the world as man made it." The rustle of the wren in the dew-covered reeds thus races against the jingle of a cash register—and all too often the wren has lost.

Applications of traditional economic principles to wildlife and wildlife habitats have been fraught with difficulties. Such applications indeed involve relatively new economic concepts. Some argue, as Chief Seattle did with the Great White Father, that monetary values cannot be assigned to "the warmth of the land." The aesthetic and, to some, the religious appeal of the natural world defy expression in dollars and cents. Although there may be a fundamental and appealing truth in those positions, the realities of the world today dictate that important decisions are reached on the basis of relative economic values. Many different approaches have been undertaken to evaluate the monetary worth of wildlife resources. Those subjects, in part, are discussed in this chapter.

NONCONSUMPTIVE VALUE OF WILDLIFE

People visiting Aransas National Wildlife Refuge in Texas annually spend large sums for transportation, tours, and souvenirs associated with whooping cranes (*Grus americana*). In 1993, 78,000 tourists visited the refuge and spent a total of nearly $4 million (B. Fletcher, personal communication). Whooping cranes are an important source of revenue for the regional economy. Likewise, gray whales (*Eschrichtius robustus*) migrating along the California coast attract thousands of visitors each year. Kaza (1982) conservatively estimated a total gross income of about $2.2 million generated from some 235,000 whale-watchers in California during the 1981 season.

Whale-watching has become a popular activity in recent years, and dozens of charter boats on both coasts of Canada and the United States now cater to that market (Fig. 20-1). The economic value of whales therefore has shifted from the sale of whale products to the sale of whale-watching tours. In the United States, the three main whale-watching centers—Hawaii, California, and New England—annually attract some 1.5 million tourists, who pay about $25 million in fares each year (Tilt 1987). Along with land transportation, meals, and lodging, the whale-watching industry is probably worth about $1 billion per year. Whale-watching requires a measure of management, however, because the feeding, mating, and nursing of calves may be disrupted by too close attention. Repeated but unintentional harassment from tour boats may cause the abandonment of otherwise choice feeding or breeding areas. Although operators of whale-watching vessels have guidelines designed to prevent undue disturbances, competition from other tour boats and pressure from passengers occasionally may cause violations of the guidelines as well as of the Marine Mammal Protection Act. Despite these problems, however, there is little doubt that whale-watching is, on balance, not only beneficial for whales, but also of significant economic benefit to communities situated on coastlines frequented by whales.

Whooping cranes and whales are resources that are used nonconsumptively; that is, these animals are not removed from their habitat and consumed for meat, hides, oil, plumage, or other goods. There are, of course, many other species of wildlife in North America that have economic value in their natural habitat. Based on a national survey conducted in 1996 by the U.S. Fish and Wildlife Service (1997), an estimated 112.1 million Americans over the age of 16—more than half of all adults—participated in wildlife-associated recreation and spent more than $87.8 billion in such activities as hunting, fishing, and bird-watching.

Figure 20-1. Whale-watching has become a new-found source of enjoying wildlife. Shown here are enthusiasts photographing a humpback whale (*Megaptera novaeangliae*) off Cape Cod. The curious whale approached this boat while its motor was shut off. The economic worth of whales has shifted from whale products to whale tours. (Photo courtesy of Center for Coastal Studies.)

According to a worldwide economic review, Kenya emerged as the leading country for tourism based on wildlife, with 263,000 foreign visitors arriving annually (Eltringham 1984). Zambia attracted 229,000 visitors per year. Figures for other countries are not readily available, but countries with national parks that feature wildlife are most likely to attract foreign visitors. Ecuador has developed a profitable tourist trade featuring cruises to the Galapagos Islands, where the unique fauna inspired the concepts of Charles Darwin. Even remote Antarctica is becoming an increasingly popular tourist area because of the continent's seal and penguin populations. The lure of Antarctica and its wildlife is unique because there are no land-based facilities for tourists; a well-provisioned ship must serve as transportation, hotel, and nature center for travelers.

Money spent on nonconsumptive wildlife activities can amount to considerable sums. The large number of tourists in Kenya spend about $130 million each year; only the export of coffee earns more foreign money for Kenya. The Kenyan people spend about $43 million providing for tourism, thereby realizing an annual profit of about $87 million. Without wildlife, the national parks in Kenya undoubtedly would attract far fewer tourists. Zambia, on the other hand, has experienced a net loss in tourist trade; an annual expenditure of about $3 million for facilities brought an annual return of only about $300,000. Factors other than wildlife affect tourism; these include political considerations, distance, reputation and tradition, and attractions such as scenery.

In North America, tourists seeking experiences with wildlife come primarily from Canada and the United States; that is, the North American fauna by itself does not attract many tourists from Europe and elsewhere in the world. Residents of the United States spend about $4 billion annually on trips where wildlife is viewed. Of that total, lodging (49 percent) and transportation (45 percent) are the major expenditures (Shaw and Mangun 1984). However, in contrast with Africa, national parks are not the primary places where tourists view wildlife. Instead, state parks and state and national wildlife refuges are the primary sites, along with numerous types of landmarks such as promontories, marshes, or mountain ridges where wildlife frequently is encountered.

Aside from traveling, 89 million U.S. residents were engaged in some type of nonconsumptive activity related to wildlife near their homes. These activities involved identifying, photographing, and feeding animals. People in the United States spent more than $600 million on field guides, birdhouses, and birdbaths in 1980, of which $500 million was spent just for birdseed (Shaw and Mangun 1984).

The contributions of people of the United States to state agencies offer another means of assessing the economic worth of wildlife on a nonconsumptive basis. By the end of 1992, 36 states were collecting funds earmarked for wildlife projects from voluntary checkoffs taken from income-tax refunds. Under such systems, taxpayers dedicate part of their refund to programs concerned with the research and management of nongame species (see Chapter 19). Some states collect more than $500,000 each year from a checkoff plan (e.g., Minnesota, $570,000, and Colorado, $692,000 in 1981). These are substantial amounts, but they are far less than the $18 million and $25 million generated, respectively, by the sales of hunting licenses in Minnesota and Colorado. Also, funds returned from the federal government to Minnesota and Colorado from fish and wildlife restoration programs amounted to $4.3 and $3.6 million, respectively, in 1982. Therefore, funds stemming from licenses and taxes associated with hunting and fishing exceed by about fortyfold the revenues voluntarily contributed for nongame programs.

A federally imposed tax on birdseed, field guides, binoculars, and similar items has been proposed as a means of generating revenue for research and management programs associated with nongame wildlife (Wildlife Management Institute 1975). Such a measure would function in ways paralleling the Pittman-Robertson excise tax on sporting arms and ammunition. The attitudes of nonconsumptive users of wildlife toward such a tax were examined in a poll conducted by Shaw and Mangun (1984). The results were that 42 percent opposed and about the same percentage favored a tax dedicated for nongame programs. Eighty percent favored voluntary purchases of conservation stamps, 61 percent supported the existing state income-tax checkoff, and 53 percent favored using revenues from the general tax fund for nongame management. On the basis of such attitudes toward taxes, nongame management programs in the United States apparently will remain severely underfinanced compared with programs for game species.

ECONOMICS OF FISHERIES

Fishing has existed as a commercial venture at least since biblical times. Assessing the value of a commercial fishery is a relatively simple task. One may simply use the average price paid per kilogram of commercial species and multiply by the number of kilograms sold per year. That figure then may be augmented by a multiplier that considers the number of exchanges of the money brought in by the fish harvest. With sport fishing, however, the market value of the catch is not an adequate measure of the worth of the fishery. Weithman (1986) noted that the income-multiplier method is useful for determining the value of recreational fishing to a local or regional economy. With this method, the total expenditures of anglers is multiplied by a factor representing the exchanges of this money in the local community. As an example comparing values, commercial fishing on the Great Lakes in 1985 brought in $15 million (U.S. Bureau of Census 1987). In Michigan waters of the Great Lakes in 1984, the value of the commercial catch was $6 million, whereas the sport fishery in the same waters was worth $500 million (Office of the Great Lakes 1985). It has been further estimated that a lake trout (*Salvelinus namaycush*) caught on hook-and-line is worth six times the amount as the same fish caught in a commercial net. A few individual costs that anglers pay are as follows: Atlantic salmon (*Salmo salar*) anglers in Quebec may pay $10–$60 per day, whereas more-elaborate guided canoe trips may cost $500 per day (M. W. Coulter, personal communication).

FARMING WILDLIFE

Domestication of oxen (*Bos bovis*), hogs (*Sus scrofa*), sheep (*Ovis aries*), and goats (*Capra hircus*) began as "farming" of wild animals. Some animals, however, resist domestication and remain intractable to human handling, although such species may be kept in enclosures. Roots (1970) reviewed the husbandry of wildlife, as follows.

A large and sometimes curious variety of wildlife and wildlife products are farmed. Civets (*Viverra civetta*) are raised in Ethiopia for musk. From each, about 10 g of musk—a base for perfume—is extracted three times a month from pouches near the anal glands. In Taiwan, sambar deer (*Cervus unicolor*) are raised for antlers in the belief that aphrodisiac and other medicinal properties accrue from the velvet. Antlers in velvet are cut into thin slices and then are steeped in wine for a week before being used. A pair of 20-cm antlers in velvet commands about $250, and there are about 150 deer farms in Taiwan. Late in the 19th century, when plumes on hats became fashionable, severe stresses were placed on wild ostriches (*Struthio camelus*) and great egrets (*Casmerodius albus*). The demand brought forth ostrich farms, and 350,000 ostriches were pastured in South Africa and elsewhere at the turn of the century. By 1935, after plumes became less fashionable, about 35,000 ostriches remained in farm flocks. At the peak of marketing plumes, a few egret farms in India and Tunisia were profitable, but egret farming never reached the proportions of ostrich farming.

Mink (*Mustela vison*), chinchilla (*Chinchilla* spp.), and nutria (*Myocastor coypus*) have been raised in captivity for fur. Chinchillas, Andean rodents, have exceptionally soft, light, but durable fur. Early in the 20th century, millions of pelts were shipped from Chile, Peru, and Bolivia. Such exploitation might have caused the extinction of chinchillas had not an American mining engineer brought a few pairs to California and started a fur farm. Chinchillas were extirpated in Peru, and only small populations remain in Bolivia and Chile. Nutria, another South American rodent, also were imported and farmed in North America and Europe. Nutria later escaped or were released in the southern United States and, in the wild, often became unwelcome additions to the native fauna (Chapter 18). Commercial production of hand-reared mink—mink farming—was a natural outgrowth of high prices and the scarcity of wild animals. Selective breeding eventually produced various colors and qualities of fur. Mink farming remains a profitable enterprise for skilled handlers. Nonetheless, trappers sell the pelts of more than 300,000 wild mink in the United States each year.

Fur farming raises two points of interest for wildlife managers. First, fur farming may relieve wild populations of undue exploitation. Second, captive animals may escape and establish troublesome feral populations. Besides the example of nutria, mink—a North American species—have escaped from fur farms in Britain and Scandinavia. Wild mink populations in Europe may compete with otters (*Lutra lutra*), although the relationship is not conclusive. Nonetheless, European otter populations have declined where mink have flourished. About 35,000 wild mink are trapped annually in Sweden and Norway.

Some animals are not confined, but they are closely managed in ways that resemble farming. Lapps herd free-ranging reindeer (*Rangifer tarandus tarandus*) for milk, hides, and meat. Other examples include the commercial cropping of ungulates in Africa and the highly regulated harvest of nests in the Orient for bird's nest soup. The nests are not a mixture of muds, sticks, and grass as constructed by most birds, but instead are built entirely from the saliva of cave swiftlets (*Collocalia brevirostra*). The swifts nest in the caves of western Borneo, and the nests have been cropped for at least 1000 years by the Chinese. Strict rules governed by the curator of the Sarawak Museum regulate the harvest of the nests. Each nesting area is registered to specific collectors, but poaching—sometimes accompanied by bloodshed—remains a profitable risk. That the harvest of nests has continued for a millennium without apparent effect on the swiftlet population bears witness to the soundness of the ancient management system. Such careful management of this unusual resource undoubtedly was initiated and perpetuated because of the nests' proven economic worth.

In recent years, proposals have outlined the potential economic and ecological advantages of managing native African mammals for meat. Darling (1960) first proposed that cropping of free-ranging game was the most ecologically efficient way of providing protein for humans. The traditional husbandry of domestic livestock, especially those stocks originating in other lands, was not well matched with the environmental conditions in Africa. Dasmann (1964) and Cole and Ronning (1974) further developed convincing arguments in favor of game ranching. Protein production, for example, in wild animals equals or exceeds the amounts found in cattle (Table 20-1).

Soil conditions in much of Africa are not well suited for domestic livestock operations—heavy grazing pressure encourages erosion or the concretelike hardening of the surface soil. Typical results include degraded ranges and cattle in poor condition. Conversely, African mammals are adapted to local conditions, especially in terms of feeding on the rich variety of native plants. Virtually all of the available feeding niches have been filled during the evolution of native grazers and browsers. Water conservation also is well developed in native wildlife, as shown in the following ecological setting. The leaves of some plants in semiarid areas absorb water from the atmosphere at night. Leaves of *Disperma* spp. contain 1–5 percent water at noon and 40 percent water at night (Taylor 1968). The water in this case is brought to the surface and then is released into the air by other deep-rooted plants and absorbed by *Disperma*. Thus, wildlife feeding at night on *Disperma* and similar plants obtain more water from their food than do daytime feeders such as cattle. Moreover, the deep roots of shrubs on the African savannas bring nutrients as well as water to the surface, and browsing animals thereby gain nutritional benefits not available to grazing species. Because of their narrow mouths and long dexterous tongues, giraffes (*Giraffa camelopardalis*), elands (*Tragelaphus oryx*), and kudus (*T. strepsiceros*) can strip leaves and fruit from branches bearing thorns "capable of puncturing a Land Rover tire" (Cole and Ronning 1974). Cattle, with their wide mouths and short tongues, are unsuited for such feeding. See Chapter 14 for more about the ecological advantages of game ranching in Africa; our focus here returns to the economic setting.

Eltringham (1984) reviewed attempts at regulated cropping of free-ranging African mammals and noted several shortcomings that invariably have produced economic failures. The problems—involving harvesting, processing, and marketing—are (1) opposition from the established meat trade toward competition from the meat of wildlife; (2) opposition from wildlife preservation groups; (3) low profit margins, largely because of the low monetary value of meat and high costs of meeting hygienic regulations; (4) lack of local markets; (5) poor communication with distant markets; (6) poaching; (7) shortfalls in quotas for maximum yields because of the elusiveness and migratory habits of some species, whereas populations of resident species may be too small for an economically worthwhile harvest (also, difficulty of harvesting in rough terrain); and (8) drought and other ecological changes leading to unpredictable trends in animal numbers.

A profitable compromise may be reached, however, when game animals are cropped along with cattle (Eltringham 1984). A well-managed ranch in Zimbabwe with 1261 animal units of cattle and 5700 wild ungulates of 10 species has been operating effectively. One section of the ranch was devoted only to wildlife, and the other to cattle and 1300 wild animals

TABLE 20-1. Analysis of Meat of Wild and Domestic Animals[a]

	Ash	Fat[b]	Protein	Calories
Eland[c]	1.1	1.9	23	125
Range cattle	1.1	2.0	22	120
Intensive fat stock	0.9	15.0	20	230

Source: Cole and Ronning (1974).

[a]Ash, fat, and protein are expressed as g/100 g of fresh tissue, energy as cal/100 g fresh tissue.

[b]The carcass-fat content from free-living species will vary with seasons. In general, fat increases in the rainy season when fresh grass is available.

[c]Hartebeest (*Alcelaphus* spp.), topi (*Damaliscus lunatus*), giraffe (*Giraffa camelopardalis*), buffalo (*Syncerus caffer*), and warthog (*Phacochoerus aethiopicus*) have similar meat.

of 3 species. Each year, 300 cattle and 730 impala (*Aepyceros melampus*) and other species are harvested. Although the marketability of the wild animals was not mentioned, this experience indicates that the income of a ranch managed for beef can be supplemented by exploiting wild ungulates.

Managers of Pilanesburg Wildlife Park in Bophuthatswana, South Africa, reached an economically feasible compromise between the opposing philosophies of strict preservation and the consumptive use of wildlife (Arms 1994). With rigorous protection, the park's population of endangered white rhinoceros (*Ceratotherium simum*) recovered from the brink of extirpation to local prosperity. Thereafter, the privilege of shooting five or six rhinos per year was sold to American and German hunters for $25,000 per animal. Other hunters paid $10,000 to shoot a rhino with a tranquilizer gun, after which they were photographed beside their drugged "trophies" (for management reasons, these rhinos were scheduled to be moved elsewhere within the park). From the funds paid by the hunters, 10 percent was distributed to villages near the park; hence local people benefited from the program and now regard the park as an asset rather than as a liability. Naturally, such a program has opponents, especially among well-fed preservationists living far off in the Western world—and their voices are sometimes heeded. For example, a wealthy trophy hunter was willing to pay $100,000 to shoot an aging black rhinoceros (*Diceros bicornis*), but the park authorities yielded to the complaints of animal rights advocates, and the old rhino later provided what one observer called "expensive hyena food" (Arms 1994).

In New Zealand, farmers have achieved commercially profitable harvests of free-ranging red deer (*Cervus elaphus elaphus*); the animals originated from stock released late in the 19th century (Chapter 18). Linear (1977) reported that New Zealand harvested 75,000 red deer in 1 year and sold the venison to the Federal Republic of Germany for $2.7 million. Recently, red deer have been farmed in New Zealand (Eltringham 1984). Indeed, more than 1000 members belong to the New Zealand Deer Farmers Association, an organization formed in 1975. Management techniques have been standardized among the farmers and, by 1981, the nationwide herd of captive red deer numbered 180,000 animals. Profits initially were spurred more by sales of antlers than of meat. In 1979, Koreans paid as much as $220/kg for antlers in velvet because of the alleged medicinal properties in the soft bone, and farmers removed and dried the antlers for sale. However, China entered the antler

market in the 1980s and the competition reduced the price to $100–$150/kg. Thereafter, farmers in New Zealand turned to venison as a more stable and marketable product, earning $3.50 to $5.00/kg for dressed carcasses of red deer.

Great Britain also has developed an industry based on deer farming and venison production. The British Deer Farmers Association consists of 250 members. The association is concerned only with meat production because, for humane reasons, laws prohibit the removal of antlers in velvet from living stags. The Oriental market for velvet antlers therefore is closed to British deer farmers. As in New Zealand, red deer form the nucleus of the British enterprise (Eltringham 1984).

The economics associated with red deer farming are important elsewhere as well (Eltringham 1984). Germany, Sweden, Austria, and Switzerland have numerous deer farms, and Denmark is beginning such ventures. In continental Europe, however, the fallow deer (*Cervus dama*) is the principal species. About 10 deer per ha, averaging 27 kg each, can be sold for meat each year at about $5.00/kg. The sales produce a return of $13.50/ha per year. In China, where antlers are the major product, about 270,000 red, fallow, and sika (*Cervus nippon*) deer are husbanded. Marketing data are available for only two provinces, but these nonetheless reflect the significance of the industry to local economies: 11.5 tons of velvet antlers worth about $7.5 million were produced in 1980.

In contrast with the struggling industry in Africa, wildlife farming seems economically feasible in temperate regions with more sedentary species, established markets, and organized transportation. Nonetheless, the ecological arguments favoring the husbandry of African mammals appear convincing and, in time, perhaps similar success may be achieved in Africa. If increased production (i.e., kilograms of protein per hectare) can be accomplished, the economic benefits would seem to follow, provided that the increased production is not offset by higher costs. Elands reach the same weight in the same amount of time as Holstein cattle, and elands accordingly have been domesticated in some parts of Africa (Crawford 1970).

ECONOMICS OF SPORT HUNTING

Sport hunting varies from a ritualized and aristocratic pursuit in many central European countries to a common form of recreation for people from all walks of life in North America. In all cases, however, money is spent for hunting, whether in large amounts for hunting leases on private land for trophy-class animals or in smaller sums of a few dollars for a license and a box of shotgun shells for quail. A brief look at history, followed by an update, will illustrate the willingness of hunters to pay for their sport.

Early in the 20th century, *Forestry Quarterly*, in response to complaints that wildlife was damaging commercial forests, published some figures for the economic value of game (Anon. 1908). These data form a backdrop for comparing present circumstances. At that time, 6.4 million kg of venison worth more than $1 million were marketed in Austria each year, and the market value of venison exceeded $5 million in Germany. Hunting privileges in Germany cost between $0.25 and $7.50/ha, and a "shooting district" in Scotland rented for as much as $25,000. In Prussia, hunting permits that sold for 15 marks each brought in more than $500,000 in 1903, and in Hungary, the forest administration sold permits for shooting stags of a given description at "astonishing prices" for the time: an 8-point stag for $100 and a 20-point stag for $400. The article also mentioned that the "privilege of shooting" a large species of European grouse, the capercaillie (*Tetrao urogallus*), often commanded $50. Modern prices reflect the magnitude of the changes that have occurred in 80 years. In 1994 dollars, the value of the capercaillie would be $485, and the two sizes of stag would increase to $1000 and $3800, respectively.

The cost of hunting in Europe therefore has increased far in excess of normal inflation. The prestige and interest in shooting trophy animals now commands high premiums in the marketplace. In Hungary, trophy stags bring an average fee of $14,000 per head, and one hunter reportedly paid $56,000 in 1979 for shooting a particularly outstanding stag. These animals are shot on hunting reserves, where individuals are cropped by age and sex in keeping with a management plan rigorously designed for high-quality trophy hunting (see also Chapter 21). In the United States, one hunter in 1986 paid $52,000 for a permit to hunt bighorn sheep (*Ovis canadensis*) in Wyoming and $33,000 to hunt desert bighorns in Utah. Still another paid $64,000 for a bighorn permit in Arizona (Foundation for North American Sheep, no date).

The money spent by hunters, like money spent by nonconsumptive users of wildlife, accrues from numerous kinds of goods and services. There are various ways of evaluating the economics of hunting. The broadest way is simply to assume that the value of

wild game animals is equivalent to the total spent on hunting. On that basis, some 14 million adult hunters in the United States spent $20.6 billion in 1996 (or about $1470 each), and 35.2 million adult anglers spent $38 billion (about $1080 each) on equipment, transportation, lodging, food, and license fees. Such data, compiled every 5 years by the U.S. Fish and Wildlife Service (1997), clearly reflect the economic worth of just one segment—game species—of the wildlife resources in the United States. Another 62.9 million people in the United States spent $29.2 billion for activities associated with watching wildlife in 1996 (e.g., film, birdseed, travel, and other items associated with the nonconsumptive uses of wildlife).

For some game species, the value of each harvested animal may be calculated by dividing the money spent by hunters per species by the number of animals killed. A value of $11,200 thus was determined for bighorn sheep in Colorado (Norman et al. 1976). Some other values from the same study are as follows: pronghorn (*Antilocapra americana*), $234; black bear (*Ursus americanus*), $6400; mule deer (*Odocoileus hemionus*), $709; blue grouse (*Dendragapus obscurus*), $20.79; bobwhite (*Colinus virginianus*), $16.21; muskrat (*Ondatra zibethicus*), $1.21; and cottontail (*Sylvilagus floridanus*), $13.66.

Another way of assessing the economic worth of game animals is to examine the annual income realized by landowners from hunting. Eltringham (1984) cited figures showing that the cost of producing a red deer for hunting was about $35 on certain infertile wildlands in Scotland. The costs covered labor, carcass transportation, and supplemental feeding in winter. At a stocking of about 12 deer per ha, annual profits amounted to about $2/ha. That figure may seem low, but the amount gains significance because the land was unsuitable for other crops. On a mixed sheep-deer enterprise of 10,000 ha in Scotland, an annual harvest of 45 stags and 50 hinds realized a profit of about 12 percent from hunting.

In North America, hunters have long enjoyed a tradition of free hunting. Because wildlife is publicly owned and the regulatory authority for its harvest rests with state, provincial, and federal governments, hunting seemed a *right* vis-à-vis a *privilege* (see Chapter 21). Public and private lands have been hunted without charge for many years. However, in recent years, ranchers, farmers, and other private landowners have become less willing to permit unrestricted hunting. The reality of "slob hunters"—those who shoot livestock, create litter, disregard closed gates, and damage property—gave landowners little peace of mind and no compensation. Moreover, with or without responsible hunters, free hunting produces no economic return for the forage consumed by wildlife. In Utah, for example, big game consume up to $9 of forage per ha on rangelands and as much as $23/ha on croplands. Such considerations, together with unstable farm and ranch economies, produced an initiative for fee-hunting. In Texas and other western states with large ranches, lease-fee hunting thus has become a profitable business. Wildlife still remains publicly owned and regulated in these systems, but the *access* of hunters to wildlife stays under the direct control of landowners. A newsletter entitled *Hunting Ranch Business,* published in Houston, Texas, keeps landowners apprised of current topics related to lease-fee hunting. Knight (1985) published a guide describing how hunting might provide additional income for landowners in New Mexico, and Nielsen et al. (1985, 1986) presented economic analyses of big-game management on privately owned rangelands in Utah.

Ramsey (1965) conducted a thorough study of lease-fee hunting for white-tailed deer (*Odocoileus virginianus*) in the Edwards Plateau of Texas; some highlights follow. Livestock grazing began in the 1800s and soon changed the range from grassland to woody vegetation. Deer flourished in the altered habitat and, by the 1940s, reached densities of 82 animals per km^2. Competition for forage increased, however, particularly between deer and the combined pressure from sheep and goats. Costs associated with the management of hunting were $2 per deer, and the value of the forage was estimated at $3 per deer per year. Against these costs were the gross returns of $42 per harvested deer, based on various types of leases; the rates varied from $5 per day to $150 per hunter for 2 weeks. Additional charges were made for the type of kill, ranging from $10 per doe to $150 for an 8-point buck. Economic returns from deer hunting are shown in Table 20-2, and for livestock and deer on the Kerr Wildlife Management Area in Table 20-3. Whereas the economics of ranching fluctuated in keeping with the production of kids and calves and the prices for meat, lamb, and mohair, a deer harvest of 20 to 30 percent can be maintained without long-term reduction of the deer population. Indeed, those data from the Kerr Area indicate that net returns from deer hunting can exceed those from livestock, provided that deer are harvested adequately. For the 6-year period, deer produced an average return of $38.60 per animal unit versus $28.82 return for livestock. On private ranches, however, hunting claimed only 5 to 8 percent of the autumn deer population; such a light harvest did not offset the value of forage consumed by deer. The implications of this study seem clear: Deer hunting yields profits, but only when landowners allow adequate harvests of the deer population.

Rates for hunting rights vary considerably by region. In North Carolina, hunting clubs leasing land for deer hunting paid an average of $0.50/ha (Franklin and Allen 1985), whereas leases for big game in Utah commanded only $0.17/ha (Nielson et al. 1986). Individuals in hunting clubs in Utah paid $50–$250 per year, with some clubs requiring an initiation fee of up to $2000. Hunters shooting on a waterfowl refuge in Illinois spent $45 per goose in the bag; yet the man-

TABLE 20-2. Economic Returns from Deer Hunting on Selected Ranches in the Edwards Plateau, Texas

Item	1955	1956	1957	1958	1959
Ranches surveyed	8	12	24	20	17
Total area in ranches (ha)	6,248	5,360	30,652	16,406	13,423
Animal units in livestock on ranches	1,049	617	4,636	3,376	2,404
Estimated deer on ranches	2,051	1,725	10,164	7,079	8,266
Antlered deer harvested	128	154	521	487	615
Antlerless deer harvested	50	75	343	124	495
Proportion of total deer population harvested (percent)	9	13	9	9	13
Total gross income from deer (dollars)[a]	8,280.00	10,475.00	47,250.00	23,705.00	30,635.00
Average net return per animal unit in deer based on total population (dollars)	16.78	18.50	25.02	15.52	9.19
Average gross return per deer killed (dollars)	46.51	45.74	54.63	38.79	27.59
Average net return per deer killed (dollars)	22.65	23.27	31.93	17.95	8.65

Source: Ramsey (1965).

[a]A total of $18 per animal unit of deer was figured as pasture expense; a total of $2 per deer killed was figured for harvesting expenses.

TABLE 20-3. Comparative Net Returns from Deer Hunting and Livestock on a 2240-ha Ranching Operation, Kerr Wildlife Management Area, Texas

Year	Deer Population No.	Deer Harvested[a]		Deer Harvested (percent)	Return per AU Deer[b] ($)	Return per AU Livestock ($)
		Antlered No.	Antlerless No.			
1957	463	24	10	7	−1.94	1.87
1958	731	37	87	17	20.54	12.86
1959	713	60	112	24	40.96	28.82
1960	625	63	115	24	41.72	11.58
1961	927	69	166	25	44.03	−3.50
1962	452	64	132	43	79.25	5.35

Source: Ramsey (1965).
[a]Antlered—male deer with forked antlers; antlerless—female deer and male deer without forked antlers.
[b]Six deer equivalent to 1 Animal Unit.

agement costs were only $6 per bird per year, thereby producing a sevenfold return on the investment (Arthur 1968). More than 30 years ago, prime waterfowl habitat on private lands in the United States was valued up to $247/ha (Bolle and Taber 1962). Depending on habitat quality and aesthetic considerations, duck hunters on the Southern High Plains of Texas placed a value of $400–$600 for hunting privileges on the small prairie wetlands known as playa lakes (Moore *in* Bolen 1982). In California, Teague (1971) reported fees of $75–$500 per duck blind, with deluxe accommodations commanding up to $10,000. One thereby can envision the financial returns potentially available at current dollar values from leasing well-managed wetlands to waterfowl hunters. Indeed, waterfowl hunters in Texas paid fees of $1500–$2500 per guided hunt on leased areas of 10–100 ha (Chamberlain and Davis 1986).

The commercialization of wildlife products and hunting rights in North America is not proceeding without ecological dangers. An outspoken critic, Geist (1985, 1986b, 1988) argues that the history of success of wildlife conservation in Canada and the United States is threatened by the movement to buy and sell wildlife and the rights to hunt. According to Geist, wildlife conservation in North America has depended on three main policies: (1) the absence of a market in meat and other parts of game animals; (2) allocation by law of material benefits of wildlife (hunting, eating, and viewing animals) equally to all people and not by marketplace economics, birthright, land ownership, or social position; and (3) the view that wildlife is a subsistence-food resource rather than a commodity. Geist notes that the first of

these policies is being eroded by "game ranching," the marketing of hunting as well as meat, antlers, claws, teeth, and so on. The second is eroded by hunting leases and other private hunting fees, which allocate wildlife only to those who can afford to pay. This, he notes, progressively alienates the general public as interested "stockholders" in conservation as fewer and fewer citizens retain access to wildlife. The third policy is eroded by the growing concept of killing for "sport." For example, trophy hunting is often viewed by the public and critics of hunting as a frivolous and cruel pastime. Geist further suggests that the successes and contributions of wildlife management in North America in recent decades stem from its ecological approach, in which game animals are seen as members of an entire community of life. In this view, assigning economic values to game animals is incompatible with the maintenance of intact ecological communities. Predators are eliminated, and vegetation is manipulated for maximum benefit for the cash crop. The potential for impoverishing the land with a wildlife monoculture has precedent in land abuses associated with intensive agriculture.

In Illinois, the costs and benefits of establishing and maintaining herbaceous vegetation along roadsides illustrated the economic value of supporting a huntable population of ring-necked pheasants (*Phasianus colchicus*) (Warner et al. 1992). Normally, farmers mow roadsides to prevent weeds from spreading into adjacent fields where cash crops are grown. Mowing in late spring and summer, however, severely reduces pheasant production by eliminating large areas of nesting and brood cover (see Chapter 13). To counter this situation, the Illinois Department of

Conservation planted smooth brome grass (*Bromus inermis*) and alfalfa (*Medicago sativa*) along roadsides selected for study. This vegetation established long-lasting cover but did not threaten crop production; hence farmers were willing to stop mowing. In response, pheasant numbers increased by 300 percent in the planted areas. Costs for planting the roadsides included labor, seed, fuel, and administration, and based on these expenses, each pheasant produced by the program and bagged by hunters cost an average of $5.22. Because each pheasant harvested in Illinois generated $64.00 in income from ammunition sales, lodging, meals, travel costs, and other expenditures, each bird attributable to the roadside program realized a twelvefold return on the state's investment in roadside habitat. Moreover, the roadside management offered farmers an annual savings in mowing costs of about $100/km of roadway.

Feltus and Langenau (1984) used a conservative method of assessing the value of wildlife on public lands. The analysis, the *travel cost method*, was used to determine the value of deer to firearm deer hunters in Michigan. This method has the advantage of describing what hunters actually pay (above the cost of staying at home) for deer hunting. It does not include expenditures for guns, licenses, ammunition, food, or lodging. Instead, the estimate is based on travel costs using the price of fuel and the distances hunters travel. The results, although conservative, placed an annual value of about $215/km^2 for deer hunting on state-owned management areas. By contrast, the annual revenue from timber sales in the same areas averaged $83/km^2. Because timber sales and deer hunting are not mutually exclusive activities, the revenues from each enterprise can be regarded as additive rather than competitive.

Waterfowl produced in the prairie provinces of Canada offer another example in which those who support a wildlife resource do not share proportionately in the resulting economic benefits. Ducks are produced instead of grain or pasture, thereby costing landowners agricultural production. Adamowicz et al. (1986) analyzed how the economic benefits are distributed from production of ducks in prairie potholes in Alberta. Their method entailed calculating the "extramarket value" of a hunting day; that is, duck hunters were asked how much they would be willing to pay above normal hunting expenditures to continue shooting ducks in the future. Duck hunters in the Alberta study were divided into five groups based on the proximity of their residences to waterfowl breeding areas. The results showed that hunters living in the duck-breeding areas received about $6 million (Canadian dollars) worth of benefits (Table 20-4). Other residents of Alberta received about $11 million, while U.S. hunters received about $104 million in benefits. Therefore, it is in the interest of all beneficiaries to maintain breeding habitat in the Canadian prairies. The study also pointed out that the costs of producing waterfowl are not derived solely on the basis of breeding areas in Canada. Winter habitat in the United States also must be included in the economic equations for migratory wildlife.

ECONOMICS OF MITIGATION

Since 1970, the environmental effects of every project requiring any type of federal permit in the United States must be evaluated in an Environmental Impact Statement (EIS). Canadian regulations require similar statements. Dams, oil developments,

TABLE 20-4. Distribution of Total Hunting Days and Direct Economic Benefits in Canada and the United States Generated by Ducks Produced in Alberta, 1973–1982

Hunter Residence	Average Total Annual Days of Duck Hunting 1973–1982	Hunting days/1000 Ducks in the May Breeding Population	Total Annual Direct Benefits, 1984 Dollars (Canadian)
Local Alberta	157,400	18.4	6,183,600
Nonlocal Alberta	266,300	31.4	11,122,700
Nonresident Canadian hunting in Alberta	10,400	1.2	805,000
Nonresident alien hunting in Alberta	6,100	0.7	323,300
United States resident hunting in the United States	2,578,500	301.5	103,956,800

Source: Adamowicz et al. (1986).

airports and highways, and open-pit mines are among the many projects requiring an EIS. Frequently, the project described in an EIS passes public scrutiny and review boards because the economic benefits of the project seemingly outweigh the values of the environmental losses. The results were piecemeal losses of wildlife habitat without equivalent replacements.

Later, however, the concept of *mitigation* was incorporated into the review process as means of offsetting the net loss of habitat. Mitigation options include (1) disapproving the proposed project, thereby avoiding the loss of habitat altogether; (2) minimizing the impact by limiting the extent of the proposed project; (3) rectifying the impact by repairing or restoring the affected environment; (4) reducing the impact by preservation or maintenance operations during the lifetime of the proposed project; or (5) compensating for the impact by replacing or substituting resources or environments. Of the five alternatives, most are rather obvious actions; that is, the public normally expects that destructive projects will be halted, modified, or repaired. The fifth action, however, is less traditional, and when applied in conjunction with other mitigation actions, it holds promise for maintaining or even increasing wildlife habitat (Soileau et al. 1985; Zagata 1985).

In mitigation, the economic units are not dollars but habitat units, which are calculated using a system known as Habitat Evaluation Procedure (HEP) employed by the U.S. Fish and Wildlife Service (Schamberger and Krohn 1982). One habitat unit (HU) is obtained by multiplying the number of acres—metric units are not used at present—by a qualitative measure known as the Habitat Suitability Index (HSI). HSI is calculated only for those acres of habitat that are of actual importance to each species at the site. The HSI ranges from 0.0 (poor) to 1.0 (excellent) for each species of wildlife, depending on the occurrence and quality of feeding, breeding, and resting sites in the area. For example, on a 10-acre (4-ha) marsh with a few small, vegetated islands, abundant waterfowl food, and 5 acres (2 ha) of meandering open water, the HSI for mallards (*Anas platyrhynchos*) might be 1.0. The HU therefore would be $10 \times 1.0 = 10$ for mallards because all of the area is of excellent value. For white-tailed deer, the marsh is less important, both in quantity and quality. Accordingly, the HU for deer might be $3 \times .03 = 0.9$. Similar calculations are completed for all vertebrate species using all or parts of the marsh. So the total for the marsh biota—both resident and transient animals—might amount to a few hundred HUs.

At this point, suppose the marsh is selected as a site for a power plant; no other area is feasible. The second, third, and fourth options for mitigation offer little help because construction of the power plant requires total elimination of the marsh. Only two choices remain; either cancel the project or enhance wildlife habitat elsewhere with a number of HUs equal to those destroyed by construction of the power plant. The latter option offers concurrent benefits: corporations and developers gain opportunities for offsetting losses of wildlife habitat and, at the same time, the public can expect positive action of an appropriate magnitude. Corporations, in fact, may undertake mitigation *before* beginning new construction projects, thereby reducing or eliminating the vacuum in habitat availability occurring between the completion of the project and the development of the new habitat.

A pilot study of mitigation is underway in the coastal marshes of Lousiana (Soileau et al. 1985; Zagata 1985). In 1982, Tenneco Oil Company signed a 25-year management agreement with the U.S. Fish and Wildlife Service for the enhancement of wildlife on 2900 ha of coastal wetlands. Tenneco will install a system of dams and dikes that will generate a net benefit of 33,500 HUs for each of the next 25 years of the agreement. A unique provision in the agreement allows Tenneco to "bank" the HUs as credits for any 1 year. An "account" thereby is established against which "debits" may be made later. The system is complex, but the net effect will be the preservation of more wildlife habitat than is lost in the course of oil-field development. Such losses otherwise would be of major proportions because transport canals must be dredged through the marshes; emplacement of an average drilling rig requires a canal 366 m long and 21 m wide. Each drilling site also claims about 0.5 ha of marsh vegetation. The loss of vegetation and the intrusion of salt water destroys about 900 HUs for fish and other marshland wildlife. As many as 100 drilling sites might be developed during the 25 years covered by the mitigation agreement (Zagata 1985).

Mitigation is a system of ecological barter. HUs form the currency instead of money, and haggles over the dollar value of a fish or bird are avoided. Although the system may hold great promise for the preservation of wildlife habitat, Soileau et al. (1985) cautioned that the credits accumulated in mitigation banking must not be "cashed in" until all other means have been exhausted for avoiding or minimizing the adverse impacts of development. Nor should developers mistakenly believe that mitigation banking will guarantee the approval of all future requests for permits. The project in Louisiana and several smaller mitigation projects elsewhere in the United States bear watching.

BEYOND DOLLARS AND CENTS

Even though the economic analysis and mitigation processes seem practical and perhaps effective, there remain some arguments against total reliance on economic values as means of protecting living resources. What price can be placed on a species that might be expected to exist for several million years in the future but is brought to extinction by human activities?

Ehrenfeld (1976) discussed the value of plants and animals that lack any known economic worth—those organisms that seemingly represent a "nonresource"—in a society built on exploitation of natural resources. An example of a nonresource is the Houston toad (*Bufo houstonensis*), an unobtrusive amphibian found only in the vicinity of Houston, Texas, which lacks direct or indirect commercial value. The snail darter (*Percina tanasi*) is another example of a nonresource. Just as the endangered snail darter held up construction of the Tellico Dam, the tenuous existence and preservation of the Houston toad were advanced as reasons for preventing oil drilling in a public park. A traditional economic argument in favor of the toad seems difficult to substantiate. Ehrenfeld (1976) described several ways in which nonresources might emerge as resources:

1. *Recreational and aesthetic values.* Widespread interest in nature generates large sums of money spent for travel, film, field guides, and other goods and services. As noted earlier, whooping cranes and whales are important economic assets.
2. *Undiscovered or undeveloped values.* Some species may have unique, but undiscovered, properties. For example, in 1975, scientists discovered that beans from jojoba plants (*Simmondsia chinensis*), a desert shrub of the American Southwest, contain oil much like the prized oil from sperm whales (*Physeter macrocephalus*). Jojoba beans suddenly became a valuable resource and, not incidentally, evidence against the continued killing of an endangered species of whale.
3. *Ecosystem stabilization values.* A well-known ecological hypothesis states that each species adds stability to ecosystems. Therefore, as species are removed, stability is lessened and the chances increase for major disruptions. This hypothesis, however, has both supporting and contradicting evidence in nature (see Chapter 4).
4. *Examples of survival.* Natural communities and species provide information about survival. Human survival may depend on knowledge gained from the study of natural ecosystems.
5. *Environmental baseline and monitoring values.* Natural communities and species may serve as standards for comparisons with artificially modified systems (i.e., they are biological indicators of environmental health). The presence or absence of certain species of aquatic organisms are measures of water quality.
6. *Scientific research values.* Some organisms have certain properties or characteristics that are valuable for research. For example, because armadillos (*Dasypus novemcinctus*) bear genetically identical quadruplets, the offspring are valuable research media for studies of leprosy and other diseases.
7. *Conservation values.* A natural community covered by concrete is essentially irretrievable, and extinction of a species is irreversible. Perhaps human survival may be threatened without our continued association with a full range of fellow organisms. Although this argument may have merit, hundreds of extinctions and severe artificial modifications of the environment have not demonstrably affected the overall success of the human species. However, we do not know what opportunities or discoveries were lost with each extinction.

"Regardless of the truth of these explanations, they are not as convincing as those that are backed by a promise of short-term economic gain. In a capitalist society, any private individual or corporation that treats nonresources as if they were resources is likely to go bankrupt. In a socialist society, the result will be nonfulfillment of growth quotas" (Ehrenfeld 1976). The difficulty with the seven arguments just given is that each might be refuted under certain circumstances by an informed adversary. For example, if a nonresource indeed were destroyed and no disaster followed, the argument for conservation would lose credibility. If a disaster did follow, it might be too late to do anything about it, or it might be difficult to prove a cause-and-effect relationship.

Instead of attempting to assign resource values to nonresources, the noneconomic values inherent in all species and natural communities should be weighed equally with economic values. Elton (1958) identified the ultimate reason for conservation—one that cannot be compromised:

The first [reason for conservation] which is not usually put first, is really religious. There are some millions of people in the world who think that animals have a right to exist and be left alone, or at any rate that they should not be persecuted or made extinct as a species. Some people will believe this even when it is quite dangerous to themselves.

Eltringham (1984) suggested that placing dollar values on wildlife to justify its preservation is to play a dangerous game. If the economic argument fails, justification for saving wildlife is lost. He contends that wildlife, like historic buildings and areas of scenic beauty, should be saved because people want them saved, regardless of measurable economic worth—and it is the role of governments to pass and enforce laws to do so.

Ehrenfeld (1976) therefore concluded that species and communities should be conserved simply because they exist and have existed for a long time. There are weaknesses and strengths in such arguments. The assumption that a resource is, by definition, something that must be bought and sold is not necessarily valid. The interest of the public in the preservation of wilderness areas, for example, constitutes a cost that can be measured in dollars. If the public deems that the wilderness area is more valuable than the minerals that might be extracted from the site—an action that might destroy the scenic and pristine landscape—then the wilderness area is a resource that has a measurable value. On the other hand, the strength of these arguments may convince the public that organisms such as the Houston toad are indeed resources—a value that emerges when society regards the survival of a toad as more important than the oil beneath the land where generations of toads have lived for millennia.

SUMMARY

Some have questioned whether nature should be treated as a commodity to be bought and sold. Native Americans, who regarded the earth as their mother, equated the sale of land with the selling of one's family. Economic values of wildlife have been calculated, however, in terms of how much money is spent by people in activities focusing on wildlife. Although expenditures for nonconsumptive uses of wildlife have increased in recent years, the amount lags well behind the funds spent by hunters and anglers.

In North America, lease-fee hunting on private lands is an expanding industry; hunting fees sometimes generate more dollars per area than livestock. Marketing limitations and other shortcomings have reduced the profits from cropping free-ranging wildlife in Africa for meat and other products. Elsewhere, large profits have been reaped from raising captive red deer for meat and antlers. Extensive selling of wildlife products and hunting rights in North America may undermine established ecological management of wildlife for the broad public interest.

A danger in relying on economic values for the conservation of wildlife is that the ecological value of certain nonresources (e.g., inconspicuous endangered species) cannot be calculated; that is, we lack knowledge of the role some species play in their ecosystems and the potential of such species as human resources. Lack of a dollar value should not be mistaken for worthlessness. Economic arguments are further complicated by differing ethical and religious attitudes of humans toward their environment. Wildlife may require integrated evaluations of economic and noneconomic factors for the fullest achievement of management goals.

CHAPTER 21

CONSERVATION BIOLOGY AND WILDLIFE MANAGEMENT

Conservation biology differs from most other biological sciences in one important way: it is often a crisis discipline. Its relation to biology, particularly ecology, is analogous to that of surgery to physiology and war to political science.

Michael E. Soulé (1985)

For most of the 20th century, the term *conservation* has been associated with managing a rather limited range of natural resources—those materials regarded as either useful or of necessity to humans. Justification for conservation therefore rested largely in the monetary values assigned to timber, water, and other natural resources. For wildlife, the principal focus for many years remained on animals sought by hunters, anglers, and trappers. Consequently, those species and communities that lacked either direct or potential economic value have not fared well in industrially based societies (i.e., those societies that have evolved an exploitative relationship with nature).

In the 1970s, however, an increasing number of conservationists expressed interest in the welfare of other species—organisms that have no obvious impact on human welfare and therefore not ordinarily regarded as "resources." Additional attention also was directed toward the conservation of biotic communities or entire ecosystems. Ehrenfeld (1976) advocated the protection of these components of our natural world for several reasons, which included holding species and communities with the same regard as we consider works of art, as things to be conserved for their inherent beauty; also for ethical

reasons, out of respect for continuing—not summarily ending—the track of evolution leading to their existence. With this broadened perspective in mind, we now turn to the role played by wildlife management, beginning with a condensed review of its history.

The realm of wildlife management has expanded gradually since the enactment of a few game laws marked its humble beginnings. As time passed, more and more game species gained protection because of legislation granting closed seasons, bag limits, and the designation of refuges. Then, early in the 1930s, Aldo Leopold championed the application of science— most notably biology and ecology—to the management of game animals in his seminal work, *Game Management* (Leopold 1933a). Formation of The Wildlife Society followed in 1937. This organization devoted itself to the development of management practices based on sound biological grounds and in keeping with high professional standards (Bennitt et al. 1937). In the middle decades of the 20th century, the focus of wildlife management spread from game animals to the more inclusive categories of birds and mammals, with emphasis on the management of populations and their habitats. By the 1970s and early 1980s, another step included entire communities, as illustrated in the management of "life-form associations," thereby enlarging the emphasis from populations of a single species to groups of populations occupying similar habitats (Thomas et al. 1979a).

The most recent broadening in the scope of managing wildlife and their habitats encompasses the subjects of biodiversity, including the preservation of threatened and endangered organisms and their genetic diversity, the conservation of whole faunas (e.g., birds of Hawaii), and the maintenance of ecosystems (e.g., rain forests of South America).

Many biologists felt that the distinction between these broader approaches and the traditional focus of wildlife management warranted a new discipline known as *conservation biology* (Soulé 1985). As a result, the Society for Conservation Biology was organized in 1986, followed in 1987 with the publication of the society's journal, *Conservation Biology*. The domain of the new society was described by its founders as crisis oriented, more theoretical, and more global than the realm of the The Wildlife Society. Nonetheless, some wildlife biologists expressed doubts about the need for such an organization, claiming that the goals of the new society were already being met, or could be met, by The Wildlife Society, an established organization then with more than 50 years of experience (Teer 1988). Others agreed, citing that a new awareness for conservation (e.g., maintaining the integrity of ecosystems) is growing within agencies once concerned with a narrower view of resource management (Aplet et al. 1992).

Temple et al. (1988) attempted to distinguish between the two closely related fields. The central goal of wildlife management was identified as the manipulation of animal populations—primarily vertebrates—whereas conservation biology stresses the maintenance of biological diversity (Table 21-1). Additionally, wildlife management tends to deal with selected species of birds and mammals, whereas conserva-tion biology addresses the full range of plant and animal taxa.

Even so, the policy of The Wildlife Society, as stated in the first issue of the *Journal of Wildlife Management*, "embraces the practical ecology of all vertebrates and their plant and animal associates," and "while emphasis may often be placed on species with special economic importance, wildlife management along sound biological lines is also a part of the greater movement for conservation of our entire native fauna and flora" (Bennitt et al. 1937). Soulé (1985) nevertheless asserted his belief that wildlife management in practice still concentrates on a relatively small number of "valuable" organisms, particularly fishes, deer, and waterfowl.

Differences in background and training represent another area where wildlife biologists may be distinguished from conservation biologists. For the most part, wildlife biologists graduate from colleges and universities where applied sciences such as forestry and agriculture have a long academic history (Bolen 1989). The curriculum is rather uniform among these institutions; hence, students graduating with a major in wildlife management enjoy similar training in such courses as mammalogy, dendrology, and soil science. In contrast, conservation biology is a synthetic discipline with a highly diversified membership (e.g., persons with degrees in genetics, physiology, oceanography, and geography, as well as numerous wildlife biologists). Data from anthropological (e.g., materials used for Native American clothing) and archaeological (e.g., rock art) sources, for example, were employed to examine the original structure of predator-prey populations in the Great Basin, thereby bringing into question just how "natural" the faunas of some "pristine" areas actually are today (Berger and Wehausen 1991). Courses dealing with conservation biology are now in the curricula of many universities, and textbooks appeared in support of those efforts (e. g. Primack 1993).

Table 21-1. Some Fundamental Differences Between Wildlife Management and Conservation Biology

Basis for Comparison	Wildlife Management	Conservation Biology
Central goal	Manipulation of populations	Maintain biodiversity
Outlook	Mostly practical	Mostly theoretical
Educational background	More uniform	More diverse
Taxonomic bias	Selected species of higher vertebrates	All taxa
Professional affiliation	Primarily state and federal agencies	Primarily academic institutions

Source: Temple et al. (1988).

Despite what may seem to be large differences, many of the dissimilarities between the two disciples are rather superficial when viewed in a modern context. We therefore regard wildlife management and conservation biology as complementary disciplines whose dual existence broadens and enhances human understanding of our biological surroundings. The dawn of the 21st century reveals no shortage of concerns, whether for deer or ducks, butterflies or orchids, prairie or rain forest.

"ON THE GROUND" WITH CONSERVATION BIOLOGY

The primary goal of conservation biology addresses the active protection of biological diversity, a term often shortened to *biodiversity*. Simply stated, this means halting the unprecedented loss of genes, species, and ecosystems from a planet still fabulously enriched with each. But where might the ideas of conservation biology be put into practice? Thomas and Salwasser (1989) addressed this issue in terms of resource management, as follows.

North America, and particularly the United States, represents one of the prime locations for conserving biodiversity. Among the reasons are (1) a relatively low human population, (2) the relatively short period of significant environmental alteration since settlement, (3) an affluent population, of which an influential segment is sensitive to environmental concerns, (4) the presence of a well-trained and motivated cadre of professional biologists and resource managers, (5) the presence of agencies and academic institutions attuned to the biology and management of wildlife and other natural resources, (6) the enactment of relatively strong and effective environmental laws, and (7) the state and federal ownership of wildlife refuges, parks, forests, and other large areas of land.

The availability of large areas of public land is especially important for the protection of biodiversity. More than one-third of the area of the United States is held in federal ownership, most of which is in large blocks under management by professional biologists, foresters, hydrologists, and others concerned with natural resources. Further, legislation governing much of this area requires some degree of conservation that favors biodiversity (e.g., Endangered Species Act of 1973, National Forest Management Act of 1976). Unfortunately, few large reserves— national parks and wilderness areas, for example— suitable for protecting biodiversity are likely to be added to the current holdings of public land (i.e., the 3 percent currently held in reserves such as wilderness areas is unlikely to reach more than 4–5 percent of the total area of public land). The existing reserves by themselves are not sufficient to protect biodiversity in a large part of the United States, which means that biodiversity must become a management goal on other types of public lands (e.g., national forests). Accordingly, land managers *already* "on the ground," namely men and women trained in fisheries, forestry, wildlife, watershed, range management, outdoor recreation, and related areas, will bear the responsibility for the conservation of biological resources for some time to come. Few such persons were schooled in conservation biology until the late 1980s, and those who received their degrees 15 or more years ago probably know little about the subject. Fortunately, however, most land managers are accepting to the importance of biodiversity, and many seem willing to apply the tenets of conservation biology to the resources in their care. Conservation biologists should interact with resource managers in ways that promote the useful transfer of their knowledge into the policies and procedures by which public land is managed.

The first targets of conservation biology should be those where there is a strong likelihood of success. Public lands managed for a variety of purposes— grazing lands, for example—offer prime settings for testing whether biodiversity can be maintained and established as a goal of national importance. Interagency cooperation forms an essential key toward achieving that aspiration (Salwasser et al. 1987; see also Salwasser et al. 1984). Conservation biologists, however, should not overlook the importance of privately owned lands as sites where biodiversity might become a dimension of the management program. Lands in the ownership of forest products industries, for example, would seem prime candidates as sites where the biodiversity might be improved.

LEVELS OF CONSERVATION BIOLOGY

Wildlife management typically directs much of its attention to such broad categories as population manipulation, in which seasons and bag limits are regulated, or by the improvement of habitat conditions for selected species. Although by no means fully exclusive of these activities, the focus of conservation biology is directed toward three other levels, each dealing with diversity, within the broad realm of biology: the preservation of genetic diversity, the preservation of species diversity, and the preservation of ecosystem diversity (Fig. 21-1).

GENETIC DIVERSITY.   The genetic variation occurring within some species has been of interest to wildlife managers. A notable example concerns the various races of Canada goose (*Branta canadensis*), whose genotypes produce birds of various sizes and habits (e.g., Hanson 1965). But although wildlife managers are well aware of these differences, their level of interest largely concerns the outward appearance of the geese as a means of gross identification and thereafter applying an appropriate management strategy to each racial group.

Conversely, certain issues require much greater definition and expertise in genetics. One case involves a small population of lions (*Panthera leo*) isolated in the Ngorongoro Crater in Tanzania where, in 1962, just 9 females and 1 male survived a disastrous epizootic of biting flies (*Stomoxys calcitrans*). Shortly thereafter, 7 additional males ventured into the seclusion of the crater, but the small population subsequently received no additional immigrants—or genes (Packer et al. 1991). The lions thus experienced a *genetic bottleneck,* a situation in which only a few individuals provide the total pool of genes for a population. Even when such a population later expands, as indeed happened when the Ngorongoro population grew to 125 lions, the genetic structure of the population reflects the genes present in the limited number of original individuals. In contrast, greater genetic diversity (i.e., more heterozygosity) is found in lions examined elsewhere on the plains of Tanzania.

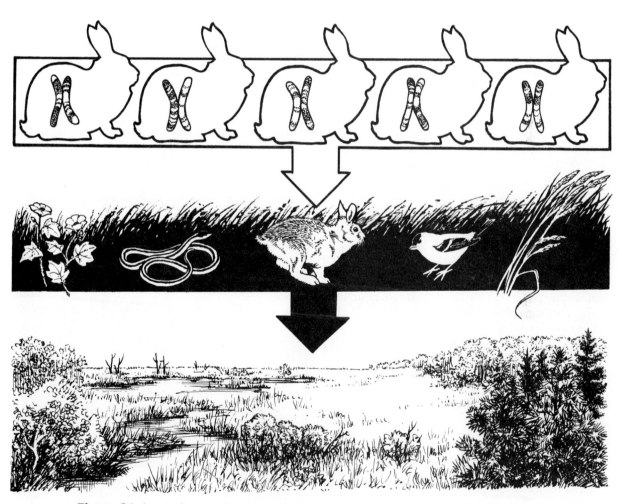

Figure 21-1.   As shown in this diagram, biological diversity—the chief concern of conservation biology—operates at three levels: genetic diversity (top), species diversity (middle), and ecosystem diversity (bottom). (From Temple [1991], drawn by T. Sayre.)

The term *founder effect* is used to characterize these circumstances, namely, when a small gene pool peculiar to the descendants of just a few "founders" occurs in a relatively large population. The founder effect also occurs when isolated areas such as islands are colonized by only a few individuals or even by a single, gravid female. In the case of the Ngorongoro lions, the consequences of the genetic bottleneck lessened heterozygosity in the population. Reproductive success declined, and males developed high levels of sperm abnormalities (e.g., sperm with two-heads; Fig. 21-2). Both conditions seem correlated with the reduced herterozygosity resulting from the physical and genetic isolation of this unique lion population.

Genetic differences also were revealed among isolated populations of the wild turkeys (*Meleagris gallopavo*). In this case, fragmented distributions and the small number of turkeys in each population—each the result of human activities—were associated with genetic variation (Leberg 1991). Many turkey populations initially were reduced as a consequence of European settlement in North America. Later, small numbers of turkeys were released to reestablish populations in areas where the species had been extirpated (e.g., a release of 22 turkeys in Connecticut included only 5 males). The proportion of genetic variation attributable to differences among turkey populations therefore was one of the highest recorded for birds of any species. Similar alterations in genetic structure may be expected whenever animal populations become artificially fragmented as a result of human activities (see Harris and Silva-Lopez 1992).

SPECIES DIVERSITY. The number of species naturally occurring within a community may have implications for the traditional species-specific focus of wildlife management. The governing features of predator-prey relationships, for example, depend in large measure on the number of prey species available to predators. As shown in Chapter 9, the relatively simple community on Isle Royale offers an instructive setting for examining the ways predators and prey interact (Allen 1979). Elsewhere, in a southern deciduous forest, for example, food webs are far more complex and the dynamics for the several predators and many species of prey are far more difficult to comprehend.

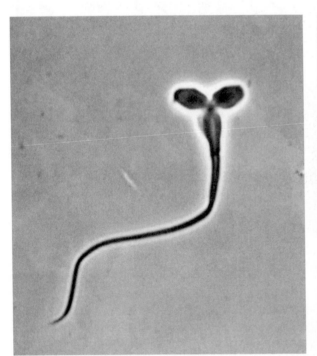
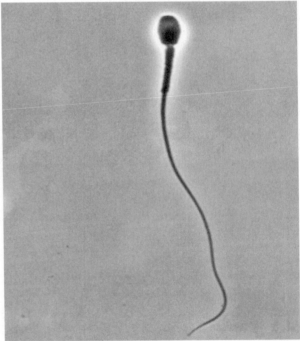

Figure 21-2.   Inbreeding depression occurring in a small population of African lions isolated in the Ngorongoro Crater in Tanzania produced reproductive dysfunctions, including the frequent occurrence of sperm cells with two heads (left). A normal sperm cell is shown for comparison (right). (Photos courtesy of D. E. Wildt and JoGayle Howard, National Zoological Park.)

Important as these relationships are to wildlife management, conservation biology takes an even broader view of the composition of a natural community. A good deal of effort may be directed toward relatively inconspicuous species of animals, many of which play crucial roles in pollination, herbivory, seed dispersal, and nutrient cycling. The component species within some communities, or segments of communities, may be unusually complex and of immense numbers. Even the most generous estimates for the numbers of species of insects and other arthropods were revised upward, based on work conducted in the rain forests of Central and South America. Most photosynthesis in forests occurs in the canopy, where the arthropod community is correspondingly rich. However, the exceptional height of the vegetation in rain forests limited the ways biologists might determine the numbers of arthropods living in the canopy. Erwin (1982) solved the sampling problem with "bug bombs," which blew a fog of quick-acting insecticide into the treetops—the dead insects fell into 1-m-wide funnels on the ground below. His analyses indicated that no fewer than 163 species of beetles lived in the canopies of a single species of tree. By extrapolation, some 8.15 million species of beetles may live in the canopies of the 50,000 species of tropical trees. Furthermore, because beetles constitute about 40 percent of all arthropods, the estimated number of arthropods living in the rain-forest canopy may total 20 million species. Finally, because the ratio of canopy to ground-dwelling arthropods is about 2:1, the estimated size of the arthropod community living in all strata of tropical rain forests almost defies imagination: 30 million, well in excess of the 1.5 million species usually proposed for arthropods. Large errors no doubt color extrapolations of this sort so that a figure of 5 to 10 million may be more realistic (see Gaston 1991a,b; also Erwin 1991), but the fact remains that an immense number of invertebrates has evolved in just one segment of the world's several ecosystems—and each species plays a role in the workings of its environment.

Investigations of species diversity, such as the study of the arthropod community in the forest canopy, fall squarely into the realm of conservation biology. Nonetheless, numerous examples might be cited in which wildlife managers also maintain a strong interest in certain species or groups of arthropods, just two of which are noted here: (1) as crucial food for breeding ducks (Krapu and Swanson 1975)

and (2) as pathogens or as vectors for pathogens (Jacobson et al. 1979; Herman 1968).

ECOSYSTEM DIVERSITY. The type of work just described offers a compelling argument for the preservation of tropical forests—and their associated biota—as a natural component of our world's ecosystems. Another reason is the role of tropical forests as winter habitat for many species of migratory birds (Hagan and Johnston 1992).

Regrettably, tropical forests are not the only ecosystems experiencing major and perhaps permanent perturbations. Virtually all of our planet's ecosystems are currently subject to some degree of inimical disturbance, whether these be off-road vehicles, the theft of cacti in deserts, overgrazing of grasslands, or erosion-causing logging in temperate forests. Wildlife managers and conservation biologists share these and other concerns for the preservation of ecosystems, not the least of which has been the unrelenting fight for the conservation of wetlands.

The disciplines of wildlife management and conservation biology therefore have similar goals, but these may be aimed at different levels in the structure of ecosystems. One thrust of conservation biology is to preserve ecosystems as a primary means of protecting natural diversity, whereas wildlife management seeks to maintain ecystems as habitat for selected species of wildlife. In many cases, management amounts to the simple "technique" of leaving the ecosystem alone. Little more than preservation is required for such environments as Alpine or Arctic Tundra. Nonetheless, both disciplines endorse treatments that preserve the integrity of some ecosystems (e.g., prescribed burns on grasslands), albeit for somewhat different reasons. To date, however, wildlife managers have assumed most of the leadership in the development of techniques associated with habitat intervention.

In summary, conservation biology is concerned with three levels of diversity—genetic, species, and ecosystem—whereas wildlife management currently places greater emphasis on the latter two levels. The differences between the two disciplines lie primarily in the academic training and outlook associated with each group. Wildlife management includes regulated harvest of selected species and therefore devotes considerable attention to manipulating the numbers and habitats of those species, whereas conservation biology devotes its efforts toward the full range of organisms and their habitats. Both groups nonetheless advocate wildlife conservation.

EXAMPLES OF CONSERVATION BIOLOGY INTERACTING WITH WILDLIFE MANAGEMENT

POPULATION GENETICS

Conservation biologists recently have examined the genetic makeup of certain small populations to determine why they fail to increase in an otherwise apparently suitable environment. The working hypotheses in this rapidly developing field suggest that high genetic variation favors survival, whereas low genetic variation does not. Hence, in a group of plants or animals that lack genetic diversity, the occurrence of a single adverse factor (e.g., disease or unfavorable weather) may prove so disastrous as to rapidly eliminate the entire population. One cause for the lack of genetic diversity in such a population is *inbreeding depression,* which results from matings between siblings or other members from the same family group and the occurrence of extreme homozygosity (many similar genes).

Conversely, a population with considerable genetic variability is more likely to include some members with genes that offer resistance to certain diseases or other mortality factors, thereby permitting greater fitness in its members. However, an exception may occur when individuals of different subspecies (= races) interbreed, with the result that the offspring experience diminished fitness; that is, the offspring in this case, although having considerable heterozygosity (many dissimilar genes), nonetheless may not be as well adapted to local conditions as are less heterozygous individuals from the local population. Therefore, matings among individuals with diverse genetic backgrounds also may lead to reduced fitness, a condition known as *outbreeding depression* (Templeton 1986). For example, in Chapter 18, we described how crosses of ibex (*Capra ibex*) led to an ill-timed rutting season and the production of kids during the coldest time of the year—an example of outbreeding depression that caused the loss of the ibex population in the Tatra Mountains of Czechoslovakia.

Biologists not long ago discovered that the genetic structure of cheetahs (*Acinonyx jubatus*) is remarkably uniform throughout the species' geographic distribution (O'Brien et al. 1985). That discovery led to the conclusion that a severe genetic bottleneck had occurred in the recent history of the species. Whereas the cause of the bottleneck remains obscure, the entire cheetah population apparently once had been reduced to only a few individuals, which became the ancestors of all the present-day cheetahs. Wayne et al. (1986) likened the genetics of cheetahs to an inbred population of laboratory mice!

The lack of heterozygosity in cheetahs produces several negative effects. Among these are a tenfold reduction in sperm count in comparison with other species of felines, which severely reduces reproductive success in captive cheetahs, and a 71 percent frequency of abnormally shaped sperm cells, compared with 29 percent in domestic cats (Wildt et al. 1983). Cheetahs also experience much greater rates of infant mortality than other cats. Finally, inbreeding has produced a high rate of asymmetry in the structure of cheetah skulls (Wayne et al. 1986). Caro and Laurenson (1994) nonetheless challenged the importance of reduced heterozygosity as the primary explanation for poor reproductive success in cheetahs. Instead of genetic limitations, they cited predation on cubs as "clearly more important" in regulating natural populations of cheetahs.

Several reproductive features of male Florida panthers (*Felis concolor coryi*), an endangered subspecies of mountain lion, are greatly dissimilar from those of other subspecies whose populations are abundant and therefore lack evidence of inbreeding depression (Barone et al. 1994). Florida panthers, whose remaining population numbers less than 50 individuals, have lower semen volume, more abnormal sperm, and reduced sperm motility (Table 21-2). Crytorchidism—the presence of only one testicle in the scrotum—also is unusually high in Florida panthers (44 percent), compared with other mountain lions (4 percent). Moreover, the average heterozygosity in Florida panthers was only 0.028, far less than the values determined for other subspecies of mountain lions in Texas (0.042) and Arizona (0.069), which further suggests inbreeding depression (Roelke et al. 1993).

The importance of genetic information to wildlife management is apparent. Without this knowledge, for example, wildlife biologists might continue seeking other explanations for poor reproductive performance (e.g., poor food supply, disease, or other environmental factors), thereby overlooking the real cause of the problem and wasting a good deal of time and other resources in the process. We see in this case a clear connection between conservation biology and wildlife management.

Table 21-2.   Means of Testicular Volume and Seminal Traits of Free-Ranging Florida Panthers Compared to Mountain Lions from Texas, Colorado, Latin America, and North American Zoos

Testicular and Semen Characteristics	Population				
	Florida	*Texas*	*Colorado*	*Latin America*	*Zoos*
Total testicular volume (cm³)	9.6	17.8	21.5	23.3	19.0
Ejaculate volume (ml)	0.7	2.0	3.9	2.1	2.9
Sperm motility (%)	38.2	49.4	80.0	71.1	53.0
Sperm concentration/ml ($\times 10^6$)	4.8	15.4	10.3	22.5	21.5
Structurally normal spermatozoa (%)	6.5	14.0	16.3	39.4	16.5

Source: Barone et al. (1994).

THE WOLVES OF ISLE ROYALE

Despite long-term research, biologists still have not satisfactorily explained why wolves (*Canis lupus*) on Isle Royle recovered so slowly from several years of low density, even with the apparent availability of abundant prey on this island-park in Lake Superior (Wayne et al. 1991). Three hypotheses were proposed to account for the apparent stagnation in the wolf population at 11–14 animals, when at one time there were more than 50: (1) low food availability, (2) lack of genetic variability and the consequent inability of the population to recover from epizootics of such diseases as canine parvovirus and Lyme disease, and (3) low genetic heterozygosity leading to poor fitness (e.g., juvenile survival).

To investigate these alternatives, Wayne et al. (1991) compared the genetic variability between the wolves on Isle Royale and those in populations on the mainland (Fig. 21-3). Blood samples were collected from the Isle Royale population and from 144 wolves at various other locations. The samples were analyzed to determine (1) the frequency of heterozygotes at several locations (loci) on the chromosomes, (2) variations in mitochrondrial DNA, and (3) multilocus hypervariable minisatellite DNA, which is popularly known as "genetic fingerprinting." The results of these tests indicated that the wolves on Isle Royale indeed had low genetic variability. Moreover, they probably were the descendants of a single female, possibly a member of a small pack that had emigrated across the ice to the island about 1949. The study did not conclude that poor reproductive success was associated with the occurrence of low genetic variability, but the authors nonetheless suggested that such a relationship probably exists. In the spring of 1993, however, after several years with essentially no successful reproduction, the wolves on Isle Royale produced 7 pups that survived their first winter, thereby increasing the population to 16 animals. (R. Peterson, personal communi-

cation). This rebound in numbers, whether temporary or relatively permanent, demonstrates the fallibility of accepting simple theory to explain complex phenomena such as population dynamics.

The genetic study of the Isle Royale wolf population is a good illustration of the interdisciplinary cooperation among scientists trained in a special area of conservation biology—genetics, in this case—and traditionally trained wildlife biologists. Indeed, 11 authors from both disciplines contributed to the publication resulting from this study. As analytical techniques improve, wildlife biologists undoubtedly will continue calling on specialists such as endocrinologists, bacteriologists, nutritionists, immunologists, and specialists from a multitude of other disciplines to help answer increasingly complex questions about wildlife populations and their environments.

NORTHERN SPOTTED OWL

Few issues in the realm of natural resources have proven as controversial in the past decade as the matter of the northern spotted owl (*Strix occidentalis caurina*) and its primary habitat, the old-growth forests of the Pacific Northwest (Fig. 21-4). The debate pits the welfare of the spotted owl population against the economic importance of logging in large areas of northern California, Oregon, and Washington. Old-growth forests—those with trees at least 200 years of age whose diameters are 80 cm or more at breast height—include, among other species, western hemlock (*Tsuga heterophylla*), Pacific silver fir (*Abies amabilis*), Douglas-fir (*Pseudotsuga menziesii*), and Sitka spruce (*Picea sitchensis*). Structurally, old-growth forests are multilayered, with a relatively closed canopy of tall, overstory trees, often with broken tops, and a high incidence of large snags and heavy accumulations of logs and other woody debris on the ground. Such settings support a rich community of wildlife (see Chapter 15).

Figure 21-3. Wolves isolated on Isle Royale exhibit more limited genetic variability than those in larger populations on the mainland. Study of their blood samples suggests that all members of the current population on Isle Royle are descendents of a single female. (Photo courtesy of R. O. Peterson.)

Because of their value as timber, old-growth forests have been heavily cut in recent decades, and today virtually none remain on privately owned lands. With this harvest—clear-cutting is typical—the prime habitat for spotted owls has steadily diminished to the point that the birds are gone from many areas and are apparently disappearing from others. As a result, spotted owls are variously listed as threatened, endangered, or of special concern by state and federal agencies. Both old-growth forests and spotted owls, however, remain on publicly owned land, much of it under the stewardship of the U.S.D.A. Forest Service and the Bureau of Land Management. These same lands are subject to logging by privately owned companies who pay the government for the timber, and as might be expected, these companies and other economic interests in the region resent interference with the timber harvest by people concerned about spotted owls. Plans for protecting the birds, in fact, are challenged by heated claims of local and regional economic disaster. However, the National Forest Management Act of 1976 requires the U.S.D.A. Forest Service to maintain viable populations of all native vertebrates on the lands it manages. This mandate, of course, includes the spotted owl, which is also considered by the Forest Service as a "sensitive species" and an "indicator species" for old-growth ecosystems.

According to Thomas et al. (1990), pairs of spotted owls remain in the same territories year after year so long as suitable habitat is available. However, even when paired and in possession of a territory, spotted owls may not nest every year. The nesting season occurs in March and April, when females usually lay two eggs per clutch. Females alone incubate the eggs and brood the young, whereas males provide most of the food (small mammals) for each family. Young owls leave the nest in May or June, although they rely on their parents for food for the remainder of the summer. By October, the young owls disperse from their parents' nesting area, but they typically do not acquire their own territories for 2 to 3 years thereafter. Unlike other species of owls, spotted owls are not wary of humans and may be approached quite closely before flushing. Unfortunately, about 60 percent of the species' habitat has disappeared, most of it in the 20th century.

The controversy—economic gains from clear-cutting old-growth forests weighed against the welfare of spotted owl populations—quickly became a rallying point for those concerned with wildlife, including conservation biologists. Numerous studies, among

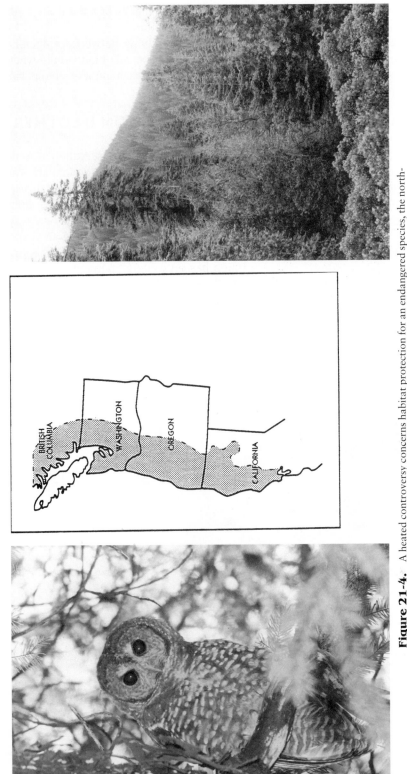

Figure 21-4. A heated controversy concerns habitat protection for an endangered species, the northern spotted owl (left), and the economic interests of the lumber industry in the Pacific Northwest. The distribution of the species (center) coincides with the occurrence of old-growth forest (right). Map from Thomas et al. (1990). (Photos courtesy of E. D. Forsman and J. F. Franklin.)

them Forsman et al. (1984), did much to describe the biology and distribution of northern spotted owls using the traditional tools and methods of wildlife management. However, the ultimate basis for managing the situation relies on island biogeography (see following) and other concepts generally associated with conservation biology (Wilcove and Murphy 1991). The conservation plan, prepared by an interagency task force, considered five key principles (Thomas et al. 1990):

• Species that are well distributed across their range are less prone to extinction than species confined to small parts of their range.
• Large blocks of habitat containing many individuals are superior to small blocks with a few individuals of a species.
• Patches of habitat near each other are better than widely dispersed patches of the same type of habitat.

• Contiguous blocks of habitat are better than fragmented habitat.
• Those sites lying between protected habitats enhance the dispersal of individuals when they closely resemble suitable habitat for the species in question.

CONSERVATION IN OTHER LANDS

An article entitled "Conservation in Action: Past, Present, and Future of the National Park System of Costa Rica" illustrates the international thrust of conservation (Boza 1993). The author, as vice-minister of Natural Resources, reviewed the progress and problems of developing and maintaining an aggressive conservation program in Costa Rica. This progressive nation has set aside 12 percent of its land for national parks and wildlife reserves (Fig. 21-5). As a result, protection is provided for most of the 205 species of mammals, 845 species of birds, 160 amphibians, 218 reptiles,

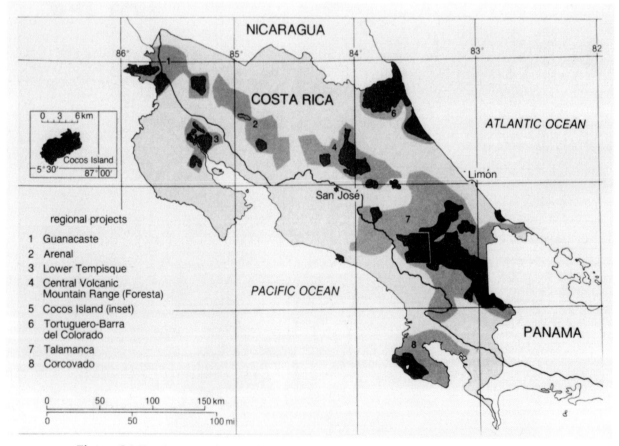

Figure 21-5.   Costa Rica devotes a larger percentage of its area to national parks (dark areas) and other kinds of protected areas (gray) than any other nation in the world. (Photo courtesy of D. Janzen.)

1013 fishes, and some 10,000 species of vascular plants found within the borders of Costa Rica.

In doing so, Costa Rica obtained national and international financial support and a good deal of backing from its internal political network, including groups such as the National Teachers' Association and the Students' Federation of the University of Costa Rica as well as the National Biologists' Association. Articles by advocates of conservation published in newspapers promoted the designation of parks. Nature conservation was added to the curriculum for the final 2 years of high school, and a university initiated a graduate program in ecotourism. Government planning included the development of corridors to allow the exchange of animals between parks, thereby lessening the chances of genetic isolation, roadkills, and other undesirable events associated with fragmented habitats.

ISLAND BIOGEOGRAPHY

MacArthur and Wilson (1967) set in motion ideas concerning the number of species that might exist on sites of various sizes (see also MacArthur 1972). The resulting concept—island biogeography—has at its heart the species-area relationship, which proposes the logical notion that larger sites contain more species than smaller locations. Larger locations are (1) more likely to be colonized by new species and (2) less likely to experience extinctions in their biota. Wilson (1989) illustrated one such relationship using the reptilian and amphibian fauna occurring on a series of islands in the West Indies (Fig. 21-6).

Initially, the rate of colonization is high, but in time the rate diminishes as more and more of the immigrants belong to species already present on the island. Conversely, species disappear (local "extinction") more rapidly with the passage of time. This results, in part, from the simple matter of chance; that is, more species are at risk as the pool of species increases. From the biological point of view, however, each additional species may increase competition, which reduces the average size of each population and thereby sets the stage for extinction. Strong competition may be inferred, for instance, when two species occur in a group of islands but never occupy the same island. Moreover, predators often are lacking in the vertebrate community on many islands; their populations necessarily occur at much lower densities than their prey. Thus, predators are far less likely to flourish in island ecosystems.

In addition to size, a second geographical factor plays a major role in the nature of an island's biota: distance from the mainland; that is, the mainland provides the source of colonists, so the number of species reaching an island decreases as the distance from the mainland to the island increases. The interaction between the size of an island and its distance from the mainland thereby forms the nucleus of island biogeography (Fig. 21-7). In general, smaller islands well offshore contain fewer species than larger islands located near the mainland. Because of the dynamics between the rates of colonization and extinction, a state of equilibrium eventually is reached in the richness (number of species) of an island's biota.

We interject here the importance of an organism's pioneering ability, which concerns its powers for reaching an island and, once arrived, its capacity for establishing a successful colony. Because of their means of dispersal, species with windborne propagules often are successful pioneers, as are those that float and can survive long periods in water (e.g., coconuts). The nature of the barrier between the "island" and its "mainland" source of immigrants, of course, significantly influences the rate of colonization. Thereafter, myriad local factors such as soil types, water resources, and topography influence the composition of an island's biota.

The sites subject to the concepts of island biogeography may be true islands or "islands" represented by such features as the remaining patches of tallgrass prairies (Sampson 1980). Even mountaintops and caves illustrate "islands," lending themselves to analysis (Vuilleumier 1970, 1973). Studies of island biogeography also have taken place in artificial settings such as woodlots separated by fields or urban development. Galli et al. (1976) recorded a strong positive correlation between the number of bird species and the size of woodlots in New Jersey. The number of species steadily increased on woodlots up to at least 24 ha in area. Forest-interior species (e.g., wood thrush, *Hylocichla mustelina*) began to appear in woodlots of 0.8 ha, but only forest-edge species (e.g., common flicker, *Colaptes auratus*) could find suitable habitat in woodlots of 0.2 ha or less. Cemeteries (Lussenhop 1977) and parks (Gavareski 1976) are among the developed areas where the concepts of island biogeography may be applied.

Some critics of island biogeography theory challenge the idea that equilibrium results from the counterbalancing forces of colonization and extinction (e.g., Gilbert 1980; Simberloff 1983). One shortcoming, for example, lies in the assumption that all species

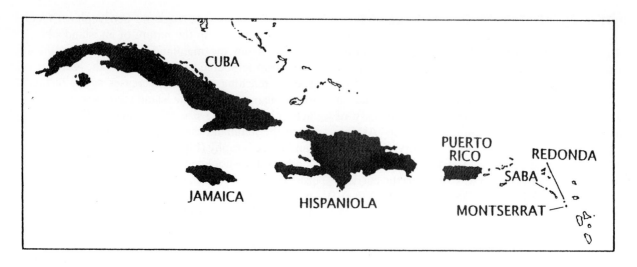

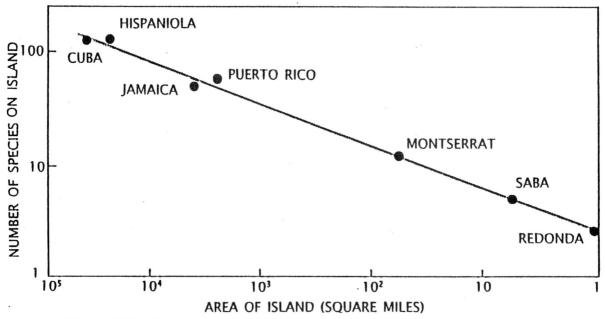

Figure 21-6.   The number of species on an island corresponds to the area of an island, as illustrated here for reptiles and amphibians occurring on an archipelago in the West Indies (top). (From Wilson 1989.)

are treated equally as units in the computations (that is, some species in fact have greater competitive abilities than others). Another is the assumption that the environment on an island remains constant throughout the processes of colonization and extinction. Such a view discounts the possibilities that succession and other types of ecological or physical change will influence the presence or absence of species. Also, no allowance is made for increases in the number of species as the result of evolution rather than by continued colonization.

Nonetheless, one of the presumed uses of island biogeography involves the minimum area required for nature reserves (Shafer 1990); that is, what size block ("island") of habitat will assure protection for the greatest number of species representative of a threatened community or ecosystem? A sanctuary therefore may fail in its intended purpose if it is too small. A case in point concerns Barro Colorado, a 15.6 -km² hilltop island formed in Lake Gatun when the Panama Canal was constructed and later declared a reserve for tropical fauna. Despite this protection, about 17 species of

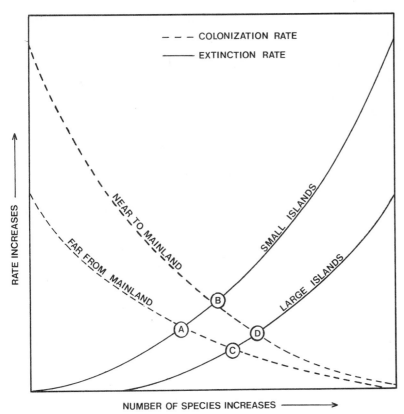

Figure 21-7. Island biogeography developed from a theory predicting the effects of distance and area on species richness (biodiversity). As shown here, the point of equilibrium—the balance between rates of colonization and extinction—reflects progressively more richness, left to right, beginning with small islands far from the mainland (A), small islands near the mainland (B), large islands far from the mainland (C), ending with large islands near the mainland (D). Drawn by Eva Lowe, from Wilcox (1980), based on MacArthur and Wilson (1967).

forest birds subsequently disappeared from Barro Colorado without replacement from the mainland (Willis 1974). Later, the decrease was reassessed upward to about 50–60 species, or three times greater than the previous estimate (Karr 1982; see also Karr 1990). Conversely, given the practical constraints of funding and politics, knowledge about the optimal upper range for the area of a nature reserve also is of obvious advantage. If, for example, 15 ha will provide adequate, long-term protection for an area surrounding the opening to a cave harboring a unique biota, then there seems little reason to press for a larger refuge (see McCoy 1983).

Wetlands are among the environments where the concepts of island biogeography might be applied to wildlife management. Regrettably, enormous numbers of wetlands across North America have been drained or otherwise rendered useless for wildlife; hence, those remaining are highly valuable resources. Brown and Dinsmore (1986) accordingly examined the importance of size and isolation of 30 wetlands, ranging from 0.2 and 182.0 ha in area in relation to the diversity of birds breeding in Iowa. Based on their results, size and isolation explained most (75 percent) of the variation in the richness of

the avifauna. However, the study also demonstrated that "bigger is not always better." Marshes 20–30 ha in size were more efficient in preserving species richness than were larger sites of up to 180 ha, and although a single large marsh may be less expensive to manage than several smaller wetlands, a cluster of smaller marshes increases habitat heterogeneity and thereby better fulfills the various needs of breeding ducks and other wildlife. Also, a complex of wetlands generally contains more species than a single large marsh, although some area-dependent species (e.g., Canada goose) may be excluded from smaller, isolated wetlands.

The importance of clustered wetlands also was illustrated in a model based on more than 350 sites in a 600-km^2 area in Maine (Gibbs 1993). When the computer simulation removed small (less than about 4 ha) wetlands, the overall area of wetlands in the study area was reduced by 19 percent, but the number of wetlands was reduced by 62 percent. Moreover, the distance between the habitats—wetlands in this case—increased by 67 percent. Increased distance between habitats often impairs the stability of *metapopulations,* which are nearby, but separated, groups of the same species between which individuals may migrate. Metapopulations may experience significant fluctuations in their numbers, including dropping to the point of extirpation, after which migrants from nearby metapopulations typically reestablish another population at the site (see Hanski and Gilpin 1991). However, when units of habitat within the cluster are eliminated, fewer individuals are exchanged and the chances for the continued survival of other metapopulations diminishes accordingly. For the wetlands in Maine, the computer simulation indicated that metapopulations of turtles and small species of birds and mammals, which were otherwise stable, indeed would be at risk if the smaller wetlands were removed from the landscape (Gibbs 1993). In other words, despite their lack of size, small wetlands may provide important sources of stock for reestablishing temporarily extirpated metapopulations.

CORRIDORS

Corridors, in ecological terms, are connections between separate areas of similar habitats. Rights-of-way for pipelines, roadways, and powerlines cutting through forested areas often are regarded as corridors, but these usually do not connect similar habitat; in fact, they generally increase fragmentation and thereby create barriers, filters, and predator lanes (Gates 1991; see also Forman and Godron 1986). Riparian forests (see Chapter 11) or hedgerows (see Chapter 13) may extend the ranges of some animals,

Photo courtesy of Edward O. Wilson

Edward O. Wilson
A Naturalist at Harvard
(1929–)

Dr. Edward O. Wilson pursued his academic training in the field of entomology, becoming a world authority in the study of ants (Formicidae); his coathored book *The Ants* (1990) won a Pulitzer Prize. However, his ideas and interests in other areas, including the controversial subject of sociobiology, brought him to even greater prominence in scientific circles. In 1967, with Robert H. MacArthur (1930–1972), Wilson enlarged upon their theory of island biogeography, which provided much of the intellectual foundation for conservation biology (i.e., factors influencing species richness and therefore of importance for maintaining Earth's rich biota). This concept, encapsulated in a simple diagram (see Fig. 21-7), stimulated a number of research projects conducted in various types of "islands" (e.g., city parks, lakes, and mountaintops, as well as real is-

lands). Wilson later published an acclaimed book, *The Diversity of Life* (1992), in which the role of biological diversity—more popularly, biodiversity—in our environment and the evolution of species are described in a semitechnical fashion.

Wilson also developed the concept of *biophilia*, the inherent connections humans subconsciously seek with other forms of life; that is, given an opportunity, humans generally search for outdoor experiences at seashores, in forests, or elsewhere where there is contact with what we call nature (e.g., vacation sites, but also year-round living places near water and woods). National parks thus are crowded with visitors, and as Wilson notes, more people in the United States and Canada visit parks and zoos than attend all professional athletic events combined. Biophilia thus provides a link between humans and their stewardship of the environment, especially in terms of preserving ecosystems and preventing further extinctions. In Wilson's words, "An enduring environmental ethic will aim to preserve not only the health and freedom of our species, but access to the world in which the human spirit was born."

Wilson's scientific contributions and far-reaching ideas have earned him numerous distinctions, including the National Medal of Science (1977), the 1990 gold medal of the Worldwide Fund for Nature, the International Prize for Biology (1990), and the Crafoord Prize (1990) of the Royal Swedish Academy of Sciences, which was established for those fields of science not covered by the Nobel Prize. Most of his career has been spent at Harvard University. In 1994, he wrote an autobiography, *Naturalist,* in which he described his humble boyhood life in Alabama, his worldwide fieldwork, and his professional associations with colleagues and students.

but this role is independent of whether they also function as travel lanes between areas of similar habitat (Simberloff et al. 1992).

Here we are interested in those corridors that indeed link "islands" of desirable wildlife habitat. With suitable corridors, plants and animals may travel between islands, thereby (1) preventing inbreeding depression by maintaining gene flow between various segments of a larger population and (2) enhancing species richness in keeping with the ideas of island biogeography. In other words, such corridors may offset the negative consequences of fragmented habitats. Ideally, corridors of this type would be planned in the design of landscapes *before* the existing vegetation was altered (e.g., greenbelts connecting urban parks in the design of new towns). More often, however, corridors are planned *after* perturbations occur in the landscape, which usually leads to the expense of acquiring land as well as additional costs for restoring the appropriate structural components (e.g., vegetative cover).

Corridors have been proposed for connecting fragmented areas of habitat for animals as large as mountain lions, as Beier (1993) suggested for an area in southern California undergoing rapid development. Based on a computer model, a population of 15–20 mountain lions might persist on an area of 2200 km², but if a corridor permitted the immigration of up to 3 males and 1 female per decade, the population could persist on an even smaller area (600–1600 km²) without experiencing a significant 100-year risk of extirpation. Immigration of this sort—that which augments the size of a local population—is sometimes characterized as the "rescue effect" (Brown and Kodric-Brown 1977). Similarly, corridors in the southeastern United States could unite three national wildlife refuges, two national forests, and other lands in a single management unit suitable for the release of Florida panthers raised in a captive breeding program (Noss and Harris 1986; Harris 1988). One link in this plan is a corridor—Pinhook Swamp—between Okefenokee National Wildlife Refuge and Osceola National Forest (Fig. 21-8). Elsewhere in Florida, highway underpasses designed to reduce collisions with wildlife also may reduce the effects of habitat fragmentation on Florida panthers and other species (Foster and Humphrey 1995).

In 1997, a particularly ambitious proposal emerged for a 2800-km corridor connecting Yellowstone National Park with Canada's Yukon Territory. Popularly known as "Y2Y," the Yellowstone to Yukon Conservation Initiative is designed to facilitate genetic exchanges among wilderness species (e.g., grizzly bears, *Ursus arctos horribilis,* which are relatively isolated in Yellowstone Park). Besides Yellowstone, Y2Y will link Glacier National Park with several parks in Canada, among them Banff and Jasper, along the spine of the northern Rocky Mountains. To implement the plan, the corridor will require large areas of additional land in nine state, provincial, and territorial jurisdictions—ranging from 8712 km² in Washington State to 446,295 km² in British Columbia. Overall, about 38 percent of Y2Y lies within the United States, with the balance in Canada. Other costs for the project include animals-only highway overpasses (Fig. 21-9), but an economic and demographic analysis also suggests the Y2Y region will receive important benefits (Rasker and Alexander 1997). To date, more than 80 organizations have endorsed Y2Y as a strategy to promote biodiversity in North America's longest chain of mountains.

There is less than full agreement about the merits of corridors as a tool of conservation biology. Introduction of pathogens, predators, or exotic species are among the potential drawbacks to corridors (Table 21-3). However, the overriding criticism of corridors concerns the lack of field data proving the success of their intended purposes (Simberloff and Cox 1987, Simberloff et al. 1992); that is, few data currently are available to show if individuals, as believed, actually move through corridors and at what rates. Even less is known about the effects, if any, on the gene pool of insular populations connected by corridors. The presumed benefits of corridors thus cannot provide justification for establishing reserves that are too small for their intended purposes. Additionally, whether a corridor represents safety or a threat to dispersing animals depends upon the ecological factors specific to the species and the site (e.g., predation may increase in corridors because of their greater amounts of edge). If corridors attract numbers of organisms but unduly increase mortality, they become population "sinks," which are more harmful than beneficial. Finally, there is the issue of cost, as noted earlier. Precious dollars committed to the expense of establishing a corridor perhaps may be better spent elsewhere in the cause of conservation.

Translocation may offer an alternate strategy for corridors. Managers could periodically move animals from one isolated population to another, which would be less expensive than establishing corridors between the populations. Bird eggs, for example, might be switched from nests at one site to those at another. Translocation may be useful for several purposes: to bolster genetic heterogeneity of small populations, to establish satellite populations as a means of reducing risk to losses from catastrophes, or to hasten recovery of species following restoration of their habitat (Griffith et al. 1989). Translocation also may be necessary when habitat fragmentation endangers the composition of natural communities, especially for those organisms whose abilities to disperse are limited. Nonetheless, the best argument for corridors may lie in the assertion that much of the original landscape was interconnected, whereas today, because of human modifications, many species are confined to isolated patches of habitat (Noss 1987). Corridors therefore represent an attempt to restore a degree of connectivity to a highly fragmented landscape. Moreover, Harris and Scheck (1991) and other advocates of corridors suggest that corridors encourage biodiversity, the fundamental goal of conservation biology.

SONGBIRDS AND FRAGMENTED FORESTS

Harris (1984) was among the first to warn of the biological implications of a highly divided landscape in his benchmark study, *The Fragmented Forest*. This book focused on island biogeography theory as it applied to old-growth forests in the western Cascades. Elsewhere, the plight of neotropical migrant songbirds—including wood warblers (Parulidae) and tanagers (Thraupidae)—has been linked not only to the destruction of tropical rain forests where these species overwinter, but also to the fragmentation of

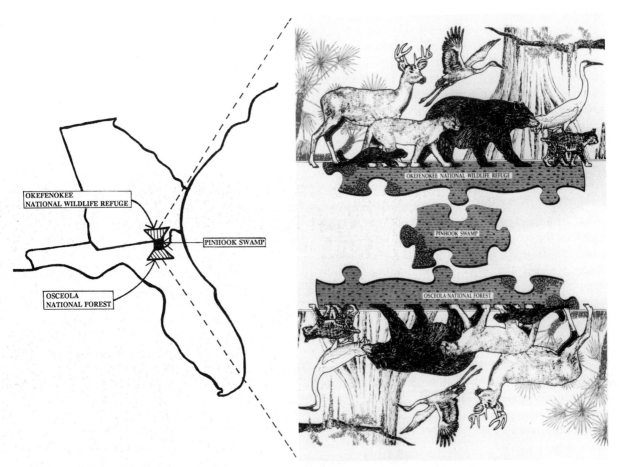

Figure 21-8. Pinhook Swamp represents a strategic corridor filling a 24-km gap between two established conservation areas, Okefenokee National Wildlife Refuge in Georgia and Osceola National Forest in Florida. The corridor, which was authorized by Congress in 1989, is of potential importance to the management of endangered species, including Florida panthers, whooping cranes, and red wolves. Map (left) drawn by Eva Lowe. (From Harris and Gallagher 1989.)

forests in North America where these species nest (Hagan and Johnston 1992; Fig. 21-10). The number of nesting species diminishes when the blocks ("islands") of forest are too small, and their populations

Figure 21-9. An experimental overpass designed for wildlife crosses Highway 1 in Banff National Park, Alberta, Canada. This structure offers a prototype for other overpasses planned as part of an immense corridor—popularly known as Y2Y—running from Yellowstone National Park to the Yukon Territory. (Photo courtest of Ben Alexander.)

thereafter drop. Hence, the preservation of suitably sized blocks of breeding habitat is a major concern, especially in the eastern deciduous and coniferous forests of North America, where the continent's largest breeding concentrations of migratory birds occur (Terborgh 1989, 1992). Thus, in keeping with the canons of island biogeography, the area—from 0.1 to more than 3000 ha—and the degree to which blocks of forest are isolated were significant in predicting the richness of the avifauna in the mid-Atlantic States (Robbins et al. 1989). Nest parasitism by brown-headed cowbirds (*Molothrus ater*) is another concern associated with forest fragmentation and the concurrent declines in several species of songbirds (Brittingham and Temple 1983). Cowbirds evolved in grasslands and savannas and rarely penetrate far into wooded areas. The newly opened woodlands thereby offer cowbirds additional habitat in areas once covered with unbroken forest. As nest parasites, brown-headed cowbirds lay their eggs in the nests of other birds, often species of wood warblers, which thereafter care for the cowbird eggs and young as if they were their own. Because of their short incubation period, cowbird eggs hatch before the host's eggs, and thereafter the large and aggressive nestlings easily

Table 21-3. Potential Benefits and Drawbacks of Corridors as a Means of Connecting Fragmented Habitats

Benefits	Drawbacks
Increase immigration, which could • favorably influence species richness, as predicted by the theory of island biogeography • increase population sizes and decrease chances for extirpation (produces a "rescue effect") or reestablish extirpated populations • prevent inbreeding depression and maintain genetic variation within populations	Increase immigration, which could • aid spread of diseases, weeds and pests, exotic species, and other undesirable species into reserves and across the landscape • affect genetic events (e.g., outbreeding depression)
Increase foraging areas for wide-ranging species	Aid spread of abiotic disturbances, especially fire
Provide escape cover for animals moving between fragmented habitats	Increase exposure of wildlife to hunters, poachers, and predators
Provide diversity for species requiring various cover types and successional stages	May favor some species and not others (e.g., riparian habitat may not serve as corridors for upland species)
Provide refuges from large disturbances (a "fire escape")	Expense, and conflicts with other conservation strategies
Provide "greenbelts" in urban areas for recreation, better scenery, improved land values, and other benefits	

Source: Noss (1987)

Figure 21-10. Forest fragmentation is clearly indicated in this aerial view of a rural area in the northeastern United States. Only small areas of forest (light colored) remain scattered among the fields in this landscape. (Photo courtesy of D. Muchoney, National Zoological Park.)

outcompete the host's offspring for food carried to the nest.

These features in the breeding biology of brown-headed cowbirds gain additional significance in light of forest fragmentation. Wood warblers and other birds, once secure in the forest interior, now nest nearer the edge of the woods where their nests are vulnerable to cowbird parasitism. In sum, cowbirds have (1) invaded areas where the original forest is now a patchwork of woodlands and (2) gained access to many more nests of host species. As a result, populations of the wood warblers and other host species steadily decrease, whereas those of brown-headed cowbirds increase.

Populations of widely distributed, abundant species, as well as rarer species such Kirtland's warbler (*Dendroica kirtlandii*), also may suffer from cowbird parasitism. Nuttall's white-crowned sparrows (*Zonotrichia leucophrys nuttalli*) in a metropolitan area in California currently experience a 40–50 percent rate of parasitism, compared to only 5 percent 15 years ago, and for this reason their long-term survival in this part of their range may be in doubt (Trail and Baptista 1993).

The geometry of the remaining blocks of forest also bears on the increase in cowbird parasitism. For example, one square 4-km² block of forest has 8 km of edge, for 2:1 ratio of edge to area. In comparison,

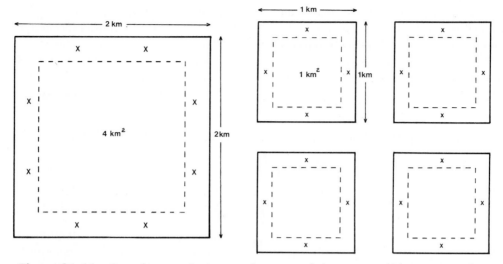

Figure 21-11. Forest fragmentation increases the amount of edge per area, which creates more sites where brown-headed cowbirds can find and parasitize the nests of other birds. As shown in this diagram, the total area (4 km²) is the same for the large block and for the four smaller blocks, but the smaller blocks collectively have 8 km more linear edge than the single large block. Twice as many opportunities for parasitism therefore occur along the borders of the smaller blocks (i.e., 16 nests compared with 8 nests). Nests (*X*) are equally spaced in all blocks. Dotted lines represent the limit to which brown-headed cowbirds penetrate into a block of forest habitat in search of nests; nests farther to the interior remain free from parasitism. (Drawn by Eva Lowe.)

four smaller blocks of 1 km² each have a 4:1 ratio. So even though the total area in the two forests remains unchanged, the amount of edge in the fragmented forest has doubled. For some species of wildlife, more edge is desirable (see Chapter 7), but for wood warblers the increased exposure of their nests to cowbirds—and poor breeding success—is a direct result of forest fragmentation (Fig. 21-11).

We have focused on songbirds to illustrate concerns about fragmented habitats, but other species also are affected. Nor are such concerns limited to animals. Menges (1991) demonstrated how habitat fragmentation limited the germination of seeds from royal catchfly (*Silene regia*), a prairie plant dependent on hummingbirds for pollination. The distribution of this plant has been severely limited by fragmentation, and only seeds from the larger populations (those of more than 150 individuals) experience uniformly high germination (i.e., 85 percent or more). Two explanations for this relationship seem possible: (1) inbreeding depression has reduced the viability of the seeds from the smaller populations, or (2) the smaller populations are not visited as often by hummingbirds, resulting in reduced pollination.

ECOLOGICAL BASELINES

Wildlife management usually involves manipulations of animal populations and/or their environments. Although clearly well intentioned, management efforts involve some degree of uncertainty about their effect on plant and animal communities because we do not yet understand perfectly how biological communities respond to human intervention, environmental changes, or natural internal processes. Therefore, in the absence of "controls," we cannot be certain whether ensuing changes in biological communities resulted from the management or occurred naturally. Arcese and Sinclair (1997) accordingly proposed setting aside protected areas throughout the world to serve as ecological baselines. Such sites, they argue, enhance our understanding of ecological processes by serving as "controls" for the large "experiment" that Earth is experiencing from deforestation, agriculture, urban development, and other human manipulations of ecosystems. In other words, we need baseline areas to evaluate the effects of humans on biodiversity, the state of biotic communities in the absence of humans, or the range of variation we might expect in these communities in the absence of human intervention.

Accordingly, the presence of baseline areas allow management to be conducted as an experimental science and thereby avoid the slower process of learning by trial and error (i.e., we increase the rate at which we learn how our interventions—or noninterventions—influence communities.)

MINIMUM VIABLE POPULATIONS

The term *minimum viable population* (MVP) gained currency when Congress enacted the National Forest Management Act of 1976, which charged the U.S. Forest Service to maintain "viable populations" for all species of vertebrates living in the nation's national forests (Gilpin and Soulé 1986). This legislation added a firm commitment to plant and animal conservation to the mandate of the Forest Service. Unfortunately, the concept of MVP may imply the existence of a "magic number" that will ensure the continued presence of wildlife populations, when in fact such a threshold may vary not only by species, but also because of other circumstances such as location. An MVP is not a population that simply maintains itself under average conditions, for it also must be large enough to survive various types of random events (see following).

Despite the dangers of suggesting a number for such a threshold, however, an MVP of at least 50 (at least for large mammals) was proposed as sufficient to cope with the short-term effects of inbreeding, but an MVP of 500 represents a target for long-term maintenance (Franklin 1980). We emphasize again, however, that these numbers were early attempts to define a complex issue in absolute terms, and the concept of MVP ultimately may defy across-the-board generalizations.

In addition to considerations of size, time should become a dimension in the definition of an MVP. Standards for flood-control structures offer an useful analogy (Shaffer 1981). Such structures, if designed to control a once-in-50-years flood, cannot handle a once-in-100-years deluge. Similarly, an MVP should be perceived in terms of its probabilities of experiencing various types of adversities during some time period. In 1989, for example, Hurricane Hugo killed 63 percent of the red-cockaded woodpeckers (*Picoides borealis*) and destroyed 87 percent of the active nest cavities at the Francis Marion National Forest in South Carolina, the site of the densest, second-largest, and only increasing population of this endangered species (Hooper et al. 1990; Fig. 21-12). In terms of probability, a hurricane of similar magnitude (category IV on the Saffire-Simpson scale of I–V) will hit the same area of South Carolina about

once every 100 years. So, the frequency of severe hurricanes should be considered in determining the MVP for the woodpeckers in this region (i.e., in this case, the MVP must be large enough to survive large-scale habitat destruction every 100 years).

Shaffer (1981) offered a tentative definition of an MVP:

A MVP for any given species in any given habitat is the smallest isolated population having a 99 percent chance of remaining extant for 1000 years despite the foreseeable effects of demographic, environmental, and genetic stochasticity and natural catastrophes.

Thomas (1990) reviewed field data from various populations to determine estimates of MVPs needed for medium- to long-term persistence. For example, studies of bird populations on the California Channel Islands offered insight concerning persistence (Jones and Diamond 1976). At least 39 percent of the populations with fewer than 10 pairs was lost within 80 years. Of those with 10 to 100 pairs, the number of populations declined by at least 10 percent, but of those with 100 to 1000 pairs, only one population was lost during the same 80-year period. No losses occurred in populations of more than 1000 pairs. Elsewhere, bird populations with average populations of 2.1–5.0 pairs lasted about 3.5 years, and those with 5.1–12.0 pairs were lost within an average of 7.5 years (Pimm et al. 1988). Such evidence indicated that a population of 10 individuals is far too small and 100 is usually inadequate, whereas 1000 may be acceptable as an MVP for species whose numbers vary normally (Thomas 1990). Because insects typically exhibit much larger variability in their numbers, their MVP is on the order of 10,000.

Any definition for an MVP should include an explicit set of performance criteria (Shaffer 1981); that is, an MVP might be defined in terms of, say, a 90 percent chance of persistence for 500 years. Such criteria vary by species and circumstances, rather than by some "magic number." Therefore, wildlife refuges or other conservation areas should be evaluated on the basis of their ability to meet MVP criteria. In general, the area required for the conservation of vertebrates increases proportionately with increases in the size of MVP and the length of the period of persistence. Some circumstances demand large areas for just a few individuals. It seems necessary, for example, to protect a large foraging area, perhaps as much as 100,000 ha, for the successful reintroduction of a MVP of California condors (*Gymnogyps californianus*) raised from captive stock (see Thiollay 1989 for the area require-

Figure 21-12. Mature longleaf pines (*Pinus paulustris*) provide prime habitat for colonies of red-cockaded woodpeckers, as shown by a 100-year-old stand at Francis Marion National Forest in South Carolina (top). In 1989, the strong winds of Hurricane Hugo—a "storm of the century"—destroyed large areas of this habitat (bottom). (Photo courtesy of U.S.D.A. Forest Service.)

ments of hawks and other raptors in rain forests and the comments of Robinson and Wilcove 1989).

Populations are jeopardized by many circumstances, most of which are related to human activities (see Table 19-2). Overhunting and pollution are examples of these hazards, which have been variously called *systemic* or *deterministic* factors. The outcome—extinction—is predictable if these conditions are severe and remain unchanged. Stochastic events—chance or random occurrences—represent another threat to the existence of organisms. These events do not immediately destroy populations; instead, they thin populations to levels at which the chances of survival are greatly diminished, especially if another unfavorable random event should occur. Obviously, stochastic events are of great importance to populations already reduced by deterministic factors. Simply stated, the smaller the population, the greater its vulnerability to stochastic events; as might be expected, the shorter the interval between the occurrence of such events, the less likely a population will survive. Shaffer (1981) identified four types of stochastic events, any one of which might bring a species or population to the point of no return:

- Demographic stochasticity, which produces a chance occurrence of an unfavorable death rate or other feature of population ecology; for example, a sex ratio heavily skewed toward one sex in the litters of adult red wolves (*Canis rufus*) released in a reestablishment program, thereby limiting the population's reproductive effectiveness (i.e., number of pairs).
- Genetic stochasticity, which results from the chance occurrence of unfavorable genetic circumstances; for example, the bottleneck in cheetahs, previously described.
- Environmental stochasticity, which occurs with a chance occurrence of unusual levels of predation, parasitism, disease, or other decimating factors; for example, the extreme infestation of biting flies on lions in Ngorongoro Crater, previously described.
- Natural catastrophes, of which random fires, floods, and droughts are representative; for example, Hurricane Hugo striking a population of red-cockaded woodpeckers, previously described.

Unfortunately, the fate of the eastern subspecies of prairie chicken known as the heath hen (*Tympanuchus cupido cupido*) was sealed by a sequence of such events, but it is uncertain which of these, singly or in combination, was most important in causing extinction of the birds (Shaffer 1981). Although once widespread on the eastern seaboard, the range of heath hens steadily declined in the last century until only a small population survived on Martha's Vineyard. By 1900, fewer than 100 heath hens remained, but under management their numbers increased to more than 800 birds by 1916. Then a fire (natural catastrophe) destroyed vital habitat and most of the year's nests; then followed a winter in which unusually heavy predation (environmental stochasticity) reduced the population to 100–150 birds. After that, their numbers increased slowly to about 200 by 1920,

when disease (more environmental stochasticity) struck, and the population plummeted to fewer than 100 birds. Thereafter, increasing sterility (genetic stochasticity) and an imbalanced sex ratio (demographic stochasticity) further diminished the remaining population; by 1932, heath hens were gone forever (see also Greenway 1967). The compound effect of such stochastic events on diminished populations produces what Gilpin and Soulé (1986) call "an extinction vortex," a train of random, normal events that jeopardize small populations, possibly to the point of their complete elimination.

An analysis of 122 fragmented populations of bighorn sheep (*Ovis canadensis*) illustrates the concept of MVP for a large game mammal (Berger 1990). The study involved isolated populations of bighorns in the American Southwest for which records spanning up to 70 years were available (reintroduced populations were not included in the study). Notably, no populations of fewer than 50 sheep persisted for more than 50 years, whereas those with more than 100 animals persisted for at least 70 years. The results strongly suggest a threshold of 50 animals, below which populations of bighorns cannot survive, and an MVP of at least 100 animals is necessary for long-term persistence. Because the analysis revealed no single cause (e.g., food shortages or predation) for the rapid elimination of the small populations, the conjecture arises that these populations were predisposed to elimination solely because of their small size. Moreover, it seems evident that the small populations of bighorn sheep still surviving will require prompt management if they are to persist much longer. Suchy et al. (1985) similarly estimated that a MVP of 125 was necessary for the maintenance of grizzly bears (*Ursus arctos*) in the Yellowstone ecosystem; at such a size, the MVP was calculated as having a 95 percent chance of persisting for 100 years. See Lehmkuhl (1984), Shaffer and Sampson (1985), and Harris et al. (1987) for more about determining MVP.

Some species have recovered even when reduced to a handful of individuals. Thousands of northern elephant seals (*Mirounga angustirostris*) once occurred in rookeries along the coast of California and Baja California, but excessive hunting between 1820 and 1880 eliminated all but about 20 individuals on Isla de Guadalupe, a remote island off the Mexican coast. With federal protection, however, this small population slowly increased and established new breeding colonies, which eventually included more than 30,000 animals. Even so, the renewed population lacks heterozygosity, possibly to the point that northern elephant seals are now entirely monogenic (i.e., the species possesses only one set of genes). Much like the cheetahs, northern elephant seals now lack a pool of genetic variability with which they might cope with changing environmental conditions (Bonnell and Selander 1974). Recently, newspaper headlines reported some recovery in the greatly diminished population of blue whales (*Balaenoptera musculus*), and although this is encouraging news, little is known about their previous or current levels of heterozygosity, which will influence their fitness and chances for long-term survival.

HUMAN POPULATION

Articles about the rampant growth of human populations seldom appear in journals devoted to either conservation biology or wildlife management; yet a perceptive editorial in *Conservation Biology* points out that human growth represents a major factor affecting the impact of conservation biology (Meffe et al. 1993; see also Chapter 23). Journals often address the effects of human disturbance or behavior on various species of wildlife, but they ordinarily do not consider the sheer numbers of humans burdening our globe. Allen (1968) likewise warned of the difficulties of managing wildlife populations in the face of the continued increase in the number of humans. In the interval since his paper appeared just a generation ago, the global population of humans has grown from 3.5 billion to 5.7 billion—an increase of 63 percent.

Meffe et al. (1993) noted that the world's human population, which is increasing at about 260,000 people per day, is probably "the most severe problem faced in human history, and the one most likely to result in breakdown of both normal ecosystem function and social structure." Although most biologists are aware of the seriousness of the rapid surge of human numbers, the general public is far less so. Consequently, biologists have the professional responsibility to call public attention to the perils of unrestrained growth in the human population. Without stabilization of the human population, Meffe et al. (1993) continue, "the biodiversity we work so hard to maintain will continue to be degraded at the hands of profiteers, short-term thinkers, the uninformed, and those blameless unfortunates who are simply struggling to stay alive in an overcrowded world." They suggest that conservation biologists respond in two ways: (1) seek more interactions with demographers, soci-

ologists, and epidemiologists, and (2) strongly advocate controlling human populations.

SUMMARY

A discipline known as conservation biology emerged in the latter half of the 1980s. Some goals of conservation biology overlap with those of wildlife management, but its foundation rests primarily on the maintenance of diversity at three levels of biological organization: genetic, species, and ecosystem. Wildlife managers and conservation biologists strive toward a common goal of creating a world in which humans and an almost unimaginably rich biota might share a fruitful and common presence for generations to come. Much will be gained if the practitioners of one discipline learn from the other.

Genetic bottlenecks and inbreeding depression limit the heterozygosity of plant and animal populations, thereby impairing their reproductive success or diminishing their ability to respond to changing environmental conditions.

Island biogeography represents a fundamental concept in the theory and practice of conservation biology. Initially based on the number of species established on small or large islands found at varying distances from the mainland, the concept of island biogeography later was applied to the size of blocks of habitat required for the maintenance of such ecosystems as wetlands or rain forest.

Other concerns include the fragmentation of habitat and the resulting increase in the ratio of edge to area, which may harm some kinds of wildlife, as shown for songbirds breeding in the forests of eastern North America. In some cases, corridors may help unite fragmented landscapes into functional management units for species as large as Florida panthers.

Some ecologists propose to establish worldwide ecological baselines as control sites for comparisons with the "grand experiment" humans are conducting on the Earth's biological communities.

Small populations face elimination from stochastic events occurring in the physical or biological environment (e.g., hurricanes, fire, and diseases) or within the population itself (e.g., imbalanced sex ratios and other features associated with population ecology). Hence, an important thrust of conservation biology is to determine the minimum viable population (MVP) for populations or species threatened with extinction. Considerations of MVP include the probability of its persistence for a given period of time.

CHAPTER 22

WILDLIFE AS A PUBLIC TRUST

When the Supreme Court in the nineteenth century enunciated the principle that wildlife was not the private property of any individual or group of individuals, but was instead the collective property of all the people, it established the paramount role of the government, as public trustee, in the task of wildlife conservation.

Michael J. Bean (1983)

Policies governing the management of natural resources developed gradually. Some skeptics might venture that these evolved as haphazard reactions to then-current events rather than to thoughtful planning for the future. In any case, myriad far-reaching laws and policies now govern the destiny of virtually every natural resource in the United States. Unfortunately, the administration of these is scattered throughout a maze of state and federal agencies, where overlap and conflicting interests sometimes hamper decisions and actions.

More than 60 years ago, an "American Game Policy" was published as a guideline for the restoration, management, and preservation of wildlife resources in North America (Leopold 1930). In its introduction, this policy noted that the demand for hunting was outstripping the supply and asked "where, how, and who" might increase game production if hunting were to continue as recreation. The policy proposed a seven-point program designed to blend the interests of the land and landowners, game and hunters and the public into a feasible means for restoring game as a natural resource. The seven basic actions were these:

1. Extend as much as possible public ownership and management of "game lands," including those lands whose primary function might be for other purposes, such as forestry.
2. Recognize landowners as custodians of public game on privately owned lands, including protecting owners from damage and providing them with compensation for labor and use of the land.
3. Experiment with ways to bring the hunters, the landowner, and the public into a productive relationship with one another, and then encourage the relationships that promise the most for game management.
4. Train people in the skills of game management, thereby establishing a profession like forestry, agriculture, and other forms of applied science.
5. Determine the facts about the ways and means of making land produce more game.
6. Recognize nonhunters and scientists as partners with hunters and landowners for the conservation of wildlife, including joint sponsorship of management activities and their funding requirements.
7. Provide funding from general taxation for the bettering of all kinds of wildlife, with the hunters

paying for those activities that alone serve game species. Private funding should help carry the costs of wildlife education and research.

The core of the policy was directed toward the conflict between the tradition that hunting was free of economic laws and the opposing view that land must produce a return on its owner's investment. Without a reward, why should owners of farms or forest produce game, especially if game production was costly or interfered with the primary crop? The 1930 policy offered new approaches to the dilemma. Hunting on public land was one avenue; compensation for the landowner was another.

The success of the 1930 policy was checkered. Some of the seven points blossomed into fruition, but others have not fared as well. Wildlife management indeed became a profession of trained men and women. Journal papers, books, and monographs now herald the scientific findings about animals and their habitats, and the revenues from special stamps, taxes, and other sources subsidize purchase of public lands and management programs, including those for game and nongame species. Less certainty surrounds the points about relationships with private owners and the availability of wildlife habitat. Numerous examples might be cited—among them wetland destruction and clean farming—but the trend clearly shows that there is less suitable habitat than ever.

Since 1930, urbanization, dramatic changes in agriculture, and perhaps most of all, burgeoning population pressures have produced new forces that influence wildlife management. The public's tastes also have changed in large measure. Many segments now disapprove of hunting and trapping, and the trend may continue. Antihunting sentiments seem sure to remain influential in the formulation and prosecution of modern wildlife policy. Wildlife photography, bird-watching, and other nonconsumptive outdoor activities are more popular than ever before. Membership in national environmental organizations has mushroomed in recent years, reflecting the surging interest in conservation among citizens from all walks of life. Time for leisure has increased steadily as has the affluence to pay for outdoor recreation. All told, managers can no longer isolate concerns simply for "fish and game." Managers are now in the business of working with *all* living things in a variety of contexts and settings, and consumptive uses of wildlife must bear the burden of proof that these uses are not damaging other interests (Allen 1972).

With these changes came the need for a new statement building upon the historic policy of 1930. Accordingly, an updated statement, prepared by a committee headed by Durward Allen, was published in 1973 under the auspices of the Wildlife Management Institute. The new statement called for developing an awareness of ecology's universal application in managing the affairs of the Earth's human population; reversing the trend of diminishing species, populations, and life communities; and preventing a cultural loss in the diversity that enriches human experiences (Allen 1973). Wildlife was proposed as an important component in the living standard that Americans should preserve now and in the future. Furthermore, the policy acknowledged the biological functions of communities and stressed that destruction of one part inexorably impairs other parts. To that end, the 1973 policy addressed wildlife management in the context of modern farming, grazing, forestry, water conservation, and wilderness preservation. Social considerations also were part of the new policy, as shown by a call for dealing with those "slob hunters" who despoil hunting by wanton shooting, by disrespecting laws and landscapes, and by violating the sensibilities of true hunters. Environmental education was endorsed as an element in the training of children. In all, the 1973 statement echoed some problems of the past (e.g., relationships with landowners) while tracking new dimensions for the future.

This chapter outlines some of the factors and influences bearing on policy. We highlight a few of the important steps leading to today's platform of wildlife management. The chapter concludes with a look at the attitudes and contrasting views of management in Europe.

SOURCES

Whereas a broad range of pressures may be responsible for initiating wildlife policies, we can assign these to two general categories. First, there are those that stem from organized citizens' groups, a more general public upwelling, court decisions, or even from a highly visible individual. Among the jurisdictions of the Environmental Protection Agency is the approval of commercially produced pesticides, a role triggered in large measure by the foreboding message of *Silent Spring* (Carson 1962). We can view these and similar sources of policy as a more or less normal evolutionary pathway developing in concert with new ideas, new discoveries, and changing times.

Political fence-jumping represents a second source of policy development. This is a more volatile process than the other and is best witnessed when the reins of government are taken in hand by a new administration. This may happen as often as every 4 years in the United States. A greater potential for policy shifts—for better or worse—develops when the philosophies of a new president and his appointees differ from those of their predecessors. New policies may be stated openly as replacements for old ones, but less direct means also shape these changes. Budget controls may starve agencies whose programs are not favored, whereas those reflecting the current administration's point of view receive strong financial support. In effect, some policies flourish while others wither under a regime of selective funding.

In more historical terms, the earliest policies concerning wildlife were legislation protecting game species; often these rules dealt only secondarily with enforcement. Hunting regulations were prescribed by nearly all of the American colonies; yet almost another century passed before a regular warden corps enforced these and subsequent restrictions. Some colonial townships appointed special persons to enforce game laws of the day, but statewide warden systems had their beginnings in the 1850s (Leopold 1933a; Trefethen 1975). A modicum of federal sovereignty over wildlife first developed with passage of the Lacey Act in 1900. State and federal authority thereafter grew continuously, though not always in agreement, until today's complex bureaucracies grapple daily with issues reaching far beyond hunting and fishing regulations.

AGENCY STRUCTURE AND DEVELOPMENT

State agencies responsible for wildlife management are designated by diverse titles. Even a small sampling of these reveals the wide-ranging nomenclature for these units.

Department of Environmental Management (Rhode Island).
Wildlife and Marine Resources Department (South Carolina).
Game and Fresh Water Fish Commission (Florida).
Game, Fish, and Parks Department (South Dakota).

Department of Conservation (Illinois).
Department of Fish and Wildlife Resources (Kentucky).
Department of Environmental Conservation (New York).
Department of Wildlife Conservation (Mississippi).
Department of Conservation and Natural Resources (Nevada).
Department of Natural Resources (Georgia).

The titles of wildlife agencies now or in the past may reflect special interests or situations. A concern for marine organisms is seen in South Carolina, clearly reflecting the state's coastal geography. Texas once authorized a Game, Fish, and Oyster Commission but now, like South Dakota, has incorporated wildlife and park management into a single department. Wildlife programs in Kansas once were directed by a Forestry, Fish, and Game Commission. A special interest is shown in Massachusetts, where the Department of Fisheries, Wildlife, and Recreational Vehicles administers boating and snowmobiles along with wildlife resources. In some agencies, we find curious distinctions in the biology or status of wildlife. A "fish and game" department seems to mean that fishes are not game or that game represents only birds and mammals that are harvested, whereas "fish and wildlife" implies that fishes are not wild. These matters aside, a large number of agencies once known as fish and game departments now have adopted broader titles that recognize a full range of responsibilities for natural resources. Departments of conservation or natural resources evolved from the change. But more than just assuming a new name, these agencies presumably gained greater effectiveness and public recognition for dealing with ecosystems where nongame, game, endangered species, and their habitats became parts of integrated management programs.

Attempts at wildlife administration began late in the 19th century. However, these early thrusts were weakly conceived and often represented only token efforts. Thereafter, the administration of Theodore Roosevelt registered its influence by championing conservation from the highest office in the land. This message was not lost, and by 1910 nearly every state had an administrative office responsible for the art of wildlife management as then practiced (Trefethen 1975).

Virtually every state agency now is directed by a commission. Originally, it was common practice for governors to appoint a single person who directed the agency's activities. Such appointments typically were political in nature and usually represented no more than a reward for support of the governor's election campaign. The single-commissioner system rapidly became the epitome of political patronage. This form of structure, at best, created considerable instability. Whereas individual commissioners sometimes proved surprisingly capable, the leadership of the department changed each time a new governor assumed office. Other commissioners left office if they fell into the governor's disfavor. In these early days, the commissioner controlled a limited staff of game wardens and perhaps the manager of a game farm or fish hatchery. Wildlife management was in its infancy, and professional managers and biologists simply were nonexistent. Furthermore, issues of the time did not seem complex, and except for hunting and fishing regulations, there was no impetus for advancing either public policy or administration. The single-commissioner system largely faded away during the 1930s after a few decades of existence as standard operation procedure.

By 1934, it was abundantly clear that other means were needed to operate state wildlife agencies. In that year, the Model State Game and Fish Administrative Law was designed as a prototype for a multimember commission. This document, adopted at the annual convention of the International Association of Game, Fish, and Conservation Commissioners, heralded a new era of wildlife administration. A commission of several people now replaced a single commissioner. Members still were appointed by the governor, but their terms of office were staggered, leading to more continuity, stability, and freedom from immediate political pressures. In fact, the model law expressly forbade any political involvement by officers or employees of the commission beyond that of voting rights and private opinions. Such commissions oversee an administrative organization headed by a director who in turn manages the agency's administration and field operations. The model law carefully outlined the selection and qualifications of the director. The term of office was indefinite, and experience with wildlife resources was required. Furthermore, the model prohibited political associations or allegiances, and an oath and bond were required as guarantees of the director's performance. By 1971, all or part of the model administrative law had been adopted by 45 states (Gutermuth 1971).

Ideally, commissions limit their activities to forming policies, ratifying legal matters, and interacting with both the legislative and executive branches of state government (Fig. 22-1). In practice, separation between policy formulation and the day-to-day operations of the agency sometimes weakens so that the commission's influence may reach deep into the ranks of the professional staff. Nonetheless, the multimember commission overcame many of the shortcomings of the single-commissioner system. Understandably, it works best when its members equitably represent the entire state—not merely a region of the state—remaining open and responsive to staff recommendations and listening to landowners, hunters, conservation groups, and other sectors of the public. Nominating procedures for commissioners vary widely from state to state, but the appointments are made by the governor, sometimes requiring ratification by the state senate (Wildlife Management Institute 1977). In Kentucky and Tennessee, candidates for the commission are selected by the governor from a list submitted by outdoor groups. Regardless of the process, the structure of multimember commissions encourages appointment of highly qualified individuals, especially if the commission is relatively small. According to Gutermuth (1971), the size of the commissions varies from 3 to 15 in the United States, with those of less than 6 members functioning more efficiently than larger bodies. In larger commissions, the membership may divide into cliques representing only parts of the state or special interest groups. Commissions averaged 7.4 members in 1968 but expanded to an average of 8.6 members by 1976, largely in response to broadened responsibilities when diversified natural resource agencies within many states were consolidated under one governing commission (Wildlife Management Institute 1977).

In cases where the state legislature exercises some or total control of bag limits and season lengths, the usefulness of the commission often is impaired. Legislative processes usually are tedious, and laws, once enacted, are repealed only with great effort. Therefore, the comparatively rapid changes in fish and wildlife populations may be unmatched by equally rapid legislative actions. In the past, a number of state legislatures reacted to the normal cyclic lows in ruffed grouse (*Bonasa umbellus*) populations by closing the hunting season. By the time these restrictions were repealed, the grouse often had reached and passed their next cyclic peak (Gutermuth 1971). By comparison, small, multimember commissions can act quickly on such matters if they, and not legislative bodies, are empowered to do so.

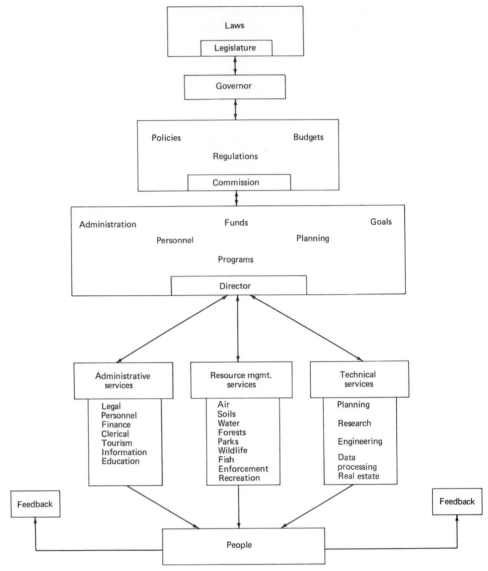

Figure 22-1.   Generalized organizational chart of natural resource management at the state level. (From Wildlife Management Institute 1977.)

Even though the multimember commission seems to represent the best way of guiding state wildlife programs, there nonetheless are some pitfalls to overcome (Towell 1971). These include: (1) commissioners delving into routine administrative matters better left to the director and his or her staff; (2) over-representation of regional interests that may deny sound management on a statewide basis (i.e., provincialism); (3) repetition of earlier mistakes leading to continuation or renewal of outdated practices such as bounty systems, artificial feeding, and stocking because of a commissioner's well-intentioned but misin-

formed preconceptions; and (4) partisan political pressures placed on individual commissioners.

More or less concurrent with adoption of the multimember commission was a change in funding sources. License fees previously disappeared into general revenue, and with this loss of identity, the fortunes of state wildlife agencies depended solely on legislative appropriations. The willingness of elected officials to respond with adequate funding was, of course, highly unpredictable. Too often, wildlife agencies foundered on meager budgets. With the multimember commission and a broader political base at hand, revenues

On The Job With...
Ray B. Owen, Jr.
Commissioner

Ray "Bucky" Owen recently stepped down as Commissioner for the Maine Department of Inland Fisheries and Wildlife. He earned an undergraduate degree in biology and then completed M.S. and Ph.D. degrees in ecology. Before assuming his current position in state government, Bucky pursued an academic career and served for many years as a professor of wildlife management at the University of Maine, where he conducted research on the ecology of woodcock, waterfowl, and other game species. Among his specific interests, he and his co-workers have studied the distribution of earthworms, the woodcock's principal food, in relation to forest communities, soil types, and land use.

As commissioner, much of his time was spent developing policies that guide wildlife management in Maine.

He functioned as liaison between the public and law-makers, using his experience as a wildlife biologist to translate and guide biological information into state law. He thus worked closely with the state legislature and legislative committees and attended numerous public meetings and hearings. Bucky's in-house administrative duties included hiring personnel, allocating positions, setting priorities for department projects, and overseeing the operating budget.

As an example of a typical day, his schedule recently included participating in a meeting of the Atlantic Salmon Recovery Plan Committee, chairing a meeting of the Baxter State Park Authority, preparing performance measures for the department's Strategic Plan (which will be presented to the governor's cabinet), and speaking to a private association of camp owners about fish management.

Bucky has gained satisfaction in his job for at least three reasons: He was able to institute better fiscal management for the department, he has expanded the program for nongame wildlife, and he has enhanced hunting and fishing opportunities. On the other hand, political interests often contributed to his frustrations, especially when these affect management decisions that instead should have been based on sound biology.

His advice to undergraduate students who hope to become wildlife biologists is to study decision-making procedures on which wildlife policy is based. These include the political and legislative processes, as well as administrative procedures. In doing so, biological information can be better transferred into appropriate management policies.

---

generated directly from hunting and fishing activities now generally remain at the command of the commission, providing a much-needed infusion of dollars to supplement legislative appropriations. In recent times, some states have required stamps for special types of hunting and fishing that generate additional funding. Boat and snowmobile registrations also provide revenues for wildlife agencies in some states.

However, passage of the Federal Aid in Wildlife Restoration Act in 1937 (the Pittman-Robertson Act; see following) clearly provided the largest continuing funding base for state wildlife agencies. In its first year of operation, this legislation provided about $0.9 million for wildlife management; in 1986, it contributed $107.4 million (Kallman 1987). With these funds, state wildlife agencies gained a reliable means to initiate and maintain programs addressing an ever-increasing array of issues.

Beyond the commission and director, a department of professional personnel responds to the policies of the commission. Salaried biologists, managers, and other employees work in appropriate divisions under the overall guidance of a director. The director is the chief executive who coordinates the programs of the divisions and forms a link between the commission and the daily activities of the department. Divisions within departments vary from state to state; a single model cannot illustrate all internal organizations. Generally, however, these divisions include (1) game management, (2) fisheries management, (3) nongame and/or endangered species management, (4) law enforcement, and (5) information and education. These divisions in turn are supported collectively by budgetary, acquisition, personnel, and planning services. Research generally is assigned within each of the resource divisions, although in

TABLE 22-1. Partial Listing of Federal Agencies Having Responsibilities for Wildlife Management

Cabinet-Level Department	Agency Name	Activities and Responsibilities for Wildlife
Interior[a]	U.S. Fish and Wildlife Service	Lead agency for conservation of migratory birds, certain mammals, and sport fishes; manages refuges and hatcheries; coordinates endangered species programs; administers federal aid to states; negotiates international agreements; works primarily from regional offices.
	U. S. Geological Survey (Biological Resources Division)	Research functions once conducted by U.S. Fish and Wildlife Service (e.g., Patuxent Wildlife Research Center); also Cooperative Fish and Wildlife Research Units on university campuses.
	National Park Service	Research and management of wildlife on national parks and monuments; coordinates Wild and Scenic Rivers System.
	Bureau of Land Management	Lead agency for managing lands in the public domain, primarily in western states; supervises multiple use, including wildlife, grazing, mining, recreation, timber, and watershed; about 55 percent of all federal lands are under Bureau of Land Management.
	Bureau of Indian Affairs	Trust for grazing, timber, water, and other resource management, including wildlife.
	Bureau of Reclamation	Leads programs for water development in western states; wildlife management and recreation considered in reclamation projects.
Agriculture[b]	U.S.D.A. Forest Service	Administers national forests and grasslands and wildlife thereon; research and management of all forest resources; fire protection and timber harvests are major concerns; regional experiment stations are activity centers.
	Natural Resources Conservation Service	Publishes soil surveys; provides data and technical assistance for soil and water conservation; no research activities; funds small watershed projects and assists with habitat development on private lands; works primarily with organized districts.
Commerce[c]	National Marine Fisheries Service	Provides management, research, and other services for living marine resources, including mammals and invertebrates as well as marine fishes; lead agency in managing offshore development as component of National Oceanic and Atmospheric Administration.
Defense[d]	Army Corps of Engineers	Major responsibilities for dredging, stream stabilization, and other developments of navigable rivers and coastal wetlands; issues dredge and fill permits authorized by Section 404 of the Clean Water Act.

Source: National Wildlife Federation (1987).

[a]Other Interior agencies in one way or another involved with wildlife management include: Bureau of Mines, Office of Surface Mines, and the Office of Water Research and Technology.

[b]Other Agriculture agencies with wildlife-related activities include Animal and Plant Health Inspection Service, Agricultural Stabilization and Conservation Service, and Economic Research Service.

[c]Other Commerce agencies with wildlife-related activities include the Office of Coastal Zone Management and the National Sea Grant College program; both are components of the National Oceanic and Atmospheric Administration.

[d]Other Defense agencies (e.g., Department of the Air Force) manage wildlife and other natural resources on military lands under provisions of the Sikes Act (see text).

Illinois an altogether separate state agency largely is responsible for this activity.

At the federal level, a multitude of agencies administers wildlife or wildlife-related activities in one way or another (Table 22-1). Most of these occur within two cabinet-level departments, Interior and Agriculture, but the National Marine Fisheries Service is housed in the Department of Commerce. Of these, we will focus on the U.S. Fish and Wildlife Service in the Department of the Interior, as it is the lead federal agency for wildlife management.

The beginnings of what today is the U.S. Fish and Wildlife Service were modest indeed. In 1885, the Division of Entomology within the U.S. Department of

Figure 22-2. The newspaper cartoons of J. N. "Ding" Darling cut to the heart of environmental issues during the 1930s. Land abuses, and especially the loss of waterfowl habitat, were popular subjects of his cartoons. His skilled pen brought Darling to the attention of President Franklin D. Roosevelt, who appointed "Ding" director of the Bureau of Biological Survey in 1934. The bureau later was renamed the U.S. Fish and Wildlife Service. (Drawing courtesy of the J. N. "Ding" Darling Foundation, Inc.)

Agriculture subsidized the efforts of the American Ornithologists' Union to determine the status of bird distributions and migrations; a federal grant of $5000 supported the project. A year later, the work was transferred to the Division of Economic Ornithology and Mammalogy, also in the Department of Agriculture. The first document issued by the infant agency reflected its associations with agriculture. Bulletin 1 was entitled "The English Sparrow (*Passer domesticus*) in North America, Especially in Relation to Agriculture" (Barrows 1889). In 1896, the division was renamed the Bureau of Biological Survey and was charged with the investigation of food habits, distribution, and migration of birds and mammals in relation to agriculture, horticulture, and forestry. The bureau gained new responsibilities for law enforcement with passage of the Lacey Act in 1900.

Waterfowl suffered extensive habitat losses during the long drought of the 1930s. The Great Depression also was wrecking national economies during the same period. Faced with this double problem, the newly appointed director of the then Bureau of Biological Survey, Jay N. "Ding" Darling, conceived the "duck stamp" as a funding measure for restoring depleted waterfowl habitat. Congress had earlier passed the Migratory Bird Conservation Act in 1929 to ac-

quire land for waterfowl refuges, but it did not include a stable source of funds for the program. With Darling's persuasion, the Migratory Bird Hunting Stamp Act was enacted by Congress in 1934. All waterfowl hunters 16 years of age or older had to obtain a stamp; the stamps had to be signed and affixed to the backs of hunting licenses. (A Migratory Bird Hunting Stamp is required only for hunting ducks, geese, and swans, and not for other migratory game birds, as its name might imply.) Darling was a clever artist whose political cartoons in newspapers attracted wide attention and probably had much to do with his appointment as director (Fig. 22-2). The first Migratory Bird Hunting Stamp, designed by Darling, was issued in 1934 and sold for $1 (Fig. 22-3). In the years between 1934 and 1980, more than 81 million stamps were sold; prices for the stamps rose gradually and reached $15 in 1991. About $226 million was brought to bear on refuges and other needs of waterfowl management as a direct result of Darling's innovation.

The first national wildlife refuge was acquired in 1903 when Pelican Island off the coast of Florida was declared a sanctuary for wading birds by President Theodore Roosevelt. The federal refuge system thereafter expanded rapidly, becoming one of the Bureau of Biological Survey's more visible activities.

Figure 22-3. The first "duck stamp," issued in 1934, was drawn by J. N. "Ding" Darling. Funds from the sale of the stamps are dedicated to the restoration of waterfowl habitat. About 635,000 of the original stamps were sold at a cost of $1 each. Today, the Migratory Bird Hunting and Conservation Stamp is issued by the Department of the Interior and costs $15 each. Several million stamps are sold each year to waterfowl hunters, nonhunting conservationists, and stamp collectors. (Photo courtesy of U.S. Fish and Wildlife Service.)

More than 6.8 million ha of refuge lands were at hand when the bureau was transferred, on June 30, 1940, to the Department of the Interior and renamed the U.S. Fish and Wildlife Service. The original plan created a "U.S. Wildlife Service," but President Franklin Roosevelt responded to those who felt that fisheries management might take second place in the new organization; the president thus expanded the name to U.S. Fish and Wildlife Service, in effect creating the artificial biological distinctions mentioned earlier. Since then, numerous administrative changes have occurred in the organization, including a short-lived designation as the U.S. Bureau of Sport Fisheries and Wildlife. The responsibilities of the agency steadily became enmeshed in an ever-wider realm of wildlife and wildlife-related activities.

NATIONAL BIOLOGICAL SERVICE

In 1993, Secretary of Interior Bruce Babbitt formed the National Biological Survey (NBS), a new unit comprised of scientists from the U.S. Fish and Wildlife Service, the National Park Service, and the Bureau of Land Management. The purpose was to bring together scientists from various disciplines to determine the distribution and abundance of various wildlife, plant, and mineral resources—a basic mission of research, inventory, and monitoring. In 1994, a newly elected conservative Congress—pledged to reduce the size of government—saw NBS as an agency designed to obstruct economic development. In the House of Representatives, the speaker threatened to lead a campaign to abolish the agency. Such

a move, if successful, would have removed a few thousand employees from the government rolls and, in turn, reduced public access to information—and experts—concerned with natural resources. However, a political compromise was reached in 1996: NBS would remain intact, with Survey changed to Service, but the unit would be housed in the U.S. Geological Survey. The effectiveness of the NBS under these circumstances remains to be tested.

Some of the several components now included in NBS are listed below, along with their previous affiliations.

Transferred from the U.S. Fish and Wildlife Service:

All campus-based Cooperative Fish and Wildlife Research Units
    Northern Prairie Wildlife Research Center
    Patuxent Wildlife Research Center
    National Wildlife Health Center
    National Wetlands Research Center

Transferred from the Bureau of Land Management:

Raptor Research and Technical Assistance Center
Desert Tortoise Research Project

Transferred from the National Park Service:

Field stations at Yellowstone National Park, Isle Royale National Park, Everglades National Park, and Padre Island National Seashore
    Water Resources Branch
    Air Quality Branch
    Center for Urban Ecology

NEW TRAINING, NEW PROFESSION

Wildlife management struggled for many years without professionally trained personnel. At first, political appointees staffed the game farms and other installations then associated with conservation agencies. Next, persons with scientific backgrounds began to enter the field, but they were trained solely in zoology and other basic sciences. They were as dedicated as today's managers, but their formal education lacked a base of resource management familiar to latter-day students.

Such circumstances handicapped the fledgling activities of wildlife agencies until 1935. At that time, J. N. "Ding" Darling again displayed his dynamic leadership as the newly appointed director of what is now the U.S. Fish and Wildlife Service. Darling seized on the idea that a nationwide, cooperative venture housed on university campuses would produce a new generation of field-trained personnel (earlier, in

1932, Darling had pledged $9000 of his own money to start a state-sponsored unit at the Iowa State University). He secured the financial backing of leading sporting arms-and-ammunition manufacturers (Trefethen 1975). Simultaneously, the American Wildlife Institute became a financial partner in the unit program. That organization became the Wildlife Management Institute in 1946 and continued its vigorous support of unit activities. This nucleus immediately attracted state and federal funding. States participating in the unit program contributed funds from their wildlife departments so that a combination of private, state, and federal sources fueled the research projects. A federal biologist was assigned to each campus as the local program supervisor and concurrently was given faculty status. The universities provided laboratory and office space and other kinds of support in addition to establishing a curriculum specializing in wildlife biology and management. The academic work,

Jay N. "Ding" Darling
Conservationist and Cartoonist
(1876–1962)

Darling's career started as a newspaper reporter in Iowa. He illustrated his stories with clever cartoons that gained him national attention and, eventually, two Pulitzer prizes. Darling signed his cartoons with the distinctive pen name "Ding," a contraction of his last name. Most of his career was spent with the *Des Moines Register,* but his political cartoons were syndicated through the *New York Herald Tribune* and appeared in 130 daily newspapers. His fame produced associations with Herbert Hoover, Will Rogers, and Henry Ford among other celebrities of the early 20th century and included a visit to Russia at the invitation of Stalin.

Many of Darling's cartoons were political satire, some of which mocked the relief programs of Presi-

dent Franklin D. Roosevelt's New Deal, but many others addressed a spectrum of conservation issues, including poaching and poor sportsmanship. Soil erosion, pollution, and other land abuses were frequent targets of his drawings (see Fig. 22-2). During the 1930s, Darling assailed the loss of waterfowl habitat, which prompted Roosevelt to appoint him in 1934 to the position of Chief, Bureau of Biological Survey, predecessor to the U.S. Fish and Wildlife Service (the appointment also was an adroit means of ending Darling's political assault on the New Deal). As chief, Darling fought successfully for the expansion of the federal wildlife refuge system. He also used his drawing skills to design (in 1934) the first "duck stamp" that waterfowl hunters have since been required to purchase (see Fig. 22-3). Although his tenure as chief lasted just 18 months, his spirited leadership brought new vigor to the bureau and its conservation activities. Darling subsequently organized and became the first president of the National Wildlife Federation, which today is the largest citizen-based conservation organization in the United States.

Darling is remembered as an energetic advocate of wildlife conservation at the national level during the 1930s—the troubled era of the Dust Bowl and the Great Depression—and the farsighted legacy of the duck stamp continues today. The J. N. (Ding) Darling National Wlidlife Refuge at Sanibel Island, Florida, is named in his honor. A foundation dedicated, in part, to conservation education in grade schools also was created in his name. For a full biography, see Lendt (1979).

to be sure, still drew heavily from zoology, botany, and other basic sciences, but these disciplines were meshed in new courses stressing their application to wildlife populations. Fieldwork became an integral part of the training experience. As a result, a four-way partnership—federal, state, university, and private—established the Cooperative Wildlife Research Units. Nine units were created in 1935–36 in Virginia, Oregon, Iowa, Connecticut, Alabama, Texas, Maine, Utah, and Ohio. Graduate students immediately began to train under the auspices of the unit program, essentially enrolling for the first time in a national educational effort geared specifically to wildlife biology and management (Jackson 1937).

In 1962, the first Cooperative Fishery Research Unit—at Utah State University—was added to the program. Today, 56 fish and/or wildlife units operate on university campuses in 41 states from Florida to Alaska; most of the units now have combined missions (i.e., fish and wildlife). In 1993, the units were incorporated into the National Biological Survey, later renamed the National Biological Service, and administered by the U.S. Geological Survey. The unit program superbly fulfilled Darling's dream for college graduates trained in wildlife management and also produced new knowledge for its practice, and in the process, a profession was born.

POLICY AND WILDLIFE LAW

The legal foundation for wildlife management is formulated by court decision and legislation. Policy thereafter is set forth in administrative regulations guiding actions at the field level. In the legislative phase, laws frequently establish the framework for regulations governing the harvest of wildlife. Season lengths, bag limits, and ways and means of harvesting wildlife usually are established by commissions in the United States. Similar procedures exist in many governments around the world. Nonetheless, policies and regulations must consider constraints on administrative flexibility. Wildlife resources regularly face new management challenges and, on occasion, emergencies demanding speedy action. We should now look at some of the basic constraints and requirements influencing conservation law, policy, and administration.

BIOLOGICAL FOUNDATION

Above all, wildlife policy must be based on the best available biological information. Commissions and other groups charged with regulatory authority must have at hand a reasonable summary of cogent data in order to act responsibly. In Chapter 8, we noted how a modest change in the opening date of the hunting season caused hunters to overcome certain psychological responses to parasitized squirrels. So facts rather than preconceptions, guesswork, or tradition are necessary if policies are truly to benefit the management of wildlife.

Changing conditions and new knowledge, of course, should prompt changes in policy. Preconceptions that bounty systems or buck laws, among others, actually favor wildlife populations sometimes fostered poor wildlife management. Large-mouth bass (*Micropterus salmoides*) populations in some southern states were governed by strict size limits for years. When research determined that the fish populations could be harvested safely without these restrictions, policies and regulations were changed appropriately. A "catch-it-and-keep-it" policy liberalized the fishing regulations in Alabama, increasing recreation without harming the welfare of fish populations (Sport Fishing Institute 1982). In Arizona, the varmint status of the mountain lion (*Felis concolor*) was changed to that of a game animal in 1970 when the trophy value of these animals was recognized. As a game species, mountain lions fell subject to management, just as do deer or any other species of traditional big game. A bag limit of one lion per year was established, and a tag was required for each of the big cats killed by sport hunters. Funds for bounties were eliminated by the Arizona Legislature at the same time (although the bounty act for mountain lions itself was not repealed). Since 1981, sport hunters have been required to report their kills, as were ranchers who killed mountain lions preying on their livestock (Shaw 1982). In some instances, the lengths of hunting seasons may not have as much bearing on game populations as once was supposed. Hunting seasons in Iowa and Minnesota have little relationship to pheasant (*Phasianus colchicus*) numbers; instead, farming patterns have more influence on pheasant populations (George et al. 1980). The effectiveness of policy-making groups accordingly rests on their objectiveness and on their desire for current biological information on which to act.

New ideas can be developed even from "old" biological data. Waterfowl biologists long knew of the early fall migration of blue-winged teal (*Anas discors*) in relation to other species of waterfowl. Bluewings move southward weeks or even months before other game ducks, thereby completing much of their migration through North America before waterfowl seasons opened in many states. Analyses of their migration schedule and hunting mortality suggested that blue-winged teal largely escaped hunting pressure. In 1965, an experimental 9-day early season was opened expressly for teal as a means of increasing hunting recreation. The season was monitored closely for the consequences, if any, the special harvest might have on continental teal populations. The effect was negligible on the continental population of teal or of any other species, including those shot by mistake (Martinson et al. 1966). The early teal season thereafter became an established part of waterfowl hunting for those states wishing to adopt it, but it was discontinued in 1988 when waterfowl populations plummeted because of widespread habitat deterioration.

At times, some biological information also has produced misdirected conclusions. Studies of predation on pheasant nests in South Dakota indicated that losses were large, suggesting that predators should be controlled (Trautman et al. 1974). Nonetheless, the real question of whether predation truly limited the *harvest* of pheasants by hunters was not addressed. In the published discussion following Trautman's paper, various speakers noted the importance of this question but left the answer unsettled. Predation may or may not markedly reduce wildlife populations, but as a basis of policy and cost effectiveness, predation should be tested against its effects on the hunters' bag before control measures gain acceptance as a part of wildlife management. In sum, research may offer useful insights about the workings of nature—predation in this case—but unless the appropriate questions are addressed, policies cannot be established correctly.

Sound biological information clearly remains a fundamental ingredient in the formulation of wildlife policies, but public support is necessary as well. In fact, some wildlife administrators feel that "people management" represents a large part of their duties. With this in mind, we now turn to other influences on wildlife policy.

ENFORCEMENT

Effective enforcement of conservation laws is based on (1) the public's acceptance and ability to comply, (2) risk and severity of punishment, and (3) personnel. These factors act in concert, rather than independently.

In considering compliance, imagine a hypothetical policy protecting hen bobwhites (*Colinus virgini-*

anus) from hunters, just as hen pheasants usually are protected. If this policy were transformed into law, the administrative phase would be faced with an enforcement problem that defies public compliance. Unlike pheasants, female quail are difficult to distinguish from males when the birds burst forth in front of hunters. The same problems would confront a sex-specific policy governing fishing. Early attempts to limit or completely protect the harvest of certain waterfowl were in fact frustrated because hunters, on the average, could not properly identify waterfowl in the marginal light of predawn or because of a duck's rapid flight. A point system, adopted in many states, later relied on identifying waterfowl *after* each bird is shot. Each species and/or sex carries separate point valuations so that hunters can keep a tally toward their total allocation of 100 points. The point system still faces the dilemma of reordering the sequence in which birds are taken in the daily bag, but it does represent an appropriate example of species management and realistic public compliance (Geis and Crissey 1973; Nelson and Low 1977).

The special season for blue-winged teal described earlier also included green-winged (*Anas crecca*) and cinnamon (*A. cyanoptera*) teal because of the gross or actual similarity of the three species. Because greenwings generally migrate later than bluewings, relatively few birds of that species become vulnerable to harvest in the total bag of teal. Legal measures designed to protect wild populations also must recognize the legal exploitation of the same species raised in captivity. Trapping seasons regulate the harvest of wild mink (*Mustela vison*), but similar limitations are not placed on commercial mink ranches. Enforcement of trapping regulations requires that the skins from one source not be attributed to the other (i.e., skins of wild mink trapped out of season might be claimed as coming from a mink ranch).

Legal safeguards afforded to some endangered animal populations also may unintentionally involve others that are not in need of the same degree of protection. This problem arises when an endangered population has a more secure counterpart so similar in appearance that the two cannot be distinguished. Two subspecies of leopard cat (*Felis bengalensis*) illustrate the concept. These subspecies, one from India and the other from China, are extremely difficult to separate, especially when their hides appear in fur markets. The Chinese subspecies is heavily involved in the New York fur trade, but it is illegal to buy or sell hides of the endangered Indian subspecies in the United States. Despite this constraint, legal sales of the Chinese subspecies could easily include illegal hides from India. Compliance and enforcement virtually are impossible under these circumstances. Furthermore, continued sales of hides from India promote further reductions of a population already facing extinction. To overcome these difficulties, it may be necessary to list the Chinese leopard cat as endangered (*fide* U.S. Fish and Wildlife Service). These look-alike situations are covered by the Similarity of Appearance passages in the Endangered Species Act of 1973. Policies for the management of American alligators (*Alligator mississipiensis*) also are enforced, in part, by these regulations. Alligators were listed as endangered in 1967 after years of poaching and overhunting. Some segments of the alligator population have since recovered to the point where hunting again is possible. However, at the same time, other segments still require protection of the Endangered Species Act. The look-alike regulations allow for compliance under these circumstances, and a statewide hunting season is now authorized in Louisiana (Bender 1981). In this case, a healthy segment of the alligator population remains classified as "Threatened by Similarity of Appearance" only because these animals look just like those for whom full protection still is required.

Personnel for enforcement is a function of funding, which represents the commitment of state and federal appropriations. There undoubtedly is a point of diminishing returns in law enforcement (W. F. Sigler, personal communication). The addition of one more officer might save 5 or 10 deer annually from poachers, but the cost per deer would be prohibitive. Therefore, the price of law enforcement has direct bearing on the degree of compliance if threat of punishment—and not sporting ethics—is the basis of regulating wildlife harvests. An idealist might argue that game laws are enforced for the sake of law and order. Perhaps so, but if resources—whether money or wildlife—did not need protection, then enforcement would have vanished long ago. Wildlife, like jewels or money, has an inherent need for protection because it is valuable. Furthermore, many game violations represent lost revenues, although it is difficult to determine the amounts involved. Hunters or anglers without licenses or with the wrong kind of licenses or permits may "cost" many thousands of dollars. Without enforcement, the treasuries of wildlife departments surely would be impoverished by considerable sums. A good deal of variation occurs among the states in the size of their conservation law-enforcement corps. In some cases, aircraft have increased the ca-

pacity for surveillance, thereby reducing the numbers of field officers responsible for large areas with sparse human populations. In others, relatively large numbers of enforcement personnel are needed where denser habitation contributes more hunting and fishing activities per unit of area. There also are regions where dense cover or other conditions conceal poaching activities. Both of these situations may be present within a single state, and the enforcement forces should be apportioned accordingly.

Today's conservation officers must be well trained. Far more than mere license inspection often is involved. Hen mallards (*Anas platyrhynchos*) cannot be confused with gadwalls (*A. strepera*) when a hunter's bag is inspected under the jurisdiction of the point system. Whether a deer was shot illegally at night sometimes may be determined by the body temperature of a carcass hanging in camp the next day (Woolf et al. 1983; see also Oates et al. 1984). By using biochemical analyses of blood and meat tissues, wildlife crime laboratories may identify closely related species subject to separate hunting regulations. Conservation officers must be aware of these possibilities when inspecting harvests of such species as bears (Wolfe 1983).

An even broader range of subject matter complements traditional training programs for law enforcement personnel in some states. In New York, classwork addresses air, water, and radiation pollution and similar topics relevant to a modern technological environment (Benoit 1973). Morse (1987) noted that about 50 percent of the enforcement effort and 25 percent of the citations issued in New York concerned environmental issues. Beyond these considerations lie the complexities of forensics. Legal procedures for proper search and arrest, collection and presentation of evidence, and many other matters must be followed precisely for successful prosecution (see Sigler 1982). For these reasons, some states require college degrees for their conservation officers. Almost all trainees attend special training sessions about wildlife in their state, and many attend police academies as well. Finally, conservation officers often are in the forefront of relationships with the public. They deal with people—not just when inspecting hunters and anglers in the field, but also when speaking to scouting groups and other organizations about wildlife conservation. In this role, public relations become a major part of the officer's duties, again requiring training and finesse (Fig. 22-4). Conservation officers now have their own professional organization, The North American Wildlife Enforce-

Figure 22-4. Wildlife enforcement officers—"game wardens"—today shoulder a broad range of responsibilities. Besides inspecting hunters and anglers for licenses and compliance with game laws (top left), officers in many states also enforce boating safety regulations (top right), participate in civic programs such as "Kid Fish" events (bottom left), and instruct clubs, schoolchildren, and other groups about wildlife conservation (bottom right). Environmental monitoring (e.g., water quality) is another responsibility in some states. (Photos courtesy of Texas Parks and Wildlife Department and Illinois Department of Conservation.)

ment Officers Association, which issues a quarterly publication entitled *The International Game Warden* (Morse 1987).

The question arises as to whether poaching detracts from the legal bag of law-abiding hunters. Unfortunately, no simple answer is at hand. However, the population turnover rate of the species influences the matter in large measure (W. F. Sigler, personal communication). Animals such as moose (*Alces alces*) indeed may be overharvested when poaching adds to the legal kill, thereby affecting not only the welfare of the moose population, but also the ability of hunters to fulfill reasonable expectations for a successful hunt. Conversely, poaching of rapidly reproducing species such as quail or rabbits likely has little effect in either regard. Furthermore, the degree with which

legal harvests are realized is related to the significance of poaching. If the full surplus of elk (*Cervus elaphus canadensis*) is taken by legal harvest, then any additional kill by poachers likely will be felt in the following hunting season. However, if only 60 percent of the surplus of elk is harvested legally, then the impact of poaching is less likely to be reflected in the next year's hunting success.

Wildlife management, in terms of game species, often has been assessed by the success of hunters or anglers in bagging their quarry. Nonetheless, broader measures of satisfaction have been suggested (Potter et al. 1973; Brown et al. 1977; Gilbert 1977). For example, a hunter filling a bag limit with recently stocked, hand-reared pheasants may not derive the same satisfaction as one bagging fewer, but completely wild, birds. In short, hunting and fishing experiences include more than bagging game (Hendee 1974); yet hunters are governed by how, when, and what to hunt as well as by bag limits. This suggests that game laws may influence, in part, the degree of satisfaction hunters experience. Beattie (1981) addressed this influence in a survey of Virginia hunters. The results indicated that hunters derived satisfaction from the belief that the laws regulated game populations. Hunters also felt that game laws promoted safe hunting and the preservation of wildlife. In all, the survey suggested that satisfactions of hunting are enhanced by game laws.

Despite the long tenure of law enforcement as a part of wildlife management, little research exists to substantiate the effectiveness of enforcement activities. Because the degree to which patrols actually deter violations has not been measured, it is not possible to test enforcement strategies or even to justify patrols knowledgeably on the basis of reduced violations (Beattie et al. 1977). Thus, it is often difficult to incorporate the full dimensions of law enforcement into many kinds of wildlife policies. Nonetheless, enforcement remains a major emphasis in the programs of state wildlife agencies, in part because of political influence and public demand. About one-third (or 7180 employees) of the personnel and 30 percent (or $275.5 million) of the budgets of state wildlife agencies are allocated to enforcement activities (Morse 1987).

A suitably sized and well-trained corps of conservation officers, coupled with a realization by hunters of punishment appropriate to the violation, present the first line of enforcement. Habitual violators, in particular, must be confronted with significant enforcement efforts. For others, rational game policies generally are enough to promote compliance. Indeed, the following scenario suggests two ways a hunter crouched in a goose blind might view the regulations governing the sport—regulations imposing bag limits, the number of shells per gun, and the shooting hours. First, such a hunter may see these laws purely as legal restrictions, impinging upon the self-imposed mission to kill geese, with violation of these laws leading to arrest, loss of equipment, a fine, and/or a jail sentence. Compliance with the regulations in this case is based simply on fears of apprehension, conviction, and a penalty.

The alternate view is to see game regulations as a set of ethical guidelines, based upon the biological welfare of goose populations and fairness to other hunters and, indeed, to other sectors of the public as well. For a hunter who views regulations this way, a violation becomes something not so much illegal as unethical—a betrayal of the public trust placed upon him or her when the hunter purchased a license and received a copy of the hunting regulations.

Because we always shall have people whose self-interest exceeds their social conscience, it remains necessary that conservation regulations be given the strength of law and the strong backing of the courts. Obviously, habitual violators act under a cloak of secrecy, avoiding the possibility of witnesses, so their chances of apprehension remain slim. Even an army of enforcement officers could not cope with poaching and other violations if the majority of hunters chose to disobey conservation laws. For this reason, most states now have developed systems for reporting violations of wildlife law. These are known by several names, among them "Operation Game Thief" and "Help Our Wildlife," but each encourages the public to report violations confidentially, using toll-free phone numbers. These systems offer rewards when convictions result. In Michigan, "Report All Poachers" (RAP) provides increased personnel concentrated in an investigative unit, a $80,000 budget for rewards, a 24-hour hotline for reporting violations, and an educational program (Nelson and Verbyla 1984). RAP is funded with a surcharge of $0.25 on each hunting, fishing, and trapping license. These programs not only assist conservation officers, but also instill among citizens an awareness of their responsibilities for protecting wildlife resources.

It appears, then, that the most productive approach toward achieving compliance with fish-and-game regulations is to establish a sense of ethics leading to the truest meaning of sportsmanship. Ethics involve personal responsibility and conscience and are motivated more by pride and opinion than by the threat of punishment (Fig. 22-5). They may be instilled in many ways, although most produce only slow results. Articles in magazines emphasizing ethical behavior, television programs showing people releasing legal-size fish and stopping short of their limits, educational programs in schools, and bumper stickers all may have a slow but cumulative effect in bringing about ethical behavior among anglers and hunters. Organizations such as Ducks Unlimited, the Boone and Crockett Club, and the Izaak Walton League each do much to promote sporting ethics. Many states now require courses in gun safety before they issue hunting licenses. It seems equally reasonable that these same

Figure 22-5. Roadside sign stressing sporting ethics may have as much impact on curtailing game violations as several wardens. (Photo courtesy of D. M. Paris.)

courses might expand their mission and uniformly include the fundamentals of sporting ethics, as is the practice in Europe.

POLICY: SOCIAL AND ECONOMIC FACTORS

Living standards and moral and religious issues may be entwined with environmental and conservation policies. Local and even regional living standards clearly are impaired when industrial facilities are closed because of the pollution they generate. However, costs are passed through to consumers when manufacturers modify their production methods to conform with new environmental regulations. Smaller, fuel-efficient cars are often in demand, but these have expensive devices for controlling toxic emissions. The controversies stirred by thoughts of regulating human populations as a means of reducing environmental pressures involve deeply entrenched religious and moral considerations.

For wildlife, social and economic factors often have played a prominent role in conservation policy. Federal policies for protecting endangered species were threatened by construction of the Tellico Dam in Tennessee. At the time, the only known population of snail darters (*Percina tanasi*) was threatened by an impoundment on the Little Tennessee River. Construction of the dam was halted, despite the contention that regional economies and living standards would suffer. Later, Congress responded to the social and economic arguments, exempting the Tellico pro-

ject from the constraints of the Endangered Species Act. Fortunately, additional numbers of snail darters were discovered in other rivers and these, together with transplants from the Little Tennessee population, may assure perpetuation of the species.

Social influences often are closely linked with political reactions. Whereas wildlife managers may have technically feasible procedures at hand to achieve desirable biological objectives, failure to consider the sociopolitical climate may negate their application to management programs. In Vermont, the Fish and Game Department sponsored antlerless deer seasons as a means of regulating the state's burgeoning deer herd. The procedure had been used successfully elsewhere for the same objective, but in 1969 it was new in Vermont. During the next two years (1969 and 1970), a social backlash formed among hunters and landowners, and the legislature responded by revoking all responsibilities (except research) previously assigned to the department for deer management. Had the social responses been considered beforehand, the legislature's political reaction likely would have been avoided and the department would not have lost its functions for deer management (Gilbert 1977). The decision to shoot antlerless deer was biologically sound and necessary, but its acceptance by the public lagged too far behind.

Political ferment, including outright warfare, remains a social phenomenon with some bearing on wildlife. European bison (*Bison bonasus*) living in the Bialowieza Forest in Poland experienced losses coincident with rebellions in 1830 and again in 1863. Both revolts were thwarted ventures for Polish independence from Russian dominance (Wren 1979). While other factors perhaps were involved, the revolutionary turmoil nonetheless contributed to periods of excessive poaching. Bison numbers decreased sharply from 1900 to 500 animals during the decade bracketing the 1863 uprising (Lydekker 1898). Later, the bison herd recovered to about 725 animals, but fell to zero in 1914 when the Bialowieza Forest became a World War I battleground for German and Russian armies. The entire herd was consumed by hungry soldiers. Fortunately, meager stocks in zoos provided a nucleus for reestablishing the herd (Boyle 1961). The new population of European bison is protected in Bialowieza National Park in a part of the forest excluded from commercial lumbering, with a free-roaming herd maintained by an intensive breeding program.

Early in the history of the New World, proprietary concern for wildlife resources influenced, in part, hostilities between European interests that were then dominating North America (Garraty 1975). French, Dutch, and, later, English claims to fur-rich lands and the commerce therefrom began as early as 1609 in what is now New York. Overtrapping quickly exhausted fur resources, leading to contested ownership of new and untrapped territories. These battles—often fought with Native American allies—moved to the Mississippi watershed as the frontier pushed westward. Beaver (*Castor canadensis*) was the principal species in most disputes, but other furbearers contributed to the bonanza of pelts entering the colonial market. Similarly, the fisheries of the North Atlantic were sources of conflict between colonists in New England and Canada. Fishermen needed land bases for processing their catches, and French and English interests struggled over harbors in Maine, Nova Scotia, and Newfoundland. In more recent times, fishing rights brought forth a brief "codfish war" between Iceland and England. Wars also seem to reduce public interest in natural resources. Kellert and Westervelt (1982) cited evidence that the U.S.'s role in World Wars I and II shifted the country's attention away from wildlife and the general subject of conservation. This relationship apparently recurred modestly during the Korean conflict and to a greater extent when fighting raged in Vietnam (Fig. 22-6).

POLICY: SCIENTIFIC AND TECHNOLOGICAL FACTORS

The rapid advance of science often outpaces our understanding of potential environmental hazards. This is amply illustrated by the application of nuclear energy without simultaneous development of adequate means for the safe disposal of radioactive wastes. Widespread use of persistent pesticides without knowledge of their residual effects on avian reproduction is another example, as was the aftermath of industrial pollution on the embryonic development of young birds (Hays and Risebrough 1972). Whereas scientific discovery has produced sophisticated equipment and procedures for human betterment, even critical evaluations and testing may not uncover all of the real or potential hazards. New medicines, for example, may cure specific maladies but produce unacceptable side effects in some patients. Thalidomide reduced nervous anxieties but, when taken by pregnant women, the drug produced deformed babies.

The effect of technology on wildlife management is not a new consideration. Many years ago, electrical recording devices as a means of calling waterfowl were outlawed, as were repeating shotguns holding more than three shells. Then, as now, this equipment

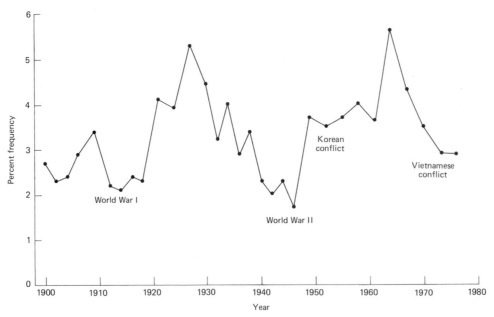

Figure 22-6.   Relationship of social unrest (in this instance, warfare) to human attitudes about wildlife and conservation affairs based on the frequency of newspaper articles about animal-related subjects. (From Kellert and Westervelt 1982.)

represented "gadget" advancements that exceeded the elements of an ethical "fair chance" policy (see Kozicky 1977). Kellert and Westervelt (1982) linked the emergence of the automobile with a heightened interest in wildlife during the 1920s. Today, snowmobiles, all-terrain vehicles, and aircraft give hunters access to wildlife where remoteness or impassable road conditions once precluded outdoor recreation. These same conveyances, of course, have become tools of poachers and conservation law-enforcement officers alike. Advances in materials and design have greatly increased the effectiveness of archery as a means of hunting. Some states now permit deer hunting with compound bows and crossbows. Hunters using compound bows may achieve success rates similar to those achieved using firearms, thereby threatening the recreational status of archery as a primitive form of hunting (Gladfelter et al. 1983). As we will mention, the development of high-powered speedboats also influenced recreational policies on at least one national wildlife refuge. New metallurgical techniques led to steel, alloy, or copper-coated shot as replacements for toxic lead shot, permitting new policies and regulations for waterfowl hunting in the United States (see Chapter 3).

LEGAL JURISDICTION

Environmental perturbations seldom are confined to political or geographical boundaries (Skillern 1981). Some, such as the killing of whales and seals, may occur in international locations. In North America, acid precipitation is generated largely in the United States, but much of it falls to Earth in Canada (see Chapter 11). Agricultural chemicals applied in the heartlands of North America drastically reduced brown pelican (*Pelecanus occidentalis*) populations on the Gulf Coast. In addition, the extinction of an organism remains a loss for the world, not just for the locality where it once lived. These and other situations underscore the legal issue of unconfined jurisdiction.

In his review of *Who Owns Wildlife?*, Matthews (1986) noted that federal power in the United States must be authorized by a constitutional clause. For issues regarding wildlife, the main clauses are the commerce clause (Art. I, Sec. 8)—probably the most far reaching—the treaty clause (Art. I, Sec. 10), and the property clause (Art. IV, Sec. 3).

The issue of jurisdiction over wildlife faced its first U.S. Supreme Court test in 1842, but in *Geer v. Connecticut* the doctrine of state ownership of wildlife

reached its zenith (Bean 1983). In 1889, Geer was convicted of violating a Connecticut statute that made the killing or possession of certain species of upland game birds unlawful "for the purpose of conveying the same beyond the limits of this state." The birds had been lawfully obtained, but Geer had violated the law only because he intended to ship the birds out of state. Connecticut thereby allowed hunting of game birds and permitted the sale, purchase, and use of birds obtained legally, but *only if the birds remained in the state.* In other words, the statute regulated wildlife within the state in a way that forbade the transport of game outside of Connecticut. Such a statute seemed to violate the commerce clause, and on that basis, the case went to the U.S. Supreme Court in 1895. The Court agreed with the conviction in a decision reached in 1896, and *Geer v. Connecticut* emerged as the bulwark for the doctrine of state ownership of wildlife (Bean 1983: see also Skillern 1985).

The Geer decision, however, was overruled in 1979 when the Supreme Court examined *Hughes v. Oklahoma.* In this case, an Oklahoma statute prohibited transporting native minnows caught within the state for sale out of state. Hughes owned a licensed bait business in Texas and bought the minnows from a licensed dealer in Oklahoma but was convicted for taking the minnows from Oklahoma for sale in Texas. The Supreme Court decided that the 1896 Geer ruling was in error and thus ruled against Oklahoma.

In legal terms, state legislation cannot conflict with a power delegated to Congress by the Constitution—interstate commerce in this case—even if Congress has not exercised its power by enacting its own legislation (Skillern 1985). The implicit limitation of state authority to act in such cases is the so-called dormant aspect of the commerce clause. Therefore, within their borders states properly may conserve and manage wildlife, but use of the animals cannot be limited to the citizens of one state to the exclusion of citizens in other states. With the overthrow of the Geer decision, the fiction of state ownership could no longer be used as a basis for inhibiting the development of federal wildlife law (Bean 1983).

In a few instances, jurisdiction has been resolved by international treaties. The best known of these concerns migratory birds. Enactment of the initial treaty between Great Britain (on behalf of Canada) and the United States in 1916 resulted in the Migratory Bird Treaty Act two years later. This legislation initiated much of the federal involvement with wildlife, then and now. Among the provisions of the act were the protection of whooping cranes (*Grus americana*), swans (*Cygnus* spp.), most shorebirds, and wood ducks (*Aix sponsa*), protection of bird nests and eggs, establishment of closed seasons for waterfowl between March 10 and September 1, definitions of *game* and *nongame birds,* and authorization for the states to adopt and enforce regulations that were not inconsistent with the federal provisions (i.e., the states could be more restrictive but not more liberal than federal limits in setting season lengths and bag limits). In the Migratory Bird Treaty Act, we find the seeds of an expanding wildlife refuge program, endangered species management, and federal law enforcement. In assuming these responsibilities, however, jurisdictional conflicts arose that continued long afterward. Baiting ducks during the hunting season years ago and, more recently, the mandated use of steel shot initiated heated controversies between state and federal governments.

Interestingly, Congress previously had enacted a Migratory Bird Law in 1913 that sparked a state-sponsored test of its constitutionality. However, passage of the 1916 treaty led to repeal of the 1913 law before the Supreme Court addressed the issue. Subsequently, both the treaty and the Migratory Bird Treaty Act were tested on the same constitutional grounds, again in litigation initiated by the states (e.g., *Missouri v. Holland*). Both were upheld by the Supreme Court in 1920. So what might have been federal intrusion into state jurisdiction had been overcome by the treaty-making authority expressly reserved by the Constitution for the federal government.

The property clause empowers the federal government to control its own land, but the government also can control activities that occur elsewhere if federal lands are affected (Matthews 1986). For example, when a Nevada rancher drilled and pumped from a well on private land, the water level dropped on nearby Devil's Hole National Monument—the home of an endangered fish—and the rancher was required to reduce the pumping rate so that federal property was not harmed.

NATIONAL POLICIES

Policy statements that affect natural resources are commonplace. Virtually every level of government pays some attention to one or more components of ecosystems. Although interest in ecological matters has been popular in many circles, especially since the 1960s, the range of commitment may vary from simple lip service to complete dedication. Unfortunately, not all policies act in concert among the agencies responsible for their implementation. A historic example concerns the efforts of the U.S. Fish and Wildlife Service to protect wetlands valuable as waterfowl habitat while, at the same time, the Soil Conservation Service aided farmers with technical and financial assistance for drainage programs. With revenues generated from the purchase of duck stamps, one arm of government was attempting to preserve and manage the same wetland resources that another federal agency was draining for farmland. All the while, taxpayers were paying for a government-backed program that guaranteed prices for grain in an era of surplus production.

We can expect inefficiencies of this sort so long as the functions of government remain compartmentalized, each serving specialized and sometimes politically potent segments of the public. The political arena and the policies it spawns are not necessarily consistent, nor do they possess far-reaching wisdom. More than one conservationist has claimed that Democrats and Republicans are the worst enemies of wildlife. Nevertheless, the management and protection of wildlife and other natural resources have benefited from far-reaching legislation, and we here highlight a selection of these policies. Other policies appear in chapters aligned with specific subject matter (e.g., see Chapter 13 for the "Swampbuster" and "Sodbuster" provisions of the Food Security Act of 1985).

LACEY ACT

Congressman John F. Lacey justifiably earned the distinction of "the father of conservation legislation" (Trefethen 1975). He championed the protection of wildlife throughout his public life, beginning with the Yellowstone Park Protection Act of 1894. Until then, only military regulations governed Yellowstone, and these contained no provisions that addressed the illicit practices affecting the park's wildlife. Passage of the act stemmed the poaching of buffalo and other violations of Yellowstone's sanctity. However, Lacey's mark in conservation history was assured when he sponsored the Game and Wild Birds Preservation and Disposition Act of 1900. Known since as the Lacey Act, this legislation prohibited the transportation of illegally killed game across state lines, curbed trafficking in plumage and other wildlife products, and initiated permit requirements and controls for the introduction of mongooses (*Herpestes auropunctatus*), starlings (*Sturnus vulgaris*), and certain other exotic animals.

Unfortunately, an enforcement arm barely existed at the time and only modest penalties were specified, even when violators were apprehended and convicted. The original legislation in fact limited fines to no more than $200 per violation. In 1976, the Lacey Act was strengthened. Penalties for importing harmful animals—now expanded to cover mollusks, amphibians, and other taxa—were increased, including imprisonment for up to 6 months. Violators of the interstate transportation provisions concerning illegally killed wildlife or the sale of wildlife products now can be assessed fines of $10,000 and/or 1 year's imprisonment.

In the Lacey Act, we find the genesis of all subsequent federal policies governing wildlife. Until its passage, only the states had addressed wildlife policies, and virtually all legislation concerned season lengths and bag limits for game species. Such measures then constituted the sole basis of wildlife management. The Lacey Act, however, established the federal role and initiated concerns for a broader spectrum of interest. Wildlife products, especially those from nongame birds, interstate commerce, and exotics, formed the substance of the Lacey Act, bringing the regulatory authority of wildlife management into areas other than those involving the simple control of hunting and fishing activities. With its now-stiffened penalties, the act continues as a strong deterrent to the unwarranted exploitation of wildlife in the United States.

RESTORATION ACTS

Perhaps the most important legislative measure promoting wildlife management in the United States was enacted in 1937 after several years of aborted treatment in the halls of Congress. Formally known as the Federal Aid in Wildlife Restoration Act and popularly known as the Pittman-Robertson Act after its Congressional sponsors from Nevada and Virginia, this legislation originally levied an excise tax of 10 percent on the sales of sporting arms and ammunition; the tax later was increased to 11 percent. Revenues from the tax are pooled in the national treasury for redistribution to the states. A formula based on the land area and the number of licensed hunters determines the amounts of money available to each state. Therefore, large, thinly populated western states such as Nevada compete evenly with small, heavily populated states such as Maryland for "P-R" funds. Once the apportionment is determined, it remains for each state to match the federal dollars

on a 1:3 ratio (i.e., for every $1 of state funding, the P-R provides another $3). An important point: no state could receive P-R dollars if, as in the past, revenues from hunting licenses were diverted to unrelated activities such as highway construction.

P-R funds were administered by federal coordinators who approve projects proposed by the states. Known as federal-aid projects, they originally supported research, land acquisition, and construction. Several amendments to the original legislation later broadened the coverage. Maintenance of completed projects was added in 1946 so that funding could continue the day-to-day operations once construction work was finished. In 1955, the scope was expanded to include management of state wildlife areas, although law enforcement and public relations programs were excluded. Two additional changes occurred in 1970. First, federal taxes on handguns were added to the pool of P-R funds, but half of this new revenue was reserved for hunter safety programs (half of the federal taxes on archery equipment were added for the same purpose in 1975). The second change gave each state an option for requesting federal money. Instead of proposing individual federal-aid projects, states could submit a comprehensive plan covering a 5-year period. Federal allocations were based on the needs to fulfill the goals of the plan. P-R generated nearly $166 million in 1997, thereby bringing the total apportioned to the states since 1937 to more than $3 billion. The act has "done its work in silent woods and fields and waters, in research laboratories and modest offices . . . and our lives are richer for it" (Hodel, *in* Kallman 1987).

The success of the Pittman-Robertson Act led to passage of a similar program in 1950 known as the Federal Aid in Fish Restoration Act. Again, the bill is popularly known by the initials of its Congressional sponsors, Dingell of Michigan and Johnson of Colorado, with "D-J" funding created by a 10 percent excise tax on certain kinds of fishing tackle. The state-federal ratio also is 1:3, and the maximum allocation is based on the number of licensed anglers and the area of water within each state. In 1984, the scope of D-J was expanded by legislation sponsored by Senator Wallop of Wyoming and Congressman Breaux of Louisiana. "Wallop-Breaux" extended the current tax on rods, reels, and lures to include all sport-fishing equipment, and a 3 percent excise tax was imposed on electric trolling motors and fish finders. Duties collected on imported boats and fishing tackle also were added to D-J funds. By far the largest source of revenue, however, comes from taxes collected on motorboat fuels. The first $45 million of the fuel tax—which totals about $90 million annually—is obligated for boating safety, but all revenues above that amount are directed into fish management. States with coastlines must use the new revenues for the management of marine as well as freshwater fisheries, and all states must obligate at least 10 percent of the new money for the construction of boating facilities. In all, including Wallop-Breaux, the total for the D-J program in fiscal 1996 was more than $273 million.

Both acts profoundly influenced the course of wildlife and fisheries management. Support for research was an important feature, as few states were involved with scientific investigations before the new sources of funding became available (Allen 1974a). Today, with these sources, all states are committed to active research programs. Language in the acts required that the federal-aid projects meet certain standards in character and design, with the implication that hunters and anglers would benefit in ways justifying the allocations (Bean 1983). Among these was the acquisition of lands for management areas that were otherwise too expensive for the states to purchase alone. Fish hatcheries were built, as were a multitude of other structures needed in support of management activities. The list of federal-aid projects would be awesome to reconstruct; any state's conservation program would include hundreds of earlier projects and dozens more currently underway. It is worth repeating that the attractive matching ratios built into each measure forced the states once and for all into stable financial positions for managing wildlife. License fees could no longer be diverted from the state wildlife agencies for other purposes if the states were to remain eligible for the 75 percent federal contribution, and all states quickly adapted their laws to meet this requirement (Bean 1983). So beginning in 1937, wildlife management gained a sure footing, free of the financial manipulations of state legislatures.

LAND AND WATER POLICIES

Passage of the Fish and Wildlife Coordination Act in 1934 and its amendments in 1946 and 1953 authorized the Department of the Interior to assure that wildlife received equal consideration with other features of water-development programs initiated or licensed by federal agencies. Specifically, the act required future developers (public or private agencies) of rivers or other bodies of water scheduled for impoundment, diversion, or channelization to consult with the state's wildlife and fisheries agency and the U.S. Fish and Wildlife Service before action would be taken on the project. The goals of the act are, first, to prevent damage or loss of fish and wildlife resources and, second, to improve fish and wildlife conditions affected by water development. All costs associated with fish and wildlife conservation, including subsequent maintenance, are borne by the federal government, the developer, or some combination of the two, as part of the overall budget. Only those impoundments exceeding 4 ha of surface area fall under jurisdiction of the act. For larger impoundments and other water-development projects, the act authorized acquisition of additional land or water expressly for wildlife as compensation or mitigation for habitat losses associated with the project; these may not be exchanged or sold later but must remain managed for wildlife resources.

The thrust of the Fish and Wildlife Coordination Act is twofold. First, massive dislocations of water supplies no longer can be achieved legally without first considering fish and wildlife values. Dams planned for rivers with migratory fish populations must be equipped with fish ladders, and the cost of such modifications becomes part of the cost of the project. Second, additional land or water acquisitions of equal value can be required if the project jeopardizes existing wildlife habitat. Many large reservoirs today are adjoined by forests or other lands acquired solely as fish and wildlife management areas. Management of these acquisitions, including recreational activities, is assured by the act.

The Wetland Loan Act recognized the plight of marshes and other wetlands under the pressures of agricultural and other developments of an expanding society. Not only was wetland drainage continuing, but it was doing so at an accelerated rate. Based on estimates made in 1955, more than 35 percent of North America's original wetland area of 51 million ha already had been drained (Shaw and Fredine 1956). The pressure from agriculture varies with regional economies so that rates of drainage are uneven in their geographical distribution. Therefore while the national average was 36 percent in 1955, wetland losses in the heavily farmed prairie states were more than 50 percent (Schrader 1955). Wooded wetlands in Missouri total no more than 4 percent of their original area (Korte and Fredrickson 1977b), and drainage has claimed all but 6 percent of Iowa's marshes (Bishop 1981). In recent decades, the national loss has been estimated at 233,000 ha per year

(National Research Council 1982). Clearly, a means of acquiring wetlands en masse was needed immediately instead of the slower year-by-year acquisitions then possible using the relatively small funds annually available from duck stamp sales. By 1982, the price of duck stamps had increased 750 percent but, in effect, that percentage adds only about $0.40 per stamp more in purchasing power compared with the $1 stamps of the 1930s. In the same interval, the average price of farm real estate also rose, but by 2600 percent (Poole 1982). All the while, wetland habitats were being lost.

This alarming pattern finally was recognized by Congress in 1961. Under terms of the Wetland Loan Act, Congress initially loaned the U.S. Fish and Wildlife Service $105 million for accelerated acquisition of wetlands crucial to waterfowl and other wildlife. The loan spanned a 7-year period, and future revenues from duck stamp sales were pledged against repayment. The act was reauthorized in 1967 and 1976, when the amount of the loan was increased to a total of $200 million. Most recently, the Emergency Wetlands Resources Act of 1986 extended the 1976 reauthorization. The 1986 legislation increased the funds available for wetlands acquisition by raising the price of duck stamps from $7.50 to $15.00 during a 5-year period. Other revenues for wetland conservtion and acquisition also are authorized in the new act; these include the transfer of import duties—estimated at $10 million annually—collected on firearms and ammunition, entry fees from selected national wildlife refuges, and access to the Land and Water Conservation Fund (see following). Moreover, loans made by the previous wetland acts were forgiven. An important provision of the 1986 act required that the secretary of the interior conduct a study for Congress on the extent to which subsidies from other federal programs encourage wetland destruction.

In 1964, the departments of the Interior and Agriculture became responsible for administering the National Wilderness Preservation System. Known as The Wilderness Act, this legislation proposed to maintain the pristine nature of land where "man himself is a visitor who does not remain."

A wilderness area was subject to four criteria: (1) only the forces of nature had affected the land, with the imprint of human work substantially unnoticeable; (2) the land presented outstanding opportunities for primitive or unconfined recreation in a climate of solitude; (3) at least 2024 ha must be available, or if less, it must be practical to preserve and use it in an unimpaired condition; and (4) the area must include ecological, geological, or other features leading to educational, scientific, scenic, or historical value. No permanent or temporary roads, motorized vehicles, boats, or other machinery, commercial activities, or structures were to intrude into wilderness areas, except by special provisions. Mining and oil claims were authorized until 1984, but only if the surface thereafter was restored and the wilderness character of the area was maintained. In 1982, Congress rejected proposals to open some wilderness areas for commercial activities. The act left wildlife under the jurisdiction of the state. In 1968, Congress approved The Wild and Scenic Rivers Act, essentially adding to the concept of The Wilderness Act. Together, these acts assured that large blocks of basically pristine landscape, including wildlife habitat, will remain unspoiled by future developments.

The implications of The Wilderness Act are severalfold. First and foremost, it established the values of pristine landscapes unfettered by human encroachment. National parks, to be sure, often provide similar benefits, but many of these are developed to some degree with roads, hotels, and other marks of modern life. The act assured that qualified sites would remain protected from these or other developments inimical to their status. Parts of existing wildlife refuges, parks, and national forests were designated as wilderness, preventing future developments that otherwise might have occurred under their previous classification. For wildlife, wilderness areas function as pristine showcases, and here larger species such as mountain lions and bears may enjoy large tracts where many of their needs are met. Wilderness areas remain free of those developments that historically forced these animals into ever more restricted ranges. In addition, the opportunities for conflict with humans were reduced, thereby lessening the persecution of large carnivores. Hunting, fishing, and trapping may continue under state authority, but the state laws must regulate wildlife populations in ways consistent with goals of wilderness status. Habitat preservation, not hunting, remains the paramount concern of the act. Wilderness areas also present unique opportunities to study wildlife in situations where the influences of humans remain negligible. The pristine nature of primitive areas covered by The Wilderness Act permits incisive research on animal social systems, predator-prey relationships, movements, and other facets of wildlife ecology for several species of wildlife (see Hornocker 1969, 1970 and Seidensticker et al. 1973 for pertinent examples with mountain lions).

The Land and Water Conservation Act also was approved in 1964. The act created the Land and Water Conservation Fund from admission fees to publicly owned land, surplus property sales, fines, and a share of the federal excise tax on motorboat fuels. Additionally, Congress might appropriate up to $900 million annually to the fund, incorporating, as needed, monies from federal oil leases authorized by the Outer Continental Shelf Lands Act. These revenues were dedicated to the acquisition of privately owned land and waters having scenic, recreational, or other values as part of the public trust. Once acquired, these parcels are added to existing refuges, parks, or other public units in federal, state, or local ownership.

Wildlife benefited from the Land and Water Conservation Act not only by the acquisition of habitat, but also because part of the funds were dedicated to endangered species. However, the impact of the act was lessened from its full potential when the monies it generated, in part, were sidetracked into recreational facilities in urban settings (Trefethen 1975). Instead of expanding desperately needed environments where both city dwellers and others might enjoy wildlife and unique outdoor experiences, powerful forces in the government directed that playgrounds and skating rinks be built with some of the money, even though other federal programs already supported such activities. Such raids on the Land and Water Conservation Fund seem likely to continue. Economic constraints and budget realignments of the early 1980s created pressures to redirect the fund's assets for still other purposes.

Legislation known as the Sikes Act was enacted in 1960. The legislation directed the U.S. Fish and Wildlife Service to cooperate in the management of wildlife resources on military bases and other lands—more than 10 million ha—administered by the Department of Defense (see Ramsey 1986). In 1974, the act was expanded to include other federally administered lands (Owen 1986). The Sikes Act was reauthorized in 1986, when the position of fish and wildlife management was strengthened. Base commanders now must improve wildlife habitat as an integral part of the other resource management activities and plans already in force (e.g., grazing). Net revenues from timber sales may be applied to fish and wildlife management instead of reverting to the general treasury. The new legislation also directs that professional biologists must be involved in the management of fish and wildlife resources on lands falling under jurisdiction of the act.

The North American Waterfowl Management Plan (NAWMP) addresses the protection and management of key wetland habitats in Canada, Mexico, and the United States. Initiated in 1986, the goals of NAWMP are (1) to protect 2.4 million ha of wetlands and (2) to attain a continental breeding population of 62 million waterfowl, with specific targets set for each of 32 species. A fall population in excess of 100 million waterfowl by the year 2000 is the anticipated outcome of these efforts, for which the habitat component alone will require more than $1.5 billion. NAWMP currently includes 15 joint ventures, each representing a project involving the partnership of public and private conservation agencies. Some of the joint ventures are in such key areas as the Prairie Pothole Region, Central Valley of California, Gulf Coast, and the lower Mississippi Valley. American black ducks (*Anas rubripes*) and arctic-nesting geese also are the species-specific targets of two joint ventures. Under NAWMP, wetlands will be purchased or protected with leases or easements, and owners of private land will receive economic incentives for implementing soil and water practices that favor waterfowl (e.g., flooding fields after crops are harvested). Additionally, the habitat protected by NAWMP benefits a large number of other wetland-related species of wildlife (e.g., shorebirds).

NATIONAL ENVIRONMENTAL POLICY ACT

National policies were directed toward settlement and development during much of U.S. history. The Homestead Act (1862), the Pacific Railway Act (1862), and the Timber and Stone Act (1878) identify, in part, the national effort once enhancing economic expansion in the United States. Industrial development proceeded without much regulation, and attractive tax benefits actually encouraged expansion. World War II later heightened industrial development to levels never before imagined.

So for more than 200 years little thought or inquiry was given to the impacts of human encroachment on the American landscape. New technologies—everything from nuclear energy to snowmobiles—added further disruptions to an environment supporting an ever-increasing human population. At first, nature seemed forgiving. Streams somehow absorbed simple chemical wastes and the atmosphere apparently disposed of the acrid smoke billowing from industrial chimneys. In any case, there was the general perception that "more" of everything could be developed someplace. For the most part, the public remained unaware or unconcerned about environmental matters, nor had the scientific community marshaled much evidence of environmental degradation. By the 1960s, however, these circumstances changed dramatically.

The National Environmental Policy Act (NEPA), passed by Congress in 1969 and signed by President Nixon early in 1970, established a new era seeking to maintain environmental quality. Previously, it remained for each federal agency, independently, to address environmental issues from its particular point of view. With NEPA, a broad charge was mandated. All federal agencies now were required to respond to the same policy. Compliance with NEPA is excused only when it conflicts with another statutory requirement, as when the length of reporting periods might differ from NEPA requirements. Otherwise, NEPA brought uniformity to the ways and means the array of federal agencies address environmental issues. Federal agencies responsible for sanitation, power, housing, or public transportation now treat the environmental aspects of their actions with the same requirements as natural resource agencies like the U.S.D.A. Forest Service. Actions by independent regulatory agencies such as the Interstate Commerce Commission—once concerned primarily with licensing and rate controls—also fall under NEPA policy. In this and other instances, NEPA forces open an agency's actions to public and judicial consideration of environmental issues rather than just economic questions (Skillern 1981). NEPA thus came to bear on a wide range of matters, such as endangered species, scenic values, and wetland preservation, that otherwise might have been impaired or ignored by federal activities.

Section 101 of NEPA sets the tone of the congressional mandate. Federal planning, actions, and functions were charged to use all practical means to enhance and protect environmental quality. The government, in short, became trustee for the future. People in the United States, while enjoying a wide range of beneficial uses of the environment, would also be assured safe, healthful, productive, and aesthetically and culturally pleasing surroundings. Perhaps the greatest challenge lies in the goal that the nation achieve a balance between population and resource use consistent with a reasonable standard of living.

NEPA also established the Council on Environmental Quality (CEQ). Activities of the council lie entirely in the executive branch of the federal government. This body reports to the president in an advisory capacity. Under the authority of executive orders, CEQ promulgates rules that govern compliance with NEPA. CEQ also monitors trends and reviews programs, publicly reporting its findings annually to the president, along with recommendations for environmental policies.

At the core of NEPA lies the Environmental Impact Statement (EIS). As required by Section 102, proposed "major" actions by the government potentially having significant impacts on the environment shall be accompanied by an EIS that discloses fully the effects of the action on the environment. In addition, an EIS must propose alternative actions and describe the environmental consequences of the proposed project. When proposing a project, most agencies prepare a preliminary study known as an Environmental Assessment (EA). If the EA determines that there is no clear indication that environmental damage can or will occur, the agency then may file "A Finding of No Significant Impact," leaving the matter resolved unless this conclusion is challenged. On the other hand, if the EA determines that there is likelihood of environmental repercussions, it triggers preparation of a more thorough EIS and provides the background for further review.

In many cases, an EIS requires field studies properly identifying the status of wildlife and other resources at the site. Biological and ecological surveys focus particularly on endangered species, if any are present, but also include information about other types of wildlife. All species of wildlife, whether game or nongame, enjoy equal status in this regard. Whereas the economic prominence of game or commercial wildlife generally is more easily documented, aesthetic values are given no less weight. An EIS thus becomes a powerful tool by which legal challenges might be issued to projects potentially degrading important ecological or biological settings.

Social as well as ecological dimensions of the action are included in the scope of an EIS. A controversy at Ruby Lake National Wildlife Refuge in Nevada illustrates social concerns in the form of recreational values. This refuge, an important breeding area for canvasbacks (*Aythya valisineria*) and other ducks, also offers opportunities for public recreation. Fishing and boating activities highlight recreation at Ruby Lake and increased dramatically between 1955 and 1980 (Table 22-2). Both activities have social values as public recreation, but

TABLE 22-2. History of Public Visits for Boating-Related Recreation at Ruby Lake National Wildlife Refuge

Year	Boating and Waterskiing[a]	Fishing	Total Refuge Visits
1955		700	1,500
1960		8,000	18,249
1965	4,030	15,000	20,100
1970	1,700	31,450	34,205
1975	2,945	41,575	45,680
1980		57,698	65,568

Source: Bouffard (1982).

[a]Recreational boating data exclude fishing from boats.

waterskiing, in particular, demands high-powered boats. A survey indicated that the average motor size on the boats exceeded 90 hp and that most of the public visits occurred between May and September (Bouffard 1982). Because of the disturbance caused by the speedboats, waterfowl nesting and brooding success suffered at Ruby Lake. Among other data collected, field studies for an EIS showed that canvasback broods reared coincident with heavy boating averaged 3.8 ducklings, whereas those raised in a boat-free area averaged 5.3 ducklings, for a difference of nearly 30 percent (U.S. Fish and Wildlife Service 1976). Propellers cut aquatic vegetation, changing the composition of the plants in some areas and vastly reducing the amount in others. The weight of aquatic vegetation per area was reduced by nearly 77 percent where boating activities were highest, compared with a boat-free zone. With less vegetation, the wakes of the speedboats caused bank erosion and siltation, especially where waterskiing occurred (Bouffard 1982).

After 3 years of study, the EIS reported that boating (numbers, size, and penetration into the marsh) interfered with waterfowl production. Public recreation, although authorized by the Refuge Recreation Act of 1962, remains secondary to the primary purpose of any national wildlife refuge. At Ruby Lake, this meant that waterfowl production outweighed the importance of fishing and waterskiing. Refuge managers thereafter limited boating activities at Ruby Lake, but the public reacted against these management efforts. New regulations were adopted, and the matter seemed settled until an animal-rights group challenged these arrangements in court. The group won the case, and today all waterskiing is prohibited at Ruby Lake. Outboard motors are regulated by size

and type, and the refuge is zoned with a schedule timed to minimize the conflicts between boating and waterfowl production.[1]

POLICIES OF SENTIMENT

Passage of the Wild Free-Roaming Horse and Burro Act in 1971 culminated a long-standing emotional issue on the part of persons and organizations championing animal rights. Unfortunately, the act did not resolve the ecological situation, and indeed, it vastly complicated wise land management. A dynamic political setting emerged (Behan 1978). On one side, the policy of shooting feral equines on public lands brought strong public reactions on behalf of the animals. In particular, burros (*Equus asinus*) summon images of tranquil beasts, not lacking in either charm or delight among children and others with sensitive feelings toward animals. On the other side, managers responsible for the stewardship of public lands well know of the damage burros cause to delicate semi-arid environments and their endemic biota (see Chapter 14). Burro populations were degrading the Grand Canyon National Park, Bandelier National Monument, and Death Valley National Monument. On these and other public lands where feral equines are numerous, they compete with other animals for food and water, degrade habitats for native wildlife, and initiate extensive erosion. At Bandelier National Monument, burros also damaged the ancient Indian ruins representing the cultural basis for originally protecting the site.

The dilemma continues. The act still protects feral burros and horses on most publicly owned land. Because national parks were exempted, however, rangers continued shooting burros, but these efforts were short-lived. Public sentiment and lawsuits quickly stopped the shooting program, and other controls were proposed. Among these were sterilization and fencing, but these measures either

---

[1]In another study, Titus and VanDruff (1981) concluded that heightened recreational pressure in a major wilderness area only slightly reduced the breeding success of common loons (*Gavia immer*) where motor-powered boats were restricted to specified routes. Overall, the adult loon population on the Boundary Waters Canoe Area of northern Minnesota increased by 35 percent during a period when recreational use of the area jumped by 800–900 percent. Loons, although generally considered indicators of wilderness conditions, may be adapting or habituating to some forms of human intrusion.

were not adopted, were not feasible, or were of little success. The most publicized method—and one that appealed to the public—involved using the animals as a source of pets. Under the auspices of The Fund for Animals, more than 500 burros were captured and removed from Grand Canyon National Park (Fig 22-7). The "Adopt a Burro" program was expensive, partly because individual animals were transported by helicopter from the canyon floor; estimates placed the cost of removal at $1000 per animal. In reporting these circumstances to the secretary of the interior, Allen et al. (1981) noted that such methods may have objectionable features, namely, the public

may be conditioned to believe that expensive operations are the appropriate way to control excessive numbers of nuisance animals. Further, the report lamented the irony that public contributions can be raised to rescue an abundant population of feral animals at a time when nearly all of the world's wild equines are faced with extinction (except for plains zebra, *Equus burchelli*). Nonetheless, a new policy was proposed in 1987 for disposing of the excess burros and horses on lands administered by the Bureau of Land Management. Along with representatives from humane organizations, resource specialists recommended (1) continuing the "Adopt" program with

Figure 22-7. Round-up of feral burros posed a humane but expensive means of lessening damage they caused to native biological communities in Grand Canyon National Park. After roping (upper left), the burros were transported over rough terrain by helicopter to holding pens (upper right and lower left). The project was financed and operated by a private organization, The Fund for Animals, which eventually distributed the burros for pets. (Photos courtesy of The Fund for Animals.)

a fee of $125 per horse and $75 per burro; (2) altering fees for "special" adoptions; (3) using prison inmates to train horses before adoption; (4) placing horses on private refuges; and (5) humanely destroying unadoptable animals. Wagner (1983), Williams (1985), and Berger (1986) further reviewed the implications and management of feral equines on North American rangelands.

By law, national wildlife refuges may be open to hunting when the primary purpose of the refuge is not compromised by this form of outdoor recreation. Up to 40 percent of a refuge may be authorized for hunting waterfowl and 100 percent for other kinds of hunting. Some refuges are closed entirely to hunting, but many others are not. Refuge functions are, of course, viewed differently by the public. Some regard them only as inviolate sanctuaries for wildlife, whereas other persons view refuges as sites of multiple use, including consumptive recreation. These conflicting views spurred another public controversy.

In New Jersey, the Great Swamp National Wildlife Refuge became a battleground in a long-standing controversy about deer hunting. Great Swamp was established in 1964 to protect a large, near-pristine wetland in a densely populated state; at that time, the site was proposed for development as a large airport servicing New York City's metropolitan area. Prior to 1964, deer were hunted in the area. Estimates of the summer deer herd then were placed at 120 animals and were projected to increase to 590 during the next decade in the absence of hunting or other control measures.

Deer hunting on the refuge stopped completely in 1968. As predicted, the herd increased under full protection, concurrently producing all the symptoms of an overloaded environment. Measurements of breeding success indicated reduced productivity. Yearling does produced 0.67 fawns, whereas the normal expectation would be 1.5 fawns per yearling doe. Indices of body condition also declined. The fat content of femur marrow fell below 25 percent, the normal minimum for winter survival. Furthermore, infections of abomasal parasites increased in the malnourished herd; counts of helminths reached more than 1300 per animal, whereas parasite loads elsewhere in poor habitat numbered 840 per deer. Finally, the vegetation developed a pronounced browse line on both the refuge and adjacent properties. The first known deaths from malnutrition were discovered in the spring of 1973. Outside the refuge, auto collisions and damage to ornamental and farm vegetation increased as the deer herd expanded.

Hunting seemed the only practical tool for managing the overabundant deer population. Refuge officials therefore sanctioned a deer hunt on the refuge in 1973, but they were quickly restrained by a lawsuit sponsored by an animal rights group. Because an EIS had not been prepared, a federal district court ruled that the U.S. Fish and Wildlife Service had not complied with the National Environmental Policy Act. Subsequently, an EIS was prepared, describing the biological and ecological maladies befalling the refuge and its deer herd, as outlined earlier, and proposing a highly controlled hunting season to thin the herd (U.S. Fish and Wildlife Service 1974). When the case was concluded, the court held that public hunting is not inconsistent with sound wildlife management and that hunting is compatible with the purposes for which the refuge was established. A hunt was held in 1974 despite efforts again to enjoin this action. The special hunting season remains in effect under the rigorous management of refuge personnel. Since hunting resumed in 1974, there have been no documented cases of starvation, and perhaps of greater importance, delicate habitat in the Great Swamp now shows recovery from overbrowsing. Antihunting groups nonetheless still keep the issue very much alive, annually continuing public demonstrations and media coverage against the Great Swamp deer hunt.

These controversies clearly were stirred by public opinion in spite of serious efforts by professional managers to fulfill their assigned responsibilities. What can be learned from these experiences? First, the public's concept of management may not always coincide with the professional judgment of wildlife managers supposedly trained and hired to do the job. Most wildlife managers are public employees. They therefore walk a fine line when their management decisions and recommendations are contrary to the public viewpoint. The dilemma is not new. Managers have wrestled for years with bounty systems, artificial feeding, and other issues popular with the public, and they will do so again in the future.

Second, managers sometimes face difficult legal constraints. Burros were given legal protection on the same public lands where they often were degrading other natural resources also protected by legal mandate. Restraining orders and other legal actions thus can stall management activities for long periods of time. Management decisions—for better or worse— are sometimes made in the courtroom where fine points of law rule the day.

Third, some traditional management tools for removing overabundant animals no longer may be acceptable to the public. Shooting, no matter how efficient and inexpensive, lacks public support in many instances. Managers may seem to represent only the interests of special groups, particularly hunters, instead of a broader public constituency. In 1982, deer were confronted with deteriorating habitat in Florida when the Everglades flooded. The deer faced starvation, and a special hunt was authorized to thin the herd. However, many sectors of the public disagreed, wishing instead to live-trap and remove the excess deer and/or feed the herd artificially. This event in many ways was not unlike the controversy at Great Swamp more than a decade earlier, suggesting that managers have made little progress in shaping public attitudes about controlling animal populations. Indeed, Applegate (1975) found that the margin between New Jersey residents approving or disapproving of deer hunting decreased from 16 percent to 6 percent in 2 years' time; approving residents still outnumbered disapproving residents, but the gap was closing rapidly.

Finally, we again see that wildlife management does not just consist of banding ducks or aging deer. It is very much *people* management as well. Management policies are subject to judicial, legislative, *and* public review. Each is a force with which managers must reckon.

PUBLIC ATTITUDES

Wildlife policy in the United States is based on the concept of public ownership of wild animals. Privately owned land may be zoned against trespass or hunting, but wildlife residing on that land itself is not private property—it remains subject to public jurisdiction. This fundamental premise has influenced U.S. actions and attitudes about wildlife. City dwellers might oppose measures adopted in rural areas or even in other states. Since 1970, NEPA has provided the public with a formal process for influencing policy decisions affecting wildlife. In the late 1970s, citizen organizations in the "lower 48" used NEPA requirements to challenge the removal of wolves (*Canis lupus*) in certain areas of Alaska where caribou (*Rangifer tarandus*) numbers had diminished. We previously mentioned that segments of the public vigorously objected to deer hunting at Great Swamp National Wildlife Refuge because of perceptions about refuges and wildlife management.

Public reactions are by no means always opposed to new management policies. Numerous state legislatures responded to public wishes for nongame pro-

grams, including specific appropriations derived from state sales and income taxes. But what are the overall perceptions of people in the United States about wildlife, and do different segments of the public view these in the same ways? Any answers may be somewhat speculative, but statistically based samples of public attitudes offer at least as much insight as other public opinion polls concerning television programs or political affairs. In recent years, questionnaires and their analyses have offered conservationists, wildlife managers, and other concerned with wildlife new opportunities to assess public attitudes.

WHO AND WHAT?

Kellert (1976a,b) initiated surveys of public attitudes toward wildlife in the United States and thereafter described nine basic attitudes toward animals. These included a range from *naturalistic*—a profound attraction to wildlife and the outdoor environment—to *negativistic*—an avoidance or even fear of animals along with a desire for a separation from any association with a nonhuman environment (Table 10-1, Chapter 10). The classifications are not completely exclusive, as one person may experience more than one attitude, but they do serve as useful indicators of human perceptions. Highlights are summarized below for each of four population variables.

1. *Ascriptive.* Persons between ages 18 and 29 contrasted significantly with those 65 and older. The younger group was more naturalistic about wildlife and the outdoors, whereas the older group was more utilitarian in its outlook. Older citizens supported predator control, hunting, and commercial activities using wildlife and preferred work animals to pets. Women showed greater compassion for wildlife than men, as measured by the moralistic and humanistic dimensions of the survey. Racial differences also were prominent, with African Americans having more negative attitudes about wildlife than European Americans. African American men, in particular, held strongly dominionistic attitudes about animals.

2. *Socioeconomic.* Persons with less than an eighth-grade education were negativistic, utilitarian, and dominionistic, whereas persons with some college education were more sensitive to ecological and natural values. Among occupations, farmers were utilitarian in their outlook, whereas students, executives, and professionals were far more naturalistic. Unskilled and clerical workers showed a

tendency toward negativistic attitudes. Income in itself, however, had surprisingly little relationship to attitudes about wildlife. Other components of socioeconomic status (e.g., education and occupation) were better predictors of public values.

3. *Geographic.* Measurements of the geographical variables revealed that childhood and current residence had a bearing on attitudes. Persons raised in rural settings or who currently were living in rural settings were more utilitarian and less moralistic than those living in large cities. Surprisingly, people from towns of 10,000 to 50,000 in population scored highest in naturalistic attitudes. Regional differences were not great, but, in general, citizens of the western states were more utilitarian and dominionistic in their attitudes, whereas those from eastern states were more humanistically oriented. Attitudes in the Rocky Mountain states were more naturalistic than elsewhere in the United States.

4. *Familial.* Age influenced the significance of familial variables so that marital status and other characteristics did not gain statistical prominence as indicators of public attitudes. In general, however, single persons were more humanistic, whereas married persons showed stronger utilitarian attitudes toward wildlife. Childless persons also tended to be more humanistic than those with five or more children; for the latter, utilitarian attitudes were dominant. Pets apparently influenced these results, as single and/or childless people seemingly substituted pets for human associations, thereby adopting closer ties with animal-oriented attitudes.

Some inferences about policy may be drawn from these data. First, the striking contrast between well-educated and less-educated citizens in their regard for wildlife implies that educational processes hold great promise in the future of wildlife management. If education beyond the eighth grade steadily increases the positive aspects of appreciation for wildlife in a broad context (e.g., outdoor experiences and ecological awareness), then wildlife agencies and organizations might direct many of their information programs toward students still in grade school. Special materials about aesthetics and ecological relationships would seem appropriate for such groups. Conversely, information about the role of hunting and other consumptive uses of wildlife would be better aimed toward citizens who have more advanced education. As a group, they tended to disapprove of utilitarian values.

In sum, perceptions of wildlife values vary so greatly among the U.S. public—in this case, by educational background—that information systems should be tailored for specific groups, with each carrying a well-researched mission.

Second, new programs dealing with more-or-less nontraditional aspects of wildlife likely will be supported if the appropriate constituency is informed. Nongame management, for example, may attract little interest among farmers or other groups holding high utilitarian values about wildlife. Conversely, some programs may have more universal appeal. The Missouri Conservation Department initiated "Eagle Days" by offering public field trips and interpretative talks and movies about eagles (Witter et al. 1980). About 1600 people preregistered for the programs, and of these, 1000 attended; some 300 eagles were observed on the 3 days the programs were held. The visitors included family and youth groups and substantial numbers of hunters and anglers. Most of the visitors were impressed, with 32 percent stating that it was the most enjoyable wildlife experience they ever had. Hunters and nonhunters responded with equal favor about the merits of the experience, and many wished for additional opportunities (e.g., for programs such as "Waterfowl Days"). The appeal of Eagle Days was enhanced by low operating costs and high visibility and suggested that such efforts promote broad-based support for wildlife management from all segments of the public. Such programs are equally effective for promoting the protection and conservation of prairies and other types of dwindling habitat (Thom et al. 1986).

Finally, formulators of wildlife policy should maintain continual contact with demographic patterns. Older age classes are expanding while the last of the World War II "baby boom" already has entered the workforce. Fewer students will enter college in the decades ahead, and those that do seemingly will select technological and professional training rather than a liberal education. Smaller families, more single people, and a higher percentage of working wives now are part of the demographic pattern in the United States. The attitudes of these and other groups need consideration if policies governing wildlife are to be accepted and maintained.

AMERICANS AND ISSUES

From the foregoing, it is clear that several variables influence the attitudes that the people of the United States hold toward wildlife. These permit some analyses of the support or nonsupport for key issues confronting the U.S. public (Kellert 1980b).

The issue of protecting endangered species was weighed against commercial development, which sharpened the issue more or less into tangible rather than idealistic attitudes. Significant numbers of the people responding to the survey were willing to underwrite additional expenses for commercial developments in order to protect some endangered species such as bald eagles (*Haliaeetus leucocephalus*) or eastern mountain lions. Conversely, endangered plants, snakes, and insects listed in the survey gained less than majority support. Further, when given the choice of protecting an unknown fish from a water-development project, people distinguished between "essential" human needs (e.g., drinking water) and "nonessential" needs (e.g., recreation). They approved of essential projects in spite of the consequences to the endangered fish but favored protection if the project was nonessential to humans.

For predator control, particularly concerning coyotes (*Canis latrans*), the public generally opposed measures that indiscriminately reduced predator populations. Poisons were opposed specifically in spite of their inexpensive applications. However, the public favored programs that selectively controlled individual animals causing damage instead of wholesale population reduction. Among the options receiving strong support were the capture and relocation of problem animals. The popularity of this approach is an indication of the public's preference for nonlethal methods of predator control. The public also disapproved of using tax monies to compensate ranchers for livestock losses.

Habitat preservation was supported consistently but at various levels of strength. Only slight majorities disapproved of housing projects that spoiled wetlands, the development of resources in wilderness areas, or grazing practices on public land that harmed wildlife. Stronger support was indicated for forestry practices that helped wildlife, even if higher timber prices resulted. Oil development in Yellowstone National Park and limited off-road activities for vehicles, if harmful to wildlife, were opposed strongly.

Taxes for funding wildlife management gained significant public support. These included excise taxes on clothing made from pelts of wild animals, on off-road vehicles, and on backpacking and bird-watching equipment, but not on wildlife-related literature and art. Entrance fees to wildlife refuges and similar areas were supported by the majority of people in the United States. Overall, 57 percent of the sample approved of expending a larger proportion of general tax revenues for wildlife management.

These and other results of the issue-oriented survey suggest that the majority of the people value wildlife and seem willing to make substantial social and economic sacrifices to maintain wildlife resources (Kellert 1980b). The results do not reflect the strong opinions of a philosophical minority, but instead represent a widespread appreciation for wildlife as an integral part of U.S. lifestyles.

PUBLIC AWARENESS

Despite the empathies generally expressed for wildlife and wildlife-related issues outlined earlier, the public concurrently displayed little or modest knowledge about animals or about specific situations confronting wildlife in recent times (Kellert 1980b). About half of the public did not know that spiders have eight legs or that insects lack backbones. Perhaps most disturbing was the finding that 75 percent of the public believed that coyotes are endangered, a surprising result in light of the publicity given to predator control by the livestock industry.

If the results of this survey indeed represent public awareness, then the complex and controversial issues facing professional managers may lack a much-needed base of support. For example, 58 percent of the public cares more about the suffering of individual animals than about the population levels of these same species (Kellert 1980b). This factor alone suggests the magnitude of translating wildlife policies into publicly acceptable activities, as managers largely address the ills or strengths of animal populations, not individuals within those populations. Clearly, better communication between professional managers and the public they serve is needed if wildlife conservation is to remain viable and effective in the future. At the same time, communication may be difficult unless new approaches are developed. Kellert and Westervelt (1982) noted the relative ease with which newspapers can publish pictorial messages about the plight of animals. Photographs of fur hunters clubbing juvenile harp seals (*Phoca groenlandica*) attract instant—and compassionate—public attention, as do those of individual animals suffering unnecessary injuries (Fig. 22-8). Conversely, it is difficult to convey with equal force pictures of habitat losses experienced almost daily by entire communities of wildlife.

The public's general lack of knowledge about wildlife, coupled with its tendency for personal

Figure 22-8.   This mallard hen attracted national attention in 1981 when she was discovered impaled by an arrow on a golf course in Las Vegas. Dubbed "Donna the Duck," she quickly became the darling of the public as efforts to capture her and remove the arrow headlined the news media. Donna eventually was tranquilized and the arrow was removed, leading to full recovery from the senseless injury. (Photo courtesy of Patricia Mortati.)

perceptions of animals, discloses still other difficulties. Those issues gaining greater recognition among the public sector generally are of lesser importance to professional biologists and managers. Conversely, issues of greatest concern among biologists were little known by the public. The U.S. Fish and Wildlife Service once ranked lead poisoning among its top priorities for management; yet at the time only 14 percent of the public was aware of the problem (Kellert 1980c).

CANADIAN WILDLIFE SERVICE

Canada is the world's second-largest country; yet most of its relatively small population—about 30 million in 1996—is concentrated along its southern border. Canada thus is one of the few countries where large areas of natural ecosystems remain largely intact, particularly its vast regions of Tundra and Boreal

Forest. Lakes and rivers in Canada also contain more fresh water than any other country. Understandably, then, wildlife resources have long been an integral and vital component of Canada's heritage for its native peoples and European settlers from at least 1670, when the Hudson's Bay Company formally began its fur trade (Newman 1989).

Federal involvement with wildlife in Canada emerged in 1916 as a result of the Migratory Bird Convention. Canada enacted the Migratory Bird Convention Act in 1917, the counterpart of the Migratory Bird Treaty Act enacted by the United States in 1918. So at the federal level, waterfowl and most other species of migratory birds became the early focus of wildlife management in Canada; regulated hunting and habitat protection were paramount in these activities. Because the majority of North America's waterfowl breed in Canada—ducks in the prairie pothole regions of Manitoba, Saskatchewan, and Alberta and geese in subarctic regions—field surveys of waterfowl populations and habitat conditions remain a major function of CWS. These surveys are conducted jointly with personnel from the U.S. Fish and Wildlife Service and represent one of the activities in which the two federal organizations regularly cooperate with each other as well as with state and provincial agencies. Wildlife in Canada's national park system also falls under federal jurisdiction. In 1947, these endeavors became the formal responsibility of the newly formed Dominion Wildlife Service, later renamed the Canadian Wildlife Service (CWS).

Other needs also required federal responses, including protection for mammals that cross international boundaries (e.g., polar bears, *Ursus maritimus*), international trade in wildlife, the growing threat of species that face extinction, and protection of nationally significant habitat (now more than 11 million ha in Canada). CWS today is one of the directorates organized within the federal department known as Environment Canada, which was established in 1971. In 1973, the Canada Wildlife Act authorized CWS to conduct wildlife research and, in cooperation with provincial governments, to undertake a wide range of activities concerned with wildlife conservation. Among these are comprehensive studies of birds breeding on Canada's immense coastline and northern regions (e.g., Reed, et al. 1996; Vermeer and Morgan 1997; Dickson 1997).

CWS is currently involved with several programs and projects, including implementing the North American Waterfowl Management Plan, monitoring toxic materials and other environmental contaminants, and sustaining biodiversity. The agency works closely with Canada's large population of Inuit and other Native Americans on matters related to the management of wildlife resources, of which subsistence hunting is a major issue. CWS also provides information to other agencies in the Canadian government about issues such as acid precipitation, global warming, and integrating wildlife values into plans for economic development (e.g., mineral extraction).

RESOURCE MANAGEMENT IN MEXICO: SEMARNAP

SEMARNAP is an acronym for *Secretaría de Medio Ambiente, Recursos Naturales y Pesca*—the Ministry of the Environment, Natural Resources and Fisheries—and serves as Mexico's foremost administrative body for wildlife management at the federal level. However, as its name implies, the scope of SE-MARNAP's responsibilities includes forestry, air and water quality, and—of course—fish and wildlife management. For example, SEMARNAP must approve reservoir and other projects that may impact the environment; it also enforces laws concerned with environmental protection.

Administrative units within SEMARNAP, known as directorates, deal more specifically with each of the ministry's various responsibilities. One of these, *Dirección General de Vida Silvestre*—the General Directorate for Wildlife—is responsible for wildlife research and management, along with conservation and educational programs—tasks that more or less coincide with the duties assigned to the U.S. Fish and Wildlife Service and the Canadian Wildlife Service. These responsibilities include oversight of both resident and migratory wildlife, including endangered species such as the jaguar (*Panthera onca*), the gray whale (*Eschrichitius robustus*), and various macaws (Psittacidae).

SEMARNAP also regulates and administers all protected areas owned by the federal government, and many of these have field stations. The ministry also maintains a delegate in the capital city of each of Mexico's 31 states; other representatives are assigned specifically to other large cities as well. SEMARNAP, as does the Canadian Wildlife Service, utilizes the data-processing services of the Bird Banding Laboratory at Patuxent Wildlife Research Center in Laurel, Maryland.

In 1992, the president of Mexico created CONABIO, the National Commission for the Knowledge and Use of Biodiversity, which falls under the administrative jurisdiction of SEMARNAP. The commission's operations are directed by a national coordinator, assisted by a staff of about 50 officers, analysts, and support personnel. The fundamental task of CONABIO is to promote and coordinate current efforts by institutions and groups in Mexico along three lines: (1) knowledge of biodiversity, primarily with databases and inventories, of which atlases for Mexico's birds, mammals, and herptiles are noteworthy; (2) sustainable use, such as addressing the legal and illegal traffic in Mexican vertebrates and biodiversity for medicinal purposes; and (3) public awareness, which includes TV programs on coral reefs and other ecosystems in Mexico. CONABIO maintains close contact with key foreign institutions such as the Canadian Museum of Nature, the Smithsonian Institution, and the Missouri Botanical Garden. A private trust fund supports CONABIO and receives grants and contributions from the World Wildlife Fund and similar organizations; the estimated 1996 budget was $2.5 million in U.S. dollars.

A CONTRAST IN EUROPE

Policies in Europe evolved from long centuries of tradition and custom that are without parallel in the United States. Despite the European origins of most Americans, policies and attitudes about wildlife generally were not transferred to the United States by our migrant ancestors. Indeed, most sought refuge from the political and economic strife of their original cultures in a land blessed with new freedoms and unfettered opportunities. Still, we might benefit by tracking the roots of European attitudes and conditions as a means of understanding how others today approach wildlife management. Our discussion also will encompass influences from other parts of the Old World (see Leopold 1933a for more details).

The origins of hunting regulations probably began with the earliest evolution of social history. Tribes that adopted conservative taboos about hunting at certain times or certain types of animals (e.g., females with young) effectively preserved wildlife. In doing so, such tribes prospered, whereas others that indiscriminately exploited animals failed when their food was no longer available. In this way, the taboos were reinforced from generation to generation by the survival of those human societies practicing even these primitive restrictions. Later in history, Moses, in delivering God's charge to the Israelites, decreed that females (in this case, birds) should not be killed while incubating so that breeding stock might be preserved. Only the offspring should be taken "unto thyself" for food. Because this restriction occurs in the Bible, in Deuteronomy 22:6, it probably is the first "game policy" to appear in writing.

An imperial era followed, exerting the greatest influence still found in European attitudes. Although the emperor Justinian during his reign (A.D. 527–565) forbade hunting without permission of the landowner, conservation policies otherwise are unknown in the Greco-Roman culture. During the 13th century, however, Marco Polo found more sophisticated measures enforced in Asia by the Mongol Empire of Kublai Khan. Game animals, including stags, hares, and any large birds, were not killed between March and October, effectively protecting wildlife during the breeding season. In addition to placing restrictions on hunting, the Khan also provided wildlife with winter food and cover.

The earliest policies in feudal Europe—particularly in England—were based largely on custom. Holidays determined the opening and closing of the hunting seasons. Certain classes of animals (e.g., does and fawns) were protected, except when royalty was part of the hunting party. The policies of the day were made by the ruling lords and monarchs and decidedly favored their own purposes. Conflicts between the hungry peasantry and the sport-minded nobility often focused on hunting privileges. Killing of the king's stags was a capital offense. Bows in England were the sporting weapons of the aristocracy and, in a climate where the nobles were fanatically jealous of their hunting rights, a peasant owning a bow thereby labeled himself as a poacher (Howarth 1978). In legend, protection from such villainous public hunting fell to the Sheriff of Nottingham pitted against the likes of Robin Hood and his Merry Men. In fact, the legend portrays the cultural realities of hunting during the Middle Ages.

Still, some measure of conservation policies was present in this period. Henry VIII and later monarchs protected waterfowl, pheasants, and partridges during the breeding season, and the ethical conduct of hunting was enforced. Predator control also was authorized by monarchs of the Middle Ages. Edward, the second Duke of York, wrote that otters should be hunted because of the great harm they cause to fish populations. Bounties were placed on a wide array of predators and nuisance species by a succession of medieval monarchs. James I functionally initiated a refuge for herons when he decreed that weapons should not be fired within 600 paces of their breeding grounds. Artificial propagation of game birds is recorded in the 16th century. Mallards and pheasants were pen-reared for falconry but perhaps also directly for the royal table. The Middle Ages witnessed little in the way of habitat management, although royal policies sometimes favored maintaining forests for the benefit of wildlife. Nonetheless, the forests remained the king's possessions for his hunting pleasures alone. All other privately owned tracts of forests were governed by common law. These eventually were protected by trespass rights so that incentives came forth for private game management. William and Mary, at the end of the 17th century, protected nesting cover from burning in the spring, but little else of habitat management seems evident in this era.

The contrasts we see today between the ways wildlife is regarded and managed in the United States and Europe thus evolved during the course of history. Europe developed under a long regime of monarchs, replete with many lesser ranks of nobility. The nobles held privileges not available to the common person, not the least of which was ownership of land and hunting privileges. In addition, the history of Europe is rich with examples of repeated warfare, so that modern-day Europe is divided into a score of countries, each considerably smaller than the United States. France, the largest nation in western Europe, is smaller than Texas; Poland is about the size of New Mexico; and the area of England equals that of Florida. Georgia alone is the size of the combined areas of Ireland, Switzerland, and the Netherlands. Nonetheless, human population densities in most European nations are far greater than occur in the United States. The former state of West Germany, about the size of Oregon, contained a human population 30 times greater than that state and 10 times the average density of the United States (Gottschalk 1972). On a per capita basis, this means that land is much less available in Europe, and those few large estates remaining have been in the hands of a few families for centuries. However, the advent of socialist governments in eastern Europe led to the confiscation and redistribution of the larger estates; in Poland, some 4500 hunting districts were established after World War II (Taber 1961). Under socialism, some of the larger estates in eastern Europe became government-owned game-management areas.

The overriding results of these situations and events—past and present—are that hunting remains connected with ownership of land, that hunting is more an individual privilege than a traditional right of the public, and that land management is extremely intensive. Hunting and, in turn, wildlife management in Europe therefore are shaped by responsibilities remaining largely in the private domain—not in the public trust. Indeed, hunting is the primary management tool, and, in many ways, wildlife management in Europe can be thought of as "harvest management" instead of as a broader concept based on an ecological footing. As we shall see, European hunters are better thought of as "gamekeepers" because of the role they assume in European wildlife management (Newman 1979).

High standards are imposed in most European countries before hunters can participate in their sport. Rigorous tests are administered by hunting societies and include not only marksmanship, ballistics, and gun safety, but also field identification of wildlife and knowledge of natural history, including diseases, aging techniques, and management practices

(Helminen 1977; Haas 1977). In Czechoslovakia, 60 hours of coursework are followed by a 1-year gamekeeping internship before a prospective hunter is tested (Newman 1979). In 1 year, more than one-third of those Germans taking these tests failed to pass, and less than 0.4 percent of the population are licensed hunters (Gottschalk 1972).

Hunting societies set bag limits on the lands they manage, sometimes adopting individual limits for each member based on that person's participation in management activities (Salo 1976). In Germany and other European countries, the societies file shooting plans (*Abschusspläne*) that determine in advance the annual harvest; once approved the plans must be fulfilled or the surplus of big game may starve the following winter (Webb 1960; Haas 1977). Big game hunters in Poland are issued specific instructions for the species, age, sex, and numbers of animals they are to harvest. At the end of the hunt, each person's success is weighed against these instructions, and points are awarded accordingly. Penalty points also are assessed for poor performances and can lead to the suspension of hunting privileges or payment of fines. This system openly displays each hunter's skills among peers and exerts substantial social pressure for improvement (Taber 1961). In all, private hunting societies in Europe function more like public agencies in the United States. The societies undertake responsibilities for nearly all aspects of wildlife management on the lands they hunt.

Because land parcels generally are small and the human population is large, land management in Europe is highly intensive. Farming and forestry are included, as well as wildlife management. For wildlife, big game management is especially intensive, often focusing on the harvest of individual animals and artificial feeding. Resident species are highly valued. Conversely, migrant birds may be exploited as they are not permanently associated with the local habitat. Ducks and other migrant species usually are subject to "taking while the taking is good," whereas resident game is guarded judiciously against overexploitation (Errington 1961; Isakovic 1970).

Wildlife management perhaps is as intensive in Germany as anywhere (Gottschalk 1972). About 91 percent of the usable land is divided into some 37,000 hunting districts, or *reviers*, but less than 5 percent of these are publicly owned. Most either are owned entirely by a single person or may be the collective properties of several owners. About one-third of the licensed hunters own or lease reviers, with the balance hunting as guests or subleasers. Hunting is highly selective because few persons have access to reviers or have mastered the license requirements. Likewise, the harvest is selective. Trophy animals are prized, and, at its extreme, animals with inferior development are removed, whereas those promising to reach regal proportions are rigorously protected until they reach perfection. This practice also gives the better-quality animals opportunities to breed before they are harvested. Responsibilities for management fall under the direction of the forest master (*Forstmeister*), who also may serve as the hunting master (*Jaegermeister*). Included in their duties are the census of big game with an eye toward composition of the herd (Webb 1960). Age and sex ratios are monitored so that productivity may be assessed each year. Barren females may be singled out for removal whenever possible, as are males of poor trophy value. Careful tallies also are kept of the annual harvest, with some records of this kind dating to the Middle Ages.

In keeping with the overtones of private management, German hunters may not retain more than the head, lungs, liver, and heart of big game they shoot. Instead, the meat is sold in public markets, with the revenues reverting to landowners for further management of the revier. In addition to management costs, each revier employs a professional gamekeeper and assumes responsibilities for any damage wildlife may inflict on forests or farmlands (Nagy and Bencze 1973; Haas 1977). Similarly, pheasant hunters shooting on Scottish estates are awarded only two cocks at the end of the hunt even though each hunter may have bagged many more birds; the rest, other than what is kept by the landowner, is shipped for sale in the larger cities (Peterle 1958).

Such rigorous practices coincide with equally intensive forest management. German forests were managed in a tradition of even-aged stands of a single species, thereby promoting a monoculture. Spruce (*Picea abies*) usually was the desired species, and beginning about 1820, virtually all German forests were converted to this species in a craze known as *Fichtenomania* (Leopold 1936a). Plantings also were made outside the normal range of spruce in the conversion. At maturity, the trees were clear-cut and replaced with rows of seedlings. Forests of mixed species and natural reproduction ceased in Germany under *Fichtenomania*. Such an environment brought forestry into direct conflict with wildlife management. Plant diversity diminished under the closed canopies of the spruce forests so that nearly all of the forbs and hardwood browse for deer no longer existed (Fig. 22-9). Deer, in fact, damaged the spruce, either by eating the bark or by browsing on the tender seedlings. Even so, deer herds were maintained continuously with artificial feeding programs.

Early in the 20th century, the folly of the spruce monoculture became apparent even to foresters who had promoted it. Timber production dropped off in the second and third generations of spruce plantings. The excessive litter of needles did not decay and piled up on the forest floor in a sterile blanket that smothered all natural undergrowth. Without hardwoods, the topsoil increased in acidity and the trees' roots no longer penetrated deeply enough to prevent windfalls. Insects, too, severely damaged the uniform forests of spruce. All told, *Fichtenomania* proved an ecological disaster for forest and wildlife alike (Leopold 1936a).

Dauerwald, or continual woods, has gradually replaced the unnatural forests of spruce in Germany. This practice allows the forest to revert to natural reproduction and a mixture of species—an environment far more suitable for browsing animals as well as for songbirds and other wildlife. But the abuses of the past remain. Deer populations of several species still are maintained by artificial feeding but nonetheless outstrip the browse generated by *Dauerwald*. Preferred species of browse quickly disappear, replaced by those of lesser quality. Permanent fencing prevents deer from entering nearby fields or new plantings so that the animals are confined inside the nearly foodless forests. Artificial feeding is mandatory under these constraints. High-intensity management thereby continues to mark the mixture of forestry and wildlife management in Germany. Winter feeding is often coordinated with installations of temporary fencing (Johnson and Adams 1955). Fences protect small areas of young trees near the feeding stations. After the new growth is established in the protected areas, the fences are removed and relocated elsewhere in the forest for the same purpose. No longer vulnerable to damage, the 6- to 8-year-old trees thereafter offer new sources of browse as well as ensure future timber production. Other means also protect trees from the high densities of browsing animals. Young trees are tarred and older ones are wrapped individually with bundles of sticks so that deer are inhibited from stripping bark (Leopold 1936a). More recently, inexpensive plastic netting has been used as wrapping for the same purpose, especially on the sapling-size ash (*Fraxinus* spp.)

Figure 22-9. Contrasts in German forestry. Monocultures of evergreens, particularly spruce, once were widespread during the 19th and early 20th centuries. Known as *Fichtenomania,* the practice left little diversity in German forests and scarce browse for wildlife (left). Later in the 20th century, the benefits of diversity and uneven-aged trees were recognized (*Dauerwald*), leading to more and better browse in the forest community (right). (Photos courtesy of Wolfgang Schiffer.)

preferred by red deer (*Cervus e. elaphus*) (D. A. Klebenow, personal communication). As another indication of high-intensity management, Gottschalk (1972) witnessed breeding experiments with red deer designed toward reducing body size, thereby lessening consumption of browse and intraspecific competition while retaining maximum antler development. The European quest for red deer with huge antlers is scarcely new, however, as enormous racks were exchanged as princely gifts during medieval times (Geist 1986a).

German policies toward orderliness also include adoption of strong measures against predators. Larger mammalian predators such as wolves and lynx (*Lynx lynx*) are exterminated whenever encountered, with the blessing of the *Jaegermeister.* Even the larger hawks are shot. Methods of predator control often are part of the test for hunting licenses (Gottschalk 1972). Leopold (1936b) estimated that predators were killed in a ratio of 1 for every 2–15

small-game animals harvested. Ecological balance within German forests thus is wanting even with the return to a more natural mixture of vegetation. With artificial feeding, restrictive fencing, and rigorous control of predators, deer populations are subject only to the hunting pressure of the privileged few with access to reviers.

The final contrast we shall mention concerns the traditions of hunting in Germany and other European countries. Many of these originated centuries ago and reflect the modern European hunters' ties with the past. In some instances, garb is traditional, as with the tweed costume worn by British grouse shooters or the dress code of mounted fox hunters (the latter is one of the few customs transferred from Europe to the United States, where it likewise remains a tradition associated with aristocratic U.S. society). German hunters religiously wear green woolens, including ties, and decorate their hats with the "brush" of a previous trophy (Fig. 22-10). They

Figure 22-10.   European hunting is rich in tradition. Hunts are preceded by rituals (left). Note the traditional hunting garb, including the formality of neckties. After the hunt, game is displayed with respect, reflecting the privilege and responsibility associated with hunting in Europe (right). Note that the pheasants and rabbits are displayed over a bier of evergreen boughs. (Photos courtesy of Wolfgang Schiffer.)

also have traditional greetings for one another and a specialized vocabulary when discussing wildlife and hunting (e.g., blood is steadfastly referred to as "sweat") (Webb 1960). Further indication of traditions associated with wildlife in Europe include the ritual of "swan upping" each year on the River Thames. Working from skiffs emblazoned with banners, costumed representatives of the crown and two trade guilds capture young swans with crooks. Each cygnet is marked on its bill as a means of determining ownership, becoming the property of one of the three proprietors according to its parentage. Swan upping—the term has obscure origins—stems from the 15th century when the two guilds received rights from the crown to own swans, making necessary the marking process.

In Hungary, Germany, and elsewhere in central Europe, hunters place fresh twigs in the mouths of big game they kill (Fig. 22-11). This tradition dates to the year 700 when, in a vision, St. Hubertus spoke with an albino stag bearing a crucifix between its antlers. The stag asked why just the best deer were shot and warned that the herd would suffer if only

the inferior animals were left to breed. Believing that God had sent the albino stag to him for a special purpose, St. Hubertus thereafter worked as a priest for better hunting ethics, including the selective removal of weaker animals. He also promoted an attitude toward wildlife as "beings" rather than mere targets for hunters. On his deathbed, he asked that his vestments be placed over his mouth to return to God the soul he had received. St. Hubertus became the patron saint of hunting, and the tradition of the twig recognizes the return to God of the slain animal's soul (Knapp 1977). This and other traditions (e.g., playing of hunting horns) deemphasize the aspect of killing and instead stress the privilege of hunting and generate respect and responsibility for wildlife in European society.

The mixture of historical and present attitudes in Europe and the intensive management it spawned might confuse people of the United States of every persuasion (viz., hunters, antihunters, and nonhunters). Much of the European approach clearly is artificial when compared to the deer herds roaming on public lands in the American West. It also is

Figure 22-11. As part of the hunting traditions in Germany and other central European nations, hunters place evergreen twigs in the mouths of boars and other big game (inset). The ceremony dates to the year 700 and the vision of St. Hubertus. (Photo courtesy of Horst Niesters.)

highly selective if not aristocratic. But the point must be made that the harmony between land, tradition, and harvest in Europe is far removed from the crowds of hunters that scramble afield on opening days in the United States. The quality of hunting ex-

periences and of the game itself is a crucial matter in any assessment of these comparisons. Furthermore, recent demographic evidence indicates that the proportion of hunters in the United States is declining, along with habitat suitable for wildlife or hunting

(Peterle 1977).[2] However, hunting may bring the place of humans in nature to the fore of our national conscience, making hunters effective agents of environmental affairs (Shepard 1959). U.S. hunters may be more responsive than supposed to participation in management if the attitudes of those associated with groups such as Ducks Unlimited, National Field Archery Association, and others are considered (Rich 1977). Bossenmaier (1976) extended these thoughts with the suggestion that hunting in North America should become institutionalized as an ecological experience. The idea seems not far removed from the traditions and mores dominating hunting in Europe. As the human population expands, hunting in the United States may become less a right and more a privilege. Will a gradual evolution occur, leading to systems similar to those in Europe? Indeed, will hunting continue in the United States as a socially accepted institution without establishing more mature ethical foundations like those surviving for centuries in Europe? These questions are posed easily—arriving at answers may be more difficult.

SUMMARY

In 1930, the public policy for wildlife management in the United States emphasized game species. Among other issues, the policy called for reconciliations between the right of free hunting and land economics. Several points in the 1930 policy later were realized, but those dealing with the preservation of wildlife habitat have not been as successful. Changing attitudes and concerns of the U.S. public produced a

new policy in 1973 stressing an ecological ethic for the continued enrichment of humans. A broader base of social responsiveness, not just for hunters, was developed in the new statement.

State and federal agencies managing wildlife resources today have broad responsibilities for many other aspects of environmental quality and outdoor recreation. State conservation agencies now are governed by multimember commissions instead of the politically expedient single commissioner of earlier times, and funding sources for wildlife management have improved greatly as the structure of state agencies has matured. Several federal agencies deal, in part, with wildlife management, but prime responsibilities at the national level lie with the U.S. Fish and Wildlife Service. This agency maintains a widespread system of federal refuges and oversees the management of migratory birds as part of its many other activities.

Conservation policies are influenced by current biological information, law enforcement and compliance, social and economic factors, new technology, and legal jurisdiction. National policies include the Lacey Act, which established the first federal involvement with wildlife management. Two restoration acts, one for wildlife and the other for fisheries, were landmarks establishing reliable funding for research and management in those states sharing 25 percent of the cost. Several land and water policies govern wildlife management on publicly owned lands and/or provide for the purchase of habitat. The National Environmental Policy Act (NEPA) brought sweeping awareness for environmental quality, including concerns for wildlife and wildlife habitat. Virtually all activities proposed by the federal government fell under the jurisdiction of NEPA, requiring an Environmental Impact Statement for any activity influencing the environment. The Wild Free-Roaming Horse and Burro Act illustrates conflicts between sentimentality and environmental management in the course of conducting public affairs.

Public attitudes toward wildlife and wildlife management have changed in recent years. The U.S. public includes many segments, each with varying characteristics and backgrounds and each with varying interests in animals and their welfare. Many people, however, are uninformed about wildlife or wildlife habitat, even though they generally support wildlife management. Some segments of the public treat wildlife on a personal basis, and the concerns of many people are not always the same issues that are addressed by wildlife managers.

[2]The number of hunters is determined in various ways. One of these is a census of persons who purchase hunting licenses, which has remained at about 15.5 million in recent years and therefore represents a declining percentage of the ever-larger U.S. population. In 1996, the U.S. Fish and Wildlife Service reported sales of 15.2 million hunting licenses, which represents 5.7 percent of a U.S. population of 265,282,000. Between 1961 and 1979, hunters represented between 7.3 percent and 7.8 percent of the U.S. population (Williamson 1981). In 1980 and 1981, the percentages were 7.2 and 7.3, respectively, but they began to decline in 1982 and dropped to about 6 percent by 1993. These data, however, include only hunters who buy licenses and omit those who, because of their age, can hunt legally without purchasing licenses. Also, some states sell various types of licenses (e.g., big game, small game, etc.), and additional bias results in the census when a single hunter purchases more than one type of license. Other types of surveys produced estimates of as many as 22 million hunters in 1996 (Ference 1997).

Wildlife management in Europe focuses on the privileges of hunting and employs a self-imposed system of intensive gamekeeping on relatively small areas of private land. Hunters represent a small proportion of the European population; yet because they undergo rigorous training and are guided by strong traditions, European hunters remain an esteemed segment of society. The European system of hunting and land management emphasizes an abiding responsibility for wildlife among those individuals worthy of harvesting its rewards.

CHAPTER 23

CONCLUSION

*Though leaves are many,
The root is one.*

**William Butler Yeats,
"The Coming of Wisdom with Time" (1910)**

In this book, we have described what we believe are the main ideas, concepts, and aspirations of wildlife management. Successes and failures have colored the history of this young profession, and much of what has been presented as fact today may become fancy tomorrow.

New discoveries, based on careful research, are steadily improving our knowledge of wildlife populations and their habitats. We hope that remedies will be forthcoming soon for epizootics of botulism, crop and livestock depredations by wildlife, and the plight of threatened and endangered species. Perhaps, too, rain forests—the most species-rich environments on Earth—may be saved from further destruction before they become only memories of a squandered biological treasure. For that matter, ecosystems of every kind will require conservation and integrated management, else life-support systems for a multitude of organisms surely will dissolve despite our best efforts at single-species management. Our goal must be the maintenance of ecological diversity—the full web of life, not just single strands. Without such a broad focus, we surely will find ourselves in a world with neither deer nor butterflies. Therefore, maturing management policies will continue recognizing the merits of nongame species in communities where game animals constitute only a small component of

our wildlife heritage. We also trust that wildlife will become a greater part of the human experience for all segments of society, no less in suburbia than in the wilderness.

For the short term, wildlife managers will continue their husbandry of whooping cranes and tabulation of game harvests. Others will continue debating the significance of predator-prey interactions and the ecological worth of exotic wildlife. Deer and ducks still will be inventoried, no doubt with better accuracy than now can be imagined, and managers will make further evaluations of crucial wildlife habitats; despite burgeoning costs, some valuable lands surely will enter into the public trust.

Fortunately, some significant and positive changes have occurred in the several years since the first edition of this book appeared. In the United States, the "Sodbuster" and "Swampbuster" provisions of the Food Security Act of 1985 now protect some of our most important resources—soil and wetlands—while concurrently providing many kinds of wildlife with crucial environments. Wolves have been reestablished in some parts of their former range, and alligators are no longer threatened with extinction in many areas. In the fall of 1991, after years of study and controversy, nontoxic shot finally was required for all waterfowl hunting, thereby thwarting further losses of ducks and geese from lead poisoning. Although not without controversy, a government policy of no net loss of wetlands in the United States was initiated in the early 1990s. Our hope, of course, is that these events portend a brightening trend for the immediate future.

We are less sanguine, however, about controlling the growth of our own species and the resulting implications of unrestrained growth on natural resources. For millennia—from the dawn of human presence to the eve of the Industrial Revolution—human populations increased slowly, probably reaching about 500 million by 1650. Then, within two centuries, the number doubled, following the geometric expansion described long ago by Thomas Malthus (1798). Since 1850, the number has increased to 5.9 billion in 1998 (Fig. 23-1). Such growth creates enormous pressures on the natural resources of the Earth.

For humans, carrying capacity has an importance beyond a load based only on numbers. Human culture generates expectations greatly exceeding our needs for survival. Whereas the per capita rate of food consumption for humans and other animals remains unchanging, unlike other animals, each new generation of humans consumes more energy than its predecessor. Eventually it will become difficult enough to feed an ever larger number of humans, but the demand for food pales against supplying the global population of humans with cars, toasters, electric blankets, and other modern conveniences. Attempts at fulfilling such desires turns *Homo sapiens* into what Catton (1987) has dubbed "*Homo colossus*"—a resource-demanding organism that evolved from the cultural pressures of modern society. Just as tractors replaced horses for U.S. farmers two generations ago, so have snowmobiles recently replaced dogsleds for arctic peoples. Air conditioners alone in the affluent United States consume more electrical energy than the *total* amount of electricity used in China (Owen 1985).

The impact of such consumption therefore extends far beyond meeting minimal levels of food and fiber production for the survival of an expanding human population. In short, the resources of our planet cannot provide a broader niche—a U.S. lifestyle—for all humans who now exist, let alone those of the next generation (the present population will have surged by another billion humans by the year 2000). Continued expansion of the human population at some point certainly will extract a price from the quality of life.

The increased consumption of energy per capita can be illustrated by comparisons with the energetic budgets of various species of dolphins and whales (Fig. 23-2). In prehistoric times, with 3 million

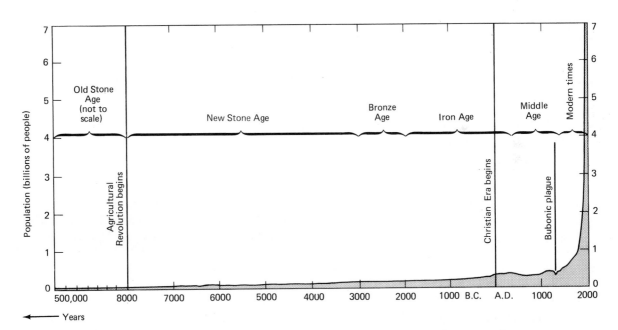

Figure 23-1. Growth of the human population. As shown, human numbers have irrupted into several billions in the last 200 years after a long history of advancing slowly. Because of the geometric progression, the numbers double in progressively smaller periods of time even if the rate of growth remains unchanged. (Compiled from various sources.)

hunter-gatherers, the energy used by a human coincided with that of a dolphin, whereas now each person in the United States consumes the energy equivalent to that used by a large whale (Catton 1987). So we see how the total load on the world's ecosystems has changed because of the immense energy consumption of humans *in addition* to the intake of food by an expanding human population.

What has been the role of the wildlife biologist in this arena? To date, largely complacency. At the 1993 North American Wildlife and Natural Resources Conference, seven papers were presented in session entitled Human Population—The Unblamed Factor. But only two of these suggested—and rather obliquely—that wildlife managers strike at the core of the problem: reducing growth of the human population. Coleman (1993) stated it succinctly:

Today, we fish and wildlife managers are expected to sustain all of our natural resources despite continued loss of habitat quantity and quality to develop-

ment. We manage the remaining natural habitats and fauna intensively in an effort to accommodate as much of the area's original biodiversity as possible. Our management focus usually is on making the best of the situation, while we refrain from addressing the core issue, the population explosion.

Wildlife biologists are generally respected by the tax-paying citizenry, in part because of their understanding of population ecology. Biologists are trained in the fundamental relationships between births and deaths, immigration and emigration, and on a day-to-day basis, they apply these concepts to the management of wildlife populations. Whereas these same principles also affect human populations, they are not well understood by the public at large. Nor does the public always understand how human populations are governed by growth rates, which at first glance may seem quite small. Growth rates of "only" 3–4 percent per year occur in some nations in Africa and Central America, for example, but these seem-

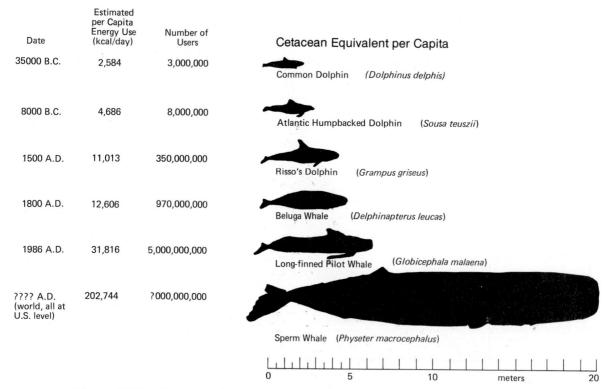

Date	Estimated per Capita Energy Use (kcal/day)	Number of Users
35000 B.C.	2,584	3,000,000
8000 B.C.	4,686	8,000,000
1500 A.D.	11,013	350,000,000
1800 A.D.	12,606	970,000,000
1986 A.D.	31,816	5,000,000,000
???? A.D. (world, all at U.S. level)	202,744	?000,000,000

Cetacean Equivalent per Capita

Common Dolphin   *(Dolphinus delphis)*

Atlantic Humpbacked Dolphin   *(Sousa teuszii)*

Risso's Dolphin   *(Grampus griseus)*

Beluga Whale   *(Delphinapterus leucas)*

Long-finned Pilot Whale   *(Globicephala malaena)*

Sperm Whale   *(Physeter macrocephalus)*

0   5   10   meters   20

Figure 23-2. The changing human load measured in equivalents of dolphins and whales per capita. The load has increased dramatically in energy consumption per capita as well as in population size. Each human once used energy equivalent to a common dolphin (top), but today the average resident of the United States uses as much energy as a sperm whale. So if all humans become as industrialized as are people of the United States, the untenable result is the equivalent of a world populated by 5+ billion sperm whales (bottom). How many such colossal humans can Earth support? (From Catton 1987.)

ingly small rates nonetheless contribute significantly to the *addition* of nearly 100 million each year to the global human population.

Given these circumstances, it seems appropriate for wildlife biologists to take part in the public debate, especially when they are asked to produce more wildlife from an ever-shrinking base of suitable habitat. In other words, because a continuously expanding human population steadily erodes the quantity and quality of land and water resources available for wildlife, questions about the abundance of wildlife and wildlife habitat offer wildlife biologists a prime opportunity to address the overarching issue of human population ecology.

We also must avoid environmental traps wherein consumption exceeds replacement. We have seen the result of maintaining overabundant deer populations with hay in winter. The carrying capacity is temporarily increased and more deer survive in the short run, but eventually the hay is no longer sufficient (i.e., the deer population continues expanding because of the artificial food source and therefore requires more and more hay each winter, hence the trap). The deer population exceeds the capacity of browse to replace itself, and each year, the larger number of deer places progressively more pressure on the remaining browse. Human populations may experience a similar fate when, for example, there no longer is enough water for the irrigation of crops grown on otherwise arid and semiarid lands.

The ecological principles of limiting factors, carrying capacity, energy flow, and diversity and stability clearly apply to humans as well as to wildlife. Human populations in most African and Central and South American nations are doubling every 25–30 years. The world population increases by 83 million people per year—230,000 each day—on a planet with finite dimensions and resources. There are, however, a few bright spots. Since the mid-1980s, several European countries have attained population growth rates close to zero (Population Reference Bureau 1998). The Chinese, recognizing the critical need for balancing their population—now more than 1.2 billion—with agricultural production initiated a national policy of one child per family (Luoma 1984). The population policy has been accompanied by a revived interest in managing and restoring the integrity of those forests and wetlands remaining in China (Ables et al. 1982).

About 58 years ago Leopold (1949) eloquently advocated the adoption of a conservation and land ethic (see also Tanner 1987). With such an ethic, humans exist in harmony with the land, recognize their dependence on the land, and show a respect for the integrity of natural communities. Adoption of such an ethic, however, has been attained only modestly—human demands for resources still take precedence over conservation.

The future of wildlife management, we believe, depends on a matter of far larger dimensions: *whether society shall reconcile the limitations of our planet's natural resources against the eroding pressures of human population.* That issue, more than any other, faces all of us. Indeed, it will be civilization's ultimate test of the conservation ethic.

GLOSSARY

Abiotic: Term describing the nonliving part of an ecosystem; consists of soil, water, air, sunlight, and climate. Producers (green plants) largely depend on the quality and quantity of abiotic conditions (e.g., soil fertility), thereby indirectly influencing higher trophic levels. *See* Ecosystem; Producer; Trophic level.

Abundance: A quantitative measure of resources, particularly food, normally expressed as biomass. Abundance usually should be evaluated concurrently with the availability of the resource to specific organisms. *See* Availability; Biomass.

Acid precipitation ("acid rain"): Rain, fog, or snow with a pH of less than 5.6; forms when emissions (SO_4 and NO_x) from the combustion of fossil fuels, mainly coal, mix with atmospheric moisture and produce sulfuric and nitric acids. *See* pH.

Acid shock: The surge of acid into aquatic systems produced by the melting of winter-long accumulations of acid-bearing snow.

Additive mortality: A concept that the effect of one kind of mortality is added to those of other sources of mortality. For example, if predation takes 10 percent of a population and an ice storm takes 20 percent, the total mortality for the year is 30 percent. If in the next year predation takes 20 percent and an ice storm takes 20 percent for a total mortality of 40 percent, the effects of the two factors are said to be additive. In this concept, hunting mortality adds to the total natural mortality rate of a population. *See* Compensatory mortality.

Adult: An animal, or age class of animals, that has reached breeding age.

Age ratio: The relative proportions of various age groups in a population. May be expressed in several ways: number of juveniles per adult, juveniles per adult female, juveniles per 100 females, or juveniles per pair of adults. Age ratios often are determined from examinations of hunter-killed animals (e.g., wings of wood-

cock, quail, or ducks) and may be used as a measure of breeding success in animal populations. *See* Juvenile; Adult.

Animal Unit (AU): The combined weight of a cow and calf, assumed to be 454 kg. Used to equate grazing pressure by different kinds of grazing animals under the assumption that the same kinds of forage are consumed. Therefore, 1 AU = 7.7 white-tailed deer or 5.8 mule deer.

Annual: In plants, a species that completes its life cycle in one growing season. Many common weeds are annuals. *For comparison, see* Perennial.

Annual increment: *See* Production.

Annulus (plural, Annuli): A mark, usually circular, indicating one year of growth, and therefore a useful measure of age. Accuracy increases or decreases in relation to the distinctiveness of seasonal growth patterns (i.e., north temperate versus tropical environments). Examples include rings on fish scales, trees, and horns.

Antler: One of paired bony structures protruding from the skulls of deer, elk, moose, and caribou (Cervidae). Covered with velvet during development. Shed annually and usually branched; on males only, except caribou. *For comparison, see* Horn; Velvet.

Arboreal: Term referring to trees; for wildlife, a lifestyle adapted to trees. For example, most squirrels are arboreal species.

Arthropod: Any species in the phylum Arthropoda, whose meaning is "jointed legs." Insects, crustaceans, spiders, and ticks are typical arthropods.

Availability: With abundance, part of the ecological equation for measuring food and other resources. For example, earthworms often are abundant beneath logs, but under those conditions, such worms are not available as food for robins or woodcock. Similarly, twigs on tall shrubs and small trees, although abundant, may not be available as browse until deep snows bring the upper branches within the reach of deer and rabbits. *See* Abundance; Browse.

Aversive conditioning: Learning that relies on the stimuli of undesirable experiences (i.e., negative reinforcement). For example, some experiments suggest that coyotes fed mutton treated with distasteful substances may avoid killing sheep.

Avifauna: A term referring to all species of birds occupying a designated area or time. For example, about 540 species of birds make up the current avifauna of Texas. Other examples include Miocene avifauna, Neotropical avifauna, and preglacial avifauna.

Backyard management: A part of urban wildlife management in suburban zones; installation of birdhouses, feeders, and layered vegetation are common techniques.

Bag limit: The legal kill per hunter per day. Bag limits for the same species may vary from year to year, depending on habitat and other conditions. *For fish, see* Creel limit.

Benthic: Term referring to the deep or bottom parts of aquatic environments. Most clams are benthic organisms.

Biological control: Any means of control that takes advantage of a pest's natural vulnerability. For example, increased contact with a species-specific disease or predator, as when rabbits in Australia were inoculated with the disease myxomatosis. *See* Juvenile hormone; Pheromone.

Biological diversity (biodiversity): The variety of life, typically expressed in terms of species richness, but also may be applied to genes and ecosystems. The preservation of biodiversity is the primary goal of conservation biology. *See* Conservation biology; Species richness.

Biomagnification: The accumulation of matter with each succeeding trophic level, bottom to top, in an ecosystem. Harmless levels of DDT applied at the lower trophic levels thereby were amplified to harmful levels in the bodies of bald eagles and other carnivores. *See* Chlorinated hydrocarbon.

Biomass: The weight (usually dry weight) in a defined part of an organism, population, or community. Often used for comparing trophic levels (e.g., the biomass of herbivores per unit of area is greater than that of carnivores). Dry weight better indicates the magnitude of biomass, but wet weight is more convenient to determine.

Biome: A worldwide grouping of similar communities, commonly described by dominant vegetation. The Grassland biome includes the Great Plains (North America), steppes (Eurasia), and pampas (South America). Deserts, Tundra, and Forests are, among other biomes, each subject to refinements (e.g., Deciduous Forest biome, Alpine Tundra biome, and Hot Desert biome).

Biosphere: The zone of life on Earth, a region usually considered to range from a few hundred meters above ground occupied by high-flying birds to a few meters below ground level where burrowing animals occur. Species living on the ocean floor extend the biosphere by several kilometers.

Biota: The term combining flora (plants) and fauna (animals).

Biotic community: *See* Community.

Bottleneck: *See* Genetic bottleneck. Also used to describe periods, typically in winter, when food or other resources are limiting, usually to the point of markedly increasing mortality in wildlife populations.

Brood: A family of young birds from a single mother; sometimes applied to fishes and reptiles. *For comparison, see* Litter.

Brood parasite: *See* Nest parasite.

Browse: Woody vegetation consumed as food. Twigs and buds are important winter browse for deer in many places.

Buffer species: A food species of secondary preference that gains importance under adverse conditions. Buffer species usually are harder to obtain and/or are less palatable than primary foods but form an alternate diet when major foods are scarce or absent. Ruffed grouse become a buffer species when lynx can no longer find snowshoe hares. A rarer meaning of the term is applied to a less-prized game animal. For example, when mallards are not abundant or have a small bag limit, duck hunters instead may shoot blue-winged teal as a buffer species.

C: Celsius, the metric unit of temperature (0.01 of the heat interval between freezing and boiling water); 0°C equals 32°F; 100°C equals 212°F.

Carbamate: A chemical belonging to one "family" of insecticides; contains carbon, hydrogen, and nitrogen. These vary in their toxic effects on vertebrates. Temik is a well-known—and highly toxic—carbamate. Most carbamates are not persistent and therefore do not biomagnify in food chains. *See* Biomagnification.

Carnivore: In general usage, a meat-eating animal or, rarely, plant. Mountain lions, owls, and pitcher plants are examples based on diet. The term is ambiguous because it also may refer specifically to members of the mammalian order Carnivora, some of which (e.g., bears, raccoons, and coyotes) at times eat plant foods and so are functional omnivores in spite of their taxonomic affinities. Cats, dogs, weasels, and seals are included in Carnivora. *For comparison, see* Herbivore; Omnivore.

Carrying capacity: The maximum population an environment can sustain without causing damage such as overbrowsing. Measured in terms of biomass or number of animals per species per unit area.

Cecum (plural, ceca): A dead-end outpouching of the digestive tract. Prevalent in fishes, birds, and mammals with high-fiber diets (e.g., spruce grouse, with a winter diet of evergreen needles). The human appendix probably is a rudimentary nonfunctional cecum.

Census: A complete count of a population, as when winter flocks of waterfowl are counted from aircraft. Censuses are not possible for most species. *See* Estimate; Index.

Chill factor (wind-chill index):   A measure of cold stress that combines temperature and wind velocity. For example, at 16 km/hr and 0°C, the chill factor is the equivalent of − 8°C.

Chlorinated hydrocarbon: A chemical belonging to one "family" of insecticides; contains carbon, hydrogen, and chlorine. DDT is the best known. Often persists for years in the environment and, as a fat-soluble compound, may accumulate in tissues. Although only mildly toxic to vertebrates at the time of application, these compounds often undergo biomagnification and thereby can harm nontarget organisms at the top of food chains (e.g., bald eagles). *See* Biomagnification; DDT.

Circadian rhythm: The regular fluctuations in bodily functions (e.g., temperature) and behavior (e.g., sleeping) during a cycle approximating 24 hours.

Clean farming: The elimination of diversity and interspersion on farmland and substitution of a monoculture. Hedgerows, roadside vegetation, and similar areas are eliminated. Instead, fields of single crops are cultivated from border to border without cover for wildlife. *See* Monoculture; Interspersion.

Clear-cut: A forestry practice that cuts all trees at one time from a block of forest.

Climax: The ultimate expression of vegetational development under prevailing local or regional conditions. Climax communities therefore are the last stage of succession and remain in dynamic equilibrium unless disturbed; that is, individual plants of the same species come and go, but the composition of the climax community remains the same. Oak-hickory forests represent one kind of climax community in the Deciduous Forest biome. *See* Community; Biome; Succession.

Clutch: The eggs laid by a bird or reptile in a single nesting attempt.

cm: Centimeter, a metric unit of length (.01 meter); 2.54 cm equals 1 inch.

Cohort: Animals born or hatched in the same year. Members of the same cohort therefore experience the same risks and benefits, which may be different from those experienced by earlier or later cohorts.

Community: The living part of an ecosystem. Communities change with succession, thereby forming distinctive ecological units both in time and space. The plant community and the animal community together form the biotic community. Size is not implied (i.e., organisms associated with a decaying log or with an entire forest each represent communities). *See* Ecosystem; Succession.

Compensatory mortality: The concept that one kind of mortality largely replaces another kind of mortality in animal populations. In simple logic, an animal dying from one cause (e.g., hunting or disease) cannot die from another cause (e.g., predation or starvation), so one source of mortality compensates for the other. Therefore, the total mortality rate normally is not greatly influenced by changes for any single cause of death. For example, if predation takes 10 percent of a population and disease takes 20 percent, total natural mortality for the year is 30 percent. If in the next year predation takes 25 percent and disease takes 5 percent, the effect of predation is said to be compensatory. Of importance, the total annual mortality rate may remain essentially unchanged with or without legal hunting. *See* Additive mortality.

Compensatory population growth: Equivalent to density-dependent population growth; that is, growth of a population is rapid when numbers are low and less rapid when numbers are high. Such growth may result from decreased mortality, increased natality, or both.

Compound 1080:   Sodium monofluoroacetate, commonly used as a coyote poison before restrictions were imposed in the United States in 1973. Considered dangerous because of nontarget victims. *See* Secondary poisoning.

Condition index: A measure of an animal's well-being, usually expressed in terms of fat content. Most condition indices are adjusted for size differences among individuals by dividing fat weight by some other physical feature (e.g., wing length for birds). In mammals, fat deposits in bone marrow or around kidneys are common condition indices.

Conifer: A cone-bearing tree. Pine, spruce, and cedar are common examples. Most are evergreens, but larch (tamarack) is a coniferous species that drops its needles in autumn. *For comparison, see* Deciduous tree.

Conservation biology: A field of many disciplines united with the common goal of preserving biodiversity. Genetics, physiology, geography, population biology, wildlife management, forestry, and veterinary science are among the basic and applied disciplines contributing to conservation biology; the professional staffs at many zoos also provide crucial expertise. *See* Biological diversity.

Conservation tillage: An all-encompassing term for minimum tillage, no tillage, zero tillage, and other kinds of farming practices that eliminate or reduce plowing. Newly planted croplands are protected by at least 30 percent cover formed by residue from the previous crop. Besides enhancing soil and water conditions, conservation tillage improves habitat, especially cover, for several species of wildlife. *See* Tillage.

Consumer: A member of the animal community. Consumers occupy the higher trophic levels in an ecosystem. *See* Primary consumer; Secondary consumer; Trophic level.

Consumptive use: Removal and alteration of natural resources by humans, as when wildlife is hunted, trees are cut, or ores are mined. *For comparison, see* Nonconsumptive use.

Coprophagy: The practice of eating feces. Coprophagy enables rabbits to recover nutrients from their droppings that escaped initial digestion.

Corpora lutea: *See* Corpus luteum.

Corpus luteum (plural, **Corpora lutea**): A structure formed in the mammalian ovary from the follicle that once contained an ovum (egg); functions as an endocrine gland by secreting hormones to maintain pregnancy. Corpora lutea, which are visible when an ovary is sectioned, indicate the number of ova (eggs) shed by the ovary, thereby providing data on fertility. *See* Fertility.

Corridor: A strip or block of habitat connecting otherwise isolated units of similar habitat that allows the dispersal of organisms and the consequent mixing of genes.

Coyote getter: A gunlike device designed to fire cyanide into the mouth of a coyote; triggered when the animal bites on a scented lure.

Creel census: A means of determining statistics about fish and fishing based on examinations of catches and interviews of anglers in the field. Species and size of the catch are recorded, and scale samples often are collected for analysis of age structure. Fish caught per fishing hour is a common statistic calculated from a creel census. (A creel is a wicker basket in which an angler carries a catch of trout or other species.)

Creel limit: The legal fishing kill per day. Creel limit in fishing is the equivalent of bag limit in hunting.

Crop: Anatomically, an expandable part of a bird's esophagus used for food storage and, perhaps, a small amount of digestion. Analysis of crop contents helps determine the diets of birds because foods in the gizzard (stomach) often are ground beyond recognition. Crop also refers to the harvested part of a plant or animal population. *See* Gizzard.

Culling: Removal of animals from a population, usually individuals at risk. For example, predators normally remove sick and injured individuals from a prey population. In some cases, culling may apply to the general thinning of a population, as when the density of foxes or raccoons is reduced for control of rabies.

Cutting cycle: A forestry term reflecting the number of years between cuts. The frequency varies in relation to species, growth rate, and management objectives.

Cycle: In animal populations, the more-or-less regular fluctuations in numbers. Some species of wildlife show 3-year cycles (e.g., lemmings) and 10-year cycles (e.g., ruffed grouse).

DDT: Chemical abbreviation for dichloro-diphenyl-trichloroethane, perhaps the most familiar pesticide in the "family" of chlorinated hydrocarbons; banned in the United States in 1972, in part because of its harmful effects on wildlife. *See* Biomagnification; Chlorinated hydrocarbon.

Deciduous tree: A tree that drops its leaves each autumn. Usually broad-leaved trees of which maple, oak, birch, and hickory are examples. *For comparison, see* Conifer.

Decomposer: One of many organisms, principally bacteria, that reduce animal wastes and the carcasses of complex organisms into elemental components. Decomposers form a trophic level in the ecosystem that makes the remains of dead organisms available to green plants. *See* Ecosystem; Trophic level.

Deer yard: *See* Yard.

Den tree: A tree with a cavity or hollow in which mammals (e.g., raccoons) sleep and/or rear their litters. *See also* Snag.

Density-dependent factor: A factor that acts in proportion to the density of animals. Some diseases are density dependent because a higher percentage of the population becomes infected as density increases. Natality and mortality often fluctuate with changes in density.

Density-independent factor: A factor that acts independently of population density. Weather is often considered density independent. A flood, for example, may kill an entire population regardless of its density.

Dispersal: Movement of organisms into unfamiliar locations; behavior usually associated with younger animals upon leaving natal areas. *Pioneering* is a synonym. *For comparison, see* Dispersion.

Dispersion: The spatial pattern of organisms. For example, many desert plants are spaced in ways that reduce competition for water. *For comparison, see* Dispersal.

Drawdown management: A practice that alternates wet and dry cycles in wetland habitats to improve food and cover for wildlife. Exposed soils undergo oxygenation, thereby promoting renewed fertility when water levels are restored. *For a related concept, see* Reservoir effect.

Ecological amplitude: *See* Range of tolerance.

Ecological equivalents: Organisms occupying the same niche but living in different communities. Examples include bison and wildebeest occupying grazing niches in the plains of North America and Africa, respectively. Tigers and jaguars are another pair of ecological equivalents. *See* Niche.

Ecology: The study of interactions between organisms and their environment. Term coined by German zoologist Ernst Haeckel in 1866 based on the Greek root *oikos*, meaning "home," and *logos*, meaning "study."

Ecosystem: Any area or volume in which there is an exchange of matter and energy between living and nonliving parts; that is, the biotic community together with soil, air, water, and sunlight form an ecosystem. Ecosystems are the best units for studying the flow of energy and matter. Trophic levels are the functional parts of an ecosystem. *See* Abiotic; Community; Trophic level.

Ecotone: In terms of space, the area where adjacent communities blend. Some change rapidly, but most ecotones blend gradually, thereby forming species-rich zones (e.g., marsh edges, forest openings, and fencerows). In terms of time, a midpoint in succession when species of the outgoing and incoming seral stages temporarily mingle. *See* Interspersion.

Edaphic: An adjective pertaining to soil (e.g., concerning texture, drainage, and fertility). Edaphic factors often are of direct importance to burrowing animals, but also exert major indirect influences on all wildlife because of ecological links with vegetation (e.g., recall the expression "good soil, good browse, good deer").

Edge: The contact zone between two different types of habitat. Edge increases as interspersion increases. To illustrate, there is proportionately more edge between the red and black squares on a checkerboard than if the same-size board were divided into halves of solid red and black. The total area of black and red is the same in either case, but the checkerboard has a far greater edge because the colors are interspersed. For many species of wildlife, additional edge improves the quality and quantity of habitat because each edge forms an ecotone. *See* Interspersion; Ecotone; Edge effect.

Edge effect: The ecological result of increasing edges in homogeneous habitats, principally the increased abundance and diversity of species. A benefit in the management of some animals (bobwhite and other "edge species"), although increased predation or nest parasitism may result in other cases. *See* Edge.

Endangered species: Legal status declared by governments for taxa or populations facing extinction. The designation provides such groups with enhanced protection and management. *See* Threatened species; Taxon.

Environmental Impact Statement (EIS): In the United States, a report mandated by the National Environmental Policy Act (NEPA) that reviews the potential for environmental damage and, if necessary, proposes alternative actions. Required for all projects using federal funds or requiring federal permits (e.g., airports, power plants, and highways).

Enzootic: The chronic level of disease frequency, that is, a low but constant occurrence of a disease in a population. *For comparison, see* Epizootic.

Epizootic: An outbreak of disease. Large numbers of animals die in a short period. *For comparison, see* Enzootic.

Epizootiology: The study of disease ecology. Addresses the "how" and "why" of diseases at either enzootic or epizootic levels.

Esophagus: A tube in the digestive system of vertebrates that connects the mouth with the stomach. Expands in some birds for temporary food storage. *See* Crop.

Estimate: A result of a statistical sample, often for determining population size. Used instead of a census or index. To obtain an estimate, animals or inanimate objects (e.g., nests or dens) may be counted on one or more sample areas known as plots. *See* Census; Index.

Ethology: The study of animal behavior.

Eury-: A prefix used to identify wide tolerances for specific components of the environment. White-tailed deer are eurythermal because of their wide tolerance of temperature extremes. *For comparison, see* Steno-.

Eutrophic: Term describing the enriched nature of some freshwater lakes and rivers. Organic materials and nutrients accumulate, leading to increased biological productivity. The enrichment process is known as eutrophication. Human activities may affect the process abnormally. *For comparison, see* Oligotrophic.

Even-aged management: A forestry practice in which trees of the same age and height form the forest, often the result of clear-cutting. *For comparison, see* Uneven-aged management.

Exotic species: An organism introduced—intentionally or accidentally—from its native range into an area where the species did not previously occur. Asian animals such as pheasants, sika deer, and carp are exotic species in North America.

Extinction: The complete loss—forever—of a unique constellation of genes known as a species (e.g., passenger pigeon). *For comparison, see* Extirpation.

Extirpation: The elimination of a species from one or more specific areas, but not from all areas. Extirpation should not be confused with extinction. Bison have been extirpated from most of their former range in North America, but many hundreds still exist on private ranches, zoos, and federal refuges. *For comparison, see* Extinction.

Fauna: A combined term for all animal life in an area. *For comparison, see* Flora; *also* Biota.

Featured-species management: A management policy keyed to a single species, perhaps at the expense of others. Often used for endangered species (e.g., forest management for red-cockaded woodpeckers) but also applied to more-abundant species.

Fecundity: A measure of reproduction. The number of eggs or sperm produced per adult. In practice, usually refers only to eggs.

Feral: A term referring to normally domestic animals that have escaped and become wild. Pigeons (rock doves), burros, hogs, cats, and dogs are examples.

Fertility: A measure of reproduction. The percentage of eggs that are fertile.

Fitness: The competence of an organism to pass on its genes to the next generation. All else being equal, some individuals within a population are of greater fitness than others. Fitness may be influenced by many factors, including physical, physiological, and social conditions.

Flora: A term for all plant life in an area. *For comparison, see* Fauna; *also* Biota.

Food chain: The step-by-step passage of matter and energy (food) through an ecosystem, beginning with vegetation and ending with carnivores. For example, aquatic plant–mayfly larva–minnow–trout–bald eagle represents a food chain. *See* Food web.

Food habits: A term for the diet of wildlife. Thorough studies of food habits include feeding behavior and food availability as well as quantifying items in the diet.

Food web: A network of interlocking food chains. Most ecosystems are complex, and matter and energy (food) move through the trophic levels in any of several food chains. *See* Food chain.

Forb: A broad-leaved herbaceous plant, often of importance as food and cover for wildlife. Dandelions, milkweed, sunflowers, and thistles are forbs. Grasses, while also herbaceous, have linear leaves and other features that contrast with forbs. *See* Herbaceous.

Fossorial species: Animals adapted for living underground. Moles and pocket gophers are examples.

Founder effect: The peculiar genetic composition of a population resulting, by chance, from the restricted gene pool of the population's founders. For example, if only blue-eyed individuals happened to reach an uninhabited island or other isolated patch of habitat, then the resulting population would reflect this trait of its founders.

g: Gram, a metric unit for the weight of 1 cm^3 of water; 1 g equals 0.035 ounce.

Gallinaceous bird: A bird of the order Galliformes. These are chickenlike birds, including pheasants, quail, grouse, prairie chickens, ptarmigan, and turkeys.

Gallinaceous guzzler: A structure that collects drinking water for wildlife in arid environments. Quail initially were the target species, but later designs serve many other kinds of desert wildlife.

Genetic bottleneck: The temporary reduction of a population to only a few individuals, thereby limiting the gene pool and increasing inbreeding. When such a population later increases in size, it will still have a limited gene pool.

Gene pool: The combined reservoir of genetic material, usually expressed for a specified population at a given time and place. A normally heterozygous pool may be greatly diminished by repeated inbreeding. *See* Heterozygosity; Inbreeding depression.

Gizzard: The muscular stomach of birds that grinds food, usually with the aid of grit. Gizzards serve as the functional equivalent of teeth in mammals. *See* Crop.

Gravid: Term describing females containing ripening eggs. Major physiological differences aside, the gravid condition in egg-laying species is somewhat analogous to pregnancy in mammals.

Group-selection cut: A forestry practice that cuts small groups of trees. *For comparison, see* Single-tree selection cut.

ha: Hectare, a metric unit of area (10,000 m^2) equaling 2.47 acres.

Habitat: The physical and biological resources required by an organism for its survival and reproduction; these requirements are species-specific. Food and cover are major components of habitat.

Habitat Evaluation Procedure (HEP): A method of the U.S. Fish and Wildlife Service that documents the quality and quantity of resources available for selected species of wildlife; involves estimation of a Habitat Suitability Index (HSI) and the amount of such habitat available. HEP numerically predicts the effects of altering habitats by development or by natural events. *See* Habitat Suitability Index; Habitat Unit.

Habitat Suitability Index (HSI): Part of habitat evaluation procedure. A value ranging from 0.0 to 1.0, assigned on the basis of known food and cover requirements of a species. HSI is multiplied by the area of habitat (in acres) to obtain the Habitat Units (HUs) available for each species. *See* Habitat Evaluation Procedure; Habitat Unit.

Habitat Unit (HU): Part of habitat evaluation procedure. The product resulting from Habitat Suitability Index (HSI) × the number of acres available with that HSI. Used for assigning a comparative value for habitat of various species. In a 150-acre marsh, for example, 100 acres of marginal habitat (HSI of 0.2) for blue-winged teal equal 20 HUs (100 × 0.2) for teal. The other 50 acres represent good teal habitat (HSI of 1.0), thereby equal to 50 HUs. Overall, the marsh represents 70 HUs for blue-winged teal. Summation of such values for all wildlife species in the marsh constitutes the Habitat Evaluation Procedure. *See* Habitat Evaluation Procedure; Habitat Suitability Index.

Habituation: A behavioral tendency wherein animals become accustomed to unnatural components in their environment. Deer feeding at the edges of highways often become habituated to traffic.

Halophyte: A plant adapted to saline conditions (e.g., salt flats or coastal areas subject to ocean spray). Sea oats and saltmarsh hay are halophytic grasses.

Heat capacity: The property of absorbing heat without becoming much warmer. Water has a very high heat capacity, which helps to stabilize body temperatures in animals facing extreme environmental conditions.

Helminth: A general term referring to any of several kinds of worms (e.g., roundworms, flatworms, and segmented worms).

Herb: Term for forbs and other nonwoody plants. *See* Forb.

Herbaceous: Term descriptive of nonwoody plants. Herbaceous vegetation includes forbs and grasses (except bamboo). Also, the nonwoody parts of trees and shrubs (e.g., leaves). *See* Forb.

Herbicide: A chemical that kills unwanted vegetation, widely used to protect crops from competition.

Herbivore: A plant-eating animal. Rabbits, deer, and carp are examples. *See* Primary consumer.

Herpetofauna: Term combining the amphibian and reptilian members of a community. *Herptiles* (or *herps*) likewise is a term combining amphibians and reptiles.

Heterozygosity: The presence of both forms (known as *alleles*)—dominant and recessive—of a single gene in an individual or population. Heterozygosity generally conveys an advantage for coping with new situations. An individual with a genetic composition (genotype) represented as AaBbCc is heterozygous, whereas another individual of the same species with AAbbCC or aaBBCC is homozygous. In a population, heterozygosity provides a variety of genotypes, thereby increasing the likelihood that some individuals are capable of surviving stresses acting on the population and thereby increasing fitness. *See* Fitness; Gene pool. *For comparison, see* Homozygosity.

Hibernaculum (plural hibernacula): The winter den of reptiles, particularly snakes.

Homeothermic: Term indicating maintenance of stable body temperatures independently of environmental temperatures. Homeotherms—mammals and birds—can produce metabolic heat to meet the stress of low ambient temperatures. "Warm blooded" is a popular description of homeothermic species. *For comparison, see* Poikilothermic.

Home range: The area included in the daily, seasonal, and annual travels of an individual animal.

Homozygosity: The limited presence or total absence of either the dominant or recessive form (allele) of a gene in an individual or population. The genes of a homozygous individual may be represented as AAbbCC or AABBcc. A population composed mainly of homozygous individuals lacks genetic diversity, thereby reducing the likelihood that some individuals are genetically more capable of surviving stresses than those in a heterozygous population. Therefore, a homozygous population is generally less able to survive over a long period of time than a heterozygous population. *See* Fitness; Gene pool. *For comparison, see* Heterozygosity.

Hormone: An internal chemical secretion of plants and animals that regulates a variety of functions (e.g., growth, sexual activity, and incubation). In animals, hormones originate from ductless glands (e.g., thyroid, ovaries, testes, and pituitary).

Horn: A structure protruding from the skulls of goats, sheep, and bovines (antelope, cows, and bison) consisting of a keratin sheath surrounding a core of bone. Usually paired, except in rhinoceros. There is no velvet covering during development. Rarely branched or shed, but pronghorns are notable exceptions. Horns occur in both sexes. *For comparison, see* Antlers; *also* Velvet.

Immature: An animal, or age class, incapable of breeding and not looking like an adult; usually young-of-the-year. *Juvenile* is a synonym.

Imprinting: A type of permanent learning acquired during a critical period in the behavioral development of an animal. Especially well-developed in some species of birds. Young waterfowl imprint on their mothers during the first few hours after hatching.

Inbreeding depression: The undesirable result of repeated matings within a small population of related individuals, typically reducing the gene pool and producing abnormalities and lessened fitness in the offspring. *See* Fitness; Gene pool. *For comparison, see* Outbreeding depression.

Index: A means of comparing the relative size of an animal population, usually from year to year, but it does not provide a census or estimate of actual numbers. Roadside drumming counts of ruffed grouse provide a useful population index. *See* Census; Estimate.

Indicator species: A *key* organism, usually a plant, that serves as an indicator of ecological conditions. For example, some plants grow only on disturbed soils, others only where the water table is high, and still others only after a fire. Range managers often use the abundance of indicator species to determine grazing pressure.

Inherent: In ethology, a behavioral characteristic that is not learned. *Innate* and *instinctive* are synonyms.

Innate: *See* Inherent.

Instinctive: *See* Inherent.

Integrated Pest Management (IPM): A combination of chemical, cultural, and biological methods designed to minimize pest damage. For example, IPM recognizes the beneficial role of predaceous insects in the control of herbivorous insects.

Interspersion: The mixture of habitats in a patchwork pattern. High interspersion means a high ratio of edge to area, producing ecotones and therefore a situation favored by many species of wildlife. Monocultures offer little or no interspersion. *See* Ecotone; Edge; Edge effect; Monoculture.

Invertebrate: Any animal without a vertebral column (i.e., backbone). Insects and crustaceans are particularly abundant groups of invertebrates.

Irruption: The sudden and massive increase in the size of an animal population. Locust irruptions ("plagues") are well-known examples. Synonymous with *eruption* (as with volcanos).

Island biogeography: The study of natural communities on islands, with an emphasis on species diversity as related to an island's area and its distance from the mainland. Principles apply not only to actual islands, but also to fragmented forests and other patches of habitat.

Juvenile: *See* Immature.

Juvenile hormone: A naturally produced hormone associated with the immature, or larval, stages of insect development. When synthesized, however, juvenile hormones present a powerful biological tool for arresting populations of harmful insects and so may substitute for toxic insecticides; that is, when sprayed on crops, the hormone prevents larval insects from maturing into breeding adults (without threatening nontarget species).

kg: Kilogram, a metric unit of weight (1000 g) equaling 2.2 pounds.

km: Kilometer, a metric unit of length (1000 m) equaling 0.6 miles.

K-selected species: Species adapted to low birthrates, low death rates, and long life spans. Depressed populations of such species recover slowly and often have specialized habitat needs. As a result, overkill and other disasters easily jeopardize *K*-selected species. Whales, elephants, and whooping cranes are examples. *For comparison, see r*-selected species.

l: Liter, a metric unit of volume (1000 cm^3) equaling about 1.1 quarts.

Lagomorph: Any member of the mammalian order Lagomorpha, which includes rabbits (cottontails), hares, and pikas. Often mistaken for rodents, lagomorphs have two pairs of upper incisors, one small set directly behind the larger front pair.

Landscape: An aggregate of different but interacting landforms, sometimes united by a cultural attribute (e.g., a mosaic of farmland, including tilled fields, woodlots, stock ponds, swales, and fencerows). Landscape ecology generally operates at a scale of at least many hectares or, more often, several square kilometers.

Laterite: An iron-rich soil that can harden into bricklike pavement when denuded of vegetation and exposed to sunlight; commonly develops in association with tropical rain forests.

Lek: The display ground for certain birds, typically prairie chickens. *Booming* or *dancing ground* is synonymous.

Lichen: A symbiotic plant consisting of part alga, part fungus. Some lichens live on rocks, where they appear as paintspots, whereas others are fibrous and drape from trees or form dense ground cover. Several of the fibrous species are staples in the diet of caribou. *See* Mutualism; Symbiosis.

Life form: The mode of existence of an organism, including its shape, size, feeding, and shelter-seeking behavior. Among plants, life forms include annuals, biennials, and herbaceous and woody perennials (e.g., trees have a common life form, as do grasses). Among animals are sessile forms such as corals and various locomotory means such as swimming, burrowing, leaping, and flying.

Life-form association: A group of organisms whose requirements are satisfied by similar successional stages in the development of communities.

Litter: A family of young mammals from a single mother. Also, the ground cover of fallen leaves and other organic matter. *For comparison, see* Brood; Mast.

m: Meter, the basic metric unit of length, equaling 39.4 inches.

Mast: A collective term for woody fruits—most often acorns and other nuts—and especially those fallen to the ground where they then become available for turkeys, deer, and other wildlife. *See* Availability; Litter.

Maximum yield: The largest number of animals or plants that can be harvested from a population and maintain the population at a stable level; usually refers to the harvest of fishes, furbearers, birds, and mammals. *For comparison, see* Optimum yield.

Metapopulation: Naturally or, more commonly, artificially isolated units of a larger population. These occur in patches separated by uninhabitable areas across which movements and exchanges, if any, are generally limited to corridors. A roadway dividing a wetland may create metapopulations of frogs and turtles.

Micronutrient: One of several minerals necessary for plant and animal growth but required in minute amounts. Primarily functions as a catalyst in enzyme systems. Zinc, copper, cobalt, and iron are examples. *Trace element* is a largely outdated synonym.

Migration: The periodic movement to and from a breeding area, but the term often is used loosely for other types of movements. Migration recurs each year in individuals of many species, but other species make only one round trip before dying (e.g., Pacific salmon). Other patterns occur (e.g., monarch butterflies) that complicate a universal definition. Altitudinal migration involves changes in elevation (e.g., some grouse), whereas latitudinal migration involves changes in latitude (e.g., the typical north-south migration of geese in North America).

Migrational homing: The return to a site of previous experience, usually a breeding area. Female ducks of many species, for example, return to the same breeding area where they nested in the previous year. The term is often shortened to *homing.*

Minimum tillage: *See* Conservation tillage.

Mitigation: The replacement, usually by substitution, of wildlife habitat lost to reservoirs or other large-scale land and water developments.

ml: Milliliter, a metric unit of volume (0.001 l); 30 ml equals 1 fluid ounce.

mm: Millimeter, a metric unit of length (0.001 m); 25 mm equals 1 inch.

Model: An abstract representation of a real situation; may be graphic, mathematical, or both. A population model depicts numbers, birthrates, death rates, immigration, and emigration as influenced by environmental factors.

Monoculture: A term describing unbroken expanses committed to a single crop, whether in forest or farmland; that is, interspersion is lacking. Highly urbanized areas ("downtown") are paved monocultures. *See* Clean farming; Interspersion.

Monogamy (monogamous): A type of mating behavior in which one male unites with one female, forming either seasonal (e.g., most ducks) or lifetime (e.g., geese and swans) pair bonds. *For comparison, see* Polygamy.

Multiple-use management: A concept of land management that integrates various activities and natural resources. For example, many national forests provide a combination of skiing, hunting, fishing, hiking, canoeing, camping, mining, and grazing in addition to lumbering and watershed protection. Wildlife represents a major component in multiple-use management.

Mutualism: A common type of symbiosis whereby two species share a mutually advantageous association. Tick-birds, for example, feed on parasites infecting the skin of rhinos without fear of disturbance. *See* Symbiosis; Lichen.

Mycorrhiza: The association of fungi with roots of trees, grasses, and crops. The fungi increase the availability of several essential nutrients (e.g., phosphorus, iron, and nitrogen) even in poor soils, whereas the higher forms of plants apparently supply the fungi with sugars and other foods. *See* Symbiosis.

Natality: Birthrate. Expressed in several ways, but often as the number of offspring per female per year or per 100 females per year.

Nest parasitism: The act or result of one species (the parasite) of bird laying its eggs in the nest of another (the host), after which the host raises the parasite's young at the expense of its own offspring. The brown-headed cowbird is a common nest parasite in North America.

Niche: The functional role of an organism considered in the multidimensional environment (i.e., time, space, food, water, and cover) in which it lives. In North America, for example, pronghorn, jackrabbits, and bison fill separate grazing niches in grassland communities.

No tillage: *See* Conservation tillage.

Nonconsumptive use: Use, without removal or alteration, of natural resources. Bird-watching and nature photography are typical examples. Catch-and-release fishing is a special case. *For comparison, see* Consumptive use.

Nonpasserine: Term applied collectively to all birds other than those in the order Passeriformes. *See* Passeriformes.

Odd area: A farm site where poor drainage or another shortcoming prevents tillage and crop production, thereby providing potential habitat for wildlife.

Old-growth forest: A virgin forest; a climax forest that has not been cut and contains dead and decaying trees. Most old-growth forests in North America occur in the Pacific Northwest.

Oligotrophic: A term describing the relative infertility of some freshwater lakes and rivers. Geologically immature, such waters are deep, clear, and cold and lack organic materials and nutrient accumulations. Biological productivity is low. *For comparison, see* Eutrophic.

Omnivore: An animal that eats both plant and animal foods. In season, the diet of grizzly bears includes berries and salmon, and coyotes eat rabbits and cactus fruits. In some species, juveniles eat animal foods but switch as adults to plant foods (e.g., mallards and bobwhite). Omnivores are not limited to any taxonomic group. *See* Carnivore; Herbivore.

Optimum yield: The number of animals or plants that can be harvested from a population, taking into account not only biological potential of the population (i.e., production), but also human demands, including aes-

thetics, ethics, and preservation. *For comparison, see* Maximum yield; Sustained yield.

Organophoshate: A chemical in one "family" of insecticides; contains carbon, hydrogen, and phosphorus. Parathion and Malathion are examples. Most are highly toxic but degrade rapidly and do not biomagnify. *See* Biomagnification.

Outbreeding depression: The undesirable result of crossings between individuals from separate populations with incompatible traits; the offspring lack fitness (e.g., young ibex born in winter instead of spring; see text). *See* Fitness. *For comparison, see* Inbreeding depression.

Passeriformes: The largest order of birds, commonly known as perching birds or "songbirds," which includes such diverse groups as crows, wrens, swallows, thrushes, warblers, blackbirds, and sparrows. *See* Nonpasserine.

Pathogen: A disease-causing agent; includes bacteria, viruses, and parasites.

Perennial: For plants, species that live two or more growing seasons. Trees and other woody plants are examples, but cattails, sod-forming grasses, and many other herbaceous species also are perennials. *For comparison, see* Annual.

Pesticide: Any of several chemicals that kill pests. Includes insecticides, herbicides, rodenticides, and fungicides. *See* Herbicide; Carbamate; Chlorinated hydrocarbon; Organophosphate.

pH: A measure of acidity, based on the concentration of hydrogen ions in solution. pH ranges from 14 (most alkaline) to 0 (most acid). Rainwater normally is slightly acid, with an average pH of 5.7. Because the pH scale is logarithmic, rainwater with pH 4.7 is 10 times more acid than normal, and rain with pH 3.7 is 100 times more acid than normal rainwater. *See* Acid precipitation.

Pheromone: A naturally produced chemical secretion for olfactory communication (i.e., smell), usually between individuals of the same species. When synthesized, pheromones acting as sex attractants offer a means of luring insects into lethal traps, thereby reducing the need for toxic insecticides. *See* Biological control.

Photoperiod: The proportion of light versus dark during a 24-hour period; influences flowering and fruiting in many species of plants and breeding and migration in many animals. An environmental constant for a given latitude and date.

Photosynthesis: A fundamental chemical reaction occurring in green plants whereby solar energy (light) converts carbon dioxide and water into oxygen and sugars. All food energy flowing through an ecosystem originates from photosynthesis (the sole exception occurs in some deep-sea communities uniquely dependent on bacteria adapted to sulfur-rich environments surrounding marine vents).

Pioneer: The first species or community in succession. On bare soil, annual weeds usually are the pioneer species, but some kinds of woody plants are also pioneers (e.g., jack pine after fires). Together with its associated animal life, such vegetation forms pioneer communities. *See* Succession.

Pioneering behavior: The dispersal of animals, often juveniles, into new habitats. Varies by species (e.g., strong in mallards, but weak in Canada geese). Important when new habitat is created (e.g., artificial wetlands). *See* Dispersal.

Poikilothermic: Term indicating the inability to maintain stable body temperatures when environmental temperatures change. Poikilotherms—fishes, amphibians, reptiles, and invertebrates—cannot generate metabolic heat to meet the stress of low ambient temperatures nor cool themselves at high ambient temperatures. "Cold blooded" is a popular description for poikilothermic species. *For comparison, see* Homeothermic.

Polyandry (polyandrous): *See* Polygamy.

Polygamy (polygamous): A type of mating behavior in which an individual has more than one mate at the same time. Subdivided into two types: polygyny, in which one male maintains concurrent pair bonds with several females (e.g., pheasants, walrus, and elk), and polyandry, in which one female maintains pair bonds with several males (rare in vertebrates, but known in phalaropes and ostriches). *For comparison, see* Monogamy.

Polygyny (polygynous): *See* Polygamy.

Precipitation: Any form of atmospheric moisture reaching the Earth's surface, including rain, snow, fog, and hail.

Predator: An organism that depends in total or part on killing another animal for its food. Bobcats, owls, and bass are well-known predators, but so are shrews, robins, bullfrogs, and dragonflies. A few species of plants also are predaceous (e.g., Venus flytrap). *See* Scavenger; Prey.

Predator lane: A narrow strip of cover in which predators easily can find prey. For example, a predator lane is created when clean farming leaves only a small ring of vegetation around a pond edge; in this way skunks, raccoons, and other predators easily locate and destroy duck nests.

Prescribed burning: The use of fire as a management tool for improving plant and animal habitats. *Controlled burning* is a synonym.

Prey: An organism killed and eaten by a predator. Most prey species are herbivores (e.g., rabbits). *See* Predator; Herbivore.

Primary consumer: A trophic level; consists of herbivores. *See* Producer; Secondary Consumer; Trophic level.

Primary production: The amount of energy incorporated into green plants by photosynthesis. Gross primary production is the total amount of solar energy converted into chemical energy (e.g., sugars). Net primary production is the energy left in plants after respiration and represents the amount available to herbivores. Production also may be expressed in terms of biomass. *See* Biomass; *for comparison, see* Production.

Producer: A trophic level; consists of green plants. *See* Abiotic; Trophic level.

Production (productivity): The number of surviving offspring produced during a specific period of time, usually expressed per year. Net productivity is the annual gain or loss between births and deaths in a population. *Recruitment* and *annual increment* are synonyms. *For comparison, see* Primary production.

Promiscuous (promiscuity): A type of mating behavior in which pair bonds are formed for only a brief period of sexual contact; individuals of either sex may breed with several other individuals during a single season. Common in mountain lions, lynx, and other members of the cat family (Felidae); also in domestic dogs. *For comparison, see* Monogamy; Polygamy.

Proventriculus: The glandular prestomach of birds, located between the esophagus and gizzard where chemical digestion softens foods. *See* Crop; Esophagus; Gizzard.

r-selected species: Species adapted to high reproductive rates and short life spans, often with wide ranges of tolerance to environmental conditions. Bobwhites, mourning doves, and cottontails are examples. *For comparison, see* K-selected species.

Rain forest (tropical): A multicanopied forest in humid tropical zones. Although lush, rain-forest vegetation actually is highly susceptible to human disturbances and, once cut, can seldom recover. More than half of the world's biota probably occurs in tropical rain forests. Rain forests also occur in some temperate zones (e.g., Olympic Peninsula, Washington). *See* Biota.

Randomized sampling: A research method whereby each individual within a population has an equal chance of being selected as a sample. "Picking numbers from a hat" is a popular explanation. *For comparison, see* Standardized sampling.

Range of tolerance: The range of environmental conditions (e.g., temperature) that limits the abundance and distribution of organisms. *Ecological amplitude* is a synonym. *See* Eury-; Steno-.

Raptor: A bird of prey, typically a hawk or owl. Based on a Latin verb meaning "to seize," the term has only a functional meaning and lacks taxonomic standing.

Recruitment: *See* Production

Renest: A new clutch of eggs laid in the same breeding season after an earlier clutch or (rarely) brood was destroyed (i.e., single-brooded species try nesting again). Usually located in the vicinity of the first nest (circa 100–300 m). An important adaptation for breeding success in waterfowl and most gallinaceous birds (e.g., mallards and quail). Differs from the second nest in species that normally produce more than one clutch per year (e.g., mourning doves, a multibrooded species). *Replacement clutch* is a synonym. *See* Gallinaceous.

Reservoir: In the context of diseases, the pool represented by animal populations where pathogens may be harbored as a source of infection for other populations. The reservoir may be wildlife or livestock populations. Also, an artificial impoundment of water formed by a dam. *See* Pathogen.

Reservoir effect: The rapid increase in biological production when land is flooded, resulting from the sudden infusion of nutrients. The effect is temporary unless the area undergoes repeated wet-dry fluctuations. *See* Drawdown management.

Riparian: Term describing the ecological zone bordering bodies of water, usually rivers. Typically, a zone enriched when seasonal floods deposit nutrients. Pin oak forests in the Mississippi River bottomlands are classic examples of riparian habitats.

Rotation time: The number of years between the cutting of a tree or forest and the regrowth of that tree or forest to an equivalent size.

Rumen: The first of four chambers in the stomach of ruminants (e.g., deer). Food in the rumen undergoes fermentation, after which the contents are returned to the mouth for further chewing ("chewing the cud"). *See* Ruminant.

Ruminant: A cud-chewing animal with a complex stomach that is often divided into four chambers. Deer, goats, sheep, and bovines (antelopes, cattle, and bison) are examples. *See* Rumen.

Sampling bias: An error in collecting or analyzing samples representing a population. Fish populations sampled with nets may not include the younger age classes if smaller individuals escape through the mesh, thereby introducing sampling bias based on equipment. Does and fawns usually are underrepresented in the deer harvest simply because of sampling bias introduced by hunter selection for mature bucks. *See* Sampling theory.

Sampling theory: The assumption that samples reflect the same attributes (e.g., age or sex ratios, stomach contents, or parasite loads) as the remainder of the population from which the samples were collected. *See* Randomized sampling; Standardized sampling; Sampling bias.

Scavenger: An organism feeding on dead organisms. Vultures and hyenas are examples, but many other species also scavenge at times (e.g., crows, golden eagles, and coyotes).

Second law of thermodynamics: One of three fundamental principles governing energy. As applied to ecology, the law states that large amounts of energy are lost with the transfer of energy from one trophic level to the next. In essence, the reason why a relatively small

biomass of predators requires a far larger biomass of prey. *See* Biomass; Predator; Prey; Trophic level.

Secondary consumer: A trophic level; consists of predators that feed on primary consumers. A bobcat is a secondary consumer. *See* Primary consumer; Trophic level.

Secondary poisoning: Unintended contamination of organisms (nontarget species) feeding on previously poisoned animals. Vultures and other scavengers are particularly susceptible (e.g., feeding on the carcasses of poisoned coyotes). *See* Scavenger.

Seed-tree cut: A forestry practice whereby several mature parent trees are left as seed sources for blocks of cutover forest.

Selection cut: A forestry practice in which certain trees, or small groups of trees, are chosen for harvest on the basis of size, shape, growth potential, or competition with neighboring trees. The result is an uneven-aged stand. *Selective cut* is a synonym.

Selective cut: *See* Selection cut.

Sex ratio: The ratio of males to females in animal populations, with the percentage of males expressed first (i.e., in a population with a 45:55 sex ratio, 45 percent are males). Sex ratios are subdivided by age: primary (fertilization); secondary (birth or hatching); tertiary (juvenile); and quaternary (adult).

Shelterwood cut: A forestry practice whereby several large trees are left temporarily as a source of shade for a new crop of shade-tolerant seedlings.

Snag: The upright trunk of a dead or dying tree; important as feeding, perching, and/or nesting sites for many birds (e.g., nuthatches, kestrels, and most woodpeckers). Snags are key sources of beetle larvae and other invertebrates in the diets of insectivorous birds; sometimes also may be a den tree. Synonymous with *stub. See* Den tree.

Species: The basic taxonomic unit. A population whose members freely interbreed and share a common gene pool. Species are identified scientifically in Latin with a two-part name known as a binomial, which includes both genus and species names (e.g., *Bubo virginianus* is the binomial for the great horned owl). *See* Taxonomy; Taxon.

Species diversity: A measure combining richness and evenness. Diversity increases when both the number of species increases and the number of individuals of each species are more evenly distributed. *See* Species evenness; Species richness.

Species evenness: The relative abundance of individuals among those species present in a specified area and/or time. Some species, although present, may have only a few individuals, whereas others are abundant; such conditions diminish evenness. *See* Species richness.

Species richness: The number of species present in a specified area and/or time; the term may be applied generally or to groups such as beetles, rodents, or grasses at selected sites. Globally, species richness decreases from the tropics toward the poles. *See* Species evenness.

Standardized sampling: A research method whereby an individual within a population is selected on a predetermined basis. For example, every 20th oak tree along a line transect might be measured for acorn production. *For comparison, see* Randomized sampling.

Steno-: A prefix used to identify narrow tolerances for specific components of the environment. Frogs are stenohaline because their eggs and larvae have almost no tolerance to salt water. *For comparison, see* Eury-.

Stochastic: Random or expected by chance. Wildfires, storms, and flooding are stochastic events, each with a statistical probability of happening. For example, rain may fall only 10 times per year in a desert, but the date of the next rainfall cannot be predicted accurately, whereas the *probability* of its occurrence on a given day can be calculated rather precisely. Rainfall therefore is a stochastic event. Some features of population ecology also may occur stochastically (e.g., chance variation in a sex ratio or birthrate).

Subadult: An animal, or age class, that resembles an adult but does not breed because of behavioral and/or sexual immaturity.

Subspecies: A taxon identifying subpopulations of a species. Names of subspecies are formed by adding a third term to the species binomial. For example, the elk (*Cervus elaphus*) includes, among several others, a North American subspecies (*C. e. canadensis*) and a European subspecies known as the red deer (*C. e. elaphus*). White-tailed deer and Canada geese each have a large number of subspecies. *Race* is a synonym. *See* Species; Taxon; Taxonomy.

Succession: The orderly progression through time of changes in community composition, usually described in terms of plant life. Unless interrupted, succession passes through intermediate stages from pioneer to climax communities. Managers often manipulate succession—with fire, grazing, or other means—thereby creating habitat conditions favoring selected species of wildlife. *See* Climax; Pioneer.

Sustained yield: The concept of steady-state management of timber, wildlife, and many other natural resources, but most often applied to forest management. Consumption is matched by production. *For comparison, see* Optimum yield.

Symbiosis: A biological relationship between two species, often benefiting both parties. *See* Lichen; Mutualism.

Taxon (plural, taxa): Any level of the hierarchy of taxonomy. Species, genus, and family are each a taxon. *See* Taxonomy; Species.

Taxonomy: The classification of organisms into units known as taxa (singular, taxon). Carolus Linnaeus, a Swedish biologist, developed the currently used taxonomic system in 1753 (for plants) and 1758 (for animals). The species is the basic taxonomic unit. After the species level, taxa progress into larger units: genus, family, order, class, and phylum. For some plants and animals, scientists identify a smaller taxon, the subspecies. *See* Taxon; Species; Subspecies.

Territory: Any area defended by an individual against intrusion by others of the same species; apparently assures spacing in keeping with availability of food and other resources (the classical explanation). Typical territorial behavior (by males) simultaneously attracts females while warning other males.

Threatened species: Legal status declared by governments for species or subspecies that may become endangered. The designation provides such taxa with enhanced protection and management. *See* Endangered species; Taxon.

Tillage: The practice of cultivating soil with plow or disk. *See* Conservation tillage.

Trace elements: *See* Micronutrients.

Trophic level: One of the various feeding levels in a community and ecosystem, classified on the basis of function and flow of energy. From bottom to top, producers (green plants); primary consumers (herbivores); secondary consumers (carnivores); and tertiary consumers (larger carnivores). Each trophic level depends on the quality and quantity of the underlying level, but producers generally depend on abiotic conditions. Another trophic level, decomposers (bacteria and fungi), breaks down dead matter from the other trophic levels. *See* Abiotic; Community; Ecosystem; Producer; Primary consumer; Secondary consumer.

Uneven-aged management: A forestry practice in which individuals or small groups of trees are cut (e.g., single-tree selection cut; group selection cut), thereby producing a forest with trees of different ages and heights as cutting and regrowth continue. *For comparison, see* Even-aged management.

Ungulate: A hooved mammal. Ungulates include members of the mammalian orders Artiodactyla (eventoed ungulates; e.g., deer, bison, true antelope, pronghorn, and peccaries) and Perissodactyla (oddtoed ungulates; e.g., horses, tapirs, and rhinos). All are herbivores, often with specialized stomachs. *See* Rumen; Ruminant.

Urbanization: The process or degree of human concentration in relatively small areas (cities), where huge amounts of basic materials are consumed from stocks produced elsewhere (e.g., food from farms, ranches, and oceans). At its extreme, urbanization covers the soil with a monoculture of buildings and pavement.

Vector: An organism that transmits a disease within and between populations, commonly an insect (e.g., mosquito) or other arthropod (e.g., tick). Fleas are vectors for sylvatic plague.

Velvet: The soft, highly vascularized tissues that cover developing antlers; normally rubbed and sloughed off when antlers reach full growth and harden. *See* Antler.

Ventriculus: The muscular, grit-bearing stomach of birds, designed for grinding food; also called the gizzard. May be modified in some species (e.g., hummingbirds).

Vertebrate: Any animal with a backbone, including cartilaginous fishes (e.g., sharks). A taxon known scientifically as Vertebrata includes fishes, amphibians, reptiles, birds, and mammals. Among the vertebrates, birds and mammals are most often considered in the narrow meaning of "wildlife," although such a limited view is not in keeping with modern wildlife management. *See* Taxon.

Weathering: The physical and chemical breakdown of parent material (rock) into mineral soil.

Wind-chill index: *See* Chill factor.

Yard: An area where deer in northern regions congregate in winter. Usually sheltered by coniferous trees; in this way, temperatures, wind chill, and snow depth are reduced. Same areas often are used year after year, and overbrowsing may be severe. Useful sites for population estimates (e.g., herd size, winter mortality, and health).

Yearling: An animal, or age class, between 12 and 24 months of age. An age class of biological importance in some species (e.g., deer and geese), but not in others (e.g., most insects, mice, and humans, among others). In many deer populations, for example, few yearlings breed successfully when range conditions are poor, but contribute a large percentage of the fawn crop under better conditions. In contrast, a yearling mouse is an adult and yearling human is a juvenile (i.e., in these cases, the term is overshadowed by another designation of greater biological significance).

LITERATURE CITED

ABELL, D. H., AND F. F. GILBERT. 1974. Nutrient content of fertilized deer browse in Maine. J. Wildl. Manage. 38:517–524.

ABLES, E. C., S. SHEN, AND X. QUIAN-ZHU. 1982. Wildlife education in China. Wildl. Soc. Bull. 10:282–285.

ACHA, P. N., AND A. M. ALBA. 1988. Economic losses due to *Desmodus rotundus*. Pp. 207–214 *in* The natural history of vampire bats (A. M. Greenhall and V. Schmidt, eds.). CRC Press, Boca Raton, Fla. 246 pp.

ADAMOWICZ, W. L., W. E. PHILLIPS, AND W. S. PATTISON. 1986. The distribution of economic benefits from Alberta duck production. Wildl. Soc. Bull. 14:396–398.

ADAMS, C. 1982. Hunting: An American tradition. The Am. Hunter. May:16–17.

ADAMS, L. W., L. E. DOVE, AND T. M. FRANKLIN. 1985. Mallard pair and brood use of urban stormwater-control impoundments. Wildl. Soc. Bull. 13:46–51.

ADAMS, L. W., L. E. DOVE, AND D. L. LEEDY. 1984. Public attitudes toward urban wetlands for stormwater control and wildlife enhancement. Wildl. Soc. Bull. 12:299–303.

ADAMS, L. W., AND D. L. LEEDY (EDS.). 1987. Integrating man and nature in the metropolitan environment. Nat. Instit. Urban Wildl., Columbia, Md. 249 pp.

ADAMS, L. W., AND D. L. LEEDY (EDS.). 1991. Wildlife conservation in metropolitan environments. Nat. Instit. Urban Wildl., Columbia, Md. 264 pp.

ADAMS, L. W., D. L. LEEDY, AND W. C. McCOMB. 1987. Urban wildlife research and education in North American colleges and universities. Wildl. Soc. Bull. 15:591–595.

ADAMS, L. W., T. M. FRANKLIN, L. E. DOVE, AND J. M. DUFFIELD. 1986. Design considerations for wildlife in urban stormwater management. Trans. N. Am. Wildl. and Nat. Resour. Conf. 51:249–259.

ADAMS, N. E. 1983. Deer harvest management, Welder and McCan Ranch, Texas. Abstract *in* Symposium on game harvest management, Caesar Kleberg Wildl. Res. Inst. Kingsville, Tex. 37 pp.

ADDISON, P., AND K. JENSEN. 1987. Long-range air pollution effects on the forest ecosystem. Trans. N. Am. Wildl. and Nat. Resour. Conf. 52:665–676.

ADDY, C. E., AND L. G. MacNAMARA. 1948. Waterfowl management on small areas. Wildl. Manage. Inst., Washington, D.C. 80 pp.

AGRICULTURE LAND USE AND WILDLIFE RESOURCES, COMMITTEE ON. 1970. Land Use and Wildlife Resources. National Acad. Sciences, National Resour. Council, Washington, D.C. 262 pp.

AHLGREN, C. E. 1974. Introduction. Pp. 1–5 *in* Fire and ecosystems (T. T. Kozlowski and C. E. Ahlgren, eds.). Academic Press, New York. 542 pp.

AID, C. S., G. G. MONTGOMERY, AND D. W. MOCK. 1985. Range extension of the Peruvian booby to Panama during the 1983 El Niño. Colonial Waterbirds 8:67–68.

AIR FORCE, DEPARTMENT OF. 1974. Weather for aircrews. AF Manual 51-12. Department of the Air Force, Washington, D.C. 235 pp.

ALBERS, P. H. 1977. Effects of external applications of fuel oil on hatchability of mallard eggs. Pp. 158–163 *in* Fate and effects of petroleum hydrocarbons in marine ecosystems and organisms (P. A. Wolfe, ed.). Pergamon Press, New York. 478 pp.

ALBRECHT, W. A. 1944. Soil fertility and wildlife—cause and effect. Trans. N. Am. Wildl. Conf. 9:19–28.

ALDOUS, S. E. 1941. Deer management suggestions for northern white cedar types. J. Wildl. Manage. 5:90–94.

ALDOUS, S. E. 1952. Deer browse clipping study in the lake states region. J. Wildl. Manage. 16:401–409.

ALDRICH, J. W. 1946. The United States races of the bob-white. Auk 63:493–508.

ALDRICH, J. W. 1967. Taxonomy, distribution, and present status. Pp. 17–44 *in* The wild turkey and its management (O. H. Hewitt, ed.). The Wildlife Society, Washington, D.C. 589 pp.

ALEXANDER, L. T., AND J. G. CADY. 1962. Genesis and hardening of laterite in soils. U.S.D.A. Soil Cons. Serv. Tech. Bull. 1282. Washington, D.C. 90 pp.

ALFORD, J. R. III, AND E. G. BOLEN. 1977. Differential responses of male and female pintail ducks to decoys. J. Wildl. Manage. 41:657–661.

ALIEV, F. F., AND G. C. SANDERSON. 1966. Distribution and status of the raccoon in the Soviet Union. J. Wildl. Manage. 30:497–502.

ALKON, P. U. 1961. Nutritional and acceptability values of hardwood slash as winter deer browse. J. Wildl. Manage. 25:77–81.

ALLAN, P. F., F. E. GARLAND, AND R. F. DUGAN. 1963. Rating northeastern soils for their suitability for wildlife habitat. Trans. N. Am. Wildl. and Nat. Resour. Conf. 28:247–261.

ALLAWAY, W. H., AND J. F. HODGSON. 1964. Symposium on nutrition, forage, and pastures: selenium in forages as related to the geographic distribution of muscular dystrophy in livestock. J. Animal Sci. 23:271–277.

ALLEN, C. R., S. DEMARAIS, AND R. S. LUTZ. 1994. Red imported fire ant impact on wildlife: an overview. Texas J. Sci. 46:51–59.

ALLEN, C. R., S. DEMARAIS, AND R. S. LUTZ. 1997. Effects of red imported fire ants on recruitment of white-tailed deer fawns. J. Wildl. Manage. 61:911–916.

ALLEN, C. R., R. S. LUTZ, AND S. DEMARAIS. 1993. What about fire ants and northern bobwhites. Wildl. Soc. Bull. 21:349–351.

ALLEN, C. R., R. S. LUTZ, AND S. DEMARAIS. 1995. Red imported fire ant impacts on northern bobwhites populations. Ecol. Applic. 5:632–638.

ALLEN, D. L. 1968. Population, resources, and the great complexity. Trans. N. Amer. Wildl. and Nat. Resour. Conf. 34:450–461.

ALLEN D. L. 1972. The need for a new North American wildlife policy. Trans. N. Am. Wildl. and Nat. Resour. Conf. 37:46–54.

ALLEN, D. L. (CHAIRMAN). 1973. Report of the Committee on North American Wildlife Policy. Trans. N. Am. Wildl. and Nat. Resour. Conf. 38:152–181.

ALLEN, D. L. 1974a. Our wildlife legacy (rev. ed.). Funk and Wagnalls, New York. 422 pp.

ALLEN, D. L. 1974b. Philosophical aspects of urban wildlife. Pp. 9–12 *in* Wildlife in an urbanizing environment (J. H. Noyes and D. R. Progulske, eds.). Holdsworth Nat. Resour. Ctr., Planning and Resour. Dev. Series No. 28, Amherst, Massachusetts. 182 pp.

ALLEN, D. L. 1979. Wolves of Minong: their vital role in a wild community. Houghton-Mifflin, Boston. 499 pp.

ALLEN, D. L., L. ERICKSON, E. R. HALL, AND W. M. SCHIRRA. 1981. A review and recommendations on animal problems and related management needs in units of the National Park System. Rept. to Secretary of the Interior. 18 pp. Mimeo.

ALLEN, L. J. (ED.). 1984. Proc. urban fishing symposium. Am. Fisheries Soc., Bethesda, Md. 297 pp.

ALLEN, R. P. 1952. The whooping crane. Res. Rept. 3, National Audubon Soc., New York. 246 pp.

ALLEN, R. P. (ED.). 1956. A report on the whooping crane's northern breeding grounds. Res. Rept. 3 (supplement), National Audubon Soc., New York. 60 pp.

ALLEN, S. H., AND A. B. SARGEANT. 1975. A rural mail-carrier index of North Dakota red foxes. Wildl. Soc. Bull. 3:74–77.

ALLEN, S. L., R. A. LANCIA, AND C. W. BETSILL. 1982. Habitat preferences of cottontail rabbits on an intensive farm and a traditional farm. Ann. Conf. SE Fish and Wildl. Agencies 36:614–626.

ALLEN, S. W., AND G. W. SHARPE. 1960. An introduction to American forestry, 3rd ed. McGraw-Hill, New York. 460 pp.

ALLRED, D. M., AND D. E. BECK. 1963. Range of movement and dispersal of some rodents at the Nevada atomic test site. J. Mamm. 44:190–200.

AMES, C. R. 1977. Wildlife conflicts in riparian management: grazing. Pp. 49–51 *in* Importance, preservation, and management of riparian habitat: a symposium. U.S.D.A. Forest Serv. Gen. Tech. Rept. RM-43. Rocky Mountain Forest and Range Exp. Sta., Fort Collins, Colo.

AMORY, C. 1974. Man kind? Our incredible war on wildlife. Harper & Row, New York. 372 pp.

ANDELT, W. F., AND B. R. MAHAN. 1977. Ecology of an urban coyote. Proc. Neb. Acad. Sci. 87:5 (abstract).

ANDERSON, B. L., R. D. PIEPER, AND V. W. HOWARD. 1974. Growth response and deer utilization of fertilized browse. J. Wildl. Manage. 38:525–530.

ANDERSON, D. C. 1982. Observations on *Thomomys talpoides* in the region affected by the eruption of Mount St. Helens. J. Mamm. 63:652–655.

ANDERSON, D. R., AND K. P. BURNHAM. 1978. Effect of restrictive and liberal hunting regulations on annual survival rates of the mallard in North America. Trans. N. Am. Wildl. and Nat. Resour. Conf. 43:181–186.

ANDERSON, D. W., J. R. JEHL, JR., W. RISEBROUGH, L. A. WOODS, JR., L. R. DEWEESE, AND W. G. EDGECOMB. 1975. Brown pelicans: improved reproduction off the southern California coast. Science 190:806–808.

ANDERSON, E. W., AND R. J. SCHERZINGER. 1975. Improving quality of winter forage for elk by cattle grazing. J. Range Manage. 28:120–125.

ANDERSON, R. C. 1963. The incidence, development, and experimental transmission of *Pneumostrongylus tenuis* Dougherty (Metastrongyloidae: Protostrongylidae) of the meninges of the white-tailed deer in Ontario. Can. J. Zool. 41:775–792.

ANDERSON, R. C. 1965. An examination of wild moose exhibiting neurologic signs in Onatario. Can. J. Zool. 43:635–639.

ANDERSON, R. C., AND U. R. STRELIVE. 1967. The penetration of *Pneumostrongylus tenuis* into the tissues of white-tailed deer. Can. J. Zool. 45:285–289.

ANDERSON, R. K. 1969a. Prairie chicken responses to changing booming-ground cover type and height. J. Wildl. Manage. 33:636–643.

ANDERSON, R. K. 1969b. Mating and interspecific behavior of greater prairie chicken. Unpubl. Ph.D. Thesis, Univ. Wis., Madison. 118 pp.

ANDERSON, R. K. 1991. Frederick N. Hamerstrom, 1910–1990. Wildl. Soc. Bull. 19:119–122.

ANDERSON, R. M., H. C. JACKSON, R. M. MAY, AND A. M. SMITH. 1981. Population dynamics of fox rabies in Europe. Nature 289:765–771.

ANDERSON, R. M., AND R. M. MAY (EDS.). 1982. Population biology of infectious diseases. Springer-Verlag, New York. 315 pp.

ANDERSON, R. S. (ED.). 1975. Pet animals and society. Williams and Wilkins, Baltimore. 165 pp.

ANDERSON, S. H. 1982. Effects of the 1976 Seney National Wildlife Refuge wildfire on wildlife and wildlife habitat. U.S. Fish and Wildl. Serv. Res. Publ. 146, Washington, D.C. 28 pp.

ANDERSON, W. L. 1978. Waterfowl collisions with power lines at a coal-fired power plant. Wildl. Soc. Bull. 6:77–83.

ANDERSON, W. L., AND L. V. COMPTON. 1971. More wildlife through soil and water conservation. Agric. Inform. Bull. 172 (revised). U.S.D.A Soil Cons. Serv., Washington, D.C. Unnumbered pages.

ANDERSON, W. L., S. P. HAVERA, AND R. A. MONTGOMERY. 1987. Incidence of ingested shot in waterfowl in the Mississippi Flyway, 1977–1979. Wildl. Soc. Bull. 15:181–188.

ANDERSON, W. L., AND P. L. STEWART. 1969. Relationships between inorganic ions and the distribution of pheasants in Illinois. J. Wildl. Manage. 33:254–270.

ANDERSON, W. L., AND P. L. STEWART. 1973. Chemical elements and the distribution of pheasants in Illinois. J. Wild. Manage. 37:142–153.

ANKNEY, C. D. 1996. An embarrassment of riches: too many geese. J. Wildl. Manage. 60:217–223.

ANKNEY, C. D., D. G. DENNIS, AND R. C. BAILEY. 1987. Increasing mallards, decreasing black ducks: coincidence or cause and effect? J. Wildl. Manage. 51:523–529.

ANKNEY, C. D., AND J. HOPKINS. 1985. Habitat selection of roof-nesting killdeer. J. Field Ornith. 56:284–286.

ANNISON, E. F., AND D. LEWIS. 1959. Metabolism in the rumen. Methuen, London. 184 pp.

ANON. 1908. Value of game. Forestry Quarterly 6:104.

ANON. 1956. Influence of man on vegetation and environment 2,300 years ago. Ecology 37:394.

ANON. 1971. International conference on the conservation of wetlands and waterfowl, Ramsar, Iran. Wildfowl 22:122–125.

ANON. 1981. Llamas—are they successful guards? The Predator 8:3.

ANON. 1983. "Rescued" deer die. Mich. Audubon 31:15.

ANON. 1985. Chernobyl and migratory birds. Wingtips 2:122–125.

ANON. 1986a. Chernobyl: new estimates of deaths, concerns for food chain. Environment 28:22.

ANON. 1986b. Palm trees in the sky. Urban Wildl. News. 3(2):2.

ANON. 1987. Treaty signed to protect porcupine caribou herd. Audubon Activist 2(1):9.

ANTHONY, H. E. 1957. But it's instinctive. Saturday Review 17:9–10, 40.

APLET, G. H., R. D. LAVEN, AND P. L. FIEDLER. 1992. The relevance of conservation biology to natural resource management. Conserv. Biol. 6:298–300.

APPLEGATE, J. E. 1975. Attitudes toward deer hunting in New Jersey: a second look. Wildl. Soc. Bull. 3:3–6.

APPLEGATE, J. E. 1984. Nongame tax check-off programs: a survey of New Jersey residents following the first year of contributions. Wildl. Soc. Bull. 12:122–128.

APPLEGATE, J. E., AND J. R. TROUT. 1976. Weather and the harvest of cottontails in New Jersey. J. Wildl. Manage. 40:658–662.

APPLEGATE, J. E, AND J. R. TROUT. 1984. Factors related to revenue yield in state tax checkoffs. Trans. N. Am. Wildl. and Nat. Resour. Conf. 49:199–204.

APPLEGATE, V. C. 1950. Natural history of the sea lamprey, *Petromyzon marinus*, in Michigan. U.S. Fish and Wildl. Serv. Spec. Sci. Rept., Fisheries, No. 55. 237 pp.

APPLEGATE, V. C., AND E. L. KING, JR. 1962. Comparative toxicity of 3-trifluormethyl-4-nitrophenol (TFM) to larval lampreys and eleven species of fishes. Trans. Am. Fisheries Soc. 91:342–345.

ARBIB, R. S., O. S. PETTINGILL, JR., AND S. H. SPOFFORD. 1966. Enjoying birds around New York City. Houghton Mifflin, Boston; Riverside Press, Cambridge, Mass. 171 pp.

ARCESE, P., AND A. R. E. SINCLAIR. 1997. The role of protected areas as ecological baselines. J. Wildl. Manage. 61:587–602.

ARMSTRONG, J. T. 1965. Breeding home range in the nighthawk and other birds: its evolutionary and ecological significance. Ecology 46:619–629.

ARMSTRONG, W. E., AND D. E. HARMEL. 1981. Exotic mammals competing with the natives. Texas Parks and Wildl. 39:6–7.

ARMSTRONG, W. E., AND S. WARDROUP. 1980. Statewide census of exotic big game animals. Fed. Aid Proj. W-109-R-3. Texas Parks and Wildl. Dept., Austin. 33 pp.

ARNER, D. H. 1963. Production of duck food in beaver ponds. J. Wildl. Manage. 27:76–81.

ARTHUR, G. C. 1968. Farming for Canada geese. Pp. 113–115 *in* Canada goose management (R. L. Hine and C. Schoenfeld, eds.). Dembar Educational Res. Serv., Madison, Wis. 195 pp.

ARTMANN, J. W., AND E. M. MARTIN. 1975. Incidence of ingested lead shot in sora rails. J. Wildl. Manage. 39:514–519.

ASH, A. N., AND J. F. BENDELL. 1979. Trials of nitrogen fertilizer on foods of blue grouse. J. Wildl. Manage. 43:503–508.

ASHBROOK, F. G. 1948. Nutrias grow in United States. J. Wildl. Manage. 12:87–95.

ATKESON, T. Z., AND L. GIVENS. 1951. Gray squirrel parasitism by heel fly larvae. J. Wildl. Manage. 15:105–106.

ATKINS, T. D., AND R. L. LINDER. 1967. Effects of dieldrin on reproduction of penned hen pheasants. J. Wildl. Manage. 31: 746–753.

ATKINSON, R. 1975. A Canadian landowner views waterfowl production. Pp. 58–59 *in* First international waterfowl symposium. Ducks Unlimited, Chicago. 224 pp.

ATTEBURY, J. T., J. C. KROLL, AND M. H. LEE. 1977. Operational characteristics of commercial exotic big game hunting ranches. Wildl. Soc. Bull. 5:179–184.

ATWOOD, E. L. 1950. Life history studies of nutria, or coypu, in coastal Louisiana. J. Wildl. Manage. 14:249–265.

AUEL, J. M. 1980. The clan of the cave bear. Bantam Books, New York. 495 pp.

AUMANN, G. D., AND J. T. EMLEN. 1965. Relation of population density to sodium availability and sodium selection by microtine rodents. Nature 208:198–199.

AUSTIN, O. L. 1951. The mourning dove on Cape Cod. Bird-banding 22:149–174.

AVARIA, S. 1985. Effects of "El Niño" on southeast Pacific fisheries. Pp. 1–15, Sec. VI *in* Proc. internat. conf. tropical ocean and global atmosphere scientific programme. World Climate Res. Programme Series No. 4, Paris.

AVERY, M. L. 1985. Application of mimicry theory to bird damage control. J. Wildl. Manage. 49:1116–1121.

BABCOCK, K. M., AND E. L. FLICKINGER. 1977. Dieldrin mortality of lesser snow geese in Missouri. J. Wildl. Manage. 41:100–103.

BACON, P. J. (ED.). 1985. Population dynamics of rabies in wildlife. Academic Press, Orlando, Fla. 358 pp.

BAILEY, T. N., AND A. W. FRANZMANN. 1983. Mortality of resident vs. introduced moose in confined populations. J. Wildl. Manage. 47:520–523.

BAILEY, W. M. 1974. An evaluation of striped bass introductions in the southeastern United States. Proc. SE Assoc. Fish and Game 28:54–68.

BAKER, B. W. 1978. Ecological factors affecting wild turkey nest predation on south Texas rangelands. Proc. Ann. Conf. SE Assoc. Fish and Wildl. Agencies 32:126–136.

BAKER, D. 1986. Virgin forests under fire. Nat. Wildl. 24:4–11.

BAKER, R. R. 1978. The evolutionary ecology of animal migration. Holmes and Meier, New York. 1012 pp.

BALCOMB, R. 1983. Secondary poisoning of red-shouldered hawks with carbofuran. J. Wildl. Manage. 47:1129–1132.

BALCOMB, R., C. A. BOWEN, D. WRIGHT, AND M. LAW. 1984. Effects on wildlife of at-planting corn applications of granular carbofuran. J. Wildl. Manage. 48:1353–1359.

BALDASSARRE, G. A., R. J. WHYTE, E. E. QUINLAN, AND E. G. BOLEN. 1983. Dynamics and quality of waste corn available to post-breeding waterfowl in Texas. Wildl. Soc. Bull. 11:25–31.

BALLARD, W. B., L. A. AYRES, P. R. KRAUSMAN, D. J. REED, AND S. G. FANCY. 1997. Ecology of wolves in relation to a migratory caribou herd in northwest Alaska. Wildl. Monogr. 135. 47 pp.

BALSER, D. S., H. H. DILL, AND H. K. NELSON. 1968. Effect of predator reduction on waterfowl nesting success. J. Wildl. Manage. 32:669–682.

BANFIELD, A. W. F. 1954. The role of ice in the distribution of mammals. J. Mamm. 35:104–107.

BANFIELD, A. W. F. 1961. A revision of the reindeer and caribou, genus *Rangifer*. Bull. 177, Biol. Serv. 66, Nat. Mus. Can., Ottawa. 137 pp.

BANKO, W. E. 1960. The trumpeter swan. North American Fauna, No. 63. U.S. Fish and Wildl. Serv., Washington, D.C. 214 pp.

BANKO, W. E., AND R. H. MACKAY. 1964. Our native swans. Pp. 155–164 *in* Waterfowl tomorrow (J. P. Linduska, ed.). U.S.

Dept. of the Interior, Bureau of Sport Fisheries and Wildlife, U.S. Govt. Printing Office, Washington, D.C. 770 pp.

BANKS, R. C. 1976. Reflective plate glass—a hazard to migrating birds. BioScience 26:414.

BANKS, R. C. 1979. Human related mortality of birds in the United States. U.S. Fish and Wildl. Serv. Spec. Sci. Rept. 215:1–16.

BANKS, W. A., D. F. WILLIAMS, AND C. S. LOFGREN. 1988. Effectiveness of fenoxycarb for control of red imported fire ants (Hymenoptera: Formicidae). J. Econ. Entomol. 81:83–87.

BAPTISTE, M. E., J. B. WHELAN, AND R. B. FRARY. 1979. Visitor perception of black bear problems at Shenandoah National Park. Wildl. Soc. Bull. 7:25–29.

BARBER, J. C., AND M. M. BARBER. 1983. Prey of an urban peregrine falcon. Maryland Birdlife 39:108–110.

BARBER, R. T., AND F. P. CHAVEZ. 1983. Biological consequences of El Niño. Science 222:1203–1210.

BARBOUR, A. G., AND D. FISH. 1993. The biological and social phenomenon of Lyme disease. Science 260:1610–1616.

BARDACH, J., AND R. M. SANTERRE. 1981. Climate and the fish in the sea. BioScience 31:206–215.

BARDACH, J. E., J. H. RYTHER, AND W. O. MCLARNEY. 1972. Aquaculture, the farming and husbandry of freshwater and marine organisms. John Wiley & Sons, New York. 868 pp.

BARKALOW, F. S., JR., AND R. F. SOOTS, JR. 1965. An analysis of the effect of artificial nest boxes on a gray squirrel population. Trans. N. Am. Wildl. and Nat. Resour. Conf. 30:349–360.

BARNES, A. M., AND L. KARTMAN. 1960. Control of plague vectors in diurnal rodents in the Sierra Nevada of California by use of insecticide bait-boxes. J. Hyg. 58:347–355.

BARNES, I. R. 1966. Amid brick and asphalt. Pp. 414–424 *in* Birds in our lives (A. Stefferud and A. L. Nelson, eds.). U.S.D.I., Bur. Sport Fisheries and Wildlife. U.S. Govt. Printing Office, Washington, D.C. 561 pp.

BARONE, M. A., M. E. ROELKE, J. HOWARD, J. L. BROWN, A. E. ANDERSON, AND D. E. WILDT. 1994. Reproductive characteristics of male Florida panthers: comparative studies from Florida, Texas, Colorado, Latin America, and North American zoos. J. Mamm. 75:150–162.

BARRETT, G. W., AND R. M. DARNELL. 1967. Effects of dimethoate on small mammal populations. Am. Midl. Nat. 77:164–175.

BARRETT, M. W. 1979. Evaluation of fertilizer on pronghorn winter range in Alberta. J. Range Manage. 32:55–59.

BARRETT, R. H. 1980. History of the Hearst Ranch Barbary sheep herd. Pp. 46–50 *in* Proc. Symposium on Ecology and Management of Barbary Sheep (C. D. Simpson, ed.). Dept. Range and Wildlife Management, Texas Tech Univ., Lubbock. 112 pp.

BARROWS, W. B. 1889. The English sparrow (*Passer domesticus*) in North America, especially in relation to agriculture. U.S.D.A. Div. Economic Ornith. and Mamm. Bull. 1, Washington, D.C. 405 pp.

BARRY, T. W. 1962. Effect of late seasons on Atlantic brant reproduction. J. Wildl. Manage. 26:19–26.

BARSNESS, L. 1985. Heads, hides and horns, the compleat buffalo book. Texas Christian Univ. Press, Forth Worth. 233 pp.

BARSOM, G. 1973. Lagoon performance and the state of lagoon technology. U.S. E.P.A. Environmental Protection Tech. Ser. Govt. EPA-R2-73-144. U.S. Govt. Printing Office, Washington, D.C. 214 pp.

BARTMANN, R. M. 1969. Pheasant nesting on soil bank land in northern Utah. J. Wildl. Manage. 33:1020–1023.

BARTSCH, A. F., AND W. M. INGRAM. 1959. Stream life and the pollution environment. Public Works: City, County, and State 90:104–110.

BASORE, N. S., L. B. BEST, AND J. B. WOOLEY, JR. 1986. Bird nesting in Iowa no-tillage and tilled cropland. J. Wildl. Manage. 50:19–28.

BASORE, N. S., L. B. BEST, AND J. B. WOOLEY, JR. 1987. Arthropod availability to pheasant broods in no-tillage fields. Wildl. Soc. Bull. 12:229–233.

BATT, B. D. J. (ED.). 1997. Arctic ecosystems in peril: report of the Arctic Goose Habitat Working Group. Arctic Goose Joint Venture Spl. Pub., U.S. Fish and Wildl. Serv., Washington D.C. and Canadian Wildl. Serv., Ottawa, Ontario. 120 pp.

BAUER, O. N., AND G. L. HOFFMAN. 1976. Helminth range extension by translocation of fish. Pp. 163–172 *in* Wildlife diseases (L. A. Page, ed.). Plenum Press, New York. 686 pp.

BAXTER, W. L., R. L. LINDER, AND R. B. DAHLGREN. 1969. Dieldrin effects in two generations of penned hen pheasants. J. Wildl. Manage. 33:96–102.

BEALS, E. W., G. COTTAM, AND P. J. VOGL. 1960. Influence of deer on vegetation of the Apostle Islands, Wisconsin. J. Wildl. Manage. 24:68–80.

BEAMISH, R. 1972. Lethal pH for the white sucker (*Catostomus commersoni* (Lacepede). Trans. Am. Fisheries Soc. 10:355–358.

BEAMISH, R. 1976. Acidification of lakes in Canada by acid precipitation and resulting effects on fishes. Water Air Soil Pollut. 6:5401–5514.

BEAMISH, R., AND H. HARVEY. 1972. Acidification of the La-Cloche Mountain lakes, Ontario, and resulting fish mortalities. J. Fish. Res. Bd. Can. 29:1131–1143.

BEAMISH, R., W. LOCKHART, J. VAN LOON, AND H. HARVEY. 1975. Long-term acidification of a lake and resulting effects on fishes. Ambio 4:98–102.

BEAN, M. J. 1983. The evolution of national wildlife law (rev). Praeger Publishers, New York. 449 pp.

BEARD, E. B. 1953. The importance of beaver in waterfowl management at the Seney National Wildlife Refuge. J. Wildl. Manage. 17:398–436.

BEASOM, S. L. 1974. Relationship between predator removal and white-tailed deer net productivity. J. Wildl. Manage. 38:854–859.

BEASOM, S. L., AND O. H. PATTEE. 1980. The effect of selected climatic variables on wild turkey productivity. Trans. Nat. Wild Turkey Symposium 4:127–135.

BEASOM, S. L., AND C. J. SCIFRES. 1977. Population reactions of selected game species to aerial herbicide applications. J. Range Manage. 30:138–142.

BEASOM, S. L., J. M. INGLIS, AND C. J. SCIFRES. 1982. Vegetation and white-tailed deer responses to herbicide treatment of a mesquite drainage habitat type. J. Range Manage. 35:790–794.

BEATTIE, K. H. 1981. The influence of game laws and regulations on hunting satisfaction. Wildl. Soc. Bull. 9:229–231.

BEATTIE, K. H., R. H. GILES, AND C. J. COWLES. 1977. Lack of research in wildlife law enforcement. Wildl. Soc. Bull. 5:170–174.

BECHTOLD, J-P. 1996. Chemical characterization of natural mineral springs in northern British Columbia, Canada. Wildl. Soc. Bull. 24:649–654.

BECK, A. M. 1973. The ecology of stray dogs: a study of free-ranging urban animals. York Press, Baltimore, Md. 98 pp.

BECK, A. M. 1984. An epizootic of rabies. Nat. Hist. 93(7):7–11.

BECK, W. A. 1942. Effect of drought on the production of plant pigments. Plant Physiol. 17:487–491.

BEGLEY, S. 1987. On the trail of acid rain. Nat. Wildl. 25(2):8–12.

BEGON, M., J. L. HARPER, AND C. R. TOWNSEND. 1986. Ecology: individuals, populations, and communities. Sinauer Assoc., Inc., Sunderland, Mass. 876 pp.

BEHAN, R. W. 1978. Political dynamics of wildlife management: the Grand Canyon burros. Trans. N. Am. Wildl. and Nat. Resour. Conf. 43:424–433.

BEIER, P. 1987. Sex differences in quality of white-tailed deer diets. J. Mamm. 68:323–329.

BEIER, P. 1993. Determining minimum habitat areas and habitat corridors for cougars. Conserv. Biol. 7:94–108.

BEISSINGER, S. R., AND D. R. OSBORNE. 1982. Effects of urbanization on avian community organization. Condor 84:75–83.

BELCHER, J. W., AND S. D. WILSON. 1989. Leafy spurge and the species composition of a mixed grass prairie. J. Range Manage. 42:172–175.

BELL, H. 1971. Effect of low pH on the survival and emergence of aquatic insects. Water Res. 5:313–319.

BELL, H. L. 1985. Seasonal changes in abundance and pond preferences of waterbirds at Moitaka sewerage works, Port Moresby, Papua New Guinea. Corella 9:108–113.

BELL, J. F., W. BURDORFER, AND G. J. MOORE. 1957. The behavior of rabies virus in ticks. J. Infect. Dis. 100:278.

BELL, J. F., G. W. SCIPLE, AND A. A. HUBERT. 1955. A microenvironment concept of the epizoology of avian botulism. J. Wildl. Manage. 19:352–357.

BELLROSE, F. C. 1941. Duck food plants of the Illinois River Valley. Ill. Nat. Hist. Surv. Bull. 21:237–280.

BELLROSE, F. C. 1955. Housing for wood ducks. Ill. Nat. Hist. Surv. Bull. 55 (2nd printing, rev.). 48 pp.

BELLROSE, F. C. 1959. Lead poisoning as a mortality factor in waterfowl populations. Ill. Nat. Hist. Surv. Bull. 27:235–288.

BELLROSE, F. C. 1968. Waterfowl migration corridors east of the Rocky Mountains in the United States. Ill. Nat. Hist. Surv. Biol. Notes 61. 23 pp.

BELLROSE, F. C. 1975. Impact of ingested lead pellets on waterfowl. Pp. 163–167 in First international waterfowl symposium. Ducks Unlimited, Chicago. 224 pp.

BELLROSE, F. C. 1976a. Comeback of the wood duck. Wildl. Soc. Bull. 4:107–110.

BELLROSE, F. C. 1976b. Ducks, geese, and swans of North America. Stackpole Books, Harrisburg, Penn. 544 pp.

BELLROSE, F. C., H. C. HANSON, AND P. D. BEAMER. 1945. Aspergillosis in wood ducks. J. Wildl. Manage. 9:325–326.

BELLROSE, F. C., AND D. J. HOLM. 1994. Ecology and management of the wood duck. Stackpole Books, Mechanicsburg, Penn. 588 pp.

BELLROSE, F. C., F. L. PAVEGLIO, AND D. W. STEFFECK. 1979. Waterfowl populations and the changing environment of the Illinois River Valley. Ill. Nat. Hist. Surv. Bull. 32. 54 pp.

BELLROSE, F. C., T. G. SCOTT, A. S. HAWKINS, AND J. B. LOW. 1961. Sex ratios and age ratios in North American ducks. Ill. Nat. Hist. Surv. Bull. 27, (Art. 6:391–474).

BELTHOFF, J. R., AND S. A. GAUTHREAUX. 1991. Partial migration and differential winter distribution of house finches in the eastern United States. Condor 93:374–382.

BEMENT, R. E. 1969. A stocking-rate guide for beef production on blue-grama range. J. Range Manage. 22:83–86.

BENDELL, B. E., P. J. WEATHERHEAD, AND R. K. STEWART. 1981. The impact of predation of red-winged blackbirds on European corn borer populations. Can. J. Zool. 59:1535–1538.

BENDELL, J. F. 1974. Effects of fire on birds and mammals. Pp. 73–138 in Fire and ecosystems (T. T. Kozlowski and C. E. Ahlgren, eds.). Academic Press, New York. 542 pp.

BENDER, M. 1981. Service recognizes recovery of alligator in Louisiana. Endangered Species Tech. Bull. 6:7.

BENDT, R. H. 1962. The Jackson Hole elk herd in Yellowstone and Grand Teton National Parks. Trans. N. Am. Wildl. and Nat. Resour. Conf. 27:191–201.

BENNET, G. F. 1960. On some ornithophilic blood-sucking diptera in Algonquin Park, Ontario, Canada. Can. J. Zool. 38:377–389.

BENNETT, C. L., JR., L. A. RYEL, AND L. J. HAWN. 1966. A history of Michigan deer hunting. Res. and Dev. Rept. 85, Michigan Dept. Nat. Resour. Lansing, Mich. 66 pp.

BENNETT, J. W., AND E. G. BOLEN. 1978. Stress response in wintering green-winged teal. J. Wildl. Manage. 42:81–86.

BENNETT, L. J. 1938. The blue-winged teal: its ecology and management. Collegiate Press, Inc., Ames, Ia. 144 pp.

BENNETTS, R. E., AND R. L. HUTTO. 1985. Attraction of social fringillids to mineral salts: an experimental study. J. Field Ornith. 56:187–189.

BENNITT, R., J. S. DIXON, W. H. CAHALANE, W. W. CHASE, AND W. L. MCATEE. 1937. Statement of policy. J. Wildl. Manage. 1:1–2.

BENOIT, P. J. 1973. From fish and wildlife officer to environmental conservation officer. Wildl. Soc. Bull. 1:128–130.

BENSON, D., AND D. FOLEY. 1956. Waterfowl use of small, man-made wildlife marshes in New York State. N.Y. Fish and Game J. 3:217–224.

BENSON, S. B. 1933. Concealing coloration among some desert rodents of the southwestern United States. Univ. Calif. Publ. in Zoology 40:1–70.

BENT, A. C. 1938. Life histories of North American birds of prey, Part 2. Smithsonian Inst. U.S. Nat. Mus. Bull. 170, Washington, D.C. 482 pp.

BENT, A. C. 1963. Life histories of North American diving birds. Dover ed. New York. 239 pp.

BERG, W. E., AND D. W. KUEHN. 1982. Ecology of wolves in north-central Minnesota. Pp. 4–11 in Wolves of the world (F. H. Harrington and P. C. Paquet, eds.). Noyes, Park Ridge, N.J. 474 pp.

BERGER, J. 1986. Wild horses of the Great Basin. Univ. of Chicago Press, Chicago. 326 pp.

BERGER, J. 1990. Persistence of different-sized populations: an empirical assessment of rapid extinctions in bighorn sheep. Conserv. Biol. 4:91–98.

BERGER, J., AND J. D. WEHAUSEN. 1991. Consequences of a mammalian predator-prey disequilibrium in the Great Basin desert. Conserv. Biol. 5:244–248.

BERRYMAN, J. H. 1972. The principles of predator control. J. Wildl. Manage. 36:395–400.

BESHEARS, W. W., JR., AND A. O. HAUGEN. 1953. Muskrats in relation to farm ponds. J. Wildl. Manage. 17:450–456.

BESSER, J. F. 1973. Protecting seeded rice from blackbirds with methiocarb. Int. Rice Comm. Newsl. 22:9–14.

BESSER, J. F., AND R. W. HANSON. 1985. 4-aminopyridine hydrochloride baits on baiting lanes protect ripening corn from blackbirds. Wildl. Soc. Bull. 13:546–551.

BEST, L. B. 1983. Bird use of fencerows: implications of contemporary fencerow management practices. Wildl. Soc. Bull. 11:343–347.

BEST, L. B. 1986. Conservation tillage: ecological traps for nesting birds? Wildl. Soc. Bull. 14:308–317.

BEZZEL, E. 1985. Birdlife in intensively used rural and urban environments. Ornis Fennica 62:90–95.

BIGLER, W. J., AND G. L. HOFF. 1976. Urban wildlife and community health: gray squirrels as environmental monitors. Proc. Ann. Conf. SE Assoc. Fish and Wildl. Agencies 30:536–540.

BINDERNAGEL, J. A., AND R. C. ANDERSON. 1972. Distribution of meningeal worm in white-tailed deer in Canada. J. Wildl. Manage. 36:1349–1353.

BioScience. 1981. Raven warns of rapid tropical deforestation. BioScience 31:633.

Birge, E. A., and C. Juday. 1929. Transmission of solar radiation by the waters of inland lakes. Trans. Wis. Acad. Sci. Arts. Lett. 24:509–580.

Birkhead, M. 1982. Causes of mortality in the mute swan (*Cygnus olor*) on the River Thames. J. Zool. (London) 198:15–25.

Bishop, R. A. 1981. Iowa's wetlands. Proc. Iowa Acad. Sci. 88:11–16.

Bishop, R. A., R. D. Andrews, and R. J. Bridges. 1979. Marsh management and its relationship to vegetation, waterfowl and muskrats. Proc. Ia. Acad. Sci. 86:50–56.

Blair, W. F., and J. D. Kilby. 1936. The gopher mouse— *Peromyscus floridanus*. J. Mamm. 17:421–422.

Blaisdell, J. P., and W. F. Mueggler. 1956. Effects of 2,4-D on forbs and shrubs associated with big sagebrush. J. Range Manage. 9:38–40.

Blancher, P. J., and D. G. McAuley. 1987. Influence of wetland acidity on avian breeding success. Trans. N. Am. Wildl. and Nat. Resour. Conf. 52:528–635.

Bleich, V. C., L. J. Coombes, and J. H. Davis. 1982. Horizontal wells as a wildlife habitat improvement technique. Wildl. Soc. Bull. 10:324–328.

Bleich, V. C., R. T. Bowyer, and J. D. Wehausen. 1997. Sexual segregation in mountain sheep: resources or predation? Wildl. Monogr. 134. 50 pp.

Block, B. C. 1966. Williamsport, Pennsylvania, tries starling control with distress calls. Pest Control 34:24–30.

Blockstein, D. E. and H. B. Tordoff. 1985. Gone forever, a contemporary look at the extinction of the passenger pigeon. Am. Birds 39:845–851.

Blokpoel, H., and G. D. Tessier. 1984. Overhead wires and monofilament lines exclude ring-billed gulls from public places. Wildl. Soc. Bull. 12:55–58.

Blus, L. J., R. K. Stroud, B. Reising, and T. McEnearney. 1989. Lead poisoning and other mortality factors in trumpeter swans. Environ. Toxicol. and Chem. 8:263–271.

Boddicker, M. L. 1981. Profiles of American trappers and trapping. Pp. 1918–1949 *in* Proc. worldwide furbearer conf., Vol. III (J. A. Chapman and D. Pursley, eds.). Frostburg, Md. 2056 pp.

Boggess, E.K., and F. R. Henderson. 1981. Trapper/hunter education and implications for furbearer management. Pp. 1950–1969 *in* Proc. worldwide furbearer conf., Vol. III (J. A. Chapman and D. Pursley, eds.). Frostburg, Md. 2056 pp.

Bogucki, D. J., G. K. Gruendling, E. B. Allen, K. B. Adams, and M. M. Remillard. 1986. Photointerpretation of historical (1948–1985) beaver activity in the Adirondacks. Pp. 299–308 *in* Technical papers, 1986 ASPRS–ACSM fall convention, Am. Soc. Photogram. and Remote Sensing, Falls Church, Va. 415 pp.

Bohn, C. 1986. Biological importance of streambank stability. Rangelands 8:55–56.

Bolen, E. G. 1971. Some views on exotic waterfowl. Wilson Bull. 83:430–434.

Bolen, E. G. 1982. Playa wetlands of the U.S. Southern High Plains: their wildlife values and challenges for management. Pp. 9–20 *in* Wetlands ecology and management, B. Gopal, R. E. Turner, R. G. Wetzel, and D. F. Whigham (eds.). Nat. Inst. Ecol. and Internat. Sci. Publ., New Delhi, India. 514 pp.

Bolen, E. G. 1989. Conservation biology, wildlife management, and Spaceship Earth. Wildl. Soc. Bull. 17:351–354.

Bolen, E. G. 1991. Analogs: a concept for the research and management of urban wildlife. Landscape and Urban Planning 20:285–289.

Bolen, E. G. 1992. The Hamerstroms, a memoir. J. Raptor Res. 26:200–201.

Bolen, E. G., M. M. Wilson, M. W. Weller, L. R. Jahn, C. S. Robbins, and B. R. Chapman. 1980. Nongame conservation and management. Wilson Bull. 92:142–148.

Bolle, A. W., and R. D. Taber. 1962. Economic aspects of wildlife abundance on private lands. Trans. N. Am. Wildl. and Nat. Resour. Conf. 27:255–267.

Bollinger, E. K., and J. W. Caslick. 1985. Factors influencing blackbird damage to field corn. J. Wildl. Manage. 49:1109–1115.

Bonnell, M. L., and R. K. Selander. 1974. Elephant seals: genetic variation and near extinction. Science 184:908–909.

Borrell, J. 1987. Trouble ahead for the canal? Time 129:63.

Bortner, J. B. 1987. American woodcock harvest and breeding population status, 1987. Office of Migratory Bird Manage., U.S. Fish and Wildl. Serv., Laurel, Md. 11 pp.

Bossenmaier, E. F. 1976. Ecological awareness and sport hunting— a viewpoint. Wildl. Soc. Bull. 4:127–128.

Bossenmaier, E. F., and W. H. Marshall. 1958. Field-feeding by waterfowl in southwestern Manitoba. Wildl. Monogr. No. 1. 32 pp.

Botkin, D. B. 1990. Discordant harmonies: a new ecology for the twenty-first century. Oxford Univ. Press, New York, N.Y. 241 pp.

Bouffard, S. H. 1982. Wildlife values versus human recreation: Ruby Lake National Wildlife Refuge. Trans. N. Am. Wildl. and Nat. Resour. Conf. 47:553–558.

Bourn, W. S., and C. Cottam. 1950. Some biological effects of ditching tidewater marshes. U.S. Fish and Wildl. Serv. Res. Rept. 19, Washington, D.C. 30 pp.

Bowen, J. T. 1970. A history of fish culture as related to the development of fishery programs. Pp. 71–93 *in* A century of fisheries in North America (N. G. Benson, ed.). Am. Fisheries Soc. Spec. Publ. No. 7, Washington, D.C. 330 pp.

Bowers, W., B. Hosford, A. Oakley, and C. Bond. 1979. Wildlife habitats in managed rangelands—the Great Basin of southeastern Oregon. Native trout. U.S.D.A. Forest Serv. Gen. Tech. Rept. PNW-84. 16 pp.

Bowman, G. B., and L. D. Harris. 1980. Effect of spatial heterogeneity on ground-nest depredation. J. Wildl. Manage. 44:806–813.

Bowman, T. D., P. F. Schempf, and J. I. Hodges. 1997. Bald eagle population in Prince William Sound after the *Exxon Valdez* oil spill. J. Wildl. Manage. 61:962–967.

Box, T. W. 1961. Relationships between plants and soils of four range plant communities in south Texas. Ecology 42:794–810

Box, T. W. 1964. Changes in wildlife habitat composition following brush control practices in south Texas. Trans. N. Am. Wildl. and Nat. Resour. Conf. 29:432–438.

Box, T. W., and J. Powell. 1965. Brush management techniques for improved forage values in Texas. Trans. N. Am. Wildl. and Nat. Resour. Conf. 30:285–296.

Boyce, M. S., and R. S. Miller. 1985. Ten-year periodicity in whooping crane census. Auk 103:658–660.

Boyle, C. L. 1961. Nature conservation in Poland. Oryx 6:6–26.

Boyle, S. A., and F. B. Samson. 1985. Effects of nonconsumptive recreation on wildlife: a review. Wildl. Soc. Bull. 13:110–116.

Boza, M. A. 1974. Costa Rica: a case study of strategy in the setting up of national parks in a developing country. Pp. 183–192 *in* Second world conference on national parks (S. H. Elliott, ed.). Internat. Union for Cons. of Nature and Nat. Resour., Morges, Switzerland. 504 pp.

BOZA, M. A. 1993. Conservations in action: past, present, and future of the National Park system of Costa Rica. Conserv. Biol. 7:239–247.

BRAITHWAITE, L. W., AND H. J. FRITH. 1969. Waterfowl in an inland swamp in New South Wales, III, Breeding. CSIRO Wildl. Res. 14:65–109.

BRAKHAGE, G. K. 1966a. Management of mast crops for wood ducks. Pp. 75–80 in Wood duck management and research: a symposium. Wildl. Manage. Inst., Washington, D.C. 212 pp.

BRAKHAGE, G. K. 1966b. Tub nests for Canada geese. J. Wildl. Manage. 30:851–853.

BRAMBELL, M. R. 1972. Mammals: their nutrition and habitat. Pp. 613–648 in Biology of nutrition (R.N.T.-W-Fiennes, ed.). Pergamon Press, New York. 681 pp.

BRANDLE, J. R., B. B. JOHNSON, AND D. D. DEARMONT. 1984. Windbreak economics: the case of winter wheat production in eastern Nebraska. J. Soil and Water Cons. 39:339–343.

BRANSON, F. A. 1953. Two new factors affecting resistance of grasses to grazing. J. Range Manage. 6:165–171.

BRAUN, C. E., T. BRITT, AND R. O. WALLESTAD. 1977. Guidelines for maintenance of sage grouse habitats. Wildl. Soc. Bull. 5:99–106.

BRAYTON, D. S. 1984. The beaver and the stream. J. Soil and Water Cons. 39:108–109.

BRIAND, F., AND J. E. COHEN. 1987. Environmental correlates of food chain length. Science 238:956–960.

BRITISH COLUMBIA MINISTRY OF ENVIRONMENT AND PARKS. 1985. Wolf management in British Columbia protecting predator and prey. B.C. Ministry Environ. and Parks. Victoria, B.C. Canada. 18 pp.

BRITTINGHAM, M. C., AND S. A. TEMPLE. 1983. Have cowbirds caused forest songbirds to decline? BioScience 33:31–35.

BRITTINGHAM, M. C., AND S. A. TEMPLE. 1986. A survey of avian mortality at winter feeders. Wildl. Soc. Bull. 14:445–450.

BROCKE, R. H. 1979. The name of the nongame. Wildl. Soc. Bull. 7:279–282.

BRODRICK, A. H. 1972. Introduction. Five unnumbered pages in Animals in archeology (A. H. Brodrick, ed.). Praeger Publishers, New York. 180 pp.

BRODSKY, L. M., AND P. J. WEATHERHEAD. 1984. Behavioral and ecological factors contributing to American black duck–mallard hybridization. J. Wildl. Manage. 48:846–852.

BRODY, S. 1945. Bioenergetics and growth. Reinhold Publishing, New York. 1023 pp.

BROOKS, A. C., AND I. O. BUSS. 1962. Past and present status of the elephant in Uganda. J. Wildl. Manage. 26:38–50.

BROOKS, P. 1972. The house of life, Rachel Carson at work. Houghton Mifflin Co., Boston, Mass. 350 pp.

BROOKS, R. P., AND W. J. DAVIS. 1987. Habitat selection by breeding belted kingfishers (Ceryle alcyon). Am. Midl. Nat. 117:63–70.

BROOKS, R. T., AND C. T. SCOTT. 1983. Quantifying land-use edge from aerial photographs. Wildl. Soc. Bull. 11:389–391.

BROWER, L. P. 1985. New perspectives on the migration biology of the monarch butterfly, Danaus plexippus L. Univ. Texas Contrib. Marine Sci. Suppl. 27:748–785.

BROWN, B. T., AND M. W. TROSSET. 1989. Nesting-habitat relationship of riparian birds along the Colorado River in Grand Canyon, Arizona. Southwestern Nat. 34:260–270.

BROWN, D. E. 1978. Grazing, grassland cover, and gamebirds. Trans. N. Am. Wildl. and Nat. Resour. Conf. 43:477–485.

BROWN, E. R., AND J. H. MANDERY. 1962. Planting and fertilization as a possible means of controlling distribution of big game animals. J. Forestry 60:33–35.

BROWN, F. C. III. 1985. National parks stagger as difficulties grow while budget shrinks. Wall Street J. 67(34):1, 6.

BROWN, J. H., AND A. KODRIC-BROWN. 1977. Turnover rates in insular biogeography: effect of immigration on extinction. Ecology 58:445–449.

BROWN, L. N. 1985. Elimination of a small feral swine population in an urbanizing section of central Florida. Florida Scientist 48:120–123.

BROWN, M., AND J. J. DINSMORE. 1986. Implications of marsh size and isolation for marsh bird management. J. Wildl. Manage. 50:392–397.

BROWN, M. K., AND G. R. PARSONS. 1979. Waterfowl production on beaver flowages in a part of northern New York. N.Y. Fish and Game J. 26:142–153.

BROWN, P. J., G. E. HAAS, AND B. L. DRIVER. 1980. Value of wildlife to wilderness users. Proc. 2nd Conf. on Sci. Res. in National Parks 6:168–179.

BROWN, P. J., M. E. HAUTALUOMA, AND S. M. MCPHAIL. 1977. Colorado deer hunting experiences. Trans. N. Am. Wildl. and Nat. Resour. Conf. 42:216–255.

BROWN, R. G. 1985. Effects of an urban wetland on sediment and nutrient loads in runoff. Wetlands 4:147–158.

BROWN, R. L. 1982. Effects of livestock grazing on Mearns quail in southeastern Arizona. J. Range Manage. 35:727–732.

BROWN, T. L., C. P. DAWSON, AND R. L. MILLER. 1979. Interests and attitudes of metropolitan New York residents about wildlife. Trans. N. Am. Wildl. and Nat. Resour. Conf. 44:289–297.

BROWN, T. L., D. J. DECKER, AND C. P. DAWSON. 1978. Willingness of New York farmers to incur white-tailed deer damage. Wildl. Soc. Bull. 6:235–239.

BROWN, W. 1978. The role of the Endangered Species Scientific Authority in management of rare and endangered wildlife. Pp. 13–15 in Proc. rare and endangered wildl. symposium. Ga. Dept. Nat. Resour., Athens. 184 pp.

BROWN, W. S. 1987. Hidden life of the timber rattler. Nat. Geograph. 172:128–138.

BROWNLEE, S. 1987. Fostering hope for the whooper. Nat. Wildl. 25:40–43.

BRUEMMER, F. 1987. Life upon the permafrost. Nat. Hist. 96(4):30–39.

BRUGGERS, R. L., J. E. BROOKS, R. A. DOLBEER, P. P. WORONECKU, R. K. PANDIT, T. TARIMO, ALL-INDIA COOR. RESOUR. PROJ. ON ECON. ORNITH., AND M. HOQUE. 1986. Responses of pest birds to reflecting tape in agriculture. Wildl. Soc. Bull. 14:161–170.

BRUGGINK, J. G. 1997. American woodcock harvest and breeding status, 1997. U.S. Fish and Wildl. Serv. Patuxent Res. Center, Laurel, Md. 12 pp.

BRYANT, F. C., F. S. GUTHERY, AND W. WEBB. 1982. Grazing management in Texas and its impact on selected wildlife. Pp. 94–112 in Wildlife-livestock relationships symposium: proceedings 10 (J. M. Peek and P. D. Dalke, eds.). Univ. of Idaho Forest, Wildl., and Range Exp. Sta., Moscow, Id. 614 pp.

BRYANT, F. C., M. M. KOTHMANN, AND L. B. MERRILL. 1979. Diets of sheep, angora goats, Spanish goats, and white-tailed deer under excellent range conditions. J. Range Manage. 32:412–417.

BRYANT, F. C., C. A. TAYLOR, AND L. B. MERRILL. 1981. White-tailed deer diets from pastures in excellent and poor range condition. J. Range Manage. 34:193–200.

BRYANT, J. P., AND P. J. KUROPAT. 1980. Selection of winter forage by subarctic browsing vertebrates: the role of plant chemistry. Ann. Rev. Ecol. Syst. 11:261–285.

BRYANT, L. D. 1982. Response of livestock to riparian zone exclusion. J. Range Manage. 35:780–785.

BRYNILDSON, O. M., V. A. HACKER, AND T. A. KLICK. 1977. Brown trout: life history, ecology and management. Wis. Dept. Nat. Resour., Madison. 16 pp.

BRYSON, H. R. 1931. The interchange of soil and subsoil by burrowing insects. J. Kans. State Ent. Soc. 4:17–24.

BUCHHOLTZ, K. P., AND D. E. BAYER. 1960. Establishment of wildlife food patches in sod without tillage. J. Wildl. Manage. 24:412–418.

BUE, I. G., L. BLANKENSHIP, AND W. H. MARSHALL. 1952. The relationship of grazing practices to waterfowl breeding populations and production on stock ponds in western South Dakota. Trans. N. Am. Wildl. Conf. 17:396–414.

BUECH, R. R. 1980. Vegetation of a Kirtland's warbler breeding area and 10 nest sites. Jack-Pine Warbler 58:59–72.

BUECHNER, H. K. 1942. Interrelationships between the pocket gopher and land use. J. Mamm. 23:346–348.

BUECHNER, H. K. 1950. Life history, ecology, and range use of the pronghorn antelope in Trans-Pecos Texas. Am. Midl. Nat. 43:257–354.

BUECHNER, H. K. 1960. The bighorn sheep in the United States, its past, present, and future. Wildl. Mongr. No. 4. 174 pp.

BUERGER, T. T., R. E. MIRARCHI, AND M. E. LISANO. 1986. Effects of lead shot ingestion on captive mourning dove survivability and reproduction. J. Wildl. Manage. 50:1–8.

BULL, E. L., AND E. C. MESLOW. 1977. Habitat requirements of the pileated woodpecker in northeastern Oregon. J. Forestry 75:335–337.

BULL, J. 1964. Birds of the New York [City] area. Harper and Row, New York. 540 pp.

BULL J. 1974. Birds of New York State. Doubleday-Natural History Press, Garden City, New York. 655 pp.

BULL, L. B., AND C. G. DICKINSON. 1937. The specificity of the virus of rabbit myxomatosis. J. Council Sci. and Indust. Res. Australia. 10:291–293.

BUMP, G. 1951. Game introductions—when, where, and how. Trans. N. Am. Wildl. Conf. 16:316–325.

BUMP, G. 1963a. Status of the Foreign Game Introduction Program. Trans. N. Am Wildl. and Nat. Resour. Conf. 28:240–247.

BUMP, G. 1963b. History and analysis of tetraonid introductions into North America. J. Wildl. Manage. 27:855–867.

BUMP, G. 1968. Exotics and the role of the State-Federal Foreign Game Introduction Program. Pp. 5–8 in Symposium on introduction of exotic animals: ecological and socioeconomic considerations. Caesar Kleberg Res. Pgm. in Wildl. Ecology, Tex. A&M Univ., College Station. 25 pp.

BUMP, G., R. W. DARROW, F. C. EDMINSTER, AND W. F. CRISSEY. 1947. The ruffed grouse: life history, propagation, management. N.Y. State Dept. Cons., Albany. 915 pp.

BUREAU OF RECLAMATION. 1995. Operation of Glen Canyon Dam, Colorado River Storage Project, Arizona. Final Environ. Impact Statement. U.S. Dept. Interior, Washington, D.C. 337 pp. plus appended material.

BURGER, G. V. 1969. Response of gray squirrels to nest boxes at Remington Farms, Maryland. J. Wildl. Manage. 33:796–801.

BURGER, G. V. 1978. Agriculture and wildlife. Pp. 89–107 in Wildlife and America (H. P. Brokaw, ed.). Council on Environmental Quality, U.S. Fish and Wildl. Serv., Forest Service, and Nat. Oceanic and Atmospheric Admin. U.S. Govt. Printing Office, Washington, D.C. 532 pp.

BURGER, J., AND R. T. ZAPPALORTI. 1986. Nest site selection by pine snakes, *Pituophis melanoleucus*, in the New Jersey Pine Barrens. Copeia 1986:116–121.

BURKHART, R. L., AND L. A. PAGE. 1971. Chlamydiosis (Ornithosis-Psittacosis). Pp. 118–140 in Infectious and parasitic diseases of wild birds (J. W. Davis, R. C. Anderson, L. Karstad, and D. O. Trainer, eds.). Ia. State Univ. Press, Ames. 344 pp.

BURNHAM, K. P., D. R. ANDERSON, AND J. L. LAAKE. 1980. Estimation of density from line transect sampling of biological populations. Wildl. Monogr. No. 72. 202 pp.

BURNHAM, K. P., D. R. ANDERSON, AND J. L. LAAKE. 1985. Efficiency and bias in strip and line transect sampling. J. Wildl. Manage. 49:1012–1018.

BURNS, R. J., G. E. CONNOLLY, AND I. OKUNO. 1986. Secondary toxicity of coyotes killed by 1080 single-dose baits. Proc. Vert. Pest Conf. 12:324–329.

BURROUGHS, J. 1906. Camping and tramping with Roosevelt. Houghton Mifflin, Boston, Mass. 110 pp.

BURZLAFF, D. F., G. W. FICK, AND L. R. RITTENHOUSE. 1968. Effect of nitrogen fertilization on certain factors of a western Nebraska range ecosystem. J. Range Manage. 21:21–24.

BUSH, G. W. 1993. Whaling activities of Norway. Letter to the Norwegian Government. U.S. Govern. Printing Office, Washington, D.C. 2 pp.

BUTCHART, D. 1988. Give a bird a bone, a brief account of vulture "restaurants" in southern Africa. African Wildl. 42:316–322.

BYERS, S. M., AND G. V. BURGER. 1979. Evaluation of three partridge species for put-and-take hunting. Wildl. Soc. Bull. 7:17–20.

CADE, T. J., AND G. L. MACLEAN. 1967. Transport of water by adult sandgrouse to their young. Condor 69:323–343.

CAESAR KLEBERG WILDLIFE RESEARCH INSTITUTE. 1983. Symposium on game harvest management. Caesar Kleberg Wildl. Res. Inst., Kingsville, Tex. 37 pp.

CAHALANE, V. H. 1948. Predators and people. Nat. Parks Magazine 22:5–12.

CAHN, A. R. 1937. The fisheries program in the Tennessee Valley. Trans. Am. Fisheries Soc. 66:398–402.

CAHOON, W. G. 1953. Commercial carp removal at Lake Mattamuskeet, North Carolina. J. Wildl. Manage. 17:312–317.

CAIN, S. A., J. A. KADLEC, D. L. ALLEN, R. A. COOLEY, M. G. HORNOCKER, A. S. LEOPOLD, AND F. H. WAGNER. 1972. Predator control—1971. Rept. to the Council on Environmental Quality and the U.S.D.I. by the Advisory Comm. on Predator Control. Inst. for Environ. Quality, Univ. Mich., Ann Arbor. 207 pp.

CALHOON, R. E., AND C. HASPEL. 1989. Urban cat populations compared by season, subhabitat, and supplemental feeding. J. Animal Ecol. 58:321–328.

CALLICOTT, J. B. (ED.).1987. Companion to *A Sand County Almanac*. Univ. Wisconsin Press, Madison, Wis. 308 pp.

CAMPBELL, H. 1950. Quail picking up lead shot. J. Wildl. Manage. 14:243–244.

CAMPBELL, R. S., AND W. KELLER (EDS.). 1973. Range resources of the southeastern United States. Am. Soc. Agron. Spec. Publ. 21, Madison, Wis. 78 pp.

CANNON, R. W., AND F. L. KNOPF. 1979. Lesser prairie chicken responses to range fires at the booming ground. Wildl. Soc. Bull. 7:44–46.

CANNON, R. W., AND F. L. KNOPF. 1981. Lesser prairie chicken densities on shinnery oak and sand sagebrush rangelands in Oklahoma. J. Wildl. Manage. 45:521–524.

CAPEL, S. W. 1988. Design of windbreaks for wildlife in the Great Plains of North America. Agric. Ecosystems & Environ. 22/23:337–347.

CAPEN, D. E., W. J. CRENSHAW, AND M. W. COULTER. 1974. Establishing breeding populations of wood ducks by relocating wild broods. J. Wildl. Manage. 38:253–256.

CAREY, A. B., AND J. D. GILL. 1980. Firewood and wildlife. U.S.D.A. Forest Serv. Res. Note 299. NE For. Expt. Sta., Broomall, Penn. 5 pp.

CARL, L. M., J. R. RYCKMAN, AND W. C. LATTA. 1976. Management of trout fishing in a metropolitan area. Mich. Dept. Nat. Resour. Fisheries Res. Rept. 1836, Lansing. 29 pp.

CARLETON, A. R., AND O. T. OWRE. 1975. The red-whiskered bulbul in Florida: 1960–71. Auk 92:40–57.

CARLSON, P. C., G. W. TANNER, J. M. WOOD, AND S. R. HUMPHREY. 1993. Fire in Key deer habitat improves browse, prevents succession, and preserves endemic herbs. J. Wildl. Manage. 57:914–928.

CARO, T. M., AND M. K. LAURENSON. 1994. Ecological and genetic factors in conservation: a cautionary tale. Science 263:485–486.

CAROTHERS, K., AND K. B. KNIGHT. 1984. A profile of contributors to the West Virginia nongame wildlife program. Trans. N. Am. Wildl. and Nat. Resour. Conf. 49:190–198.

CAROTHERS, S. W., AND B. T. BROWN. 1991. The Colorado River through the Grand Canyon, natural history and human change. Univ. Arizona Press, Tucson, Ariz. 235 pp.

CAROTHERS, S. W., M. E. STITT, AND R. R. JOHNSON. 1976. Feral asses on public lands: an analysis of biotic impact, legal considerations, and management alternatives. Trans. N. Am. Wildl. and Nat. Resour. Conf. 41:396–406.

CARPENTER, L. H., AND G. L. WILLIAMS. 1972. A literature review on the role of mineral fertilizers in big game range improvement. Colo. Div. Fish and Game Spec. Rept. 28, Fort Collins. 25 pp.

CARR, A. 1965. The navigation of the green turtle. Sci. Am. 212:79–86.

CARR, A. 1982. Tropical forest conservation and estuarine ecology. Biol. Cons. 23:247–259.

CARRUTHERS, D. R., AND R. D. JAKIMCHUK. 1987. Migratory movements of the Nelchina caribou herd in relation to the Trans-Alaska pipeline. Wildl. Soc. Bull. 15:414–420.

CARSON, R. 1951. The sea around us. Oxford Univ. Press, New York, N.Y. 230 pp. [Revised in 1961.]

CARSON, R. 1962. Silent spring. Houghton Mifflin, Boston. 368 pp.

CARTER, J. G., R. A. VALDEZ, R. J. RYEL, AND V. A. LAMARRA. 1985. Fisheries habitat dynamics in the upper Colorado River. J. Freshwater Ecol. 3:249–264.

CARTER, V. G., AND T. DALE. 1974. Topsoil and civilization (rev. ed.) Univ. Okla. Press, Norman. 292 pp.

CASLICK, J. W., AND D. J. DECKER. 1979. Economic feasibility of a deer-proof fence for apple orchards. Wildl. Soc. Bull. 7:173–175.

CASTO, S. D. 1984. Captain M. B. Davis' war to save the birdlife of Texas. Bull. Texas Ornith. Soc. 17:2–12.

CASTRALE, J. S. 1985. Responses of wildlife to various tillage conditions. Trans. N. Am. Wildl. and Nat. Resour. Conf. 50: 142–156.

CATT, J. 1950. Some notes on brown trout with particular reference to their status in New Brunswick and Nova Scotia. Can. Fish Culturist 7:25–27.

CATTON, W. R., JR. 1987. The world's most polymorphic species. BioScience 37:413–419.

CAUFIELD, C. 1984. In the rainforest. Alfred A. Knopf, New York. 304 pp.

CAUGHLEY, G. 1970. Eruption of ungulate populations, with emphasis on Himalayan thar in New Zealand. Ecology 51:53–72.

CAUGHLEY, G., G. C. GRIGG, AND L. SMITH. 1985. The effect of drought on kangaroo populations. J. Wildl. Manage. 49: 679–685.

CAULEY, D. L., AND J. R. SCHINNER. 1973. The Cincinnati raccoons. Nat. Hist. 82:58–60.

CAVIEDES, C. N. 1984. El Niño 1982–83. Geograph. Review 74: 267–290.

CENTERS FOR DISEASE CONTROL AND PREVENTION. 1996. Lyme disease—United States. Morbidity and Mortality Weekly Rept. 46:531–535.

CHAMBERLAIN, E. B. 1948. Ecological factors inflencing the growth and management of certain waterfowl food plants on Back Bay National Wildlife Refuge. Trans. N. Am Wildl. Conf. 13:347–356.

CHAMBERLAIN, P. A. 1980. Armadillos: problems and control. Proc. Vertebrate Pest Conf. 9:163–169.

CHAMBERLAIN, P. A., M. CAROLINE, AND W. A. WRIGHT. 1982. Urban vertebrate pest management: a practical approach. Proc. Great Plains Wildl. Damage Control Workshop 5:78–96.

CHAMBERLAIN, P. A., AND B. DAVIS. 1986. Marketing waterfowl hunting lease opportunities. Pp. 219–225 in Proc. 1986 internat. ranchers roundup, Kerrville, Texas (D. E. Guynn and T. R. Troxel, eds.). Texas A&M Univ., Texas Agric. Res. and Ext. Ctr., Uvalde. 316 pp.

CHAMRAD, A. D., AND T. W. BOX. 1968. Food habits of white-tailed deer in south Texas. J. Range Manage. 21:158–164.

CHAMRAD, A. D., AND J. D. DODD. 1972. Prescribed burning and grazing for prairie chicken habitat manipulation in the Texas coastal prairie. Proc. Tall Timbers Fire Ecol. Conf. 12:257–276.

CHANCELLOR, R. J. 1978. The long-term effects of herbicides on weed populations. Ann. Appl. Biol. 91:141–144.

CHAPMAN, F. D. 1939. The whitetail deer and its management in southeastern Ohio. Trans. N. Am. Wildl. Conf. 4:257–267.

CHAPMAN, H. H. 1932. Is the longleaf type a climax? Ecology 13:328–334.

CHAPMAN, H. H. 1936. Effects of fire in propagation of seedbed for longleaf pine seedlings. J. Forestry 34:852–854.

CHAPMAN, H. H. 1944. Fire and pines. Am. Forestry 50:62–64, 91–93.

CHASE, A. 1986a. The grizzly and the juggernaut. Outside 11(1):29–34, 55–65.

CHASE, A. 1986b. Playing God in Yellowstone: the destruction of America's first national park. Atlantic Monthly Press, Boston. 446 pp.

CHEATUM, E. L. 1951. Disease in relation to winter mortality of deer in New York. J. Wildl. Manage. 15:216–220.

CHEATUM, E. L., AND C. W. SEVERINGHAUS. 1950. Variations in fertility of white-tailed deer related to range conditions. Trans. N. Am. Wildl. Conf. 15:170–189.

CHERNIN, E. 1952. The relapse phenomenon in the Leucocytozoon simondi infection of the domestic duck. Am. J. Hyg. 56:101–118.

CHILDS, J. E., KSIAZEK, T. G., SPIROPOULOU, C. F., ET AL. 1994. Serologic and genetic identification of Peromyscus maniculatus as the primary rodent reservoir for a new hantavirus in the southeastern United States. J. Infect. Dis. 169:1271–1280.

CHITTY, D. 1959. A note on shock disease. Ecology 40:728–731.

CHOATE, J. R., AND K. M. REED. 1986. Historical biogeography of the woodchuck in Kansas. Prairie Nat. 18:37–42.

CHOULES, G. L., W. C. RUSSELL, AND D. A. GAUTHIER. 1978. Duck mortality from detergent-polluted water. J. Wildl. Manage. 42:410–414.

CHRISTENSEN, G. C. 1970. The chukar partridge. Nev. Dept. Fish and Game Biol. Bull. No. 4, Reno. 82 pp.

CHRISTIANSEN, E. 1976. Pheromones in small rodents and their potential use in pest control. Proc. Vertebrate Pest Conf. 7:185–195.

CHRISTISEN, D. M. 1985. The greater prairie chicken and Missouri's land-use patterns. Terrestrial Ser. No. 15. Mo. Dept. Cons., Jefferson City. 51 pp.

CHRUCHER, P. B., AND J. H. LAWTON. 1987. Predation by domestic cats in an English village. J. Zool. (London) 212:439–455.

CHUPP, N. R., AND P. D. DALKE. 1964. Waterfowl mortality in the Coeur D'Alene River Valley, Idaho. J. Wildl. Manage. 28: 692–702.

CHURA, N. J. 1961. Food availability and preference of juvenile mallards. Trans. N. Am. Wildl. and Nat. Resour. Conf. 26: 121–134.

CHURCH, D. C. 1971. Trace (or micro) minerals. Pp. 453–504 *in* Digestive physiology and nutrition of ruminants. Vol. 2. O&B Books, Corvallis, Ore. 693 pp.

CLAPHAM, W. B., JR. 1973. Natural ecosystems. Macmillan, New York. 248 pp.

CLARK, D. R., JR. 1979. Lead concentrations: bats vs. terrestrial small mammals collected near a major highway. Environ. Sci. Technol. 13:338–341.

CLARK, R. G., H. GREENWOOD, AND L. G. SUGDEN. 1986. Estimation of grain wasted by field-feeding ducks in Saskatchewan. J. Wildl. Manage. 50:184–189.

CLARKE, C. H. D. 1958. Autumn thoughts of a hunter. J. Wildl. Manage. 32:430–437.

CLARKE, F. C. 1913. Preliminary report upon the disease occurring among the ducks of the southern San Joaquin Valley during the fall of 1913. Condor 15:214–226.

CLARKE, J. A., B. A. HARRINGTON, T. HRUBY, AND F. E. WASSERMAN. 1984. The effect of ditching for mosquito control on salt marsh use by birds in Rowley, Massachusetts. J. Field Ornith. 55:160–180.

CLOUDSLEY-THOMPSON, J. L. 1977. Animal life and desertification. Environ. Cons. 4:199–204.

CLOUGH, G. 1965. Lemmings and population problems. Am. Scientist 53:199–212.

COADY, J. W. 1974. Influence of snow on behavior of moose. Nat. Can. (Que.) 101:417–436.

COADY, J. W. 1982. Moose. Pp. 902–922 *in* Wild mammals of North America (J. A. Chapman and G. A. Feldhamer, eds.). Johns Hopkins Univ. Press, Baltimore, Md. 1147 pp.

COBB, E. 1987. Notes from the field: victim of technology. Am. Heritage, Invent. & Technol. 2:6–7.

COGBILL, C. 1976. The history and character of acid precipitation in eastern North America. Pp. 363–370 *in* Proc. first internat. symposium on acid precipitation and the forest ecosystem (L. Dochinger and T. Seliga, eds.). U.S.D.A. Forest Serv. Gen. Tech. Rept. NE-23. 1074 pp.

COHEN, J. A. 1977. Why do you hunt? The Humane Soc. News 22:25.

COLE, G. F. 1970. Managing the Yellowstone elk. Nat. Parks and Cons. Magazine 44:20–22.

COLE, G. F. 1974. Management involving grizzly bears and humans in Yellowstone National Park, 1970–73. BioScience 24: 335–338.

COLE, H. H., AND M. RONNING. 1974. Animal agriculture: the biology of domestic animals and their use by man. W. H. Freeman and Co., San Francisco, Calif. 788 pp.

COLE, S. 1972. Animals of the new stone age. Pp. 15–41 *in* Animals in archeology (A. H. Brodrick, ed.). Praeger Publishers, New York. 180 pp.

COLEMAN, J. S., AND S. A. TEMPLE. 1993. Rural residents' free-ranging domestic cats: a survey. Wildl. Soc. Bull. 21:381–390.

COLEMAN, R. A. 1993. San Francisco Bay—an urban/wildlife shuffle. Trans. N. Am. Wildl. and Nat. Resour. Conf. 58:137–142.

COLINVAUX, P. 1986. Ecology. John Wiley & Sons, New York. 725 pp.

COLLINS, H. L., A-M. CALLCOTT, T. C. LOCKLEY, AND A. LADNER. 1992. Seasonal trends in effectiveness of mydramethylnon (AMDRO) and fenoxycarb (LOGIC) for control of red improted fire ants (Hymenoptera: Formicidae). J. Econ. Entomol. 85: 2131–2137.

COLLINS, J. M. 1974. The relative abundance of ducks breeding in southern Ontario in 1951 and 1971. Pp. 32–44 *in* Waterfowl studies (H. Boyd, ed.). Can. Wildl. Serv. Rept. Series 29. 106 pp.

COLLS, D. G. 1951. The conflict between waterfowl and agriculture. Trans. N. Am. Wildl. Conf. 16:89–93.

COLVIN, B. A. 1985. Common barn-owl population decline in Ohio and the relationship to agricultural trends. J. Field Ornith. 56:224–235.

COMMITTEE ON AGRICULTURE, LAND USE, AND WILDLIFE RESOURCES. 1970. Land Use and Wildlife Resources. Nat. Acad. of Sci.–Nat. Resour. Council. Washington, D.C. 262 pp.

COMMITTEE ON MERCHANT MARINE AND FISHERIES, U.S. HOUSE OF REPRESENTATIVES. 1993. Prince William Sound after *Exxon Valdez* oil spill. Serial No. 103–10. U.S. Gov. Printing Office, Washington, D.C. 389 pp.

CONNELLY, N. A., D. J. DECKER, AND T. L. BROWN. 1985. New opportunities with a familiar audience: where esthetics and harvest overlap. Wildl. Soc. Bull. 13:399–403.

CONNOLLY, G. E., AND W. M. LONGHURST. 1975. The effects of control on coyote populations: a simulation model. Univ. Calif. Div. Agric. Sci. Bull. 1872. 37 pp.

CONOVER, M. R. 1982. Behavioral techniques to reduce bird damage to blueberries: methiocarb and a hawk-kite predator model. Wildl. Soc. Bull. 10:211–216.

CONOVER, M. R. 1984. Effectiveness of repellents in reducing deer damage in nurseries. Wildl. Soc. Bull. 12:399–404.

CONOVER, M. R. 1985. Alleviating nuisance Canada goose problems through methiocarb-induced aversive conditioning. J. Wildl. Manage. 49:631–636.

CONOVER, M. R., AND G. G. CHASKO. 1985. Nuisance Canada goose problems in the eastern United States. Wildl. Soc. Bull. 13:228–233.

CONSERVATION TILLAGE INFORMATION CENTER. 1987. Executive summary, 1986 national survey of conservation tillage practices. Cons. Tillage Inform. Center, Ft. Wayne, Ind. 8 pp.

CONSTANTINE, D. G. 1993. Chiroptera: bat medicine, management, and conservation. Pp. 310–321 *in* Zoo & wild animal medicine, current therapy 3 (M. E. Fowler, ed.). W. B. Saunders, Philadelphia, Penn. 617 pp.

COOCH, F. G. 1964. A preliminary study of the survival value of a functional salt gland in prairie Anatidae. Auk 81:380–393.

COOCH, F. G. 1969a. The current state of the art. Proc. Fed.-Prov. Wildl. Conf. 33:39–50. Can. Wildl. Serv., Edmonton.

COOCH, F. G. 1969b. Waterfowl-production habitat requirements. Pp. 5–10 *in* Trans. Saskatoon wetlands seminar. Dept. of Indian Affairs and Northern Development, Can. Wildl. Serv. Rept. Series No. 6. 262 pp.

COOK, R. S. 1971. *Leucocytozoon* Danilewsky 1890. Pp. 291–299 *in* Infectious and parasitic diseases of wild birds (J. W. Davis, R. C. Anderson, L. Karstad, and D. O. T, eds.). Ia. State Univ. Press, Ames. 344 pp.

COOPER, C. F. 1961. The ecology of fire. Sci. Am. 204:150–160.

COOPER, E. L. 1951. Rate of exploitation of wild eastern brook trout and brown trout populations in the Pigeon River, Otsego County, Michigan. Trans. Am. Fisheries Soc. 81:224–234.

COOPER, J. A. 1978. The history and breeding biology of the Canada geese of Marshy Point, Manitoba. Wildl. Monogr. No. 61. 87 pp.

COOPER, J. A. 1987. The effectiveness of translocation control of Minneapolis–St. Paul Canada goose populations. Pp. 169–171 *in* Integrating man and nature in the metropolitan environment (L. W. Adams and D. L. Leedy, eds.). Nat. Instit. Urban Wildl., Columbia, Md. 249 pp.

COOPER, J. A., AND M. A. JOHNSON. 1977. Wintering waterfowl in the Twin Cities. Loon 49:212–238.

COOPER, J. A., AND T. KEEFE. 1997. Urban Canada goose management: policies and procedures. Trans. N. Amer. Wildl. and Nat. Resour. Conf. 62:412–430.

COPPINGER, L., AND R. COPPINGER. 1982. Livestock-guarding dogs that wear sheep's clothing. Smithsonian 13:65–73.

CORINTH, R. L. AND W. L. ROBINSON. 1997. The status of vegetation and white-tailed deer in Beaver Basin, Pictured Rocks National Lakeshore. Paper presented *at* Conf. on Research in National Parks in the Lake Superior Region. Nat. Park Serv., Munising, Mich.

CORLISS, J. B., J. DYMOND, L. I. GORDON, J. M. EDMOND, R. P. VON HERZEN, R. D. BALLARD, K. GREEN, D. WILLIAMS, A. BAINBRIDGE, K. CRANE, AND T. H. VANANDEL. 1979. Submarine thermal springs on the Galápagos Rift. Science 203:1073–1083.

COTTAM, C. 1969. Pesticide pollution. Nat. Parks and Cons. Magazine 43:4–9.

COTTAM, C., AND P. KNAPPEN. 1939. Food of some uncommon North American birds. Auk 56:138–169.

COTTAM, C., J. J. LYNCH, AND A. L. NELSON. 1944. Food habits and management of American sea brant. J. Wildl. Manage. 8:36–56.

COTTAM, C., A. L. NELSON, AND L. W. SAYLOR. 1940. The chukar and Hungarian partridges in America. U.S.D.A. Wildl. Leaflet BS-159. 6 pp.

COULTER, M. W., AND W. R. MILLER. 1968. Nesting biology of black ducks and mallards in northern New England. Vt. Fish and Game Dept. Bull. 68-2. Montpelier. 73 pp.

COUNCIL ON ENVIRONMENTAL QUALITY. 1979. Environmental quality, the Tenth Annual Report of the Council on Environmental Quality. Washington, D.C. 816 pp.

COURTENAY, W. R., JR., AND J. R. STAUFFER, JR. (EDS.). 1984. Distribution, biology, and management of exotic fishes. Johns Hopkins Univ. Press, Baltimore, Md. 430 pp.

COWAN, I. McT. 1936. Distribution and variation in deer (genus *Odocoileus*) of the Pacific coastal region of North America. Calif. Fish and Game 22:155–246.

COWAN, I. McT. 1947. The timber wolf in the Rocky Mountain National Parks of Canada. Can. J. Res. 25:139–174.

COWAN, I. McT., AND V. C. BRINK. 1949. Natural game licks in the Rocky Mountain National Parks of Canada. J. Mamm. 30:379–387.

COWAN, W. F. 1982. Waterfowl production on zero tillage farms. Wildl. Soc. Bull. 10:305–308.

COWARDIN, L. M. 1969. Use of flooded timber by waterfowl at the Montezuma National Wildlife Refuge. J. Wildl. Manage. 33:829–842.

COWARDIN, L. M., AND D. H. JOHNSON. 1979. Mathematics and mallard management. J. Wildl. Manage. 43:18–35.

COWELL, D. W. 1984. The Canadian beaver, *Castor canadensis*, as a geomorphic agent in karst terrain. Can. Field-Nat. 98:227–230.

COWLES, H. C. 1899. The ecological relations of the vegetation on the sand dunes of Lake Michigan. Bot. Gaz. 27:95–117; 167–202; 281–308; 361–391.

COWLING, E. B., AND R. A. LINTHURST. 1981. The acid precipitation phenomenon and its ecological consequences. BioScience 31:649–654.

CRAIGHEAD, F. C. 1968. The role of the alligator in shaping plant communities and maintaining wildlife in the southern Everglades. Florida Nat. 41:2–7; 69–74, 94.

CRAIGHEAD, F. C., JR. 1979. Track of the grizzly. Sierra Club Books, San Francisco. 261 pp.

CRAIGHEAD, J. J., AND F. C. CRAIGHEAD. 1971. Grizzly bear–man relationships in Yellowstone National Park. BioScience 21:845–857.

CRAWFORD, B. T. 1946. Wildlife sampling by soil types. Trans. N. Am. Wildl. Conf. 11:357–364.

CRAWFORD, B. T. 1950. Some specific relationships between soils and wildlife. J. Wildl. Manage. 14:115–123.

CRAWFORD, J. A., AND E. G. BOLEN. 1976. Effects of land use on lesser prairie chickens in Texas. J. Wildl. Manage. 40:96–104.

CRAWFORD, S. M. 1970. Wild protein: a vital role for Africa. Animals 12:540–543.

CREWS, K. A. 1987. Human needs and nature's balance: population, resources, and the environment. Population Reference Bureau, Inc. Washington, D.C. 12 pp.

CRICHTON, V. 1963. Autumn and winter foods of the spruce grouse in central Ontario. J. Wildl. Manage. 27:597.

CRINGAN A. T. 1957. History, food habits, and range requirements of the woodland caribou of continental North America. Trans. N. Am. Wildl. Conf. 22:485–501.

CRINGAN, A. T. 1970. Reproductive biology of ruffed grouse in southern Ontario. J. Wildl. Manage. 34:756–761.

CRISLER, L. 1956. Observations of wolves hunting caribou. J. Mamm. 37:337–346.

CRISSEY, W. F. 1969. Prairie potholes from a continental viewpoint. Pp. 161–171 *in* Trans. Saskatoon wetlands seminar. Dept. of Indian Affairs and Northern Development, Can. Wildl. Serv. Rept. Series No. 6. 262 pp.

CROFT, B. A., AND A. W. A. BROWN. 1975. Responses of arthropod natural enemies to insecticides. Ann. Rev. Ent. 20:285–335.

CROME, F. H. J. 1986. Australian waterfowl do not necessarily breed on a rising water level. Aust. Wildl. Res. 13:461–480.

CRONEMILLER, F. P. 1955. Soil surveys for game-range development. Trans. N. Am. Wildl. Conf. 20:532–539.

CRONIN, E. W. 1979. Doe harvest managing the deer herd. Blair and Ketchum's Country J. 6:100–109.

CULLEY, D. E., JR., AND D. E. FERGUSON. 1969. Patterns of insecticide resistance in the mosquito fish, *Gambusia affinis*. J. Fish. Res. Bd. Can. 26:2395–2401.

CUMBERLIDGE, N. 1987. Gorilla thriller in Rwanda. Internat. Living 7(6):9.

CURATALO, J. A., AND S. M. MURPHY. 1986. The effects of pipelines, roads, and traffic on the movements of caribou, *Rangifer tarandus*. Can. Field-Nat. 100:218–224.

CURRY-LINDAHL, K. 1972. Let them live. W. Morrow, New York. 394 pp.

CURTIS, J. T. 1959. The vegetation of Wisconsin, an ordination of plant communities. Univ. Wis. Press, Madison. 657 pp.

CURTIS, P. 1981. Animals are good for the handicapped, perhaps all of us. Smithsonian 12:49–57.

CUSTER, T. W., AND P. H. ALBERS. 1980. Response of captive, breeding mallards to oiled water. J. Wildl. Manage. 44:915–918.

CUTRIGHT, P. R. 1956. Theodore Roosevelt, the naturalist. Harper & Brothers, New York, N.Y. 297 pp.

CUTRIGHT, P. R. 1985. Theodore Roosevelt, the making of a conservationist. Univ. Illinois Press, Urbana, Ill. 285 pp.

DAHL, B. E., AND D. N. HYDER. 1977. Developmental morphology and management implications. Pp. 257–290 *in* Rangeland and plant physiology (R. E. Sosebee, ed.). Soc. for Range Manage. Rangeland Sci. Series, No. 4. 290 pp.

DAHLGREN, R. B. 1967. The pheasant decline. S.D. Dept. Game, Fish, and Parks. 44 pp.

DAHLGREN, R. B., AND R. L. LINDER. 1970. Eggshell thickness in pheasants given dieldrin. J. Wildl. Manage. 34:226–228.

DAHLGREN, R. B., AND R. L. LINDER. 1971. Effects of polychlorinated biphenyls on pheasant reproduction, behavior, and survival. J. Wildl. Manage. 35:315–319.

DALE, F. H. 1942. Influence of rainfall and soil on Hungarian partridges and pheasants in southeastern Michigan. J. Wildl. Manage. 6:17–18.

DALE, F. H. 1943. History and status of the Hungarian partridge in Michigan. J. Wildl. Manage. 7:368–377.

DALE, F. H. 1954. Influence of calcium on the distribution of the pheasant in North America. Trans. N. Am. Wildl. Conf. 19:316–323.

DALE, F. H. 1955. The role of calcium in reproduction of the ring-necked pheasant. J. Wildl. Manage. 19:325–331.

DALKE, P. D., R. D. BEEMAN, F. J. KINDEL, R. J. ROBEL, AND T. R. WILLIAMS. 1965. Use of salt by elk in Idaho. J. Wildl. Manage. 29:319–332.

DALKE, P. D., A. S. LEOPOLD, AND D. L. SPENCER. 1946. The ecology and management of the wild turkey in Missouri. Tech. Bull. No. 1. Mo. Cons. Comm., Jefferson City. 86 pp.

DALKE, P. D., D. P. PYRAH, D. C. STANTON, J. E. CRAWFORD, AND E. F. SCHLATTERER. 1963. Ecology, productivity, and management of sage grouse in Idaho. J. Wildl. Manage. 27:810–841.

DAMBACH, C. A., AND D. L. LEEDY. 1948. Ohio studies with repellent materials with notes on damage to corn by pheasants and other wildlife. J. Wildl. Manage. 12:392–398.

DANA, S. T. 1968. Remarks in Goddard, M. K., and R. R. Widner. The job ahead. Pp. 562–565 in Forests and forestry in the American states (R. R. Widner, ed.). Nat. Assoc. of State Foresters. Washington, D.C. 594 pp.

DANIELS, T. J., AND R. C. FALCO. 1989. The Lyme disease invasion. Natural Hist. 98(7):4, 6, 8, 10.

DANSEREAU, P. 1957. Biogeography: an ecological perspective. Ronald Press, New York. 394 pp.

DANTZLER, W. H. 1982. Renal adaptations of desert vertebrates. BioScience 32:108–113.

DARLING, F. F. 1960. An ecological reconnaissance of the Mara Plains in Kenya Colony. Wildl. Monogr. No. 5. 41 pp.

DARR, G. W., AND D. A. KLEBENOW. 1975. Deer, brush control, and livestock on the Texas Rolling Plains. J. Range Manage. 28:115–119.

DARWIN, C. 1845. Journal of researches into the natural history and geology of the countries visited during the voyage of the H.M.S. Beagle round the world, under the command of Capt. FitzRoy, R. N. 2nd ed., corrected, with additions. John Murray, Colonial and Home Library, London. 519 pp.

DARWIN, C. 1859. The origin of species by means of natural selection. John Murray, London. Reprinted in 1960 by Harvard Univ. Press, Cambridge, Mass. 502 pp.

DASMANN, R. F. 1964. African game ranching. Macmillan, New York. 76 pp.

DASMANN, R. F. 1968. Game ranching potentials in North America. Pp. 11–12 in Symposium on introduction of exotic animals: ecological and socioeconomic considerations. Caesar Kleberg Res. Pgm. in Wildl. Ecology, Texas A&M Univ., College Station. 25 pp.

DASMANN, R. F. 1975. The conservation alternative. John Wiley & Sons, New York. 164 pp.

DASMANN, R. F. 1981. Wildlife biology, 2nd ed. John Wiley & Sons, New York. 212 pp.

DASMANN, W. 1981. Deer range improvement and management. McFarland and Co., Jefferson, N. C. 168 pp.

DAUBENMIRE, R. 1968. Plant communities: a textbook of plant synecology. Harper and Row, New York. 300 pp.

D'AULAIRE, P. O., AND E. D'AULAIRE. 1986. Lessons from a ravaged jungle. Internat. Wildl. 16:36–40.

DAVEY, S. P. 1967. The role of wildlife in an urban environment. Trans. N. Am. Wildl. and Nat. Resour. Conf. 32:50–60.

DAVIDSON, W. R., J. L. BLUE, L. B. FLYNN, S. M. SHEA, R. L. MARCHINTON, AND J. A. LEWIS. 1987. Parasites, diseases and health status of sympatric populations of sambar deer and white-tailed deer in Florida. J. Wildl. Dis. 23:267–272.

DAVIDSON, W. R., AND C. B. CROW. 1983. Parasites, diseases, and health status of sympatric populations of sika deer and white-tailed deer in Maryland and Virginia. J. Wildl. Dis. 19:345–348.

DAVIDSON, W. R., J. M. CRUM, J. L. BLUE, D. W. SHARP, AND J. H. PHILLIPS. 1985. Parasites, diseases, and health status of sympatric populations of fallow deer and white-tailed deer in Kentucky. J. Wildl. Dis. 2:153–159.

DAVIDSON, W. R., F. A. HAYES, V. F. NETTLES, AND F. E. KELLOGG (EDS.). 1981. Diseases and parasites of white-tailed deer. Tall Timbers Res. Sta., Univ. Ga., Athens. 458 pp.

DAVIS, D. E., AND R. L. WINSTEAD. 1980. Estimating the numbers of wildlife populations. Pp. 221–245 in Wildlife management techniques manual, 4th ed. (S. D. Schemnitz, ed.). The Wildlife Society, Washington, D.C. 686 pp.

DAVIS, D. S. 1986. Preliminary bison/brucellosis report. P. 4 in Buffalo World (newsletter), Am. Buffalo Assoc. 24 pp.

DAVIS, J. W., L. H. KARSTAD, AND D. O. TRAINER (EDS.). 1970. Infectious diseases of wild mammals. Ia. State Univ. Press, Ames. 421 pp.

DAVIS, J. W., L. H. KARSTAD, AND D. O. TRAINER (EDS.). 1981. Infectious diseases of wild mammals, 2nd ed. Ia. State Univ. Press, Ames. 446 pp.

DAVIS, R. B., AND C. K. WINKLER. 1968. Brush vs. cleared range as deer habitat in southern Texas. J. Wildl. Manage. 32:321–329.

DAVIS, W. B., AND J. L. ROBERTSON. 1944. The mammals of Culberson County, Texas. J. Mamm. 25:254–273.

DAY, A. M. 1949. North American waterfowl. Stackpole and Heck, Inc., New York. 329 pp.

DAY, A. M. 1964. Developing fish and wildlife resources through Public Law 566 projects. Trans N. Am. Wildl. and Nat. Resour. Conf. 29:112–118.

DAY, B. W., JR., L. E. GARLAND, F. L. MCLAUGHLIN, C. H. WILEY, T. R. MYERS, AND J. W. ARTMANN. 1971. Vermont's game annual. Vt. Fish and Game Dept. Bull. 71-1, Montpelier. 42 pp.

DAY, D. 1981. The doomsday book of animals. Viking Press, New York. 288 pp.

DAY, R. H., S. M. MURPHY, J. A. WIENS, G. D. HAYWARD, E. J. HARNER, AND B. E. LAWHEAD. 1997. Effects of the *Exxon Valdez* oil spill on habitat use by birds along the Kenai Peninsula, Alaska. Condor 99:728–742.

DAYE, P. 1980. Attempts to acclimate embryos and alevins of Atlantic salmon, *Salmo salar*, and rainbow trout, *S. gairdneri*, to low pH. Can. J. Fish Aquat. Sci. 27:1035–1038.

DAYE, P. G., AND GARSIDE, E. T. 1977. Lower lethal levels of pH for embryos and alevins of Atlantic salmon (*Salmo salar*). Can. J. Zool. 55:1504–1508.

DEARMENT, R. 1972. When disease strikes livestock, don't blame big game. Texas Parks and Wildl. 30:12–15.

DEBUS, S. J. S. 1985. Cuckoo-shrikes as urban scavangers. Australian Birds 20:28–29.

DEBYLE, N. V. 1981. Clearcutting and fire in the larch/Douglas-fir forests of western Montana—a multifaceted research summary. U.S.D.A. Forest Serv. Intermountain Forest and Range Expt. Sta. Gen. Tech. Rept. INT-99. 73 pp.

DEBYLE, N. V. 1985. Managing wildlife habitat with fire in the aspen ecosystem. Pp. 73–82 *in* Proc. Symposium on fire's effects on wildlife habitat. U.S.D.A. Forest Serv. Gen. Tech. Rept. INT 186. Missoula, Mont. 96 pp.

DECALESTA, D. S. 1983. Influence of regulation on deer harvest. Abstract *in* Symposium on game harvest management. Caesar Kleberg Wildl. Res. Inst., Kingsville, Tex. 37 pp.

DECALESTA, D. S., AND D. B. SCHWENDEMAN. 1978. Characterization of deer damage to soybean plants. Wildl. Soc. Bull. 6:250–253.

DECKER, D. J., AND T. A. GAVIN. 1987. Public attitudes toward a suburban deer herd. Wildl. Soc. Bull. 15:173–180.

DEEVEY, E. S., JR. 1947. Life tables for natural populations of animals. Quarterly Rev. Biol. 22:283–314.

DEFENDERS OF WILDLIFE. 1984. Changing U.S. trapping policy: a handbook for activists. Defenders of Wildlife, Washington, D.C. 56 pp.

DEGABRIELE, R. 1980. The physiology of the koala. Sci. Am. 243:110–117.

DEGRAAF, R. M., AND B. R. PAYNE. 1975. Economic values of non-game birds and some urban wildlife research needs. Trans. N. Am. Wildl. and Nat. Resour. Conf. 40:281–287.

DEGRAAF, R. M., AND J. M. WENTWORTH. 1981. Urban bird communities and habitats in New England. Trans. N. Am. Wildl. and Nat. Resour. Conf. 46:396–413.

DEGRAAF, R. M., AND G. M. WITMAN. 1979. Trees, shrubs, and vines for attracting birds. Univ. Mass. Press, Amherst. 194 pp.

DEGRAZIO, J. W., J. F. BESSER, T. J. DECINO, J. L. GUARINO, AND E. W. SCHAFER. 1972. Protecting ripening corn from blackbirds by broadcasting 4-aminopyridine baits. J. Wildl. Manage. 36:1316–1320.

DEKKER, D., AND G. ERICKSON. 1986. Releases of peregrine falcons in southern and central Alberta, 1976–1985. Alberta Nat. 16:1–3.

DE KOCK, W. C., AND C. T. BOWMER. 1993. Bioaccumulation, biological effects, and food chain transfer of contaminants in the zebra mussel (*Dreissena polymorpha*). Pp. 503–533 *in* Zebra mussels: biology, impacts, and control (T. F. Nalepa and D. W. Schloesser, eds.). Lewis Publ., Boca Raton, Fla. 810 pp.

DELACOUR, J. 1951. The pheasants of the world. Charles Scribner's Sons, New York. 347 pp.

DELACOUR, J. 1954. The waterfowl of the world. Vol. I. Country Life Ltd., London. 284 pp.

DELACOUR, J. 1959. The waterfowl of the world. Vol. IV. Country Life Ltd., London. 364 pp.

DEL RIO, C., S. R. GRANTER, AND P. H. DURAY. 1997. Lyme borreliosis. Pp. 269–283 *in* Pathology of emerging infections (C. R. Horsburgh, Jr. and A. M. Nelson, eds.). ASM Press, Washington, D.C. 332 pp.

DEMARAIS, S., H. A. JACOBSON, AND D. C. GUYNN. 1983. Abomasal parasites as a health index for white-tailed deer in Mississippi. J. Wildl. Manage. 47:247–252.

DEMENT, S. H., J. J. CHISOLM, JR., J. C. BARBER, AND J. D. STRANDBERG. 1986. Lead exposure in an "urban" peregrine falcon and its avian prey. J. Wildl. Dis. 22:238–244.

DEMENT, S. H., J. J. CHISOLM, JR., M. A. ECKHAUS, AND J. D. STRANDBERG. 1987. Toxic lead exposure in the urban rock dove. J. Wildl. Dis. 23:272–278.

DENNEY, A. H. 1944. Wildlife relationships to soil types. Trans. N. Am. Wildl. Conf. 9:316–323.

DERKSEN, D. V., AND W. D. ELDRIDGE. 1980. Drought-displacement of pintails to the Arctic Coastal Plain, Alaska. J. Wildl. Manage. 44:224–229.

DESAI, J. R. 1974. The Gir forest reserve: its habitats, faunal, and social problems. Pp. 193–198 *in* Second world conference on national parks (S. H. Elliott, ed.). Internat. Union for Cons. of Nature and Natural Resources, Morges, Switzerland. 504 pp.

DESGRANGES, J-L., AND M. L. HUNTER, JR. 1987. Duckling response to lake acidification. Trans. N. Am. Wildl. and Nat. Resour. Conf. 52:636–644.

DEVINE, T., AND T. J. PETERLE. 1968. Possible differentiation of natal areas of North American waterfowl by neutron activation analysis. J. Wildl. Manage. 32:274–279.

DEVOS, A. 1979. The need for an action programme on wildlife conservation in arid and semi-arid regions of Africa. J. Arid Environ. 2:369–372.

DEVOS, A., R. H. MANVILLE, AND R. G. VAN GELDER. 1956. Introduced mammals and their influence on native biota. Zoologica 41:163–194.

DEXTER, R. W. 1955. The vertebrate fauna on the campus of Kent State University. The Biologist 37:84–88.

DIAMOND, J. 1987. The worst mistake in the history of the human race. Discover 8:64–66.

DICE, L. R. 1947. Effectiveness of selection by owls of deer-mice (*Peromyscus maniculatus*), which contrast in color with their background. Lab. Vertebrate Biol. Contr. 34, Univ. Mich. Press, Ann Arbor. 20 pp.

DICE, L. R., AND P. M. BLOSSOM. 1937. Studies of mammalian ecology in southwestern North America with special attention to the colors of desert mammals. Carnegie Inst. of Washington Publ. No. 485. Waverly Press, Baltimore, Md. 129 pp.

DICKSON, D. L. (ED.). 1997. King and common eiders of the western Canadian arctic. Occas. Paper 94, Canadian Wildl. Serv., Ottawa, Ontario. 79 pp.

D'ITRI, F. M. (ED.). 1985. A systems approach to conservation tillage. Lewis Publ., Chelsea, Mich. 384 pp.

DIXON, K. R., AND M. C. SWIFT. 1981. The optimal harvesting concept in furbearer management. Pp. 1524–1551 *in* Proc. worldwide furbearer conf., Vol. II. (J.A. Chapman and D. Pursley, eds.). Frostburg, Md. 1552 pp.

DODGE, R. I. 1883. Our wild Indians. A. D. Worthington and Co., Hartford, Conn. 653 pp.

DOERR, T. B., AND F. S. GUTHERY. 1983. Effects of tebuthiuron on lesser prairie-chicken habitat and foods. J. Wildl. Manage. 47:1138–1142.

DOLBEER, R. A. 1981. Cost-benefit determination of blackbird damage control for cornfields. Wildl. Soc. Bull. 9:44–51.

DOLBEER, R. A., P. P. WORONECKI, AND R. L. BRUGGERS. 1986a. Reflecting tapes repel blackbirds from millet, sunflowers, and sweet corn. Wildl. Soc. Bull. 14:418–425.

DOLBEER, R. A., P. P. WORONECKI, AND R. A. STEHN. 1986b. Resistance of sweet corn to damage by blackbirds and starlings. J. Am. Hort. Soc. 111:306–311.

DOLBEER, R. A., P. P. WORONECKI, R. A. STEHN, G. J. FOX, J. J. HANZEL, AND G. M. LINZ. 1986c. Field trails of sunflower resistant to bird depredation. North Dakota Farm Res. 43:21–24, 28.

DOMAN, E. R., AND D. I. RASMUSSEN. 1944. Supplemental winter feeding of mule deer in northern Utah. J. Wildl. Manage. 8:317–338.

DORNBUSH, J. N., AND J. R. ANDERSON. 1964. Ducks on the wastewater pond. Water and Sewage Works 111:271–276.

DOWNHOWER, J. F., AND E. R. HALL. 1966. The pocket gopher in Kansas. U. Kans. Mus. Nat. Hist. and State Biol. Surv. Misc. Publ. 44. Lawrence, Kans. 32 pp.

DOWNING, R. L. 1987. Success story: white-tailed deer. Pp. 45–57 *in* Restoring America's wildlife (H. Kallman, ed.). U.S. Dept. Interior. Washington, D.C. 394 pp.

DRAWE, D. L. 1968. Midsummer diet of deer on the Welder Wildlife Refuge. J. Range Manage. 21:164–166.

DRAWE, D. L., A. D. CHAMRAD, AND T. W. BOX. 1978. Plant communities of the Welder Wildlife Refuge. Contrib. 5, Series B, 2nd ed. Welder Wildl. Foundation, Sinton, Tex. 38 pp.

DREGNE, H. E. 1983. Desertification of arid lands. Harwood Academic Publ., New York. 246 pp.

DREWIEN, R. C., AND E. G. BIZEAU. 1978. Cross-fostering whooping cranes to sandhill crane foster parents. Pp. 201–222 *in* Endangered birds: management techniques for preserving threatened species (S. A. Temple, ed.). Univ. Wis. Press, Madison. 466 pp.

DROBNEY, R. D., AND L. H. FREDRICKSON. 1979. Food selection by wood ducks in relation to breeding status. J. Wildl. Manage. 43:109–120.

DROZE, W. H. 1977. Trees, prairies, and people. Texas Women's Univ. Press, Denton. 313 pp.

DUEBBERT, H. F. 1969. High nest density and hatching success of ducks on South Dakota CAP land. Trans. N. Am. Wildl. and Nat. Resour. Conf. 34:218–229.

DUEBBERT, H. F., AND H. A. KANTRUD. 1974. Upland duck nesting related to land use and predator reduction. J. Wildl. Manage. 38:257–265.

DUEBBERT, H. F., AND J. T. LOKEMOEN. 1976. Duck nesting in fields of undisturbed grass-legume cover. J. Wildl. Manage. 40:39–49.

DUEBBERT, H. F., AND J. T. LOKEMOEN. 1980. High duck nesting success in a predator-reduced environment. J. Wildl. Manage. 44:428–437.

DUEBBERT, H. F., E. T. JACOBSON, K. F. HIGGINS, AND E. B. PODOLL. 1981. Establishment of seeded grasslands for wildlife habitat in the prairie pothole region. U.S. Fish and Wildl. Serv. Spec. Sci. Rept.—Wildl. 234. 21 pp.

DURANT, A. J. 1956. Impaction and pressure necrosis in Canada geese due to eating dry hulled soybeans. J. Wildl. Manage. 20:399–404.

DUSI, J. L.1978. Impact of cattle egrets on an upland colony area. Proc. 1977 Conf. Colonial Waterbird Group 1:128–130.

DUSI, J. L. 1979. Stabilities of heron colonies in swamp and upland sites. Proc. 1978. Conf. Colonial Waterbird Group 2:38–40.

DUTTWEILER, M. W. 1975. Urban sport fishing: a review of literature and programs. N.Y. Coop. Fish Res. Unit, Cornell Univ. Press, Ithaca. 52 pp.

DVORAK, D. F. 1980. A brief history and status of aoudad sheep in Palo Duro Canyon, Texas. P. 23 *in* Proc. symposium on ecology and management of Barbary sheep (C. D. Simpson, ed.). Dept. of Range and Wildlife Management, Texas Tech Univ., Lubbock. 112 pp.

DWERNYCHUK, L. W., AND D. A. BOAG. 1972. How vegetative cover protects duck nests from egg-eating birds. J. Wildl. Manage. 36:955–958.

DYKSTERHUIS, E. J. 1949. Condition and management of range land based upon quantitative ecology. J. Range Manage. 2:104–115.

DYKSTERHUIS, E. J., AND E. M. SCHMUTZ. 1947. Natural mulches or "litter of grasslands": with kinds and amounts on a southern prairie. Ecology 28:163–179.

EATON, R. L. 1986. Zen and the art of hunting. Carnivore Press, Reno, Nev. 74 pp.

EBEDES, H. 1976. Anthrax epizootics in wildlife in the Etosha National Park, South West Africa. Pp. 519–526 *in* Wildlife diseases (L. A. Page, ed.). Plenum Press, New York. 686 pp.

EBERHARD, T. 1954. Food habits of Pennsylvania house cats. J. Wildl. Manage. 18:284–286.

EDMINSTER, F. C. 1954. American game birds of field and forest. Charles Scribner's Sons. New York. 490 pp.

EDWARDS, R. Y. 1954. Fire and the decline of a mountain caribou herd. J. Wildl. Manage. 18:521–526.

EDWARDS, R. Y. 1956. Snow depths and ungulate abundance in the mountains of western Canada. J. Wildl. Manage. 20:159–168.

EDWARDS, R. Y., AND R. W. RITCEY. 1956. The migrations of a moose herd. J. Mamm. 37:486–494.

EDWARDS, W. R., AND K. E. SMITH. 1984. Exploratory experiments on the stability of mineral profiles of feathers. J. Wildl. Manage. 48:853–866.

EHRENFELD, D. W. 1976. The conservation of nonresources. Am. Scientist 64:648–656.

EHRLICH, P. R., AND J. ROUGHGARDEN. 1987. The science of ecology. Macmillan, New York. 710 pp.

EIDE, S. H., S. D. MILLER, AND M. A. CHIHULY. 1986. Oil pipeline crossing sites utilized in winter by moose, *Alces alces*, and caribou, *Rangifer tarandus*, in southcentral Alaska. Can. Field-Nat. 100:197–207.

EINARSEN, A. S. 1946. Crude protein determination of deer food as an applied management technique. Trans. N. Am. Wildl. Conf. 11:309–312.

EINARSEN, A. S. 1948. The pronghorn antelope and its management. Wildl. Manage. Inst., Washington, D.C. 238 pp.

EISENBERG, J. F., AND R. W. THORINGTON. 1973. A preliminary analysis of a neotropical mammal fauna. Biotropica 5:150–161.

EKBALD, S., AND J. CROCKFORD. 1983. A no-freeze watering trough. Rangelands 5:15.

ELDER, J. B. 1956. Watering patterns of some desert game animals. J. Wildl. Manage. 20:368–378.

ELGMORK, K., A. HAGEN, AND A. LANGELAND. 1973. Polluted snow in southern Norway during the winters 1968–1971. Environ. Pollut. 4:41–52.

ELLIS, D. H., AND J. W. CARPENTER. 1981. A technique for vasectomizing birds. J. Field Ornith. 52:69–71.

ELLIS, D. H., S. T. DOBROTT, AND J. G. GOODWIN, JR. 1977. Reintroduction techniques for masked bobwhites. Pp. 345–354 *in* Endangered birds, management techniques for preserving threatened species (S. A. Temple, ed.). Univ. Wis. Press, Madison. 466 pp.

ELLIS, M. M. 1937. Some fishery problems in impounded waters. Trans. Am. Fisheries Soc. 66:63–75.

ELLIS, M. M. 1942. Freshwater impoundments. Trans. Am. Fisheries Soc. 71:80–93.

ELMAN, R. 1982. America's pioneering naturalists. Winchester Press, Tulsa, Okla. 231 pp.

ELTON, C. 1927. Animal Ecology. Macmillan, New York. 207 pp.

ELTON, C. 1939. On the nature of cover. J. Wildl. Manage. 3:332–338.

ELTON, C. 1958. The ecology of invasions by animals and plants. Methuen & Co., London. 181 pp.

ELTRINGHAM, S. K. 1984. Wildlife resources and economic development. John Wiley & Sons, New York. 325 pp.

EMMERICH, J. M., AND P. A. VOHS. 1982. Comparative use of four woodland habitats by birds. J. Wildl. Manage. 46:43–49.

ENDERSON, J. H., AND D. D. BERGER. 1970. Pesticides: eggshell thinning and lowered production of young in prairie falcons. BioScience 20:355–356.

ENDERSON, J. H., G. R. CRAIG, W. A. BURNHAM, AND D. D. BERGER. 1982. Eggshell thinning and organochlorine residues in Rocky Mountain peregrines, *Falco peregrinus*, and their prey. Can. Field-Nat. 96:255–264.

ENGBRING, J., AND T. H. FRITTS. 1988. Demise of an insular avifauna: the brown tree snake on Guam. Trans. Western Sect. Wildl. Soc. 24:31–37.

ENNIS, H. J. 1981. Agricultural crop losses and migratory bird management—a Manitoba government perspective. Pp. 155–158 *in* Fourth internat. waterfowl symposium. Ducks Unlimited, Chicago. 265 pp.

ENNIS, W. B., JR. 1964. Selective toxicity for herbicides. Weed Res. 4:93–104.

ENVIRONMENTAL PROTECTION AGENCY. 1994. Lead fishing sinkers: response to citizens' petition and proposed ban. 40 CFR Part 745:11122–11145.

ERICKSON, R. E., AND J. E. WIEBE. 1973. Pheasants, economics, and land retirement programs in South Dakota. Wildl. Soc. Bull. 1:22–27.

ERICKSON, R. E., R. L. LINDER, AND K. W. HARMON. 1979. Stream channelization (P. L. 83-566) increased wetland losses in the Dakotas. Wildl. Soc. Bull. 7:71–78.

ERIKSSON, M. 1976. Food and feeding habits of downy goldeneye *Bucephala clangula* (L.) ducklings. Ornis Scand. 7:159–169.

ERIKSSON, M. 1979. Competition between freshwater fish and goldeneyes *Bucephala clangula* (L.) for common prey. Oecologia 41:99–107.

ERIKSSON, M. O. G. 1986. Fish delivery, production of young, and nest density of osprey (*Pandion haliaetus*) in southwest Sweden. Can. J. Zool. 64:1961–1965.

ERLINGE, S., G. GORANSSON, G. HOGSTEDT, G. JANSSON, O. LIBERG, J. LOMAN, I. N. NILSSON, T. VON SCHANTZ, AND M. SILVEN. 1984. Can vertebrate predators regulate their prey? Am. Nat. 123:125–133.

ERRINGTON, P. L. 1945. Some contributions of a fifteen-year local study of the northern bobwhite to a knowledge of population phenomena. Ecol. Monogr. 15:1–34.

ERRINGTON, P. L. 1946. Predation and vertebrate populations. Quarterly Rev. Biol. 21:144–177, 221–245.

ERRINGTON, P. L. 1956. Factors limiting higher vertebrate populations. Science 124:304–307.

ERRINGTON, P. L. 1961. An American visitor's impressions of Scandinavian waterfowl problems. J. Wildl. Manage. 25:109–130.

ERRINGTON, P. L. 1967. Of predation and life. Ia. State Univ. Press, Ames. 277 pp.

ERWIN, T. L. 1982. Tropical forests: their richness in Coleoptera and other arthropod species. Coleopterists' Bull. 36:74–75.

ERWIN, T. L. 1991. How many species are there?: revised. Conserv. Biol. 5:330–333.

ESTABROOKS, S. R. 1987. Ingested lead shot in northern red-billed whistling ducks (*Dendrocygna autumnalis*) and northern pintails (*Anas acuta*) in Sinaloa, Mexico. J. Wildl. Dis. 23:169.

EULER, D., AND H. SMITH. 1983. Calculating kill quotas for deer with minimal data. Abstract *in* Symposium on game management. Caesar Kleberg Wildl. Res. Inst., Kingsville, Tex. 37 pp.

EVANS, A. C., AND W. J. McL. GUILD. 1947. Studies on the relationships between earthworms and soil fertility. I. Biological studies in the field. Ann. Appl. Biol. 34:307–330.

EVANS, J. 1970. About nutria and their control. U.S. Fish and Wildl. Serv. Res. Publ. 86:1–65.

EVANS, K. E., AND R. N. CONNER. 1979. Snag management. Pp. 214–225 *in* Proc. workshop on management of north central and northeastern forests for nongame birds. U.S.D.A. Forest Serv., North Central Forest Exp. Sta., St. Paul, Minn. 268 pp.

EVE, J. H., AND F. E. KELLOGG. 1977. Management implications of abomasal parasites in southeastern white-tailed deer. J. Wildl. Manage. 41:169–177.

EVENDEN, F. G. 1974. Wildlife as an indicator of a quality environment. Pp. 19–21 *in* A symposium on wildlife in an urbanizing environment (J. H. Noyes and D. R. Progulske, eds.). Univ. Mass. Coop. Ext. Serv., Amherst. 182 pp.

FAANES, C. A. 1987. Bird behavior and mortality in relation to power lines in prairie habitats. Fish Wildl. Tech. Rept. 7. U.S. Fish and Wildl. Serv., Washington, D.C. 24 pp.

FALK, N. W., H. B. GRAVES, AND E. D. BELLIS. 1978. Highway right-of-way fences as deer deterrents. J. Wildl. Manage. 42:646–650.

FALLIS, A. M., AND G. F. BENNETT. 1966. On the epizootiology of infections caused by *Leucocytozoon simondi* in Algonquin Park, Canada. Can. J. Zool. 44:101–112.

FANCY, S. G., AND R. G. WHITE. 1985. Energy expenditures by caribou while cratering in snow. J. Wildl. Manage. 49:987–993.

FARNER, D. S. 1945. Age groups and longevity in the American robin. Wilson Bull. 57:56–74.

FARNWORTH, E. G., AND F. B GOLLEY (EDS.). 1974. Fragile ecosystems. Springer-Verlag, New York. 258 pp.

FASSETT, N. C. 1957. A manual of aquatic plants. Univ. Wis. Press, Madison. 405 pp.

FAY, L. D., A. P. BOYCE, AND W. G. YOUATT. 1956. An epizootic in deer in Michigan. Trans. N. Am. Wildl. Conf. 21:173–184.

FAYRER-HOSKEN, R. A., P. BROOKS, H. J. BERTSCHINGER, J. F. KIRKPATRICK, J. W. TURNER, AND I. K. M. LIU. 1997. Management of African elephant populations by immunocontraception. Wildl. Soc. Bull. 25:18–21.

FEARNSIDE, P. M. 1983. Land-use trends in the Brazilian Amazon region as factors in accelerating deforestation. Environ. Cons. 10:141–148.

FEATHERLY, H. I. 1940. Silting and forest succession on Deep Fork in southwestern Creek County, Oklahoma. Proc. Okla. Acad. Sci 24:63–65.

FEIERABEND, J. S. 1983. Steel shot and lead poisoning in waterfowl: an annotated bibliography of research 1976–1983. Sci. and Tech. Series 8, Nat. Wildl. Fed., Washington, D.C. 62 pp.

FELBECK, H. 1981. Chemoautotrophic potential of the hydrothermal vent tube worm, *Riftia pachyptila* Jones (Vestimentifera). Science 213:336–338.

FELDHAMER, G. A., J. A. CHAPMAN, AND R. L. MILLER. 1978. Sika deer and white-tailed deer on Maryland's eastern shore. Wildl. Soc. Bull. 6:155–157.

FELDMAN, B. M. 1974. The problem of urban dogs. Science 185:903.

FELTUS, D. G., AND E. E. LANGENAU, JR. 1984. Optimization of firearm deer hunting and timber values in northern lower Michigan. Wildl. Soc. Bull. 12:6–12.

FENDERSON, C. N. 1954. The brown trout in Maine. Maine Dept. Inland Fisheries and Game, Fishery Res. and Manage. Div. Bull. No. 2, Augusta. 16 pp.

FENNER, F., AND F. N. RATCLIFFE. 1965. Myxomatosis. Univ. Press, Cambridge, England. 379 pp.

FENNER, F., AND G. M. WOODROOFE. 1953. The pathogenesis of infectious myxomatosis: the mechanism of infection and the immunological response in the European rabbit (*Oryctolagus cuniculus*). Brit. J. Exptl. Pathol. 34:400–411.

FENNER, F., M. F. DAY, AND G. M. WOODROOFE. 1952. The mechanisms of the transmission of myxomatosis in the European rabbit (*Oryctolagus cuniculus*) by the mosquito *Aedes aegypti*. Australian J. Exptl. Biol. Med. Sci. 30:139–152.

FERBER, A. E. 1974. Windbreaks for conservation. U.S.D.A. Soil Cons. Serv. Agric. Inform. Bull. 339 (reprinted). U.S. Govt. Printing Office, Washington, D.C.

FERENCE, L. 1997. Profile of the shooting sports, 1996. Nat. Shooting Sports Found., Newton, Conn. 16 pp.

FERGUSON, D. E. 1963. Notes concerning the effects of heptachlor on certain poikilotherms. Copeia 1963:441–443.

FERGUSON, D. E. 1967. The ecological consequences of pesticide resistance in fishes. Trans. N. Am. Wildl. and Nat. Resour. Conf. 32:103–107.

FERGUSON, D. E., AND C. R. BINGHAM. 1966. Endrin resistance in the yellow bullhead, *Ictalurus natalis*. Trans. Am. Fisheries Soc. 95:325–326.

FIGLEY, W. K., AND L. W. VANDRUFF. 1982. The ecology of urban mallards. Wildl. Monogr. No. 81. 40 pp.

FINDHOLT, S. L., AND C. H. TROST. 1985. Organochlorine pollutants, eggshell thickness, and reproductive success of black-crowned night-herons in Idaho, 1979. Colonial Waterbirds 8:32–41.

FINDLEY, J. S., AND S. ANDERSON. 1956. Zoogeography of the montane mammals of Colorado. J. Mamm. 37:80–82.

FINKELSTEIN, D. 1983. Deer as industry. Audubon 84:38–39.

FISCHER, J. R., D. E. STALLKNECHT, M. P. LATTRELL, A. A. DHONDT, AND K. A. CONVERSE. 1997. Mycophasmal conjunctivitis in wild songbirds: the spread of a new contagious disease in a mobile host population. Emerg. Infect. Dis. 3:69–72.

FISH, D., AND T. J. DANIELS. 1990. The role of medium-sized mammals as reservoirs of *Borrelia burgdorferi* in southern New York. J. Wildl. Disease 26:339–345.

FISHBEIN, D. B., A. J. BELOTTO, R. E. PACER, J. S. SMITH, W. G. WINKLER, S. R. JENKINS, AND K. M. PORTER. 1986. Rabies in rodents and lagomorphs in the United States, 1971–1984: increased cases in the woodchuck (*Marmota monax*) in mid-Atlantic states. J. Wildl. Dis. 22:151–155.

FISHER, A. R., AND R. G. ANTHONY. 1980. The effect of soil texture on distribution of pine voles in Pennsylvania orchards. Am. Midl. Nat. 104:39–46.

FISHER, J., N. SIMON, AND J. VINCENT. 1969. Wildlife in danger. Viking Press, New York. 368 pp.

FISHERIES AND ENVIRONMENT, DEPARTMENT OF. 1977. The seal hunt. Dept. Fisheries and Environ., Ottawa, Canada. 24 pp.

FISK, E. J. 1978. The growing use of roofs by nesting birds. Bird-Banding 49:135–141.

FITTKAU, E. J. 1970. Role of caimans in the nutrient regime of mouth-lakes of Amazon affluents (an hypothesis). Biotropica 2:138–142.

FITTKAU, E. J. 1973. Crocodiles and the nutrient metabolism of Amazonian waters. Amazoniana 4:103–133.

FITZGERALD, B. M. 1988. Diet of domestic cats and their impact on prey populations. Pp. 123–144, plus appendix, *in* The domestic cat, the biology of its behaviour (D.C. Turner and P. Bateson, eds.). Cambridge Univ. Press, New York. 222 pp.

FLADER, S. L. 1974. Thinking like a mountain: Aldo Leopold and the evolution of an ecological attitude toward deer, wolves, and forests. Univ. Neb. Press. Lincoln. 284 pp.

FLADER, S. L., AND J. B. CALLICOTT (EDS.). 1991. The river of the mother of God and other essays by Aldo Leopold. Univ. Wisconsin Press, Madison, Wis. 384 pp.

FLEMING, C. A. 1951. Sea lions as geological agents. J. Sedimentary Petrology 21:22–25.

FLICKINGER, E. L. 1979. Effects of aldrin exposure on snow geese in Texas rice fields. J. Wildl. Manage. 43:94–101.

FLICKINGER, E. L. 1981. Wildlife mortality at petroleum pits in Texas. J. Wildl. Manage. 45:560–564.

FLICKINGER, E. L., AND C. M. BUNCK. 1987. Number of oil-killed birds and fate of bird carcasses at crude oil pits in Texas. Southwestern Nat. 32:377–381.

FLICKINGER, E. L., K. A. KING, W. F. STOUT, AND M. M. MOHN. 1980. Wildlife hazards from furadan 3G applications to rice in Texas. J. Wildl. Manage. 44:190–197.

FLICKINGER, E. L., C. A. MITCHELL, D. H. WHITE, AND E. J. KOLBE. 1986. Bird poisoning from misuse of the carbamate Furadan in a Texas rice field. Wildl. Soc. Bull. 14:59–62.

FLICKINGER, E. L., D. H. WHITE, C. A. MITCHELL, AND T. G. LAMONT. 1984. Monocrotophos and dicrotophos residues in birds as a result of misuse of organophosphates in Matagorda County, Texas. J. Assoc. Off. Anal. Chem. 67:827–828.

FLYGER, V. 1959. A comparison of methods for estimating squirrel populations. J. Wildl. Manage. 23:220–223.

FLYGER, V. 1960. Sika deer on islands in Maryland and Virginia. J. Mamm. 41:140.

FLYGER, V. 1970. Urban gray squirrels—problems, management, and comparisons with forest populations. Trans. NE Fish and Wildl. Conf. 27:107–113.

FLYGER, V. 1974. Tree squirrels in urbanizing environments. Pp. 121–124 *in* A symposium on wildlife in an urbanizing environment (J. H. Noyes and D. R. Progulske, eds.). Univ. Mass. Coop. Ext. Serv., Amherst. 182 pp.

FOGG, G. E. 1975. Park planning guidelines. Nat. Soc. for Park Resour. Spec. Publ. Series 15001. 151 pp.

FORD, J. 1971. The role of the trypanosomiases in African ecology: a study of the tsetse-fly problem. Clarendon Press, Oxford. 568 pp.

FOREYT, W. J., AND W. M. SAMUEL. 1979. Parasites of white-tailed deer of the Welder Wildlife Refuge in southern Texas: a review. Pp. 105–132 *in* Proc. first Welder Wildlife Foundation symposium (D. L. Drawe, ed.). Sinton, Tex. 275 pp.

FOREYT, W. J., AND A. C. TODD. 1972. The occurrence of *Fascioloides magna* and *Fasciola hepatica* together in the liver of naturally infected cattle in south Texas, and the incidence of the flukes in cattle, white-tailed deer, and feral hogs. J. Parasitol. 58:1010–1011.

FOREYT, W. J., AND A. C. TODD. 1976. Development of the large American liver fluke, *Fascioloides magna*, in white-tailed deer, cattle, and sheep. J. Parasitol. 62:26–32.

FOREYT, W. J., A. C. TODD, AND K. FOREYT. 1975. *Fascioloides magna* (Bassi, 1875) in feral swine from southern Texas. J. Wildl. Dis. 11:554–559.

FORMAN, R. T. T., AND J. BAUDRY. 1984. Hedgerows and hedgerow networks in landscape ecology. Environ. Manage. 8:495–510.

FORMAN, R. T. T., AND M. GODRON. 1986. Landscape ecology. John Wiley & Sons, New York. 619 pp.

FORMOZOV, A. N. 1946. The snow cover as an environment factor and its importance in the life of mammals and birds. Trans. from Russian. Boreal Inst., Univ. Alberta, Edmonton, Canada. 176 pp.

FORSMAN, E. D., E. C. MESLOW, AND H. M. WIGHT. 1984. Distribution and biology of the spotted owl in Oregon. Wildl. Monogr. No. 87. 64 pp.

FOSBURGH, W. 1986. Spotted owl report opens new round in old-growth fight. Audubon Action 1:1, 15.

FOSTER, M. L., AND S. R. HUMPHREY. 1995. Use of highway underpasses by Florida panthers and other wildlife. Wildl. Soc. Bull. 23:95–100.

FOSTER, N. M., R. D. BRECKTON, A. J. LUEDKE, R. H. JONES, AND H. E. METCALF. 1977. Transmission of two strains of epizootic hemorrhagic disease virus in deer by *Culicoides variipennis*. J. Wildl. Dis. 13:9–16.

FOWLER, L. J., AND R. W. DIMMICK. 1983. Wildlife use of nest boxes in eastern Tennessee. Wildl. Soc. Bull. 11:178–181.

FOX, G. A. 1971. Recent changes in the reproductive success of the pigeon hawk. J. Wildl. Manage. 35:122–128.

FRANCIS, W. J. 1967. Prediction of California quail populations from weather data. Condor 69:405–410.

FRANKLIN, C., AND J. A. ALLEN. 1985. Hunting leases on private nonindustrial forestland in North Carolina. Proc. Ann. Conf. SE Assoc. Fish and Wildl. Agencies 39:344–350.

FRANKLIN, I. R. 1980. Evolutionary change in small populations. Pp. 135–149 *in* Conservation biology, an evolutionary-ecological perspective (M. E. Soulé and B. A. Wilcox, eds.). Sinauer Assoc., Sunderland, Mass. 395 pp.

FRANKLIN, J. F., J. A. MACMAHON, F. J. SWANSON, AND J. R. SEDELL. 1985. Ecosystem responses to the eruption of Mount St. Helens. Nat. Geograph. Res. 1:198–216.

FRANKLIN, T. M., AND K. B. REIS. 1996. Teaming with wildlife: an investment in the future of wildlife management. Wildl. Soc. Bull. 24:781–782.

FRANZMANN, A. W., A. FLYNN, AND P. D. ARNESON. 1975. Levels of some mineral elements in Alaskan moose hair. J. Wildl. Manage. 39:374–378.

FRASER, D. 1985. Mammals, birds, and butterflies at sodium sources in northern Ontario Forests. Can. Field-Nat. 99:365–367.

FRASER, D., AND E. R. THOMAS. 1982. Moose-vehicle accidents in Ontario: relation to highway salt. Wildl. Soc. Bull. 10:261–265.

FRASER, D., E. REARDON, F. DIEKEN, AND B. LOESCHER. 1980. Sampling problems and interpretation of chemical analysis of mineral springs used by wildlife. J. Wildl. Manage. 44:623–631.

FRASER-DARLING, F. 1960. Wildlife husbandry in Africa. Sci. Am. 203:123.

FREDA, J. 1986. The influence of acidic pond water on amphibians: a review. Water Air Soil Pollut. 30:439–450.

FREDRICKSON, L. F., R. L. LINDER, R. B. DAHLGREN, AND C. G. TRAUTMAN. 1978. Pheasant reproduction and survival as related to agricultural fertilizer use. J. Wildl. Manage. 42:40–45.

FREDRICKSON, L. H. 1978. Lowland hardwood wetlands: current status and habitat values for wildlife. Pp. 296–306 *in* Wetland functions and values: the state of our understanding (P. E. Greeson, J. R. Clark, and J. E. Clark, eds.). Am. Water Resour. Assoc., Bethesda, Md. 674 pp.

FREDRICKSON, L. H., AND J. L. HANSEN. 1983. Second broods in wood ducks. J. Wildl. Manage. 47:320–326.

FREDRICKSON, L. H., AND T. S. TAYLOR. 1982. Management of seasonally flooded impoundments for wildlife. U.S.D.I. Fish and Wildl. Serv. Res. Publ. 148. Washington, D.C. 29 pp.

FREIHERR, G. 1986. Vaccine sandwich to go. Science 86:68–69.

FRENCH, C. E., L. C. McEWEN, N. D. MAGRUDER, R. H. INGRAM, AND R. W. SWIFT. 1955. Nutritional requirements of white-tailed deer for growth and antler development. Penn. State Univ. Agric. Expt. Sta. Bull. 600, State College. 50 pp.

FRIEDMAN, G. M., AND J. E. SANDERS. 1978. Principles of sedimentology. John Wiley & Sons, New York. 792 pp.

FRIEDMAN, R., AND P. MUNDY. 1983. The use of "restaurants" for the survival of vultures in South Africa. Pp. 345–355 *in* Vulture

biology and management (S. R. Wilbur and J. A. Jackson, eds.). Univ. Calif. Press, Berkeley. 550 pp.

FRIEDMANN, H. 1963. Host relations of the parasitic cowbirds. U.S. Nat. Mus. Bull. 233, Washington, D.C. 276 pp.

FRIEND, M. 1976. Diseases: a threat to our waterfowl. Ducks Unlimited 40:36–37.

FRIEND, M. 1981. Waterfowl management and waterfowl disease: independent or cause and effect relationships? Trans. N. Am. Wildl. and Nat. Resour. Conf. 46:94–103.

FRIEND, M. 1984. Meeting migratory bird management needs by integrated disease control. Trans. N. Am. Wildl. and Nat. Resour. Conf. 49:480–488.

FRIEND, M. (ED.). 1987. Field guide to wildlife diseases. U.S. Fish and Wildl. Serv., Resour. Publ. 167. Washington, D.C. 225 pp.

FRINGS, H., AND J. JUMBER. 1954. Preliminary studies on the use of a specific sound to repel starlings (*Sturnus vulgaris*) from objectionable roosts. Science 119:318–319.

FRITH, H. J. 1959. The ecology of wild ducks in inland New South Wales: IV. Breeding, CSIRO Wildl. Res. 4:156–181.

FRITH, H. J. 1967. Waterfowl in Australia. East-West Center Press, Honolulu. 328 pp.

FRITH, H. J. 1973. Wildlife conservation. Angus and Robertson, Sydney. 414 pp.

FRITTS, S. H. 1983. Record dispersal by a wolf from Minnesota. J. Mamm. 64:166–167.

FRITTS, S. H., AND L. D. MECH. 1981. Dynamics, movements, and feeding ecology of a newly protected wolf population in northwestern Minnesota. Wildl. Monogr. No. 80. 79 pp.

FRITTS, T. H. 1988. The brown tree snake, *Boiga irregularis*, a threat to Pacific islands. U.S. Fish and Wildl. Serv. Biol. Rept. 88(31). 36 pp.

FRITTS, T. H., M. J. McCOID, AND R. L. HADDOCK. 1990. Risks to infants on Guam from bites of the brown tree snake (*Boiga irregularis*). Am. J. Trop. Med. Hyg. 42:607–611.

FRITTS, T. H., N. J. SCOTT, JR., AND J. A. SAVIDGE. 1987. Activity of the arboreal brown tree snake (*Boiga irregularis*) on Guam as determined by electrical outages. Snake 19:51–58.

FULBRIGHT, T. E., AND S. L. BEASOM. 1987. Long-term effects of mechanical treatments on white-tailed deer browse. Wildl. Soc. Bull. 15:560–564.

FULLER, T. K., AND D. M. HEISEY. 1986. Density-related changes in winter distribution of snowshoe hares in northcentral Minnesota. J. Wildl. Manage. 50:261–264.

FULLER, T. K., AND L. B. KEITH. 1981. Woodland caribou population dynamics in northeastern Alberta. J. Wildl. Manage. 45:197–213.

FULLER, T. K., AND W. L. ROBINSON. 1982. Winter movements of mammals across a large northern river. J. Mamm. 63:506–510.

FULLER, W. A. 1969. National parks and nature preservation. Pp. 99–110 *in* Canadian parks in perspective (J. G. Nelson, ed.). Harvest House Ltd., Montreal, Canada. 343 pp.

FURNISS, O. C. 1935. The duck situation in the Prince Albert District, Central Saskatchewan. Wilson Bull. 47:111–118.

GAFFNEY, W. S. 1941. The effects of winter elk browsing, South Fork of the Flathead River, Montana. J. Wildl. Manage. 5:427–453.

GAGE, L. 1987. Rescue attempt at Hubbard Glacier. Cetus 7(1):28–29.

GALLI, A. E., C. F. LECK, AND R. T. T. FORMAN. 1976. Avian distribution patterns in forest islands of different sizes in central New Jersey. Auk 93:356–364.

GALVIN, J. W., T. J. HOLLIER, K. D. BODINNAR, AND C. M. BUNN. 1985. An outbreak of botulism in wild waterbirds in southern Australia. J. Wildl. Dis. 21:347–350.

GARD, R. 1961. Effects of beaver on trout in Sagehen Creek, California. J. Wildl. Manage. 25:221–242.

GARRATY, J. A. 1975. The American nation, a history of the United States to 1877, Vol. 1, 3rd ed. Harper & Row, New York. 490 pp.

GARTNER, F. R. 1986. Horizontal wells—an economical water development option. Rangelands 8:8–11.

GASAWAY, W. C., AND I. O. BUSS. 1972. Zinc toxicity in the mallard duck. J. Wildl. Manage. 36:1107–1117.

GASAWAY, W. C., R. O. STEPHENSON, J. L. DAVIS, P. E. K. SHEPHERD, AND O. E. BURRIS. 1983. Interrelationships of wolves, prey, and man in interior Alaska. Wildl. Monogr. No. 84. 50 pp.

GASTON, K. J. 1991a. The magnitude of global insect species richness. Conserv. Biol. 5:283–296.

GASTON, K. J. 1991b. Estimates of the near-imponderable: a reply to Erwin. Conserv. Biol. 5:564–566.

GATES, J. E. 1991. Powerline corridors, edge effects, and wildlife inforested landscapes of the central Appalachians. Pp. 15–32 *in* Wildlife and habitats in managed landscapes (J. E. Rodiek and E. G. Bolen, eds.). Island Press, Washington, D.C. 219 pp.

GATES, J. E., AND L. W. GYSEL. 1978. Avian nest dispersion and fledging success in field-forest ecotones. Ecology 59:871–883.

GATES, J. M., AND G. E. OSTROM. 1966. Feed-grain program related to pheasant production in Wisconsin. J. Wildl. Manage. 30:612–617.

GAVARESKI, C. A. 1976. Relation of park size and vegetation to urban bird populations in Seattle, Washington. Condor 78:375–382.

GEBHARDT, M. R., T. C. DANIEL, E. E. SCHWEIZER, AND R. R. ALLMARAS. 1985. Conservation tillage. Science 230:625–630.

GEHLBACH, F. R. 1986. Odd couples of suburbia. Nat. Hist. 95(6):56–66.

GEIER, A. R., AND L. B. BEST. 1980. Habitat selection by small mammals of riparian communities: evaluating effects of habitat alterations. J. Wildl. Manage. 44:16–24.

GEIS, A. D. 1974. Effects of urbanization and type of urban development on bird populations. Pp. 97–105 *in* A symposium on wildlife in an urbanizing environment (J. H. Noyes and D. R. Progulske, eds.). Univ. Mass. Coop. Ext. Serv., Amherst. 182 pp.

GEIS, A. D., AND W. F. CRISSEY. 1973. 1970 test of the point system for regulating duck harvests. Wildl. Soc. Bull. 1:1–21.

GEIST, J. M., P. J. EDGERTON, AND A. W. ADAMS. 1983. Elk pellets aren't all alike. Rangelands 5:28–29.

GEIST, V. 1971. Mountain sheep—a study in behavior and evolution. Univ. Chicago Press. 383 pp.

GEIST, V. 1985. Game ranching: threat to wildlife conservation in North America. Wildl. Soc. Bull. 13:594–598.

GEIST, V. 1986a. Super antlers and pre-World War II European research. Wildl. Soc. Bull. 14:91–94.

GEIST, V. 1986b. Conservation unravelling: three threats to wildlife. Probe Post 9(2):26–29.

GEIST, V. 1988. How markets in wildlife meat and parts, and the sale of hunting privileges, jeopardize wildlife conservation. Conserv. Biol. 2:15–26.

GEORGE, J. L., S. S. DUBIN, AND B. M. NEAD. 1974. Continuing education needs of wildlife managers. Wildl. Soc. Bull. 2:59–62.

GEORGE, J. L., A. P. SNYDER, AND G. HANLEY. 1982. An initial survey of the value of the wild bird products industry. Penn. State Univ., University Park. 13 pp.

GEORGE, R. R., A. L. FARRIS, C. C. SCHWARTZ, D. D. HUMBURG, AND J. C. COFFEY. 1979. Native prairie grass pastures as nest cover for upland birds. Wildl. Soc. Bull. 7:4–9.

GEORGE, R. R., J. B. WOOLEY, JR., J. M. KIENZLER, A. L. FARRIS, AND A. H. BERNER.1980. Effect of hunting season length on ring-necked pheasant populations. Wildl. Soc. Bull. 8:279–283.

GEORGE, W. G. 1974. Domestic cats as predators and factors in winter shortages of raptor prey. Wilson Bull. 86:384–396.

GEORGHIOU, G. P., AND C. E. TAYLOR. 1977. Pesticide resistance as an evolutionary phenomenon. Proc. Internat. Congress of Ent. 15:759:785.

GERSTELL, R. 1942. The place of winter feeding in practical wildlife management. Penn. Game Comm. Res. Bull. 3, Harrisburg. 121 pp.

GESSEL, S. P., AND G. H. ORIANS. 1967. Rodent damage to fertilized Pacific silver fir in western Washington. Ecology 48:694–697.

GHIGLIERI, M. P. 1984. Realm of the mountain gorillas. J. East African Wildl. Soc. 7(2):24–27.

GIBBS, H. L., S. C. LATTA, AND J. P. GIBBS. 1987. Effects of the 1982–83 El Niño event on blue-footed and masked booby populations on Isla Daphne Major, Galápagos. Condor 89:440–442.

GIBBS, J. P. 1993. Importance of small wetlands for the persistence of local populations of wetland-associated animals. Wetlands 13:25–31.

GILBERT, A. H. 1977. Influence of hunter attitudes and characteristics on wildlife management. Trans. N. Am. Wildl. and Nat. Resour. Conf. 42:226–236.

GILBERT, D. C. 1971. Natural resources and public relations. The Wildl. Soc., Washington, D.C. 320 pp.

GILBERT, F. F. 1974. *Parelaphostrongylus tenuis* in Maine: II—Prevalence in moose. J. Wildl. Manage. 38:42–46.

GILBERT, F. F. 1976. Impact energy thresholds for anaesthetized raccoons, mink, muskrats, and beavers. J. Wildl. Manage. 40:669–676.

GILBERT, F. F. 1981. Maximizing the humane potential of traps—the Vital and the Conibear 120. Pp. 1630–1646 *in* Proc. worldwide furbearer conf., Vol. III. (J. A. Chapman and D. Pursley, eds.). Frostburg, Md. 2056 pp.

GILBERT, F. F. 1982. Public attitudes toward urban wildlife: a pilot study in Guelph, Ontario. Wildl. Soc. Bull. 10:245–253.

GILBERT, F. S. 1980. The equilibrium theory of island biogeography: fact or fiction? J. Biogeogr. 7:209–235.

GILBERT, P. F., O. C. WALLMO, AND R. B. GILL. 1970. Effect of snow depth on mule deer in Middle Park, Colorado. J. Wildl. Manage. 34:15–23.

GILES, R. H. 1978. Wildlife management. W. H. Freeman and Co., San Francisco. 416 pp.

GILL, A. E., AND E. M. RASMUSSON. 1983. The 1982–83 climate anomaly in the equatorial Pacific. Nature 306:229–234.

GILL, D. 1966. Coyote and man in Los Angeles. British Columbia Geograph. Series 7:69–84.

GILL, D. 1970. The coyote and the sequential occupants of the Los Angeles basin. Am. Anthropol. 72:821–826.

GILL, D., AND P. BONNETT. 1973. Nature in the urban landscape: a study of city ecosystems. York Press, Baltimore, Md. 209 pp.

GILLESPIE, G. D. 1985. Feeding behavior and impact of ducks on ripening barley crops in Otago, New Zealand. J. Applied Ecol. 22:347–356.

GILLESPIE, G. D. 1985a. Hybridization, introgression, and morphometric differentiation between mallard (*Anas platyrhynchos*) and grey duck (*Anas superciliosa*) in Otago, New Zealand. Auk 102:459–469.

GILMER, D. S., AND R. E. STEWART. 1984. Swainson's hawk nesting ecology in North Dakota. Condor 86:12–18.

GILPIN, M. E., AND M. E. SOULÉ. 1986. Minimum viable populations: process of species extinction. Pp. 19–34 *in* Conservation biology, the science of scarcity and diversity (M. E. Soulé, ed.). Sinauer Assoc., Sunderland, Mass. 584 pp.

GILTNER, L. T., AND J. F. COUCH. 1930. Western duck sickness and botulism. Science 72:660.

GINSBERG, H. S. (ED.). 1993. Ecology and environmental management of Lyme disease. Rutgers Univ. Press, New Brunswick, N.J. 224 pp.

GJERSING, F. M. 1975. Waterfowl production in relation to rest-rotation grazing. J. Range Manage. 28:37–42.

GLADFELTER, H. L., J. M. KIENZLER, AND K. J. KOEHLER. 1983. Effects of compound bow use on deer hunter success and crippling rates in Iowa. Wildl. Soc. Bull. 11:7–12.

GLADING, B. 1943. A self-filling quail watering device. Calif. Fish and Game 29:157–164.

GLADING, B. 1947. Game-watering devices for the arid Southwest. Trans. N. Am. Wildl. Conf. 12:286–292.

GLASS, B. P. 1970. Feeding mechanisms of bats. Pp. 84–92 *in* About bats, a chiropteran biology symposium (B. H. Slaughter and D. W. Walton, eds.). S. Meth. Univ. Press, Dallas. 339 pp.

GLAZENER, W. C. 1967. Management of the Rio Grande turkey. Pp. 453–492 *in* The wild turkey and its management (O. H. Hewitt, ed.). The Wildlife Society, Washington, D.C. 589 pp.

GLINSKI, R. L., AND R. D. OHMART. 1983. Breeding ecology of the Mississippi kite in Arizona. Condor 85:200–207.

GOLDSTEIN, M. I., B. WOODBRIDGE, M. E. ZACCAGNINI, AND S. B. CANAVELLI. 1996. An assessment of mortality of Swainson's hawks on wintering grounds in Argentina. J. Raptor. Res. 30:106–107.

GOLIGHTLY, R. T., JR., AND R. D. OHMART. 1984. Water economy of two desert canids: coyote and kit fox. J. Mamm. 65:51–58.

GOMEZ-POMPA, A., C. VAZQUEZ-YANES, AND S. GUEVARA. 1972. The tropical rain forest: a nonrenewable resource. Science 177:762–765.

GOODLAND, R. J. A., AND H. S. IRWIN. 1975. Amazon jungle: green hell to red desert? Elsevier Sci. Publ. Co., New York. 155 pp.

GOODRUM, P. 1949. Status of bobwhite quail in the United States. Trans. N. Am. Wildl. Conf. 14:359–369.

GORDON, K. 1966. Mammals and the influence of the Columbia River Gorge on their distribution. Northwest Science 40:142–146.

GORE, H. G., AND W. F. HARWELL. 1983. White-tailed deer age, weight, and antler development survey. Job 14, Federal Aid Proj. W-109-R-6, Texas Parks and Wildlife Dept., Austin. 43 pp. Processed.

GORE, H. G., W. F. HARWELL, M. D. HOBSON, AND W. J. WILLIAMS. 1983. Buck permits as a management tool in south Texas. Abstract in Symposium on Game Harvest Management. Caesar Kleberg Wildl. Res. Inst., Kingsville, Tex. 37 pp.

GORHAM, E. 1955. On the acidity and salinity of rain. Geochim. et Cosmochim., Acta 7:231–239.

GOTTSCHALK, J. S. 1972. The German hunting system, West Germany, 1968. J. Wildl. Manage. 36:110–118.

GOULDING, M. 1980. The fishes and the forest. Univ. Calif. Press, Berkeley. 280 pp.

GOWATY, P. A. 1984. House sparrows kill eastern bluebirds. J. Field Ornith. 55:378–380.

GRABER, R. R., AND J. W. GRABER. 1963. A comparative study of bird populations in Illinois, 1906–1909 and 1956–1958. Ill. Nat. Hist. Surv. Bull. 28:383–528.

GRABILL, B. A. 1977. Reducing starling use of wood duck boxes. Wildl. Soc. Bull. 5:69–70.

GRAF, W. 1955. Cottontail rabbit introductions and distribution in western Oregon. J. Wildl. Manage. 19:184–188.

GRAHAM, F. 1970. Since Silent Spring. Houghton Mifflin Co., Boston, Mass. 333 pp.

GRAHAM, F., JR. 1976. Blackbirds, a problem that won't fly away. Audubon 78:118–125.

GRANDY, J. W. 1986. Providing humane stewardship for wildlife: the case against sport hunting. Pp. 295–300 *in* Advances in animal welfare science 1986/87 (M. W. Fox and L. D. Mickley, eds.). Humane Soc. of the United States, Washington, D.C. 302 pp.

GRANDY, J. W., L. N. LOCKE, AND G. E. BAGLEY. 1968. Relative toxicity of lead and five proposed substitute shot types to pen-reared mallards. J. Wildl. Manage. 32:483–488.

GRANGE, W. B. 1948. Wisconsin grouse problems. Wis. Cons. Dept. Publ. No. 328, Madison. 318 pp.

GRANT, W. E., E. C. BIRNEY. N. R. FRENCH, AND D. M. SWIFT. 1982. Structure and productivity of grassland small mammal communities related to grazing-induced changes in vegetative cover. J. Mamm. 63:248–260.

GRAUL, W. D., AND G. C. MILLER. 1984. Strengthening ecosystem management approaches. Wildl. Soc. Bull. 12:282–289.

GRAUL, W. D., J. TORRES, AND R. DENNEY. 1976. A species-ecosystem approach for nongame programs. Wildl. Soc. Bull. 4:79–80.

GREELEY, F., R. F. LABISKY, AND S. H. MANN. 1962. Distribution and abundance of pheasants in Illinois. Ill. Nat. Hist. Biol. Notes No. 47. 16 pp.

GREEN, J. S., AND R. A. WOODRUFF. 1990. Livestock guarding dogs: protecting sheep from predators. Agric. Inform. Bull. 588, U.S. Dept. Agric., Washington, D.C. 31 pp.

GREEN, R. G., AND C. A. EVANS. 1940. Studies on a population cycle of snowshoe hares in the Lake Alexander area. J. Wildl. Manage. 4:220–238, 247–258, 267–278.

GREEN, R. G., AND C. L. LARSON. 1938a. A description of shock disease in the snowshoe hare. Am. J. Hyg. 28:190–212.

GREEN, R. G., C. L. LARSON, AND D. W. MATHER. 1938b. The natural occurrence of shock disease in hares. Trans. N. Am. Wildl. Conf. 3:877–881.

GREEN, R. G., C. L. LARSON, AND J. F. BELL. 1939. Shock disease as the cause of the periodic decimation of the snowshoe hare. Am. J. Hyg. 30B:83–102.

GREEN, W. C. H. 1986. Age-related differences in nursing behavior among American bison cows (*Bison bison*). J. Mamm. 67:739–741.

GREENE, R. A., AND G. H. MURPHY. 1932. The influence of two burrowing rodents, *Dipodomys spectabilis spectabilis* (kangaroo rat) and *Neotoma albigula albigula* (pack rat), on desert soils in Arizona. II. Physical effects. Ecology 13:359–363.

GREENE, R. A., AND C. REYNARD. 1932. The influence of two burrowing rodents, *Dipodomys spectabilis spectabilis* (kangaroo rat) and *Neotoma albigula albigula* (pack rat), on desert soils in Arizona. Ecology 13:73–80.

GREENHALL, A. M. 1972. The problem of bat rabies, migratory bats, livestock and wildlife. Trans. N. Am. Wildl. and Nat. Resour. Conf. 37:287–293.

GREENWAY, J. C., JR. 1967. Extinct and vanishing birds of the world, 2nd ed. Dover, New York. 520 pp.

GREER, D. M. 1982. Urban waterfowl population: ecological evaluation of management and planning. Environ. Manage. 6:217–229.

GRICE, D., AND J. P. ROGERS. 1965. The wood duck in Massachusetts. Bull. Mass. Div. Fisheries and Game, Westboro. 96 pp.

GRIFFIN, D. R. 1940. Migrations of New England bats. Bull. Mus. Comp. Zool. Harvard Univ. 86:217–246.

GRIFFIN, D. R. 1945. Travels of banded cave bats. J. Mamm. 26: 15–23.

GRIFFITH, B., J. M. SCOTT, J. W. CARPENTER, AND C. REED. 1989. Translocation as a species conservation tool: status and strategy. Science 245:477–480.

GRIFFITH, P. 1976. Introduction of the problems. Pp. 3–7 in Shelterbelts on the Great Plains: proceedings of the symposium (R. W. Tinus, ed.). Great Plains Agric. Council Publ. 78. Lincoln, Neb. 218 pp.

GRIZZELL, R. A. 1960. Fish and wildlife management on watershed projects. Trans. N. Am. Wildl. and Nat. Resour. Conf. 25:186–192.

GROSS, B. D., J. L. HOLECHEK, D. HALLFORD, AND R. D. PIEPER. 1983. Effectiveness of antelope pass structures in restriction of livestock. J. Range Manage. 36:22–24.

GRUBB, T. G. 1976. A survey and analysis of bald eagle nesting in western Washington. Unpubl. M.S. Thesis, Univ. Washington, Seattle. 87 pp.

GRUE, C. E., W. J. FLEMING, D. G. BUSBY, AND E. F. HILL. 1983. Assessing hazards of organophosphate pesticides to wildlife. Trans. N. Am. Wildl. and Nat. Resour. Conf. 48:200–220.

GRUE, C. E., T. J. O'SHEA, AND D. J. HOFFMAN. 1984. Lead concentrations and reproduction in highway-nesting barn swallows. Condor 86:383–389.

GRZIMEK, B. 1975. Grzimek's Animal Life Encyclopedia. Vol. 12. Mammals III. Van Nostrand Reinhold, New York. 657 pp.

GUARINO, J. L. 1972. Methiocarb, a chemical bird repellent: a review of its effectiveness on crops. Proc. Vertebrate Pest Conf. 5:108–111.

GUILIANO, W. M., C. R. ALLEN, R. S. LUTZ, AND S. DEMARAIS. 1996. Effects of red imported fire ants on northern bobwhite chicks. J. Wildl. Manage. 60:309-313.

GULDEN, N. A., AND L. L. JOHNSON. 1968. History, behavior, and management of a flock of giant Canada geese in southeastern Minnesota. Pp. 58–71 in Canada goose management (R. L. Hine and C. Schoenfeld, eds.). Dembar Educational Resour. Serv., Madison, Wis. 195 pp.

GULJAS, E. 1975. Managing woodlands for wildlife. Indiana Dept. Nat. Resour. Manage. Series 3. 9 pp.

GULLION, G. W. 1965. A critique concerning foreign game bird introductions. Wilson Bull. 77:409–414.

GULLION, G. W. 1966. A viewpoint concerning the significance of studies of game bird food habits. Condor 68:372–376.

GULLION, G. W. 1970. Factors influencing ruffed grouse populations. Trans. N. Am. Wildl. and Nat. Resour. Conf. 35:93–105.

GULLION, G. W. 1972. Improving your forested lands for ruffed grouse. The Ruffed Grouse Soc., Coraopolis, Penn. 36 pp.

GULLION, G. W. 1977. Forest manipulation for ruffed grouse. Trans. N. Am. Wildl. and Nat. Resour. Conf. 42:449–458.

GULLION, G. W. 1982. Michigan Out-of-Doors interviews Gordon Gullion. Michigan Out-of-Doors 36:42–46.

GUNDERSON, D. R. 1968. Floodplain use related to stream morphology and fish populations. J. Wildl. Manage. 32:507–514.

GURCHINOFF, S., AND W. L. ROBINSON. 1972. Chemical characteristics of jackpine needles selected by feeding spruce grouse. J. Wildl. Manage. 36:80–87.

GUTERMUTH, C. R. 1971. Role of policy-making boards and commissions. Proc. 7th Ann. Short Course Game and Fish Management, Colo. State Univ., Fort Collins, 11 pp.

GUTH, R. W. 1981. Wildlife in the Chicago area: the interaction of feeding and vegetation. Trans. N. Am. Wildl. and Nat. Resour. Conf. 46:432–438.

GUTHERY, F. S., T. E. ANDERSON, AND V. W. LEHMANN. 1979. Range rehabilitation enhances cotton rats in south Texas. J. Range Manage. 32:354–356.

GUTHERY, F. S., J. CUSTER, AND M. OWEN. 1980. Texas Panhandle pheasants: their history, habitat needs, habitat development opportunities, and future. U.S.D.A. Forest Serv. Gen. Tech. Rept. RM-74, Rocky Mountain Forest and Range Exp. Sta., Fort Collins, Colo. 11 pp.

HAAPANEN, A. 1965. Bird fauna of the Finnish forests in relation to forest succession. J. Ann. Zool. Fenn. 2:153–196.

HAAS, G. H. 1977. German hunting: it's the tradition that counts. Internat. Wildl. 7:38–43.

HAGAN, J. M., III, AND D. W. JOHNSTON (EDS.). 1992. Ecology and conservation of Neotropical migrant landbirds. Smithsonian Instit. Press, Washington, D.C. 609 pp.

HAGAR, D. C. 1960. The interrelationships of logging, birds, and timber regeneration in the Douglas-fir region of northwestern California. Ecology 41:116–125.

HAHN, H. C., JR. 1949. A method of censusing deer and its application in the Edwards Plateau of Texas. Texas Game, Fish and Oyster Comm. 24 pp.

HAILEY, T. L., J. W. THOMAS, AND R. M. ROBINSON. 1966. Pronghorn die-off in Trans-Pecos Texas. J. Wildl. Manage. 30: 488–496.

HAINES, T. A. 1981. Acid precipitation and its consequences for aquatic ecosystems: a review. Trans. Am. Fisheries Soc. 110: 669–707.

HAINSWORTH, F. R. 1981. Animal physiology: adaptations in function. Addison-Wesley, Reading, Mass. 600 pp.

HALEY, J. E. 1936. Charles Goodnight, cowman and plainsman. Univ. Okla. Press, Norman. 485 pp.

HALL, J. S. 1962. A life history and taxonomic study of the Indiana bat, *Myotis sodalis*. Reading Pub. Mus. Art Gal. Sci. Publ. 12, Reading, Penn. 68 pp.

HALL, L. S., P. R. KRAUSMAN, AND M. L. MORRISON, 1997. The habitat concept and plea for standard terminology. Wildl. Soc. Bull. 25:173–182.

HALLIDAY, T. R. 1980. The extinction of the passenger pigeon, *Ectopistes migratorius*, and its relevance to contemporary conservation. Biol. Cons. 17:157–162.

HALLORAN, A. F., AND O. V. DEMING. 1958. Water development for desert bighorn sheep. J. Wildl. Manage. 22:1–9.

HAMERSTROM, F. 1970. An eagle to the sky. Ia. State Univ. Press, Ames. 142 pp.

HAMERSTROM, F. 1979. Effect of prey on predator: voles and harriers. Auk 96:370–374.

HAMERSTROM, F. 1980. Strictly for the chickens. Iowa State Univ. Press, Ames, I. 174. [An expanded version was published in 1994, retitled My double life: memoirs of a naturalist. Univ. Wisconsin Press, Madison, Wis. 316 pp.]

HAMERSTROM, F. 1986. Harrier, hawk of the marshes: the hawk that is ruled by a mouse. Smithson. Instit. Press, Washington, D.C. 171 pp.

HAMERSTROM, F., F. N. HAMERSTROM, AND C. J. BURKE. 1985. Effect of voles on mating systems in a central Wisconsin population of harriers. Wilson Bull. 97:332–346.

HAMERSTROM, F. N., JR. 1939. A study of Wisconsin prairie chicken and sharp-tailed grouse. Wilson Bull. 51:105–120.

HAMILTON, G., A. D. RUTHVEN, E. FINDLAY, K. HUNTER, AND D. A. LINDSAY. 1981. Wildlife deaths in Scotland resulting from misuse of agricultural chemicals. Biol. Cons. 21:315–326.

HAMILTON, K. C., AND K. P. BUCHHOLTZ. 1953. Use of herbicides for establishing food patches. J. Wildl. Manage. 17:509–516.

HAMILTON, S. J., A. N. PALMISANO, G. A. WEDEMEYER, AND W. T. YASUTAKE. 1986. Impacts of selenium on early life stages and smoltification of fall Chinook salmon. Trans. N. Am. Wildl. and Nat. Resour. Conf. 51:343–356.

HAMMACK, J., AND G. M. BROWN, JR. 1974. Waterfowl and wetlands: toward bioeconomic analysis. Washington Resources for the Future. Distrib. by Johns Hopkins Univ. Press, Baltimore, Md. 95 pp.

HAMMAN, R. L. 1982. Spawning and culture of humpback chub. Prog. Fish Cult. 44:213–216.

HAMMERQUIST-WILSON, M. M., AND J. A. CRAWFORD. 1981. Response of bobwhites to cover changes within three grazing systems. J. Range Manage. 34:213–215.

HAMPY, D. B., B. PENCE, AND C. D. SIMPSON. 1979. Serological studies on sympatric Barbary sheep and mule deer from Palo Duro Canyon, Texas. J. Wildl. Dis. 15:443–446.

HANLEY, T. A. 1983. Black-tailed deer, elk, and forest edge in a western Cascades watershed. J. Wildl. Manage. 47:237–242.

HANLEY, T. A., AND W. W. BRADY. 1977a. Seasonal fluctuations in nutrient content of feral burro forages, lower Colorado River Valley, Arizona. J. Range Manage. 30:370–373.

HANLEY, T. A., AND W. W. BRADY. 1977b. Feral burro impact on a Sonoran Desert range. J. Range Manage. 30:374–377.

HANSEN, C. S. 1983. Costs of deer-vehicle accidents in Michigan. Wildl. Soc. Bull. 11:161–164.

HANSEN, H. A. 1961. Loss of waterfowl production to tide floods. J. Wildl. Manage. 25:242–248.

HANSEN, P. 1987. Acid rain and waterfowl: the case for concern in North America. Izaak Walton League of America, Minneapolis, Minn. 39 pp.

HANSEN, R. M., AND I. K. GOLD. 1977. Blacktail prairie dogs, desert cottontails and cattle trophic relations on shortgrass range. J. Range Manage. 30:210–214.

HANSEN, R. M., AND P. S. MARTIN. 1973. Ungulate diets in the lower Grand Canyon. J. Range Manage. 26:380–381.

HANSKI, I., AND M. E. GILPIN. 1991. Metapopulation dynamics: brief history and conceptual domain. Biol. J. Linnean Soc. 42:3–16.

HANSON, H. C. 1965. The giant Canada goose. Southern Ill. Univ. Press, Carbondale. 226 pp.

HANSON, H. C., AND R. L. JONES. 1976. The biogeochemistry of blue, snow, and Ross's geese. Ill. Nat. Hist. Surv. Spec. Publ. No. 1. 281 pp.

HANSON, H. C., AND C. W. KOSSACK. 1963. The mourning dove in Illinois. Tech. Bull. 2, Ill. Dept. Cons., Springfield. 133 pp.

HANSON, H. C., AND R. H. SMITH. 1950. Canada geese of the Mississippi Flyway. Bull. Ill. Nat. Hist. Surv. 25(Art. 3):67–210.

HANSON, L. E., AND D. R. PROGULSKE. 1973. Movements and cover preferences of pheasants in South Dakota. J. Wildl. Manage. 37:454–461.

HANSON, W. C. 1967. Cesium-137 in Alaskan lichens, caribou, and Eskimoes. Health Physics 13:383–389.

HARDIN, J. W., AND L. W. VANDRUFF. 1978. Characteristics of human visitors at urban waterfowl ponds in the vicinity of Syracuse, New York. Trans. NE Fish and Wildl. Conf. 35:130–142.

HARGER, E. M. 1956. Behavior of a ring-necked pheasant on a prairie chicken booming ground, Wilson Bull. 68:70–71.

HARMON, K. W. 1976. The economies of wetland values. U.S. Fish and Wildl. Wetlands Symposium, Watertown, S.D. 11 pp. Mimeo.

HARMON, K. W., AND M. W. NELSON. 1973. Wildlife and soil considerations in land retirement programs. Wildl. Soc. Bull. 1:28–38.

HARMON, W. H. 1944. Notes on mountain goats in the Black Hills. J. Mamm. 25:149–151.

HARPER, J. A. 1963. Calcium in grit consumed by juvenile pheasants in east-central Illinois. J. Wildl. Manage. 27:362–367.

HARPER, J. A. 1964. Calcium in grit consumed by hen pheasants in east-central Illinois. J. Wildl. Manage. 28:264–270.

HARPER, J. A., AND R. F. LABISKY. 1964. The influence of calcium on the distribution of pheasants in Illinois. J. Wildl. Manage. 28:722–731.

HARPMAN, D. A., AND C. F. REULER. 1985. Economic aspects of the nongame checkoff. Nongame Newsl. 3(5):4–8.

HARRINGTON, F. H., AND L. D. MECH. 1979. Wolf howling and its role in territory maintenance. Behaviour 68:207–249.

HARRIS, L. D. 1984. The fragmented forest, island biogeography theory and the preservation of biotic diversity. Univ. Chicago Press, Chicago. 211 pp.

HARRIS, L. D. 1988. Landscape linkages: the dispersal corridor approach to wildlife conservation. Trans. N. Am. Wildl. and Nat. Resour. Conf. 53:595–607.

HARRIS, L. D., AND P. B. GALLAGHER. 1989. New initiatives for wildlife conservation, the need for movement corridors. Pp. 11–34 in In defense of wildlife: preserving communities and corridors (G. Mackintosh, ed.). Defenders of Wildl., Washington, D.C. 96 pp.

HARRIS, L. D., AND I. H. KOCHEL. 1981. A decision-making framework for population management. Pp. 221–239 in Dynamics of large mammal populations (C. W. Fowler and T. D. Smith, eds.). John Wiley & Sons, New York. 477 pp.

HARRIS, L. D., AND J. SCHECK. 1991. From implications to applications: the dispersal corridor principle applied to the conservation of biological diversity. Pp. 189–220 in Nature conservation 2: the role of corridors (D. A. Saunders and R. J. Hobbs, eds.). Surrey Beatty & Sons, Chipping Norton, NSW, Australia. 442 pp.

HARRIS, L. D., AND G. SILVA-LOPEZ. 1992. Forest fragmentation and the conservation of biological diversity. Pp. 197–237 in Conservation biology, the theory and practice of nature conservation preservation and management (P. L. Fiedler and S. K. Jain, eds.). Chapman and Hall, New York. 507 pp.

HARRIS, L. D., C. MASER, AND H. J. ANDREWS. 1982. Patterns of old-growth liquidation and implications for Cascades wildlife. Trans. N. Am. Wildl. and Nat. Resour. Conf. 47:374–392.

HARRIS, M. T., W. L. PALMER, AND J. L. GEORGE. 1983. Preliminary screening of white-tailed deer repellents. J. Wildl. Manage. 47:516–519.

HARRIS, P., AND R. CRANSTON. 1979. An economic evaluation of control methods for diffuse and spotted knapweed in western Canada. Can. J. Plant Sci. 59:375–382.

HARRIS, R. B., L. A. MCGUIRE, AND M. L. SHAFFER. 1987. Sample sizes for minimum viable population estimation. Conserv. Biol. 1:72–76.

HARRIS, V. T. 1952. An experimental study of habitat selection by prairie and forest races of the deermouse, *Peromyscus maniculatus*. Contr. Lab. Vert. Biol., Univ. Mich. 56:1–53.

HARRIS, V. T. 1956. The nutria as a wild fur animal in Louisiana. Trans. N. Am. Wildl. Conf. 21:474–485.

HARRISON, G. H. 1992. Is there a killer in your house? Nat. Wildl. 30(6):10–13.

HARRISON, P. D., AND M. I. DYER. 1984. Lead in mule deer forage in Rocky Mountain National Park, Colorado. J. Wildl. Manage. 48:510–517.

HARRISON, R. L. 1997. A comparison of gray fox ecology between residential and undeveloped rural landscapes. J. Wildl. Manage. 61:112–122.

HARTMAN, D. S. 1979. Ecology and behavior of the manatee (*Trichechus manatus*) in Florida. Am. Soc. Mammalogists Spec. Publ. 5. 153 pp.

HARTUNG, R. 1965. Some effects of oiling on reproduction of ducks. J. Wildl. Manage. 29:872–874.

HARTUNG R. 1967. Energy metabolism in oil-covered ducks. J. Wildl. Manage. 31:798–804.

HARTUNG, R., AND G. S. HUNT. 1966. Toxicity of some oils to waterfowl. J. Wildl. Manage. 30:564–570.

HASLER, A. D. 1960. Guideposts of migrating fish. Science 132: 785–792.

HASSINGER, J., C. E. SCHWARZ, AND R. G. WINGARD. 1981. Timber sales and wildlife. Penn. Game Comm., Harrisburg. 14 pp.

HATHAWAY, M. B. 1973. Ecology of city squirrels. Nat. Hist. 82: 61–62.

HAUGEN, A. O. 1952. Trichomoniasis in Alabama mourning doves. J. Wildl. Manage. 16:164–169.

HAUGEN, A. O., AND J. KEELER. 1952. Mortality of mourning doves from trichomoniasis in Alabama during 1951. Trans. N. Am. Wildl. Conf. 17:141–151.

HAVERA, S. P., AND F. C. BELLROSE. 1985. The Illinois River: a lesson to be learned. Wetlands 4:29–41.

HAWKINS, A. S., R. C. HANSON, H. K. NELSON, AND H. M. REEVES (EDS.). 1984. Flyways: pioneering waterfowl management in North America. U.S. Fish and Wildl. Serv., Washington, D.C. 517 pp.

HAY, K. G. 1975. The status of oiled wildlife: research and planning. Pp. 249–253 *in* Proc. on prevention and control of oil spills, Am. Petroleum Inst., Washington, D.C. 612 pp.

HAYS, C. G. 1984. The Humboldt penguin (*Spheniscus humboldti*) in Peru and the effects of the 1982–1983 El Niño. M.S. Thesis, Univ. Florida, Gainesville. 86 pp.

HAYS, H., AND R. W. RISEBROUGH. 1972. Pollutant concentrations in abnormal young terns from Long Island Sound. Auk 89:19–35.

HEADY, H. F. 1960. Range management in the semi-arid tropics of east Africa according to principles developed in temperate climates. Proc. Internat. Grassland Congress 8:223–226.

HEBERT, D. M., AND I. M. COWAN. 1971a. White muscle disease in the mountain goat. J. Wildl. Manage. 35:752–756.

HEBERT, D. M., AND I. M. COWAN. 1971b. Natural salt licks as a part of the ecology of the mountain goat. Can. J. Zool. 49:605–610.

HEIMER, W. E. 1972. 1972 annual report. P. R. Projects W-17-3 and W-17-4, Jobs 6.1R and 6.2R. Alaska Dept. Fish and Game, Juneau.

HEIMSTRA, N. W., D. K. DAMKOT, AND N. G. BENSON. 1969. Some effects of silt turbidity on behavior of juvenile largemouth bass and green sunfish. U.S. Fish and Wildl. Serv. Bur. Sport Fisheries Wildl. Tech. Paper No. 20. Washington, D.C. 9 pp.

HEINZ, G. H. 1976a. Methylmercury: Second-year feeding effects on mallard reproduction and duckling behavior. J. Wildl. Manage. 40:82–90.

HEINZ, G. H. 1976b. Methylmercury: Second-generation reproductive and behavioral effects on mallard ducks. J. Wildl. Manage. 40:710–715.

HEINZ, G. H. 1976c. Behavior of mallard ducklings from parents fed 3 ppm DDE. Bull. Environ. Contam. and Toxicol. 16:640–645.

HELMINEN, M. 1977. Hunter qualification and education programs in Finland. Trans. N. Am. Wildl. and Nat. Resour. Conf. 42:490–492.

HENDEE, J. C. 1974. A multiple-satisfaction approach to game management. Wildl. Soc. Bull. 2:104–113.

HENDRICKSON, G. O. 1947. Cottontail management in Iowa. Trans. N. Am. Wildl. Conf. 12:473–479.

HENDRICKSON, J., W. L. ROBINSON, AND L. D. MECH. 1975. Status of the wolf in Michigan, 1973. Am. Midl. Nat. 94:226–232.

HENDRIX, L. B. 1981. Post-eruption succession on Isla Fernandina, Galápagos. Madrono 28:242–254.

HENDRY, L. C., T. M. GOODWIN, AND R. F. LABISKY. 1980. Florida's vanishing wildlife. Florida Coop. Ext. Serv. Circular 485, Gainesville. 69 pp.

HENNY, C. J., L. J. BLUS, E. J. KOLBE, AND R. E. FITZNER, 1985. Organophosphate insecticide (Famphur) topically applied to cattle kills magpies and hawks. J. Wildl. Manage. 49:648–658.

HENSLEY, M. M., AND J. B. COPE. 1951. Further data on removal and repopulation of the breeding birds in a spruce-fir forest community. Auk 68:483–493.

HEPP, G. R., R. L. BLOHM, R. E. REYNOLDS, J. E. HINES, AND J. D. NICHOLS. 1986. Physiological condition of autumn-banded mallards and its relationship to hunting vulnerability. J. Wildl. Manage. 50:177–183.

HERBERT, R. A., AND K. G. S. HERBERT. 1965. Behavior of peregrine falcons in the New York City region. Auk 82:62–94.

HERBERT, R. A., AND K. G. S. HERBERT. 1969. The extirpation of the Hudson River peregrine falcon population. Pp. 133–154 *in* Peregrine falcon populations, their biology and decline (J. J. Hickey, ed.). Univ. Wis. Press, Madison. 596 pp.

HERMAN, C. M. 1968. Blood parasites of North American waterfowl. Trans. N. Am. Wildl. and Nat. Resour. Conf. 33:348–359.

HERMAN, C. M., J. H. BARROW, AND I. B. TARSHIS. 1975. Leucocytozoonosis in Canada geese at the Seney National Wildlife Refuge. J. Wildl. Dis. 11:404–411.

HERMAN, S. G., AND J. B. BULGER. 1979. Effects of a forest application of DDT on nontarget organisms. Wildl. Monogr. No. 69. 62 pp.

HERRERO, S. 1970. Human injury inflicted by grizzly bears. Science 170:593–598.

HERZOG, P. W., AND D. A. BOAG. 1977. Seasonal changes in aggressive behavior of female spruce grouse. Can. J. Zool. 55:1734–1739.

HESTER, F. E. 1965. Survival, renesting, and return of adult wood ducks to previously used nest boxes. Proc. Ann. Conf. SE Assoc. Game & Fish Comm. 16:67–70.

HEUSMANN, H. W. 1974. Mallard–black duck relationships in the Northeast. Wildl. Soc. Bull. 2:171–177.

HEUSMANN, H. W., AND R. G. BURRELL. 1974. Park mallards. Pp. 77–86 *in* A symposium on wildlife in an urbanizing environment (J. H. Noyes and D. R. Progulske, eds.). Univ. Mass. Coop. Ext. Serv., Amherst. 182 pp.

HIBLER, C. P., C. O. METZGER, T. R. SPRAKER, AND R. E. LANGE. 1974. Further observations on *Protostrongylus* sp. infection by transplacental transmission in bighorn sheep. J. Wildl. Dis. 10:39–41.

HICKEY, J. J. 1952. Survival studies of banded birds. U.S. Fish and Wildl. Serv. Spec. Sci. Rept.: Wildl. No. 15. Washington, D.C. 177 pp.

HICKEY, J. J. 1955. Is there a scientific basis for flyway management? Trans. N. Am. Wildl. Conf. 20:126–150.

HICKEY, J. J. (ED.). 1969. Peregrine falcon populations, their biology and decline. Univ. Wis. Press, Madison. 596 pp.

HICKEY, J. J. 1974. Some historical phases in wildlife conservation. Wildl. Soc. Bull. 2:164–170.

HICKEY, J. J., AND D. W. ANDERSON. 1968. Chlorinated hydrocarbons and eggshell changes in raptorial and fish-eating birds. Science 162:271–273.

HICKEY, J. J., AND L. B. HUNT. 1960. Initial songbird mortality following a Dutch elm disease control program. J. Wildl. Manage. 24:259–265.

HICKS, R. E. 1979. Guano deposition in an Oklahoma crow roost. Condor 81:247–250.

HIGGINS, K. F. 1977. Duck nesting in intensively farmed areas of North Dakota. J. Wildl. Manage. 41:232–242.

HILE, R., P. H. ESCHMEYER, AND G. F. LUNGER. 1951. Decline of the Lake Trout Fishery in Lake Michigan. U.S. Fish and Wildl. Serv. Fishery Bull. 52:77–95.

HILL, E. F., AND W. J. FLEMING. 1982. Anticholinesterase poisoning of birds: field monitoring and diagnosis of acute poisoning. Environ. Toxicol. Chem. 1:27–38.

HILL, E. F., AND V. M. MENDENHALL. 1980. Secondary poisoning of barn owls with famphur, an organophosphate insecticide. J. Wildl. Manage. 44:676–681.

HILL, E. F., R. G. HEATH, J. W. SPANN, AND J. D. WILLIAMS. 1975. Lethal dietary toxicities of environmental pollutants to birds. U.S. Fish and Wildl. Serv. Spec. Sci. Rept. No. 191. Washington, D.C. 61 pp.

HILL, E. P. 1972. Litter size in Alabama cottontails as influenced by soil fertility. J. Wildl. Manage. 36:1199–1209.

HILL, H. R. 1976. Feeding habits of the ring-necked pheasant chick, *Phasianus colchicus*, and the evaluation of available foods. Ph.D. Thesis. Mich. State Univ., East Lansing. 70 pp.

HILLMAN, C. N. 1968. Field observations of black-footed ferrets in South Dakota. Trans. N. Am. Wildl. and Nat. Resour. Conf. 33:433–443.

HILLMAN, C. N., R. L. LINDER, AND R. B. DAHLGREN. 1979. Prairie dog distribution areas inhabited by black-footed ferrets. Am. Midl. Nat. 102:185–187.

HOBAUGH, W. C., AND J. G. TEER. 1981. Waterfowl use characteristics of flood-prevention lakes in north-central Texas. J. Wildl. Manage. 45:16–26.

HOCHBAUM, H. A. 1944. The canvasback on a prairie marsh. The Stackpole Company, Harrisburg, Penn., and The Wildlife Management Institute, Washington, D.C. 207 pp.

HOCHBAUM, H. A. 1955. Travels and traditions of waterfowl. Charles T. Branford, Newton, Mass. 301 pp.

HOCHBAUM, H. A., AND E. F. BOSSENMAIER. 1965. Waterfowl research—accomplishments, needs, objectives. Trans. N. Am. Wildl. and Nat. Resour. Conf. 30:222–229.

HODDER, J., AND M. R. GRAYBILL. 1985. Reproduction and survival of seabirds in Oregon during the 1982–1983 El Niño. Condor 87:535–541.

HODGDON, K. W., AND J. H. HUNT. 1955. Beaver management in Maine. Maine Dept. of Inland Fisheries and Game Final Rept. Federal Aid to Wildl. Resour. Project 9-R, Augusta. 102 pp.

HOFF, G. L., AND D. O. TRAINER. 1981. Hemorrhagic diseases of wild ruminants. Pp. 45–53 *in* Infectious diseases of wild mammals (J. W. Davis, L. H. Karstad, and D. O. Trainer, eds.). Ia. State Univ. Press, Ames, 446 pp.

HOFF, G. L., S. H. RICHARDS, AND D. O. TRAINER. 1973. Epizootic of hemorrhagic disease in North Dakota deer. J. Wildl. Manage. 37:331–335.

HOFFMAN, C. O., AND J. L. GOTTSCHANG. 1977. Numbers, distribution, and movements of a raccoon population in a suburban residential community. J. Mamm. 58:623–636.

HOFFMAN, D. A., AND A. R. JONEZ. 1973. Lake Mead, case history. Pp. 220–233 *in* Man-made lakes: their problems and environmental effects (W. C. Ackerman et al., eds.). Am. Geophysical Union, Washington, D.C. 847 pp.

HOFFMAN, D. J. 1978. Embryotoxic effects of crude oil in mallard ducks and chicks. Toxicol. and Appld. Pharmacol. 46:183–190.

HOFFMAN, D. J., AND W. C. EASTIN. 1982. Effects of lindane, paraquat, toxaphene, and 2,4,5-trichlorophenoxyacetic acid on mallard embryo development. Arch. Environ. Contam. Toxicol. 11:79–86.

HOFFMAN, D. M. 1963. The lesser prairie chicken in Colorado. J. Wildl. Manage. 27:726–732.

HOHF, R. S., J. T. RATTI, AND R. CROTEAU. 1987. Experimental analysis of winter food selection by spruce grouse. J. Wildl. Manage. 51:159–167.

HOLBROOK, S. H. 1945. Burning an empire: the story of American forest fires. Macmillan, New York. 229 pp.

HOLECHEK, J. L. 1982. Managing rangelands for mule deer. Rangelands 4:25–28.

HOLECHECK, J. L., R. VALDEZ, S. D. SCHEMNITZ, R. D. PIEPER, AND C. A. DAVIS. 1982. Manipulation of grazing to improve or maintain wildlife habitat. Wildl. Soc. Bull. 10:205–210.

HOLEMAN, J. N. 1968. The sediment yield of major rivers of the world. Water Resour. Res. 4:737–747.

HOLL, S. A., AND V. C. BLEICH. 1987. Mineral lick use by mountain sheep in the San Gabriel Mountains, California. J. Wildl. Manage. 51:383–385.

HOLLING, C. S. 1959. The components of predation as revealed by a study of small-mammal predation of the European pine sawfly. Can. Ent. 91:293–320.

HOLMES, J. C. 1982. Impact of infectious disease agents on the population growth and geographical distribution of animals. Pp. 37–51 *in* Population biology of infectious diseases (R. M. Anderson and R. M. May, eds.). Springer-Verlag, New York. 315 pp.

HOLT, S. J., AND L. M. TALBOT. 1978. New principles for the conservation of wild living resources. Wildl. Monogr. No. 59. 33 pp.

HONESS, R. F. 1942. Lungworms of domestic sheep and bighorn sheep in Wyoming. Wyo. Agric. Expt. Sta. Bull. 255, Laramie. 24 pp.

HOOPER, R. G., A. F. ROBINSON, JR., AND J. A. JACKSON. 1980. The red-cockaded woodpecker: notes on life history and management. U.S.D.A. Forest Serv. Gen. Rept. SA-GR 9, Southeastern Area. 8 pp.

HOOPER, R. G., E. F. SMITH, H. S. CRAWFORD, B. S. MCGINNES, AND V. J. WALKER. 1975. Nesting bird populations in a new town. Wildl. Soc. Bull. 3:111–118.

HOOPER, R. G., J. C. WATSON, AND R. E. F. ESCANO. 1990. Hurricane Hugo's initial effects on red-cockaded woodpeckers in the Francis Marion National Forest. Trans. N. Am. Wildl. and Nat. Resour. Conf. 55:220–224.

HOOPES, D. T. 1982. Old-growth timber and wildlife management in southeast Alaska: a question of balance. Trans. N. Am. Wildl. and Nat. Resour. Conf. 47:588–604.

HOOVER, R. L., C. E. TILL, AND S. OGILVIE. 1959. The antelope of Colorado: a research and management study. Colo. Dept. Fish and Game Tech. Bull. 4, Denver. 110 pp.

HOPKINS, D. D., AND R. B. FORBES. 1979. Size and reproductive patterns of the Virginia opossum in northwestern Oregon. Murrelet 60:95–98.

HOPKINS, D. D., AND R. B. FORBES. 1980. Dietary patterns of the Virginia opossum in an urban environment. Murrelet 61:20–30.

HOPKINS, D. D., AND R. A. GRESBRINK. 1982. Surveillance of sylvatic plague in Oregon by serotesting carnivores. Am. J. Publ. Health 72:1295–1297.

HORKEL, J. D., AND W. D. PEARSON. 1976. Effects of turbidity on ventilation rates and oxygen consumption of green sunfish, *Lepomis cyanellus*. Trans. Am. Fisheries Soc. 105:107–113.

HORN, E. 1949. Waterfowl damage to agricultural crops and its control. Trans. N. Am. Wildl. Conf. 14:577–585.

HORNADAY, W. T. 1913. Our vanishing wild life, its extermination and preservation. N.Y. Zool. Soc. 411 pp.

HORNOCKER, M. G. 1969. Winter territoriality in mountain lions. J. Wildl. Manage. 33:457–464.

HORNOCKER, M. G. 1970. An analysis of mountain lion predation upon mule deer and elk in the Idaho Primitive Area. Wildl. Monogr. No. 21. 39 pp.

HORST, R. 1972. Bats as primary producers in an ecosystem. Bull. Nat. Speleol. Soc. 34:49–54.

HOTCHKISS, N., AND R. H. POUGH. 1946. Effect on forest birds of DDT used for gypsy moth control in Pennsylvania. J. Wildl. Manage. 10:202–207.

HOWARD R. J., AND J. S. LARSON. 1985. A stream habitat classification system for beaver. J. Wildl. Manage. 49:19–25.

HOWARD, R. P., AND M. L. WOLFE. 1976. Range improvement practices and ferruginous hawks. J. Range Manage. 29:33–37.

HOWARD, W. E. 1953. Nutria (*Myocastor coypus*) in California. J. Mamm. 34:512–513.

HOWARD W. E. 1960. Innate and environmental dispersal of individual vertebrates. Am. Midl. Nat. 63:152–161.

HOWARD, W. E. 1964. Introduced browsing mammals and habitat stability in New Zealand. J. Wildl. Manage. 28:421–429.

HOWARD, W. E. 1966. Control of introduced mammals in New Zealand. New Zealand Dept. of Sci. and Industrial Res. Animal Ecology Div. Inform. Series. No. 45. Whitcombe and Tombs Ltd., Wellington. 96 pp.

HOWARD, W. E. 1974. Predator control: whose responsibility? BioScience 24:359–363.

HOWARD, W. E., R. L. FENNER, AND H. E. CHILDS, JR. 1959. Wildlife survival in brush burns. J. Range Manage. 12:230–234.

HOWARTH, D. 1978. 1066 The year of the conquest. Viking Press, New York. 207 pp.

HOY, D. R. 1978. Geography and development, a world regional approach. Macmillan, New York. 728 pp.

HUBBARD, D. R. 1985. A descriptive epidemiological study of raccoon rabies in a rural environment. J. Wildl. Dis. 21:105–110.

HUBBS, C. L. 1940. Speciation of fishes. Am. Nat. 74:198–211.

HUBERT, W. A., AND J. N. KRULL. 1973. Seasonal fluctuations of aquatic macroinvertebrates in Oakwood Bottoms Greentree Reservoir. Am. Midl. Nat. 90:177–185.

HUDSON, M. S. 1983. Waterfowl production in three age-classes of stock ponds in Montana. J. Wildl. Manage. 47:112–117.

HUEY, W. S., AND W. H. WOLFRUM. 1956. Beaver-trout relations in New Mexico. Prog. Fish-Culturist 18:70–74.

HUFFAKER, C. B. 1958. Experimental studies on predation: dispersion factors and predator-prey oscillations. Hilgardia 27:343–383.

HUFFAKER, C. B. (ED.). 1980. New technology of pest control. John Wiley & Sons, New York. 500 pp.

HUGHES, J. M., C. J. PETERS, M. L. COHEN, AND B. W. J. MAHY. 1993. Hantavirus pulmonary syndrome: an emerging infectious disease. Science 262:850–851.

HUMPHREY, S. R. 1974. Zoogeography of the nine-banded armadillo (*Dasypus novemcinctus*) in the United States. BioScience 24:457–462.

HUNGERFORD, C. R. 1964. Vitamin A and productivity in Gambel's quail. J. Wildl. Manage. 28:141–147.

HUNT, E. G., AND A. I. BISCHOFF. 1960. Inimical effects on wildlife of periodic DDD applications to Clear Lake. Calif. Fish and Game 46:91–106.

HUNT, G. S. 1961. Waterfowl losses on the lower Detroit River due to oil pollution. Pp. 10–26 *in* Proc. 4th conf. on Great Lakes research, Great Lakes Res. Div., Inst. Sci. and Tech. Publ. 7. Univ. Mich., Ann Arbor. 213 pp.

HUNT, G. S., AND A. B. COWAN. 1963. Causes of death of waterfowl on the Lower Detroit River—winter 1960. Trans. N. Am. Wildl. and Nat. Resour. Conf. 28:150–163.

HUNT, G. S., AND R. W. LUTZ. 1959. Seed production by curly-leaved pondweed and its significance to waterfowl. J. Wildl. Manage. 23:405–408.

HUNT, L. B. 1960. Songbird breeding populations in DDT-sprayed Dutch elm disease communities. J. Wildl. Manage. 24:139–146.

HUNTER, B. F., AND M. N. ROSEN. 1965. Occurrence of lead poisoning in a wild pheasant (*Phasianus colchicus*). Calif. Fish and Game 51:207.

HUNTER, B. F., W. E. CLARK, P. J. PERKINS, AND P. R. COLEMAN. 1970. Applied botulism research including management recommendations. Wildl. Manage. Branch, Calif. Dept. Fish and Game, Sacramento. 87 pp.

HURCOMB, L. 1969. Protection of wildlife in London and its outskirts by public authorities. Biol. Cons. 1:166–169.

HUTCHINS, M., AND V. STEVENS. 1981. Olympic mountain goats. Nat. Hist. 90:58–69.

HUTCHINS, M., V. STEVENS, AND N. ATKINS. 1982. Introduced species and the issue of animal welfare. Internat. J. Stud. Anim. Prob. 3:318–336.

HUTCHINS, M., AND C. WEMMER. 1986. Wildlife conservation and animal rights: are they compatible? Pp. 111–137 *in* Advances in animal welfare science 1986/87 (M. W. Fox and L. D. Mickley, eds.). Humane Soc. of the United States, Washington, D.C. 302 pp.

HUTCHINSON, G. E. 1978. An introduction to population ecology. Yale Univ. Press, New Haven, Conn. 260 pp.

HYDER, D. N. 1972. Defoliation in relation to vegetative growth. Pp. 304–317 *in* Biology and utilization of grasses (V. B. Younger and C. M. McKell, eds.). Academic Press, New York. 426 pp.

HYDER, D. N., AND F. A. SNEVA. 1959. Growth and carbohydrate trends in crested wheatgrass. J. Range Manage. 12:271–276.

HYDER, D. N., AND F. A. SNEVA. 1963. Studies of six grasses seeded on sagebrush-bunchgrass range—yield, palatability, carbohydrate accumulation, and developmental morphology. Ore. Agric. Exp. Sta. Tech. Bull. No. 71. Corvallis. 20 pp.

HYLTON, A. R., G. R. SAVAGE, AND C. D. REESE. 1972. The use and effects of pesticides for rangeland sagebrush control. Environmental Protection Agency Office of Water Programs, Pesticides Study Series No. 3. Environmental Protection Agency, Washington, D.C. 170 pp.

IDYLL, C. P. 1973. The anchovy crisis. Sci. Am. 228:22–29.

IKEDA, A. Y. 1971. A study of the 1970 urban fishing program in the city of St. Louis, Missouri. M.S. Thesis, Univ. Mo., Columbia. 93 pp.

INTERAGENCY GRIZZLY BEAR COMMITTEE. 1986. Interagency Grizzly Bear Guidelines. U.S.D.A. Forest Service, Rocky Mountain Region, Lakewood, Colo. 99 pp.

INTERIOR, DEPARTMENT OF. 1980. Service plan maps whooping crane recovery. Endangered Species Tech. Bull. 5:1,4.

IRBY, H. D., L. N. LOCKE, AND G. E. BAGLEY. 1967. Relative toxicity of lead and selected substitute shot types to game farm mallards. J. Wildl. Manage. 31:253–257.

IRWIN, L. L. 1975. Deer-moose relationships on a burn in northeastern Minnesota. J. Wildl. Manage. 39:653–662.

ISAKOVIC, I. 1970. Game management in Yugoslavia. J. Wildl. Manage. 34:800–812.

ISHMAEL, W. E., AND O. J. RONGSTAD. 1984. Economics of an urban deer-removal program. Wildl. Soc. Bull. 12:394–398.

JACKSON, A. S. 1969. A handbook for bobwhite quail management in the west Texas Rolling Plains. Texas Parks and Wildl. Dept. Bull. No. 48, Austin. 77 pp.

JACKSON, A. S., AND R. DEARMENT. 1963. The lesser prairie chicken in the Texas panhandle. J. Wildl. Manage. 27:733–737.

JACKSON, H. H. T. 1937. Some accomplishments of the cooperative research units—a summary to January 31, 1937. Trans. N. Am. Wildl. Conf. 2:108–118.

JACKSON, H. O., AND W. C. STARRETT. 1959. Turbidity and sedimentation at Lake Chautauqua, Illinois. J. Wildl. Manage. 23:157–158.

JACKSON, J. J. 1982. Public support for nongame and endangered wildlife management: which way is it going? Trans. N. Am. Wildl. and Nat. Resour. Conf. 47:432–440.

JACKSON, P. 1983. The tragedy of our tropical rainforests. Ambio 12:252–254.

JACOBSON, H. A., D. C. GUYNN, AND E. J. HACKETT. 1979. Impact of the botfly on squirrel hunting in Mississippi. Wildl. Soc. Bull. 7:46–48.

JAEGER, E. C. 1949. Further observations on the hibernation of the poor-will. Condor 51:105–109.

JAENKE, E. A. 1966. Opportunities under the cropland adjustment program. Trans. N. Am. Wildl. and Nat. Resour. Conf. 31:323–331.

JAHNKE, H. E. 1976. Tsetse flies and livestock development in east Africa. Weltforum Verlag, Munchen. 180 pp.

JAMES, D., AND K. B. DAVIS. 1965. The effect of sublethal amounts of DDT on the discrimination ability of the bobwhite, *Colinus virginianus* (Linnaeus). Am. Zool. 5:229. (Abstract.)

JAMES, P. C., A. R. SMITH, L. W. OLIPHANT, AND I. G. WARKENTIN. 1987. Northward expansion of the wintering range of Richardson's merlin. J. Field Ornith. 58:112–117.

JANZEN, D. H. 1974. Tropical blackwater rivers, animals, and mast fruiting by the Diptercarpaceae. Biotropica 6:69–103.

JANZEN, D. H. 1978. New horizons in the biology of plant defenses. Pp. 331–350 *in* Herbivores: their interaction with secondary plant metabolites (G. A. Rosenthal and D. H. Janzen, eds.). Academic Press, New York. 718 pp.

JAQUETTE, D. S. 1948. Copper sulfate as a treatment for subclinical trichomoniasis in pigeons. Am. J. Vet. Res. 9:206–209.

JARVIS, R. J. 1976. Soybean impaction in Canada geese. Wildl. Soc. Bull. 4:175–179.

JEFFERIES, D. J. 1967. The delay in ovulation produced by pp'-DDT and its possible significance in the field. Ibis 109:266–272.

JEFFERIES, D. J., AND M. C. FRENCH. 1972. Changes induced in the pigeon thyroid by p,p'-DDE and dieldrin. J. Wildl. Manage. 36:24–30.

JEFFERSON, T. 1785. Letter to James Madison, President of William and Mary College *in* A Jefferson profile, as revealed in his letters (S. K. Padover, ed., 1956). John Day Co., New York. 359 pp.

JENKINS, D., A. WATSON, AND G. R. MILLER. 1963. Population studies on red grouse, *Lagopus lagopus scoticus* (Lath.) in north-east Scotland. J. Animal Ecol. 32:317–376.

JENKINS, R. M., AND D. I. MORAIS. 1971. Reservoir sport fishing effort and harvest in relation to environmental variables. Pp. 371–384 *in* Reservoir fisheries and limnology (G. E. Hall, ed.). Am. Fisheries Soc. Spec. Publ. 8. Washington, D.C. 511 pp.

JENNINGS, W. S., AND J. T. HARRIS. 1953. The collared peccary in Texas. Texas Game and Fish Comm., Austin. 31 pp.

JENSEN, C. H., A. D. SMITH, AND G. W. SCOTTER. 1972. Guidelines for grazing sheep on rangelands used by big game in winter. J. Range Manage. 25:346–352.

JENSEN, R. W. 1980. Effects of winter navigation on waterfowl and raptors in the St. Mary's River area. M.A. Thesis, Northern Mich. Univ., Marquette. 105 pp.

JENSEN, W. F. 1982. Vegetational analysis and an analytical model of the Beaver Basin deeryard. Unpubl. M.A. Thesis, Northern Mich. Univ., Marquette. 156 pp.

JOHNSGARD, P. A. 1968. Waterfowl: their biology and natural history. Univ. Neb. Press, Lincoln. 138 pp.

JOHNSGARD, P. A. 1975. Waterfowl of North America. Ind. Univ. Press, Bloomington. 575 pp.

JOHNSGARD, P. A. 1978. Ducks, geese, and swans of the world. Univ. Neb. Press, Lincoln. 404 pp.

JOHNSON, D. H., AND A. B. SARGEANT. 1977. Impact of red fox predation on the sex ratio of prairie mallards. Wildl. Res. Rept. 6, U.S. Fish and Wildl. Serv., Washington, D.C. 56 pp.

JOHNSON, F. A., AND F. MONTALBANO III. 1987. Considering waterfowl habitat in hydrilla control policies. Wildl. Soc. Bull. 15:466–469.

JOHNSON, F. W., AND L. ADAMS. 1955. Some lessons from Europe in forest-big game management. J. Forestry 53:436–438.

JOHNSON, K. P., AND C. H. NELSON. 1984. Side-scan sonar assessment of gray whale feeding in the Bering Sea. Science 225:1150–1152.

JOHNSON, M. K. 1979. Review of endangered species: policies and legislation. Wildl. Soc. Bull. 7:79–93.

JOHNSON, M. K. 1980. Management involvement lacking at recent international meeting. Wildl. Soc. Bull. 8:65–69.

JOHNSON, R. J., AND M. M. BECK. 1988. Influences of shelterbelts on wildlife management and biology. Agric. Ecosystems & Environ. 22/23:301–335.

JOHNSON, R. J., P. H. COLE, AND W. W. STROUP. 1985. Starling response to three auditory stimuli. J. Wildl. Manage. 49:620–625.

JOHNSON, R. J., A. E. KOEHLER, O. C. BURNSIDE, AND S. R. LOWRY. 1985. Response of thirteen-lined ground squirrels to repellents and implications for conservation tillage. Wildl. Soc. Bull. 13:317-324.

JOHNSON, R. R., L. T. HAIGHT, AND J. M. SIMPSON. 1977. Endangered species versus endangered habitats: a concept. U.S.D.A. Forest Serv. Gen. Tech. Rept. RM-43:68–79.

JOHNSON, T. K., AND C. D. JORGENSEN. 1981. Ability of desert rodents to find buried seeds. J. Range Manage. 34:312–314.

JOHNSON, W. M. 1953. Effect of grazing intensity upon vegetation and cattle gains on ponderosa pine–bunchgrass ranges of the Front Range of Colorado. U.S.D.A. Circ. 939.

JOHNSON, W. M. 1969. Life expectancy of a sagebrush control in central Wyoming. J. Range Manage. 22:177–182.

JOHNSTON, A., S. SMOLIAK, A. D. SMITH, AND L. E. LUTWICK. 1967. Improvement of southeastern Alberta range with fertilizers. Can. J. Plant Sci. 47:671–678.

JONES, H. L., AND J. M. DIAMOND. 1976. Short-time-base studies of turnover in breeding bird populations on the California Channel Islands. Condor 78:526–549.

JONES, J. K., JR. 1964. Distribution and taxonomy of mammals of Nebraska. Univ. Kans. Pubs. Mus. Nat. Hist. 16, Lawrence. 356 pp.

JONES, J. K., JR., D. M. ARMSTRONG, R. S. HOFFMAN, AND C. JONES. 1983. Mammals of the northern Great Plains. Univ. Neb. Press, Lincoln. 379 pp.

JONES, K. B. 1981. Effects of grazing on lizard abundance and diversity in western Arizona. Southwestern Nat. 26:107–115.

JONES, L. H. P., AND K. HANDRECK. 1967. Silica in soils, plants, and animals. Adv. Agronomy 19:107–149.

JONES, R. H., R. D. ROUGHTON, N. M. FOSTER, AND B. M. BANDO. 1977. *Culicoides*, the vector of epizootic hemorrhagic disease in white-tailed deer in Kentucky in 1971. J. Wildl. Dis. 13:2–8.

JONES, R. L., AND H. C. HANSON. 1985. Mineral licks, geophagy, and biogeochemistry of North American ungulates. Ia. State Univ. Press, Ames. 302 pp.

JONES, V. E., AND R. D. PETTIT. 1984. Low rates of Tebuthiuron for control of sand shinnery oak. J. Range. Manage. 37:488–490.

JORDAN, C. F. 1982. Amazon rain forests. Am. Scientist 70: 394–401.

JORDAN, C. F., AND C. UHL. 1978. Biomass of a "tierra firme" forest of the Amazon Basin. Oecologia Plantarum 13:387–400.

JORDAN, J. S. 1953a. Consumption of cereal grains by migratory waterfowl. J. Wildl. Manage. 17:120–123.

JORDAN, J. S. 1953b. Effects of starvation on wild mallards. J. Wildl. Manage. 17:304–311.

JORDAN, P. A., D. B. BOTKIN, AND M. L. WOLFE. 1971. Biomass dynamics in a moose population. Ecology 52:147–152.

JORDAN, P. A., D. B. BOTKIN, A. S. DOMINSKI, H. S. LOWENDORF, AND G. E. BELOVSKY. 1973. Sodium as a critical nutrient for the moose of Isle Royale. Proc. N. Am. Moose Conf. Workshop 9:1–28.

JOSELYN, G. B., AND G. I. TATE. 1972. Practical aspects of managing roadside cover for nesting pheasants. J. Wildl. Manage. 36:1–11.

JOSELYN, G. B., AND J. E. WARNOCK. 1964. Value of federal feed-grain program to production of pheasants in Illinois. J. Wildl. Manage. 28:547–551.

JOSELYN, G. B., J. E. WARNOCK, AND S. L. ETTER. 1968. Manipulation of roadside cover for nesting pheasants—a preliminary report. J. Wildl. Manage. 32:217–233.

JOYNER, D. E., J. D. SOMERS, F. F. GILBERT, AND R. J. BROOKS. 1980. Use of methiocarb as a blackbird repellent in field corn. J. Wildl. Manage. 44:672–676.

JUENEMANN, B. G. 1973. Habitat evaluations of selected bald eagle nest sites in the Chippewa National Forest. M.S. Thesis, Univ. Minnesota, Minneapolis. 170 pp.

JUGENHEIMER, R. W. 1976. Corn: improvement, seed production, and uses. John Wiley & Sons, New York. 670 pp.

KABAT, C., N. E. COLLIAS, AND R. C. GUETTINGER. 1953. Some winter habits of white-tailed deer and the development of census methods in the Flag Yard of northern Wisconsin. Wisconsin Conserv. Dept. Tech. Bull. 7. Madison. 32 pp.

KADLEC, J. A. 1962. Effects of a drawdown on a waterfowl impoundment. Ecology 43:267–281.

KAEDING, L. R. AND M. A. ZIMMERMAN. 1983. Life history and ecology of the humpback chub in the Little Colorado and Colorado rivers of the Grand Canyon. Trans. Amer. Fisheries Soc. 112:577–594.

KAHL, M. P. 1964. Food ecology of the wood stork (*Mycteria americana*) in Florida. Ecol. Monogr. 34:97–117.

KAHL, R. B., AND F. B. SAMSON. 1984. Factors affecting yield of winter wheat grazed by geese. Wildl. Soc. Bull. 12:256–262.

KALLAND, A., AND B. MOERAN. 1992. Japanese whaling, end of an era? Curzon Press, London. 228 pp.

KALLMAN, H. (ED.). 1987. Restoring America's wildlife, 1937–1987. U.S. Fish and Wildl. Serv., Washington, D.C. 394 pp.

KALMBACH, E. R. 1939. American vultures and the toxin of *Clostridium botulinum*. J. Am. Vet. Med. Assoc. 94:187–191.

KALMBACH, E. R. 1968. Type C botulism among wild birds—a historical sketch. U.S. Fish and Wildl. Serv. Spec. Sci. Rept. No. 110, Washington, D.C. 8 pp.

KALMBACH, E. R., AND M. F. GUNDERSON. 1934. Western duck sickness: a form of botulism. U.S.D.A. Tech. Bull. No. 411, Washington, D.C. 81 pp.

KANTRUD, H. A., AND R. L. KOLOGISKI. 1982. Effects of soil and grazing on breeding birds of uncultivated upland grasslands of the northern Great Plains. U.S.D.I. Fish and Wildl. Serv. Wildl. Res. Rept. No. 15, Washington, D.C. 33 pp.

KAPLAN, C. 1977. The world problem. Pp. 1–21 *in* Rabies, the facts (C. Kaplan, ed.). Oxford Univ. Press, Oxford, England. 116 pp.

KARAS, N. 1974. The complete book of the striped bass. Winchester Press, New York. 367 pp.

KARL, D. M., C. O. WIRSEN, AND H. W. JANNASCH. 1980. Deep-sea primary production at the Galápagos hydrothermal vents. Science 207:1345–1347.

KARNS, P. D. 1967. *Pneumostrongylus tenuis* in deer in Minnesota and implications for moose. J. Wildl. Manage. 31:299–303.

KARR, J. R. 1982. Avian extinction on Barro Colorado Island, Panama: a reassessment. Am. Nat. 199:220–239.

KARR, J. R. 1990. Avian survival rates and the extinction process on Barro Colorado Island, Panama. Conserv. Biol. 4:391–397.

KARR, J. R., L. A. TOTH, AND D. R. DUDLEY. 1985. Fish communities of midwestern rivers: a history of degradation. BioScience 35:90–95.

KARSTAD, L. 1971. Diseases of wildlife. Pp. 79–81 *in* A manual of wildlife conservation (R. D. Teague, ed.). The Wildlife Society, Washington, D.C. 206 pp.

KARSTAD, L., A. WINTER, AND D. O. TRAINER. 1961. Pathology of epizootic hemorrhagic disease of deer. Am. J. Vet. Res. 22: 227–235.

KAUFMAN, D. W. 1974. Adaptive coloration in *Peromyscus polionotus*: experimental selection by owls. J. Mamm. 55:271–283.

KAZA, S. 1982. Recreational whalewatching in California: a profile. Whalewatcher 16(1):6–8.

KEANE, C. 1926. The epizootic of foot-and-mouth disease in California. Calif. Dept. Agric. Spec. Publ. No. 65, Sacramento. 54 pp.

KEAR, J. 1972a. Feeding habits of birds. Pp. 471–503 *in* Biology of nutrition (R.N.T-W-Fiennes, ed.). Pergamon Press, Elmsford, N.Y. 681 pp.

KEAR, J. 1972b. The blue duck of New Zealand. The Living Bird 11:175–192.

KEAR, J., AND A. J. BERGER. 1980. The Hawaiian goose. Buteo Books, Vermillion, S. D. 154 pp.

KEAR, J., AND P. J. K. BURTON. 1971. The food and feeding apparatus of the Blue duck (*Hymenolaimus*). Ibis 113:483–493.

KEENLEYSIDE, M. H. A. 1967. Effects of forest spraying with DDT in New Brunswick on food of young Atlantic salmon. J. Fish. Res. Bd. Can. 24:807.

KEIPER, R. R. 1985. Are sika deer responsible for the decline of white-tailed deer on Assateague Island, Maryland? Wildl. Soc. Bull. 13:144–146.

KEIPER, R. R., J. STEPHENS, AND D. BALDWIN. 1984. Sex, age, and dressed weights of hunter-killed sika and white-tailed deer from Assateague Island, Maryland. Proc. Penn. Acad. Sci. 58: 101–102.

KEITER, R. B. 1997. Greater Yellowstone's bison: unraveling of an early American wildlife conservation achievement. J. Wildl. Manage. 61:1–11.

KEITH, L. B. 1961. A study of waterfowl ecology on small impoundments in southeastern Alberta. Wildl. Monogr. No. 6. 88 pp.

KEITH, L. B. 1963. Wildlife's ten-year cycle. Univ. Wis. Press, Madison. 201 pp.

KEITH, L. B. 1974. Some features of population dynamics in mammals. Proc. Internat. Congress Game Biol. 11:17–58.

KEITH, L. B., J. R. CARY, O. J. RONGSTAD, AND M. C. BRITTINGHAM. 1984. Demography and ecology of a declining snowshoe hare population. Wildl. Monogr. No. 90. 43 pp.

KEITH, L. B., A. W. TODD, C. J. BRAND, R. S. ADAMCIK, AND D. H. RUSCH. 1977. An analysis of predation during cyclic fluctuations of snowshoe hares. Proc. Internat. Congress Game Biol. 13:151–175.

KEITH, L. B., AND L. A. WINDBERG. 1978. A demographic analysis of the snowshoe hare cycle. Wildl. Monogr. No. 58. 70 pp.

KELLERT, S. R. 1976a. A study of American attitudes toward animals, Part II. U.S. Fish and Wildl. Serv., Washington, D.C. 453 pp.

KELLERT, S. R. 1976b. Perceptions of animals in American society. Trans. N. Am. Wildl. and Nat. Resour. Conf. 41:533–546.

KELLERT, S. R. 1978. Attitudes and characteristics of hunters and antihunters. Trans. N. Am. Wildl. and Nat. Resour. Conf. 43:412–423.

KELLERT, S. R. 1980a. Needed research on people-animal interactions in National Parks. Proc. 2nd Conf. on Sci. Res. in National Parks, 6:165–167.

KELLERT, S. R. 1980b. Americans' attitudes and knowledge of animals. Trans. N. Am. Wildl. and Nat. Resour. Conf. 45:111–124.

KELLERT, S. R. 1980c. What do North Americans expect of wildlife management agencies? Proc. Internat. Assoc. Fish and Wildl. 70:29–38.

KELLERT, S. R. 1984. Comprehensive wildlife management: an approach for developing a nongame program in Connecticut. Trans. N. Am. Wildl. and Nat. Resour. Conf. 49:215–221.

KELLERT, S. R. 1985. Public perceptions of predators, particularly the wolf and coyote. Biol. Cons. 31:167–189.

KELLERT, S. R., AND M. O. WESTERVELT. 1982. Historical trends in American animal use and perception. Trans. N. Am. Wildl. and Nat. Resour. Conf. 47:649–664.

KELLY, D. 1986. The decadent forest. Audubon 88:46–73.

KELLY, S. T., AND M. E. DeCAPITA. 1982. Cowbird control and its effect on Kirtland's warbler reproductive success. Wilson Bull. 94:363–364.

KELSALL, J. P., AND J. R. CALAPRICE. 1972. Chemical content of waterfowl plumage as a potential diagnostic tool. J. Wildl. Manage. 36:1088–1097.

KELSALL, J. P., AND W. PRESCOTT. 1971. Moose and deer behavior in snow in Fundy National Park, New Brunswick. Can. Wildl. Serv. Rept. No. 15, Ottawa. 26 pp.

KENDALL, R. J., AND P. F. SCANLON. 1979. Lead concentrations in mourning doves collected from middle Atlantic game management areas. Proc. Ann. Conf. SE Assoc. Fish and Wildlife Agencies 33:165–172.

KENNAMER, R. A., AND G. R. HEPP. 1987. Frequency and timing of second broods in wood ducks. Wilson Bull. 99:655–662.

KENNEDY, J. J. 1974. Motivations and rewards of hunting in a group versus alone. Wildl. Soc. Bull. 2:3–7.

KERBES, R. H., P. M. KOTANEN, AND R. L. JEFFERIES. 1990. Destruction of wetland habitats by lesser snow geese; a keystone species on the west coast of Hudson bay. J. Appl. Ecol. 27:242–258.

KERLEY, G. I. H. 1986. Jackass penguins harmed in South Africa oil spills. World Birdwatch 8(2):4.

KESSEL, B. 1953. Distribution and migration of the European starling in North America. Condor 55:49–67.

KESSLER, W. B., AND J. D. DODD. 1978. Responses of coastal prairie vegetation and Attwater's prairie chicken to range management practices. Proc. Internat. Rangeland Congress 1:473–476.

KIE, J. G., M. WHITE, AND F. K. KNOWLTON. 1979. Effects of coyote predation on population dynamics of white-tailed deer. Pp. 65–82 in Proc. first Welder Wildlife Foundation symposium (D.L. Drawe, ed.). Sinton, Tex. 275 pp.

KIEL, W. H., JR. 1976. Bobwhite quail population characteristics and management implications in south Texas. Trans. N. Am. Wildl. and Nat. Resour. Conf. 41:407–420.

KILGEN, R. H. 1978. Growth of channel catfish and striped bass in small ponds stocked with grass carp and water hyacinths. Trans. Am. Fisheries Soc. 107:176–180.

KILGORE, B. M., AND D. TAYLOR. 1979. Fire history of a sequoia-mixed conifer forest. Ecology 60:129–142.

KIMERLE, R. A., AND W. R. ENNS. 1968. Aquatic insects associated with midwestern water stabilization lagoons. Water Pollut. Control Fed. J. 40:R31–R41.

KING, J. A. 1955. Social behavior, social organizations, and population dynamics in a black-tailed prairie dog town in the Black Hills of South Dakota. Univ. Mich. Lab. Vertebrate Biol. Contr. No. 67, Ann Arbor. 128 pp.

KING, J. R. AND D. S. FARNER. 1961. Energy metabolism, thermoregulation and body temperature. Pp. 215–288 in Biology and comparative physiology of birds, Vol. II (A. J. Marshall, ed.). Academic Press, New York. 468 pp.

KINGSCOTE, B. F., AND J. G. BOHAC. 1986. Antibodies to bovine bacterial and viral pathogens in pronghorns in Alberta, 1983. J. Wildl. Dis. 22:511–514.

KINGSCOTE, B. F., W. D. G. YATES, AND G. B. TIFFIN. 1987. Diseases of wapiti utilizing cattle range in southwestern Alberta. J. Wildl. Dis. 23:86–91.

KIRKPATRICK, J. F., J. W. TURNER, JR., I. K. M. LIU, AND R. A. FAYRER-HOSKEN. 1996. Applications of pig zona pellucida immunocontraception to wildlife fertility control. J. Reprod. Fert. Suppl. 51:183–189.

KIRSCH, L. M. 1969. Waterfowl production in relation to grazing. J. Wildl. Manage. 33:821–828.

KISHIDA, K. 1934. The mammal fauna of the great city of Tokyo, Japan. Lansania (J. Arach. and Zool.) 6:17–30.

KLEBENOW, D. A. 1970. Sage grouse versus sagebrush control in Idaho. J. Range Manage. 23:396–400.

KLEBENOW, D. A. 1972. The habitat requirements of sage grouse and the role of fire in management. Proc. Tall Timbers Fire Ecol. Conf. 12:305–315.

KLEBENOW, D. A., AND G. M. GRAY. 1968. The food habits of juvenile sage grouse. J. Range Manage. 21:80–83.

KLEIN, D. R. 1968. The introduction, increase, and crash of reindeer on St. Matthew Island. J. Wildl. Manage. 32:350–367.

KLEIN, D. R. 1979. The Alaska oil pipeline in retrospect. Trans. N. Am. Wildl. and Nat. Resour. Conf. 44:235–246.

KLEIN, D. R. 1982. Fire, lichens, and caribou. J. Range Manage. 35:390–395.

KLEIN, D. R. 1987. Vegetation recovery patterns following overgrazing by reindeer on St. Matthew Island. J. Range Manage. 40:336–338.

KLEIN, D. R., AND S. T. OLSON. 1960. Natural mortality patterns of deer in southeast Alaska. J. Wildl. Manage. 24:80–88.

KLEIN, D. R., AND H. STRANDGAARD. 1972. Factors affecting growth and body size of roe deer. J. Wildl. Manage. 36:64–79.

KLEIN, D.R., AND R. G. WHITE. 1978. Grazing ecology of caribou and reindeer in tundra systems. Proc. Internat. Range Congress 1:469–472.

KLEM, D., JR. 1981. Avian predators hunting birds near windows. Proc. Penn. Acad. Sci. 55:90–92.

KLEM, D., JR. 1990a. Bird injuries, cause of death, and recuperation from collisions with windows. J. Field Ornithol. 61:115–119.

KLEM, D., JR. 1990b. Collisions between birds and windows: mortality and prevention. J. Field Ornithol. 61:120–128.

KLEM, D., JR. 1991. Glass and bird kills: an overview and suggested planning and design methods of preventing a fatal hazard. Pp. 99–103 in Wildlife conservation in metropolitan environments.

(L. W. Adams and D. L. Leedy, eds.). Nat. Instit. Urban Wildl., Columbia, Md. 264 pp.

KLETT, A. T., H. F. DUEBBERT, AND G. L. HEISMEYER. 1984. Use of seeded native grasses as nesting cover by ducks. Wildl. Soc. Bull. 12:134–138.

KLIMSTRA, W. D., AND J. L. ROSEBERRY. 1975. Nesting ecology of the bobwhite in southern Illinois. Wildl. Monogr. No. 41. 37 pp.

KLIMSTRA, W. D., AND T. G. SCOTT. 1957. Progress report on bobwhite nesting in southern Illinois. Proc. SE Assoc. Game and Fish Comm. 11:351–355.

KLOPFER, P. H. 1963. Behavioral aspects of habitat selection: the role of early experience. Wilson Bull. 75:15–22.

KLOPFER, P. H., D. K. ADAMS, AND M. S. KLOPFER. 1964. Maternal "imprinting" in goats. Proc. Nat. Acad. Sci. U.S.A. 52:911–914.

KLOS, H. G., AND A. WUNSCHMANN. 1972. The wild and domestic oxen. Pp. 331–398 *in* Grzimek's animal life encyclopedia, vol. 13. Van Nostrand Reinhold, New York. 657 pp.

KNAPP, F. 1977. St. Hubertus: a conservation legacy. Internat. Wildl. 7:42.

KNICK, S. T. 1990. Ecology of bobcats relative to exploitation and a prey decline in southeastern Idaho. Wildl. Monogr. 108. 42 pp.

KNIGHT, J. E. 1985. Potential returns for landowner management of wildlife. Range Improvement Taskforce Rept. 19, Agric. Exper. Sta. Coop. Ext. Serv., New Mexico State Univ., Las Cruces. 19 pp.

KNIGHT, R. L., AND S. K. KNIGHT. 1984. Responses of wintering bald eagles to boating activity. J. Wildl. Manage. 48:999–1004.

KNIGHT, R. L., D. J. GROUT, AND S. A. TEMPLE. 1987. Nest-defense behavior of the American crow in urban and rural areas. Condor 89:175–177.

KNIGHT, R. R., AND L. L. EBERHARDT. 1985. Population dynamics of Yellowstone grizzly bears. Ecology 66:223–334.

KNIGHT, R. R., AND M. R. MUDGE. 1967. Characteristics of some natural licks in the Sun River area, Montana. J. Wildl. Manage. 31:293–299.

KNIPE, T. 1944. The status of the antelope herds of northern Arizona. Ariz. Game and Fish Comm. Fed. Aid Div. P-R Project 9-R, Phoenix. 40 pp.

KNOWLTON, F. F. 1972. Preliminary interpretations of coyote population mechanics with some management implications. J. Wildl. Manage. 36:369–382.

KNOX, K. L., J. G. NAGY, AND R. D. BROWN. 1969. Water turnover in mule deer. J. Wildl. Manage. 33:389–393.

KNUDSEN, G. J. 1962. Relationship of beaver to forests, trout, and wildlife in Wisconsin. Wis. Cons. Dept. Tech. Bull. No. 25, Madison. 53 pp.

KNUDSON, D. M. 1980. Outdoor recreation. Macmillan, New York. 655 pp.

KOCAN, R. M. 1969. Various grains and liquid as potential vehicles of transmission for *Trichomonas gallinae*. Bull. Wildl. Dis. Assoc. 5:148–149.

KOCAN, R. M., AND C. M. HERMAN. 1971. Trichomoniasis. Pp. 282–290 *in* Infectious and parasitic diseases of wild birds (J. W. Davis, R. C. Anderson, L. Karstad, and D. O. Trainer, eds.). Ia. State Univ. Press, Ames. 344 pp.

KOERTH, B. H., W. M. WEBB, F. C. BRYANT, AND F. S. GUTHERY. 1983. Cattle trampling of simulated ground nests under short duration and continuous grazing. J. Range Manage. 36:385–386.

KOFORD, C. B. 1958. Prairie dogs, whitefaces, and blue grama. Wildl. Monogr. No. 3. 78 pp.

KOLOSOV, A. M. 1935. Sur la biologie du Korsak et du renard des steppes. Bull. Soc. Nat. Moscou. Sect. Biol. 44:165–177.

KORSCHGEN, L. J. 1948. Late-fall and early-winter food habits of bobwhite quail in Missouri. J. Wildl. Manage. 12:46–57.

KORSCHGEN, L. J. 1970. Soil-food-chain pesticide wildlife relationships in aldrin-treated fields. J. Wildl. Manage. 34:186–199.

KORSCHGEN, L. J. 1971. Disappearance and persistence of Aldrin after five annual applications. J. Wildl. Manage. 35:494–500.

KORSCHGEN, L. J. 1980. Procedures for food-habits analyses. Pp. 113–127 *in* Wildlife management techniques manual (S. D. Schemnitz, ed.). The Wildl. Society, Washington, D.C. 686 pp.

KORTE, P. A., AND L. H. FREDRICKSON. 1977a. Swamp rabbit distribution in Missouri. Trans. Mo. Acad. Sci. 10–11:72–77.

KORTE, P. A., AND L. H. FREDRICKSON. 1977b. Loss of Missouri's lowland hardwood ecosystem. Trans. N. Am. Wildl. and Nat. Resour. Conf. 42:31–41.

KOTANEN, P., AND R. L. JEFFERIES. 1997. Long-term destruction of sub-arctic wetland vegetation by lesser snow geese. EcoScience 4:179–182.

KOTHMANN, M. M. 1974. Grazing management terminology. J. Range Manage. 27:326–327.

KOTHMANN, M. M. 1980. Integrating livestock needs to the grazing system. Pp. 65–83 *in* Proc. grazing management systems for southwest rangelands symposium (K. C. McDaniel and C. D. Allison, eds.). N. M. State Univ., Las Cruces. 183 pp.

KOTHMANN, M. M., AND G. W. MATHIS. 1974. Calf production from ten management systems. Proc. West. Sect. Am. Soc. Animal Sci. 25:185–188.

KOTT, E., AND W. L. ROBINSON. 1963. Seasonal variation in litter size of the meadow vole in southern Ontario. J. Mamm. 44: 467–470.

KOZICKY, E. L. 1975. The steel shot issue. Pp. 168–175 *in* First internat. waterfowl symposium. Ducks Unlimited, Chicago. 224 pp.

KOZICKY, E. L. 1977. Tomorrow's hunters—gadgeteers or sportsmen? Wildl. Soc. Bull. 5:175–178.

KRAMER, R. J. 1971. Hawaiian land mammals. Charles E. Tuttle Co., Rutland, Vt. 347 pp.

KRAPU, G. L. 1974a. Feeding ecology of pintail hens during reproduction. Auk 91:278–290.

KRAPU, G. L. 1974b. Foods of breeding pintails in North Dakota. J. Wildl. Manage. 38:408–417.

KRAPU, G. L., AND G. A. SWANSON. 1975. Some nutritional aspects of reproduction in prairie-nesting pintails. J. Wildl. Manage. 39:156–162.

KRAPU, G. L., AND G. A. SWANSON. 1977. Foods of juvenile, brood, hen, and post-breeding pintails in North Dakota. Condor 79:504–507.

KRAUSMANN, P. R., AND J. A. BISSONETTE. 1977. Bone-chewing behavior of desert mule deer. Southwestern Nat. 22:149–150.

KREBS, C. J. 1978. Ecology: The experimental analysis of distribution and abundance. 2nd ed. Harper & Row, New York. 678 pp.

KREFTING, L. W. 1962. Use of silvicultural techniques for improving deer habitat in the Lake States. J. Forestry 16:40–42.

KREFTING, L. W., AND H. L. HANSEN. 1969. Increasing browse for deer by aerial applications of 2, 4-D. J. Wildl. Manage. 33: 784–790.

KREFTING, L. W., H. L. HANSEN, AND M. H. STENLUND. 1956. Stimulating regrowth of mountain maple for deer browse by herbicides, cutting, and fire. J. Wildl. Manage. 20:434–441.

KREITLOW, K. W., AND R. H. HART (EDS.). 1974. Plant morphogenesis as the basis for scientific management of range resources. Pp. 138–146 *in* U.S.D.A. Misc. Publ. 1271. Washington, D.C. 232 pp.

KROODSMA, R. L. 1984. Ecological factors associated with degree of edge effect in breeding birds. J. Wildl. Manage. 48:418–425.

KRULL, J. M. 1970. Aquatic plant-macroinvertebrate associations and waterfowl. J. Wildl. Manage. 34:707–718.

KRUTCH, J. W. 1957. A damnable pleasure. Saturday Review 17:8–9, 39–40.

KRYSL, L. J., C. D. SIMPSON, AND G. G. GRAY. 1980. Dietary overlap of sympatric Barbary sheep and mule deer in Palo Duro Canyon, Texas. Pp. 97–103 *in* Proc. of the symposium on ecology and management of Barbary sheep (C. D. Simpson, ed.). Dept. Range and Wildlife Management, Texas Tech Univ., Lubbock. 112 pp.

KUBOTA, J., S. RIEGER, AND V. A. LAZAR. 1970. Mineral composition of herbage browsed by moose in Alaska. J. Wildl. Manage. 34:565–569.

KUHN, L. W., AND E. P. PELOQUIN. 1974. Oregon's nutria problem. Proc. Vertebrate Pest Conf. 6:101–105.

KURZEJESKI, E. W., L. W. BURGER, JR., M. J. MONSON, AND R. LENKNER. 1992. Wildlife conservation attitudes and land use intentions of Conservation Reserve Program participants in Missouri. Wildl. Soc. Bull. 20:253–259.

KUSHLAN, J. A. 1974. Observations on the role of the American alligator *(Alligator mississipiensis)* in the southern Florida wetlands. Copeia 1974:993–996.

KUSHLAN, J. A. 1976. Wading bird predation in a seasonally fluctuating pond. Auk 93:464–476.

KUSHLAN, J. A. 1979. Temperature and oxygen in an Everglades alligator pond. Hydrobiologist 67:267–271.

KUSHLAN, J. A., AND B. P. HUNT. 1979. Limnology of an alligator pond in south Florida. Florida Scientist 42:65–84.

KYLE, R. 1972. Will the antelope recapture Africa? New Scientist 53:640–643.

LABASTILLE, A. 1974. Ecology and management of the Atitlan Grebe, Lake Atitlan, Guatemala. Wildl. Monogr. No. 37. 66 pp.

LABISKY, R. F. 1957. Relation of hay harvesting to duck nesting under a refuge-permittee system. J. Wildl. Manage. 21:194–200.

LABISKY, R. F. 1961. Report of attempts to establish Japanese quail in Illinois. J. Wildl. Manage. 25:290–295.

LABISKY, R. F. 1975. Illinois pheasants: their distribution and abundance, 1958–1973. Ill. Nat. Hist. Surv. Biol. Notes No. 94. 11 pp.

LABISKY, R. F. 1976. Midwest pheasant abundance declines. Wildl. Soc. Bull. 4:182–183.

LABISKY, R. F., AND R. W. LUTZ. 1967. Responses of wild pheasants to solid-block applications of aldrin. J. Wildl. Manage. 21:13–24.

LABISKY, R. F., C. V. BURGER, R. S. ELLARSON, D. J. FORSYTH ET AL. 1975. Ecopolicies of The Wildlife Society. Wildl. Soc. Bull. 3:36–43.

LABOV, J. B. 1977. Minireview-phytoestrogens and mammalian reproduction. Comp. Biochem. Physiol. 57A(1A):3–9.

LACK, D. 1947. Darwin's finches: an essay on the general biological theory of evolution. Harper and Brothers, New York. 204 pp.

LACK, D. 1951. Population ecology of birds: a review. Pp. 409–448 *in* Proc. Xth Internat. Ornith. Congress. 622 pp.

LACK, D. 1954. The natural regulation of animal numbers. Clarendon Press, Oxford. 343 pp.

LACK, D. 1966. Population studies of birds. Clarendon Press, Oxford. 341 pp.

LACOMBE, D., P. MATTON, AND A. CYR. 1987. Effect of Ornitrol on spermatogenesis in red-winged blackbirds. J. Wildl. Manage. 51:596–601.

LAETSCH, W. M. 1979. Plants, Basic concepts in botany. Little, Brown, Boston. 510 pp.

LAGLER, K. F. 1956a. Freshwater fishery biology, 2nd ed. Wm. C. Brown Co., Dubuque, Ia. 421 pp.

LAGLER, K. F. 1956b. The pike, *Esox lucius* Linnaeus, in relation to waterfowl on the Seney National Wildlife Refuge, Michigan. J. Wildl. Manage. 20:114–124.

LAMB, D. W., R. L. LINDER, AND Y. A. GREICHUS. 1967. Dieldrin residues in eggs and fat of penned pheasant hens. J. Wildl. Manage. 31:24–27.

LAMBRECHT, F. L. 1966. Principles of tsetse control and land use with emphasis on animal husbandry. E. African Wildl. J. 1:63.

LAMBRECHT, F. L. 1983. Game animals: a substitute for cattle? Rangelands 5:22–24.

LAMPREY, H. F. 1974. Management of flora and fauna in national parks. Pp. 237–257 *in* Second world conference on national parks (S. H. Elliott, ed.). Internat. Union for Cons. of Nature and Nat. Resour. Morges, Switzerland. 504 pp.

LANE, R. S., J. PIESMAN, AND W. BURGDORFER. 1991. Lyme borreliosis: relation of its causative agent to its vectors and hosts in North America and Europe. Ann. Rev. Entomol. 36:587–609.

LANGENAU, E., JR., R. J. MORAN, J. R. TERRY, AND D. C. CUE. 1981. Relationship between deer kill and ratings of the hunt. J. Wildl. Manage. 45:959–964.

LANKESTER, M. W., AND R. C. ANDERSON. 1968. Gastropods as intermediate hosts of *Pneumostrongylus tenuis* Dougherty of white-tailed deer. Can. J. Zool. 46:373–383.

LANTIS, M. 1954. Problems of human ecology in the North American Arctic. Arctic 7:307–320.

LARIMORE, R. W., AND P. W. SMITH. 1963. The fishes of Champaign County, Illinois, as affected by 60 years of stream changes. Ill. Nat. Hist. Surv. Bull. 28:299–382.

LARSON, F. D. 1966. Cultural conflicts with the cattle business in Zambia, Africa. J. Range Manage. 19:367–370.

LARSON, J. S., AND R. D. TABER. 1980. Criteria of sex and age. Pp. 143–202 *in* Wildlife management techniques manual (S. D. Schemnitz, ed.). The Wildl. Society, Washington, D.C. 686 pp.

LAUNDRE, J. W., AND N. K. APPEL. 1986. Habitat preferences for burrow sites of Richardson's ground squirrels in southwestern Minnesota. Prairie Nat. 18:235–239.

LAY, D. W. 1938. How valuable are woodland clearings to wildlife? Wilson Bull. 50:254–256.

LAY, D. W. 1954. Quail management on pastures. Proc. SE Assoc. Game and Fish Comm. 8:11–12.

LAYCOCK, G. 1966. The alien animals. Nat. Hist. Press, Garden City, New York. 240 pp.

LEACH, H. R., AND E. G. HUNT. 1974. Coyotes and people. Pp. 117–119 *in* A symposium on wildlife in an urbanizing environment (J. H. Noyes and D. R. Progulske, eds.). Univ. Mass. Coop. Ext. Serv., Amherst. 182 pp.

LEAR, L. 1997. Rachel Carson: witness for nature. Henry Holt and Co., New York, N.Y. 634 pp.

LEBERG, P. L. 1991. Influence of fragmentation and bottlenecks on genetic divergence of wild turkey populations. Conserv. Biol. 5:522–530.

LEE, D.-P., AND T.-H. KOO. 1986. A comparative study of the breeding density of magpies, *Pica pica sericea* Gould, between urban and rural areas. Bull. Inst. Ornithol., Kyung Hee Univ. 1:39–51.

LEE, K. E. 1985. Earthworms: their ecology and relationships with soils and land use. Academic Press, Orlando, Fla. 441 pp.

LEEDY, D. L., AND E. H. DUSTMAN. 1947. The pheasant decline and land-use trends, 1941–1946. Trans. N. Am. Wildl. Conf. 12:479–490.

LEEDY, D. L., T. M. FRANKLIN, AND R. M. MAESTRO. 1981. Planning for urban fishing and waterfront recreation. U.S.D.I. Rept. FWS/OBS 80/35, Washington, D.C. 108 pp.

LEEDY, D. L., R. M. MAESTRO, AND T. M. FRANKLIN. 1978. Planning for wildlife in cities and suburbs. Am. Soc. of Planning Officials, Chicago, and U.S. Fish and Wildl. Serv. Rept. FWX/OSB-77/66, Washington, D.C. 64 pp.

LEEGE, T. A., AND W. O. HICKEY, 1977. Elk/snow habitat relationships in the Pete King Drainage, Idaho. Idaho Dept. Fish and Game Wildl. Bull. No. 6, Boise. 23 pp.

LEHMANN, V. W. 1946. Bobwhite quail reproduction in southwestern Texas. J. Wildl. Manage. 10:111–123.

LEHMANN, V. W. 1953. Bobwhite population fluctuations and vitamin A. Trans. N. Am. Wildl. Conf. 18:199–246.

LEHMANN, V. W., AND R. G. MAUERMANN. 1963. Status of Attwater's prairie chicken. J. Wildl. Manage. 27:714–725.

LEHMKUHL, J. F. 1984. Determining size and dispersion of minimum viable populations for land management planning and species conservation. Envir. Manage. 8:167–176.

LEHRMAN, D. S. 1964. The reproductive behavior of ringed doves. Sci. Am. 211:48–54.

LEIF, A. P. 1994. Survival and reproduction of wild and pen-reared ring-necked pheasant hens. J. Wildl. Manage. 58:501–506.

LEISTER, C. W. 1932. The pronghorn of North America. Bull. N.Y. Zool. Soc. 35:182–205.

LEITCH, W. G. 1951. Saving, maintaining, and developing waterfowl habitat in western Canada. Trans. N. Am. Wildl. Conf. 16:94–103.

LEMAHO, Y., P. DELCLITTE, AND J. CHATONNET. 1976. Thermoregulation in fasting emperor penguins under natural conditions. Am. J. Physiol. 231:913–922.

LENNON, R. E., J. B. HUNN, R. A. SCHNICK, AND R. M. BURRESS. 1970. FAO Fisheries Tech. Paper 100. Reprinted (1971) by Bur. Sport Fisheries and Wildl., Washington, D.C. 99 pp.

LEOPOLD, A. (CHAIRMAN). 1930. Report to the American Game Conference on an American game policy. Trans. Am. Game Conf. 17:284–309.

LEOPOLD, A. 1931. Report on a game survey of the north central states. Sporting Arms and Ammunition Manufacturers' Institute, Madison, Wis. 299 pp.

LEOPOLD, A. 1933a. Game management. Scribner, New York. 481 pp.

LEOPOLD, A. 1933b. The conservation ethic. J. Forestry 31: 634–643.

LEOPOLD, A. 1936a. Deer and dauerwald in Germany. J. Forestry 34: 366–375; 460–466.

LEOPOLD, A. 1936b. Naturschutz in Germany. Bird-Lore 38: 102–111.

LEOPOLD, A. 1938. Chukaremia. Outdoor America 3:3.

LEOPOLD, A. 1943. Deer irruptions. Wis. Cons. Bull. (reprinted in Wis. Cons. Dept. Publ. 321:1–11).

LEOPOLD, A. 1949. A sand county almanac. Oxford Univ. Press, New York. 226 pp.

LEOPOLD, A., L. K. SOWLS, AND D. L. SPENCER. 1947. A survey of overpopulated deer ranges in the United States. J. Wildl. Manage. 11:162–177.

LEOPOLD, A. S. 1963. Study of wildlife problems in national parks. Trans. N. Am. Wildl. and Nat. Resour. Conf. 28:28–45.

LEOPOLD, A. S. 1969. Adaptability of animals to habitat changes. Pp. 51–63 in Readings in conservation ecology (G. W. Cox, ed.). Appleton-Century-Crofts, New York. 595 pp.

LEOPOLD, A. S. 1977. The California quail. Univ. Calif. Press, Berkeley. 281 pp.

LEOPOLD, A. S., AND F. F. DARLING. 1953a. Wildlife in Alaska. Ronald Press, New York. 129 pp.

LEOPOLD, A. S., AND F. F. DARLING. 1953b. Effects of land use on moose and caribou in Alaska. Trans N. Am. Wildl. Conf. 18:553–560.

LEOPOLD, A. S., S. A. CAIN, C. M. COTTAM, I. N. GABRIELSON, AND T. I. KIMBALL. 1964. Predator and rodent control in the United States. Trans. N. Am. Wildl. and Nat. Resour. Conf. 29:27–49.

LEOPOLD, A. S., C. COTTAM, I. McT. COWAN, I. N. GABRIELSON, AND T. L. KIMBALL. 1968. The National Wildlife Refuge System. Trans. N. Amer. Wildl. and Nat. Resour. Conf. 33:30–54.

LEOPOLD, A. S., AND M. K. DEDON. 1983. Resident mourning doves in Berkeley, California. J. Wildl. Manage. 47:780–789.

LEOPOLD, A. S., M. ERWIN, J. OH, AND B. BROWNING. 1976. Phytoestrogens: adverse effects on reproduction in California quail. Science 191:98–100.

LEOPOLD, L. B. 1953. Round river, from the journals of Aldo Leopold. Oxford Univ. Press, N.Y. 173 pp.

LESLIE, D. M., AND C. L. DOUGLAS. 1980. Human disturbance at water sources of desert bighorn sheep. Wildl. Soc. Bull. 8:284–290.

LEWIN, R. 1986. Damage to tropical forests, or why were there so many kinds of animals. Science 234:149–150.

LEWIN, V., AND G. LEWIN. 1984. The Kalij pheasant, a newly established game bird on the island of Hawaii. Wilson Bull. 96:634–646.

LEWIS, J. B. 1961. Wild turkeys in Missouri, 1940–1960. Trans. N. Am. Wildl. and Nat. Resour. Conf. 26:505–513.

LEWIS, J. C. 1966. The fox rabies control program in Tennessee, 1956–66. Trans. N. Am. Wildl. and Nat. Resour. Conf. 31: 269–280.

LEWIS, J. C., AND E. LEGLER, JR. 1968. Lead shot ingestion by mourning doves and incidence in soil. J. Wildl. Manage. 32: 476–482.

LEWIS, J. C., G. W. ARCHIBALD, R. C. DREWIEN, C. R. FRITH, E. A. GLAESING, R. D. KLATASKE, C. D. LITTLEFIELD, J. SANDS, W. J. D. STEPHEN, AND L. E. WILLIAMS, JR. 1977. Sandhill crane (*Grus canadensis*). Pp. 5–43 in Management of migratory shore and upland game birds in North America (G. C. Sanderson, ed.). Internat. Assoc. of Fish and Wildl. Agencies in Coop. with Fish and Wildl. Serv. U. S. D. I., Washington, D.C. 358 pp.

LEWIS, S. J., AND R. A. MALECKI. 1984. Effects of egg oiling on larvid productivity and population dynamics. Auk 101:584–592.

LEWIS, T. A. 1987. How did the giants die? Internat. Wildl. 17(5):4–10.

LIGON, J. S. 1952. The vanishing bobwhite. Condor 54:48–50.

LIKENS, G., AND F. BORMANN. 1974. Acid rain; a serious regional environmental problem. Science 184:1176–1179.

LINCOLN, F. C. 1935. The waterfowl flyways of North America. U.S.D.A. Circ. No. 342. 12 pp.

LINDEMANN, W. 1956. Transplantation of game in Europe and Asia. J. Wildl. Manage. 20:68–70.

LINDUSKA, J. P. 1948. Insecticides vs. wildlife. Trans. N. Am. Wildl. Conf. 13:121–128.

LINEAR, M. 1977. The economics of wildlife: marginal land often yields profits when left alone. Ceres: the FAO Rev. on Agric. and Dev. 3:51–54.

LINHART, S. B., J. D. ROBERTS, AND G. J. DASCH. 1982. Electric fencing reduces coyote predation on pastured sheep. J. Range Manage. 35:276–281.

LINHART, S. B., R. T. STERNER, G. J. DASCH, AND J. W. THEADE. 1984. Efficacy of light and sound stimuli for reducing coyote predation upon pastured sheep. Protection Ecol. 6:75–84.

LIPSKE, M. 1985. Night stalker. Nat. Wildl. 23(4):20–24.

LITTRELL, E. E. 1986. Mortality of American wigeon on a golf course treated with the organophosphate, Diazinon. Calif. Fish and Game 72:117–126.

LITVAITIS, J. A., AND R. VILLAFUERTE. 1996. Factors affecting the persistence of New England cottontail metapopulations: the role of habitat management. Wildl. Soc. Bull. 24:686–693.

LIVEZEY, B. C. 1981. Duck nesting in retired croplands at Horicon National Wildlife Refuge, Wis. J. Wildl. Manage. 45:27–37.

LLOYD, H. G. 1977. Wildlife rabies: prospects for Britain. Pp. 91–103 *in* Rabies, the facts (C. Kaplan, ed.). Oxford Univ. Press, Oxford, England. 116 pp.

LOCKE, L. N., AND G. E. BAGLEY. 1967. Lead poisoning in a sample of Maryland mourning doves. J. Wildl. Manage. 31:515–518.

LOCKE, L. N., H. D. IRBY, AND G. E. BAGLEY. 1967. Histopathology of mallards dosed with lead and selected substitute shot. Bull. Wildl. Dis. Assoc. 3:143–147.

LONGCORE, J. R., AND F. B. SAMSON. 1973. Eggshell breakage by incubating black ducks fed DDE. J. Wildl. Manage. 37:390–394.

LONGCORE, J. R., P. O. CORR, AND H. E. SPENCER. 1982. Lead shot incidence in sediments and waterfowl gizzards from Merrymeeting Bay, Maine. Wildl. Soc. Bull. 10:3–10.

LONGCORE, J. R., R. K. ROSS, AND K. L. FISHER. 1987. Wildlife resources at risk through acidification of wetlands. Trans. N. Am. Wildl. and Nat. Resour. Conf. 52:608–618.

LONGHURST, W. M., H. K. OH, M. B. JONES, AND R. E. KEPNER. 1968. A basis for the palatability of deer forage plants. Trans. N. Am. Wildl. and Nat. Resour. Conf. 34:181–192.

LORENZ, K. Z. 1952. King Solomon's ring. Harper & Row, New York. 202 pp.

LOTKA, A. J. 1925. Elements of physical biology. Williams & Wilkins, Baltimore, Md. 460 pp.

LOUCKS, O. L. 1980. Acid rain: living resource implications and management needs. Trans. N. Am. Wildl. and Nat. Resour. Conf. 45:24–31.

LOVAT, LORD (S. J. FRASER, 14TH BARON LOVAT, ED.). 1911. Moor management. Pp. 372–391 *in* The grouse in health and disease, Vol. I. Smith, Elder & Co., London. 512 pp.

LOVE, D., AND F. L. KNOPF. 1978. The utilization of tree plantings by Mississippi kites in Oklahoma and Kansas. Pp. 70–74 *in* Trees, a valuable Great Plains multiple use resource. Proc. 13th Ann. Meeting of the Forestry Comm., Great Plains Agric. Council Publ. No. 87, Lincoln, Neb. 111 pp.

LOVE, D., J. A. GRZYBOWSKI, AND F. L. KNOPF. 1985. Influences of various land uses on windbreak selection by nesting Mississippi kites. Wilson Bull. 97:561–565.

LOVEJOY, T. 1976. We must decide which species will go forever. Smithsonian 7:52–58.

LOW, J. B., AND F. C. BELLROSE. 1944. Seed and vegetative yield of waterfowl food plants in the Illinois River Valley. J. Wildl. Manage. 8:7–22.

LOWERY, G. H., JR. 1974a. The mammals of Louisiana and its adjacent water. La. State Univ. Press, Baton Rouge. 565 pp.

LOWERY, G. H., JR. 1974b. Louisiana birds, 3rd ed. La. State Univ. Press, Baton Rouge. 651 pp.

LOWRY, D. A., AND K. L. MCARTHUR. 1978. Domestic dogs as predators on deer. Wildl. Soc. Bull. 6:38–39.

LUDYANSKIY, M. L., D. MCDONALD, AND D. MACNEILL. 1993. Impact of the zebra mussel, a bivalve invader. BioScience 43:533–544.

LUNT, H. A., AND H. G. M. JACOBSON. 1944. The chemical composition of earthworm casts. Soil Sci. 58:367–375.

LUOMA, S. N. 1984. Introduction to environmental issues. Macmillan, New York. 548 pp.

LUSSENHOP, J. 1977. Urban cemeteries as bird refuges. Condor 79:456–461.

LUTTRELL, M. P., J. R. FISCHER, D. E. STALLKNECHT, AND S. H. KLEVEN. 1996. Field investigation of *Mycoplasm gallisepticum* infections in house finches (*Carpodacus mexicanus*) from Maryland and Georgia. Avian Dis. 40:335–341.

LYDEKKER, R. 1898. Wild oxen, sheep, and goats of all lands. Rowland Ward, Ltd., London. 318 pp.

LYE, B. L. 1981. Letter in The Predator Forum. The Predator 8:3.

LYNCH, J. J., C. D. EVANS, AND V. C. CONOVER. 1963. Inventory of waterfowl environments of prairie Canada. Trans. N. Am. Wildl. and Nat. Resour. Conf. 28:93–109.

LYON, L. J. 1961. Evaluation of the influence of woody cover on pheasant hunting success. J. Wildl. Manage. 25:421–428.

LYONS, J. R., AND D. L. LEEDY. 1984. The status of urban wildlife programs. Trans. N. Am. Wildl. and Nat. Resour. Conf. 49: 233–251.

MABBUTT, J. A. 1981. Review and retrospect. Pp. 1–14 *in* The threatened drylands (J. A. Mabbutt and S. M. Berkowicz, eds.). School of Geography, Univ. New South Wales, Sydney, Australia. 153 pp.

MACARTHUR, R. H. 1972. Geographical ecology, patterns in the distribution of species. Harper & Row, New York. 269 pp.

MACARTHUR, R. H., AND E. O. WILSON. 1967. The theory of island biogeography. Princeton Univ. Press, Princeton. N.J. 203 pp.

MACCRACKEN, J. G. 1982. Coyote foods in a southern California suburb. Wildl. Soc. Bull. 10:280–281.

MACCRACKEN, J. G., D. W. URESK, AND R. M. HANSEN. 1985. Vegetation and soils of burrowing owl nest sites in Conata Basin, South Dakota. Condor 87:152–154.

MACCRIMMON, H. R., AND T. L. MARSHALL. 1968. World distribution of brown trout, *Salmo trutta*. J. Fish. Res. Bd. Can. 25: 2527–2548.

MACCURDY, E. 1939. The notebooks of Leonardo da Vinci. Reynal and Hitchcock, New York. 1247 pp.

MACINNES, C. D., R. A. DAVIS, R. N. JONES, B. C. LIEFF, AND A. J. PAKULAK. 1974. Reproductive efficiency of McConnell River small Canada geese. J. Wildl. Manage. 38:686–707.

MACK, G. D., AND L. D. FLAKE. 1980. Habitat relationships of waterfowl broods on South Dakota stock ponds. J. Wildl. Manage. 44:695–700.

MACKIE, R. J. 1987. Mule deer. Pp. 265–271 *in* Restoring America's wildlife (H. Kallman, ed.). U.S. Dept. Interior. Washington, D.C. 394 pp.

MACLEAN, G. L. 1985. Sandgrouse: models of adaptive compromise. S. Afr. J. Wildl. Res. 15:1–6.

MACLENNAN, R. R. 1978. Solutions to waterfowl depredation. Pp. 59–66 *in* Third internat. waterfowl symposium. Ducks Unlimited, Chicago. 244 pp.

MACLULICH, D. A. 1937. Fluctuations in the numbers of the varying hare *(Lepus americanus)*. Univ. Toronto Stud. Biol. Series No. 43. 136 pp.

MACPHERSON, A. H. 1969. The dynamics of Canadian arctic fox populations. Can. Wildl. Serv. Rept. Series No. 8, Ottawa. 52 pp.

MAHONEY, J. J. 1975. DDT and DDE effects on migratory condition in white-throated sparrows. J. Wildl. Manage. 39:520–527.

MAITLAND, P. S., AND C. E. PRICE. 1969. *Urocleidus principalis* (Mizelle, 1936), a North American monogenetic trematode new to the British Isles, probably introduced with the largemouth bass *Micropterus salmoides* (Lacepede, 1802). J. Fish Biol. 1:17–18.

MALCOLM, J. M. 1982. Bird collisions with a power transmission line and their relation to botulism at a Montana wetland. Wildl. Soc. Bull. 10:297–304.

MALECKI, R. A., B. BLOSSEY, S. D. HIGHT, D. SCHROEDER, L. T. KOK, AND J. R. COULSON. 1993. Biological control of purple loosestrife. BioScience 43:680–686.

MALOIY, G. M. O., AND H. F. HEADY. 1965. Grazing conditions in Kenya Masailand. J. Range Manage. 18:269–272.

MALTHUS, T. R. 1798. An essay on the principle of population. Reprinted in 1976 as An essay on the principle of population: text, sources, and background criticism. (P. Appleman, ed.). W. W. Norton, New York. 260 pp.

MANFREDO, M. J., AND B. HAIGHT. 1986. Oregon's nongame tax checkoff: a comparison of donors and nondonors. Wildl. Soc. Bull. 14:121–126.

MANGUN, W. R. 1986. Fiscal constraints to nongame management programs. Proc. Midwest. Fish and Wildl. Conf. 47:23–32.

MANNING, A. 1967. An introduction to animal behavior. Addison-Wesley, Reading, Mass. 208 pp.

MANSKI, D. A., L. W. VANDRUFF, AND V. FLYGER. 1981. Activities of gray squirrels and people in a downtown Washington, D.C. park: management implications. Trans. N. Am. Wildl. and Nat. Resour. Conf. 46:439–454.

MAPSTON, R. D., R. S. ZOBELL, K. B. WINTER, AND W. D. DOOLEY. 1970. A pass for antelope in sheep-tight fences. J. Range Manage. 23:457–459.

MARBURGER, R. G., AND J. W. THOMAS. 1965. A die-off of white-tailed deer of the central mineral region of Texas. J. Wildl. Manage. 29:706–716.

MARCH, G. L., AND R. M. SADLEIR. 1970. Studies on the band-tailed pigeon *(Columba fasciata)* in British Columbia. I. Seasonal changes in gonadal development and crop gland activity. Can. J. Zool. 48:1353–1357.

MARCH, G. L., AND R. M. SADLEIR. 1972. Studies on the band-tailed pigeon *(Columba fasciata)* in British Columbia. II. Food resource and mineral-gravelling activity. Syesis 5:279–284.

MARCY, B. C., AND R. C. GALVIN. 1973. Winter-spring sport fishery in the heated discharge of a nuclear power plant. J. Fisheries Biol. 5:541–547.

MARSH, H. 1938. Pneumonia in Rocky Mountain bighorn sheep. J. Mamm. 19:214–219.

MARSH, P. C. 1985. Effects of incubation temperature on survival of embryos of native Colorado River fishes. Southwestern Nat. 30:129–140.

MARTI, R. 1993. White-headed duck [Red data bird]. World Bird-watch 15(3):18–19.

MARTIN, A. C. 1951. Identifying pondweed seeds eaten by ducks. J. Wildl. Manage. 15:253–258.

MARTIN, A. C., AND F. M. UHLER. 1951. Food of game ducks in the United States and Canada. U.S. Fish and Wildl. Serv. Res. Rept. 30, Washington, D.C. 308 pp.

MARTIN, A. C., H. S. ZIM, AND A. L. NELSON. 1951. American wildlife and plants, a guide to wildlife food habits. McGraw-Hill, New York. 500 pp.

MARTIN, N. S. 1970. Sagebrush control related to habitat and sage grouse occurrence. J. Wildl. Manage. 34:313–320.

MARTIN, P. S. 1971. Prehistoric overkill. Pp. 612–624 in Man's impact on environment (T. R. Detwyler, ed.). McGraw-Hill, New York. 731 pp.

MARTIN, P. S. 1973. The discovery of America. Science 179:969–974.

MARTIN, P. S. 1987. Clovisia the beautiful! Nat. Hist. 96:10, 12–13.

MARTIN, P. S., AND R. G. KLEIN (EDS.). 1984. Quaternary extinctions, a prehistoric revolution. Univ. Ariz. Press, Tucson. 892 pp.

MARTIN, T. E., AND P. A. VOHS. 1978. Configuration of shelterbelts for optimum utilization by birds. Pp. 79–88 in Trees, a valu-able Great Plains multiple use resource. Proc. 13th Ann. Meeting of the Forestry Comm., Great Plains Agric. Council Publ. No. 87, Lincoln, Neb. 111 pp.

MARTINKA, C. J. 1976. Fire and elk in Glacier National Park. Proc. Tall Timbers Fire Ecol. Conf. 14:377–389.

MARTINKA, C. J. 1982. Rationale and options for management in grizzly bear sanctuaries. Trans. N. Am. Wildl. and Nat. Resour. Conf. 47:470–475.

MARTINSON, R. K., M. E. ROSASCO, E. M. MARTIN, M. G. SMART, S. M. CARNEY, C. F. KACZYNSKI, AND A. D. GEIS. 1966. 1965 experimental September hunting season on teal. U.S. Fish and Wildl. Serv. Spec. Sci. Rept. Wildl. No. 95, Washington, D.C. 36 pp.

MASER, C., R. G. ANDERSON, K. CROMARK, JR., J. T. WILLIAMS, AND R. E. MARTIN. 1979. Dead and down woody material. Pp. 78–95 in Wildlife habitats in managed forests: the Blue Mountains of Oregon and Washington (J. W. Thomas, ed.). U.S.D.A. Forest Serv. Agric. Handbook No. 553, 512 pp.

MASTER, L. L. 1979. Some observations on great gray owls and their prey in Michigan. Jack-Pine Warbler 57:215–217.

MATTHEWS, J. W., AND D. E. MCKNIGHT. 1982. Renewable resource commitments and conflicts in southeast Alaska. Trans. N. Am. Wildl. and Nat. Resour. Conf. 47:573–582.

MATTHEWS, O. P. 1986. Who owns wildlife? Wildl. Soc. Bull. 14:459–465.

MAXSON, G. D. 1981. Waterfowl use of a municipal sewage lagoon. Prairie Nat. 13:1–12.

MAY, J. F. 1978. Design and modification of windbreaks for better winter protection of pheasants. Unpubl. Doctoral Diss., Ia. State Univ., Ames. 179 pp.

MAY, R. M. 1983. Parasitic infections as regulators of animal populations. Am. Scientist 71:36–45.

MAYER, W. V. 1955. The protective value of the burrow system to the hibernating arctic ground squirrel, *Spermophilus undulatus.* The Anatomical Record 122:437–438 (abstract).

MAYFIELD, H. F. 1960. The Kirtland's warbler. Cranbrook Inst. Sci. Bull. 40, Bloomfield Hills, Mich. 242 pp.

MAYFIELD, H. F. 1961. Cowbird parasitism and the population of the Kirtland's warbler. Evolution 15:174–179.

MAYFIELD, H. F. 1972. Third decennial census of Kirtland's warbler. Auk 89:263–268.

MAYFIELD, H. F. 1973. Census of Kirtland's warbler in 1972. Auk 90:684–685.

MAYFIELD, H. F. 1977. Brood parasitism: reducing interactions between Kirtland's warblers and brown-headed cowbirds. Pp. 85–91 in Endangered birds, management techniques for preserving threatened species (S. A. Temple, ed.). Univ. Wis. Press, Madison. 466 pp.

MAYO, L. R. 1986. Breakup of the Hubbard Glacier ice dam and outburst of Russell Lake. Memo., Oct. 19, 1986. U.S. Geol. Surv., Fairbanks, Alaska. 8 pp.

MAYR, E. 1942. Systematics and the origin of species. Columbia Univ., Press, New York. 334 pp.

MAZAK, E. J., H. J. MACISAAC, M. R. SERVOS, AND R. HESSLEIN. 1997. Influence of feeding habits on organochlorine contaminant accumulation in waterfowl on the Great Lakes. Ecol. Applic. 7:1133–1143.

MCADOO, J. K., C. C. EVANS, B. A. ROUNDY, J. A. YOUNG, AND R. A. EVANS. 1983. Influence of heteromyid rodents on *Oryzopsis hymenoides* germination. J. Range Manage. 36:61–64.

MCARTHUR, K. L. 1980. Methods in the study of grizzly bear behavior in relation to people in Glacier National Park. Proc. 2nd Conf. on Sci. Res. in National Parks 6:234–247.

McBee, R. H., and G. C. West. 1969. Cecal fermentation in the willow ptarmigan. Condor 71:54–58.

McCabe, R. A. 1947. The homing of transplanted young wood ducks. Wilson Bull. 59:104–109.

McCabe, R. A. 1987. Aldo Leopold: the professor. Rusty Rock Press, Madison, Wis. 172 pp.

McCabe, R. A. 1991. Frederick N. Hamerstrom, 1910–1990, an addendum. Wildl. Soc. Bull. 19:378–379.

McCloskey, M. 1982. At last—the IWC bans whaling. Sierra 67:17–20.

McColloch, J. W. 1926. The role of insects in soil deterioration. J. Am. Soc. Agron. 18:143–150.

McCormack, D. E. 1974. Soil potentials: a positive approach to urban planning. J. Soil and Water Cons. 29:258–262.

McCorquodale, S. M. 1997. Cultural context of recreational hunting and native subsistence and ceremonial hunting: their significance for wildlife management. Wildl. Soc. Bull. 25: 568–573.

McCorquodale, S. M., and R. F. DeGiacomo. 1985. The role of wild North American ungulates in the epidemiology of bovine brucellosis: a review. J. Wildl. Dis. 21:351–357.

McCoy, E. D. 1983. The application of island-biogeographic theory to patches of habitat: how much land is enough? Biol. Conserv. 25:53–61.

McCullough, D. R. 1982. Behavior, bears, and humans. Wildl. Soc. Bull. 10:27–33.

McDougle, H. C., and R. W. Vaught. 1968. An epizootic of aspergillosis in Canada geese. J. Wildl. Manage. 32:415–417.

McEwen, L. C., and R. L. Brown. 1966. Acute toxicity of dieldrin and malathion to wild sharp-tailed grouse. J. Wildl. Manage. 30:604–611.

McFadden, J. T. 1961. A population study of the brook trout, *Salvelinus fontinalis*. Wildl. Monogr. No. 7. 73 pp.

McGinnies, W. J. 1968. Effects of nitrogen fertilizer on an old stand of crested wheatgrass. Agron. J. 60:560–562.

McKell, C. M., J. Major, and E. R. Perrier. 1959. Annual-range fertilization in relation to soil moisture depletion. J. Range Manage. 12:189–193.

McKelvey, J. J., Jr. 1973. Man against tsetse: struggle for Africa. Cornell Univ. Press, Ithaca, N.Y. 306 pp.

McKenzie, D. W. 1985. Range water pumping systems: State-of-the-art-review. U.S.D.A. Forest Service Proj. Rept. 8522-1201, San Dimas, Calif. 22 pp.

McKinley, D. 1960. A history of the passenger pigeon in Missouri. Auk 77:399–420.

McKinnon, J. G., G. L. Hoff, W. J. Bigler, and E. C. Prather. 1976. Heavy metal concentrations in kidneys of urban gray squirrels. J. Wildl. Dis. 12:367–371.

McKnight, D. E., and J. B. Low. 1969. Factors affecting waterfowl production on a spring-fed salt marsh in Utah. Trans. N. Am. Wildl. and Nat. Resour. Conf. 34:307–314.

McKnight, T. 1964. Feral livestock in Anglo-America. Univ. Calif. Publ. Geog. Vol. 16. 87 pp.

McKnight, T. L. 1958. The feral burro in the United States: distributions and problems. J. Wildl. Manage. 22:163–179.

McLaughlin, C. L., and D. Grice. 1952. The effectiveness of large-scale erection of wood duck boxes as a management procedure. Trans. N. Am. Wildl. Conf. 17:242–259.

McLouglin, D. K. 1966. Observations on the treatment of *Trichomonas gallinae* in pigeons. Avian Dis. 10:288.

McMahan, C. A. 1964. Comparative food habits of deer and three classes of livestock. J. Wildl. Manage. 28:798–808.

McMartin, W., A. B. Frank, and R. H. Heintz. 1974. Economics of shelterbelt influence on wheat yields in North Dakota. J. Soil Water Cons. 29:87–91.

McMillan, I. I. 1964. Annual population changes in California quail. J. Wildl. Manage. 28:702–711.

McMurry, F. B., and C. C. Sperry. 1941. Food of feral housecats in Oklahoma, a progress report. J. Mamm. 22:185–190.

McNicol, D. K., B. E. Bendell, and D. G. McAuley. 1987. Avian trophic relationships and wetland acidity. Trans. N. Am. Wildl. and Nat. Resour. Conf. 52:619–627.

McNulty, F. 1966. The whooping crane, the bird that defies extinction. E. P. Dutton, New York. 190 pp.

McNulty, F. 1971. Must they die? Doubleday, Garden City, New York. 86 pp.

McNulty, F. 1978. Last days of the condor. Audubon 80:53.

McQuilkin, R. A., and R. A. Musbach. 1977. Pin oak acorn production on green tree reservoirs in southeastern Missouri. J. Wildl. Manage. 41:218–225.

Meade, G. M. 1942. Calcium chloride—a death lure for crossbills. Brit. Birds 59:439–440.

Meagher, M. M. 1973. The bison of Yellowstone National Park. Sci. Monogr. Series 1, Nat. Park Serv. Dept. of Interior, Washington, D.C. 161 pp.

Means, D. B. 1982. Responses to winter burrow flooding of the gopher tortoise *(Gopherus polyphemus* Daudin). Herpetologica 38:521–525.

Mech, L. D. 1966. The wolves of Isle Royale. U.S. Nat. Park Serv. Faunal Series No. 7. 210 pp.

Mech, L. D. 1970. The wolf: the ecology and behavior of an endangered species. Nat. Hist. Press, Garden City, New York. 384 pp.

Mech, L. D. 1973. Wolf numbers in the Superior National Forest of Minnesota. U.S.D.A. Forest Serv., N. Central Forest Exp. Sta. Res. Paper NC-97. 10 pp.

Mech, L. D. 1977a. Productivity, mortality and population trends of wolves in northeastern Minnesota. J. Mamm. 58:559–574.

Mech, L. D. 1977b. Wolf-pack buffer zones as prey reservoirs. Science 198:320–321.

Mech, L. D. 1977c. A recovery plan for the eastern timber wolf. Nat. Parks and Cons. Magazine 51:17–21.

Mech, L. D., and L. D. Frenzel, Jr. 1971. An analysis of the age, sex, and condition of deer killed by wolves in northwestern Minnesota. U.S.D.A. Forest Serv., N. Central Forest Exp. Sta. Res. Paper NC-52:35–51.

Mech, L. D., and P. D. Karns. 1978. Role of the wolf in a deer decline in the Superior National Forest. U.S.D.A. Forest Serv., N. Central Forest Exp. Sta. Res. Paper NC-148. 23 pp.

Mech, L. D., L. D. Frenzel, Jr., and P. D. Karns. 1971a. The effect of snow conditions on the vulnerability of white-tailed deer to wolf predation. U.S.D.A. Forest Serv., N. Central Forest Exp. Sta. Res. Paper NC-52:51–62.

Mech, L. D., L. D. Frenzel, Jr., R. R. Ream, and J. W. Winship. 1971b. Movements, behavior, and ecology of timber wolves in northeastern Minnesota. U.S.D.A. Forest Serv., N. Central Forest Exp. Sta. Res. Paper NC-52:1–35.

Mech, L. D., and S. M. Goyal. 1993. Canine parvovirus effect on wolf population change and pup survival. J. Wildl. Dis. 29: 330–333.

Mech, L. D., S. M. Goyal, C. M. Bota, and U.S. Seal. 1986. Canine parvovirus infection in wolves *(Canis lupus)* from Minnesota. J. Wildl. Dis. 22:104–106.

Medin, D. E., and A. E. Anderson. 1979. Modeling the dynamics of a Colorado mule deer population. Wildl. Monogr. No. 68. 77 pp.

MEEHAN, W. R., AND W. S. PLATTS. 1978. Livestock grazing and the aquatic environment. J. Soil and Water Cons. 33:274–278.

MEEKS, R. L. 1968. The accumulation of ^{36}Cl ring-labeled DDT in a freshwater marsh. J. Wildl. Manage. 32:376–398.

MEFFE, G. K., A. H. EHRLICH, AND D. EHRENFELD. 1993. Human population control: the missing agenda. Conserv. Biol. 7:1–4.

MEINE, C. 1988. Aldo Leopold, his life and work. Univ. Wisconsin Press., Madison, Wis. 638 pp.

MEIER, K. E. 1983. Habitat use by opossums in an urban environment. Unpubl. M.S. Thesis, Oregon State Univ., Corvallis. 69 pp.

MENASCO, K. A. 1986. Stocktanks: an underutilized resource. Trans. N. Am. Wildl. and Nat. Resour. Conf. 51:304–409.

MENDALL, H. L. 1958. The ring-necked duck in the Northeast. Univ. Maine Bull. Vol. LX, No. 16, Orono. 317 pp.

MENDALL, H. L., AND C. M. ALDOUS. 1943. The ecology and management of the American woodcock. Maine Coop. Wildl. Res. Unit, Orono. 201 pp.

MENDELSSOHN, H., AND U. PAZ. 1977. Mass mortality of birds of prey caused by Azodrin, an organophosphorous insecticide. Biol. Cons. 11:163–170.

MENGES, E. S. 1991. Seed germination percentage increases with population size in a fragmented prairie species. Conserv. Biol. 5:158–164.

MERESZCZAK, I. M., W. C. KRUEGER, AND M. VAVRA. 1981. Effects of range improvement on Roosevelt elk winter nutrition. J. Range Manage. 34:184–187.

MERLEN, G. 1984. The 1982–83 El Niño: some of its consequences for Galápagos wildlife. Oryx 18:210–214.

MERRILL, L. B. 1975. Effects of grazing management practices on wild turkey habitat. Pp. 108–112 in Proc. third nat. wildl. turkey symposium (L. K. Halls, ed.). Texas Chapter—The Wildlife Society, Austin. 227 pp.

MERRILL, L. B. 1980. Considerations necessary in selecting and developing a grazing system—what are the alternatives? Pp. 29–35 in Proc. grazing management systems for southwest rangelands symposium (K. C. McDaniel and C. D. Allison, eds.). N. M. State Univ. Press, Las Cruces. 183 pp.

MERRILL, L. B., AND C. A. TAYLOR, JR. 1975. Advantages and disadvantages of intensive grazing management systems near Sonora and Barnhart. Pp. 10–11 in Texas Agric. Exp. Sta. Rept. PR-3341, College Station. 70 pp.

MERZ, R. W., AND G. K. BRAKHAGE. 1964. The management of pin oak in a duck shooting area. J. Wildl. Manage. 28:233–239.

MESLOW, E. C., AND L. B. KEITH. 1968. Demographic parameters of snowshoe hare population. J. Wildl. Manage. 32:812–834.

MESLOW, E. C., C. MASER, AND J. VARNER. 1981. Old-growth forests as wildlife habitat. Trans. N. Am. Wildl. and Nat. Resour. Conf. 46:329–335.

MESSMER, T. A., C. A. LIVELY, D. D. MACDONALD, AND S. A. SCHROEDER. 1996. Motivating landowners to implement wildlife conservation practices using calendars. Wildl. Soc. Bull. 24: 757–763.

MICHAEL, V. C., AND S. L. BECKWITH. 1955. Quail preference for seed of farm crops. J. Wildl. Manage. 19:281–296.

MILLER, G. R. 1968. Evidence for selective feeding on fertilized plots by red grouse, hares, and rabbits. J. Wildl. Manage. 32:849–853.

MILLER, G. S., R. J. SMALL, AND E. C. MESLOW. 1997. Habitat selection by spotted owls during natal dispersal in western Oregon. J. Wildl. Manage. 61:140–150.

MILLER, G. T., JR. 1982. Living in the environment, 3rd ed. Wadsworth, Belmont, Calif. 500 pp.

MILLER, H. W. 1976. Crop depredations as a limiting factor. Pp. 35–39 in Second internat. waterfowl symposium. Ducks Unlimited, Chicago. 192 pp.

MILLER, K. 1982. Parks under siege. Internat. Wildl. 12:32–34.

MILLER, M. R. 1975. Gut morphology of mallards in relation to diet quality. J. Wildl. Manage. 39:168–173.

MILLER, R. R. 1948. The cyprinodont fishes of the Death Valley system of eastern California and southwestern Nevada. Univ. Mich. Mus. Zool. Misc. Publ. 68, Ann Arbor. 155 pp.

MILLER, R. S. 1964. Ecology and distribution of pocket gophers (Geomyidae) in Colorado. Ecology 45:256–272.

MILLER, R. S., AND D. B. BOTKIN. 1974. Endangered species: models and predictions. Am. Scientist 62:172–181.

MILLER, W. R., AND F. E. EGLER. 1950. Vegetation of the Wequetequock-Pawatuck tidal marshes. Ecol. Monogr. 20: 143–172.

MILONSKI, M. 1958. The significance of farmland for waterfowl nesting and techniques for reducing losses due to agricultural practices. Trans. N. Am. Wildl. Conf. 23:215–228.

MINCKLEY, W. L., AND J. E. DEACON. 1968. Southwestern fishes and the enigma of "endangered species." Science 159: 1424–1432.

MIQUELLE, D. G., J. M. PEEK, AND V. VAN BALLENBERGHE. 1992. Sexual segregation in Alaskan moose. Wildl. Monogr. 122. 57 pp.

MITCHELL, C. A., AND J. CARLSON. 1993. Lesser scaup forage on zebra mussels at Cook Nuclear Plant, Michigan. J. Field Ornithol. 64:219–222.

MITCHELL, J. G. 1982. The trapping question, soft skins and sprung steel. Audubon 84:64–89.

MITZNER, L. 1978. Evaluation of biological control of nuisance aquatic vegetation by grass carp. Trans. Am. Fisheries Soc. 107:135–145.

MOEN, A. N. 1978. Seasonal changes in heart rates, activity, metabolism, and forage intake of white-tailed deer. J. Wildl. Manage. 42:715–738.

MOMENT, G. B. 1968. Bears: the need for a new sanity in wildlife conservation. BioScience 8:1105–1108.

MOMENT, G. B. 1970. Man-grizzly problems—past and present. BioScience 20:1142–1144.

MONK, C. E. 1981. History and present status of fur management in Ontario. Pp. 1501–1521 in Proc. worldwide furbearer conf., vol. II (J.A. Chapman and D. Pursley, eds.). Frostburg, Md. 1552 pp.

MOODY, J. D., AND C. D. SIMPSON. 1977. Population status, habitat, and movement of elk in the Guadalupe Mountains National Park, Texas. Proc. Ann. Conf. SE Assoc. Fish and Wildl. Agencies 31:151–158.

MOORE, J. 1983. Responses of an avian predator and its isopod prey to an acanthocephalan parasite. Ecology 64:1000–1015.

MOORE, J. C. 1956. Observations of manatees in aggregations. Am. Mus. Novit. 1811:1–24.

MOORE, W. H., AND W. S. TERRY. 1979. Short-duration grazing may improve wildlife habitat in southeastern pinelands. Proc. Ann. Conf. SE Assoc. Fish and Wildl. Agencies 33:279–287.

MOOREHEAD, B. B., J. AHO, P. CRAWFORD, D. HOUSTON, R. OLSON, AND E. SCHREINER. 1981. An environmental assessment of the management of introduced mountain goats in Olympic National Park. Olympic National Park, Port Angeles, Wash. 49 pp.

MORRISON, B. L. 1980. History and status of Barbary sheep in New Mexico. Pp. 15–16 in Proc. of the symposium on ecology and management of Barbary sheep (C. D. Simpson, ed.). Dept. of Range and Wildlife Management, Texas Tech Univ. Lubbock. 112 pp.

MORRISON, M. L., AND E. C. MESLOW. 1984. Response of avian communities to herbicide-induced vegetation changes. J. Wildl. Manage. 48:14–22.

MORSE, D. H. 1980. Behavioral mechanisms in ecology. Harvard Univ. Press, Cambridge, Mass. 383 pp.

MORSE, W. B. 1987. Conservation law enforcement: a new profession is forming. Trans. N. Am. Wildl. and Nat. Resour. Conf. 52:169–175.

MOSHER, L. 1985. Unknown quantities of toxic chemicals find their way to wildlife refuges. National J. 17:905–908.

MOSS, A. E. 1940. The woodchuck as a soil expert. J. Wildl. Manage. 4:441–443.

MOSS, M. B., J. D. FRASER, AND J. D. WELLMAN. 1986. Characteristics of nongame fund contributors vs. hunters in Virginia. Wildl. Soc. Bull. 14:107–115.

MOULTON, D. W., W. I. JENSEN, AND J. B. LOW. 1976. Avian botulism epizootiology on sewage oxidation ponds in Utah. J. Wildl. Manage. 40:735–742.

MOWAT, F. 1963. Never cry wolf. Dell Publ., New York. 176 pp.

MUNDY, P. 1983. The conservation of the Cape griffon vulture of southern Africa. Pp. 57–74 *in* Vulture biology and management (S. R. Wilbur and J. A. Jackson, eds.). Univ. Calif. Press, Berkeley. 550 pp.

MUNDY, P., D. BUTCHART, J. LEDGER, AND S. PIPER. 1992. The vultures of Africa. Academic Press, New York. 460 pp.

MUNDINGER, J. G. 1976. Waterfowl response to rest-rotation grazing. J. Wildl. Manage. 40:60–68.

MURIE, A. 1944. The wolves of Mount McKinley. U.S. Nat. Park Serv. Fauna Series No. 5. 238 pp.

MURIE, O. J. 1959. Fauna of the Aleutian Islands and Alaska Peninsula. U. S. Fish and Wildl. Serv. N. Am. Fauna No. 61. 406 pp.

MURPHY, E. C., AND K. R. WHITTEN. 1976. Dall sheep demography in McKinley Park and a re-evaluation of Murie's data. J. Wildl. Manage. 40:597–609.

MURPHY, R. K., N. F. PAYNE, AND R. K. ANDERSON. 1985. White-tailed deer use of an irrigated agriculture-grassland complex in central Wisconsin. J. Wildl. Manage. 49:125–128.

MURPHY, S. M., R. H. DAY, J. A. WIENS, AND K. R. PARKER. 1997. Effects of the *Exxon Valdez* oil spill on birds: comparisons of pre- and post-spill surveys in Prince William Sound, Alaska. Condor 99:299–313.

MURRAY, R. W. 1958. The effect of food plantings, climatic conditions, and land use practices upon the quail population of an experimental area in northwest Florida. Proc. SE Assoc. Game and Fish Comm. 12:269–274.

MUSGROVE, J. W. 1949. Iowa. Pp. 193–223 *in* Wildfowling in the Mississippi Flyway (E. V. Connett, ed.). D. Van Nostrand Co., Inc., New York. 387 pp.

MUTH, O. H. 1963. White muscle disease, a selenium-responsive myopathy. J. Am. Vet. Med. Assoc. 142:272–277.

MYERS, J. P., R. I. G. MORRISON, Z. ANTAS, B. A. HARRINGTON, T. E. LOVEJOY, M. SALLABERRY, S. E. SENNER, AND A. TURAK. 1987. Conservation strategy for migratory species. Am. Scientist 75:19–26.

MYERS, N. 1976. An expanded approach to the problem of disappearing species. Science 193:198–202.

MYERS, N. 1979. The sinking ark. Pergamon Press, Elmsford, N.Y. 307 pp.

MYERS, N. 1980. Conversion of tropical moist forests. Nat. Acad. Sci., Washington, D.C. 205 pp.

NAGY, J. G., AND L. BENCZE. 1973. Game management, administration, and harvest in Hungary. Wildl. Soc. Bull. 1:121–127.

NAGY, J. G., H. W. STEINHOFF, AND G. M. WARD. 1964. Effects of essential oils of sagebrush on deer rumen microbial function. J. Wildl. Manage. 28:785–790.

NAIMAN, R. J., AND J. M. MELILLO. 1984. Nitrogen budget of a subarctic stream altered by beaver *(Castor canadensis)*. Oecologia 62:150–155.

NAIMAN, R. J., J. M. MELILLO, AND J. E. HOBBIE. 1986. Ecosystem alteration of boreal forest streams by beaver *(Castor canadensis)*. Ecology 67:1254–1269.

NASH, R. 1982. Wilderness and the American mind, 3rd ed. Yale Univ. Press, New Haven. 425 pp.

NASH, R. G., AND E. A. WOOLSON. 1967. Persistence of chlorinated hydrocarbon insecticides in soils. Science 157:924–926.

NASON, G. W. 1982. Ecofallow and wildlife in Nebraska. Pp. 13–14 *in* Proc. Great Plains Agric. Council, Lincoln, Neb. 81 pp.

NASS, R. D., G. LYNCH, AND J. THEADE. 1984. Circumstances associated with predation rates on sheep and goats. J. Range Manage. 37:423–426.

NATIONAL RESEARCH COUNCIL. 1982. Impacts of emerging agricultural trends on fish and wildlife habitat. Comm. on Impacts of Emerging Agric. Trends on Fish and Wildl. Habitat, Bd. on Agric. and Renewable Resour., and Comm. on Nat. Resour. National Academy Press, Washington, D.C. 303 pp.

NATIONAL SHOOTING SPORTS FOUNDATION. No date. Un-endangered species. The success of wildlife management in North America. National Shooting Sports Foundation, Riverside, Conn. 48 pp.

NATIONAL WILDLIFE FEDERATION. 1987. Conservation Directory, 32nd ed. National Wildlife Federation, Washington, D.C. 306 pp.

NATIONS, J. D., AND D. I. KOMER. 1983. Central America's tropical rainforests: positive steps for survival. Ambio 12:232–238.

NATIONS, J. D., AND D. I. KOMER. 1984. Chewing up the jungle. Internat. Wildl. 14:14–16.

NCTRH. No date. People and plants, for therapy and rehabilitation. National Council for therapy and rehabilitation through horticulture, Alexandria, Virginia. Brochure, unnumbered pp.

NEESS, J. C. 1946. Development and status of pond fertilization in central Europe. Trans. Am. Fisheries Soc. 76:335–358.

NEFF, D. J. 1957. Ecological effects of beaver habitat abandonment in the Colorado Rockies. J. Wildl. Manage. 21:80–84.

NEFF, J. A. 1955. Outbreak of aspergillosis in mallards. J. Wildl. Manage. 19:415–416.

NELSON, C., AND D. VERBYLA. 1984. Characteristics and effectiveness of state anti-poaching campaigns. Wildl. Soc. Bull. 12:117–122.

NELSON, H. K., AND R. B. OETTING. 1981. An overview of management of Canada geese and their recent urbanization. Pp. 128–133 *in* Fourth internat. waterfowl symposium. Ducks Unlimited, Chicago. 265 pp.

NELSON, J. S. 1965. Effects of fish introductions and hydroelectric development on fishes in the Kananaskis River system, Alberta. J. Fish. Res. Bd. Can. 22:721–753.

NELSON, L., AND J. B. LOW. 1977. Acceptance of the 1970–71 point system season by duck hunters. Wildl. Soc. Bull. 5:52–55.

NELSON, M. E., AND L. D. MECH. 1981. Deer social organization and wolf predation in northeastern Minnesota. Wildl. Monogr. No. 77. 53 pp.

NELSON, M. E., AND L. D. MECH. 1986. Relationship between snow depth and gray wolf predation on white-tailed deer. J. Wildl. Manage. 50:471–474.

NERNEY, N. J. 1958. Grasshopper infestations in relation to range condition. J. Range Manage. 11:247.

NESTLER, R. B. 1946. Vitamin A, vital factor in the survival of bob-whites. Trans. N. Am. Wildl. Conf. 11:176–192.

NESTLER, R. B., T. B. DEWITT, AND T. V. DERBY, JR. 1949. Vitamin A storage in wild quail and its possible significance. J. Wildl. Manage. 13:265–271.

NEWMAN, J. R. 1979. Hunting and hunter education in Czechoslovakia. Wildl. Soc. Bull. 7:155–161.

NEWMAN, J. R., W. H. BRENNAN, AND M. L. SMITH. 1977. Twelve-year changes in nesting bald eagles *(Haliaeetus leucocephalus)* on San Juan Island, Washington. Murrelet 58:37–39.

NEWMAN, P. C. 1989. Empire of the bay, an illustrated history of the Hudson's Bay Company. Madison Press, Toronto, Ontario. 232 pp.

NICE, M. M. 1937. Studies in the life history of the song sparrow. Vol. I. Trans. Linnaean Soc., New York. 246 pp.

NICE, M. M. 1943. Studies in the life history of the song sparrow. Vol. II. Trans. Linnaean Soc., New York. 328 pp.

NICHOLS, J. D., M. J. CONROY, D. R. ANDERSON, AND K. P. BURNHAM. 1984. Compensatory mortality in waterfowl populations: a review of evidence and implications for research and management. Trans. N. Am. Wildl. and Nat. Resour. Conf. 49:535–554.

NIELSEN, D. B., D. D. LYTLE, AND F. WAGSTAFF. 1985. Big game on private lands: who benefits (or loses)? Utah Sci. 46:48–50.

NIELSEN, D. B., F. J. WAGSTAFF, AND D. LYTLE. 1986. Big-game animals on private range. Rangelands 8:36–38.

NIELSON, C. K., W. F. PORTER, AND H. B. UNDERWOOD. 1997. An adaptive management approach to controlling suburban deer. Wildl. Soc. Bull. 25:470–477.

NOBLE, G. K. 1939. The role of dominance in the social life of birds. Auk 56:263–273.

NOON, B. R., V. P. BINGHAM, AND J. P. NOON. 1979. The effects of changes in habitat on northern hardwood bird communities. Pp. 33–48 *in* Management of north central and northeastern forests for non-game birds. U.S.D.A. Forest Serv., N. Central For. Exp. Sta., St. Paul, Minn. 268 pp.

NORD, W. H. 1963. Wildlife in the small watershed program. Trans. N. Am. Wildl. and Nat. Resour. Conf. 28:118–124.

NORMAN, R. L., L. A. ROPER, P. D. OLSON, AND R. L. EVANS. 1976. Using wildlife values in benefit/cost analysis and mitigation of wildlife losses. Colorado Div. of Wildlife, Denver. 18 pp.

NOSS, R. F. 1987. Corridors in real landscapes: a reply to Simberloff and Cox. Conserv. Biol. 1:159–164.

NOSS, R. F., AND L. D. HARRIS. 1986. Nodes, networks, and MUMs: preserving diversity at all scales. Environ. Manage. 10:299–309.

NOYES, J. H. 1986. Consumption of largemouth bass eggs by redhead ducks at Ruby Lake, Nevada. Auk 103:219–221.

OATES, D. W., J. L. COGGIN, F. E. HARTMAN, AND G. I. HOILIEN. 1984. A guide to the time of death in selected wildlife species. Neb. Tech. Series 14, Neb. Game and Parks Comm., Lincoln. 72 pp.

OBERHOLSER, H. C. 1974. Bird life of Texas. Univ. Texas Press, Austin. 1069 pp.

O'BRIEN, M., AND R. A. ASKINS. 1985. The effects of mute swans on native waterfowl. Connecticut Warbler, J. Conn. Ornithol. 5:27–39.

O'BRIEN, S. J., M. E. ROELKE, L. MARKER, A. NEWMAN, C. A. WINKLER, D. MELTZER, L. COLLY, J. F. EVERMAN, M. BUSH, AND D. E. WILDT. 1985. A genetic basis for species vulnerability in the cheetah. Science 227:1428–1434.

ODEN, S. 1976. The acidity problem—an outline of concepts. Water Air Soil Pollut. 6:317–366.

ODUM, E. P. 1975. Ecology: the link between social and natural sciences, 2nd ed. Holt, Rinehart and Winston, Orlando, Fla. 244 pp.

ODUM, E. P. 1983. Basic ecology. Saunders, New York. 613 pp.

OETTING, R. B., AND J. F. CASSEL. 1971. Waterfowl nesting on interstate highway right-of-way in North Dakota. J. Wildl. Manage. 35:774–781.

OFFICE OF TECHNOLOGY ASSESSMENT. 1984. Technologies to sustain tropical forest resources. OTA-F-214, U.S. Congress, Washington, D.C. 344 pp.

OFFICE OF TECHNOLOGY ASSESSMENT. 1987. Technologies to maintain biological diversity: summary. Congress of the United States, Washington, D.C. 47 pp.

OFFICE OF THE GREAT LAKES. 1985. State of the Great Lakes. Office of the Great Lakes, Lansing, Michigan. 36 pp.

O'GARA, B. W. 1982. Let's tell the truth about predation. Trans. N. Am. Wildl. and Nat. Resour. Conf. 47:476–484.

OGBUEHI, S. N., AND J. R. BRANDLE. 1981. Influence of windbreak-shelter on soybean production under rainfed conditions. Agron. J. 73:625–628.

OGDEN, E. C. 1943. The broad-leaved species of *Potamogeton* of North America north of Mexico. Rhodora 45:57–106; 119–163; 171–214.

OGDEN, E. C. 1953. Key to the North American species of *Potamogeton*. N.Y. State Mus. Circ. 31. 11 pp.

OGREN, H. A. 1965. Barbary sheep. N. M. Dept. Fish and Game Bull. No. 13, Santa Fe. 117 pp.

OHLENDORF, H. M., R. L. HOTHEM, C. M. BUNCK, T. W. ALDRICH, AND J. F. MOORE. 1986a. Relationships between selenium concentrations and avian reproduction. Trans. N. Am. Wildl. and Nat. Resour. Conf. 51:330–342.

OHLENDORF, H. M., D. J. HOFFMAN, M. K. SAIKI, AND T. W. ALDRICH. 1986b. Ebryonic mortality and abnormalities of aquatic birds: apparent impacts of selenium from irrigation drainwater. Sci. Total Environ. 52:49–63.

OHLENDORF, H. M., R. W. LOWE, P. R. KELLY, AND T. E. HARVEY. 1986c. Selenium and heavy metals in San Francisco Bay diving ducks. J. Wildl. Manage. 50:64–70.

OHLENDORF, H. M., R. W. RISEBROUGH, AND K. VERMEER. 1978. Exposure of marine birds to environmental pollutants. U.S. Fish and Wildl. Serv. Wildl. Res. Rept. 9, Washington, D.C. 40 pp.

OHMART, R. D., AND B. W. ANDERSON. 1982. North American desert riparian ecosystems. Pp. 433–467 *in* Reference handbook on the desert of North America (G. L. Bender, ed.). Greenwood Press, Westport, Conn. 594 pp.

OKA, I. N., AND D. PIMENTEL. 1974. Corn susceptibility to corn leaf aphids and common corn smut after herbicide treatment. Environ. Ent. 3:911–915.

OLINDO, P. M. 1974. Discussion. P. 254 *in* Second world conference on national parks (S. H. Elliott, ed.). Internat. Union for Cons. of Nature and Natural Resources, Morges, Switzerland. 504 pp.

OLSEN, O. W. 1949. White-tailed deer as a reservoir host of the large American liver fluke. Vet. Med. 44:26–30.

OLSEN, P. F. 1981. Sylvatic plague. Pp. 232–243 *in* Infectious diseases of wild mammals, 2nd ed. (J. W. Davis, L. H. Karstad, and D. O. Trainer, eds.). Ia. State Univ. Press, Ames. 446 pp.

OLSON, D. 1980. When a naturalist goes to kill, he raises some hard questions. Milwaukee Journal, January 6, 1980, Discover Section, pp. 1 and 9.

OLSON, D. P. 1965. Differential vulnerability of male and female canvasbacks to hunting. Trans. N. Am. Wildl. and Nat. Resour. Conf. 30:121–135.

OLSON, T. A. 1964. Blue-greens. Pp. 349–356 *in* Waterfowl tomorrow (J.P. Linduska, ed.). U.S.D.I. Bur. of Sport Fisheries and Wildl., Fish and Wildl. Serv., U.S. Govt. Printing Office, Washington, D.C. 770 pp.

O'MEARA, T. E., J. B. HAUFLER, L. H. STELTER, AND J. G. NAGY. 1981. Nongame wildlife responses to chaining of pinyon-juniper woodlands. J. Wildl. Manage. 45:381–389.

O'NEIL, T. 1949. The muskrat in the Louisiana coastal marshes. Louisiana Dept. Wildl. and Fish, New Orleans. 152 pp.

O'NEILL, D. H., R. J. ROBEL, AND A. D. DAYTON. 1983. Lead contamination near Kansas highways: implications for wildlife enhancement programs. Wildl. Soc. Bull. 11:152–160.

O'NEILL, E. J. 1947. Waterfowl grounded at the Muleshoe National Wildlife Refuge, Texas. Auk 64:457.

ONSTAD, C. A. 1972. Soil and water losses as affected by tillage practices. Trans. Am. Soc. Agric. Eng. 15:287–289.

OPPENHEIMER, E. 1980. *Felis catus* population densities in an urban area. Carnivore Genet. Newsletter 4:72–80.

OOSTING, H. J. 1958. The study of plant communities: an introduction to plant ecology. W. H. Freeman, San Francisco. 440 pp.

ORMEROD, W. E. 1978. The relationship between economic development and ecological degradation: how degradation has occurred in West Africa and how its progress might be halted. J. Arid Environ. 1:357–379.

O'ROKE, E. C. 1934. A malaria-like disease of ducks caused by *Leucocytozoon anatis* Wickware. Univ. Mich. School Forest Cons. Bull. 4. Ann Arbor. 44 pp.

O'ROKE, E. C., AND F. N. HAMERSTROM, JR. 1948. Productivity and yield of the George Reserve deer herd. J. Wildl. Manage. 12:78–86.

ORR, R. T. 1970. Animals in migration. Macmillan, New York. 303 pp.

OSBORN, B. 1942. Prairie dogs in shinnery (oak shrub) savannah. Ecology 23:110–115.

OSBORN, B., AND P. E. ALLAN. 1949. Vegetation of an abandoned prairie dog town in tall grass prairie. Ecology 30:322–332.

OSTFELD, R. S., M. C. MILLER, AND J. SCHNURR. 1993. Ear tagging increases tick (*Ixodes dammini*) infestation rates of white-footed mice (*Peromyscus leucopus*). J. Mamm. 74:651–655.

OWEN, M. 1975. Cutting and fertilizing grassland for winter goose management. J. Wildl. Manage. 39:163–167.

OWEN, M. 1980. Wild geese of the world. B.T. Batsford Ltd., London. 236 pp.

OWEN, M., AND M. NORDERHAUG. 1977. Population dynamics of barnacle geese *Branta leucopsis* breeding in Svalbard, 1948–1976. Ornis Scand. 8:161–174.

OWEN, O. S. 1985. Natural resource conservation: an ecological approach, 4th ed. Macmillan, New York. 657 pp.

OWEN, R. B., JR. 1970. The bioenergetics of captive blue-winged teal under controlled and outdoor conditions. Condor 72:153–163.

OWEN, R. B., JR., J. M. ANDERSON, J. W. ARTMANN, E. R. CLARK, T. G. DILWORTH, L. E. GREGG, F. W. MARTIN, J. D. NEWSOM, AND S. R. PURSGLOVE, JR. 1977. American woodcock (*Philohela minor=Scolopax minor* of Edwards 1974). Pp. 149–186 *in* Management of migratory shore and upland game birds in North America (G. C. Sanderson, ed.). Internat. Assoc. of Fish and Wildl. Agencies in Coop. with the Fish and Wildl. Serv. U. S. D. I., Washington, D.C. 358 pp.

OWEN, R. M. 1986. Idaho's cooperative Sikes Act wildlife management program. Trans. N. Am. Wildl. and Nat. Resour. Conf. 51:132–140.

OZOGA, J. J., AND E. M. HARGER. 1966. Winter activities and feeding habitats of northern Michigan coyotes. J. Wildl. Manage. 30:809–818.

OZOGA, J. J., AND L. J. VERME. 1986. Relation of maternal age to fawn-rearing success in white-tailed deer. J. Wildl. Manage. 50:480–486.

OZOGA., J. J., L. J. VERME, AND C. S. BIENZ. 1982. Parturition behavior and territoriality in white-tailed deer: impact on neonatal mortality. J. Wildl. Manage. 46:1–11.

PACKER, C., A. E. PUSEY, H. ROWLEY, D. A. GILBERT, J. MARTENSON, AND S. J. O'BRIEN. 1991. Case study of a population bottleneck: lions of the Ngorongoro Crater. Conserv. Biol. 5:219–230.

PACKER, P. E. 1963. Soil stability requirements for the Gallatin elk winter range. J. Wildl. Manage. 27:401–410.

PAGE, L. A. (ED.). 1976. Wildlife diseases. Plenum Press, New York. 686 pp.

PAGE, R. D., AND J. F. CASSEL. 1971. Waterfowl nesting on a railroad right-of-way in North Dakota. J. Wildl. Manage. 35:544–549.

PALMER, W. L., G. M. KELLY, AND J. L. GEORGE. 1982. Alfalfa losses to white-tailed deer. Wildl. Soc. Bull. 10:259–261.

PALMER, W. L., R. G. WINGARD, AND J. L. GEORGE. 1983. Evaluation of white-tailed deer repellents. Wildl. Soc. Bull. 11:164–166.

PALMER, W. L., J. M. PAYNE, R. G. WINGARD, AND J. L. GEORGE. 1985. A practical fence to reduce deer damage. Wildl. Soc. Bull. 13:240–245.

PARKER, J. C. 1968. Parasites of the gray squirrel in Virginia. J. Parasitol. 54:633–634.

PARKER, R. L. 1968. Quarantine and health problems associated with introductions of animals. Pp. 21–22 *in* Introduction of exotic animals: ecological and socioeconomic considerations. Caesar Kleberg Res. Pgm. in Wildl. Ecology, Texas A&M Univ., College Station. 25 pp.

PARNELL, R., A. SPRINGER, L. STEVENS, J. B. BENNETT, T. HOFFINAGLE, T. MELIS, AND D. STANITSKI-MARTIN. 1997b. Flood-induced backwater rejuvenation along the Colorado River in Grand Canyon, Arizona. Pp. 41–51 *in* Glen Canyon Dam beach-habitat building flow, abst. and exec. summaries. Grand Canyon Monit. and Res. Center, U.S. Dept. Interior, Flagstaff, Ariz. 129 pp.

PARNELL, R. A., L. STEVENS, J. B. BENNETT, J. E. HAZEL, JR., M. KAPLINSKI, M. MALONE, AND A. DALE. 1997a. Impacts of controlled flooding on physical and chemical characteristics of riparian ecosystems along the Colorado River, Grand Canyon, Arizona. Abst., Geol. Soc. Amer. 29, No. 6.

PATTEE, O. H., AND S. K. HENNES. 1983. Bald eagles and waterfowl: the lead shot connection. Trans. N. Amer. Wildl. and Nat. Resour. Conf. 48:230–237.

PATTERSON, J. C. 1974. Planting in urban soils. Nat. Park Serv. Ecol. Serv. Bull. No. 1. 23 pp.

PAULUS, S. L. 1982. Gut morphology of gadwalls in Louisiana in winter. J. Wildl. Manage. 46:483–489.

PAVEGLIO, F. L., K. M. KILBRIDE, AND C. M. BUNCK. 1997. Selenium in aquatic birds from central California. J. Wildl. Mange. 61:832–839.

PAYNE, W. J. A., AND H. G. HUTCHISON. 1963. Water metabolism in cattle in East Africa. I. The problem and the experimental procedure. J. Agric. Sci. 61:255–266.

PEACOCK, D. 1997. The Yellowstone massacre. Audubon 99(3):41–49, 102–103, 106–110.

PEARSON, T. G. 1915. Cemeteries as bird sanctuaries. Audubon Soc. Circ. 2. 4 pp.

PEHRSSON, O. 1979. Feeding behavior, feeding habitat utilization, and feeding efficiency of mallard ducklings *Anas platyrhynchos,* domestic duck. Viltrevy 10:193–218.

PENCE, D. B. 1980. Diseases and parasites of the Barbary sheep. Pp. 59–62 *in* Proc. of the symposium on ecology and management of Barbary sheep (C. D. Simpson, ed.). Dept. Range and Wildlife Management, Texas Tech Univ., Lubbock. 112 pp.

PENDERGAST, B. A., AND D. A. BOAG. 1971. Nutritional aspects of the diet of spruce grouse in central Alberta. Condor 73:437–443.

PENDERGAST, B. A., AND D. A. BOAG. 1973. Seasonal changes in the internal anatomy of spruce grouse in Alberta. Auk 90:307–317.

PENGELLY, W. L. 1972. Clearcutting: detrimental aspects for wildlife resources. J. Soil and Water Cons. 27:255–258.

PERRY, M. C., F. FERRIGNO, F. H. SETTLE. 1978. Rehabilitation of birds oiled on two mid-Atlantic estuaries. Proc. Ann. Conf. SE Assoc. Fish and Wildl. Agencies 32:318–325.

PERSSON, C., H. RODHE, AND L. DeGEER. 1987. The Chernobyl accident—a meteorological analysis of how radionuclides reached and were deposited in Sweden. Ambio 16:20–31.

PETERLE, T. J. 1958. Game management in Scotland. J. Wildl. Manage. 22:221–231.

PETERLE, T. J. 1977. Hunters, hunting, anti-hunting. Wildl. Soc. Bull. 5:151–161.

PETERLE, T. J., AND K. C. SADLER. 1965. An analysis of the effect of artificial nest boxes on a gray squirrel population. Trans. N. Am. Wildl. and Nat. Resour. Conf. 30:349–360.

PETERS, J. H. 1967. Effects on a trout stream of sediment from agricultural practices. J. Wildl. Manage. 31:805–812.

PETERS, R. P., AND L. D. MECH. 1975. Scent marking in wolves. Am. Scientist 63:628–637.

PETERSEN, L. 1979. Ecology of great horned owls and red-tailed hawks in southeastern Wisconsin. Wis. Dept. Nat. Resour. Tech. Bull. III, Madison. 63 pp.

PETERSON, J. G. 1970a. Gone with the sage. Montana Outdoors 5:1–3.

PETERSON, J. G. 1970b. The food habits and summer distribution of juvenile sage grouse in central Montana. J. Wildl. Manage. 34:147–155.

PETERSON, R. O. 1977. Wolf ecology and prey relationships on Isle Royale. U.S. Nat. Park Serv. Sci. Monogr. Series, No. 7. 210 pp.

PETERSON, R. O. 1980. Ecological studies of wolves on Isle Royale, Annual Report 1979–80. Mich. Tech. Univ., Houghton. 19 pp.

PETERSON, R. O. 1995. The wolves of Isle Royale: a broken balance. Willow Creed Press, Minocqua, Wis. 189 pp.

PETERSON, R. O. 1997. Ecological studies of wolves on Isle Royale. Annual Rept. 1996-97. Michigan Tech. Univ., Houghton, Mich. 16 pp.

PETERSON, R. O., AND D. L. ALLEN. 1974. Snow conditions as a parameter in moose-wolf relationships. Nat. Can. 101:481–492.

PETERSON, R. O., AND R. J. KRUMENAKER. 1989. Wolves approach extinction on Isle Royle: a biological and policy conundrum. George Wright Forum 6:10–15.

PETERSON, R. O., R. E. PAGE, AND K. M. DODGE. 1984. Wolves, moose, and the allometry of population cycles. Science 224:1350–1352.

PETERSON, R. P. 1987. A new bluebird nesting structure for highway rights-of-way. Wildl. Soc. Bull. 15:200–204.

PETERSON, T. L., AND J. A. COOPER. 1987. Impacts of center pivot irrigation systems on birds in prairie wetlands. J. Wildl. Manage. 51:238–247.

PETRABORG, W. H., E. G. WELLEIN, AND V. E. GUNVALSON. 1953. Roadside drumming counts, a spring census method for ruffed grouse. J. Wildl. Manage. 17:292–295.

PETRIDES, G. A., AND D. L. LEEDY. 1948. The nutria in Ohio. J. Mamm. 29:182–183.

PETRIDES, G. A., AND W. G. SWANK. 1958. Management of the big game resource in Uganda, East Africa. Trans. N. Am. Wildl. Conf. 23:461–477.

PFITSCH, W. A. 1980. The effect of mountain goats on the subalpine plant communities of Klahhane Ridge, Olympic National Park, Washington. Unpubl. M.S. Thesis, Univ. Washington, Seattle. 113 pp.

PFLIEGER, W. L. 1975. The fishes of Missouri. Mo. Dept. Cons., Columbia. 343 pp.

PHILLIPS, J. C. 1928. Wild birds introduced or transplanted in North America. U.S.D.A. Tech. Bull. No. 61. 63 pp.

PHILLIPS, R. L., R. D. ANDREWS, G. L. STORM, AND R. A. BISHOP. 1972. Dispersal and mortality of red foxes. J. Wildl. Manage. 36:237–248.

PHINNEY, H. K. 1959. Turbidity, sedimentation, and photosynthesis. Proc. Symposium Pacific Northwest: siltation—its sources and effects on the aquatic environment. U.S. Dept. Health, Education and Welfare 5:4–12.

PIANKA, E. R. 1983. Evolutionary ecology, 2nd ed. Harper & Row, New York. 397 pp.

PIELOWSKI, Z. 1959. Studies on the relationship: predator (goshawk) and prey (pigeon). Bull. Acad. Polish Sci. Serv. Sci. Biol. 7:401–403.

PIERCE, B. A. 1983. Grass carp status in the United States: a review. Environ. Manage. 7:151–160.

PIERCE, U. D. 1981. Brown trout management plan. Pp. 177–221 *in* Planning for Maine's inland fish and wildlife, Vol. 2, Pt. 1. Maine Dept. Inland Fisheries and Wildl., Augusta. 445 pp.

PIKE, D. K. 1981. Effects of mountain goats on three plant species unique to the Olympic Mountains, Washington. Unpubl. M.S. Thesis, Univ. Washington, Seattle. 188 pp.

PIMENTEL, D. 1961. Animal population regulation by the genetic feedback mechanism. Am. Nat. 95:65–79.

PIMENTEL, D., AND C. A. EDWARDS. 1982. Pesticides and ecosystems. BioScience 32:595–600.

PIMENTEL, D., D. ANDOW, D. GALLAHAN, I. SCHREINER, T. E. THOMPSON, R. DYSON-HUDSON, S. N. JACOBSON, M. A. IRISH, S. F. KROOP, A. M. MOSS, M. D. SHEPARD, AND B. G. VINZANT. 1980. Pesticides: environmental and social costs. Pp. 99–158 *in* Pest control: cultural and environmental aspects (D. Pimentel and J. H. Perkins, eds.). Westview Press, Boulder, Colo. 243 pp.

PIMLOTT, D. H. 1972. Wolves around the world. Nat. Parks and Conserv. Magazine 46:18–22.

PIMLOTT, D. H., AND W. J. CARBERRY. 1958. North American moose transplantations and handling techniques. J. Wildl. Manage. 22:51–62.

PIMLOTT, D. H., AND P. W. JOSLIN. 1968. The status and distribution of the red wolf. Trans. N. Am. Wildl. and Nat. Resour. Conf. 33:373–389.

PIMM, S. L., H. L. JONES, AND J. DIAMOND. 1988. On the risk of extinction. Amer. Nat. 132:757–785.

PINCHOT, G. 1947. Breaking new ground. Harcourt, Brace and Co., New York, N.Y. 522 pp.

PINKETT, H. T. 1970. Gifford Pinchot, private and public forester. Univ. Illinois Press, Urbana, Ill. 167 pp.

PINSHOW, B., M. A. FEDAK, D. R. BATTLES, AND K. SCHMIDT-NIELSEN. 1976. Energy expenditure for thermoregulation and locomotion in emperor penguins. Am. J. Physiol. 231:903–912.

PITMAN, W. D., AND E. C. HOLT. 1983. Herbage production and quality of grasses with livestock and wildlife values in Texas. J. Range Manage. 36:52–54.

PLATTS, W. S. 1981. Effects of sheep grazing on a riparian-stream environment. U.S.D.A. Forest Serv. Res. Note INT-307. 6 pp.

PLATTS, W. S. 1982. Livestock and riparian-fishery interactions: what are the facts? Trans. N. Am. Wildl. and Nat. Resour. Conf. 47:507–515.

PLETSCHER, D. H. 1987. Nutrient budgets for white-tailed deer in New England with special reference to sodium. J. Mamm. 68: 330–336.

POCHE, R. M. 1980. Elephant management in Africa. Wildl. Soc. Bull. 8:199–207.

PODOLL, E. B. 1979. Utilization of windbreaks by wildlife. Pp. 121–127 in Windbreak management. Great Plains Agric. Council Publ. 92, Lincoln, Neb. 132 pp.

POKRAS, M. A., AND R. CHAFEL. 1992. Lead toxicosis from ingested fishing sinkers in adult common loons *(Gavia immer)* in New England. J. Zoo and Wildl. Medicine 23:92–97.

POLLEY, H. W., AND L. L. WALLACE. 1986. The relationship of plant species heterogeneity to soil variation in buffalo wallows. Southwestern Nat. 31:493–501.

POOLE, D. A. 1982. Where have all the dollars gone? Ducks Unlimited 46:17–18, 20.

POOLE, K. G. 1997. Dispersal patterns of lynx in the Northwest Territories. J. Wildl. Manage. 61:497–505.

POPULATION REFERENCE BUREAU. 1988. 1988 world population data sheet. Pop. Ref. Bur., Washington, D.C. 1 p.

POPULATION REFERENCE BUREAU. 1998. 1998 World population data sheet. Pop. Ref. Bur., Washington, D.C. 2 pp.

PORTER, S. D., A. BHATKAR, R. MULDER, S. B. VINSON, AND J. D. CLAIR. 1991. Distribution and density of polygyne fire ants (Hymenoptera: Formicidae) in Texas. J. Econ. Entomol. 84: 866–874.

PORTER, W. F. 1983. A baited electric fence for controlling deer damage to orchard seedlings. Wildl. Soc. Bull. 11:325–327.

POSITION STATEMENTS COMMITTEE. 1975. Ecopolitics of The Wildlife Society. The Wildlife Society, Washington, D.C. 19 pp.

POSPAHALA, R. S., D. R. ANDERSON, AND C. J. HENNY. 1974. Population ecology of the mallard. II. Breeding habitat conditions, size of the breeding populations, and production indices. U. S. Fish and Wildl. Serv. Res. Publ. 115. Washington, D.C. 73 pp.

POSSARDT, E. E., AND W. E. DODGE. 1978. Stream channelization impacts on songbirds and small mammals in Vermont. Wildl. Soc. Bull. 6:18–24.

POST, G. 1971. The pneumonia complex in bighorn sheep. Trans. N. Am. Wildl. Sheep Conf. 1:98–106.

POTTER, D. R., J. C. HENDEE, AND R. N. CLARK. 1973. Hunting satisfaction: games, guns, or nature? Trans. N. Am. Wildl. and Nat. Resour. Conf. 38:220–229.

POTTS, G. R. 1977. Population dynamics of the grey partridge: overall effects of herbicides and insecticides on chick survival rates. Proc. Internat. Congress Game Biol. 13:203–211.

POTTS, G. R. 1980. The effects of modern agriculture, nest predation and game management on the population ecology of partridges *(Perdix perdix* and *Alectoris rufa).* Adv. Ecol. Res. 11:2–79.

POULSON, T. L., AND W. B. WHITE. 1969. The cave environment. Science 165:971–981.

POWELL, J., AND T. W. BOX. 1966. Brush management influences preference values of south Texas woody species for deer and cattle. J. Range Manage. 19:212–214.

PRESIDENT'S COMMISSION ON AMERICANS OUTDOORS. 1987. Americans outdoors: the legacy, the challenge. Island Press, Covelo, Calif. 426 pp.

PRESIDENT'S COUNCIL ON RECREATION AND NATURAL BEAUTY. 1968. From sea to shining sea. A report on the American environment—our natural heritage. U.S. Govt. Printing Office, Washington, D.C. 304 pp.

PRESNALL, C. C. 1958. The present status of exotic mammals in the United States. J. Wildl. Manage. 22:45–50.

PRESTWOOD, A. K., AND J. F. SMITH. 1969. Distribution of meningeal worm *(Pneumostrongylus tenuis)* in deer in the southeastern United States. J. Parasitol. 55:720–725.

PRICE, J. I. 1985. Immunizing Canada geese against avian cholera. Wildl. Soc. Bull. 13:508–515.

PRIMACK, R. B. 1993. Essentials of conservation biology. Sinauer Assoc., Sunderland, Mass. 564 pp.

PRITCHARD, P. C. 1987. Parks and conservation: a vision for the future. Renewable Resour. J. 5:10–12.

PROULX, G. 1986. Civic-service clubs—a new horizon for wildlife professionals. Wildl. Soc. Bull. 14:94–97.

PROVOST, M. W. 1948. Marsh-blasting as a wildlife management technique. J. Wildl. Manage. 12:350–387.

PRUITT, W. O., JR. 1959. Snow as a factor in the winter ecology of the barren-ground caribou. Arctic 12:159–179.

QUINTON, D. A., R. G. HOREJSI, AND J. T. FLINDERS. 1979. Influence of brush control on white-tailed deer diets in north-central Texas. J. Range Manage. 32:93–97.

RADABAUGH, B. E. 1974. Kirtland's warbler and its Bahama wintering grounds. Wilson Bull. 86:374–383.

RAMSEY, C. 1986. Natural resources management in support of national defense. Trans. N. Am. Wildl. and Nat. Resour. Conf. 51:125–131.

RAMSEY, C. W. 1965. Potential economic returns from deer as compared with livestock in the Edwards Plateau region of Texas. J. Range Manage. 18:247–250.

RAMSEY, C. W. 1968. Texotics. Bull. No. 49. Texas Parks and Wildl. Dept., Austin. 46 pp.

RAMSEY, C. W. 1970. Exotics. Bull. No. 49 (revised). Texas Parks and Wildl. Dept., Austin. 46 pp.

RANDS, M. R. W. 1985. Pesticide use on cereals and the survival of grey partridge chicks: a field experiment. J. Applied Ecol. 22:49–54.

RANDS, M., AND N. SOTHERTON. 1985. Pesticides threaten British wildlife. New Scientist 107:32.

RANGEL-WOODYARD, E., AND C. D. SIMPSON. 1980. Status of Barbary sheep in Mexico. Pp. 30–32 in Proc. of the symposium on ecology and management of Barbary sheep (C. D. Simpson, ed.). Dept. of Range and Wildlife Management, Texas Tech Univ., Lubbock. 112 pp.

RAPPOLE, J. H., AND E. S. MORTON. 1985. Effects of habitat alteration on a tropical avian forest community. Ornith. Monogr. 36:1013–1021.

RAPPOLE, J. H., AND D. W. WARNER. 1978. Migratory bird population ecology: conservation implications. Trans. N. Am. Wildl. and Nat. Resour. Conf. 43:235–240.

RAPPOLE, J. H., AND D. W. WARNER. 1980. Ecological aspects of migrant bird behavior in Veracruz, Mexico. Pp. 353–393 in Migrant birds in the Neotropics: ecology, behavior, distribution, and conservation (A. Keast and E. S. Morton, eds.). Smithsonian Inst. Press, Washington, D.C. 576 pp.

RAPPOLE, J. H., E. S. MORTON, T. E. LOVEJOY III, AND J. L. RUOS. 1983. Nearctic avian migrants in the Neotropics. U.S. Fish and Wildl. Serv., Washington, D.C. 646 pp.

RASKER, R., AND B. ALEXANDER. 1997. The new challenge: people, commerce, and the environment in the Yellowstone to Yukon region. The Wilderness Soc., Bozeman, Mont. 59 pp. (plus appended data).

RASMUSSEN, O. I. 1941. Biotic communities of Kaibab Plateau, Arizona. Ecol. Monogr. 3:229–275.

RATCLIFFE, D. A. 1967. Decrease in eggshell weight in certain birds of prey. Nature 215:208.

RATCLIFFE, D. A. 1969. Population trends of the peregrine falcon in Great Britain. Pp. 239–269 *in* Peregrine falcon populations: their biology and decline (J. J. Hickey, ed.). Univ. Wis. Press, Madison. 596 pp.

RAU, G. H. 1981. Hydrothermal vent clam and tube worm $13_c/12_c$: further evidence of nonphotosynthetic food sources. Science 213:338–340.

RAUZI, F., AND C. L. HANSON. 1966. Water intake and runoff as affected by intensity of grazing. J. Range Manage. 19:351–356.

RAVELING, D. G. 1978. Dynamics of distribution of Canada geese in winter. Trans. N. Am. Wildl. and Nat. Resour. Conf. 43:206–225.

RAVEN, P. H. 1981. Tropical rain forests: a global responsibility. Nat. Hist. 90:28–32.

RAVEN, P. H. 1986. The urgency of tropical conservation. Nat. Conserv. News 36:7–11.

REAM, C. H. 1980. Managing interactions of people and wildlife in backcountry. Proc. 2nd Conf. on Sci. Res. in National Parks 6:212–220.

REARDON, J. 1986. The Alaska sportsman. Alaska 52(8):58.

REARDON, P. O., L. B. MERRILL, AND C. A. TAYLOR, JR. 1978. White-tailed deer preferences and hunter success under various grazing systems. J. Range. Manage. 31:40–42.

REDFORD, P. 1962. Raccoon in the U.S.S.R. J. Mamm. 43:541–542.

REED, A., AND A. BOURGET. 1977. Distribution and abundance of waterfowl wintering in southern Quebec. Can. Field-Nat. 91:1–7.

REED, A., R. BENOID, M. JULIEN, AND R. LALUMIERE. 1996. Goose use of the coastal habitats of northeastern James Bay. Occas. Paper 92, Canadian Wildl. Serv., Ottawa, Ontario. 60 pp.

REED, D. F., T. M. POJAR, AND T. N. WOODARD. 1974. Use of one-way gates by mule deer. J. Wildl. Manage. 38:9–15.

REED, D. F., T. N. WOODARD, AND T. M. POJAR. 1975. Behavioral response of mule deer to highway underpass. J. Wildl. Manage. 39:361–367.

REESE, J. G. 1975. Productivity and management of feral mute swans in Chesapeake Bay J. Wildl. Manage. 39:280–286.

REESE, J. G. 1980. Demography of European mute swans in Chesapeake Bay. Auk 97:449–464.

REGELIN, W. L., AND O. C. WALLMO. 1978. Duration of deer forage benefits after clearcut logging of subalpine forest in Colorado. U.S.D.A. Forest Serv. Rocky Mountain Forest and Range Exp. Sta. Note RM-356. Fort Collins, Colo. 4 pp.

REID, H. W., M. G. BURRIDGE, N. B. PULLAN, R. W. SUTHERST, AND E. W. WAIN. 1966. A survey of trypanosomiasis in the domestic cattle of South Busoga. Internat. Sci. Comm. Trypanosomiasis No. 11. 11 pp.

REID, V. H. 1953. Multiple land use: timber, cattle, and bobwhite quail. Trans. N. Am. Wildl. Conf. 18:412–420.

REID, W. H., AND G. R. PATRICK. 1983. Gemsbok *(Oryx gazella)* in White Sands National Monument. Southwestern Nat. 28:97–99.

REINECKE, K. 1979. Feeding ecology and development of juvenile black ducks in Maine. Auk 96:737–745.

RENOUF, R. N. 1972. Waterfowl utilization of beaver ponds in New Brunswick. J. Wildl. Manage. 36:740–744.

RESLER, R. A. 1972. Clearcutting: beneficial aspects for wildlife resources. J. Soil and Water Conserv. 27:250–254.

REYNOLDS, H. G. 1966. Use of openings in spruce-fir forests of Arizona by elk, deer, and cattle. U.S.D.A. Forest Serv. Rocky Mountain Forest and Range Exp. Sta. Note RM-66. Fort Collins, Colo. 4 pp.

RICH, M. E. 1977. Hunter education programs in the United States. Trans. N. Am. Wildl. and Nat. Resour. Conf. 42:502–506.

RICHARDSON, P. R. K., P. J. MUNDY, AND I. PLUG. 1986. Bone crushing carnivores and their significance to osteodystrophy in griffon vulture chicks. J. Zool. (London) 210: 23–43.

RICKLEFS, R. E. 1979. Ecology. Chiron Press, New York. 966 pp.

RILEY, G. A., AND R. T. MCBRIDE. 1972. A survey of the red wolf *(Canis rufus)*. U.S. Fish and Wildl. Serv. Spec. Sci. Rept. Wildl. 162, Washington, D.C. 15 pp.

RINEY, T. 1967. Conservation and management of African wildlife. Food and Agriculture Organization of the United Nations, Rome. 35 pp.

RINGELMAN, J. K., M. W. MILLER, AND W. F. ANDELT. 1993. Effects of ingested tungsten-bismuth-tin shot on captive mallards. J. Wildl. Manage. 57:725–732.

RITCEY, R. W., AND R. Y. EDWARDS. 1958. Parasites and diseases of the Wells Gray moose herd. J. Mamm. 39:139–145.

RIVNAY, E. 1964. The influence of man on insect ecology in arid zones. Ann. Rev. Ent. 9:41–62.

ROBBINS, C. S. 1979. Effect of forest fragmentation in bird populations. Pp. 198–212 *in* Management of north central and northeastern forests for nongame birds. Workshop Proc. U.S.D.A. Forest Serv. North Central Forest Exp. Sta., St. Paul, Minn. 268 pp.

ROBBINS, C. S., AND R. E. STEWART. 1949. Effects of DDT on bird populations of scrub forest. J. Wildl. Manage. 13:11–16.

ROBBINS, C. S., D. K. DAWSON, AND B. A. DOWELL. 1989. Habitat requirements of breeding forest birds of the middle Atlantic states. Wildl. Monogr. No. 103. 34 pp.

ROBBINS, C. S., P. F. SPRINGER, AND C. G. WEBSTER. 1951. Effects of five-year DDT application on breeding bird population. J. Wildl. Manage. 15:213–216.

ROBBINS, C. T. 1983. Wildlife feeding and nutrition. Academic Press, New York. 343 pp.

ROBEL, R. J. 1961a. The effects of carp populations on the production of waterfowl food plants on a western waterfowl marsh. Trans. N. Am. Wildl. and Nat. Resour. Conf. 26:147–159.

ROBEL, R. J. 1961b. Water depth and turbidity in relation to growth of sago pondweed. J. Wildl. Manage. 25:436–438.

ROBEL, R. J. 1962. Changes in submersed vegetation following a change in water level. J. Wildl. Manage. 26:221–224.

ROBEL, R. J. 1982. Testimony presented at EPA Administrative Hearings on Registration of 1080 as a Pesticide, July 16, 1982, Washington, D.C. 28 pp.

ROBERTS, R. F. 1977. Big Game guzzlers. Rangeman's J. 4:80–82.

ROBINETTE, W. L., O. JULANDER, J. S. GASHWILER, AND J. G. SMITH. 1952. Winter mortality of mule deer in Utah in relation to range condition. J. Wildl. Manage. 16:289–299.

ROBINSON, R. M., T. L. HAILEY, C. W. LIVINGSTON, AND J. W. THOMAS. 1967. Bluetongue in the desert bighorn sheep. J. Wildl. Manage. 31:165–168.

ROBINSON, R. M., T. L. HAILEY, R. G. MARBURGER, AND L. WEISHUHN. 1974. Vaccination trials in desert bighorn sheep against blue-tongue virus. J. Wildl. Dis. 10:228–231.

ROBINSON, S. K., AND D. S. WILCOVE. 1989. Conserving tropical raptors and game birds. Conserv. Biol. 3:192–193.

ROBINSON, T. S. 1956. Climate and bobwhites in Kansas—1955. Trans. Kan. Acad. Sci. 59:206–212.

ROBINSON, W. L. 1960. Test of shelter requirements of penned white-tailed deer. J. Wildl. Manage. 24:364–371.

ROBINSON, W. L. 1969. Habitat selection by spruce grouse in Michigan. J. Wildl. Manage. 33:113–120.

ROBINSON, W. L. 1980. Fool hen: the spruce grouse on the Yellow Dog Plains. Univ. Wis. Press, Madison. 244 pp.

ROBINSON, W. L. 1986. The case for hunting. Pp. 273–281 *in* Advances in animal welfare science 1986/87 (M. W. Fox and L. D. Mickley, eds.). Humane Soc. of the United States, Washington, D.C. 302 pp.

ROBINSON, W. L., AND G. J. SMITH. 1977. Observations on recently killed wolves in upper Michigan. Wildl. Soc. Bull. 5:25–26.

ROBINSON, W. L., L. H. FANTER, A. G. SPALDING, AND S. L. JONES. 1980. Biological impacts of political mismanagement of white-tailed deer in Pictured Rocks National Lakeshore. Proc. 2nd Conf. on Sci. Res. in National Parks 7:283–292.

ROCKE, T. E., C. J. BRAND, AND J. G. MENSIK. 1997. Site-specific lead exposure from lead pellet ingestion in sentinel mallards. J. Wildl. Manage. 61:228–234.

RODDA, G. H., AND T. H. FRITTS. 1992. The impact of the introduction of the colubrid snake *Boiga irregularis* on Guam's lizards. J. Herpetol. 26:166–174.

RODDA, G. H., T. H. FRITTS, AND P. J. CONRY. 1992. Origin and population growth of the brown tree snake, *Boiga irregularis,* on Guam. Pacific Sci. 46:46–47.

RODGERS, J. A., JR., AND H. T. SMITH. 1997. Buffer zone distances to protect foraging and loafing waterbirds from human disturbance in Florida. Wildl. Soc. Bull. 25:139–145.

RODGERS, R. D. 1983. Reducing wildlife losses to tillage in fallow wheat fields. Wildl. Soc. Bull. 11:31–38.

ROELKE, M. E., J. S. MARTENSON, AND S. J. O'BRIEN. 1993. The consequences of demographic reduction and genetic depletion in the endangered Florida panther. Current Biol. 3:340–350.

ROGERS, D. J., AND H. J. McALLISTER. 1969. Preferential grazing by game on a fungus "fairy ring." J. Wildl. Manage. 33:1034–1036.

ROGERS, J. G. 1974. Responses of caged red-winged blackbirds to two types of repellents. J. Wildl. Manage. 38:418–423.

ROGERS, J. G., JR. 1978. Some characteristics of conditioned taste aversions in red-winged blackbirds. Auk 95:362–369.

ROGERS, J. P. 1964. Effect of drought on reproduction of the lesser scaup. J. Wildl. Manage. 28:213–222.

ROGERS, L. 1981. A bear in its lair. Nat. Hist. 90:64.

ROGOFF, M. J., E. DERSCH, D. L. MOKMA, AND E. P. WHITESIDE. 1980. Computer-assisted ratings of soil potentials for urban land uses. J. Soil and Water Cons. 35:237–241.

ROLETTO, J. 1987. The Russell Lake rescue effort. CMMC News 7(1):1–3. Calif. Marine Mammal Center, Fort Cronkhite, Calif.

ROLLINS, H. B., D. H. SANDWEISS, AND J. C. ROLLINS. 1986. Effect of the 1982–1983 El Niño on bivalve mollusks. Nat. Geograph. Res. 2:106–112.

ROOSEVELT, T. 1893. The wilderness hunter. G. P. Putnam's Sons, New York, N.Y. 296 pp.

ROOSEVELT, T. 1905. Outdoor pastimes of an American hunter. Scribner, New York. 409 pp.

ROOTS, C. 1970. Wild harvest: A look at man's association with wild animals. Lutterworth Press, London. 199 pp.

ROOTS, C. 1976. Animal invaders. Universe Books, New York. 203 pp.

ROSATTE, R. C., C. D. MacINNES, R. T. WILLIAMS, AND O. WILLIAMS. 1997. A proactive prevention strategy for raccoon rabies in Ontario, Canada. Wildl. Soc. Bull. 25:110–116.

ROSEN, M. N., AND A. L. BISCHOFF. 1953. A new approach toward botulism control. Trans. N. Am. Wildl. Conf. 18:191–199.

ROSENBERG, K. V., S. B. TERRILL, AND G. H. ROSENBERG. 1987. Value of suburban habitats to desert riparian birds. Wilson Bull. 99:642–654.

ROSENE, W. 1969. The bobwhite quail: its life and management. Rutgers Univ. Press, New Brunswick, N.J. 418 pp.

ROUSH, G. J. 1982. On saving diversity. Nature Conservancy News 32:4–10.

ROYCE-MALMGREN, C. H., AND W. H. WATSON III. 1987. Modification of olfactory-related behavior in juvenile Atlantic salmon by changes in pH. J. Chem. Ecol. 13:533–546.

ROZE, U. 1985. How to select, climb, and eat a tree: lessons from the porcupine school of forestry. Nat. Hist. 94:62–69.

RUBINOFF, I. 1983. A strategy for preserving tropical rainforests. Ambio 12:255–258.

RUDD, R. L. 1964. Pesticides and the living landscape. Univ. Wis. Press, Madison. 320 pp.

RUDEBECK, G. 1950. The choice of prey and modes of hunting of predatory birds with special reference to their selective effect. Oikos 2:65–88.

RUDEBECK, G. 1951. The choice of prey and modes of hunting of predatory birds with special reference to their selective effect. Oikos 3:200–231.

RUDOLPH, R. R., AND C. G. HUNTER. 1964. Green trees and greenheads. Pp. 611–618 *in* Waterfowl tomorrow (J. P. Linduska, ed.). U.S. Govt. Printing Office, Washington, D.C. 770 pp.

RUFFNER, G. A., AND S. W. CAROTHERS. 1982. Age structure, condition, and reproduction of two *Equus asinus* (Equidae) populations from Grand Canyon National Park, Arizona. Southwestern Nat. 27:403–411.

RUNN, P., N. JOHANSSON, AND G. MILBRINK. 1972. Some effects of low pH on the hatchability of eggs of perch, *Perca fluviatilis* L. Zoon. 5:115–125.

RUPP, R. S. 1955. Beaver-trout relationship in the headwaters of Sunkhaze Stream, Maine. Trans. Am. Fisheries Soc. 84: 75–85.

RUPPRECHT, C. E., AND J. S. SMITH. 1994. Raccoon rabies: the re-emergence of an epizootic in a densely populated area. Seminars in virology, new and emerging diseases 5:155–164.

RUPPRECHT, C. E., C. A. HANLON, H. KOPROWSKI, AND A. HAMIR. 1992. Oral wildlife rabies vaccination: development of a recombinant virus vaccine. Trans. N. Am. Wildl. and Nat. Resour. Conf. 57:439–452.

RUSCH, D. H., AND L. B. KEITH. 1971. Ruffed grouse-vegetation relationships in central Alberta. J. Wildl. Manage. 35:417–429.

RUSH, G. 1973. The hen-brood release as a restoration technique. Pp. 192–197 *in* Wild turkey management: current problems and programs (G. C. Sanderson and H. C. Schultz, eds.). Univ. Mo. Press. Columbia. 355 pp.

RUSSELL, E. W. 1966. Soil conditions and plant growth, 9th ed. John Wiley & Sons, New York. 688 pp.

RUSZ, P. J., H. H. PRINCE, R. D. RUSZ, AND G. A. DAWSON. 1986. Bird collisions with transmission lines near a power plant cooling pond. Wildl. Soc. Bull. 14:441–444.

RUTHERFORD, W. H. 1955. Wildlife and environmental relationships of beavers in Colorado forests. J. Forestry 53:803–806.

RUWALDT, J. J., JR., L. D. FLAKE, AND J. M. GATES. 1979. Waterfowl pair use of natural and man-made wetlands in South Dakota. J. Wildl. Manage. 43:375–383.

RYAN, D. A., AND J. S. LARSON. 1976. Chipmunks in residential environments. Urban Ecol. 2:173–178.

RYDER, J. P. 1967. The breeding biology of Ross' goose in the Perry River region, Northwest Territories. Can. Wildl. Serv. Rept. Series No. 3, Ottawa. 56 pp.

SADLER, K.C. 1961. Grit selectivity by the female pheasant during egg production. J. Wildl. Manage. 25:339–341.

SAFONOV, V. G. 1981. The status and reestablishment of fur resources in the U.S.S.R. Pp. 65–110 *in* Proc. worldwide furbearer conf. (J. A. Chapman and D. Pursley, eds.). Frostburg, Md. 652 pp.

SALO, L. J. 1976. History of wildlife management in Finland. Wildl. Soc. Bull. 4:167–174.

SALWASSER, H., S. P. MEALEY, AND K. JOHNSON. 1984. Wildlife population viability: a question of risk. Trans. N. Am. Wildl. and Nat. Resour. Conf. 49:421–439.

SALWASSER, H., C. SCHONEWALD-COX, AND R. BAKER. 1987. The role of interagency cooperation in managing for viable populations. Pp. 159–173 *in* Viable populations for conservation (M. Soulé, ed.). Cambridge Univ. Press, New York. 189 pp.

SAMPSON, F. B. 1980. Island biogeography and the conservation of nongame birds. Trans. N. Am. Wildl. and Nat. Resour. Conf. 45:245–251.

SAMUEL, W. M., G. A. CHALMERS, J. G. STELFOX, A. LOEWEN, AND J. J. THOMSEN. 1975. Contagious ecthyma in bighorn sheep and mountain goats from western Canada. J. Wildl. Dis. 11:26–31.

SANDERSON, G. C. 1976. Conservation of waterfowl. Pp. 43–58 *in* Ducks, geese, and swans of North America. Stackpole Books, Harrisburg, Penn. 544 pp.

SANDERSON, G. C., AND F. C. BELLROSE. 1986. A review of the problem of lead poisoning in waterfowl. Ill. Nat. Hist. Surv. Spec. Pub. 4. 34 pp.

SANFORD, R. L., JR. 1987. Apogeotropic roots in an Amazon rain forest. Science 235:1062–1064.

SANKHALA, K. S. 1974. Discussion. P. 255 *in* Second world conference on national parks (S. H. Elliott, ed.). Internat. Union for Cons. of Nature and Natural Resources, Morges, Switzerland. 504 pp.

SARGEANT, A. B., S. H. ALLEN, AND R. T. EBERHARDT. 1984. Red fox predation on breeding ducks in midcontinent North America. Wildl. Monogr. No. 89. 41 pp.

SAUMIER, M. D., M. E. RAU, AND D. M. BIRD. 1986. The effect of *Trichinella pseudospiralis* infection on the reproductive success of captive American kestrels *(Falco sparverius)*. Can. J. Zool. 64:2123–2125.

SAUNDERS, B. P. 1973. Meningeal worm in white-tailed deer in northwestern Ontario and moose population densities. J. Wildl. Manage. 37:327–330.

SAVIDGE, J. A. 1987. Extinction of an island forest avifauna by an introduced snake. Ecology 68:660–668.

SAVORY, A. 1978. A holistic approach to ranch management using short duration grazing. Pp. 555–559 *in* Proc. 1st internat. rangeland congress (D. H. Hyder, ed.). Soc. Range Manage., Denver. 742 pp.

SCHAFER, E. W., AND R. B. BRUNTON. 1971. Chemicals as bird repellents: two promising agents. J. Wildl. Manage. 35:569–572.

SCHALLER, G. B. 1986. Secrets of the wild panda. Nat. Geogr. 169:284–309.

SCHALLER, G. B., AND P. G. CRAWSHAW. 1980. Movement patterns of jaguar. Biotropica 12:161–168.

SCHAMBERGER, M., AND W. B. KROHN. 1982. Status of the habitat evaluation procedures. Trans. N. Am. Wildl. and Nat. Resour. Conf. 47:154–164.

SCHEFFER, V. B. 1976. The future of wildlife management. Wildl. Soc. Bull. 4:51–54.

SCHEMNITZ, S. D. 1964. Comparative ecology of bobwhite and scaled quail in the Oklahoma panhandle. Am. Midl. Nat. 71: 429–433.

SCHEUHAMMER, A. M., AND S. I. NORRIS. 1995. A review of the environmental impacts of lead shotshell ammunition and lead fishing weights in Canada. Occas. Paper 88, Canadian Wildl. Serv., Hull, Quebec, Canada. 54 pp.

SCHINNER, J. R., AND D. L. CAULEY. 1974. The ecology of urban raccoons in Cincinnati, Ohio. Pp. 125–130 *in* A symposium on wildlife in an urbanizing environment (J. H. Noyes and D. R. Progulske, eds.). Univ. Mass. Coop. Ext. Serv., Amherst. 182 pp.

SCHITOSKEY, F., JR., J. EVANS, AND G. K. LAVOIE. 1972. Status and control of nutria in California. Proc. Vertebrate Pest Conf. 5:15–17.

SCHLADWEILER, J. L., AND J. R. TESTER. 1972. Survival and behavior of hand-reared mallards in the wild. J. Wildl. Manage. 36:1118–1127.

SCHLESINGER, W., W. REINERS, AND D. KNOPMAN. 1974. Heavy metal concentrations and deposition in bulk precipitation in montane ecosystems of New Hampshire, U.S.A. Environ. Pollut. 6:39–47.

SCHMIDT, R. L., C. P. HIBLER, T. R. SPRAKER, AND W. H. RUTHERFORD. 1979. An evaluation of drug treatment for lungworm in bighorn sheep. J. Wildl. Manage. 43:461–467.

SCHMIDT-NIELSEN, K. 1964. Desert animals: physiological problems of heat and water. Oxford Univ. Press, London. 277 pp.

SCHMIDT-NIELSEN, K. 1979. Animal physiology: adaptation and environment. Cambridge Univ. Press, New York. 560 pp.

SCHMIDT-NIELSEN, K., AND Y. T. KIM. 1964. The effect of salt intake on the size and function of the salt gland in ducks. Auk 81:160–172.

SCHMITZ, R. A., A. A. AGUIRRE, R. S. COOK, AND G. A. BALDASSARRE. 1990. Lead poisoning of Caribbean flamingos in Yucatán, Mexico. Wildl. Soc. Bull. 18:399–404.

SCHNURRENBERGER, P. R., J. R. BECK, AND D. PEDEN. 1964. Skunk rabies in Ohio. Public Health Rept. 79:161.

SCHOEN, J. W., O. C. WALLMO, AND M. D. KIRCHHOFF. 1981. Wildlife-forest relationships: is a reevaluation of old growth necessary? Trans. N. Am. Wildl. and Nat. Resour. Conf. 46:531–544.

SCHOFIELD, C. 1976. Acid precipitation: effects on fish. Ambio 5:228–230.

SCHOFIELD, C., AND J. TROJNAR. 1980. Aluminum toxicity to fish in acidified waters. Pp. 341–366 *in* Polluted rain (T. Toribara, M. Miller, and P. Morrow, eds.). Plenum Press, New York. 502 pp.

SCHORGER, A. W. 1937. A great Wisconsin passenger pigeon nesting in 1871. Proc. Linn. Soc., New York 48:1–26.

SCHORGER, A. W. 1955. The passenger pigeon: its natural history and extinction. Univ. Wis. Press, Madison. 424 pp.

SCHORGER, A. W. 1960. The crushing of *Carya* nuts in the gizzard of the turkey. Auk 77:337–340.

SCHORGER, A. W. 1966. The wild turkey: its history and domestication. Univ. Okla. Press, Norman. 625 pp.

SCHRADER, T. A. 1955. Waterfowl and the potholes of the North Central States. Pp. 596–604 *in* Yearbook of agriculture, 1955. U.S.D.A., Washington, D.C. 751 pp.

SCHRADER, T.A. 1960. Does soil bank aid pheasants? Conserv. Volunteer 23:34–37.

SCHRANCK, B. W. 1972. Waterfowl nest cover and some predation relationships. J. Wildl. Manage. 36:182–186.

SCHREIBER, R. W., AND E. A. SCHREIBER. 1984. Central Pacific seabirds and the El Niño Southern Oscillation: 1982 to 1983 perspectives. Science 225:713–716.

SCHREINER, C. III. 1968. Uses of exotic animals in a commercial hunting program. Pp. 12–16 *in* Introduction of exotic animals: ecological and socioeconomic considerations. Caesar Kleberg Res. Pgm. in Wildl. Ecology, Texas A&M Univ., College Station. 25 pp.

SCHROEDER, R. L. 1982. Habitat suitability index models: pileated woodpecker. FWS/OBS-82/10.39, U.S. Fish and Wildl. Serv. Fort Collins, Colo. 15 pp.

SCHULTZ, V. 1948. Vitamin A as a survival factor of the bobwhite quail *(Colinus v. virginianus)* in Ohio during the winter of 1946–47. J. Wildl. Manage. 12:251–263.

SCHUTZ, F. 1965. Sexuelle pragung bei Anatiden. Z. Tierpsychologie 22:50–103.

SCHWEITZER, A. 1965. The teaching of reverence for life. Translated by R. and C. Winston. Holt, Rinehart and Winston, Orlando, Fla. 63 pp.

SCIFRES, C. J., AND D. B. POLK, JR. 1974. Vegetation response following spraying of light infestation of honey mesquite. J. Range Manage. 27:462–465.

SCOTT, J. D., AND T. W. TOWNSEND. 1985a. Characteristics of deer damage to commercial tree industries of Ohio. Wildl. Soc. Bull. 13:135–143.

SCOTT, J. D., AND T. W. TOWNSEND. 1985b. Methods used by selected Ohio growers to control damage by deer. Wildl. Soc. Bull. 13:234–240.

SCOTT, M. E. 1982. Seasonal use of clearcuts by white-tailed deer in Vermont. M.S. Thesis, Univ. Vermont, Burlington. 63 pp.

SCOTT, T. G. 1963. Paul L. Errington, 1902-1962. J. Wildl. Manage. 27:321–324.

SCOTT, T. G., AND W. L. DEVER. 1940. Blasting to improve wildlife environment in marshes. J. Wildl. Manage. 4:373–374.

SCOTT, T. G., Y. L. WILLIS, AND J. A. ELLIS. 1959. Some effects of a field application of dieldrin on wildlife. J. Wildl. Manage. 23:409–427.

SCOTT, V. E. 1979. Bird response to snag removal in ponderosa pine. J. Forestry 77:26–28.

SCOTT, V. E., G. L. CROUCH, AND J. A. WHELAN. 1982. Responses of birds and small mammals to clearcutting in a subalpine forest in central Colorado. U.S.D.A. Forest Serv. Res. Note RM—422. Rocky Mountain Forest and Range Expt. Sta., Fort Collins, Colo. 6 pp.

SCOTTER, G. W. 1964. Effects of forest fires on the winter range of barren-ground caribou in northern Saskatchewan. Can. Wildl. Serv. Wildl. Manage. Bull. Series 1, No. 18, Ottawa. 111 pp.

SCOTTER, G. W. 1970a. Wildfires in relation to the habitat of barren-ground caribou in the taiga of northern Canada. Proc. Tall Timbers Fire Ecol. Conf. 10:85–105.

SCOTTER, G. W. 1970b. Reindeer husbandry as a land use in Northern Canada. Pp. 159–169 *in* Proc. of the conf. on productivity and conservation in northern circumpolar lands. Internat. Union for Cons. of Nature and Natural Resources, Morges, Switzerland. 344 pp.

SEAL, U. S. 1979. Assessment of habitat condition by measurement of biochemical and endocrine indicators of the nutritional, reproductive, and disease status of free-ranging animal populations. Pp. 305–329 *in* Proc. symposium on classification, inventory, and analysis of fish and wildlife habitat, U.S. Fish and Wildl. Serv. Off. Biol. Surv., Washington, D.C. 604 pp.

SEAMANS, R., J. WILSON, A. KENNEDY, G. MOORE, C. GILCHRIST, F. SCHMIDT, AND E. B. CHAMBERLAIN. 1959. An illustrated small marsh construction manual based on standard designs. Atlantic Waterfowl Council, Comm. on Habitat Manage. and Development. 60 pp.

SEATTLE, CHIEF. 1854. Speech. Reprinted in 1975 in Michigan Out-of-Doors 29:24–25.

SEBER, G. A. F. 1973. The estimation of animal abundance and related parameters. Griffin, London. 506 pp.

SEEGMILLER, R. F., AND R. D. OHMART. 1981. Ecological relationships of feral burros and desert bighorn sheep. Wildl. Monogr. No. 78. 58 pp.

SEEGMILLER, R. F., AND C. D. SIMPSON. 1979. The Barbary sheep: some conceptual implications of competition with desert bighorns. Trans. Desert Bighorn Council 23:47–49.

SEGELQUIST, C. A., AND M. J. ROGERS. 1975. Response of Japanese honeysuckle to fertilization. J. Wildl. Manage. 39:769–775.

SEIDENSTICKER, J. C. IV, M. G. HORNOCKER, W. V. WILES, AND J. P. MESSICK. 1973. Mountain lion social organization in the Idaho primitive area. Wildl. Monogr. No. 35. 60 pp.

SENNER, S. E. 1997. [Review of] *Exxon Valdez* oil spill: fate and effects in Alaskan waters. Wilson Bull. 109:549–555.

SEPIK, G. F., R. B. OWEN, AND M. W. COULTER. 1981. A landowner's guide to woodcock management in the northeast. Moosehorn National Wildl. Refuge, U.S. Fish and Wildl. Serv., Univ. Maine Misc. Rept. 253, Orono. 23 pp.

SETON, E. T. 1929. Lives of game animals, Vol. 4. Doubleday, Doran, New York. 780 pp.

SEVERINGHAUS, C. W. 1947. Relationship of weather to winter mortality and population levels among deer in the Adirondack region of New York. Trans. N. Am. Wildl. Conf. 12:212–223.

SEVERINGHAUS, C. W., AND L. W. JACKSON. 1970. Feasibility of stocking moose in the Adirondacks. N.Y. Fish Game J. 17:18–32.

SHAFER, C. L. 1990. Nature reserves: island theory and conservation practice. Smithsonian Insti. Press, Washington, D.C. 189 pp.

SHAFFER, M. L. 1981. Minimum population sizes for species conservation. BioScience 31:131–134.

SHAFFER, M. L., AND F. B. SAMPSON. 1985. Population size and extinction: a note on determining critical population sizes. Amer. Nat. 125:144–152.

SHALAWAY, S. D. 1985. Fencerow management for nesting birds in Michigan. Wildl. Soc. Bull. 13:302–306.

SHANE, S. H. 1984. Manatee use of power plant effluents in Brevard County, Florida. Florida Scientist 47:180–188.

SHAPIRO, A. E., F. MONTALBANO III, AND D. MAGER. 1982. Implications of construction of a flood control project upon bald eagle nesting activity. Wilson Bull. 94:55–63.

SHARP, M. S., AND R. L. NEILL. 1979. Physical deformities in a population of wintering blackbirds. Condor 81:427–430.

SHARP, W. M. 1957. Social and range dominance in gallinaceous birds—pheasants and prairie grouse. J. Wildl. Manage. 21:242–244.

SHAW, H. 1982. Looking back at the mountain lion. Arizona Game and Fish Dept. Wildl. Views 25:4–5.

SHAW, S. P., AND C. G. FREDINE. 1956. Wetlands of the United States. U.S. Fish and Wildl. Serv. Circ. 39, Washington, D.C. 67 pp.

SHAW, W. W., AND T. COOPER. 1980. Managing wildlife in national parks for human benefits. Proc. 2nd Conf. on Sci. Res. in National Parks 6:189–198.

SHAW, W. W., AND W. R. MANGUN. 1984. Nonconsumptive use of wildlife in the United States. Resour. Publ. 154, U.S. Fish and Wildl. Serv., Washington, D.C. 20 pp.

SHAW, W. W., W. R. MANGUN, AND J. R. LYONS. 1985. Residential enjoyment of wildlife resources by Americans. Leisure Sci. 7:361–375.

SHEFFIELD, W. J., JR., E. D. ABLES, AND B. A. FALL. 1971. Geographic and ecologic distribution of Nilgai antelope in Texas. J. Wildl. Manage. 35:250–257.

SHELDON, W. G. 1950. Denning habits and home range of red foxes in New York State. J. Wildl. Manage. 14:33–42.

SHELDON, W. G. 1967. The book of the American woodcock. Univ. Mass. Press, Amherst. 227 pp.

SHELFORD, V. E. 1963. The ecology of North America. Univ. Ill. Press, Urbana. 610 pp.

SHEPARD, P. 1959. The theory of the value of hunting. Trans. N. Am. Wildl. Conf. 24:504–512.

SHERIDAN, D. 1981. Desertification of the United States. Council on Environmental Quality, U.S. Govt. Printing Office, Washington, D.C. 142 pp.

SHEWELL, G. E. 1955. Identity of the black fly that attacks ducklings and goslings in Canada (Diptera: Simuliidae). Can. Ent. 87:345–349.

SHIRLEY, M. L., AND J. A. RAGSDALE. 1966. Deltas in their geologic framework. Houston Geological Society. 251 pp.

SHOESMITH, M. W., AND W. H. KOONZ. 1977. The maintenance of an urban deer herd in Winnipeg, Manitoba. Trans. N. Am. Wildl. and Nat. Resour. Conf. 42:278–285.

SHOPE, R. E., L. G. MACNAMARA, AND R. MANGOLD. 1955. Report on deer mortality: epizootic hemorrhagic disease of deer. N. J. Outdoors 6:16–21.

SHORTEN, M. 1946. A survey of the distribution of the American gray squirrel *(Sciurus carolinensis)* and the British red squirrel *(S. vulgaris leucourus)* in England and Wales in 1944–5. J. Animal Ecol. 15:82–92.

SHORTEN, M. 1953. Notes on the distribution of the gray squirrel *(Sciurus carolinensis)* and the red squirrel *(Sciurus vulgaris leucourus)* in England and Wales from 1945 to 1952. J. Animal Ecol. 22:134–140.

SIGLER, W. F. 1982. Wildlife law enforcement, 3rd ed. Wm. C. Brown Co., Dubuque, Ia. 403 pp.

SIGLER, W. F., AND R. R. MILLER. 1963. Fishes of Utah. Utah Dept. Fish and Game, Salt Lake City. 203 pp.

SIKES, R. K., SR. 1981. Rabies. Pp. 3–17 *in* Infectious diseases of wild mammals, 2nd ed. (J. W. Davis, L. H. Karstad, and D. O. Trainer, eds.). Ia. State Univ. Press, Ames. 446 pp.

SILVER, H., AND N. F. COLOVOS. 1957. Nutritive evaluation of some forage rations of deer. Tech. Circ. 15, N. H. Fish and Game Dept., Concord. 56 pp.

SILVER, H., N. F. COLOVOS, J. B. HOTER, AND H. H. HAYES. 1969. Fasting metabolism of white-tailed deer. J. Wildl. Manage. 33:490–498.

SIMBERLOFF, D. S. 1983. When is an island community in equilibrium? Science 220:1275–1277.

SIMBERLOFF, D. S., AND J. COX. 1987. Consequences and costs of conservation corridors. Conserv. Biol. 1:63–71.

SIMBERLOFF, D. S., J. A. FARR, J. COX, AND D. W. MEHLMAN. 1992. Movement corridors: conservation bargains or poor investments? Conserv. Biol. 6:493–504.

SIMPSON, C. D., L. J. KRYSL, D. B. HAMPY, AND G. G. GRAY. 1978. The Barbary sheep: a threat to desert bighorn survival. Trans. Desert Bighorn Council 22:26–31.

SIMPSON, P. W., J. R. NEWMAN, M. A. KEIRN, R. M. MATTER, AND P. A. GUTHRIE. 1982. Manual of stream channelization impacts on fish and wildlife. FWS/OBS-82/24. Office of Biological Service, Washington, D.C. 155 pp.

SIMPSON, V. R., A. E. HUNT, AND M. C. FRANK. 1979. Chronic lead poisoning in a herd of mute swans. Environ. Pollut. 18:187–202.

SINGER, F. J., W. T. SWANK, AND E. E. C. CLEBSCH. 1984. Effect of wild pig rooting in a deciduous forest. J. Wildl. Manage. 48:464–473.

SKILLERN, F. F. 1981. Environmental protection: the legal framework. Shepard's/McGraw-Hill, Colorado Springs. 384 pp.

SKILLERN, F. F. 1985. Constitutional and statutory issues of federalism in the development of energy resources. Nat. Resources Lawyer 17:533–622.

SKOVLIN, J. M. 1971. Ranching in east Africa: a case study. J. Range Manage. 24:263–270.

SKOVLIN, J. M., R. W. HARRIS, G. S. STRICKLER, AND G. A. GARRISON. 1976. Effects of cattle grazing methods on ponderosa pine-bunchgrass range in the Pacific Northwest. U.S.D.A. Forest Serv. Tech. Bull. 1531. 40 pp.

SLONEKER, L. L., AND W. C. MODENHAUER. 1977. Measuring the amounts of crop residue remaining after tillage. J. Soil and Water Cons. 32:231–236.

SMITH, A. G., J. H. STOUDT, AND J. B. GOLLOP. 1964. Prairie potholes and marshes. Pp. 39–50 *in* Waterfowl tomorrow (J. P. Linduska, ed.). U.S.D.I. Bur. of Sport Fisheries and Wildl., U.S. Fish and Wildl. Serv., Washington, D.C. 770 pp.

SMITH, C. A., R. E. WOOD, L. BEIER, AND K. BOVEE. 1987. Wolf-deer habitat relationships in southeast Alaska. Final Rept. Proj. W-22-4, W-22-5, W-22-6, Alaska Dept. Fish and Game, Juneau. 20 pp.

SMITH, C. C. 1940. The effect of overgrazing and erosion upon the biota of the mixed-grass prairie of Oklahoma. Ecology 21:381–397.

SMITH, C. C. 1970. The coevolution of pine squirrels *(Tamiasciurus)* and conifers. Ecol. Monogr. 40:349–371.

SMITH, D. C. 1975. Rehabilitating oiled aquatic birds. Pp. 241–247 *in* Proc. 1975 conf. on prevention and control of oil spills. American Petrol. Instit., Washington, D.C. 612 pp.

SMITH, G. F., J. R. CARY, AND O. J. RONGSTAD. 1981. Sampling strategies for radio-tracking coyotes. Wildl. Soc. Bull. 9:88–93.

SMITH, H. N., AND C. A. RECHENTHIN. 1964. Grassland restoration: the Texas brush problem. U.S.D.A. Soil Cons. Serv. Publ., Temple, Tex. 33 pp.

SMITH, J. S. 1989. Rabies virus epitopic variation: use in ecologic studies. Advan. Virus Res. 36:215–253.

SMITH, L. C. 1975. Urban wildlife—is it wanted and needed? Can. Field-Nat. 89:351–353.

SMITH, L. M., AND I. L. BRISBIN, JR. 1984. An evaluation of total trapline captures as estimates of furbearer abundance. J. Wildl. Manage. 48:1452–1455.

SMITH, M. A., J. C. MALECHEK, AND K. O. FULGHAM. 1979. Forage selection by mule deer on winter range grazed by sheep in spring. J. Range Manage. 32:40–45.

SMITH, R. E. 1955. Natural history of the prairie dog in Kansas. Univ. Kans. Mus. Nat. Hist. and State Biol. Surv. Publ. 16, Lawrence. 36 pp.

SMITH, R. I. 1970. Response of pintail breeding populations to drought. J. Wildl. Manage. 34:943–946.

SMITH, R. I. 1981. Update on steel shot regulations. Pp. 112–113 *in* Fourth internat. waterfowl symposium. Ducks Unlimited, Chicago. 265 pp.

SMITH, R. I., AND T. W. TOWNSEND. 1981. Attitudes of Ohio hunters toward steel shot. Wildl. Soc. Bull. 9:4–7.

SMITH, S. H. 1968. Species succession and fishery exploitation in the Great Lakes. J. Fish. Res. Bd. Can. 25:667–693.

SMITH, S. H. 1972a. Factors of ecologic succession in oligotrophic fish communities of the Laurentian Great Lakes. J. Fish. Res. Bd. Can. 29:717–730.

SMITH, S. H. 1972b. The future of salmonid communities in the Laurentian Great Lakes. J. Fish. Res. Bd. Can. 29:951–957.

SMUTS, G. L. 1982. Interrelations between predators, prey, and their environment. Pp. 207–211 *in* Environment 82/83 (J. Allen, ed.). Dushkin Publishing Group, Inc., Guilford, Conn. 240 pp.

SNELL, G. P., AND B. D. HLAVACHICK. 1980. Control of prairie dogs—the easy way. Rangelands 2:239–240.

SNYDER, W. D. 1985. Survival of radio-marked hen ring-necked pheasants in Colorado. J. Wildl. Manage. 49:1044–1050.

SOIL CONSERVATION SERVICE. 1980. Save fuel . . . use conservation tillage. U.S.D.A. Soil Cons. Serv. Pgm. Aid 1263, Washington, D.C.

SOILEAU, D. M., J. D. BROWN, AND D. W. FRUGÉ. 1985. Mitigation banking: a mechanism for compensating unavoidable fish and wildlife habitat losses. Trans. N. Am. Wildl. and Nat. Resour. Conf. 50:465–474.

SOKAL, R. R., AND F. J. ROHLF. 1981. Biometry, the principles and practice of statistics in biological research. W. H. Freeman and Co., San Francisco. 859 pp.

SOLMAN, V. E. F. 1945. The ecological relations of pike, *Esox lucius* L., and waterfowl. Ecology 26: 157–170.

SOLMAN, V. E. F. 1968. Bird control and air safety. Trans. N. Am. Wildl. and Nat. Resour. Conf. 33:328–336.

SOMERS, J. D., F. F. GILBERT, D. E. JOYNER, R. J. BROOKS, AND R. G. GARTSHORE. 1981. Use of 4-aminopyridine in cornfields under high foraging stress. J. Wildl. Manage. 45:702–709.

SOULÉ, M. E. 1985. What is conservation biology? BioScience 35:727–734.

SOUTHWOOD, T. R. E. 1972. Farm management in Britain and its effect on animal populations. Proc. Tall Timbers Ecol. Animal Cont. Habitat Manage. Conf. 3:29–51.

SOUTIERE, E. C., AND E. G. BOLEN. 1972. Role of fire in mourning dove nesting ecology. Proc. Tall Timbers Fire Ecol. Conf. 12:277–288.

SOUTIERE, E. C., AND E. G. BOLEN. 1976. Mourning dove nesting on tobosa grass-mesquite rangeland sprayed with herbicides and burned. J. Range Manage. 29:226–231.

SOWLS, L. K. 1955. Prairie ducks, a study of their behavior, ecology, and management. Stackpole Books, Harrisburg, Penn. 193 pp.

SOWLS, L. K. 1960. Results of a banding study of Gambel's quail in southern Arizona. J. Wildl. Manage. 24:185 –190.

SOWLS, L. K. 1961. Hunter-checking stations for collecting data on the collared peccary *(Pecari tajacu)*. Trans. N. Am. Wildl. and Nat. Resour. Conf. 26:496–505.

SPALDING, D. 1873. Instinct: with original observations on young animals. Macmillan's Magazine 27:282–293.

SPENCER, D. L., AND E. F. CHATELAIN. 1953. Progress in the management of the moose of south-central Alaska. Trans. N. Am. Wildl. Conf. 18:539–552.

SPENCER, D. L., U. C. NELSON, AND W. A. ELKINS. 1951. America's greatest goose-brant nesting area. Trans. N. Am. Wildl. Conf. 16:290–295.

SPENCER, S. R., G. N. CAMERON, B. D. ESHELMAN, L. C. COOPER, AND L. R. WILLIAMS. 1985. Influence of pocket gopher mounds on a Texas coastal prairie. Oecologia 66:111–115.

SPEVAK, T. A. 1983. Population changes in a Mediterranean scrub rodent assembly during drought. Southwestern Nat. 28:47–52.

SPILLETT, J. J., J. B. LOW, AND D. SILL. 1967. Livestock fences—how they influence pronghorn antelope movements. Utah Agric. Exp. Sta. Bull. 470, Logan, Utah. 80 pp.

SPINAGE, C. A. 1962. Rinderpest and faunal distribution patterns. African Wildl. 16:55–60.

SPITZER, P. R., R. W. RISEBROUGH, W. WALKER, R. HERNANDEZ, A. POOLE, D. PULESTON, AND I. C. T. NISBET. 1978. Productivity of ospreys in Connecticut–Long Island increases as DDE residues decline. Science 202:333–335.

SPORT FISHING INSTITUTE. 1982. The view from Alabama on size limits. SFI Bull. 334:2–4.

SPRAGUE, M. A., AND G. B. TRIPLETT (EDS.). 1986. No-tillage and surface tillage agriculture, the tillage revolution. John Wiley & Sons, New York. 472 pp.

SPRUNT, A. 1955. The spread of the cattle egret. Pp. 259–276 *in* Publ. 4190, Ann. Rept. of the Bd. of Regents of the Smithsonian Inst., showing the operations, expenditures, and condition of the Institution for the year ended June 30, 1954. U.S. Govt. Printing Office, Washington, D.C. 455 pp.

SPRUNT, A., AND E. B. CHAMBERLAIN. 1949. South Carolina bird life. Charleston Mus. Contr. No. 11, Columbia, S.C. 585 pp.

SSEMWEZI, P. 1974. Discussion. P. 254 *in* Second world conference on national parks (S. H. Elliott, ed.). Internat. Union for Cons. of Nature and Natural Resources, Morges, Switzerland. 504 pp.

STABLER, R. M. 1954. *Trichomonas gallinae:* a review. Experimental Parasitol. 3:368–402.

STABLER, R. M., AND C. M. HERMAN. 1951. Upper digestive tract trichomoniasis in mourning doves and other birds. Trans. N. Am. Wildl. Conf. 16:145–163.

STABLER, R. M., AND R. W. MELLENTIN. 1953. Effect of 2-amino-5-nitro-thiazole (Enheptin) and other drugs on *Trichomonas gallinae* infections in the domestic pigeon. J. Parasitol. 39:637.

STAFFORD, K. 1997. Migrating to safer solutions. Americas 49(5):6–15.

STALMASTER, M. V., AND J. R. NEWMAN. 1978. Behavioral responses of wintering bald eagles to human activity. J. Wildl. Manage. 42:506–513.

STANFORD, J. A. 1957. A progress report of coturnix quail investigations in Missouri. Trans. N. Am. Wildl. Conf. 22:316–359.

STANFORD, J. A. 1972. Bobwhite quail population dynamics: relationships of weather, nesting, production patterns, fall population characteristics, and harvest in Missouri quail. Pp. 115–139 *in* Proc. first national bobwhite quail symposium (J. A. Morrison and J. C. Lewis, eds.). Okla. State Univ., Stillwater. 390 pp.

STANLEY, J. G., AND W. M. LEWIS (EDS.). 1978. Grass carp in the United States. Trans. Am. Fisheries Soc. 107:104–224.

STANLEY, W. C. 1963. Habits of the red fox in northeastern Kansas. Univ. Kans. Mus. Nat. Hist. Misc. Publ. No. 34. Lawrence. 31 pp.

STARK, N., AND M. SPRATT. 1977. Root biomass and nutrient storage in rain forest oxisols near San Carlos de Rio Negro. Tropical Ecol. 18:1–9.

STAUFFER, D. F., AND L. B. BEST. 1980. Habitat selection by birds of riparian communities: evaluating effects of habitat alterations. J. Wildl. Manage. 44:1–15.

STAUFFER, R. C. 1937. Changes in the invertebrate community of a lagoon after disappearance of the eel-grass. Ecology 18:427–431.

STEARNS, F. W. 1967. Wildlife habitat in urban and suburban environments. Trans. N. Am. Wildl. and Nat. Resour. Conf. 32: 61–69.

STEBBINS, G. L., AND B. CRAMPTON. 1961. A suggested revision of the grass genera of temperate North America. Rec. Adv. Bot. 1:133–145.

STEERE, A. C. 1994. Lyme disease: a growing threat to urban populations. Proc. Nat. Acad. Sci. 91:2378–2383.

STEGER, R. E. 1982. Rapid rotation grazing programs in Texas. Rangelands 4:75–77.

STENBERG, K., AND W. W. SHAW (EDS.). 1986. Wildlife conservation and new residential developments. Proc. Nat. Symp. Urban Wildlife, Tucson, Ariz. 203 pp.

STEPHEN, J. W. D. 1975. Status of crop depredation problems and programs in Canada. Pp. 120–126 *in* First internat. waterfowl symposium. Ducks Unlimited, Chicago. 224 pp.

STEPHENS, S. 1987. Lapp life after Chernobyl. Nat. Hist. 96(12): 33–40.

STEUTER, A. A., AND H. A. WRIGHT. 1980. White-tailed deer densities and brush cover on the Rio Grande plain. J. Range Manage. 33:328–331.

STEVENS, C. 1980. The whales versus Japan. Living Wilderness 44:18–20.

STEVENS, R. E. 1957. The striped bass of the Santee-Cooper Reservoir. Proc. SE Assoc. Game and Fish Comm. 11:253–264.

STEVENS, R. O. 1944. Talk about wildlife. Byrum Printing Co., Raleigh, N.C. 229 pp.

STEWART, P. A. 1952. Dispersal, breeding behavior, and longevity of banded barn owls in North America. Auk 69:227–245.

STICKEL, W. H. 1974. Effects on wildlife of newer pesticides and other pollutants. Proc. Ann. Conf. Western Assoc. State Game and Fish Comm. 53:484–491.

STICKEL, W. H., D. W. HAYNE, AND L. F. STICKEL. 1965. Effects of heptachlor-contaminated earthworms on woodcocks. J. Wildl. Manage. 29:132–155.

STICKLEY, A. R., AND J. L. GUARINO. 1972. A repellent for protecting corn seed from blackbirds and crows. J. Wildl. Manage. 36:150–152.

STICKLEY, A. R., R. T. MITCHELL, R. G. HEATH, C. R. INGRAM, AND E. L. BRADLEY. 1972. A method for appraising the bird repellency of 4-aminopyridine. J. Wildl. Manage. 36:1313–1316.

STICKLEY, A. R., JR., D. J. TWEDT, J. F. HEISTERBERG, D. F. MOTT, AND J. F. GLAHN. 1986. Surfactant spray system for controlling blackbirds and starlings in urban roosts. Wildl. Soc. Bull. 14: 412–418.

STIEGLITZ, W. O., AND R. L. THOMPSON. 1967. Status and life history of the Everglade kite in the United States. U.S. Fish and Wildl. Serv. Spl. Sci. Rept.:—Wildl. No. 109. Washington, D.C. 21 pp.

STOCKSTAD, D. S., M. S. MORRIS, AND E. C. LORY. 1953. Chemical characteristics of natural licks used by big game animals in western Montana. Trans. N. Am. Wildl. Conf. 18:247–258.

STODDARD, H. L. 1931. The bobwhite quail: its habits, preservation, and increase. Scribner, New York. 559 pp.

STODDARD, H. L. 1939. The use of controlled fire in southeastern quail management. Coop. Quail Study Assoc. 21 pp.

STODDARD, H. L. 1962. Use of fire in pine forests and game lands of the deep southeast. Proc. Tall Timbers Fire Ecol. Conf. 1:31–42.

STODDART, L. A., AND A. D. SMITH. 1955. Range management, 2nd ed. McGraw-Hill, New York. 433 pp.

STODDART, L. A., A. D. SMITH, AND T. W. BOX. 1975. Range management, 3rd ed. McGraw-Hill, New York. 532 pp.

STOLL, G. D. 1977. Michigan's experience with violation simulation. Paper presented at Assoc. of Midwest Fish and Game Law Enforcement Officers, Okoboji, Ia. 2 pp.

STONE, C. P., D. F. MOTT, J. F. BESSER, AND J. W. DEGRAZIO. 1972. Bird damage to corn in the United States in 1970. Wilson Bull. 84:101–105.

STONE, C. P., W. F. SHAKE, AND D. J. LANGOWSKI. 1974. Reducing bird damage to highbush blueberries with a carbamate repellent. Wildl. Soc. Bull. 2:135–139.

STONE, R. 1993. The mouse-pinon nut connection. Science 262:833.

STONE, W. B., AND P. B. GRADONI. 1985. Recent poisonings of wild birds by diazinon and carbofuran. Northeastern Environ. Sci. 4:160–164.

STONE, W. B., S. R. OVERMANN, AND J. C. OKONIEWSKI. 1984. Intentional poisoning of birds with parathion. Condor 86:333–336.

STORMER, F. A., AND G. L. VALENTINE. 1981. Management of shelterbelts for wildlife. Proc. Ann. Meeting of the Forestry Comm. 33:169–181. Great Plains Agric. Council Publ. 102, Lincoln, Neb.

STRANAHAN, S. Q. 1987. Many happy returns for wildlife. Nat. Wildl. 25(3):50–51.

STREATOR, C. P. 1930. Commercial fertilizer from wood rat nests. J. Mamm. 11:318.

STRELL, W. K., AND J. G. DICKSON. 1980. Effect forest clear-cut edge on breeding birds in east Texas. J. Wildl. Manage. 44:559–567.

STRESEMANN, E. 1934. (1926–1934). Aves. Vol. 7, Part 2 *in* Handbuch der Zoologie (W. Kukenthal and T. Krumbach, eds.). Walter de Gruyter and Co., Berlin. 899 pp.

STROHMEYER, D. L., AND L. H. FREDRICKSON. 1967. An evaluation of dynamited potholes in northwest Iowa. J. Wildl. Manage. 31:525–532.

STROUD, R. K., C. O. THOEN, AND R. M. DUNCAN. 1986. Avian tuberculosis and salmonellosis in a whooping crane (*Grus americana*). J. Wildl. Dis. 22:106–110.

STUEWER, F. W. 1948. Artificial dens for raccoons. J. Wildl. Manage. 12:296–301.

SUCHY, W. J., L. L. MCDONALD, M. D. STRICKLAND, AND S. H. ANDERSON. 1985. New estimates of minimum viable population size for grizzly bears of the Yellowstone ecosystem. Wildl. Soc. Bull. 13:223–228.

SUETSUGU, H. Y., AND K. E. MENZEL. 1963. Wild turkey introductions in Nebraska. Trans. N. Am. Wildl. and Nat. Resour. Conf. 28:297–307.

SUGDEN, L. G., AND G. W. BEYERSBERGEN. 1984. Farming intensity on waterfowl breeding grounds in Saskatchewan parklands. Wildl. Soc. Bull. 12:22–26.

SUGDEN, L. G., W. T. THURLOW, R. D. HARRIS, AND K. VERMEER. 1974. Investigations of mallards overwintering at Calgary, Alberta. Can. Field-Nat. 88:303–311.

SULLIVAN, R. A. 1980. Fisheries issues in the 80s (editorial). Fisheries 5:1, 6–8.

SUN, M. 1987. Trouble ahead for exotic Mono Lake. Science 237:716–717.

SURBER, E. W. 1957. Results of striped bass (*Roccus saxatilis*) introductions into freshwater impoundments. Proc. SE Assoc. Game and Fish Comm. 11:273–276.

SWANK, W. G. 1955. Nesting and production of the mourning dove in Texas. Ecology 36:495–505.

SWANK, W. G., AND S. GALLIZIOLI. 1954. The influence of hunting and of rainfall upon Gambel's quail populations. Trans. N. Am. Wildl. Conf. 19:283–296.

SWANK, W. G., AND G. A. PETRIDES. 1954. Establishment and food habits of the nutria in Texas. Ecology 35:172–176.

SWANSON, G. A. 1977. Diel food selection by Anatinae on a waste stabilization system. J. Wildl. Manage. 41:226–231.

SWANSON, G. A., AND J. C. BARTONEK. 1970. Bias associated with food analysis in gizzards of blue-winged teal. J. Wildl. Manage. 34:739–746.

SWANSON, G. A., M. I. MEYER, AND J. R. SERIE. 1974. Feeding ecology of breeding blue-winged teals. J. Wildl. Manage. 38:396–407.

SWENSON, J. E. 1985. Compensatory reproduction in an introduced mountain goat population in the Absaroka Mountains, Montana. J. Wildl. Manage. 49:837–843.

SWIHART, R. K., AND R. H. YAHNER. 1982. Habitat features influencing use of farmstead shelterbelts by eastern cottontails. Am. Midl. Nat. 107:411–414.

SZARO, R. C., P. H. ALBERS, AND N. C. COON. 1978. Petroleum: effects on mallard egg hatchability. J. Wildl. Manage. 42:404–406.

TABER, R. D. 1961. Wildlife administration and harvest in Poland. J. Wildl. Manage. 25:353–363.

TABER, R. D., AND R. F. DASMANN. 1957. The dynamics of three natural populations of the deer *Odocoileus hemionus columbianus*. Ecology 38:233–246.

TABOR, S. P., AND R. E. THOMAS. 1986. The occurrence of plague *(Yersinia pestis)* in a bobcat from the Trans-Pecos area of Texas. Southwestern Nat. 31:135–136.

TAKEKAWA, J. Y., E. O. GARTON, AND L. A. LANGELIER. 1982. Biological control of forest insect outbreaks with avian predators. Trans. N. Am. Wildl. and Nat. Resour. Conf. 47:393–409.

TALBOT, L. M. 1957. The lions of Gir: wildlife management problems of Asia. Trans. N. Am. Wildl. Conf. 22:570–579.

TALBOT, L. M., AND M. H. TALBOT. 1963. The wildebeest in western Masailand, East Africa. Wildl. Monogr. No. 12. 88 pp.

TANNER, O. 1975. Urban wilds. Time-Life Books, New York. 184 pp.

TANNER, T. (ED.). 1987. Aldo Leopold, the man and his legacy. Soil Conserv. Soc. Amer., Ankeny, Ia. 175 pp.

TANSEY, M. R., AND T. D. BROCK. 1973. *Dactylaria gallopava*, a cause of avian encephalitis, in hot spring effluents, thermal soils, and self-heated coal waste piles. Nature 242:202–203.

TATE, J. 1980. Aggression in human-bear interactions: the influence of setting. Proc. 2nd Conf. on Sci. Res. in National Parks 6:221–233.

TAUTIN, J., P. H. GEISSLER, R. E. MUNRO, AND R. S. POSPAHALA. 1983. Monitoring the population status of American woodcock. Trans. N. Am. Wildl. and Nat. Resour. Conf. 48:376–388.

TAYLOR, C. R. 1968. Hygroscopic food: a source of water for desert animals. Nature 219:181–182.

TAYLOR, M. W., C. W. WOLFE, AND W. L. BAXTER. 1978. Land-use change and ring-necked pheasants in Nebraska. Wildl. Soc. Bull. 6:226–230.

TAYLOR, W. P. 1935. Some animal relations to soils. Ecology 16:127–136.

TAYLOR, W. P., C. T. VORHIES, AND P. B. LISTER. 1935. The relation of jackrabbits to grazing in southern Arizona. J. Forestry 33:490–498.

TEAGUE, R. D. 1971. Wildlife enterprises on private land. Pp. 140–143 *in* A manual of wildlife conservation (R. D. Teague, ed.). The Wildlife Society, Washington, D.C. 206 pp.

TECHNAU, G. 1936. Die Nasendruse der Vogel. J. für Ornith. 84:511–617.

TEER, J. G. 1975. Commercial uses of game animals on rangelands of Texas. J. Animal Sci. 40:1000–1008.

TEER, J. G. 1988. Conservation biology the science of scarcity and diversity. Book review. J. Wildl. Manage. 52:570–572.

TEER, J. G., AND N. K. FOREST. 1968. Bionomic and ethical implications of commercial game harvest programs. Trans. N. Am. Wildl. and Nat. Resour. Conf. 33:192–204.

TEER, J. G., J. W. THOMAS, AND E. A. WALKER. 1965. Ecology and management of white-tailed deer in the Llano Basin of Texas. Wildl. Monogr. No. 15. 62 pp.

TEETER, J. W. 1965. Effects of sodium chloride on the sago pondweed. J. Wildl. Manage. 29:838–845.

TELFAIR, R. C. II. 1983. The cattle egret: a Texas focus and world view. Kleberg Studies in Natural Resources. Texas Agric. Exp. Sta. Texas A&M Univ., College Station. 144 pp.

TELFER, E. S. 1967. Comparison of moose and deer winter range in Nova Scotia. J. Wildl. Manage. 31:418–425.

TELFER, E. S., AND J. P. KELSALL. 1984. Adaptation of some large North American mammals for survival in snow. Ecology 65:1828–1834.

TEMPLE, S. A. 1977. Plant-animal mutualism: coevolution with dodo leads to near extinction of plant. Science 197:885–886.

TEMPLE, S. A. (ED.). 1977a. The concept of managing endangered species. Pp. 3–8 *in* Endangered birds, management techniques for preserving threatened species (S. A. Temple, ed.). Univ. Wis. Press, Madison. 466 pp.

TEMPLE, S. A. (ED.). 1977b. Endangered birds, management techniques for preserving threatened species. Univ. Wis. Press, Madison. 466 pp.

TEMPLE, S. A. 1991. Conservation biology: new goals and new partners for managers of biological resources. Pp. 45–54 *in* Challenges in the conservation of biological resources: a practioner's guide (D. J. Decker, M. E. Krasny, G. R. Goff, C. R. Smith, and D. W. Gross, eds.). Westview Press, Boulder. Colo. 402 pp.

TEMPLE, S. A., E. G. BOLEN, M. E. SOULÉ, P. F. BRASSARD, H. SALWASSER, AND J. G. TEER. 1988. What's new about conservation biology? Trans. N. Amer. Wildl. and Nat. Resour. Conf. 53:609–612.

TEMPLETON, A. R. 1986. Coadaptation and outbreeding depression. Pp. 105–116 *in* Conservation biology, the science of scarcity and diversity (M. E. Soulé, ed.). Sinauer Assoc., Sunderland, Mass. 584 pp.

TERBORGH, J. 1989. Where have all the birds gone? Princeton Univ. Press, Princeton, N.J. 207 pp.

TERBORGH, J. 1992. Perspectives on the conservation of Neotropical migrant landbirds. Pp. 7–12 *in* Ecology and conservation of Neotropical migrant landbirds (J. M. Hagan and D.W. Johnston, eds.). Smithsonian Instit. Press, Washington, D.C. 609 pp.

THAYER, V. G., AND R. T. BARBER. 1984. At sea with El Niño. Nat. Hist. 93:4–12.

THIOLLAY, J. M. 1989. Area requirements for the conservation of rain forest raptors and game birds in French Guiana. Conserv. Biol. 3:128–137.

THOM, R. H., D. T. CATLIN, AND D. J. WITTER. 1986. Missouri's prairie day: a targeted interpretive event. Proc. N. Am. Prairie Conf. 9:175–178.

THOMAS, C. D. 1990. What do real population dynamics tell us about minimum viable population sizes? Conserv. Biol. 4: 324–327.

THOMAS, G. W., S. E. CURL, AND W. F. BENNETT. 1982. Food and fiber for a changing world: third-century challenge to American agriculture. Interstate Publishers, Danville, Ill. 258 pp.

THOMAS, J. R., H. R. COSPER, AND W. BEVER. 1964. Effects of fertilizers on the growth of grass and its use by deer in the Black Hills of South Dakota. Agron. J. 56:223–226.

THOMAS, J. W. 1979. Preface. P. 6 *in* Wildlife habitats in managed forests—the Blue Mountains of Oregon and Washington (J. W. Thomas, ed.). U.S.D.A. Forest Serv. Agric. Handbook No. 553. U.S. Govt. Printing Office, Washington, D.C. 512 pp.

THOMAS, J. W., AND R. A. DIXON. 1974. Cemetery ecology. Pp. 107–110 *in* A symposium on wildlife in an urbanizing environment (J. H. Noyes and D. R. Progulske, eds.). Univ. Mass. Coop. Ext. Serv., Amherst. 182 pp.

THOMAS, J. W., AND H. SALWASSER. 1989. Bringing conservation biology into a position of influence in natural resource management. Conserv. Biol. 3:123–128.

THOMAS, J. W., R. G. ANDERSON, C. MASER, AND E. L. BULL. 1979b. Snags. Pp. 60–77 *in* Wildlife habitats in managed forests—the Blue Mountains of Oregon and Washington (J. W. Thomas, ed.). U.S.D.A. Forest Serv. Agric. Handbook No. 553. U.S. Govt. Printing Office, Washington, D.C. 512 pp.

THOMAS, J. W., E. D. FORSMAN, J. B. LINT, E. C. MESLOW, B. P. NOON, AND J. VERNER. 1990. A conservation strategy for the northern spotted owl. Interagency Scientific Committee (U.S. Department of Agriculture and U.S. Department of the Interior), Washington, D.C. 427 pp.

THOMAS, J. W., C. MASER, AND J. E. RODIEK. 1979. Riparian zones. Pp. 40–47 in Wildlife habitats in managed forests—the Blue Mountains of Oregon and Washington (J. W. Thomas, ed.). U.S.D.A. Forest Serv. Agric. Handbook No. 553. U.S. Govt. Printing Office, Washington, D.C. 512 pp.

THOMAS, J. W., R. M. ROBINSON, AND R. G. MARBURGER. 1964a. Velvet-horn investigation, Part I. Texas Game and Fish 22:4–7.

THOMAS, J. W., R. M. ROBINSON, AND R. G. MARBURGER. 1964b. Velvet-horn investigation, Part II. Texas Game and Fish 22:5–8.

THOMAS, J. W., R. M. ROBINSON, AND R. G. MARBURGER. 1964c. Hypogonadism in white-tailed deer of the central mineral region in Texas. Trans. N. Am. Wildl. and Nat. Resour. Conf. 29:225–236.

THOMAS, J. W., R. M. ROBINSON, AND R. G. MARBURGER. 1970. Studies in hypogonadism in white-tailed deer of the central mineral region of Texas. Texas Parks and Wildl. Dept. Tech. Series No. 5, Austin. 50 pp.

THOMAS, J. W., R. J. MILLER, C. MASER, R. G. ANDERSON, AND B. E. CARTER. 1979a. Plant communities and successional stages. Pp. 23–39 in Wildlife habitats in managed forests—the Blue Mountains of Oregon and Washington (J. W. Thomas, ed.). U.S.D.A. Forest Serv., Agric. Handbook No. 553. U.S. Govt. Printing Office, Washington, D.C. 512 pp.

THOMPSON, B. C. 1987. Attributes and implementation of nongame and endangered species programs in the United States. Wildl. Soc. Bull. 15:210–216.

THOMPSON, D. Q., P. B. REED, G. E. CUMMINGS, AND E. KIVISALU. 1968. Muck hardwoods as green-timber impoundments for waterfowl. Trans. N. Am. Wildl. and Nat. Resour. Conf. 33:142–159.

THOMPSON, D. Q., R. L. STUCKEY, AND E. B. THOMPSON. 1987. Spread, impact, and control of purple loosestrife (Lythrum salicaria) in North American wetlands. U.S. Fish and Wildl. Serv., Fish and Wildl. Res. 2. 55 pp.

THOMPSON, J. D., B. J. SHEFFER, AND G. A. BALDASSARRE. 1989. Incidence of ingested lead shot in waterfowl harvested in Yucatán, Mexico. Wildl. Soc. Bull. 17:189–191.

THOMPSON, L. 1977. Behavior and ecology of burrowing owls on the Oakland Municipal Airport. Condor 73:177–192.

THORNE, E. T., AND S. L. ANDERSON. 1985. Immune response of elk vaccinated with a reduced dose of strain 19 Brucella vaccine. Job Perform. Rept., Job Nos. BDSWCBF551 and BDGACBF551. Pp. 48–62 in Game and Fish Research, Wyo. Game and Fish Dept., Cheyenne. 68 pp.

THORNE, E. T., J. K. MORTON, AND W. C. RAY. 1979. Brucellosis, its effect and impact on elk in western Wyoming. Pp. 212–220 in North American elk: ecology, behavior, and management (M. S. Boyce and L. O. Hayden-Wing, eds.). Univ. Wyo., Laramie. 294 pp.

THORNE, E. T., T. J. WALTHALL, AND H. A. DAWSON. 1981. Vaccination of elk with strain 19 Brucella abortus. Proc. U.S. Animal Health Assoc. 85:359–374.

THORP, J. 1949. Effects of certain animals that live in soils. Sci. Month. 68:180–191.

THREINEN, C. W., AND W. T. HELM. 1954. Experiments and observations designed to show carp destruction of aquatic vegetation. J. Wildl. Manage. 18:247–251.

TILT, W. C. 1987. From whaling to whalewatching. Trans. N. Am. Wildl. and Nat. Resour. Conf. 52:567–585.

TIME. 1981. Bad news for the birds. Time, the weekly newsmagazine. 117:52 (October 5, 1981).

TIME. 1986. A proud river runs red. Time, the weekly newsmagazine. 128:36–37 (November 24, 1986).

TITUS, J. R., AND L. W. VANDRUFF. 1981. Response of the common loon to recreational pressure in the Boundary Waters Canoe Area, northeastern Minnesota. Wildl. Monogr. No. 79. 60 pp.

TOLBA, M. K. 1979. What could be done to combat desertification. Pp. 5–22 in Advances in desert and arid land technology and development, Vol. 1 (A. Bishay and W. G. McGinnies, eds.). Harwood Academic Publishers, New York. 618 pp.

TOMIALOJC, L. 1980. The impact of predation on urban and rural wood pigeon (Columba palumbus L.) populations. Polish Ecol. Studies 5:141–220.

TOMLINSON, R. E. 1972. Current status of the endangered masked bobwhite quail. Trans. N. Am. Wildl. and Nat. Resour. Conf. 37:294–311.

TONER, G. C. 1956. House cat predation on small mammals. J. Mamm. 37:199.

TOTH, S. J., F. TOURINE, AND S. J. TOTH, JR. 1972. Fertilization of smartweed. J. Wildl. Manage. 36:1356–1363.

TOWELL, W. E. 1971. Role of policy-making boards and commissions. Pp. 2–5 in A manual of wildlife conservation (R. D. Teague, ed.). The Wildlife Society, Washington, D.C. 206 pp.

TRAIL, P. W., AND L. F. BAPTISTA. 1993. The impact of brown-headed cowbird parasitism on populations of the Nuttall's white-crowned sparrow. Conserv. Biol. 7:309–315.

TRAINER, D. O. 1970. Bluetongue. Pp. 55–59 in Infectious diseases of wild mammals (J. W. Davis, L. H. Karstad, and D. O. Trainer, eds.). Ia. State Univ. Press, Ames. 421 pp.

TRAINER, D. O., AND M. M. JOCHIM. 1969. Serologic evidence of bluetongue disease in wild ruminants of North America. Am. J. Vet. Res. 30:2007–2011.

TRAINER, D. O., AND L. H. KARSTAD. 1970. Epizootic hemorrhagic disease. Pp. 50–54 in Infectious diseases of wild mammals (J. W. Davis, L. H. Karstad, and D. O. Trainer, eds.). Ia. State Univ. Press, Ames. 421 pp.

TRAINER, D. O., C. S. SCHILDT, R. A. HUNT, AND L. R. JAHN. 1962. Prevalence of Leucocytozoon simondi among some Wisconsin waterfowl. J. Wildl. Manage. 26:137–143.

TRAUTMAN, C. G., L. F. FREDRICKSON, AND A. V. CARTER. 1974. Relationship of red foxes and other predators to populations of ring-necked pheasants and other prey, South Dakota. Trans. N. Am. Wildl. and Nat. Resour. Conf. 39:241–255.

TRAUTMAN, M. B. 1957. The fishes of Ohio, with illustrated keys. Ohio Div. Wildl., Ohio State Univ. Press, Columbus. 683 pp.

TRAWEEK, M. A. 1985. Statewide census of exotic big game animals. Fed. Aid. Proj. W-109-R-8, Job 21, Texas Parks and Wildl. Dept., Austin. 40 pp.

TREFETHEN, J. B. 1975. An American crusade for wildlife. Winchester Press, New York. 409 pp.

TRIPPENSEE, R. E. 1948. Wildlife management: upland game and general principles. McGraw-Hill, New York. 479 pp.

TUCHMAN, B. W. 1978. A distant mirror: the calamitous 14th century. Alfred A. Knopf, New York. 677 pp.

TUCKER, R. D., AND G. W. GARNER. 1983. Habitat selection and vegetational characteristics of antelope fawn bedsites in west Texas. J. Range Manage. 36:110–113.

TURCEK, F. J. 1951. Effect of introductions on two game populations in Czechoslovakia. J. Wildl. Manage. 15:113–114.

TURNER, R. W. 1974. Mammals of the Black Hills of South Dakota and Wyoming. Univ. Kansas Mus. Nat. Hist. Misc. Publ. 60. Lawrence, Kans. 178 pp.

TURNER, W. R. 1971. Sport fish harvest from Rough River, Kentucky, before and after impoundment. Pp. 321–329 *in* Reservoir fisheries and limnology (G. E. Hall, ed.). Am. Fisheries Soc. Spec. Publ. No. 8, Washington, D.C. 511 pp.

TWISS, R. H. 1967. Wildlife in the metropolitan landscape. Trans. N. Am. Wildl. and Nat. Resour. Conf. 32:68–74.

TWOMEY, A. C. 1936. Climographic studies of certain introduced and migratory birds. Ecology 17:122–132.

TYNDALE-BISCOE, H. 1973. Life of marsupials. American Elsevier, New York. 254 pp.

TYRON, C. A. 1954. The effect of carp exclosures on growth of submerged aquatic vegetation in Pymatuning Lake, Penn. J. Wildl. Manage. 18:251–254.

UHAZY, L. S., J. C. HOLMES, AND J. G. STELFOX. 1973. Lungworms in the Rocky Mountain bighorn sheep of western Canada. Can. J. Zool. 51:817–824.

UHLER, F. M. 1956. New habitats for waterfowl. Trans. N. Am. Wildl. Conf. 21:453–469.

UHLER, F. M. 1964. Bonus from waste places. Pp. 643–653 *in* Waterfowl tomorrow (J. P. Linduska, ed.). U.S.D.I., Bur. of Sport Fisheries and Wildl., Fish and Wildl. Serv., U.S. Govt. Printing Office, Washington, D.C. 770 pp.

UHLIG, H. G., AND R. W. BAILEY. 1952. Factors influencing the distribution and abundance of the wild turkey in West Virginia. J. Wildl. Manage. 16:24–32.

ULLREY, D. E., W. G. YOUATT, H. E. JOHNSON, L. D. FAY, B. E. BRENT, AND K. E. KEMP. 1968. Digestibility of cedar and balsam fir browse for the white-tailed deer. J. Wildl. Manage. 32:162–171.

ULRICH, R. S. 1984. View through a window may influence recovery from surgery. Science 224:420–421.

UNDERWOOD, B., AND W. F. PORTER. 1983. The use of sex ratios for determining harvest quotas in white-tailed deer. Abstract *in* Symposium on game harvest management. Caesar Kleberg Wildl. Res. Inst., Kingsville, Tex. 37 pp.

UNDERWOOD, G. 1994. Brief history of the sacred ibis colony at Healesville Sanctuary. Unpubl. rep., Healesville, Victoria, Australia. 7 pp.

UNITT, P. 1987. *Empidonax traillii extimus*: an endangered subspecies. Western Birds 18:137–162.

URESK, D. W., J. G. MACCRACKEN, AND A. J. BJUGSTAD. 1982. Prairie dog density and cattle grazing relationships. Proc. Great Plains Wildlife Damage Control Workshop 5:199–201.

URNESS, P. J., AND C. Y. MCCULLOCH. 1973. Deer nutrition in Arizona chaparral and desert habitats. Ariz. Dept. Game and Fish Resour. Div. and U.S.D.A. Forest Serv. Rocky Mountain Forest and Range Exp. Sta. Spec. Rept. 3. 68 pp.

U.S. BUREAU OF THE CENSUS. 1981. Statistical abstract of the United States: 1981, 102nd ed. Washington, D.C. 1031 pp.

U.S. BUREAU OF THE CENSUS. 1987. Statistical abstract of the United States: 1987. U.S. Government Printing Office, Washington, D.C. 668 pp.

U.S. DEPARTMENT OF AGRICULTURE. 1981. Soil and Water Resources Conservation Act: 1981 program report and environmental impact statement (revised). U.S.D.A., Washington, D.C. 134 pp.

U.S. ENVIRONMENTAL PROTECTION AGENCY. 1979. Fish kills caused by pollution: fifteen year summary 1961–1975. Office of Water Planning and Standards. Washington, D.C. 8 pp.

U.S. FISH AND WILDLIFE SERVICE. 1974. White-tailed deer hunting program, Great Swamp National Wildlife Refuge. U.S. Fish and Wildl. Serv. Final Environ. Statement. FES-74-58, Washington, D.C. 58 pp.

U.S. FISH AND WILDLIFE SERVICE. 1976. Effect of boating on management of Ruby Lake National Wildlife Refuge. U.S. Fish and Wildl. Serv. Environ. Impact Assessment. Portland, Ore. 182 pp.

U.S. FISH AND WILDLIFE SERVICE. 1978. Predator damage in the west: a study of coyote management alternatives. U.S.D.I., Washington, D.C. 168 pp.

U.S. FISH AND WILDLIFE SERVICE. 1980. Regional policy for management of nonmigratory nuisance Canada geese. Release 5-1, Administrative Manual, Region 5, U.S. Fish and Wildl. Serv., Newton Corner, Mass. 6 pp. Mimeo.

U.S. FISH AND WILDLIFE SERVICE. 1982a. Spraying of treflan may pose a hazard to reproduction in birds. Research Information Bull. 82-20, Patuxent Wildl. Res. Center, Laurel, Md. 1 p. (used with permission).

U.S. FISH AND WILDLIFE SERVICE. 1982b. Half of America fishes, hunts, or enjoys wildlife, 1980 survey reveals. U.S. Dept. Int. news release. 3 pp.

U.S. FISH AND WILDLIFE SERVICE. 1984. Summary of the report and recommendations on funding sources to implement the Fish and Wildlife Conservation Act of 1980. U.S. Dept. of Interior, Washington, D.C. 54 pp.

U.S. FISH AND WILDLIFE SERVICE. 1987. National survey of fishing, hunting, and wildlife-associated recreation (preliminary results). U.S. Fish and Wildl. Serv., Washington, D.C. 3 pp. + tables.

U.S. FISH AND WILDLIFE SERVICE. 1994. Whooping crane recovery plan. U.S. Fish and Wildl. Serv., Albuquerque, N. M. 92 pp.

U.S. FISH AND WILDLIFE SERVICE. 1997. 1996 national survey of fishing, hunting, and wildlife-associated recreation. U.S. Fish and Wildl. Serv., Washington, D.C.

USHER, M. B. 1972. Developments in the Leslie matrix model. Pp. 29–60 *in* Mathematical models in ecology (J. N. R. Jeffers, ed.). Symposium of the British Ecol. Soc. 12. Blackwell Scientific Publ., Oxford. 398 pp.

VALDEZ, R. 1997. Effects of an experimental flood on fish and backwaters in the Colorado River, Grand Canyon. Pp. 60–62 *in* Glen Canyon Dam beach-habitat building flow, abst. and exec. summaries. Grand Canyon Monit. and Res. Center, U.S. Dept. Interior, Flagstaff, Ariz. 129 pp.

VALENTINE, J. M., J. R. WALTHER, K. M. MCCARTNEY, AND L. M. IVEY. 1972. Alligator diets on the Sabine National Wildlife Refuge, Louisiana. J. Wildl. Manage. 36:809–815.

VALLE, C. A., AND M. C. COULTER. 1987. Present status of the flightless cormorant, Galápagos penguin and greater flamingo populations in the Galápagos Islands, Ecuador, after the 1982–83 El Niño. Condor 89:276–281.

VANCE, D. R. 1976. Changes in land use and wildlife populations in southeastern Illinois. Wildl. Soc. Bull. 4:11–15.

VANCE, D. R., AND R. L. WESTEMEIER. 1979. Interactions of pheasants and prairie chickens in Illinois. Wildl. Soc. Bull. 7:221–225.

VANDEL, G. M., AND R. L. LINDER. 1981. Pheasants decline but covertype acreages unchanged on South Dakota study area. Wildl. Soc. Bull. 9:299–302.

VANDRUFF, L. W. 1979. Urban wildlife—neglected resource. Pp. 184–190 *in* Wildlife conservation: principles and practices (R. Teague and E. Decker, eds.). The Wildlife Society, Washington, D.C. 280 pp.

VANGILDER, L. D., L. M. SMITH, AND R. K. LAWRENCE. 1986. Nutrient reserves of premigratory brant during spring. Auk 103: 237–241.

VAN VELZEN, A. C., W. B. STILES, AND L. F. STICKEL. 1972. Lethal mobilization of DDT by cowbirds. J. Wildl. Manage. 36: 733–739.

VAUGHT, R. W., H. C. MCDOUGLE, AND H. H. BURGESS. 1967. Fowl cholera in waterfowl at Squaw Creek National Wildlife Refuge, Missouri. J. Wildl. Manage. 31:248–253.

VERME, L. J. 1965a. Reproduction studies on penned white-tailed deer. J. Wildl. Manage. 29:74–79.

VERME, L. J. 1965b. Swamp conifer deeryards in northern Michigan, their ecology and management. J. Forestry 63:523–529.

VERME, L. J. 1968. An index of winter severity for northern deer. J. Wildl. Manage. 32:566–574.

VERME, L. J. 1969. Reproductive patterns of white-tailed deer related to nutritional plane. J. Wildl. Manage. 33:881–887.

VERME, L. J. 1973. Movements of white-tailed deer in Upper Michigan. J. Wildl. Manage. 37:545–552.

VERMEER, K., AND K. H. MORGAN (EDS.). 1997. The ecology, status, and conservation of marine and shoreline birds of the Queen Charlotte Islands. Occas, Paper 93, Canadian Wildl. Serv., Ottawa, Ontario. 150 pp.

VERTS, B. J., AND L. N. CARRAWAY. 1980. Natural hybridization of *Sylvilagus bachmani* and introduced *S. floridanus* in Oregon. Murrelet 61:95–98.

VERTS, B. J., AND L. N. CARRAWAY. 1981. Dispersal and dispersion of an introduced population of *Sylvilagus floridanus*. Great Basin Nat. 41:167–175.

VINSON, S. B., AND A. A. SORENSEN. 1986. Imported fire ants: life history and impact. Texas Dept. Agric., Austin Tex. 28 pp.

VOLTERRA, V. 1926. Variations and fluctuations of the number of individuals in animal species living together. Pp. 409–448 *in* Animal ecology (R. N. Chapman, ed.). McGraw-Hill, New York. 464 pp.

VORHIES, C. T., AND W. P. TAYLOR. 1933. The life histories and ecology of jack rabbits, *Lepus alleni* and *Lepus californicus* spp., in relation to grazing in Arizona. Univ. Ariz. Agric. Exp. Sta. Bull. 49:478–587.

VUILLEUMIER, F. 1970. Insular biogeography in continental regions. I. The northern Andes of South America. Amer. Natl. 104:593–595.

VUILLEUMIER, F. 1973. Insular biogeography in continental regions. II. Cave faunas from Tessin, southern Switzerland. Syst. Zool. 22:64–76.

WAGNER, F. H. 1972. Coyotes and sheep: some thoughts on ecology, economics, and ethics. Forty-fourth Honor Lecture, Utah State Univ., Logan. 59 pp.

WAGNER, F. H. 1983. Status of wild horse and burro management on public rangelands. Trans. N. Am. Wildl. and Nat. Resour. Conf. 48:116–133.

WAGNER, F. H., C. D. BESADNY, AND C. KABAT. 1965. Population ecology and management of Wisconsin pheasants. Wis. Dept. Cons. Tech. Bull. 34. Madison. 168 pp.

WAGNER, R. H. 1978. Environment and man. W. W. Norton, New York. 591 pp.

WAKELEY, J. S., AND R. C. MITCHELL. 1981. Blackbird damage to ripening field corn in Pennsylvania. Wildl. Soc. Bull. 9:52–55.

WALCOTT, C. F. 1974. Changes in bird life in Cambridge, Massachusetts, from 1860 to 1964. Auk 91:151–160.

WALLACE, G. J. 1963. An introduction to ornithology, 2nd ed. Macmillan, New York. 491 pp.

WALLEN, I. E. 1951. The direct effect of turbidity on fishes. Oklahoma Agric. and Mech. Coll. Bull. 48, Biol. Series 2, Stillwater. 27 pp.

WALLESTAD, R. 1975. Male sage grouse responses to sagebrush treatment. J. Wildl. Manage. 39:482–484.

WALLMO, O. C. 1969. Response of deer to alternate-strip clearcutting of lodgepole pine and spruce-fir timber in Colorado. U.S.D.A. Forest Serv. Rocky Mountain Forest and Range Exp. Sta. Res. Note RM—141. Fort Collins, Colo. 4 pp.

WALLMO, O. C., AND J. W. SCHOEN. 1980. Response of deer to secondary forest succession in southeast Alaska. Forest Sci. 26:448–462.

WALSH, J. 1987. Bolivia swaps debt for conservation. Science 237:596–597.

WALTERS, C. J. 1971. Systems ecology: the systems approach and mathematical models in ecology. Pp. 276–292 *in* Fundamentals of ecology, 3rd ed. (E. P. Odum, ed.). W. B. Saunders, Philadelphia. 574 pp.

WANDELL, W. N. 1949. Status of ring-necked pheasants in the United States. Trans. N. Am. Wildl. Conf. 14:370–390.

WARNER, R. E. 1968. The role of introduced diseases in the extinction of the endemic Hawaiian avifauna. Condor 70:101–120.

WARNER, R. E. 1979. Use of cover by pheasant broods in east-central Illinois. J. Wildl. Manage. 43:334–346.

WARNER, R. E. 1981. Illinois pheasants: populations, ecology, distribution, and abundance, 1900–1978. Ill. Nat. Hist. Surv. Biol. Notes No. 115. 22 pp.

WARNER, R. E. 1984. Effects of changing agriculture on ring-necked pheasant brood movements in Illinois. J. Wildl. Manage. 48:1014–1018.

WARNER, R. E., L. M. DAVID, S. L. ETTER, AND G. B. JOSELYN. 1992. Costs and benefits of roadside management for pheasants in Illinois. Wildl. Soc. Bull. 20:270–285.

WARNER, R. E., AND G. B. JOSELYN. 1978. Roadside management for pheasants in Illinois: acceptance by farm cooperators. J. Wildl. Manage. 6:128–134.

WARNER, R. E., AND G. B. JOSELYN. 1986. Responses of Illinois ring-necked pheasant populations to block roadside management. J. Wildl. Manage. 50:525–532.

WARNER, R. E., S. L. ETTER, G. B. JOSELYN, AND J. A. ELLIS. 1984. Declining survival of ring-necked pheasants in Illinois agricultural ecosystems. J. Wildl. Manage. 48:82–88.

WARNER, R. E., S. P. HAVERA, AND L. M. DAVID. 1985. Effects of autumn tillage systems on corn and soybean harvest residues in Illinois. J. Wildl. Manage. 49:185–190.

WARNER, R. E., G. B. JOSELYN, AND S. L. ETTER. 1987. Factors affecting roadside nesting by pheasants in Illinois. Wildl. Soc. Bull. 12:221–228.

WARRELL, D. A. 1977. Rabies in man. Pp. 32–52 *in* Rabies, the facts (C. Kapan, ed.). Oxford Univ. Press, Oxford, England. 116 pp.

WARREN, R. J., AND L. J. KRYSL. 1983. White-tailed deer food habits and nutritional status as affected by grazing and deer-harvest management. J. Range Manage. 36:104–109.

WARRICK, G., AND P. R. KRAUSMANN. 1986. Bone-chewing by desert bighorn sheep. Southwestern Nat. 31:414.

WATSON, A., AND R. MOSS. 1970. Dominance, spacing behavior, and aggression in relation to population limitation in vertebrates. Pp. 167–220 *in* Animal populations in relation to their food resources (A. Watson, ed.). Blackwell Sci., Oxford, U.K. 477 pp.

WATSON, A., AND P. J. O'HARE. 1979. Red grouse populations on experimentally treated and untreated Irish bog. J. Applied Ecol. 16:433–452.

WATT, B. K., AND A. L. MERRILL. 1963. Composition of foods . . . raw, processed, prepared. Agric. Handbook No. 8, U.S. Dept. Agric., Washington, D.C. 190 pp.

WATT, K. E. F. 1982. Understanding the environment. Allyn & Bacon, Boston. 431 pp.

WATT, W. D., C. D. SCOTT, AND W. J. WHITE. 1983. Evidence of acidification of some Nova Scotian rivers and its impact on Atlantic salmon *(Salmo salar)*. Can. J. Fish. Aqua. Sci. 40:462–473.

WATT, W., D. SCOTT, AND S. RAY. 1979. Acidification and other chemical changes in Halifax County lakes after 21 years. Limnol. Oceanogr. 24:1154–1161.

WATTS, P. D., N. A. ORITSLAND, C. JONKEL, AND K. RONALD. 1981. Mammalian hibernation and the oxyen consumption of a denning black bear. Comp. Biochem. Physiol. 69: 121–123.

WATTS, T. J., AND S. D. SCHEMNITZ. 1985. Mineral lick use and movement in a remnant desert bighorn sheep populations. J. Wildl. Manage. 49:994–996.

WAYNE, R. K., N. HELMAN, D. GIRMAN, P. J. P. GOGAN, D. A. GILBERT, K. HANSE, R. O. PETERSON, U. S. SEAL, A. EISENHAWER, L. D. MECH, AND R. J. KRUMENAKER. 1991. Conservation genetics of the endangered Isle Royale gray wolf. 1991. Conserv. Biol. 5:41–51.

WAYNE, R. K., W. S. MODI, AND S. J. O'BRIEN. 1986. Morphological variability and asymmetry in the cheetah *(Acinonyx jubatus)*, a genetically uniform species. Evolution 40:78–85.

WEAVER, J. E., AND F. E. CLEMENTS. 1929. Plant ecology. McGraw-Hill, New York. 520 pp.

WEAVER, J. K., AND H. S. MOSBY. 1979. Influence of hunting regulations on Virginia wild turkey populations. J. Wildl. Manage. 43:128–135.

WEAVER, R. A. 1973. California's bighorn management plan. Trans. Desert Bighorn Council 17:22–42.

WEBB, R. E., AND F. HORSFALL, JR. 1967. Endrin resistance in the pine mouse. Science 156:1762.

WEBB, R. E., R. W. HARTGROVE, W. C. RANDOLPH, V. J. PETRELLA, AND F. HORSFALL, JR. 1973. Toxicity studies in endrin-susceptible and resistant strains of pine mice. Toxicol. Appl. Pharmacol. 25:42–47.

WEBB, W. L. 1960. Forest wildlife management in Germany. J. Wildl. Manage. 24:147–161.

WEBER, B., AND A. VEDDER. 1984. Forest conservation in Rwanda and Burundi. J. East African Wildl. Soc. 7(6):32–35.

WEBER, M. 1982. Quixotic victory for whales. Oceans 15:68–69.

WEBER, W. 1979. Pigeon-associated people diseases. Proc. Bird Control Seminar 8:156–158.

WEBSTER, E. B. 1925. Status of mountain goats introduced into the Olympic Mountains, Washington. Murrelet 6:10.

WECKER, S. C. 1963. The role of early experience in habitat selection by the prairie deer mouse, *Peromyscus maniculatus bairdii*. Ecol. Monogr. 33:307–325.

WEEKS, H. P., AND C. M. KIRKPATRICK. 1976. Adaptations of white-tailed deer to naturally occurring sodium deficiencies. J. Wildl. Manage. 40:610–625.

WEEKS, H. P., JR., AND C. M. KIRKPATRICK. 1978. Salt preferences and sodium drive phenology in fox squirrels and woodchucks. J. Mamm. 59:531–542.

WEIGAND, J. P. 1980. Ecology of the Hungarian partridge in north-central Montana. Wildl. Monogr. No. 74. 106 pp.

WEISE, T. F., W. L. ROBINSON, R. A. HOOK, AND L. D. MECH. 1975. An experimental translocation of the eastern timber wolf. Audubon Conserv. Rept. No. 5. The Audubon Society, New York. 28 pp.

WEITHMAN, A. S. 1986. Measuring the value and benefits of reservoir fisheries programs. Pp. 11–17 *in* Reservoir fisheries management: strategies for the 80's (G. E. Hall and M. J. Van Den Avyle, eds.). Southern Div. Am. Fisheries Soc., Bethesda, Md. 327 pp.

WELCH, B. L., AND J. C. PEDERSON. 1981. In vitro digestibility among accessions of big sagebrush by wild mule deer and its relationship to monoterpenoid content. J. Range Manage. 34: 497–500.

WELCHERT, W. T., AND B. N. FREEMAN. 1973. "Horizontal" wells. J. Range Manage. 26:253–256.

WELLER, M. W. 1969. Potential dangers of exotic waterfowl introductions. Wildfowl 20:55–58.

WELLER, M. W. 1980. The island waterfowl. Ia. State Univ. Press, Ames. 121 pp.

WELLS, P. G., J. N. BUTLER, AND J. S. HUGHES (EDS.). 1995. *Exxon Valdez* oil spill: fate and effects in Alaskan waters. ASTM Spl. Tech. Publ. 1219, Amer. Soc. Testing and Materials, Philadelphia, Penn. 955 pp.

WELTY, J. C. 1982. The life of birds, 3rd ed. Saunders College Publ., Philadelphia. 754 pp.

WENDT, J. S., AND J. A. KENNEDY. 1991. Policy considerations regarding the use of lead pellets for waterfowl hunting in Canada. Pp. 61–67 *in* Lead poisoning in waterfowl (D. J. Pain, ed.). Int. Waterfowl and Wetlands Res. Bur. Spec. Publ. 16. Slimbridge, U.K. 105 pp.

WESELOH, D. V., AND R. T. BROWN. 1971. Plant distribution within a heronry. Am. Midl. Nat. 86:57–64.

WESLEY, D. E., K. L. KNOX, AND J. G. NAGY. 1970. Energy flux and water kinetics in young pronghorn antelope. J. Wildl. Manage. 34:908–912.

WEST, R. R., R. B. BRUNTON, AND D. J. CUNNINGHAM. 1969. Repelling pheasants from sprouting corn with a carbamate insecticide. J. Wildl. Manage. 33:216–219.

WEST, S. 1980. Acid from heaven. Sci. News 117:76–78.

WESTEMEIER, R. L. 1966. Apparent lead poisoning in a wild bobwhite. Wilson Bull 78:471–472.

WESTEMEIER, R. L. 1972. Prescribed burning in grassland management for prairie chickens in Illinois. Proc. Tall Timbers Fire Ecol. Conf. 12:317–338.

WETMORE, A. 1915. Mortality among waterfowl around Great Salt Lake, Utah. U.S.D.A. Bull. No. 217. 10 pp.

WETMORE, A. 1918. The duck sickness in Utah. U.S.D.A. Bull. No. 672. 26 pp.

WHITE, D. H. 1985. Secondary poisoning of Franklin's gulls in Texas by monocrotophos. J. Wildl. Dis. 21:76–78.

WHITE, D. H., C. A. MITCHELL, L. D. WYNN, E. L. FLICKINGER, AND E. J. KOLBE. 1982. Organophosphate insecticide poisoning of Canada geese in the Texas panhandle. J. Field Ornith. 53: 22–27.

WHITE, J. 1981a. Trouble on Angel Island. Outdoor California (Sept.–Oct.) 42:9–10.

WHITE, J. 1981b. Largest deer capture in DFG history. Outdoor California (Nov.–Dec.) 42:11–13.

WHITE, L., JR. 1967. The historical roots of our ecologic crisis. Science 155:1203–1207.

WHITE, P. T. 1983. Tropical rain forests: nature's dwindling treasures. Nat. Geogr. 163:2–9, 20–46.

WHITE, S. B., R. A. DOLBEER, AND T. A. BOOKHOUT. 1985. Ecology, bioenergetics, and agricultural impacts of a winter-roosting population of blackbirds and starlings. Wildl. Monogr. No. 93. 42 pp.

WHITFORD, W. G., D. SCHAEFER, AND W. WISDOM. 1986. Soil movement by desert ants. Southwestern Nat. 31:273–274.

WHITMORE, R. W., K. P. PRUESS, AND R. E. GOLD. 1986. Insect food selection by 2-week-old ring-necked pheasant chicks. J. Wildl. Manage. 50:223–228.

WHITTAKER, R. H., AND G. E. LIKENS. 1975. The biosphere and man. Pp. 305–328 *in* Primary productivity of the biosphere (H. Lieth and R. H. Whittaker, eds.). Ecological Studies 14. Springer-Verlag, New York. 339 pp.

WHITTAKER, R. H., AND G. M. WOODWELL. 1969. Structure, production, and diversity of the oak-pine forest at Brookhaven, New York. J. Ecology 57:155–174.

WHYTE, R. J., AND B. W. CAIN. 1981. Wildlife habitat on grazed or ungrazed small pond shorelines in south Texas. J. Range Manage. 34:64–68.

WHYTE, R. J., N. J. SILVY, AND B. W. CAIN. 1981. Effects of cattle on duck food plants in southern Texas. J. Wildl. Manage. 45:512–515.

WIENER, J. G. 1987. Metal contamination of fish in low-pH lakes and potential implications for piscivorous wildlife. Trans. N. Am. Wildl. and Nat. Resour. Conf. 52:645–657.

WIENS, J. A. 1973. Pattern and process in grassland bird communities. Ecol. Monogr. 43:237–270.

WIESE, J. H. 1978. Heron nest-site selection and its ecological effect. Pp. 27–34 *in* Wading birds. National Audubon Soc. Res. Rept. 7. The Audubon Society, New York. 381 pp.

WILCOVE, D., AND D. MURPHY. 1991. The spotted owl controversy and conservation biology. Conserv. Biol. 5:261–262.

WILCOX, B. A. 1980. Insular ecology and conservation. Pp. 95–117 *in* Conservation biology (M. E. Soulé and B. A. Wilcox, eds.). Sinauer Assoc., Sunderland, Mass. 395 pp.

WILDE, S. A. 1946. Soil-fertility standards for game food plants. J. Wildl. Manage. 10:77–81.

WILDE, S. A. 1958. Forest soils, their properties and relation to silviculture. Ronald Press, New York. 537 pp.

WILDE, S. A., C. T. YOUNGBERG, AND J. H. HOVIND. 1950. Changes in composition of ground water, soil fertility, and forest growth produced by the construction and removal of beaver dams. J. Wildl. Manage. 14:123–128.

WILDLIFE MANAGEMENT INSTITUTE. 1975. Current investments, projected needs, and potential new sources of income for nongame fish and wildlife programs in the United States. Wildl. Manage. Inst., Washington, D.C. 93 pp.

WILDLIFE MANAGEMENT INSTITUTE. 1977. Organization, authority, and programs of state fish and wildlife agencies. Wildl. Manage. Inst., Washington, D.C. 5 pp.

WILDLIFE SOCIETY, THE. 1986. Position statement of The Wildlife Society: management and conservation of brown bears. The Wildlifer 216:28.

WILDT, D. W., M. BUSH, J. G. HOWARD, S. J. O'BRIEN, D. MELTZER, A. VANDYK, H. EBEDES, AND D. J. BRAND. 1983. Diminished ejaculate quality of the South African cheetah. Biol. Reprod. 21:1019–1025.

WILES, G. J. 1987. Current research and future management of Marianas fruit bats (Chiroptera: Pteropodidae) on Guam. Aust. Mammal. 10:93–95.

WILES, G. J., AND H. P. WEEKS, JR. 1986. Movements and use patterns of white-tailed deer visiting natural licks. J. Wildl. Manage. 50:487–496.

WILLIAMS, A. S. 1977. Current methods of bird rehabilitation. Pp. 125–134 *in* Proc. in the 1977 oil spill response workshop (P. L. Fore, ed.). U.S. Fish and Wildl. Serv., Biol. Serv. Pgm. FWS/OBS/77-24, Washington, D.C. 153 pp.

WILLIAMS, A. S. 1978. Saving oiled seabirds: a manual for cleaning and rehabilitating oiled waterfowl. American Petroleum Inst., Washington, D.C. 7 pp.

WILLIAMS, C. E. 1964. Soil fertility and cottontail body weight: a reexamination. J. Wildl. Manage. 28:329–338.

WILLIAMS, E. S., E. T. THORNE, M. J. G. APPEL, AND D. W. BELITSKY. 1988. Canine distemper in black-footed ferrets (*Mustela nigripes*) from Wyoming. J. Wildl. Dis. 24:385–398.

WILLIAMS, J. D., AND C. K. DODD, JR. 1978. Importance of wetlands to endangered and threatened species. Pp. 565–575 *in* Wetland functions and values: the state of our understanding (P. E. Greeson, J. R. Clark, and J. E. Clark, eds.). Am. Water Resour. Assoc., Minneapolis. 674 pp.

WILLIAMS, J. G. 1963. Freeing flamingoes from anklets of death. Nat. Geogr. 124:934–944.

WILLIAMS, R. E., B. W. ALLRED, R. M. DENIO, AND H. A. PAULSEN, JR. 1968. Conservation, development, and use of the world's rangelands. J. Range Manage. 21:355–360.

WILLIAMS, R. N., AND L. V. GIDDINGS. 1984. Differential range expansion and population growth of bulbuls in Hawaii. Wilson Bull. 96:647–655.

WILLIAMS, T. 1985. Horses, asses, and asininities. Audubon 87(5):20, 22–23.

WILLIAMS, T. 1996. Seeking refuge, Audubon 98(3):36–45, 90–94.

WILLIAMSON, L. 1981. Hunting—an American tradition. The American Hunter 10:12–13.

WILLIAMSON, L. 1985. Whither goes the Park Service? Outdoor Life 176(9):29, 33. Dec. 1985.

WILLIAMSON, R. D. 1973. Bird—and people—neighborhoods. Nat. Hist. 82:55–57.

WILLIAMSON, S. J., AND D. H. HIRTH. 1985. An evaluation of edge use by white-tailed deer. Wildl. Soc. Bull. 13:252–257.

WILLIS, E. O. 1974. Populations and local extinctions of birds on Barro Colorado Island, Panama. Ecol. Monogr. 44:153–169.

WILLMS, W. D. 1971. The influence of forest edge, elevation, aspect, site index, and roads on deer use of logged and mature forest, northern Vancouver Island. M.S. Thesis, Univ. British Columbia, Vancouver. 184 pp.

WILLNER, G. R. 1982. Nutria. Pp. 1059–1076 *in* Wild mammals of North America (J. A. Chapman and G. A. Feldhamer, eds.). Johns Hopkins Univ. Press, Baltimore, Md. 1147 pp.

WILLNER, G. R., J. A. CHAPMAN, AND D. PURSLEY. 1979. Reproduction, physiological responses, food habits, and abundance of nutria on Maryland marshes. Wildl. Monogr. No. 65. 43 pp.

WILSON, D. E., AND S. M. HIRST. 1977. Ecology and factors limiting roan and sable antelope populations in South Africa. Wildl. Monogr. No. 54. 111 pp.

WILSON, E. B., AND J. M. HUNT (EDS.). 1975. Petroleum in the marine environment. Nat. Acad. of Sci., Washington, D.C. 107 pp.

WILSON, E. O. 1975. Sociobiology, the new synthesis. Belknap Press, Cambridge, Mass. 697 pp.

WILSON, E. O. 1984. Biophilia. Harvard Univ. Press, Cambridge, Mass. 157 pp.

WILSON, E. O. 1989. Threats to biodiversity. Sci. Am. 261(3):108–116.

WILSON, E. O. 1992. The diversity of life. Harvard Univ. Press, Cambridge, Mass. 424 pp.

WILSON, E. O. 1994. Naturalist. Island Press, Washington, D.C. 380 pp.

WILSON, E. O., AND B. HOLLDOBLER. 1990. The ants. Harvard Univ. Press, Cambridge, Mass. 732 pp.

WILSON, E. O., AND S. R. KELLERT (EDS.). 1993. The biophilia hypothesis. Island Press, Washington, D.C. 484 pp.

WILSON, M. M., AND J. A. CRAWFORD. 1979. Response of bobwhites to controlled burning in south Texas. Wild. Soc. Bull. 7:53–56.

WING, L. W. 1951. Practice of wildlife conservation. John Wiley & Sons, New York. 412 pp.

WINKLER, W. G., AND K. BOGEL. 1992. Control of rabies in wildlife. Sci. Am. 266(6):86–92.

WINKLER, W. G., R. G. MCLEAN, AND J. C. COWART. 1975. Vaccination of foxes against rabies using ingested baits. J. Wildl. Dis. 11:382–388.

WINSTON, F. A. 1954. Status, movement, and management of the mourning dove in Florida. Florida Game and Fresh Water Fish Comm. Tech. Bull. No. 2, Tallahassee. 86 pp.

WINTER, W. R., AND J. L. GEORGE. 1981. The role of feeding stations in managing nongame bird habitats in urban and suburban areas. Trans. N. Am. Wildl. and Nat. Resour. Conf. 46:414–423.

WISCONSIN DEPT. OF NATURAL RESOURCES. 1985. Better lagoons. Wis. Nat. Resour. Mag. 9(2):29.

WISHART, R. A., AND J. R. BIDER. 1976. Habitat preferences of woodcock in southwestern Quebec. J. Wildl. Manage. 40:523–531.

WISHART, W. 1978. Bighorn sheep. Pp. 161–171 in Big game of North America: ecology and management (J. L. Schmidt and D. L. Gilbert, eds.). Stackpole Books, Harrisburg, Penn. 494 pp.

WITHAM, J. J., AND J. M. JONES. 1987a. Deer-human interactions and research in the Chicago Metropolitan area. Pp. 155–159 in Integrating man and nature in the metropolitan environment (L. W. Adams and D. L. Leedy, eds.). Nat. Instit. Urban Wildl., Columbia, Md. 249 pp.

WITHAM, J. J., AND J. M. JONES. 1987b. Chicago urban deer study. Ill. Nat. Hist. Surv. Rept. 265. 4 pp.

WITTER, D. J., D. L. TYLKA, AND J. E. WERNER. 1981. Values of urban wildlife in Missouri. Trans. N. Am. Wildl. and Nat. Resour. Conf. 46:424–431.

WITTER, D. J., J. D. WILSON, AND G. T. MAUPIN. 1980. "Eagle Days" in Missouri: characteristics and enjoyment ratings of participants. Wildl. Soc. Bull. 8:64–65.

WOBESER, G., AND C. J. BRAND. 1982. Chlamydiosis in two biologists investigating disease occurrences in wild waterfowl. Wildl. Soc. Bull. 10:170–172.

WOBESER, G., AND J. HOWARD. 1987. Mortality of waterfowl on a hypersaline wetland as a result of salt encrustation. J. Wildl. Dis. 23:127–134.

WOBESER, G., AND W. RUNGE. 1975. Rumen overload and rumenitis in white-tailed deer. J. Wildl. Manage. 39:596–600.

WOBESER, G. A. 1981. Diseases of wild waterfowl. Plenum Press, New York. 300 pp.

WOLFE, M. L., N. V. DEBYLE, C. S. WINCHELL, AND T. R. MCCABE. 1982. Snowshoe hare cover relationships in northern Utah. J. Wildl. Manage. 46:662–670.

WOLFE, J. R. 1983. Electrophoretic differentiation between Alaskan brown and black bears. J. Wildl. Manage. 47:268–271.

WOOD, D. H. 1980. The demography of a rabbit Oryctolagus cuniculus population in an arid region of New South Wales, Australia. J. Animal Ecol. 49:55–80.

WOOD, G. W., AND R. H. BARRETT. 1979. Status of wild pigs in the United States. Wildl. Soc. Bull. 7:237–246.

WOOD, G. W., AND J. S. LINDZEY. 1980. The effects of forest fertilization on the crude protein, calcium, and phosphorus content of deer browse in a mixed oak forest. Can. Field Nat. 94:335–346.

WOOD, G. W., AND T. E. LYNN, JR. 1977. Wild hogs in southern forests. So. J. Applied Forestry 1:12–17.

WOOD, G. W., AND L. J. NILES. 1978. Effects of management practices on nongame bird habitat in longleaf-slash pine forests. Pp. 40–49 in Proc. of workshop management of southern forests for nongame birds, Atlanta, Georgia. U.S.D.A. Forest Serv. Gen. Tech. Rept. SE-14, Southeastern Forest and Range Exp. Sta., Asheville, N.C. 176 pp.

WOOD, G. W., M. K. CAUSAY, AND R. M. WHITING. 1985. Perspectives on American woodcock in the southern United States. Trans. N. Am. Wildl. and Nat. Resour. Conf. 50:573–585.

WOODALL, P. F. 1982. Botulism outbreak in waterbirds at Sevenmile Lagoon in south-east Queensland. Aust. Wildl. Res. 9:533–539.

WOODARD, T. N., R. J. GUTIERREZ, AND W. H. RUTHERFORD. 1974. Bighorn lamb production, survival, and mortality in south-central Colorado. J. Wildl. Manage. 38:771–774.

WOODBRIDGE, B., K. K. FINLEY, AND S. T. SEAGER. 1995. An investigation of the Swainson's hawk in Argentina. J. Raptor Res. 29:202–204.

WOODSTREAM CORPORATION. 1977. Trapping and wildlife management, 11th ed. Woodstream Corp., Lititz, Penn. 22 pp.

WOODWARD, S. L., AND R. D. OHMART. 1976. Habitat use and fecal analysis of feral burros (Equus asinus), Chemeheuvi Mountains, California, 1974. J. Range Manage. 29:482–485.

WOOLEY, J. B., JR., L. B. BEST, AND W. R. CLARK. 1985. Impacts of no-till row cropping on upland wildlife. Trans. N. Am. Wildl. and Nat. Resour. Conf. 50:157–168.

WOOLF, A., AND J. D. HARDER. 1979. Population dynamics of a captive white-tailed deer herd with emphasis on reproduction and mortality. Wildl. Monogr. No. 67, 53 pp.

WOOLF, A., J. L. ROSEBERRY, AND J. WILL. 1983. Estimating time of death of deer in Illinois. Wildl. Soc. Bull. 11:47–51.

WORMINGTON, A., AND J. H. LEACH. 1992. Concentrations of migrant diving ducks at Point Pelee National Park, Ontario, in response to invasion of zebra mussels, Dreissena polymorpha. Can. Field-Nat. 106:376–380.

WORSTER, D. 1979. Dust bowl: the southern plains in the 1930s. Oxford Univ. Press, New York. 277 pp.

WREN, C. D., P. M. STOKES, AND K. L. FISCHER. 1986. Mercury levels in Ontario mink and otter relative to food levels and environmental acidification. Can. J. Zool. 64:2854–2859.

WREN, M. C. 1979. The course of Russian history, 4th ed. Macmillan, New York. 598 pp.

WRIGHT, B. S. 1960a. Predation on big game in East Africa. J. Wildl. Manage. 24:1–15.

WRIGHT, B. S. 1960b. Woodcock reproduction in DDT-sprayed areas of New Brunswick. J. Wildl. Manage. 24:419–420.

WRIGHT, B. S. 1965. Some effects of heptachlor and DDT on New Brunswick woodcocks. J. Wildl. Manage. 29:172–185.

WRIGHT, H. A., AND A. W. BAILEY. 1982. Fire ecology: United States and southern Canada. John Wiley & Sons, New York. 501 pp.

WRIGHT, H. A., AND K. J. STINSON. 1970. Response of mesquite to season of top removal. J. Range Manage. 23:127–128.

WRIGHT, R., AND E. SNEKVIK. 1978. Acid precipitation: chemistry and fish populations in 700 lakes in southernmost Norway. Verh. Int. Verein. Limnol. 20:765–775.

WRIGHT, R., T. DALE, E. GJESSING, G. HENDREY, A. HENRIKSEN, M. JOHANNESSEN, AND I. MUNIZ. 1976. Impact of acid precipitation on freshwater ecosystems in Norway. Pp. 459–476 *in* Proc. of the first internat. conf. on acid precipitation and the forest ecosystem (L. Dochinger and T. Seliga, eds.). U.S.D.A. Forest Serv. Gen. Tech. Rept. NE-23. 1074 pp.

WYNNE-EDWARDS, V. C. 1962. Animal dispersion in relation to social behavior. Hafner, New York. 653 pp.

YAHNER, R. H. 1982a. Avian nest densities and nest-site selection in farmstead shelterbelts. Wilson Bull. 94:156–175.

YAHNER, R. H. 1982b. Microhabitat use by small mammals in farmstead shelterbelts. J. Mamm. 63:440–445.

YAHNER, R. H. 1982c. Avian use of vertical strata and plantings in farmstead shelterbelts. J. Wildl. Manage. 46:50–60.

YAHNER, R. H. 1983a. Small mammals in farmstead shelterbelts: habitat correlates of seasonal abundance and community structure. J. Wildl. Manage. 47:74–84.

YAHNER, R. H. 1983b. Seasonal dynamics, habitat relationships, and management of avifauna in farmstead shelterbelts. J. Wildl. Manage. 47:85–104.

YEAGER, L. E. 1949. Effects of permanent flooding in a river-bottom timber area. Ill. Nat. Hist. Surv. Bull. 25:33–65.

YEATTER, R. E. 1934. The Hungarian partridge in the Great Lakes region. Bull. No. 5. Univ. Michigan School of Forestry and Cons. 92 pp.

YEATTER, R. E., AND D. H. THOMPSON. 1952. Tularemia, weather, and rabbit populations. Bull. Ill. Nat. Hist. Surv. 25:351–382.

YEOMANS, J. A., AND J. S. BARCLAY. 1981. Perceptions of residential wildlife programs. Trans. N. Am. Wildl. and Nat. Resour. Conf. 46:390–395.

YOAKUM, J., W. P. DASMANN, H. R. SANDERSON, C. M. NIXON, AND H. S. CRAWFORD. 1980. Habitat improvement techniques. Pp. 329–403 *in* Wildlife habitat management techniques manual, 4th ed. (S. D. Schemnitz, ed.). The Wildlife Society, Washington, D.C. 686 pp.

YOAKUM, J. D. 1980. Barbary sheep in the United States: past, present, and future, Pp. 9–15 *in* Proc. of the symposium on ecology and management of Barbary sheep (C. D. Simpson, ed.). Dept. Range and Wildlife Management, Texas Tech Univ., Lubbock. 112 pp.

YOCOM, C. F. 1970. The giant Canada goose in New Zealand. Auk 87:812–814.

YONCE, F. J. 1983. Forest ownership. Pp. 222–229 *in* Encyclopedia of American forest and conservation history, Vol. 1 (R. C. Davis, ed.). Macmillan, New York. 400 pp.

YOUATT, W. G., L. J. VERME, AND D. E. ULLREY. 1965. Composition of milk and blood in nursing white-tailed does and blood composition of their fawns. J. Wildl. Manage. 29:79–84.

YOUNG, J. A. 1984. Rabies in rangeland environments. Rangelands 6:51–53.

YOUNG, V. A. 1936. Edaphic and vegetational changes associated with injury of a white pine plantation by roosting birds. J. Forestry 34:512–523.

YUILL, T. M. 1970. Myxomatosis and fibromatosis of rabbits, hares, and squirrels. Pp. 104–130 *in* Infectious diseases of wild mammals (J. W. Davis, L. H. Karstad, and D. O. Trainer, eds.). Ia. State Univ. Press, Ames. 421 pp.

ZAGATA, M. D. 1985. Mitigation by "banking" credits: a Louisiana pilot project. Trans N. Am. Wildl. and Nat. Resour. Conf. 50:475–484.

ZAHM, G. R. 1986. Kesterson Reservoir and Kesterson National Wildlife Refuge: history, current problems and management alternatives. Trans. N. Am. Wildl. and Nat. Resour. Conf. 51:324–329.

ZAPPALORTI, R. T., AND H. K. REINERT. 1986. Final report on habitat utilization by the timber rattlesnake, *Crotalus horridus* (Linnaeus), in southern New Jersey with notes on hibernation. Herpetol. Associates, Inc., Beachwood, N.J. 170 pp.

ZARET, T. M., AND R. T. PAINE. 1973. Species introduction in a tropical lake. Science 182:449–455.

ZELENY, L. 1976. The bluebird. Indiana Univ. Press, Bloomington. 170 pp.

ZERN, E. 1972. I am a hunter. Audubon 74:16–19.

ZINKL, J. G., J. RATHERT, AND R. R. HUDSON. 1978. Diazinon poisoning in wild Canada geese. J. Wildl. Manage. 42:406–408.

ZISWILER, V., AND D. S. FARNER. 1972. Digestion and the digestive system. Pp. 343–430 *in* Avian biology, Vol. 2 (D. S. Farner and J. R. King, eds.). Academic Press, New York. 612 pp.

ZOBELL, R. S. 1968. Field studies of antelope movements on fenced ranges. Trans. N. Am. Wildl. and Nat. Resour. Conf. 33:211–217.

INDEX